Important Dates, Periods, and People in the Development of Probability and Statistics

Mid-1500s   Girolamo Cardano (1501-1576) publishes *The Gambler's Handbook*, the first known suggestion of the theory of probability.

1654   Letters between Blaise Pascal (1623-1662) and Pierre de Fermat (1601-1665) further develop the theory of gambling based on probability concepts.

1713-1718   Jacques Bernoulli (1654-1705) publishes the first formal text on probability in 1713. This is followed by the publication of *Doctrine of Chance* in 1718 by Abraham Demoivre (1667-1754). These texts, together with published work printed in 1713 by Ars Conjectandi (unknown) establish the notion of statistics, the laws of large numbers, the binomial distribution (due to Bernoulli), and the transition from the binomial to the normal distribution (due to Demoivre).

1812   Pierre Simon-Laplace (1749-1827) publishes *Théorie Analytique des Probabilités*. It is the first comprehensive treatment of probability.

1809   Karl Friedrich Gauss (1777-1855) extends and develops the least squares criterion, first introduced by Adrien Legendre (1752-1833) in what becomes known as regression analysis. Gauss applies the normal distribution to regression analysis. The term, "regression" is first used by Sir Francis Galton (1822-1911).

1901   Initial publication of the journal, *Biometrika*. Through publications in this journal, biometrical methods are developed due to Karl Pearson (1857-1936), Sir Ronald Fisher (1890-1962), and later by Jerzy Neyman (1894-1981). Pearson develops the chi-square statistic, first introduced by J. J. Bienayme (unknown) in 1858.

1933   Sir Ronald Fisher publishes *Design of Experiments* which introduces the concepts of experimental design and the analysis of variance.

1950   Abraham Wald (1902-1950) publishes *Statistical Decision Functions*, a comprehensive treatment of statistical decision theory.

Statistical Methods
For Business and Economics

Statistical Methods
For Business and Economics

Roger C. Pfaffenberger
Texas Christian University

James H. Patterson
Indiana University

Third Edition

1987

IRWIN

Homewood, Illinois   60430

The Irwin Series in Quantitative Analysis for Business

Consulting Editor

Robert B. Fetter
Yale University

PHOTO CREDITS
pp. 115 and 116: Photos of 1963 Corvettes courtesy of Automobile
Quarterly Publications, Princeton, NJ. From *Corvette: America's
Star-Spangled Car* by Karl Ludvigen. Photos by Rick Lenz.

p. 115: Photo of Corvette seats courtesy of Motorbooks International,
Illustrated Corvette Buyer's Guide by Michael Antonick. Photo by
Ed Olson.

ISBN 0-256-03664-0

Library of Congress Catalog Card No. 86–82595

Printed in the United States of America

1 2 3 4 5 6 7 8 9 0 V 4 3 2 1 0 9 8 7

Preface

In most disciplines, including business and economics, there is an increasing awareness of the need to comprehend basic statistical methodology and its language for successful careers in industry, government, and self-employed positions. Today, with the widespread use of computers (mainframes and now microcomputers), the collection and analysis of data for decision making are, for many enterprises, essential activities for efficient operations. In a typical organization, perhaps only a few individuals are directly involved in the collection, tabulation, and the analysis of data, but many others must have a sufficient understanding of statistics to *interpret* the analyses in order to make reasonable and timely decisions.

Because statistical methods are now a part of numerous non-mathematics curricula, it is no longer possible, or indeed appropriate, to assume that students who study introductory statistics are well trained mathematically. The demand for statistical training has increased so dramatically in the past several years and has affected so many disciplines that most introductory courses *cannot* require more than a knowledge of algebra on the part of the student. We believe that most students taking an introductory course do not require the level of knowledge that would enable them to pursue training for a career in statistics. Rather, they need to understand the fundamental ideas of statistical analysis presented in a competent but not necessarily mathematically rigorous fashion. This is the essential characteristic of this text. By suggesting the potential importance that statistical analysis has in various careers, we hope to further motivate students in their study of statistics. Furthermore, we have emphasized the development of the logic of statistical procedures without resorting to mathematical derivations or proofs. We firmly believe that the extensive use of examples, the careful discussion of assumptions required to implement statistical procedures, and the proper interpretation of results can competently and effectively replace a mathematically rigorous development of statistics in an introductory course.

The text is designed to be quite flexible. A typical one-semester course would cover Chapters 1 through 10, the core of classical statistical inference, and perhaps two or three additional chapters that may be selected to satisfy specific course objectives (e.g., the regression chapters—13, 14, and 15). The accompanying chart indicates the interdependence of the chapters.

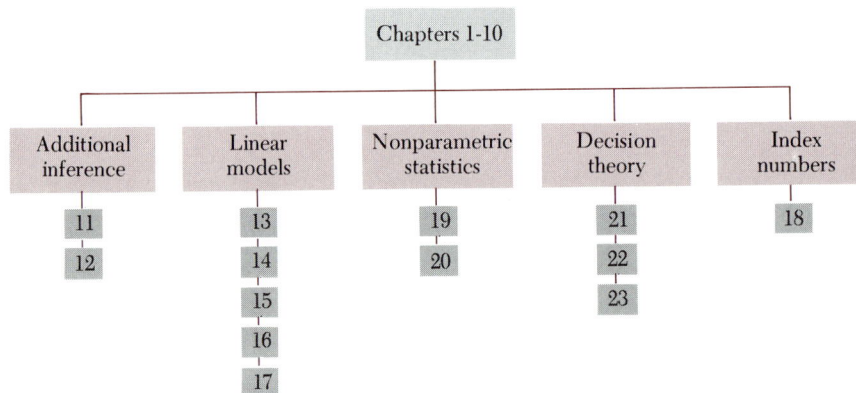

Chapters 1-10				
Additional inference	Linear models	Nonparametric statistics	Decision theory	Index numbers
11	13	19	21	18
12	14	20	22	
	15		23	
	16			
	17			

In a two-semester sequence, we suggest Chapters 1 through 10, the linear models chapters (Chapters 13 through 17), including analysis of variance (Chapter 12), and a selection of two or three additional chapters. We have included what we feel are the most important special statistical methods in this text. Nonparametric statistical methods (Chapters 19 and 20) are becoming increasingly important in business and social science applications, where, all too often, the data do not support the classical statistical method assumptions. Forecasting (Chapter 17) also has become an increasingly used statistical tool in business and economic applications in recent years. Most financial institutions depend heavily on the development of accurate forecasts for their success. Statistical decision theory (Chapters 21–23) is finding an important place in business applications of statistics, where prior information relevant to a current problem is available and may be combined with current information for decision-making purposes.

The third edition contains several significant changes:

- The number of problems at the ends of the chapters has been increased from approximately 600 to 1,075. In addition to the increase in the number of problems, many realistic examples have been added to the problem sets.

- MINITAB is used throughout to illustrate the use and interpretation of statistical software packages to perform statistical analyses.

- The descriptive statistics chapter (Chapter 2) has been rewritten to include new material on the use of graphical techniques to present statistical information and to illustrate the use of MINITAB in describing data sets.

- The regression chapters (Chapters 13–15) have been extensively rewritten. The multiple regression chapter (Chapter 15) has new sections on multicollinearity, model development, and an extensive case example. The statistical software packages MINITAB and SAS are used to illustrate the computer analysis of regression data.

- The hypothesis testing chapter (Chapter 10) has been rewritten to improve the flow of the presentation and readability.

- Advanced material has been placed in separate sections. These sections may be skipped by the instructor without loss of continuity in the presentation of the material.

- Margin notes have been used to indicate key definitions, concepts, and terms to students.

- The use of calculus has been relegated to the advanced sections and to appendixes at the end of chapters. By doing so, the instructor can choose to cover or not to cover this material, as desired to meet course objectives.

The text retains important characteristics of the second edition. Most notably, the concept of a random variable is emphasized by distinguishing among a population parameter (usually denoted by a Greek letter such as μ, σ, π, and β), a random variable or an *estimator* of a population parameter (denoted in boldface such as **X**, **S**, and **P**), and the *estimate* of the parameter obtained in a specific sample (denoted by lowercase letters such as x, s, and p). This notation emphasizes the notion of a random variable being a *function*, and that what is obtained after sampling is an estimate, which is a *value* of the function or random variable. This distinction is used throughout the text after its development in Chapter 5. For those who do not wish to emphasize random variables in the teaching of statistics at the introductory level, this distinction, for the most part, can be ignored. But, we have found that several instructors desire to highlight this distinction. Therefore, we continue to use this notation to facilitate the teaching of the difference between a random variable and its value. In addition, p-values or probability values continue to be used throughout the text after their introduction in Chapter 10.

An instructor's manual is available which gives complete solutions to all problems at the end of the chapters and provides helpful guidance to the instructor on how to present the text material. The instructor is encouraged to use a statistical software package to minimize the computational burden on the student while adding realism to the analysis of data sets. We strongly recommend MINITAB, a statistical programming language that has proved over time to be easy for students to use and very effective as an aid in illustrating statistical concepts. Alternatives are SAS (Statistical Analysis

System), SPSS (Statistical Procedures for the Social Sciences), or any of the other fine software statistical systems available. If microcomputers are available for student use, then several additional packages now are available such as SYSTAT, ABSTAT, and P-STAT. These packages, as well as others, are referenced at the end of Chapter 2.

■
A note to the student

The competent interpretation of statistical data and analyses is possible only with an understanding of basic statistical concepts. All research that involves the collection of data inevitably leads to a statistical analysis of these data. A well-known research effort that has involved considerable statistical analysis in recent years is the study conducted under the authority of the Surgeon General of the United States to assess the effect of smoking on the health of an individual. While this study has received notoriety because of various interpretations of the results reported, it is in no way unique in the sense of using *statistical analysis to interpret data*. Today, the applications and reports of statistical data abound, and we are frequently called upon to draw *our own* conclusions based on the analysis of the reported results. It is not surprising, therefore, to find that many disciplines have added introductory statistics courses to their undergraduate and graduate curricula.

It is the authors' view that statistics is best learned on first exposure by presenting an example of the application of the statistical concept immediately following its introduction. Accordingly, most discussions of statistical techniques and concepts are followed by at least one example (with solution) of an application of the technique or concept introduced. Although the inclusion of these example problems and solutions has made the text somewhat longer than several introductory statistics texts, we believe that the increased exposure to the many actual problem solutions warrants the additional length. The student is encouraged to attempt to solve several of the examples given in the various chapters and then compare results with those presented. This "self-test" feature of the text should provide students rapid feedback on their understanding of the material presented and therefore lessen the time required for the mastery of the concepts involved. We have also highlighted important definitions, theorems, rules, and so on in boxes and with margin notes. In addition, where possible, summary computational formulas have been provided at the end of selected chapters.

To assist the student in doing the exercises at the end of each chapter, we have provided the answers to selected end-of-chapter problems in the back of the text. Consistent with our belief that statistics can be better learned "by doing," the student is encouraged to work as many problems at the end of each chapter as possible.

There is a workbook prepared by Dr. Jack Hayya, Pennsylvania State University, that accompanies this text. The workbook presents many solved problems using the concepts and techniques introduced in the text. Since reviewing solved problems is an excellent learning device, students are encouraged to consider purchasing the workbook. This is particularly true for students who may consider themselves deficient in mathematics or have anxiety over taking a quantitative course. There is additionally a Student's Solution Manual that may be purchased at your college or university bookstore. This manual provides complete solutions with explanations to every other even-numbered problem at the end of the chapters in the text. The answers to all even-numbered problems at the end of the chapters are given in the back of the text.

Finally, students are encouraged to use computers and statistical software in solving text problems. Your instructor most likely will provide you with the information needed to use statistical software on your college or university mainframe computers or microcomputers. Statistics and computers are inseparable. The analysis of data, particularly large data sets, is not feasible generally without computers. In most business situations, answers are needed quickly in "real time." Computers make this possible. It has been predicted that the typical manager or executive will have a microcomputer at his or her desk as standard equipment. In many corporations, this is already true today. The ready access to data enabled by management information systems and microcomputers increases the interest and demand to analyze these data statistically for decision making. It would appear in the future that all managers and executives in business must be "literate" in computers and in statistical analysis and interpretation to ensure professional success.

Acknowledgments

The authors wish to thank the many students and instructors whose comments about the second edition aided significantly the development of this third edition of the text. Prior to developing the revised manuscript for the third edition, the second edition was carefully and thoroughly evaluated by four reviewers. We wish to express our appreciation for these competent and constructive reviews to: Michael Broida, Miami University; Timothy Ryan and Philip Jeffress, University of New Orleans; and Warren Boe, University of Iowa. Six reviewers evaluated the third edition manuscript during its development. We are grateful to these reviewers for their thoughtful and thorough suggestions for the improvement of the third edition manuscript. We took these suggestions seriously, but could not of course follow all of them. Our special thanks go to these reviewers: Martin Puterman, University of British Columbia; Ralph St. John, Bowling Green State University; Peter Rossi, University of Chicago; Ralph Miller, California State Polytechnic University-Pomona; Edna White, Texas A&M University; and Robert Fetter of Yale University.

Roger Pfaffenberger
James Patterson

Contents

Statistical Methods
For Business and Economics

1 Introduction

1.1 Introduction to statistical problems

Statistics can be retrieved instantly. . . .

Latest government statistics show. . . .

The top 20 percent of Fortune 500 companies agree. . . .

The average annual return to investors is. . . .

The tables provide 32 columns of data, which. . . .

Comparative annual operating costs indicate. . . .

Median housing costs for Washington, D.C., are. . . .

Surveyed companies responded. . . .

The phrases listed above were all found in a recent edition of a popular weekly business periodical. They illustrate the extent to which we are bombarded with numbers. More important, the use of such phrases demonstrates that we are constantly asked to meaningfully relate these numbers to our lives and business activities. Given this fact, we are faced with some significant questions:

1. How are these numbers collected?

2. How are they processed into neat little graphs and tables?

3. What do they mean?

4. How can we use them?

The science of answering these and many other questions related to the use of numerical data is called "statistics."

Used in the context of its original meaning, statistics generally refers to information about an activity or a process that is expressed in numbers listed in tables or illustrated in figures. But, since its early connotation, statistics has grown to encompass a larger role than presenting us with charts, graphs, and tables or figures. In a modern setting, statistics refers to the science of collecting, presenting, and analyzing data. A statistician is a person who engages in one or more of the following tasks: (1) the clerical activities of tabulating, summarizing, and displaying statistical data; (2) analyzing data by using statistical methods, usually for the purposes of decision making; or (3) advancing the science of statistics by developing new and better analysis methods. The level of expertise required by statisticians ranges from mastering simple clerical operations with data to advanced training in applied mathematics, and statisticians are needed at all levels.

Governments, businesses, and individuals collect statistical data required to carry out their activities efficiently and effectively. The rate at which statistical data are being collected is staggering and is primarily due to the realization that better decisions are possible with more information and, perhaps more importantly, to technological advances that have enabled the efficient collection and analysis of large bodies of data. The most important technological advance in this area has been the development of the electronic digital computer. More recently, of course, the technology of microcomputers has advanced to the point that many desk-top micros are capable of performing functions that only a mainframe could have been used to do a decade ago. Statistical concepts and methods, and the use of computers in statistical analyses, have affected virtually all disciplines—physics, engineering, economics, sociology, psychology, business, and others. In business and economics, the development and application of statistical methods have led to greater production efficiency, better forecasting techniques, and better management practices. It is becoming increasingly apparent that some knowledge of statistics and computers is essential for careers in economics, business, administration, and many other fields as well. To gain an appreciation for the breadth of applications of statistics to business and economic problems in particular, let us consider five examples.

Example 1.1 Many newspapers provide daily and/or weekly summaries of the New York Stock Exchange transactions. For example, on a particular day, we may find that 3,369,000 shares of Procter and Gamble were traded, with an average high for the past 52 weeks of 77⅝ and an average low for that period of 51⅝ (dollars per share). The value of a share is currently listed as 75½, with a change of +1½ from the closing value of the previous day. This information is illustrated in an excerpt from *The Wall Street Journal* for Friday, May 23, 1986, in Figure 1.1. Since it would not be feasible to list the values of all 3,369,000 shares traded during the day, values of the daily high, low, closing price, and net change statistics are used to summarize the entire set of transactions. The values of the statistics presented in these tables provide essential information for investment decisions.

FIGURE   1.1   Excerpt from a summary of daily New York Stock
Exchange transactions

52 weeks			Volume	Daily			
High	Low	Stock	(in 100s)	High	Low	Close	Change
28	14½	PrimeC	2860	21⅛	20⅜	20⅞	+⅜
42⅞	21	PrimM	355	40	39¾	40	+⅛
77⅝	51⅝	ProctG	3369	75½	74⅜	75½	+1½
17¼	11¼	PrdRs	400	15¾	15⅛	15¾	+⅝
43½	33½	Proler	92	36	34	36	+1¾

Source: *The Wall Street Journal,* May 23, 1986.

Example 1.2   Annually, the Bureau of the Census publishes the *Statistical Abstract of the United States,* which contains economic and business data on a variety of subjects. The data are compiled from the U.S. census of the population (conducted every 10 years), annual samples drawn from the U.S. population by the bureau, surveys conducted by other government agencies, and government-supported research projects conducted by private organizations.

In Table 1.1, extracted from the 1986 *Statistical Abstract,* statistics relating to the receipts and outlays of the federal government are presented. Although the sources and functions are listed only by major sources, the table contains valuable information for business decision making. For example, the increase in corporate income taxes from 37.0 billion dollars in 1983 to 66.4 billion dollars (estimated) in 1985 may dictate how a company transacts its business. With the new tax laws on the horizon (perhaps in place during this edition) that further increase corporate tax liability, the increasing trend suggests that companies must pursue new accounting devices to minimize tax liability.

A table such as Table 1.1 conveys a substantial amount of information in a very useful and concise format. Note the steady increase in the budget deficit, from $2.8 billion in 1970 to $222.2 billion (estimated) in 1985. The outlay for national defense is interesting, for example. The outlay rose from $81.7 billion in 1970 to $253.8 billion (estimated) in 1985. However, in 1970 the outlay for national defense represented 41.77 percent of the budgeted outlays, whereas in 1985 the percentage dropped to 26.46 percent (estimated). Notice the increase in net interest to service the budget deficit—it rose from $14.4 billion in 1970 to $130.4 billion (estimated) in 1985. Thus 13.60 percent (estimated) of the budget outlay in 1985 went to pay the *interest* on the money borrowed by the federal government to cover the budget deficit!

Example 1.3   In operations management, a primary concern is controlling the quality of the items being produced. If the product is a transistor radio battery, for example, we may be concerned with the longevity of the

TABLE 1.1 | Federal receipts by source and outlays by function: 1970 to 1985 (in $ billions)

Source or function	1970	1975	1978	1979	1980	1981	1982	1983	1984	1985, est.	Percent distribution 1970	1985
Surplus or deficit (−)	**−2.8**	**−53.2**	**−59.0**	**−40.2**	**−73.8**	**−78.9**	**−127.9**	**−207.8**	**−185.3**	**−222.2**	(×)	(×)
BY SOURCE												
Total receipts	**192.8**	**279.1**	**399.7**	**463.3**	**517.1**	**599.3**	**617.8**	**600.6**	**666.5**	**736.9**	**100.00**	**100.00**
Individual income taxes	90.4	122.4	181.0	217.8	244.1	285.9	297.7	288.9	296.2	329.7	46.89	44.74
Corporation income taxes	32.8	40.6	60.0	65.7	64.6	61.1	49.2	37.0	56.9	66.4	17.01	9.01
Social insurance taxes and contributions	44.4	84.5	121.0	138.9	157.8	182.7	201.5	209.0	241.7	268.4	23.03	36.42
Employment taxes and contributions	39.1	75.2	103.9	120.1	138.7	163.0	180.7	185.8	211.9	238.1	20.28	32.31
Unemployment insurance	3.5	6.8	13.9	15.4	15.3	15.8	16.6	18.8	25.1	25.6	1.82	3.47
Contributions for other insurance and retirement	1.8	2.6	3.2	3.5	3.7	4.0	4.2	4.4	4.6	4.7	.93	.64
Excise taxes	15.7	16.6	18.4	18.7	24.3	40.8	36.3	35.3	37.4	37.0	8.14	5.02
Estate and gift taxes	3.6	4.6	5.3	5.4	6.4	6.8	8.0	6.1	6.0	5.6	1.87	.76
Customs duties	2.4	3.7	6.8	7.4	7.2	8.1	8.9	8.7	11.4	11.8	1.24	1.60
Miscellaneous receipts	3.4	6.7	7.4	9.3	12.7	13.8	16.2	15.6	17.0	18.0	1.76	2.44
Federal Reserve earning deposits	**3.3**	**5.8**	**6.6**	**8.3**	**11.8**	**12.8**	**15.2**	**14.5**	**15.7**	**16.4**	**1.71**	**2.23**
BY FUNCTION												
Total outlays	**195.6**	**332.3**	**458.7**	**503.5**	**590.9**	**678.2**	**745.7**	**808.3**	**851.8**	**959.1**	**100.00**	**100.00**
National defense	**81.7**	**86.5**	**104.5**	**116.3**	**134.0**	**157.5**	**185.3**	**209.9**	**227.4**	**253.8**	**41.77**	**26.46**
International affairs	**4.3**	**7.1**	**7.5**	**7.5**	**12.7**	**13.1**	**12.3**	**11.8**	**15.9**	**19.6**	**2.20**	**2.04**
Income security	**15.6**	**50.2**	**61.5**	**66.4**	**86.5**	**99.7**	**107.7**	**122.6**	**122.7**	**127.2**	**7.98**	**13.26**

Category												
Health	5.9	12.9	18.5	20.5	23.2	26.9	27.4	28.6	30.4	33.9	3.02	3.53
Social Security and medicare	36.5	77.5	116.6	130.6	150.6	178.7	202.5	223.3	235.8	257.4	18.66	26.84
Veterans benefits and services	8.7	16.6	19.0	19.9	21.2	23.0	24.0	24.8	25.6	26.9	4.45	2.80
Education, training, employment, and social services	8.6	16.0	26.7	30.2	31.8	33.7	27.0	26.6	27.6	30.4	4.40	3.17
Commerce and housing credit	2.1	9.9	6.3	4.7	9.4	8.2	6.3	6.7	6.9	6.0	1.07	.63
Transportation	7.0	10.9	15.5	17.5	21.3	23.4	20.6	21.3	23.7	27.0	3.58	2.82
Natural resources and environment	3.1	7.4	11.0	12.1	13.9	13.6	13.0	12.7	12.6	13.0	1.58	1.36
Energy	1.0	2.9	8.0	9.2	10.2	15.2	13.5	9.4	7.1	8.2	.51	.85
Community and regional development	2.4	4.3	11.8	10.5	11.3	10.6	8.3	7.6	7.7	8.6	1.23	.90
Agriculture	5.2	3.0	11.4	11.2	8.8	11.3	15.9	22.9	13.6	20.2	2.66	2.11
Net Interest	14.4	23.2	35.4	42.6	52.5	68.7	85.0	89.8	111.1	130.4	7.36	13.60
General purpose fiscal assistance	.5	7.2	8.4	8.4	8.6	6.9	6.4	6.5	6.8	6.6	.26	.69
General science, space, and technology	4.5	4.0	4.9	5.2	5.8	6.5	7.2	7.9	8.3	8.7	2.30	.91
General government	1.8	3.2	3.6	3.9	4.4	4.6	4.5	4.8	5.1	5.8	.92	.60
Administration of justice	1.0	3.0	3.8	4.2	4.6	4.8	4.7	5.1	5.7	6.7	.51	.70
Undistributed offsetting receipts	-8.6	-13.6	-15.7	-17.5	-19.9	-28.0	-26.1	-34.0	-32.0	-32.3	-4.40	-3.36
Off-budget Federal entities	—	8.1	10.4	12.5	14.2	21.0	17.3	12.4	10.0	12.5	(×)	1.30

Source: U.S. Bureau of the Census, *Statistical Abstract of the United States: 1986*, 106th ed. (Washington, D.C.: U.S. Government Printing Office, 1986).

batteries. Suppose it is desired that at least 95 percent of the batteries last through at least 20 hours of continuous use. The actual percentage of batteries lasting more than 20 hours could be determined by inserting every battery produced into a transistor radio and recording its time to failure, but then there would be no batteries to sell. Rather, a manager may decide in a day's production to pull every 100th battery off the production line, insert the sample batteries in electric test circuits, and record their times to failure. The percentage of these batteries lasting through more than 20 hours of continuous use could be used to estimate the percentage of all batteries produced during that day that will last more than 20 hours. Moreover, if this estimated percentage drops much below 95 percent (say, to 80 percent), the manager may wish to stop the production line until she can determine why the percentage of bad batteries appears to be greater than the tolerated 5 percent. The manager is using the value of a percentage statistic computed from a sample of all batteries produced to arrive at a decision regarding the quality of the set of all batteries produced on a given day.

Destructive sampling

This example illustrates a common phenomenon in quality control: *destructive sampling*. It is impossible to test the quality (longevity) of each battery produced because the test for longevity will ordinarily involve its destruction. The manager has little recourse but to sacrifice a small number of batteries (the *sample*) in order to gain information about the entire set of batteries composing the daily production (the *population*).

Example 1.4 Determining the marketability of a new product is a constant problem posed to many marketing research groups. To determine whether or not a new kitchenware product will sell, the marketers might conduct a house-to-house survey of 1,000 households selected randomly in the product target areas, during which they present the product to the homemaker for evaluation. The percentage of the homemakers willing to buy the product at its listed price, together with other information obtained from the interviews, could be used to decide whether or not the new item should undergo full-scale production.

Example 1.5 Politicians and their supporters are immensely interested in knowing their prospects of winning an election as the campaign heads toward final balloting. By sampling 1,000 registered voters prior to the election, the percentage who claim they will vote for a given candidate may be used to estimate the percentage of the votes the candidate will receive in the election. The estimated percentage could be used to decide, for example, whether or not a greater campaign effort (more money) is required to assure the candidate's election.

There are many more examples in business and other areas that might be cited, but the preceding five should indicate the many ways in which statistics can be used. In the first two examples, statistics is used to *describe* large bodies of data. In this application, the word *statistic* is used to describe the value of a specific numerical measure, such as an average or a total, and the collection of the values of the statistics is used to summarize or condense a large set of numbers. These compiled values of the statistics may in turn be used to assist in decision making. In the last three examples, statistics may be interpreted in a much broader sense—namely, the process of drawing conclusions about an entire *population* or collection of things based on a *sample,* a subset of the population or collection.

Most students probably view statistics in the context of the first two examples—that is, as tables of figures, charts, and graphs (batting averages, pie charts illustrating the sources of government revenue, and so on). This concept is called *descriptive statistics* and was at one time the principal use of statistics in business. Currently, there is an increasing interest in the methods and uses of *inferential statistics*—the process of drawing inferences about the whole (the *population*) from a subset of it (the *sample*), as exemplified in Examples 1.3–1.5. Schematically, the process of drawing inferences about an unknown population numerical quantity (the proportion of defectives in a production lot, mean income of a class of laborers, etc.) is illustrated in Figure 1.2. Units are selected from the population to form the

| Descriptive statistics |

| Inferential statistics |

FIGURE 1.2 | The statistical inference process

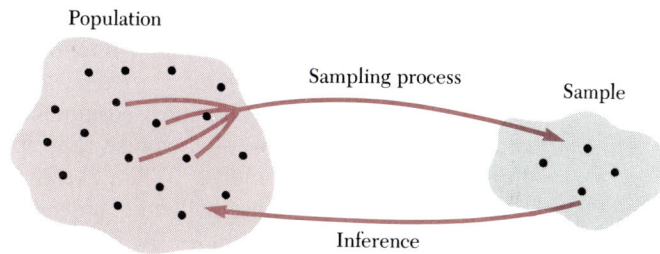

sample, which in turn is used to draw inferences about the population characteristic of interest. Much of this text is devoted to the study of statistical inference. In subsequent sections of this chapter, we will focus attention on the sources of data, the methods of obtaining data, and data measurement considerations.

■ **1.2**

Sources of data

| Internal and external data |

| Primary data |

In many applications of statistics, businesses use *internal* data—that is, data arising from bookkeeping practices, standard operating business procedures, or planned experiments by research divisions within the company. Examples are profit and loss statements, employee salary information, production data and economic forecasts. Occasionally, it may be necessary or desirable to use sources of *external* data. By external, we mean sources of data outside the firm. External data may be of two types: *primary* data and *secondary* data. By primary data, we mean data obtained from the organization that originally collected them. An example is the population data collected by and available from the U.S. Bureau of the Census. Secondary data come from a source other than the one that originally collected them. The federal budget information used in Example 1.2 was published by the Bureau of the Census, but originated from the Bureau of the Budget. Ordinarily, if external data must be used, it is recommended that primary data be sought since it will not have undergone any "refining" by the secondary source.

Secondary data

In the election survey in Example 1.5, the *Statistical Abstract* provides numerous tables of both primary and secondary data, such as past voting records in districts and numbers of registered Democrats and Republicans, which may supply important information in conjunction with the internal sampled data on estimating a candidate's probability of being elected.

Sources of data

There are many excellent sources of published (primary and secondary) data compiled by the state and federal government, by business and economic associations, and by commercial sources (periodicals). Some examples are: *The Statistical Abstract of the United States* (published annually by the Bureau of the Census), *Survey of Current Business* (published by the Department of Commerce), *Monthly Labor Review* (published by the Bureau of Labor Statistics), *Harvard Business Review* (periodical), *Business Week* (periodical), *The Wall Street Journal* (periodical), *Dun's Review* (periodical), and *The Journal of Management Science* (association journal). Additional sources of external data that are available in most reference libraries are *The Economic Almanac, Federal Reserve Bulletin, Life Insurance Fact Book, International Financial Statistics,* and *Business Conditions Digest.*

Caution must always be exercised in using external sources of data, particularly secondary sources, because they may contain errors in transcription from the primary source. When external data are used, the conditions under which the data were collected and summarized must be determined to assure that they are relevant for the intended use. This determination usually requires identifying and locating the primary source, which typically will discuss any restrictions placed on the data due to the process of their collection. Thus, although secondary sources of data are convenient, it usually is prudent to seek out and use primary sources of external data.

■ 1.3
Statistical terminology

When statistical data are collected and analyzed, it is usually in the context of *populations* and their *characteristics*.

Population

Definition 1.1
Population and population characteristic

A *population* is the totality of units under study. A *population characteristic* is an attribute of a population unit.

Population characteristic

We may be interested, for example, in the salaries of workers in a particular industry. If so, the population is the totality of these workers and the characteristic of interest is each worker's salary. In collecting the salary data, we may be interested in other population characteristics as well, including sex, age, educational level, and other information. In general, a population unit may have one or more characteristics of interest in a particular study.

As another illustration of a population, a firm may be interested in the proportion of defective units of a certain product that it has produced in a large lot stored in a warehouse. The population is the totality of units in the warehouse, and the characteristic is the acceptability of each unit of the product—it is either defective or nondefective.

Suppose we are interested in information about a population characteristic. There are two ways we can collect data to study the population characteristic of interest: inspect all the population units or inspect a portion of the population units. When the data are produced by measuring the population characteristic for *each and every* unit in the population, we say that a *census* of the population has been taken.

Population census

Definition 1.2
A population census

A *population census* is the evaluation of each and every unit in the population under study.

In some situations it is possible to take a complete census of the population. This rarely occurs in business unless the population size is very small, due to cost and time considerations. A census of the U.S. population is undertaken every 10 years, and it is truly a Herculean effort, paid for, naturally, by the taxpayers. The U.S. census produces a wealth of data of considerable importance to the federal government and to firms and institutions, many of whom view the census as an important source of external data.

Of course the U.S. census taken every 10 years is not a *true* census—not everyone in the U.S. population is contacted! In recent years the Bureau of the Census has turned to sampling techniques to *estimate* U.S. population statistics and no longer attempts to take a true census.

In most instances, it is impossible to take a census of a population. It may be too costly, too time-consuming, not feasible (such as a true U.S. census), or the evaluation process may destroy the population unit as in Example 1.3.

Sample

> **Definition 1.3**
> A sample
>
> A *sample* is a part of a population in which the population characteristic is studied so that inferences may be made from the sample study about the entire population.

A classic example of a situation in which samples must be used rather than a census taken is *destructive sampling,* in which the process of evaluating a unit of the population destroys or irrevocably damages that unit. For example, suppose a tire manufacturer wishes to claim its new radial tire will last 40,000 miles or more. To support this claim, a sample from all tires produced (the population) is selected for testing to determine how many miles the tires will last. Since testing destroys the tires, a complete census of the population is impossible.

The advantages of sampling over taking a census are rather obvious. A sample is less expensive than a census; it can produce data more quickly; and the data are often more reliable because more time can be spent studying each sampled unit. But there clearly is a price to be paid as well. By looking at only a portion of the population, we are subject to committing errors because the sample may not be representative of the whole population.

Sampling error

> **Definition 1.4**
> Sampling error
>
> *Sampling error* is the difference between (1) studying a sample and inferring a *result* about a population characteristic and (2) determining the *result* by taking a census of the population and evaluating the population characteristic.

As an illustration of sampling error, suppose we are interested in the average salary (the characteristic) of unionized workers in a specific industry and we know from union membership lists that there are 1,000 workers in the industry (the population). If we had taken a census, we might have found that the average salary is, let us say, $25,000 (the result by taking the population census). Based on a sample of, say, 100 workers, we may find that the sample average salary is $27,200 (the result by taking the sample). The difference between these two figures—$2,200—is the sampling error, if nonsampling errors have not occurred.

Errors in acquiring and tabulating statistical data can arise in other ways as well, and these errors are called *nonsampling errors*.

Nonsampling error

> **Definition 1.5**
> Nonsampling error
>
> *Nonsampling errors* are errors that occur in acquiring, recording, or tabulating statistical data that cannot be ascribed to sampling error. They may arise in either a census or a sample.

Nonsampling errors are usually more difficult to control and detect than sampling errors. Suppose we are acquiring data on the 1,000 unionized workers mentioned earlier. If we approached a particular worker and asked for his or her income, we could be lied to—a troublesome and frequent source of nonsampling error when a sensitive question is asked directly of a person. (What is your grade-point average?) In some instances, a person may give a false response out of ignorance rather than by design. Another source of nonsampling error is in recording that data. A "7" may be written as a "9," the decimal point may be incorrectly placed, and so on. Errors may also occur in tabulating the data—keypunching errors in preparing computer input and typing errors in transcribing data, for instance. It is always necessary to edit data carefully to minimize the chance of nonsampling errors adversely affecting the statistical analysis of the data.

When referring to numbers, such as the average salary of the union workers in a certain industry, to avoid confusion it is necessary to use terminology that indicates whether the number refers to a population or to a sample. If a number or numerical measure refers to a population, then we call it the value of a population *parameter*. If a number or numerical measure refers to a sample, then we call it the value of a sample *statistic*.

Parameter

> **Definition 1.6**
> A parameter
>
> A *parameter* is a numerical measure of a population characteristic.

Statistic

Definition 1.7
A statistic

A *statistic* is a numerical measure calculated from a set of sample observations.

In the union worker illustration above, the average salary of the population of 1,000 union workers is a value of a *parameter*. The average salary of a sample of 100 of these workers represents a value of a *statistic*. In the next chapter, we will describe numbers (numerical measures) calculated for population and sample sets of data such as averages and medians. The discussion of the concepts of a parameter and of a statistic will then be continued in Chapter 8.

The identification of the units in a population under study can often be a surprisingly difficult task. We refer to a listing of population units as a *frame*.

Population frame

Definition 1.8
Population frame

The listing of all units in the population under study is called the *population frame*.

If the population is a production lot of units stored in a warehouse, production records will give us a listing of the serial numbers of the units from which each unit may be identified. If the population is the 1,000 unionized workers in a specific industry, union membership records may serve as a frame. But what about a frame for all persons who will vote in a particular election? A listing of registered voters is not appropriate, because in many elections less than 50 percent of the registered voters actually vote. Some classic errors have been made in identifying the frame. In the 1948 Dewey-Truman presidential race, one survey prior to the election used telephone directories as the frame. Random telephone numbers were selected and called; the person was asked whether he or she was a registered voter and intended to vote on Election Day. As it turned out, not everyone owned a telephone in 1948 (incredible as that may seem today), and a preponderance of those who did were Republicans. The survey came to the incorrect conclusion that Dewey would be elected president of the United States.

■ 1.4

**The acquisition
of data: Surveys
and experiments**

When internal or external data are not readily available or are incomplete in a study attempting to answer questions about a population characteristic, a *survey* or an *experiment* may be conducted to provide the required information.

Statistical survey

Definition 1.9
Statistical survey

A *survey* is a process of collecting data from existing population units, with no particular control over factors that may affect the population characteristics of interest in the study.

Most of us are very familiar with surveys. As students, we are asked about our opinions regarding dining hall food, impending tuition hikes, teaching effectiveness, and so on. Filling out survey questionnaires or answering an interviewer's questions has become a routine occurrence in most of our lives. To better understand our definition of a survey, suppose we are interested in acquiring data on the salaries of 1,000 unionized workers in a specific industry. The population characteristic "salary" may be affected by a host of factors—age, race, sex, educational level, etc. As we elicit a particular worker's salary by a survey, we have no control over educational level, age, and so on—these are existing attributes of the worker.

In contrast to a survey is a statistical *experiment* in which we do exercise control over factors that may affect the population characteristics of interest.

Statistical experiment

Definition 1.10
Statistical experiment

An *experiment* is a process of collecting data about population characteristics when control is exercised over some or all factors that may affect the characteristics of interest in the study.

We may be interested, for example, in the yield of a chemical process that is affected by temperature and pressure. A variety of settings for temperature and pressure could be selected, and the chemical process run for each setting to determine the yield. In this way, the joint effect of temperature and pressure on yield is studied in a controlled manner.

In management, we may be interested in the effects of a training program on the first-year performance of new employees. A set of new employees may be split into two groups such that both groups are approximately alike in terms of age, sex, education, and other factors. The training program could be administered to one group and not to the other (the *control* group). At the end of the first year, performance characteristics could be measured to assess the effects of the training program, accounting for factors other than the training program that may affect performance.

Experiments almost always provide better information than do surveys, but both are extremely important and useful tools for acquiring data. Though an experiment is preferred to a survey, much of the data used in statistical analyses in business and economics are survey data. There are a number of reasons for this. First, most internal and external data are collected by surveys. Second, it is not always possible to conduct an experiment to acquire the needed information. An interesting example of this is the effect of smoking on health. Virtually all data on the relationship between smoking and health are survey data; other factors that may affect health, such as age, race, sex, and physiological properties, are not under the control of those collecting the data. To run an experiment in this case would involve controlling people's lives. Some people in the experiment would be required to smoke, and others would not. It is neither feasible nor desirable to approach the acquisition of the data for the study of the relationship between smoking and health in this way.

| Steps in planning a survey or an experiment |

The planning of a survey or an experiment is essential to ensure that the resulting information will be useful. A good plan usually involves the following steps (these steps are applied to the quality control problem in Section 1.1):

1. **A clear and detailed statement of the problem.** The statement of the problem should clearly indicate that we are interested in determining whether or not the percentage of good batteries (those lasting through 20 hours) exceeds a specified number (95 percent). The population is composed of all batteries produced during a chosen period of time (a day or a week), and the characteristic in the population of interest is the number of hours the battery will continuously operate before failure.

2. **A decision to survey or to experiment.** In this problem it is possible to answer the question about the population by experimentation. We may test each selected battery under the set of conditions in which it was designed to operate.

3. **A decision to take a census or a sample.** This is a case of destructive sampling. In determining the proportion of batteries that will last 20 hours or more, the tested batteries are "spent." We must therefore take a sample of batteries.

4. **Designing the survey or experiment.** The experiment must be designed so that we isolate the characteristic of interest—the lifetime of the bat-

tery. Test circuits must be constructed and carefully monitored when the selected batteries are inserted for testing.

5. **Collecting and analyzing the data.** For each battery, the time to failure is recorded, and the proportion of batteries lasting 20 hours or more is calculated.

6. **Reaching conclusions about the population characteristics.** The sample proportion of batteries surviving 20 or more hours is used as an estimate of the population proportion that survives 20 or more hours.

7. **Reporting the results.** The report should include a thorough description of the problem, the sampling design, the testing method, and the inferences. Sufficient monies should be allocated for a competent writing of the report. Indeed, many companies employ technical writers to put into ''laymen's'' words the experimental results.

The manner in which a sample is drawn, the methods of analyzing statistical data, and the kinds of inferences that may be drawn from the analysis (steps 4, 5, and 6) are major topics in this text. It is important not to minimize the other steps, particularly steps 1 and 7. A clear and detailed statement of the problem is essential in planning a survey or an experiment. And the best analysis of survey or experimental results is meaningless unless the analysis can be accurately and understandably reported.

■ **1.5**

Obtaining data

Once it has been determined that a survey or experiment is required, there are a variety of methods that may be used. The most difficult problems arise when gathering information from people in surveys, and the methods most relevant to this situation will be emphasized.

1.5.1 | Self-enumeration

Self-enumeration is probably the most common method of acquiring data from people in a survey or experiment. Questionnaires are usually distributed to selected individuals by mail, although the distribution mechanism depends to a large extent on the purpose and nature of the questionnaire. For example, if the purpose of the questionnaire is to survey the attitudes of those using public transportation, the questionnaire may be distributed to people while commuting to and from work on buses, subways, and trains.

The use of questionnaires suffers from two serious drawbacks. First, if the respondent has difficulty in interpreting the questions, no one is available for assistance. If this situation arises, the information received may contain a high degree of nonsampling error, or the respondent may become frustrated and not bother to complete or return the questionnaire. Furthermore, if a questionnaire is mailed to a household, it is often not clear who in the household responded to it. Second, questionnaires typically have an extremely poor response rate. It is not uncommon to have less than 30 percent returned on the first mailing of a questionnaire. The principal advantage of a questionnaire is the low cost relative to the other means of obtaining information. Most mail questionnaires may be bulk mailed at a reasonable rate. But it is almost always necessary to contact nonrespondents to the first mailing by subsequent mailings, telephone calls, or personal interviews, and these costs must be planned for in a well-designed self-enumeration survey or experiment. In most instances, those who do respond to the first mailing of a questionnaire are not representative of the entire population. To use only their responses would tend to bias the analytical results. Some self-enumeration questionnaires do enjoy high initial response rates. Examples are questions asked on warranty cards that must be returned to the manufacturer for warranty coverage of a new product and Census Bureau questions asked on federal income tax returns.

The federal government, particularly the Census Bureau, frequently uses mail questionnaires to acquire information necessary for economic forecasts and planning. The response rate to government questionnaires tends to be much greater than to those distributed by the private sector.

1.5.2 | Personal interview

In most situations, the best method of eliciting information from individuals is by a personal interview. The interviewer personally contacts individuals who are selected to participate in the survey or experiment. Responses are recorded on a *schedule* (a questionnaire form filled out by the interviewer).

Schedule

The personal interview method produces a higher response rate than self-enumeration questionnaires and allows the interviewer to clear up any misunderstandings about any questions on the schedule. But personal interviews are generally very expensive. Interviewers must be carefully selected and trained, and sufficient remuneration must be provided to ensure that the interviewer is competent and dedicated to the chore. It is always prudent in a personal interview survey to call a selected set of respondents to ensure that they were in fact contacted (as opposed to the interviewer filling fake responses), to ascertain if the interviewer's demeanor was appropriate, and to determine whether or not the interviewer may have biased responses by making gestures or comments when asking the questions or recording the responses.

Overall, the personal interview method of conducting an experiment or survey is the best way to acquire data when the population units are people, when the process is properly planned and executed, and if it can be afforded.

1.5.3 | Telephone interview

Occasionally, it is possible to conduct an interview over the telephone with the interviewer working from a schedule as in a personal interview. Polls to determine the most popular programs on television are frequently conducted in this manner. Telephone interviews are usually less expensive than personal interviews, but the response rate is lower, and fewer questions may be asked before the respondent tires of the proceedings. And not everyone owns a phone—even today.

■ 1.6

Constructing questionnaires and schedules

There are three steps in constructing and analyzing a questionnaire or a schedule: (1) designing the instrument, (2) conducting the pretest, and (3) editing the results. The construction of a questionnaire or schedule instrument is time-consuming and difficult. There is a natural tendency to rush through the construction of the instrument so that the data collection process can commence. But time spent in this stage of a well-planned survey or experiment is invariably found to be extremely valuable in retrospect.

The proper construction of questionnaires is a skill that is generally developed only by experience in the use of research methodology or by on-the-job training. We will discuss only some of the basic concepts concerning the construction of a questionnaire. For further information on this subject, see the Hansen et al. and Kish references listed at the end of this chapter.

1.6.1 | The design

There are basically three kinds of questions that may be asked: dichotomous, multiple choice, or free answer. In the *dichotomous* question, the respondent is asked to select one of two responses, usually yes and no. For example, in a transportation study, a worker may be asked:

Dichotomous question

> Did you drive a car to work this morning?
> Yes ( )   No ( )

The dichotomous question is simple and straightforward, and perhaps comes closest to decisions that respondents are used to making.

In the multiple-choice question, the respondent is asked to select one of a number of responses:

Multiple choice question

> What is the likelihood of your using the following services for preventive health care purposes in the next two years? (*a*) Dental checkup, (*b*) eye exam, (*c*) general physical.

	a	*b*	*c*
Extremely unlikely	( )	( )	( )
Unlikely	( )	( )	( )
Slightly unlikely	( )	( )	( )
Not certain	( )	( )	( )
Slightly likely	( )	( )	( )
Likely	( )	( )	( )
Extremely likely	( )	( )	( )

The multiple-choice question gives the respondent a greater range of responses to choose from, but it may also request a more qualified response than the respondent is prepared to make. For instance, a respondent may answer yes to the dichotomous question, "Will you have a physical this year?'' when it is not a certain event; good intentions are not always realized. Yet the respondent may not be able to conceptualize properly the assignment of a *likelihood* (slightly likely, likely, extremely likely) to the event, "I will have a physical this year." All too often, responses in situations like this lead to "end loading"—selecting the response that most closely approaches a simple yes-no response. In this case, the respondent would select the "extremely likely" response in place of yes if he or she were given the multiple-choice response format, for instance.

Free answer form

In the free answer form, the respondent is asked to answer a question in his or her own words in essay form. For example:

> What is your opinion of the dining hall food and service?

The difficulty with the free answer question is in classifying the responses. This not only may be difficult and somewhat arbitrary, but it is also time-consuming.

In most instruments, it usually is necessary to use all three types of questions to elicit the information required.

The order of the questions in the instrument can be extremely important. The questionnaire or survey should begin slowly with easily answered questions to develop rapport with the respondent. Respondents tend to "tie" questions together, and one particular ordering of questions may produce a different set of responses than another set for this reason.

The degree of directness of the questions is also important. If sensitive questions are asked directly, respondents may distort their answers. This invariably happens when a person is asked for her/his income. To elicit information about sensitive questions, indirect questions may be used. For example, we may ask the respondent to indicate a salary range among a set of ranges. Later, we may ask what proportion of the monthly income is spent on food, and much later, ask for average monthly expenditure for food. We may be able to determine a person's salary indirectly in this way better than by directly asking for income. At the very least, we have a consistency check to determine how reliable the responses are.

It is important that the questions are stated clearly and do not bias the results. Ideally, the question should have the same meaning to every respondent in the survey or experiment. And the questions should be relatively short. Bias may arise when leading questions or statements are used, such as:

> The food in the dining hall is rotten.
> Agree ( )   Uncertain ( )   Disagree ( )

Given to the typical college student, the response will invariably be, "Agree." A less biased statement might be, "The food in the dining hall is of acceptable quality."

1.6.2 | The pretest

The pretest is an essential step in constructing a questionnaire or schedule instrument. The instrument is given to a small number of respondents to determine whether or nor there are any problems with it. Almost always there are. There may be ambiguous questions, the ordering may require changing, and some questions may have to be asked in different forms. The time to identify difficulties with the instrument is before the full-scale survey or experiment is conducted—not after.

Furthermore, the information gathered during the pretest phase may be used to estimate quantities required for the proper planning of the statistical design of the experiment. We will return to the discussion of this aspect of the pretest in Chapter 9.

1.6.3 | Editing

The completed questionnaires or schedules must be carefully checked and edited for errors. Often it is possible to design questions that represent internal consistency checks for the respondent's answers. Finding recording, transcription, or clerical errors can be very tedious work, but it is necessary if the data are going to be of value in decision making.

Today, the computer is used extensively to edit data. Various computer-assisted techniques have been developed to identify "outliers"—responses that are greatly different from the majority of the responses. Many outliers result from recording, transcription, or clerical errors, or from false information provided by the respondent.

■ 1.7

Variables and scales of measurement

| Variables |

| Dependent and independent variables |

| Quantitative and qualitative variables |

| Nominal scale |

The characteristic of the population under study is called a *variable* if it can take on two or more different values among the population units. For instance, if we are interested in the incomes of workers in a particular industry, we may record other characteristics about the worker as well: age, race, level of education, and sex. In this instance the five characteristics—income, age, race, level of education, and sex—are variables in the survey or experiment.

Furthermore, we would call income a *dependent variable* and call the other four *independent variables* if we are concerned with how sex, age, level of education, and race affect income. Income is the basic variable of interest, and our interest in the other variables is in their influence on income.

If we are measuring a set of variables from a population, then determining which are dependent and which are independent variables is a function of the purpose of the survey or experiment. An independent variable in one study may be a dependent variable in another.

A *quantitative* variable is one that can be measured numerically, such as income and age. A *qualitative* variable is one that is nonnumeric, such as sex, race, and level of education (high school, college, graduate school, etc.).

In preparing data for analysis, we must be familiar with the four numerical scales of measurement: *nominal, ordinal, interval,* and *ratio*. The *nominal* scale applies whenever we use numbers only to categorize values of a variable. For instance, we could let a male be 1 and a female be 0, but this

Ordinal scale

numerical assignment is clearly arbitrary; a female could be assigned 100 and a male 0. The *ordinal* scale differs from the nominal scale in that the ordering of the numbers has meaning. An example is the responses to a multiple-response question:

Strongly disagree	Disagree	Uncertain	Agree	Strongly agree
−2	−1	0	+1	+2

Interval scale

Ratio scale

The numerical assignments of −2, −1, 0, 1, and 2 indicate the degree of agreement, but they could just as easily have been 0, 10, 100, 200, and 500, respectively. The key here is that although a +2 indicates stronger agreement than a +1, the difference in the strength of the respondent's conviction between +1 and +2 may not be the same as between 0 and +1. In the *interval* scale, the relative order of the numbers is important, but so is the difference between them. This scale uses the concept of unit distance such that the difference between any two numbers may be expressed as some number of units. The interval scale requires a zero point, but its location may be arbitrary. Good examples of interval scales are the Fahrenheit and Celsius temperature scales. Both have different zero points and unit distances. The principle of an interval scale is not violated by a change in scale or location or both. The *ratio* scale is used when the interval size is important and also the ratio between two numbers has meaning. By this we mean it is appropriate to speak of one number being, say, twice as big as another. This is clearly not possible with an interval scale, where, for instance, 80°F is not twice as "hot" as 40°F; measured on the Celsius scale, these two temperatures are 27°C and 4°C, respectively, and 27°C is not twice 4°C. Examples of instances when ratio scales are appropriate are measurements of heights, weights, and age.

Section 20.2 in Chapter 20 contains additional information about and examples of the four numerical scales of measurement.

Most of the statistical methods we will develop in this book require that the variable be measured at least on the interval scale.

■ 1.8
Text orientation

In most practical applications of statistics, it is not feasible to take a census of the population under study. Therefore we will emphasize the development and proper use of statistical methods that allow us to infer results about population characteristics from *sample* information. This process is called *inferential statistics*. In effect we are defining inductive reasoning, for it is reasoning from the part to the whole. This is the very essence of statistics as it is used today in research and decision making. Considerable space is devoted in this text to describing the inductive reasoning process of inferring from the part (the *sample*) to the whole (the *population*).

■ 1.9
Summary

The science of statistics encompasses the methodologies that are used for the collection, presentation, and analysis of quantitative data, and the interpretations of these data. Descriptive statistics refers to methods used to describe data numerically or graphically, whereas inferential statistics refers to the methods of inferring results about one or more characteristics in the population from a sample. A sample is a subset of the population units, and a census is a complete enumeration and use of all population units in a survey or experiment. A listing of all population units is called a *frame*. In a survey, factors that may affect the population characteristics of interest are not controlled, but in an experiment they are. When the population units are people, personal interviews using schedules are usually the most effective method of acquiring data, although self-enumeration questionnaires and telephone interviews are more commonly used and are less expensive than interviews. In constructing a questionnaire, there are three important steps: the design, the pretest, and editing the responses.

A population characteristic under study is called a *variable* if it can take on two or more values among the population units. If the value of a variable is numeric, it is called a *quantitative* variable; otherwise, it is called a *qualitative* variable. There are four scales of measurement: nominal, ordinal, interval, and ratio.

With the extensive use of inferential statistics in business operations and the power of modern computers to do great amounts of computation in a very short time, there is a tendency for students to assume that their training in statistics will provide them with the *only* tools needed for successful decision making. Nothing could be further from the truth. Appropriately applied, inferential statistics will provide the effective manager with *one* source of information which, in conjunction with other sources (the manager's intuition for one), enables him to make a good decision. Inferential statistics should always be used with common sense; indeed, it *is* the use of common sense by definition (inductive reasoning).

■ References

Cochran, W. G. *Sampling Techniques*. 3rd ed. New York: John Wiley & Sons, 1977.

Ferber, R.; P. Sheatsley; A. Turner; and J. Wakesbury. *What Is a Survey?* Washington, D.C.: American Statistical Association, 1980.

Hansen, M. H.; N. W. Hurwitz, and W. G. Madow. *Sample Survey Methods and Theory*. Vols. 1 and 2. New York: John Wiley & Sons, 1953.

Kish, L. *Survey Sampling*. New York: John Wiley & Sons, 1965.

Ostle, B., and R. W. Mensing. *Statistics in Research*. 3d ed. Ames: Iowa State University Press, 1976.

Scheaffer, R., W. Mendenhall; and R. Ott. *Elementary Survey Sampling*. 2nd ed. North Scituate, Mass.: Duxbury Press, 1979.

Snedecor, G., and W. Cochran. *Statistical Methods*. 7th ed. Ames: Iowa State University Press, 1980.

Taeuber, C. "Information for the Nation from a Sample Survey." In *Statistics: A Guide to the Unknown,* 2nd ed., ed. J. Tanur et al. San Francisco: Holden-Day, 1978.

Tanur, J. et al., eds. *Statistics: A Guide to the Unknown*. 2nd ed. San Francisco: Holden-Day, 1978.

■ Problems

Section 1.2 Problems

1.1 Briefly describe each of the following terms:
  a. Primary data.
  b. Secondary data.
  c. External data.
  d. Internal data.

1.2 Distinguish between *primary* and *secondary* data. Which are the most reliable? Why?

1.3 For each of these listed sources of statistical data, determine the publisher, frequency of publication, and nature of the data— monthly, quarterly, semiannually, or yearly. Cite two statistics from each source. Example: The number of persons in the United States in 1980 who were 20 years of age or less was: _____ .
  a. Statistical Abstract of the United States.
  b. Survey of Current Business.
  c. Monthly Labor Review.
  d. Federal Reserve Bulletin.
  e. Life Insurance Fact Book.
  f. International Financial Statistics.
  g. Business Conditions Digest.
  h. Dun's Review.
  i. The Economic Almanac.
  j. Automobile Facts and Figures.

Section 1.3 Problems

1.4 Briefly describe each of the following terms:
  a. Population.
  b. Population characteristic.
  c. Census.
  d. Sample.
  e. Sampling error.
  f. Nonsampling error.
  g. Parameter.
  h. Statistic.
  i. Population frame.

1.5 Distinguish between *sampling* and *nonsampling* error. Which can occur in a census? Which can occur in a sample?

1.6 A manufacturer buys electronic parts from a supplier with the understanding that 1 per-cent or fewer of the parts are defective. In a particular shipment of 5,000 parts, the supplier finds in a sample of 100 parts that none is defective. The manufacturer decides to check the parts as well and, in another sample of 100 parts, finds that four are defective. On this basis, the manufacturer decides to reject the lot.
  a. How is it possible that one sample produced 0 percent defectives, and another produced 4 percent defectives?
  b. Is it possible that the manufacturer is making a mistake by not accepting the shipment? If you were the supplier, how would you defend the shipment as possibly being acceptable (1 percent or less defectives)?

1.7 How can sampling error be minimized? Discuss. How can nonsampling error be reduced? Discuss. Is it possible for sample results to be more accurate than census results? Explain.

1.8 In the following situations, indicate whether a sample or a census should be taken and explain why.
  a. A firm that employs 500 persons wishes to determine the acceptability of subscribing to a new employee insurance program.
  b. A car manufacturer wishes to obtain information on customer preferences with respect to size of cars.
  c. The Internal Revenue Service wishes to obtain data on the proportion of income tax returns that contain arithmetic mistakes.

1.9 Discuss the following statement: "A census will always give better information than does a sample."

Section 1.4 Problems

1.10 Briefly describe each of the following terms:
  a. Statistical survey.
  b. Statistical experiment.

1.11 Distinguish between a *survey* and an *experiment*. Which is preferred and why?

1.12 In the following situations, indicate whether

a survey or an experiment is more suitable, and explain why.

a. Collecting data on the gas mileage of a specific model and make of car.

b. Collecting data on the number of sick days per month used by workers in a large company.

c. Collecting data on the interest paid for home mortgages in a large metropolitan area.

1.13 List and briefly describe the steps in conducting a survey or an experiment

Section 1.5 Problems

1.14 There are three ways of obtaining information from persons in a survey or an experiment: self-enumeration, personal interview, and a telephone interview. Briefly describe each method.

1.15 Distinguish between a *schedule* and a *questionnaire*. What is each used for?

1.16 There are three kinds of questions that may be used in a schedule or questionnaire. Describe each, and discuss its advantages and disadvantages.

1.17 In the following situations, which method of data collection—self-enumeration, personal interview, or telephone interview—would you select and why? Keep in mind the cost of the method, the response rate, and the time necessary to obtain the information, as well as other relevant factors.

a. Consumer acceptance of a new camera model before it is placed on the market.

b. National survey to determine, in the public's view, the best way to fund the social security system in future years.

c. Data on preventive health care behavior of persons in the United States.

d. The determination of the national ranking of a television show in a specific time slot.

e. Data on electricity rates nationally.

Section 1.6 Problems

1.18 Why is pretesting a questionnaire or a schedule a necessary step in conducting a survey or an experiment that uses either instrument?

1.19 In constructing a schedule or questionnaire, there are three primary steps: *design, pretest,* and *editing*. Describe each step.

Section 1.7 Problems

1.20 Briefly describe each of the following terms:

a. *Variable*.

b. *Dependent variable*.

c. *Independent variable*.

d. *Quantitative variable*.

e. *Qualitative variable*.

1.21 There are four measurement scales: *nominal, ordinal, interval,* and *ratio*. Describe each, and give an example of a survey question that may use measurements of each type.

1.22 For each of the following, indicate the scale of measurement:

a. Red (1), Blue (0), Yellow (−1).

b. Extremely likely (5), Likely (4), Indifferent (3), Unlikely (2), and Extremely unlikely (1).

c. Pressure in pounds per square inch; from 0 to ∞.

d. Volume in cubic centimeters; from 0 to ∞.

e. Age in years; 0 to ?

f. Salary in dollars, 0 to ?

g. Rank of a state in population; 1 to 50.

1.23 For each of the following, indicate whether it is a *quantitative* or *qualitative* variable.

a. Hair color.

b. Sales volume of an automotive firm.

c. Sex of an individual.

d. Number of persons unemployed in the United States.

1.24 A corner drugstore is losing business and wishes to determine the causes for the loss. Customers entering the store are given a questionnaire that inquires about their satisfaction with the store, the service, and so on, and asks for suggestions on improvements.

a. What is the population in this survey?

b. Is the sample biased? That is, are the respondents representative of the population?

c. How would you suggest collecting data to determine the causes for the loss in business?

Additional Problems

1.25 Athletic teams generally wear numbers on the backs of their uniforms. These numbers represent which measurement scale? Explain.

1.26 The *Seal Point Daily Tribune* conducted a telephone survey asking the question, "Should the proposed nuclear power plant at Seal Point be built?" Of the 514 people contacted, 183 said yes, 307 said no, and the rest said they had no opinion on the matter.
  a. What is the target population of the survey?
  b. What is the sampled population?
  c. The survey produces three statistics. What are they?
  d. On what measurement scale are the three statistics measured?
  e. What are the values of the statistics?

1.27 For each of the following, indicate an appropriate scale of measurement.
  a. Temperature (in degrees Centigrade).
  b. Stock prices per share.
  c. Ranking of the horses finishing a horse race (first, second, etc.).
  d. Bond ratings (AAA, AA, A, etc.).

1.28 Give an example of each of the following.
  a. Descriptive statistics.
  b. Inferential statistics.

1.29 In each of the following statements, indicate whether descriptive or inferential statistics is being used.
  a. Last year, the inflation rate rose by 5 percent.
  b. Based on a sample of 233 students at Tech University, 72 percent of all university students at least occasionally drink alcoholic beverages.
  c. Based on an average of 50 selected stocks, the stock market's average value of a share rose by 5.2 points at the end of the trading today.
  d. The average grade-point average of all business students at Tech University is 3.12, based on current enrollment data supplied by the registrar's office.

1.30 Is it possible to "solve" the potential problem of sample bias due to a low response rate by simply sending out a sufficiently large number of questionnaires to produce a sizable sample? Explain.

1.31 There are 5,000 members in a community group health association, and 500 are selected for a mail survey on level of satisfaction with the services provided by the association. Thirty percent reply to the initial mailing of the questionnaire.
  a. Is the sampled population the same as the target population (the 5,000 members)? Explain why the sample is likely to be biased in this instance.
  b. If 90 percent responded in the sample of 500, would you be less concerned about bias? Why?

1.32 In the following set of questions, find at least one fault in each. Also, suggest an improved rewording of the question.
  a. Do you agree that too much money is being spent on national defense?
  b. How many tubes of toothpaste did you purchase in the past year?
  c. Does the name "Frontier" come to mind when hi-fidelity equipment is mentioned?
  d. It is a waste of money to send people into outer space.
     (Strongly agree, Agree, Undecided, Disagree, Strongly disagree)

1.33 Criticize or defend the following statements.
  a. Only 23 percent of the Tech University students who had flu shots last winter caught the flu. This provides evidence that the flu shots were effective in preventing a flu attack.
  b. The responses to an interviewer's questions are free from nonsampling error if they are accurate and honest.
  c. When mail questionnaires are used, it must be kept in mind that more responses tend to be from persons with special interests in the outcome of the survey than from others.

2

Descriptive statistics

■ **2.1**

Introduction

When a survey or an experiment has produced a body of data, the original state of the data will not generally convey much information about the characteristics of interest. Typically, there will simply be too many observations to give one an insight into the nature of the data. Whether a data set represents a sample or a population, it is necessary to organize and reduce the data into such meaningful forms as graphs and charts or such numerical quantities as averages, totals, and percentages. The resulting statistical summaries of the data can be used as a convenient and meaningful framework for data analysis and interpretation.

> Graphical and
> numerical methods

There are basically two methods to describe data: *graphical* methods and *numerical* methods. We shall discuss both in this chapter. As the methods are being discussed, it is important to keep in mind that they may be applied to either population or sample data sets. If it is possible to take a complete census, then these methods may be used to summarize information about the population characteristics. A company, for example, may be interested in the number or proportion of its employees who make more than $50,000 per year. The information on salaries is maintained for all employees, so a census can be taken to determine this number or proportion.

If the data set represents a sample, the statistical summaries, particularly the numerical measures, may be used to draw inferences about the population characteristics.

Mainframe and microcomputers are used extensively to analyze data. Once the data constituting a population or a sample has been encoded on computer disk or tape, the analysis of these data is facilitated enormously by the use of computers. Very recently, numerical and graphical software packages have become readily available for microcomputers. This software can be used to prepare impressive and effective statistical summaries of data sets. It seems clear that in the future most if not all business reports will be expected to have microcomputer-generated statistical summaries of data including graphical presentations as integral parts of the reports.

■ **2.2**

Frequency distributions

As a means of explaining concepts in this chapter, we shall make use of the data in the following example.

Example 2.1   A medium-size manufacturing firm produces metal furniture for business offices as its main product. The firm currently has 643 employees, approximately 50 percent of whom are laborers, 30 percent of whom are in sales, and 20 percent of whom are in managerial/executive positions. The firm maintains data files that include, among other things, the following information on each employee:

1. Sex (1 = Male, 0 = Female).

2. Age in years.

3. Number of dependents.

4. Years working for the firm.

5. Employee rating in most recent evaluation (measured on a scale from 0 to 100).

6. Whether or not employee holds a college degree (1 = College degree; 0 = No college degree).

7. Whether or not employee participates in a stock profit sharing plan (1 = Employee participates; 0 = Employee does not participate).

8. Job classification (1 = Laborer, 2 = Sales, 3 = Managerial/Executive).

9. Annual wage, not including benefits or bonuses.

The data for 100 of these 643 employees is given in Table 1, Appendix C. For example, the first row of the table corresponds to the first employee in this *sample* of 100 employees. This individual is a 34-year-old male who has two dependents and has been working for the firm in sales for eight years. His most recent employee rating is 85; he has a college degree; he participates in the profit sharing plan, and his annual wage is $36,250.

Suppose the firm is interested in summarizing the annual wage data for these 100 employees. An efficient means of doing so is to form intervals for the annual wage characteristic and record the number of annual wages that fall into each interval. An example using this approach is shown in Table 2.1.

The grouping of data in Table 2.1 is called a *frequency distribution*. A frequency distribution is a convenient way of grouping data so that the important aspects of the raw data are more readily apparent. In this instance the 100 employee annual wages have been reduced to seven "numbers"—

Frequency distribution

TABLE 2.1 | Frequency distribution of the annual wages for the 100 employees of the manufacturing firm

Class number	Class limit	Frequency	Relative frequency
1	$15,000 but less than $25,000	30	30/100 = 0.30
2	$25,000 but less than $35,000	27	27/100 = 0.27
3	$35,000 but less than $45,000	21	21/100 = 0.21
4	$45,000 but less than $55,000	9	9/100 = 0.09
5	$55,000 but less than $65,000	9	9/100 = 0.09
6	$65,000 but less than $75,000	2	2/100 = 0.02
7	$75,000 but less than $85,000	2	2/100 = 0.02
	Totals:	100	1.00

Source: Personnel data files for the manufacturing firm; data given in Table 1, Appendix C.

the frequencies in each class, which have been tallied and tabulated in Table 2.1. The relative frequency in each class gives the ratio of the annual wages in each class to the total number of wages in the data set. So, for example, 27 percent of these employees make $25,000 but less than $35,000, and only 2 percent make $75,000 but less than $85,000.

When data are grouped into a frequency distribution, some information is lost. For example, we cannot determine from Table 2.1 the number of employees who have exactly a $25,000 annual wage or the number of employees who have an annual wage of $30,000 but less than $35,000.

The selection of the number of classes and the class limits that define the intervals of the frequency distribution is somewhat arbitrary. A useful series of steps for forming a frequency distribution is given below. [In brackets, these steps are applied to the Example 2.1 annual wage data.]

Steps in forming a frequency distribution	**Steps in forming a frequency distribution**

Range	1. Determine the *range* of the ungrouped numbers by finding the difference between the largest and the smallest numbers.

$$[\$82,500 - \$18,960 = \$63,540]$$

Number of classes	2. Select the *number of classes* into which the range will be divided. As a rule the number of classes should be between 4 and 20. The idea is to have a sufficient number of classes to show details and properties of the data, but not too many classes so that the data set has not been reduced to a manageable form. For the wage data, selecting seven classes gives a

reasonable balance between too much or too little data reduction (choosing five, six or eight classes would be fine, as well).

[7 classes]

Class width

3. Divide the number of classes into the range. This ratio represents the *minimum class width* for each class. Then choose a number equal or greater than this ratio that produces a *convenient class width* with which to work.

[63,540/7 = 9,077.14 (minimum class width)]

[Choose 10,000 for a convenient class width]

Class limits

4. Select the *class limits* by beginning with a convenient whole number less than the minimum value in the data set and constructing classes with the width determined in step 3 and stopping when the range of numbers has been covered. The minimum annual wage for the 100 employees is $18,960. Rather than starting at $18,000, we will start with $15,000. Using multiples of 5s and 10s generally produces a frequency distribution that is easy to read and understand.

[$15,000 but less than $25,000; $25,000 but less than $35,000;
. . . ; $75,000 but less than $85,000]

Frequencies

5. Form the column headings for the class number, the class limit, the frequency, and the relative frequency. Count the frequencies in each class, and record the frequencies in the table. Be certain to total the frequencies to ensure that all numbers have been accounted for.

Relative frequencies

6. Form the column of relative frequencies by dividing the frequency in each class by the total number of observations.

As suggested by the above six steps, forming a frequency distribution is as much a matter of art as it is a science. As such, there is not a unique "correct" frequency distribution for a given data set. Choosing the class width and the class limits is a matter of taste, provided that certain rules and guidelines are followed:

Each number must belong to one and only one class

1. A hard and fast rule is that every number in the data set must belong to one and only one class. We could not, for example, use such class limits as $15,000 to $25,000 and $25,000 to $35,000. By doing so an annual wage of $25,000 would belong to two classes (the $15,000 to $25,000 class and the $25,000 to $35,000 class). Nor could we begin the first class with a lower limit of $20,000, for it would exclude one of the numbers (an annual wage of $18,960). That is why we use the "but less than" statement for the class limits. This ensures that an annual wage will not belong to two or more classes.

Use simple class width and limits

2. Attempt to select the class width and class limits so the resulting frequency distribution is appealing and understandable. Selecting such limits as $18,882 but less than $27,443 defeats the purpose of a frequency

distribution to reduce the data set to a convenient and readily understandable form.

| Goal: Classes with equal widths |

3. Attempt to select the class limits so that each class has the same width. Sometimes this is difficult to do if the numbers in the data set are very spread out over a large range. We will return to this point in Section 2.9 entitled, "More on Descriptive Statistical Techniques."

| Avoid open-ended classes if possible |

4. Avoid open-ended classes if at all possible. Sometimes the data set is so spread out that there is no alternative to the use of open-ended classes. For example, suppose one of the executives for the manufacturing firm in Example 2.1 has an annual wage of $225,000, and this wage is included in the data set containing 100 annual wages. If we used the six steps given above for forming a frequency distribution, then the resulting frequency distribution would contain either several empty classes if we chose a large number of classes or a few classes that contain most of the annual wages if we chose a small number of classes. The only reasonable recourse in this case is to use an open-ended class for the last class, such as "$75,000 or more."

| Cumulative frequency |

Frequently, two additional columns are added to the frequency distribution table: *cumulative frequency* and *cumulative relative frequency*. These two columns are shown for the Example 2.1 frequency distribution in Table 2.2.

| Cumulative relative frequency |

The cumulative frequencies accumulate the frequencies in each successive class. Thus the cumulative frequency in class 2 is $30 + 27 = 57$, and the cumulative frequency in class 5 is $30 + 27 + 21 + 9 + 9 = 96$, for example. The cumulative relative frequencies accumulate the relative frequencies for successive classes. Thus, for example, the cumulative relative frequency in

TABLE 2.2 | Complete frequency distribution table for the Example 2.1 data including the cumulative frequencies and the cumulative relative frequencies.

Class number	Class limit	Frequency	Cumulative frequency	Relative frequency	Cumulative relative frequency
1	$15,000 but less than $25,000	30	30	0.30	0.30
2	$25,000 but less than $35,000	27	57	0.27	0.57
3	$35,000 but less than $45,000	21	78	0.21	0.78
4	$45,000 but less than $55,000	9	87	0.09	0.87
5	$55,000 but less than $65,000	9	96	0.09	0.96
6	$65,000 but less than $75,000	2	98	0.02	0.98
7	$75,000 but less than $85,000	2	100	0.02	1.00
	Totals:	100		1.00	

Source: Personnel records for the manufacturing firm; data given in Table C.1.

class 4 is 0.30 + 0.27 + 0.21 + 0.09 = 0.87. The cumulative relative frequencies are particularly useful, for they tell us the proportion of employees whose wages are less than an upper class limit for each class. For example, in class 6, the cumulative relative frequency tells us that a proportion of 0.98 of the 100 employees, or 98 percent, make less than $75,000 per year.

■ 2.3

Graphical methods of data presentation

The frequency distribution discussed in Section 2.2 conveniently places the data in a form for a graphical display called a *histogram*. The following steps are required in constructing a histogram:

| Histogram |

1. Mark off the x (horizontal) axis with the class limits taken from the frequency distribution.

2. Construct rectangles over the class limits so that the area of each rectangle is proportional to the class frequency with a constant coefficient of proportionality for all classes. If the class widths are equal, choosing the height proportional to class frequency is equivalent to choosing area proportional to class frequency.

In applying these steps to the frequency distribution in Table 2.2, we see that the height of each rectangle may be taken as the class frequency, since each class is of equal width. The resulting histogram is shown in Figure 2.1. The histogram clearly shows the "slide" effect of the annual wages for the 100 sampled employees of the manufacturing firm—most of the wages are to

FIGURE 2.1 | Histogram for the annual wages of the 100 sampled employees of the manufacturing firm

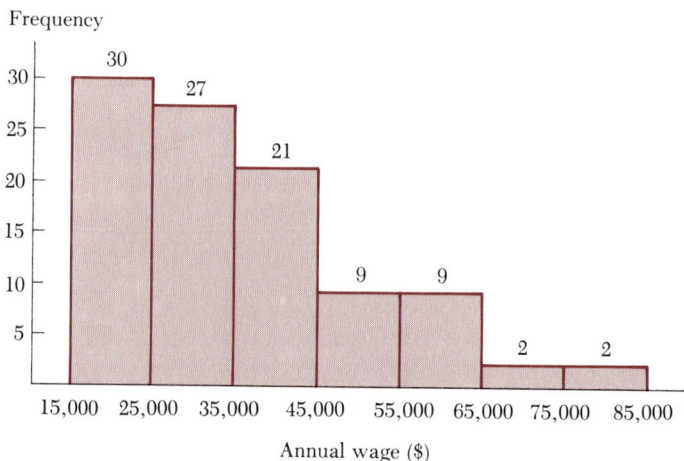

the left at the lower end of the wage range, and a few of the wages are to the right at the very upper end of the salary range.

From a frequency distribution, it is possible to construct numerous other forms for graphical displays of data sets. A *frequency polygon* is constructed by plotting the class frequency and the class *midpoint* for each class and joining the resulting points by straight lines. A frequency polygon for the frequency distribution in Table 2.2 is shown in Figure 2.2. Notice that the polygon is "tied down" at the ends by starting with a frequency of 0 at $10,000 and ending with a frequency of 0 at $90,000. The number $10,000 represents the midpoint of a class with the limits "$5,000 but less than

Frequency polygon

Class midpoint

FIGURE 2.2 | Frequency polygon for the frequency distribution in Table 2.2

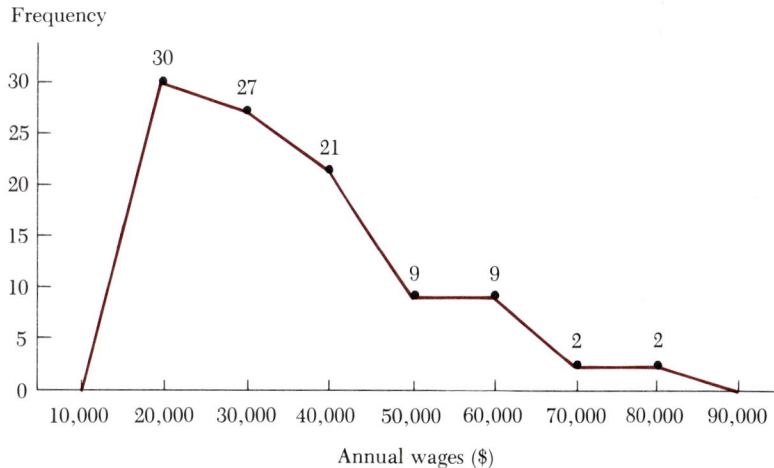

$15,000." This class precedes the first class in the frequency distribution, "$15,000 but less than $25,000." Similarly, the $90,000 represents the midpoint of a class with the limits "$85,000 but less than $95,000." This class immediately follows the last class in the frequency distribution, "$75,000 but less than $85,000." By tying down the ends of the polygon, it has a neater appearance than if we had just connected the dots corresponding to the class midpoints in the frequency distribution alone. The frequency polygon is another means of shedding light on the nature of a data set. As with the histogram, the polygon in Figure 2.2 indicates that most of the annual wages are at the left of the range of wages, and a few of the annual wages are at the right of the range of wages.

The cumulative relative or absolute frequencies may also be used to construct a graph that is called an *ogive* (pronounced oh-jive, as in five). A "less than or equal to" cumulative relative frequency polygon (ogive) for the frequency distribution in Table 2.2 is illustrated in Figure 2.3.

Ogive

FIGURE 2.3 | A "less than" ogive for the frequency distribution in Table 2.2

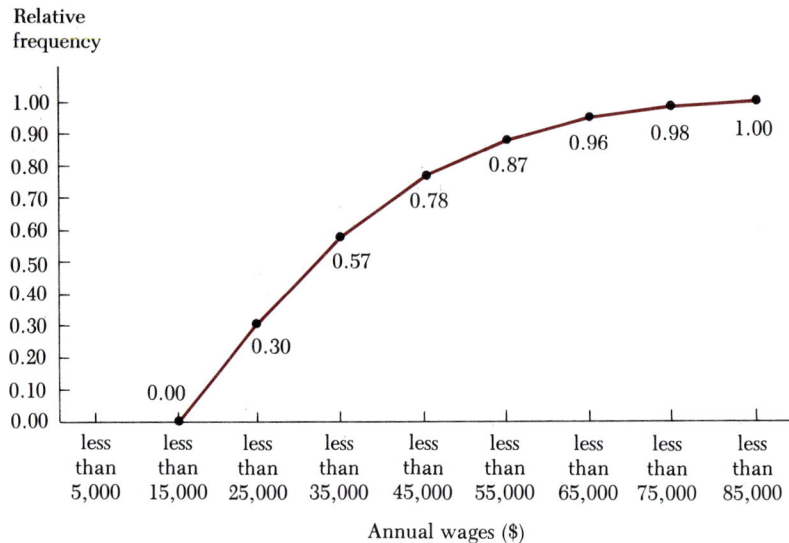

Notice that the ogive plots the cumulative relative frequencies for the upper class limit in each class. Therefore, in the first class, "$15,000 but less than $25,000" the cumulative relative frequency is 0.30, and we plot this number at the upper class limit of "less than $25,000." We cannot use exactly $25,000 because this may include numbers belonging to the next class.

There are several forms of graphical displays that are not usually associated with a frequency distribution. Examples are *bar charts, line charts,* and *pie charts*. In Figure 2.4a, a *pie chart* summarizes the sales of different models of Porsche automobiles for the year 1983, and in Figure 2.4b a pie chart summarizes the distribution of Porsche sales to different geographic regions during this same period. In pie charts like these, an area or "slice" of the pie is allocated to each class or grouping in proportion to the relative frequency or percentage of the total. Production managers can use such visual statistical displays in depicting gross and item production during a

Pie chart

FIGURE **2.4a** | Distribution of product sales for Porsche, Inc. for the year 1983

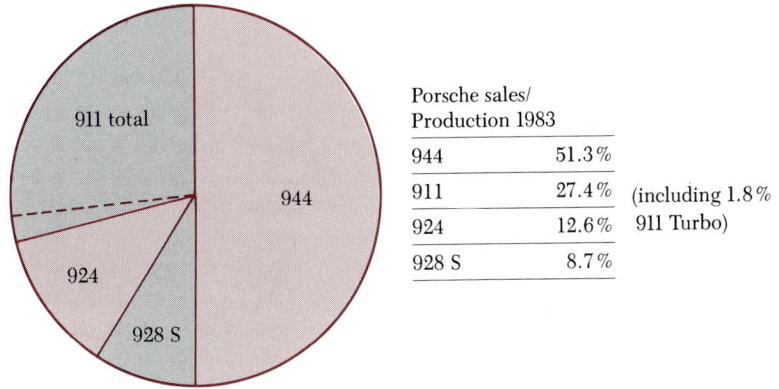

Porsche sales/ Production 1983		
944	51.3%	
911	27.4%	(including 1.8%
924	12.6%	911 Turbo)
928 S	8.7%	

Source: Company records.

FIGURE **2.4b** | Geographic distribution of Porsche, Inc. sales for 1983

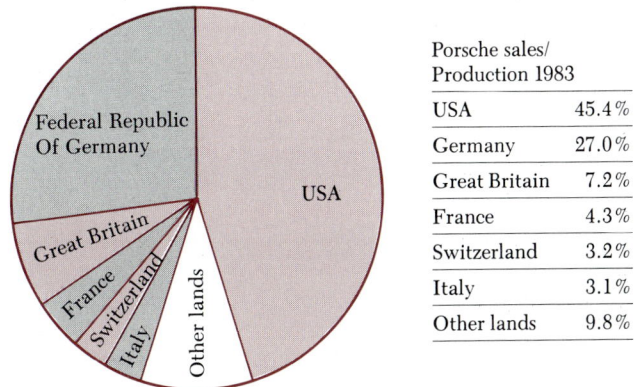

Porsche sales/ Production 1983	
USA	45.4%
Germany	27.0%
Great Britain	7.2%
France	4.3%
Switzerland	3.2%
Italy	3.1%
Other lands	9.8%

Source: Company records.

yearly period. The second pie chart is useful for those in logistics in indicating gross distribution requirements for company output.

Artists with statistical training can enhance and improve upon the presentation of such charts as those shown in Figure 2.4. In Figure 2.5, for exam-

FIGURE 2.5 | Enhanced display of Porsche, Inc. sales by product for the year 1983

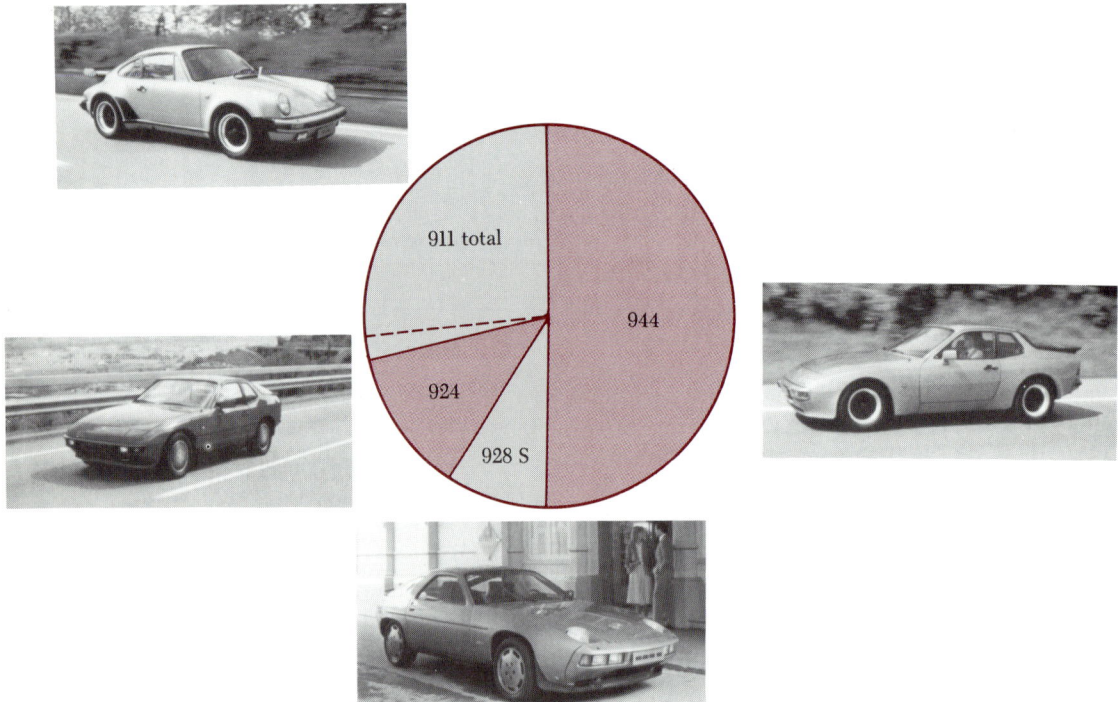

Source: Company records.

ple, the pie chart has been augmented with a drawing of each of the products involved, giving an even clearer and more appealing display of the statistical data on sales for Porsche during 1983.

Pie charts can take imaginative forms, as illustrated in Figure 2.6. In this figure the basic circle form of the pie chart has been replaced by a form representing the state of Texas to show the distribution of Texas nonagricultural employment in 1980.

Line chart

Figure 2.7 illustrates a *line chart*. The numbers of units produced yearly by Porsche from 1962 to midyear 1983 are shown by drawing a line through the output figures for each year. Notice the change in 1972 from a calender year to a fiscal year for accounting purposes.

FIGURE   2.6 | Distribution of Texas nonagricultural employment in 1980

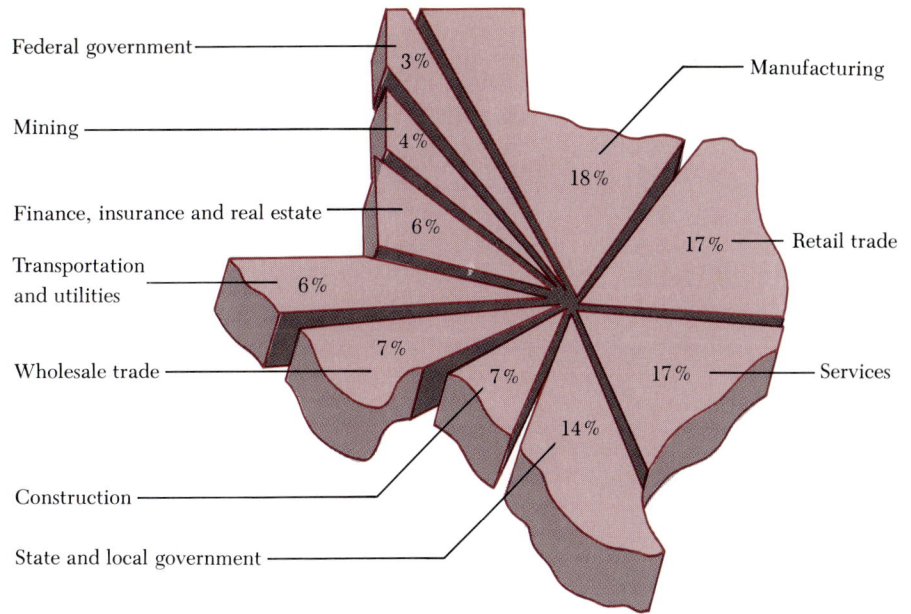

Federal government — 3%

Mining — 4%

Finance, insurance and real estate — 6%

Transportation and utilities — 6%

Wholesale trade — 7%

Construction — 7%

State and local government — 14%

Manufacturing — 18%

Retail trade — 17%

Services — 17%

Source: Texas Employment Commission.

FIGURE   2.7 | Units produced annually by Porsche

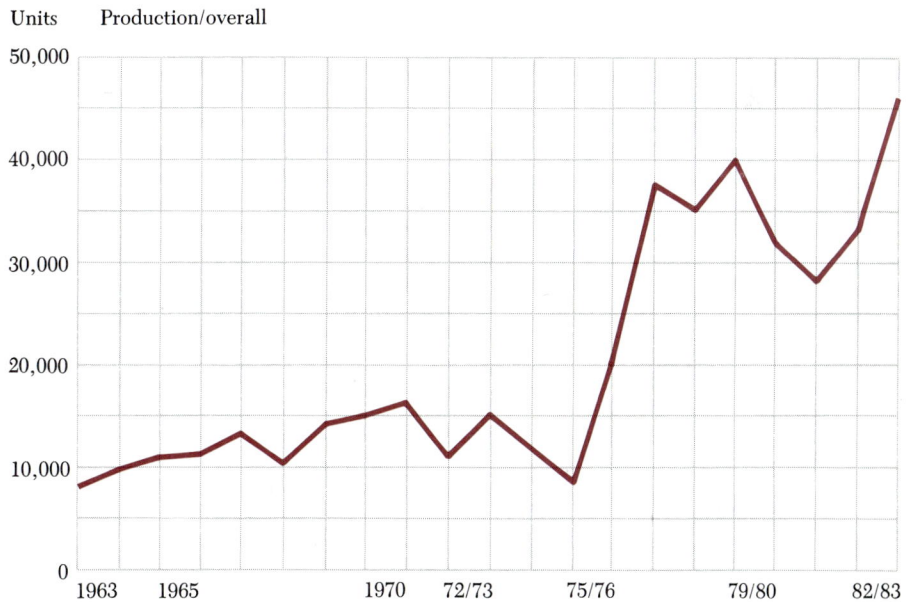

Units   Production/overall

Note: Following a lull during the 1980/81 business year, caused by the overall market climate, the past two years have again shown steep improvement in production numbers.

Source: Company records.

Figure 2.8 combines the use of a pie chart with a *bar chart*. The chart on the right uses bars to show the maintenance and operating budgets for the academic years 1967–68 through 1985–86 for the Tarrant County Junior College (TCJC) system. The use of the vertical bars dramatically illustrates the growth of the TCJC system during this period. However, these dollar figures have not been adjusted for inflation, so that the chart is somewhat

FIGURE 2.8 | Growth in maintenance and operating budgets (bar chart) and income sources and expenditures (pie charts)

Maintenance and operations income and expenditures, 1984-85

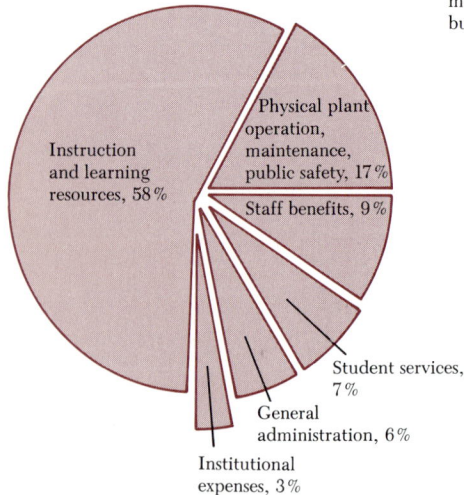

Income

State aid, 65%
Local maintenance tax, 16%
Tuition and fees, 13%
Other income 6%

As shown on the accompanying chart, the local maintenance tax accounts for 16 percent of the College's annual income, and tuition and fees for 13 percent.

The size of TCJC's fiscal operation and an idea of the growth of the College are seen in the increase of the total maintenance and operations budget from year to year:

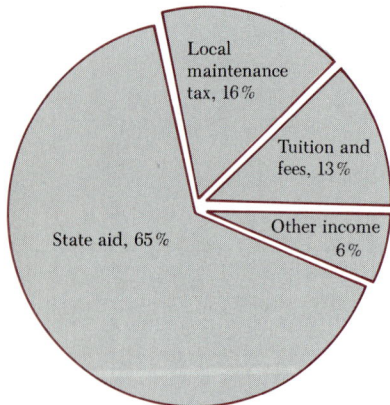

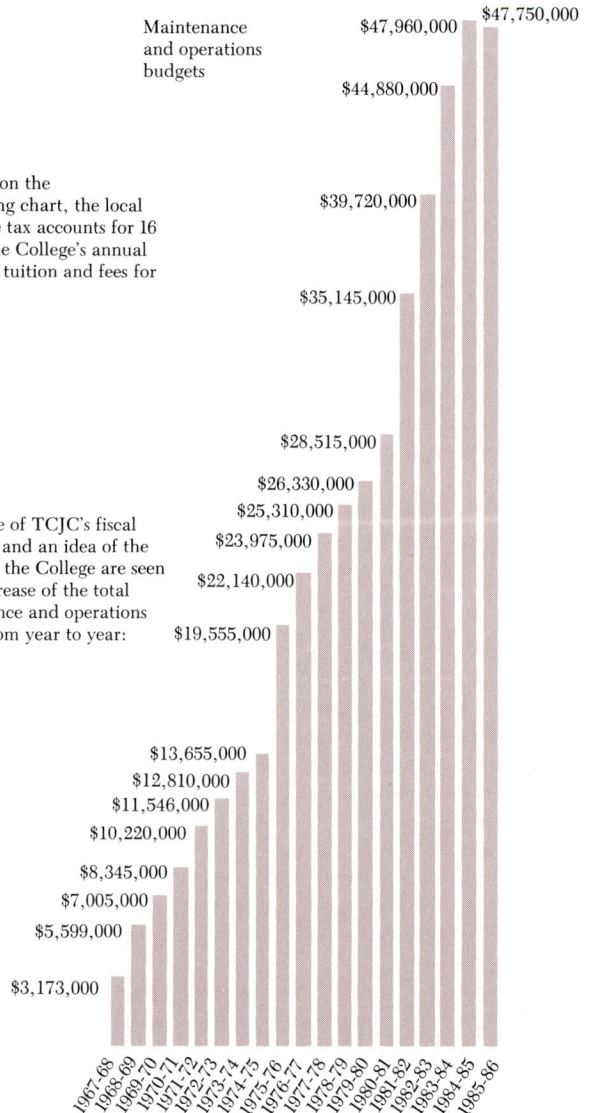

Expenditures

Instruction and learning resources, 58%
Physical plant operation, maintenance, public safety, 17%
Staff benefits, 9%
Student services, 7%
General administration, 6%
Institutional expenses, 3%

Maintenance and operations budgets

$47,960,000
$47,750,000
$44,880,000
$39,720,000
$35,145,000
$28,515,000
$26,330,000
$25,310,000
$23,975,000
$22,140,000
$19,555,000
$13,655,000
$12,810,000
$11,546,000
$10,220,000
$8,345,000
$7,005,000
$5,599,000
$3,173,000

1967-68 1968-69 1969-70 1970-71 1971-72 1972-73 1973-74 1974-75 1975-76 1976-77 1977-78 1978-79 1979-80 1980-81 1981-82 1982-83 1983-84 1984-85 1985-86

Source: Tarrant County Junior College District records; Tarrant County, Texas, 1986.

deceiving. We will return to the use of charts with dollar figures not adjusted for inflation in Section 2.9 ("How to Lie with Statistics"). The pie charts on the left of Figure 2.8 show the income sources as percentages and the expenditures for the TCJC system for the 1984–85 academic year.

Figure 2.9 uses a bar chart to illustrate the rapid depreciation of a new car. As an illustration, a 1986 Chevrolet Cavalier Z24 hatchback costing $12,411 is used to show the value of the car at the age of 1 year through 10 years. The bar chart shows the percent of the original value that the car is worth each year together with the actual dollar amount.

The use of imaginative graphical presentations of statistical data is clearly on the rise. Software is now readily available for color graphics on microcomputers. The color images on screen can be transferred to transparencies or to film for use in reports and presentations. For a thorough and fascinating discussion of the use of graphics for transmitting information, see the Tufte reference at the end of this chapter. We will return to a discussion of graphical methods using illustrations from this reference in Section 2.9 of this chapter.

■ 2.4

Representative measures

Numerical measures

Notation for population and sample numerical measures

Representative measures

The frequency distribution and its graphical display, the histogram, can provide considerable insights into the nature of the distribution of a data set. From the histogram, we can quickly tell the form or shape of the distribution—the degree of spread in the observations, their tendency to cluster about a "middle" value, the existence of one or few very large or very small observations, and so forth. It is also possible to characterize this information by computing *numerical measures* such as averages or quantities that measure the spread of the observations. Although these numbers may not contain as much information as a frequency distribution or its histogram, when they are taken collectively, the statistician can readily construct a visual image of how the frequency distribution must look. More important, these numerical measures are easy to work with mathematically and are used extensively in the inference-making process if they have been calculated from sample data.

The formulas for the numerical measures will depend on whether or not the data set represents a population or a sample. Throughout this text, Greek letters (μ, σ, ρ, etc.) will be used to denote *population* numerical measures, and roman letters (s, r, t, etc.) will be used to denote *sample* numerical measures. For each numerical measure presented, the formulas for the population numerical measure and the sample numerical measure will be given.

The first numerical measures we will consider are called *representative measures*. By this we mean measures that are "representative" of all the observations. These measures may attempt to locate the "center" or "middle" value of the distribution of observations. But in some instances it may not make much sense to talk about a central value, because of a wide spread in the observations or concentrations in the ends of the distribution rather than in the middle.

The most commonly used representative measure is the simple arithmetic mean.

What's your car worth?

Here's what depreciation would cost the owner of the car pictured above—a well-equipped 1986 Chevrolet Cavalier Z24 hatchback costing $12,411. The Chevrolet Cavalier was a top selling car in 1985.

[Note: The percentage of depreciation will vary with different sources. The percentages used in this figure do not specifically reflect the formula of any particular source.]

Estimated value of car and percentage of new-car price retained

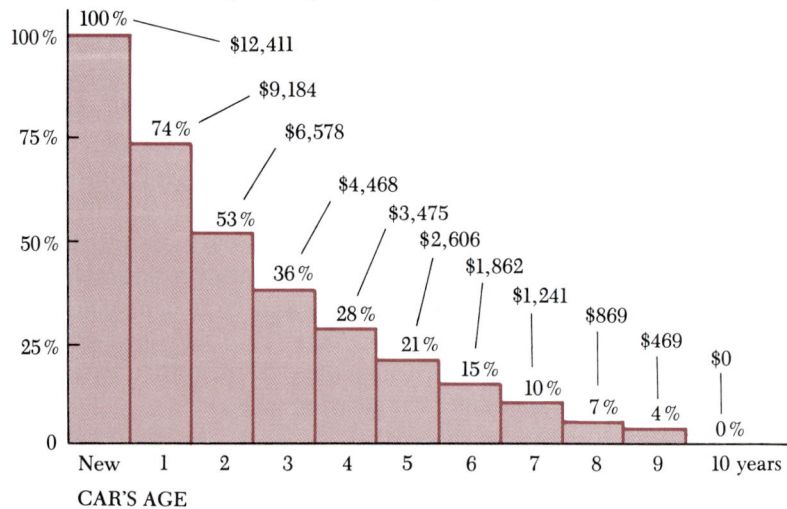

Source: Reprinted from the *Star-Telegram,* Fort Worth Texas, Sunday, November 3, 1985.

Arithmetic mean

> **Definition 2.1**
> The arithmetic mean
>
> The *population arithmetic mean*, denoted by μ, of a set of N population measurements $x_1, x_2, \ldots, x_N$, is given by the formula:
>
> $$\mu = \frac{\sum\limits_{i=1}^{N} x_i}{N} \qquad (2.1)$$
>
> The *sample arithmetic mean*, denoted by $\bar{x}$, of a set of n sample measurements $x_1, x_2, \ldots, x_n$, is given by the formula:
>
> $$\bar{x} = \frac{\sum\limits_{i=1}^{n} x_i}{n} \qquad (2.2)$$

The arithmetic mean of a set of measurements is the sum of the measurements divided by the number of measurements. This rule applies to both the population and the sample arithmetic means.

Example 2.2   Find the population arithmetic mean of the following five population measurements:

$$x_1 = 10 \quad x_2 = 15 \quad x_3 = 6 \quad x_4 = 12 \quad x_5 = 11$$

Solution

$$\mu = \frac{\sum\limits_{i=1}^{5} x_i}{5} = \frac{10 + 15 + 6 + 12 + 11}{5} = \frac{54}{5} = 10.8$$

Arithmetic mean as the center of gravity

The arithmetic mean gives the "center of gravity" of a set of numbers as indicated in Figure 2.10 for the data set of Example 2.2. The mean of 10.8 is the "balancing point" of the five numbers.

FIGURE 2.10 | Arithmetic mean as the center of gravity

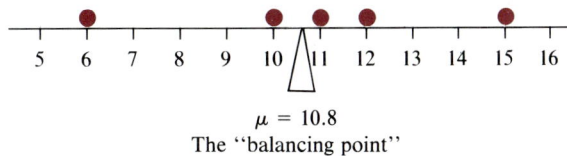

$\mu = 10.8$
The "balancing point"

Example 2.3 The following eight *sample* measurements represent the gain or loss in the daily closing price of a commodity on eight consecutive market days:

$$3.375 \quad -1 \quad 0 \quad 0 \quad 2.25 \quad 2.5 \quad 1.75 \quad -1.875$$

Find the average loss or gain in the commodity price.

Solution

$$\bar{x} = \frac{\sum_{i=1}^{8} x_i}{8} = \frac{3.375 + (-1) + 0 + 0 + 2.25 + 2.5 + 1.75 + (-1.875)}{8}$$
$$= 0.875 \quad \text{(the average gain)}$$

Median

Definition 2.2

The median

The *population median*, denoted by $\mathfrak{M}$, of a set of N population measurements $x_1, x_2, \ldots, x_N$ is the value x such that it falls in the middle of the array of the N population measurements when they have been ordered from the least numerical value to the greatest numerical value.

The *sample median*, denoted by M, of a set of n sample measurements $x_1, x_2, \ldots, x_n$ is the value x such that it falls in the middle of the array of the n sample measurements when they have been ordered from the least numerical value to the greatest numerical value.

Example 2.4 Find the population median of the $N = 5$ *population* measurements in Example 2.2.

Solution We first order the measurements from the smallest to the largest: 6, 10, 11, 12, and 15. The "middle" value is 11. Thus the population median is $\mathfrak{M} = 11$.

Example 2.5 Find the sample median of the $n = 8$ *sample* gains and losses in Example 2.3.

Solution The measurements when ordered are: $-1.875, -1, 0, 0, 1.75, 2.25, 2.5,$ and 3.375. In this case, there is no "middle" number. When the number of measurements is even in a data set (population or sample), we define the median to be the value halfway between the two "middle" numbers in the ordered array. Thus, in this example, the sample median is:

$$M = \frac{0 + 1.75}{2} = 0.875$$

Mode

Definition 2.3
The mode

The *population mode*, denoted by $\mathfrak{M}_o$, of a set of N population measurements $x_1, x_2, \ldots, x_N$ is the value x that occurs with the greatest frequency.
The *sample mode*, denoted by M_o, of a set of n sample measurements $x_1, x_2,$ $\ldots, x_n$ is the value that occurs with the greatest frequency.

Example 2.6   Find the mode of the following $N = 6$ *population* measurements:

$$x_1 = 3, x_2 = 4, x_3 = 3, x_4 = 5, x_5 = 2, x_6 = 1$$

Solution $\mathfrak{M}_o = 3$, since the value 3 occurs with the greatest frequency (twice).

Example 2.7   Find the mode of the following *sample* measurements:

$$2 \quad 3 \quad 2 \quad 5 \quad -1 \quad -2 \quad -1 \quad 3 \quad 2 \quad -1 \quad 5$$

Solution   Both -1 and 2 occur three times. Thus there are two sample modes: $M_o = -1$ and $M_o = 2$. In this case, we say that the sample set of measurements is *bimodal* (two modes).

Example 2.8   Find the mode of the following *sample* measurements:

$$1 \quad 3 \quad -1 \quad -2 \quad 4 \quad 0 \quad 5$$

Solution   Each number occurs once. In this case, we say that the set of sample measurements *does not have a mode*.

In reviewing Definitions 2.1, 2.2, and 2.3, it is clear that there is no difference in the way the population numerical measure and the sample numerical measure are computed in each case. The arithmetic mean is the sum of the measurements divided by the number of measurements, the median is the middle value in the ordered array of measurements, and the mode is the most frequently occurring measurement—*regardless* of whether the measurements constitute a population or a sample. Although there is not a difference between how the population numerical measure and the sample numerical measure is computed for the mean, median, and mode, there will be differences for other numerical measures as we shall see later in this chapter.

A reasonable question to raise at this point is: Which numerical measure—the mean, the median, or the mode—should be used as the representative measure of a data set?

The arithmetic mean is most often used, and for some very good reasons. All three measures are functions of all the observations, but the mean seems intuitively to take all the observations into account "better" than does the median or the mode. More important, the mean lends itself "best" among the representative measures for the purpose of statistical inference when it has been computed over a sample data set.

However, in most instances all three representative measures should be computed for a data set to help describe its properties. Differences among their values for a given data set contain useful information regarding how the measurements in the data set are distributed. For example, consider the frequency distribution polygon in Figure 2.11.

Most of the measurements are at the left end of the polygon in Figure

FIGURE 2.11 | A frequency distribution polygon indicating a skewed distribution of measurements

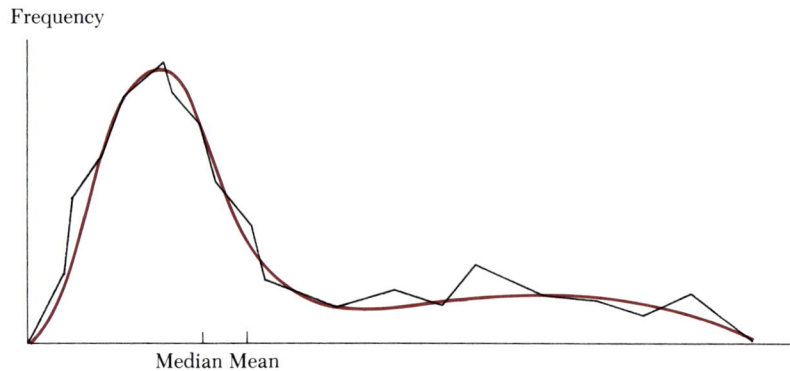

2.11, and only a few measurements are distributed over the larger values at the right end of the polygon. In this instance, the mean will be greater in value than the median due to the center of gravity (the mean) being "pulled" to the right by the few large values at the right end of the polygon. When this occurs (mean > median), we say that the polygon is *skewed to the right*—the long tail is at the right end of the polygon. When the mean is less than the median (mean < median), then we say that the polygon is *skewed to the left*—the long tail is at the left end of the polygon.

The smooth curve drawn through the frequency distribution polygon in Figure 2.11 is for convenience only. For simplicity of drawing polygons to illustrate concepts, we will straighten out the line segments in the polygons by using smooth curves. We will refer to the smooth curve approximation of the line segment polygon as a representation of the *distribution of the data set measurements*. In fact, the smooth curve is simply a convenient way of drawing the frequency distribution polygon for the measurements to illustrate concepts in this chapter.

A distribution is called *symmetric* if there is a value such that the portion of the distribution to the left of this value is a mirror image of the portion of the distribution to the right of this value. In Figure 2.12, examples of sym-

> Skewed right:
> Mean > Median

> Skewed left:
> Mean < Median

> Symmetric:
> Mean = Median

FIGURE 2.12 | Examples of symmetric and skewed distributions

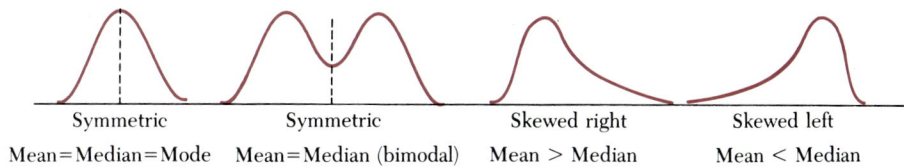

| Symmetric | Symmetric | Skewed right | Skewed left |
| Mean=Median=Mode | Mean=Median (bimodal) | Mean > Median | Mean < Median |

metric, skewed right, and skewed left distributions are illustrated. If the distribution is unimodal and symmetric, the three representative measures (the mean, the median, and the mode) have identical numerical values. In Section 2.6, numerical measures for determining the degree of skewness will be discussed. However, as the above discussion suggests, the three representative measures frequently provide a clue as to the symmetry or skewness of the distribution of a set of measurements.

When the distribution is skewed, there is a problem in selecting the representative measure that best represents the data set. Unfortunately, in some cases the decision is made on the basis of which measure best supports an individual's argument, and this is a clear abuse of statistics. For example, suppose that a large insurance company has a branch in Arlington, Texas, that is staffed by a branch manager and 10 insurance agents. The annual salaries (including commissions) of the 10 insurance agents are listed in Table 2.3. Insurance agent 10 is experienced and very successful with an annual salary (including commissions) of $280,000. The *median* of these annual salaries is $40,000 [($40,000 + $40,000)/2 = $40,000], and the *arithmetic mean* is $63,200. The manager of the Arlington branch office may use the arithmetic mean of $63,000 to argue that the salaries of the agents in her branch are high when compared with other branches of the insurance com-

TABLE 2.3 | Annual salaries (including commissions) of the 10 insurance agents in the Arlington, Texas, branch of a large insurance company.

Insurance agent	Annual salary	Insurance agent	Annual salary
1	$40,000	6	$ 40,000
2	36,000	7	52,000
3	42,000	8	35,000
4	27,000	9	48,000
5	32,000	10	280,000

Source: Branch office records.

pany. However, the median of $40,000 may be a better representative numerical measure of the "typical" salary of the agents in her branch in this case. The arithmetic mean is heavily influenced by the one large salary that pulls the center of gravity to the right. The distribution of the 10 salaries is skewed to the right due to the one very high salary ($280,000). The other 9 agents certainly may feel that the median is a better representative measure of the typical salary! The best solution in cases such as this is to report both the arithmetic mean and the median in describing the 10 measurements. The fact that the mean is greater than the median indicates that the frequency distribution polygon of the salaries is skewed to the right. This would suggest that there is at least one very large salary in the set of 10 salaries.

In business applications, the mode has been commonly used in demand analysis. A retail shoe store might be concerned, for example, about the modal sizes of shoes that are demanded by customers. The median has been used extensively in the social sciences and is a favorite government statistic as a representative measure in skewed distributions (median income, for instance). However, for the purposes of inference making, the mean will be used extensively in this text as a measure of a representative value for a data set. Its advantages outweigh its disadvantages. Additionally, in many practical applications of statistical inference, the distribution of the values is nearly symmetric, so that the three representative measures are approximately equal numerically in the sample selected.

■ 2.5

Measures of variability

Dispersion or variability

Although representative measures provide certain information about a distribution, more is needed before a clear picture of the shape of the distribution can be formulated. In Figure 2.13, for example both distributions have the same mean value but obviously differ in another respect—the amount of *dispersion* or *variability* of the values. The concept of variability is very important in statistics. For example, in production management, a major concern is the variability of the quality of a product being produced or

FIGURE 2.13 | Two dissimilar distributions with identical population means

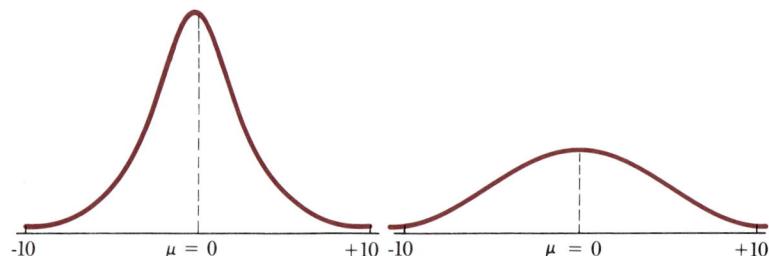

the variability of a crucial measurement of a product, such as a bearing diameter. More important, in statistical inference, we use the concept of variability to determine how good our inferences are. For the moment, it is sufficient to recognize the need for measuring the variability of a set of values to get a better idea of the shape of the distribution of the values.

The first and simplest measure of variability is the *range*.

Range

> **Definition 2.4**
> The range
>
> The *population range*, denoted by $\mathcal{R}$, of a set of N population measurements $x_1, x_2, \ldots, x_N$ is the algebraic difference between the largest and the smallest values.
> The *sample range*, denoted by R, of a set of n sample measurements $x_1, x_2, \ldots, x_n$ is the algebraic difference between the largest and the smallest values.

Example 2.9   Given the following six *population* measurements, determine their range.

$$x_1 = 0 \quad x_2 = 5 \quad x_3 = 6 \quad x_4 = 2 \quad x_5 = -1 \quad x_6 = 10$$

Solution   The largest number is 10, and the smallest is -1. The range is $10 - (-1) = 11$. Thus $\mathcal{R} = 11$.

Although the range is easy to compute, it is not a very satisfactory measure of variability, as Figure 2.14 illustrates. The distribution on the left

FIGURE   2.14 | Two distributions with equal ranges but unequal variance or dispersion of the measurements

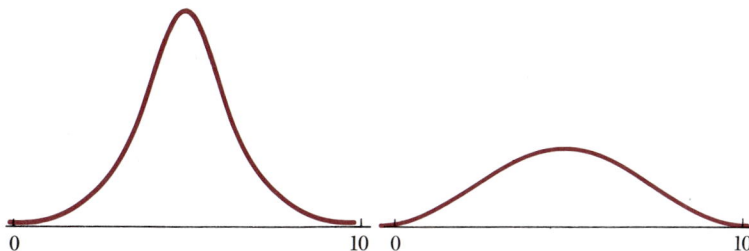

clearly is less variable than the distribution on the right, yet the ranges for the two distributions are identical $(10 - 0 = 10)$.

A better measure of variability is provided by using the *deviations* of the measurements from their mean or average value. For example, suppose we have a set of six *population* measurements:

$$x_1 = 1 \quad x_2 = 4 \quad x_3 = 6 \quad x_4 = 7 \quad x_5 = 4 \quad x_6 = 8$$

The mean of these six population measurements is:

$$\mu = \frac{\sum_{i=1}^{6} x_i}{6} = \frac{1 + 4 + 6 + 7 + 4 + 8}{6} = \frac{30}{6} = 5$$

In Figure 2.15 the six values are placed on the x axis, and the deviation $(x_i - \mu)$ of each value from the mean is indicated by a horizontal line. The "lengths" of these horizontal lines contain valuable information about the dispersion of the values. However, if we attempt to form a numerical measure based on the sum of these deviations, we find that it is not very helpful in measuring dispersion—the sum is 0 as the computations in Table 2.4 verify. Indeed, it can be shown that for *any* set of N measurements, $\sum_{i=1}^{N} (x_i - \mu) = 0$. We will not prove this result algebraically. Rather, the result can be appreciated on intuitive grounds. Since μ is the center of gravity, the positive and negative deviations about μ must balance out. Thus the sum of the positive deviations corresponding to the x_i values that are greater than μ equal the sum of the negative deviations corresponding to the x_i values that are less than μ.

One way to use the deviations of the x_i about their mean μ is to take their

FIGURE 2.15 | Deviations of the six population measurements from their population mean μ

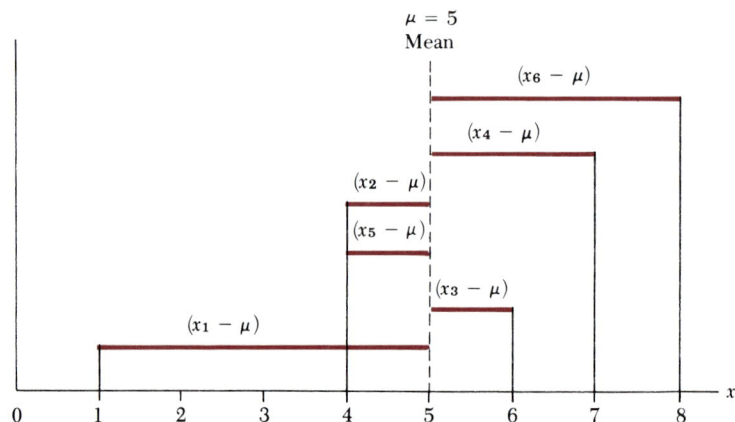

TABLE 2.4 | Computation of $\sum_{i=1}^{N} (x_i - \mu)$ for the data associated with Figure 2.15

x_i	$x_i - \mu$
1	$1 - 5 = -4$
4	$4 - 5 = -1$
6	$6 - 5 = 1$
7	$7 - 5 = 2$
4	$4 - 5 = -1$
8	$8 - 5 = 3$
	0

absolute values—the absolute values will eliminate any negative signs. Although the sum of the absolute values of the deviations has been used in statistics to measure dispersion, it is not a favored statistic because of the mathematical problems involved in working with the absolute value function. A more attractive statistic is the "average" of the squared deviations.

Variance

Definition 2.5

The variance

The *population variance*, denoted by σ^2, of a set of N population measurements $x_1, x_2, \ldots, x_N$ is given by the formula:

$$\sigma^2 = \frac{\sum_{i=1}^{N} (x_i - \mu)^2}{N} \tag{2.3}$$

The *sample variance*, denoted by s^2, of a set of n sample measurements $x_1, x_2, \ldots, x_n$ is given by the formula:

$$s^2 = \frac{\sum_{i=1}^{n} (x_i - \bar{x})^2}{n - 1} \tag{2.4}$$

Notice that this is the first numerical measure where there is a difference between how the population and sample measures are computed. The numerators of both measures are computed in the same fashion: The sum of the squared deviations of the measurements about the mean of the measurements. However, the denominators are different. In the population variance formula, the divisor is equal to the number of measurements (N); and in the sample variance formula, the divisor is equal to *one less than the number of measurements* ($n - 1$). This is an important difference that will be explained in Chapters 8 and 9. For the moment, it is important to select the correct formula: If we have a *population* set of N measurements, then we calculate σ^2 for the variance measure with a divisor of N, and if we have a *sample* set of n measurements, then we calculate s^2 for the variance measure with a divisor of $n - 1$.

Example 2.10 Compute the population variance for the six *population* measurements: 1, 4, 6, 7, 4, and 8.

Solution We can compute σ^2 by forming a table of deviations and their squares, as in Table 2.5.

TABLE 2.5 | Computation of σ^2 for data in Example 2.10

x_i	$(x_i - \mu)$	$(x_i - \mu)^2$
1	$(1 - 5) = -4$	$(-4)^2 = 16$
4	$(4 - 5) = -1$	$(-1)^2 = 1$
6	$(6 - 5) = 1$	$(1)^2 = 1$
7	$(7 - 5) = 2$	$(2)^2 = 4$
4	$(4 - 5) = -1$	$(-1)^2 = 1$
8	$(8 - 5) = 3$	$(3)^2 = 9$
		32

$\sigma^2 = \dfrac{32}{6} = 5.33$

Example 2.11 Compute the *population* variance for the six *population* measurements: 0, -10, 20, 30, -15, and 5.

Solution

$$\mu = \frac{\sum_{i=1}^{6} x_i}{6} = \frac{0 - 10 + 20 + 30 - 15 + 5}{6} = \frac{30}{6} = 5$$

$$\begin{aligned}
\sigma^2 &= \frac{\sum_{i=1}^{6} (x_i - \mu)^2}{6} \\
&= \frac{(0 - 5)^2 + (-10 - 5)^2 + (20 - 5)^2 + (30 - 5)^2 + (-15 - 5)^2 + (5 - 5)^2}{6} \\
&= \frac{(-5)^2 + (-15)^2 + (15)^2 + (25)^2 + (-20)^2 + (0)^2}{6} \\
&= \frac{25 + 225 + 225 + 625 + 400 + 0}{6} = \frac{1,500}{6} = 250
\end{aligned}$$

Examples 2.10 and 2.11 indicate that the larger the variance is, the more variable the distributions of values are. In these two examples, both sets of population measurements have the same mean value, but the larger variance in the second example clearly indicates that the second set of population measurements is more variable.

Example 2.12   Compute the sample variance for the following set of six *sample* measurements: 20, −10, 0, 30, −15, and 5.

Solution

$$\bar{x} = \frac{\sum_{i=1}^{n} x_i}{n} = \frac{20 - 10 + 0 + 30 - 15 + 5}{6} = \frac{30}{6} = 5$$

$$
\begin{aligned}
s^2 &= \frac{\sum_{i=1}^{n} (x_i - \bar{x})^2}{n - 1} \\
&= \frac{(20 - 5)^2 + (-10 - 5)^2 + (0 - 5)^2 + (30 - 5)^2 + (-15 - 5)^2 + (5 - 5)^2}{6 - 1} \\
&= \frac{225 + 225 + 25 + 625 + 400 + 0}{5} = \frac{1,500}{5} = 300
\end{aligned}
$$

Notice that the sample data set in Example 2.12 is the same as the population data set in Example 2.11 (only the order of the measurements is different). The means are the same in these two examples ($\mu = 5$ and $\bar{x} = 5$), *but the variances differ* ($\sigma^2 = 250$ and $s^2 = 300$). The difference in the variances is due to the different divisors in the two variance formulas in Definition 2.5.

Standard deviation

Definition 2.6

The standard deviation

The *population standard deviation,* denoted by σ, of a set of N population measurements $x_1, x_2, \ldots, x_N$, is the *positive* square root of the population variance, σ^2:

$$\sigma = \sqrt{\frac{\sum_{i=1}^{N} (x_i - \mu)^2}{N}} \tag{2.5}$$

The *sample standard deviation,* denoted by s, of a set of n sample measurements $x_1, x_2, \ldots, x_n$, is the *positive* square root of the sample variance, s^2:

$$s = \sqrt{\frac{\sum_{i=1}^{n} (x_i - \bar{x})^2}{n - 1}} \tag{2.6}$$

For example, the population standard deviation of the set of *population* measurements in Example 2.11 is:

$$\sigma = \sqrt{250} = 15.81$$

Standard deviation
is the square root
of the variance

It is usually more convenient to work with the standard deviation than the variance, since the standard deviation is in the same units (inches, for example) as the measurements, and the variance is measured in the original units squared (inches², for example). Furthermore, interpretations about the amount of variability in a set of data are more easily drawn using the standard deviation rather than the variance, as we shall see in Section 2.7.

There is an easier way to calculate the variance (and consequently the standard deviation) than the preceding formulas in Definition 2.5. We may use the following *computing formulas* for the population and the sample variances.

Computing formulas for the variance

Definition 2.7

Computing formulas for the population and sample variances

Computing formulas for the population variance σ^2:

$$\sigma^2 = \frac{\sum\limits_{i=1}^{N} x_i^2 - \dfrac{\left(\sum\limits_{i=1}^{N} x_i\right)^2}{N}}{N} = \frac{\sum\limits_{i=1}^{N} x_i^2 - N\mu^2}{N} \qquad (2.7)$$

Computing formulas for the sample variance s^2:

$$s^2 = \frac{\sum\limits_{i=1}^{n} x_i^2 - \dfrac{\left(\sum\limits_{i=1}^{n} x_i\right)^2}{n}}{n-1} = \frac{\sum\limits_{i=1}^{n} x_i^2 - n\bar{x}^2}{n-1} \qquad (2.8)$$

The first formula above for the population variance and the sample variance usually produces a more accurate answer for the variance (less numerical roundoff). In Example 2.13 below, the population variance is computed using the first formula for the population variance in Definition 2.7 above for the set of *population* measurements in Example 2.11.

Example 2.13 Compute the population variance σ^2 using the first formula in Equation 2.7 for the set of *population* measurements in Example 2.11.

Solution First, form a table to sum the values and their squares (see Table 2.6).

At this stage, your primary concern should be to compute the variance and to verify that the formulas in Definitions 2.5 and 2.7 give the same results (apart from numerical roundoff) for the population and sample vari-

TABLE 2.6 | Calculations required for σ^2 (data from Example 2.11)

x_i	x_i^2
0	0
−10	100
20	400
30	900
−15	225
5	25
30	1,650

$$\sigma^2 = \frac{\sum\limits_{i=1}^{6} x_i^2 - \dfrac{\left(\sum\limits_{i=1}^{6} x_i\right)^2}{6}}{6} = \frac{1,650 - \dfrac{(30)^2}{6}}{6}$$

$$= \frac{1,650 - \dfrac{900}{6}}{6} = \frac{1,650 - 150}{6} = \frac{1,500}{6} = 250$$

ances. The interpretation and use of the variance are very important, but these considerations will be reserved for later sections of this chapter and later chapters.

2.6

Other numerical measures

There are numerous representative measures other than the mean, median, and mode. The *geometric mean* is commonly used in business problems to describe the "average" of ratios. Although the geometric mean is not as important as the three principal representative measures (mean, median, and mode), we shall introduce it now while reserving its application until the discussion of Fisher's ideal price index in Chapter 18.

Geometric mean

Definition 2.8

The geometric mean

The *population geometric mean,* denoted by Γ (gamma), of a set of N population measurements $x_1, x_2, \ldots, x_N$ is given by the formula:

$$\Gamma = (x_1 \cdot x_2 \cdot x_3 \cdots x_N)^{1/N} = \left(\prod_{i=1}^{N} x_i\right)^{1/N} \qquad (2.9)$$

where Π represents the product function.

The *sample geometric mean,* denoted by GM, of a set of n sample measurements $x_1, x_2, \ldots, x_n$ is given by the formula:

$$GM = (x_1 \cdot x_2 \cdot x_3 \cdots x_n)^{1/n} = \left(\prod_{i=1}^{n} x_i\right)^{1/n} \qquad (2.10)$$

where Π represents the product function.

For a set of N population measurements, the geometric mean Γ is the Nth root of the product of all N measurements. It is not as easily computed as the arithmetic mean. The computation is eased somewhat by taking logarithms of both sides of equation (2.9):

$$\log(\Gamma) = (1/N)[\log x_1 + \log x_2 + \cdots + \log x_N] = \frac{\sum_{i=1}^{N} \log x_i}{N} \quad (2.11)$$

Thus the *log* of the population geometric mean Γ is the *arithmetic mean* of the *logs* of the measurements. Using equation 2.11, we can compute the arithmetic mean of the logs of the measurements and take the antilog of the result to produce the population geometric mean Γ. The sample geometric mean GM can be computed in the same way:

$$\log(GM) = (1/n)[\log x_1 + \log x_2 + \cdots + \log x_n] = \frac{\sum_{i=1}^{n} \log x_i}{n} \quad (2.12)$$

Bankers, brokers, and others in the financial community use the geometric mean to compute the return on certain classes of investments. For example, the computation of the geometric mean is used frequently in bond valuation. These computations are illustrated in the following example.

Example 2.14 We will suppose for the moment that we are in a period of interest rate decline (true in late 1985 and early 1986). Based upon government projections, we estimate the return on a certain class of investments to be as follows:

Year	Interest rate earned
1987	14%
1988	12
1989	10
1990	8
1991	7

Let us further suppose that an investor has $1,000 to invest at the beginning of 1987 and that she wishes to accumulate and reinvest her funds until December 31, 1991, when she begins college. What constant rate of interest would the investor have to earn on a bond or other financial instrument in order to equal the accumulation from the above investment alternative? Assume that all accumulated funds can be reinvested at the determined rate of interest.

Solution First we will determine the accumulation from the variable rate alternative (remember that all funds are to be reinvested at the "current" rate of interest).

Year 1: $1,000.00 + 14\%(\$1,000.00) = \$1,140.00$
Year 2: $1,140.00 + 12\%(\$1,140.00) = \$1,276.80$
Year 3: $1,276.80 + 10\%(\$1,276.80) = \$1,404.48$
Year 4: $1,404.48 + 8\%(\$1,404.48) = \$1,516.84$
Year 5: $1,516.84 + 7\%(\$1,516.84) = \$1,623.02$

Hence, at the end of the fifth year, the investor will have accumulated $1,623.02 for her college education.

The astute reader will recognize that this same quantity can be more easily determined by using the compound interest formula:

$$\text{Accumulation} = P\left[\prod_{t=1}^{n}(1 + I_t)\right]$$

where Π represents the product function, P represents the principal amount invested, and I_t represents the interest rate available during time period t. Using the data provided,

$$\begin{aligned}\text{Accumulation} &= \$1,000[(1.14)(1.12)(1.10)(1.08)(1.07)] \\ &= \$1,000[1.62302] \\ &= \$1,623.02\end{aligned}$$

In order to determine the constant rate of interest that will yield $1,623.02 at the end of the fifth year on a $1,000 investment at the given annual interest rates, we need to compute the *geometric mean* of the quantities 1.14, 1.12, 1.10, 1.08, and 1.07 in the brackets above, and then subtract 1 to get the interest rate:

$$\begin{aligned}[(1.14)(1.12)(1.10)(1.08)(1.07)]^{1/5} - 1 &= 1.1017032 - 1 \\ &= 0.1017032\end{aligned}$$

A bond with a constant annual interest rate of 10.17032 percent and a maturity of five years will thus produce the same return as the above investment. We can verify this as follows:

Year 1: $1,000.00 + 10.17032\%(\$1,000.00) = \$1,101.70$
Year 2: $1,101.70 + 10.17032\%(\$1,101.70) = \$1,213.75$
Year 3: $1,213.75 + 10.17032\%(\$1,213.75) = \$1,337.19$
Year 4: $1,337.19 + 10.17032\%(\$1,337.19) = \$1,473.19$
Year 5: $1,473.19 + 10.17032\%(\$1,473.19) = \$1,623.02$

Hence the geometric mean is important in finance for determining yields on alternative financial instruments with differing rates of interest and different timings of these interest payments.

Example 2.15   Find the arithmetic mean and the geometric mean of the *population* set of measurements: 100, 100, 100, and 1,000.

Solution   Population arithmetic mean:

$$\mu = \frac{100 + 100 + 100 + 1,000}{4} = 325$$

Population geometric mean:

$$\log\Gamma = (1/4)[\log100 + \log100 + \log100 + \log1{,}000]$$
$$= (1/4)[2 + 2 + 2 + 3] = (1/4)[9] = 2.25$$
$$\Gamma = 10^{2.25} = 177.8$$

by using logs to the base 10.

This latter example illustrates that the geometric mean is less affected by one (or a few) extremely large (or small) measurements in a data set than is the arithmetic mean. Unfortunately, it is neither simple to compute, nor is it amenable to use for making statistical inferences. Still it is useful in analyzing financial payment streams as indicated in Example 2.14, and it is useful in analyzing ratios—a process that frequently arises in computing a cost of living or other index numbers as described in Chapter 18.

2.6.2 | Quartiles and percentiles

The mean, median, and mode can be thought of as measures of location—they attempt to locate the most representative value. Other measures of location are *quartiles* and *percentiles*.

100*p*th percentile

Definition 2.9
The 100*p*th percentile

The *100pth percentile* of either a *population* or a *sample* set of measurements that have been ordered from the smallest to the largest in value is a value so that at least 100*p* percent of the measurements are at or to the left of (less than or equal to) this value *and* at least 100(1 − *p*) percent of the measurements are at or to the right of (greater than or equal to) this value in the ordered array of measurements, ordered from the smallest to the largest.

Example 2.16 Find the 20th percentile for the data set:

20 34 17 18 28 33 12 15 17 12 41
45 18 19 16 21 26 14 26 13 29

Solution Ordered from the smallest to the largest, the 21 measurements are:

12 12 13 14 15 16 17 17 18 18 19
20 21 26 26 28 29 33 34 41 45

To determine the 20th percentile, we look for a value so that at least $100(0.20) = 20$ percent of the measurements are at or to the left of this value and at least $100(1 − 0.20) = 80$ percent of the measurements are at or to the right of this value. The measurement, 15, satisfies these conditions: at least 20 percent of the measurements [(0.20)(21) = 4.2] are at or to the left of 15 (exactly 5 measurements are at or to the left of 15), and at least 80 percent of the measurements [(0.80)(21) = 16.8] are at or to the right of 15 (exactly 17 measurements are at or to the right of 15). Therefore, the 20th percentile is 15.

Definition 2.10

The lower, middle, and upper quartiles

For either a *population* or a *sample* data set, the *lower quartile* Q_1 of a set of measurements is the 25th percentile. The *middle quartile* Q_2 is the 50th percentile. The *upper quartile* Q_3 is the 75th percentile.

Quartiles

Example 2.17 Find Q_1, Q_2, and Q_3 for the data set in Example 2.16.

Solution For Q_1 at least 25 percent of the ordered measurements should be at or to the left of Q_1 and at least 75 percent of the ordered measurements should be at or to the right of Q_1. Thus at least (0.25)(21) = 5.25 measurements are at or to the left of Q_1, and at least (0.75)(21) = 15.75 measurements are at or to the right of Q_1. Therefore Q_1 is 16.

For Q_2 at least 50 percent of the ordered measurements should be at or to the left of Q_2, and at least 50 percent of the ordered measurements should be at or to the right of Q_2. Thus at least (0.50)(21) = 10.5 measurements are at or to the left of Q_2, and at least (0.50)(21) = 10.5 measurements are at or to the right of Q_2. Therefore Q_2, the median, is equal to 19.

For Q_3 at least 75 percent of the ordered measurements should be at or to the left of Q_3, and at least 25 percent of the ordered measurements should be at or to the right of Q_3. Thus at least (0.75)(21) = 15.75 measurements are at or to the left of Q_3, and at least (0.25)(21) = 5.25 measurements are at or to the right of Q_3. Therefore Q_3 is 28.

By calculating the difference between the lower and upper quartiles, we can get some feel for the variability of the data. The difference, $Q_3 - Q_1$, is called the *interquartile range*. The larger the interquartile range is for a data set, the more variable (spread out) the set of measurements is.

2.6.3 | Skewness

As suggested earlier in this chapter, one possible measure of the *skewness* of a distribution of a set of measurements is the difference between its mean and its median. We use a function of this difference as a measure of skewness.

> **Definition 2.11**
> The skewness measure
> (Pearson's second coefficient of skewness)
>
> The *population skewness measure*, denoted by γ, of a set of N population measurements $x_1, x_2, \ldots, x_N$ is given by the formula:
>
> $$\gamma = \frac{3(\mu - \mathfrak{M})}{\sigma} \qquad (2.13)$$
>
> where μ = Population mean, $\mathfrak{M}$ = Population median, and σ = Population standard deviation.
> The *sample skewness measure*, denoted by SK, of a set of n sample measurements $x_1, x_2, \ldots, x_n$ is given by the formula:
>
> $$SK = \frac{3(\bar{x} - M)}{s} \qquad (2.14)$$
>
> where $\bar{x}$ = Sample mean, M = Sample median, and s = Sample standard deviation.

If the distribution is skewed to the right, the mean will be larger than the median, and the skewness measure (population or sample) will be positive. If the distribution is skewed to the left, the mean will be smaller than the median, and the skewness measure will be negative.

The effect of dividing the skewness measure by the standard deviation is to provide a measure that is not dependent on the unit of measurement. The mean, median, and the standard deviation are all measured in the same units for a given data set.

The skewness measure can be used in two ways. First, the sign of the skewness measure indicates the direction of skewness: +, skewed right; and −, skewed left. Second, if the skewness measure is larger in magnitude in one data set than another, the first data set distribution is more skewed than the other. So it can be used in a relative sense to compare the skewness in the distributions of two or more data sets.

Example 2.18 Compute the skewness measure for the following set of $n = 5$ *sample* measurements: 10, 4, 4, 6, and 1.

Solution The sample mean, median, and standard deviation for this sample data set are $\bar{x} = 5$, $M = 4$, and $s = 3.32$. Therefore, the sample skewness measure is given by

$$SK = \frac{3(5 - 4)}{3.32} = 0.90$$

Since SK is positive, the distribution of the five sample measurements is skewed to the right.

There are other properties of a set of measurements and their corresponding frequency distributions than representative measure, variance, and skewness. But these three numerical measures provide the foundation for almost all the basic methods of statistical inference. We will therefore discuss no other kinds of numerical measures than those presented here.

2.7

The Chebychef theorem: The use of the standard deviation

The representative measures convey an intuitive idea of a property of the distribution of a set of measurements—the location of a representative or "middle" value of the distribution. The variance and its positive square root, the standard deviation, are not so easily interpreted; they measure dispersion, but a numerical value of the variance does not help us in visualizing the distribution of the measurements. A theorem by the Russian mathematician Chebychef does provide a means of using the standard deviation (either population or sample) to gain insights into the dispersion of a set of measurements.

Theorem 2.1
Chebychef theorem

Given a set of measurements and a number k greater than or equal to 1 ($k \geq 1$), at least $(1 - 1/k^2)$ of the measurements will lie within k standard deviations of their mean value.

The remarkable property of the Chebychef theorem is that it is applicable to *any* set of measurements. The result of the theorem is illustrated in Figure 2.16 for a *population* set of measurements with population mean μ and

FIGURE 2.16 | Illustration of the Chebychef theorem

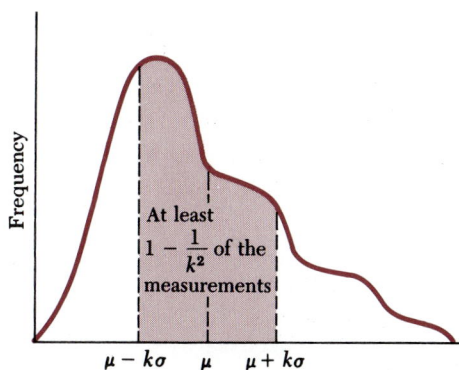

population standard deviation σ. In the interval $\mu - k\sigma$ to $\mu + k\sigma$, at least $(1 - 1/k^2)$ of the measurements will reside. To see the effect of selecting different values of k, in Table 2.7 a few values of k are selected and the quantity $(1 - 1/k^2)$ calculated. When $k = 1$, at least 0 of the measurements lie in the interval from $\mu - \sigma$ to $\mu + \sigma$; thus selecting $k = 1$ in the Chebychef theorem does not produce a useful result. But, say, when $k = 3$, at least 8/9 of the measurements lie in the interval from $\mu - 3\sigma$ to $\mu + 3\sigma$.

TABLE 2.7 | Values of k and $(1 - 1/k^2)$ for the Chebychef theorem

k	$1 - 1/k^2$
1	0
3/2	5/9
2	3/4
5/2	21/25
3	8/9

Example 2.19 Suppose a set of *population* measurements has a mean of $\mu = 50$ and a standard deviation of $\sigma = 10$. Apply the Chebychef theorem to these population data to describe their dispersion.

Solution In Table 2.8 three values of k (3/2, 2, and 3) are chosen and the corresponding intervals computed with $\mu = 50$ and $\sigma = 10$. These results are illustrated in Figure 2.17.

TABLE 2.8 | Application of the Chebychef theorem to data from Example 2.19

k	$1 - 1/k^2$	Interval $\mu - k\sigma$ to $\mu + k\sigma$	
3/2	5/9	$50 - (3/2)10$ to $50 + (3/2)10$;	35–65
2	3/4	$50 - (2)10$ to $50 + (2)10$;	30–70
3	8/9	$50 - (3)10$ to $50 + (3)10$;	20–80

The mean of the measurements is 50. At least 5/9 of the measurements fall between 35 and 65, at least 3/4 between 30 and 70, and at least 8/9 between 20 and 80. Thus 1/9 or less of the measurements are less than 20 or greater than 80. It is not necessarily true that half the 1/9 of the measurements outside the interval are to the left of 20 and the other half are to the right of 80, because the distribution may be skewed either to the right or to the left.

FIGURE 2.17 | Illustration of the Chebychef theorem applied to Example 2.19

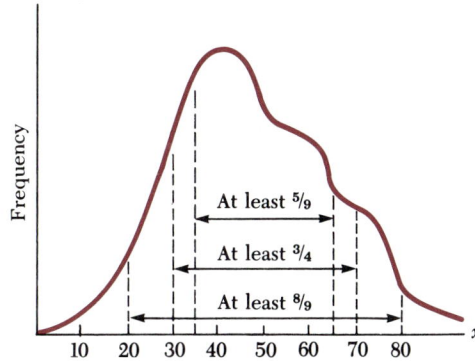

Example 2.20 In Example 2.1 reference was made to the personnel files for a *sample* of 100 employees in a manufacturing firm with 643 total employees. The data for these 100 employees are given in Table 1, Appendix C. Suppose the company is interested in describing the dispersion of the annual wages for these 100 employees. The sample mean wage and the standard deviation of the wages are $\bar{x} = \$35,484$ and $s = \$14,074$, respectively. Using the Chebychef theorem, answer the following:

a. Determine an interval that contains at least 80 percent of the wages of these 100 employees.

b. At least what percentage of the wages are contained in the interval $299 to $70,669?

Solution

a. From the Chebychef theorem, we know that the interval $\bar{x} \pm ks$ contains at least a proportion $1 - 1/k^2$ of the measurements in a data set. We wish to construct an interval containing at least a proportion of 0.80 (80 percent) of the wages. Thus we need to find the value of k. This can be done by setting the proportion $1 - 1/k^2$ equal to the desired proportion 0.80 and solving algebraically for k:

$$1 - 1/k^2 = 0.80$$
$$1/k^2 = 0.20$$
$$k^2 = 1/0.20 = 5$$
$$k = \sqrt{5} = 2.24$$

Thus the interval is $\bar{x} \pm 2.24s = \$35,484 \pm 2.24(\$14074)$, or $3958.24 to $67,009.76. Thus the Chebychef theorem guarantees that at least 80 percent of the annual wages are between $3958.24 and $67,009.76.

In general, if we let p represent the minimum proportion of measurements contained in the desired interval, then k can be found from the following formula:

$$k = \sqrt{\frac{1}{1-p}}$$

For example, in part a we wish an interval that contains at least 80 percent of the wages. Therefore, $p = 0.80$ and

$$k = \sqrt{\frac{1}{1-0.80}} = \sqrt{\frac{1}{0.20}} = \sqrt{5} = 2.24$$

b. In this part we know the interval ($299 to $70,669) and wish to find at least what proportion of the wages for the 100 employees falls in this interval. The Chebychef interval is $\bar{x} \pm ks$—this interval is symmetric about the mean value $\bar{x}$. The constant k represents the number of standard deviations to move below the mean and above the mean to produce the endpoints of the interval. Thus we first subtract the lower limit from the mean and the mean from the upper limit:

$$\$35,484 - \$299 = \$35,185$$
$$\$70,699 - \$35,484 = \$35,185$$

If the resulting numbers are not the same, then the interval is not symmetric about the mean, and we cannot answer the question by the Chebychef theorem. Since the numbers are the same ($35,185), we can now divide this number by the standard deviation of $14,074 to determine how many standard deviations (k) above and below the mean the endpoints of the interval are:

$$k = \frac{\$35,185}{\$14,074} = 2.5$$

Now the desired proportion is:

$$1 - 1/k^2 = 1 - 1/(2.5)^2 = 1 - 1/6.25 = 1 - 0.16 = 0.84$$

Thus at least 84 percent of the 100 employee wages are contained in the interval between $299 and $70,669.

From Table C.1 the actual percentage of annual wages contained in this interval is 97 percent (97 out of 100). Thus the Chebychef theorem produces a very conservative percentage in this case.

The Chebychef theorem is used when we have no idea of the nature of the distribution of a set of measurements, since it applies to *any* data set (population or sample). Though it tends to give conservative estimates of the percentage of measurements contained in a given interval, it is very useful in providing an initial idea of the spread of the measurements in a data set.

■ 2.8

Descriptive statistics via computer statistical software analysis

The widespread availability of computer statistical software packages has greatly enhanced the analysis of data sets. Until recently, most of these packages were available only on large mainframe computers. Now there are several packages designed to operate on microcomputers. Their existence has made it much easier to use statistical software for data analysis. In the reference section at the end of this chapter, we list several of the available software packages and indicate the types of machines (mainframe or micro-computer) for which they are designed. In this section, we will demonstrate the use of MINITAB, a very popular statistical software package available for both mainframe computers and microcomputers.

For the purpose of demonstrating how MINITAB can assist in the use of descriptive statistics, let us return to the data described in Example 2.1. In Example 2.1 data from the personnel files of a medium-size manufacturing plant were introduced. These data are given in Table C.1, in the back of the book. The following characteristics were measured for a *sample* of 100 employees (total company size: 643 employees):

1. Sex (1 = Male, 0 = Female).

2. Age in years.

3. Number of dependents.

4. Years working for the company.

5. Employee rating in most recent evaluation.

6. Whether or not employee holds a college degree (1 = College degree, 0 = No college degree).

7. Whether or not employee participates in a stock profit sharing plan (1 = Employee participates, 0 = Employee does not participate).

8. Job classification (1 = Laborer, 2 = Sales, 3 = Managerial/Executive).

9. Annual wage, not including benefits or bonuses.

The company is interested in summarizing these characteristics for the sample of 100 employees for the purpose of developing an employee profile.

The data listed in Table 1, Appendix C, are typed into a disk file labeled PERSONNEL.DATA on the computer. The software package MINITAB reads the data by using the following command:

```
READ 'PERSONNEL.DATA' INTO COLUMNS C1-C9
```

The columns C1 through C9 now contain the measurements for the nine characteristics listed above. MINITAB allows the columns to be named for easy reference. The following names are used:

1. Sex   MF

2. Age   AGE

3. Number of dependents   NODEP

4. Years with company   YEARS

5. Employee rating   RATING

6. College degree   COLDEG

7. Profit sharing plan   STOCK

8. Job classification   JCLASS

9. Annual wage   WAGE

As a first step in describing these characteristics, we use MINITAB to find the sample mean $\bar{x}$ and the sample standard deviation s for each of the 9 characteristics by using the DESCRIBE command:

```
                  DESCRIBE COLUMNS C1-C9

    MF       N = 100   MEAN =    0.83000   ST.DEV. =     0.378
    AGE      N = 100   MEAN =     34.070   ST.DEV. =      10.7
    NODEP    N = 100   MEAN =     1.3400   ST.DEV. =      1.25
    YEARS    N = 100   MEAN =     9.7800   ST.DEV. =      7.77
    RATING   N = 100   MEAN =     85.550   ST.DEV. =      13.6
    COLDEG   N = 100   MEAN =    0.44000   ST.DEV. =     0.499
    STOCK    N = 100   MEAN =    0.72000   ST.DEV. =     0.451
    JCLASS   N = 100   MEAN =     1.6500   ST.DEV. =     0.809
    WAGE     N = 100   MEAN =     35484.   ST.DEV. =    14074.
```

From the output of the DESCRIBE command, we see among other things that there are 83 percent males (the average of the 1s—males—and 0s—females—is 0.83), the average age of the employees is 34.07 years, with a standard deviation of 10.7 years; the average number of years with the company is 9.78 years; 44 percent of the employees hold a college degree; 72 percent participate in the stock profit sharing plan; and the average annual wage is $35,484, with a standard deviation of $14,074.

MINITAB also produces simple graphical displays of the information. The HISTOGRAM command automatically establishes the class width and class midpoint. Rather than using rectangles for the histogram, it uses asterisks to indicate the frequencies in each class. Further, the histogram is printed on its side for printing convenience. For example, the command

```
                  HISTOGRAM COLUMN C9
```

produces the following graph of the annual wages:

```
WAGE

  MIDDLE OF   NUMBER OF
  INTERVAL    OBSERVATIONS
    20000.       11     **********
    25000.       31     ******************************
    30000.       13     *************
    35000.        8     ********
    40000.       13     *************
    45000.        3     ***
    50000.        6     ******
    55000.        4     ****
    60000.        6     ******
    65000.        2     **
    70000.        1     *
    75000.        1     *
    80000.        0
    85000.        1     *
```

Although this "histogram" is not as effective as the one in Example 2.1 at the beginning of this chapter, it nevertheless captures the important properties of the annual wages. The frequency distribution of the wages is skewed to the right, and most of the wages occur toward the lower end of the wage distribution. Histograms from MINITAB for the remaining eight employee characteristics are shown below.

```
MF
 EACH * REPRESENTS 2 OBSERVATIONS

 MIDDLE OF   NUMBER OF
 INTERVAL    OBSERVATIONS
     0.        17     ********
     1.        83     *****************************************

AGE

 MIDDLE OF   NUMBER OF
 INTERVAL    OBSERVATIONS
    20.        10     *********
    25.        27     **************************
    30.        15     **************
    35.        16     ***************
    40.         8     ********
    45.         8     ********
    50.        10     *********
    55.         4     ****
    60.         2     **
```

```
NODEP

  MIDDLE OF   NUMBER OF
  INTERVAL    OBSERVATIONS
      0.        31    ******************************
      1.        31    ******************************
      2.        18    ******************
      3.        14    **************
      4.         5    *****
      5.         1    *

YEARS

  MIDDLE OF   NUMBER OF
  INTERVAL    OBSERVATIONS
      0.         5    *****
      4.        34    **********************************
      8.        22    **********************
     12.        13    *************
     16.         5    *****
     20.        12    ************
     24.         4    ****
     28.         3    ***
     32.         1    *
     36.         1    *

RATING

  MIDDLE OF   NUMBER OF
  INTERVAL    OBSERVATIONS
     50.         1    *
     55.         3    ***
     60.         5    *****
     65.         4    ****
     70.         8    ********
     75.         5    *****
     80.         9    *********
     85.         5    *****
     90.        21    *********************
     95.        18    ******************
    100.        21    *********************
```

```
COLDEG
 EACH * REPRESENTS 2 OBSERVATIONS

 MIDDLE OF   NUMBER OF
 INTERVAL    OBSERVATIONS
      0.        56     ***************************
      1.        44     *********************
STOCK
 EACH * REPRESENTS 2 OBSERVATIONS

 MIDDLE OF   NUMBER OF
 INTERVAL    OBSERVATIONS
      0.        28     **************
      1.        72     ************************************
JCLASS
 EACH * REPRESENTS 2 OBSERVATIONS

 MIDDLE OF   NUMBER OF
 INTERVAL    OBSERVATIONS
      1.        56     ***************************
      2.        23     ************
      3.        21     **********
```

Notice that the age distribution of the employees is also skewed to the right (AGE), as is the distribution of the number of years the employee has been with the company (YEARS). The distribution of most current employee rating (RATING) is skewed to the left—most of the employees sampled have received high ratings. The histogram for job classification (JCLASS) readily shows that there are 56 laborers, 23 employees in sales, and 21 manager/executive types.

If we wished to know the minimum, maximum, and median for each characteristic, say wages, we could use the following commands:

```
MINIMUM OF WAGES IN C9
MINIMUM = 18960

MAXIMUM OF WAGES IN C9
MAXIMUM = 82500

MEDIAN OF WAGES IN C9
MEDIAN = 29808
```

Thus the minimum wage is $18,960; the maximum wage is $82,500; and the median of the wages is $29,808. Noting that the mean of the wages is $35,484, since the median is less than the mean, the distribution of wages is skewed to the right (toward the higher wages).

The MINITAB PLOT command produces plots of two variables using vertical and horizontal axes. These plots allow us to understand quickly the relationships between pairs of characteristics. For example, invoking the command

```
PLOT WAGES IN C9 AGAINST AGE IN C2
```

produces the following plot:

The numbers in the plot indicate x, y coordinate points where two or more pairs of wages and ages fall. Not surprisingly, as the age of the employee increases, the annual wage increases as well in a very definite linear fashion. Four additional plots are shown below: wage against years with the firm, wage against employee rating, wage against college degree, and wage against job classification (note: a + on the plot indicates that more than nine points fall at the x, y coordinate).

As the number of years with the firm increases, the first plot shows that the annual wage also increases in a linear fashion. The plot of wages against the employee rating shows that there is a definite relationship between higher annual wages and higher ratings, although the relationship is not as strong as the firm may hope for. From the plot of wages against college degrees, it is clear that those with college degrees have higher annual wages than those without college degrees. Finally, the plot of wages against job classification clearly indicates that it is better to be a manager/executive type (3) than a sales type (2) or a laborer (1)!

MINITAB can also be used to produce average wages for specified classes of employees.[1] For example, the output below shows mean wage and the standard deviation of wages for employees by job classification (1 = Laborer, 2 = Sales, and 3 = Manager/executive). Notice that the manager/executive types (3) have both a much larger mean wage and a standard deviation of wages ($53,857, $13,699) than do the laborers ($27,136, $5,896).

LEVEL	N	MEAN	ST. DEV.
1	56	27136.	5896.
2	23	39032.	11003.
3	21	53857.	13699.

Below are listed other group comparisons of wages that are of interest. The first output table gives the means and standard deviations of wages for those with (1) and without (0) a college degree.

LEVEL	N	MEAN	ST. DEV.
0	56	27228.	5930.
1	44	45990.	14469.

The next output table gives the means and standard deviations of wages for males (1) and females (0). Clearly, there is quite a difference in the mean annual wages for males and females in this sample of 100 employees.

LEVEL	N	MEAN	ST. DEV.
0	17	26651.	5265.
1	83	37293.	14637.

The use of such computer software packages as MINITAB make it very easy to quickly and effectively describe large data sets containing measurements of several characteristics. In this illustration, these descriptive statis-

---

[1] The average wages printed below are part of the output from the MINITAB ONEWAY analysis of variance command. We will discuss the use of this command in Chapter 12.

tics techniques provide useful information in preparing a profile of the employees of the manufacturing firm.

■ **2.9**

More on descriptive statistical techniques

In this section we introduce relatively new graphical procedures beyond pie charts, line and bar graphs, and histograms. In addition, examples illustrating the correct procedures for drawing histograms are reviewed with examples of histograms that are distorted to emphasize a point in violation of these procedures. Finally, examples of distortions in the use of words to represent and to describe counts of things are presented.

2.9.2 | Stem and leaf plots

Stem and leaf plot

A *stem and leaf* plot is an alternate way of producing a histogram-type graph. Instead of using asterisks to indicate frequencies as in the MINITAB histogram plots, the actual data are used. The basic idea of a stem and leaf plot is illustrated in the following example.

Example 2.21 Consider the following set of measurements (they may represent either a population or a sample set):

24, 56, 64, 33, 27, 22, 42, 48, 21, 36, 61, 26, 33, 30
23, 29, 25, 25, 37, 41, 55, 30, 21, 40, 32, 22, 20, 48

To form a stem and leaf plot for these numbers, we treat the first digit (the 10's place) as the stem and the second digit as the leaf (the 1's place) to produce the following plot:

```
2 | 011223455679
3 | 0023367
4 | 01288
5 | 56
6 | 14
```

For example, in the row beginning with 4   01288, the number 4 represents the first digit of the two-digit numbers, and numbers 01288 represent the second digit of the two-digit numbers. Thus this row corresponds to the numbers 40, 41, 42, 48, and 48. Notice within each of the two-digit sets, the numbers are ordered from the smallest to the largest. It is not necessary to do so in a stem and leaf plot, but it does help in making the plot more appealing to the eye. Notice that the stem and leaf plot is essentially equivalent to the MINITAB histogram laid on its side. However, the stem and leaf plot replaces the asterisks in MINITAB's histogram with the second digit in the two-digits numbers, thus providing useful additional information about the data set. In this stem and leaf plot, it is evident immediately that the distribution of the numbers is skewed to the right (long tail toward the higher values).

Now let's turn to the Example 2.1 data set (the data are given in Table 1, Appendix C). Given below is the stem and leaf plot of the ages of the 100 sampled employees of the manufacturing firm:

```
STEM-AND-LEAF DISPLAY OF AGE
LEAF DIGIT UNIT = 1.0000
1 2 REPRESENTS 12.

    1     1. 9
   21     2* 0012222223344444444
   46     2. 5555556666667777888888999
  (12)    3* 001122333444
   42     3. 5666666677888
   29     4* 00011334
   21     4. 566668889
   12     5* 0222224
    5     5. 5668
    1     6* 0
```

The minimum age of these employees is 19 years, and the maximum is 60. The median age is 31.5 years. These numerical measures can be verified from the data set in Table 1, Appendix C, for the age measure. Since these ages are two-digit numbers, the stem is taken as the first digit and the leaf as the second digit. Notice that the MINITAB stem and leaf plot splits each 10's group into two groups to show more of the distribution of the ages. For example, the stem 2* represents the ages from 20 to 24, and the stem 2. represents the ages from 25 to 29. This stem and leaf plot has some extra features beyond the plot in Example 2.21 above. The label of the plot includes the statement "Leaf digit unit = 1.0000." This indicates that the number in each row represents one unit (one year, in this case). The second statement in the label is "1 2 represents 12." This tells us that the stem of 1 together with the leaf of 2 represents the number 12 (twelve). Finally, the left-most column contains useful information. The row corresponding to the number in parentheses contains the *median* of the set of numbers. We know that the median of these 100 ages is 31.5. Since the size of the data set is even, the median is the average of the "two middle" numbers, which turn out to be 31 and 32 in the ordered array of ages from the smallest to the largest. Since both the numbers 31 and 32 satisfy the definition of the 50th percentile (the median), either of these two numbers can be viewed as representing the median. The number in the parentheses tells us how many numbers are in the row containing the median (12). The numbers above the number in parentheses tell us how many numbers are in or above the specific row, and the numbers below the number in parentheses tell us how many numbers are in or below the specific row. For example, in the second row (a stem of 2*), the number 21 tells us that there are 21 ages in the first and second row. In the third row (a stem of 2.), the number 46 tells us that there are 46 numbers in the first, second, and third rows. In the 7th row (a stem of 4.), the number 21 tells us that there are 21 ages in the 7th, 8th, 9th, and 10th rows.

The stem and leaf plot of the annual wages for the 100 sampled employees is given below:

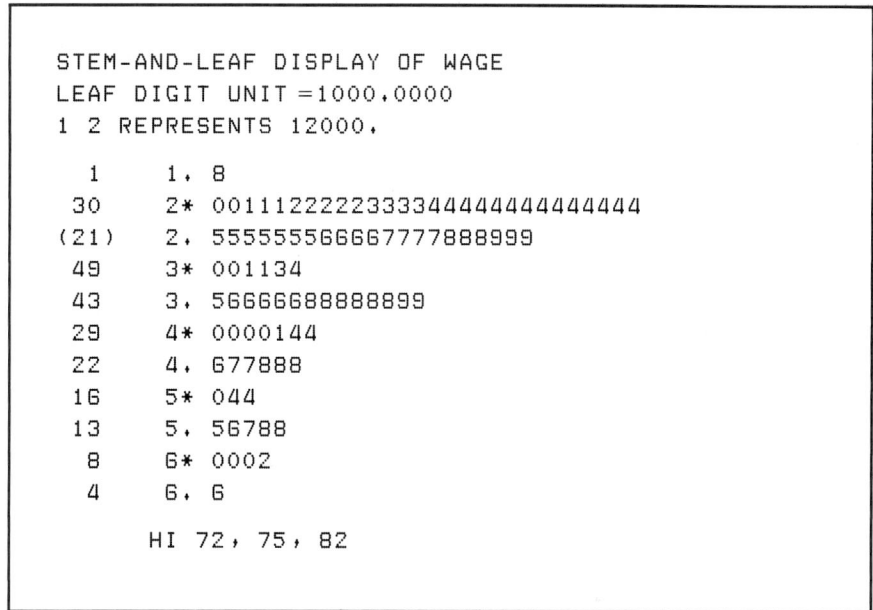

```
STEM-AND-LEAF DISPLAY OF WAGE
LEAF DIGIT UNIT =1000.0000
1 2 REPRESENTS 12000.

    1      1. 8
   30      2* 0011122222333344444444444444
  (21)     2. 555555566667777888999
   49      3* 001134
   43      3. 56666688888899
   29      4* 0000144
   22      4. 677888
   16      5* 044
   13      5. 56788
    8      6* 0002
    4      6. 6

         HI 72, 75, 82
```

In the label, the statement "leaf digit unit = 1,000.0000" tells us that each leaf is equivalent to 1,000 (dollars). The statement "1   2 represents 12,000" tells us that a stem of 1 and a leaf of 2 represents 12,000 (dollars). Therefore in the first row [1.   8] represents a wage of $18,000. Actually, the minimum wage is $18,960. In this stem and leaf plot, only the first two digits of the numbers are used. Thus for a wage of $18,960, the stem is 1, and the leaf is 8—the 960 is not used in the plot. The median wage is contained in the third row (a stem of 2.) and the last three wages (72, 75, and 82) are not directly included in the stem and leaf plot. When there are a very few big or small numbers, the convention is to not include them in the plot so as not to stretch the plot out too much. Rather, the numbers are listed above or below the plot so that they may be checked for accuracy. These numbers represent "outliers"—numbers that are quite a bit removed from the rest of the numbers. Often, outliers represent data encoding errors that are easily identified by using this convention. In this case the three highest wages are $72,475, $75,000, and $82,500. Although these numbers are unusually high when compared with the rest of the wages, they are not encoding mistakes—these wages belong to high-ranking executives in the manufacturing firm.

Outlier

2.9.2 | Box plots

Box plot

Another interesting way to summarize numerical measures in a data set is a *box plot*. A box plot (produced by MINITAB) for the ages of the 100 employees of the manufacturing firm is shown below:

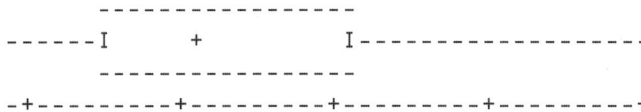

```
 -- BOXPLOT AGES

                  -----------------
          ------I      +           I--------------------
                  -----------------
          -+---------+---------+---------+---------+
 ONE HORIZONTAL SPACE = 0.10E + 01
 FIRST TICK AT      20.000
```

The label below the plot tells us the spacing and at what value the plot begins. The statement "One horizontal space = 0.10E + 01" tells us that each horizontal space (−) is equal to 1 unit. The number 0.10 E + 01 uses scientific notation: E stands for exponent, and +01 means to raise 10 to the 01 power and multiply it times the number to the left of E—$10^1(0.10) = 1$. An easier way to deal with numbers in scientific notation is as follows: If the number to the right of E is positive (+), then move the decimal point in the number to the *left* of E to the *right* by this number of places. In the number 0.10E + 01, we would move the decimal place in 0.10 one place to the right, giving 1.0. The number 0.10E + 04 is 1,000 (the decimal place in 0.10 is moved four places to the right). If the number to the right of E is negative (−), then move the decimal point in the number to the *left* of E to the *left* by this number of places. Thus the number 0.10E − 02 is 0.001.

Returning to the boxplot, the second statement in the label, "First tick at 20.000," indicates that the first + on the bottom line corresponds to 20 (years). Since each tick is 1 year, the next tick (+) corresponds to 30 years, and so forth. The + in the box represents the location of the median, and the ends of the box represent the first quartile (on the left) and the third quartile (on the right). The lines extending from the left of the box and from the right of the box represent the range of the numbers. Since the minimum age is 19, and the maximum age is 60, these lines extend to the left to 19 and to the right to 60.

The boxplot enables us to quickly form an impression of where the "center" of the numbers are (the median + in the box), where most of the numbers are (the interquartile range—from the first quartile (25th percentile) to the third quartile (75th percentile), and how spread out the numbers are (the range given by the lines from each side of the box). Further, we can see immediately if the distribution of the numbers (ages) is skewed or symmetric. Here, it is apparent that the distribution of the ages is skewed to the right, since the box is moved toward the left range of the ages, and the long line is to the right of the box.

The boxplot for the annual wages is shown below:

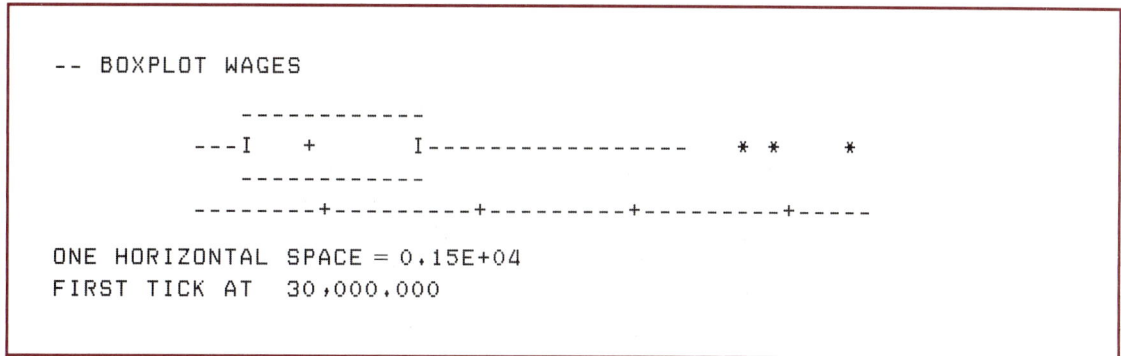

```
--  BOXPLOT WAGES

                  - - - - - - - - - - -
        - - - I     +         I - - - - - - - - - - - - - - - -     * *     *
                  - - - - - - - - - - -
        - - - - - - - + - - - - - - - - - - + - - - - - - - - - + - - - - - - - - - + - - - - -
ONE HORIZONTAL SPACE = 0.15E+04
FIRST TICK AT   30,000,000
```

The label tells us that the first tick on the axis is at 30,000 ($30,000) and that each horizontal space is equal to 0.15E + 04 (which equals 1,500—moving the decimal place in 0.15 four places to the right). The three asterisks at the right end of the plot denote outliers—these represent the three very high wages of $72,475, $75,000, and $82,500. The boxplot shows that the annual wages are skewed to the right and that most of the wages are at the lower end of the range.

2.9.3 | More on histograms

Fundamental rule in constructing histograms

A fundamental rule in constructing a histogram is that the ratio of the area of the rectangle to the frequency in each class must be the same for all classes. This rule is violated frequently in histograms published in newspapers, magazines, and corporation annual reports. To illustrate this rule, we will begin with the following example.

Example 2.22 Consider the frequency distribution of gasoline prices for regular unleaded gasoline (dollars per gallon) for 100 sampled gas stations nationally in mid-1985 in Table 2.9.

The histogram for the frequency distribution in Table 2.9 is given in Figure 2.18. Since each class has an equal class width (0.10), the heights of the rectangles can be set by the frequencies in each of the classes. By doing so, the ratio of the *area* of each rectangle to the frequency in each class will be the same for all classes. For example, the area of the rectangle in the first class is (40)(0.10) = 4. The ratio of this area to the frequency in the first class is 4/40 = 0.10, the class width. This ratio holds for all classes. Consider, for example, the fourth class. The area of the rectangle is (10)(0.10) = 1. The ratio of this area to the class frequency is 1/10 = 0.10.

Now suppose that the first and second classes of the frequency distribution in Table 2.9 are combined into one class with a width of 0.20. The resulting frequency distribution is given in Table 2.10.

TABLE 2.9 | Frequency distribution for the price per gallon in dollars regular unleaded gasoline at 100 gas stations sampled nationally in June 1985

Class number	Class	Frequency
1	$1.10 but less than $1.20	40
2	$1.20 but less than $1.30	30
3	$1.30 but less than $1.40	15
4	$1.40 but less than $1.50	10
5	$1.50 but less than $1.60	5
Total:		100

FIGURE 2.18 | Histogram of the frequency distribution in Table 2.9

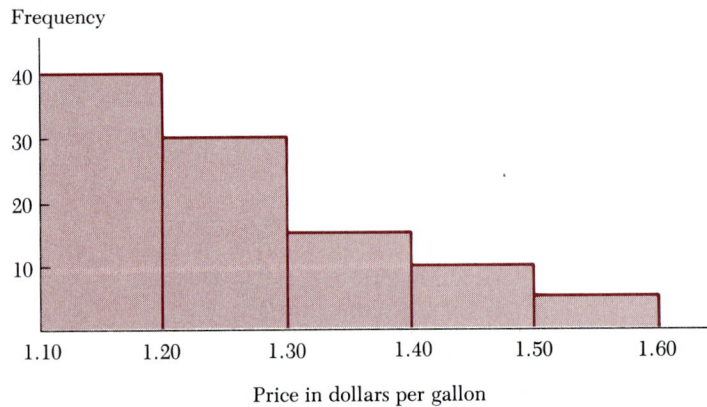

Price in dollars per gallon

TABLE 2.10 | Frequency distribution of the unleaded regular gasoline prices by combining the first two classes of the frequency distribution in Table 2.9 into one class

Class number	Class	Frequency
1	$1.10 but less than $1.30	70
2	$1.30 but less than $1.40	15
3	$1.40 but less than $1.50	10
4	$1.50 but less than $1.60	5
Total:		100

FIGURE 2.19 | Two histograms constructed for the frequency distribution in Table 2.10

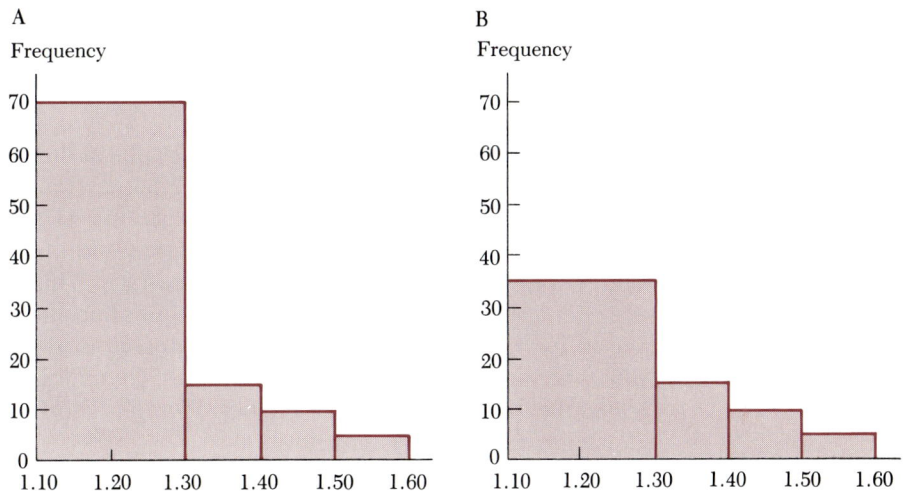

A histogram constructed by setting the height of the rectangle equal to the frequency in each class is shown in Figure 2.19A.

Notice that the frequency distribution in Table 2.10 has classes with unequal widths—the width of the first class is 0.20, and the widths of the remaining classes are 0.10. For the histogram in Figure 2.19A, consider the ratio of the area of each rectangle to its class frequency. In the first class, this ratio is $(70)(0.20) = 0.20$, the class width for this class. In the remaining three classes, the ratio of the area of the rectangle to the frequency in each class is 0.10. The consequence of the different ratios is that the rectangle in the first class has twice as much area as it should have. The ratio of its area to its frequency should also be 0.10. To achieve the proper ratio, the height of the rectangle must be adjusted. Let h represent the height of this rectangle. Then the area is $(h)(0.20)$. The ratio of the area to the class frequency must be 0.10. Thus

$$[(h)(0.20)]/70 = 0.10$$

Solving for h:

$$h = \frac{(70)(0.10)}{(0.20)} = 35$$

Thus the height of the rectangle should be 35, not 70 as it is in the histogram in Figure 2.19A. The correct histogram is shown in Figure 2.19B. The histogram in Figure 2.19A clearly overemphasizes the first class, making it appear that more of the gasoline prices of unleaded regular gasoline are less than $1.30 than there are. The histogram in Figure 2.19B has the first class rectangle in the correct proportion when compared with the other three rectangles.

The type of error exhibited in the histogram in Figure 2.19A can be extended beyond the basic histogram. Consider, for example, Figure 2.20. The shrinkage in the value of the dollar from $1.00 in 1958 to $0.44 in 1978 is equal to the *heights* of the dollars, *not* the areas of the dollar bills. Yet the visual impression is two dimensional. To fairly represent the purchasing power change, the Carter dollar should be approximately twice its present area so that its area is 44 percent of the area of the Eisenhower dollar at the top.

Figure 2.21 shows another example of graphical distortion. The percentage increase from the 1973 price of $2.41 per barrel to the 1979 price of $13.34 per barrel is 454 percent $[(13.34 - 2.41)/2.41 = 10.93/2.41 = 4.54$; 4.54×100 percent $= 454$ percent]. The increase in the *areas* of the barrels from 1973 to 1979 is approximately 4,300 percent. Thus the use of the two-dimensional oil barrels gives the visual impression that oil prices have increased by 4,300 percent, but in fact the increase is 454 percent. Edward Tufte, the author of *The Visual Display of Quantitative Information*, has created the *lie factor* ratio for the kinds of graphs shown in Figures 2.20 and 2.21. Tufte's lie factor is given by:

Lie factor

$$\text{Lie factor} = \frac{\text{Percent increase in area}}{\text{Percent increase in amount}}$$

For Figure 2.21, the lie factor is:

$$\text{Lie factor} = \frac{4,300\%}{454\%} = 9.47$$

If the *volume* of the barrels is considered, then the Lie factor is:

$$\text{Lie factor} = \frac{27,000\%}{454\%} = 59.47!!$$

A well designed graph should have a lie factor of 1.00. When the lie factor is less than 0.95 or greater than 1.05, the distortion in the graph becomes serious in Tufte's view.

There is another source of distortion in Figure 2.21. The oil prices are given in nominal dollars, not adjusted for inflation. An increase from $2.41 per barrel in 1973 to $13.34 per barrel in 1979 is large, but it must be remembered that some of this difference is attributable to inflation. A method to correct for the inflation effects over time is discussed in Chapter 18, Index Numbers. Using 1972 as the base year with an index of 100, the nominal and real prices per barrel of oil are graphed in Figure 2.22. The real prices are adjusted for inflation using methods described in Chapter 18. The increase in the real price of oil per barrel from 1972 to 1979, although large, is not nearly as dramatic as the increase in the nominal price per barrel.

For further discussion of the lie factor and graphical design, see the reference *The Visual Display of Quantitative Information*, written by Edward Tufte. It is a delightful, fascinating, and informative treatment of graphical displays, describing the correct construction of graphs with many examples

FIGURE 2.20 | Purchasing power of the diminishing dollar

1958 — EISENHOWER: $1.00

1963 — KENNEDY: 94¢

1968 — JOHNSON: 83¢

1973 — NIXON: 64¢

1978 — CARTER: 44¢
(August)

Source: *The Visual Display of Quantitative Information,* p. 70. (Originally taken from the *Washington Post,* October 25, 1978; page 1.)

IN THE BARREL...

Price per bbl. of
light crude, leaving
Saudi Arabia
on Jan. 1

April 1
$14.55

$13.34

$12.70

$12.09

$11.51

$10.46

$10.95

$2.41

'73 '74 '75 '76 1977 1978 1979

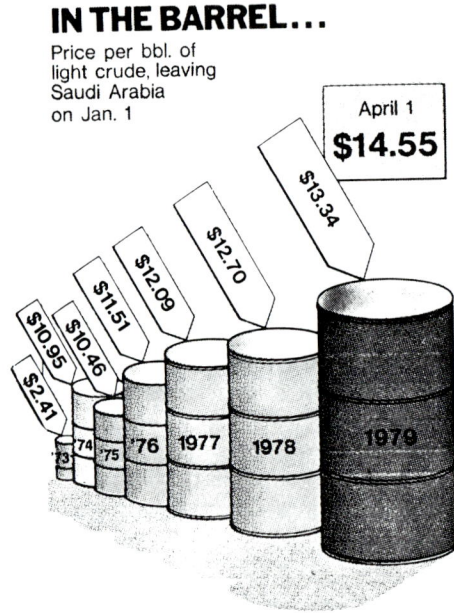

Source: *The Visual Display of Quantitative Information,* p. 62. (Originally taken from *Time Magazine,* April 9, 1979; p. 57.)

FIGURE   2.22 | Prices of oil per barrel in nominal and real dollars, 1972–79

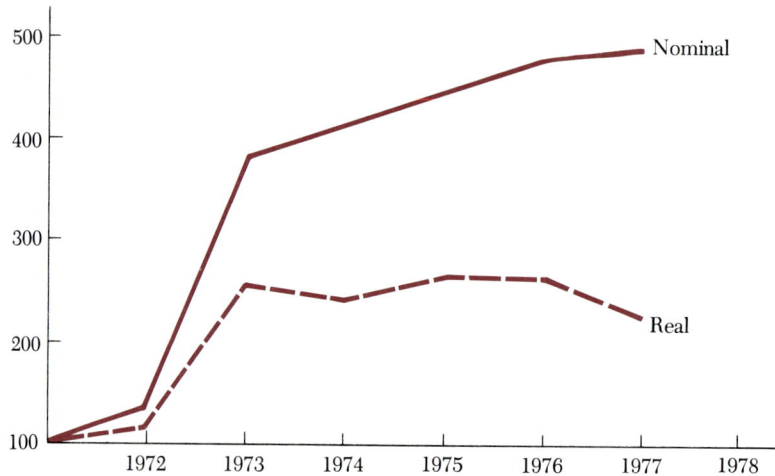

Nominal

Real

500

400

300

200

100

1972   1973   1974   1975   1976   1977   1978

and describing several poorly constructed graphs, such as those in Figures 2.20 and 2.21 above.

2.9.4 | Using words to describe counts of things

Often words are used to describe counts of objects, people, and so forth. These words are subject to varying interpretations. Don Reid, a technical writer, prepared a table of values for common phrases denoting counts of things or nonspecific quantities. This table is commonly referred to as Reid's table and is duplicated in Table 2.11.

TABLE 2.11 | Reid's table

Common phrase	Absolute value(s)
One	1
Only one	1
A couple	2 to 4
A few	3 to 5
Quite a few	3 to 6
Several	3 to 9
Many	3 to 8
Most (as in "most authorities")	4 to 6
Half a dozen	5 to 7
About a half dozen	4 to 8
A lot	6 to 10
Quite a lot	7 to 11
A whole lot	8 to 17
Ten*	9 to 11
Around 10	7 to 13
A dozen	11 to 13
About a dozen	9 to 15
A bunch	8 to 15
A whole bunch	9 to 19
Two dozen	22 to 26
About two dozen	21 to 27
A few hundred	75 to 125
A couple of hundred	99 to 139
Half a million (as in a promoter's estimate of a crowd size at a rock show)	90,000 to 125,000
Most (when expressed as a percentage)	10 percent to 20 percent
A majority	50 percent + 1
A clear majority	51 percent
A vast majority	52 percent to 60 percent
An overwhelming majority	61 percent to 70 percent
Almost everyone/all	71 percent to 75 percent
Practically everyone/all	76 percent to 80 percent
Everyone/all	81 percent to 85 percent
Absolutely everyone/all	86 percent to 90 percent
100 percent of those surveyed	91 percent to 95 percent

* Applies to any round number, and ranges from −10 percent to +10 percent. Thus 20 people means 18, 19, 20, 21, or 22; 100 is any number between 90 and 110, and so on.

Although Reid's table is humorous, these interpretations are frequently accurate. Therefore, in describing counts of things, we must be very careful when we use such terms as *several* and *most*. In this subsection and the previous one, we have described ways in which graphs and words representing counts of things can be misleading. The illustrations used should convince you that a great deal of care is necessary in describing things and that a critical view should always be taken when reviewing descriptions of numerical quantities by words, graphs, and numerical measures.

■ 2.10 Summary

In this chapter we have been concerned with describing sets of data by numerical and graphical methods. Both methods are extremely useful in describing either population or sample data sets. When the data set represents a sample, numerical measures lend themselves to inference making better than do graphical methods. As a consequence, numerical measures will be used extensively in the inference-making process, a major topic in this text.

Among the graphical methods, the frequency distribution histogram is very useful in summarizing a set of data, for it provides at a glance information about the shape of the data set. The other graphical methods can be useful if they supply an easily understood and accurate summary of the data.

Among the representative measures, the arithmetic mean is generally preferred in inference making, although the median, mode, geometric mean, quartiles, and percentiles can be helpful in determining a representative value for the data set and in providing information concerning the location of most values in the set.

The variance and the standard deviation are the preferred measures of variability, although determining differences of quartiles and percentiles may also be helpful in assessing the spread of the measurements. Used in conjunction with the Chebychef theorem, the standard deviation provides meaningful information about the dispersion of the data set.

The frequency distribution can be used to compute the numerical measures approximately, but the measures can easily be computed from ungrouped data by computers. The most important reason for constructing a frequency distribution is, therefore, to generate a histogram that gives a pictorial display of the distribution of the measurements.

■ References

Bell, P. C., and E. F. P. Newson. *Statistics for Business with MINITAB*. Palo Alto, Calif.: The Scientific Press, 1985.

Campbell, S. K. *Flaws and Fallacies in Statistical Thinking*. Englewood Cliffs, N.J.: Prentice-Hall, 1974.

Finkelstein, M. *Statistics at Your Fingertips*. Belmont, Calif.: Wadsworth Publishing, 1985.

Freedman, D.; R. Pisani; and R. Purves. *Statistics*. New York: W. W. Norton, 1978.

Huff, D. *How to Lie with Statistics*. New York: W. W. Norton, 1954.

Iman, R. L., and W. J. Conover. *Modern Business Statistics*. New York: John Wiley & Sons, 1983.

Kenkel, J. L. *Introductory Statistics for Management and Economics*. Boston: Prindle, Weber, and Schmidt, 1981.

Kohler, H. *Statistics for Business and Economics*. Glenview, Ill.: Scott, Foresman, 1985.

Reichman, W. J. *Use and Abuse of Statistics*. Baltimore: Penguin Books, 1971 (paperback).

Ryan, T.; B. Joiner; and B. Ryan. *MINITAB Handbook*. 2nd ed. Boston: Duxbury Press, 1985.

Tufte, E. R. *The Visual Display of Quantitative Information*. Cheshire, Conn.: Graphics Press, 1983.

Velleman, P. F., and D. C. Hoaglin. *Applications, Basics, and Computing of Exploratory Data Analysis*. Boston: Duxbury Press, 1981.

Wonnacott, T. H., and R. J. Wonnacott. *Introductory Statistics for Business and Economics*. 3rd ed. New York: John Wiley and Sons, 1984.

Computer statistical software

Name of software	Mainframe computer	Microcomputer	Vendor
ABstat		X	Anderson-Bell TEC Building 11479 S. Pine Drive Suite 407 Parker, Colo. 80134
BMDP	X	X	BMDP Software 1964 Westwood Blvd. Los Angeles, Calif. 90025
DASY		X	Statistical Software Resources 20355 Seaboard Road Malibu, Calif. 90265
MINITAB	X	X	MINITAB, Inc. 3081 Enterprise Drive State College, Pa. 16801
MSUSTAT		X	Research and Development, Inc. Montana State University Bozeman, Mont. 59717
PRODAS		X	Effort Rockefeller Center P. O. Box 3012 New York, N.Y. 10185
P-STAT	X	X	P-STAT Princeton, N.J.
SAS	X	X	SAS Institute P. O. Box 8000 Cary, N.C. 27511
SPSS	X	X	SPSS, Inc. 444 North Michigan Avenue Chicago, Ill. 60611
STATPAC		X	Walonick Associates 6500 Nicollet Ave. So. Minneapolis, Minn. 55423
SYSTAT		X	SYSTAT, Inc. 603 Main Street Evanston, Ill. 60202

Contact vendors for current prices (usually academic discounts are given). See recent issues of *The American Statistician* for software reviews. Reviews in 1985 include DASY, SYSTAT, and MSUSTAT.

◼ Problems

Section 2.2 Problems

2.1 The following 25 measurements represent the number of business trips taken annually by 25 claim adjusters of the Acme Insurance Company:

33 17 2 10 12 15 22 18 20 24 8 27 8
 5 28 12 17 15 21 38 16 18 10 12 9

Construct a frequency distribution with five classes for these data. Include in the frequency distribution columns for the relative frequency, the cumulative frequency, and the cumulative relative frequency.

2.2 The number of sales made by the Spring Plumbing Supply Company in 30 consecutive working days are:

63 47 41 72 65 61 27 33 49 21 45 26 31 48 18
74 20 29 49 44 26 19 78 64 36 47 52 50 24 35

Construct a frequency distribution with five classes for these data. In the frequency distribution, include columns for the relative frequency, the cumulative frequency, and the cumulative relative frequency.

2.3 An investment firm handles small investment portfolios for customers in a large metropolitan area. As a part of their annual review, they have sampled 60 customer accounts to determine the *net* (after taxes) return on investment for each account. The net returns are given as the following percentages:

5.81	6.39	5.67	5.96	6.54	5.85	4.94
4.98	6.30	6.14	5.04	5.01	6.02	4.06
5.58	5.98	5.85	5.26	4.84	5.96	5.30
3.96	5.96	5.24	5.92	4.55	4.98	6.02
5.28	4.20	4.46	5.08	5.71	5.27	4.89
4.92	4.34	5.62	6.18	4.30	4.64	6.18
6.14	6.08	6.18	4.18	6.01	5.92	6.04
5.70	5.87	5.74	5.66	6.13	3.75	6.37

6.12 5.04 4.72 5.84

Using the classes given in the table below, construct a frequency distribution giving both the frequency and the cumulative relative frequency for each class.

Net return	Frequency	Cumulative relative frequency
3.50–3.99		
4.00–4.49		
4.50–4.99		
5.00–5.49		
5.50–5.99		
6.00–6.49		
6.50–6.99		

2.4 The board of directors of a large hospital in Fort Worth, Texas, are considering a proposal to construct a $5 million fitness center to be built next to the existing hospital building. As a part of the study to determine the feasibility of building such a facility, a telephone survey is conducted of residences located within three miles of the location of the proposed fitness center to determine the frequency of use for various activities. The number of times per month that 100 sampled respondents stated that they would use specific facilities in the proposed fitness center if they became members are given below:

Nautilus equipment:
0 0 12 10 0 18 0 0 0 5 12 15 0 20 25
0 10 8 15 0 0 5 2 4 0 10 12 16 0 0

```
0   3  12 14 25  0  6  0 30  0  5 14 24  0  8
6   9  30 24  0  8  4 12 15  0 16 12 20 20 10
0   8   4 20  0  0  0  8 11 14 20 15  0  9  0
9  24  12 15 12  6  8  9  0  0 22 16 30  0  8
8   0   0  0 28 25 12  0  0 24
```

Aerobic exercise:
```
25 20  0 24 30 12  0  0 30 20 20 12  4  0
25  0 12 15  0  0 15 18 24  0  8  5 14  0
 0 10  8  0  0 12  0  0 24  0  0 20  8  0
 0  0 12 12 24  0  0  8 10  0 14 10 10  0
30  8 24 20 20 10 12 18  0  0  6  8 18  0
 0  0  8 20 12  0  0 10 15  0 20 25 15 16
20 24 10 12  0  0  0 12 12 14  0 18 20  8
 6 10
```

Jogging track:
```
20  0 10 10 12 12 15  0  0 20  0 12 10  8
24  0  8  6  0  5 12 16  5  4  0  0 10 12
 8  0 12 10 12 15  0  0  0  8  6 20  9 14
16  0 20 10  0  8 10  6  5  2  0  0  0  0
 0 12 10 24 28  0  0 12  0  0  4  2  0  5
12 15  8  0  0 20 24  0 10  0  8  0  1 24
 9 10  0  0  0  0 14 10 12  0 20  0 12 12
20 15
```

a. Form a frequency distribution for the number of times per month that potential members would use the Nautilus equipment. Include in your frequency distribution columns for the relative frequency, the cumulative frequency, and the cumulative relative frequency.

b. Form a frequency distribution for the number of times per month that potential members would use aerobic exercise. Include in your frequency distribution columns for the relative frequency, the cumulative frequency, and the cumulative relative frequency.

c. Form a frequency distribution for the number of times per month that potential members would use the jogging track. Include in your frequency distribution columns for the relative frequency, the cumulative frequency, and the cumulative relative frequency.

2.5 A survey is conducted based on a sample of 50 lending institutions in the state of Texas in January 1986 to determine the mortgage interest rate for a conventional loan on home purchases. The resulting interest rates in percentages are:

```
10.500  10.250  10.750  11.000  11.000  10.125  10.875
10.750  10.750  11.125  11.500  10.625  10.375  10.500
10.750  10.750  10.250  10.875  11.375  11.250  11.000
11.125  10.875  11.125  10.000  11.875  11.250  10.750
10.750  10.875  10.750  10.500  11.000  11.375  10.750
10.500  10.375  10.875  10.750  11.000  10.750  10.500
10.750  11.000  11.125  10.500  10.375  11.375  10.750
11.000
```

Form a frequency distribution with five classes for these interest rates. Include in the frequency distribution columns for the relative frequency, the cumulative frequency, and the cumulative relative frequency.

Section 2.3 Problems

2.6 In the Complaints Department of a large department store on a given day, the lengths (in seconds) of the first 100 telephone calls were recorded (rounded to the nearest second), and the following frequency distribution was constructed:

Class	Class limits	Frequency
1	0 but less than 60	5
2	60 but less than 120	20
3	120 but less than 180	40
4	180 but less than 240	25
5	240 but less than 300	10

a. Construct a histogram from this frequency distribution.

b. Construct a polygon from this frequency distribution.

2.7 In its annual report to shareholders, Soybean International, Inc., provides the annual net earnings per common share for each year in the 10-year period from 1977 through 1986. The data are:

1977	1978	1979	1980	1981
$0.79	0.50	0.60	0.74	1.42

1982	1983	1984	1985	1986
4.73	3.50	2.20	1.26	1.86

Draw a line chart with the years on the horizontal axis and net earnings per common share on the vertical axis.

2.8 Polyglas, Inc., reports at its annual stockholders meeting that the net revenue for 1986 has been expended as follows:

Expenditure	Percent of total revenue
Raw materials	56.0
Supplies and services	28.7
Employee benefits	6.9
Depreciation	2.4
Taxes	1.8
Advertising and promotion	1.3
Other (interest, dividends, reinvestment)	2.9
Total	100.0

Form a pie chart to show how the net revenue was expended.

2.9 For the frequency distribution constructed in Problem 2.3,
a. Draw a histogram.
b. Draw a polygon.
c. Draw a "less than" ogive.

2.10 For the frequency distribution that you constructed in Problem 2.1,
a. Draw a histogram.
b. Draw a polygon.
c. Draw a "less than" ogive.

2.11 For the frequency distribution that you constructed in Problem 2.2,
a. Draw a histogram.
b. Draw a polygon.
c. Draw a "less than" ogive.

2.12 In Problem 2.4, you were asked to construct frequency distributions for the numbers of times per month that potential members of the proposed fitness center state that they would use the Nautilus equipment, aerobics exercise, and the jogging track. Construct a histogram for *each* of the frequency distributions that you constructed.

2.13 For the frequency distribution constructed in Problem 2.5,
a. Draw a histogram.
b. Draw a polygon.
c. Draw a "less than" ogive.

2.14 Sixty-four students who were admitted to an MBA program in the fall semester, 1985 complete the fall semester 1985 and the spring semester 1986 with at least 24 credits (eight three-hour-credit courses). The frequency distribution of these students' grade-point average (GPA) is given below:

Class number	Class limit	Frequency	Relative frequency	Cumulative relative frequency
1	1.00 but less than 1.50	1	0.0156	0.0156
2	1.50 but less than 2.00	5	0.0781	0.0937
3	2.00 but less than 2.50	12	0.1875	0.2812
4	2.50 but less than 3.00	22	0.3438	0.6250
5	3.00 but less than 3.50	20	0.3125	0.9375
6	3.50 but less than 4.00	4	0.0625	1.0000

a. What percentage of these students received a GPA of less than 3.00?
b. What percentage of these students received a GPA of 2.50 or more?
c. Draw a histogram for this frequency distribution.
d. Draw a polygon for this frequency distribution.
e. Draw a "less than" ogive for this frequency distribution.

2.15 The Carnegie Corporation of New York conducted a study on the age distribution in the

U.S. population. In the year 1982 the percentages of ages in the U.S. population for the stated age intervals were given as:

Age interval	Percentage
Less than 18	29%
18–34	28%
35–54	22%
More than 54	21%

Based on a statistical analysis, the following forecasts of the percentages in the year 2010 were made:

Age interval	Percentage
Less than 18	25%
18–34	23%
35–54	27%
More than 54	25%

a. Form a pie chart for the 1982 actual percentages.
b. Form a pie chart for the forecasted 2010 percentages.
c. Compare the pie charts in parts *a* and *b*. What do these two charts indicate is happening with the distribution of ages for persons in the United States from 1982 to 2010?

Section 2.4 Problems

2.16 The dean of a leading business school is interested in the starting salaries of graduates of the BBA program offered by the school. Over the past two years, a random sample of graduates is taken and the starting salaries are recorded in thousands of dollars. The starting salaries are separated by undergraduate major. Below are given 25 starting salaries for accounting majors and 20 starting salaries for management majors:

Accounting majors:
18.6 22.3 20.6 16.5 24.3 25.5 20.8 18.0 26.7 23.0
18.0 21.0 25.0 24.6 17.9 20.0 18.5 16.4 28.4 26.0
19.8 20.0 25.5 22.8 28.2

Management majors:
14.4 20.0 16.6 15.0 15.6 18.5 21.8 16.0 15.5 14.0
13.9 23.5 18.7 16.0 15.0 16.4 14.2 13.9 20.0 18.8

a. For the accounting graduates, calculate the minimum salary, the maximum salary, and the mean, mode, and median of the salaries.
b. For the management graduates, calculate the minimum salary, the maximum salary, and the mean, mode, and median of the salaries.
c. Are the distributions of the salaries for the accounting and management majors skewed? Explain. If the distributions are skewed, in what direction are they skewed?

2.17 The price to earnings ratios for a sample of 20 companies in the pharmaceutical/cosmetic industry in the first quarter of 1985 are:

2.24 3.36 0.50 4.45 10.00 3.60 8.87 2.21 4.45 12.60
4.44 10.60 2.21 5.60 22.36 9.90 2.20 3.23 12.50 4.45

a. Calculate the mean price to earnings ratio for these sample data.
b. Calculate the median price to earnings ratio for these sample data.
c. Compare the mean to the median. Is the distribution of these sample values skewed? Explain.

2.18 Consider the data in Problem 2.1 representing the number of business trips taken annually by 25 claim adjusters of the Acme Insurance Company. Treating these data as a population set of data, determine the following:
a. The mean number of business trips for the 25 adjusters.
b. The median number of business trips for the 25 adjusters.
c. The mode of the number of business trips for the 25 adjusters.

2.19 Consider the data in Problem 2.2 representing the number of sales made by the Spring Plumbing Supply Company in 30 consecutive working days. Treating this data set as a sample, determine the following:
a. The mean number of sales.
b. The median number of sales.
c. Is the distribution of the number of sales skewed? Explain.

2.20 Consider the net return on investment data

for the sample of 60 returns in Problem 2.3. For these sample net return on investment data, determine:

a. The mean net return on investment.
b. The median net return on investment.

2.21 In Problem 2.4 the numbers of times per month that individuals would use Nautilus equipment, aerobic exercise, and a jogging track were recorded in a survey of 100 sampled individuals who stated that they would use the fitness center. For these data, determine:

a. The mean, median, and mode of the number of times per month that the Nautilus equipment would be used by the 100 sampled individuals.
b. The mean, median, and mode of the number of times per month that the aerobic exercise would be used by the 100 sampled individuals.
c. The mean, median, and mode of the number of times per month that the jogging track would be used by the 100 sampled individuals.

2.22 For the Problem 2.5 sample data, determine the following:

a. The mean mortgage interest rate.
b. The median mortgage interest rate.
c. The mode of the 50 sample interest rates.

Section 2.5 Problems

2.23 In finance the standard deviation is frequently used to measure the risk of an investment. An investor is considering two possible investments, A and B. A sample of 10 rates of return are recorded for each investment. These rates of return given as percentages are as follows.

Investment A:
    4   6   6   5   4   8   5   5   6   7

Investment B:
    0   −2   6   12   9   16   −4   20   2   5

a. Calculate the mean rate of return for each of these investments.
b. Calculate the standard deviation of the rate of return for each of these investments.
c. Which of these investments has the larger mean rate of return? Which of these investments is riskier? Explain.

2.24 A metal part used in the Space Shuttle is designed to be 60 inches long. The contractor who produces this part has had difficulty in the fabrication process with the result that the length tends to vary around the targeted 60 inches. NASA has set an acceptable tolerance of 0.005 inches for the length; that is, the length may vary between 59.995 inches and 60.005 inches. The contractor produces a sample of 10 of these metal parts for evaluation by NASA. The lengths of these 10 parts are:

60.001	60.000	59.996	59.999	60.007	59.998	59.992
60.003	60.005	59.996				

a. What percentage of these parts fail to meet NASA's tolerance interval?
b. What is the mean length of these 10 sample parts?
c. What is the median length of these ten sample parts?
d. What is the range of the 10 lengths?
e. What is the standard deviation of the lengths of these 10 sampled parts?

2.25 Consider the data in Problem 2.1. These data represent the number of business trips taken annually by 25 claim adjusters of the Acme Insurance Company. Treating these 25 measurements as a population data set, determine the following:

a. The range of the 25 measurements.
b. The population variance of the 25 measurements.
c. The population standard deviation of the 25 measurements.

2.26 Consider the data in Problem 2.2. Treating these 30 days of sales as a sample, determine the following:

a. The range of the numbers of sales.
b. The sample variance of the numbers of sales.
c. The sample standard deviation of the numbers of sales.

2.27 Consider the data in Problem 2.5. Treating these 50 mortgage interest rates as a sample, determine the following:

a. The range of the mortgage interest rates.
b. The sample variance of the mortgage interest rates.

c. The sample standard deviation of the mortgage interest rates.

2.28 Compute the variance and the standard deviation for the following set of 20 population measurements:

12 36 10 20 22 14 11 13 14 15
22 31 12 10 28 19 17 12 11 10

Section 2.6 Problems

2.29 The following 20 measurements represent the number of calls received by a receptionist of the Eastern Petroleum Company per hour for 20 sampled hours during a two-week period:

12 22 30 10 40 25 28 26 44 36
10 22 29 19 41 37 32 33 34 40

a. Determine the first and the third quartiles for these 20 measurements.
b. Determine the second quartile (median) for these 20 measurements.
c. Determine the 90th percentile for these 20 measurements. Interpret the 90th percentile for these data.

2.30 Consider the Problem 2.1 data. For these 25 measurements, determine the following:
a. The 20th percentile. Interpret this number.
b. The 70th percentile. Interpret this number.
c. The interquartile range.

2.31 For the following set of 10 sample measurements, calculate the mean, geometric mean, median, and mode:

10 10 10 20 20 50 80 100 100 1,000

Which numerical measure(s)—the mean, geometric mean, median, or mode—do you believe best represents these 10 measurements? Why?

2.32 For the following 10 sample measurements, calculate the mean, median, and the skewness measure:

0 2 0 1 5 2 5 2 0 15

Is the distribution of these measurements skewed? If so, in which direction? Explain.

2.33 You are given the choice of the following two investments. You may invest any amount you like in either one. Assume also that any earnings are reinvested at the current rate.

Investment 1: This investment is a fixed interest rate instrument that pays 9 percent compounded yearly on the invested funds. Funds must remain invested for six years, and interest is compounded each year. That is, interest is earned on the total amount accumulated and thus includes earnings on previous earnings.

Investment 2: This instrument yields the following rates for the years indicated. (Total accumulations earn interest at the stipulated rate each year the funds are on deposit.) All earnings must be reinvested the next year. All funds are withdrawn at the end of six years.

Year	*Interest Rate*
1	14%
2	10
3	9
4	8
5	5
6	8

Which of these investments do you prefer, and why?

2.34 The American Chamber of Commerce Researchers Association (ACCRA) has developed a cost of living index for U.S. cities. The index is based on 59 items that include the cost of a two-liter bottle of Coca-Cola, a new home of 1,800 square feet, teeth cleaning by a dentist, a McDonald's quarter-pound hamburger, and one line of bowling (evening rate). In the April 9th, 1985, issue of *The Wall Street Journal,* the 1984 cost of living index was reported for a sample of 20 cities taken from the set of all participating cities. The average of the index for all participating cities is 100. The sample values are:

San Francisco	145.9	Des Moines, Iowa	102.1
Anchorage, Alaska	136.5	Birmingham, Ala.	101.7
New York	134.4	Tucson, Ariz.	99.8
Fairfield, Conn.	122.9	Oshkosh, Wis.	99.3

Boca Raton, Fla.	118.2	Greensboro, N.C.	96.6
San Diego	113.4	Huntsville, Ala.	94.3
Seattle	110.5	Fort Smith, Ark.	92.4
Denver	109.6	Joplin, Mo.	90.9
Kalamazoo	105.1	Pueblo, Colo.	88.4
Pittsburgh	104.4	Clinton, Mo.	88.3

a. Determine the mean index value for the 20 sampled cities.

b. Determine the 50th percentile (median) of the index value for the 20 sampled cities.

c. Determine the geometric mean for the 20 sampled cities.

d. Compare the mean, median, and geometric mean computed in parts a, b and c above. Which of these would you choose as a representative measure for the sampled 20 cities? Explain.

2.35 For the 20 sample price to earnings ratios in Problem 2.17, determine the following:

a. The 20th percentile.

b. The 75th percentile (third quartile).

c. The skewness measure. Based on the value of the skewness measure, is the distribution of the 20 price to earnings ratios skewed? If so, in which direction?

2.36 It is known that the average annual salary among workers in a particular industry is $20,000, with a standard deviation of $2,000. The distribution of salaries is known to be skewed to the right. Using Chebychef's theorem, within what interval are contained at least 84 percent of the salaries?

2.37 The Painful Dentist Clinic, Inc., has a receptionist who makes appointments by telephone for patients who wish to see one of the four dentists in the clinic. If the phone is busy, the computerized system automatically answers up to seven calls, putting the callers on hold and providing them with musical entertainment. The computer also records how long each person must wait before being served by the receptionist (or before hanging up in disgust!). The average waiting time for service when the system is busy is 2.66 minutes, with a standard deviation of 1.82 minutes.

a. Use the Chebychef theorem to determine an interval that contains at least 90 percent of the waiting times.

b. Since no waiting times can be negative, what problem does this present, if any,

with the interpretation of the Chebychef interval in part a? Explain.

2.38 Most income measures are skewed to the right, which requires the use of the median as a representative measure rather than the mean in many instances. In a study of union salaries for garment workers in California, the average salary is found to be $18,642, with a standard deviation of $3,613. The median salary is $16,021.

a. Calculate the skewness measure for these data. Is the distribution skewed, and if so, in which direction?

b. Using the Chebychef theorem, determine an interval that contains at least 80 percent of the salaries of the garment workers.

2.39 The average attendance at a major league baseball park for the 81 home games in 1985 was 22,300, with a standard deviation of 3,800. Using the Chebychef theorem, determine the following:

a. An interval that contains at least 90 percent of the attendances for the 81 home games.

b. At least what proportion of the 81 home games had attendances between 11,660 and 32,940?

Section 2.9 Problems

2.40 In a sample of 100 audits, the percentage of errors are recorded and are placed in the following frequency distribution:

Class	Class limits	Frequency
1	0% but less than 2%	60
2	2% but less than 4%	20
3	4% but less than 7%	10
4	7% but less than 10%	5
5	10% but less than 15%	5
	Total	100

Draw a histogram for this frequency distribution. Notice that there are three different class widths.

2.41 For the 60 sample measurements in Problem 2.3, construct a stem and leaf plot. In this plot, use stems of 3* for leafs 0, 1, 2, 3, and 4, and 3 for leafs of 5, 6, 7, 8, and 9. Do the same for stems 4, 5, and 6. For the leafs, use only the first decimal place of the net return on investment, disregarding the second decimal place.

2.42 For the sample data in Problem 2.4, construct stem and leaf plots for the number of times per month each of the three fitness activities (Nautilus equipment, aerobic exercise, and jogging track) would be used by potential members. For each stem and leaf plot, split the 10's digit at the midpoint. For example, use stem 1* with leafs 0, 1, 2, 3, 4 for the numbers 10, 11, 12, 13, 14, and use stem 1. with leafs 5, 6, 7, 8, 9 for the numbers 15, 16, 17, 18, 19.

2.43 For the sample data in Problem 2.29, construct a box plot. Describe in words the information the box plot contains about the data set of 20 measurements.

2.44 The union shop manager keeps records on the number of complaints per week issued by employees of a manufacturing firm that represent employee grievances against management. In a sample of 20 weeks, the following numbers of complaints were received:

10	24	0	4	12	22	9	0	10	2
0	6	3	11	7	0	12	6	9	8

a. Find the first, second, and third quartiles for these data.

b. Draw a box plot for these data. Based on the box plot, is the distribution of these data skewed? Explain.

2.45 For the sample data sets representing starting salaries of BBA graduates in accounting and in management in Problem 2.16, determine the following:

a. The first, second, and third quartiles of the starting salaries in accounting.

b. The first, second, and third quartiles of the starting salaries in management.

c. Draw a box plot of the starting salaries in accounting.

d. Draw a box plot of the starting salaries in management.

2.46 For a national U.S. airline, the average salary of a pilot is $54,000. The median salary, the first quartile, and the third quartile are: $50,000, $38,000, and $70,000, respectively. The minimum salary is $32,000, and the maximum salary is $120,000.

a. Is the distribution of these salaries skewed? Explain.

b. Form a box plot of these salaries.

2.47 Eighty students were recently admitted to the fall 1986 MBA class in a business school of a southwestern university. Treating these 80 students as a population, the numerical measures for the GMAT (Graduate Management Admissions Test) scores of these 80 students are: Mean = 510; Median = 500; Minimum = 425; Maximum = 700; First quartile = 460; Third quartile = 600.

a. Is the distribution of these GMAT scores skewed? Explain.

b. Draw a box plot for these 80 GMAT scores. Explain what the box plot tells us about these scores.

2.48 The following graph was used to illustrate the growth in sales of compact disk players used in stereo systems by an audio equipment retailer in a metropolitan area. The graph begins in 1982 with annual sales of 20 units and ends in 1986 with annual sales of 600 units. Circles representing disks are used to represent the number of units sold per year.

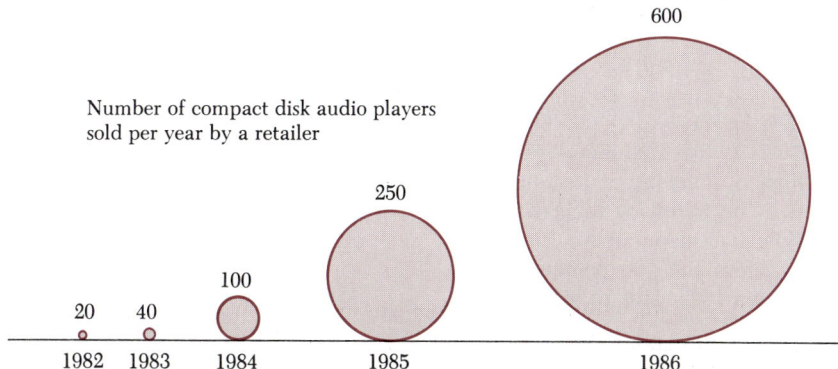

Number of compact disk audio players sold per year by a retailer

600
250
100
20 40
1982 1983 1984 1985 1986

Note: In the original graphic, the 1986 circle had a diameter (height) of 2 inches, (thus, a radius of 1 inch), and the 1982 circle had a diameter (height) of 0.067 inch (thus, a radius of 0.033 inch).

a. Is this a well-constructed graph? Explain.
b. Noting that the area of a circle is πr^2, what approximately is the lie factor for this graph?

Additional Problems

2.49 Compute the mean, median, mode, range, and variance for the following set of six measurements:

$$6\ \ 6\ \ 4\ \ 3\ \ 8\ \ 5$$

2.50 Compute the mean, median, mode, range, and variance for the following set of 10 measurements:

$$-8\ \ \ 0\ \ \ 0\ \ \ 4\ \ \ 5\ \ -10\ \ \ 6\ \ \ 10\ \ \ 4\ \ \ 2$$

2.51 For the data in Problem 2.50, find the percentage of the values lying in the intervals $\mu \pm \sigma$, $\mu \pm 2\sigma$, and $\mu \pm 3\sigma$. Apply the Chebychef theorem to these data, and compare the percentages of the values contained in these intervals.

2.52 The average daily high temperature recorded at a weather station is 80°F during June. On six consecutive days, the difference between the recorded high and this average high were:

$$-10\ \ -8\ \ -2\ \ 6\ \ 4\ \ 10$$

Compute the average difference and the standard deviation.

2.53 For the following five measurements, compute the arithmetic mean, geometric mean, median, and mode:

$$10\ \ 10\ \ 50\ \ 100\ \ 1,000$$

Which number(s) would you select as the best representative measure, and why?

2.54 The Nitro-Fusion Company, in its annual report, states, "The following table supports our contention that shares of NF stock are widely held—more than 90 percent of the stockholders own 100 shares or less."

Do you agree with this statement? Discuss.

Number of shares held	Percent of total number of stockholders
1–20	48.76
21–50	25.44
51–100	18.53
101–500	5.81
More than 500	1.46

2.55 Under what conditions may the arithmetic average be a poor representative measure? Explain.

2.56 Compute the mean, median, mode, range, and variance for the following 10 numbers:

$$-60\ \ \ 10\ \ \ 10\ \ \ 20\ \ -75$$
$$5\ \ \ 0\ \ \ 10\ \ \ 20\ \ \ \ 0$$

2.57 For the data in Problem 2.56, apply the Chebychef theorem to determine the minimum proportion of observations lying in the intervals $\mu \pm 2\sigma$ and $\mu \pm 3\sigma$. What proportions of the observations actually lie in these intervals?

2.58 For the following 10 numbers, compute the arithmetic mean, the geometric mean, and the median. Is the distribution of the numbers skewed? Explain.

$$2\ \ 6\ \ 1\ \ 2\ \ 5\ \ 6\ \ 4\ \ 15\ \ 6\ \ 2$$

2.59 A coffee machine dispenses coffee in 8-ounce cups. The machine is supposed to fill the cups with 7.75 ounces of liquid. In 25 recent purchases from the machine, the amounts of the fills were:

7.2	7.2	7.6	7.2	7.0	7.7	7.3
7.2	7.6	7.3	7.0	7.8	7.6	7.6
7.1	7.2	7.5	7.4	7.2	7.8	7.1
7.4	7.2	7.5	7.4			

a. Calculate the mean, mode, median, and standard deviation for these data.
b. Construct a frequency distribution using the following five classes: 7.00 but less than 7.20; 7.20 but less than 7.40; 7.40 but less than 7.60; 7.60 but less than 7.80; and 7.80 but less than 8.00.

2.60 The average net profit on weekly sales invoices at the Maui Gift Shop on Kaanapali

Beach on the Island of Maui, Hawaii, is $15.60, with a standard deviation of $6.62. Using the Chebychef theorem, determine the following:

a. An interval that contains at least 75 percent of the profits on the invoices.

b. An interval that contains at least 90 percent of the profits on the invoices.

c. At least what percentage of the invoices produce a profit between $0 and $31.82? In determining this percentage, what assumption needs to be made regarding the lower end of this interval?

2.61 In Snake, Texas, a small residential community in Ragweed County, 916 residential properties paid real estate taxes last year. The amounts paid are summarized in the following frequency distribution:

Class	Class limits	Frequency
1	$0 but less than $300	125
2	$300 but less than $600	463
3	$600 but less than $900	206
4	$900 but less than $1,200	122

a. Draw a histogram for this frequency distribution.

b. Draw a polygon for this frequency distribution.

c. Draw a "less than" ogive for this frequency distribution.

2.62 The average prices of homes sold in Fort Worth, Texas, on a quarterly basis from 1980 to 1983 are shown in the table below.

Average price of homes in Fort Worth (sold through Multiple Listing Service)

	Number of homes sold	Average price
1st quarter '80	1,108	49,400
2nd quarter '80	962	59,400
3rd quarter '80	1,663	53,800
4th quarter '80	1,243	51,100
1st quarter '81	915	67,400
2nd quarter '81	1,184	59,000
3rd quarter '81	1,145	80,000
4th quarter '81	905	59,800

Average price of homes in Fort Worth (sold through Multiple Listing Service)

	Number of homes sold	Average price
1st quarter '82	915	68,700
2nd quarter '82	1,250	61,600
3rd quarter '82	1,140	63,000
4th quarter '82	768	65,200
1st quarter '83	887	65,300
2nd quarter '83	1,190	68,300
3rd quarter '83	1,465	66,000
4th quarter '83	1,220	65,000

Source: Greater Fort Worth Board of Realtors.

a. Calculate the mean of these 16 quarterly average prices for homes sold in Fort Worth.

b. Calculate the median of these 16 quarterly average prices for homes sold in Forth Worth.

c. Draw a line chart for these 16 quarterly averages.

2.63 The monthly consumer price index values in the United States in 1984 are listed below.

Month	CPI
January	305.2
February	306.6
March	307.3
April	308.8
May	309.7
June	310.7
July	311.7
August	313.0
September	314.5
October	315.3
November	315.3
December	315.5

Form a line plot of these monthly CPI values in 1984.

2.64 In the table below is given data on a random sample of 20 MBA students enrolled in an MBA program at a major southwestern university.

Student number	GMAT score	Under-graduate grade-point average	Years of work experience	Age	Sex (1 = M, 0 = F)
1	604	3.78	2	24	0
2	474	3.08	6	27	1
3	505	3.24	0	22	1
4	480	3.12	4	23	0
5	510	2.88	8	30	0
6	560	3.67	0	23	1
7	524	3.22	2	24	1
8	495	2.44	2	23	1
9	460	2.56	4	25	0
10	540	3.31	0	22	0
11	520	3.60	1	23	0
12	477	3.06	4	24	1
13	440	2.45	0	22	1
14	486	2.90	5	27	1
15	590	3.20	10	30	0
16	450	2.62	2	25	1
17	492	2.98	5	26	1
18	515	3.45	0	24	1
19	475	2.32	8	29	1
20	604	3.88	5	25	0

a. For the GMAT scores, find the mean, median, first quartile, third quartile, range, and standard deviation of the scores.

b. For the undergraduate grade-point average, find the mean, median, first quartile, range, and standard deviation of the GPAs.

c. Plot the GMAT score and the undergraduate grade-point average on an x–y graph with GMAT on the vertical axis and undergraduate grade-point average on the horizontal axis. Does this graph suggest that there is a relationship between an MBA student's GMAT score and undergraduate grade-point average? Explain.

d. Form a box plot of the GMAT scores for the 20 students.

e. Form a box plot of the undergraduate grade-point average of the 20 students.

f. Calculate the mean, median, and mode of the ages of the 20 students. Is the distribution of the ages skewed? Explain.

g. What is the mean number of years of work experience for these 20 students?

h. Plot the GMAT score (vertical axis) against the age (horizontal axis). Does this plot suggest any relationship between the GMAT score and age? Explain.

i. Plot the GMAT score (vertical axis) against the number of years of work experience. Does this plot suggest any relationship between the GMAT score and the number of years of work experience? Explain.

j. Calculate the mean GMAT score and the mean undergraduate grade-point average of males in the program. Do the same for the females in the program. Based on the means of these two characteristics, do females have higher GMAT and undergraduate GPAs than do males? Make this judgment strictly on the basis of whether or not the means for the males exceed or do not exceed the means for the females for the GMAT scores and the undergraduate grade-point average.

3

Introduction to probability

Probability is a commonly used concept that most of us encounter on a regular basis. For instance, on the local evening newscast, we may hear such statements as: "There is a 60 percent chance of rain tomorrow," "Federal economists believe that there is a 50 percent chance that interest rates will rise in the next quarter," and "A survey indicates that 70 percent of the drivers are using their seat belts as required by state law, with an error of no more than ±10 percent." In the sports page of the daily paper, we may read that "The odds that the Dallas Cowboys will beat the Washington Redskins (in a professional football game) are 2 to 1." These statements convey information about the probabilities that certain events—such as rising interest rates or wearing seat belts—will happen. In predicting the occurrence of rain tomorrow ("There is a 60 percent chance of rain tomorrow"), the event or outcome is "rain tomorrow," and the 60 percent is a statement of the probability that the event will occur. One possible interpretation of the probability statement "60 percent" for the occurrence of rain tomorrow is the following: In the past when the weather conditions that we believe will exist tomorrow have prevailed, on 60 percent of those days, it has rained. The weather reporter in this instance would be using an analysis of past conditions to formulate the forecast. She would first decide on the most likely weather conditions for tomorrow: temperature, humidity, wind velocity at various altitudes, pressures, the location of weather fronts that may enter the area, and so forth. With computers, this information is readily available and is being constantly updated by the National Weather Service. Then, from past data, the reporter would find a set of days during which these

93

weather conditions existed. Finding, for example, that it rained on 60 percent of these days, she would obtain her forecast for the chance of rain tomorrow—60 percent. If she chose not to go to the computer files that contained information on the weather in the past, she might use her memory, knowledge, and experience to produce the forecast of a 60 percent chance of rain tomorrow.

As we will learn in this chapter, probability is a number that can vary between 0 and 1 (or, converted to a percentage, from 0 percent to 100 percent). If an event, such as ''rain tomorrow,'' has a probability of 0, then it means that the event is *certain not to happen*. If an event has a probability of 1, then it means that the event is *certain to happen*. For a probability between 0 and 1—say 0.60, such as in the forecast-for-rain-tomorrow illustration—the probability gives us a number that can be used relative to the endpoints of 0 and 1 to tell us how likely it is that an event will occur. The closer the number is to 0, the less likely it is that the event will occur; the closer the number is to 1, the more likely it is that the event will occur.

In this chapter, we will formalize the concepts concerning probability. In the above discussion, we have used such terms as, *event, probability, likely, odds,* and *certain to occur*. Through the use of definitions and illustrations, these expressions (and others) will be described and explained. The study of probability is very important in business applications. Most firms must continually forecast the chance of events occurring in the future for decision-making purposes. Forecasting borrowing rates for short-term cash needs, the money supply, amount of tax liability, and the unemployment rate as well as other economic quantities is essential to the efficient operation of modern businesses. Probability establishes the bases for the forecasting process. Probability also plays a fundamental role in inferential statistics by building a ''bridge'' between the population and the sample taken from it. Our initial applications of probability in this connection will be to make deductions about a sample from a known population, whereas in Chapter 9 the ''traffic'' will flow in the opposite direction on this bridge. Thus, until Chapter 9, the use of probability is as indicated in Figure 3.1: Probability

FIGURE 3.1 | Role of probability in the statistical inference process

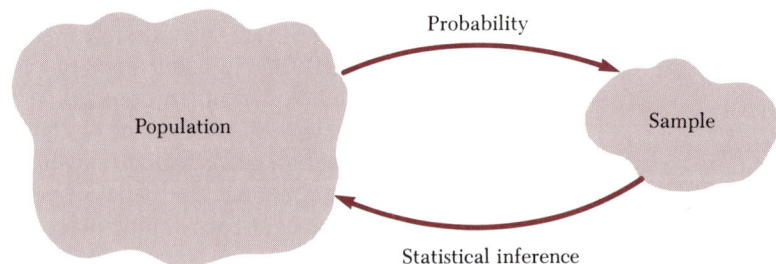

reasons from the population to the sample, while statistical inferences are drawn about the population from the sample.

As an example of the use of probability in this context, consider the next national election for the office of president of the United States. Let us suppose that only two candidates are listed on the ballot for the presidency, the Democratic candidate (A) and the Republican candidate (B). Furthermore, suppose it is believed that in the population of registered persons who will vote on election day that 60 percent will vote for candidate A, and 40 percent will vote for candidate B. If we now randomly select one person from this population, what is the probability that he or she will vote for candidate A? Since we believe that 60 percent of the persons will vote for candidate A, it seems reasonable that the probability that one sampled person will vote for candidate A is 0.60 (or, as a percentage, 60 percent). Knowledge of the probabilities of the two possible outcomes (voting for candidate A or voting for candidate B) of the experiment (randomly selecting one person from the population) enables us to deduce the probability of the outcome in our sample of size one.

Indeed, by using this knowledge, we could deduce the probabilities of zero, one, or two persons voting for candidate A in a sample of size two, and so on for larger sample sizes. The methods for deducing these probabilities will be developed in this chapter and in the next four chapters. Thus, if the population is known in the sense that the probabilities associated with the values in it are known, then this knowledge can be used to deduce the probabilities of the outcomes in the sample.

To illustrate the use of probability in making inferences from a sample to the population, suppose that candidate A conjectures *before* the election that 60 percent of the people in the voting population will vote for her. To demonstrate that this conjecture is reasonable, her campaign manager randomly selects 10 individuals from the population and finds that all 10 intend to vote for candidate B. If the probability of a randomly selected person voting for candidate A is really 0.60, it is extremely unlikely that 10 randomly selected people will all vote for candidate B. It is more likely that the true percentage of people who will vote for candidate A is considerably less than 60 percent. Hence knowledge of this experiment (sample outcome) may suggest to candidate A that more resources (campaigning, advertising, etc.) may be necessary if she is to have a chance of winning the election. Candidate A is interested, of course, in testing whether this sample is representative of the population characteristic (voting pattern), whether more sample information should be obtained, or whether the election is likely to go to candidate B in any case (so that she would be wasting her time and money by campaigning further).

In this chapter, we begin the development of the principles and concepts of probability. This development is continued in Chapter 4. In the remainder of the text, these concepts and principles are applied to infer measures of population characteristics from samples taken from the population as in the voting illustration above.

■ **3.2**

The sample and event spaces

In the presidential election example discussed in the previous section, we defined a population consisting of registered persons who will vote on election day. Suppose we assign a 1 to an individual if he or she intends to vote for candidate A and assign 0 if he or she intends to vote for candidate B. The population can then be thought of as a collection of 1s and 0s. How are these 1s and 0s generated? Each person in the population must be contacted and represented by a 1 or a 0. The process of contacting each person to determine the outcome (1 or 0) is called an *experiment*.

Experiment

Definition 3.1

Experiment

An *experiment* is a process that results in one of two or more distinct outcomes, where the specific outcome that occurs cannot be predicted with certainty.

In the voting illustration above, there are only two possible outcomes and a sample of size one. They are distinct, meaning that one outcome can be clearly distinguished from the other. In the sample of size one, the selected person either votes for candidate A or votes for candidate B. Selecting a sample of one person and determining who the person will vote for thus represents an experiment as in Definition 3.1. The experiment results in one of two distinct outcomes (the person either votes for candidate A or votes for candidate B), and it is not possible to determine beforehand who the person will vote for with certainty. Other examples of experiments are:

1. A professor at a large university is selected and his salary is recorded.

2. A unit of a product is selected from an assembly line and is analyzed to determine whether or not it is defective.

3. A light bulb is randomly selected from the day's production and its time to failure measured.

By repeating an experiment many times, a population of outcomes can be generated. For example, if we repeated the experiment in 3 until each and every light bulb in the day's production run had been tested to failure, the population of all times to failure of this set of light bulbs would have been generated. In the process of doing this, however, the entire day's production of light bulbs (the population) would have been destroyed! We can also think of the sample being generated by repeated trials of an experiment. For example, if we wanted to sample 10 light bulbs, we could repeat the experiment 10 times.

<table>
<tr><td>

Elementary out-
comes

</td><td>

> **Definition 3.2**
> **Elementary outcome**
>
> Each distinct outcome of an experiment is called an *elementary outcome* of the experiment.

</td></tr>
</table>

Simple events

The elementary outcomes of an experiment are sometimes referred to as the *simple events* of the experiment. We will, however, refer to the distinct outcomes of an experiment as the *elementary outcomes* throughout this chapter. The elementary outcomes of an experiment will be denoted by the capital letter E. A subscript is added to associate it with a distinct outcome of the experiment.

Example 3.1 A dice game is played by tossing a pair of dice and recording the numbers of dots on the top two faces. There are 36 elementary outcomes when the pair of dice is tossed, and the numbers of dots appearing on the top two faces are recorded. The experiment in this case is the process of tossing the dice and recording the number of dots on the two top faces of the pair. The 36 elementary outcomes for this experiment are illustrated in Figure 3.2, and the totals of the dots showing on the top faces of the pair of dice are given beneath each pair of top faces. We will be referring to these totals in a later example.

The elementary outcomes of the dice game experiment are denoted by the double-subscripted term E_{ij}, where i gives the number of dots on the top face of the first die and j gives the number of dots on the top face of the second die. Thus the elementary outcomes of this experiment are labeled E_{11}, E_{12}, . . . , E_{66}; 36 elementary outcomes in all.

Outcome tree

The 36 elementary outcomes of this experiment can be generated by an *outcome tree,* as illustrated in Figure 3.3. The first set of branches in the tree represents the possible outcomes for the first die. The second set of branches in the tree represents the possible outcomes for the second die. Thus there are $(6)(6) = 36$ elementary outcomes of this experiment.

The outcome tree is a logical way to list the elementary outcomes of an experiment. It is very practical and efficient if the number of elementary outcomes is not too large.

Example 3.2 Suppose three persons, A, B, and C, are interviewing for a job. Two will be hired. The experiment is the selection of two of the three individuals interviewed for the job. List the elementary outcomes of this experiment using an outcome tree.

Solution The outcome tree is given in Figure 3.4.

Notice in Example 3.2 that the six elementary outcomes listed specify not only the two individuals selected, but also the *order* in which they are selected. If the order in which the two individuals are selected is not impor-

FIGURE 3.2 | The 36 elementary outcomes $E_{11}, E_{12}, \ldots, E_{66}$ corresponding to the dice experiment

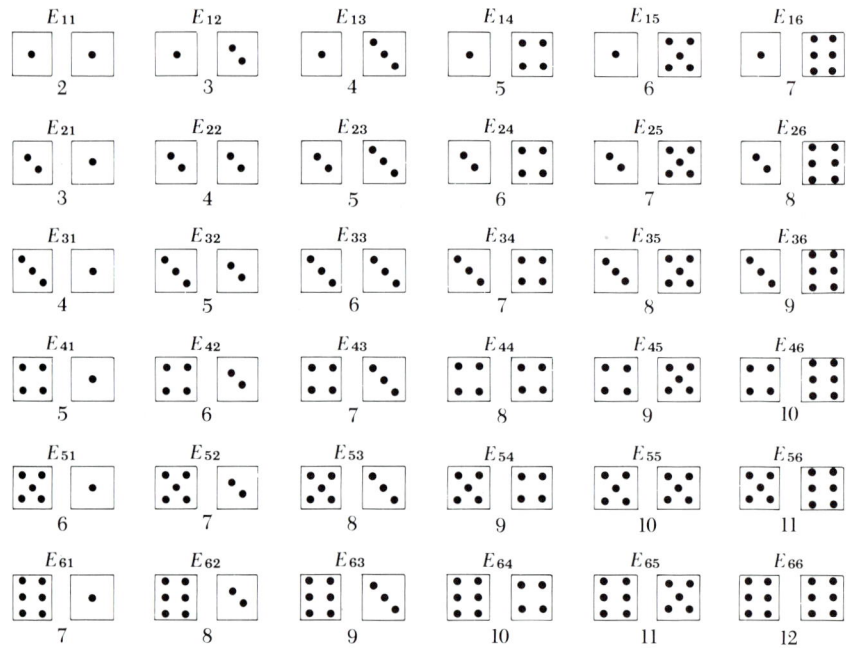

tant, then we need not distinguish between E_1 and E_3, E_2 and E_5, and E_4 and E_6. In this case, a "simpler" set of outcomes would be:

E_1^*: A and B are selected[1]
E_2^*: A and C are selected
E_3^*: B and C are selected

As suggested, it is often possible to define the elementary outcomes differently in the same experiment. To gain a better understanding of how to define the elementary outcomes of an experiment, consider the following example.

Example 3.3 Suppose that a coin is tossed twice by an individual. He tells you that there are three possible outcomes for this experiment:

E_1^*: Two heads (no tails)
E_2^*: One head   (one tail)
E_3^*: No heads   (two tails)

---

[1] The asterisk(*) differentiates between two sets of elementary outcomes, E_i and E_i^*, defined for the same experiment.

FIGURE 3.3 | Outcome tree for the dice experiment in Example 3.1

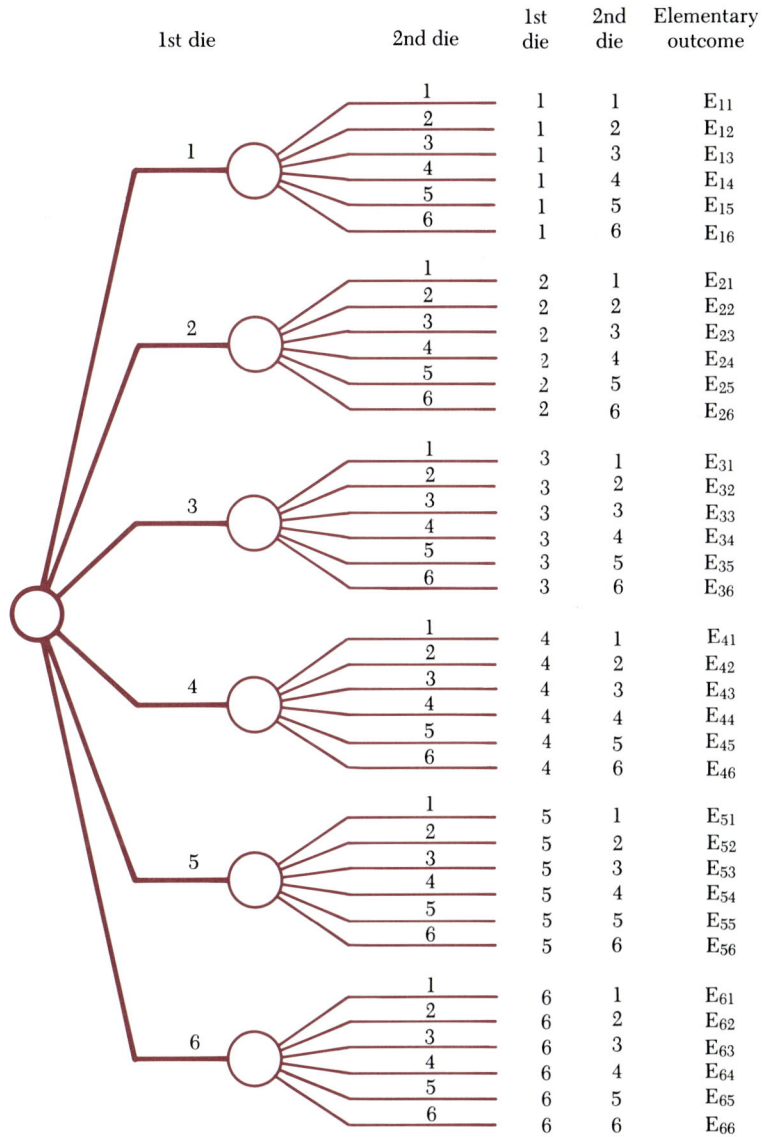

1st die	2nd die	1st die	2nd die	Elementary outcome
	1	1	1	E_{11}
	2	1	2	E_{12}
1	3	1	3	E_{13}
	4	1	4	E_{14}
	5	1	5	E_{15}
	6	1	6	E_{16}
	1	2	1	E_{21}
	2	2	2	E_{22}
2	3	2	3	E_{23}
	4	2	4	E_{24}
	5	2	5	E_{25}
	6	2	6	E_{26}
	1	3	1	E_{31}
	2	3	2	E_{32}
3	3	3	3	E_{33}
	4	3	4	E_{34}
	5	3	5	E_{35}
	6	3	6	E_{36}
	1	4	1	E_{41}
	2	4	2	E_{42}
4	3	4	3	E_{43}
	4	4	4	E_{44}
	5	4	5	E_{45}
	6	4	6	E_{46}
	1	5	1	E_{51}
	2	5	2	E_{52}
5	3	5	3	E_{53}
	4	5	4	E_{54}
	5	5	5	E_{55}
	6	5	6	E_{56}
	1	6	1	E_{61}
	2	6	2	E_{62}
6	3	6	3	E_{63}
	4	6	4	E_{64}
	5	6	5	E_{65}
	6	6	6	E_{66}

FIGURE 3.4 | Outcome tree for Example 3.2

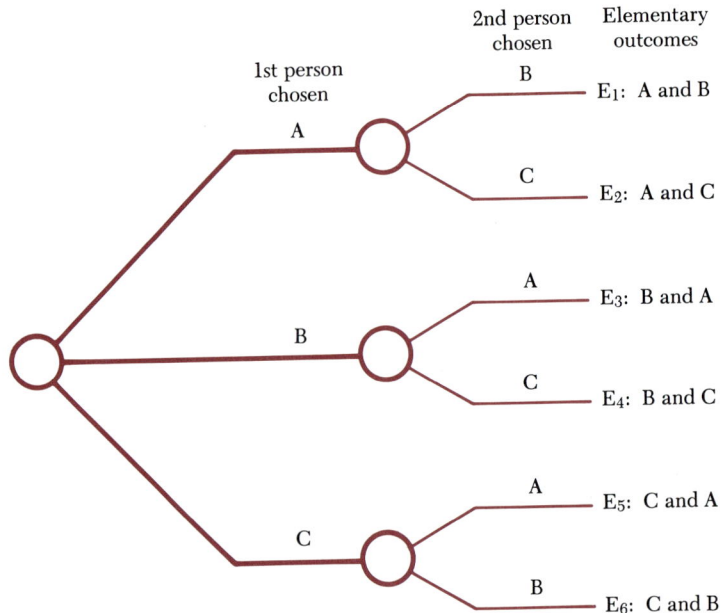

Construct an outcome tree for this experiment to determine whether or not his claim that there are three outcomes for the experiment is valid.

Solution   The outcome tree for this experiment is given in Figure 3.5.

From the outcome tree, it is clear that it is possible to define another set of elementary outcomes for this experiment:

$$E_1: \quad (H,H) \qquad E_3: \quad (T,H)$$
$$E_2: \quad (H,T) \qquad E_4: \quad (T,T)$$

Outcome E_2^* (one head) is equivalent to the two outcomes E_2 and E_3 in the outcome tree in Figure 3.5. Although his claim that there are three outcomes for this experiment is valid if one considers the number of heads for each outcome, a complete set of outcomes is generated by the outcome tree. The tree shows that there are two ways to get one head: a head on the first toss of the coin and a tail on the second toss *and* a tail on the first toss of the coin and a head on the second toss. A danger in using his set of outcomes (E_1^*, E_2^*, and E_3^*) is assuming that each of these outcomes is equally likely to occur in the experiment. That is clearly not the case, since there are two ways to get the outcome "one head," and there is only one way to get the outcome "no heads" and to get the outcome "two heads." We will return to this example when we begin our discussion on assigning probabilities to the outcomes of an experiment.

FIGURE 3.5 | Outcome tree for a coin-tossing experiment

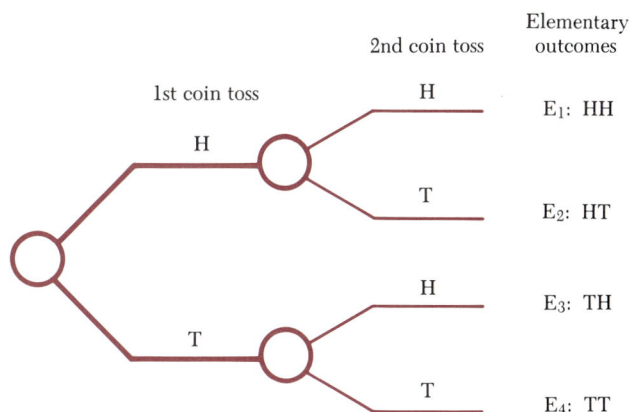

As Examples 3.2 and 3.3 suggest, typically there is more than one way to define a set of elementary outcomes of an experiment. Indeed this is the case in most experiments. In Section 3.3 we will learn that probability questions formulated about an experiment can be correctly answered using two (or more) elementary outcome definition sets, provided that the probabilities have been correctly associated with the elementary outcomes of the experiment.

After the elementary outcomes of an experiment have been defined, collections of the elementary outcomes, called *events*, may be defined.

Event

Definition 3.3
Event

An *event* is a collection of one or more elementary outcomes of an experiment.

In Example 3.1, we might define the event A to be "A total of seven dots appear on the top two faces of the pair of dice." The event A is composed of the elementary outcomes E_{16}, E_{25}, E_{34}, E_{43}, E_{52}, and E_{61}. In each of these elementary outcomes, the total of the dots on the top faces of the dice is 7. Note that *each* of these elementary outcomes could be an event as well. For example, we may define the event F to be "a total of two dots on the top two faces of the dice." Event F is composed of the single elementary outcome E_{11}. Hence E_{11} is both an elementary outcome and an event in this case.

Null event

> **Definition 3.4**
> Null event
>
> A *null event* is an event that contains no elementary outcomes of the experiment. It is denoted by ϕ.

In Example 3.1, an example of a null event is "a total of one dot on the two top faces of the dice." It is impossible for this event to happen because at least one dot appears on the top face of each die. In this instance the event set is empty; it does not contain any of the elementary outcomes of the experiment.

Venn diagram

The elementary outcomes of an experiment and events defined to be collections of one or more of these elementary outcomes can be illustrated graphically by a *Venn diagram*. The Venn diagram associated with the elementary outcomes in Example 3.1 is shown in Figure 3.6. Each elementary outcome in a Venn diagram is shown as a "point" with its corresponding subscripted letter E. The collection of *all* elementary outcomes in an experiment is called the *sample space of the experiment*. The sample space of an experiment is denoted by S and is shown in the Venn diagram as the collection of all elementary outcomes. The event A, "a total of seven dots appearing on the top two faces of the dice," is illustrated in the Venn diagram by enclosing the elementary outcomes belonging to it in red as illustrated in Figure 3.6. The resulting enclosed region is the *event space*, A.

Sample space

> **Definition 3.5**
> Sample space
>
> A *sample space* of an experiment is the collection of *all* elementary outcomes of the experiment.

Event space

> **Definition 3.6**
> Event space
>
> An *event space* of an experiment is the collection of the elementary outcomes constituting the event.

FIGURE 3.6 | Venn diagram for the elementary outcomes in Example 3.1

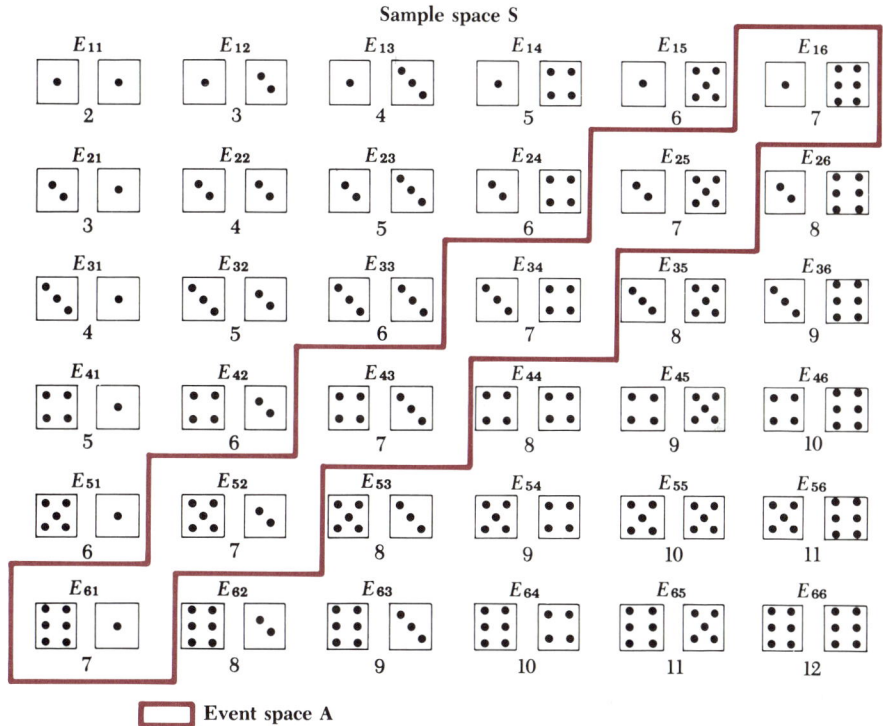

Sample space S

Event space A

When referring to the sample space S of an experiment, the elementary outcomes are also called the *sample points* in S.

Sample point

Definition 3.7
Sample point

A *sample point* in a sample space S of an experiment is an elementary outcome of the experiment.

Example 3.4 In Example 3.2, suppose we define the following events:

F: A is chosen first.
G: A is chosen without regard to selection order.
H: A and B are chosen without regard to selection order.

Assume that we define the following elementary outcomes:

E_1: A and B $\quad$ E_3: B and A $\quad$ E_5: C and A
E_2: A and C $\quad$ E_4: B and C $\quad$ E_6: C and B

Draw a Venn diagram showing the sample space S, the sample points in the sample space, and the event spaces F, G, and H for this experiment.

   Solution The Venn diagram showing the sample space S, the sample points, and the event spaces F, G, and H is illustrated in Figure 3.7.

FIGURE 3.7 | Venn diagram for the events defined in Example 3.4

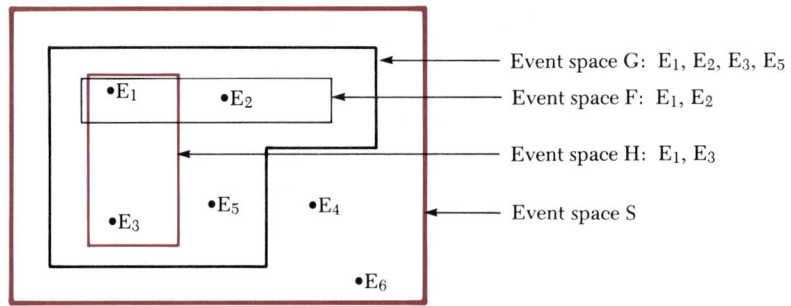

Event space G: E_1, E_2, E_3, E_5
Event space F: E_1, E_2
Event space H: E_1, E_3
Event space S

■ **3.3**

Computing probabilities from the sample space

   Suppose an event A is defined over a sample space S. To determine the probability that event A occurs in a single trial of the experiment, it is necessary to know the probability of each and every sample point in the sample space S occurring. Therefore, to answer the question, "What is the probability that an event, say event A, occurs in the experiment?" we first must assign probabilities to the elementary outcomes of the experiment (or equivalently, to the sample points in the sample space S of the experiment). We will denote by $P(E_i)$ the probability assigned to the elementary outcome E_i of an experiment. The assigned probabilities $P(E_1)$, $P(E_2)$, . . . , $P(E_N)$, where there are N elementary outcomes or sample points in the sample space S, must satisfy three probability axioms for experiments that have a finite number of outcomes.

Axiom 1

$$0 \leq P(E_i) \leq 1 \quad \text{for} \quad i = 1, 2, \ldots, N$$

The first axiom requires that every elementary outcome be assigned as its probability a nonnegative number between 0 and 1, inclusive.

Axiom 2

$$P(S) = 1$$

If $E_1, E_2, \ldots, E_N$ are disjoint events, then the condition $P(S) = 1$ is equivalent to the condition

$$\sum_{i=1}^{N} P(E_i) = 1$$

The second axiom requires that the sample space probability equals one. Since the elementary outcomes are disjoint (one and only one elementary outcome can occur on each trial of the experiment), the condition $P(S) = 1$ is equivalent to the sum of the probabilities of all the elementary outcomes being equal to one. The dice experiment is an example of disjoint elementary outcomes. On each trial of the experiment (tossing the pair of dice and recording the number of dots on each of the top faces), one and only one of the 36 elementary outcomes composing the sample space S can occur.

Axiom 3

For elementary outcomes $E_1, E_2, E_3, \ldots,$

$$P(E_1 \text{ or } E_2 \text{ or } E_3 \text{ or } \ldots) = P(E_1) + P(E_2) + P(E_3) + \ldots$$

The third axiom requires that the probability of one or more members of a set of elementary outcomes occurring in an experiment is the sum of their respective probabilities.

Example 3.5 Suppose in Example 3.1 that we assign a probability of $\frac{1}{36}$ to each of the 36 elementary outcomes of the experiment. This would seem to be reasonable if the dice are "fair." By fair, we mean that each die, as a six-sided cube, will have any one of its faces upward with equal probability on each toss. If this is so for each die, then each of the 36 elementary outcomes in tossing the pair of dice should be equally likely. Thus the assignment of a probability of $\frac{1}{36}$ to each elementary outcome is established.

Elementary outcome	Definition	Probability
E_{ij}	i dots on the top face of the first die j dots on the top face of the second die $i = 1, 2, 3, 4, 5, 6;\ \ j = 1, 2, 3, 4, 5, 6$	$P(E_{ij}) = \frac{1}{36}$

These probability assignments to the 36 elementary outcomes satisfy the three axioms of probability: (1) each probability is between 0 and 1, inclusive; (2) the probabilities sum to 1; and (3) the probability of one or more members of a set of elementary outcomes occurring in an experiment is the sum of their respective probabilities.

The satisfaction of the third axiom by these probability assignments may be argued as follows. By the definition of the elementary outcomes (E_{ij} values), it is obvious that one and only one of the elementary outcomes can occur in any single trial of the experiment. Thus, for example, the probability of, say, either E_{11} or E_{12} occurring in a single trial of the experiment is simply the sum of the probability that E_{11} occurs and the probability that E_{12} occurs.

These axioms are intuitive; most of us understand them before taking any formal training in probability. If an event is certain to happen, its probability of occurrence is 1, and if an event is certain not to happen, its probability of occurrence is 0.

How do we, in fact, formally define probability? We will consider three ways.

Relative frequency

Definition 3.8

Relative frequency definition of probability

If an event E is defined in an experiment, the experiment is repeated a very large number of times, say, N, and the event E is observed to occur in n of these N experimental trials, then

$$P(E) = \frac{n}{N}$$

The ratio n/N represents the proportion of the time that event E occurs in repeated experiments. Suppose we wish to assign a probability to the elementary outcome E_1: a head, in a coin-tossing experiment that consists of tossing a coin and observing whether it lands heads or tails. If we toss the coin a large number of times, say 10,000, and if the coin is fair, we expect the proportion of heads that occurs in the 10,000 trials to be about ½. The larger the number of trials, the closer we should expect the proportion of heads to approach ½. The relative frequency definition of probability defines probability in a limiting sense; it is the limit of the ratio n/N as the total number of trials N approaches infinity.

Logic

Definition 3.9
Deductive logic definition of probability

The probability of an event E is determined logically from symmetry or geometric considerations associated with the experiment.

An example of applying Definition 3.9 is the assignment of probability ½ to the elementary outcome E: a head, in a coin-tossing experiment, on the basis of the symmetry of a fair coin. Another example is the assignment of probability ⅟₃₆ to each elementary outcome in the dice game in Example 3.5.

Subjective

Definition 3.10
Subjective definition of probability

The probability of an event E is determined subjectively, reflecting a person's "degree of belief" that an event will occur.

For example, a person who is knowledgeable about an experiment and its outcomes subjectively assigns probabilities to the elementary outcomes. This is generally the case in preparing forecasts of such economic variables as interest rates and the GNP. For instance, an economist might define the following elementary outcomes concerning the prime interest rate in the next quarter: The interest rate will decrease; the interest rate will remain the same as it is this quarter; or the interest rate will increase. The economist may subjectively assign probabilities that satisfy Axioms 1, 2, and 3 to these three elementary outcomes that compose the sample space.

Objective proba-
bility

When probabilities are assessed in ways that are consistent with the relative frequency or the symmetry determination of probability, we call them *objective* probabilities. Objective and subjective probabilities are fundamentally different. If we asked a number of people to determine the probability of an event objectively, they would all arrive at the same answer, provided they did the calculations properly. But, if we asked them to determine the probability subjectively, each individual would arrive at his or her own answer. As a consequence, not all probability theory and methods that can be applied to objective probabilities can be applied to subjective ones.

In this chapter, we shall assume that the probabilities of the elementary outcomes in an experiment have been determined objectively. In Chapter 4, we shall again discuss how one assigns objective probabilities to events. Subjective probabilities will be considered in Chapters 21, 22, and 23.

Now we will consider methods to determine the probability that an event occurs in an experiment.

Event occurrence

Definition 3.11
Event occurrence

An event, defined over a sample space S in an experiment, *occurs* if one or more of the sample points in its event space occur.

Once the probabilities have been assigned to the elementary outcomes of an experiment, we are able to compute the probabilities of events occurring that are defined over the sample space for the experiment by using the following theorem.

Theorem 3.1[2]

Suppose event A is an event defined over a sample space. The probability of event A occurring is equal to the sum of the probabilities of the elementary outcomes composing event A.

The following three examples illustrate the use of the sample space–event space framework for answering probability questions.

Example 3.6 In the dice game introduced in Example 3.1, suppose we have defined the following events:

A: Sum of the dots appearing on the top two faces of the dice is seven.
B: Sum of the dots appearing on the top two faces of the dice is three or less.

What are the probabilities of events A and B occurring?

---

[2] A theorem is a statement of a result that is a logical consequence of definitions and axioms.

FIGURE 3.8 | Venn diagram of the events in Example 3.6

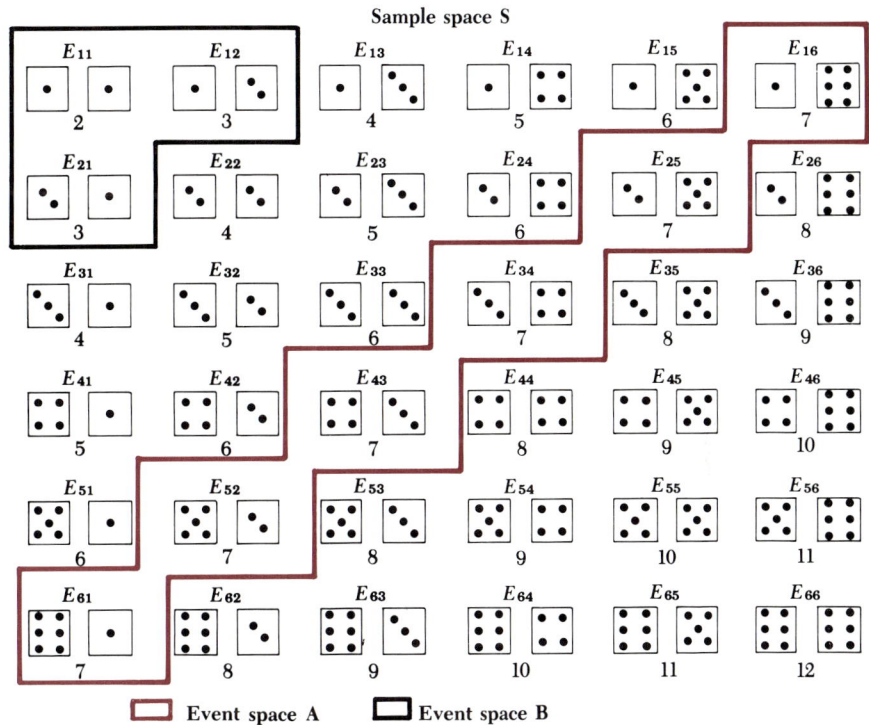

Sample space S

Event space A   Event space B

Solution   The Venn diagram in Figure 3.8 shows the two event spaces, A and B. The elementary outcomes composing event A are: E_{16}, E_{25}, E_{34}, E_{43}, E_{52}, and E_{61}. Thus

$$P(A) = P(E_{16}) + P(E_{25}) + P(E_{34}) + P(E_{43}) + P(E_{52}) + P(E_{61})$$
$$= \tfrac{1}{36} + \tfrac{1}{36} + \tfrac{1}{36} + \tfrac{1}{36} + \tfrac{1}{36} + \tfrac{1}{36} = \tfrac{6}{36} = \tfrac{1}{6}$$

The elementary outcomes composing event B are: E_{11}, E_{12}, and E_{21}. Thus

$$P(B) = P(E_{11}) + P(E_{12}) + P(E_{21}) = \tfrac{1}{36} + \tfrac{1}{36} + \tfrac{1}{36} = \tfrac{3}{36} = \tfrac{1}{12}$$

Example 3.7   A construction manager is in charge of preparing bids on construction jobs for submission to a county commission, which awards contracts for these jobs. Let us suppose that during the current time period, there are three construction jobs available in the county for which the manager wishes to bid. What is the probability that the manager will win at least two of the three submitted bids if he has a 50 percent chance of winning each of the three bids?

Solution   The experiment consists of placing bids on three construction jobs. The elementary outcomes of this experiment are generated by using the

FIGURE 3.9 | Outcome tree for the experiment in Example 3.7

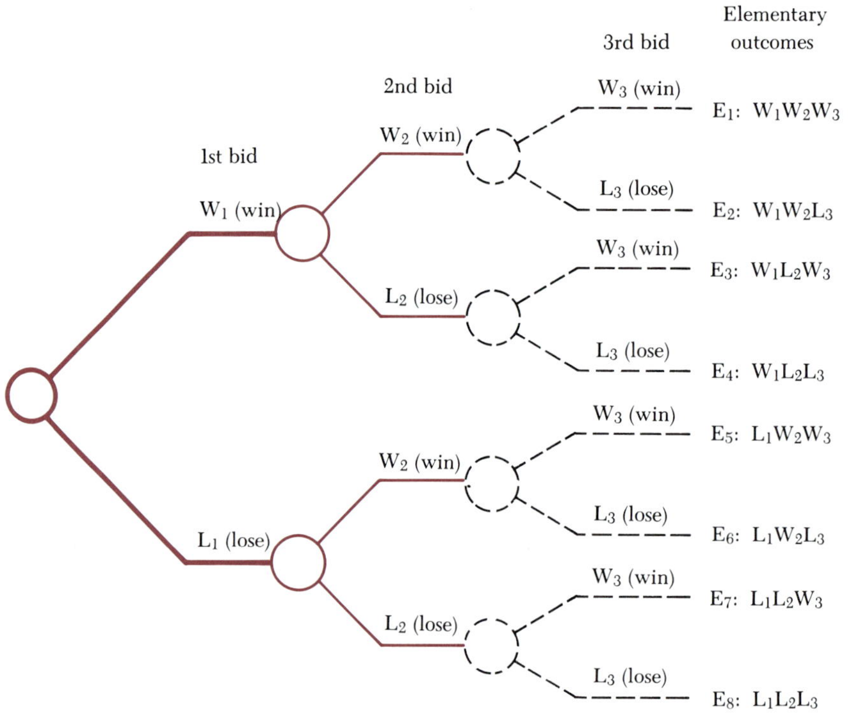

outcome tree illustrated in Figure 3.9. Thus there are eight points in the sample space for this experiment. The assumption that the manager has a probability of ½ of winning each of the bids will result in each of the eight elementary outcomes being equally likely to occur if the outcome of any one bid will not affect the outcomes of the other two. We will show in Chapter 4 how to arrive at this conclusion. For the moment, it seems reasonable, and we shall assume in this problem that

$$P(E_i) = \frac{1}{8}, \quad i = 1, 2, \ldots, 8$$

Now, define the event A, "at least two bids are won." Since the elementary outcomes E_1, E_2, E_3, and E_5 belong to A,

$$P(A) = P(E_1) + P(E_2) + P(E_3) + P(E_5) = \frac{1}{8} + \frac{1}{8} + \frac{1}{8} + \frac{1}{8} = \frac{1}{2}$$

Example 3.8 Consider the problem introduced in Example 3.2: there are three persons, A, B, and C, interviewing for a job and two will be hired. If the recruiter views all three candidates as being equally qualified, what is the probability that A is among the two hired?

Solution The elementary outcomes of the experiment are:

E_1: A and B are selected.
E_2: A and C are selected.
E_3: B and C are selected.

Since the recruiter considers all three to be equally qualified, each pair is equally likely to be chosen. Thus $P(E_i) = \frac{1}{3}$, $i = 1, 2, 3$. Let the event D be "A is among the two hired." Since D is composed of E_1 and E_2,

$$P(D) = P(E_1) + P(E_2) = \frac{1}{3} + \frac{1}{3} = \frac{2}{3}$$

The rule for computing probabilities using the sample space–event space method can be modified if each sample point in the sample space is equally likely to occur.

Theorem 3.2

If each of the elementary outcomes in an experiment has the same probability of occurring, and A is an event in the sample space, then

$$P(A) = \frac{n}{N}$$ where N = Number of points in sample space S.

n = Number of points in event space A.

Example 3.9 The sample space S for the dice game introduced in Example 3.1 together with the event space for the event A, "a total of seven dots on the top two faces of the dice," are shown in Figure 3.6. In Example 3.5, a probability of $\frac{1}{36}$ was assigned to each of the sample points (E_{ij}) in the sample space S. What is the probability of Event A, "a total of seven dots appear on the top two faces," occurring?

Solution Since each of the sample points in sample space S has the same probability of occurring, the probability that event A occurs in a single trial of the experiment is given by:

$$P(A) = \frac{\text{Number of points in event space A}}{\text{Number of points in sample space S}}$$

The number of points in sample space S is 36, and the number of points in event space A is 6 (E_{16}, E_{25}, E_{34}, E_{43}, E_{52}, and E_{61}). Thus $P(A) = \frac{6}{36} = \frac{1}{6}$. The probability of tossing a 7 (the dots total seven on the two top faces) in a single roll of the dice is $\frac{1}{6}$. Therefore, rather than summing the probabilities of the elementary outcomes composing A, we can simply *count the number of sample points* in event space A and in sample space S to compute $P(A)$.

Since in Examples 3.6–3.8 the elementary outcomes have equal probabilities assigned to them, Theorem 3.2 could have been used to answer the probability questions. In most problems if Theorem 3.2 can be applied, it will usually be easier than summing probabilities of elementary outcomes to determine the probability of an event occurring in a sample space. We will return to this point later, but first let's see how a sample space can be formed to take advantage of Theorem 3.2 in a probability problem.

Example 3.10 Herman purchases a new four-cylinder compact automobile. After one week, the car begins to run poorly. It is difficult to start, and the engine misses and quits at stoplights. Herman suspects that something is wrong. Indeed, unknown to Herman, two of the four spark plugs in the engine are defective. Herman takes the car to the garage and issues a complaint. On the basis of the symptoms Herman describes, the mechanic immediately suspects spark plug difficulties. He randomly selects two of the four spark plugs and pulls them out to be checked. What is the probability that he selected the two defective spark plugs?

Solution A possible sample space is:

E_1^*: Two defective spark plugs are selected.
E_2^*: One defective and one good spark plug are selected.
E_3^*: Two good spark plugs are selected.

These three elementary outcomes are not equally likely to occur. There are more ways that E_2^* can occur than E_1^*. To see this, label the good spark plugs as G_1 and G_2 and the defective spark plugs as D_1 and D_2. The possible extracted pairs are:

$$E_1: \ G_1G_2 \qquad E_3: \ G_1D_2 \qquad E_5: \ G_2D_2$$
$$E_2: \ G_1D_1 \qquad E_4: \ G_2D_1 \qquad E_6: \ D_1D_2$$

These pairs are equally likely to occur, because the mechanic is selecting two spark plugs at random. Let A be the event "The two defectives are selected." Since $P(E_i) = \frac{1}{6}$, $i = 1, \ldots, 6$, Theorem 3.2 may be applied:

$$P(A) = \frac{1}{6}$$

The sample space composed of the points E_1^*, E_2^*, and E_3^* certainly could have been used to solve this problem. In this space, $P(E_1^*) = \frac{1}{6}$, $P(E_2^*) = \frac{4}{6}$, and $P(E_3^*) = \frac{1}{6}$. Thus by using Theorem 3.1,

$$P(A) = P(E_1^*) = \frac{1}{6}$$

It would have been very easy, for example, to mistakenly assign equal probabilities of $\frac{1}{3}$ to E_1^*, E_2^*, and E_3^*. That is, if Theorem 3.2 had been applied to the sample space composed of E_1^*, E_2^*, and E_3^*, the probability assigned to event A would have incorrectly been $\frac{1}{3}$—1 [the number of points in the event space (E_1^*)] divided by 3 [the number of points in the sample space (E_1^*, E_2^*, and E_3^*)].

Example 3.11 Reconsider now the problem in Example 3.3—a fair coin is tossed twice. The outcome tree in Figure 3.5 gives the four outcomes of the experiment: E_1 (H, H), E_2 (H, T), E_3 (T, H), and E_4 (T, T). By using the deductive logic definition of probability, we assign a probability of ¼ to each of the four outcomes, arguing that if the coin is fair, then a head or a tail on each toss should be equally likely, and therefore each of the four outcomes should be equally likely to occur. Determine the probability that one head occurs in the two tosses.

Solution   Define the event, A, to be "One head occurs in the two tosses of the coin." Then

$$P(A) = \frac{\text{Number of points in event space A}}{\text{Number of points in sample space S}} = \frac{2}{4} = 0.50$$

since each of the four points in sample space S is equally likely to occur. Note that if the alternate sample space in Example 3.3 had been used, then each of the experimental outcomes—E_1^* (no heads), E_2^* (one head), and E_3^* (two heads)—is not equally likely to occur (E_2^* can occur in two ways, but E_1^* and E_3^* can occur only one way each). Therefore, Theorem 3.2 does not apply. A common mistake in probability is to assume that outcomes in a sample space are equally likely to occur when in fact they are not. If we mistakenly made this assumption with the sample space composed of E_1^*, E_2^*, and E_3^*, then applying Theorem 3.2 would produce $P(E_2^*) = P(\text{one head}) = \frac{1}{3}$, clearly an incorrect result. Notice the similarity between this example and the previous example, Example 3.10.

The reason for trying to construct a sample space so that the sample points have equal probabilities assigned to them becomes more pronounced as the number of sample points N in S increases. It is common for a sample space to contain thousands of points. Identifying them and adding the probabilities of a subset of the points composing an event become rather troublesome and time-consuming. In the next section, we shall present various counting rules that enable us to compute probabilities using Theorem 3.2 when it becomes difficult to list the entire set of elementary outcomes corresponding to an experiment due to the sheer number of them.

■ **3.4**

Permutations, combinations, and other counting rules

Numerous counting rules can be used to count the number of points in sample and event spaces. We shall consider four of the most important counting rules. Each will be presented without proof and followed by examples.

Product rule

Rule 3.1

Product rule

Suppose there are m groups of objects. The first group contains k_1 distinguishable objects, the second group contains k_2 distinguishable objects, and so forth, and the mth and last group contains k_m distinguishable objects. If the experiment is to draw *one* object from *each* of the m groups, then there are $(k_1)(k_2) \cdots (k_m)$ elementary outcomes in the experiment.

The basis for the product rule can be illustrated by an outcome tree. Suppose that there are $m = 2$ groups. In the first group there are $k_1 = 2$ distinguishable items, and in the second group there are $k_2 = 3$ distinguishable items. Then, by the product rule, there are $(k_1)(k_2) = (2)(3) = 6$ elementary outcomes to the experiment where one object is drawn from each of the two groups. These six outcomes are shown in Figure 3.10. The first set of

FIGURE 3.10 | The product rule for $m = 2$ groups, $k_1 = 2$, and $k_2 = 3$, illustrated by an outcome tree

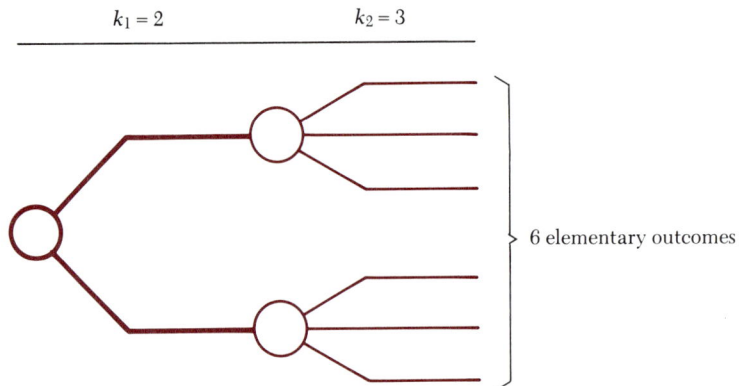

two branches of the tree represents the two objects in the first group. The second set of branches represents the three objects in the second group. Since each of the two initial branches is followed by three branches, there are $(2)(3) = 6$ possible pairs of branch pathways in moving from the left to the right in the tree.

Example 3.12 Herman has decided to purchase a new hi-fi system with the money he saved by buying a compact car instead of a large sedan. His hi-fi system will be composed of a receiver, a pair of speakers, a turntable, and a tape deck. In the store where he will make the purchase, there are 10 different kinds of receivers, five kinds of speakers, four kinds of turntables, and eight kinds of tape decks. How many different systems can Herman choose from? Assume that all combinations are compatible.

Solution There are $m = 4$ groups (types of equipment) with $k_1 = 10$ (receivers), $k_2 = 5$ (speakers), $k_3 = 4$ (turntables), and $k_4 = 8$ (tape decks). Since he must select one object from each group, he can choose from $(10)(5)(4)(8) = 1,600$ possible systems.

Example 3.13 In 1953, its first year of production, the Chevrolet Corvette was offered in the following color schemes and with the indicated options:

1953 Chevrolet Corvette Ordering Information

Exterior Body Color: Polo White
Interior Body Color: Red
Options:
  Heater
  Signal-Seeking AM Radio

By 1963 (10 years later), the choices of colors, body styles, options, and so on had increased as indicated below:

1963 Corvette Ordering Information
(Approximate)

Body Style:
  Coupe
  Roadster
Exterior Body Color:
  Tuxedo Black
  Ermine White
  Riverside Red
  Silver blue
  Daytona Blue
  Saddle Tan
  Sebring Silver
Interior Body Colors:
  Black
  Red
  Saddle
  Dark Blue
Options:
  Genuine leather seat trim
  Soft ray tinted glass, windshield
  Soft ray tinted glass, all windows
  Electric power windows
  Heater and defroster deletion (credit)
  Air conditioning
  Positraction rear axle, all ratios
  Special highway 3.08 : 1 rear axle
  Power brakes
  Metallic brakes
  Optional 300 horsepower engine
  Optional 340 horsepower engine
  Optional 370 horsepower engine (fuel injected)
  Four-speed transmission

Powerglide automatic transmission
Thirty-six gallon fuel tank
Off-road exhaust system
Woodgrained plastic steering wheel
Power steering
Special cast aluminum knock-off wheels
Blackwall nylon tires, 6.70 × 15
Whitewall rayon tires, 6.70 × 15
Backup lights
Signal-seeking AM radio
AM–FM radio

Determine the number of different Corvettes that could be produced/ordered in 1953 and 1963. Assume that all color combinations, engine choices, options, and so on could be ordered or produced independently of any other choice. That is, for example, that a 1963 Sebring Silver Coupe could be ordered with any of the engine choices, interior colors, and so on.

Solution   Since the 1953 Corvette came in only one exterior color, one interior color, and with two options, the number of different 1953 Corvettes that could be produced was:

> Number of radio options (includes no radio)
>> times Number of heater options (includes no heater)
>> = 2 × 2 = 4.

The 1963 Corvette, on the other hand, could come in a variety of configurations:

> Number of body styles (coupe or roadster) times Number of exterior colors times Number of interior colors times Number of interior fabrics times Number of glass-tinting possibilities (including clear) times Power window choice (or not) times Heater delete (or not) times Air conditioning (or not) times Standard or 3.08 ratio axle times Positraction rear axle (or not) times Power brakes (or not) times Metallic brakes (or not) times number of engine choices (including standard engine) times Number of transmission choices (including standard three-speed transmission) times Standard or optional fuel tank times Standard or off-road exhaust times Standard or woodgrained steering wheel times Power steering (or not) times Knock-off or standard wheels times Number of tire choices times Number of radio choices (includes radio delete) times Backup lights (or not)
>> = (2)(7)(4)(2)(3)(2)(2)(2)(2)(2)(2)(2)(4)(3)(2)(2)(2)(2)(2)(2)(3)(2)
>> = 198,180,864!

Since 21,513 1963 Corvettes were produced by General Motors, it is easy to see that it would have been impossible to produce every conceivable different variety that could have been produced. Actually, the quantity 198,180,864 is just slightly inflated—the reason being that it was impossible to order every option/color independently of one another. For example, the 36-gallon fuel tank was only optional on the coupe model; the leather trim option was only available in saddle (although leather trim was offered in every interior color choice the following year); a car could not be ordered both with fuel injection (high-performance engine) and air conditioning. Still, the number of possibilities remaining after considering the dependent choices is still in the 10s of millions, far surpassing the number of cars produced. The use of outcome trees and counting rules discussed in this section would allow us to precisely determine the number of different automobiles that could have been produced (it is large!).

The rather large number of different types of automobiles that could be produced each year is not restricted to Corvettes, obviously. To convince yourself of this, simply read a showroom brochure to see the many ways in which several American manufacturers currently offer their models. These early Corvette models are offered simply as examples of the variety of ways American manufacturers were configuring their product.

The trend toward more individualization (believe it or not) with Corvette and with many other American automobiles continued throughout the 1970s, far surpassing the number of possibilities reported above. Then, fearing competition from abroad, manufacturers in the 1980s gradually began to cut back on the number of different possibilities for each model. This was accomplished in a variety of ways, one of which was to make more of the "options" standard equipment (e.g., air conditioning, power steering, power brakes) on certain models. The number of options available has also declined on certain models, partially through the influence of the federal government. For example, there is only one engine choice available presently for the Chevrolet Corvette due to meeting pollution control requirements. These same requirements mandated the choice of only one transmission type (automatic) on the 1984 Corvette. Manufacturers have also reduced the number of possibilities in production scheduling by assembling many of the models with a nearly fixed set of the options. For example, although the 1953 Corvette listed two options, all 300 cars produced were actually assembled that first year with both available options included with each car. This was accomplished by producing each automobile for inventory rather than by special order. Allowing the dealers to install many of the options has been another method of reducing the number of different types of automobiles assembled. This has worked particularly well with many of the import models.

Reducing the number of ways in which an automobile can be assembled has had a dramatic impact on quality (it has raised quality!). It has had a significant impact on reducing inventories (one hopes there has been a concomitant reduction in price) and has resulted in the inclusion of many more items as standard equipment on automobiles. The use of statistical analysis alerts managers to problems potentially caused by product variety that may interfere with assuring high-quality and effectively produced end products. Reducing the number of ways to configure each automobile assembled has been one of many reasons for the remarkable turnaround for the American automobile manufacturer, in the 1980s. Compare some of the imported cars with their American counterparts, for example, for variety both in the 1970s, and at present.

The next two counting rules apply to a different type of experiment as indicated in the following example.

Example 3.14   Suppose three persons, A, B, and C, are competing for two job positions. How many ways can two people be selected for employment from the three?

Solution   In this problem, it is easy to list the possible outcomes of the experiment. There are three: AB, AC, and BC. However, we may be interested in the order of the selection as well as the content of the resulting pairs. If this is the case, there are six possible outcomes: AB, BA, AC, CA, BC, and CB.

In Example 3.14 there are two different ways to view the pairs formed by selecting two persons out of three. Suppose there are n distinguishable objects from which we are selecting a subset of size r. If we are concerned about the number of groups of size r that can be formed from the n where one group of size r is different from another if its content is different, then we

| Combinations |

want to determine the number of *combinations* of r things selected from n. If, on the other hand, we want to compute the number of groups of size r that can be formed from the n where one group of size r is different from another in terms of *both* its content and the order in which the r things were drawn,

| Permutations |

then we want to determine the number of *permutations* of r things drawn from n.

In Example 3.14, the number of combinations of two persons drawn from three is three, and the number of permutations of two persons drawn from three is six.

In conjunction with the rules for computing the number of permutations and combinations, the complete definitions follow.

Permutations

Definition 3.12
Permutation

An ordered arrangement of r distinguishable objects is called a *permutation*. The number of permutations of r objects taken from n distinguishable objects is denoted by P_r^n.

Rule 3.2
Permutations

$$P_r^n = \frac{n!}{(n-r)!}$$

Combinations

Definition 3.13
Combination

A set of r distinguishable objects is called a *combination*. The number of combinations of r objects taken from n distinguishable objects is denoted by C_r^n.

Rule 3.3
Combinations

$$C_r^n = \frac{n!}{r!(n-r)!} = \frac{1}{r!}\, P_r^n$$

$n! = n(n-1)(n-2)$ $\cdots (2)(1)$

$0! = 1$
$1! = 1$

The symbol $n!$ is called "n factorial"; $n! = n(n-1)(n-2)\cdots(2)(1)$. Thus $4! = (4)(3)(2)(1) = 24$, and $6! = (6)(5)(4)(3)(2)(1) = 720$; $1! = 1$ and, by definition, $0! = 1$.

Example 3.15 A committee of three is to monitor the activities of the local lonely hearts club. The committee is to be formed by selecting three people from a group of five persons. How many different committees could be formed?

Solution Nothing is mentioned about the order or arrangement of the three selected individuals. Thus one committee will be different from another if it has one or more different people in it. We are only concerned about the *content* of each group, and therefore we want to determine the number of combinations of three things taken from five.

$$C_3^5 = \frac{5!}{3!(5-3)!} = \frac{5 \cdot 4 \cdot 3 \cdot 2 \cdot 1}{(3 \cdot 2 \cdot 1)(2 \cdot 1)} = 10$$

Thus it is possible to develop 10 different committees of three people selected from five. You can check this result by listing all possible groups of three drawn from five, where the five people are labeled A, B, C, D, and E.

Example 3.16 The lonely hearts club committee of three is to be formed by selecting three people from a group of five. One of the selected people will be chosen as chair of the committee, another secretary, and the third person simply a member of the committee. How many different committees can be formed?

Solution Suppose the three people (denoted by A, B, and C) have been chosen from among the five. Once we have this combination of three people, we must assign them to the three positions: chair, secretary, and member. We can view this process as ordering the three persons. That is, let the first position in the ordering (or *permutation*) be the chairperson, the second the secretary, and the third the member. The possible permutations are: ABC, ACB, BAC, BCA, CAB, and CBA. Each of these permutations is a different committee, although each contains the same three people, A, B, and C. Thus we want to compute the number of *permutations of* $r = 3$ people chosen from $n = 5$:

$$P_3^5 = \frac{5!}{(5-3)!} = \frac{5 \cdot 4 \cdot 3 \cdot 2 \cdot 1}{2 \cdot 1} = 60$$

In this case 60 different committees can be formed. Notice that this is six times as many committees as could be formed in Example 3.15 because for each combination in Example 3.15, there are six permutations in Example 3.16.

Example 3.17 The Universal Motors Company has decided to manufacture a new automobile named the Jupiter. A new division has been created within the company for the production of the Jupiter car. Universal decides to build two production plants initially. There will be a primary plant and a secondary plant. The primary plant will be much larger than the secondary plant. Universal must decide where to locate the two plants. After considerable analysis and debate, the list of possible states has been reduced to 10. Universal decides to locate the primary plant in one of these states and the secondary plant in another of these states. In how many ways can the two states be chosen, where one of the states selected is to be the location of the primary plant and the other state the location of the secondary plant?

Solution There are

$$C_2^{10} = \frac{10!}{2!(10 - 2)!} = \frac{10!}{2!8!} = \frac{10 \cdot 9 \cdot 8 \cdot 7 \cdot 6 \cdot 5 \cdot 4 \cdot 3 \cdot 2 \cdot 1}{2 \cdot 1 \cdot 8 \cdot 7 \cdot 6 \cdot 5 \cdot 4 \cdot 3 \cdot 2 \cdot 1} = \frac{10 \cdot 9}{2 \cdot 1} = 45$$

combinations of 2 states chosen from the 10 states. Then there are $2! = (2)(1)$ permutations or orderings of the two selected states where one is chosen for the primary plant location and the other is chosen for the secondary plant location. Therefore, there are $(45)(2) = 90$ ways that 2 states from the 10 can be selected where one of the states receives the primary plant and the other state receives the secondary plant. This number can be found directly from the permutation formula:

$$P_2^{10} = \frac{10!}{(10 - 2)!} = \frac{10!}{8!} = \frac{10 \cdot 9 \cdot 8 \cdot 7 \cdot 6 \cdot 5 \cdot 4 \cdot 3 \cdot 2 \cdot 1}{8 \cdot 7 \cdot 6 \cdot 5 \cdot 4 \cdot 3 \cdot 2 \cdot 1} = 10 \cdot 9 = 90$$

Hypergeometric rule

Rule 3.4

Hypergeometric rule

Suppose there are n objects in a group, and that n_1 are of one type and n_2 are of another type. The number of groups of r objects, where r_1 are of the first type and r_2 are of the second type, that can be formed by drawing r objects from the n is given by

$$C_{r_1}^{n_1} \cdot C_{r_2}^{n_2}, \text{ where } n_1 + n_2 = n; \, r_1 + r_2 = r$$

Example 3.18 A college recruiter for the Acme Corporation has set up interviews with 10 college seniors. Six students are fraternity members and four are not. The recruiter plans to hire 5 of the 10 interviewed students. How many possible groups of the five individuals hired will contain exactly three fraternity students?

Solution Let $n_1 = 6$ and $n_2 = 4$ in the hypergeometric rule. In the subgroup of size $r = 5$, we want $r_1 = 3$ fraternity students and $r_2 = 2$ nonfraternity students. The number of groups of size five with this property are:

$$C_3^6 \cdot C_2^4 = \frac{6!}{3!(6 - 3)!} \cdot \frac{4!}{2!(4 - 2)!} = \frac{6 \cdot 5 \cdot 4 \cdot 3 \cdot 2 \cdot 1}{3 \cdot 2 \cdot 1(3 \cdot 2 \cdot 1)} \cdot \frac{4 \cdot 3 \cdot 2 \cdot 1}{2 \cdot 1(2 \cdot 1)}$$
$$= (20)(6) = 120$$

Notice that the hypergeometric rule is "putting together" the product and the combination rules. In Example 3.18, we first compute the number of ways of selecting three fraternity students from six individuals [$C_3^6 = 20$]; then we compute the number of ways of selecting two nonfraternity students from four individuals [$C_2^4 = 6$]. A particular set of three fraternity students and two nonfraternity students is formed by selecting one of the 20 groups of three fraternity students and one of the 6 groups of nonfraternity students. Thus, by the product rule, the number of "pairs" of groups is $(20)(6) = 120$. The reader is *not* encouraged to list all 120 elementary outcomes of this experiment!

Example 3.19   The new Jupiter Division of the Universal Motors Company needs to establish a board of directors. It is decided that there will be eight board members, three of whom will be selected from automotive-related industries and five of whom will be selected from other industries. The division president compiles a list of 8 names from automotive-related industries and 10 names from other industries. In how many ways can the board of eight members be formed?

Solution   The hypergeometric rule applies in this case with $n_1 = 8$, $r_1 = 3$ (automotive-related industries) and $n_2 = 10$, $r_2 = 5$ (other industries). The number of possible boards of eight members each is given by:

$$
\begin{aligned}
C_3^8 \cdot C_5^{10} &= \frac{8!}{(3!)(8-3)!} \cdot \frac{10!}{(5!)(10-5)!} \\
&= \frac{8 \cdot 7 \cdot 6 \cdot \cancel{5 \cdot 4 \cdot 3 \cdot 2 \cdot 1}}{3 \cdot 2 \cdot 1 \cdot \cancel{5 \cdot 4 \cdot 3 \cdot 2 \cdot 1}} \cdot \frac{10 \cdot 9 \cdot 8 \cdot 7 \cdot 6 \cdot \cancel{5 \cdot 4 \cdot 3 \cdot 2 \cdot 1}}{5 \cdot 4 \cdot 3 \cdot 2 \cdot 1 \cdot \cancel{5 \cdot 4 \cdot 3 \cdot 2 \cdot 1}} \\
&= (56)(252) = 14{,}112
\end{aligned}
$$

Thus there are a lot of ways to form the eight-member board of directors (14,112 ways, to be exact)!

Counting rules are used to determine probabilities with the aid of Theorem 3.2. From Theorem 3.2, if the elementary outcomes of an experiment are equally likely to occur in any trial, then the probability that an event A occurs is the ratio of the number of points in the event space to the number of points in the sample space. The counting rules can be used to count the number of points in the event and sample spaces as the following four example problems illustrate.

Example 3.20   A corporation has expanded its business by opening an office in a new location. The office manager must select a cleaning service to care for the office and a delivery service to make local deliveries. In the Yellow Pages, she finds listed four cleaning services (A, B, C, and D) and three delivery services (I, II, and III). The services in each category charge similar rates, and all receive a good rating by the Better Business Bureau. Therefore, the manager decides to select one cleaning service and one delivery service at random. Unknown to her, cleaning service B and delivery service III are the poorest in each category. What is the probability that she will select the poorest service in each category?

Solution   The number of elementary outcomes of the experiment can be determined from the product rule because one service will be chosen from the first group of four cleaning services, and one service will be selected from the second group of three delivery services. The selection is made at random, so each elementary outcome of the experiment will occur with the same probability as any other. Thus Theorem 3.2 may be used. Let G denote the event "cleaning service B and delivery service III are selected." Then

$$P(G) = \frac{\text{Number of sample points in event space G}}{\text{Number of sample points in sample space S}} = \frac{n}{N}$$

By using the product rule, the total number of points in the sample space S is $N = (4)(3) = 12$. The number of points in event space G is $n = 1$, since for event G to occur, the pair (B, III) must be chosen. Thus

$$P(G) = \frac{n}{N} = \frac{1}{12}$$

Example 3.21 Suppose we take a well-shuffled deck of 52 playing cards and deal a five-card hand. That is, five cards are drawn randomly without replacement from the deck. What is the probability that our hand is a straight flush? A *straight flush* is five cards in one suit in order; A, K, Q, J, 10 of spades or 8, 7, 6, 5, 4 of hearts, for example. An ace can be counted either as a low card or as a high card.

Solution Define A to be the event "a five-card straight flush is dealt." The number of points N in the sample space is $C_5^{52} = 2{,}598{,}960$—the total number of possible different five-card hands. The number of points n in the event space can be counted as follows. In any particular suit, there are 10 possible straight flushes: (A, K, Q, J, 10), . . . , (6, 5, 4, 3, 2), (5, 4, 3, 2, A)—the ace may be used as the highest card or the lowest. There are four suits (hearts, diamonds, clubs, spades). Thus there are $n = (4)(10) = 40$ possible straight flushes by using the product rule. Hence

$$P(A) = \frac{n}{N} = \frac{40}{2{,}598{,}960} = 0.000015$$

A straight flush is very unlikely to occur, and a player holding this hand is in an ideal betting situation! The quality of a poker hand is determined by probability. Hand A is "better" than hand B if it is less likely to occur.

Notice that it would be virtually impossible to list all the points in the sample space in this problem. However, the combination rule makes it quite easy to count these points.

Example 3.22 The manager of a large accounting firm decides to send four employees in the tax division to a workshop on "New Ways to Shelter Income from Federal Income Taxes" presented by the local state university. There are 10 employees in the tax division, three of whom are women. The manager decides to select the four employees at random from the 10. What is the probability that two of the three women are selected to attend the workshop?

Solution Each possible sample point (set of four employees) in the experiment has the same probability of occurring as any other sample point. Thus we can use Theorem 3.2 to solve this problem. Let the event "two women and two men are selected'' be denoted by A. The probability $P(A)$ is given by:

$$P(A) = \frac{\text{Number of points in event space A}}{\text{Number of points in the sample space S}} = \frac{n}{N}$$

Let us first consider the denominator in the probability calculation. The experiment is drawing 4 employees from the 10 in the tax division. The set of elementary outcomes composes the sample space S. The total number of points in S is the number of combinations of 4 things taken from 10 because the order in which the 4 employees are chosen is unimportant. Thus

$$N = C_4^{10} = \frac{10 \cdot 9 \cdot 8 \cdot 7 \cdot 6 \cdot 5 \cdot 4 \cdot 3 \cdot 2 \cdot 1}{4 \cdot 3 \cdot 2 \cdot 1(6 \cdot 5 \cdot 4 \cdot 3 \cdot 2 \cdot 1)} = 210$$

Now, how many points are in the event space A? The 10 employees are divided into two groups (male and female), so we can use the hypergeometric rule to determine the number of sample points in A. Let n_1 and r_1 denote the number of women in the division and group selected, respectively. Then $n_1 = 3$, $r_1 = 2$, and $n_2 = 7$, $r_2 = 2$. Thus

$$n = C_2^3 C_2^7 = \left[\frac{3 \cdot 2 \cdot 1}{2 \cdot 1(1)}\right]\left[\frac{7 \cdot 6 \cdot 5 \cdot 4 \cdot 3 \cdot 2 \cdot 1}{2 \cdot 1(5 \cdot 4 \cdot 3 \cdot 2 \cdot 1)}\right] = (3)(21) = 63$$

Therefore

$$P(A) = \frac{C_2^3 C_2^7}{C_4^{10}} = \frac{63}{210} = 0.3$$

Example 3.23 American Gas and Oil Company leases the use of a pipeline to ship oil between its Texas plant and a processing center in Oklahoma. On a particular day, there are five shipments waiting to be sent through the pipeline, including American's shipment. Only three shipments can be sent on that day. The three to be sent are selected at random from the five so that each of the five shipments has an equal chance of being selected. The management of American is hoping that its shipment will be sent, preferably first or second, so that the oil arrives in the Oklahoma processing plant on the same day.

a. What is the probability that American's shipment will be one of the three chosen for shipment that day?

b. What is the probability that the first shipment sent will be American's?

c. What is the probability that the first or second shipment sent will be American's?

Solution

a. Let M be the event that American's shipment is selected to be sent on that day. Since Theorem 3.2 applies,

$$P(M) = \frac{\text{Number of points in event space M}}{\text{Number of points in sample space S}}$$

There are

$$C_3^5 = \frac{5!}{3!(5-3)!} = \frac{(5)(4)\cancel{(3)}\cancel{(2)}\cancel{(1)}}{\cancel{(3)}\cancel{(2)}\cancel{(1)}(2)(1)} = 10 \text{ ways}$$

to select the three shipments to be sent from the five. Thus there are 10 points in sample space S. The number of points in event space M can be determined from the hypergeometric rule. Let $n_1 = 1$ and $r_1 = 1$ (American forms the first group of one shipment) and $n_2 = 4$ and $r_2 = 2$ (the second group is composed of the remaining four shipments, two of which will be selected). Therefore

$$\text{Number of points in event space M} = C_1^1 C_2^4$$

$$= \frac{1!}{1!(1-1)!} \frac{4!}{2!(4-2)!}$$

$$= (1)(6) = 6$$

Therefore $P(M) = 6/10 = 0.60$.

b. Let N be the event that American's shipment is selected and sent first among the three selected shipments. Now, the order in which the shipments are selected is relevant. The number of points in sample space S is

$$P_3^5 = \frac{5!}{(5-3)!} = \frac{(5)(4)(3)\cancel{(2)}\cancel{(1)}}{\cancel{(2)}\cancel{(1)}} = 60$$

There are 60 ways in which the three shipments can be selected and ordered first, second, and third for shipment from the five shipments. The number of points in the event space can be determined by using the result in part a. In each of the six ways in which American's shipment can be selected, there are $3! = (3)(2)(1) = 6$ orderings of the three selected shipments. For example, suppose that American's shipment is denoted by A, and the other two shipments are denoted by B and C. The possible orderings are: ABC, ACB, BAC, BCA, CAB, CBA. In two of these orderings, A is first. Therefore

$$\text{Number of points in event space N} = (2)(6) = 12$$

and

$$P(N) = 12/60 = 0.20$$

Note that the product rule could be used as well. The number of permutations composing sample space S is given by $\boxed{5} \cdot \boxed{4} \cdot \boxed{3} = 60$. That is, any of the five shipments can be selected first to be sent, any one of the remaining four shipments can be selected to be sent second, and any one of the remaining three shipments can be selected to be sent third. In the event space, we have $\boxed{1} \cdot \boxed{4} \cdot \boxed{3} = 12$. That is, American's shipment must be selected first (there is only one way to do that); then one of the remaining four shipments must be selected to be sent second; and finally one of the remaining three shipments must be selected to be sent third. Thus

$$P(N) = 12/60 = 0.20$$

There is even an easier way to determine $P(N)$. We can think of the problem as follows. Suppose that there are five shipments—A, B, C, D, and E—and each has an equal chance of being the first shipment selected to be sent on that day. Then the probability that American's shipment, say A, is sent is $1/5 = 0.20$; that is, the sample space S contains five points, and the event space contains 1 point (A is selected). Thus $P(N) = 0.20$.

c. Let O be the event that American's shipment is selected either first or second to be sent. We will use the product rule to determine this probability:

$$P(O) = \frac{\boxed{1} \cdot \boxed{4} \cdot \boxed{3} + \boxed{4} \cdot \boxed{1} \cdot \boxed{3}}{\boxed{5} \cdot \boxed{4} \cdot \boxed{3}} = \frac{12 + 12}{60} = \frac{24}{60} = 0.40$$

That is, American's shipment can be selected first with four ways to select the second shipment sent and three ways to select the third shipment sent; or American's shipment can be selected second with four ways to select the first shipment sent and three ways to select the third shipment sent.

Example 3.23 illustrates an important result concerning probability problems (particularly part *b*). There are usually several ways to solve a particular problem. Of course, all we need to do is find *one* way, and in some problems that is a sufficient challenge!

■ 3.5
Summary

In this chapter, we have introduced the basic concepts of probability—the experiment, elementary outcomes, events, the sample and event spaces, and the probability axioms. Three ways of determining the probability of an event were discussed—the relative frequency definition, probability deduced logically, and subjective probability. Counting rules—the product rule, permutations, combinations, and the hypergeometric rule—were shown to be helpful in determining the probability of an event occurring in an experiment.

■ References

Introductory

Anderson, D.; D. Sweeney; and T. Williams. *Statistics for Business and Economics*. 2nd ed. St. Paul, Minn.: West Publishing, 1984, Chapter 4.

Harnett, D. *Statistical Methods*. 3rd ed. Reading, Mass.: Addison-Wesley, 1982, Chapter 2.

Lapin, L. *Statistics for Modern Business Decisions*. 3rd ed. New York: Harcourt, Brace, Jovanovich, 1982, Chapter 5.

Wonnacott, R., and T. Wonnacott. *Introductory Statistics*. 4th ed. New York: John Wiley & Sons, 1985, Chapter 3.

Advanced

Feller, W. *An Introduction to Probability Theory and Its Applications*. 3rd ed. New York: John Wiley & Sons, 1968.

Parzen, E. *Modern Probability Theory and Its Applications*. New York: John Wiley & Sons, 1960.

Ross, S. *Introduction to Probability Models*. 2nd ed. New York: Academic Press, 1980, Chapter 1.

■ Problems

Section 3.2 Problems

3.1 Give the sample space of each of the following experiments in the form of a Venn diagram. Be certain to define the elementary outcomes corresponding to the sample points in the sample space.
  a. A fair coin is tossed three times.
  b. A coin and a die are tossed together.
  c. A student receives his score on a multiple-choice exam containing 20 questions.
  d. A student receives his grade on an exam.
  e. The number of telephone calls received at a switchboard during a five-minute interval is recorded.
  f. A child is selected in a first grade class, and his or her weight (to the nearest pound) is recorded.

3.2 A committee is composed of two men and two women. One member of the committee is selected to serve as chairperson and another to serve as secretary.
  a. Define the elementary outcomes composing the sample space of this experiment.
Identify which sample points in the sample space belong to the following event spaces:
  b. The younger man is selected as chairperson: event A.
  c. A man is selected as chairperson: event B.
  d. A woman is selected as secretary: event C.
  e. Events A and C occur: event D.
  f. Events B or C or both occur: event E.
  g. Show the five event spaces in a Venn diagram.

3.3 Two college job recruiting officers, Herman and Bill, come to the University of Truth campus to fill positions in their organizations. Each officer is attempting to fill three positions. Three students qualify for the positions described, and each will be interviewed by the two officers. If a sample point is defined as a specific number of students hired by Herman or Bill, define the following events as specific collections of sample points. Note: There are six jobs and three students—three jobs (or possibly more!) will not be filled. (Hint: The sample space is two-dimensional.)
  a. The sample space S that consists of all elementary outcomes defining the number of students hired by each officer.
  b. Event A: Herman hires at least two students.
  c. Event B: All three students are hired by Bill.
  d. Event C: Exactly one student is hired by Bill.
  e. Event D: Bill hires two students and so does Herman.
  f. Event E: Events A or C or both occur.
  g. Event F: Events A and C occur.

3.4 An investment specialist is to rate a bond issue. The rating is on an eight-point scale: A+, A, A−, B+, B, B−, C, F.
  a. Develop three alternative sample spaces for this experiment. [Hint: One possible space is E_1 = high rating (A+, A, A−), E_2 = Medium rating (B+, B, B−), E_3 = poor rating (C, F).]
  b. Is any one of the three sample spaces "best"? Discuss.
  c. Describe a situation wherein each sample space developed in part *a* is appropriate.

Section 3.3 Problems

3.5 In Problem 3.2, suppose each individual committee member has an equal chance of being selected as either chairperson or secretary. Compute the probabilities of the events A, B, C, D, and E by summing the probabilities of the appropriate sample points.

3.6 In Problem 3.3, suppose all the sample points of the sample space S are equally likely to occur. Compute the probability of the events A, B, C, D, E, and F by summing the probabilities of the appropriate sample points.

3.7 A piece of equipment is assembled in three separate operations, which may be performed in any sequence. Herman must decide which order to use. He places three slips of paper marked #1, #2, and #3 in a hat and draws them out one at a time without looking.

    a. Define the elementary events composing the sample space of this experiment. For each event described below, identify its sample points and compute the probability of its occurrence. Assume all the sample points in the sample space are equally likely to occur.

    b. The operation marked #1 comes first in the implemented production sequence.

    c. The operation marked #2 comes either second or third in the implemented production sequence.

    d. The operation marked #1 comes first, *and* the operation marked #2 comes either second or third in the implemented production sequence.

3.8 For each example below, discuss whether it may be viewed as an *objective* (frequency or deductive logic definition) probability or a *subjective* (personal) probability, or both.

    a. The probability that the 'Skins will beat the Cowboys in an upcoming football game.

    b. The probability that a new electric toaster will survive its warranty period (six months) without failure or defect.

    c. The probability that a newly purchased lot of 100 calculators contains no defectives.

    d. The probability that Herman will be fired from his job in the next year.

    e. The probability that Green will win the presidential nomination of her party.

Section 3.4 Problems

3.9 A sales representative is traveling from New York to Los Angeles with a stop at Detroit. From New York to Detroit there are five possible flights, and from Detroit to Los Angeles there are six. In how many ways can the sales representative make the trip from New York to Los Angeles with a stop-over in Detroit?

3.10 Suppose auto license plates in a particular state are designed so that the first three characters are letters of the alphabet, and the second three characters are numbers (e.g., ABC 127). How many different license plates can be produced by following this convention?

3.11 A brief market research questionnaire requires the respondent to answer each of eight questions. The first four questions are to be answered either *yes* or *no,* and the remaining four questions have three possible answers (*always, sometimes, never*). The sequence of responses to all eight questions is defined as the respondent's ''profile.'' How many different profiles are there?

3.12 Six brands of soft drinks are ranked in order of preference by independent taste testers.

    a. How many distinct rankings (using all six brands and ruling out ties) are possible?

    b. In how many of these possible rankings can brand A be ranked best?

3.13 A group is composed of six men and four women. How many committees of four members can be formed from this group if:

    a. There are no restrictions placed on the composition of the committee?

    b. The committee must contain two men and two women?

    c. The committee must contain at least one woman?

3.14 Herman is taking four courses during the present term. In each course, he can receive one of five possible grades: A, B, C, D, and F.

    a. At the end of the term, how many different combinations of grades can he receive in the four courses?

    b. Suppose in his first course he knows he

can make no better than a C. How many combinations of grades can he receive?

c. Suppose by midterm it is apparent that he will flunk botany (he quits going to class), but he knows he has As in two other courses. How many grade combinations are possible?

3.15 A customer considering the purchase of a new car is told that the car comes in five body styles, six colors, two transmissions, and five engines. How many choices does the customer have?

3.16 A firm has 10 sales representatives. In how many ways can they be assigned to two territories, with seven sales representatives in one territory and three in the other?

3.17 A university telephone system is established with the prefix 921. How many campus numbers with this prefix of the form 921–XXXX are possible?

3.18 A combination lock has 40 numbers on it—0, 1, . . . , 39. The combination is a sequence of three numbers. How many combinations are there? Suppose that you own such a lock and decide to try all of the combinations. What is the probability that the first combination that you try works?

3.19 The computer center in a corporation has instituted a password system to minimize the unauthorized use of its computer system. The password is six characters in length; and each character may be one of the 26 letters in the alphabet, one of the 10 digits 0, 1, 2, . . . , 9, or one of the 10 characters ! @ # $ % ^ & * ( ). How many passwords are possible? If a person randomly types in six characters, what is the probability that he or she will have typed in the proper password for a particular account?

3.20 A puzzle in a newspaper presents a matching problem. The names of 10 U.S. presidents are listed in one column, and their vice presidents are listed in a random order in the second column. The puzzle asks the reader to match each president with his vice president. If someone makes the matches randomly, what is the probability that all 10 matches are correct?

3.21 A department in a company has a system by which certain information—brochures, no-

tices, and the like—are distributed to the eight individuals in the office. The names of the eight individuals are placed on a sheet attached to the information, and it is sent to the first person on the list. That person passes it on to the second person on the list, and so on until it gets to the eighth (and last) person on the list. How many ways can the eight names be listed on the sheet?

3.22 The cafeteria at the Acorn corporation provides lunches for free to its employees. The employee is allowed to select from the following: four salads, three vegetable dishes, four main dishes, seven drinks, and five desserts. The employee can select one or none of the offerings in each category: salad, vegetable, main dish, drink, and dessert. For example, the employee may decide to have a salad, a vegetable, and a drink but no main dish or dessert. How many different lunches are possible?

3.23 The Computer Store sells microcomputer systems to small businesses. Each system is composed of a central processing unit, one or more disk drives, a printer, a specified amount of memory, and a modem for communications. The store offers five microcomputer central processing options; either one or two disk drives, either of which may be a floppy disk drive or a hard disk drive; six printer options; and four memory options (256K, 384K, 512K, 640K). How many different systems are possible?

3.24 A businessperson has the option of selecting one of five lines to fly from Washington, D.C., to Paris, France. Once in Paris, she can select one of seven rental car agencies to rent a car from, and she can stay in one of 10 hotels in Paris. How many travel combinations are available to her (airline, rental car, and hotel)?

3.25 Define the experimental conditions under which each of the following counting rules is appropriate:
a. The product rule.
b. Combinations.
c. Permutations.
d. Hypergeometric rule.

3.26 In a certain region of the country, radio station call letters are restricted to either three-letter or four-letter sequences beginning with the letter *W*. How many possible call letters are there?

3.27 A machine shop has six work orders that require lathe work and four work orders that require milling work. If the shop has three lathes and two milling machines, in how many different ways can the work orders be assigned to the machines?

3.28 How many different ways can five horses in a five-horse race finish if there are no ties?

Additional Problems

3.29 How many ways can six people be seated in a row if:
   a. There are no restrictions on the seating arrangement?
   b. There are three married couples, and each couple must sit together?

3.30 A corporate information division of a large multinational company subscribes to national newspapers to keep up with news in the various regions of the country. To simplify matters, suppose the division divides the nation into four regions: West, North, South, and East. In the West, they have identified 10 potential papers for subscription, 6 in the North, 8 in the South, and 12 in the East. It is decided to subscribe to five papers in each region. How many choices are possible?

3.31 In a market survey, customers are asked to taste and rank three breakfast cereals from best to worst. The cereals are labeled A, B, and C.
   a. Develop a sample space for this experiment.
   b. If each elementary outcome in the sample space is equally likely to occur, what is the probability that either cereal A or cereal B ranks first?

3.32 Distinguish between objective and subjective probability. Is it possible to apply both definitions of probability to a specific event? Under what conditions is it impossible to determine an objective probability?

3.33 Five men are scheduled to participate in a 100-meter race at a track meet. What is the probability that a sportswriter picks the first-,

second-, and third-place winners if he makes the selections at random?

3.34 In the dice game, a player wins on the first toss of the pair of dice if the total number of dots on the two top faces of the dice is either 7 or 11 and loses if the total number of dots on the two top faces is either 2 or 12.
   a. What is the probability that the player wins on the first toss?
   b. What is the probability that the player loses on the first toss?

3.35 A division minicomputer is used by four departments in a company. Jobs may be submitted as "priority" or "nonpriority" jobs. A job sent to the computer room is selected at random.
   a. Develop a sample space for this experiment.
   b. Is it likely that the events in this sample space would be equally likely to occur in a typical company setting? Explain.

3.36 Four persons carpool together each morning to work. Two of them are women; three are more than 30 years old; and all are either women or more than 30, or both. Suppose that one of these people is randomly selected.
   a. Develop a sample space for this experiment.
   b. Assuming that each elementary outcome in the sample space is equally likely to occur, what is the probability that the person selected is more than 30 and is a man?

3.37 In the first semester of the freshman year at a large university, every student must register for six courses. Two of the courses are fixed: P.E. and English. For the remaining four courses, the student must choose one course from each of four groups: (*a*) social sciences (10 courses available for selection), (*b*) sciences (12 courses available for selection), (*c*) mathematics (3 courses available for selection), and (*d*) arts and humanities (15 courses available for selection). Assuming that all courses are available when a student registers, how many schedules may the student choose from?

3.38 There are 10 horses in a horse race, including Stewball. The bets are placed for win, place, or show (finishing first, second, or third). Suppose that each horse is equally likely to win the race.

a. What is the probability that Stewball finishes the race in the money (finishes either first, second, or third)?

b. What is the probability that Stewball places or shows?

3.39 The game of Keno is played as follows: A sheet contains 80 numbers, 1, 2, . . . , 80. In each game 20 of these numbers are randomly selected. The player may select one, two, three, . . . up to 10 numbers from 1 to 80. The payoff depends on how many numbers the player selected in his or her set of numbers that are included among the 20 numbers randomly selected for each game.

a. Suppose that the player selects three numbers. What is the probability that all three of his numbers will be selected among the 20 selected in a particular game? What is the probability that two of his three numbers will be selected?

b. Suppose that the player selects 10 numbers. What is the probability that all 10 numbers will be among the 20 selected in a particular game?

3.40 The president and chief executive officer of a large corporation randomly visits several of the offices of the corporation located around the country each quarter. The company has 16 offices from which she selects her visits.

a. Suppose that she decides to visit 4 of the 16 offices in a given quarter. How many ways can she order the visits to the four offices?

b. If you are the director of one of these offices, what is the probability that your office will be one of the four visited?

c. What is the probability that your office will not be visited first, but will be one of the four offices visited?

4

Conditional probability, the probability of compound events, and Bayes' theorem

■ 4.1

Introduction

In Chapter 3, the basic concepts of probability were developed. In this chapter, we will continue the development of probability concepts by first introducing the compound event—an event that is the combination of other events. Second, we will develop the notion of conditional probability—the probability that an event occurs given (or conditioned upon) the knowledge that one or more other events have occurred. Finally, ways of describing the relationships between two events are presented and are used in stating probability laws concerning compound events.

Compound events

■ 4.2

Compounding events: Unions and intersections

In many experiments, it is convenient or necessary to combine or to *compound* events to answer probability questions formulated from the experiment. In many problems, the compounding of events may lead to a solution of a probability problem without the need of listing all points in the sample space.

The compounding of events can be manifested in two ways: *unions* and *intersections*.

Example 4.1   To illustrate the union, suppose there are six points in the sample space S; E_1, E_3, and E_6 belong to the event space A; and E_1, E_2, and E_6 belong to the event space B. Construct a Venn diagram indicating the union of these two events.

Solution   The union of events A and B is the set of sample points E_1, E_2, E_3, and E_6. The compound event A $\cup$ B is illustrated in Figure 4.1.

Union of events

Definition 4.1

Union of events

Suppose A and B are two events defined in a sample space S. The *union* of A and B, denoted by A ∪ B, is the set of all sample points in the event space A *or* in event space B *or* in both event spaces A and B.

FIGURE 4.1 | Venn diagram of the union of two events, A and B

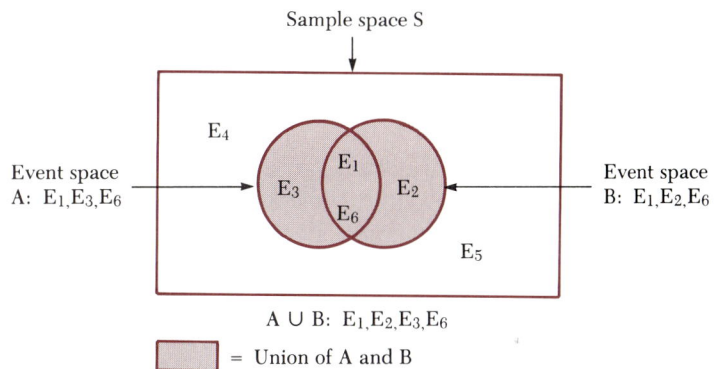

Sample space S

E_4

Event space
A: E_1, E_3, E_6

E_1

E_3 E_2

E_6

Event space
B: E_1, E_2, E_6

E_5

A ∪ B: E_1, E_2, E_3, E_6

☐ = Union of A and B

Intersection of events

Definition 4.2

Intersection of events

Suppose A and B are two events defined in a sample space S. The *intersection* of A and B, denoted by A ∩ B, is the set of sample points in the event space A *and* in the event space B.

Example 4.2   Suppose, as in the previous example, events A and B are defined as the following collections of elementary outcomes: A: E_1, E_3, and E_6, and B: E_1, E_2, and E_6. Construct a Venn diagram illustrating the intersection of these two events.

Solution   The intersection of events A and B is illustrated in the shaded area in the Venn diagram in Figure 4.2.

FIGURE 4.2 | Venn diagram of the intersection of two events, A and B

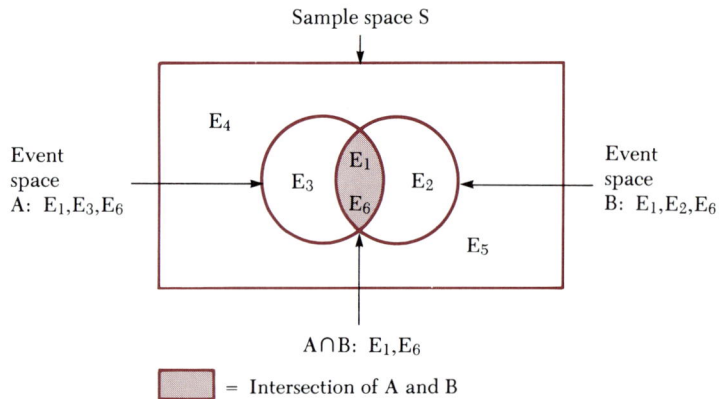

Sample space S

Event
space
A: E_1, E_3, E_6

E_4

E_3 E_1 E_2

E_6

E_5

Event
space
B: E_1, E_2, E_6

$A \cap B$: E_1, E_6

= Intersection of A and B

It may well be that the two events A and B do not overlap, as illustrated in Figure 4.3. In this instance, the compound event A ∩ B is empty—it contains no common elementary outcomes—and is denoted by ϕ, the null event.

FIGURE 4.3 | Venn diagram of the intersection of two nonoverlapping events, A and B

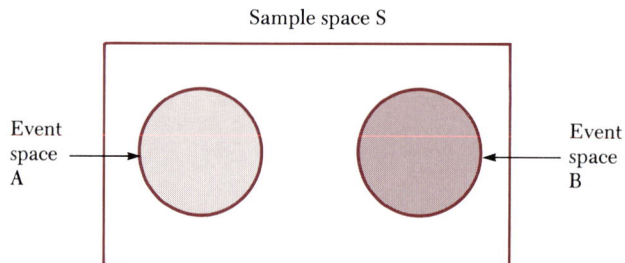

Sample space S

Event
space
A

Event
space
B

Example 4.3 Recently, there has been considerable interest in computer literature searches. The method works as follows: A data base is formed of the titles or key words in articles appearing in magazines, journals, or government documents. If we were interested in doing research on, say, the effects of a recession on unemployment, we would instruct the computer literature search specialist to search the data base for papers, articles, or documents on this topic. The logic of the search usually uses the intersection logic of sets. For example, the specialist would enter into the computer on a terminal the request to search for and to print out the citations for all papers, articles, or documents that have the words *recession* and (intersection) *unemployment* in their titles or key words (many articles now contain a set of three to five key words that attempt to capture the main ideas in the article). Draw a Venn diagram illustrating the logic of the search process.

Solution   Figure 4.4 shows the intersection logic used. Set A contains all citations that have the word *recession* in their titles or lists of key words, and set B contains all citations with the word *unemployment* in their titles or lists of key words. For example, there may be 512 citations in set A and 783 citations in set B, as illustrated in Figure 4.4. What we want is the *intersec-*

FIGURE 4.4 | The intersection of the title words or key words *recession* and *unemployment*

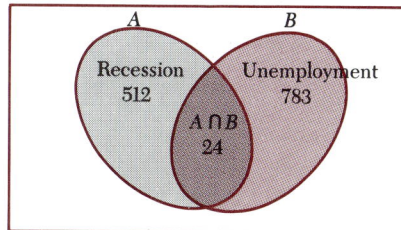

tion of the two sets, which contains, in this example, 24 citations. The computer search specialist would then instruct the computer to print out the citations for the 24 "hits."

Computer bibliographic literature searching is a relatively new field that is bound to grow dramatically in the next decade as more written work is stored in computer files. For anyone writing a term paper or a research paper, this innovation is significant, for it can save many hours of "searching the stacks" for references.

Example 4.4   The dice game was introduced in Example 3.1 in Chapter 3. The experiment is tossing a pair of dice and noting the number of dots appearing on each of the top faces of the dice. There are 36 possible elementary outcomes in the experiment and, if the dice are "fair," each elementary outcome has an equal probability of $\frac{1}{36}$ of occurring in a single trial of the experiment.

Suppose we define the following events in this experiment:

A: Total of the dots on the top two faces of the dice is seven.
B: Total of the dots on the top two faces of the dice is even.
C: Total of the dots on the top two faces of the dice is less than or equal to four.

What are $P(A \cup B)$, $P(A \cup C)$, $P(B \cup C)$, $P(A \cap B)$, $P(A \cap C)$, and $P(B \cap C)$?

Solution   The event spaces A, B, and C and the sample space S are illustrated in Figure 4.5. Table 4.1 gives the elementary outcomes contained in each compound event. For the intersection compound events, $A \cap B$, $A \cap C$, and $B \cap C$, it is evident that A and B do not overlap (no intersection) and A and C do not overlap, as Figure 4.5 illustrates.

Now, since each sample point in the sample space is equally likely to occur, Theorem 3.2 in Chapter 3 can be used:

$$P(\text{event}) = \frac{\text{Number of points in the event space}}{\text{Number of points in the sample space}}$$

Therefore

$$P(A \cup B) = {}^{24}\!/_{36} = {}^{2}\!/_{3} \qquad P(A \cap C) = {}^{0}\!/_{36} = 0 \qquad P(A \cap B) = {}^{0}\!/_{36} = 0$$
$$P(B \cup C) = {}^{20}\!/_{36} = {}^{5}\!/_{9} \qquad P(A \cup C) = {}^{12}\!/_{36} = {}^{1}\!/_{3} \qquad P(B \cap C) = {}^{4}\!/_{36} = {}^{1}\!/_{9}$$

■ **4.3**

Conditional probability

Suppose in the dice game that the first die is black and the second die is white. The experiment, as before, is tossing the pair of dice and noting the total number of dots on the two top faces. The sample space is repeated in Figure 4.6. Suppose we define the following two events for this experiment:

A: Dots on the top two faces total 11.
B: Six dots show on the top of the first (black) die.

Since there are two points in the event space for A (E_{56} and E_{65}), $P(A) = {}^{2}\!/_{36} = {}^{1}\!/_{18}$. Now, suppose we have the following situation. The dice are tossed, but one of them rolls under a table so that its top face cannot be seen. The other die, the black one, shows six dots on its top face. Now, what is the probability that event A has occurred? This probability is called a *conditional probability* and is denoted by $P(A/B)$.

The notation $P(A/B)$ is read, "the probability that event A occurs, given (slash) that event B has occurred."[1] The conditional probability, $P(A/B)$, can be determined from the sample space. Since we know that event B has occurred (the black die shows six dots on its top face), the only possible elementary outcomes in the sample space are E_{61}, E_{62}, E_{63}, E_{64}, E_{65}, and E_{66}. These elementary outcomes represent the boxed region in the sample space illustrated in Figure 4.6. These sample points compose the *reduced sample space*—it is the portion of the sample space that still remains, given

Conditional probability

---

[1] Notice that $P(A/B)$ does not mean $P(A) \div P(B)$.

FIGURE 4.5 | Sample space and event spaces for Example 4.4

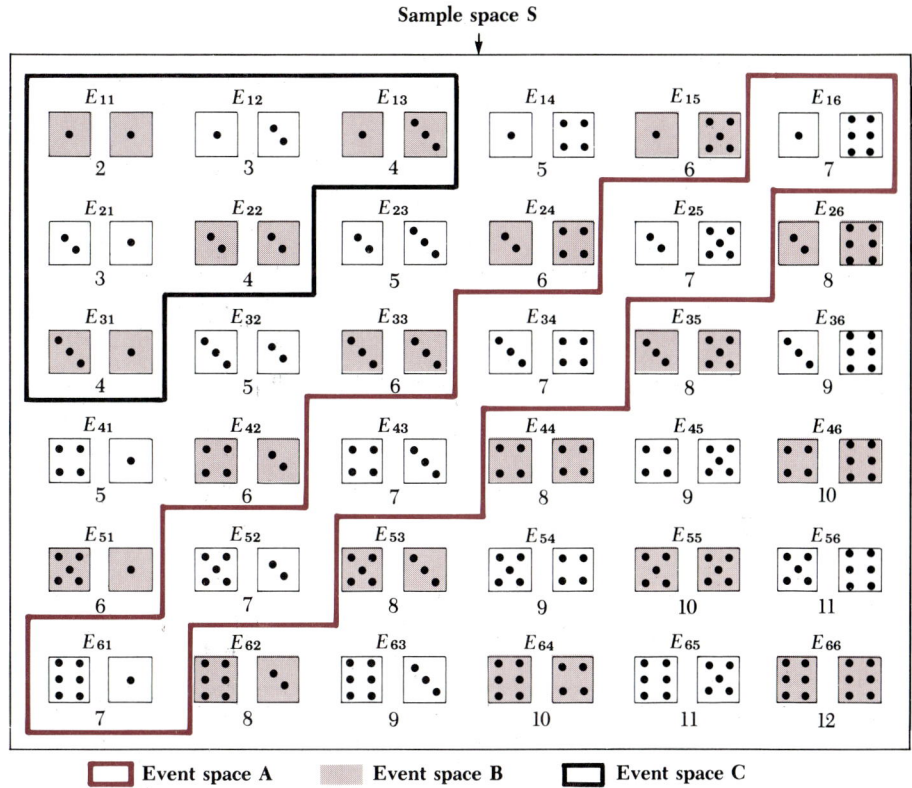

Sample space S

Event space A   Event space B   Event space C

TABLE 4.1 | Elementary outcomes contained in the compound events in Example 4.4

Compound event	Elementary outcomes in compound events	Number of elementary outcomes
$A \cup B$	E_{16}, E_{25}, E_{34}, E_{43}, E_{52}, E_{61}, E_{11}, E_{13}, E_{22}, E_{31}, E_{15}, E_{24}, E_{33}, E_{42}, E_{51}, E_{26}, E_{35}, E_{44}, E_{53}, E_{62}, E_{46}, E_{55}, E_{64}, E_{66}	24
$A \cup C$	E_{16}, E_{25}, E_{34}, E_{43}, E_{52}, E_{61}, E_{11}, E_{12}, E_{21}, E_{13}, E_{22}, E_{31}	12
$B \cup C$	E_{11}, E_{31}, E_{22}, E_{13}, E_{51}, E_{42}, E_{33}, E_{24}, E_{15}, E_{26}, E_{35}, E_{44}, E_{53}, E_{62}, E_{46}, E_{55}, E_{64}, E_{66}, E_{21}, E_{12}	20
$A \cap B$	ϕ	0
$A \cap C$	ϕ	0
$B \cap C$	E_{11}, E_{31}, E_{22}, E_{13}	4

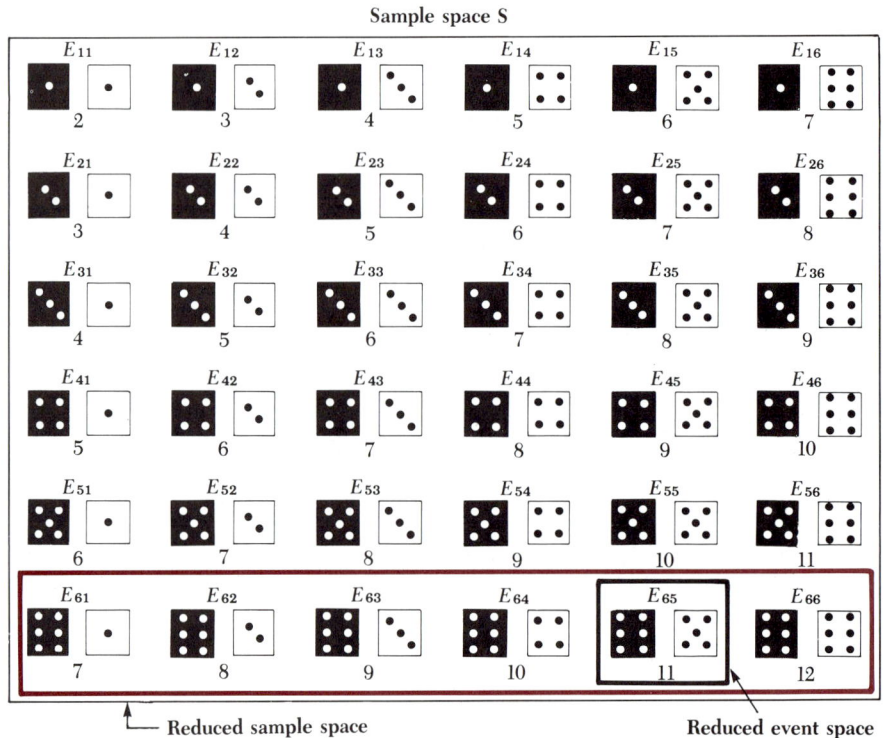

Sample space S

Reduced sample space — Reduced event space

that event B has occurred (six dots on the black die). In the reduced sample space, the only sample point that produces a total of 11 dots on the top faces of the dice is E_{65}. Since all the sample points are equally likely to occur, $P(A/B) = \frac{1}{6}$. In essence, we are merely reusing Theorem 3.2:

$$P(A/B) = \frac{\text{Number of points in the reduced event space}}{\text{Number of points in the reduced sample space}}$$

The number of points in the reduced sample space is six (E_{61}, E_{62}, E_{63}, E_{64}, E_{65}, and E_{66}). *Within* the reduced sample space, the reduced event space has one element, E_{65}. Thus $P(A/B) = \frac{1}{6}$.

Notice that knowing that the first (black) die is showing six dots on its top face substantially changes the probability that the dots total 11 on the top faces of the pair of dice (event A). Without knowing that event B has occurred, $P(A) = \frac{2}{36} = \frac{1}{18}$, but if we know B has occurred, $P(A/B) = \frac{1}{6}$. Thus the knowledge that event B has occurred *triples* the probability that event A occurs in this example.

Often two events are related in such a way that the probability of the occurrence of one event, say event A, is affected by the occurrence of another event, say event B, as in the dice game illustration. If we know that event B has occurred, then we can compute the *conditional probability* of A, *given* B.

Although it is possible to calculate conditional probabilities from the sample space by finding the appropriate reduced sample space and the appropriate reduced event space, there is a formula that may be used as well. It is given in Definition 4.3.

Definition 4.3

Conditional probability

The *conditional probability* of event A, given event B has occurred, is the joint probability of events A and B occurring (the intersection event) divided by the probability that event B occurs:

$$P(A/B) = \frac{P(A \cap B)}{P(B)} \quad \text{provided } P(B) \neq 0;$$

similarly, $\quad P(B/A) = \dfrac{P(A \cap B)}{P(A)} \quad$ provided $P(A) \neq 0$.

The conditional probability $P(A/B) = \dfrac{P(A \cap B)}{P(B)}$ can be illustrated in a Venn diagram. In Figure 4.7, let the area in the circles corresponding to event spaces A and B represent $P(A)$ and $P(B)$, respectively. Thus the probability that event A occurs can be thought of as the ratio of the area in the circle representing event space A to the area in the square representing the sample space S. *Given* that event B occurs, then the sample space S is reduced to event space B—only elementary outcomes of the experiment in event space B can now occur. Then the probability that A occurs *given* that B has occurred is the ratio of the area representing the overlapping elementary outcomes belonging to both event space A and event space B (A ∩ B) to the area representing event space B.

Example 4.5 In the dice game illustrated earlier, events A and B were defined as follows:

A: Dots on the top faces of the pair of dice sum to 11.
B: Dots on the top face of the first (black) die number 6.

Use Definition 4.3 to compute the probability of event A occurring given that event B has occurred.

FIGURE 4.7 | The conditional probability, $P(A/B)$, illustrated by a Venn diagram

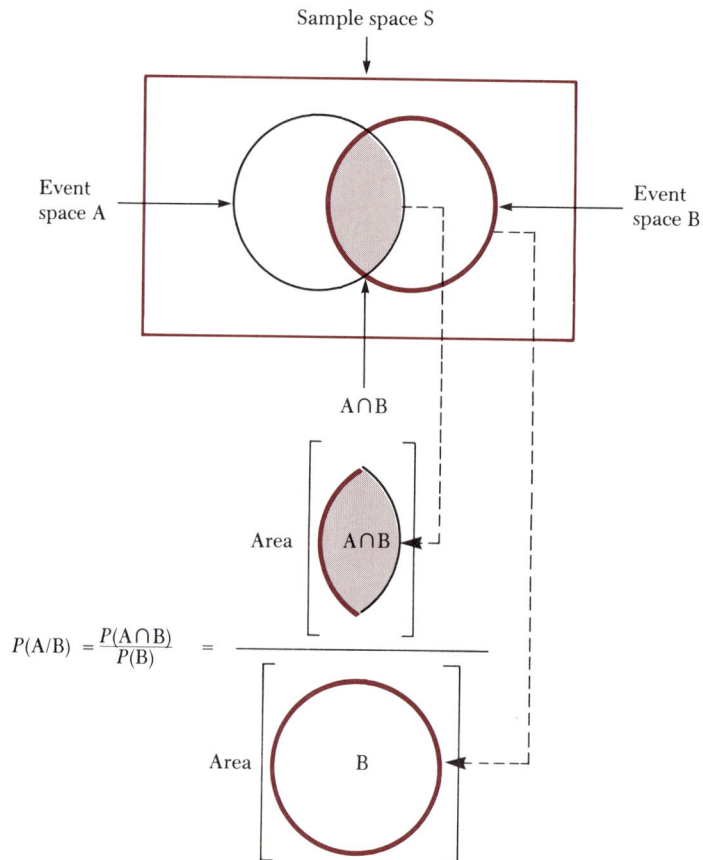

$$P(A/B) = \frac{P(A \cap B)}{P(B)} = \frac{\text{Area } A \cap B}{\text{Area } B}$$

Solution From Definition 4.3,

$$P(A/B) = \frac{P(A \cap B)}{P(B)}$$

The intersection event space $A \cap B$ contains the sample point E_{65}. That is, E_{65} is the *only* sample point that has six dots showing on the top face of the first (black) die *and* a total of 11 dots on the top faces of the pair of dice. Thus $P(A \cap B) = 1/36$. The event space B contains the following sample points: E_{61}, E_{62}, E_{63}, E_{64}, E_{65}, and E_{66}. Thus $P(B) = 6/36$. Therefore

$$P(A/B) = \frac{P(A \cap B)}{P(B)} = \frac{1/36}{6/36} = \frac{1}{6}$$

It should be apparent that these formulas need not always be used in computing conditional probabilities. The conditional probability $P(A/B)$, can be determined from the sample space associated with the experiment by reducing the space using the knowledge that event B has occurred, and then forming the ratio of the number of points in reduced event space A to the total number of points in the reduced sample space. This method of computing the conditional probability will work if each sample point in the original sample space is *equally likely to occur*.

The next example further illustrates the notion of conditional probability and the methods used to compute it.

Example 4.6   A study to determine the (possible) connection between smoking and lung cancer involved the survey of 100 patients at a local Veterans Administration hospital. Each patient was classified as a smoker or a nonsmoker and as having lung cancer or not having lung cancer. The

| Contingency table |

results of this study are shown in the *contingency table* in Table 4.2. For

TABLE 4.2 | Contingency table for Example 4.6

Cancer / *Smoking*	*Lung cancer* A	*No lung cancer* $\overline{A}$	*Totals*
Smokers B	15	25	40
Nonsmokers $\overline{B}$	5	55	60
Totals	20	80	100

example, it was found that 20 patients had lung cancer, 40 were smokers, and 15 had lung cancer *and* smoked. As indicated in Table 4.2, the event "Lung cancer" is denoted by A, "No lung cancer" by $\overline{A}$, "Smoker" by B, and "Nonsmoker" by $\overline{B}$. The entries in the cells of the contingency table give the numbers of patients who belong to the intersections of the corresponding row and column events. The contingency table is a convenient way to summarize the elementary outcomes of an experiment and tabulate the sample points in the sample space S. The experiment in this problem consists of contacting each patient and recording whether or not he or she has lung cancer and/or smokes. The experiment was repeated 100 times, generating the 100 sample points summarized in the contingency table. There are, for example, 40 points in event space B and 80 points in event space $\overline{A}$.

Suppose that one patient is drawn at random from the 100 patients in this survey. Determine $P(B)$, $P(A \cap B)$, $P(A/B)$, and $P(A/\overline{B})$.

Solution Since each of the 100 sample points is equally likely to occur, we can use Theorem 3.2 (in Chapter 3) to determine each of these probabilities.

$$P(B) = \frac{\text{Number of points in event space B}}{\text{Number of points in sample space S}} = \frac{40}{100} = 0.4$$

Marginal probability

The probability $P(B)$ is often referred to as a *marginal probability*, since it is computed from the contingency table by dividing the row margin 40 by the grand total 100. In a similar way, the other marginal probabilities are $P(\overline{B}) = 60/100 = 0.60$, $P(A) = 20/100 = 0.20$, and $P(\overline{A}) = 80/100 = 0.80$.

$$P(A \cap B) = \frac{\text{Number of points in compound event space } (A \cap B)}{\text{Number of points in sample space S}}$$

$$= \frac{15}{100} = 0.15$$

Joint probability

The probability $P(A \cap B)$ is often referred to as a *joint probability*, since it gives the probability of the joint occurrence of events A and B. It is determined from the contingency table by dividing the cell total 15 representing the intersection of events A and B by the grand total of 100. In a similar fashion, the other joint probabilities are $P(A \cap \overline{B}) = 5/100 = 0.05$, $P(\overline{A} \cap B) = 25/100 = 0.25$, and $P(\overline{A} \cap \overline{B}) = 55/100 = 0.55$.

$$P(A/B) = \frac{\begin{array}{c}\text{Number of points in reduced event space A} \\ \text{within reduced sample space R}\end{array}}{\text{Number of points in reduced sample space R}}$$

$$= \frac{15}{40} = 0.375$$

The computation of $P(A/B)$ is illustrated in Figure 4.8.

Given that event B has occurred, we know that the one patient we selected is a smoker. Therefore, he must be one of the 40 patients who smokes out of the total of 100 patients—the reduced sample space R has 40 points in it.

FIGURE 4.8 | Computation of $P(A/B)$

	A	$\overline{A}$	
B	15	25	40
$\overline{B}$	5	55	60
	20	80	100

Reduced sample space R (40 points)

$$P(A/B) = \frac{15}{40}$$

$$= 0.375$$

Now, what is the probability that this patient has lung cancer? There are 15 sample points in the reduced space containing the 40 points that belong to event A "lung cancer." Thus

$$P(A/B) = \frac{15}{40}$$

Notice that the conditional probability, $P(A/B)$, could also have been determined from the definitional formula:

$$P(A/B) = \frac{P(A \cap B)}{P(B)} = \frac{15/100}{40/100} = \frac{15}{40} = 0.375$$

From the contingency table, we can write the probability $P(A/\overline{B})$ directly (alternatively, the formula for conditional probability could be used):

$$P(A/\overline{B}) = \frac{5}{60} = 0.0833.$$

Comparing the conditional probabilities $P(A/B)$ and $P(A/\overline{B})$ with $P(A)$ demonstrates rather dramatically the idea of conditioning. The probability $P(A) = 0.20$ gives the probability that a randomly selected patient has lung cancer with *no knowledge about whether or not the patient smokes*. If we know the selected patient smokes, then $P(A/B) = 0.375$ gives us the probability that the patient has lung cancer. The probability that the patient has lung cancer—$P(A) = 0.20$—has been *revised upward* to $P(A/B) = 0.375$, given that the patient is a smoker. If we know that the selected patient does not smoke, then $P(A/\overline{B}) = 0.0833$ gives us the probability that the patient has lung cancer. The probability that the patient has lung cancer—$P(A) = 0.20$—has been *revised downward* to $P(A/\overline{B}) = 0.0833$, given that the selected patient does not smoke.

In comparing $P(A/B)$ and $P(A/\overline{B})$ with $P(A)$, we can certainly appreciate why the surgeon general warns smokers on cigarette packages about the health risks associated with smoking! The interpretation of conditional prob-

Revised probabilities

abilities as *revised probabilities* is common and easy to understand. The probability $P(A)$ is the probability that event A occurs with no knowledge about the occurrence of another (or other) events. The probabilities $P(A/B)$ and $P(A/\overline{B})$ *revise* the probability $P(A)$, depending on whether or not event B has occurred.

■ **4.4**

Event relationships

In most probability problems, the methods used in determining the probabilities associated with two or more events depend on the form of relationships among the events.

The statement, $P(A/B) = P(A)$, says that the occurrence of event B does not affect the probability of the occurrence of event A. Similarly, the statement $P(B/A) = P(B)$ states that the occurrence of event A does not affect the probability of the occurrence of event B. As it turns out, if one of these conditions $[P(A/B) = P(A)$ or $P(B/A) = P(B)]$ is satisfied, then the other

Independent events

Definition 4.4

Independent events

Two events, A and B, are said to be *independent* if either

$$P(A/B) = P(A) \quad \text{or} \quad P(B/A) = P(B)$$

Dependent events

If either of these conditions is not satisfied, then the two events are said to be *dependent*.

---

will be also. Therefore we need to check only one of these conditions to determine if the two events A and B are independent.

Example 4.7   In Example 4.6, are the two events, lung cancer (A) and smoker (B), independent events?

Solution   Since $P(A) = {}^{20}/_{100} = 0.20$, $P(A/B) = {}^{15}/_{40} = 0.375$, and $P(A) \neq P(A/B)$, the two events are *dependent*.

Example 4.8   For the two events, F and G, it is known that $P(F) = 0.6$, $P(G) = 0.4$, and $P(F \cap G) = 0.10$. Are the two events independent?

Solution   Since $P(F \cap G) = 0.10$, $P(F/G) = P(F \cap G)/P(G) = 0.10/0.40 = \frac{1}{4} = 0.25$. Since $P(F/G) = 0.25 \neq P(F) = 0.6$, the two events are *dependent*.

Dependency versus cause-effect relationships

The fact that two events are dependent does not imply a cause-effect relationship. For instance, in Example 4.7, we can state that the occurrence of the event "smoking" affects the probability of the occurrence of the event "lung cancer." This does not mean that smoking *causes* lung cancer, but rather that the act of smoking is related to the *probability* of contracting lung cancer. The causal agent may be a physiological factor, for example, that increases a person's likelihood of both smoking and contracting lung cancer. This is a rather salient point concerning the dependence of two events that will be dealt with in detail in later chapters.

Definition 4.5

Mutually exclusive events

Mutually exclusive events

Two events, A and B, are said to be *mutually exclusive* if their intersection, $A \cap B$, is the null set ϕ.

Thus if A and B are mutually exclusive events, then

$$P(A \cap B) = 0$$

That is, in words, two events A and B are mutually exclusive if they cannot both occur in a single trial of the experiment. In Figure 4.9, two mutually exclusive events are illustrated; the event spaces A and B do not intersect.

Example 4.9   Suppose the experiment is flipping a coin once. Let events A and B be getting a head and a tail, respectively. In the single coin toss, are the two events mutually exclusive?

FIGURE 4.9 | Venn diagram of two mutually exclusive events

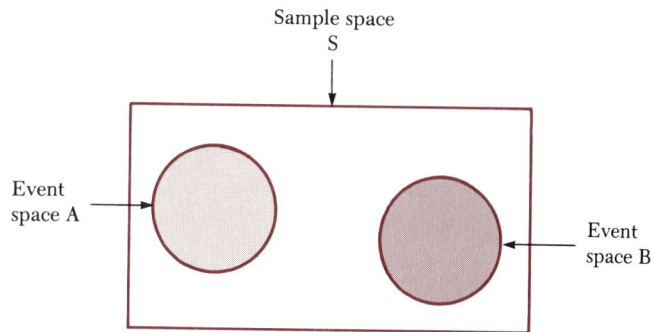

Solution   Since, if a head occurs, a tail cannot, and vice versa, events A and B are mutually exclusive.

Example 4.10   In Example 4.6 where the experiment is randomly selecting one of the 100 hospital patients, are events A, "lung cancer," and B, "smoking," mutually exclusive?

Solution   Since $P(A \cap B) = {}^{15}/_{100} \neq 0$, the two events are not mutually exclusive. They can occur jointly in a single trial of the experiment.

The two event relationships, *independence* and *mutual exclusivity,* are the most important ones in probability! In most probability problems, the nature of the questions asked require determining the independence or dependence of events or whether or not the events are mutually exclusive. These determinations can be made separately by applying the definitions, but it is interesting to ask whether or not these two event relationships are related. Can a Venn diagram be used to show independent events as was done to illustrate mutually exclusive events. Let us suppose, for the moment, that two events, A and B, are mutually exclusive. This means that the occurrence of event A precludes the occurrence of event B, since by definition, the two events cannot occur jointly in a single trial of an experiment. This certainly ensures the *dependence* of the two events, because the occurrence of A, for example, means that $P(B/A) = 0 \neq P(B)$. If two events are not mutually exclusive, then they *may* be independent events, but it is not a certainty. In Example 4.6, the events A (lung cancer) and B (smoker) were not mutually exclusive, but they were not independent either. The property

FIGURE 4.10 | Venn diagrams of dependence and independence of two events

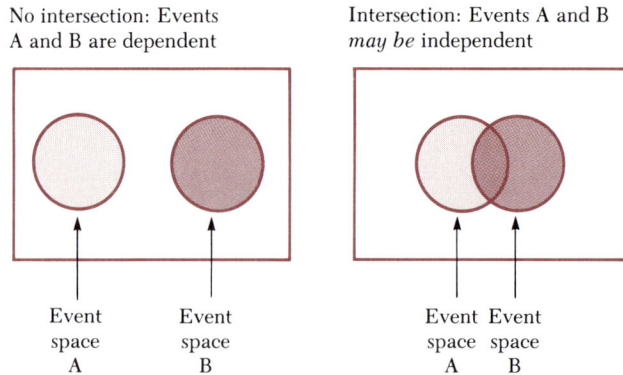

No intersection: Events
A and B are dependent

Intersection: Events A and B
may be independent

Event
space
A

Event
space
B

Event Event
space space
A B

of event independence may be illustrated by Venn diagrams as in Figure 4.10.

Definition 4.6

Collectively exhaustive events

A set of events $A_1, A_2, \ldots, A_r$ is said to be *collectively exhaustive* if

$$A_1 \cup A_2 \cup A_3 \cup \ldots \cup A_r = S$$

Collectively ex-
haustive events

Collectively exhaustive events "fill" the sample space. Figure 4.11 illustrates a set of five collectively exhaustive events. Notice that the events can overlap.

In Figure 4.11, A_1 and A_2 are not mutually exclusive events; nor are A_2 and A_4. However, A_1, A_2, A_3, A_4, and A_5 do fill the sample space S. That is, the union of the five events is S. We now extend Definition 4.5, the definition of a pair of mutually exclusive events, to several events.

A set of r events, $A_1, A_2, \ldots, A_r$ is said to be *mutually exclusive and collectively exhaustive* if both Definitions 4.6 and 4.7 are satisfied for the events. Thus a set of events is mutually exclusive and collectively exhaustive if every possible pair of events is mutually exclusive and if the union of all events is the sample space S. A set of five mutually exclusive and collectively exhaustive events is illustrated in Figure 4.12.

Mutually exclusive events

Definition 4.7
Mutually exclusive events

A set of events $A_1, A_2, \ldots, A_r$ is said to be *mutually exclusive* if

$$A_i \cap A_j = \phi \quad \text{for all } i \text{ and } j; \, i \neq j$$

FIGURE 4.11 | A set of five collectively exhaustive events

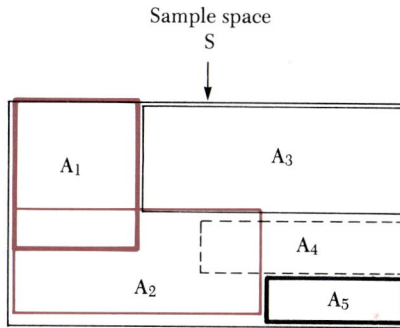

FIGURE 4.12 | A set of five mutually exclusive and collectively exhaustive events

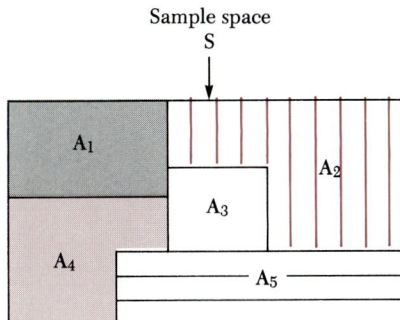

The implication of a set of mutually exclusive and collectively exhaustive events is that an outcome in one and only one of the events in the set will occur in a single trial of the experiment.

Example 4.11 An experiment consists of tossing a die a single time. Events are defined to be the number of dots that appear on the top face of the die. Is this set of events mutually exclusive and collectively exhaustive?

Solution The events are mutually exclusive, because if one occurs, none of the others may occur. If the die is fair, then each event has a probability of ⅙ of occurring, and the sum of the probabilities assigned to the individual events is 1. Thus the set of events is mutually exclusive and collectively exhaustive.

Complement of an event

Definition 4.8
Complement of an event

The *complement* of an event A is the collection of all points in the sample space that are *not* in the event space A. The complement of A is denoted by $\overline{A}$.

As illustrated by the Venn diagram in Figure 4.13, an event and its complement are simply a special case of a set of mutually exclusive and collectively exhaustive events.

From Figure 4.13, it is apparent that

$$P(A) + P(\overline{A}) = 1 \qquad \text{so that} \qquad P(A) = 1 - P(\overline{A})$$

FIGURE 4.13 | An event A and its complement $\overline{A}$

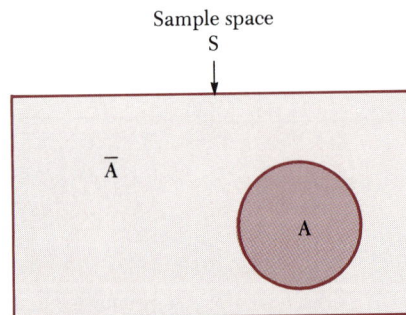

■ **4.5**

Additive and multiplicative probability laws

In addition to answering probability questions by the sample space–event space approach, we can approach probability questions by defining events, determining event relationships, and then applying the following two probability laws.

Additive law of probability

Law 1
Additive law of probability

The probability of the union of two events A and B (A ∪ B) is given by:

$$P(A \cup B) = P(A) + P(B) - P(A \cap B)$$

If A and B are mutually exclusive, then

$$P(A \cup B) = P(A) + P(B)$$

The additive law can be used to compute the probability of the union of two events. If the two events are not mutually exclusive, then the law tells us to add the probability of event A to the probability of event B and subtract the probability of the intersection of A and B because it has been counted twice when $P(A)$ is added to $P(B)$, as Figure 4.14 illustrates.

If the two events are mutually exclusive, then the event spaces A and B do not overlap, and the probability of the union is determined by simply summing the probabilities assigned to event A and event B.

Example 4.12   In Example 4.2, determine the probability that the randomly sampled patient is a smoker (B) or has lung cancer (A).

Solution   This problem may be solved using the sample space–event space method simply by summing the number of sample points in the event space A ∪ B (5 + 15 + 25) and dividing this sum by the total number of points in the sample space (100); thus

$$P(A \cup B) = \frac{5 + 15 + 25}{100} = \frac{45}{100} = 0.45$$

Alternatively, $P(A \cup B)$ may be determined by the event composition method and the additive probability law. Since $P(A) = 0.20$, $P(B) = 0.40$, $P(A \cap B) = 0.15$, and A and B are not mutually exclusive events,

$$P(A \cup B) = P(A) + P(B) - P(A \cap B) = 0.20 + 0.40 - 0.15 = 0.45$$

It should be apparent that both methods perform the same operations, but in slightly different ways. In many problems, the sample space may contain too many points to enumerate. In others, the only information we are given may

FIGURE 4.14 | Computation of $P(A \cup B)$

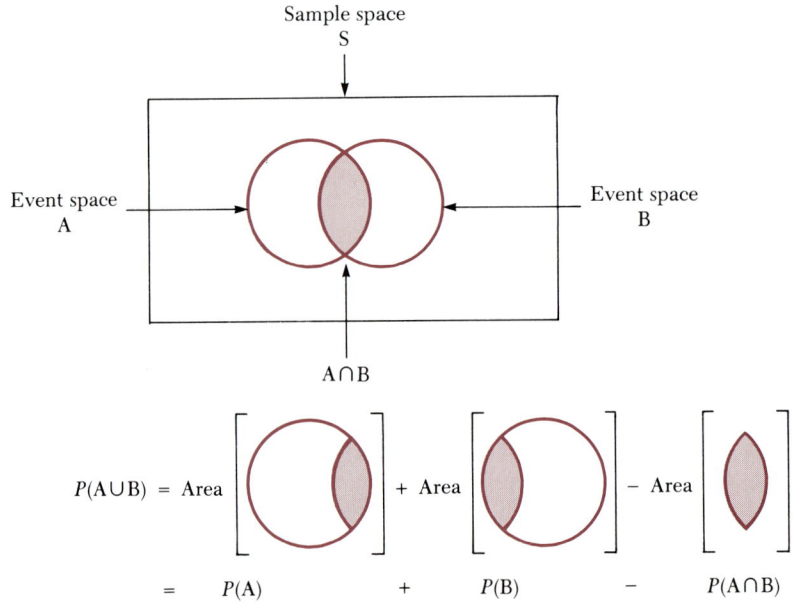

already be in the form of the probabilities. Thus we may have no choice but to use the event composition method.

Multiplicative law of probability

Law 2
Multiplicative law of probability

The probability of the intersection (A ∩ B) of two events A and B is given by:

$$P(A \cap B) = P(A/B)P(B) = P(B/A)P(A)$$

If A and B are independent events, then

$$P(A \cap B) = P(A)P(B)$$

Example 4.13 An applicant for a quality control position in a manufacturing firm is given the following test: Twenty units of a product are placed in a large bin. Two of the units are defective. Inspect the units and select the two that are defective. What is the probability the applicant will select the two defective units if he randomly selects any pair of units from the bin?

Solution   Define the two events: A, the first unit selected is defective, and B, the second unit selected is defective. Then we wish to find the probability of the intersection of the two events; that is, $P(A \cap B)$. The events are clearly dependent. The probability that the second unit is defective depends on whether or not the first unit selected is defective. Thus we must use the first form of the multiplicative law:

$$P(A \cap B) = P(A)P(B/A) = \left(\frac{2}{20}\right)\left(\frac{1}{19}\right) = \frac{1}{190} = 0.00526$$

The conditional probability $P(B/A)$ equals $\frac{1}{19}$, because if A occurs (a defective unit is selected in the first draw), there will be 19 units left to select from on the second draw, one of which is defective. Hence it is very unlikely that the applicant would select the two defective units by chance alone.

Example 4.14   In the dice game, a total of seven dots on the top faces of the pair of dice wins. What is the probability that a player tosses two consecutive sevens?

Solution   The experiment in this case is tossing the pair of dice twice. Let event A be "a total of seven dots appear on the top faces of the dice on the first toss," and let event B be "a total of seven dots appear on the top faces of the dice on the second toss." Then we wish to find $P(A \cap B)$. It is reasonable to assume that the two tosses are independent; that is, the outcome in the second toss in no way is affected by or affects the outcome in the first toss. Thus the second form of the multiplicative law may be used:

$$P(A \cap B) = P(A)P(B) = \left(\frac{6}{36}\right)\left(\frac{6}{36}\right) = \frac{1}{36}$$

The event composition approach using the probability laws to solve probability problems is more direct than the sample space–event space approach. Indeed, as suggested by Examples 4.13 and 4.14, it may not be feasible to use the latter approach due to the sheer number of points in the sample space. The sample space–event space method could have been used to solve the two example problems, but the work involved would be considerably more time-consuming. In Example 4.13, it would have been necessary to "list out" or count the 380 (20 × 19) points (pairs of units) in the sample space, and in Example 4.14 to work with the sample space containing 1,296 (36 × 36) points.

Although the event composition method of determining probabilities may be appealing in many problems, it takes considerable creativity and experience in using this method to select a proper set of event descriptions and to determine the proper relationships between events. The importance of event relationships should now be apparent; if we are finding $P(A \cap B)$, the determination that the events are independent will simplify computations considerably. Similarly, if we are finding $P(A \cup B)$, the determination that A and B are mutually exclusive will simplify computations.

In the next section, example problems are presented and solved using the event composition technique in conjunction with the additive and multiplicative probability laws. It will be seen that the use of probability trees can be particularly helpful in solving problems by the event composition method.

4.6

Computing probabilities: Probability laws and probability trees

Example 4.15   The workers in a large assembly plant must check out and check in tools required during the day as the need for a particular tool arises. There is a centrally located tool bin operated by two employees who get the required tools for the workers and return them to the bin shelves when the workers are finished. The availability of one bin employee is assumed to be independent of the other. The probability that a specific employee in the bin is available to get a tool is 0.8. What is the probability that if a worker requires a tool, she or he can get it from the bin without waiting?

Solution   Define the two events:

A: First bin employee is available for service.
B: Second bin employee is available for service.

If the worker requiring a tool does not have to wait in line at the bin, then events A *or* B *or* both have occurred. Thus we are interested in computing $P(A \cup B)$. Since we have determined that the question may be answered by using the additive probability law, the next step is to determine whether or not A and B are mutually exclusive events. Since both bin employees can be busy at the same time, they are clearly not mutually exclusive events. Then

$$P(A \cup B) = P(A) + P(B) - P(A \cap B) = (0.8) + (0.8) - (0.8)(0.8)$$
$$= 1.6 - 0.64 = 0.96$$

Notice that $P(A \cap B)$ has been computed by the multiplicative law, using the fact that events A and B are independent (the availability of one bin employee is independent of the availability of the other employee). This problem can also be solved using a *probability tree* as illustrated in Figure 4.15.

Probability tree

The first branch on the tree depicts the availability of employee A; he is either available [with probability $P(A) = 0.8$] or not available [with probability $P(\overline{A}) = 0.2$]. On the upper branch A, we now branch on B. These two branches represent conditional events—employee B is available given that employee A is available (B/A), and employee B is not available given that employee A is available ($\overline{B}$/A). The conditional probabilities reflect that two events, A and B, are independent; that is, $P(B/A) = P(B) = 0.8$, and $P(\overline{B}/A) = P(\overline{B}) = 0.2$. By multiplying the probabilities along the branches, the probability of the intersection of events can be determined. For example, on the upper branch, $(0.8)(0.8) = 0.64$ is the probability of $(A \cap B)$. [By the multiplicative rule, $P(A)P(B/A) = P(A \cap B)$.] The probability tree gives us the required components to use in conjunction with the additive probability law to find $P(A \cup B)$: $P(A) = 0.8$, $P(B) = 0.8$, and $P(A \cap B) = 0.64$. Notice that we can determine $P(A \cup B)$ directly from the tree by summing the probabilities of the three mutually exclusive outcomes $A \cap B$, $A \cap \overline{B}$, and $\overline{A} \cap B$ $(0.64 + 0.16 + 0.16 = 0.96)$.

FIGURE 4.15 | Probability tree for Example 4.15

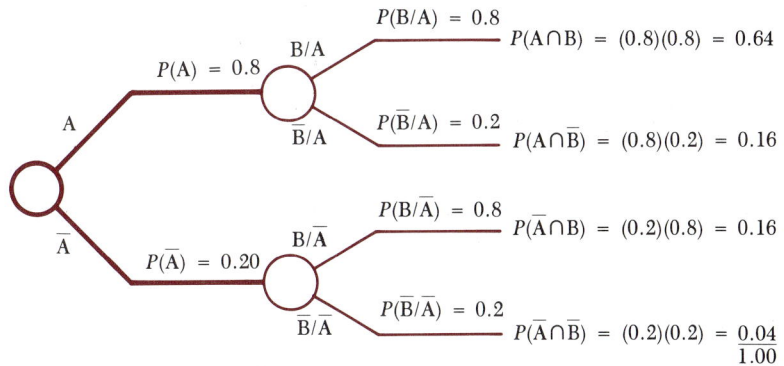

Further, notice that the "ends" of the probability tree list the four mutually exclusive and collectively exhaustive compound events associated with the problem—$(A \cap B)$, $(A \cap \overline{B})$, $(\overline{A} \cap B)$, and $(\overline{A} \cap \overline{B})$. The question could have been answered somewhat more easily had we recognized that at least one employee being available is the complement event to the event neither employee is available. Thus

$$P(A \cup B) = P(A \cap B) + P(A \cap \overline{B}) + P(\overline{A} \cap B) = 1 - P(\overline{A} \cap \overline{B})$$
$$= 1 - 0.04 = 0.96$$

Example 4.16 Three fire detectors are placed in an office building. The probability that *each* will signal a fire when the temperature reaches or exceeds a specific value is 0.9. The signaling by one detector is independent of the signaling of the other two detectors when this minimum temperature is reached. If there is a fire in the office and the critical temperature is reached, what is the probability that at least one detector will signal a fire?

Solution If we define the event A, "At least one detector signals the fire," then $P(A) = 1 - P(\overline{A})$, where $\overline{A}$ is the complementary event, "None of the detectors signals a fire." Let F_1, F_2, and F_3 denote the events, "The first detector signals," "The second detector signals," and "The third detector signals" when the critical temperature has been reached, respectively. Then $P(\overline{A}) = P(\overline{F}_1 \cap \overline{F}_2 \cap \overline{F}_3)$, since the probability that none will signal is the probability of the intersection $(\overline{F}_1 \cap \overline{F}_2 \cap \overline{F}_3)$. Since the signals of the detectors are independent events and $P(\overline{F}_i) = 1 - P(F_i) = 1 - 0.9 = 0.1$ for $i = 1, 2,$ and 3,

$$P(\overline{F}_1 \cap \overline{F}_2 \cap \overline{F}_3) = P(\overline{F}_1)P(\overline{F}_2)P(\overline{F}_3) = (0.1)(0.1)(0.1) = 0.001$$

Thus

$$P(A) = 1 - P(\overline{A}) = 1 - 0.001 = 0.999$$

Example 4.17 A local Ford dealer received a shipment of 50 new ½-ton pickup trucks and sold two of them to the Cabot Construction Company, which will use them on construction jobs. A week after the trucks were delivered, the dealer received a notice that 10 of the 50 trucks had faulty rear axles. What is the probability that the Cabot Company bought two defective trucks?

Solution Let event A be defined as "Both trucks purchased by Cabot are defective." The event A can be written as the compound event, $A_1 \cap A_2$, where $A_i = i$th purchased truck is defective and $i = 1, 2$.

To stress that there are frequently many ways to approach a probability problem, this problem will be solved in two ways.

1. Sample space–event space method. Assuming that each of the total number of pairs of trucks that could have been selected by Cabot were equally likely to have been chosen, then $P(A)$ can be determined by counting the number of points in the event space A (n) and dividing this by the number of points in the sample space S (N).

Since two trucks are selected from the 50, the number of points in the sample space is $N = C_2^{50} = (50 \cdot 49)/2 = 1,225$. Since the set of 50 trucks from which the two were selected contains two types, defectives (faulty rear axles) and nondefectives, the hypergeometric counting rule is used to count the number of points in the event space A:

$$n = C_2^{10} C_0^{40} = \left(\frac{10 \cdot 9}{2}\right)(1) = 45$$

Thus

$$P(A) = \frac{45}{1,225} = \frac{9}{245} = 0.037$$

2. Multiplicative rule. Since $P(A) = P(A_1 \cap A_2)$, the multiplicative rule can be used to determine $P(A)$. The two events A_1 and A_2 are not independent because if the first truck purchased is defective (event A_1 occurs), this affects the probability that the second event A_2 occurs. Thus

$$P(A) = P(A_1)P(A_2/A_1) = \frac{10}{50} \cdot \frac{9}{49} = \frac{9}{245} = 0.037$$

The conditional probability, $P(A_2/A_1)$, is determined by simple reasoning: If event A_1 occurs, then a defective truck was chosen as the first truck selected. This leaves 49 trucks, 9 of which are defective. Thus the probability that the second purchased truck is defective given that the first one is defective is 9/49.

4.7

Bayes' theorem

Bayes' theorem provides a formula for the conditional probability, $P(A/B)$, that may be used in place of the formula in Definition 4.3 [$P(A/B) = P(A \cap B)/P(B)$]. To produce the formula for Bayes' theorem, we first must present a very useful law that results in a formula used in the theorem.

Law of total proba-
bility

Law 3
Law of total probability

$$P(B) = P(A \cap B) + P(\overline{A} \cap B)$$

That is, the probability that event B occurs is equal to the probability that events B and A occur jointly plus the probability that events B and $\overline{A}$ occur jointly. Since either event A occurs or its complementary event $\overline{A}$ occurs, event B *must* occur jointly with either event A or its complement event $\overline{A}$. This relationship is illustrated in Figure 4.16. The intersection $(A \cap B)$ plus the intersection $(\overline{A} \cap B)$ is the event space B.

FIGURE 4.16 | Venn diagram illustrating $P(B) = P(A \cap B) + P(\overline{A} \cap B)$

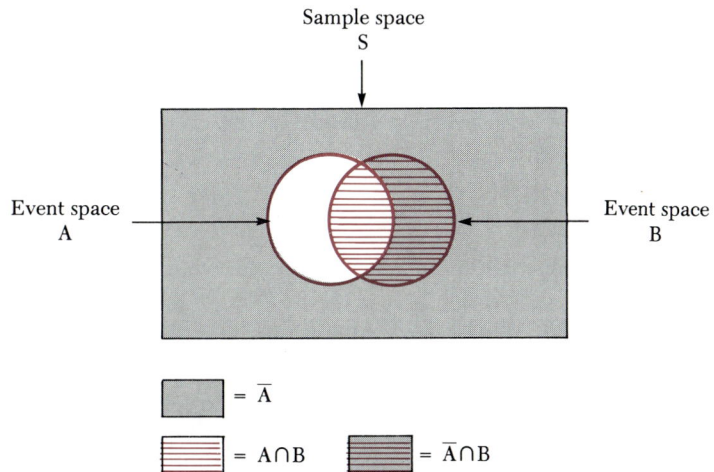

It is possible to write $P(B)$ in Law 3 in another form as well. The multiplicative law provides the following formula for the intersection of two events, A and B:

$$P(A \cap B) = P(B/A)P(A)$$

Therefore in Law 3 we can replace $P(A \cap B)$ with $P(B/A)P(A)$. Also, by replacing A with $\overline{A}$ in the multiplicative law above, we have

$$P(\overline{A} \cap B) = P(B/\overline{A})P(\overline{A})$$

Therefore we can rewrite the law of total probability for event B as:

$$P(B) = P(B/A)P(A) + P(B/\overline{A})P(\overline{A})$$

This expression is a key part of Bayes' theorem.

Theorem 4.1

Bayes' theorem

Let A be an event. If B is another event such that $P(B)$ is not 0, then

$$P(A/B) = \frac{P(B/A)P(A)}{P(B/A)P(A) + P(B/\overline{A})P(\overline{A})}$$

From the definition of the conditional probability $P(A/B)$, we know that

$$P(A/B) = \frac{P(A \cap B)}{P(B)}$$

However, from the multiplicative law, $P(A \cap B) = P(B/A)P(B)$ and from above, $P(B) = P(B/A) P(A) + P(B/\overline{A})P(\overline{A})$. Therefore, Bayes' theorem is simply an alternate formula for calculating the conditional probability $P(A/B)$. In many probability problems where it is necessary to calculate the conditional probability $P(A/B)$, the probabilities $P(A \cap B)$ and $P(B)$ are not given directly. Therefore, the formula in the definition for $P(A/B)$ cannot be used. Rather, the problem statement gives the probabilities $P(B/A)$, $P(B/\overline{A})$, $P(A)$, and $P(\overline{A})$. With these probabilities, the conditional probability $P(A/B)$ can be found by Bayes' theorem.

Posterior probability

In application of Bayes' theorem, the computed probability $P(A/B)$ is often called the *posterior probability* of event A, given that event B has occurred. The probabilities $P(A)$ and $P(\overline{A})$ are called the *prior probabilities* of the event A and its complement $\overline{A}$. The prior probability $P(A)$ is thus *revised* using the information that event B has occurred. The probability of the intersection of events A and B, $[P(A \cap B)]$, is the *joint probability* of events A and B; and the divisor in Bayes' theorem, $[P(B)]$, is called the *unconditional* or the *marginal probability* of event B. These relationships between events are developed in detail in Chapter 23.

Prior probability

Probability revision

Bayes' theorem can be very useful when we wish to determine a conditional probability, say $P(A/B)$, but we have the probabilities $P(A \cap B)$ and $P(B)$ given in decomposed forms, as the next example illustrates.

Example 4.18   The Superior Research Hospital has publicized the discovery of a new diagnostic test for cancer. If the person has cancer, the test will be *positive* (indicating the person has cancer) 99 percent of the time. If the person does not have cancer, the test is *negative* 95 percent of the time. From past medical histories and population census files, the probability that a person has cancer is 0.005. If the test is administered to a randomly selected individual in this population and it is positive, what is the probability that the person has cancer?

Solution   Define the events S, "Test is positive," and C, "Person has cancer." Using the given information, we can assign the following probabilities:

$$P(C) = 0.005 \qquad P(S/C) = 0.99 \qquad P(\overline{S}/\overline{C}) = 0.95$$

We wish to determine $P(C/S)$.

Bayes' theorem may be used to determine $P(C/S)$. Associating the event C with A and S with B in Bayes' theorem above,

$$P(C/S) = \frac{P(S/C)P(C)}{P(S/C)P(C) + P(S/\overline{C})P(\overline{C})}$$

In the statement of the problem, we have all the pieces required to calculate $P(C/S)$ by using the above formula except $P(S/\overline{C})$ and $P(\overline{C})$. But recall that

$$P(S/\overline{C}) + P(\overline{S}/\overline{C}) = 1$$

That is, given that the person does not have cancer, the test will be either positive or negative. The probabilities of these two conditional events $(S/\overline{C})$ and $(\overline{S}/\overline{C})$ must total 1. Since $P(\overline{S}/\overline{C}) = 0.95$, it follows that

$$P(S/\overline{C}) = 1 - P(\overline{S}/\overline{C}) = 1 - 0.95 = 0.05$$

Similarly, $P(\overline{C}) = 1 - P(C) = 1 - 0.005 = 0.995$. Now,

$$P(C/S) = \frac{(0.99)(0.005)}{(0.99)(0.005) + (0.05)(0.995)} = \frac{0.00495}{0.00495 + 0.04975} = 0.091$$

Thus the probability that the person has cancer has been *revised* from 0.005 to 0.091 by knowing that the test for cancer was positive. Notice that the denominator $(0.00495 + 0.04975 = 0.0547)$ gives us the probability that the test will be positive $[P(S)]$.

This problem can also be solved by using a probability tree as illustrated in Figure 4.17. Notice that the first branching of the tree was performed on the given event C. The tree may be used to compute $P(C/S)$ from the definitional form, $P(C/S) = P(S \cap C)/P(S)$. To determine $P(S)$, add the joint probabilities at the end of the tree with the event letter S in them—$0.00495 + 0.04975 = P(S)$.

FIGURE   4.17   │   Solution to Example 4.18 using a probability tree

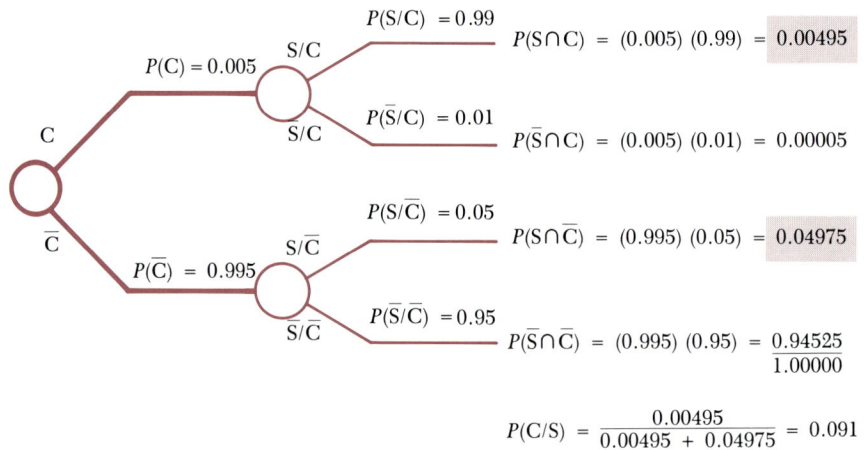

$$P(S/C) = 0.99$$

$$P(C) = 0.005$$

$$P(S\cap C) = (0.005)(0.99) = 0.00495$$

$$P(\bar{S}/C) = 0.01$$

$$P(\bar{S}\cap C) = (0.005)(0.01) = 0.00005$$

$$P(S/\bar{C}) = 0.05$$

$$P(S\cap \bar{C}) = (0.995)(0.05) = 0.04975$$

$$P(\bar{C}) = 0.995$$

$$P(\bar{S}/\bar{C}) = 0.95$$

$$P(\bar{S}\cap \bar{C}) = (0.995)(0.95) = \underline{0.94525}$$
$$1.00000$$

$$P(C/S) = \frac{0.00495}{0.00495 + 0.04975} = 0.091$$

Bayes' theorem can be extended to more than two events that are conditioned by an event B. We will state the general form of Bayes' theorem and then apply it in an example.

Theorem 4.2
Bayes' theorem—general form

Let $A_1, A_2, \ldots, A_K$ be a set of K mutually exclusive and collectively exhaustive events. If B is another event, such that $P(B)$ is not 0, then

$$P(A_i/B) = \frac{P(B/A_i)P(A_i)}{\sum_{i=1}^{K} P(B/A_i)P(A_i)}, \qquad i = 1, 2, \ldots, K$$

If we let $K = 2$, $A_1 = A$, and $A_2 = \bar{A}$, Theorem 4.1 is a special case of the general form of Bayes' theorem.

Example 4.19   A manufacturing firm produces units of a product in four plants. Define event A_i, "A unit is produced in plant i, where $i = 1, 2, 3, 4$," and event B, "A unit is defective." From past records of the proportions of defectives produced at each plant, the following conditional probabilities are set: $P(B/A_1) = 0.10$, $P(B/A_2) = 0.05$, $P(B/A_3) = 0.02$, and $P(B/A_4) = 0.15$.

The first plant produces 40 percent of the units of the product, the second plant 20 percent, the third 5 percent, and the fourth 35 percent. A unit of the product made at one of these plants is tested and found to be defective. What is the probability that the unit was produced in plant 3? That is, find $P(A_3/B)$, the probability that the unit came from plant 3, given that it is defective.

Solution   We wish to determine $P(A_3/B)$. From the general form of Bayes' theorem, this probability is given by

$$P(A_3/B) = \frac{P(B/A_3)P(A_3)}{\sum\limits_{i=1}^{4} P(B/A_i)P(A_i)}$$

The computation of $P(A_3/B)$ by Bayes' theorem is eased somewhat by applying the tabular format as given in Table 4.3. From the third row, fifth

TABLE 4.3 | Application of Bayes' theorem to more than two events

Plant (event) i	$P(A_i)$	$P(B/A_i)$	$P(A_i)P(B/A_i)$	$\dfrac{P(A_i)P(B/A_i)}{\sum\limits_{i=1}^{4} P(A_i)P(B/A_i)}$
1	0.40	0.10	0.0400	0.04/0.1035 = 0.3865
2	0.20	0.05	0.0100	0.01/0.1035 = 0.0966
3	0.05	0.02	0.0010	0.001/0.1035 = 0.0097
4	0.35	0.15	0.0525	0.0525/0.1035 = 0.5072
	1.00		0.1035 = $P(B)$	1.0000

column of this table, $P(A_3/B) = 0.0097$. The fourth column, when summed, gives the denominator of Bayes' theorem. From Table 4.3 $P(B) = 0.1035$; that is, the probability that a defective part is produced by this firm is 0.1035. An advantage to using a table similar to Table 4.3 with Bayes' theorem is that summing of probabilities in the last column to 1.0 (or close to 1.0, allowing for some numerical roundoff) gives some assurance that the required probability has been correctly determined.

■ **4.8**

Additive and multiplicative laws for more than two events

Law 1 gives the additive law of probability, and Law 2 gives the multiplicative law of probability for two events, A and B. These laws can be extended to apply to more than two events. We will not give the laws for the general case with more than two events (see, for example, the Feller reference at the end of this chapter for these laws). However, we will give a special case for each of the two laws. The special cases are very important and occur frequently in the analysis of problems by the use of probability.

First, consider the additive law for determining the probability of the union of a set of events. Suppose that we have r events denoted by A_1, A_2, . . . , A_r. *If these r events are mutually exclusive,* then the probability of the union of these events is given by:

$$P(A_1 \cup A_2 \cup \ldots \cup A_r) = P(A_1) + P(A_2) + \ldots + P(A_r)$$

That is, the probability of the union of r events is the sum of the probabilities of the r events. If the r events are not mutually exclusive, then this result does not hold (see problem 4.44 at the end of the chapter: By using a Venn diagram, you are asked to derive the formula for $P(A \cup B \cup C)$ when the three events are *not* mutually exclusive).

Second, consider the multiplicative law for determining the probability of the intersection of a set of events. Suppose that we have r events denoted by A_1, A_2, . . . , A_r. *If these r events are independent,* then the probability of the intersection of these events is given by:

$$P(A_1 \cap A_2 \cap \ldots \cap A_r) = P(A_1)P(A_2) \ldots P(A_r)$$

That is, if each pair of the r events is independent, then the probability of the intersection of these events is the product of the probabilities $P(A_1)$, $P(A_2)$, . . . , $P(A_r)$. This formula *does not* apply if the r events are dependent.

Example 4.20 The space shuttle, Liberty, has four redundant systems which act independently for a certain electronic task. The probability that each system operates properly is 0.96. If the first system fails, then the second system is tried and so on until one is found to work or it is found that none of the four work.

a. What is the probability that none of the four systems work?

b. What is the probability that at least one of the four systems will work?

Solution In part a, let A_i denote the event that the ith system works. Then we want $P(\overline{A}_1 \cap \overline{A}_2 \cap \overline{A}_3 \cap \overline{A}_4)$. Since $P(\overline{A}_i) = 1 - P(A_i)$, and since the four systems are independent, this probability is given by $(0.04)(0.04)(0.04)(0.04) = 0.00000256$. In part b, the probability that at least one of the four work is the complement to the event that none of the four systems work. Therefore

$$P(\text{at least one system works}) = 1 - P(\text{none of the four work})$$
$$= 1 - 0.00000256 = 0.99999744$$

This is the kind of reliability figure that astronauts like to see when they are flying around the earth in the shuttle!

Example 4.21   Executives traveling on business for the Bleecker Company rent cars when needed from one of several car rental agencies. The proportion of time agency A is used is 0.20, agency B is used is 0.15, agency C is used is 0.25, and some other agency is used is 0.40. Assuming that these four events are mutually exclusive, what is the probability that an executive rents a car on a particular trip from agency A, B, or C?

Solution   We want to find the probability of the union of the three events A, rents from agency A; B, rents from agency B; and C, rents from agency C. Since the three events are mutually exclusive,

$$P(A \cup B \cup C) = P(A) + P(B) + P(C)$$
$$= 0.20 + 0.15 + 0.25$$
$$= 0.60$$

Therefore the probability that she rents from one of the big three—agencies A, B, and C—is 0.60.

Example 4.22: The birthday problem   An interesting application of conditional probabilities for a set of events that are *not* pairwise independent is the birthday problem. Suppose you are in a group of n people $(n > 1)$. What is the probability that at least two people in the group have the same birthday (day and month)? In answering this question, we will treat the leap year date of February 29 as March 1 to simplify computations.

Solution   It is easier to find the probability that no one in the group of n persons has the same birthday as anyone else. Then

$$P\begin{bmatrix}\text{At least two people} \\ \text{have the same birthday.}\end{bmatrix} = 1 - P\begin{bmatrix}\text{No one has the same} \\ \text{birthday as anyone else.}\end{bmatrix}$$

Let the event "No one has the same birthday as anyone else in a group of n persons" be A. Then from above, $P(\overline{A}) = 1 - P(A)$, where the event $\overline{A}$ is the complement of event A and is defined as, "At least two people have the same birthday." The formula for $P(A)$ is given by

$$P(A) = \frac{\text{Number of points in event space A}}{\text{Number of points in sample space S}}$$

$$= \frac{[365][364][363] \; \ldots \; [365 - (n-1)]}{[365]^n}$$

assuming that there are 365 days in the year. This formula may be explained by an illustration. Suppose there are $n = 3$ people forming the group. The number of points in the sample space is $365 \cdot 365 \cdot 365 = 365^3$; that is, each person's birthday can be any one of the 365 days in the year. Therefore, the total number of sets of three birthdays is 365^3. The number of points in the event space A can be determined as follows. Let the three boxes below represent the birthdays of the $n = 3$ people.

Person 1   Person 2   Person 3

$\boxed{365}$ · $\boxed{364}$ · $\boxed{363}$ = [365][364][363]

[365 − (3 − 1)]

n

If there are no matches, then the first person's birthday can be any one of the 365 days in the year. The second person's birthday may be any one of the remaining 365 − 1 = 364 days. The third person's birthday may be any one of the remaining 365 − 2 = 363 days. Now,

$$P(\overline{A}) = 1 - P(A)$$
$$= 1 - \left[\frac{[365][364][363] \ . \ . \ . \ [365 - (n - 1)]}{[365]^n} \right]$$
$$= 1 - \left[\frac{365!}{(365 - n)![365]^n} \right]$$

Table 4.4 shows the probabilities of events A and $\overline{A}$ for several group sizes. Thus in a group of $n = 64$ people, it is *almost certain* that at least two people will have the same birthday! When the group size becomes $n = 23$, the probability of at least two matches is greater than ½. These probabilities surprise most people. It is commonly thought that $P(\overline{A}) = n/365$—why? In actuality, $P(\overline{A})$ is much larger than $n/365$, as can be seen from Table 4.4.

TABLE 4.4 | The probabilities of the events A—"There is no one in a group of n people with the same birthday as anyone else"—and $\overline{A}$—"There are at least two people in the group of n persons with the same birthday"—for selected values of n, the number of persons in the group

n	A	$\overline{A}$
4	0.984	0.016
8	0.926	0.074
12	0.833	0.167
16	0.716	0.284
22	0.524	0.476
23	0.493	0.507
24	0.462	0.538
28	0.346	0.654
32	0.247	0.753
40	0.109	0.891
48	0.039	0.961
56	0.012	0.988
64	0.003	0.997

Source: Parzen, *Modern Probability Theory and Its Applications* (New York: John Wiley & Sons, 1960), p. 47.

■ 4.9
On the assignment of probability

An often quoted advantage of following the relative frequency or the symmetry definition of probability is that all people assessing probabilities of various events would arrive at the same probabilities (if calculations are performed correctly). For example, two different individuals would likely assess the probability of a head on the single toss of a fair coin at one half, or $P(\text{H}) = 0.50$. On the other hand, these same two individuals might assess the probability of Penn State beating Texas at the Cotton Bowl quite differently, depending on subjective beliefs concerning the outcome, which in turn are influenced by a host of other factors.

The interpretation of various statistical tests varies considerably, depending on how one uses subjective probabilities. Not all probability theory and procedures that can be applied to objective probabilities, for example, can be applied and interpreted similarly when using subjective probabilities. A statistician who uses subjective probabilities is called a *Bayesian* statistician. The topics of Bayesian statistics and Bayesian decision theory are developed in Chapters 21, 22, and 23.

■ 4.10
Summary

Probability measures the likelihood of an event occurring. It must be a number between 0 and 1, inclusive. If the probability of an event is 0, it means that the event is certain not to occur; if the probability is 1, then the event is certain to occur in a single trial of an experiment.

A convenient framework for computing probabilities is the use of sample spaces to list, at least conceptually, all the elementary outcomes that can occur in the experiment. The elementary outcomes should form a set of mutually exclusive and collectively exhaustive outcomes of the experiment. Additionally, they should be chosen in such a way that each elementary outcome is equally likely to occur in a single trial of the experiment.

The assignments of probabilities to the elementary outcomes may be accomplished in three ways: experimentally, using the frequency definition of probability; by symmetrical arguments; or subjectively. In any case, the assigned probabilities must all be numbers between 0 and 1, inclusive; and the sum of the probabilities associated with all the elementary outcomes in an experiment must be 1.

Once the elementary outcomes have been identified and probabilities assigned to them, the probabilities associated with more complex events in the experiment can be determined by one or more of the methods presented in this chapter, among which are: the sample space–event space method, the event composition method, and the use of probability trees. The latter method is suggested for most problems, because it conveniently illustrates the events associated with an experiment and minimizes the risk of defining events incorrectly or misassigning probabilities from the wording of the problem.

Knowledge of the relationships among events is important in solving probability problems, because it indicates which probability law should be used to solve the problem. The two most important event relationships are *mutual exclusiveness* and *independence*.

■ References

Introductory

Anderson, D.; D. Sweeney; and T. Williams. *Statistics for Business and Economics*. 2nd ed. St. Paul, Minn.: West Publishing, 1984, Chapter 4.

Harnett, D. *Statistical Methods*. 3rd ed. Reading, Mass.: Addison-Wesley, 1982, Chapter 2.

Lapin, L. *Statistics for Modern Business Decisions*. 3rd ed. New York: Harcourt, Brace, Jovanovich, 1982, Chapter 5.

Wonnacott, R., and T. Wonnacott. *Introductory Statistics*. 4th ed. New York: John Wiley & Sons, 1985, Chapter 3.

Advanced

Feller, W. *An Introduction to Probability Theory and Its Applications*. 3rd ed. New York: John Wiley & Sons, 1968.

Parzen, E. *Modern Probability Theory and Its Applications*. New York: John Wiley & Sons, 1960.

Ross, S. *Introduction to Probability Models*. 2nd ed. New York: Academic Press, 1980, Chapter 1.

■ Problems

Section 4.2 Problems

4.1 Suppose there are six points in a sample space S: E_1, E_2, E_3, E_4, E_5, and E_6. Events A, B, and C defined over this sample space contain the following points in the sample space:

  A: E_1, E_2, and E_5.
  B: E_2, E_3, and E_4.
  C: E_6.

Find the points in the sample space that belong to the following compound events:
a. A ∪ B
b. A ∩ B
c. B ∩ C
d. A ∪ C

4.2 Assume that the six points in the sample space in Problem 4.1 are equally likely to occur; that is,

$$P(E_1) = P(E_2) = P(E_3) = P(E_4)$$
$$= P(E_5) = P(E_6) = \frac{1}{6}$$

a. Determine $P(A \cup B)$.
b. Determine $P(A \cap B)$.
c. Determine $P(B \cap C)$.
d. Determine $P(B \cup C)$.

4.3 In the dice game example (Example 4.4 in this chapter), suppose that the following three events are defined:

  A: Total of dots on the top two faces of the dice is seven.
  D: Total of dots on the top two faces of the dice is 11.
  E: Total of dots on the top two faces is 10 or more.

For the following compound events, list the elementary outcomes belonging to each.
a. A ∪ E
b. A ∪ D
c. D ∪ E
d. D ∩ E
e. A ∩ D
f. A ∩ E

4.4 For the compound events listed in parts *a* through *f* in Problem 4.3, find the probability of each compound event occurring, assuming that each sample point in the sample space shown in Figure 4.5 is equally likely to occur.

4.5 Suppose that there are five sample points E_1, E_2, E_3, E_4, and E_5 in a sample space S. Further, suppose that the probability each occurs in a single trial of the experiment is: $P(E_1) = 0.15$; $P(E_2) = 0.25$; $P(E_3) = 0.20$; $P(E_4) = 0.05$; and $P(E_5) = 0.35$. Suppose now that the following events are defined over this sample space:

A: E_1, E_3, and E_5.
B: E_2 and E_4.
C: E_1, E_2, and E_3.

Find the probability that each of the following compound events occur [Note: The elementary outcomes are not equally likely to occur in a single trial of the experiment. Therefore, Theorem 3.1 in Chapter 3 must be used to find these probabilities.]
a. $P(A \cup B)$
b. $P(A \cup C)$
c. $P(B \cup C)$
d. $P(A \cap B)$
e. $P(A \cap C)$
f. $P(B \cap C)$

Section 4.3 Problems

4.6 Sales of an automobile dealer over the past month can be classified by method of pay-

ment and type of car sold, as indicated in the following table:

Type of car	Methods of payment	
	Cash	Installments
New	5	95
Used	25	25

A sales record is selected at random from the previous month's sales.
a. What is the probability that it represents a used car purchase?
b. If the purchase is of a new car, what is the probability that it was paid by installments?

4.7 A poll taken among 50 employees of an industrial firm on the question of changing the work week from five days (8 A.M.–5 P.M.) to four days (7 A.M.–6 P.M.) yielded the following results:

	Classification of employees				
Response	Executive	Sales-person	Office worker	Plant worker	Total
Favor change	2	8	10	10	30
Oppose change	1	3	3	8	15
No opinion	0	2	2	1	5
Totals	3	13	15	19	50

One employee is selected at random from among the 50.
a. What is the probability that an employee is a plant worker and favors the change?
b. What is the probability that the employee will oppose the change given that he is a salesperson?
c. If the employee is either an executive or a salesperson, what is the probability that he expressed no opinion?

d. Given that the employee is *not* a plant worker, what is the probability that he favors the change?
e. What is the probability that the employee either favors the change or is an office worker?

4.8 The data at the top of page 166 pertain to a survey taken of new car dealers in the United States. If a dealer is selected at random from this survey, determine:

a. $P(A_3)$
b. $P(A_1 \cap B_4)$
c. $P(A_4/B_1)$
d. $P(A_3 \cup B_4)$
e. $P[(B_3 \cup B_4)/A_1]$

f. $P[(A_1 \cup A_3)/B_5]$
g. $P(A_1 \cup A_2 \cup A_3)$
h. $P[B_1/(A_3 \cup A_4)]$
i. $P[A_3/(B_3 \cup B_5)]$

		Region of dealer				
		A_1	A_2	A_3	A_4	
Event	Type of dealership	Northeast	Northwest	Southeast	Southwest	Totals
B_1:	Fjord Motors	200	100	50	150	500
B_2:	Chysler Motors	150	50	30	70	300
B_3:	National Motors	50	50	20	30	150
B_4:	Chevylay	300	150	150	200	800
B_5:	Toysalot	10	40	10	90	150
	Totals	710	390	260	540	1,900

4.9 The probability that an employee selected at random from a work force at a certain factory is a male is 0.65; the probability that the employee is married is 0.75; and the probability that the employee is both married and male is 0.50. If an employee in this group is chosen at random, calculate the probability that the employee is:
a. Female.
b. Single.
c. Single and female.
d. Female or married.

4.10 A town has both a morning and an evening newspaper. Of the families in the town, 50 percent buy the morning paper, 40 percent buy the evening paper, and 25 percent buy both papers. Define the events: A_1: Buys a morning paper. A_2: Does not buy a morning paper. B_1: Buys an evening paper. B_2: Does not buy an evening paper.
a. List all pairs of events from the set that are mutually exclusive.
b. List all pairs of events that are statistically dependent.
c. Find $P(B_2/A_1)$.
d. Find the probability that a family buys either one paper or the other *but not both*.

4.11 If $P(B) = 0$, does the statement $P(A/B)$ have any meaning? Discuss.

4.12 For two events A and B, $P(A) = 0.10$, $P(B) = 0.40$, and $P(A \cap B) = 0.05$.
a. Determine $P(A/B)$.
b. Determine $P(B/A)$.

Section 4.3 Problems

4.13 For the events defined in Problem 4.7, answer the following:
a. Are the events "office worker" and "plant worker" statistically independent? Why or why not?
b. Are the events "office worker" and "favor change" statistically independent? Why or why not?
c. Are the events "executive" and "plant worker" mutually exclusive? Why or why not?
d. Are the events "executive" and "no opinion" mutually exclusive? Why or why not?

4.14 Consider the statement: If two events are mutually exclusive, they must be dependent events. Is this statement true? Discuss. If two events are not mutually exclusive, does this imply that they are independent events? Discuss.

4.15 The following table has been prepared from a survey of 100 companies in three industries

Industry	IBM	Apple	IBM compatible	Total
Banking	25	5	5	35
Electronics	10	20	10	40
Accounting	25	0	0	25
Total	60	25	15	100

Computer type

to determine the primary type of microcomputer used by employees in the companies. Suppose one of these 100 surveyed companies is chosen at random, and answer the following questions:

a. Are the events "use IBM microcomputer" and "Accounting industry" independent for these 100 surveyed companies? Explain.

b. Are the events "use IBM microcomputer" and "Accounting industry" mutually exclusive events? Explain.

c. Are the events "use IBM microcomputer" and "use Apple microcomputer" mutually exclusive events? Explain.

d. Are the events "use IBM-compatible microcomputer" and "Electronics industry" independent events? Explain.

4.16 For the three events A, B, and C, we know the following: $P(A) = 0.20$, $P(B) = 0.30$, $P(C) = 0.40$, $P(A \cap B) = 0.06$, $P(A \cap C) = 0.10$, and $P(B \cup C) = 0$.

a. Are events A and B independent? Explain.

b. Are events A and C independent? Explain.

c. Are events A and B mutually exclusive? Explain.

d. Are events A and C mutually exclusive? Explain.

e. Are events B and C mutually exclusive? Explain.

4.17 On a given day, either it rains during the 24-hour period, or it does not rain. Are these events mutually exclusive and collectively exhaustive? Explain.

4.18 The probability that the solid booster rocket used in launches of the space shuttle fails is 0.001. What is the probability that the solid booster rocket does not fail?

4.19 Suppose for three events, A, B, and C, we know the following: $P(A) = 0.50$, $P(B) = 0.30$, $P(C) = 0.10$, $P(A \cap B) = 0$, $P(A \cap C) = 0$, and $P(B \cap C) = 0$.

a. Is this a set of $r = 3$ mutually exclusive events? Explain.

b. Is this a set of $r = 3$ collectively exhaustive events? Explain.

Section 4.5 Problems

4.20 It is known that the probability is 0.10 that an item from a specific production line is defective. Three items are randomly selected from this line. Determine:

a. The probability that all three are defective.

b. The probability that at least one item is not defective.

4.21 Given events A and B, where $P(A) = 0.4$, $P(B) = 0.3$, and $P(B/A) = 0.5$, find:

a. $P(A/B)$

b. $P(A \cup B)$

4.22 Is $P(A \cup B)$ greater when $P(A \cap B) = 0$ or when $P(A \cap B) > 0$? Discuss.

4.23 If $P(A \cup B) = 0$, can the event $A \cap B$ occur? Discuss.

4.24 A gambler is playing the following game: A coin is tossed and if it comes up heads, he wins $100. If the coin comes up tails, he loses $100. After five tosses of the coin, he has lost $500. Assume that the coin is fair.

a. What is the probability of five straight tails?

b. What is the probability that he will lose $100 on the next coin toss?

4.25 For two events A and B, $P(A) = 0.20$, $P(B) = 0.40$, and $P(B/A) = 0.60$,
a. Determine $P(A \cap B)$.
b. Determine $P(A \cup B)$.

Section 4.6 Problems

4.26 A shipment of four automobile air conditioners is received by an automobile dealer. All four are identical in appearance, but one of them is defective. One unit is selected at random from the four and tested by mechanic A. If she does not find it defective, a second unit is randomly selected from the remaining three and tested by mechanic B. What is the probability that one of the two mechanics will discover the defective unit? Assume that a mechanic always finds a defective part.

4.27 From company records it is known that 80 percent of all applicants pass the American Trust Company's Managerial Trainee Examination on the first attempt. An applicant who fails the exam on the first try has a probability of 0.90 of passing the exam on the second try.
a. Three candidates take the examination for the first time. What is the probability that two will pass? What is the probability that at least one will pass?
b. What is the probability that a candidate fails the exam twice?
c. If a candidate can only take the exam twice, what is the probability of passing the exam in one of the two trials?

4.28 A psychological experiment is designed to test whether or not intelligence of the type measured by IQ tests can be detected from photographs. Photographs of five individuals (whose IQs are known only by the experimenter) were presented to a subject. The subject was asked to arrange the photographs in order according to IQ. Suppose the subject could not detect IQ from photographs.
a. What is the probability that the subject would by chance achieve the correct ordering of the five pictures?
b. What is the probability that this subject would by chance succeed in placing the photograph of the individual having the highest IQ in one of the first two positions?

4.29 Suppose two defective color TV sets have been included in a shipment of six. The buyer begins to test the six TVs one at a time.
a. What is the probability that the second defective is found on the fourth test?
b. What is the probability that no more than four TV sets need to be tested before locating both defectives?

4.30 An article in the newspaper claims that if the probability of destroying an attacking airplane is 0.3 at each of three independent defense barriers, and if the attacking plane has to pass all three barriers to reach its target, the probability that the plane would not reach its target is 0.90. Correct this statement.

4.31 A portable calculator salesperson finds that the probability of selling a unit to a prospective buyer on the first contact is 0.30 but improves to 0.60 on the second contact. The salesperson will not contact a prospective buyer more than twice. If the salesperson contacts Martha, a prospective client, determine:
a. The probability that Martha will buy a calculator.
b. The probability that Martha will not buy a calculator.

4.32 Each of two packages of six flashlight batteries contains exactly two inoperable batteries. If two batteries are selected from each package, what is the probability that all four batteries will operate?

4.33 Herman awakens in the middle of the night with a headache and stumbles into the bathroom without turning on the light. In the dark, he grabs one of three bottles containing pills and takes a pill from the selected bottle. One hour later, he is feeling much worse and recalls that one of the three bottles contained tranquilizing pills and the other two aspirin. Herman quickly consults his handy medical text and finds that 80 percent of normal individuals show his symptoms after taking tranquilizing pills and only 5 percent have these symptoms after taking aspirin. Assume that each of the three bottles had an equal probability of being chosen in the dark.

a. Ignoring Herman's symptoms, what is the probability that he took a tranquilizing pill?
b. Find the probability that Herman took a tranquilizing pill, given his symptoms.
c. Find the probability that Herman took an aspirin, given his symptoms.

4.34 Construct a probability tree to determine $P(A)$ in Example 4.16 in the text, where event A is defined to be, "At least one detector signals the fire."

4.35 Construct a probability tree to determine $P(A)$ in Example 4.17 in the text, where event A is defined to be, "Both trucks purchased by Cabot are defective."

Section 4.7 Problems

4.36 A company has three factories producing the same item. Factory A produces 20 percent of the total output, factory B, 50 percent of the output, and factory C, 30 percent of the total output. The proportion of defectives found at the three factories is 1 percent, 2 percent, and 3 percent, respectively. An item is selected at random from the combined output of the three factories and is found to be defective. What is the probability that the item came from factory B?

4.37 An oil-well drilling company must decide whether or not to drill at a particular site the company has under lease. On the basis of a geological survey, it is decided that there is a 0.45 probability that a formation of type 1 lies beneath the well site, a 0.30 probability that a formation of type 2 lies beneath the site, and a 0.25 probability of a type 3 formation. Records indicate that oil is discovered 30 percent of the time in type 1 formations, 40 percent of the time in type 2 formations, and 20 percent in type 3 formations. What are the following probabilities?
a. If the well produces oil, it comes from a type 1 or type 2 formation.
b. If the well is not productive, it comes from a type 3 formation.

4.38 A compositor makes at least one typesetting error on each page printed with probability 0.01. Assume a six-page brochure is printed.

a. What is the probability that the brochure contains at least one page with at least one typesetting error?
b. What is the probability that exactly three pages contain at least one typesetting error?

4.39 Parts used in manufacturing microcomputers are rated as "acceptable," "unacceptable, but repairable," or "unacceptable—scrap." From recent experience, the probability that a part is acceptable is 0.85, the probability that a part is unacceptable, but repairable is 0.10, and the probability that a part is unacceptable and must be scrapped is 0.05. Five of these parts are randomly selected from the production line.
a. What is the probability that all five are acceptable?
b. What is the probability that at least three parts are acceptable?
c. What is the probability that no more than two of the five parts are unacceptable and must be scrapped?

4.40 A corporation has become very concerned that travel expense statements submitted for executive travel are inaccurate. Specifically, the corporation believes that some of these statements contain expenses that were not incurred by the executive on a business trip. These expense statements are defined as being "fraudulent." The corporation classifies travel into two categories for all business trips: a trip within the continental United States and a trip outside the continental United States. Recent records show that 75 percent of the business trips are made in the continental United States. Further, inspection of a sample of expense statements shows: (1) The proportion of fraudulent statements for trips within the continental United States is 0.05; and (2) the proportion of fraudulent statements for trips not made within the continental United States is 0.25. What is the proportion of fraudulent statements for business trips submitted by executives in this corporation?

4.41 A firm receives a product it uses in manufacturing from three different suppliers: supplier A, supplier B, and supplier C. The probability that a unit of this product is received from each of the three suppliers is as follows:

(supplier A) = 0.40
(supplier B) = 0.30
(supplier C) = 0.30
─────
1.00

The firm is concerned about the number of defective units of this product that it receives from these three suppliers. Based on past experience, the firm estimates the following conditional probabilities (likelihoods):

(unit of product is defective/supplier A) = 0.10
(unit of product is defective/supplier B) = 0.05
(unit of product is defective/supplier C) = 0.01

Given that a sampled unit of the product is defective, what is the probability that it came from supplier A?

4.42 The Acme Corporation determines that their sensitive and heavily protected financial computer data file has been penetrated illegally. The corporation believes that the penetration could have resulted in three ways. These ways and their associated probabilities are given in the table below:

Event	Type of penetration	Probability
A	Use of computer terminal from outside the corporate offices (by telephone modem)	0.40
B	Use of computer terminal from within the corporate offices (direct cable connection)	0.50
C	Corporate computer department employee	0.10
	Total	1.00

The type of penetration is of extreme concern to the corporation. Specifically, the corpora-

tion is concerned with the manipulation of the financial data base (as opposed to merely printing out information from the data base). Label the event, the manipulation of the data base, S. Based on previous experience with illegal entries to its financial data base, the corporation makes the following conditional probability assignments:

$$P(S/A) = 0.20$$
$$P(S/B) = 0.50$$
$$P(S/C) = 0.90$$

For example, $P(S/A)$ represents the probability that the data base has been manipulated, *given* that the access occurred from an outside terminal.

The corporation finds that indeed the data base has been manipulated (event S has occurred). Given that the event S has occurred, what is the probability that event C occurred (a computer department employee committed the illegal penetration)?

Section 4.8 Problems

4.43 If three events, A_1, A_2, and A_3, are mutually exclusive and collectively exhaustive and B is not a null event, show that

$$P(B/A_1)P(A_1) + P(B/A_2)P(A_2) + P(B/A_3)P(A_3) = P(B)$$

using a Venn diagram.

4.44 Using a Venn diagram, develop a general formula for $P(A \cup B \cup C)$.

4.45 An engineer is designing a system based on two components, component A and component B. The probability that component A fails is 0.10, and the probability that component B fails is 0.05. The system requires that both components work for the system to operate as shown below:

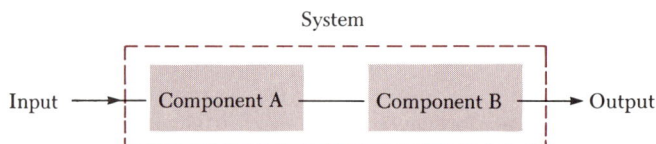

System

Input → Component A → Component B → Output

a. What is the probability that the system operates?
b. Suppose the engineer reconfigures the system as shown below.
   The system will now operate if at least one unit of component A works and if at least one unit of component B works. What is the probability that this system will operate?
c. Compare the two systems in parts a and b. Which produces the higher probability that the system will operate?

System

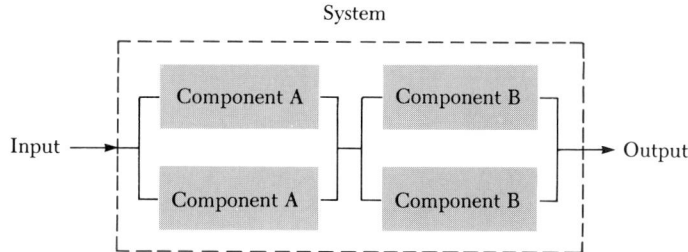

Additional Problems

4.46 Given $P(A) = \frac{1}{3}$, $P(B) = \frac{3}{4}$, and $P(A \cap B) = \frac{1}{6}$
a. Find $P(A \cup B)$.
b. Find $P(B/A)$.
c. Are A and B independent? Explain.

4.47 A survey is taken on drivers in a certain city to determine whether or not they use seat belts regularly. In the survey, data is taken on the age of the driver sampled as well. The contingency table below gives the survey results.

Age	Seat belts		
	Use	Do not use	Total
Younger than 40	10	40	50
40 or older	20	30	50
Total	30	70	100

Suppose that one person is randomly selected from among these 100 surveyed individuals.
a. What is the probability that the person sampled is younger than 40 *and* wears a seat belt?
b. What is the probability that the person sampled is 40 or older *or* wears a seat belt?
c. Given that the person sampled is 40 or older, what is the probability that this person wears a seat belt?
d. Are the events "use a seat belt" and "younger than 40" independent? Explain.

4.48 A family owns a house in town and one at the beach. In any one year, the probability of the town house being burglarized is 0.01, and the probability that the beach house is burglarized is 0.05. Assuming that the two events are independent, what is the probability that, in any given year, both will be burglarized?

4.49 An arts and crafts store decides to record, over its next 100 customer purchases, sales on a cash basis and on a charge basis as a function of the size of the sale. The results are shown in the table on page 172.
a. Given that a purchase is charged, what is the probability that the size of the purchase is more than $100?
b. Given that a purchase is charged, what is the probability that the size of the purchase is less than $25?
c. Determine $P(A_1 \cup B_1)$.
d. Determine $P(A_1 \cap \overline{B}_3)$.
e. Are the compound events "type of purchase" and "size of purchase" independent? Explain.

Size of purchase / Type of purchase	Less than $25 B_1	$25.01–$100 B_2	Over $100 B_3	Totals
A_1 (cash)	25	20	5	50
A_2 (charge)	5	20	25	50
Totals	30	40	30	100

4.50 An insurance agency employs two salespeople. Sally contacts 75 percent of the prospective clients and signs a sales contract with 60 percent of the clients she contacts. Hal contacts 25 percent of the prospective clients and signs 40 percent of those contacted. If a sales contract has just been signed, what is the probability that it was signed by Sally?

4.51 Suppose the last three male customers out of a restaurant have all lost their hat checks so that the checker hands back their hats in random order. What are the following probabilities?

a. No man will get the right hat.
b. Exactly one man will get the right hat.
c. Exactly two men will get the right hats.

4.52 A department store surveyed 200 customer purchases to determine whether or not each purchase was charged, paid for in cash, or paid for by check. Define these events A_1: cash used, A_2: check used, A_3: charged; B_1: purchase less than $10, B_2: purchase more than $10 but less than $50, and B_3: purchase $50 or more. The survey results are:

Payment / Purchase	B_1	B_2	B_3	Totals
A_1	25	15	10	50
A_2	10	40	10	60
A_3	65	15	10	90
Totals	100	70	30	200

If one purchase is selected at random from this survey, determine:

a. $P(B_3)$
b. $P(A_3 \cap B_2)$
c. $P[(A_1 \cup A_2)/B_3]$
d. $P[(A_1 \cup A_3)/B_2]$
e. $P[A_1/(B_1 \cup B_2)]$
f. $P(A_1/B_3)$
g. Are events A_1 and B_1 independent? Why or why not?

4.53 At the AppleWorks Company, 30 percent of the executives are women, 60 percent of the executives have MBAs, and 80 percent of the male executives have MBAs.

a. What is the probability that an executive is a woman and has an MBA degree?

b. Given that the executive is a woman, what is the probability that she has an MBA degree?

4.54 A large corporation is interested in the relationship between terminal degree (B.S., M.S., and Ph.D.) and starting salaries at various jobs within the company. The starting salaries are placed into three ranges: $15,000 to $24,999, $25,000 to $34,999, and $35,000 or more. The starting salaries of 100 employees of the corporation are arranged in a contingency table by terminal degree (highest degree earned):

		Degrees		
Starting salary	B.S. A_1	M.S. A_2	Ph.D. A_3	Total
$15,000 to $24,999 B_1	42	18	0	60
$25,000 to $34,999 B_2	28	2	0	30
$35,000 or more B_3	0	5	5	10
Total	70	25	5	100

If one of the employees is chosen at random from this group, find the following:

a. $P(A_1)$
b. $P(A_1 \cap A_2)$
c. $P(A_1/B_1)$
d. $P(B_3/A_1 \cup A_2)$
e. $P(B_1 \cup B_2/A_2)$
f. $P(\bar{A_3})$
g. Are A_1 and B_1 independent events?
h. Are A_1 and B_3 independent events?
i. Are A_1 and B_1 mutually exclusive events?
j. Are A_1 and B_3 mutually exclusive events?
k. Are B_1 and B_3 mutually exclusive events?

4.55 Mike becomes ill in the evening and calls his family physician. The physician listens carefully to Mike's description of his ailments and determines that he has one of three diseases, denoted by D_1, D_2, and D_3. On the basis of the telephone conversation, the physician assigns the following prior probabilities that Mike has each of the following diseases:

$$P(D_1) = 0.50 \quad P(D_2) = 0.40 \quad P(D_3) = 0.10$$

The physician instructs Mike to meet her at the emergency room of the local hospital.

When he arrives, the physician administers a certain test and finds that the test is positive. Let this event be denoted as follows:

T: Test is positive when administered to a patient.

Based on past experience with the test and the three diseases, the physician chooses the following conditional probabilities:

$$P(T/D_1) = 0.10 \quad P(T/D_2) = 0.10$$
$$P(T/D_3) = 0.90$$

That is, $P(T/D_1)$ is the conditional probability that the test is positive, given that the patient has disease D_1. With this new information (the test is positive when administered to the patient), find the revised (posterior) probabilities that the patient has disease D_1, D_2, or D_3.

4.56 In a certain population of individuals, the probability that an individual in the population has AIDS is 0.20. A new test has been devised for the detection of AIDS. The test is applied to a general population of individuals and it is found to produce a positive result (indicating the presence of the AIDS virus) with a probability of 0.10. The test is also applied to a set of individuals who are known to have AIDS and the test is positive 25 percent of the time. That is, given that a person has AIDS, the test will be positive with a probability of 0.25.

a. Label the events in the above statement and, using your labels and probability notation, write the appropriate probability statements for the three probabilities 0.20, 0.10, and 0.25 in the above statement.

b. What is the probability that an individual in this population has AIDS and the test is positive when administered to this individual?

c. Given that a person is tested in this population and the test proves to be positive, what is the probability that the person has AIDS?

d. Are the events "Person has AIDS" and "The test is positive" independent events? Explain.

e. Are the events "Person has AIDS" and "The test is positive" mutually exclusive events? Explain.

4.57 A large corporation receives express mail deliveries at its main office in New York daily. The express mail is delivered by three firms: Fad Express, Percolator, and Emerald. From recent corporate data, 50 percent of the deliveries are made by Fad Express, 30 percent by Percolator, and 20 percent by Emerald. All three companies guarantee delivery by 11 A.M. Also, based on recent data, given that Fad Express made the delivery, the probability it was late (i.e., arrived after 11 A.M.) is 0.10; given that Percolator made the delivery, the probability that it was late is 0.20; and given that Emerald made the delivery, the probability that it was late is 0.50.

a. What is the probability that an express mail delivery arrives late at the corporate office in New York?

b. Given that an express mail delivery arrives late at the corporate office in New York, what is the probability that Emerald delivered it?

c. Given that an express delivery arrives on time (i.e., at or before 11 A.M.), what is the probability that it was *not* delivered by Fad Express?

4.58 A survey conducted to analyze TV viewing habits turned up the following result concerning the most popular comedy show on TV in 1985 ("The Bill Cosby Show," of course!): The probability of one or more adults in households with children watching the show is 0.50. The probability of children in these households watching the show is 0.60. Given that the children in these households watch the show, the probability that the adults watch the show is 0.80.

a. What is the probability that both the adults and the children watch the show in these households?

b. What is the probability that either the adults or the children watch the show in these households?

c. What is the probability that neither the adults nor the children watch the show in these households?

d. Given that the children do not watch the show in one of these households, what is the probability that the adults do watch the show?

5

Random variables and their distributions

In Chapters 3 and 4, the basic concepts of probability were developed in detail. Although the notion of chance events (e.g., head or tail in the flip of a coin, or winning or losing a bet) is familiar to most of us, the specific elements of probability theory, such as the concepts of independence, mutual exclusiveness, and a statistical experiment, are frequently difficult to grasp on first exposure. Indeed, the student may wonder why it is necessary to spend so much time on probability in an introductory statistics course. In this chapter, we shall answer this query. Probability provides a framework upon which the process of making inferences about a population characteristic from a sampled portion of the population is based.

In a typical population, it is usually possible to identify more than one characteristic of the units that constitute it. For example, suppose the population is the *collection* of all full-time students registered at your university or college during the present academic term. In this instance, it is possible to identify numerous characteristics of the population unit—earned income (if any!), height, weight, sex, hair color, number of parking tickets accumulated during the term, grade point average, and so on. In a statistical study of the units in this population, we may be interested in just one characteristic or in a collection of such characteristics, such as sex and grade point average or earned income and grade point average and so on.

In Chapter 1, we referred to a population characteristic as a *variable* if it can assume one or more values in the population. We must now define

175

variable more specifically when we use the word to mean a measure of a population characteristic.

> **Definition 5.1**
> Random variable
>
> A *random variable* is a numerically valued function defined over a sample space.

Suppose we consider the population of full-time students registered at your university or college. If we are interested in the population characteristic "grade point average" and we select one student at random from this population, then the characteristic may be viewed as a random variable. Its value is clearly numeric (usually measured on the interval between 0.00 and 4.00). In addition, it is a function because it defines a correspondence between members of one set (the elementary outcomes in the sample space which are all the full-time students at your college or university) and members of another set (the set of all possible grade point averages, from 0.00 to 4.00). For each student, the random variable defines one and only one grade point average, although more than one student may have the same grade point average.

To appreciate the concept of a random variable being a function, let us consider an example.

Example 5.1 Suppose the random experiment is tossing a coin twice and we define the random variable **X** as the number of heads in the two tosses. Give the correspondence between members of the experimental elementary outcomes and possible values of the random variable.

Solution Figure 5.1 gives the correspondence between members of the experimental elementary outcomes and possible values of the random variable.

Notice that the random variable **X** is a function. To each member in the first set, the elementary outcomes in the experimental sample space, there corresponds one and only one member in the second set, the values of the random variable. Each value of the random variable may correspond to one or more elementary outcomes, however. Notice that we have written the random variable in functional notation in Figure 5.1 to emphasize its meaning.

Example 5.2 To appreciate more fully the notion that a random variable is a function (that the *value* of the random variable may correspond to one or more elementary outcomes of an experiment, but to each elementary outcome there corresponds only one value of the random variable), consider the die-tossing experiment of Example 3.1. Suppose we wish to establish a

FIGURE 5.1 | Random variable **X** = number of heads in two tosses of a coin

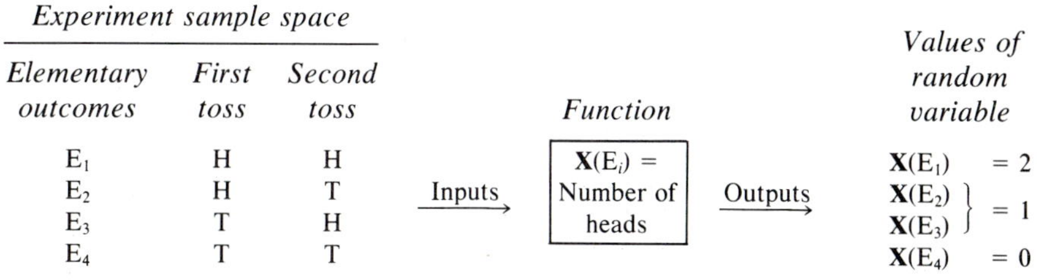

Experiment sample space					*Values of random variable*
Elementary outcomes	*First toss*	*Second toss*	*Function*		
E_1	H	H	$X(E_i) =$ Number of heads	$X(E_1)$ = 2	
E_2	H	T	Inputs →	Outputs →	$X(E_2)$ ⎱ = 1
E_3	T	H			$X(E_3)$ ⎰
E_4	T	T			$X(E_4)$ = 0

correspondence between the elementary outcomes of the experiment and the sum of the number of dots appearing on the top face of each die in the toss. Define the random variable in this experiment, and list the elementary outcomes corresponding to the values the random variable can assume.

Solution The random variable **X** is defined to be the *number* of dots that appear on the top faces of the two dice in a random roll. Thus to each of the elementary outcomes of the experiment, we associate the number of dots on the two top faces. This is the random variable. Values of the random variable and elementary outcomes that produce the values of the random variable are given in Table 5.1.

In most random experiments, we are interested in measuring more than one characteristic of the population. Suppose we randomly select two students from the population of full-time students at your university or college. We are interested in three population characteristics: grade point average,

TABLE 5.1 | Values of the random variable and elementary outcomes corresponding to values of the random variable for the die-tossing experiment

Values of the random variable	*Elementary outcomes*
2	E_{11}
3	E_{12}, E_{21}
4	E_{13}, E_{22}, E_{31}
5	E_{14}, E_{23}, E_{32}, E_{41}
6	E_{15}, E_{24}, E_{33}, E_{42}, E_{51}
7	E_{16}, E_{25}, E_{34}, E_{43}, E_{52}, E_{61}
8	E_{26}, E_{35}, E_{44}, E_{53}, E_{62}
9	E_{36}, E_{45}, E_{54}, E_{63}
10	E_{46}, E_{55}, E_{64}
11	E_{56}, E_{65}
12	E_{66}

number of parking tickets received, and sex. If we define the three random variables as **X** = grade point average, **Y** = number of parking tickets received, and **Z** = sex (1 if male, 0 if female), then the measurements for the three students can be determined by using these functions as illustrated in Figure 5.2.

FIGURE 5.2 | Experiment involving three random variables

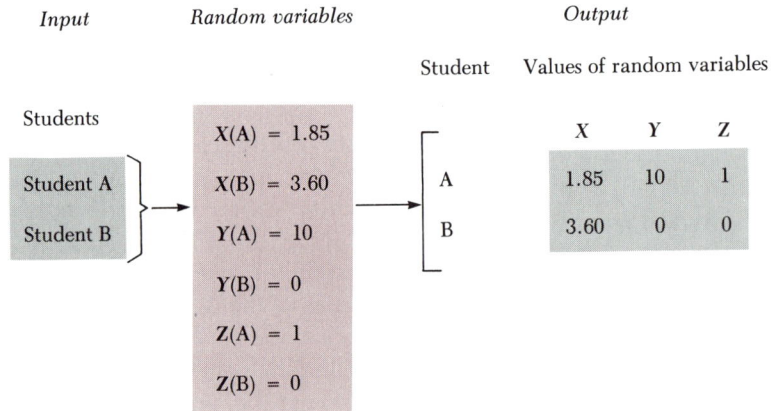

In this experiment, we can think of the sample space as being composed of all possible pairs of students that could be selected from the population of full-time students at your college or university. The random variables **X, Y,** and **Z,** then, establish a functional correspondence between sample space points (pairs of students) and possible values of the three random variables ($0.00 \leq \mathbf{X} \leq 4.00$, $0 \leq \mathbf{Y} \leq ?$, and $\mathbf{Z} = 0, 1$). This may seem like a rather abstract way to say that student A's grade point average is 1.85 (he is in trouble), the number of parking tickets he has received at the time of the experiment is 10 (he is in *big* trouble), and he is a male (what more can be said!). But, we will come to appreciate the need to establish a mathematical foundation for measuring population characteristics, as we have done here.

Note the treatment of the random variable **Z.** We could not let the values of this random variable be represented by "male" and "female" because these values are not numerical and a random variable is a *numerically* valued function. Whenever we are dealing with a qualitative population characteristic such as sex, we must assign to its outcomes numerical values if we are to treat the characteristic as a random variable.

When the probabilities of the elementary outcomes in the experimental sample space are known, it is possible to determine the probabilities of the values of the random variable by using the fact that the random variable is a function that establishes a correspondence between sample space points and the values of the random variable, as the next example illustrates.

Example 5.3 A production lot of 100 transistor radios contains 10 defectives. A retailer decides to select two of the radios randomly and, by extensive testing, determine how many are defective. If neither radio is defective, he will accept the lot. Define the random variable **X** = number of defective radios (**X** = 0, 1, or 2). Determine the probability that the random variable **X** assumes each of its three possible values.

Solution In the experiment of selecting two radios at random, define the elementary outcomes:

R_1: First radio is defective. $\overline{R}_1$: First radio is nondefective.
R_2: Second radio is defective. $\overline{R}_2$: Second radio is nondefective.

The four elementary outcomes of the experiment are given in Table 5.2. The probability of each elementary outcome occurring is determined by using the multiplicative law for events (see Chapter 4). For example,

$$P(R_1 \cap R_2) = P(R_1)P(R_2/R_1) = \frac{10}{100} \cdot \frac{9}{99} = 0.0091$$

Notice that the two events are *not independent;* the outcome of the first selection affects the chance of the second radio being defective. Notice that the probabilities in Table 5.2 sum to 1—we have specified the four mutually exclusive and collectively exhaustive elementary outcomes of the experiment.

TABLE 5.2 | Elementary outcomes and probabilities of the experiment described in Example 5.3

Elementary outcomes	Probability	Value of random variable **X**
1: $R_1 \cap R_2$	$(^{10}/_{100})(^{9}/_{99}) = 0.0091$	2
2: $R_1 \cap \overline{R}_2$	$(^{10}/_{100})(^{90}/_{99}) = 0.0909$	1
3: $\overline{R}_1 \cap R_2$	$(^{90}/_{100})(^{10}/_{99}) = 0.0909$	1
4: $\overline{R}_1 \cap \overline{R}_2$	$(^{90}/_{100})(^{89}/_{99}) = 0.8091$	0

Probability distribution

Since the elementary outcomes are mutually exclusive and collectively exhaustive, $P(\mathbf{X} = 2) = 0.0091$, $P(\mathbf{X} = 1) = 0.0909 + 0.0909 = 0.1818$, and $P(\mathbf{X} = 0) = 0.8091$. The values of the random variable **X** and its probabilities of occurrence are given in Table 5.3. This table represents the *probability distribution* of the random variable **X**—a list of each random variable value and its probability of occurrence.

In order not to confuse a random variable with values that the random variable might assume, we will adopt the notation that random variables are expressed as boldface capital letters, and *values* of the random variable are expressed as lowercase letters. Thus, for example, the random variable "X" will be denoted by **X**, and values that the random variable **X** may assume will be denoted by x. Similarly, $P(\mathbf{X})$ will denote the probability distribution of the random variable **X**, while $P(x) = P(\mathbf{X} = x)$ will denote the probability that the random variable **X** assumes the particular value x. From Example 5.3, $P(1)$ represents the probability that the random variable **X** assumes the value of 1, or $P(1) = 0.1818$. The probability that the retailer will accept the lot is (from Table 5.3) $P(\mathbf{X} = 0) = P(0) = 0.8091$.

TABLE 5.3 | Probability distribution of **X**

x	$P(x)$
0	0.8091
1	0.1818
2	0.0091

The probabilities in Table 5.3 are *relative frequency* probabilities. Since the population is composed of 100 distinguishable units, there are 4,950 possible pairs of different radios that could be drawn from this population[1]— the percentage of the total of 4,950 radio pairs in which both are nondefective is 80.91 percent. Thus, by experimentally forming every different pair of radios that could be randomly drawn from this population, we would find that the proportion of pairs that contained no defectives would be 0.8091.

Most statistical inference methods of inducing information about a population characteristic from a sample of units drawn from the population are based on defining a random variable and determining its probability distribution.

It is important to distinguish between two types of random variables— those that are discrete and those that are continuous. A *discrete* random variable is one for which the number of its values is finite (as in Example 5.3) or countably infinite. By *countably infinite,* we mean that the values of the random variable can be placed in one-to-one correspondence with the positive integers.[2] An example of a random variable that has a countably infinite

---

[1] The number of possible pairs of radios that can be selected from 100 is given by the combination rule of Chapter 3 (Rule 3.3). Since the order of selection is not important and we are selecting 2 items from 100, the number of possible different pairs selected is given by:

$$C_2^{100} = \frac{100!}{2!(100 - 2)!} = 4,950$$

[2] An alternative definition of a discrete random variable is: the set of values is such that there is at most a finite number of values in every finite interval on the real number line.

number of values is a *counting variable,* a random variable with values of 0 and the positive integers 1, 2, . . . , ∞. There are an infinite number of values, but these values are restricted to specific real numbers—the set of positive integers plus 0. The counting random variable typically is appropriate when we are counting the number of occurrences of a particular event in a fixed interval of time, where there is no upper bound on the maximum number of occurrences of the event in the specified time interval.

Definition 5.2
Discrete random variable

A random variable is *discrete* if its set of values is finite or countably infinite in number.

Examples of discrete random variables are the number of parking tickets issued to a student during a term, the number of dots appearing on the top faces of a pair of dice, the number of heads occurring in a specified number of coin flips, and the sampling problem of Example 5.3. In each instance, there is only a finite number of possible outcomes in a simple statistical experiment (selection of a student to determine the number of tickets received, roll of a die, flip of a coin, etc.). Some additional examples of discrete random variables are given in Table 5.4.

TABLE 5.4 | Examples of discrete random variables

Random variable X definition	Values of the random variable	Number of values in the set
1. Let X be the number of trucks available to make a delivery in a firm with a fleet of six trucks.	0, 1, 2, 3, 4, 5, 6	Finite (7)
2. Let X be the number of days per week that an employee works overtime.	0, 1, 2, 3, 4, 5	Finite (6)
3. Let X be the number of persons entering Sears department stores nationally each month.	0, 1, 2, . . .	Countably infinite
4. Let X be the number of telephone mail orders received by a mail order firm per month.	0, 1, 2, . . .	Countably infinite

In the last two examples in Table 5.4, the largest value of the random variables defined can be arbitrarily large, at least conceptually. Thus the sets of values in these two examples are countably infinite. In practice, a countably infinite discrete random variable can be converted to a finite discrete random variable by placing an upper bound on the largest value that the random variable can assume.

Continuous random variable

A *continuous* random variable assumes values that occur over an interval or on an intersection of intervals on the real number line. The number of values that a continuous random variable may assume is infinite. Examples of continuous random variables are the height and weight of individuals and the diameter of ball bearings produced by a certain machine. Although each variable is bounded (for example, the weight of individuals is bounded between 0 and, say, 500 pounds), the variable can assume any of an infinite number of values between these bounds. Other examples of continuous random variables are given in Table 5.5.

TABLE 5.5 | Examples of continuous random variables

Random variable X *definition*	*Range of values of* X
1. Let X be the diameter of ½″ bolts produced by a machine in a machine shop.	$X \geq 0$
2. Let X be the weight of a student in your class.	$X \geq 0$
3. Let X be the amount of rainfall in inches recorded at a weather station on a given day.	$X \geq 0$
4. Let X be the number of ounces of coffee contained in one fill from the school cafeteria vending machine, where the cup can contain at most 8 ounces before it overflows.	$0 \leq X \leq 8$

Definition 5.3
Continuous random variable

A random variable is *continuous* if it can assume all the real number values in an interval on the real line (in theory).

In the following discussion of the probability distribution of a random variable, it is essential to keep in mind the difference between these two basic types of random variables. In the next two sections, our attention will be focused on *discrete* random variables. The discussion of the probability distribution of a *continuous* random variable will be deferred until Sections 5.5 and 5.6.

Before leaving this section, let us focus our attention on the distinction between quantitative and qualitative population characteristics, since qualitative characteristics present a problem when we attempt to associate them with a random variable.

Qualitative charac-
teristics

A difficulty arises because a random variable is a *numerically* valued function, whereas qualitative characteristics by definition are nonnumerical. One way to circumvent this problem is to assign numbers to the categories of the qualitative characteristic. For example, we might let a 2 denote a male and a 1 denote a female if the qualitative characteristic is sex. However, this may create additional problems *unless* the qualitative characteristic can be quantified on a *meaningful* scale. To illustrate this point, consider the quantification of the qualitative characteristic sex: whereas $10,000 has twice the value of $5,000, being a male (2) does not have twice the value of being female (1).

This difficulty also arises, for example, in ratings. We may have four ratings of a new product—excellent, good, fair, and poor. To treat the rating characteristic as a random variable, we may assign the numbers 4, 3, 2, and 1 to the ratings excellent, good, fair, and poor, respectively. But is excellent four times the value of poor and twice the value of fair? Obviously, attention should be paid to these considerations before a qualitative characteristic is quantified. A more realistic quantification in this illustration might be 10, 8, 3, and 1 for the ratings excellent, good, fair, and poor, respectively.

In quantifying a qualitative characteristic, the problem is thus twofold: Categories must be constructed that identify understandable and acceptable gradations of the qualitative characteristic, and a meaningful scale must be identified for its categories. For example, an acceptable gradation of a new product might be excellent, very good, good, fair, poor, and very poor as opposed to the gradations of excellent, good, fair, and poor. Note that this gradation difficulty presumably would not arise in assigning different gradations to the qualitative variable sex—one usually would use two gradations to represent this qualitative random variable! Although our concern is principally with *quantitative* characteristics of the population units, some methods of dealing with *qualitative* factors as random variables will be discussed in Chapters 19 and 20, which describe *nonparametric* statistical procedures.

■ **5.3**

Discrete random variables and their distributions

Consider the problem in Example 5.3: Two radios are sampled from a population of 100 radios in which it is known that 10, or 10 percent of the radios, are defective. As a radio is withdrawn from the population to be examined, it is not replaced, so the probability of a defective unit changes on the second and succeeding trials of the experiment. The experiment is to select randomly two radios and count the number of defectives. The sample space generated by this experiment contains four mutually exclusive and collectively exhaustive events. Table 5.6 summarizes the values that the random variable **X** can assume, the corresponding elementary outcomes of the sample space of the experiment, and the probabilities corresponding to the values of **X**.

TABLE 5.6 | Probability distribution with corresponding elementary outcomes for Example 5.3 (random variable **X**, event R_i = ith radio is defective, event $\bar{R}_i$ = ith radio is nondefective)

Value of random variable **X**	Corresponding elementary outcomes	Probability $P(\mathbf{X} = x)$
2	$(R_1 \cap R_2)$	$P(2) = 0.0091$
1	$(R_1 \cap \bar{R}_2), (\bar{R}_1 \cap R_2)$	$P(1) = 0.0909 + 0.0909 = 0.1818$
0	$(\bar{R}_1 \cap \bar{R}_2)$	$P(0) = 0.8091$

Since the four elementary outcomes in the experiment form a set of mutually exclusive and collectively exhaustive events, we know that the probabilities corresponding to these events sum to 1. Thus the probabilities associated with the values of the random variable—$P(0)$, $P(1)$, and $P(2)$—must also sum to 1. Indeed, the probability distribution of a discrete random variable must have the property that the probabilities associated with the set of mutually exclusive and collectively exhaustive values of the random variable sum to 1.

The probability distribution can be described by a function $P(\mathbf{X})$, called a *probability mass function,* which assigns probabilities to the values of a discrete random variable.

Probability mass function

Definition 5.4

Probability mass function (finite case)

Let the random variable **X** assume a finite number of values, r in total, and denote these values by $x_1, x_2, \ldots , x_r$, where $x_1 < x_2 < x_3 < \cdots < x_r$. Let $P(x_i)$ be the probability that the random variable **X** assumes the value x_i [i.e., $P(\mathbf{X} = x_i)$]. A *probability mass function* is a function that assigns probabilities to the values of a discrete random variable such that the following two conditions on the function $P(\mathbf{X})$ are satisfied:

$$1. \quad 0 \le P(x_i) \le 1, \qquad i = 1, 2, \ldots , r \qquad 2. \quad \sum_{i=1}^{r} P(x_i) = 1$$

The probability mass function, $P(\mathbf{X})$, of a discrete random variable can be stated in three ways. The first method is to list the values of the random variable **X** and the probabilities corresponding to these values $P(x)$ in tabular form as in Table 5.6. Second, it may be possible to write a function in the

---

Notation

Recall from Section 5.2 the notation for a random variable. A random variable is denoted by an uppercase boldfaced letter, such as **X**. A *value* of the random variable **X** is denoted by the corresponding lowercase letter x. Thus $P(\mathbf{X})$ represents the probability mass function of the random variable **X**. $P(x_i)$ represents the probability mass function of **X** *evaluated* at the value x_i. Thus $P(x_i)$ is the probability that the random variable **X** is equal to the specific value x_i.

---

form of an equation to determine each value the random variable may assume. We give such a function in Table 5.7 for determining the probability of selecting 0, 1, or 2 defective radios from a population of 100 radios in which it is known that 10 of them are defective. It is *not* important at this time that

TABLE 5.7 | Three ways of presenting the distribution of the discrete random variable in Example 5.3

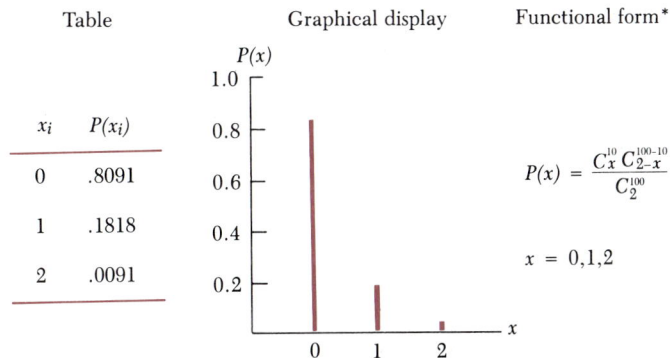

Table		Graphical display	Functional form*
x_i	$P(x_i)$		
0	.8091		$P(x) = \dfrac{C_x^{10}\, C_{2-x}^{100-10}}{C_2^{100}}$
1	.1818		
2	.0091		$x = 0,1,2$

* By substituting the values of 0, 1, and 2 into this equation, the probabilities given in the table on the left are determined. This probability mass function is the probability mass function of a discrete random variable discussed in Section 6.4.

you understand how the equation for determining $P(\mathbf{X})$ is derived in Table 5.7—indeed, it will be the topic of Section 6.4 to develop the probability mass function of this random variable. *It is important* at this time to recognize that it is possible to state the probability mass function of a discrete random variable in the form of an equation for many different discrete random variables. The functional form is an alternative to presenting the values of a discrete random variable along with their probabilities of occurrence in the form of a table or a graph, to be discussed next. It should be pointed out that for some discrete random variables, it is not possible to express the probability distribution in functional form. The distribution can only be given by a stick diagram or by a table in these instances.

Stick diagram

The third way to present a probability mass function is a graphical display called a *stick diagram*. In a stick diagram, the *x*-axis is used for the values of the random variable (**X**) and the *y*-axis is used as the probability scale (0 to 1). At each value of the random variable, say x_i, a "stick" of height $P(x_i)$ is drawn vertically, giving the probability of the occurrence of the value x_i.

In Table 5.7, the three ways of presenting a discrete random variable are illustrated for the random variable described in Example 5.3.

Certainly not all three methods of describing the distribution of a discrete random variable need to be used. In Example 5.3, the tabular form of presentation alone would normally be sufficient.

Example 5.4 Due to revelations made known in a congressional hearing on wiretapping, a certain senator became intensely concerned about securing his office from intruders, "bugs," and the like. At his request, three intrusion-sensing devices were installed in his office (at taxpayers' expense!). The devices work independently of one another. Given an intrusion, the first device will detect it with probability 0.8, the second with probability 0.9, and the third with probability 0.7.

Suppose an intrusion occurs. Let **X** be the number of sensing devices that correctly signal the intrusion. Form the probability distribution of the random variable **X**.

Solution The possible values of the random variable **X** are clearly 0, 1, 2, and 3. Define the events S_i: the *i*th device correctly triggers, and $\bar{S}_i$: the *i*th device fails, where $i = 1$, 2, and 3. The elementary outcomes of the experiment are most easily enumerated by using an outcome tree as illustrated in Figure 5.3.

Since the three sensing devices work independently, the probabilities corresponding to the elementary outcomes are computed by using the corollary of the multiplicative law of probability; for example, $P(E_1) = P(S_1 \cap S_2 \cap S_3) = P(S_1)P(S_2)P(S_3) = (0.8)(0.9)(0.7) = 0.504$.

In Table 5.8, the probability distribution of **X** is formulated by accumulating the probabilities of corresponding elementary outcomes with the various values of the random variable **X**. From the probability distribution, it is clear that the probability that at least one device signals $[P(1) + P(2) + P(3) = 1 - P(0)]$ is 0.994—the senator should feel reasonably safe about discovering an intrusion.

The distribution of the random variable **X** in the form of a stick diagram is given in Figure 5.4A. In this instance, a reasonably simple functional form for $P(\mathbf{X})$ is not obvious. Can you formulate one?

One final note regarding discrete random variables and their distributions: If **X** is a discrete random variable and we wish to determine the probability that **X** is greater than or equal to a but less than or equal to b, where $b \geq a$, we need only to sum the probabilities of the occurrence of the values a through b; that is

$$P(a \leq \mathbf{X} \leq b) = \sum_{x=a}^{x=b} P(x)$$

FIGURE 5.3 | Eight mutually exclusive and collectively exhaustive
elementary outcomes of Example 5.4

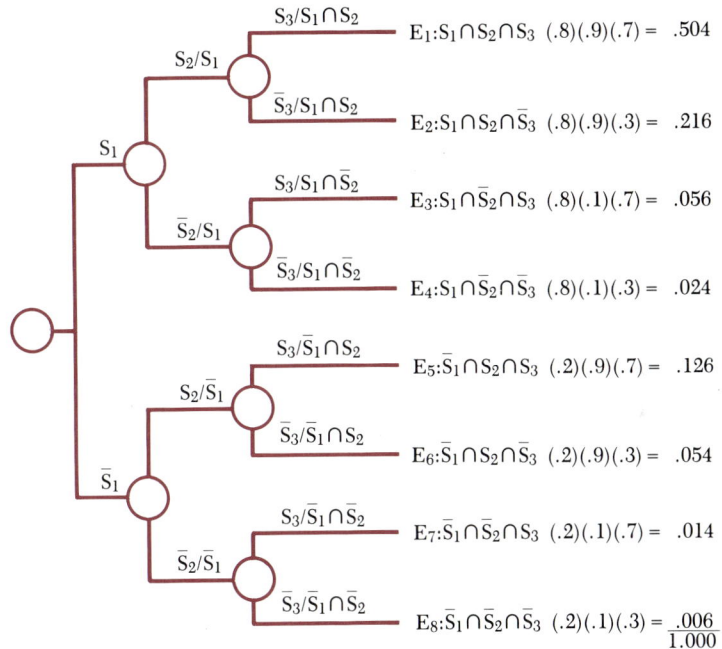

$E_1{:}S_1 \cap S_2 \cap S_3 \ (.8)(.9)(.7) = .504$

$E_2{:}S_1 \cap S_2 \cap \bar{S}_3 \ (.8)(.9)(.3) = .216$

$E_3{:}S_1 \cap \bar{S}_2 \cap S_3 \ (.8)(.1)(.7) = .056$

$E_4{:}S_1 \cap \bar{S}_2 \cap \bar{S}_3 \ (.8)(.1)(.3) = .024$

$E_5{:}\bar{S}_1 \cap S_2 \cap S_3 \ (.2)(.9)(.7) = .126$

$E_6{:}\bar{S}_1 \cap S_2 \cap \bar{S}_3 \ (.2)(.9)(.3) = .054$

$E_7{:}\bar{S}_1 \cap \bar{S}_2 \cap S_3 \ (.2)(.1)(.7) = .014$

$E_8{:}\bar{S}_1 \cap \bar{S}_2 \cap \bar{S}_3 \ (.2)(.1)(.3) = \underline{.006}$
1.000

TABLE 5.8 | Probability distribution of the discrete random variable **X** in
Example 5.4

Values of **X** x_i	Corresponding elementary outcomes E_i	Probabilities $P(x_i)$
0	E_8	0.006
1	E_4, E_6, and E_7	$0.024 + 0.054 + 0.014 = 0.092$
2	E_2, E_3, and E_5	$0.216 + 0.056 + 0.126 = 0.398$
3	E_1	$\underline{0.504}$
		1.000

For instance, in Example 5.4, the probability that **X**, the number of intrusion devices that operate properly, is 1 or more but less than 3 is given by

$$P(1 \le \mathbf{X} < 3) = P(1 \le \mathbf{X} \le 2) = \sum_{x=1}^{x=2} P(x)$$

$$= P(1) + P(2) = 0.092 + 0.398 = 0.490$$

Many important questions concerning a quantitative characteristic of a population can be answered by summing probabilities corresponding to the values of a random variable representing the characteristic. For example, an important special case of summing probabilities is the determination of the *cumulative mass function* of a discrete random variable, given in Definition 5.5. For a discrete random variable **X** with values $x_1 \leq x \leq x_r$, the cumulative mass function, $F(x)$, gives the probability that the random variable **X** assumes the value *x or less*. Thus $F(x_i)$ can be determined by summing the probabilities that **X** $= x_1, x_2, x_3, \ldots, x_i$. Or,

$$F(x_i) = \sum_{x=x_1}^{x=x_i} P(x) = P(x_1) + P(x_2) + \cdots + P(x_i)$$

Cumulative mass function

Definition 5.5

Cumulative mass function of a discrete random variable (finite case)

Let the random variable **X** assume a finite number of values $x_1, x_2, \ldots, x_r$ and have a probability mass function $P(\mathbf{X})$ as delineated in Definition 5.4. The *cumulative mass function* of the random variable **X**, denoted $F(\mathbf{X})$, is the probability that the random variable assumes a value less than or equal to $x = x_i$, where $1 \leq i \leq r$. Mathematically, it is given by:

$$F(\mathbf{X} = x_i) = P(\mathbf{X} \leq x_i) = \sum_{x=x_1}^{x=x_i} P(x)$$

We illustrate the computation of a cumulative mass function of a discrete random variable in the following example.

Example 5.5 Using the data in Example 5.4, determine the values of the cumulative mass function $F(x)$ where the random variable **X** is defined to be the number of devices that correctly signal an intrusion. Graph the resulting values.

Solution Using the data in Table 5.8, we can easily determine the required probabilities. A graph of the resulting cumulative mass function is given in Figure 5.4B, and a list of values of $F(x)$ is given in Table 5.9.

The use of the cumulative mass function of a discrete random variable is required in many of the nonparametric statistical tests described in Chapter 20.

We will not, in this text, develop the countably infinite discrete random variable case. We will, however, use an important discrete mass function in Chapter 6 that is based on a countably infinite discrete random variable. For the development of the countably infinite discrete random variable, the interested reader is directed to the Harnett reference at the end of this chapter.

TABLE 5.9 | Cumulative mass function of the data in Example 5.4

Values of **X** x	P(x)	F(x) = P(**X** ≤ x)
0	0.006	0.006 = P(**X** ≤ 0)
1	0.092	0.098 = P(**X** ≤ 1)
2	0.398	0.496 = P(**X** ≤ 2)
3	0.504	1.000 = P(**X** ≤ 3)

FIGURE 5.4 | Stick diagrams for the probability mass function and the cumulative mass function for the data in Examples 5.4 and 5.5

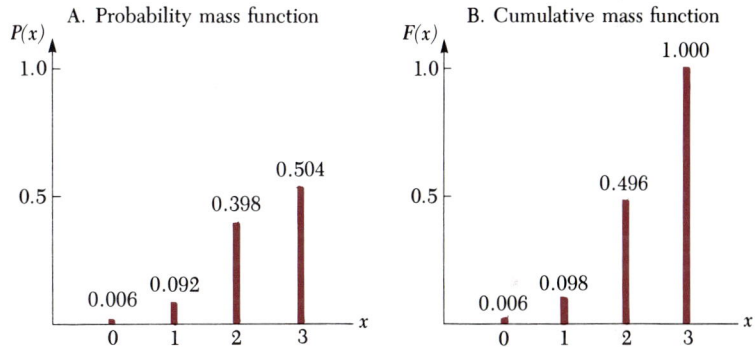

5.4

The mean and variance of a discrete random variable

In Example 5.3, we illustrated the connection between a random variable and a population of values. This relationship is shown in Figure 5.5 for the experiment in Example 5.3: selecting 2 radios at random from a production lot of 100 radios. There are 4,950 possible pairs of radios that can be randomly drawn from the population. In 4,005 of these pairs [(0.8091)(4,950)], neither radio is defective; in 900 of these pairs [(0.1818)(4,950)], one radio is defective; and in 45 of these pairs [(0.0091)(4,950)], both radios are defective. The mean of the population of values of the random variable **X** is, therefore,

$$\frac{(0)(4,005) + (1)(900) + (2)(45)}{4,950} = (0)\left(\frac{4,005}{4,950}\right) + (1)\left(\frac{900}{4,950}\right)$$
$$+ (2)\left(\frac{45}{4,950}\right) = 0.2$$

Thus the average of the 0s, 1s, and 2s is 0.2—on the average, there are $0.2 = \frac{2}{10}$ radio defective in each pair. Notice that 0.2 is 10 percent of 2, the

FIGURE 5.5 | Population set of values of the random variable in Example 5.3

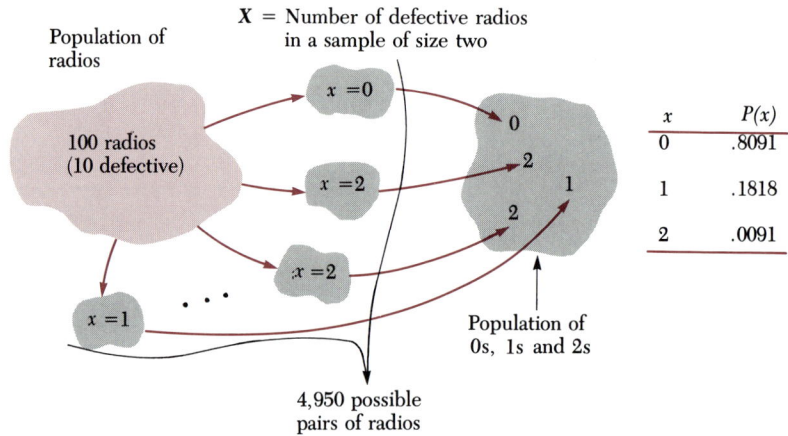

X = Number of defective radios in a sample of size two

Population of radios

100 radios (10 defective)

x = 0

x = 2

x = 2

x = 1

4,950 possible pairs of radios

Population of 0s, 1s and 2s

x	P(x)
0	.8091
1	.1818
2	.0091

sample size. Since there are 10 percent defective radios in the population, it is not surprising to find that the average number of defectives in a sample of size 2 is 10 percent of 2, or 0.2. Also, notice that the average of the 0s, 1s, and 2s can be determined by summing the products of the values of the random variable **X** and their probabilities. This calculation is shown in Table 5.10. The average of the values of a random variable is called the *expected value* of the random variable and is given for a discrete random variable by the formula:

| Expected value of a random variable |

$$E(\mathbf{X}) = \sum_{\substack{\text{all values} \\ \text{of } \mathbf{X}}} xP(x)$$

TABLE 5.10 | Expected value of the random variable in Example 5.3

x	P(x)	xP(x)
0	0.8091	0.0000
1	0.1818	0.1818
2	0.0091	0.0182
Total		0.2000

Expected value of a
discrete random
variable

Definition 5.6
Expected value of a random variable
(discrete case)

Let $\mathbf{X}$ be a discrete random variable with a finite number of values denoted by $x_1, x_2, \ldots, x_r$. The *mean* or *expected value* of the random variable, denoted by $E(\mathbf{X})$ or, equivalently by μ, is given by

Notation: $\mu = E(\mathbf{X})$

$$\mu = E(\mathbf{X}) = \sum_{i=1}^{r} x_i P(x_i)$$

Since $E(\mathbf{X})$ represents the mean of the population of values of the random variable $\mathbf{X}$, it is common to use μ, the population mean symbol, to represent the expected value of $\mathbf{X}$.

We have shown that by using the formula for $E(\mathbf{X})$ we can determine the mean of the values of the random variable for the random variable given in Example 5.3. Indeed, this formula simply "weights" the 0s, 1s, and 2s with their relative frequencies of occurrence: 0.8091, 0.1818, and 0.0091, respectively. Another interpretation of $\mu = E(\mathbf{X})$ is the center of gravity of the distribution of the random variable $\mathbf{X}$. Figure 5.6 shows the distribution of the Example 5.3 random variable in a stick diagram. Notice that the "balancing" point (the center of gravity) occurs at the value of $E(\mathbf{X})$ on the x-axis.

FIGURE   5.6 | The stick diagram for the random variable $\mathbf{X}$ in the Example 5.3 problem showing $E(\mathbf{X})$ as the center of gravity of the distribution.

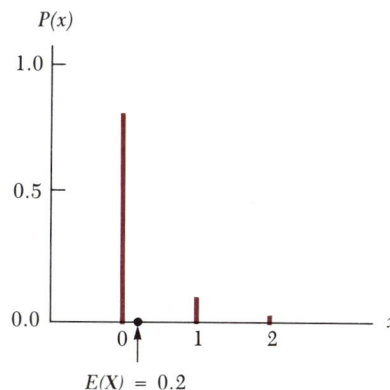

$E(X) = 0.2$

In an analogous manner, we can measure the variance of the values of a random variable—in this case, the variability of the 0s, 1s, and 2s.

Variance of a random variable

> **Definition 5.7**
> Variance of a random variable
> (discrete case)
>
> Let **X** be a discrete random variable with a finite number of values denoted by $x_1, x_2, \ldots, x_r$. The *variance* of the random variable **X**, denoted by $V(\mathbf{X})$ or, equivalently, by σ^2, is given by:
>
> $$\sigma^2 = V(\mathbf{X}) = \sum_{i=1}^{r} [x_i - E(\mathbf{X})]^2 P(x_i) = \sum_{i=1}^{r} x_i^2 P(x_i) - [E(\mathbf{X})]^2$$

Notation:
$\sigma^2 = V(\mathbf{X})$

The second form of $V(\mathbf{X})$ in Definition 5.7 is the "computing formula"—giving two expressions for $V(\mathbf{X})$ is similar to giving two expressions for the variance in Chapter 2: The first is a definitional form and the second is the computing form.

Example 5.6 Compute the variance of the random variable given in Example 5.3.

Solution The easiest way to compute the variance of a discrete random variable is by using a table similar to Table 5.11.

TABLE 5.11 | Partial computation of the variance of the random variable **X** described in Example 5.3

x_i	$P(x_i)$	x_i^2	$x_i^2 P(x_i)$
0	0.8091	0	0.0000
1	0.1818	1	0.1818
2	0.0091	4	0.0364
Total			0.2182

From Table 5.10, $E(\mathbf{X}) = 0.2$. Thus

$$V(\mathbf{X}) = \sum_{i=1}^{3} x_i^2 P(x_i) - [E(\mathbf{X})]^2$$
$$= 0.2182 - (0.2)^2 = 0.1782$$

The variance of **X** is 0.1782, and the standard deviation of **X** is $\sqrt{0.1782} = 0.422$. The population of 0s, 1s, and 2s has a standard deviation of 0.422.

Example 5.7 Compute the expected value, the variance, and the standard deviation of the random variable described in Example 5.4.

Solution In this problem, the random variable **X** is the number of sensing devices that correctly signal an intrusion into the senator's office. The computation of $E(\mathbf{X})$ and part of $V(\mathbf{X})$ is shown in Table 5.12. The expected value of **X** is $E(\mathbf{X}) = 2.4$, and the variance of **X** is

$$V(\mathbf{X}) = 6.220 - (2.4)^2 = 6.22 - 5.76 = 0.46$$

The standard deviation of **X** is thus $\sqrt{0.46} = 0.678$.

TABLE 5.12 | Computation of $E(\mathbf{X})$ and $V(\mathbf{X})$ in Example 5.7

Value of **X** x_i	Probability $P(x_i)$	$x_i P(x_i)$	x_i^2	$x_i^2 P(x_i)$
0	0.006	(0)(0.006) = 0.000	0	(0)(0.006) = 0.000
1	0.092	(1)(0.092) = 0.092	1	(1)(0.092) = 0.092
2	0.398	(2)(0.398) = 0.796	4	(4)(0.398) = 1.592
3	0.504	(3)(0.504) = 1.512	9	(9)(0.504) = 4.536
		2.400		6.220

Therefore, in repeated trials of the experiment (each trial is an intrusion into the office), on the average 2.4 sensors will correctly identify the intrusion. The standard deviation of the collection of 0s, 1s, 2s, and 3s is 0.678 units.

In Chapter 2, we introduced the Chebychef theorem as a method of interpreting the standard deviation. An alternative statement in the context of random variables is:

Chebychef theorem for random variables

Let **X** be a random variable. Then for $k \geq 1$,

$$P(\mu - k\sigma \leq \mathbf{X} \leq \mu + k\sigma) \geq 1 - \frac{1}{k^2}$$

For example, if $k = 2$ is used in the theorem, then from the theorem, at least $1 - (1/2)^2 = 3/4$ of the possible values of **X** will be within $k = 2$ standard deviations of its mean. In Example 5.7, the standard deviation of the random variable **X** is $\sigma = 0.678$. The mean μ of the random variable **X** is 2.4.

Therefore, if we want an interval that will contain at least 3/4 of the possible values of the random variable **X**, then we have:

$$\mu \pm 2\sigma \rightarrow 2.4 \pm 2(0.678) \text{ or } 1.044 \text{ to } 3.756$$

or

$$P(\mu - 2\sigma \leq \mathbf{X} \leq \mu + 2\sigma) \geq 1 - (1/2)^2 = 3/4$$

In fact, 90.2 percent of the values lie in this interval (the interval contains the values 2 and 3 of **X**). The Chebychef theorem provides a lower bound (3/4 in this case) for the number of values of **X** in the interval. In this instance, the actual number of values within the interval (90.2 percent) is appreciably higher than this lower bound. But the use of the theorem together with the expected value and standard deviation of a random variable can shed much light on the nature of a discrete probability distribution when the distribution mass function is not completely known.

5.5

Continuous random variables and their distributions

We have thus far been concerned with random variables that can assume a finite or countably infinite number of values. In many applications of statistics, the random variable can assume *any* value on an interval of the real number line. Examples are measurements such as height, weight, length, and speed. In these cases, the random variable is called *continuous* because it can take on an infinite number of values that are not countable.

A continuous random variable must be dealt with in a slightly different way than a discrete random variable for reasons that we shall now make clear. Suppose that a machine is producing ball bearings for gyroscopes in space vehicles and that these ball bearings are intended to be exactly 0.5 inch in diameter. Due to flaws in the composition of the bearings, changes in temperature and humidity in the production cycle, and faults related to the fine-tuning of the machine itself, very few bearings produced will measure exactly 0.5 inch in diameter. As a means of quality control, random samples are selected from production runs, and the bearings in these samples are measured by a caliper to an accuracy of 0.001 inch. Let us suppose that 10 bearings are randomly sampled, the diameters are measured, and the histogram shown in Figure 5.7A is formed from the data. Six classes have been chosen to form the histogram in Figure 5.7A. Now suppose that another 10 ball bearings are added to the sample. The histogram formed from using the 20 measurements is shown in Figure 5.7B. A smooth curve is drawn through the midpoints of the tops of the histogram rectangles to accentuate the form of the distribution. Naturally, as more bearings are used to form the random sample, more divisions will appear on the *x*-axis of the histogram, each representing the diameter of a selected ball bearing. If the sample is increased to include 100 ball bearings, a histogram formed from the resulting diameter measurements and by using more classes may appear as in Figure 5.7C.

There are two important features of the histograms in Figure 5.7A, B, and C that deserve special attention. First, as the sample size increases, many

FIGURE 5.7

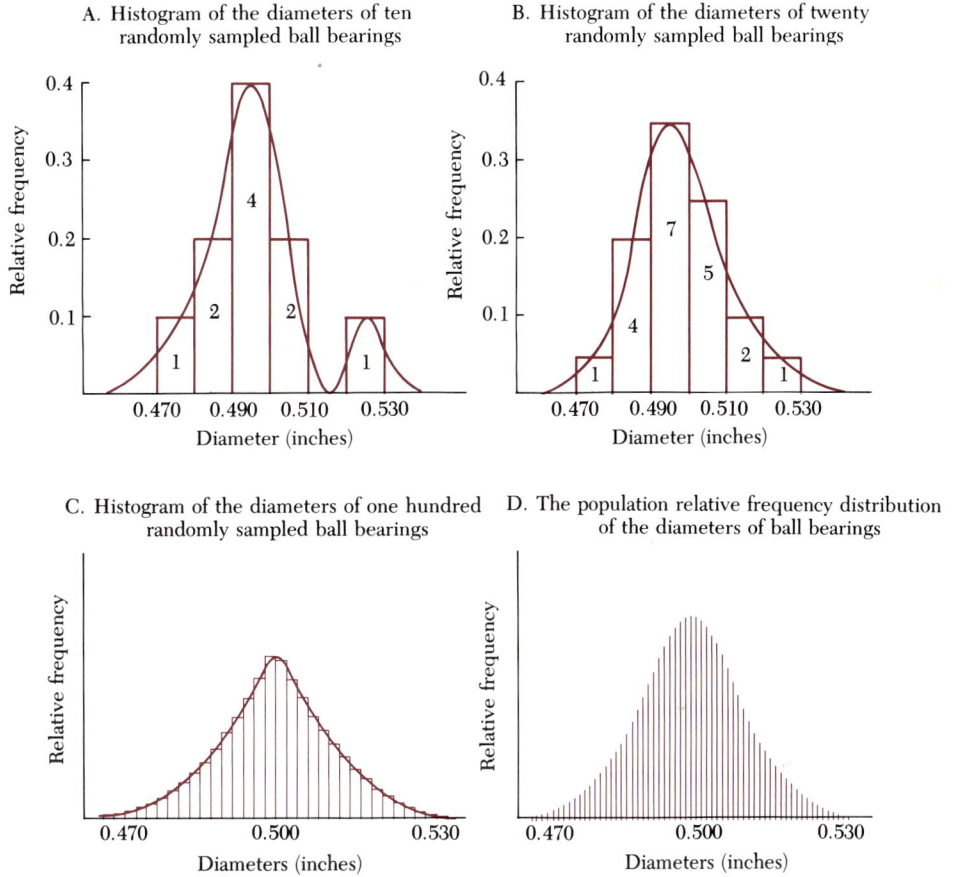

A. Histogram of the diameters of ten randomly sampled ball bearings

B. Histogram of the diameters of twenty randomly sampled ball bearings

C. Histogram of the diameters of one hundred randomly sampled ball bearings

D. The population relative frequency distribution of the diameters of ball bearings

more different values of the diameter will occur. Thus the set of marks on the histogram axis will more densely cover the interval from 0.470 to 0.530 as the number of sample ball bearings increases. We are assuming here that 0.470 and 0.530 are the minimum and maximum possible diameters that the machine can produce, respectively. Second, as the sample size increases, the values begin to cluster more densely about a central value. If the number of classes used to form the histogram is increased, the width of the histogram rectangles becomes smaller, and the midpoints on the tops of the rectangles more closely follow a smooth, continuous curve.

Suppose that we now denote the diameter of the ball bearings by the random variable **X** and continue to take larger and larger samples of ball bearings from the production line. Intuitively, it is reasonable to assume that the larger the sample, the more closely the sample frequency distribution and histogram will resemble the population frequency distribution (the probability distribution of the random variable **X**) and its stick diagram. Figure 5.7D shows the population frequency distribution (stick diagram) for the diameters of all ball bearings produced on the machine. Since the caliper can only measure with an accuracy of 0.001 inch, the population values are 0.470, 0.471, 0.472, . . . , 0.529, and 0.530 inch. This is a set of finite values (61 in all), so Figure 5.7D is the probability distribution of the *discrete* random variable **X**, where its values are given on the *x*-axis of the graph and the probabilities that **X** assumes these values are given on the *y*-axis.

The random variable **X** = diameter of ball bearings is, in fact, however, a *continuous* random variable because its values can occur *anywhere* on the real number line from 0.470 to 0.530. The values of **X** are restricted to the finite set above 0.470, 0.471, . . . , 0.529, and 0.530, in this case since our caliper can only measure with an accuracy of 0.001 inch. If the measuring instrument were accurate to 0.00001 inch, then many more values of **X** would be possible and many sticks would occur in the gaps between the sticks in Figure 5.7D. Taken to the limit, if our measuring device is "infinitely" accurate, then the probability distribution of **X** may be represented as a dense collection of sticks as shown in Figure 5.8.

FIGURE   5.8 | Probability distribution of the random variable **X** = diameter of ball bearings

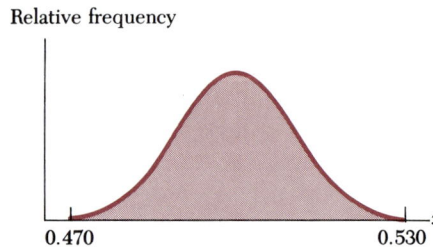

The analogy to the collection of sticks in Figure 5.7D for the discrete random variable **X** is the shaded area under the curve, denoted by $f(x)$, in Figure 5.8. Since the sum of the heights of the sticks in Figure 5.7D must be 1, by analogy, the area under $f(x)$ must also be 1 in Figure 5.8. This leads us to a formal definition of the probability distribution of a continuous random variable.

Notice the similarity between the conditions for $P(\mathbf{X})$ to be a probability mass function of a discrete random variable given in Definition 5.4 and the conditions placed on $f(\mathbf{X})$ to be a probability density function for a continuous random variable given in Definition 5.8. In the discrete case, probability is *massed* at the discrete values of the random variable, whereas in the

<table>
<tr><td>Probability density
function</td><td>

Definition 5.8

Probability density function (continuous case)

Let **X** be a continuous random variable defined over an interval of the real number line from a to b as illustrated in Figure 5.9. The *probability density function* of **X**, denoted by $f(\mathbf{X})$, must satisfy two conditions:

1. $f(x) \geq 0,\ a \leq x \leq b$
2. The area under $f(x)$ from $x = a$ to $x = b$ must be 1.

</td></tr>
</table>

FIGURE 5.9 | Probability density function $f(x)$ of a continuous random variable

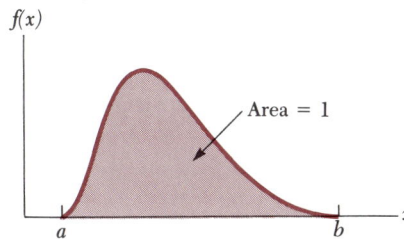

continuous case, the probability is spread *densely* over the range of values of the random variable. In the discrete case, the probability sticks must sum to 1, whereas in the continuous case, the dense set of sticks [the area under the function $f(x)$ from a to b] must have an area of 1.

Recall that if we want to compute $P(c \leq \mathbf{X} \leq d)$ where $d \geq c$ for a discrete random variable, we need only sum the probabilities of **X** taking on the values c through d; that is:

$$P(c \leq \mathbf{X} \leq d) = \sum_{x=c}^{x=d} P(x); \qquad d \geq c$$

The analogy to a continuous random variable is straightforward. The probability that a continuous random variable **X** takes on a value between c and d is equal to the *area* under $f(x)$ between c and d, as illustrated in Figure 5.10.

There is *one* significant difference in computing probabilities for a discrete and a continuous random variable. If the discrete random variable **X** can assume the value e, then the probability that **X** does assume this value is $P(e)$, the height of the stick over e on the stick diagram for **X**. However, if e

FIGURE 5.10 | Area under the curve corresponding to $P(c \le \mathbf{X} \le d)$

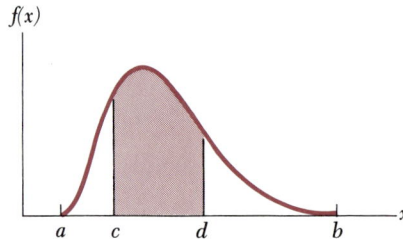

resides within the defined interval for a continuous random variable $\mathbf{X}$, the probability that $\mathbf{X}$ assumes the value of e is 0; that is, $P(\mathbf{X} = e) = 0$. This is true regardless of the numerical value of e over the defined interval of real numbers for the random variable $\mathbf{X}$. *Thus $P(\mathbf{X} = e)$ is not equal to $f(e)$, the height of the curve at the point $x = e$.*

The reason for this may be argued as follows. Let us assume that the continuous random variable $\mathbf{X}$ is defined for values on the real number line from $x = a$ to $x = b$. Assume that e is in the interval from a to b, and let $P(\mathbf{X} = e) = f(e)$. Since we have defined the probability that $\mathbf{X}$ assumes the value of e in the interval a to b in this manner, this definition must also hold for all other values, say $e_1, e_2, e_3, \ldots$ in the interval from a to b. There are an infinite number of values (not countable) between a and b, so we can see that the sum of the "probabilities" $f(e) + f(e_1) + f(e_2) + f(e_3) + \cdots$ will quickly exceed 1, which means that any particular number, say $f(e_2)$, can no longer be interpreted as a probability. Indeed, any one of the $f(e_i)$ values may exceed 1 by itself if the height of the curve at the point $\mathbf{X} = e_i$ is greater than 1.

Another way to look at this is to write $P(\mathbf{X} = e)$ as $P(e \le \mathbf{X} \le e)$ and use the definition for the probability that $\mathbf{X}$ assumes a value between two points, say c and d, as illustrated in Figure 5.10. Since there is no *area* between e and e, $P(\mathbf{X} = e) = P(e \le \mathbf{X} \le e) = 0$.

The student who has had calculus should begin to recognize some familiar notions at this point. First, the area under a continuous curve, say $f(x)$, from c to d can be determined by *integration* between the limits of c to d. Second, the condition that the area under the curve $f(x)$ from $x = a$ to $x = b$ (where the random variable is defined over the range of values, $a \le \mathbf{X} \le b$) must equal 1 can be concisely stated as: The integral of $f(x)$ from $x = a$ to $x = b$ must equal 1.

It is unnecessary to know calculus to understand the underlying concept of a probability distribution of a continuous random variable. Tables in Appendix B at the end of the text perform the integration required to determine $P(c \le \mathbf{X} \le d)$ for *all* the continuous random variables treated in the text. However, it is important to understand the basic differences between discrete and continuous random variables and that—although it is always possible (theoretically, at least) to determine the probability that a discrete random variable assumes a particular value—the probability that a continuous random variable assumes a particular value is *always* 0, regardless of the value of the random variable or the particular probability density function. *Always!* Table 5.13 summarizes the similarities and the differences between the two types of random variables.

TABLE 5.13 | Comparison of properties of discrete and continuous random variables

Property	Discrete random variable	Continuous random variable
Number of values of the random variable	Finite *or* countably infinite	Infinite (not countable)
$P(c \leq \mathbf{X} \leq d)$	$\sum_{x=c}^{x=d} P(x)$	Area under the curve $f(x)$ from $x = c$ to $x = d$
$P(\mathbf{X} = e)$	$P(e)$	*Always zero*

Example 5.8 Consider the probability density function

$$f(x) = \begin{cases} 1 & 0 \leq x \leq 1 \\ 0 & x < 0 \text{ or } x > 1 \end{cases}$$

(This density function might arise in a practical situation when the random variable **X** is representing service times that are assumed to be uniformly distributed over the interval from 0 to 1—measured in time units—for example.)

a.   Demonstrate that $f(x)$ is a probability density function.
b.   Determine $P(\mathbf{X} \leq 0.50)$.
c.   Determine $P(0.25 \leq \mathbf{X} \leq 0.60)$.
d.   Determine $P(\mathbf{X} > 0.75)$.
e.   Determine $P(\mathbf{X} \geq 0.75)$.

Solution The probability distribution of the random variable **X** is illustrated in Figure 5.11.

a.   For $f(x)$ to be a probability density function, the area under $f(x)$ must be 1, and $f(x) \geq 0$ for all x. Since $f(x)$ is equal to 1 for $0 \leq x \leq 1$ and 0 elsewhere, we only have to consider the area under $f(x)$ from 0 to 1. This area is represented by a square with height $= 1$ and base $= 1$. The area of a square is (height)(base) $= (1)(1) = 1$. Also, for all x values on the real line, $f(x) \geq 0$.

FIGURE 5.11 | The probability density function for the Example 5.8
random variable

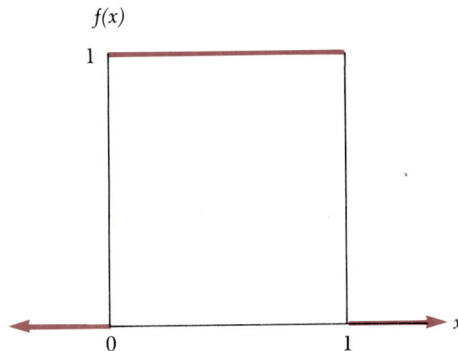

$f(x)$

b. To find $P(\mathbf{X} \leq 0.50)$, we must find the area under $f(x)$ from 0 to 0.50. Since the area of a rectangle equals base times height $\{(b)(h)\}$, this area is $(0.50)(1) = 0.50$.

$$P(\mathbf{X} \leq 0.50) = 0.50$$

c. To determine $P(0.25 \leq \mathbf{X} \leq 0.60)$, we again use the fact that the area of a rectangle $= (b)(h)$. In this case the length of the base is $(0.60 - 0.25) = 0.35$. Then the area $= (0.35)(1) = 0.35$.

$$P(0.25 \leq \mathbf{X} \leq 0.60) = 0.35$$

This area is illustrated in Figure 5.12.

d. For $P(\mathbf{X} > 0.75)$, the length of the base of the rectangle is $(1.00 - 0.75) = 0.25$. Therefore, the area of the rectangle is $(0.25)(1) = 0.25$.

$$P(\mathbf{X} > 0.75) = 0.25$$

This area is illustrated in Figure 5.12.

e. Since $P(\mathbf{X} = 0.75) = 0$, $P(\mathbf{X} \geq 0.75) = P(\mathbf{X} > 0.75) = 0.25$ from part *d*.

The definition of the cumulative density function of a continuous random variable is given in Definition 5.9 below. The cumulative density function is the continuous variable analogue of the cumulative mass function for a discrete random variable given in Definition 5.5.

Definition 5.9
Cumulative density function of a continuous random variable

Let the random variable **X** assume an infinite number of values on the real number line and have a probability density function $f(x)$ as delineated in Definition 5.8. The *cumulative density function* of the random variable **X**, denoted $F(\mathbf{X})$, is the probability that the continuous random variable assumes a value less than or equal to x. Mathematically, it is given by:

$$F(x) = P(\mathbf{X} \le x)$$

Cumulative density function

FIGURE 5.12 | The probability density function in Example 5.8 with the part c and d probabilities indicated

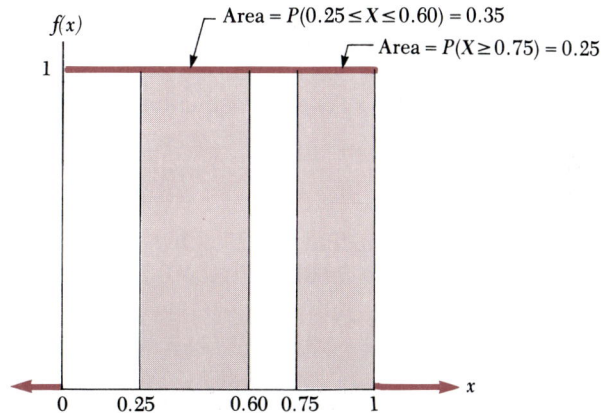

Example 5.9 The cumulative density function $F(x)$ for the probability density function $f(x)$ in Example 5.8 is given by:

$$F(x) = \begin{cases} 0 & x < 0 \\ x & 0 \le x \le 1 \\ 1 & x > 1 \end{cases}$$

The cumulative density function $F(x)$ is illustrated in Figure 5.13.

FIGURE 5.13 | The cumulative density function for the Example 5.8 probability density function

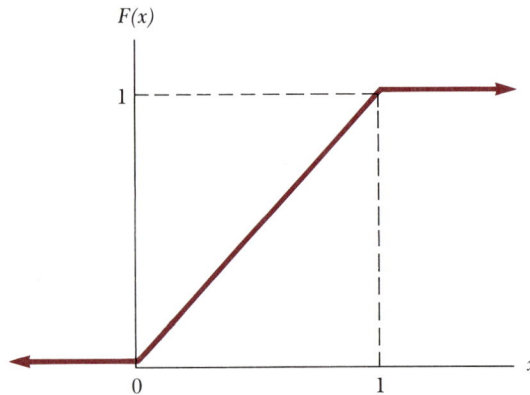

a. Using $F(x)$, find $P(\mathbf{X} \le 0.50)$.
b. Using $F(x)$, find $P(0.25 \le \mathbf{X} \le 0.60)$.
c. Using $F(x)$, find $P(\mathbf{X} > 0.75)$

   Solution Notice that these probabilities have already been determined in Example 5.8 above.

a. Since $F(x)$ gives the cumulative probabilities from the left,

$$P(\mathbf{X} \le 0.25) = F(0.25) = 0.25$$

b. We can find $P(0.25 \le \mathbf{X} \le 0.60)$ by noting that this probability is equal to $P(\mathbf{X} \le 0.60) - P(\mathbf{X} \le 0.25)$. Therefore

$$P(0.25 \le \mathbf{X} \le 0.60) = F(0.60) - F(0.25)$$
$$= 0.60 - 0.25 = 0.35$$

c. Finally, $P(\mathbf{X} > 0.75) = 1 - P(\mathbf{X} < 0.75)$. Therefore

$$P(\mathbf{X} > 0.75) = 1 - F(0.75) = 1 - 0.75 = 0.25$$

■ **5.6**

The mean and variance of a continuous random variable (advanced section)

Just as a discrete random variable has a mean and a variance, so does a continuous random variable. These are denoted by $E(\mathbf{X})$ and $V(\mathbf{X})$, respectively, just as they are in the discrete case. The definitions for the expected value and the variance of a continuous random variable are given in Definition 5.10.

Definition 5.10
Expected value and variance of a continuous random variable

Let the probability density function of the random variable **X** be given by $f(x)$. The expected value of **X**, denoted either by $E(\mathbf{X})$ or by μ is given by

$$\mu = E(\mathbf{X}) = \int_{-\infty}^{+\infty} xf(x)dx$$

The variance of **X**, denoted by either $V(\mathbf{X})$ or by σ^2 is given by

$$\sigma^2 = V(\mathbf{X}) = \int_{-\infty}^{+\infty} (x - \mu)^2 f(x)dx = \int_{-\infty}^{+\infty} x^2 f(x)dx - \mu^2$$

The standard deviation of **X**, denoted by σ, is given by $\sigma = +\sqrt{\sigma^2}$.

The formulas for $E(\mathbf{X})$ and $V(\mathbf{X})$ for a continuous random variable are analogous to the formulas for $E(\mathbf{X})$ and $V(\mathbf{X})$ for a discrete random variable—the summing operator Σ has been replaced by the integration operator $\int$.

Example 5.10 Consider the following probability density function:

$$f(x) = \begin{cases} 0 & x < 0 \\ 2x & 0 \le x \le 1 \\ 0 & x > 1 \end{cases}$$

Find the expected value and the variance of this random variable.

Solution From Definition 5.10,

$$\mu = E(\mathbf{X}) = \int_0^1 xf(x)dx = \int_0^1 x(2x)dx = \int_0^1 2x^2 dx$$

$$= 2x^3/3 \Big|_0^1 = (\tfrac{2}{3}) - (0) = \boxed{\tfrac{2}{3}}$$

$$\sigma^2 = V(\mathbf{X}) = \int_0^1 (x - \mu)^2 f(x)dx = \int_0^1 [x - (\tfrac{2}{3})]^2 f(x)dx$$

$$= \int_0^1 x^2 f(x)dx - \int_0^1 (\tfrac{4}{3})xf(x)dx + \int_0^1 (\tfrac{4}{9})f(x)dx$$

$$= \int_0^1 2x^3 dx - \int_0^1 (\tfrac{8}{3})x^2 dx + \int_0^1 (\tfrac{8}{9})x dx$$

$$= x^4/2 \Big|_0^1 - (\tfrac{8}{9})x^3 \Big|_0^1 + (\tfrac{4}{9})x^2 \Big|_0^1$$

$$= [(\tfrac{1}{2}) - (0)] - [(\tfrac{8}{9}) - (0)] + [(\tfrac{4}{9}) - (0)]$$

$$= \boxed{\tfrac{1}{18}}$$

Notice that the ranges of integration go from 0 to 1, since $f(x) = 0$ for values of x such that $x < 0$ or $x > 0$. Thus the mean of this random variable is $\frac{2}{3}$ and the variance is $\frac{1}{18}$.

Appendix A in the back of the book presents the formulas for the mean and variance of a continuous random variable, repeating Definition 5.10 above, and provides additional examples, including the mean and variance of the probability density function in Example 5.8.

■ 5.7
Mathematical expectation

Computing an expected value is a much more general tool than simply a means of finding the mean or the variance of a random variable. Let us review the process of determining the expected value of a discrete random variable. Suppose the discrete random variable **X** has the probability mass function $P(\mathbf{X})$ given in Table 5.14. The computation of $E(\mathbf{X}) = \Sigma x P(x)$ is also

TABLE 5.14 | Probability mass function of the discrete random variable **X** and the computation of its expected value $E(\mathbf{X})$

x	$P(x)$	$xP(x)$
0	0.2	$(0)(0.2) = 0.0$
1	0.3	$(1)(0.3) = 0.3$
2	0.5	$(2)(0.5) = 1.0$
	1.0	$E(\mathbf{X}) = 1.3$

shown in Table 5.14. The probability distribution of **X** tells us that 20 percent of the values of **X** are 0s, 30 percent are 1s, and 50 percent are 2s among the population of all values of **X** generated by repeated trials of the experiment. As noted earlier, by forming the products of the value of the random variable and its relative frequency in this conceptualized population and by adding these products together for each and every value of **X**, we are forming the population mean value—the mean of the random variable **X**. Thus the values of **X**, weighted by their respective relative frequencies, give us the average of **X**. We can extend this notion to form the expected value of *any* function of the random variable **X**, as Definition 5.11 describes.

First, we note that if $h(\mathbf{X}) = \mathbf{X}$, then $E[h(\mathbf{X})]$ represents the expected value of **X**. Recall the definition of the variance, $V(\mathbf{X})$, of a discrete random variable (Definition 5.7):

$$V(\mathbf{X}) = \sum_{\text{all } \mathbf{X}} [x - E(\mathbf{X})]^2 P(x)$$

Thus, if $h(\mathbf{X}) = [\mathbf{X} - E(\mathbf{X})]^2$, then $E[h(\mathbf{X})]$ represents the variance of the random variable **X**. The variance is, therefore, the *average* of the squared deviations of the values of the random variable from the mean or expected value.

Expected value of a
function of a ran-
dom variable

Definition 5.11
Expected value of a function of a random variable
(discrete case)

Let X be a discrete random variable with probability mass function $P(X)$, and consider a function of X, denoted by $h(X)$. The *average* or *expected value* of $h(X)$ is given by

$$E[h(X)] = \sum_{\text{all } X} h(x)P(x)$$

where the symbol $\sum\limits_{\text{all } X}$ means that the summation process is to be conducted over all values of the random variable X.

Example 5.11 Consider the distribution of the random variable described in Example 5.4.

x	$P(x)$
0	0.006
1	0.092
2	0.398
3	0.504
	1.000

Determine the expected value of $h(X)$, where $h(X)$ is equal to X, $[X - E(X)]^2$, and $2X$.

Solution The computations of the required expectations are shown in Table 5.15. Note that the solutions to the first two parts represent the mean

TABLE 5.15 | Computation of $E(X)$, $E[X - E(X)]^2$, and $E(2X)$

x	$P(x)$	$xP(x)$	$x - E(X)$	$[x - E(X)]^2$	$[x - E(X)]^2P(x)$	$2xP(x)$
0	0.006	0.000	−2.4	5.76	0.03456	0.000
1	0.092	0.092	−1.4	1.96	0.18032	0.184
2	0.398	0.796	−0.4	0.16	0.06368	1.592
3	0.504	1.512	+0.6	0.36	0.18144	3.024
	1.000	2.400			0.46000	4.800

and the variance of the random variable **X**, which we previously found in Table 5.12. Thus $E(\mathbf{X}) = 2.4$, $E[\mathbf{X} - E(\mathbf{X})]^2 = V(\mathbf{X}) = 0.46$, and $E(2\mathbf{X}) = 4.8$.

A similar definition holds for continuous random variables.

Occasionally, we may wish to construct a new random variable, say **Y**, from an old one, **X**, by forming the linear relationship $\mathbf{Y} = a + b\mathbf{X}$, where a and b are constants. If we know the mean $E(\mathbf{X})$ and the variance $V(\mathbf{X})$ of **X**, can we determine the mean $E(\mathbf{Y})$ and the variance $V(\mathbf{Y})$ of **Y** without knowing the probability distribution of either random variable? The answer is given by the following theorem.

Theorem 5.1

Let **X** denote a random variable (discrete or continuous) with mean and variance known to be $E(\mathbf{X})$ and $V(\mathbf{X})$, respectively. Define a new random variable $\mathbf{Y} = a + b\mathbf{X}$, where a and b are known constants. Then

$$E(\mathbf{Y}) = E(a + b\mathbf{X}) = a + bE(\mathbf{X}) \quad \text{and} \quad V(\mathbf{Y}) = V(a + b\mathbf{X}) = b^2 V(\mathbf{X})$$

Theorem 5.1 tells us that the expected value of a constant $\{E(a)\}$ is the constant $\{a\}$, whereas the expected value of a constant times a random variable $\{E(b\mathbf{X})\}$ is equal to the constant times the expected value of the random variable $\{b \cdot E(\mathbf{X})\}$. Furthermore, it indicates that the variance of a constant $\{V(a)\}$ is 0, whereas the variance of a constant times a random variable $\{V(b\mathbf{X})\}$ is the *square* of the constant times the variance of the random variable $\{b^2 \cdot V(\mathbf{X})\}$.

Example 5.12   Suppose the discrete random variable **X** has the distribution shown in Table 5.16. From the computations in Table 5.16, $E(\mathbf{X}) = 2.0$ and $V(\mathbf{X}) = 4.0$. Now suppose that we consider a new random variable **Y**, where $\mathbf{Y} = 10 + 2\mathbf{X}$. Theorem 5.1 tells us that

$$E(\mathbf{Y}) = E(10 + 2\mathbf{X}) = 10 + 2E(\mathbf{X}) = 10 + 2(2) = 14$$

and

$$V(\mathbf{Y}) = V(10 + 2\mathbf{X}) = 0 + (2)^2 V(\mathbf{X}) = (4)(4) = 16$$

Verify this by forming the probability distribution of the random variable **Y**.

Solution   Since $\mathbf{Y} = 10 + 2\mathbf{X}$, the values of **Y** are: $10 + 2(0) = 10$, $10 + 2(1) = 12$, and $10 + 2(5) = 20$. The probabilities corresponding to the values 10, 12, and 20 are 0.2, 0.5, and 0.3 (from the distribution of **X**), respectively. Table 5.17 gives the probability distribution for **Y** and shows the computation of $E(\mathbf{Y})$ and $V(\mathbf{Y})$, which are 14 and 16, respectively, as Theorem 5.1 told us they would be!

TABLE 5.16 | Computation of $E(\mathbf{X})$ and $V(\mathbf{X})$ for a discrete random variable

x	$P(x)$	$xP(x)$	$x - E(\mathbf{X})$	$[x - E(\mathbf{X})]^2$	$[x - E(\mathbf{X})]^2P(x)$
0	0.2	0.0	-2	4	0.8
1	0.5	0.5	-1	1	0.5
5	0.3	1.5	3	9	2.7
	1.00	$E(\mathbf{X}) = 2.0$			$V(\mathbf{X}) = 4.0$

TABLE 5.17 | Computation of $E(\mathbf{Y})$ and $V(\mathbf{Y})$ for the discrete random variable $\mathbf{Y} = 10 + 2\mathbf{X}$

y	$P(y)$	$yP(y)$	$y - E(\mathbf{Y})$	$[y - E(\mathbf{Y})]^2$	$[y - E(\mathbf{Y})]^2P(y)$
10	0.2	2	-4	16	3.2
12	0.5	6	-2	4	2.0
20	0.3	6	6	36	10.8
	1.00	$E(\mathbf{Y}) = 14$			$V(\mathbf{Y}) = 16.0$

The properties of the expected value (E) operator and the variance (V) operator are summarized in Table 5.18 on page 208.

We will now consider two example problems to illustrate the properties listed in Table 5.18.

Example 5.13 Suppose that for a random variable $\mathbf{X}$, $E(\mathbf{X}) = 100$ and $V(\mathbf{X}) = 25$. Determine the following:
a. $E(5\mathbf{X})$
b. $E(-50 + 2\mathbf{X})$
c. $V(20\mathbf{X})$
d. $V(5 - 10\mathbf{X})$
e. The standard deviation of $5\mathbf{X}$.

Solution
a. $E(5\mathbf{X}) = 5E(\mathbf{X}) = 5(100) = 500$ from Property 2.
b. $E(-50 + 2\mathbf{X}) = -50 + 2E(\mathbf{X}) = -50 + 2(100) = 150$ from Property 3.
c. $V(20\mathbf{X}) = (20)^2V(\mathbf{X}) = 400(25) = 10,000$ from Property 5.
d. $V(5 - 10\mathbf{X}) = V(5 + (-10)\mathbf{X}) = (-10)^2V(\mathbf{X}) = 100(25) = 2,500$ by Property 6.
e. Since the standard deviation σ is the square root of the variance $\sigma^2 = V(\mathbf{X})$, we first find the variance: $V(5\mathbf{X}) = 25V(\mathbf{X}) = 25(25) = 625$ by Property 5. Now, $\sigma = \sqrt{V(\mathbf{X})} = \sqrt{625} = 25$. Therefore, the standard deviation of $\mathbf{X}$ is 25.

Example 5.14 A microcomputer salesperson working for Computer World receives a salary of $50 per day plus a commission of $100 for every

TABLE 5.18 | Properties of the expectation and variance operators

Property number	Property	Interpretation
Property 1	$E(a) = a$	The expected value of a constant (a) is equal to the value of the constant (a).
Property 2	$E(b\mathbf{X}) = bE(\mathbf{X})$	The expected value of a constant (b) times a random variable ($\mathbf{X}$) is equal to the constant times the expected value of the random variable.
Property 3	$E(a + b\mathbf{X}) = a + bE(\mathbf{X})$	Theorem 5.1: The expected value of a constant (a) plus a constant (b) times a random variable ($\mathbf{X}$) is the constant (a) plus the constant (b) times the expected value of the random variable.
Property 4	$V(a) = 0$	The variance of a constant (a) is equal to 0.
Property 5	$V(b\mathbf{X}) = b^2V(\mathbf{X})$	The variance of a constant (b) times a random variable ($\mathbf{X}$) is the constant squared times the variance of the random variable.
Property 6	$V(a + b\mathbf{X}) = b^2V(\mathbf{X})$	Theorem 5.1: The variance of a constant (a) plus a constant (b) times a random variable ($\mathbf{X}$) is the constant b squared times the variance of the random variable.

microcomputer that she sells. Let the random variable $\mathbf{X}$ = number of micro-computers the salesperson sells per day. Suppose it is known that $E(\mathbf{X})$ = 0.80, and $V(\mathbf{X})$ = 0.25. That is, on the average, she sells 0.80 microcompu-ters per day with a variance of 0.25. What is her average daily income, and what is the standard deviation of her daily income?

Solution Let the random variable $\mathbf{Y}$ = Daily income of the salesperson. Then, $\mathbf{Y}$ = \$50 + \$100$\mathbf{X}$. That is, her daily income is the fixed salary of \$50 plus \$100 times the number of microcomputers ($\mathbf{X}$) she sells daily. Now, from Property 3 and from Property 6,

$$E(\mathbf{Y}) = E(\$50 + \$100\mathbf{X}) = \$50 + \$100E(\mathbf{X}) = \$50 + \$100(0.80)$$
$$= \$50 + \$80 = \$130$$

$$V(\mathbf{Y}) = V(\$50 + \$100\mathbf{X}) = (100)^2V(\mathbf{X}) = (10,000)(0.25) = 2,500$$

Thus, her average daily income is \$130 and the standard deviation of her daily income is $\sqrt{2500}$ = \$50. (Remember: the standard deviation is the square root of the variance.) Note that we did not need to know the probabil-ity distribution of the random variable $\mathbf{Y}$ to obtain these results—only the mean and standard deviation of the random variable $\mathbf{X}$ is required.

5.8

Bivariate random variables (advanced section)

Multivariate probability function

Bivariate probability function

Often in statistics we are interested in studying the relationship between variables. For example, we may be interested in studying the relationship between an individual's height and weight, or in exploring the statistical relationship between dollars expended on advertising effort and sales of a product. When two or more random variables are jointly involved in an experiment, their outcomes may be generated by what is called a *multivariate probability function*. In this chapter, we will discuss only *bivariate* probability functions and only the discrete case. The development of a multivariate continuous probability function follows directly from the explanation of a bivariate probability function of discrete random variables.

The joint or bivariate probability mass function of two random variables **X** and **Y**, denoted $P(\mathbf{X},\mathbf{Y})$, is given in Definition 5.12: $P(x,y) = P(\mathbf{X} = x, \mathbf{Y} = y)$. That is, $P(x,y)$ denotes the probability that the random variable **X** is equal to x *and* the random variable **Y** is equal to y. Hence the term *bivariate* or *joint* probability function.

Definition 5.12
Bivariate probability mass function and
cumulative mass function (finite case)

Let the random variable **X** assume a finite number of values $x_1, x_2, \ldots, x_r$, where $x_1 < x_2 < \cdots < x_r$, and let the random variable **Y** also assume a finite number of values $y_1, y_2, \ldots, y_s$, where $y_1 < y_2 < \cdots < y_s$, Let $P(x,y)$ denote the probability that the random variable **X** assumes the value x *and* the random variable **Y** assumes the value y.

A *bivariate probability mass function* is a function that assigns probabilities to joint values of the discrete random variables **X** and **Y** such that the following two conditions on the function $P(\mathbf{X},\mathbf{Y})$ are satisfied:

$$0 \le P(x_i,y_j) \le 1; \qquad 1 \le i \le r, 1 \le j \le s$$

$$\sum_{\text{all } \mathbf{Y}} \sum_{\text{all } \mathbf{X}} P(x,y) = 1$$

The *bivariate cumulative mass function*, denoted $F(\mathbf{X},\mathbf{Y})$, gives the probability that the random variable **X** is less than or equal to x *and* the random variable **Y** is less than or equal to y. Mathematically, it is given by:

$$F(x,y) = P(\mathbf{X} \le x, \mathbf{Y} \le y)$$

To understand the notion of a bivariate probability function, consider the data in Table 5.19. This information represents sales of walnut veneers to

TABLE 5.19 | Sales of walnut veneers

Width (inches) \ Length (inches)	24	36	48	Totals
6	12	16	22	50
12	8	24	18	50
Totals	20	40	40	100

Source: Company records.

home crafters in standard widths and lengths. For example, over the interval examined, 100 pieces of veneer were sold, and 18 of these pieces were 48 inches long and 12 inches wide. Fifty of the pieces were 6 inches wide, 40 of the pieces were 36 inches long, and so on. We can convert the data in Table 5.19 to a bivariate probability distribution for the lengths and widths of veneers sold if we divide each quantity given in Table 5.19 by the total number of occurrences, 100. This has been done, and the resulting bivariate probability function, $P(X,Y)$, where the random variable X represents the widths of veneer sold and Y represents the lengths of the veneers sold, is given in Table 5.20. A graph of the resulting bivariate probability mass function is given in Figure 5.14. Note that since two random variables (X and Y) are involved, it becomes necessary to draw the resulting mass function in

TABLE 5.20 | Probabilities of selling various lengths and widths of walnut veneers based on data in Table 5.19

x \ y	24	36	48	$P(x)$
6	0.12	0.16	0.22	0.50
12	0.08	0.24	0.18	0.50
$P(y)$	0.20	0.40	0.40	1.00

FIGURE 5.14 | Graph of bivariate probability distribution given in Table 5.20

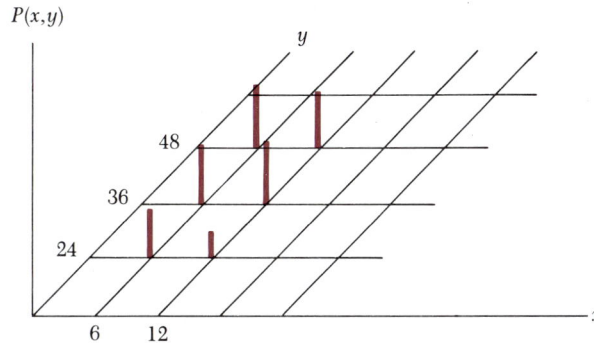

three dimensions. From Table 5.20, we can determine that the probability that a sheet of veneer 12 inches wide and 24 inches long is sold is equal to $P(12,24) = 0.08$. Other probabilities can also be read directly from the table.

A very important property concerning two random variables is the *independence* of the two variables. When two random variables are independent, the values that one random variable can assume are not affected by the values that the other random variable can assume, and vice versa. Definition 5.13 gives the formal definition of the independence of two random variables.

Independence of two random variables

Definition 5.13

Independence of two random variables **X** and **Y**

Two discrete random variables **X** and **Y** are independent if their joint probability bivariate mass function $P(x,y)$ is the product of the probability mass function for **X** and the probability mass function for **Y** for all possible pairs of x and y values. That is,

$$P(x,y) = P(x)P(y), \text{ for all } (x,y) \text{ pairs}$$

Conversely, if **X** and **Y** are independent, then the bivariate mass function $P(x,y)$ must be the product of the probability mass function for **X** and the probability mass function for **Y**. If the two random variables **X** and **Y** are not independent, then we say that they are *dependent* random variables.

To determine if two random variables are independent, we must check to see if $P(x,y)$ equals $P(x)P(y)$ for *all* pairs (x,y). If we find that $P(x,y) \neq P(x)P(y)$ for at least one pair (x,y), then the two random variables are *dependent*.

Example 5.15 Are the two random variables, **X** and **Y**, whose bivariate mass function is given in Table 5.20, independent?

Solution Consider the first (x,y) pair: $x = 6$ and $y = 24$. The bivariate mass function $P(x,y)$ evaluated at this pair is $P(6, 24) = 0.12$. The probability mass function of **X**, $P(x)$ evaluated at $\mathbf{X} = 6$, is $P(6) = 0.50$ from the first-row margin of the table. The probability mass function of **Y**, $P(y)$ evaluated at $\mathbf{Y} = 24$, is $P(24) = 0.20$ from the first-column margin. Since the product $P_\mathbf{X}(6)P_\mathbf{Y}(24) = (0.50)(0.20) = 0.10 \neq P(6, 24) = 0.12$, the two random variables **X** and **Y** are not independent. Therefore, **X** and **Y** are dependent random variables. Since the values of **X** represent the lengths, and the values of **Y** represent the widths of veneer sheets, we should not be too surprised to find that the two random variables are dependent; that is, the values that one random variable can assume are affected by the values that the other random variable can assume and vice versa.

The properties concerning the expected value and variance operators for two random variables are listed in Table 5.21.

Example 5.16 A random variable **X** has an expected value $E(\mathbf{X}) = 0.124$ and a variance $V(\mathbf{X}) = 0.0454$. A random variable **Y** has an expected value $E(\mathbf{Y}) = 0.124$ and a variance $V(\mathbf{Y}) = 0.0681$. Determine the expected value and the variance of the random variable **Z** whose values are given by:

$$z = (\tfrac{1}{2})x + (\tfrac{1}{2})y$$

Assume that the random variables **X** and **Y** are independent.

Solution From Properties 2 and 4 in Table 5.21, we have

$$
\begin{aligned}
E(\mathbf{Z}) &= (\tfrac{1}{2})E(\mathbf{X}) + (\tfrac{1}{2})E(\mathbf{Y}) \\
&= (\tfrac{1}{2})(0.124) + (\tfrac{1}{2})(0.124) \\
&= 0.124 \\
V(\mathbf{Z}) &= (\tfrac{1}{2})^2 V(\mathbf{X}) + (\tfrac{1}{2})^2 V(\mathbf{Y}) \\
&= (\tfrac{1}{2})^2(0.0454) + (\tfrac{1}{2})^2(0.0681) \\
&= 0.01135 + 0.0170 \\
&= 0.02835
\end{aligned}
$$

In Section 5.7, we examined the general form for the expectation of a function of a random variable. The counterpart to this when two random variables are involved is given in Definition 5.14. A special case of the expectation of a function of a bivariate random variable occurs when $h(\mathbf{X},\mathbf{Y}) = [\mathbf{X} - E(\mathbf{X})][\mathbf{Y} - E(\mathbf{Y})]$. In this instance, $E[h(\mathbf{X},\mathbf{Y})]$ is called the *covariance* of the random variables **X** and **Y** as described in Definition 5.15.

Covariance of two random variables

The covariance is a statistical measure that indicates how the two random variables vary together, or "co-vary," $C(\mathbf{X},\mathbf{Y})$ will tend to be a large positive number whenever large values of **X** [relative to $E(\mathbf{X})$] tend to be associated with large values of **Y** [relative to $E(\mathbf{Y})$], and small values of **X** tend to

TABLE 5.21 | Properties of the expectation and variance operators for two random variables **X** and **Y**

Property number	Property	Interpretation
Property 1	$E(\mathbf{X} \pm \mathbf{Y}) = E(\mathbf{X}) \pm E(\mathbf{Y})$	The expected value of the sum or difference of two random variables is the sum or difference of the expected values of the two random variables.
Property 2	$E(a\mathbf{X} \pm b\mathbf{Y}) = aE(\mathbf{X}) \pm bE(\mathbf{Y})$	The expected value of the sum (or difference) of two random variables each multiplied by a constant is equal to the first constant times the expected value of the first random variable plus (or minus) the second constant times the expected value of the second random variable.
Property 3	$V(\mathbf{X} \pm \mathbf{Y}) = V(\mathbf{X}) + V(\mathbf{Y})$ if **X** and **Y** are independent	If **X** and **Y** are independent, then the variance of the sum or the difference of the two random variables is equal to the sum of the variances of the two random variables.
Property 4	$V(a\mathbf{X} \pm b\mathbf{Y}) = a^2V(\mathbf{X}) + b^2V(\mathbf{Y})$ if **X** and **Y** are independent	If **X** and **Y** are independent, then the variance of the sum (or difference) of the two random variables each multiplied by a constant is equal to the square of the first constant times the variance of the first random variable plus the square of the second constant times the variance of the second random variable.
Property 5	$E(\mathbf{XY}) = E(\mathbf{X})E(\mathbf{Y})$ if **X** and **Y** are independent	If **X** and **Y** are independent, then the expected value of the product of the two random variables is equal to the product of the expected values of the two random variables.

be associated with small values of **Y**. Similarly, $C(\mathbf{X},\mathbf{Y})$ will tend to be a large negative number whenever low values of one random variable tend to be associated with high values of the other random variable, and high values of one random variable tend to be associated with low values of the other random variable. Whenever two random variables are independent (i.e., they are not statistically related), then $C(\mathbf{X},\mathbf{Y}) = 0$.

Definition 5.14

Expected value of a function of a bivariate random variable (discrete case)

Let **X** and **Y** be discrete random variables with bivariate probability mass function $P(\mathbf{X},\mathbf{Y})$ as delineated in Definition 5.11, and consider a function of both **X** and **Y** denoted by $h(\mathbf{X},\mathbf{Y})$. The *expected value of* $h(\mathbf{X},\mathbf{Y})$ is given by:

$$E[h(\mathbf{X},\mathbf{Y})] = \sum_{\text{all } \mathbf{Y}} \sum_{\text{all } \mathbf{X}} h(x,y)P(x,y)$$

Definition 5.15

Covariance

The *covariance* of two random variables **X** and **Y**, denoted by $C(\mathbf{X},\mathbf{Y})$, is the expected value of $[\mathbf{X} - E(\mathbf{X})]$ times $[\mathbf{Y} - E(\mathbf{Y})]$. Mathematically, it is given by:

$$C(\mathbf{X},\mathbf{Y}) = E\{[\mathbf{X} - E(\mathbf{X})] \cdot [\mathbf{Y} - E(\mathbf{Y})]\} = E(\mathbf{X} \cdot \mathbf{Y}) - E(\mathbf{X}) \cdot E(\mathbf{Y})$$

The covariance of two random variables is a statistical measure of how the two random variables vary together (i.e., how they "co-vary").

Experience in computing the covariance of two random variables **X** and **Y** is provided in the following example.

Example 5.17   Using the data in Table 5.20, compute $C(\mathbf{X},\mathbf{Y})$, where the random variable **X** corresponds to the widths of veneer sold and **Y** corresponds to the lengths of veneer sold. Use the first expression in Definition 5.15 for the covariance of two random variables to compute the covariance for this data set, and verify that the second expression for the covariance gives the same result using the data provided. Are the random variables **X** = widths of veneer sold and **Y** = lengths of veneer sold independent?

Solution   Relevant computations are given in Tables 5.22 and 5.23. Note that either of the definitional forms for the covariance of two random variables given in Definition 5.15 produces the result $C(\mathbf{X},\mathbf{Y}) = 0$.

We have to be extremely careful in answering the second part of Example 5.17. Although it is true that if two random variables are independent, their covariance is 0, it is not *in general* true that if the covariance of two random variables is equal to 0, the random variables are independent. To show independence between two random variables, we must demonstrate that $P(x,y) = P(x)P(y)$ for *all* possible combinations of **X** and **Y**. That is, if

TABLE 5.22 | Computation of the covariance using the data in Table 5.20 and the first definitional form in Definition 5.15

(x,y)	$P(x,y)$	$x - E(X)$	$y - E(Y)$	$[x - E(X)][y - E(Y)]$	$[x - E(X)][y - E(Y)]P(x,y)$
(6, 24)	0.12	−3	−14.4	+43.2	5.184
(6, 36)	0.16	−3	− 2.4	+ 7.2	1.152
(6, 48)	0.22	−3	9.6	−28.8	−6.336
(12, 24)	0.08	3	−14.4	−43.2	−3.456
(12, 36)	0.24	3	− 2.4	− 7.2	−1.728
(12, 48)	0.18	3	9.6	+28.8	5.184
Totals	1.00				$C(\mathbf{X},\mathbf{Y}) =$ 0.000

TABLE 5.23 | Computation of the covariance using the data in Table 5.20 and the second definitional form in Definition 5.15

x	$P(x)$	$xP(x)$	y	$P(y)$	$yP(y)$	(x,y)	$P(x,y)$	$x \cdot y$	$x \cdot yP(x,y)$
6	0.50	3.0	24	0.20	4.8	(6, 24)	0.12	144	17.28
12	0.50	6.0	36	0.40	14.4	(6, 36)	0.16	216	34.56
			48	0.40	19.2	(6, 48)	0.22	288	63.36
						(12, 24)	0.08	288	23.04
						(12, 36)	0.24	432	103.68
						(12, 48)	0.18	576	103.68
Totals	1.00	$E(\mathbf{X}) = 9.0$		1.00	$E(\mathbf{Y}) = 38.4$		1.00		$E(\mathbf{X},\mathbf{Y}) = 345.60$

$$C(\mathbf{X},\mathbf{Y}) = E(\mathbf{X} \cdot \mathbf{Y}) - E(\mathbf{X}) \cdot E(\mathbf{Y}) = 345.60 - 9.0 \cdot 38.4 = 0$$

$P(x,y) \neq P(x)P(y)$ for *any* values of **X** and **Y**, then the random variables are dependent. For example,

$$\left.\begin{array}{l} P(\mathbf{X} = 6, \mathbf{Y} = 24) = 0.12 \\ P(\mathbf{X} = 6) = 0.50 \\ P(\mathbf{Y} = 24) = 0.20 \end{array}\right\} P(6, 24) = 0.12 \neq 0.10 = P(\mathbf{X} = 6)P(\mathbf{Y} = 24)$$

Hence we would conclude that the two random variables are dependent, even though their covariance is equal to zero. We claim that the two random variables are dependent because we were able to derive at least one value x and one value y for which $P(x,y) \neq P(x)P(y)$.

Now that we have developed the notion of a covariance, we can state the general form of the variance of the sum or difference of two random variables. The general statement of the variance of the sum or difference of two random variables is given in Theorem 5.2. Note that from Theorem 5.2, whenever the covariance of two random variables is equal to zero, [i.e., $C(\mathbf{X},\mathbf{Y}) = 0$], the variance of the sum or difference of these two random variables is given by the sum of the variances of the two random variables.

Theorem 5.2
Variance of the sum or difference of two random variables
General case

Let $\mathbf{X}$ and $\mathbf{Y}$ be two random variables. Let the random variable $\mathbf{Z}$ be the sum of these two random variables:

$$\mathbf{Z} = \mathbf{X} + \mathbf{Y}$$

Then the variance of $\mathbf{Z}$ is given by:

$$V(\mathbf{Z}) = V(\mathbf{X} + \mathbf{Y}) = V(\mathbf{X}) + V(\mathbf{Y}) + 2C(\mathbf{X},\mathbf{Y})$$

Let the random variable $\mathbf{W}$ be the difference between the random variables $\mathbf{X}$ and $\mathbf{Y}$:

$$\mathbf{W} = \mathbf{X} - \mathbf{Y}$$

Then the variance of $\mathbf{W}$ is given by:

$$V(\mathbf{W}) = V(\mathbf{X} - \mathbf{Y}) = V(\mathbf{X}) + V(\mathbf{Y}) - 2C(\mathbf{X},\mathbf{Y})$$

Extension: Let a and b be two constants.

If $\mathbf{Z} = a\mathbf{X} + b\mathbf{Y}$, then $V(\mathbf{Z}) = a^2 V(\mathbf{X}) + b^2 V(\mathbf{Y}) + 2ab C(\mathbf{X},\mathbf{Y})$
If $\mathbf{W} = a\mathbf{X} - b\mathbf{Y}$, then $V(\mathbf{W}) = a^2 V(\mathbf{X}) + b^2 V(\mathbf{Y}) - 2ab C(\mathbf{X},\mathbf{Y})$

When $\mathbf{X}$ and $\mathbf{Y}$ are independent random variables, then $C(\mathbf{X},\mathbf{Y}) = 0$, and the variance of the sum or difference of these two random variables is given by the formulas in Table 5.21.

The results given in Theorem 5.2 can be combined with the properties of Table 5.21 as indicated in the following example.

Example 5.18   Rarely does an individual have all of his or her entire wealth tied up in one investment or in one asset. Portfolio managers of brokerage houses, banks, and the like recognized early that it might be possible to combine groups of assets or investments in order to reduce the risk associated with such holdings. Here, of course, risk is being assessed by the variability in returns of an asset or a group of assets. It might be possible, that is, to combine assets into a *portfolio* that would reduce the risk (variability in returns) of such holdings. Mutual funds have grown as a result of being able to diversify and reduce risk for the investor in precisely this fashion.

Table 5.24 gives the returns of two investments over a 40-year period. This table is similar to Table 5.19. For example, in three of the 40 years

TABLE 5.24 | Number of occurrences of various annual rates of return on two stocks over a 40-year period

Stock B returns \ Stock A returns	−0.10	0.05	0.15	0.38	Total
−0.15	1	1	1	9	12
−0.05	3	1	1	3	8
0.12	2	1	1	2	6
0.46	10	2	1	1	14
Total	16	5	4	15	40

examined, Stock A had a return of 38 percent (0.38) and Stock B had a return of −10 percent (−0.10)—a loss! Other quantities are derived similarly.

When we divide each of the frequencies in Table 5.24 by their total (40), we obtain the bivariate mass function for the bivariate random variable (return on Stock A, return on Stock B) as indicated in Table 5.25. Thus, for example, the probability of a return from Stock A of 38 percent *and* Stock B of 46 percent in the same year is 0.025 (low!).

Assume that a portfolio manager has decided to construct a portfolio consisting of 50 percent Stock A and 50 percent Stock B. Compute the expected returns and the variance of returns for:

a. A portfolio consisting of 100 percent Stock A.
b. A portfolio consisting of 100 percent Stock B.
c. A portfolio consisting of 50 percent Stock A and 50 percent Stock B.

Solution We will denote the return from Stock A by the random variable **X** and the return from Stock B by the random variable **Y**. Thus the bivariate mass function would be denoted by $P(\mathbf{X}, \mathbf{Y})$.

From Table 5.26 the expected return from Stock A is 12.4 percent per

TABLE 5.25 | Probabilities of various rates of return based upon data in Table 5.24.

Stock B returns \ Stock A returns	−0.10	0.05	0.15	0.38	P(y)
−0.15	0.025	0.025	0.025	0.225	0.300
−0.05	0.075	0.025	0.025	0.075	0.200
0.12	0.050	0.025	0.025	0.050	0.150
0.46	0.250	0.050	0.025	0.025	0.350
P(x)	0.400	0.125	0.100	0.375	1.000

TABLE 5.26 | Computation of $E(\mathbf{X})$ and $V(\mathbf{X})$ for Example 5.18

x	$P(x)$	$xP(x)$	$x - E(\mathbf{X})$	$[x - E(\mathbf{X})]^2$	$[x - E(\mathbf{X})]^2 P(x)$
−0.10	0.400	−0.040	−0.2238	0.0501	0.0200
0.05	0.125	0.006	−0.0738	0.0054	0.0007
0.15	0.100	0.015	0.0262	0.0007	0.0001
0.38	0.375	0.143	0.2563	0.0657	0.0246
		$E(\mathbf{X}) = 0.124$			$V(\mathbf{X}) = 0.0454$

TABLE 5.27 | Computation of $E(\mathbf{Y})$ and $V(\mathbf{Y})$ for Example 5.18

y	$P(y)$	$yP(y)$	$y - E(\mathbf{Y})$	$[y - E(\mathbf{Y})]^2$	$[y - E(\mathbf{Y})]^2 P(y)$
−0.15	0.300	−0.045	−0.2740	0.0751	0.0225
−0.05	0.200	−0.010	−0.1740	0.0303	0.0061
0.12	0.150	0.018	−0.0040	0.0000	0.0000
0.46	0.350	0.161	0.3360	0.1129	0.0395
		$E(\mathbf{Y}) = 0.124$			$V(\mathbf{Y}) = 0.0681$

year, with a variance in return of 4.54 percent. And from Table 5.27 the expected return of Stock B is 12.4 percent, with a variance of 6.81 percent.

Table 5.28 gives the computation of the covariance of $\mathbf{X}$ and $\mathbf{Y}$—$C(\mathbf{X},\mathbf{Y})$ = −0.318. From these data and the information given in Table 5.21 and

TABLE 5.28 | Computation of $C(\mathbf{X},\mathbf{Y})$ for Example 5.18

(x,y)	$P(x,y)$	$x - E(\mathbf{X})$	$y - E(\mathbf{Y})$	$[x - E(\mathbf{X})][y - E(\mathbf{Y})]$	$[x - E(\mathbf{X})][y - E(\mathbf{Y})]P(x,y)$
-0.10, -0.15	0.025	-0.2238	-0.2740	0.0613	0.0015
-0.10, -0.05	0.075	-0.2238	-0.1740	0.0389	0.0029
-0.10, 0.12	0.050	-0.2238	-0.0040	0.0009	0.0000
-0.10, 0.46	0.250	-0.2238	0.3360	-0.0752	-0.0188
0.05, -0.15	0.025	-0.0738	-0.2740	0.0202	0.0005
0.05, -0.05	0.025	-0.0738	-0.1740	0.0128	0.0003
0.05, 0.12	0.025	-0.0738	-0.0040	0.0003	0.0000
0.05, 0.46	0.050	-0.0738	0.3360	-0.0248	-0.0012
0.15, -0.15	0.025	0.0262	-0.2740	-0.0072	-0.0002
0.15, -0.05	0.025	0.0262	-0.1740	-0.0046	-0.0001
0.15, 0.12	0.025	0.0262	-0.0040	-0.0001	0.0000
0.15, 0.46	0.025	0.0262	0.3360	0.0088	0.0002
0.38, -0.15	0.225	0.2563	-0.2740	-0.0702	-0.0158
0.38, -0.05	0.075	0.2563	-0.1740	-0.0446	-0.0033
0.38, 0.12	0.050	0.2563	-0.0040	-0.0010	-0.0001
0.38, 0.46	0.025	0.2563	0.3360	0.0861	0.0022
	1.000				$C(\mathbf{X},\mathbf{Y}) = -0.0318$

Theorem 5.2, we can determine the expected return and the variance in return from a portfolio that consists of 50 percent Stock A and 50 percent Stock B. Let $\mathbf{P} = 0.5\mathbf{X} + 0.5\mathbf{Y}$ denote the return on the portfolio. Then,

$$E(\mathbf{P}) = E(0.50\mathbf{X} + 0.50\mathbf{Y}) = (0.50)E(\mathbf{X}) + (0.50)E(\mathbf{Y})$$
$$= (0.50)(0.124) + (0.50)(0.124)$$
$$= 0.124$$

$$V(\mathbf{P}) = V(0.50\mathbf{X} + 0.50\mathbf{Y}) = (0.50)^2 V(\mathbf{X}) + (0.50)^2 V(\mathbf{Y})$$
$$+ 2(0.50)(0.50)C(\mathbf{X},\mathbf{Y})$$
$$= (0.50)^2(0.0454) + (0.50)^2(0.0681)$$
$$+ 2(0.50)(0.50)(-0.0318)$$
$$= 0.012463$$

Hence a portfolio consisting of 100 percent Stock A has an expected return per year of 12.4 percent and a variance in returns of 4.54 percent; a portfolio consisting of 100 percent Stock B has an expected return per year of 12.4 percent and a variance in returns of 6.61 percent; and a portfolio P consisting of 50 percent Stock A and 50 percent Stock B has a return of 12.4 percent (approximately—there is slight round-off error here), and a variance in returns of 1.246 percent! By combining these two stocks into one portfolio, we have been able to maintain the expected returns from the portfolio (approximately) while reducing the risk or the variance associated with this return.

Portfolio managers (good ones!) analyze the movements of several stocks at a time and combine them into a portfolio such that the portfolio constructed meets the investor's needs in terms of expected return and minimal risk associated with this return. For example, a "better" combination (in terms of variance or risk reduction) is possible than given above by combining other than 50 percent of each stock into the portfolio. Most finance texts cover the determination of the "optimal" portfolio composition given a stated return and minimal risk or variability in returns.

Statistical analysis aids the portfolio manager in providing estimates of return, variability, covariability, and so on, as well as the basis for combining stocks or investment alternatives in order to reach a stated objective of each investor (to the extent that such objectives *can* be met!).

Use of the covariance operator is not restricted to financial applications alone. A producer of outboard motors for pleasure boats, for example, recently acquired a new product line (snowmobiles) in order to reduce the variability in output or production. This decision was based upon a statistical analysis of combining several product lines the manufacturer might consider in order to reduce the variability in personnel needs and the demand for physical facilities.

Unfortunately, the covariance measure is often not an easy statistical measure to interpret. Quite often, for example, the value of the covariance will change simply because the units used to measure either or both of the quantitative random variables change. Thus we may find that the covariance of two random variables differs from a value previously determined simply because the units of measurement have changed. Still the covariance of two random variables is an important statistical measure as Example 5.18 indicates. We will have an opportunity to use this statistical measure again in Chapter 13 when we study correlation—a statistical measure of the degree of linear association between two random variables. We will learn that the measure of correlation is independent of the units of measurement and is a function of the covariance of the two random variables involved.

■ 5.9 Summary

In applications of inferential statistics, random variables and their distributions play a vital role. The process may be summarized as follows. A population is described, and the numerical characteristics of the population units of interest are identified. A random variable is defined to represent the values of the numerical characteristic. The probability distribution of the random variable is formed and used to answer questions concerning the population characteristic.

There are two kinds of random variables—*discrete* and *continuous*. A discrete random variable takes on a finite or countably infinite number of values, whereas a continuous random variable takes on an infinite number of values that are not countable.

The average value, the standard deviation, and the variance of a characteristic of a random variable can be determined by computing the expected value and standard deviation of the random variable representing the values of the characteristic. The expected value, variance, and standard deviation of a random variable are computed from its probability distribution. The use of integral calculus is often required to compute these numerical measures of a continuous random variable, although tabled values are available for most density functions of a continuous random variable.

The expected value operator may be used to find the average value of any function of a random variable. Particular attention is focused in this chapter on using this notion to compute the mean and variance of a linear function of a random variable. Tables 5.18 and 5.21 summarize several important properties of the expectation operator.

■ **References**

Introductory

Anderson, D. R.; D. J. Sweeney; and T. A. Williams. *Statistics for Business and Economics.* 2nd ed. St. Paul, Minn.: West Publishing, 1984, Chapter 5.

Harnett, D. L. *Introduction to Statistical Methods.* 3rd ed. Reading, Mass.: Addison-Wesley, 1982, Chapter 3.

Neter, J.; W. Wasserman; and G. A. Whitmore. *Applied Statistics.* Boston: Allyn & Bacon, 1982, Chapter 5.

Advanced

Hogg, R. V., and A. T. Craig. *Introduction to Mathematical Statistics.* 4th ed. New York: MacMillan, 1978, Chapter 2.

Wonnacott, R. J., and T. H. Wonnacott. *Introductory Statistics.* 4th ed. New York: John Wiley & Sons, 1985, Chapter 5.

■ **Problems**

Section 5.2 Problems

5.1 Explain in your own words what is meant by a *random variable*. List several experiences you encountered recently in which reference was made to a random variable without actually calling it by that name. In each instance, try to identify the process that was generating the values of the random variable.

5.2 An experiment produces one of four elementary outcomes: E_1, E_2, E_3, and E_4 with the following probabilities $P(E_1) = 0.20$, $P(E_2) = 0.30$, $P(E_3) = 0.10$, and $P(E_4) = 0.40$. A random variable $\mathbf{X}$ is defined in the following way over the sample space S containing the four points:

$$\mathbf{X}(E_1) = 0 \qquad \mathbf{X}(E_2) = 1$$
$$\mathbf{X}(E_3) = 1 \qquad \mathbf{X}(E_4) = 2$$

a. Is $\mathbf{X}$ a random variable? Explain.

b. If $\mathbf{X}$ is a random variable, then determine its probability distribution.

5.3 Explain the difference between a discrete and a continuous random variable.

5.4 Explain the difference between a discrete

random variable with a finite number of values and a discrete random variable with a countably infinite number of values.

5.5 A coin is flipped three times. On each flip, it is noted whether the coin came up heads or tails. The elementary outcomes of this experiment are:

$$
\begin{array}{llll}
E_1: & HHH & E_5: & HTT \\
E_2: & HHT & E_6: & THT \\
E_3: & HTH & E_7: & TTH \\
E_4: & THH & E_8: & TTT
\end{array}
$$

Define the random variable, X = Number of heads in the three flips of the coin.
a. Is X a random variable? Explain.
b. If X is a random variable and each of the eight elementary outcomes is equally likely to occur in each trial of the experiment, then find the probability distribution of X.

Section 5.3 Problems

5.6 Distinguish between a probability mass function and a cumulative mass function of a discrete random variable.

5.7 In a certain population of voters, it is known that 60 percent are Democrats and 40 percent are Republicans. If a sample of three voters is extracted from this population, find the probability distribution of the random variable X = Number of Democrats in the sample. Assume that the population is growing sufficiently so that as a unit is extracted from the population, a new one enters such that the percentage of Democrats and Republicans in the population remains constant.

5.8 Assume in Problem 5.7 that the population of voters consists of 100 individuals (i.e., 60 Democrats and 40 Republicans). Find the probability distribution of the random variable X = number of Democrats sampled in a sample of size three. How does this result differ from that obtained in Problem 5.7, and how do you explain this difference?

5.9 A college recruiter wishes to select three of the five job applicants he will interview. Although all three candidates appear to be equally qualified, there is a best candidate, a second best candidate, and so on. Refer to these three unknown best candidates as successes, and let the random variable X = Number of successes hired. Determine the probability distribution of the random variable X.

5.10 Sketch in the form of a stick diagram the cumulative mass function for the distribution developed in Problem 5.9. Using the cumulative density function developed, derive the mass function for this distribution by taking the differences of appropriate cumulative values. Plot in the form of a stick diagram the mass function for this distribution.

5.11 A multiple-choice exam consists of four questions, each of which has five possible answers (a through e). If a student is forced to guess on all four questions, what is the probability distribution of the random variable X = Number of correct guesses? What is the most likely number of correct guesses?

5.12 In an organization consisting of three women and seven men, a committee of four individuals is to be selected at random from the 10 people. Find the probability distribution of the random variable X = Number of women on the committee.

5.13 An urn contains six colored balls numbered from one to six. Random samples of two balls are drawn without replacement from the urn. The random variable X is defined to be the sum of the numbers on the two balls drawn in each sample. Determine the probability distribution of the random variable X.

5.14 Assume in Problem 5.13 that the first ball is returned to the urn after recording its number and before the second ball is drawn (i.e., sampling with replacement). Determine the probability distribution of the random variable X. Compare the distribution with the one determined in Problem 5.13 by sketching each in the form of a stick diagram.

Section 5.4 Problems

5.15 Let X be a random variable with the probability distribution given in the following table. Find the following:

x	$P(x)$
0	0.60
1	0.30
2	0.05
3	0.05

a. The expected value $E(\mathbf{X})$ of $\mathbf{X}$.
b. The variance $V(\mathbf{X})$ of $\mathbf{X}$.
c. The mode of $\mathbf{X}$.
d. The range of $\mathbf{X}$.

5.16 Determine the expected value $E(\mathbf{X})$ and the variance $V(\mathbf{X})$ of the random variable defined in Problem 5.11.

5.17 Plot the cumulative mass function of the random variable described in Problem 5.11.

5.18 Each new car rolling off an assembly line is equipped with five new tires (one on each wheel plus a spare). Let $\mathbf{X}$ be the number of defective tires on each car. It is found that the probability distribution of $\mathbf{X}$ is:

x	0	1	2	3	4	5
$P(x)$	0.94	0.035	0.020	0.003	0.0015	0.0005

Find (a) the average number of defective tires per car, and (b) the most likely number of defective tires per car.

5.19 Consider the formula: $P(x) = \dfrac{x^2}{21}$, $x = 0$, 1, 2, 4. Is $P(x)$ a probability mass function? If so, show the distribution of $\mathbf{X}$ in tabular form, and compute the expected value of $\mathbf{X}$.

5.20 Consider the formula: $P(x) = \dfrac{C_x^2 C_{2-x}^3}{C_2^5}$, $x = 0, 1, 2$. Is $P(x)$ a probability mass function? If so, show the distribution of $\mathbf{X}$ in tabular form and compute the expected value of $\mathbf{X}$.

5.21 Consider the formula $P(x) = 1/5$, $x = 17, 22, 27, 33$, and 129. Is $P(x)$ a probability mass function? If it is, calculate the expected value and the variance for the random variable $\mathbf{X}$.

5.22 Determine the expected value $E(\mathbf{X})$ and the variance $V(\mathbf{X})$ of the random variable defined in Problem 5.7.

5.23 Determine the expected value $E(\mathbf{X})$ and the variance $V(\mathbf{X})$ of the random variable defined in Problem 5.8. How do the expected values and variances compare with those determined in Problem 5.22?

5.24 In a certain population of individuals, the probability that an individual has AIDS is 0.05. Five individuals are randomly taken from the population. Define the random variable $\mathbf{X}$ to be the number of individuals in the selected five who have AIDS. Assume that the trials—selecting each of the five individuals one at a time—are independent.
a. Determine the probability mass function for the random variable $\mathbf{X}$.
b. Determine the cumulative mass function for the random variable $\mathbf{X}$.
c. What is the expected number of individuals in the selected 5 who have AIDS?
d. What is the standard deviation of $\mathbf{X}$?

Section 5.5 Problems

5.25 Explain why, in general, it is unnecessary to use calculus to determine the probability that a continuous random variable assumes a value between two values of the random variable, c and d, with $c < d$.

5.26 Suppose $\mathbf{X}$ is a continuous random variable. What is the probability that $\mathbf{X}$ equals a specific real number, say a, for any value a on the real line?

5.27 Suppose that a continuous random variable $\mathbf{X}$ is defined over the range, $0 \le \mathbf{X} \le 10$. Explain in words how $P(2 \le \mathbf{X} \le 5)$ is found from the probability density function $f(x)$ of $\mathbf{X}$.

5.28 A continuous random variable $\mathbf{X}$ has the following probability density function:

$$f(x) = \begin{cases} 0 & x < 10 \\ 0.10 & 10 \le x \le 20 \\ 0 & x > 20 \end{cases}$$

a. Draw the graph of the function $f(x)$.
b. Does $f(x)$ satisfy the definition of a probability density function? Explain.
c. Determine $P(12 \le \mathbf{X} \le 18)$.

d. Determine $P(\mathbf{X} = 15)$.

e. Determine $P(10 \le \mathbf{X} \le 12)$.

5.29 The cumulative density function for the probability density function in Problem 5.28 is given by

$$F(x) = \begin{cases} 0 & x < 10 \\ 0.10x - 1.0 & 10 \le x \le 20 \\ 1 & x > 20 \end{cases}$$

Using $F(x)$, determine the following:

a. $P(\mathbf{X} \le 20)$

b. $P(\mathbf{X} \le 16)$

c. $P(10 \le \mathbf{X} \le 12)$

d. $P(12 \le \mathbf{X} \le 18)$

Section 5.6 Problems

5.30 (Calculus required.) A random variable $\mathbf{X}$ has the following density function:

$$f(x) = \begin{cases} 0 & x < 0 \\ (1/9)x^2 & 0 \le x \le 3 \\ 0 & x > 3 \end{cases}$$

a. Demonstrate that $f(x)$ is a probability density function.

b. Determine $E(\mathbf{X})$.

c. Determine $V(\mathbf{X})$.

d. What is the mode of $\mathbf{X}$?

e. What is the median of $\mathbf{X}$? (Hint: The median is the value of $\mathbf{X}$ such that one half of the area under $f(x)$ is to the left of the value and one half is to the right of the value.)

f. Find the cumulative density function of $\mathbf{X}$. Hint:

$$F(x) = \int_{-\infty}^{x} f(x)\, dx$$

5.31 (Calculus required.) The density function for a random variable $\mathbf{X}$ is given by

$$f(x) = \begin{cases} 0 & x < 0 \\ (1/2)x & 0 \le x < 1 \\ (3/2)(2 - x) & 1 \le x \le 2 \\ 0 & x > 2 \end{cases}$$

a. Demonstrate that $f(x)$ is a density function.

b. Determine $E(\mathbf{X})$.

c. Determine $V(\mathbf{X})$.

d. Determine $F(x)$, the cumulative density function. (See the hint in 5.30, part *f*.)

5.32 Using the hint in Problem 5.30, part *f*, derive the cumulative density function given in Problem 5.29 from the probability density function given in Problem 5.28.

Section 5.7 Problems

5.33 Given the following probability distribution,

x	$P(x)$
0	0.30
1	0.40
2	0.30

determine:

a. $E(\mathbf{X})$.

b. $V(\mathbf{X})$.

c. $E(10\mathbf{X})$.

d. $E(12 - \mathbf{X})$.

e. $E\{[\mathbf{X} - E(\mathbf{X})]^2\}$.

f. $E(\mathbf{Y})$, where $\mathbf{Y} = 12 + 3\mathbf{X}$.

g. $V(\mathbf{Y})$, where $\mathbf{Y} = 13 - 2\mathbf{X}$.

Are any of the results the same? If so, why? For each of your answers, list those properties from Table 5.18 you used to simplify computations.

5.34 Suppose that the random variable $\mathbf{X}$ has the following probability distribution. Find:

x	$P(x)$
0	0.10
1	0.10
2	0.40
3	0.40

a. $E(\mathbf{X})$.

b. $E(2\mathbf{X})$.

c. $E[(\mathbf{X} - 1)^2]$.

d. $E[\mathbf{X} - E(\mathbf{X})]$.

5.35 A salesperson receives orders for zero, one, two, three, or four units of a particular product each day with probability 0.45, 0.30, 0.15, 0.05, and 0.05, respectively. His salary is $60 per day plus $15 for each unit ordered. What is his expected monthly salary (based on five days/week, four weeks/month)? What properties in Table 5.18 did you use (if any) to determine the salesperson's expected salary?

5.36 A quality control expert is paid $50 a day to test complex electronic equipment units produced by a small firm. The firm produces five units per day, and the expert is instructed to check each unit thoroughly for defects. For each defective unit found, she is given a bonus of $30 (high incentive—the production workers *love* her; the $30 bonus comes out of the salaries of the production workers!). The distribution of the number of defectives is given in the table below.

x	$P(x)$
0	0.80
1	0.16
2	0.02
3	0.008
4	0.006
5	0.006

a. Find the expected number of defectives per day.
b. Find the expert's expected salary per day.
c. What is the expert's most likely daily salary?

Section 5.8 Problems

5.37 The Ajax Paperboard Company, Inc., supplies sheets of cardboard to several manufacturers, which they in turn fold into boxes for distributing their product. Shown here is the number of sheets of each size sold from the last 1,000 delivered.

Width (feet) \ Length (feet)	4	5	6
2	25	25	50
3	50	400	50
4	225	75	100

Source: Company records.

Using the table to estimate the probabilities of various sales combinations, determine the following quantities where X = the width of

cardboard sold and Y = the length of cardboard sold. Are the random variables X = width and Y = length independent?

a. The expected square feet (area = width × length) of the boxes sold.
b. $P(X = 4)$
c. $P(Y = 6)$
d. $P(X = 3, Y = 5)$
e. $C(X,Y)$

5.38 A statistics instructor is interested in the relationship between the amount of studying time for an exam and the grade earned. In a sample of 100 students, the following frequency table results.

	Grades			
Study Time	A or B	C	D or F	Total
Less than 10 hours per week	20	20	20	60
Ten hours per week or more	20	10	10	40
Total	40	30	30	100

a. Form the bivariate mass function for the random variables X and Y, where $X = 0$ if study time is less than 10 hours per week, $X = 1$ if study time is greater than or equal to 10 hours per week; $Y = 2$ if grade is A or B, $Y = 1$ if grade is C, and $Y = 0$ if grade is D or F.
b. Are the two random variables X and Y independent? Explain.

5.39 Given a random variable X with $E(X) = 10$ and $V(X) = 5$, one other random variable Y with $E(Y) = 15$ and $V(Y) = 10$, and independence assumed for the random variables X and Y, determine the following quantities:
a. $E(X \cdot Y)$
b. $E(X + 5Y)$
c. $V(X - Y)$
d. $V(15 - 5X)$
e. $C(X,Y)$

5.40 Assume that the following values of X and Y occur with equal probability:

x	y
7	3
5	3
6	3
11	3
9	3

Determine the following quantities:
a. $E(X), E(Y), V(X), V(Y)$
b. $E(X \cdot Y)$
c. $C(X,Y)$
Are **X** and **Y** independent?

Additional Problems

5.41 In an eight-cylinder automobile engine, two of the eight sparkplugs are defective. Three sparkplugs are removed at random from the engine and examined for defects. Let **X** be the number of defectives found. Find the probability distribution of **X**.

5.42 Simulate the experiment in Problem 5.41 by taking eight scraps of paper and marking on two of them the letter *D*. Place the eight scraps of paper in a box, and randomly select three of the scraps—record the number marked *D*. Repeat this experiment 100 times. Construct a frequency distribution for this sample, and compare it with the theoretical probability distribution determined in Problem 5.41. How do they compare?

5.43 Compute the expected value of the random variable **X** in Problem 5.41. Find the sample mean number of defectives recorded per trial in the Problem 5.42 simulation experiment. Does the sample mean provide a good estimate of $E(X)$? Should the sample mean provide a good estimate of $E(X)$? Why?

5.44 Consider the formula $P(x) = (2x - x^2)/2.5$, where $x = 0, 0.5, 1.5$, and 2.0. Is $P(x)$ a probability mass function? Explain.

5.45 Distinguished between a probability mass function of a discrete random variable and a probability density function of a continuous random variable. In what ways are these functions similar?

5.46 An automobile salesperson has a probability of 0.25 of selling a car to each person interested in buying a car with whom she speaks on the showroom floor. On a certain day, she talks with four persons regarding the purchase of a car. Find the probability distribution of the random variable, **X** = number of cars sold. Show the distribution in the form of a table, a stick diagram and, if possible, a formula.

5.47 A production process is producing defective units of a product at the rate of 10 percent. If six items are randomly chosen from the production line, determine the following:
a. Probability mass function for the random variable **X** = Number of defective units in the set of six units chosen.
b. The expected number of defective units in the lot of six units.
c. The most likely number of defective units in the lot of six units.
d. The standard deviation of the random variable **X**.

5.48 Let the random variable **X** denote the number of emergency calls received by the campus police office each day. The probability mass function of **X** is given by:

x	P(x)
0	0.75
1	0.05
2	0.10
3	0.05
4	0.05
	1.00

a. Find the expected value of **X**.
b. Determine the cumulative mass function of **X**.
c. What is the probability that the campus police office will receive at least one emergency call on a given day?
d. Each time an emergency call is received, it requires one hour to complete the written report related to the call. What is the average amount of time spent per day in completing these reports?

5.49 In a particular business venture, a man can make a profit of $1,000 with probability 0.2, a profit of $500 with probability 0.2, and a loss of $100 with probability 0.6. Let **X** be his profit or loss if he decides to participate in this game. *(a)* What is the expected value of **X**? *(b)* What is the most likely value of **X**?

5.50 A manufacturing firm is committed to the production of a new type of camera that produces instant-developing pictures. The production lot can be divided into three groups: group I, nondefective cameras; group II, defective cameras, but salvageable; and group III, defective cameras that cannot be salvaged, except for parts. The mathematical group predicts that 80 percent of all cameras produced will belong to I, 15 percent to II, and 5 percent to III, if the current production schedule is followed. The marketing group predicts that each camera in group I will bring a profit of $50, each camera in group II $10, and the group III cameras will each result in a loss of $100.

a. What is the expected profit (or loss) per camera?

b. If the company expects to produce 100,000 cameras in the next planning period, what is the total expected profit (or loss) during this period?

c. The mathematical group reports that if production is slowed down, the percentage of cameras in groups II and III can be cut to 10 percent and 3 percent, respectively. On the slowed-down schedule, they estimate that 88,000 cameras can be produced in the next planning period. Should the current production schedule or the slowed-down schedule be adopted?

5.51 If the probability that a man aged 50 will live another year is 0.995, how much should he pay for a $100,000 life insurance policy for the next year if the insurance company is to make an expected profit of $20?

5.52 From the most recent national census, it is found that the number of children (**X**) in American families follows the following probability distribution:

Number of children, x	0	1	2	3	4	5	6
Proportion of families, P(x)	0.48	0.20	0.15	0.08	0.05	0.03	0.01

Here it is assumed that the proportion of families with more than six children is negligible.

a. Find the expected value and variance of **X**.

b. Form a stick diagram of this distribution. Is the distribution skewed?

c. If we let **Y** represent the number of children in a family, *given* that the family has at least one child, the probability distribution of **Y** is given by:

y	1	2	3	4	5	6
P(y)	0.38	0.29	0.15	0.10	0.06	0.02

How is this distribution formed from the distribution of the random variable **X**?

d. Find the expected value and variance of **Y**.

e. Is $E(X)$ or $E(Y)$ more properly the "average family size"? Explain.

f. Define the random variable $Z = Y + 2$. Find $E(Z)$ and $V(Z)$. Interpret $E(Z)$. Might not this be a better number to call the "average family size"? Explain.

g. Use the Chebychef theorem to determine the proportion of values of **X** that are contained within two standard deviations of the mean value. Compare this proportion with the actual proportion in this interval determined directly from the probability distribution.

5.53 A group of college students conducts a survey on the quality of food served by the Goliet Corporation Food Service Unit in the campus dining room. They decide on the following measure: **X** = Number of days each week that the food is awful. Assuming that **X**

is measured only on Mondays through Fridays inclusive, the random variable **X** can take on one of six values: 0, 1, 2, 3, 4, and 5. After several weeks of survey gathering, the students determine the following probability distribution:

x	P(x)
0	0.05
1	0.20
2	0.30
3	0.25
4	0.10
5	0.10
	1.00

a. What is the expected number of days each week that the food is awful?
b. What is the standard deviation of the random variable **X**?
c. What is the mode of **X**?
d. The Goliet food manager figures that $100 is lost each week for every day that the food is awful due to the loss of customers in the future (that is, students who will not buy the meal plan in future semesters). What is the average loss per week for the Goliet Food Service due to times when they serve awful food?

5.54 The ABM corporation is concerned about the large number of its new AT microcomputers that are defective. Suppose that ABM is currently shipping the AT computer to retailers in lots of five units. Let the random variable **X** represent the number of defective AT computers in each lot of five units. Suppose it is known that **X** follows the probability mass function given in the table below:

x	P(x)
0	0.50
1	0.20
2	0.10
3	0.10
4	0.05
5	0.05

a. What is the expected number of defective units in each lot of five units?
b. What is the standard deviation of the random variable **X**?
c. What is the most likely number of defective units in a typical lot of five units?
d. For each defective unit in the lot of five units shipped to retailers, it costs ABM $250 to have the unit shipped back to its plant, repaired, and shipped back to the retailer. What is the expected cost to IBM to repair defective units in each lot of five units sent to retailers?

5.55 A survey is taken on drivers in a certain city to determine whether or not they use seat belts regularly. In the survey, data are taken on the age of the driver as well. The resulting frequency table is given below:

Age	Seat Belts		
	Use	Do Not Use	Total
Younger than 40	10	40	50
40 or older	20	30	50
Total	30	70	100

Define the following two random variables:

$$\mathbf{X} = \begin{cases} 0 \text{ if age of driver is less than 40} \\ 1 \text{ if age of driver is 40 or older} \end{cases}$$

$$\mathbf{Y} = \begin{cases} 0 \text{ if driver uses seat belts} \\ 1 \text{ if driver does not use seat belts} \end{cases}$$

a. Form the bivariate mass function for the random variables **X** and **Y**.
b. Are the two random variables **X** and **Y** independent? Explain.

5.56 The expected value and the variance of the random variable **X** are 10 and 15, respectively. The expected value and the variance of the random variable **Y** are 8 and 4, respectively. The two random variables are independent.
a. Determine E(**XY**).
b. Determine V(**X** + **Y**)
c. Determine V(**X** − **Y**)
d. Determine the covariance of **X** and **Y**.

6

Discrete probability distribution models

6.1

Introduction

Three possible representations of a probability mass function for a discrete random variable—a random variable that takes on only discrete values—are shown in Figure 6.1. In Figure 6.1A the discrete random variable can take on one of nine values, and the distribution is skewed to the right. In Figure 6.1B the discrete random variable can take on one of 13 values, and the distribution is symmetric. In Figure 6.1C the discrete random variable can take on one of 10 values, and the distribution is skewed to the left.

Skewness is an important characteristic of a probability distribution. Determining whether a probability distribution is symmetric or skewed will be useful in the selection of an appropriate inference technique as we shall see in later chapters.

In applied problems the exact form of the distribution of a discrete random variable may not be known, since only one or a few values of the random variable may be available for analysis. A clue to determining the form of the distribution of a random variable is given by a frequency distribution of the available values. Frequently, additional insight in determining the form of a probability distribution is made available by studying the nature of the experiment that produces specific values of the random variable.

In this chapter we shall study the nature of several experiments that lead to discrete random variables with distributions that frequently occur in applied problems and are of specific types. In practice the experiment producing the values of the random variable will often be similar to one of the experiments presented in this chapter. The probability distribution associated with the selected specific experiment is then used as a "model" for the

FIGURE 6.1 | Three possible representations of a probability distribution of a discrete random variable

A. Skewed right (9 values)

B. Symmetric (13 values)

C. Skewed left (10 values)

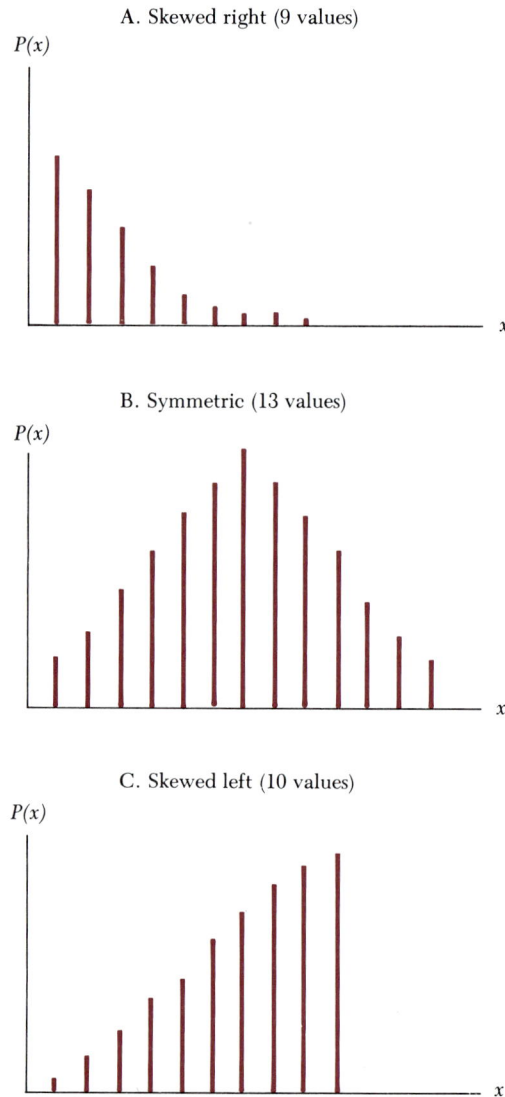

exact, but unknown, distribution of the random variable considered in the problem.

It is important to check how well the selected probability distribution model "fits" the available data—a poor fit can easily lead to erroneous inferences about the properties of the random variable. In Chapter 19 methods will be presented to help assure us that the selected probability distribution model is appropriate to represent the distribution of the random variable of interest. For the present we shall be concerned with describing the properties of several of the more important discrete probability distributions frequently used in statistics. We will then focus our attention in Chapter 7 on experiments that produce values of a continuous random variable and the resulting probability distribution models.

■ 6.2

The discrete uniform distribution

6.2.1 | Introduction

A distribution that arises frequently in practice—especially in certain gambling situations—is the discrete uniform distribution. The distribution is also used in making decisions (statistical decision theory) when we are uncertain about events that might occur, and it is being used increasingly in computer simulation experiments where "trial environments" are modeled on a computer. Figure 6.2 gives several examples of experiments that produce values of a discrete uniform random variable.

Assume that we assign the values 1, 2, . . . , N to a discrete random variable when it is in state 1, state 2, . . . , state N and the random variable is qualitative (e.g., head, tail; heart, spade, diamond, club; etc.), and that a quantitative discrete random variable assumes the values 1 through N. Then

FIGURE 6.2 | Experiments producing values of a discrete uniform random variable

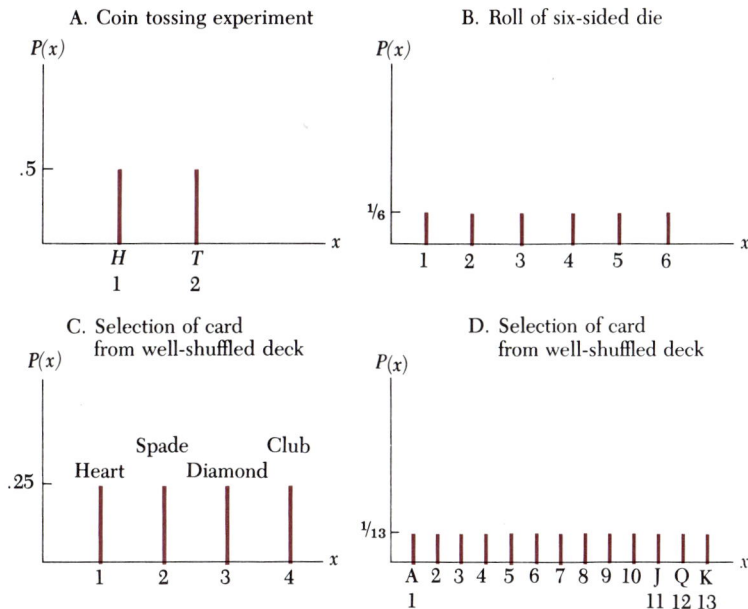

the definition of a discrete uniform random variable is as given in Definition 6.1.

Discrete uniform random variable

> **Definition 6.1**
> Discrete uniform random variable
>
> Let the random variable **X** assume the integer values 1 through N. If the probability that the random variable assumes one of the values 1 through N is $1/N$ for all values i, where $1 \leq i \leq N$, and if these probabilities do not change on repeated experiments to produce values of the random variable (independence), then **X** is a *discrete uniform random variable*.

The probability mass function of a discrete uniform random variable is given in Definition 6.2.

Discrete uniform probability mass function

> **Definition 6.2**
> Discrete uniform probability mass function
>
> Let **X** be a discrete uniform random variable. Then the *probability mass function* of **X** is given by
>
> $$P(x) = \frac{1}{N} \qquad x = 1, 2, \ldots, N$$
>
> where N = The number of states or the values the random variable can assume.

6.2.2 | The mean and variance of discrete uniform random variables

The mean and variance of a discrete uniform distribution can be computed using the formulas presented in Chapter 5.

$$\text{Mean:} \qquad E(\mathbf{X}) = \sum_{x=1}^{N} x \cdot P(x) = \sum_{x=1}^{N} x \cdot \frac{1}{N} = \frac{1}{N} \sum_{x=1}^{N} x$$

$$\text{Variance:} \quad V(\mathbf{X}) = \sum_{x=1}^{N} [x - E(\mathbf{X})]^2 P(x) = \frac{1}{N} \sum_{x=1}^{N} [x - E(\mathbf{X})]^2$$

In general, the mean and variance of a discrete uniform distribution with N states or with N values, 1, 2, . . . , N, are given by the formulas in

Theorem 6.1. The derivation of the mean and the variance formulas for the discrete uniform distribution are given in the appendix at the end of this chapter.

Mean and variance of a discrete uniform random variable

Theorem 6.1

Mean and variance of a discrete uniform random variable

If X is a discrete uniform random variable with N states or with N values, 1, 2, . . . , N, then its *mean*, $E(X)$, and *variance*, $V(X)$, are given by

$$E(X) = \frac{N + 1}{2} \quad \text{and} \quad V(X) = \frac{N^2 - 1}{12}$$

6.2.3 | Use of the discrete uniform model

We will illustrate the use of the discrete uniform distribution in the following three examples.

Example 6.1 Let the possible outcomes of the roll of a six-sided die be the *number of dots* that appear on the top side of the die, the integers 1, 2, 3, 4, 5, and 6. Determine the expected value and the variance of this random variable using Theorem 6.1.

Solution

$$E(X) = \frac{6 + 1}{2} = 3.5 \qquad V(X) = \frac{6^2 - 1}{12} = \frac{35}{12}$$

Example 6.2 In the die-tossing example, what is the probability of rolling a three *or higher* (i.e., three or more dots appear on the top side) on each roll of the die?

Solution We may use the addition law from Chapter 3 to determine:

$$P(X \geq 3) = P(3) + P(4) + P(5) + P(6) = \frac{1}{6} + \frac{1}{6} + \frac{1}{6} + \frac{1}{6} = \frac{4}{6} = \frac{2}{3}$$

Example 6.3 A large corporation has an office staffed by five people who are on call to pick up and deliver packages throughout the building. One person is always in the office to answer the telephone. Packages can only be picked up and delivered immediately if there are two or more of the five staff members in the office at the time a request is made. By experience, the CEO's secretary has learned that the number of staff members in the office at any given time follows a discrete uniform distribution. Let the random variable X = Number of staff members in the office at any given time.

a. What is the expected number of staff members in the office at any given time?

b. If the CEO's secretary calls the office for a pickup, what is the probability that the office can send a staff member immediately for the pickup of the package?

Solution   In part *a*, $E(\mathbf{X}) = (N + 1)/2 = (5 + 1)/2 = 3$ staff members. In part *b*, we want $P(\mathbf{X} \geq 2)$, and this is given by

$$\dot{P}(\mathbf{X} \geq 2) = P(2) + P(3) + P(4) + P(5)$$
$$= 0.20 + 0.20 + 0.20 + 0.20$$
$$= 0.80$$

Therefore, 80 percent of the time when the CEO's secretary calls for a package pickup, the office will be able to respond immediately.

■ **6.3**	**6.3.1** │ **Introduction**
The binomial distribution	A very important distribution of a discrete random variable is the binomial distribution. We will introduce the binomial distribution by first presenting the *Bernoulli random variable* and its distribution.

Bernoulli random
variable

Definition 6.3
Bernoulli random variable

Suppose that an experiment produces only one of two possible outcomes, labelled either a "success" or a "failure." Let the random variable **X** equal 1 if the outcome is a success and 0 if the outcome is a failure. Then the random variable **X** is a *Bernoulli random variable*.

The probability mass function of the Bernoulli random variable is quite simple. Let π = The probability of a success in the experiment. Then the probability mass function for the Bernoulli random variable **X** is given by Definition 6.4 below.

Bernoulli probabil-
ity mass function

Definition 6.4
Bernoulli probability mass function

Let the random variable **X** be a Bernoulli random variable. Then the *probability mass function* of **X** is given by

$$P(\mathbf{X} = 0) = P(0) = 1 - \pi$$
$$P(\mathbf{X} = 1) = P(1) = \pi,$$

where $\pi = P(\text{Experimental outcome is a "success"})$, and $0 \leq \pi \leq 1$.

The mean and the variance of the Bernoulli random variable can be computed by using the formulas presented in Chapter 5:

$$\text{Mean:} \quad E(\mathbf{X}) = \sum_{x=0}^{1} xP(x) = (0)(1 - \pi) + (1)(\pi) = \pi$$

$$\text{Variance:} \quad V(\mathbf{X}) = \sum_{x=0}^{1} [x - E(\mathbf{X})]^2 P(x)$$
$$= [0 - \pi]^2(1 - \pi) + [1 - \pi]^2(\pi)$$
$$= \pi^2 - \pi^3 + (1 - 2\pi + \pi^2)(\pi)$$
$$= \pi^2 - \pi^3 + \pi - 2\pi^2 + \pi^3$$
$$= \pi - \pi^2 = \pi(1 - \pi)$$

These results are summarized in Theorem 6.2 below.

Mean and variance
of a Bernoulli
random variable

> **Theorem 6.2**
>
> **Mean and variance of the Bernoulli random variable**
>
> If **X** is a Bernoulli random variable, then its *mean*, $E(\mathbf{X})$ and *variance*, $V(\mathbf{X})$, are given by
>
> $$E(\mathbf{X}) = \pi \quad \text{and} \quad V(\mathbf{X}) = \pi(1 - \pi)$$

A binomial random variable occurs when the experiment producing a Bernoulli random variable is repeated under certain conditions. When the conditions exist, then we say that a *Bernoulli process* has occurred. The definition of a Bernoulli process is given in Definition 6.5 below.

Bernoulli process

> **Definition 6.5**
>
> **Bernoulli process**
>
> A *Bernoulli process* is a process in which an experiment is repeatedly performed, yielding either a "success" or a "failure" in each trial, and where the occurrence of a success or a failure in a particular trial does not affect or is not affected by the outcomes in any previous or subsequent trial—the trials are independent.

Consider the following experiment: A coin is tossed *n* times, and the *number of heads* is counted in the *n* tosses. If the *n* coin tosses are independent events, and if we define the random variable **X** to be the *number of heads* occurring in the *n* tosses, then **X** is a *binomial random variable*. We can think of the trials of the experiment as being generated by a Bernoulli process. Each trial produces a head ("success") or a tail ("failure"), and all trials are independent.

The key elements of the experiment that produce values of a binomial random variable are:

1. There are *n* independent trials of the experiment.

2. In each trial, there are only two possible outcomes of the experiment—"success" and "failure."

3. The probability of "success" is the same for each trial.

4. The random variable is defined to be the *number of successes* in the *n* trials.

The definition of the binomial random variable is now summarized in Definition 6.6 below.

Binomial random variable

Definition 6.6
Binomial random variable

Let the random variable **X** be the number of successes in *n* independent trials of a simple experiment, where each trial results in one of two possible outcomes—"success" or "failure." Let π be the probability of a success in each and every one of the *n* trials. Then, **X** is a *binomial random variable*.

Example 6.4   Returning to the die-tossing experiment of Examples 6.1 and 6.2, suppose a net payoff is *received* if four, five, or six dots appear on the top side, and a net payout is *made* if one, two, or three dots appear on the top side. Is this experiment a Bernoulli process, and, if so, how should we define **X** so that it is a binomial random variable?

Solution   Define a "success" to be a four, five, or six, and define a "failure" to be a one, two, or three. Repeated trials of the experiment (rolls of the die) are indeed a Bernoulli process because:

1. The die is rolled independently *n* times (or at least we hope so).

2. In each trial, there are only two possible outcomes: a "success" (four, five, or six), or a "failure" (one, two, or three).

3. The probability of success is the same for each trial [$\pi = P(4) + P(5) + P(6) = 1/6 + 1/6 + 1/6 = 1/2$].

4.   The random variable **X** is defined to be the number of successes (i.e., rolls resulting in a four, five, or six) in n trials of the experiment.

The binomial random variable occurs in many other situations. Table 6.1 gives examples of random variables with distributions that are binomial *or with distributions that may be approximated* by the binomial distribution. Assuming the coin is fair and it is flipped fairly, the number of heads in example I is binomially distributed. In examples II, III, and IV, the random variables are *not* binomially distributed since the Bernoulli trials are dependent. To appreciate this, suppose in example III that there are 1,000 units of a product in a warehouse, 100 of which are defective. If we take two units from the warehouse, the probability that the second item withdrawn is defective depends on the outcome in the first trial. If the first trial (selection) produces a defective, then the probability that a second trial produces a defective is 99/999, whereas the probability of a defective in the second trial if the first trial produced a nondefective is 100/999. The appropriate probability distribution for these random variables is the hypergeometric distribution, which is considered in Section 6.4. In many real situations, the binomial distribution is used as a *model* (approximating distribution) for the distribution of a hypergeometric random variable. We will discuss this more fully in Section 6.4. For the moment, the trials in these experiments would be independent, and the binomial distribution would be the *exact* distribution if each item were replaced in the population before the next selection. For instance, in example III, we could replace each item before the next is drawn, but this would make it possible to examine the same item more than once, an inefficient sampling process.

In examples V and VI, the trials are likely to be dependent as well. In the birth example, the probability of a success (a boy, let us say) is not the same in each and every trial since some families may produce more offspring of one sex than the other (due to genetic considerations). However, it is reasonable to assume that the binomial distribution would fit well to the true distribution of the random variable due to the similarities of the experiments that produce the two random variables (the binomial random variable and the random variable aggregating the number of boy births). In example VI, the probability of winning a bid may drastically change from trial to trial. Thus the binomial distribution would not be a good model for the random variable *unless* it is reasonable to assume that the probability of winning a bid remains relatively the same for all bids placed.

The labeling of the two outcomes in each trial of the experiment as "success" and "failure" is completely arbitrary as long as the probabilities associated with each outcome are not confused. For instance, suppose in example IV that the proportion of Democrats in the population is 0.60. If the outcome "Democrat" is labeled as a success, then the probability of success in each trial of the experiment, π, is equal to 0.60. If the outcome "Republican" is labeled as a success, then $\pi = 0.40$—the definition of a successful outcome must be consistent with the assignment of the probability of a success in each trial of the simple experiment. The probability mass function of a binomial random variable is delineated in Definition 6.7.

TABLE 6.1 | Examples of binomial random variables and random variables that could possibly be modeled by the binomial distribution

Example	Trial	Success	Failure	π	n	X
I. Tossing a coin		Head	Tail	½ (if the coin is fair)	n tosses	Number of heads
II. Contacting a woman and recording whether or not she would purchase a particular consumer product		Decision to purchase	Decision not to purchase	Proportion of women who would purchase the product	Number of women contacted	Number of purchase decisions in the sample selected
III. Selecting a product from a production line		Nondefective	Defective	Proportion of nondefectives in lot	Size of sample	Number of nondefectives in the sample
IV. Selecting a registered voter in a poll		Democrat	Republican	Proportion of Democrats in population of registered voters	Size of sample in poll	Number of Democrats in sample
V. Birth of a child		Boy	Girl	Virtually ½	Number of births during April at a specific hospital	Number of boys born at the hospital during April
VI. Bidding on a construction project		Winning bid	Losing bid	Probability of a successful bid	Number of bids during the year	Number of bids won during year

Definition 6.7
Binomial probability mass function

Let **X** be a binomial random variable. Then the *probability mass function* of **X** is given by

$$P(x) = C_x^n \pi^x (1 - \pi)^{n-x} \qquad x = 0, 1, 2, \ldots, n$$

where

n = Number of trials.
π = Probability of a "success" in each trial.
$$C_x^n = \frac{n!}{x!(n - x)!}$$

In the next example problem, the derivation of the probability mass function for the binomial distribution is illustrated.

Example 6.5 The Baldwin, Rhodes, and Hilltop law firm uses microcomputers extensively in its offices for word processing and analyzing data arising from legal cases. Baldwin is very concerned about the security of the room containing the floppy disks and the microcomputers in the evenings and on weekends. The floppy disks contain sensitive information, and the firm's reputation would be seriously damaged if the disks were stolen from the room. Baldwin believes that there is adequate security during the day when people are in the offices, but not when the offices are left unattended. He contracts with a security systems company to install three movement-sensing alarms that can be turned on when no one is in the room in the evenings and on the weekends. Each alarm scans the same area, and the alarms act independently (that is, an alarm is triggered by detecting movement and not by another alarm going off). Given that there is movement in the room, each alarm has a probability of 0.90 of detecting the movement. Let the random variable **X** = Number of alarms that detect movement in the room when it occurs. Determine the probability distribution of **X**.

Solution The probability distribution of **X** can be found by the methods presented in Chapters 3, 4, and 5. In Figure 6.3 a probability tree for the experiment is shown. Each alarm either detects (D) the movement or does not detect ($\overline{D}$) the movement. Each alarm therefore represents a Bernoulli experiment. Since the alarms operate independently, a Bernoulli process generates the values of the random variable **X** = Number of alarms that detect the movement. By using the fact that the trials are independent, the probability of each outcome can be determined by multiplying the probabilities together on the tree branches as shown in Figure 6.3. The values of the random variable **X** are listed in the right-most column for each of the eight experimental outcomes. Using the probabilities of the outcomes and the values of **X**, we can determine the probability distribution of **X**:

x	$P(x)$
0	0.001
1	$0.009 + 0.009 + 0.009 = 3(0.009) = 0.027$
2	$0.081 + 0.081 + 0.081 = 3(0.081) = 0.243$
3	0.729
	1.000

FIGURE 6.3 | The probability tree for the Example 6.3 experiment

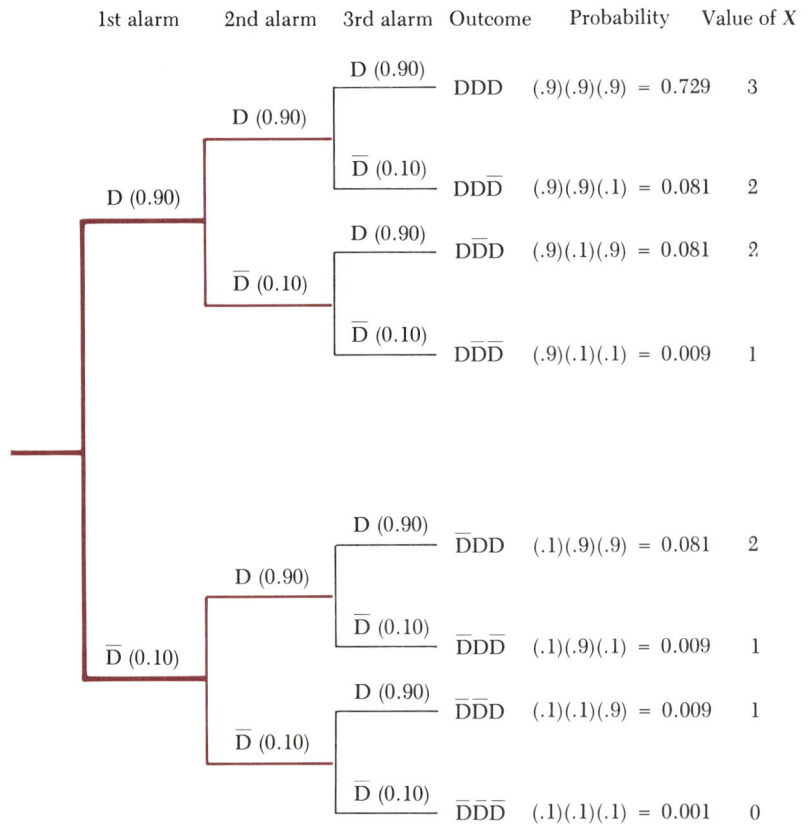

1st alarm	2nd alarm	3rd alarm	Outcome	Probability	Value of X

Notice from the table that there are three ways that one alarm can detect the movement, and there are three ways that two alarms can detect the movement. Now, consider the binomial mass function for this experiment: $n = 3$ trials, and $\pi = 0.90$ (probability of detection on each trial). The mass function is:

$$P(x) = C_x^n \pi^x (1 - \pi)^{n-x} = C_x^3 (0.90)^x (1 - 0.90)^{3-x}, \qquad x = 0, 1, 2, 3$$

Therefore,

$$P(0) = C_0^3(0.90)^0(0.10)^{3-0} = \frac{3!}{0!(3-0)!}(1)(0.001)$$
$$= (1)(1)(0.001) = 0.001$$

$$P(1) = C_1^3(0.90)^1(0.10)^{3-1} = \frac{3!}{1!(3-1)!}(0.90)(0.01)$$
$$= (3)(0.90)(0.01) = 0.027$$

$$P(2) = C_2^3(0.90)^2(0.10)^{3-2} = \frac{3!}{2!(3-2)!}(0.81)(0.10)$$
$$= (3)(0.81)(0.10) = 0.243$$

$$P(3) = C_3^3(0.90)^3(0.10)^{3-3} = \frac{3!}{3!(3-3)!}(0.729)(1)$$
$$= (1)(0.729)(1) = 0.729$$

In comparing the probabilities determined from the binomial mass function and the probabilities determined from the probability tree in Figure 6.3, notice that the term C_x^n in the mass function counts the number of ways that each value of the random variable can occur. For example, when $\mathbf{X} = 2$, two of the alarms detect the movement, and one does not. This value of the random variable corresponds to three experimental outcomes ($DD\overline{D}$, $D\overline{D}D$, and $\overline{D}DD$), each of which has a probability of $(0.90)^2(0.10)^1$ of occurring. While it would be possible to use a probability tree to answer probability questions concerning a binomial random variable, using the binomial probability mass function certainly turns out to be much easier!

The specific form of the binomial distribution depends on the values of the two *parameters*—π and n—of the distribution. After values have been assigned to the parameters π and n, probabilities of the values of the binomial random variable $\mathbf{X}$ occurring can be computed for the specific distribution. Thus the probability mass function, $P(x) = C_x^n \pi^x (1 - \pi)^{n-x}$, where $x = 0, 1, 2, \ldots, n$, represents a *family* of probability distributions, each member of which arises when numerical values are specified for the parameters π and n. Figure 6.4 illustrates the form of specific members of the binomial family for selected values of the parameters π and n. Note that the distribution is symmetric when $\pi = 0.5$, skewed to the left when $\pi > 0.5$, and skewed to the right when $\pi < 0.5$.

6.3.2 | Computing binomial probabilities

Given the values π and n, binomial probabilities may be computed in a number of ways. If n is not too large (less than 10), the probability mass function formula can be used to determine the probability that the binomial random variable $\mathbf{X}$ assumes a particular value x. If n is very large, the use of the definition of the probability mass function to determine $P(\mathbf{X} = x)$ becomes, at best, tedious. By using computers, statisticians have compiled extensive tables of binomial probabilities for a large number of values of π and n.

FIGURE 6.4 | Specific members of the binomial family of probability distributions

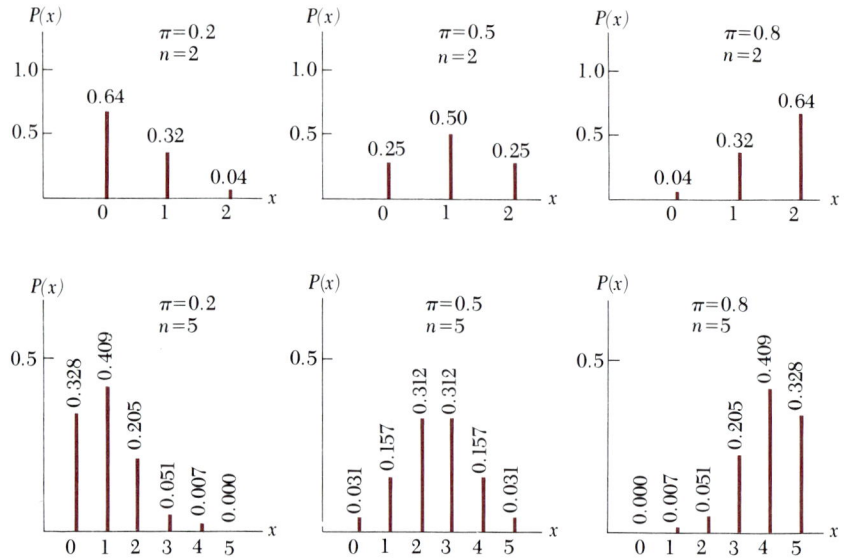

Table B.1 in Appendix B contains two binomial tables: The first table gives the binomial mass function probabilities $P(x)$, and the second gives the cumulative binomial probabilities $F(x)$, where

$$F(x) = \sum_{x=0}^{a} P(x) = P(0) + P(1) + \cdots + P(a) = P(\mathbf{X} \leq a/n, \pi)$$

The probabilities $P(x)$ and the cumulative probabilities $F(x)$ are given for the following parameter values: $n = 1, 2, \ldots , 25$ and 50; $\pi = 0.01, 0.05, 0.10,$ $0.20, 0.30, 0.40, 0.50, 0.60, 0.70, 0.80, 0.90, 0.95,$ and 0.99. More extensive tables have been compiled by statisticians and are usually available at the college library. In addition, there are several computer software programs that will compute binomial probabilities for any (n, π) parameter combination. We will consider one of these in Section 6.6 in this chapter.

A method for approximating binomial probabilities for large n will be given in Chapter 7. For the moment, we will consider several examples illustrating the use of the binomial distribution.

Example 6.6 In a family of three children, what is the probability that there will be exactly two girls, assuming that the sexes are equally likely to occur in each birth and that the "trials" are independent?

Solution Let the event "girl" represent a "success," and define $\mathbf{X}$ to be the number of successes (girls) produced in the $n = 3$ trials. The probability of a success in a single trial, π, is 0.5. Thus the specific member of the binomial family of distributions, expressed by its mass function, is:

$$P(x) = C_x^3(0.5)^x(1 - 0.5)^{3-x}, \qquad x = 0, 1, 2, 3$$

The probability that **X** assumes the value of 2 is given by:

$$P(\mathbf{X} = 2) = P(2) = C_2^3(0.5)^2(1 - 0.5)^{3-2} = \frac{3!}{2!(3 - 2)!}(0.5)^2(0.5)^1$$

$$= (3)(0.25)(0.5) = 0.375$$

In Table B.1, we can find the binomial probabilities $P(0)$, $P(1)$, $P(2)$, and $P(3)$ from the *probability mass table* (the first of the two tables in Table B.1). We first find the subtable for $n = 3$ and then locate the column heading $\pi = 0.500$. The probabilities associated with $\mathbf{X} = 0, 1, 2$, and 3 are then given in the column. The binomial probability distribution for $n = 3$ and $\pi = 0.5$ is given in three different ways (table, stick diagram, and formula) in Table 6.2.

TABLE 6.2 | Binomial distribution for $\pi = 0.5$ and $n = 3$; tabular, stick diagram, and functional forms

Table		Stick diagram	Formula
x	$P(x)$		$P(x) = C_x^3(.5)^x(1-.5)^{3-x}$
0	0.125		
1	0.375		$x = 0, 1, 2, 3$
2	0.375		
3	0.125		

Example 6.7 A large lot of items (10,000) has been produced by a certain production process. It is known by the wholesaler that 20 percent of the items are defective. Herman, the retailer, randomly samples six items from this lot. He will purchase the lot if he finds one or fewer defective items among the sampled six. Determine the probability that the wholesaler sells the lot to Herman.

Solution[1] Let a "success" be a defective item and a "failure" be a nondefective item. The probability of success, π, is 0.20, and the probability

---

[1] Actually, we are using the binomial distribution to *approximate* the true distribution in this example, for this is similar to example III in Table 6.2. The probability of a success in the first trial (selection of a defective) is 0.20. If a defective is found in the first trial, then the probability of a defective on the second trial is $1,999/9,999 = 0.19992 \approx 0.20$. If the first trial results in a nondefective, then the probability of a defective in the second trial is $2,000/9,999 = 0.20002$. Hence the probability of a success on each trial is *not independent* of what preceded it. However, because the number selected (6) is small in relation to the population size (10,000), the binomial distribution appears to be a reasonable approximation of the true distribution of defectives in this experiment.

of a failure—a nondefective item—is 0.80. The approximating probability mass function is:

$$P(x) = C_x^6(0.20)^x(1 - 0.20)^{6-x} \qquad x = 0, 1, 2, 3, 4, 5, 6$$

Herman will accept the lot if he finds one or fewer defectives in the six trials of the experiment. Therefore, the probability that he will accept the lot is given by:

$$P(\text{accept}) = P(0) + P(1) = C_0^6(0.20)^0(0.80)^6 + C_1^6(0.20)^1(0.80)^5$$
$$= 0.2621 + 0.3932 = 0.6553$$

The probabilities $P(0)$ and $P(1)$ can, of course, be found from the probability mass function table in Table B.1. With $n = 6$ and $\pi = 0.20$, we find from the table that $P(0) = 0.262$, and $P(1) = 0.393$. Notice that the table gives the probabilities only to three decimal places. Similarly, the probability that Herman will reject the lot is given by:

$$P(\text{reject}) = P(2) + P(3) + P(4) + P(5) + P(6) = 1 - [P(\text{accept})]$$
$$= 1 - [P(0) + P(1)] = 0.3447$$

A "success" in this example is defined to be a defective item so that we can work with a direct statement of the problem. If instead we define a success to be a *nondefective* item, then the appropriate probability mass function is:

$$P(x) = C_x^6(0.80)^x(1 - 0.80)^{6-x}, \qquad x = 0, 1, 2, 3, 4, 5, 6$$

(where $\pi = 0.80$), and the probability that Herman will accept the lot is given by:

$$P(\text{accept}) = P(5) + P(6) = C_5^6(0.80)^5(0.20)^1 + C_6^6(0.80)^6(0.20)^0$$
$$= 0.3932 + 0.2621 = 0.6553$$

This stresses the arbitrariness of the labeling of the two outcomes in each trial of the experiment as to which outcome constitutes a "success" and which constitutes a "failure." The probabilities $P(5)$ and $P(6)$ can also be found from the probability mass function table in Table B.1. With $n = 6$ and $\pi = 0.80$, we find from the table that $P(5) = 0.393$ and $P(6) = 0.262$.

Example 6.8 An exam containing 10 multiple-choice questions is designed so that the probability of a correct choice on any question by guessing alone is 0.20. What is the probability that a student will get no more than six questions right by guessing? What is the probability that he will get exactly five questions right? What is the probability that he will get more than two right?

Solution We wish to find $P(\mathbf{X} \leq 6)$, $P(\mathbf{X} = 5)$, and $P(\mathbf{X} > 2)$, where $\mathbf{X}$ is defined to be the number of questions answered correctly. These probabilities could be computed by using the probability mass function with $n = 10$ and $\pi = 0.20$, where "success" is correctly answering a question, and "failure" is incorrectly answering a question. However, let us use Table B.1

in Appendix B to calculate these probabilities. The first probability that we want is $P(\mathbf{X} \leq 6)$. We can use the following notation,

$$P(\mathbf{X} \leq 6/n = 10, \pi = 0.20)$$

to conveniently display the values of the parameters n and π together with the probability we wish to find. This statement is read, "the probability that $\mathbf{X}$ is less than or equal to 6, *given* that $n = 10$ and $\pi = 0.20$." We used the slash (/) in probability statements in Chapter 4 for conditional probabilities. This probability can be determined by using the *cumulative binomial probability table*, the second table contained in Table B.1 in Appendix B. With $n = 10$ and $\pi = 0.20$, the entry in the row with $a = 6$ and the column $\pi = 0.20$ in the table is 0.999. Therefore

$$P(\mathbf{X} \leq 6/n = 10, \pi = 0.20) = 0.999$$

Of course, the first table contained in Table B.1—the probability mass table—could also be used by finding the probabilities for $P(0)$, $P(1)$, $P(2)$, $P(3)$, $P(4)$, $P(5)$, and $P(6)$ and summing these probabilities. These probabilities are shown in Figure 6.5.

FIGURE 6.5 $\quad$ $P(\mathbf{X} \leq 6/n = 10, \pi = 0.20)$

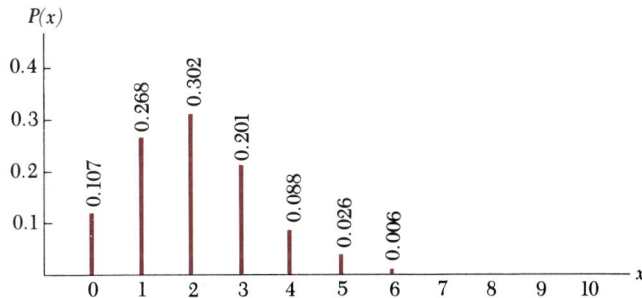

Next, we want to find $P(\mathbf{X} = 5)$, or in the more complete notation,

$$P(\mathbf{X} = 5/n = 10, \pi = 0.20)$$

This probability can be found directly from the probability mass function table—the first table in Table B.1 in Appendix B. With $n = 10$ and $\pi = 0.20$, the row entry for $a = 5$ is 0.027. Therefore

$$P(\mathbf{X} = 5/n = 10, \pi = 0.20) = 0.027$$

The final probability, $P(\mathbf{X} > 2)$, can be found from the cumulative binomial table in Table B.1 by noting two facts. First, $P(\mathbf{X} > 2) = P(\mathbf{X} \geq 3)$, since there is no binomial probability between the real numbers 2 and 3. Second, $P(X \geq 3) = 1 - P(X \leq 2)$, since the set of binomial probabilities for $n = 10$ and $\pi = 0.20$ must sum to 1. Thus the probability of "three or more" must be the complement of the probability "two or fewer." Now

$$
\begin{aligned}
P(X > 2/n = 10, \pi = 0.20) &= P(X \geq 3/n = 10, \pi = 0.20) \\
&= 1 - P(X \leq 2/n = 10, \pi = 0.20) \\
&= 1 - 0.678 \\
&= 0.322
\end{aligned}
$$

The probability $P(X \leq 2/n = 10, \pi = 0.20)$ is found from the cumulative binomial table with $n = 10$ and $\pi = 0.20$ in the row with $a = 2$.

6.3.3 | The mean and variance of binomial random variables

The mean and variance of a binomial distribution may be computed by using the formulas presented in Chapter 5:

Mean: $$E(\mathbf{X}) = \sum_{x=0}^{n} x \cdot P(x) = \sum_{x=0}^{n} x \cdot C_x^n \pi^x (1 - \pi)^{n-x}$$

Variance: $$V(\mathbf{X}) = \sum_{x=0}^{n} [x - E(\mathbf{X})]^2 P(x) = \sum_{x=0}^{n} [x - E(\mathbf{X})]^2 C_x^n \pi^x (1 - \pi)^{n-x}$$

In general, however, the mean and variance of a binomial distribution with parameters n and π can be simplified to the formulas presented in Theorem 6.3. These formulas are much simpler to work with than are the definitions given in Chapter 5, but they are equivalent.

Mean and variance of the binomial random variable	**Theorem 6.3** Mean and variance of the binomial random variable If $\mathbf{X}$ is a binomial random variable, then its *mean*, $E(\mathbf{X})$ and *variance*, $V(\mathbf{X})$, are given by $$E(\mathbf{X}) = n\pi \qquad \text{and} \qquad V(\mathbf{X}) = n\pi(1 - \pi)$$

The mean and variance of the binomial random variable are quite easy to derive by using the results on the expected value operator (E) and the variance operator (V) in Chapter 5. Let $\mathbf{Y}_i$ be a Bernoulli random variable with values 0 and 1 such that $P(\mathbf{Y}_i = 1) = \pi$, and $P(\mathbf{Y}_i = 0) = 1 - \pi$. Then, from Theorem 6.2, $E(\mathbf{Y}_i) = \pi$, and $V(\mathbf{Y}_i) = \pi(1 - \pi)$. Assume that we have n independent Bernoulli random variables $\mathbf{Y}_1, \mathbf{Y}_2, \ldots, \mathbf{Y}_n$. Then the binomial random variable $\mathbf{X}$ may be defined in terms of these n Bernoulli random variables by:

$$X = Y_1 + Y_2 + \cdots + Y_n$$

From Chapter 5,

$$
\begin{aligned}
E(X) = E(Y_1 + Y_2 + \cdots + Y_n) &= E(Y_1) + E(Y_2) + \cdots + E(Y_n) \\
&= \pi + \pi + \cdots + \pi \\
&= n\pi
\end{aligned}
$$

$$
\begin{aligned}
V(X) = V(Y_1 + Y_2 + \cdots + Y_n) &= V(Y_1) + V(Y_2) + \cdots + V(Y_n) \\
&= \pi(1 - \pi) + \pi(1 - \pi) + \cdots + \\
&\quad \pi(1 - \pi) \\
&= n\pi(1 - \pi)
\end{aligned}
$$

The computation of the variance requires that $Y_1, Y_2, \ldots, Y_n$ are independent random variables.

Example 6.9   At a large state university, 70 percent of the students have a grade point average of less than 3.00 on a 4-point scale. Suppose that 25 of these students are chosen at random. Using the binomial distribution as an approximation, answer the following:

a.   What is the expected number of students in the set of 25 who have a grade point average of less than 3.00?
b.   What is the most likely number of students in the set of 25 who have a grade point average of less than 3.00?
c.   What is the standard deviation of the number of students in the set of 25 who have a grade point average less than 3.00?

Solution   Let the random variable X = Number of students in the set of 25 who have a grade point average of less than 3.00. Assuming that the distribution of X can be approximated by a binomial distribution, the parameters of the binomial distribution are $n = 25$ and $\pi = 0.70$. We can now answer the three questions above.

a. $E(X) = n\pi = (25)(0.70) = 17.5$ students. Thus if we repeatedly took random sets of 25 students, on the average we would expect to find that 17.5 of the 25 students had grade point averages less than 3.00.
b.   The most likely number of students in the set of 25 who have grade point averages less than 3.00 is the *mode* of the random variable X. To find the mode, we must find the value of X that has the largest probability of occurring. From the probability mass function table in Table B.1, Appendix B, with $n = 25$ and $\pi = 0.70$, scanning the column containing the probabilities for this binomial distribution, we find that when $a = 18$ the largest probability occurs (0.171). Therefore, the mode is 18.
c.   From Theorem 6.3, the variance of the number of students with a grade point average less than 3.00 is given by

$$V(X) = n\pi(1 - \pi) = 25(0.70)(0.30) = 5.25$$

Since the standard deviation is the square root of the variance, the standard deviation is equal to $\sqrt{5.25} = 2.29$ students.

■ **6.4**

The hypergeometric distribution

Frequently, when sampling from a finite population, the probability of a success or a failure changes as sample units are withdrawn (examined) and are not returned to the population (i.e., sampling without replacement). From Example 6.7, the probability of a success or a defective item for the sixth item sampled is *conditional upon* the number of defectives and nondefectives encountered in the first five units examined, since units are not returned to the population after being inspected so that they can be inspected more than once. If the number of units examined without replacement is small in relation to the population size, then the binomial distribution is a reasonable approximation of the true distribution of sample results. If, on the other hand, the ratio of units examined to units in the population is large, then the binomial distribution may not be a satisfactory approximation to the true distribution. Statisticians use the rule of thumb that if the ratio of units examined without replacement (n) to units in the population (N) is less than 5 percent (i.e., $n/N < 0.05$), then the binomial distribution is a reasonable approximation of the true distribution of selected successes.

It may also be possible to divide the population of N items into k mutually exclusive categories, each containing $N_1, N_2, \ldots, N_k$ elements. However, in this chapter, we will focus our attention on the special case of $k = 2$, where N_1 units in the finite population are labeled "successes," N_2 units are labeled "failures," and $N_1 + N_2 = N$.

The experiment producing values of a random variable **X** that has a hypergeometric distribution consists of selecting a set of n elements from a finite population of N elements without replacement. The hypergeometric random variable is described in Definition 6.8.

Hypergeometric random variable

Definition 6.8

Hypergeometric random variable

Let the random variable **X** be the number of successes observed when n items are selected at random from a finite population of N elements without replacement, and the number of successes in the finite population is equal to $N_1 \leq N$. Then **X** is a *hypergeometric random variable*.

The description of the hypergeometric probability mass function is given in Definition 6.9, and the mean and variance of the hypergeometric distribution where each element in the population is labeled either a "success" or a "failure" are given in Theorem 6.4.

Probability mass function for a hypergeometric random variable

---

Definition 6.9
Hypergeometric probability mass function

Let **X** be a hypergeometric random variable. Then the *probability mass function* of **X** is given by

$$P(x) = \frac{C_x^{N_1} \cdot C_{n-x}^{N_2}}{C_n^N}, \qquad x = 0, 1, 2, \ldots, min\{n, N_1\}$$

where

$$N_1 = \text{Number of successes in the population}$$
$$N_2 = \text{Number of failures in the population}$$
$$N_1 + N_2 = N$$
$$n = \text{Number of items (trials) examined (without replacement)}$$
$$x = \text{Number of successes in the } n \text{ trials}$$
$$min\{n, N_1\} = \text{Minimum of } n \text{ and } N_1$$

Mean and variance of a hypergeometric random variable

Theorem 6.4

If **X** is a hypergeometric random variable where the population of N items consists of two mutually exclusive and collectively exhaustive groups with N_1 and N_2 items in each group, then its *mean, $E(\mathbf{X})$, and variance, $V(\mathbf{X})$* are given by

$$E(\mathbf{X}) = n \cdot \frac{N_1}{N} \qquad \text{and} \qquad V(\mathbf{X}) = \frac{N-n}{N-1} \left[n \cdot \left(\frac{N_1}{N} \right) \cdot \left(1 - \frac{N_1}{N} \right) \right]$$

The use of the hypergeometric distribution is illustrated in the following example.

Example 6.10   Assume that a population of 10 items contains four defective units. If a sample of three items is selected at random without replacement, what is the probability that zero, one, two, or three defectives will appear in the items selected? What is the expected number (mean) and the variance in the sample? How do these results compare with those that would have been obtained if the binomial distribution had been used to approximate the hypergeometric distribution?

Solution   The determination of the various probabilities of occurrence can be accomplished by the application of Definition 6.9. For example,

$$P(0) = \frac{C_0^4 C_3^6}{C_3^{10}} = \frac{\left(\frac{4!}{0!4!}\right)\left(\frac{6!}{3!3!}\right)}{\frac{10!}{3!7!}} = \frac{1 \cdot 20}{120} = 0.167$$

where $P(0)$ is the probability of selecting zero "successes" (defectives) in a sample of three without replacement. Similarly,

$$P(1) = \frac{C_1^4 C_2^6}{C_3^{10}} = 0.500 \quad P(2) = \frac{C_2^4 C_1^6}{C_3^{10}} = 0.300 \quad P(3) = \frac{C_3^4 C_0^6}{C_3^{10}} = 0.033$$

From Theorem 6.4,

$$E(\mathbf{X}) = 3 \cdot \frac{4}{10} = 1.2$$

$$V(\mathbf{X}) = \left(\frac{7}{9}\right)(3)\left(\frac{4}{10}\right)\left(1 - \frac{4}{10}\right) = 0.56$$

Using Table B.1 in Appendix B with $n = 3$ and $\pi = N_1/N = 0.4$,

$$P(0) = 0.216 \qquad P(2) = 0.288$$
$$P(1) = 0.432 \qquad P(3) = 0.064$$

From Theorem 6.3,

$$E(\mathbf{X}) = n\pi = 1.2 \qquad \text{and} \qquad V(\mathbf{X}) = n\pi(1 - \pi) = 0.72$$

These results are summarized in Table 6.3. Note that the means for the two distributions are equal, and that the variances differ by a factor of $(N - n)/(N - 1)$, or by $7/9$. If we assume that $\pi = N_1/N$ in the hypergeometric distribution, then the formulas for the mean and variance from Theorem 6.4 are

$$E(\mathbf{X}) = n\pi \quad \text{and} \quad V(\mathbf{X}) = \frac{N - n}{N - 1}[n \cdot \pi(1 - \pi)]$$

Now the relationship between the binomial and the hypergeometric distributions becomes apparent. The quantity $(N - n)/(N - 1)$, which gives the difference between the variances of the two distributions, is called the *finite population correction factor,* and is further discussed in Chapter 8.

Because of the similarity between the hypergeometric and the binomial probability distributions, the binomial distribution is often used in practice even when the hypergeometric distribution applies. This is because the differences in the two distributions are important in practice only when the sample size n, or the number of units examined, is large in relation to the number of units in the population N. Statisticians use the general rule that if $n/N < 0.05$, the binomial distribution is a reasonable approximation of the hypergeometric distribution.

Binomial used to approximate hypergeometric when $n/N < 0.05$

TABLE 6.3 | Summary computations for Example 6.10, hypergeometric distribution and binomial approximation

	Hypergeometric distribution	Binomial approximation
$P(0)$	0.167	0.216
$P(1)$	0.500	0.432
$P(2)$	0.300	0.288
$P(3)$	0.033	0.064
	1.000	1.000
$E(\mathbf{X})$	1.2	1.2
$V(\mathbf{X})$	0.56	0.72

■ 6.5

The Poisson distribution

6.5.1 | Introduction

The final discrete distribution we will discuss in this chapter is the *Poisson distribution*. To describe the Poisson distribution and, in particular, to develop the notion of a Poisson random variable, consider a production process in which insulated wire is produced. In this process, it is known that defects (bare spots) occur at random but are known to average one defect per foot of wire produced. We shall initially view the inspection of the wire as a Bernoulli process in which the inspections of lengths of wire correspond to the trials of an experiment. We will concentrate on determining the probability of two defects in each segment of the wire we examine, and label the discovery of a defect a "success."

In the 1 foot of wire we select at random, we know that the *expected number* of defects is one. If we cut the wire in half—that is, into two equal 6-inch segments—and make the assumption that no more than one defect can occur in each 6-inch segment, then the examination of the wire can be treated as a Bernoulli process. Each segment either has a defect (success) or does not have a defect (failure). Since the expected number of defects per foot of wire is one, the probability of a defect is ½ on each of the 6-inch segments, and the probability of two defectives in a foot of wire (we can have only zero, one, or two defectives, since we are assuming that each segment contains at most one defect) is

$$P(2) = \frac{2!}{2!0!} \left(\frac{1}{2}\right)^2 \left(\frac{1}{2}\right)^{2-2} = 0.25$$

We know, of course, that there is some probability that two *or more* defects could be found in each 6-inch segment. Therefore, assume we now cut our wire into six 2-inch segments for a closer approximation, and again assume that no more than one defect can occur in each 2-inch segment. Since we now have six segments and each segment is assumed to contain at most one defect, we can calculate the probabilities of zero, one, two, three, four, five, or six defects in our 1-foot segment of wire. The probability of two defects is given by:

$$P(2) = \frac{6!}{2!(6-2!)} \left(\frac{1}{6}\right)^2 \left(\frac{5}{6}\right)^{6-2} = 0.201$$

If the examination of six 2-inch segments seems more valid than the examination of two 6-inch segments because the probability of more than one defect on each segment is less in the former instance, then we might envision cutting our wire into smaller and smaller segments such that the number of defects per piece approaches zero. For example, if we cut our wire into 50 pieces and examine each for a defect (here again we are assuming that each segment can contain no more than one defect), the probability that a 1-foot length of wire contains two defects is given by:

$$P(2) = \frac{50!}{2!(50-2)!} \left(\frac{1}{50}\right)^2 \left(\frac{49}{50}\right)^{50-2} = 0.1858$$

Since 50 pieces is a better basis for approximation, we might divide our wire into still smaller segments such that the probability of a segment containing more than one defect does indeed approach 0. Specifically, we might ask the limit of the following expression as n approaches infinity:

$$P(2) = \frac{n!}{2!(n-2)!} \left(\frac{1}{n}\right)^2 \left(\frac{n-1}{n}\right)^{n-2}$$

From calculus,

$$\lim_{n \to \infty} P(2) = \frac{e^{-1}1^2}{2!} = 0.184$$

where e is the base of the natural logarithms and is approximately 2.718.

Thus our approximation of $P(2)$ becomes more accurate as the number of trials of the experiment increases:

n	$P(2)$
2	0.25
6	0.201
50	0.1858
∞	0.184

If we let λ represent the average number of defects per unit of length ($\lambda = 1$ in our example) and let x represent the number of defects encountered, then the general expression for the probability of x defects in the wire example becomes

$$P(x) = \frac{e^{-\lambda}\lambda^x}{x!}$$

which, we will see shortly, is the probability mass function of a Poisson random variable.

In a large number of applied problems, the random variable of interest can assume a countably infinite number of possible integer values, 0, 1, 2, . . . , in a continuous interval. The interval might consist of a minute, an hour, a day, etc., and hence need not be specifically tied to a physical interval, such as the length of a wire. Other examples might be the number of customers arriving at a supermarket checkout counter in a minute and the number of aircraft arriving at an airport per minute. More specifically, the experiment generating a Poisson random variable is described in the following definition.

Poisson random variable

Definition 6.10

Poisson random variable

Let the random variable **X** be the number of occurrences ($x = 0, 1, 2, . . .$) of a specific event in a given continuous interval. The random variable **X** is called a *Poisson random variable* if the following conditions on the experiment generating values of **X** are satisfied:

1. The numbers of occurrences of the specific event in two nonoverlapping intervals are independent.

2. The probability of an occurrence of the specific event in a small interval is small and is proportional to the length of the interval.

3. The probability of two or more occurrences of the specific event in a small interval is negligible or is 0.

Another classic example of an experiment that generates values of a Poisson random variable is recording the number of telephone calls received at a switchboard in some *time* interval. For example, define the random variable **X** to be the number of calls received during a 5-minute interval at the switchboard. The possible values that **X** can assume are 0, 1, 2, . . .— countably infinite in number. If we consider the 5-minute interval to be composed of many small subintervals, say 1 second long, then the conditions for **X** to be a Poisson random variable described in Definition 6.8 are satisfied. This follows, since in the small time interval of, say, 1 second,

1. The number of calls received in a second time interval is independent of all other intervals of 1 second.

2. The probability of a phone call in a 1-second time interval is small and is proportional to the length of the interval.

3. The probability of two or more phone calls in 1 second is extremely small.

Several examples of other Poisson random variables are listed in Table 6.4.

TABLE 6.4 | Examples of Poisson random variables

Example	Description of **X**
I	Number of deaths per year due to a rare disease in a large city
II	Number of machines that break down in a large plant during any one day
III	Number of ships arriving per hour at a large dock
IV	Number of typographic errors per page in a large volume of printed material

Poisson probability
mass function

Definition 6.11
Poisson probability mass function

Let **X** be a Poisson random variable. Then the *probability mass function* of **X** is given by

$$P(x) = \frac{e^{-\lambda}\lambda^x}{x!}, \qquad x = 0, 1, 2, \ldots; \qquad \lambda \geq 0$$

where $e = 2.71828$ and λ is a parameter of the distribution. Numerically, λ represents the average number of occurrences of the specific event in the given interval of measurement.

As with the binomial distribution, we can think of the Poisson probability mass function as a representation of a family of distributions, each member of which is specified by selecting a particular value of the parameter λ. Figure 6.6 illustrates the form of the Poisson distribution for several selected values of the parameter λ.

From Figure 6.6, we see that when $\lambda = 0.1$, virtually all the probability is absorbed by the first four values of **X**. Indeed, if we compute $P(0) + P(1) + P(2) + P(3)$ using the values in Figure 6.6, we find that the sum is 1. We must recognize, however, that $P(\mathbf{X} \leq 3)$ is not exactly 1; rather, this is a *rounded* solution to four decimal places. The probability distribution when $\lambda = 0.1$ assigns probabilities to *all* the positive integers, but the probabilities assigned to the values of **X** that are greater than 3 are *very* small.

We can see from Figure 6.6 that as the value of λ increases, the probabil-

FIGURE 6.6 | Some specific members of the Poisson family of probability distributions (λ = 0.1, 0.5, and 1.0)

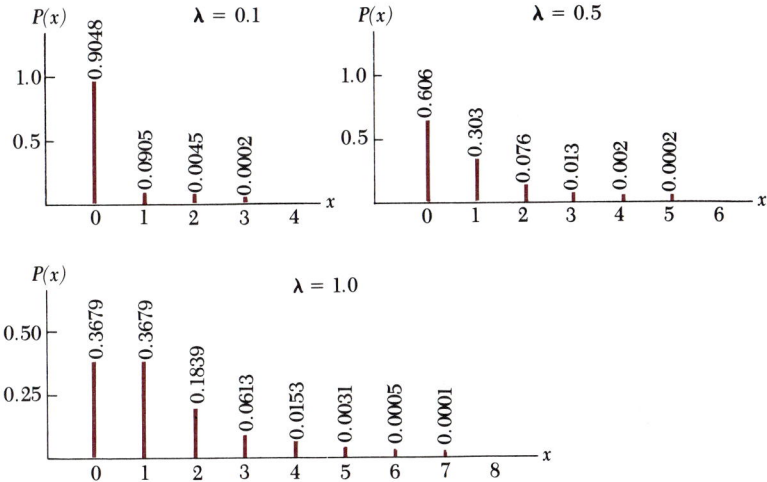

ity becomes more spread out over the values of the random variable **X**. In Table B.2 of Appendix B, Poisson probability distributions are presented in tabular form for several values of λ. In most problems, this table can be used to compute Poisson probabilities for specified values of λ rather than using the mass function given in Definition 6.11.

6.5.2 | The mean and variance of Poisson random variables

The mean and variance of the Poisson random variable may be computed from the formulas:

Mean:
$$E(\mathbf{X}) = \sum_{x=0}^{\infty} xP(x) = \sum_{x=0}^{\infty} x \cdot \frac{e^{-\lambda}\lambda^{x}}{x!}$$

Variance:
$$V(\mathbf{X}) = \sum_{x=0}^{\infty} [x - E(\mathbf{X})]^2 P(x) = \sum_{x=0}^{\infty} [x - E(\mathbf{X})]^2 \frac{e^{-\lambda}\lambda^{x}}{x!}$$

By the use of algebraic manipulation and the fact that

$$\sum_{x=0}^{\infty} P(x) = \sum_{x=0}^{\infty} \frac{e^{-\lambda}\lambda^{x}}{x!} = 1$$

it is possible to show that $E(\mathbf{X}) = \lambda$ and $V(\mathbf{X}) = \lambda$.

> **Mean and variance of a Poisson random variable**

Theorem 6.5

Mean and variance of the Poisson random variable

If **X** is a Poisson random variable, then its *mean*, $E(\mathbf{X})$, and *variance*, $V(\mathbf{X})$, are given by

$$E(\mathbf{X}) = \lambda \quad \text{and} \quad V(\mathbf{X}) = \lambda$$

6.5.3 | Use of the Poisson model

We will consider two examples that illustrate the use of the Poisson distribution.

Example 6.11 If the number of telephone calls that an operator receives in a 10-minute interval follows a Poisson distribution with $\lambda = 1$ (an average of one call each 10 minutes), what is the probability that she will receive no calls in a 10-minute interval? What is the probability of less than four calls? What is the most likely number of calls that she will receive?

Solution With $\lambda = 1$, the appropriate Poisson mass function is:

$$P(x) = \frac{e^{-1}1^x}{x!} = \frac{1}{e(x!)}, \qquad x = 0, 1, 2, \ldots$$

Therefore

$$P(\mathbf{X} = 0) = P(0) = \frac{1}{e(0!)} = \frac{1}{e} = \frac{1}{2.72} = 0.3679$$

Notice that this answer can be determined from Table B.2.

$$P(\mathbf{X} < 4) = P(\mathbf{X} \le 3) = P(0) + P(1) + P(2) + P(3)$$
$$= 0.3679 + 0.3679 + 0.1839 + 0.0613 \text{ (from Table B.2)} = 0.9810$$

The most likely value of **X** is the mode. From Figure 6.6 with $\lambda = 1.0$, we can see that the distribution is *bimodal*: The two most frequently occurring values of **X** are $x = 0$ and $x = 1$, each of which occurs with probability 0.3679.

Example 6.12 The number of admissions to the emergency ward of a hospital during the time beginning at 12:00 midnight and ending at 2:00 A.M. Saturday is found to be Poisson distributed with an average of 3.5 admissions. During this period on a particular Saturday morning, what is the probability that no one will be admitted? What is the probability that between two and five persons (inclusive) will be admitted? What is the most likely number of admissions?

Solution The appropriate Poisson mass function is:

$$P(x) = \frac{e^{-3.5}(3.5)^x}{x!}, \qquad x = 0, 1, 2, \ldots$$

Therefore

$$P(\mathbf{X} = 0) = P(0) = 0.0302 \quad \text{(from Table B.2)}$$
$$P(2 \le \mathbf{X} \le 5) = P(2) + P(3) + P(4) + P(5)$$
$$= 0.1850 + 0.2158 + 0.1888 + 0.1322 = 0.7218$$

The mode is 3, since $x = 3$ occurs with the greatest probability (0.2158).

The Poisson distribution is used extensively in applied problems to model the distribution of the number of persons joining a *queue* (a line) who wish to receive a service of some kind or purchase a product. With very little thought, it is easy to list numerous instances of this type: a grocery store checkout line, the line at your *friendly* college bookstore, dentist appointments, and so on. One must be careful to ensure that the Poisson distribution is appropriate in these applications. The experiment leading to the values of the random variable in a given applied problem should be critically compared with the Poisson random variable experiment to assess whether or not the conditions given in Definition 6.11 are satisfied. Once it has been determined that the Poisson distribution is an acceptable model for the random variable, the specific member of the Poisson family must be selected; that is, a value of λ must be chosen. Ordinarily, λ must be *estimated* from sample data (the estimation process will be discussed in Chapter 9). However, in many problems, it must be recognized that the value of λ is time-dependent. For example, in Example 6.12 the average number of emergency admissions during the 12:00–2:00 A.M. period on Saturday morning is most likely quite different from the average number of admissions during the period from 2:00 to 4:00 A.M. Tuesday. In this instance, we would have to use two different members of the Poisson family for the two periods. Care must be taken, therefore, to specify the period and its length to ensure that the average number of events (λ) occurring in the period is reasonably homogeneous throughout the entire period.

When the Poisson distribution is applied to queueing problems, it is often reparameterized as follows:

Reparameterized Poisson mass function

$$P(x) = \frac{e^{-\lambda t}(\lambda t)^x}{x!}, \qquad x = 0, 1, 2, \ldots ; \lambda t > 0,$$

where λ = Average number of events occurring per one unit of time measure, and t = Number of units of time over which the random variable $\mathbf{X}$ counts the number of events occurring. In Example 6.12, let $\lambda = 1.75$ admissions per hour with the time unit set equal to one hour (if there is an average of 3.5 admissions over the two-hour period, then there is an average of $3.5/2 = 1.75$ admissions over a one-hour period). Let the random variable $\mathbf{X}$ = Number of admissions in the two-hour period. Thus $t = 2$ in the reparameterized Poisson mass function. The mass function is given by:

$$P(x) = \frac{e^{-\lambda t}(\lambda t)^x}{x!} = \frac{e^{-(1.75)(2)}[(1.75)(2)]^x}{x!}$$
$$= \frac{e^{-3.5}(3.5)^x}{x!}$$

This is identical to the probability mass function used in the solution to Example 6.12 above. In most queueing applications, it is easier to specify λ as the number of events occurring per one unit of time measure and let t be the number of units of time over which the random variable **X** is counting events than to let λ = The average number of events occurring in the period of time directly.

■ **6.6**

Use of computer software packages

Most computer software packages will determine binomial and Poisson probabilities. One such package is MINITAB. MINITAB will produce the binomial mass function probabilities and the cumulative probabilities for specified values of the parameters n and π. For example, in an interactive session with MINITAB, we can type:

```
MTB> BINOMIAL PROBABILITIES FOR N=4 AND P=0.25
```

Note that MINITAB specifies the binomial parameter π as P. MINITAB then returns the following:

```
BINOMIAL PROBABILITIES FOR n = 4 AND P = 0.25

  K         P(X = K)        P(X less or = K)

  0          0.3164            0.3164
  1          0.4219            0.7383
  2          0.2109            0.9492
  3          0.0469            0.9963
  4          0.0039            1.0000
```

The first column lists the values of the random variable **X**, the second column gives the probabilities for the mass function, and the third column gives the cumulative probabilities. Notice that for this example Table B.1 in Appendix B cannot be used since the table gives the probabilities for $\pi = 0.20$ and $\pi = 0.30$, but skips over $\pi = 0.25$.

For the Poisson distribution, the command is:

```
MTB> POISSON PROBABILITIES FOR MEAN = 0.50
```

The mean of the Poisson distribution is λ, and in this example we have set $\lambda = 0.50$. The MINITAB output is:

```
POISSON PROBABILITIES FOR MEAN = 0.50

K          P(X = K)          P(X less or = K)

0           0.6065              0.6065
1           0.3033              0.9098
2           0.0758              0.9856
3           0.0126              0.9982
4           0.0016              0.9998
5           0.0002              1.0000
```

The first column lists the values of the random variable **X**, the second column lists the probabilities for the mass function, and the third column lists the cumulative probabilities.

The widespread use of microcomputers and software packages such as MINITAB may make many of the tables such as those contained in Appendix B obsolete, since the tables for any values of the parameters can be easily generated on the computer.

TABLE 6.5 | Summary characteristics of the discrete uniform, binomial, hypergeometric, and Poisson distributions

Distribution	Parameters	Mean $E(\mathbf{X})$	Variance $V(\mathbf{X})$	Calculated probabilities (Appendix B)
Discrete uniform	N	$\dfrac{N+1}{2}$	$\dfrac{N^2-1}{12}$	—
Binomial	n, π	$n\pi$	$n\pi(1-\pi)$	Table B.1: $P(\mathbf{X} \le a) = \displaystyle\sum_{x=0}^{a} P(x)$ and $P(\mathbf{X} = a) = P(a)$
Hypergeometric	n, N_1, N	$n\left(\dfrac{N_1}{N}\right)$	$\dfrac{N-n}{N-1}\left[n\left(\dfrac{N_1}{N}\right)\left(1-\dfrac{N_1}{N}\right)\right]$	—
Poisson	λ	λ	λ	Table B.2: $P(\mathbf{X} = a) = P(a)$

■ **6.7**

Summary

In this chapter, several examples of experiments that produce values of a discrete random variable were described. In particular, the specific discrete random variables and their probability mass functions developed were the discrete uniform, binomial, hypergeometric, and Poisson random variables and associated distributions. Table 6.5 summarizes several characteristics of these important discrete random variables.

In Chapter 7, we will describe several experiments that produce values of a continuous random variable. We will develop the associated probability distributions of the random variables described.

■ **References**

Introductory

McClave, J. T., and P. G. Benson. *Statistics for Business and Management*. 3rd ed. San Francisco: Dellen Publishing, 1985, Chapter 5.

Neter, J.; W. Wasserman; and G. A. Whitmore. *Applied Statistics*. 2nd ed. Boston: Allyn & Bacon, 1982, Chapter 6.

Wonnacott, R. J., and T. H. Wonnacott. *Introductory Statistics*. 4th ed. New York: John Wiley & Sons, 1985, Chapter 4.

Advanced

Harnett, D. L. *Introduction to Statistical Methods*. 3rd ed. Reading, Mass.: Addison-Wesley, 1982, Chapter 4.

Hogg, R. V., and A. T. Craig. *Introduction to Mathematical Statistics*. 4th ed. New York: MacMillan, 1978.

Ross, S. *Introduction to Probability Models*. 2nd ed. New York: Academic Press, 1980, Chapter 2.

■ **Appendix**

The mean and the variance of the discrete uniform distribution are given by:

$$E(\mathbf{X}) = \sum_{x=1}^{N} xP(x) = \sum_{x=1}^{N} x \frac{1}{N} = \frac{1}{N} \sum_{x=1}^{N} x$$

and

$$V(\mathbf{X}) = E(\mathbf{X} - E(\mathbf{X}))^2 = E(\mathbf{X}^2) - [E(\mathbf{X})]^2$$

where

$$E(\mathbf{X}^2) = \sum_{x=1}^{N} x^2 P(x) = \sum_{x=1}^{N} x^2 \frac{1}{N} = \frac{1}{N} \sum_{x=1}^{N} x^2$$

The $E(\mathbf{X})$ and $E(\mathbf{X}^2)$ can easily be found by using formulas for the sum of the first N integers and the sum of the squares of the first N integers. These formulas from math tables are:

$$\sum_{x=1}^{N} x = \frac{N(N+1)}{2} \qquad \sum_{x=1}^{N} x^2 = \frac{N(N+1)(2N+1)}{6}$$

Therefore

$$E(\mathbf{X}) = \frac{1}{N} \sum_{x=1}^{N} x = \frac{1}{N} \frac{N(N+1)}{2} = \frac{N+1}{2}$$

and

$$E(\mathbf{X}^2) = \frac{1}{N} \sum_{x=1}^{N} x^2 = \frac{1}{N} \frac{N(N+1)(2N+1)}{6} = \frac{(N+1)(2N+1)}{6}$$

Now,

$$V(\mathbf{X}) = E(\mathbf{X}^2) - [E(\mathbf{X})]^2 = \frac{(N+1)(2N+1)}{6} - \left[\frac{N+1}{2}\right]^2$$

$$= \frac{2(N+1)(2N+1)}{12} - \frac{3(N+1)^2}{12} = \frac{4N^2 + 6N + 2}{12} - \frac{3N^2 + 6N + 3}{12}$$

$$= \frac{N^2 - 1}{12}$$

■ **Problems**

Section 6.2 Problems

6.1 The number of times that a radio station gives the time each hour is known to be uniformly distributed between one and five, inclusive.
  a. What is the average number of times per hour that the time is given?
  b. What is the standard deviation?
  c. What is the probability that the time is given four or more times in a particular hour of broadcasting?

6.2 In Problem 6.1, use the Chebychef theorem to determine what interval contains at least 75 percent of the number of times the time is given per hour. What is the actual proportion of hourly time announcements in this interval?

6.3 The number of orders per day for a product is uniformly distributed over the values 0, 1, 2, 3, and 4. What is the average daily order, and what is the standard deviation of the daily orders? [*Hint:* Consider the uniform random variable **X** defined over the integers 1, 2, 3, 4, and 5, and the random variable **Y** = **X** − 1. Use Theorem 5.1 to determine E(**Y**) and V(**Y**).

6.4 Using the definition of the expected value and the variance of a random variable given in Chapter 5, verify that the mean and the variance are as given in Theorem 6.1 for the six-sided die experiment in Example 6.1.

6.5 During a day, the number of times a machine fails is known to be uniformly distributed over the values 0, 1, 2, . . . , 10.
  a. What is the average number of times the machine fails?
  b. What is the standard deviation of the number of times the machine fails?
  c. From the Chebychef theorem, what interval contains at least 84 percent of the numbers of times the machine fails daily?
  d. What proportion of numbers of daily failures are contained in the interval determined in part *c*?

Section 6.3 Problems

6.6 In 12 independent tosses of a fair coin, what is the probability of getting the following?
  a. Three heads.

b. At most three heads.

c. At least three heads.

d. Two to four heads.

e. What is the expected value of the number of heads in 12 tosses of the fair coin? What is the variance?

6.7 Sketch the binomial probability mass function for $n = 10$ and $\pi = 0.60$. What is the probability that the binomial random variable **X** will be equal to seven? What is the probability that the binomial random variable **X** will be less than or equal to seven? Find the mean and the variance of this distribution. Find the mode of this distribution.

6.8 In a certain location in the summer months, rain falls on two out of five days on the average. If three days are selected at random during the summer, what is the probability that rain will fall on exactly one of these three days?

6.9 The probability that a person selected to participate in a survey will respond to a mail questionnaire is 0.20. Assume the binomial distribution is applicable.

a. What is the probability that *less than* five will respond in a sample of 20 persons?

b. What is the most likely number of responses in the sample of 20 persons?

c. What is the expected number of responses in the sample of 20 persons?

6.10 Assuming that the probability of a male birth is ½, find the following in a family of five.

a. The probability of at least one boy.

b. The probability of at least one boy and at least one girl.

6.11 The probability that an entering college student will graduate from a certain college is 0.4. Determine the probability of the following out of six students.

a. None will graduate.

b. One will graduate.

c. At least one will graduate.

6.12 Twenty percent of the bolts produced by a machine are defective. If 50 bolts are chosen at random, determine the following probabilities.

a. Five or fewer are defective.

b. Exactly 10 are defective.

c. More than 15 are defective.

6.13 A library receives 95 percent of its loaned books back on time. A sample of 50 books now out on loan is taken. Assume that the binomial distribution is applicable.

a. What is the expected number of books among these 50 that will be returned on time?

b. What is the probability that *all* 50 books will be returned on time?

6.14 The probability that a certain part is defective is 0.10. Assume that the binomial distribution is applicable and that 25 parts are randomly selected from a very large lot.

a. What is the probability that five or more are defective?

b. What is the probability that exactly two are defective?

c. What is the expected number of defectives in the sample of 25?

6.15 The probability that a life insurance agent sells a policy to a prospective client is 0.20. On a particular day, the agent contacts 15 prospective clients. Let **X** = Number of policies sold.

a. Is **X** a binomial random variable? Explain.

b. What assumptions must you make, if any, to use the binomial distribution for **X** in this case?

c. Find $E(\mathbf{X})$ and $V(\mathbf{X})$ assuming that **X** is binomially distributed.

d. Using Chebychef's theorem, determine an interval that contains at least 75 percent of the values of **X**.

e. Assuming the binomial distribution is applicable, what proportion of the values of **X** actually lie in the interval constructed in part *d*?

6.16 A wholesaler will accept a shipment lot of *n* items from a producer if *d* or fewer of the items in the lot are defective. Determine, using the binomial distribution, the probability of accepting the lot, when the lot has the following proportion of defective items, given *n* and *d*.

a. 0.00; $n = 5, d = 0$

b. 0.05; $n = 5, d = 0$

c. 0.50; $n = 6, d = 1$

d. 0.10; $n = 10, d = 0$

e. 0.10; $n = 10, d = 1$

f. $0.50; n = 10, d = 0$
g. $0.50; n = 12, d = 2$
h. $0.20; n = 25, d = 2$
i. $0.10; n = 20, d = 2$
j. $0.50; n = 20, d = 3$

6.17 A large department store gives a 5 percent discount to customers who pay cash. From past experience, the store determines that 30 percent of its sales are cash sales. This experience is prior to the store's new policy on cash sale discounts. In a sample of 22 transactions taken since the new policy was enacted, it is found that there are 10 cash sales. Assume that the binomial distribution is applicable.

a. What is the probability that the number of cash sales, X, is 10 or more in a sample of size 22 if the proportion of cash sales, π, is 0.30?
b. What is the probability that X is 10 or more in a sample of 22 if $\pi = 0.40$?
c. What is the probability that X is 10 or more in a sample of 22 if $\pi = 0.50$?
d. Is it likely that π is greater than 0.30 since the enactment of the new cash sales policy? Why or why not?

6.18 If three-fourths of the students on a certain college campus are freshmen and sophomores, what is the probability that six students selected at random will contain exactly 50 percent freshmen and sophomore members?

6.19 A multiple-choice exam contains 25 questions. Each question contains five possible answers—a, b, c, d, e—only one of which is correct. Let X = Number of correctly answered questions and π = Probability that a question is answered correctly.

a. If a student randomly selects an answer to every question so that $\pi = 0.20$, what is the probability that he or she will get 10 or more questions right?
b. In part a, what is the expected number of correct responses reached by guessing?
c. A student receives a score of 15. Is it likely that he or she guessed each answer? Explain.
d. If a student must answer 15 or more questions correctly to pass the examination, what is the probability that a person will pass the examination solely by guessing?

6.20 In a certain community, 60 percent of those who are going to vote in an upcoming election favor candidate A. If a pollster samples 20 people planning to vote, prior to the election, what is the probability that he or she will predict that candidate A will receive *less than* 50 percent of the votes cast? Assume that the binomial distribution is applicable.

6.21 In Arlington, Virginia, 70 percent of all people who seek assistance in completing their federal income tax forms do so with the offices of the H. R. Clock Company. In a random sample of 20 people who have sought assistance with their federal income tax forms, what is the expected number who received assistance from the offices of the H. R. Clock Company? What is the most likely number? What is the probability that 15 or more of the 20 people received assistance from the offices of the H. R. Clock Company?

6.22 A nursery company that sells trees through the mail receives 10 percent positive responses from its mail advertisements. Assuming that the binomial distribution applies and that 20 advertisements are mailed out,

a. What is the probability that more than five people will respond to the advertisement by purchasing one or more trees?
b. What is the expected number responding positively to the advertisement?
c. What is the most likely number responding positively to the advertisement?

Section 6.4 Problems

6.23 A population consists of one dozen employment files, six of which need updating. Four files are randomly selected for inspection.

a. What is the probability that three of the four sampled files will be out of date?
b. What is the expected number of files that are out of date in the sample?

6.24 A lot of 10 calculators contains two defectives. A customer purchases three of these calculators. Let X = Number of defectives she has purchased.

a. Using the hypergeometric distribution, find the probability that **X** = 2.

b. Using the binomial distribution, find the probability that **X** = 2.

c. Which answer is "correct"? Why?

6.25 A lot of 15 transistor radios contains four defectives. If a customer buys three transistor radios for gifts and they are randomly drawn from the lot of 15, what is the probability that all three are defective?

6.26 In a statistics course with 20 students, 10 are not satisfied with the text. If a random set of five of these students show up on the day of the course evaluation,

a. What is the probability that all five are dissatisfied with the text?

b. What is the probability that at least three of the five are dissatisfied with the text?

Section 6.5 Problems

6.27 Ten percent of the units made in a production process are defective. Suppose that 10 units are chosen at random.

a. Using the binomial distribution, determine the probability that exactly two units are defective.

b. Using the Poisson distribution approximation to the binomial distribution, determine the probability that exactly two units are defective.

6.28 The probability that an individual suffers a bad reaction from taking a new drug orally is 0.001. Using the Poisson distribution, determine the following probabilities out of 2,000 individuals.

a. Three will suffer a bad reaction.

b. At least three will suffer a bad reaction.

c. Could these probabilities be determined by using the binomial distribution? Explain.

6.29 A retailer determines that the average number of orders per day for a certain product is five. Assume the Poisson distribution is applicable.

a. What is the probability that no orders are received on a given day?

b. What is the probability that between two and eight orders, inclusive, will be received on a given day?

c. What is the probability that more than 10 orders will be received on any given day?

d. What assumptions are necessary to use the Poisson distribution in parts *a–c*?

6.30 In preparing mathematical manuscripts, it is found that the average number of typographic errors made per page is 0.4. If 10 pages are randomly selected from a large manuscript, what is the probability that two or more errors are found? What assumptions did you make in calculating this probability?

6.31 The emergency squad at a metropolitan fire station receives an average of 0.20 calls per hour. Assume that the Poisson distribution is applicable.

a. What is the probability that the squad will receive *more than* one call in one hour?

b. Over the 24-hour day, what is the average number of calls the squad receives?

c. In studying the recent experience of squad calls, it is found that an average of four calls are received between the hours of 6:00 A.M. and 10:00 P.M., and an average of one call is received between the hours of 10:00 P.M. and 6:00 A.M. Is this experience consistent with part *b* and the assumptions required to determine the probability in part *a*? Explain.

6.32 Suppose that in a carpet manufacturing process, an average of two flaws occur per 10 running yards of material. What is the probability that a given 10-yard segment will have one or fewer flaws, if the number of flaws is known to be Poisson distributed?

6.33 A customer service desk receives an average of five customers per hour. Assume that the number of customers arriving per hour is Poisson distributed.

a. What is the probability that more than 10 people will request service in a particular hour?

b. What is the mean number of arrivals per hour?

c. What is the standard deviation of the number of arrivals per hour?

d. Using the Chebychef theorem, determine what interval contains at least three-fourths of the numbers of arrivals per hour.

e. What proportion of values of the Poisson random variable are actually contained in the interval determined in part *d*?

Additional Problems

6.34 In 800 trials of tossing a fair six-sided die, how many times would you expect three or fewer dots to turn up on the top face of the die?

6.35 A random variable X is uniformly distributed over the positive integers 1, 2, . . . , 10. What is the probability that X is greater than 7?

6.36 A national firm sells magazines by mail. From past experience, it is known that the success rate is 10 percent; that is, 10 percent of those receiving a mail offer subscribe to at least one magazine out of the set in the offer. In a sample of 25 recipients of the mail offer, what is the probability that five or more will subscribe to one or more magazines? Assume that the binomial distribution is applicable.

6.37 The mean and the variance of a certain binomial distribution are known to be 7.5 and 1.875, respectively. What are the values of the parameters (n and π) in this distribution?

6.38 In a manufacturing process, 10 percent of the units produced are known to be defective. A random sample of 50 units is taken from the production process. Let the random variable X denote the number of defective items in the sampled lot of 50. Assume that the binomial distribution applies in this problem.
a. What is the probability that 5 or fewer defective items will be found in the sample of 50?
b. What is the most likely number of defectives to be found in the sample of 50?
c. What is the probability that exactly 5 defectives will be found in the sample of 50?
d. What is the expected number of defectives to be found in the sample of size 50?

6.39 There are 20 stamping machines in a large plant. The machines operate independently of one another, and breakdowns are random occurrences. The probability of a breakdown for each machine in one day's operation is 0.10.

a. Using the binomial distribution, determine the probability of two or more breakdowns on any given day.
b. Using the Poisson distribution with $\lambda = (20)(0.10) = 2$ breakdowns each day on the average, determine the probability that two or more breakdowns occur on any given day.
c. Which distribution—binomial or Poisson—is more applicable in this problem? Explain.

6.40 Describe the nature of a random experiment that will produce the following.
a. A binomial random variable.
b. A hypergeometric random variable.
c. What are the differences in these two experiments?

6.41 In each of the following experiments, determine whether the random variable is binomial, hypergeometric, or neither. Explain your answers.
a. An auditor randomly selects 10 of 1,000 tax returns; X = Number of returns in which taxes are underpaid.
b. By extensive testing, it has been determined that an electronic sensing device will properly operate when exposed to a specific stimulus 90 percent of the time. Five devices are randomly selected from 1,000 and are exposed to the stimulus; X = Number of devices that properly sense the stimulus.
c. A five-person committee is to be randomly formed from a group of four men and four women; X = Number of women on the committee.
d. The probability of tossing a seven in craps (the number of dots on the two top faces of the dice sum to seven) is 1/6. The dice are tossed 10 times; X = Number of sevens in the 10 tosses.
e. A major league baseball player enters a game with a 0.330 batting average. He comes to bat four times during the game; X = Number of hits in the four at-bats.

6.42 A computer room contains five fire alarms that act independently. The probability that each alarm detects a fire of a certain magnitude is 0.95. If a fire of the specified magnitude occurs, *a.* What is the probability that all five alarms detect the fire?

b. What is the probability that at least one alarm detects the fire?

c. What is the expected number of alarms that detect the fire?

6.43 Suppose that 80 percent of the people who drink Koca-Kola prefer the old formula to the new, "lighter and sweeter" formula. If 25 Koca-Kola drinkers are randomly chosen,

a. What is the probability that 20 or more will state a preference for the old formula?

b. What is the probability that all 25 will state a preference for the old formula?

c. What is the probability that 15 or fewer will state a preference for the old formula?

6.44 A faulty switch operates correctly 95 percent of the time. Suppose that the switch is used. Either it operates correctly or it does not. Let the random variable $X = 1$ if the switch operates correctly and $X = 0$ if the switch does not operate correctly.

a. What is the probability distribution model for the random variable X?

b. Specify the probability distribution of the random variable X either in table form or by using a stick diagram.

c. What is the expected value of X? What does this value represent?

6.45 The average number of requests for tools at a tool checkout station in a manufacturing plant is 10 per hour. Assuming that these requests follow a Poisson distribution,

a. What is the probability that there will be 15 requests in a given period of one hour?

b. What is the probability that there will be less than eight requests in a given period of one hour?

c. What is the most likely number of requests in a one-hour period?

6.46 A small business is about to receive an audit from the accounting firm it uses for accounting purposes. Suppose that 5 percent of the invoices for the past quarter contain errors. If the auditor randomly selects 20 invoices,

a. What is the probability that the auditor will find four invoices that contain errors?

b. What is the probability that the auditor will find no invoices that contain errors?

c. What is the expected number of invoices that contain errors?

6.47 A manufacturer of watches knows that 5 percent of the watches of a certain type will be returned during the warranty period. A department store purchases 50 of these watches for sale. Assuming that the binomial distribution applies,

a. What is the probability that 7 of the 50 watches will be returned during the warranty period?

b. What is the probability that five or fewer of the watches will be returned during the warranty period?

c. What is the most likely number of watches that will be returned?

6.48 A customer takes her car to a mechanic for a tune-up. The tune-up will include the replacement of the six sparkplugs in the six-cylinder car. The mechanic has 10 spark plugs from which he will select six during the tune-up. Four of the spark plugs are defective—this situation is unknown to the mechanic. The mechanic randomly selects six of the spark plugs from the 10. Let the random variable $X =$ Number of nondefective spark plugs that he puts into the customer's car.

a. What is the correct probability distribution model for the random variable X?

b. What is the probability that all six of the spark plugs placed in the car are nondefective?

6.49 A certain crucial electronic component on the Space Shuttle fails 1 percent of the time. Because of this failure rate, the Shuttle contains four of these components so that the system operates correctly if at least one of these components does not fail. What is the probability that the system operates correctly?

6.50 The number of accidents per month in a large assembly plant is known to follow a Poisson distribution with a mean rate of 4.5 accidents per month.

a. What is the probability that there are eight accidents in a given month?

b. What is the probability that there will be at least two accidents in a given month?

c. What is the probability there will be 12 accidents in a quarter?

d. What is the probability there will be 20 or fewer accidents in a year?

6.51 The number of accidents per day in downtown Dallas during the morning rush period between 7:00 A.M. and 9:00 A.M. is known to follow a Poisson distribution with a mean of five accidents during the two-hour period.

a. On a given morning, what is the probability that there are 10 accidents during the morning rush period?

b. What is the probability that there are no accidents during the morning rush period on a certain day?

c. What is the most likely number of accidents during the morning rush period on any given day?

d. At what rate per hour are accidents occurring?

7

Continuous probability distribution models

■ **7.1**

Introduction

The distribution of a continuous random variable can assume many forms. Examples of representations of a probability *density function* for a continuous random variable are illustrated in Figure 7.1. Distributions A and B are skewed, while distributions C, D, and E share the property of symmetry but differ with respect to other characteristics (number of modes, dispersion, etc.).

The random variables and their associated distributions to be discussed in this chapter differ primarily from those described in Chapter 6 in that the random variable can assume any one of an infinite number of values. Examples of such random variables are the lengths of steel bars produced by a certain production process, the lifetime in hours of light bulbs, and the time between the arrivals of customers at a service counter.

We will begin our discussion of continuous distributions with the uniform distribution. The uniform distribution is the continuous analogue of the discrete uniform distribution described in Section 6.2.

■ **7.2**

The uniform (continuous) distribution

A distribution that is being used increasingly—especially in simulation experiments—is the uniform distribution. Random variates from other continuous distributions can be generated by considering special properties of the uniform distribution. For example, the *sum* of 12 uniform random variables with only slight modification can be made to approximate a normal distribution (described in Section 7.3), and random draws from an exponential distribution (described in Section 7.4) can be made by simply taking the logarithm of a uniform random variable. The simulated distributions dis-

FIGURE 7.1 | Representations of a probability distribution of a continuous random variable

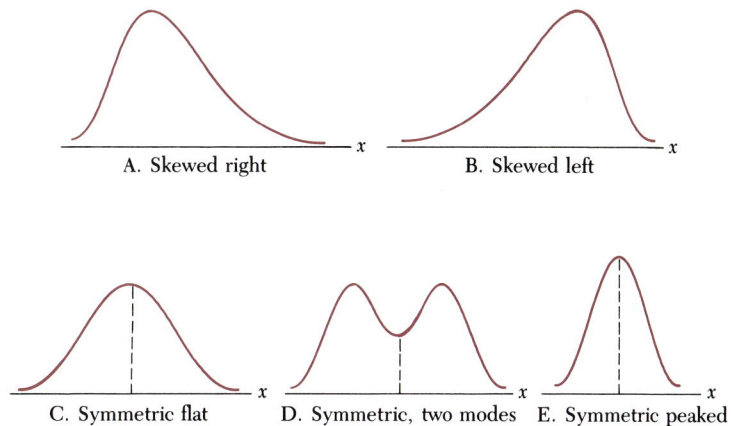

A. Skewed right   B. Skewed left

C. Symmetric flat   D. Symmetric, two modes   E. Symmetric peaked

cussed in Chapter 8 on sampling were derived by sampling random variables on a computer from a uniform distribution.

The uniform distribution is also used frequently to characterize an "almost informationless" state. That is, we *assume* that all possible values of the continuous random variable are equally likely, and thus the density function is constant over the range of values of the random variable. In this instance, the random variable is said to be *uniformly distributed*. A uniform random variable and its associated probability density function are defined in Definition 7.1.

Continuous uniform random variable

Definition 7.1

Uniform random variable and its probability density function

Let the random variable **X** be defined over the interval a to b $(a < b)$ with *probability density function* given by:

$$f(x) = \begin{cases} \dfrac{1}{b - a} & \text{if } a \le x \le b \\ 0 & \text{all other } x \end{cases}$$

Then **X** is a *uniform random variable*.

The mean and variance of a uniform random variable are given in Theorem 7.1. The derivation of the mean and variance of the uniform distribution is given in the appendix at the end of this chapter.

Mean and variance
of continuous
uniform random
variable

Theorem 7.1

Mean and variance of the uniform random variable

If **X** is a uniform random variable defined over the interval a to b ($a < b$), then its *mean*, $E(\mathbf{X})$, and *variance*, $V(\mathbf{X})$, are given by

$$E(\mathbf{X}) = \frac{b + a}{2} \quad \text{and} \quad V(\mathbf{X}) = \frac{(b - a)^2}{12}$$

Two examples of uniform random variables are given in Figure 7.2. Figure 7.2B is a special case of the uniform random variable called the *standardized uniform random variable*. The mean and the variance of a standardized uniform random variable are developed in Example 7.1.

FIGURE 7.2 | Two examples of uniform random variables

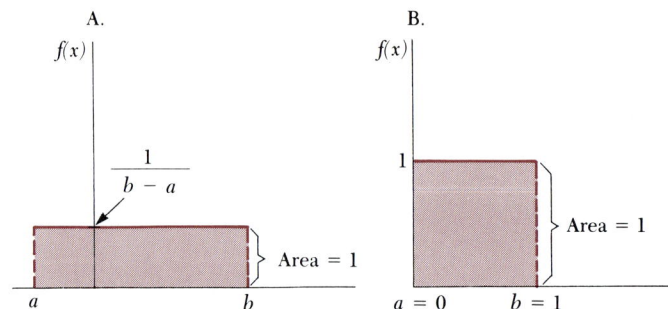

Example 7.1 A special case of the continuous uniform distribution occurs whenever $a = 0$ and $b = 1$. This distribution is called the *standardized uniform distribution*. Determine the density function, $f(x)$, the mean, $E(\mathbf{X})$, and the variance, $V(\mathbf{X})$, of a standardized uniform random variable.

Solution Definition 7.1 can be appropriately modified by substituting 0 for a and 1 for b. The density function then becomes

Standardized uni-
form distribution

$$f(x) = \begin{cases} \dfrac{1}{1 - 0} = 1 & \text{if } 0 \le x \le 1 \\ 0 & \text{all other } x \end{cases}$$

The mean and the variance of a standardized uniform random variable are

$$E(\mathbf{X}) = \frac{1 + 0}{2} = \frac{1}{2}$$

$$V(\mathbf{X}) = \frac{(1 - 0)^2}{12} = \frac{1}{12}$$

Note that the mean of this distribution is ½, which is intuitively reasonable, since this is the midpoint of the interval from 0 to 1.

Recall from Chapter 5 that because a uniform random variable is continuous, the probability that $\mathbf{X}$ equals a specific value, say $\mathbf{X} = a$, is zero regardless of the value of a. The probability that $\mathbf{X}$ assumes some value between c and d ($c < d$) is determined by finding the area under the density function from c to d. The determination of the probability that a uniform random variable assumes a value between c and d is illustrated in the following example.

Example 7.2   A uniform random variable is defined over the interval 0 to 6. What is the probability that the random variable assumes a value between 2 and 4?

Solution   The required probability is given in Figure 7.3. In essence, we are interested in the ratio of the shaded area to the total area. Since the total area is equal to 1, however, the required probability becomes

$$\frac{4 - 2}{6 - 0} = \frac{2}{6} = \frac{1}{3}$$

In general, the probability that a uniform random variable defined over the interval a to b ($a < b$) assumes a value between c and d ($c \geq a$, $d \leq b$) is

$$\left(\frac{1}{b - a}\right)(d - c) = \frac{d - c}{b - a}$$

Since the probability that the uniform random variable assumes a value between two numbers, say c and d ($c < d$) is the area of a rectangle, no

FIGURE 7.3 | Determination of $P(2 \leq \mathbf{X} \leq 4)$ from Example 7.2

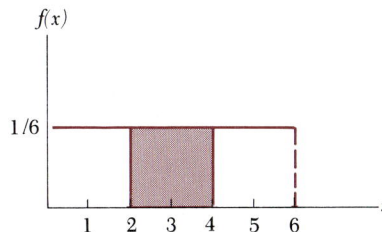

specific tables are required to find probabilities as was the case with the binomial and Poisson distributions in Chapter 6, for example.

■ 7.3

The normal distribution

7.3.1 | Introduction

The normal distribution is *probably* the most important probability distribution in statistics! It is a probability distribution of a continuous random variable, yet it is often used to model the distribution of other continuous random variables and discrete random variables.

The reason for the versatility in using the normal distribution as a probability distribution model is indicated in Figure 7.4. The basic form of the

FIGURE 7.4 | Three forms of the normal distribution

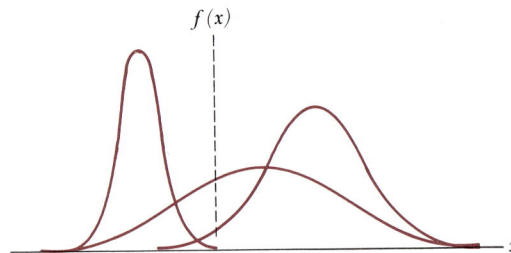

normal distribution is that of a bell—it has a single mode and is symmetric about its central value. The flexibility in using the normal distribution is due to the fact that the "bell" may be centered over any number on the real line, and it may be made flat or peaked to correspond to the amount of dispersion that the values of a random variable may assume. Many quantitative characteristics have distributions similar in form to the normal distribution's bell shape.

Examples of random variables that have been successfully modeled by the normal distribution are the height and weight of people, the diameter of bolts of a specified size produced on a machine, the IQ of people, and the lifetime in hours of batteries or light bulbs. Typically, in the type of experiment that produces a random variable that can be successfully approximated by a normal random variable, the values of the random variable are produced by a measuring process, where it is known that the measurements tend to cluster symmetrically about a central value. A random variable that is an average or a sum of values of another random variable is, under very *general* conditions, *almost always* distributed approximately as a normal random variable, *regardless* of the form of the distribution of the random variable with values that are summed or averaged. An example of such a random variable is the average grade point average of a group of students selected at random from the population of students at your university or college. The notion that a random variable that is an average is distributed as a normal random variable is discussed in Chapter 8 when we describe the central limit theorem.

For a random variable to be normally distributed, the mathematical expression describing the form of the bell must be of a specific type as described in Definition 7.2.

Normal random
variable and its
probability density
function

Definition 7.2

Normal random variable and its probability density function

A continuous random variable **X** is said to be normally distributed if its *probability density function* is

$$f(x) = \frac{1}{\sigma \sqrt{2\pi}} e^{\left[-\frac{1}{2}\left(\frac{x-\mu}{\sigma}\right)^2\right]}, \qquad -\infty < x < +\infty, \ -\infty < \mu < +\infty, \ \sigma > 0$$

where μ and σ are parameters of the distribution and π and e are mathematical constants equal to 3.14159 and 2.71828, respectively.

Recall that the probability that a continuous random variable assumes a value between two constants a and b $(a < b)$ can be determined by finding the area under its density function from a to b. Thus, for a normal random variable, $P(a \leq \mathbf{X} \leq b)$ = area under $f(x)$ above the x-axis between $x = a$ and $x = b$ as illustrated in Figure 7.5.

FIGURE 7.5 | Area representing $P(a \leq \mathbf{X} \leq b)$ for a normal random variable

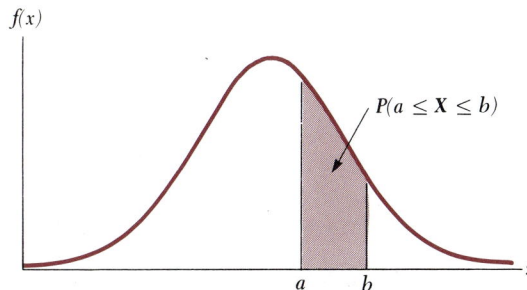

The formula giving $f(x)$ in Definition 7.2 appears to be quite imposing on first glance. But, as we shall see, it is possible to compute probabilities corresponding to the range of values a normally distributed random variable can assume using tables rather than performing mathematical computations involving $f(x)$. Before learning how to do this, we will consider the mean and the variance of a normal random variable.

7.3.2 | The mean and variance of the normal random variable

The mean and variance of the normal random variable are given in Theorem 7.2.

<table>
<tr><td>

Mean and variance
of a normal random
variable

</td><td>

Theorem 7.2
Mean and variance of the normal random variable

If **X** is a normal random variable, then its *mean*, $E(\mathbf{X})$, and *variance*, $V(\mathbf{X})$, are given by:

$$E(\mathbf{X}) = \mu \quad \text{and} \quad V(\mathbf{X}) = \sigma^2$$

</td></tr>
</table>

Notice that the mean depends *only* on the parameter μ and the variance depends *only* on the parameter σ. Thus the normal distribution may be located over its mean value on the real line separately from the amount of dispersion (σ^2) specified for the distribution! Contrast this with the binomial distribution (and others discussed thus far) in which the mean and the variance jointly depend on the *same* parameters n and π [$E(\mathbf{X}) = n\pi$; $V(\mathbf{X}) = n\pi(1 - \pi)$]. This property of the normal distribution adds immeasurably to its flexibility in modeling the distributions of nonnormal random variables, as we shall see.

We now return to the problem of computing probabilities associated with a normal random variable.

7.3.3 | The standardized normal distribution

Probabilities associated with any member of the normal distribution family can be computed from a table of probabilities compiled for the *standard normal distribution*.

The form of the standard normal distribution is illustrated in Figure 7.6. As indicated in the figure, and for *any* normal distribution, 68.27 percent of the values of **Z** lie within one standard deviation of the mean; 95.45 percent

Standard normal
distribution

Definition 7.3

Standard normal distribution

A normal distribution with $\mu = 0$ and $\sigma = 1$ is called a *standard normal distribution*. When a normal random variable **X** has a mean of 0 and a variance of 1, it is called a *standardized normal random variable* and will be denoted by **Z**. The probability density function of the standardized normal random variable **Z** is:

$$f(z) = \frac{1}{\sqrt{2\pi}}\, e^{[-\frac{1}{2}(z)^2]}, \qquad -\infty < z < +\infty$$

FIGURE 7.6 │ Standard normal distribution

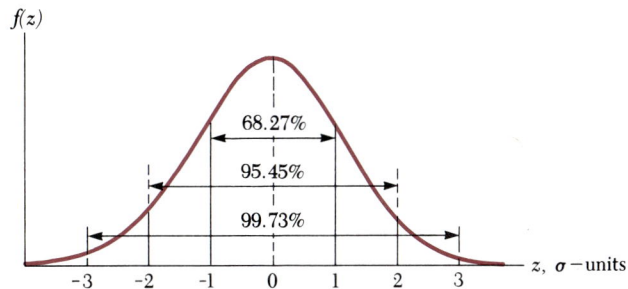

of the values lie within two standard deviations of the mean; and 99.73 percent of the values lie within three standard deviations of the mean.

Probabilities of a standardized normal random variable of the form $P(0 \leq \mathbf{Z} \leq a)$ are provided in Table B.3, Appendix B, and on the inside back cover of your text. By using the fact that the normal distribution is symmetric about its mean (0 in this case), and that the total area under the curve is 1 ($\frac{1}{2}$ to the left of 0, and $\frac{1}{2}$ to the right of 0), the probability that **Z** resides in any interval on the real line may be determined from this table as the following example indicates.

Example 7.3   Find the area under the standard normal distribution curve for each of the following intervals.

a.   Between $z = 0$ and $z = 2.0$.
b.   Between $z = -1.28$ and $z = 0.0$.
c.   Between $z = -0.58$ and $z = 2.54$.

d. Between $z = 1.20$ and $z = 2.44$.

e. Greater than $z = 2.87$.

Solution

a. In Table B.3, Appendix B, proceed downward in the leftmost column to 2.0. Select the first column marked .00 indicating that the second decimal place is 0—the area read from the table is 0.4772.

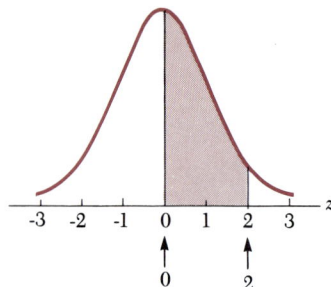

$$P[0 < \mathbf{Z} < 2] = 0.4772$$

b. Since the normal distribution is symmetric, the area between -1.28 and 0.0 is equal to the area between 0.0 and $+1.28$. Thus proceed down the leftmost column to 1.2. Select the ninth column marked .08— the resulting number in the table is 0.3997.

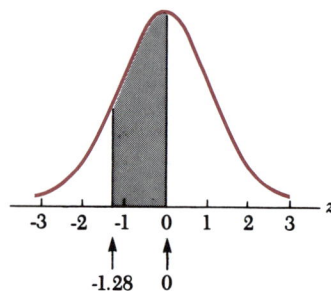

$$P[-1.28 < \mathbf{Z} < 0] = P[0 < \mathbf{Z} < 1.28] = 0.3997$$

c. We may determine this area in two parts: Total area = (Area between 0.0 and 2.54) + (Area between -0.58 and 0.0). The area between 0.0 and 2.54 is 0.4945. The area between -0.58 and 0.0 is the same as the area between 0.0 and 0.58, which is 0.2190. The answer is $0.4945 + 0.2190 = 0.7135$.

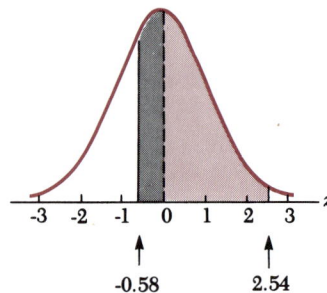

$$
\begin{aligned}
P[-0.58 < \mathbf{Z} < 2.54] &= P[-0.58 < \mathbf{Z} < 0] + P[0 < \mathbf{Z} < 2.54] \\
&= P[0 < \mathbf{Z} < 0.58] + P[0 < \mathbf{Z} < 2.54] \\
&= 0.2190 + 0.4945 \\
&= 0.7135
\end{aligned}
$$

d. We can determine this area by finding the difference between the areas from 0 to 2.44 and from 0 to 1.20. The area between 0 and 2.44 is 0.4927, and the area between 0 and 1.20 is 0.3849. Thus the area between 1.20 and 2.44 is 0.4927 − 0.3849 = 0.1078.

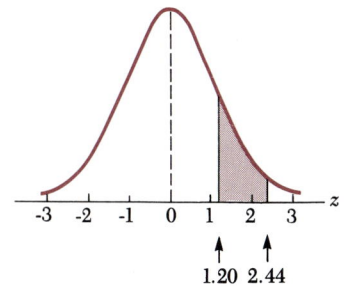

$$P[1.20 < Z < 2.44] = P[0 < Z < 2.44] - P[0 < Z < 1.20]$$
$$= 0.4927 - 0.3849$$
$$= 0.1078$$

e. Since the area between 0 and $+\infty$ is 0.5, we can determine the area from 2.87 to ∞ by subtracting the area from 0 to 2.87 (0.4979) from 0.5: 0.5000 − 0.4979 = 0.0021.

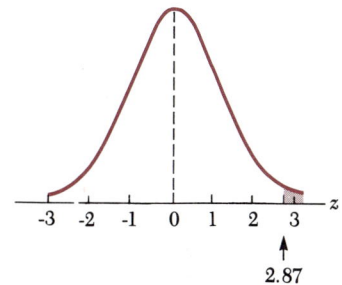

$$P[Z > 2.87] = 0.5000 - P[0 < Z < 2.87]$$
$$= 0.5000 - 0.4979$$
$$= 0.0021$$

In many problems, we are given the area in a certain interval and then asked to determine the value of **Z** that specifies the interval. This is the "reverse" of the problem solved in Example 7.3. In Example 7.4, the use of the Table B.3, Appendix B, to solve the reverse problem is demonstrated.

Example 7.4 Find the value or values of **Z** on the standard normal distribution axis when (*a*) the area between 0 and *z* is 0.3413, and (*b*) the area to the right of *z* is 0.8982.

Solution

a. Now the normal curve area table must be used in reverse. Look in the body of the table for the number 0.3413. It appears in the row marked 1.0 and the column marked 0.0. Thus the value of **Z** is 1.00.

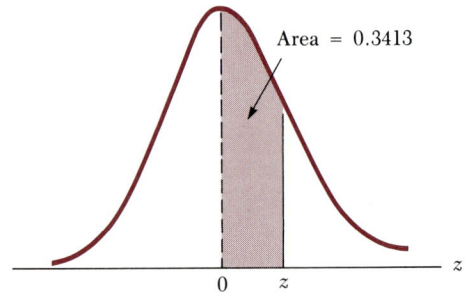

Area = 0.3413

$$P[0 < \mathbf{Z} < 1] = 0.3413$$

b. Since the area given is greater than 0.5, we know that z must be less than 0. The area between z and 0 is $0.8982 - 0.5000 = 0.3982$. Now assume that z is positive, and find z such that the area between 0 and z is 0.3982. The area of 0.3982 does not appear in the table—the closest numbers are 0.3980 and 0.3997. The exact value of z could be determined by interpolation, but we will use $z = 1.27$, since 0.3980 is closer to 0.3982 than 0.3997. Now, we must remember that z must be to the left of 0 (a negative number). Thus $z = -1.27$.

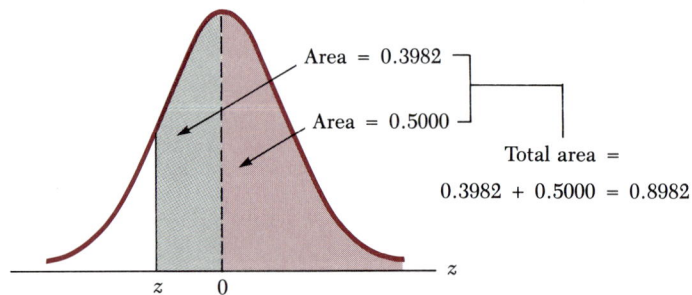

Area = 0.3982

Area = 0.5000

Total area =

0.3982 + 0.5000 = 0.8982

$$P[-1.27 < \mathbf{Z} < +\infty] = 0.8982$$

7.3.4 | Areas under the normal distribution

Probabilities associated with a normal random variable **X** that is not standardized can be determined by using the results of Theorem 7.3.

Theorem 7.3
Standardization of a normal random variable

If **X** is a normal random variable, the mean of which is μ and the standard deviation of which is σ, then

$$\mathbf{Z} = \frac{\mathbf{X} - \mu}{\sigma}$$

is a *standardized normal random variable* with a mean of 0 and a standard deviation of 1.

Standardized normal random variable

The use of this theorem is illustrated in Figure 7.7. The probability that **X** is between a and b ($b > a$) can be determined by computing the probability that **Z** is between $(a - \mu)/\sigma$ and $(b - \mu)/\sigma$; the area between a and b is preserved by the linear transformation $\mathbf{Z} = (\mathbf{X} - \mu)/\sigma$!

FIGURE 7.7 | Relationship between normal distribution with mean μ and standard deviation σ and the standard normal distribution with mean $\mu = 0$ and standard deviation $\sigma = 1$

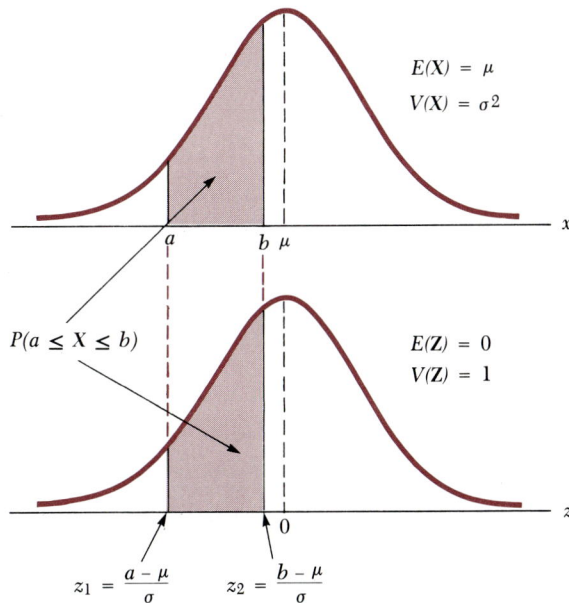

$E(X) = \mu$
$V(X) = \sigma^2$

$P(a \leq X \leq b)$

$E(Z) = 0$
$V(Z) = 1$

$z_1 = \frac{a - \mu}{\sigma}$ $z_2 = \frac{b - \mu}{\sigma}$

The following examples illustrate the use of Theorem 7.3 and, more generally, the applicability of the normal distribution model.

Example 7.5 The mean lifetime of 50-watt light bulbs produced by the Stay-Bright Light Bulb Company is 200 hours. It is known that the standard deviation is 20 hours. Assuming that the lifetimes of the light bulbs are normally distributed, what is the probability that a single 50-watt light bulb extracted from the production lot will (*a*) burn out at a time between 180 and 210 hours, and (*b*) burn out at a time greater than 250 hours?

Solution

a.   The solution on the **X**-distribution with a mean of 200 and a standard deviation of 20 is the area between $x_1 = 180$ and $x_2 = 210$. This area can be determined by first standardizing x_1 and x_2:

$$z_1 = \frac{x_1 - \mu}{\sigma}$$

$$= \frac{180 - 200}{20}$$

$$= \frac{-20}{20} = -1.00$$

$$z_2 = \frac{x_2 - \mu}{\sigma}$$

$$= \frac{210 - 200}{20}$$

$$= \frac{10}{20} = 0.50$$

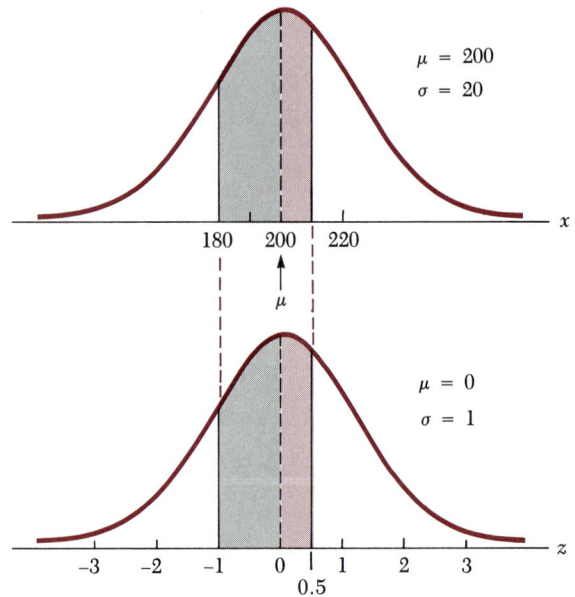

The area between -1.0 and 0.50 on the standard normal distribution will equal the area between 180 and 210 on the **X**-distribution. The area between $z_1 = -1.00$ and $z_2 = 0.50$ is:

$$\begin{aligned} \text{Area } (-1.00 \text{ to } 0.00) &= 0.3413 \\ \text{Area } (0.00 \text{ to } 0.50) &= \underline{0.1915} \\ & \; 0.5328 \end{aligned}$$

$$\begin{aligned} P[180 < \mathbf{X} < 210] &= P[-1 < \mathbf{Z} < 0.50] \\ &= P[-1 < \mathbf{Z} < 0] + P[0 < \mathbf{Z} < 0.50] \\ &= P[0 < \mathbf{Z} < 1] + P[0 < \mathbf{Z} < 0.50] \\ &= 0.3413 + 0.1915 \\ &= 0.5328 \end{aligned}$$

Thus $P(180 < \mathbf{X} < 210) = 0.5328$. This answer tells us that 53.28 percent of the 50-watt light bulbs composing the population will have times to failure between 180 and 210 hours.

  b. The solution on the **X**-distribution is the area greater than $x_1 = 250$ hours. The standardized value is:

$$
\begin{aligned}
z_1 &= \frac{x_1 - \mu}{\sigma} \\
&= \frac{250 - 200}{20} \\
&= \frac{50}{20} = 2.5
\end{aligned}
$$

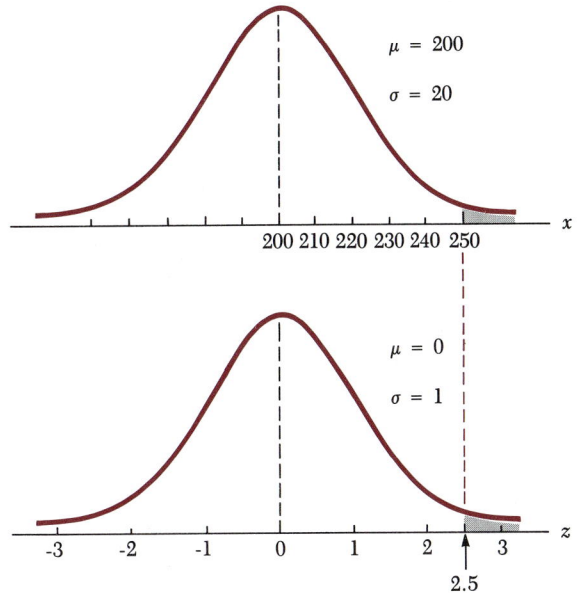

The area from $z_1 = 2.5$ to $+\infty$ is:

$$
\begin{aligned}
&0.5000 - \\
&\quad \text{Area } (0.00 \text{ to } 2.5) \\
&= 0.5000 - 0.4938 \\
&= 0.0062
\end{aligned}
$$

$$
\begin{aligned}
P[\mathbf{X} > 250] &= P[\mathbf{Z} > 2.5] \\
&= 0.5000 - P[0 < \mathbf{Z} < 2.5] \\
&= 0.5000 - 0.4938 \\
&= 0.0062
\end{aligned}
$$

Thus 0.62 percent of the 50-watt light bulbs will fail at a time exceeding 250 hours.

  Example 7.6 A statistics exam is based on 800 points. It is known that the mean of the exam is 500 and the standard deviation is 100. Assuming that the exam scores are normally distributed (i.e., the normal is a good model for this problem), what score must a student make on the exam to be placed in the upper 15 percent of all test scores?

  Solution This is the reverse problem—we are given the area under the normal curve and asked to determine the test score x. If the student is in the upper 15 percent of all students taking the exam, his score x must be such that the area to the right of x is 0.15 and the area between the mean of 500 and x is 0.35.

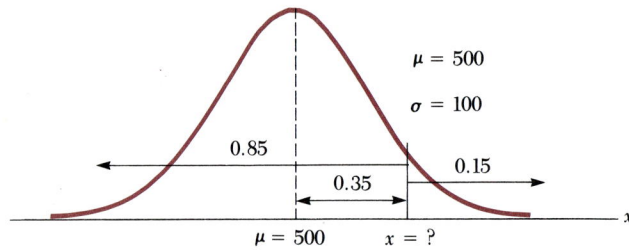

We first solve the problem on the standard normal distribution:

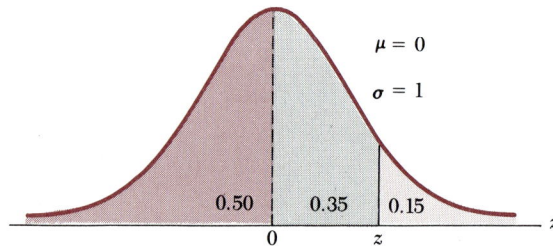

The value of **Z** such that the area between 0 and z is 0.35 is 1.04 ($z = 1.03$ corresponds to an area of 0.3485 and $z = 1.04$ corresponds to an area of 0.3508—0.3508 is closer to 0.3500). Now the standardizing formula

$$z = \frac{x - \mu}{\sigma}$$

is solved for x: $x = \sigma z + \mu$. Thus the value of **X** that corresponds to $z = 1.04$ is $x = 100(1.04) + 500 = 604$. Thus a student must score 604 or more points to be placed among the upper 15 percent of the scores.

Whenever computing probabilities associated with normal random variables, the student is advised to draw a picture of the areas to be computed as in Examples 7.4, 7.5, and 7.6—errors are usually greatly reduced if time is taken to make these sketches!

7.3.5 | The normal approximation to the binomial distribution

In Chapter 6, the binomial distribution was introduced; its probability mass function is:

$$P(x) = C_x^n(\pi)^x(1 - \pi)^{n-x}, \qquad x = 0, 1, 2, \ldots, n$$

In Chapter 6, we mentioned that a method other than using the mass function or the tables to compute binomial probabilities could be applied under certain circumstances. This third method involves fitting a normal curve to the binomial distribution and calculating areas under the curve to *approximate* the sum of binomial probabilities. To illustrate this method, suppose we flip a fair coin $n = 10$ times and record the number of heads, **X.** The appropriate binomial mass function that describes the probability distribution of **X** in this experiment is:

$$P(x) = C_x^{10}(0.5)^x(1 - 0.5)^{10-x}, \qquad x = 0, 1, 2, \ldots, 10$$

This binomial distribution is presented in tabular form in Table 7.1 and as a stick diagram in Figure 7.8. We first note from Figure 7.8 that the sticks

TABLE 7.1 | Binomial distribution when $\pi = 0.5$ and $n = 10$

x	$P(x)$	x	$P(x)$
0	0.001	6	0.205
1	0.010	7	0.117
2	0.044	8	0.044
3	0.117	9	0.010
4	0.205	10	0.001
5	0.246		

FIGURE 7.8 | Stick diagram of binomial distribution when $\pi = 0.50$ and $n = 10$

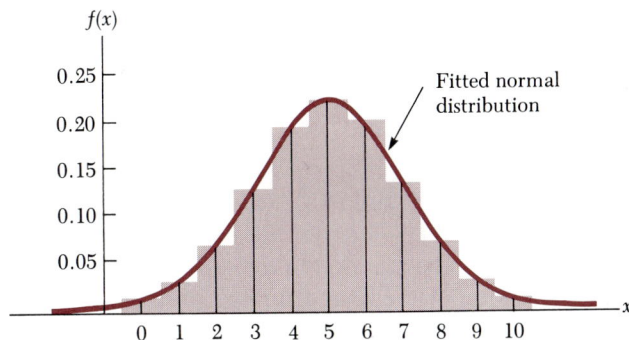

suggest the form of a bell-shaped distribution—this property is strengthened by drawing a smooth curve along the tops of the sticks. Recall that the mean and variance of the binomial distribution are given by $E(\mathbf{X}) = n\pi$ and $V(\mathbf{X}) = n\pi(1 - \pi)$, respectively. When $\pi = 0.50$ and $n = 10$, we have $E(\mathbf{X}) = 5$ and

$V(\mathbf{X}) = 2.5$. Since, in the normal distribution, the mean is μ and the variance is σ^2, we will select a normal distribution with $\mu = 5$ and $\sigma^2 = 2.5$ ($\sigma = \sqrt{2.5} = 1.58$) to fit over the sticks of this binomial distribution. Then, if we wish to determine the probability that the binomial random variable $\mathbf{X}$ assumes a value between a and b ($a < b$), it can be approximated by the area under the normal curve from a to b.

There is a problem in this approximation process, however. The normal distribution assigns probability continuously over the real number line, whereas binomial probability is massed only at integer values. For example, the probability of getting more than two heads but less than three heads in 10 tosses of a coin [$P(2 < \mathbf{X} < 3)$] is 0, but the area under the fitted normal distribution between 2 and 3 would not be 0, of course. A partial solution to this problem is indicated in Figure 7.9. Suppose we wish to approximate

FIGURE 7.9 | Normal approximation to binomial probabilities

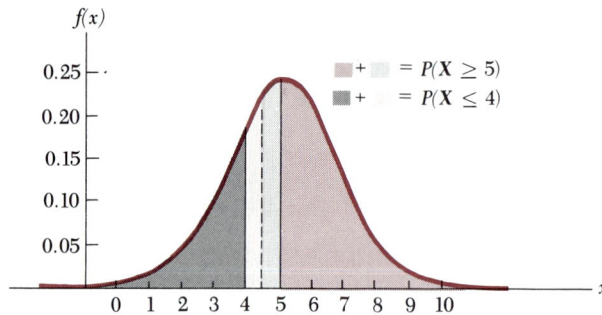

$P(\mathbf{X} \leq 4)$ and $P(\mathbf{X} \geq 5)$ using the normal distribution. If only the area to the left of $x = 4$ under the normal curve is used to approximate $P(\mathbf{X} \leq 4)$, and only the area to the right of $x = 5$ is used to approximate $P(\mathbf{X} \geq 5)$, then the sum of the approximated probabilities will not equal 1. But we know that $P(\mathbf{X} \leq 4) + P(\mathbf{X} \geq 5) = 1$ for the binomial distribution.

The problem is the "wedge" of area under the normal curve between 4 and 5. A fair allocation of this area would be to roughly split it, giving half this area to the approximation of $P(\mathbf{X} \leq 4)$ and the other half to the approximation of $P(\mathbf{X} \geq 5)$. Thus, in approximating $P(\mathbf{X} \leq 4)$, we would find the area under the fitted normal distribution to the left of $4 + 0.5 = 4.5$. The $P(\mathbf{X} \geq 5)$ would be approximated by finding the area under the fitted normal

Correction for
continuity

distribution to the right of $5 - 0.5 = 4.5$. The one-half unit (0.5) that is being added and subtracted here is called the *correction for continuity*—it accounts for the gaps between integer values where the binomial mass function is not defined.

Example 7.7 In the coin-tossing experiment described earlier, approximate the following binomial probabilities using the fitted normal distribution.

a. $P(3 \leq \mathbf{X} \leq 8)$, without the correction for continuity.
b. $P(3 \leq \mathbf{X} \leq 8)$, with the correction for continuity.
c. $P(\mathbf{X} \geq 7)$, with the correction for continuity.

Solution

a. The parameters of the fitted normal distribution are $\mu = 5$ and $\sigma = 1.58$, since the mean and standard deviation of the binomial distribution are 5 and 1.58, respectively. Without the correction for continuity, we wish to find the area under the normal distribution with $\mu = 5$ and $\sigma = 1.58$ between $x_1 = 3$ and $x_2 = 8$. The standardized values are:

$$z_1 = \frac{x_1 - \mu}{\sigma} = \frac{3 - 5}{1.58} = -1.27$$

$$z_2 = \frac{x_2 - \mu}{\sigma} = \frac{8 - 5}{1.58} = 1.90$$

The area between $z_1 = -1.27$ and $z_2 = 1.90$ on the standard normal curve is 0.8693.

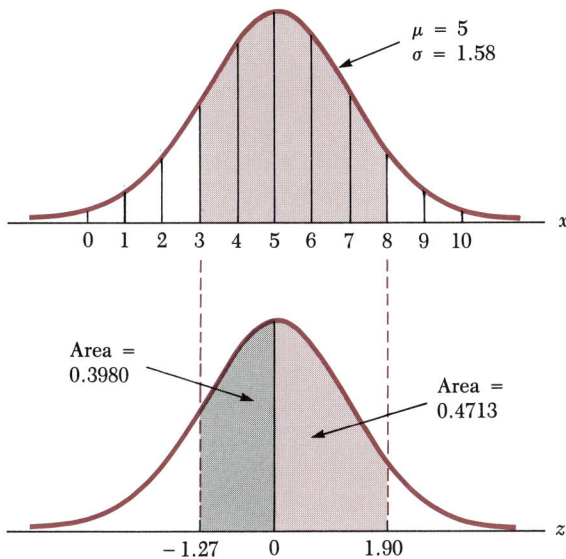

$$
\begin{aligned}
P(3) + P(4) + P(5) + P(6) + P(7) + P(8) &\doteq P[3 \leq \mathbf{X} \leq 8] \\
&= P[-1.27 \leq \mathbf{Z} \leq 1.90] \\
&= P[-1.27 \leq \mathbf{Z} \leq 0] \\
&\quad + P[0 \leq \mathbf{Z} \leq 1.90] \\
&= P[0 \leq \mathbf{Z} \leq 1.27] \\
&\quad + P[0 \leq \mathbf{Z} \leq 1.90] \\
&= 0.3980 + 0.4713 \\
&= 0.8693
\end{aligned}
$$

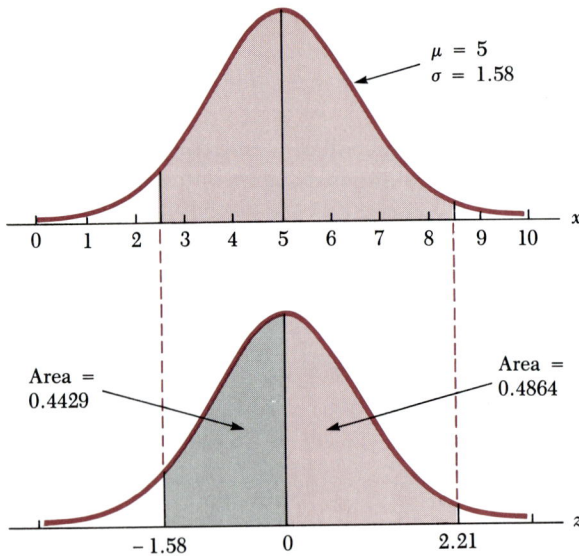

b. Using the correction for continuity, we move down 0.5 unit from 3 ($x_1 = 3 - 0.5 = 2.5$) and up 0.5 unit from 8 ($x_2 = 8 + 0.5 = 8.5$). In this way, the area between 2 and 3 is shared with $P(\mathbf{X} \le 2)$, and the area between 8 and 9 is shared with $P(\mathbf{X} \ge 9)$. Now,

$$z_1 = \frac{x_1 - \mu}{\sigma} = \frac{2.5 - 5}{1.58} = -1.58$$

$$z_2 = \frac{x_2 - \mu}{\sigma} = \frac{8.5 - 5}{1.58} = 2.21$$

The area between $z_1 = -1.58$ and $z_2 = 2.21$ on the standard normal curve is 0.9293.

$$
\begin{aligned}
P(3) + P(4) + P(5) + P(6) + P(7) + P(8) &\doteq P[2.5 \le \mathbf{X} \le 8.5] \\
&= P[-1.58 \le \mathbf{Z} \le 2.21] \\
&= P[-1.58 \le \mathbf{Z} \le 0] \\
&\quad + P[0 \le \mathbf{Z} \le 2.21] \\
&= P[0 \le \mathbf{Z} \le 1.58] \\
&\quad + P[0 \le \mathbf{Z} \le 2.21] \\
&= 0.4429 + 0.4864 \\
&= 0.9293
\end{aligned}
$$

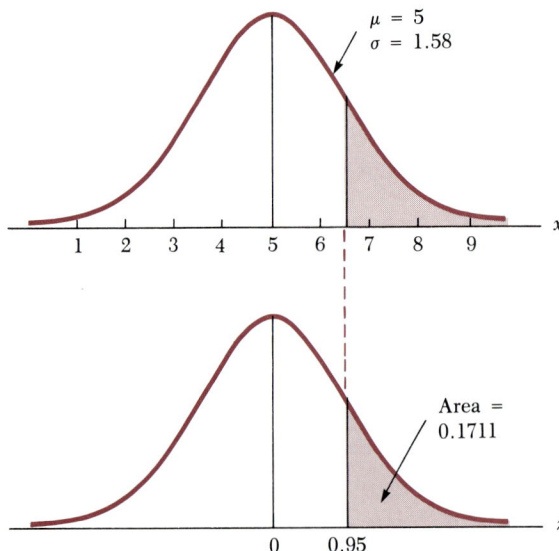

c. Using the correction for continuity, we move back 0.5 unit to share the area between 6 and 7 with $P(\mathbf{X} \le 6)$. Thus $x = 7 - 0.5 = 6.5$. Now,

$$z_1 = \frac{x_1 - \mu}{\sigma} = \frac{6.5 - 5}{1.58} = 0.95$$

The area under the standard normal curve from $z_1 = 0.95$ to $+\infty$ is 0.1711.

$$P(7) + P(8) + P(9) + P(10) \doteq P[\mathbf{X} \geq 6.5]$$
$$= P(\mathbf{Z} \geq 0.95)$$
$$= 0.5000 - P[0 \leq \mathbf{Z} \leq 0.95]$$
$$= 0.5000 - 0.3289$$
$$= 0.1711$$

From Table 7.1; we find that $P(3 \leq \mathbf{X} \leq 8) = 0.934$, and $P(\mathbf{X} \geq 7) = 0.172$. Notice that the correction for continuity gives a much better approximation to $P(3 \leq \mathbf{X} \leq 8)$—contrast the solution in part *b* to the one in part *a*.

The normal approximation to the binomial distribution should be used only if n is large. (The statistician's rule in this case is n must be sufficiently large to satisfy two conditions: first, $n\pi \geq 5$; and second, $n(1 - \pi) \geq 5$.) It works so well in Example 7.7 because when $\pi = 0.5$, the binomial distribution is symmetric. If we had chosen $\pi = 0.2$ with $n = 10$ and used the normal approximation, the approximated probabilities would not have been very close to the exact binomial probabilities, as can be seen from Figure 7.10. By

Rule for approximating binomial distribution by the normal distribution $n\pi \geq 5$ and $n(1 - \pi) \geq 5$

FIGURE 7.10 | Normal curve fitted to the binomial distribution when $\pi = 0.2$ and $n = 10$

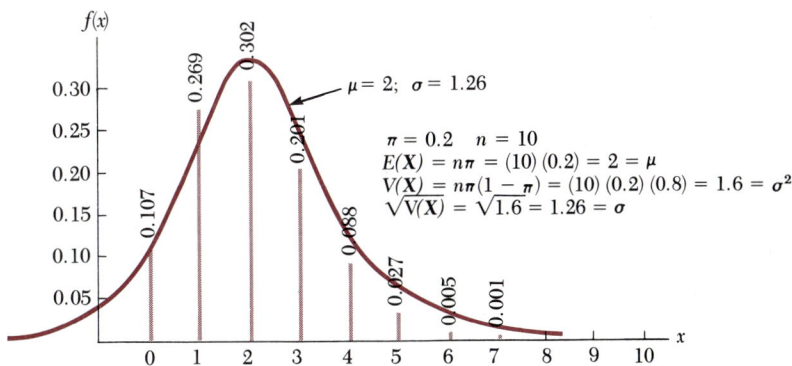

fitting the normal curve over the mean value of 2, a large portion of the area must be allocated to the region to the left of $x = 0$ where we know there is no binomial probability. As n becomes very large ($n > 100$), the area between the integers under the fitted normal curve becomes small—it is possible then not to use the correction for continuity because its effect becomes very small.

Example 7.8 In a large lot of AM/FM transistor radios produced by the Sunny Corp., it is known that 10 percent are defective. A certain retailer purchases 100 of these radios from Sunny. What is the probability that more than 15 of them are defective?

Solution Let **X** be the number of defectives in the lot of 100 purchased by the retailer. Assuming that the random variable **X** is binomially distributed, the appropriate mass function is:

$$P(x) = C_x^{100}(0.1)^x(1 - 0.1)^{100-x}, \qquad x = 0, 1, 2, \ldots , 100$$

We wish to determine $P(\mathbf{X} > 15)$ or $P(\mathbf{X} \geq 16)$. We will approximate this probability by using the normal distribution. Since $n = 100$ and $\pi = 0.1$,

$$E(\mathbf{X}) = n\pi = 100(0.1) = 10 \qquad \text{and} \qquad V(\mathbf{X}) = n\pi(1 - \pi) = 100(0.1)(0.9) = 9$$

Therefore, the fitted normal distribution has parametric values $\mu = 10$ and $\sigma = \sqrt{9} = 3$. Since we wish to approximate the probability that $\mathbf{X} \geq 16$, the value on the normal distribution is taken to be $16 - 0.5 = 15.5$; in this way, the area between $x = 15$ and $x = 16$ is shared with $P(\mathbf{X} \leq 15)$. Notice that the correction for continuity could be dropped here since n is large ($n = 100$). We include it, however, to show how the 0.50 increment is used. Now the approximated probability on the normal distribution is the area to the right of 15.5 as illustrated in Figure 7.11. The standardized value is:

$$z_1 = \frac{x_1 - \mu}{\sigma} = \frac{15.5 - 10}{3} = \frac{5.5}{3} = 1.83$$

The area to the right of $z = 1.83$ on the standard normal distribution is 0.0336. Thus the approximate probability that the retailer's lot contains more than 15 defectives is 0.0336.

$$
\begin{aligned}
P(16) + P(17) + \cdots + P(100) &\doteq P[\mathbf{X} \geq 15.5] \\
&= P[\mathbf{Z} \geq 1.83] \\
&= 0.5000 - P[0 \leq \mathbf{Z} \leq 1.83] \\
&= 0.5000 - 0.4664 \\
&= 0.0336
\end{aligned}
$$

■ 7.4

The exponential distribution

The exponential distribution is frequently used to model the distribution of a random variable that is representing service times—the time taken to complete service at a filling station, grocery store checkout stand, automobile repair shop, dentist office, and so on. It is also used in reliability applications to model the lifetimes of components that are subject to wear. Examples are the lifetimes of batteries, transistors, tubes, and bearings.

FIGURE 7.11 | Normal approximation of the binomial probability
$P(X \geq 16/n = 100, \pi = 0.10)$

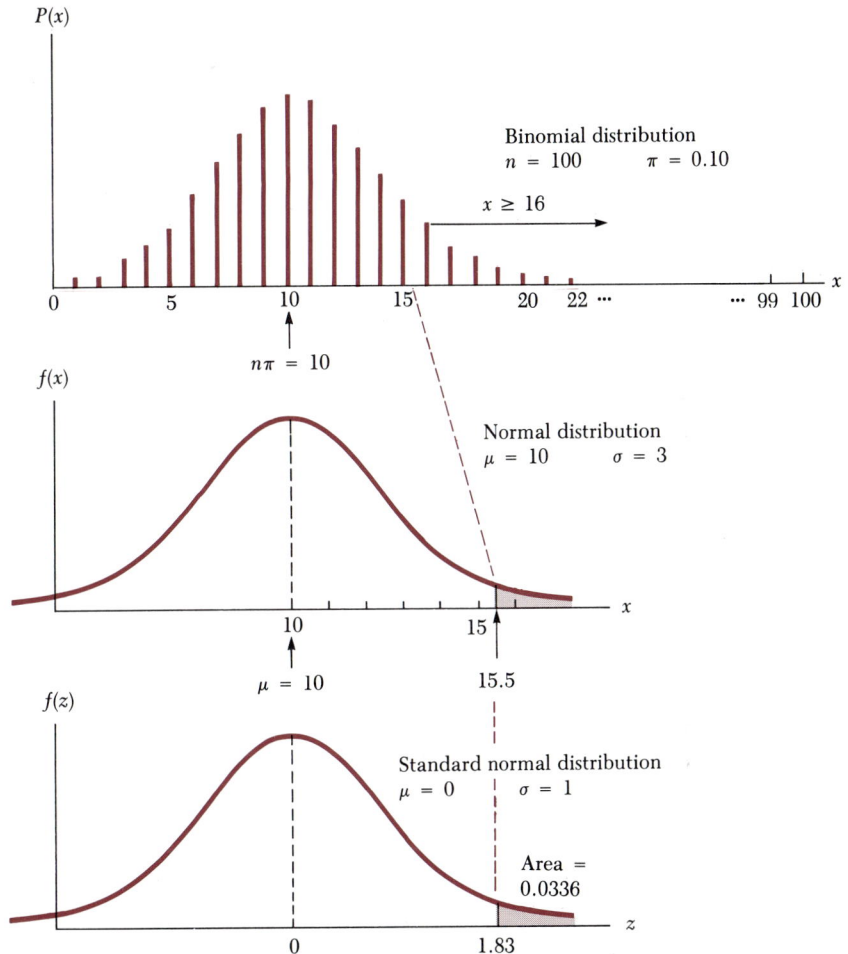

Furthermore, it has been used successfully to model the distribution of the length of time *between* successive random events—the time between the arrival of customers at a service counter, the time between breakdowns of a machine, and the time between admissions to an emergency ward of a hospital, for example. Figure 7.12 suggests why the exponential distribution has been used successfully to model these types of random variables. First, note that the distribution is defined for only positive values of the random variable **X** representing service times, times between breakdowns, and so on. Since time is a positive number, the exponential random variable is often well

FIGURE 7.12 | Two members of the exponential family of distributions

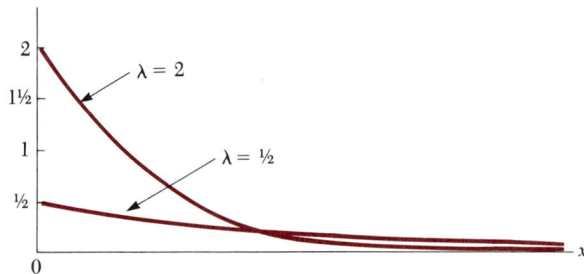

suited to represent quantitative characteristics measured in time units. (Some textbooks even go so far as to denote the exponential random variable by **T** instead of **X** to emphasize the use of this random variable in modeling events that are *time* related.) Second, the distribution is rather severely skewed to the right. Since very few service times, or times to failure of components, or times between the occurrence of successive events greatly exceed their expected values, the exponential distribution seems appropriate to model the distribution of these kinds of random variables.

It may come as a surprise, but the exponential distribution is related to a discrete distribution discussed in Chapter 6, the Poisson distribution. It can be shown that if events occur according to a Poisson process, the time *between the occurrences* of these events is exponentially distributed. We show an example of this relationship in Example 7.10.

Two members of the exponential distribution family are shown in Figure 7.12. Note from this figure that the y-intercept of the probability density function is equal to the single parameter of this distribution, λ; that is, $f(0) = \lambda$. The exponential random variable and its probability density function are described in Definition 7.4, and its mean and variance are given in Theorem 7.4. Note that unlike the normal distribution, the exponential distribution is a single-parameter distribution. Hence the mean of the distribution cannot be set separately from the amount of the dispersion. In fact, for this random variable, its variance $V(\mathbf{X})$ is always equal to the square of its expected value or its mean. That is, if $\mathbf{X}$ is an exponential random variable, $V(\mathbf{X}) = [E(\mathbf{X})]^2$, or the expected value of this random variable is equal to its standard deviation.

The parameter λ of this distribution is interpreted similarly to its use in the Poisson case—λ is the average or the expected number of occurrences of an event in a given interval of measurement. For example, $\lambda = 5$ could be interpreted as an average of five occurrences of an event in a unit of measurement, say, one minute. In this case, the mean of the distribution is $E(\mathbf{X}) = 1/\lambda = 1/5 = 0.20$, or, if events are generated according to a Poisson

Exponential ran-
dom variable and
its probability
distribution

Definition 7.4
Exponential random variable and its probability density function

A continuous random variable **X** is said to be *exponentially distributed* if its *probability density function* is

$$f(x) = \lambda e^{-\lambda x}, \qquad x \geq 0; \lambda \geq 0$$

where λ is a parameter of the distribution, and e is a mathematical constant equal to 2.71828.

Mean and variance
of an exponential
random variable

Theorem 7.4
Mean and variance of the exponential random variable

If **X** is an exponential random variable, then its *mean*, $E(\mathbf{X})$, and *variance*, $V(\mathbf{X})$, are given by

$$E(\mathbf{X}) = \frac{1}{\lambda} \qquad \text{and} \qquad V(\mathbf{X}) = \frac{1}{\lambda^2}$$

process at the rate of five per minute, then the average time between occurrences of events is 0.2 minute.

The probability that an exponential random variable assumes a value between a and b ($a < b$) is the area under the curve from a to b illustrated in Figure 7.13.

Since the computation of this area under the exponential function involves the evaluation of probability integrals between the limits of a and b, a table has been provided to find exponential distribution probabilities. The following probability formulas are used with the exponential table:

$$P[a \leq \mathbf{X} \leq b] = e^{-\lambda a} - e^{-\lambda b} \qquad (7.1)$$
$$P[\mathbf{X} \leq a] = 1 - e^{-\lambda a} \qquad (7.2)$$
$$P[\mathbf{X} \geq a] = e^{-\lambda a} \qquad (7.3)$$

FIGURE 7.13 | $P(a \le \mathbf{X} \le b)$ for the exponential distribution

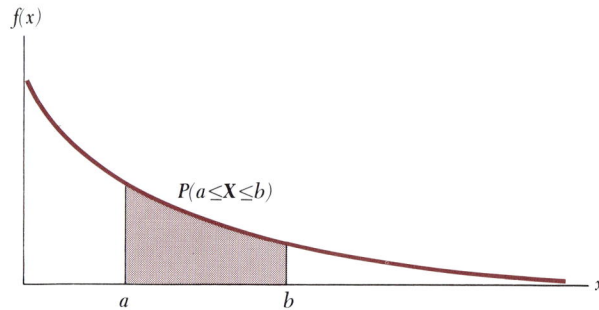

These formulas are derived in the appendix to this chapter by using integral calculus. The appendix also includes the derivation of the mean and variance of the exponential distribution by using integral calculus. Table B.4 in Appendix B contains the values of the mathematical constant e raised to the minus c power [e.g., e^{-c}], where c varies from 0.00 to 10.00 in steps of 0.01. The next three example problems will illustrate the use of the three formulas together with Table B.4.

Example 7.9 The length of time required to service customers arriving at a drive-in bank teller's window is assumed to be exponentially distributed with an expected service time per customer of 5 minutes. What proportion of customers are serviced within 1 minute? 5 minutes? Between 4 and 8 minutes?

Solution We must be careful how we define the units of measurement to be used in computing the required probabilities in this example. We are interested in the expected number of occurrences of an event in a specified length of time—in this case, in minutes. We shall use intervals of one minute to compute the required probability. Thus, since customers are served on the average in five minutes (we can think of this as being served in five continuous one-minute intervals), the expected number of services in one minute—our unit of measurement—is $\frac{1}{5} = 0.20$. That is, if we expect to service *one* customer in five minutes, then we can also expect to service *one-fifth* customer in one minute, or $\lambda = 0.20$. The random variable $\mathbf{X}$ is defined to be the time required (in minutes) to complete a service. Also, λ, the parameter of the exponential distribution, is equal to the expected number of occurrences in one minute, $\lambda = 0.20$.

We first want to determine the probability that the random variable $\mathbf{X} =$ Length of time for service is less than or equal to 1 minute [$P(\mathbf{X}) \le 1$]. From Formula 7.2 above,

$$P[\mathbf{X} \le 1] = 1 - e^{-(0.20)(1)} = 1 - e^{-0.20}$$

From Table B.4, $e^{-0.20} = 0.8187$. Therefore, $P[X \leq 1]$ is equal to $1 - 0.8187 = 0.1813$. Thus we would conclude that the percentage of customers who receive service within one minute is $100(0.1813)$ percent $= 18.13$ percent. To determine $P[X \leq 5]$, we again use Formula 7.2:

$$P[X \leq 5] = 1 - e^{-(0.20)(5)} = 1 - e^{-1.00}$$

From Table B.4, $e^{-1.00} = 0.3679$. Therefore, $P[X \leq 5] = 1 - 0.3679 = 0.6321$. That is, 63.21 percent of the customers will be served in the *average time* or less! Since the exponential distribution is skewed to the right, the mean is greater than the median. By the time we reach the mean service time of five minutes, 63.21 percent of the area has been accumulated to the left of the mean (the median is the point where 50 percent of the area under the exponential distribution is to the right and to the left of the point).

To find $P[4 < X < 8]$, we use Formula 7.1. Note first that the probabilities $P[a < X < b]$ and $P[a \leq X \leq b]$ are equal, since $P[X = a] = P[X = b] = 0$ for the continuous random variable X. Now, we have from Formula 7.1 and Table B.4,

$$
\begin{aligned}
P[4 < X < 8] &= e^{-(0.20)(4)} - e^{-(0.20)(8)} \\
&= e^{-0.80} - e^{-1.60} \\
&= 0.4493 - 0.2019 \\
&= 0.2474
\end{aligned}
$$

Thus 24.74 percent of the customers receive service in the time interval between four and eight minutes.

Example 7.10 Example 6.11 required the computation of the probability of receiving no (0) calls by an operator in a 10-minute interval, where the calls the operator receives in a 10-minute interval follow a Poisson distribution with $\lambda = 1$ (an average of one call each 10 minutes), and Example 6.12 required computing the probability that no one would be admitted to the emergency ward of a hospital during a given time interval, where the average number of admissions during this interval are Poisson distributed with an average of 3.5 admissions. Determine each of the required probabilities in these two examples using the exponential probability distribution.

Solution We will do Example 6.11 first. Since "arrivals" of phone calls are Poisson distributed, with the mean number of calls being one in each 10-minute interval, the parameter λ of the exponential distribution is also $\lambda = 1$. (They are the same!) The probability of receiving no calls in a 10-minute interval [Poisson process, $P(X = 0)$] is equivalent to the probability that the time between calls is greater than one 10-minute interval [exponential process, $P(X > 1)$].

Using Formula 7.3, we have

$$
\begin{aligned}
P[X > 1] &= e^{-(1)(1)} = e^{-1.00} \\
&= 0.3679
\end{aligned}
$$

Thus the probability is 0.3679 that $X = $ Time between calls is greater than one minute. This is the same result obtained in Example 6.11 in Chapter 6.

From Example 6.12 in Chapter 6, the average number of admittances per

the two-hour period from 12 midnight to 2:00 A.M. is 3.5. Thus $\lambda = 3.5$ admittances in the two-hour period. The probability of no admittances in the two-hour period is equivalent to the exponential probability that the time between successive admittances is greater than the length of one period (one period = 2 hours). Thus we wish to find $P[\mathbf{X} > 1]$. From Formula 7.3,

$$P[\mathbf{X} > 1] = e^{-(3.5)(1)} = e^{-3.50}$$
$$= 0.0302$$

Example 7.11 The length of the lifetime in years of a cathode ray tube used in monitors for microcomputer systems is known to be exponentially distributed with a mean of five years. Determine the following:
a. The value of the exponential distribution parameter λ.
b. The variance and the standard deviation of the lifetimes.
c. The probability that a tube will last more than 10 years.
d. The proportion of lifetimes of the tubes that fall in the interval, mean ± 2 standard deviations.

Solution Let the random variable $\mathbf{X}$ = Lifetime of the tube in years. Since the mean lifetime is five years and $E(\mathbf{X}) = 1/\lambda$, we have:

$$1/\lambda = 5 \text{ years}; \lambda = (\frac{1}{5}) \text{ per year}$$

Therefore in part a, $\lambda = \frac{1}{5} = 0.20$. For part b, since $V(\mathbf{X}) = 1/\lambda^2$, $V(\mathbf{X}) = 1/(0.20)^2 = 1/(0.04) = 25$. And the standard deviation is $\sqrt{V(\mathbf{X})} = \sqrt{25} = 5$ years. In part c we want $P[\mathbf{X} > 10]$. From Formula 7.3, we have:

$$P[\mathbf{X} > 10] = e^{-(0.20)(10)} = e^{-2.00} = 0.1353$$

Thus the probability that a tube will last more than 10 years is 0.1353. In part d we want the probability that the random variable $\mathbf{X}$ will take on a value in the interval, mean ± 2 (standard deviations). Since the mean is 5 years and the standard deviation is also 5 years, the specified interval is $5 - 2(5) = -5$ years to $5 + 2(5) = 15$ years. Thus we wish to find $P[-5 < \mathbf{X} < 15]$. Since $\mathbf{X}$ cannot be less than 0, this probability is equivalent to $P[0 < \mathbf{X} < 15]$. From Formula 7.1, we have:

$$P[0 < \mathbf{X} < 15] = e^{-(0.20)(0)} - e^{-(0.20)(15)}$$
$$= e^{0.00} - e^{-3.0}$$
$$= 1.00 - 0.0498$$
$$= 0.9502$$

Therefore the probability is 0.9502 that the tube lasts between the mean minus two standard deviations (−5 years) and the mean plus two standard deviations (15 years), or equivalently between 0 years and 15 years.

■ 7.5

Use of MINITAB to simulate distributions

In Chapter 6 we saw how the statistical software package MINITAB can be used to generate the probability distributions for the binomial and Poisson distributions. Another interesting use of MINITAB is the *simulation* of probability distributions. We will use this feature in later chapters, most notably, in the next chapter (Chapter 8).

Suppose, for example, that the random variable **X** represents the GMAT (Graduate Management Admissions Test) score. Currently, the mean test score nationally is approximately $\mu = 475$, and the standard deviation is approximately $\sigma = 75$. Further, the test scores are distributed very nearly as a normal distribution. We will assume that **X** is normally distributed with a mean of $\mu = 475$ and a standard deviation of $\sigma = 75$. Using the NRANDOM MINITAB command, we can generate a set of random (test score) values from this distribution. For example, the following command generates 50 random normally distributed values with a population mean of 475 and a population standard deviation of 75:

```
-- NRANDOM 50 OBSERVATIONS WITH MU=475 AND SIGMA=75, PUT IN
   COLUMN C1 50 NORMAL OBS, WITH MU = 475.0000 AND SIGMA = 75.0000
   437.2827    298.3938    449.4160    482.4619    517.7961    509.3606
   372.6663    503.0493    433.6160    412.4924    462.2632    633.3259
   550.8789    466.4685    434.7715    519.5305    480.0640    434.5801
   546.3479    507.3379    445.0083    666.4658    472.7166    363.4780
   477.3140    555.0347    543.7871    443.9062    399.0693    361.0024
   511.8342    456.6543    396.1030    467.5056    425.6118    508.2576
   527.3103    535.7222    514.0435    501.8816    498.6682
   584.0371    560.6646    505.8562    486.7097    570.5381
   510.6011    535.9146    350.9858    445.4648
```

For these 50 values, we can use the MINITAB DESCRIBE command to calculate the mean and standard deviation of these values:

```
-- DESCRIBE C1
   C1         N =  50       MEAN =     481.48    ST.DEV. =     70.8
```

Note that the mean and standard deviation (481.48 and 70.8) deviate from the population mean ($\mu = 475$) and the population standard deviation ($\sigma = 75$). This is because these 50 values represent a *sample* of size $n = 50$ drawn from the population of normal values with a mean of 475 and a standard deviation of 75. A histogram of these fifty values is shown below:

```
    -- HISTOGRAM C1

     C1

       MIDDLE OF      NUMBER OF
       INTERVAL       OBSERVATIONS
          300,           1      *
          350,           4      ****
          400,           3      ***
          450,          14      **************
          500,          16      ****************
          550,           9      *********
          600,           1      *
          650,           2      **
```

Notice that the histogram looks similar to a normal distribution (turned on its side) but is not perfectly symmetric as the normal distribution is.

By using the MINITAB simulation feature, we can quickly see the effects of sampling from a known probability distribution. In this illustration, a sample of size 50 produces a set of values that conform fairly well, but not exactly, to the distribution from which they are drawn. The differences (in mean, standard deviation, and shape) between the population distribution and the distribution of the sample values (the histogram) is a key element in devising methods of estimating population parameters from sample data. We will begin discussing these ideas in the next chapter, and the presentation of statistical inference methods based on these ideas will extend throughout the remainder of the book.

■ 7.6

Summary

Three important continuous probability distributions have been presented in this chapter—the uniform, normal, and exponential distributions. The situations where these distributions can be used as models for the distributions of random variables with unknown exact distribution form were mentioned. In Table 7.2, the more important characteristics of these distributions are summarized.

If the values of the parameters of the distribution are known, we have seen how to use the probability distribution to answer questions regarding properties of the random variable, such as the probability that it exceeds a specified value and the determination of its mean value.

In most cases, the values of the parameters of the probability distribution are unknown and must be estimated from sample data. The process of estimating parametric values is the primary topic of Chapters 8 and 9.

TABLE 7.2 | Summary characteristics of the uniform, normal, and exponential distributions

Distribution	Parameters	Mean $E(\mathbf{X})$	Variance $V(\mathbf{X})$	Calculated probabilities (Appendix B)
Uniform	a, b	$\dfrac{b + a}{2}$	$\dfrac{(b - a)^2}{12}$	—
Normal	μ, σ	μ	σ^2	Table B.3
Exponential	λ	$\dfrac{1}{\lambda}$	$\dfrac{1}{\lambda^2}$	Table B.4

■ **References**

Introductory

McClave, J. T., and P. G. Benson. *Statistics for Business and Management*. 3rd ed. San Francisco: Dellen Publishing, 1985, Chapter 6.

Neter, J.; W. Wasserman; and G. A. Whitmore. *Applied Statistics*. 2nd ed. Boston: Allyn & Bacon, 1982, Chapter 7.

Wonnacott, R. J., and T. H. Wonnacott. *Introductory Statistics*. 4th ed. New York: John Wiley & Sons, 1985, Chapter 4.

Advanced

Harnett, D. L. *Introduction to Statistical Methods*. 3rd ed. Reading, Mass.: Addison-Wesley, 1982, Chapter 5.

Hogg, R. V., and A. T. Craig. *Introduction to Mathematical Statistics*. 4th ed. New York: MacMillan, 1978.

■ **Appendix**

The uniform distribution

$$\text{Density function: } f(x) = \begin{cases} \dfrac{1}{b - a} & a \le x \le b \\ 0 & x < a \text{ or } x > b \end{cases}$$

$$E(\mathbf{X}) = \int_{-\infty}^{+\infty} x f(x)\, dx = \int_{-\infty}^{a} x(0)\, dx + \int_{a}^{b} x\, \frac{1}{b - a}\, dx + \int_{b}^{\infty} x(0)\, dx$$

$$= 0 + \frac{1}{b - a} \frac{x^2}{2}\Big|_{a}^{b} + 0$$

$$= \frac{1}{b - a} \left[\frac{b^2}{2} - \frac{a^2}{2} \right] = \frac{1}{2} \frac{1}{b - a} (b^2 - a^2)$$

$$= \frac{1}{2} \frac{1}{b - a} (b - a)(b + a) = \boxed{\frac{b + a}{2}}$$

$$V(\mathbf{X}) = E(\mathbf{X} - \mu)^2 = E(\mathbf{X}^2) - [E(\mathbf{X})]^2$$

$$E(\mathbf{X}^2) = \int_{-\infty}^{+\infty} x^2 f(x)dx = \int_{-\infty}^{a} x^2(0)dx + \int_{a}^{b} x^2 \frac{1}{b-a} dx + \int_{b}^{+\infty} x^2(0)dx$$

$$= 0 + \frac{1}{b-a} \frac{x^3}{3} \Big|_{a}^{b} + 0$$

$$= \frac{1}{b-a} \left[\frac{b^3}{3} - \frac{a^3}{3} \right] = \frac{1}{3} \frac{1}{b-a} (b^3 - a^3)$$

$$= \frac{1}{3} \frac{1}{b-a} (b-a)(b^2 + ab + a^2)$$

$$= \frac{1}{3} (b^2 + ab + a^2)$$

$$V(\mathbf{X}) = E(\mathbf{X}^2) - [E(\mathbf{X})]^2 = \frac{1}{3} (b^2 + ab + a^2) - \left[\frac{(b+a)}{2} \right]^2$$

$$= \frac{1}{3} (b^2 + ab + a^2) - \frac{1}{4} (b^2 + 2ab + a^2)$$

$$= \frac{4(b^2 + ab + a^2) - 3(b^2 + 2ab + a^2)}{12}$$

$$= \frac{4b^2 + 4ab + 4a^2 - 3b^2 - 6ab - 3a^2}{12}$$

$$= \frac{b^2 - 2ab + a^2}{12} = \boxed{\frac{(b-a)^2}{12}}$$

The exponential distribution

Density function: $f(x) = \begin{cases} \lambda e^{-\lambda x} & x \geq 0 \\ 0 & x < 0 \end{cases}$

$$E(\mathbf{X}) = \int_{-\infty}^{+\infty} x f(x)\, dx = \int_{-\infty}^{0} x(0)\, dx + \int_{0}^{+\infty} x\lambda e^{-\lambda x}\, dx$$

$$= 0 + \frac{1}{\lambda} e^{-\lambda x}(-\lambda x - 1) \Big|_{0}^{\infty}$$

$$= \frac{1}{\lambda} [0 - (-1)] = \boxed{\frac{1}{\lambda}}$$

$$V(\mathbf{X}) = E(\mathbf{X} - \mu)^2 = E(\mathbf{X}^2) - [E(\mathbf{X})]^2$$

$$E(\mathbf{X}^2) = \int_{-\infty}^{+\infty} x^2 f(x)\, dx = \int_{-\infty}^{0} x^2(0)\, dx + \int_{0}^{+\infty} x^2\lambda e^{-\lambda x}\, dx$$

$$= 0 + \left[-x^2 e^{-\lambda x} - \frac{2}{\lambda^2} x^{-\lambda x}(\lambda x + 1) \right] \Big|_{0}^{\infty}$$

$$= \left[0 - \left(-\frac{2}{\lambda^2} \right) \right] = \frac{2^*}{\lambda^2}$$

---

* This result uses integration by parts on the second integral.

$$V(\mathbf{X}) = E(\mathbf{X}^2) - [E(\mathbf{X})]^2 = \frac{2}{\lambda^2} - \left[\frac{1}{\lambda}\right]^2 = \frac{2}{\lambda^2} - \frac{1}{\lambda^2} = \boxed{\frac{1}{\lambda^2}}$$

$$P[a \le \mathbf{X} \le b] = \int_a^b \lambda e^{-\lambda x}\,dx = -e^{-\lambda x}\Big|_a^b = \boxed{e^{-\lambda a} - e^{-\lambda b}}$$

$$P[\mathbf{X} \le a] = \int_0^a \lambda e^{-\lambda x}\,dx = -e^{-\lambda x}\Big|_0^a = \boxed{1 - e^{-\lambda a}}$$

$$P[\mathbf{X} \ge a] = 1 - P[\mathbf{X} \le a] = 1 - [1 - e^{-\lambda a}] = \boxed{e^{-\lambda a}}$$

■ Problems

Section 7.2 Problems

7.1 A random variable $\mathbf{X}$ is uniformly distributed over the interval between 10 and 26, inclusive.
  a. Find the mean and standard deviation of $\mathbf{X}$.
  b. Using the Chebychef theorem, determine an interval that contains at least eight-ninths of the values of $\mathbf{X}$.
  c. What proportion of the values of $\mathbf{X}$ are contained in the interval determined in part b?

7.2 The flight time between two cities is known to be uniformly distributed between 60 and 75 minutes.
  a. What is the average flight time between the two cities?
  b. What is the standard deviation of the flight times between the two cities?
  c. What is the probability that a flight between the two cities takes between 65 and 70 minutes?
  d. What is the probability that a flight takes more than 70 minutes?

7.3 The time between successive arrivals of customers at a post office station is uniformly distributed between 1 and 5 minutes.
  a. Write the formula for the density function of the uniform distribution.
  b. What is the mean interarrival time?

  c. What is the standard deviation of the interarrival times?
  d. What is the probability that the time between successive arrivals is more than 3.5 minutes?

Section 7.3 Problems

The standard normal distribution

7.4 Using the standard normal distribution tables, determine the following probabilities:
  a. $P(0.00 < \mathbf{Z} < 2.50)$
  b. $P(0.00 < \mathbf{Z} < 0.40)$
  c. $P(-1.25 < \mathbf{Z} < 0.00)$
  d. $P(-2.50 < \mathbf{Z} < -0.60)$
  e. $P(2.00 < \mathbf{Z} < 2.75)$
  f. $P(\mathbf{Z} < -1.67)$
  g. $P(\mathbf{Z} > 0.80)$
  h. $P(\mathbf{Z} < 1.90)$

7.5 Given the following probabilities, find the value of $\mathbf{Z}$:
  a. $P(0.00 < \mathbf{Z} < z) = 0.4929$
  b. $P(-2.50 < \mathbf{Z} < z) = 0.4938$
  c. $P(\mathbf{Z} > z) = 0.0018$
  d. $P(\mathbf{Z} < z) = 0.6293$

7.6 Using the normal probability tables, calculate the areas under the standard normal curve for the following z values.
  a. Between $z = 0.0$ and $z = 1.2$

b.   Between $z = 0.0$ and $z = -0.9$.
c.   Between $z = 0.0$ and $z = 1.45$.
d.   Between $z = 0.0$ and $z = -1.44$.
e.   Between $z = 0.3$ and $z = 1.56$.
f.   Between $z = -1.71$ and $z = -2.03$.
g.   Between $z = -1.72$ and $z = 2.53$.
h.   Between $z = -0.02$ and $z = 3.09$.
i.   Greater than $z = 2.50$.
j.   Greater than $z = -0.60$.
k.   Less than $z = -1.22$.
l.   Less than $z = 1.66$.

7.7   Given the following areas under the standard normal distribution, find the appropriate z value.
a.   Area between 0 and z is 0.4808.
b.   Area to the left of z is 0.8621.
c.   Area to the right of z is 0.9959.
d.   Area to the right of z is 0.0020.

The general normal distribution

7.8   Given the areas under the normal distribution, find the value x_0 so that the stated probability is correct.
a. $\mu = 500$, $\sigma = 100$; $P(\mathbf{X} \leq x_0) = 0.4286$
b. $\mu = 40$, $\sigma = 10$; $P(\mathbf{X} \geq x_0) = 0.0110$
c. $\mu = 100$, $\sigma = 20$; $P(\mathbf{X} \leq x_0) = 0.9986$

7.9   Determine the following probabilities for the given normal distributions:
a. $\mu = 500$, $\sigma = 100$; $P(\mathbf{X} \geq 700)$
b. $\mu = 40$, $\sigma = 10$; $P(\mathbf{X} \leq 25)$
c. $\mu = 100$, $\sigma = 20$; $P(\mathbf{X} \geq 75)$

7.10   The mean and standard deviation for the lifetimes of a population of light bulbs are 1,200 and 150 hours, respectively. Assuming these lifetimes are normally distributed, what is the probability that a light bulb will last more than 1,500 hours?

7.11   The average length of time required to complete the civil service PACE exam is 100 minutes with a standard deviation of 15 minutes. When should the examination be terminated if the examiner wishes to allow sufficient time for 90 percent of the applicants to complete the exam, assuming that the test times are normally distributed?

7.12   Heights of individuals in a certain population are assumed to be normally distributed with a mean of 66 inches and a standard deviation of 4 inches. At what height must door frames be constructed so that no more than 0.1 percent of the individuals in this population will be too tall to pass through them without stooping?

7.13   The heights of students in a certain population are known to be normally distributed with mean $\mu = 68$ inches and standard deviation $\sigma = 3$ inches. If two students are randomly selected from this population, what is the probability that they both will be 76 inches or taller? What assumption is necessary to solve this problem?

7.14   A machine is producing ball bearings with diameters that are normally distributed with mean and standard deviation equal to 0.498 and 0.002, respectively. If specifications require that the bearing diameters equal 0.500 inch plus or minus 0.004 inch, what fraction of the production will be acceptable?

7.15   A soft drink machine can be regulated so that it discharges an average of μ ounces per cup. If the ounces of fill are normally distributed with standard deviation $\sigma = 0.3$ ounce, give the setting for μ so that 8-ounce cups will overflow only 1 percent of the time.

7.16   The Midstate Paper Company employs two sales representatives, A and B. Salesperson A averages \$20,000 in sales per month with a standard deviation of \$8,000, and salesperson B averages \$35,000 in sales per month with a standard deviation of \$5,000. During a given month, B's sales amount to \$43,000. If it is assumed that the sales volumes for both representatives are normally distributed, how great would A's sales volume have to be for it to be proportionally as great as B's sales volume?

7.17   All prospective employees at the Naval Supply Depot in Mechanicsburg, Pennsylvania, are required to pass a special civil service examination before they are hired. Over a period of time, the examination scores have been normally distributed with mean $\mu = 85$ and standard deviation $\sigma = 4$. Suppose $n = 50$ applicants for a job sit for the test.
a.   What proportion of the 50 can be expected to obtain a score of 95 or greater?
b.   If below 80 is considered failing, what is the probability that an individual will fail the exam?
c.   Of the 50 sitting for the examination,

what is the expected number who will pass?

7.18 The income in a certain community is normally distributed with a mean annual income of $35,000 and a standard deviation of $3,000.

a. What minimum income level does a member of this community have to have to be in the top 10 percent of the citizenry?

b. What is the maximum income one can have and still be in the middle 50 percent of all incomes?

c. What is the minimum income one can have and still be in the middle 50 percent of all incomes?

7.19 Boxes of a specific cereal are packaged with a mean weight of 13.1 ounces and a standard deviation of 0.1 ounce. The box states that it contains 13 ounces of cereal. Regulations allow a box to be no more than 0.2 ounce lighter than the stated weight. If an inspector randomly selects one box from the production line, what is the probability that he will find the box is too light? Assume the package weights are normally distributed.

7.20 The number of customers entering a certain store on any given day is approximately normally distributed with mean $\mu = 50$ and standard deviation $\sigma = 9$. Find the following probabilities during a given day.

a. At least 50 customers arrive.

b. At least 45 customers arrive.

c. Between 45 and 55 customers arrive.

7.21 The average time required to finish a civil service exam is normally distributed with a mean of 60 minutes and a standard deviation of 12 minutes. How many minutes should be allowed for the examination if the supervisor wishes to allow sufficient time for 90 percent of the applicants to complete the test?

7.22 At the Gold Muffler Shop, it takes a service-person an average of 20 minutes with a standard deviation of 5 minutes to put on a new muffler. A customer leaves her car and is told to come back in 30 minutes and the car will be ready. Assuming that the muffler-mounting times are normally distributed, what is the probability that her car will not be finished when she returns in 30 minutes?

The normal approximation to the binomial distribution

7.23 Consider the binomial distribution with $n = 50$ and $\pi = 0.40$. For each part, find the *exact* binomial probability and the normal approximated probability, using the correction for continuity.

a. $P(\mathbf{X} \geq 15)$

b. $P(7 \leq \mathbf{X} \leq 13)$

c. $P(\mathbf{X} = 9)$

7.24 In a certain city, it is known that 40 percent of all color TV sets bought will be made by the Xenith Co. In a given week, $n = 25$ color TVs are sold. Find the probability that at least 8 but no more than 11 of the buyers purchased a color TV produced by Xenith.

a. Use the binomial tables.

b. Use the normal approximation to the binomial distribution.

7.25 How many questions n would a statistics professor have to put on a multiple-choice examination (where there are five responses—a, b, c, d, and e—to each question) for her to be 95 percent certain that a student making random guesses on each question will miss at least 50 percent of the questions? Use the normal approximation to the binomial distribution without the correction for continuity.

7.26 A newly designed portable radio was styled on the assumption that 50 percent of all purchasers are female. If a random sample of $n = 400$ purchasers is selected, what is the probability that the number of female purchasers in the sample will be greater than 175? Use the normal approximation to the binomial distribution without the correction for continuity.

7.27 The manufacturer of an electronic part has found that 98 percent of the parts are acceptable. A random sample of 500 parts is tested. What is the probability that 12 or more defective parts will be found? Use the normal approximation to the binomial distribution without the correction for continuity.

7.28 A new serum is 90 percent effective in preventing colds. If 100 people take the serum, what is the probability that between 7 and 12 (inclusive) will catch a cold? Use the normal approximation to the binomial distribution with the correction for continuity.

7.29 Of the people entering a large department store, it is found that 50 percent will make at least one purchase. For a sample of 100 persons, what is the probability that at least 60 will make at least one purchase? Use the normal approximation to the binomial distribution without the correction for continuity.

Section 7.4 Problems

7.30 Given the exponential distribution with parameter λ equal to 0.80,
  a. Graph the probability density function for this distribution.
  b. Determine the mean and the variance of this distribution.
  c. What percentage of the area in this distribution lies within one standard deviation of the mean? Within two standard deviations of the mean?

7.31 The length of time required to service a car at a gas station is exponentially distributed with a mean service time of 4.0 minutes.
  a. What proportion of the cars are serviced within 1 minute?
  b. What proportion of the cars are serviced within 6 minutes?
  c. What is the probability that a car will be serviced within 2 to 8 minutes?

7.32 The life of an electronic component (number of hours before failure) is exponentially distributed with an average time to failure of 550 hours.
  a. What is the probability that a component fails before 375 hours?
  b. What is the probability that a component fails before 1,000 hours?
  c. What is the probability that a component fails between 500 and 1,200 hours?

7.33 Suppose a fast-food hamburger chain can serve its customers on the average at the rate of two customers in 3 minutes. Assume the service rate to be Poisson distributed.
  a. What is the probability that a restaurant in this chain will be able to serve four or more customers in a 3-minute period?
  b. What is the probability it will take longer than 3 minutes to serve a customer?
  c. What is the probability it will take between 3 and 6 minutes to serve a customer?

7.34 Ships arrive at a dock on an average of one ship every two days. What is the probability that after the departure of a ship, four or more days will elapse before another arrives?

7.35 Each 500-foot roll of fabric produced in a mill contains two flaws on the average. What is the probability that as the fabric is unrolled, the first flaw occurs within the first 50 feet?

Additional Problems

7.36 A worker submits two jobs at a service window. The average time required to complete a job is 100 minutes with a standard deviation of 20 minutes. Processing is begun on both jobs immediately, and the worker decides to return in two hours to check whether either or both jobs have been completed. Assuming the processing times are normally distributed, what is the probability that both jobs will be ready when he returns? What assumption is required to answer this question?

7.37 The diameter of ball bearings produced in a manufacturing process is known to have a mean of 0.032 inch and a standard deviation of 0.001 inch.
  a. Using the Chebychef theorem, construct an interval that contains at least 75 percent of the diameters of all ball bearings produced.
  b. If it is assumed that the diameters are normally distributed, what proportion of ball bearings is contained in the interval developed in part a?
  c. Compare parts a and b. Discuss why the proportion of diameters contained in the interval differs in parts a and b.

7.38 A production process is known to produce 10 percent defectives. A buyer purchases 225 of these items. What is the probability that 20 or less are defective? Use the normal approximation to the binomial distribution without the correction for continuity.

7.39 Indicate the type of probability distribution (normal, exponential, Poisson) that you would expect to be most applicable in each of the following instances.
  a. The pattern of variation in the number of typesetting errors per printed page when the printer averages 0.05 error per page.
  b. The pattern of variation in the tempera-

ture readings of thermometers produced under the same conditions and experimentally exposed to a temperature of 50°C.

c. The pattern of variation in the lengths of time to receive a hair style at the local barber shop.

7.40 A type C transistor has a mean lifetime of 10,000 hours. Assume that the lifetimes are exponentially distributed.

a. What is the probability that a particular type C transistor will last more than 30,000 hours?

b. Is the exponential distribution an appropriate model for the lifetimes in this case? Why or why not?

7.41 An electronic component is known to have a mean time to failure of 1,000 hours and a standard deviation of 500 hours.

a. Why is the normal distribution an inappropriate model for the distribution of this variable?

b. Is the exponential distribution appropriate in this case? Discuss.

7.42 The length of time in minutes of calls made to a toll-free number is known to be exponentially distributed with a mean time of three minutes.

a. What is the probability that the length of a call will exceed five minutes?

b. What is the probability that the length of a call will be less than two minutes?

c. What is the probability that the length of a call will be between two minutes and four minutes?

7.43 The wages of mechanics working for a major airline are known to be normally distributed with a mean of $35,000 and a standard deviation of $5,000.

a. What proportion of mechanics receive a wage greater than $50,000?

b. What is the probability that a mechanic's wage is less than $30,000?

c. What is the third quartile of the wages for the mechanics? Recall that the third quartile is the 75th percentile so that at least 75 percent of the values of **X** are at or less than the third quartile and at least 25 percent of the values of **X** are at or more than the third quartile.

7.44 The Graduate Management Admissions Test (GMAT) produces a score between 200 and 800 points with a mean score of 475 and a standard deviation of 75 points. Assuming that the normal distribution can be used to approximate the actual distribution of test scores, answer the following:

a. What is the probability that a test taker makes a score of more than 600 points?

b. What proportion of test scores are between 300 and 400 points?

c. A test taker makes a score of 650. What proportion of test takers make better scores?

d. A test taker makes a score that places her in the 90th percentile (that is, 90 percent of the test takers make her score or less, and 10 percent make her score or more). What score did the test taker make?

7.45 The useful lifetime of a specific model of a computer terminal is assumed to be normally distributed with a mean of five years and a standard deviation of one year.

a. What proportion of the terminals will have a useful lifetime between three and seven years?

b. What is the probability that a terminal will have a useful lifetime exceeding eight years?

c. Find the first and third quartiles of the useful lifetimes of the computer terminals in years. Recall that the first quartile is the 25th percentile: At least 25 percent of the values of **X** are at or less than the first quartile, and at least 75 percent of the values of **X** are at or more than the first quartile. The third quartile is the 75th percentile: At least 75 percent of the values of **X** are at or less than the third quartile, and at least 25 percent of the values of **X** are at or more than the third quartile.

7.46 A production process produces defectives at the rate of 5 defectives per 100 manufactured units of a product. One hundred units of the product are taken from the production process. Using the normal distribution approximation to the binomial distribution, determine the following:

a. The probability that more than eight of the selected items are defective.

b. The probability that between three and six of the selected items are defective.

7.47 A disk drive on a microcomputer has a mean time to failure of 8,000 hours of continuous use. Assume that the random variable **X** = Time (in hours) to failure is exponentially distributed.

    a. What is the probability that a disk drive will fail between 6,000 and 10,000 hours of continuous use?

    b. What is the probability that a disk drive will not fail through 12,000 hours of continuous use?

7.48 Boards are cut for a construction project that are 6 feet in length. The cutting machine does not cut each board so that its length is exactly 6 feet. Rather, the error in the length varies between −0.5 inch and 0.5 inch. Assume that the errors are uniformly distributed.

    a. Write the density function for the random variable **X**, where **X** = Error in inches from the specified 6 feet length.

    b. What is the mean error?

    c. What is the standard deviation of the errors?

    d. What is the probability that a board will have between a +0.02- and a +0.45-inch error?

7.49 A random variable **X** has the following density function:

$$f(x) = \begin{cases} x & 0 \le x \le 1 \\ 2 - x & 1 \le x \le 2 \\ 0 & x < 0 \text{ or } x > 2 \end{cases}$$

    a. Graph the density function.

    b. Determine $E(X)$.

    c. Determine $P[0 < X < 0.75]$ using the fact that the area of a triangle is $(1/2)(base)(height)$.

    d. Determine $P[0.50 < X < 1.50]$.

7.50 (Calculus required.) In Problem 7.49, using the integration formulas for $E(X)$ and $V(X)$ (given in the appendix), find the mean and the variance of the random variable **X** whose density function is given in Problem 7.49. What is the median of the random variable **X**?

7.51 (Calculus required.) The time to complete an assembly task is known to be uniformly distributed with a time in minutes between 10 and 15. Using integration, find the following:

    a. The probability that the assembly time takes between 12 and 14 minutes.

    b. The probability that the assembly time takes more than 13 minutes.

    c. The median of the random variable.

7.52 (Calculus required.) A battery is known to have a lifetime that is exponentially distributed with a mean lifetime of 10 hours. Using integral calculus, determine the following:

    a. The probability that the lifetime of a battery exceeds 15 hours.

    b. The probability that the lifetime of a battery will be between 8 and 12 hours.

    c. The median of the lifetimes of the batteries.

7.53 The following 50 values have been randomly drawn from an exponential distribution with a mean of 750. The random variable **X** represents the lifetime in hours of 100-watt light bulbs. Thus the random variable **X** is assumed to be exponentially distributed with a mean $(1/\lambda)$ of 750 hours.

330.34	82.16	757.23	1646.97	279.65
140.21	545.19	556.96	710.28	370.23
276.48	439.42	1377.09	855.66	799.71
1688.52	11.52	274.86	1776.95	807.16
611.89	1278.51	344.92	728.53	198.17
437.97	603.67	686.44	624.14	165.10
800.27	675.23	580.63	366.47	771.95
835.32	1240.19	135.61	162.72	550.49
1815.52	193.56	832.02	1321.53	652.84
1244.66	636.90	1917.19	1036.91	335.99

    a. Calculate the sample mean and the sample standard deviation of these 50 light bulb lifetimes. How do these sample numerical measures compare with the population mean and the population standard deviation of an exponentially distributed random variable with $1/\lambda = 750$?

    b. Using the exponential distribution with $1/\lambda = 750$, find $P(X > 1,000 \text{ hours})$. How does this probability compare with the proportion of the 50 lifetimes that exceed 1,000 hours?

8

Samples and sampling distributions

■ 8.1

Introduction

In Chapters 3–7, we developed the ideas of a random variable and the distribution of a random variable. We have seen that by designating a random variable to represent a quantitative characteristic in a population, properties of the quantitative characteristic including its mean and standard deviation may be determined from the distribution of the random variable.

In this chapter, we will draw on these ideas to lay the foundation for methods of drawing inferences about the properties of population characteristics from a sample.

■ 8.2

Parameters and statistics

When we have defined a random variable in a population to represent a quantitative characteristic of interest, we are then usually interested in the properties of the random variable—such as its mean, median, and standard deviation. For example, suppose the population is defined to be the collection of the Fortune 500 companies, and we are interested in the annual rate of return on investments in 1986 for each company. The quantitative characteristic of interest is the annual rate of return on investments, which we could represent by the random variable X. We may then be interested in the *average* or *mean* annual rate of return, the *median* rate of return, or the *standard deviation* of the rates of return for these 500 companies. The numerical measures—mean, median, standard deviation—are referred to as *parameters* when they are used to measure properties of a population characteristic, or of a random variable representing that characteristic.

A parameter

> **Definition 8.1**
> A parameter
>
> A *parameter* is a numerical measure of a population characteristic.

When the numerical measures are applied to a set of sampled observations from a population, then the numerical measures are called *statistics*.

A statistic

> **Definition 8.2**
> A statistic
>
> A *statistic* is a numerical measure calculated from a set of sample observations.

For example, suppose that we took a sample of 25 Fortune 500 companies in 1986 and calculated the mean, median, and standard deviation of the annual rate of return on investments for these 25 companies. These numerical measures—mean, median, and standard deviation—are *statistics,* since they have been computed on a set of sample observations.

Statistical inference is the process of drawing inferences about population characteristics of interest from a sample drawn from the population. Typically, this process is equivalent to estimating the values of population parameters by using sample statistics. We will develop the ideas of statistical inference in this chapter and in the following two chapters. We begin by describing how a sample is taken from the population.

■ 8.3
Simple random sampling

There are several ways that samples can be drawn from populations. We will describe the most commonly used way, a *simple random sample*. Other sampling methods are described later in this chapter (Section 8.9).

Before defining a simple random sample, it is necessary to review the notation developed in Chapters 5, 6, and 7 for a random variable and its values. A *random variable* measuring a population characteristic of interest, say **X**, is denoted by a capital **boldfaced** letter or symbol. A *value* of this random variable is denoted by the corresponding lowercase letter x.

The distinction between random variables and their values arises first in the sampling context when we speak of a random sample drawn from a population. If we are referring to a random sample of size n *before* the

Notation: **X** denotes a random variable; x denotes a value of a random variable

sample is actually drawn, the random sample of size n is designated by $\mathbf{X}_1$, $\mathbf{X}_2, \ldots, \mathbf{X}_n$, a set of *random variables*. After a specific, single sample is drawn from the population, the values of these random variables are designated by $x_1, x_2, \ldots, x_n$. The set $x_1, x_2, \ldots, x_n$ represents the collection of n *numbers* that compose a single sample of size n drawn from the population. For example, suppose we have a finite population of N values $x_1, x_2,$ $\ldots, x_N$ and consider the random sample $\mathbf{X}_1, \mathbf{X}_2, \ldots, \mathbf{X}_n$. That $\mathbf{X}_1, \mathbf{X}_2,$ $\ldots, \mathbf{X}_n$ are random variables may best be understood by considering the following fact: The first sample random variable, $\mathbf{X}_1$, may take on *any one* of the population values $x_1, x_2, \ldots, x_N$ *before* the sample is actually drawn. In a similar fashion, the second sample random variable, $\mathbf{X}_2$, may take on one of the remaining population values before the sample is drawn, and so forth for the rest of the sample random variables. Thus *before the sample is taken,* $\mathbf{X}_1, \mathbf{X}_2, \ldots, \mathbf{X}_n$ are random variables with values that are determined once a single sample is taken from the population. *After the sample is taken,* the values of the random variables $\mathbf{X}_1, \mathbf{X}_2, \ldots, \mathbf{X}_n$ are known and are denoted by $x_1, x_2, \ldots, x_n$ (note that these values are not necessarily the first n values in the population set of values $x_1, x_2, \ldots, x_N$).

Two definitions of a simple random sample are given. In the first we define a simple random sample taken from a *finite* population. In the second we define a simple random sample taken from an *infinite* population; that is, a population that has an infinite number of elements.

Simple random sample, *finite population*

Definition 8.3
Simple random sample, finite population

A *simple random sample* taken from a finite population gives each possible subset of the population (sample) of a specified size an equal probability of being selected.

This definition applies only when we are sampling without replacement; that is, when each unit drawn from the population for the sample is not returned prior to drawing the next unit. We will be concerned only with sampling without replacement in this text. Sampling with replacement has very limited and special uses in statistics.

To illustrate Definition 8.3, suppose that we have a population that contains three elements, A, B, and C. We wish to draw a simple random sample of size two from this population of three elements. There are three possible samples that could be taken: AB, AC, and BC. Note that the order in which each member of the pair is drawn is immaterial. Thus AB (drawing A first, then B) is the same sample as BA (drawing B first, then A). To draw the simple random sample of size two, we could place the letters *AB*, *AC*, and

BC on three chips and select one of the chips after mixing them thoroughly in a container, such as a hat. Although this method would produce a simple random sample satisfying Definition 8.3, it becomes impractical when the number of elements in the population is large. For example, if there are 1,000 elements in the population and we wish to draw a simple random sample of size 100 of these elements, there is an enormous number of possible samples of size 100 that could be drawn. We would not have enough tags for them all! A better way of producing a simple random sample is to use the idea in Definition 8.3 sequentially. Suppose that the population contains N elements, and we wish to sample n of them, where $n < N$. We could do the following:

1. Select the first element to be sampled by giving the N population elements an equal chance of being selected.

2. Select the second element to be sampled by giving the remaining $N - 1$ population elements an equal chance of being selected.

3. Repeat this process until all n sample elements have been selected.

Computers can be used very effectively in applying this process. In fact, almost all random samples are formed by using computers to draw the sample. For a simple random sample, we only need to "tag" the population elements by numbers from 1 to N and instruct the computer software to draw a simple random sample of size n from the numbers $1, 2, \ldots, N$.

When the population is infinite in size, we must modify our previous definition of a simple random sample, for there are an infinite number of samples that can be drawn of a specified size in this case.

To illustrate this fact, consider the following example. A machine is designed to produce steel ball bearings with a 1-inch diameter. Although the machine is set to produce 1-inch diameter bearings, there is a slight variation in the diameters of the bearings produced that is within the reliability tolerance set for the machine by the engineers. The machine produces several hundred of these bearings per day. Let the random variable $\mathbf{X}$ measure the diameter of a ball bearing. Further, let the random variable $\mathbf{X}_i$ measure the diameter of the ith bearing produced by the machine and x_i be the value (diameter) of the random variable $\mathbf{X}_i$, $i = 1, 2, 3, \ldots$. Clearly, there are an infinite number of values $x_1, x_2, x_3, \ldots$, since the machine can run forever (at least conceptually!). Therefore, there are an infinite number of samples of size n that could be taken from the population set of values, $x_1, x_2, x_3, \ldots$. If we choose n ball bearings produced by the machine, then we say that the random variables $\mathbf{X}_1, \mathbf{X}_2, \ldots, \mathbf{X}_n$ measuring the diameters of these bearings constitute a *simple random sample* if the conditions in Definition 8.4 are satisfied.

More simply, we require that the random variables $\mathbf{X}_1, \mathbf{X}_2, \ldots, \mathbf{X}_n$ are chosen so that the value each variable assumes neither is affected by nor affects the value that any other variable assumes. We further require that all variables are drawn from the probability distribution of the same population random variable $\mathbf{X}$.

Definition 8.4

Simple random sample, infinite population

The *n* random variables $X_1, X_2, \ldots, X_n$ are a *simple random sample* of size *n* if two conditions are satisfied:

1. The random variables $X_1, X_2, \ldots, X_n$ are statistically independent.
2. The random variables $X_1, X_2, \ldots, X_n$ are drawn from the same probability distribution, the probability distribution of the population random variable X.

In many business applications of inferential statistics, the population contains a finite number of elements so that Definition 8.3 is applicable. However, we may choose to approximate the discrete distribution of a population of finite elements by using a continuous distribution with an infinite number of elements. A good example is the use of the normal distribution introduced in Chapter 7 for approximating discrete distributions of finite sets of population elements. We will use the normal approximation frequently in this and the next two chapters.

In the next section, we will apply Definition 8.3 to a small finite population set of elements to investigate how sample statistics vary in value from one sample to the next.

■ **8.4**

Sampling and nonsampling errors

We cannot expect that the numerical measure of a characteristic of interest will have the same value in the sample as it does in the population. That is, the value of a statistic will rarely, if ever, equal the value of the corresponding population parameter. To see this, consider the following example.

Example 8.1 Suppose that there are $N = 5$ Fortune 500 companies whose corporate headquarters are in Dallas. Let these five companies constitute the population of interest. Further, suppose that we are interested in the annual rate of return on investments in 1986 for these five companies. Stated as percentages, suppose that these rates of return are: 7.5, 8.0, 2.2, 15.9, and 16.4 percent.

Let the random variable X equal the annual rate of return on investments for a company selected at random from this population. Then, the five rates of return above can be thought of as values of the random variable X and can be denoted by:

$$x_1 = 7.5, \quad x_2 = 8.0, \quad x_3 = 2.2, \quad x_4 = 15.9, \quad x_5 = 16.4$$

Suppose that a simple random sample of size three is taken from this population of five companies. The parameter of interest in the population is the mean annual rate of return on investments for the five companies. Calculate the value of the parameter μ, the population mean annual rate of return on investments. Form the set of samples of size three drawn from the five population elements, and calculate the value of the statistic $\overline{X}$, the sample mean.

Solution Let the five companies constituting the population be tagged by the letters A, B, C, D, and E. Table 8.1 lists the tags for the five companies and the rates of return for each.

TABLE 8.1 | Set of population values for the rate of return random variable **X** in Example 8.1

Company	Value of **X**, annual rate of return on investments
A	$x_1 = 7.5$
B	$x_2 = 8.0$
C	$x_3 = 2.2$
D	$x_4 = 15.9$
E	$x_5 = 16.4$

The value of the population parameter μ is given by:

$$\mu = \frac{\sum_{i=1}^{N} x_i}{N} = \frac{7.5 + 8.0 + 2.2 + 15.9 + 16.4}{5} = 10.0$$

The list of the possible samples of size three taken from the population of five companies and the value of the statistic $\overline{X}$ for each sample are shown in Table 8.2.

TABLE 8.2 | The possible samples of size $n = 3$ from $N = 5$ elements in the Example 8.1 population

Sample composition	Values of **X**	$\overline{x}$ sample mean
ABC	7.5, 8.0, 2.2	5.90
ABD	7.5, 8.0, 15.9	10.47
ABE	7.5, 8.0, 16.4	10.63
ACD	7.5, 2.2, 15.9	8.53
ACE	7.5, 2.2, 16.4	8.70
ADE	7.5, 15.9, 16.4	13.27
BCD	8.0, 2.2, 15.9	8.70
BCE	8.0, 2.2, 16.4	8.87
BDE	8.0, 15.9, 16.4	13.43
CDE	2.2, 15.9, 16.4	11.50

The value of the population parameter μ is 10.00. It is interesting to note that none of the 10 possible samples of size three taken from the population of five companies produced a sample mean of 10.00. This result should not be too surprising. The sample mean is based on only three of the five population values of the population random variable **X**. It is also of interest to notice how the sample mean varies from sample to sample. The largest value is 13.43, and the smallest value is 5.90. If we were to use the sample mean as an estimate of the population mean μ, then we could see from the set of all possible samples that some would produce sample means that were not very good estimates of the population mean. For example, the first sample (companies A, B, and C) produces a sample mean of 5.90, considerably under the rate of return for all five companies (10.00).

The difference between the value of a population parameter, such as the population mean in Example 8.1, and the value of the corresponding numerical measure in the sample, the statistic, is referred to as *sampling error*.

Definition 8.5

Sampling error

Sampling error is the difference between the value of the parameter in the population and the value of the statistic in the sample, excluding other errors, such as errors of measurement or technical errors in taking the sample.

Sampling error

The first sample in Example 8.1, companies A, B, and C, produce a sample mean rate of return (value of the statistic) of 5.90 percent when the population mean rate of return (value of the parameter) for all five companies is 10.00 percent. The difference between the values of the parameter and the statistic—10.00 percent − 5.90 percent = 4.10 percent—is sampling error.

It is important to understand that sampling error is a controllable, but unavoidable, part of sampling. The amount of sampling error can be diminished by increasing the sample size as the next example illustrates.

Example 8.2   Consider the Example 8.1 sampling experiment. Instead of taking samples of size $n = 3$, suppose that the sample size is increased to $n = 4$. Form all possible samples of size $n = 4$, and calculate the sample mean in each sample.

Solution   The list of all possible samples of size $n = 4$ from the population of $N = 5$ companies together with the sample mean for each sample is shown in Table 8.3.

There are only five possible samples of size $n = 4$ that can be drawn from the population of $N = 5$ companies. The first sample (ABCD) produces the smallest sample mean (8.40), and the third sample (ABDE) produces the

TABLE 8.3 | The list of all possible samples of size $n = 4$ and the sample mean for each sample for the Example 8.1 population data

Sample composition	Values of X	$\bar{x}$ sample mean
ABCD	7.5, 8.0, 2.2, 15.9	8.40
ABCE	7.5, 8.0, 2.2, 16.4	8.53
ABDE	7.5, 8.0, 15.9, 16.4	11.95
ACDE	7.5, 2.2, 15.9, 16.4	10.50
BCDE	8.0, 2.2, 15.9, 16.4	10.63

largest sample mean (11.95). Compare the smallest and largest values here with those in Example 8.1 when samples of size $n = 3$ are taken. It is clear that these samples have less sampling error, for their means tend to be closer to the parameter value of $\mu = 10.00$ than do the means for the samples of size $n = 3$.

Of course, we can eliminate the sampling error altogether by sampling *all* of the population values. In Example 8.1, if we sample all five companies in the population, then the sample is ABCDE, with values 7.5, 8.0, 2.2, 15.9, and 16.4. The sample average of these five values is 10.00 of course—the value of the parameter μ, the population mean. When we sample all of the population values, we say that a census has been taken.

Census

> **Definition 8.6**
> Census
>
> A *census* of a finite population is a procedure or a study that includes (measures) every element of the population.

Thus, if we sampled all five companies in the population of the five Fortune 500 companies in Example 8.1, we would have taken a census of the population. Perhaps the most extensive census taken is the census of the United States population every 10 years. Interestingly, it has been demonstrated that samples of the U.S. population can produce more accurate numerical measures of population characteristics than the census can. This is due in part to the fact that errors arise in measurement and recording information in the census that can be minimized in taking a sample due to having more time to study each sampled element. In addition, it is impossible to contact each and every element (person) in the U.S. population.

Hence nonreporting elements result in the population being systematically misrepresented.

Errors may occur in sampling in addition to sampling error.

Nonsampling error

> **Definition 8.7**
> Nonsampling error
>
> *Nonsampling errors* are errors that occur in acquiring, recording, or tabulating statistical data that cannot be attributed to sampling error.

Biased sample

Measurement error

Nonsampling errors can result from technical errors in taking the sample. One might want a simple random sample but might fail to get one because of a failure in meeting the conditions in either Definition 8.3 or 8.4 (whichever applies). For example, the sample may be *biased*—it may tend to favor certain population elements so that not all elements have an equal chance of being represented in the sample if the population is finite. Another common form of nonsampling error is *measurement error*—incorrectly encoding information for computer analysis, for example. Although sampling error is unavoidable, nonsampling errors can be avoided by carefully planning and executing experiments and surveys.

Examples 8.1 and 8.2 illustrate two key ideas in our development of taking samples from a population for inference purposes:

1. The value of a statistic, such as the sample mean in Examples 8.1 and 8.2, *varies* from sample to sample of the same size.

2. The amount of the sampling error can be decreased by increasing the sample size.

In the next section, we will investigate more fully the variation of a statistic from sample to sample where each sample is drawn from a population of elements.

■ 8.5

Introduction to sampling distributions

A *statistic,* such as the sample mean $\overline{X}$, is a *random variable*. Hence a statistic has an *expected value* (mean), a *variance*, and a *distribution*. To demonstrate this idea, let us return to the Example 8.1. In this problem all samples of size $n = 3$ are drawn from a population of $N = 5$ elements. Using the notation developed in section 8.3, the sample of size $n = 3$ can be represented by three random variables—X_1, X_2, and X_3. These variables assume specific values of the population random variable X in each sample. Consider the first three samples listed in Table 8.2, for example:

Sample composition	*Values of* $\mathbf{X}$	$\bar{x}$ *sample mean*
ABC	7.5, 8.0, 2.2	5.90
ABD	7.5, 8.0, 15.9	10.47
ABE	7.5, 8.0, 16.4	10.63

In the first sample (ABC), the values of $\mathbf{X}_1$, $\mathbf{X}_2$, and $\mathbf{X}_3$ are 7.5, 8.0, and 2.2, respectively. In the second sample (ABD), the values of $\mathbf{X}_1$, $\mathbf{X}_2$, and $\mathbf{X}_3$ are 7.5, 8.0, and 15.9, respectively. Before a particular sample is taken, we do not know the values of these random variables. After the sample is taken, then we know the values—the numerical values constitute the sample. The sample mean is also a random variable prior to taking a sample and determining the numerical value of the sample mean. Using the boldfaced notation for a random variable, we can write the formula for the sample mean as:

$$\overline{\mathbf{X}} = \frac{\sum_{i=1}^{n} \mathbf{X}_i}{n} = \frac{\mathbf{X}_1 + \mathbf{X}_2 + \cdots + \mathbf{X}_n}{n}$$

Once a specific sample has been taken, say the first sample above (ABC), then the random variable $\overline{\mathbf{X}}$ can be evaluated. The value of the statistic $\overline{\mathbf{X}}$ is denoted by:

$$\bar{x} = \frac{\sum_{i=1}^{n} x_i}{n} = \frac{x_1 + x_2 + \cdots + x_n}{n}$$

In the first sample (ABC), we have therefore $\overline{\mathbf{X}} = \bar{x} = 5.90$. In fact, every statistic is a random variable. That is, any numerical measure that is calculated based on a sample set of observations is a random variable. We can now restate the definition of a statistic, given in Definition 8.2, in the language of random variables.

Definition of a statistic as a random variable

> A *statistic* is a random variable that is a function of the sample random variables $\mathbf{X}_1$, $\mathbf{X}_2$, . . . , $\mathbf{X}_n$. The random variables $\mathbf{X}_1$, $\mathbf{X}_2$, . . . , $\mathbf{X}_n$ assume the values x_1, x_2, . . . , x_n when a specific random sample is drawn from the population.

Since a statistic, such as the sample mean $\overline{\mathbf{X}}$, is a random variable, it has a probability distribution, and this distribution is called the *sampling distribution* of the statistic.

> **Definition 8.8**
> The sampling distribution of a statistic
>
> A statistic is a random variable with a value that is determined by taking a random sample of size n from the population. Since a statistic is a random variable, it has a distribution, called the *sampling distribution of the statistic*.

The sampling distribution of a statistic

When the population is finite in size, we can think of the sampling distribution of a statistic as being formed by taking all possible samples of a fixed sample size n, evaluating the statistic in each sample, and constructing a distribution of these values. These ideas are illustrated in the following example. In this example (Example 8.3), there are two population characteristics measured by random variables **X** and **Y**. For each random variable, the sampling distribution of a statistic is constructed and discussed. Thus the example is used to illustrate the sampling distributions of *two* statistics.

Example 8.3 Suppose we select five students from England who are studying at a large university in the United States and consider them to constitute a population. Two quantitative characteristics are measured for each student: (1) the number of semester hour credits taken during the current semester measured by the random variable **X** and (2) a binary variable that is 1 if the student's grade point average is greater than or equal to 3.00 (on a 4.00 scale) and 0 if the student's grade point average is less than 3.00—this variable is measured by the random variable **Y**. The random variable **Y** = 1 if GPA $\geq$ 3.00, and **Y** = 0 if GPA $<$ 3.00. Table 8.4 lists the

TABLE 8.4 | Set of population values of the random variables **X** and **Y**.

Student	Value of **X** (number of credits)	Value of **Y** ($1 = GPA \geq 3.00$) ($0 = GPA < 3.00$)
A	$x_1 = 12$	$y_1 = 1$
B	$x_2 = 10$	$y_2 = 1$
C	$x_3 = 14$	$y_3 = 1$
D	$x_4 = 9$	$y_4 = 0$
E	$x_5 = 10$	$y_5 = 0$

values of the random variables **X** and **Y** for the population of five students. Thus the first student is currently taking 12 semester credit hours (**X** = 12) and has a grade point average (GPA) that is greater than or equal to 3.00 (**Y** = 1).

We will now consider taking samples of size $n = 3$ from this finite population of $N = 5$ elements (students). In each sample, we will compute the value of *two* sample statistics:

1. The sample mean number of credits taken by the three students. This is the statistic $\overline{X}$, the sample mean.

2. The proportion of the three students who have a GPA $\geq$ 3.00. We will use the symbol P to designate this statistic.

For example, suppose we selected the first three students in Table 8.4 for our sample (students A, B, and C). The average number of credits taken by these three students is:

$$\frac{12 + 10 + 14}{3} = 12.00$$

and the proportion that have a GPA $\geq$ 3.00 is 1.00 (all three do). Therefore, for this sample of three students, $\overline{X} = 12$, and $P = 1.00$. Before we compute the values of the statistics (random variables) $\overline{X}$ and P, we will compute parameters of the population random variables X and Y.

For the random variable X = Number of semester credit hours taken, we will compute two parameters: the population mean, μ_X, and the population standard deviation, σ_X. The formulas for these population numerical measures are given in Chapter 2 on pages 39 and 47.

We are using the subscript X to indicate that these are the population mean and the population standard deviation for the random variable X. These computations follow:

$$\text{Mean:} \quad \mu_X = \frac{\sum_{i=1}^{5} x_i}{5} = \frac{12 + 10 + 14 + 9 + 10}{5} = 11$$

$$\text{Standard deviation:} \quad \sigma_X = \sqrt{\frac{\sum_{i=1}^{5} (x_i - \mu_X)^2}{5}}$$

$$= \sqrt{\frac{(12 - 11)^2 + (10 - 11)^2 + (14 - 11)^2 + (9 - 11)^2 + (10 - 11)^2}{5}}$$

$$= \sqrt{\frac{16}{5}} = \sqrt{3.2} = 1.789$$

Therefore the population mean number of credits taken by the five students is $\mu_X = 11$, and the population standard deviation is $\sigma_X = 1.789$.

For the random variable $\mathbf{Y}$ (1 if GPA $\geq$ 3.00 and 0 if GPA $<$ 3.00), we will also compute the population mean, $\mu_\mathbf{Y}$, and the population standard deviation, $\sigma_\mathbf{Y}$. These computations follow:

Mean:
$$\mu_\mathbf{Y} = \frac{\sum_{i=1}^{5} y_i}{5} = \frac{1 + 1 + 1 + 0 + 0}{5} = \frac{3}{5} = 0.60$$

Standard deviation:
$$\sigma_\mathbf{Y} = \sqrt{\frac{\sum_{i=1}^{5} (y_i - \mu_\mathbf{Y})^2}{5}}$$

$$= \sqrt{\frac{(1 - 0.6)^2 + (1 - 0.6)^2 + (1 - 0.6)^2 + (0 - 0.6)^2 + (0 - 0.6)^2}{5}}$$

$$= \sqrt{\frac{1.2}{5}} = \sqrt{0.24} = 0.490$$

Notice that the population mean $\mu_\mathbf{Y} = 0.60$ represents the population proportion of students with a GPA $\geq$ 3.00. The standard deviation of these three 1s and two 0s in the population is $\sigma_\mathbf{Y} = 0.490$.

Now suppose that we take a sample of size $n = 3$ from this population of $N = 5$ students. The 10 possible samples of size $n = 3$ are given in Table 8.5. (What counting rule in Chapter 3 could be used to determine that there are 10

TABLE 8.5 | Possible samples of size $n = 3$ taken from $N = 5$ elements, values of the random variables **X** and **Y** for each sample, and values of the statistics $\overline{\mathbf{X}}$ and **P** for each sample.

Sample composition	*Values of* **X**	*Statistic* $\overline{\mathbf{X}}$ *value*	*Values of* **Y**	*Statistic* **P** *value*
ABC	12, 10, 14	12.00	1, 1, 1	1.00
ABD	12, 10, 9	10.33	1, 1, 0	0.67
ABE	12, 10, 10	10.67	1, 1, 0	0.67
ACD	12, 14, 9	11.67	1, 1, 0	0.67
ACE	12, 14, 10	12.00	1, 1, 0	0.67
ADE	12, 9, 10	10.33	1, 0, 0	0.33
BCD	10, 14, 9	11.00	1, 1, 0	0.67
BCE	10, 14, 10	11.33	1, 1, 0	0.67
BDE	10, 9, 10	9.67	1, 0, 0	0.33
CDE	14, 9, 10	11.00	1, 0, 0	0.33

possible samples of size $n = 3$ taken from a population of $N = 5$ elements?) In addition, Table 8.5 gives for each of the 10 possible samples the following:

1. The three values of the random variable **X**.

2. The value of the statistic $\overline{X}$, the sample average number of credits taken by the three students.

3. The three values of the random variable **Y**.

4. The value of the statistic **P**, the sample proportion of the three students who have a GPA ≥ 3.00.

We will first focus attention on the statistic $\overline{X}$, the sample mean. The population mean, μ_X, is 11, but only 2 of the 10 samples produce a sample mean $\overline{x}$ of 11; the other samples produce sample averages (values of the statistic $\overline{X}$) that are fairly well clustered about $\mu_X = 11$. The *sampling distribution* of the statistic $\overline{X}$ is shown in Figure 8.1.

FIGURE 8.1 | Sampling distribution of the statistic $\overline{X}$

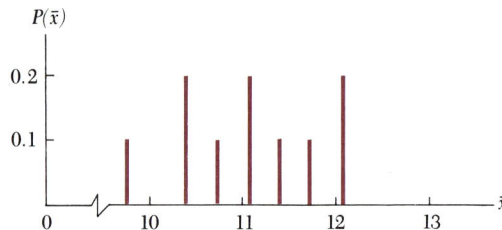

We will now compute the mean and standard deviation of the statistic $\overline{X}$. Denote the mean of the random variable $\overline{X}$ by $\mu_{\overline{X}}$:

$$\mu_{\overline{X}} = \frac{\sum_{j=1}^{10} \overline{x}_j}{10}$$

$$= \frac{12 + 10.33 + 10.67 + 11.67 + 12 + 10.33 + 11 + 11.33 + 9.67 + 11}{10}$$

$$= 11.00$$

Notice that the divisor is 10, the number of possible samples of size $n = 3$ taken from $N = 5$ population elements. The standard deviation of the 10 values of the statistic $\overline{X}$ is:

$$\sigma_{\overline{\mathbf{x}}} = \sqrt{\dfrac{\displaystyle\sum_{j-1}^{10} (\overline{x}_j - \mu_{\overline{\mathbf{x}}})^2}{10}}$$

$$= \sqrt{\dfrac{\begin{array}{l}(12 - 11)^2 + (10.33 - 11)^2 + (10.67 - 11)^2 + (11.67 - 11)^2 \\ \quad + (12 - 11)^2 + (10.33 - 11)^2 + (11 - 11)^2 + (11.33 - 11)^2 \\ \quad + (9.67 - 11)^2 + (11 - 11)^2\end{array}}{10}}$$

$$= 0.73$$

Notice that the divisor of the sum of squares under the radical is 10, not 10 − 1 = 9. It is 10, the number of possible samples of size $n = 3$ taken from $N = 5$ elements because these 10 values of the statistic $\overline{\mathbf{X}}$ represent a population set of values.

Therefore the mean and standard deviation of the random variable (statistic) $\overline{\mathbf{X}}$ are $\mu_{\overline{\mathbf{x}}} = 11.00$ and $\sigma_{\overline{\mathbf{x}}} = 0.73$, respectively. Intuitively, we might expect the values of the sample mean $\overline{\mathbf{X}}$ to be distributed about the population mean $\mu_{\overline{\mathbf{x}}}$. In fact

$$\mu_{\overline{\mathbf{x}}} = \mu_{\mathbf{X}}$$

In other words, the average of the sample means is equal to the population mean. This is an important result that holds generally. We will return to this point in the next section.

Now let's consider the other statistic in this illustration, the sample proportion $\mathbf{P}$. The sampling distribution of the random variable $\mathbf{P}$ is shown in Figure 8.2. From Table 8.5, three of the sample proportions are equal to 0.33, six proportions are equal to 0.67, and one of the proportions is equal to 1.00.

FIGURE 8.2 | The sampling distribution of the statistic **P**

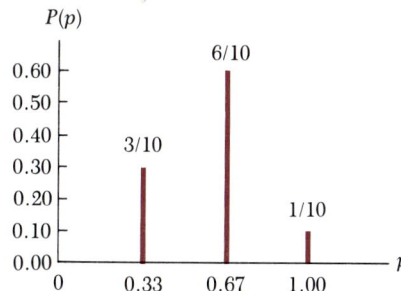

Denote the mean of the random variable **P** by $\mu_\mathbf{P}$. Then

$$\mu_\mathbf{P} = \frac{1.00 + 0.67 + 0.67 + 0.67 + 0.67 + 0.33 + 0.67 + 0.67 + 0.33 + 0.33}{10}$$

$$= 0.60$$

Thus the mean of the 10 values of the statistic **P** is equal to the population proportion 0.60 ($\mu_\mathbf{Y}$ above).

Two important outcomes of the sampling experiment in Example 8.3 are:

1. The average of the values of the statistic $\overline{\mathbf{X}}$ from all possible samples of size $n = 3$ from $N = 5$ population elements is the population average. In other words, the average of the sample means is equal to the population mean.
2. The average of the values of the statistic **P** from all possible samples of size $n = 3$ from $N = 5$ population elements is the population proportion. In other words, the average of the sample proportions is equal to the population proportion.

Suppose we decided to use the statistic $\overline{\mathbf{X}}$ to estimate the population mean. Example 8.3 shows us that a value of $\overline{\mathbf{X}}$ produced in a specific sample may not equal the population mean μ. But the average of all possible values of $\overline{\mathbf{X}}$ will equal μ. At least the values of the statistic $\overline{\mathbf{X}}$ are "aimed at the right target" (μ) if we choose to use $\overline{\mathbf{X}}$ to estimate μ. This concept is essential to the development of inferential statistics methods. We will further develop this concept in this chapter and continue the development in Chapters 9 and 10.

In the next section (section 8.6), we will more fully investigate the sampling distribution of $\overline{\mathbf{X}}$, the sample mean. In section 8.7 we will investigate the sampling distribution of **P**, the sample proportion. Although these are the only two statistics whose sampling distributions we will consider in detail, it is important to remember that every statistic is a random variable whose distribution is called its sampling distribution. Thus every numerical measure calculated from the sample observations is a random variable with a sampling distribution.

■ 8.6

Sampling distribution of the statistic $\overline{\mathbf{X}}$, the sample mean

Since the statistic $\overline{\mathbf{X}}$, the sample mean, is a random variable, it has a sampling distribution with a mean and a variance. Using the expectation operator E presented in Chapter 5, it is possible to derive the formulas for the mean and the variance of $\overline{\mathbf{X}}$. Let μ be the mean and σ^2 be the variance of the population random variable **X**. The formula for $\overline{\mathbf{X}}$ is

$$\overline{\mathbf{X}} = \frac{\sum_{i=1}^{n} \mathbf{X}_i}{n} = \frac{\mathbf{X}_1 + \mathbf{X}_2 + \cdots + \mathbf{X}_n}{n}$$

Therefore

$$E(\overline{\mathbf{X}}) = E\left[\frac{\mathbf{X}_1 + \mathbf{X}_2 + \cdots + \mathbf{X}_n}{n}\right] = \frac{E(\mathbf{X}_1) + E(\mathbf{X}_2) + \cdots + E(\mathbf{X}_n)}{n}$$

$$= \frac{\mu + \mu + \cdots + \mu}{n} = \frac{n\mu}{n} = \mu$$

This formula is the result of using the algebra of the expectation operator (E operator) given in Table 5.21 in Chapter 5 (property 2). The expectation of a constant ($1/n$) times a random variable is the constant times the expectation of the random variable and the expectation of a sum of random variables is the sum of the expectations of the random variable. Therefore the expected or average value of the random variable $\overline{\mathbf{X}}$ is μ, the population mean. Notice in the above derivation that the expectation of each $\mathbf{X}_i$ is μ, the population mean. This result follows since each sample random variable $\mathbf{X}_i$ has the same distribution as the population random variable $\mathbf{X}$ with mean μ and variance σ^2. In a similar fashion, it can be shown that when the population is infinite the variance of $\overline{\mathbf{X}}$ is given by $V(\overline{\mathbf{X}}) = \sigma^2/n$. When the population is finite, then it can be shown using more involved arguments that

$$V(\overline{\mathbf{X}}) = \frac{\sigma^2}{n}\left[\frac{N - n}{N - 1}\right]$$

where N = Population size, and n = Sample size.

Rather than using $E(\overline{\mathbf{X}})$ to represent the mean of $\overline{\mathbf{X}}$ and $V(\overline{\mathbf{X}})$ to represent the variance of $\overline{\mathbf{X}}$, we will use an alternate notation: $\mu_{\overline{\mathbf{X}}} = E(\overline{\mathbf{X}})$ and $\sigma^2_{\overline{\mathbf{X}}} = V(\overline{\mathbf{X}})$. Using this notation, we can denote the standard deviation of $\overline{\mathbf{X}}$ by $\sigma_{\overline{\mathbf{X}}}$. The results concerning the mean and standard deviation of $\overline{\mathbf{X}}$ are summarized in Theorem 8.1. Figure 8.3 summarizes these results in a graphical form.

Mean and standard deviation of the statistic $\overline{\mathbf{X}}$

Theorem 8.1

Mean and standard deviation of the statistic $\overline{\mathbf{X}}$

If we take a simple random sample of n random variables $\mathbf{X}_1, \mathbf{X}_2, \ldots, \mathbf{X}_n$ drawn from a population with mean μ and standard deviation σ, then the *mean* of $\overline{\mathbf{X}}$, denoted by $\mu_{\overline{\mathbf{X}}}$, and the *standard deviation of* the statistic $\overline{\mathbf{X}}$ denoted by $\sigma_{\overline{\mathbf{X}}}$, are given by

$$\mu_{\overline{\mathbf{X}}} = \mu$$

$$\sigma_{\overline{\mathbf{X}}} = \begin{cases} \dfrac{\sigma}{\sqrt{n}} & \text{if population is infinite} \\[2ex] \dfrac{\sigma}{\sqrt{n}} \sqrt{\dfrac{N - n}{N - 1}} & \text{if population is finite} \end{cases}$$

FIGURE 8.3 | Relationship between the mean and standard deviation of the population random variable **X** and the mean and standard deviation of the statistic $\overline{X}$

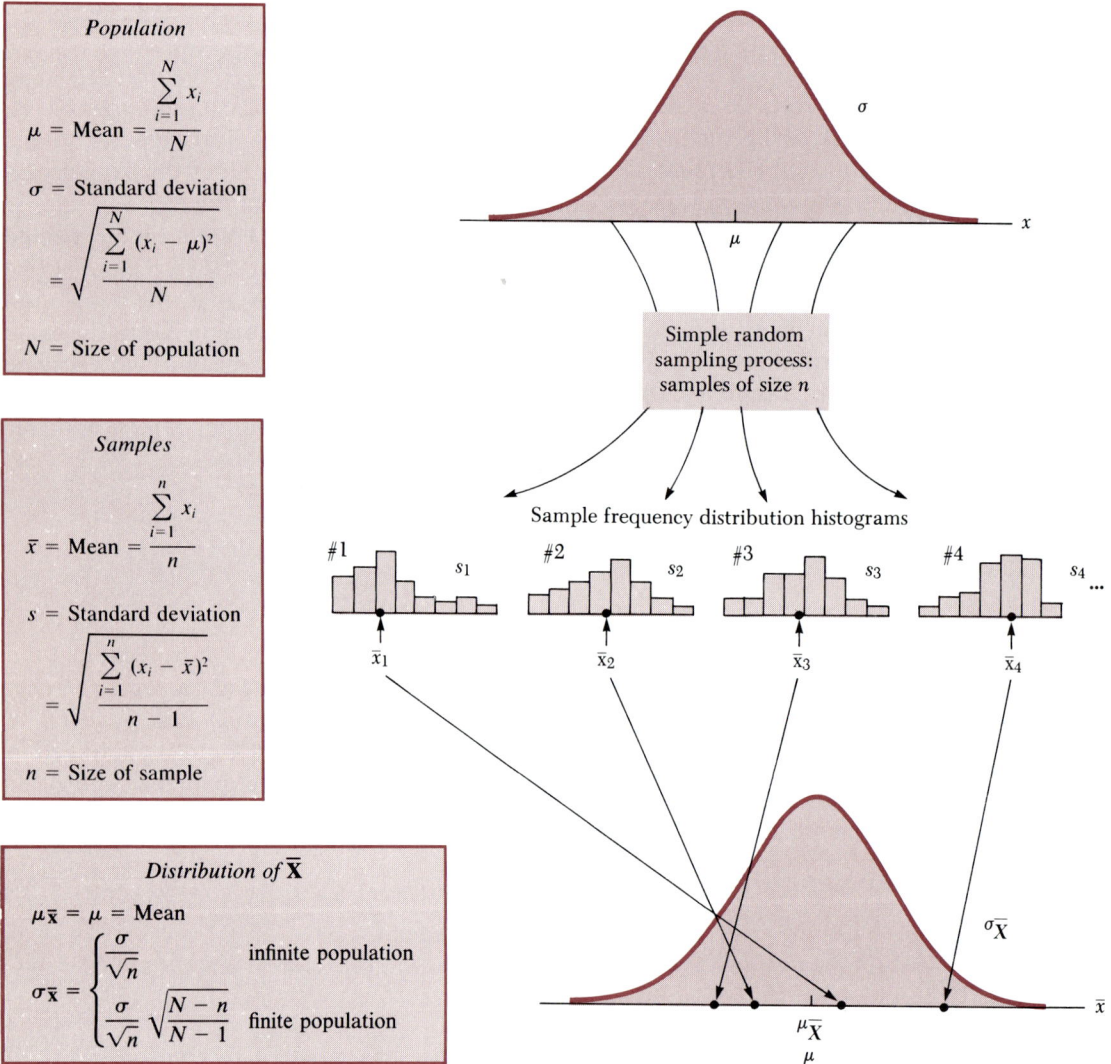

Population

$\mu = \text{Mean} = \dfrac{\sum\limits_{i=1}^{N} x_i}{N}$

$\sigma = \text{Standard deviation}$

$= \sqrt{\dfrac{\sum\limits_{i=1}^{N} (x_i - \mu)^2}{N}}$

$N = \text{Size of population}$

Samples

$\bar{x} = \text{Mean} = \dfrac{\sum\limits_{i=1}^{n} x_i}{n}$

$s = \text{Standard deviation}$

$= \sqrt{\dfrac{\sum\limits_{i=1}^{n} (x_i - \bar{x})^2}{n - 1}}$

$n = \text{Size of sample}$

Distribution of $\overline{X}$

$\mu_{\bar{x}} = \mu = \text{Mean}$

$\sigma_{\bar{x}} = \begin{cases} \dfrac{\sigma}{\sqrt{n}} & \text{infinite population} \\[2ex] \dfrac{\sigma}{\sqrt{n}} \sqrt{\dfrac{N - n}{N - 1}} & \text{finite population} \end{cases}$

Simple random sampling process: samples of size n

Sample frequency distribution histograms

An important observation from Theorem 8.1 is that the variability of the statistic $\overline{X}$ and the sample size are inversely related. If the population is finite, for example, then

$$\sigma_{\bar{x}} = \frac{\sigma}{\sqrt{n}} \sqrt{\frac{N - n}{N - 1}}$$

As the sample size increases, this quantity becomes smaller. When $n = N$, $\sigma_{\overline{X}} = 0$; there is no variability in $\overline{X}$. Indeed, when we have sampled the entire population, $\overline{X} = \mu$.

If the population is infinite, then the sample size may be increased without bound. Thus $\sigma/\sqrt{n}$ can be made arbitrarily small by making n arbitrarily large. Therefore this theorem tells us that as n increases, the sampling distribution of $\overline{X}$ converges on its mean $\mu_{\overline{X}}$, which is equal to the value of population mean μ. This is intuitive as well—the larger the sample size, the closer we should expect the value of $\overline{X}$ to be the value of μ.

Example 8.4 Consider the Example 8.3 problem. For the population random variable $\mathbf{X}$ = Number of credits taken by each student, the mean number of credits taken by the $N = 5$ students is $\mu = 11.00$, and the standard deviation is $\sigma = 1.789$. Noting that this is a finite population, determine the mean $\mu_{\overline{X}}$ and the standard deviation $\sigma_{\overline{X}}$ of the statistic $\overline{X}$, the sample mean, using Theorem 8.1. Compare these results with the computation of $\mu_{\overline{X}}$ and $\sigma_{\overline{X}}$ in Example 8.3.

Solution From Theorem 8.1

$$\mu_{\overline{X}} = \mu = 11.00$$

$$\sigma_{\overline{X}} = \frac{\sigma}{\sqrt{n}}\sqrt{\frac{N-n}{N-1}} = \frac{1.789}{\sqrt{3}}\sqrt{\frac{5-3}{5-1}} = 0.73$$

As expected, these are the same results that we obtained in Example 8.3 by computing $\mu_{\overline{X}}$ and $\sigma_{\overline{X}}$ directly from the 10 sample means. Further, the computation based on Theorem 8.1 is much simpler.

Knowing the mean and standard deviation of $\overline{X}$ is very important. The fact that $\mu_{\overline{X}} = \mu$ tells us that the sample averages $\bar{x}$ from all possible samples taken from the population of size n average out to the population mean μ. Further, the standard deviation formulas tell us that as the sample size increases, the amount of the variation of $\overline{X}$ about the population mean μ decreases. This latter result is illustrated in the next example.

Example 8.5 Suppose the population characteristic of interest is the time between telephone calls at a switchboard. We could make two assumptions regarding the distribution of the population random variable $\mathbf{X}$ = Time between calls:

Case 1. $\mathbf{X}$ is normally distributed with mean $\mu = 2.00$ minutes and standard deviation $\sigma = 0.25$.

Case 2. $\mathbf{X}$ is uniformly distributed over the range 0–4 minutes. Its mean is therefore $\mu = 2.00$ minutes.

The two distribution models are illustrated in Figure 8.4.

By using computer simulation methods, simple random samples of size $n = 5, 10, 20, 50,$ and 100 were drawn from each population for 100 repeated samples of each size. In each of the 100 repeated samples, the value of the sample mean $\overline{X}$ is calculated, and a histogram is formed. The resulting histograms for the samples sizes $n = 5, 10, 20, 50,$ and 100 and for the two

FIGURE 8.4 | Two assumed distribution models for the distribution of random variable **X** = time between calls

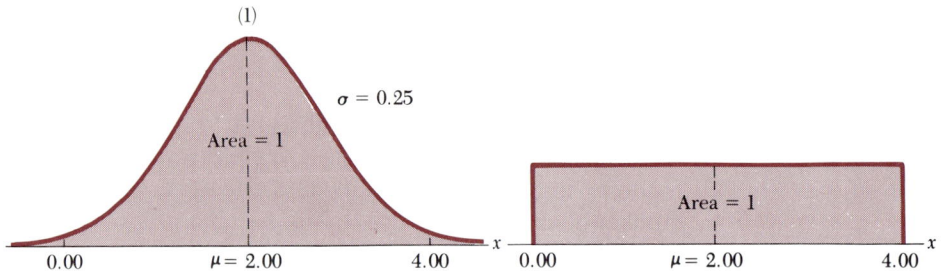

population random variables (**X** is normal and **X** is uniform) are shown in Figure 8.5.

There are two important results to observe from Figure 8.5. First, notice that as the sample size increases, the variation in the sampling distributions of $\overline{X}$ does decrease, as promised by the results on $\sigma_{\overline{x}}$ in Theorem 8.1. Second, notice the *form* of these histograms. In Case 1 where the population random variable **X** is normally distributed, the histograms look very much like a normal distribution for all sample sizes. In Case 2 where the population random variable is uniformly distributed, the histograms begin to look like the normal, bell-shaped curve by the time that n becomes 50. The form appears even more normal when $n = 100$.

The observations about the normal form of these histograms turns out to be correct. If we have the Case 1 situation where the population random variable **X** is normally distributed, then the statistic $\overline{X}$ is normally distributed for *all sample sizes*. This result is summarized in Theorem 8.2.

Sampling distribution of $\overline{X}$ when the population random variable **X** is normally distributed

Theorem 8.2
Sampling distribution of $\overline{X}$ (normal data)

If the population random variable **X** is normally distributed with mean μ and standard deviation σ, and a simple random sample of size n is to be drawn, then the sample mean statistic $\overline{X}$ is normally distributed—for all sample sizes n—with a mean $\mu_{\overline{x}} = \mu$ and a standard deviation $\sigma_{\overline{x}} = \sigma/\sqrt{n}$.

Since the normal distribution is a distribution of a continuous random variable, the population is infinite in the case where Theorem 8.2 applies. Therefore, $\sigma_{\overline{x}} = \sigma/\sqrt{n}$.

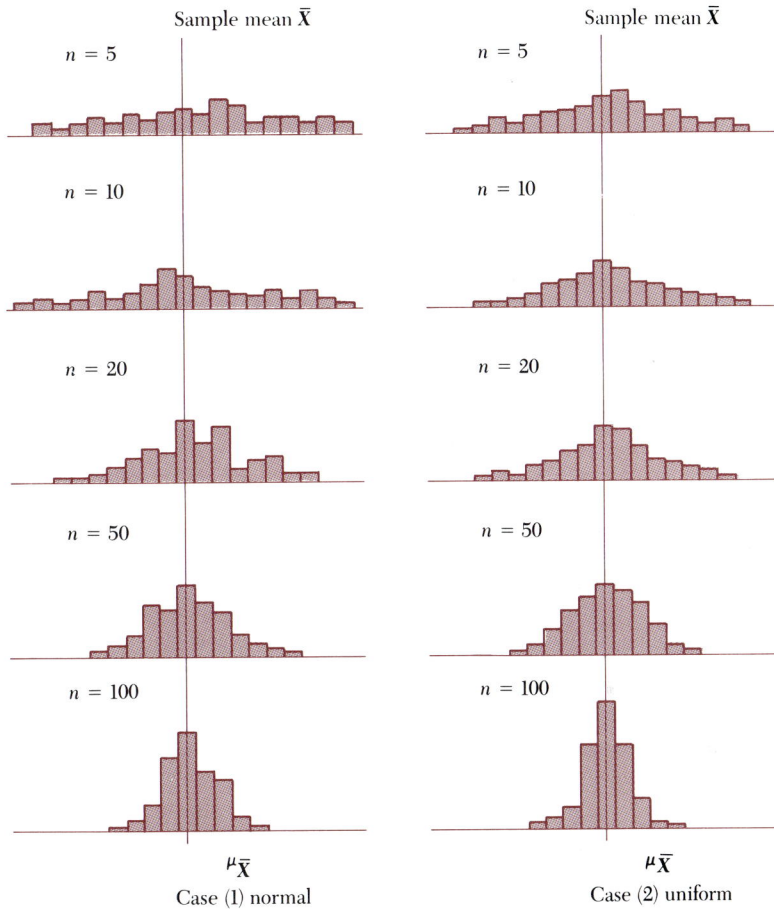

The use of Theorem 8.2 is illustrated in the following example.

Example 8.6 Suppose the time between telephone calls at a switchboard is normally distributed with a mean $\mu = 2$ minutes and a standard deviation $\sigma = 0.25$. If a random sample of size $n = 25$ is drawn, what is the probability that the sample mean statistic $\overline{X}$ will be greater than 2.10 minutes?

Solution From Theorem 8.2, the sampling distribution is formed and is shown in Figure 8.6. To find the probability that $\overline{X}$ will be greater than 2.10 minutes, we must standardize the value of the variable $\overline{X}$ by using

$$z = \frac{\overline{x} - \mu_{\overline{X}}}{\sigma_{\overline{X}}}$$

FIGURE 8.6 | Calculation of the probability in Example 8.6

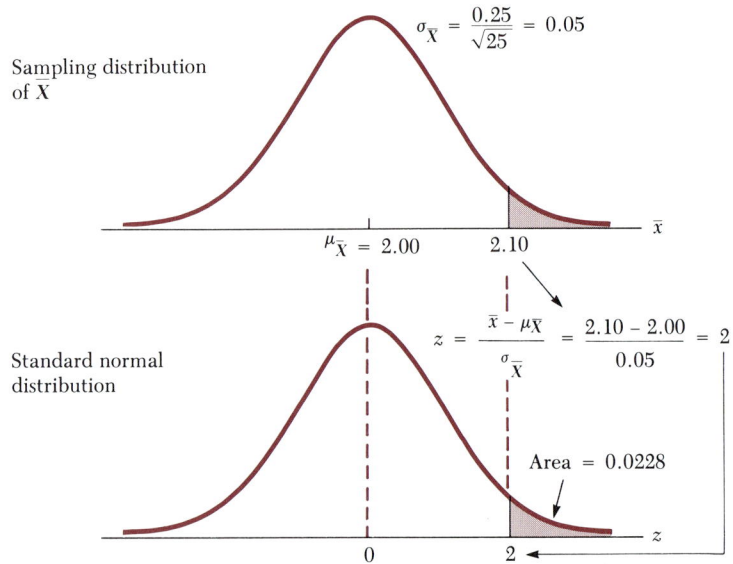

Then Table B.3 in Appendix B may be used to determine the desired probability. The answer is 0.0228. Therefore in only 2.28 percent of all possible samples of size $n = 25$ will the sample mean statistic $\overline{X}$ be 2.10 minutes or greater.

The use of Theorem 8.2 as in Example 8.6 is important conceptually. By knowing that the sampling distribution of $\overline{X}$ is normal, we can determine the probability that a sample of a specified size will produce a sample average that is more than a specified number of units from the population mean μ. In Example 8.6, the population mean $\mu = 2$, and the sample size is $n = 25$. *Before* the sample is taken, we know the probability that the one sample we select will produce a sample mean of 2.10 or more (an overestimate of 0.10 minute or more) is 0.0228. Since this probability is small, we can be confident that our sample of size $n = 25$ will not likely overestimate μ by 0.10 minutes or more.

If the distribution of the population random variable X is not normally distributed (such as Case 2, Example 8.5) or has an unknown distribution, then one of the most remarkable theorems in statistics applies—*the central limit theorem*. This theorem is stated next as Theorem 8.3.

This theorem tells us that the sampling distribution of the statistic $\overline{X}$ will become more like a normal distribution as the sample size increases; and in the limit when the sample size becomes arbitrarily large, the sampling distribution of $\overline{X}$ becomes the normal distribution. The remarkable aspect of this

Theorem 8.3
Central limit theorem

Let X_1, X_2, . . . , X_n be a simple random sample of size n drawn from an infinite population with a finite mean μ and a finite standard deviation σ. Then, for sufficiently large n, the random variable $\overline{X}$ is approximately normally distributed with mean μ and standard deviation $\sigma/\sqrt{n}$.

Central limit
theorem

theorem is that it places only two qualifications on the population distribution of X, the population random variable: (1) The population is infinite, and (2) the population has a finite mean μ and a finite variance σ^2. Notice Figure 8.5 Case 2, wherein the population random variable is uniformly distributed. For $n \geq 50$, the histograms look very much like the normal, bell-shaped curve *even though the distribution of the population random variable looks nothing like a normal curve*—it is a uniform distribution.

This theorem can be used with good results even when the population is finite, provided that the sample size is large. The theorem is summarized below.

Theorem 8.5
Sampling distribution of the statistic $\overline{X}$ (nonnormal data)

If the population random variable X is not normally distributed, but has a finite mean and variance given by μ and σ^2, respectively, and if a simple random sample is drawn, then the sample mean statistic $\overline{X}$ is approximately normally distributed for sufficiently large n with mean $\mu_{\overline{x}}$ and a standard deviation $\sigma_{\overline{x}}$, where $\mu_{\overline{x}} = \mu$, and $\sigma_{\overline{x}} = \sigma/\sqrt{n}$—provided that the population is infinite in size. If the population is finite in size, then a better approximation results by using

$$\sigma_{\overline{x}} = \frac{\sigma}{\sqrt{n}} \sqrt{\frac{N-n}{N-1}}$$

Sampling distribution of $\overline{X}$ when the population random variable X is not normally distributed

In section 8.8, rules are given for when the factor

$$\sqrt{\frac{N-n}{N-1}}$$

can be ignored, though the population is finite in size.

An issue in the use of Theorem 8.4 is how large the sample size n must be for the theorem to apply. If the population random variable is continuous, then usually a sample of size $n = 25$ or more will suffice. If the population random variable is discrete, then a sample size of 50 to 100 or more may be required. As we use this theorem in the next three chapters, we will comment on the size of the sample required in each application. We will now apply the results of Theorem 8.4 in an example.

Example 8.7   The number of people entering a service facility in a plant during 15-minute intervals during the working day is known to have a distribution with a mean rate of two arrivals and a standard deviation of 1.42 arrivals in the 15-minute intervals. If 100 of these 15-minute intervals are randomly selected and the number of arrivals in each interval is recorded, what is the probability that the sample mean number of arrivals will be less than 1.85?

Solution   Since the random variable $\mathbf{X}$ = Number of arrivals in a 15-minute interval has an unspecified distribution (the distribution is most likely Poisson; see Chapter 6), Theorem 8.4 is applicable.

Theorem 8.4 tells us that the statistic $\overline{\mathbf{X}}$ will be approximately normally distributed with $\mu_{\overline{\mathbf{X}}} = \mu = 2$, and $\sigma_{\overline{\mathbf{X}}} = \sigma/\sqrt{n} = 0.142$. Thus the probability that $\overline{\mathbf{X}}$ is less than 1.85 can be computed *approximately* by finding the area under the normal curve as indicated in Figure 8.7. The probability is approximately 0.1446 that the average number of arrivals in the sample will be less than 1.85.

FIGURE   8.7 | Calculation of the probability in Example 8.7

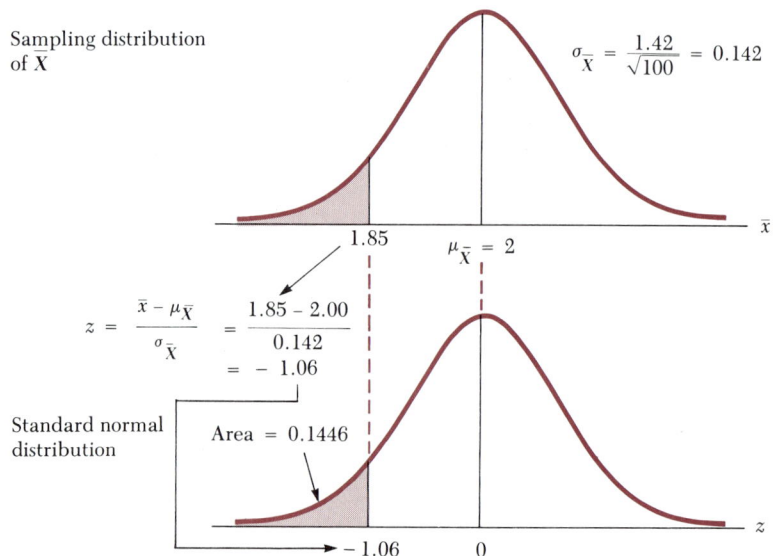

Sampling distribution of $\overline{X}$

$$\sigma_{\overline{X}} = \frac{1.42}{\sqrt{100}} = 0.142$$

1.85

$\mu_{\overline{X}} = 2$

$$z = \frac{\overline{x} - \mu_{\overline{X}}}{\sigma_{\overline{X}}} = \frac{1.85 - 2.00}{0.142} = -1.06$$

Standard normal distribution

Area = 0.1446

-1.06 0

The importance of Theorems 8.3 and 8.4 is that they allow us to deduce results concerning the outcome in the sample, based on knowledge of the population mean and standard deviation under the stated conditions. Specifically, they allow us to determine the likelihood that the sample mean statistic $\overline{\mathbf{X}}$ is larger or smaller than a stated value, or will fall into a stated interval. This result produces a framework with which the direction of inference is "turned around"—instead of _deducing_ properties of $\overline{\mathbf{X}}$, we will in subsequent chapters use values of $\overline{\mathbf{X}}$ and these theorems to _induce_ results about the population characteristic.

■ 8.7

Sampling distribution of the statistic P, the sample proportion

In Chapter 6 we considered the instance when a discrete random variable may be of a very special type—the binomial random variable. If $\mathbf{X}$ is a binomial random variable, it represents the number of successes in n independent and identical trials where each trial can result in a "success" or a "failure." If a sample of n trials of a binomial experiment is taken, the sample proportion of successes, the statistic $\mathbf{P}$, may be computed by dividing the number of successes in the n trials by n; that is,

$$\mathbf{P} = \frac{\mathbf{X}}{n}$$

Since $\mathbf{X}$ is the "total" number of successes in the n trials, $\mathbf{P}$ represents the "average" number of successes in these trials. Since $\mathbf{P}$ is in the form of an average, it is possible to apply the central limit theorem. The central limit theorem tells us that the form of the distribution of the statistic $\mathbf{P}$ is approximately normal for large sample sizes (numbers of binomial trials). To use this fact in computing probabilities concerning $\mathbf{P}$, it is necessary to first determine which member of the normal family of distributions is the appropriate one for modeling the distribution of $\mathbf{P}$. That is, the mean of $\mathbf{P}$, $\mu_{\mathbf{P}}$, and the variance of $\mathbf{P}$, $\sigma_{\mathbf{P}}^2$ must be determined.

Recall that the binomial random variable $\mathbf{X}$ has a mean of $n\pi$ and a variance of $n\pi(1 - \pi)$, where n is the number of binomial trials, and π is the probability of success on each and every trial. Since $\mathbf{P} = \mathbf{X}/n$,

$$\mu_{\mathbf{P}} = E\left(\frac{\mathbf{X}}{n}\right) = \frac{1}{n} E(\mathbf{X}) = \frac{1}{n} (n\pi) = \pi$$

$$\sigma_{\mathbf{P}}^2 = V\left(\frac{\mathbf{X}}{n}\right) = V\left[\left(\frac{1}{n}\right) \cdot \mathbf{X}\right] = \left(\frac{1}{n}\right)^2 V(\mathbf{X}) = \frac{n\pi(1 - \pi)}{n^2} = \frac{\pi(1 - \pi)}{n}$$

Therefore, the mean of the statistic $\mathbf{P}$, $\mu_{\mathbf{P}}$, is given by π; the variance of the statistic $\mathbf{P}$, $\sigma_{\mathbf{P}}^2$, is given by $[\pi(1 - \pi)]/n$; and the standard deviation of the statistic $\mathbf{P}$, $\sigma_{\mathbf{P}}$, is given by $\sqrt{[\pi(1 - \pi)]/n}$. The results for the sampling distribution of the statistic $\mathbf{P}$ are summarized in Theorem 8.5.

The fact that the sample proportions computed in repeated samples of size n trials average out to π should not be surprising—the statistic $\mathbf{P}$ is the sample proportion of successes, and π is the population proportion of successes. We observed this result in Example 8.1, our sampling experiment—the sample proportions averaged out to be the population proportion.

Sampling distribution of the statistic **P**, the sample proportion

---

Theorem 8.5

Sampling distribution of the statistic **P**

If **X** is a random variable representing the number of successes in n independent and identical trials where the probability of success in each and every trial is π, then the sampling distribution of the statistic, $\mathbf{P} = \mathbf{X}/n$ is, for sufficiently large n, approximately normally distributed with mean $\mu_\mathbf{P} = \pi$ and standard deviation

$$\sigma_\mathbf{P} = \sqrt{\frac{\pi(1 - \pi)}{n}}$$

if the population is infinite in size. If the population is finite in size, a better approximation results by using

$$\sigma_\mathbf{P} = \sqrt{\frac{\pi(1 - \pi)}{n}} \sqrt{\frac{N - n}{N - 1}}$$

The standard deviation of the statistic **P**, like the standard deviation of the statistic $\overline{\mathbf{X}}$, is a function of $1/n$; as the sample size increases, the statistic **P** tends toward the population proportion. Notice the two forms for the standard deviation $\sigma_\mathbf{P}$. If the population is finite in size, we must correct the standard deviation by the factor $\sqrt{(N - n)/(N - 1)}$.

We have already used the results of Theorem 8.5 in the text. Recall in Chapter 7 that we approximated binomial probabilities concerning the random variable **X**, the number of successes in n trials, by using the normal distribution. In these cases, we were using the central limit theorem and the results of Theorem 8.5, which ensure that the random variables **X** and **P** will be approximately normally distributed for large samples.

In Theorem 8.5 we must be somewhat careful about what is meant by a "sufficiently large sample." Chapter 7 used the following guidelines for the normal approximation to the binomial distribution: Use the normal approximation if both $n\pi \geq 5$ and $n(1 - \pi) \geq 5$. When $\pi = 0.5$, notice that $n = 10$ (or more) satisfies these conditions. However, as π becomes closer in value to either 0 or 1, the sample size n required becomes larger. Suppose, for example, that $\pi = 0.20$. Then, $n = 25$ (or more) satisfies these conditions [$(25)(0.20) = 5$ and $(25)(0.80) = 20$]. If $\pi = 0.01$, then $n = 500$ (or more) is required to satisfy the conditions. Generally speaking, it is best to take samples of $n = 50$ or more when using the sample proportion **P**, although $n = 25$ will be suitable if π is between 0.20 and 0.80.

Example 8.8 In a 1984 primary election, a candidate received 60,000 of 100,000 votes, a proportion of "successes" equal to $\pi = 0.6$. If we had taken a simple random sample of 100 voters on the eve of election day, what would have been the probability that the sample proportion would have exceeded 50 percent successes?

Solution   The first step is checking to see if the normal approximation to the binomial distribution is appropriate when $\pi = 0.60$ and $n = 100$. Checking the conditions, we have $n\pi = (100)(0.60) = 60 \geq 5$, and $n(1 - \pi) = (100)(0.40) = 40 \geq 5$. Thus the conditions are satisfied, and we can use Theorem 8.5. From Theorem 8.5 the mean of the statistic **P** is $\mu_\mathbf{P} = 0.6$, and the standard deviation of the statistic **P** is

$$\sigma_\mathbf{P} = \sqrt{\frac{\pi(1 - \pi)}{n}} \sqrt{\frac{N - n}{N - 1}} = \sqrt{\frac{0.6(1 - 0.6)}{100}} \sqrt{\frac{100{,}000 - 100}{100{,}000 - 1}}$$
$$= (0.0489)(0.9995) = 0.0489$$

Notice that the effect of the radical $\sqrt{(N - n)/(N - 1)}$ is very small, because the population is large, and the sample is only a small portion of the population. The sampling distribution of **P** is illustrated in Figure 8.8.

FIGURE 8.8   Sampling distribution of the statistic **P** in Example 8.8

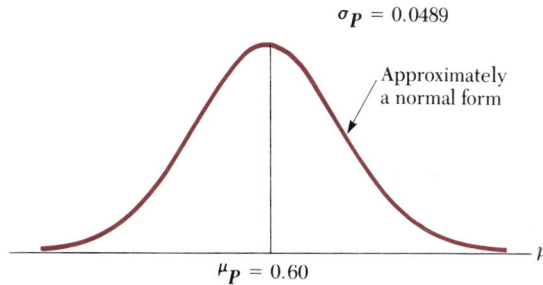

To find the probability that the statistic **P** is greater than 0.50, we first standardize the value of **P**:

$$z = \frac{p - \mu_\mathbf{P}}{\sigma_\mathbf{P}} = \frac{0.50 - 0.60}{0.0489} = -2.04$$

From the standard normal tables (Table B.3 in Appendix B), the probability that the random variable **Z** is greater than -2.04 is 0.9793. Therefore the probability that the value of the statistic **P** is greater than 0.50 in a sample is 0.9793. Another way of interpreting this answer is that if we took all possible samples of size $n = 100$ from the population of $N = 100{,}000$ voters, in approximately 97.93 percent of these samples, the sample proportion of sampled voters who say they will vote for the candidate is greater than 0.50.

Example 8.9   A production process for a high-volume product is known to produce defectives at the rate of 5 percent. A random sample of $n = 400$ units of the product is sampled from a production run. What is the probability that the sample proportion **P** will be 0.025 or less? Assume that the production run is very large.

Solution With $\pi = 0.05$, we first must check the two conditions on n to determine if the sample size taken ($n = 400$) is large enough to use Theorem 8.5. With $\pi = 0.05$ and $n = 400$, we have:

$$n\pi = (400)(0.05) = 20 \geq 5 \quad \text{and} \quad n(1 - \pi) = (400)(0.95) = 380 \geq 5$$

Thus the sample of size $n = 400$ is sufficient to use Theorem 8.5. The mean of the sampling distribution of **P** is:

$$\mu_\mathbf{P} = \pi = 0.05$$

Since the production run is "very large" (and no population size N is given), we will use the first formula in Theorem 8.5 for the standard deviation of **P**:

$$\sigma_\mathbf{P} = \sqrt{\frac{\pi(1 - \pi)}{n}} = \sqrt{\frac{(0.05)(0.95)}{400}} = 0.011$$

We want $P(\mathbf{P} \leq 0.025)$. This probability, together with the approximate normal sampling distribution of **P**, is shown in Figure 8.9.

To find $P(\mathbf{P} \leq 0.025)$, we must standardize $p = 0.025$:

$$z = \frac{p - \mu_\mathbf{P}}{\sigma_\mathbf{P}} = \frac{0.025 - 0.05}{0.011} = -2.27$$

FIGURE 8.9 | The approximate sampling distribution of **P** in Example 8.9

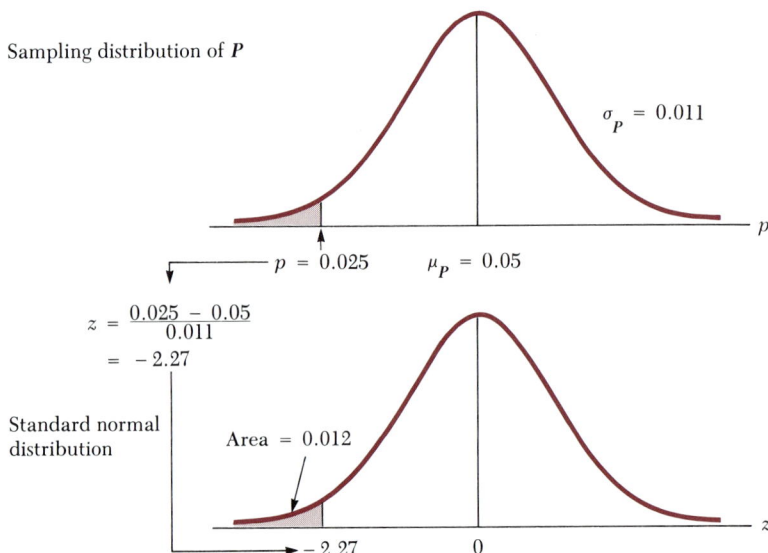

Sampling distribution of P

$\sigma_p = 0.011$

$p = 0.025$ $\mu_p = 0.05$

$z = \dfrac{0.025 - 0.05}{0.011}$
$= -2.27$

Standard normal distribution

Area $= 0.012$

-2.27 0

From Table B.3, Appendix B, the area under the standard normal distribution to the left of -2.27 as shown in Figure 8.9 is 0.0116. In Figure 8.9, this answer is rounded to three decimal places, producing 0.012. Therefore, if we select a random sample of size $n = 400$ from a population of units of this product, we have a probability of 0.012 that the sample proportion of defectives will be 0.025 (2.5 percent) or less.

8.8

Discussion

The central limit theorem applies to other sample statistics as well. For example, the sample variance statistic $\mathbf{S}^2$ is an average of sorts; it "averages" the squared deviations of the sample values about their mean. As the sample size increases, the sampling distribution of $\mathbf{S}^2$ will approach a normal distribution. But there are better approximating distributions for $\mathbf{S}^2$ than the normal distribution, so we will not apply the central limit theorem to $\mathbf{S}^2$. The exact and approximate sampling distributions of the statistic $\mathbf{S}^2$ will be given in Chapter 9.

We should also comment about the correction factor for the standard deviation of the sample statistic $\overline{\mathbf{X}}$ or $\mathbf{P}$ when the population is finite. The term,

$$FPC = \sqrt{\frac{N - n}{N - 1}}$$

Finite population correction factor

is called the *finite population correction factor (FPC)*. The reason it is necessary in finite population sampling is intuitive; as each unit is withdrawn from the finite population to be placed in the sample, the variability in the population is reduced before the next draw—there is one less unit in the population. In contrast to this, if the population is infinite, as each unit is withdrawn to be placed in the sample, the remaining population is still infinite—the variability has not been reduced by withdrawing a unit. The *FPC* factor corrects for this difference by adjusting the standard deviation formula for sample statistics drawn from a finite population.

FPC may be ignored if $n/N < 0.05$

Notice that if n is relatively small in relation to N, the value of the *FPC* will be close to 1. This situation occurred in Example 8.8 where the FPC was equal to 0.995. As a general guideline, the *FPC* may be ignored if the ratio n/N is less than 0.05 (in Example 8.8, $n/N = 100/100{,}000 = 0.001$). This guideline may be used in both Theorem 8.4 (the sampling distribution of the statistic $\overline{\mathbf{X}}$ and in Theorem 8.5 (the sampling distribution of the statistic $\mathbf{P}$). In general, if the population is described as "being large" and the population size N is not given in a problem, then we will assume that the *FPC* need not be used.

8.9

Other types of samples

A simple random sample is one kind of probability sample. Although simple random samples are being extensively used, increasingly more sophisticated probability sampling techniques are being employed. We shall consider three: *stratified sampling, cluster sampling,* and *systematic sampling*.

Definition 8.9
Stratified sample

A *stratified sample* is obtained by forming strata in the population, and from each stratum, selecting a simple random sample.

The basic idea in stratified sampling is illustrated in Figure 8.10. The population is "stratified," and from each stratum, a simple random sample is

FIGURE 8.10 | Stratified sampling

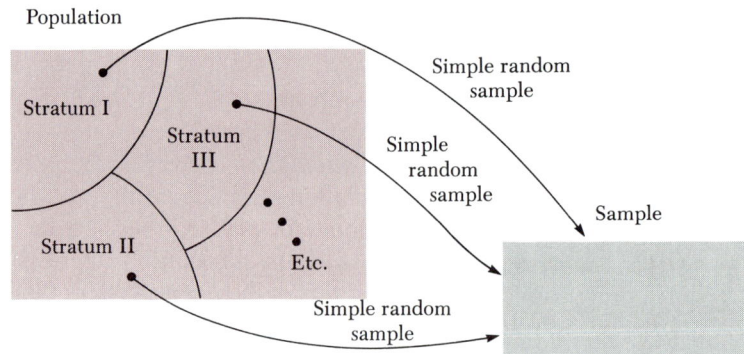

drawn. As an example, we may be planning a survey to determine the profitability of offering a lawn care service in a town. Recognizing that wealthy homeowners may be more receptive to the offer than others, we might stratify the population of homes on the basis of value, forming three strata: homes valued at $50,000 and less, more than $50,000 but less than $75,000, and homes valued at $75,000 and more. From each stratum, we then take a random sample of selected homes. Stratified sampling adds control to the sampling process by decreasing the amount of sampling error. For instance, a simple random sample may produce, by chance, proportionally too many homes in the $75,000 range, which may in turn bias our predictions of the number of homes in the town that would accept a lawn care offer.

In general, population units may be stratified on any number of characteristics. Common variables used for stratifying in "people" surveys are age, income, and location of residence.

Cluster sample

> **Definition 8.10**
> Cluster sample
>
> A *cluster sample* is obtained by identifying a set of clusters that compose the population, randomly selecting a subset of these clusters, and taking a census of each cluster selected.

The notion of cluster sampling is suggested in Figure 8.11. Clusters are identified in the population, a set of clusters is randomly sampled, and a

FIGURE   8.11 | Cluster sampling

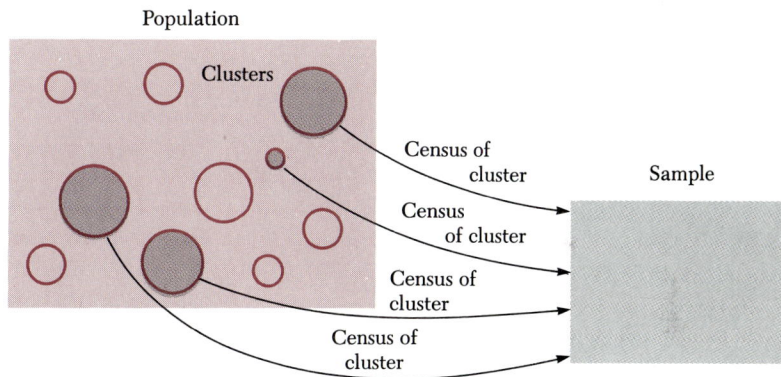

complete census is taken of each to form the sample. As an example, we may be interested in surveying the attitudes of airline passengers concerning service. Of all flights in a given week, we may select 10 and take a census of everyone aboard each flight. Each flight is a cluster in the population of flights.

Systematic samples are commonly used in reliability testing—every kth item $(k > 1)$ is selected from the assembly line and tested.

The kinds of probability samples all have one thing in common: The sampled population units are chosen according to a probability plan. Two other types of sampling, *judgment* and *convenience,* choose the sampled population units in the ways suggested by their names.

Judgment sampling is common when the sample is to be very small (usually due to the expense of sampling), the population is very heterogeneous,

Systematic sample

> **Definition 8.11**
> Systematic sample
>
> A *systematic sample* is formed by selecting one unit at random and then selecting additional units at evenly spaced intervals (every kth population unit, $k > 1$) until the sample has been formed.

or special skills are required to form a representative subset of the population.

Auditors occasionally use judgment samples to select items for study to determine whether or not a complete audit of the items may be necessary. However, auditors recently have begun to use probability samples increasingly for this purpose.

Judgment sample

> **Definition 8.12**
> Judgment sample
>
> A *judgment sample* is obtained by having an expert who is familiar with the population characteristics select units from the population.

Convenience sample

> **Definition 8.13**
> Convenience sample
>
> A *convenience sample* is obtained by selecting "convenient" population units.

Suppose we are interested in conducting a study of the attitudes of shoppers at a new shopping mall on the kinds of stores in the mall, the attractiveness of the mall, parking difficulties, and so on. To collect sample information, we ask persons to participate in the survey who happen to walk past the central area of the mall. The sample in this instance is a convenience sample—the people are not being selected according to a probability plan and, presumably, judgment is not being used in selecting those to participate in the survey.

Judgment and convenience samples have the disadvantage when compared with probability samples of being subject to sampling bias. The quality of the judgment sample depends on the competence of the expert who selects the population units to be sampled. Convenience samples are prone to bias by their very nature—selecting population elements that are convenient to choose almost always makes them special or different from the rest of the elements in the population in some way. For example, convenience sampling in attitudinal studies generally results in samples that contain too high a proportion of complainers—people who are only too happy to let their grievances be known. In almost all cases, it is desirable to take probability samples—they best guard against bias.

Among the types of probability sampling plans—which include simple random sampling, stratified random sampling, cluster sampling, and systematic sampling—methods other than simple random sampling are being used more commonly today in business applications. An excellent text on probability sampling plans is the Schaeffer and Mendenhall reference at the end of this chapter. For a more advanced treatment of these sampling plans, the Cochran text referenced at the end of this chapter is highly recommended.

■ 8.10

Computer sampling simulation experiments

Perhaps one of the best ways to understand the concept of the sampling distribution of a statistic is to use one of the many computer statistical software packages for simulating the sampling process. A good software package for this purpose is MINITAB. The following example illustrates the use of MINITAB in investigating the properties of the sampling distribution of the statistic $\overline{X}$, the sample mean.

Example 8.10 Suppose we wish to simulate 100 random samples each of size $n = 25$ drawn from a normal distribution with a mean of $\mu = 50$ and a standard deviation of $\sigma = 10$. Using MINITAB, produce the 100 samples of size $n = 25$, and describe the sampling distribution of the statistic, $\overline{X}$, the sample mean.

Solution Using MINITAB release 82.1 (the version of MINITAB released in 1982), the procedure for creating the simulated samples of size $n = 25$ is described below:

1. Create a MINITAB Macro (a set of programming steps) to produce the simple random samples of size $n = 25$ from a normal distribution with $\mu = 50$ and $\sigma = 10$. The Macro can be written by using the STORE MINITAB command:

```
MTB  > STORE 'SIMULATE'
STOR > NOPRINT
STOR > NRANDOM 100 OBSERVATIONS WITH MU = 50 AND
       SIGMA = 10, PUT INTO CK1
STOR > LET K1 = K1 + 1
STOR > END
```

The Macro is stored in the file named SIMULATE and becomes a permanent file in the account under which it is created. The first statement in the Macro, NOPRINT, tells MINITAB that we do not want the 100 observations to be stored in column CK1 to be printed out. The next statement, NRANDOM, is the random number generator, which generates normal values with a mean of 50 and a standard deviation of 10. The next statement, LET K1 = K1 + 1, is used to change the constant K1 from the initial value (which will be 1) to the next integer value. This statement allows us to ''loop'' through the Macro, reusing it several times to create the required random samples.

2. Use the EXECUTE command to loop through the Macro. The commands we will use are:

```
LET K1 = 1
EXECUTE 'SIMULATE' 25
```

The first command sets the initial value of the constant K1 equal to one. The EXECUTE command tells MINITAB to execute the Macro 25 times. The first time the Macro is executed, the constant K1 is equal to 1 so that 100 normal values with a mean of 50 and a standard deviation of 10 are placed in column CK1 = C1. Then, in the Macro, the LET statement adds 1 to K1 so the next time the Macro is executed, K1 = 2. Thus the EXECUTE command will execute the Macro 25 times, storing 100 normal values in columns C1, C2, . . . , C25. We want 100 samples of size $n = 25$, not 25 samples of size 100 (the 100 normal values stored in C1, C2, . . . , C25). However, we can use the *rows* of the columns as samples of size $n = 25$. For example, we can take the first row values in columns C1, C2, . . . , C25—these 25 values represent a simple random sample of size $n = 25$ from a normal distribution with mean 50 and standard deviation 10. Using the rows in each column as samples of size $n = 25$, we now have our 100 random samples of size $n = 25$.

3. Use MINITAB commands to calculate the sample means in each of the 100 random samples of size $n = 25$, and describe the sampling distribution of these means. Using the Macro, the set of MINITAB statements is:

```
MTB > LET K1 = 1
MTB > EXECUTE 'SIMULATE' 25
MTB > ADD C1-C25, PUT SUM IN C26
MTB > DIVIDE C26 BY 25, PUT SAMPLE AVERAGES IN C27
MTB > DESCRIBE C27
MTB > HISTOGRAM C27
```

The ADD command adds the values in each row across the 25 columns and places the sum in column C26. The DIVIDE command divides each sum by 25, thus calculating the sample mean. At this point, the columns C1, C2, . . . , C25, C26, and C27 look like the following:

```
ROW     C1       C2      . . .     C25        C26       C27
  1    55.88    63.34             59.18     1189.63    47.58
  2    41.79    53.06             54.24     1203.56    48.14
  .      .        .                 .          .          .
  .      .        .                 .          .          .
  .      .        .                 .          .          .
100    33.97    68.26             44.48     1304.71    52.19
```

Column C27 now contains the 100 sample means. The DESCRIBE command calculates the mean, standard deviation, maximum, and minimum for these 100 sample means.

```
MTB > DESCRIBE C27

                 C27
N                100
MEAN           50.36  ←
MEDIAN         50.45
TMEAN          50.36
STDEV           2.12  ←
SEMEAN          0.21
MAX            55.85
MIN            45.12
Q3             51.68
Q1             48.85
```

Thus the mean is 50.36 (the population mean is 50) and the standard deviation is 2.12 (the standard deviation of $\overline{X}$ is $\sigma/\sqrt{n} = 10/\sqrt{25} = 2.00$). The largest sample mean is 55.85, and the smallest sample mean is 45.12. The HISTOGRAM command produces the following histogram:

```
MTB > HISTOGRAM C27

   C27

      MIDDLE OF        NUMBER OF
      INTERVAL         OBSERVATIONS
           45              3     ***
           46              0
           47              7     *******
           48             10     **********
           49             11     ***********
           50             20     ********************
           51             20     ********************
           52             15     ***************
           53              6     ******
           54              5     *****
           55              2     **
           56              1     *
```

From Theorem 8.2 we know that the sampling distribution of $\overline{X}$ should be a normal distribution (with a mean of 50 and a standard deviation of 2). The histogram does indeed look very much like a normal distribution.

Example 8.11   Use MINITAB to investigate the sampling distribution of the statistic **P**, the sample proportion. Specifically, suppose that the population proportion of successes in a population of successes and failures is $\pi = 0.20$. Take 100 simple random samples of size $n = 25$ from this population, and describe the sampling distribution of the 100 sample proportions p.

Solution   We must first create the MINITAB Macro for the EXECUTE command. The Macro is named SIMP, for *sim*ulate the statistic **P**:

```
MTB > STORE 'SIMP'
STOR> NOPRINT
STOR> BTRIALS 100 RANDOM TRIALS WITH P=0.20, PUT INTO CK1
STOR> LET K1=K1+1
STOR> END
MTB >
```

Then the Macro is executed 25 times. Each time the Macro is executed, 100 0s and 1s are placed in a column with 100 rows.

```
MTB > LET K1=1
MTB > EXECUTE 'SIMP' 25
MTB > ADD C1-C25, PUT SUM IN C26
MTB > DIVIDE C26 BY 25, PUT VALUES IN C27
```

After the EXECUTE statement is completed, we add the row totals across the columns and divide by 25 to calculate the sample proportions using the ADD and DIVIDE statements. Columns C1, C2, . . . , C25, C26, and C27 look like the following:

ROW	C1	C2	. . .	C25	C26	C27
1	1	0		0	6	0.24
2	0	0		0	4	0.16
.	.	.		.	.	.
.	.	.		.	.	.
.	.	.		.	.	.
100	0	1		1	8	0.32

The DESCRIBE command produces the following results:

```
MTB > DESCRIBE C27

                C27
N               100
MEAN         0.1928  ←
MEDIAN       0.2000
TMEAN        0.1907
STDEV        0.0744  ←
SEMEAN       0.0074
MAX          0.4400
MIN          0.0400
Q3           0.2400
Q1           0.1600
```

Thus the average of the 100 sample proportions is 0.1928 (the population proportion is 0.20), and the standard deviation is 0.0744 (the standard deviation of P is $\sigma_P = \sqrt{\pi(1 - \pi)/n} = \sqrt{0.2(0.8)/25} = 0.080$).

The histogram of the 100 sample proportions is:

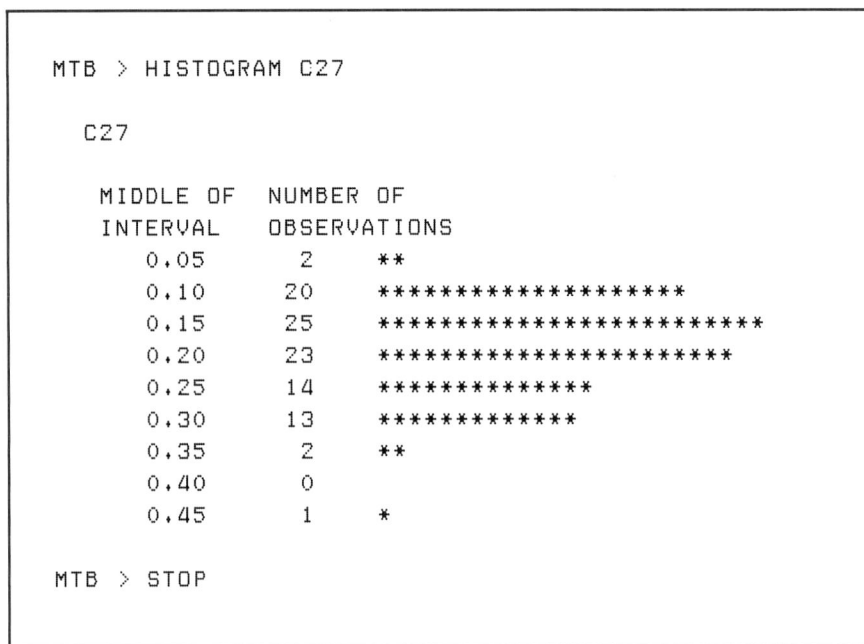

```
MTB > HISTOGRAM C27

   C27

   MIDDLE OF    NUMBER OF
   INTERVAL     OBSERVATIONS
      0.05          2     **
      0.10         20     ********************
      0.15         25     *************************
      0.20         23     ***********************
      0.25         14     **************
      0.30         13     *************
      0.35          2     **
      0.40          0
      0.45          1     *

MTB > STOP
```

From Theorem 8.5 the sampling distribution of the statistic, P, should be approximately normal for sufficiently large sample size. In this case $n = 25$, and the sampling distribution of the 100 sample proportions appears to be skewed to the right. Using the guidelines $n\pi \geq 5$ and $n(1 - \pi) \geq 5$, with $\pi = 0.20$, a sample of size $n = 25$ just qualifies for Theorem 8.5 to apply. However, in viewing this histogram, it should be clear that using the normal distribution to approximate the true sampling distribution of P will produce only approximate probability answers due to the skewed nature of the sampling distribution.

In the 1985 release of MINITAB, new commands have been created for conducting simulation experiments for the sampling distributions of P and $\overline{X}$. The new command for the normal distribution is RANDOM K OBSERVATIONS INTO EACH OF C, . . . , C with the subcommand NORMAL MU = K, SIGMA = K. (See the reference at the end of this chapter for the revised *MINITAB Handbook*.) This command eliminates the need to use the Macro presented in Examples 8.10 and 8.11. As a consequence, it is now much easier to do simulation experiments using MINITAB. Presumably, the 1985 release of MINITAB will be available to MINITAB users by early 1986.

We will consider two more examples of simulation experiments using MINITAB that illustrate the effects of increasing the sample size on statistics and sample frequency distributions.

Example 8.12 In Chapter 7 the *exponential distribution* was introduced. The continuous random variable **X** follows an exponential distribution if its *probability density function $f(x)$* is:

$$f(x) = \lambda e^{-\lambda x}, \qquad x \geq 0$$

where $\mu = E(\mathbf{X}) = 1/\lambda$, and $\sigma^2 = V(\mathbf{X}) = 1/\lambda^2$. Suppose $\mu = E(\mathbf{X}) = 2$ so that $\lambda = 1/2$. The probability density function is then

$$f(x) = (1/2)e^{-(1/2)x}, \qquad x \geq 0$$

The graph of this density function is illustrated in Figure 8.12.

FIGURE 8.12 | The graph of the exponential distribution $f(x) = (1/2)e^{-(1/2)x}$

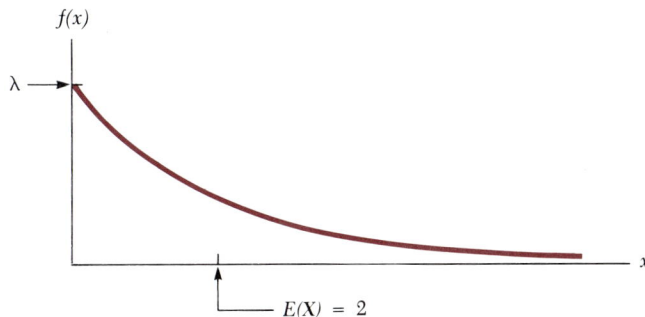

From Figure 8.12 it is evident that the distribution is very skewed to the right. Now consider the statistic $\overline{\mathbf{X}}$, the sample mean, whose value in a sample of size n is given by

$$\bar{x} = \frac{\sum\limits_{i=1}^{n} x_i}{n}$$

By the *central limit theorem,* the sampling distribution of $\overline{\mathbf{X}}$ should be approximately normal (though the distribution of **X** is skewed right) for sufficiently large sample size n. To explore this by simulation, we can use the following MINITAB commands to generate values from the exponential distribution:

```
URANDOM N, PLACE VALUES IN COLUMN C1
SUBTRACT COLUMN C1 FROM 1, PUT VALUES IN COLUMN C2
LOGE COLUMN C2, PUT IN COLUMN C3
MULTIPLY COLUMN C3 BY (-1/LAMBDA), PUT IN COLUMN
C4
```

By using these commands, column C4 will contain a simple random sample of exponential variates of size n with an expected value of $(1/\lambda)$. (See the MINITAB reference at the end of this chapter for an explanation of these commands.)

The histograms of the sampling distribution of $\overline{X}$ for 100 repeated samples of sizes $n = 2$, $n = 10$, and $n = 25$ are shown below:

n = 2

```
MIDDLE OF     NUMBER OF
INTERVAL      OBSERVATIONS
    0             12        ***********
    1             31        *******************************
    2             26        **************************
    3             13        *************
    4              9        *********
    5              4        ****
    6              2        **
    7              1        *
    8              2        **
```

n = 10

```
MIDDLE OF     NUMBER OF
INTERVAL      OBSERVATIONS
   0.5            2         **
   1.0            8         ********
   1.5           27         ***************************
   2.0           28         ****************************
   2.5           15         ***************
   3.0            8         ********
   3.5            7         *******
   4.0            4         ****
   4.5            1         *
```

n = 25

```
MIDDLE OF      NUMBER OF
INTERVAL       OBSERVATIONS
   1.2              4        ****
   1.4              6        ******
   1.6             13        *************
   1.8             26        **************************
   2.0             21        *********************
   2.2             11        ***********
   2.4             11        ***********
   2.6              5        *****
   2.8              3        ***
```

When $n = 2$ the sampling distribution of $\overline{X}$ is skewed to the right. When n is increased to 10, the distribution is still skewed but is beginning to take on a more bell-shaped, normal distribution form. When $n = 25$ the bell-shaped form of the normal distribution is becoming more apparent. The central limit theorem tells us that as n is increased without bound, the sampling distribution of $\overline{X}$ will become the normal distribution.

Example 8.13 In Chapter 5 we considered the probability distribution of the random variable X = Total number of dots on the top two faces of a pair of dice. The probability distribution of this random variable is shown in Figure 8.13.

Another interesting use of MINITAB's simulation commands is in conducting gambling simulation experiments. Suppose we repeat the experiment of

FIGURE 8.13 | The probability mass function of the random variable X = Total dots on the top two faces of a pair of dice

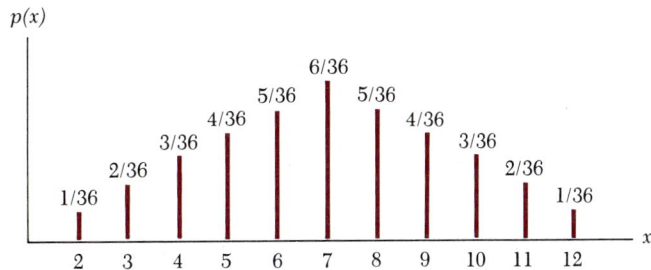

tossing a pair of dice n times. On each toss, we record the total number of dots appearing on the top two faces of the pair of dice. As the number of trials (n) of the experiment increases, we should see that the histogram of the values of **X** begins to look like the theoretical distribution of **X** shown in Figure 8.13. To investigate this using MINITAB, we can use the command:

```
DRANDOM N OBSERVATIONS, VALUES IN COLUMN C1,
AND PROBABILITIES IN COLUMN C2
```

To use this command, we first set the 11 values of **X** (2, 3, . . . , 12) into column C1 and the corresponding probabilities into column C2 (1/36, 2/36, . . . , 1/36).

The sample histograms of values of **X** for samples of size $n = 25$, 50, and 100 tosses of the pair of dice are shown below.

n = 25 trials

```
MIDDLE OF          NUMBER OF
INTERVAL           OBSERVATIONS
   3,                 2       **
   4,                 1       *
   5,                 1       *
   6,                 5       *****
   7,                 6       ******
   8,                 3       ***
   9,                 3       ***
  10,                 1       *
  11,                 2       **
  12,                 1       *
```

For $n = 25$ and $n = 50$ tosses of the pair of dice, the histograms do not conform very well to the theoretical distribution shown in Figure 8.13, although the $n = 50$ histogram is better than the $n = 25$ histogram. When $n = 100$, the shape of the histogram is beginning to look like the theoretical distribution of **X** shown in Figure 8.13.

These MINITAB simulation experiments show the effects of sampling. It is clear that we cannot expect a sample to produce the population results

n = 50 trials

```
MIDDLE OF          NUMBER OF
INTERVAL           OBSERVATIONS
   2.                  2      **
   3.                  3      ***
   4.                  3      ***
   5.                  9      *********
   6.                  6      ******
   7.                 10      **********
   8.                  5      *****
   9.                  3      ***
  10.                  7      *******
  11.                  2      **
```

n = 100 trials

```
MIDDLE OF          NUMBER OF
INTERVAL           OBSERVATIONS
   2.                  1      *
   3.                  7      *******
   4.                 11      ***********
   5.                  8      ********
   6.                 16      ****************
   7.                 15      ***************
   8.                 12      ************
   9.                 10      **********
  10.                 11      ***********
  11.                  5      *****
  12.                  4      ****
```

(parameter values or distribution) exactly (unless we sample the entire population!). However, as the sample size increases, these experiments show that the sample statistic values and sample distributions can be expected to converge on the corresponding population parameters and population distributions. The trick is to use a sufficiently large sample size so that our sample results will be representative of the population characteristics. This issue will be a primary topic over the next several chapters.

8.11
Summary

There are various ways a sample can be withdrawn from a population: *probability sampling* (simple random samples, stratified samples, cluster samples, and systematic samples), *judgment sampling,* and *convenience sampling*. The last two types must be used with care, for they are subject to sampling bias, particularly convenience samples. In this text, we will use simple random samples.

The sample statistics are random variables—their values vary from sample to sample of a fixed size drawn from a population. For sufficiently large sample size, the distribution of the sample average statistic $\overline{X}$ and the sample proportion statistic **P**, as well as other sample statistics, are approximately normally distributed. By knowing the form of the sampling distribution of a statistic, we are able to answer probability questions about it, such as the probability that the statistic is greater than a stated value. In this way, we can measure the extent of sampling error in the sampling experiment. The amount of sampling error can be reduced by increasing the sample size.

References

Introductory

McClave, J. T., and P. G. Benson. *Statistics for Business and Economics*. 3rd ed. San Francisco: Dellen Publishing, 1985, Chapters 7 and 20.

Neter, J.; W. Wasserman; and G. A. Whitmore. *Applied Statistics*. Boston: Allyn & Bacon, 1982, Chapters 8 and 9.

Ryan, T.; B. Joiner; and B. Ryan. *MINITAB Handbook*. Boston: Duxbury Press, 1976.

Ryan, T.; B. Joiner; and B. Ryan. *MINITAB Handbook*. 2nd ed. Boston: Duxbury Press, 1985.

Schaeffer, R., and W. Mendenhall. *Elementary Survey Sampling*. 3rd ed. Boston: Duxbury Press, 1986.

Advanced

Arkin, H. *Sampling Methods for the Auditor*. New York: McGraw-Hill, 1982, Chapters 1, 3, and 4.

Cochran, W. G. *Sampling Techniques*. 3rd ed. New York: John Wiley & Sons, 1977.

Kish, L. *Survey Sampling*. New York: John Wiley & Sons, 1965.

Williams, B. *A Sampler on Sampling*. John Wiley & Sons, 1978.

Problems

Section 8.2 Problems

8.1 Explain in what sense the statistic $\overline{X}$ is a random variable.

8.2 Define a parameter and a statistic. What is the difference between these two numerical measures?

8.3 For each of the cases below, indicate whether a statistic or a parameter has been computed.

a. A company calculates the *proportion* of employees who have a college degree by analyzing the personnel records of all employees.

b. A marketing research firm calculates the *proportion* of respondents who prefer brand A of a product to brand B in a survey of 1,000 sampled potential buyers of the product.

c. The National Basketball Association calculates the *mean* and the *median* salary of its current players in the league.

d. The Battery Company calculates the *mean* lifetime of its AA-size batteries based on a simple random sample of 500 batteries.

Section 8.3 Problems

8.4 Define a simple random sample taken from a population that is finite in size.

8.5 Define a simple random sample taken from a population that is infinite in size.

8.6 Suppose a finite population contains 100 elements. Discuss a procedure for taking a sample of size five from this population that satisfies the definition of a simple random sample drawn from a finite population.

Section 8.4 Problems

8.7 Define the term *sampling error*. Can sampling error be controlled or not in a sampling experiment? Explain.

8.8 Define the term *census*. Give two examples in business applications wherein it is possible to take a census rather than a sample to calculate a numerical measure of a population characteristic.

8.9 Define the term *nonsampling error*. Can nonsampling error be controlled or not in a sampling experiment? Explain.

8.10 Distinguish between sampling and nonsampling error. Can both occur in a sampling experiment? Explain.

Section 8.5 Problems

8.11 Distinguish between the *population* distribution and the *sampling* distribution of sample statistics.

8.12 A population consists of the elements listed in the table.

Item	Value
A	0
B	2
C	7
D	4
E	3
F	2

a. Compute the population mean, median, standard deviation, and proportion of values greater than or equal to 2.

b. Form a table showing all possible samples of the size $n = 2$. For each sample of size 2, compute the mean, the median, and the proportion of values greater than or equal to 2.

c. Which distribution has less variability— the population distribution or the sampling distribution of the sample mean? Why?

8.13 Repeat problem 8.12 using a sample of size $n = 3$.

8.14 A population consists of five units of a product. Each unit is either defective or nondefective. If the unit is defective, then the random variable **X** is set at 1; and if the item is nondefective, then the random variable **X** is set at 0. The population values of **X** are listed below:

Item	Value
A	0
B	0
C	0
D	1
E	0

a. Compute the population proportion π.

b. Form a table showing all possible samples of size $n = 2$. For each sample of size 2, compute the sample proportion p.

Section 8.6 Problems

8.15 A population consists of four values of a random variable **X**:

Item	Value
A	10
B	15
C	15
D	20

a. Compute the population mean μ and the population standard deviation σ.
b. Form the table showing all possible samples of size $n = 2$. For each sample of size $n = 2$, compute the sample mean $\bar{x}$ and the standard deviation s.
c. Compute the mean and standard deviation of the sampling distribution of the statistic $\bar{\mathbf{X}}$, the sample mean.

8.16 Five students are on academic probation in the Department of Finance in the School of Business at Watermark University. Let the random variable **X** = Grade point average of a student. The five students' grade point averages (values of **X**) are given in the table below:

Student	Grade point average
Besmerck	1.20
Walston	1.85
Deletory	1.66
Samuelson	1.95
Clarke	1.45

a. Compute the population mean and standard deviation of the random variable **X**.
b. Form a table showing all possible samples of size $n = 3$. For each sample of size $n = 3$, compute the sample mean grade point average $\bar{x}$ and the sample standard deviation s.
c. Compute the mean and the standard deviation of the statistic $\bar{\mathbf{X}}$, the sample mean.

8.17 Describe the sampling distribution of the statistic $\bar{\mathbf{X}}$ when
a. The population random variable is normally distributed.
b. The population random variable is not normally distributed, but the sample size is "large."

8.18 Distinguish among the population mean, μ, the sample mean, $\bar{x}$, and the mean of the sampling distribution of $\bar{\mathbf{X}}$, $\mu_{\bar{x}}$. How do these measures compare in size?

8.19 Describe in your own words the central limit theorem. Why is it such an important theorem in statistics?

8.20 A population random variable is normally distributed with a mean of 50 and a variance of 16. If a sample of size $n = 25$ is drawn from this population, what are the probabilities of the following events?
a. $\bar{\mathbf{X}}$ will exceed 51.
b. $\bar{\mathbf{X}}$ will be between 48.5 and 51.5.

8.21 Repeat Problem 8.20, but with $n = 36$. What is the effect on the probabilities in parts a and b of increasing the sample size from 25 to 36?

8.22 Repeat Problem 8.20, but with the variance of the population random variable equal to 25. What is the effect on the probabilities in parts a and b of increasing the population variance from 16 to 25?

8.23 The Sanso Petroleum Company fills steel drums with a lubricating oil in such a way that the weight of the oil in each drum is normally distributed with a mean of 250 pounds and a standard deviation of 2 pounds. If a random sample of nine drums is taken, what is the probability that the average weight per drum in the sample is more than 252 pounds?

8.24 A certain species of apricot tree produces a yield that is distributed with a mean of 2.0 bushels per tree and a standard deviation of 0.5 bushel.
a. If the yields per tree are normally distributed, what is the probability that the average yield per tree in a random sample of $n = 4$ trees will be less than 1.5 bushels?
b. If the yields per tree are *not* normally distributed, is your answer in part a a useful and meaningful one? Explain.

8.25 The mean annual income of a population of union workers in a certain local is known to be $15,000. The population standard deviation is $2,500. A sample of $n = 100$ of these workers is drawn.

a. Describe the sampling distribution of the sample mean statistic $\overline{\mathbf{X}}$.

b. What is the probability that $\overline{\mathbf{X}}$ will be within $400 of the population mean annual income?

8.26 In a population of 1,000 college students in a particular major at a large university, the mean grade point average is 2.80, and the population standard deviation is 0.20.

a. If a random sample of $n = 100$ students is taken from this population, what is the probability that the statistic $\overline{\mathbf{X}}$, the sample mean grade point average, will differ from the population mean grade point average by more than 0.01 unit?

b. Repeat part a with $n = 500$.

c. Repeat part a with $n = 1,000$.

Section 8.7 Problems

8.27 In a very large production lot, the proportion of defectives is 0.10. If a sample of 100 items in this lot is selected, what is the probability that two or fewer of the items are defective? Is this an approximate or an exact answer? Explain.

8.28 What is the exact sampling distribution of the sample proportion statistic $\mathbf{P}$ when the population is infinite in size? Under what circumstances can the normal distribution be used as an approximating sampling distribution for $\mathbf{P}$?

8.29 A population random variable $\mathbf{X}$ is a Bernoulli variable taking on values 0 or 1. The proportion of 1s (successes) in the population is $\pi = 0.80$. If a sample of size $n = 64$ is taken from this population, then answer the following probability questions concerning the statistic $\mathbf{P}$, the sample proportion:

a. $P(\mathbf{P} < 0.75)$

b. $P(0.70 < \mathbf{P} < 0.90)$

8.30 Repeat Problem 8.29, but with $n = 100$. What is the effect on the probabilities in parts a and b by increasing the sample size from $n = 64$ to $n = 100$?

8.31 In a set of 10,000 invoices, it is known that 500 contain at least one error. If 100 of the 10,000 invoices are randomly sampled, what is the probability that the sample proportion of invoices with at least one error will exceed 0.08? Is this an approximate or an exact answer? Explain.

8.32 Assume 60 percent of the voters in a large city are going to vote for the Democratic candidate. A random sample of $n = 100$ voters is chosen.

a. Determine the probability that the sample proportion of voters who will vote Democratic is within 10 percentage points of the population proportion (i.e., between 0.50 and 0.70).

b. What assumptions did you make in answering part a?

8.33 A production process is known to produce 5 percent defective items. A random sample of 100 items is taken from the process.

a. What is the probability that the sample proportion of defectives is greater than 0.10?

b. If it is known that 5 percent of the items produced by a production process in a lot of 1,000 are defective, what is the probability, in a sample of 100 drawn from the lot, that the sample proportion of defectives exceeds 0.10?

c. Compare the answers in parts a and b. Why do they differ?

8.34 A production process is known to produce defective units at a rate of 10 percent. A random sample of $n = 100$ units is to be taken from the production process. The value of the statistic $\mathbf{P}$, the sample proportion, will be computed.

a. Describe the sampling distribution of the statistic $\mathbf{P}$, the sample proportion—what is the mean and standard deviation of the sampling distribution, and what is the approximate form of the sampling distribution?

b. What is the probability that the sample of size $n = 100$ will contain more than 15 percent defective units?

c. What is the probability that six or fewer defective units will be found in the sample of size $n = 100$?

Section 8.8 Problems

8.35 Explain what the *finite population correction factor* is. When should it be used?

8.36 A sample of size $n = 25$ is to be taken from a finite population of size $N = 500$. Once the sample is taken, the value of the statistic $\overline{X}$ will be computed. Should we correct the standard deviation of $\overline{X}$ using the finite population correction factor? Explain.

8.37 A sample of size $n = 100$ is taken from a population of size $N = 2,000$. In this population, the proportion of college graduates is 0.32.

a. Calculate the standard deviation of the statistic, **P**, the sample proportion *ignoring the finite population correction factor*.

b. Calculate the standard deviation of the statistic **P** *using the finite population correction factor*.

c. Compare the answers in parts *a* and *b*. Is this a case in which the finite population correction factor needs to be used? Explain.

Section 8.9 Problems

8.38 Describe each sampling plan, and give one example where each is appropriate.

a. Simple random sample.
b. Cluster sample.
c. Stratified sample.
d. Systematic sample.
e. Judgment sample.
f. Convenience sample.

8.39 Why do convenience samples tend to be biased? Is this also true about judgment samples? Explain.

8.40 If you were asked to sample 50 households in a section of a city that contains 1,000 households, explain how you would go about determining which households to include if your sample is to be each of the following.

a. Simple random sample.
b. Stratified sample on household annual incomes with three strata: $0 to $5,000, $5,001 to $15,000, and above $15,000.
c. Convenience sample.
d. Judgment sample.

8.41 In the following situations, which type of sample—random, judgment, convenience, or a combination of sampling plans—would you recommend and why?

a. A calculator manufacturer wishes to test the reliability of a certain microcircuit used in its leading model.

b. A magazine editor selects "Letters to the Editor" to publish monthly.

c. A university motor pool must purchase a large fleet of cars from one of the leading manufacturers. The primary consideration will be economy of operation.

d. An accountant must select a sample of invoices in a large department store to determine whether the proportion of invoices with errors is within a set tolerance.

e. A reporter wishes to receive suggestions about how his daily column in the local newspaper can be improved.

Additional Problems

8.42 Consider the population of the first seven integers: 1, 2, 3, 4, 5, 6, and 7; $N = 7$. For this population, $\mu = 4$, and $\sigma = 2$.

a. How many samples of size three can be extracted from this population (sampling without replacement)?

b. Form the complete set of samples of size three and for each sample, compute the sample mean and median.

c. Form the sampling distributions for the sample mean statistic and for the sample median statistic.

d. Find the mean and standard deviation of each sampling distribution in part *c*.

e. If you had to select *at random* one of the samples of size three generated in part *b* and had to use either the sample's mean or median to estimate μ, which would you choose and why?

8.43 An automobile dealer has been chastised by the manufacturer of the cars sold by the dealership because of the high percentage of complaints received from customers about service for their cars at the dealership. To assess the seriousness of the problem, the dealer decides to sample individuals randomly who have brought their cars in for service during the past six months. Each sampled individual will receive a mail questionnaire requesting comments about the quality of service received when they brought their cars in. Discuss why the sample set of questionnaires most likely will not be representative of the attitudes toward the service received by all customers who brought their cars in for service during the past six months.

8.44 Distinguish among the population variance, σ^2, the sample variance, s^2, and the variance of the sampling distribution of $\overline{X}$, $\sigma_{\overline{X}}$. How do these measures compare in size?

8.45 Nationally, it is known that 87 percent of the secondary teachers in the United States belong to the Association of Teachers. A random sample of $n = 400$ secondary teachers is taken in a survey on salaries and benefits. Included in the survey is a question about whether or not the respondent belongs to the Association of Teachers. The statistic $\mathbf{P}$, the sample proportion of respondents who belong to the association, will be computed when the survey results are analyzed.
  a. What is the probability that this statistic will be greater than 0.90?
  b. What is the probability that fewer than 325 of the respondents will indicate that they belong to the association?

8.46 The CEO of the AXXON Corporation is concerned about the average age of the employees who work for AXXON. Suppose that the average age of all employees of AXXON is 42 years with a standard deviation of 10 years and that this information is unknown to the CEO. According to her instructions, the personnel director plans to take a random sample of 200 employees of AXXON and calculate the average age of the sampled employees.
  a. What is the probability that the average age in the sample of 200 employees will exceed 43 years?
  b. What is the probability that the average age in the sample of 200 employees will be between 41 and 44 years?

8.47 The L & R Law Firm has 400 client accounts. An auditor plans to randomly sample 64 of these accounts to determine the proportion of

accounts that are in error. Suppose that 40 of the accounts are in error.
  a. What is the probability that the auditor will find that the sample proportion $\mathbf{P}$ of accounts in error is between 0.05 and 0.15? Does the *FPC*—the finite population correction factor—have to be used in this case? Explain.
  b. What is the probability that the auditor will find that the sample proportion $\mathbf{P}$ of accounts in error will exceed 0.20?

8.48 In Problem 8.47 suppose that the auditor will certify that the accounts are in order if the sample mean balance in the sample of size $n = 64$ is within $50 of the reported mean balance. Suppose that the mean balance of the 400 accounts is $800 with a standard deviation of $250.
  a. What is the probability that the auditor will certify that the accounts are in order? Is it necessary to use the *FPC* in this case? Explain.
  b. What is the probability that the auditor will not certify that the accounts are in order?

8.49 A manufacturing firm produces pressure gauges used to determine the pressure on the bulkheads of nuclear submarines. The gauges signal an alarm when a specified average pressure is sensed by the gauge. Suppose that the standard deviation of pressure needed to trigger the alarm is 0.5 pound. A random sample of $n = 100$ gauges is taken, and the pressure required to trigger the alarm is recorded. From the 100 pressure recordings, the average pressure required to trigger the alarm is recorded. What is the probability that the average pressure needed to trigger the alarm in the sample of 100 gauges is within 0.18 pound of the population mean pressure μ required to trigger the alarm?

8.50 An automobile manufacturer claims that the average gasoline mileage of its Hawk automobile is 40 miles per gallon (MPG) with a standard deviation of 2 MPG. A consumer group samples $n = 25$ Hawk automobiles, records the MPG for each, and calculates the sample average MPG for the 25 cars.
  a. Assuming that the manufacturer's claim is correct, what is the probability that the

sample average MPG will be less than 39?

b. Assuming that the manufacturer's claim is correct, what is the probability that the sample average MPG will be less than 38.8?

c. Suppose that the consumer group finds that the sample average MPG is 38.0 MPG. Based on parts a and b, do you believe that the manufacturer's claim is justified? Explain.

8.51 The average number of semester credit hours taken by students at RSU, a university in the Midwest, is 13.2 hours with a standard deviation of 2.2 hours. If a random sample of 36 students at RSU is taken,

a. What is the probability that the average number of semester credit hours taken by the sampled students will exceed 14 hours?

b. What is the probability that the average number of semester hours taken by the sampled students will be between 12 and 13.5 hours?

8.52 According to published salary data for the labor unions at Southern Airlines, the average annual salary of all employees is $40,000, with a standard deviation of $8,000. A random sample of 100 employees of Southern Airlines is taken. The value of the statistic $\overline{X}$, the sample mean, will be calculated.

a. Describe the sampling distribution of the statistic $\overline{X}$—what is the mean and the standard deviation of the sampling distribution, and what is its approximate distributional form?

b. What is the probability that $\overline{X}$ will be less than or equal to $39,000 when the sample of size $n = 100$ is taken?

8.53 A random variable X is known to be normally distributed with a mean of 100 and a standard deviation of 100. A random sample of size n is taken from the distribution of X. It is found, based on the random sample of size n, that the probability that $\overline{X}$, the sample mean, exceeds 125 is 0.1056. What was the sample size n drawn from the distribution of X?

8.54 A large corporation has five senior vice presidents. The table below lists the annual salary including bonuses (in thousands of dollars) and whether or not the executive has an MBA degree (1 if yes, 0 if no).

Executive	Salary (000s)	MBA
Olson	215	1
Newmark	185	0
Stanley	240	0
McDonough	160	0
Carson	225	1

Let X = Annual salary including bonuses in thousands of dollars and Y = 1 (if executive has MBA) or 0 (if executive does not have MBA).

a. Compute the population mean and standard deviation for the random variable X.

b. Compute the population proportion π, where π = Proportion of executives who have an MBA degree (the average of the random variable Y).

c. Form a table showing all possible samples of size $n = 3$. For each sample of size 3, compute the sample mean salary $\bar{x}$ and the proportion of executives who have the MBA degree p.

d. Compute the mean and standard deviation of the statistic $\overline{X}$, the sample mean.

e. Compute the mean and standard deviation of the statistic P, the sample proportion.

9

Statistical inference: Estimation

In Chapter 2, we observed that a set of numbers can be described by numerical measures, such as representative measures and measures of dispersion. As noted in Chapter 8, if the set of numbers being examined comprises a population, then the numerical measures such as the population mean μ and the population standard deviation σ are called *parameters*. On the other hand, if we intend to take a sample of size n from the population, then the numerical measures $\overline{X}$ (the mean), and S (the standard deviation), and so on are functions of the random variables that constitute the sample, $X_1, X_2, \ldots, X_n$. The functions $\overline{X}$ and S are called *statistics*. The statistics $\overline{X}$ and S are random variables with values that are determined when a specific sample is taken and the numerical values $x_1, x_2, \ldots, x_n$ of the random variables $X_1, X_2, \ldots, X_n$ are known.

Since a population set of numbers can often be characterized by the values of the population parameters, attention is focused on these parameters in most statistical problems. For example, the population may consist of the lifetimes of a specific electronic component, and we may be interested in the mean lifetime, the population parameter μ. The value of the parameter μ could be computed by using each component in the population, recording the lifetime of each and every component, and averaging these lifetimes. It would clearly not be feasible to compute the value of μ in this fashion, because this process would destroy all components—there would be no components left to sell. Alternatively, a sample of components could be selected and their lifetimes used in some manner to "guess," or estimate, the value of μ. In this instance, the statistic $\overline{X}$ would seem like a reasonable function of the sample random variables $\mathbf{X}_1, \mathbf{X}_2, \ldots, \mathbf{X}_n$ to use in estimating the value of the population parameter μ. Indeed, as we shall see later in this chapter, the sample mean statistic $\overline{\mathbf{X}}$ is the "best" statistic to use in estimating μ.

In many cases, such as the electronic component lifetime example above, it is impossible to determine the values of a population parameter by analyzing the entire set of population values. The process of determining the value of the parameter may destroy the population units, or it may simply be too expensive in money and/or time to analyze each unit. In these instances, there is little choice but to use statistical inference to gain information about the values of the population parameters.

The objective of *statistical inference* is to make inferences about a population based on a subset of it—the sample drawn from the population. More specifically, *statistical inference is the process of selecting and using a sample statistic to draw inferences about a population parameter.*

To illustrate the process of drawing inferences about a population parameter, consider the example of the electronic component lifetimes mentioned above. For many electronic components used in computers, televisions, and other electronic devices, the manufacturer is interested in the "mean time to failure" (the average lifetime) of the component. Often, the manufacturer is required to state the mean time to failure on the component or on the package containing it. Suppose we are interested in a specific component's mean time to failure. If we let the random variable $\mathbf{X}$ denote the lifetime in hours of use of the component, then the mean time to failure for the population of those components is μ, the population mean.

The process of using statistical inference to draw conclusions about the value of μ based on a sample of components is illustrated in Figure 9.1.

FIGURE 9.1 | Statistical inference process for the population parameter μ

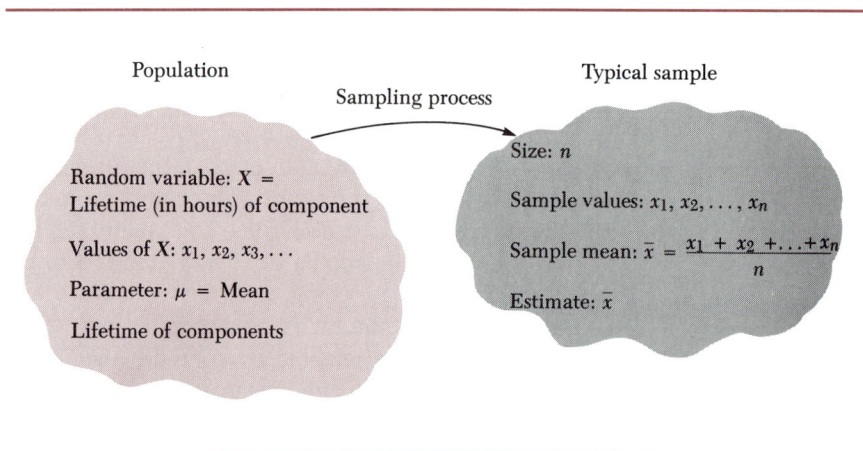

Population

Sampling process

Typical sample

Random variable: $X =$ Lifetime (in hours) of component

Values of X: $x_1, x_2, x_3, \ldots$

Parameter: $\mu =$ Mean Lifetime of components

Size: n

Sample values: $x_1, x_2, \ldots, x_n$

Sample mean: $\overline{x} = \dfrac{x_1 + x_2 + \ldots + x_n}{n}$

Estimate: $\overline{x}$

The size of the population can be either finite or infinite. Thus we show its values as x_1, x_2, x_3, . . . , indicating by the ellipses that there may be an infinite number of values. From the population, a sample of size n is taken, producing the sample values x_1, x_2, . . . , x_n. In the sample, we calculate the value of the sample mean, denoted by $\bar{x}$.

| Two methods for drawing inferences about a population parameter: Estimation and hypothesis testing |

There are *two ways* we make inferences about μ. The first way is to *estimate* the value of μ. It seems reasonable that the sample mean $\bar{x}$, calculated in a typical sample as illustrated in Figure 9.1, is a good choice to estimate μ. (In the statistical literature, this estimate is denoted by $\hat{\mu}$ where the "hat" over μ indicates that it is an estimate, not the actual value of μ. We will not use this notation in this text). For example, if $\bar{x}$ turned out to be a mean lifetime of 2,000 hours for the component based on the sample values, then we would say that the estimate of μ is 2,000 hours.

A second way to draw inferences about μ from the inference process illustrated in Figure 9.1 is *hypothesis testing*. By this method, we hypothesize a value of μ and use the sample information to make a decision as to whether or not the hypothesis is reasonable. For example, we may hypothesize that the mean lifetime of this component is 2,500 hours. If we take a sample and find that the sample mean $\bar{x}$ is 1,750 hours, then we *may* have reason to doubt that the hypothesized value of μ (2,500 hours) is true. The key idea in hypothesis testing is the construction of a *decision rule*. This rule determines how much the sample mean $\bar{x}$ can deviate from the hypothesized value of μ before we conclude that the difference is too great for μ to equal the hypothesized value.

This chapter will describe the methods of statistical estimation. The next chapter will describe the methods of hypothesis testing. As we shall see, the two ways of making inferences about a population parameter are very similar. In many instances, either approach may be used to come to conclusions about the value of a population parameter. Typically, hypothesis testing is more important in scientific studies or in designed experiments, and estimation is more important in analyzing survey data or in economic problems.

The next section will give some basic concepts regarding statistical estimation. These will be followed by concept applications for drawing inferences about the population mean (μ), the population proportion (π), and the population standard deviation (σ) in the remaining sections.

■ 9.2

Types of statistical estimation procedures

Statistical estimation is divided into two main types: *Point estimation* and *interval estimation*. In point estimation, a *single number* computed from the sample values x_1, x_2, . . . , x_n is used to estimate the value of the population parameter. It is called a point estimate because one *point* on the real number line is used to estimate the value of the population parameter. In interval estimation, two points are used to define an *interval* on the real number line, which should contain the value of the population parameter. Suppose the population parameter is μ, the mean lifetime of an electronic component. We might arrive at a point estimate of μ of 2,000 hours and an interval

| Point estimation |

estimate of μ from 1,800 hours to 2,200 hours based on a sample from the population of these components. Normally, both types of estimates are given in an estimation problem. The interval estimate is very desirable, for the *width* of the interval indicates to some degree the point estimate quality—the shorter the interval width, the more precise the point estimate.

We will now consider each type of estimation in more detail, beginning with point estimation.

9.2.1 | Point estimation

When using the point estimation method to determine a single numerical value estimate of a population parameter, we must first determine what *function* of the sample values $x_1, x_2, \ldots, x_n$ to use. For example, suppose that you wish to determine a point estimate of the population mean μ based on a typical sample of size n, whose values are $x_1, x_2, \ldots, x_n$. Several functions of the sample values could be considered. For example, consider the function f defined below:

$$f(x_1, x_2, \ldots, x_n) = \frac{x_1 + x_2 + \ldots, + x_n}{n}$$

The rule f says to sum the sample values $x_1, x_2, \ldots, x_n$ and divide the sum by the sample size n. This rule describes the *sample mean, $\bar{x}$,* but we could consider other functions of the sample values in constructing an estimate of μ. For example:

$$g(x_1, x_2, \ldots, x_n) = \frac{\max\{x_i\} + \min\{x_i\}}{2}$$

$h(x_1, x_2, \ldots, x_n) =$ The 50th percentile of the ordered sample values, ordered *(sample median)* from smallest to largest.

$m(x_1, x_2, \ldots, x_n) =$ Most frequently occurring value in the sample of values $x_1, x_2,$ *(sample mode)* $\ldots, x_n$.

The g rule says to find the largest value and smallest value in the sample, take the difference between the largest and smallest values, and divide by 2. The h rule describes the *sample median,* and the m rule describes the *sample mode.*

All four functions of $x_1, x_2, \ldots, x_n$ could be considered in selecting one for use in calculating an estimate of μ. To find which one is best, one must define desirable properties of the functions chosen to estimate the value of a population parameter. Measured against the desirable properties that will be given, the function f (the sample mean) is the "best" function to use in forming a point estimate of the population mean μ.

Before discussing the desirable properties of a function used to estimate the value of a population parameter, we need to set the notation carefully and to define several terms. The notation for estimating μ, the population mean, is illustrated in Figures 9.2 and 9.3. In Figure 9.2, **X** denotes the population random variable measuring the characteristic of interest in the population (say, the lifetime in hours of an electronic component). *Before the sample is actually taken and the sample values are known, we can represent the sample as a collection of* n *random variables,* $\mathbf{X}_1, \mathbf{X}_2, \ldots, \mathbf{X}_n$. For example, the first sample random variable $\mathbf{X}_1$ can have as its value any one of the values of $\mathbf{X}$ in the population. The *estimator* of μ, denoted by $f(\mathbf{X}_1, \mathbf{X}_2, \ldots, \mathbf{X}_n)$, is a function of the n sample random variables $\mathbf{X}_1, \mathbf{X}_2, \ldots, \mathbf{X}_n$. We do not know its value until we know the values of $\mathbf{X}_1, \mathbf{X}_2, \ldots, \mathbf{X}_n$.

Estimator of μ

FIGURE 9.2 | Notation for the point estimation of the value of the population mean μ—*the estimator of* μ

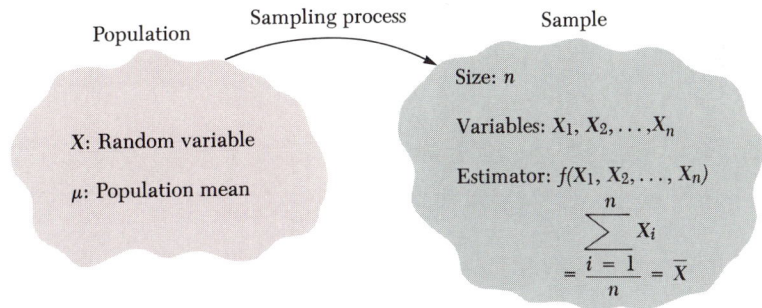

In Figure 9.3, a *typical sample* of n values of $\mathbf{X}_1, \mathbf{X}_2, \ldots, \mathbf{X}_n$ is shown. Now we have numerical values denoted by $x_1, x_2, \ldots, x_n$. The *estimate* of μ is a function of the n numerical values of $\mathbf{X}_1, \mathbf{X}_2, \ldots, \mathbf{X}_n$, denoted by $x_1, x_2, \ldots, x_n$.

Estimate of μ

To distinguish between an *estimator* and an *estimate*, we need only return to the concept of a sampling distribution, described in Chapter 8. Suppose, for example, that we decide to use the sample mean as the best function of the sample random variables, as illustrated in Figure 9.2. The sampling distribution of $\overline{\mathbf{X}}$ is shown in Figure 9.4. The sample mean $\overline{\mathbf{X}}$ is a *random variable*, often called a *statistic*, whose value varies from sample to sample.

Statistic

Its sampling distribution is approximately normal for sufficient sample size (via the central limit theorem), its mean is $\mu_{\overline{\mathbf{x}}} = \mu$, and its standard deviation is $\sigma_{\overline{\mathbf{x}}} = \sigma/\sqrt{n}$ [if the population is infinite in size; if the population is finite in size, then we must multiply by the finite population correction factor discussed in Chapter 8 so that $\sigma_{\overline{\mathbf{x}}} = (\sigma/\sqrt{n})(\sqrt{(N-n)/(N-1)})$]. Thus the notation $\overline{\mathbf{X}}$ defines the estimator of μ, and $\overline{x}$ defines the estimate of μ calculated from one of the many possible samples of size n taken from the popula-

FIGURE 9.3 | Notation for the estimation of the value of the population mean μ—*the point estimate of* μ

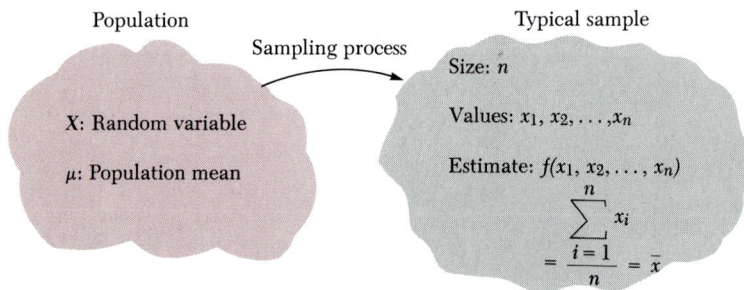

Population

X: Random variable

μ: Population mean

Sampling process

Typical sample

Size: n

Values: $x_1, x_2, \ldots, x_n$

Estimate: $f(x_1, x_2, \ldots, x_n)$

$$= \frac{\sum\limits_{i=1}^{n} x_i}{n} = \bar{x}$$

FIGURE 9.4 | Sampling distribution of $\overline{\text{X}}$

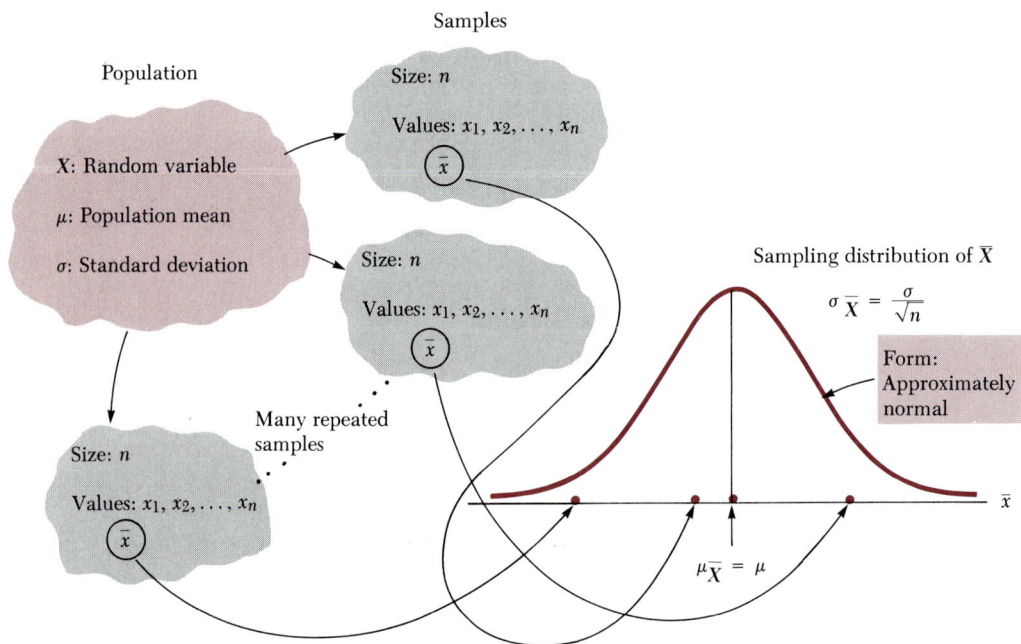

Population

X: Random variable

μ: Population mean

σ: Standard deviation

Samples

Size: n

Values: $x_1, x_2, \ldots, x_n$

$\bar{x}$

Size: n

Values: $x_1, x_2, \ldots, x_n$

$\bar{x}$

Many repeated samples

Size: n

Values: $x_1, x_2, \ldots, x_n$

$\bar{x}$

Sampling distribution of $\overline{X}$

$$\sigma\,\overline{X} = \frac{\sigma}{\sqrt{n}}$$

Form: Approximately normal

$$\mu\overline{X} = \mu$$

tion. The following definitions are now given for an estimator and an estimate.

Estimator

Definition 9.1

Estimator of a population parameter

An *estimator* of a population parameter is a function of the sample random variables X_1, X_2, . . . , X_n. It is a random variable with a sampling distribution.

Estimate

Definition 9.2

Estimate of a population parameter

An *estimate* of a population parameter is a value of the estimator of that population parameter. It has a numerical value calculated from the sample values x_1, x_2, . . . , x_n.

We now return to the idea of selecting a good estimator (function of the sample random variables X_1, X_2, . . . , X_n) to estimate a population parameter. The first measure of the "goodness" of an estimator is *bias*. It is desirable that an estimator *not* be biased.

Bias of an estimator

Definition 9.3

Bias of an estimator of a population parameter

The *bias* of an estimator is the difference between the expected value of the estimator and the value of the population parameter. An estimator is *unbiased* if this difference is zero.

Figure 9.5 illustrates a biased estimator and an unbiased estimator of a population parameter θ. The first estimator of θ is unbiased, since its expected value (center of gravity of its sampling distribution) is equal to the true value of the population parameter θ. The second estimator of θ is

FIGURE 9.5 | A biased and an unbiased estimator of a population parameter θ

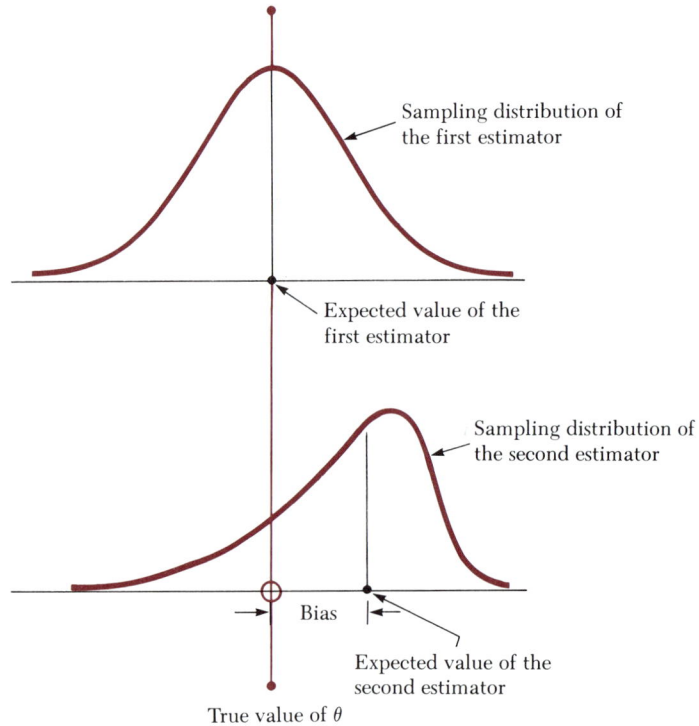

biased, since its expected value is not equal to the true value of θ. The *amount* of the bias is the difference between the expected value of the estimator and the true value of θ.

Example 9.1 Is $\overline{\mathbf{X}}$, the average of the sample random variables $\mathbf{X}_1$, $\mathbf{X}_2$, . . . , $\mathbf{X}_n$, an unbiased estimator of μ, the population mean?

Solution Yes, since from Chapter 8 on the sampling distribution of $\overline{\mathbf{X}}$, we know that $E(\overline{\mathbf{X}}) = \mu$. Thus the expected value of $\overline{\mathbf{X}}$ is equal to the true value of μ, and $\overline{\mathbf{X}}$ is therefore an unbiased *estimator* of μ.

In addition to whether or not an estimator is unbiased, there are other desirable properties of estimators that are important. For example, consider the sampling distributions illustrated in Figure 9.6. If we had to choose between these two unbiased estimators, then the second would be preferable since its sampling distribution is less variable. Because of the smaller variation in the sampling distribution of the second estimator, the chances are higher for getting a good estimate of θ (an estimate closer to the value of θ) from the one sample selected to be better than with the first estimator.

FIGURE 9.6 | Two unbiased estimators of a population parameter θ with different sampling distribution variances

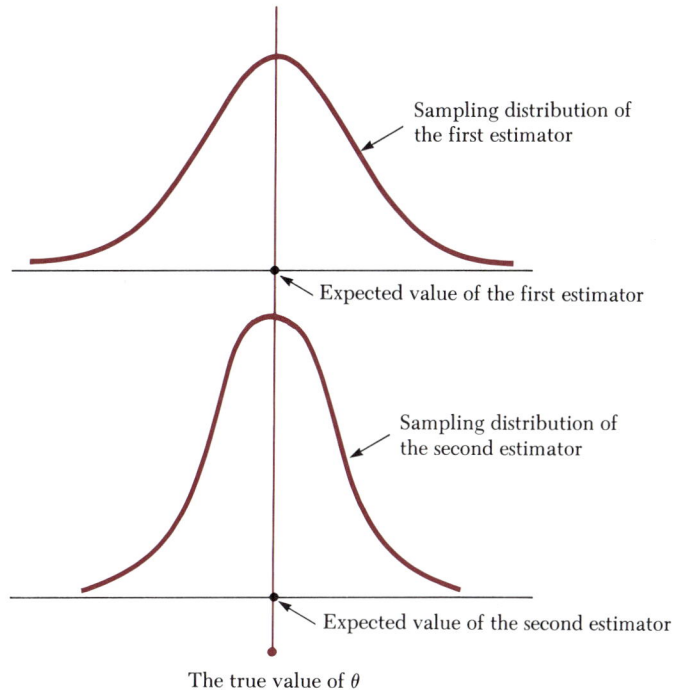

In some cases, the decision to select the best estimator from among several alternatives is not clear-cut. For example, consider the two estimators whose sampling distributions are illustrated in Figure 9.7. The first estimator is *biased* but has less variation in its sampling distribution than does the second estimator, which is unbiased. In this case, one might be inclined to use the first estimator of θ even though it is biased, since its value in the one sample selected may be closer to the value of θ than the value of the second estimator as illustrated in Figure 9.7.

In recent years, statisticians have developed estimators that may be biased but have desirable properties that compensate for their biasedness. An important measure in this regard is the *mean squared error* of an estimator.

The mean squared error measure "balances" off the variability of an estimator with its amount of squared bias. One estimator of a population parameter is said to be better than another by this measure if its *MSE is less than* that of the other estimator. In Figure 9.7, the first estimator would probably be a better estimator in mean squared error than the second estimator.

Definition 9.4
Mean squared error of an estimator

Mean squared error

The *mean squared error* of an estimator, denoted by *MSE*, is given by

$$MSE = E \text{ (estimator} - \text{parameter)}^2$$
$$= \text{(variance of estimator)} + \text{(its bias)}^2$$

FIGURE 9.7 | Two estimators of a population parameter θ with different bias and variance properties

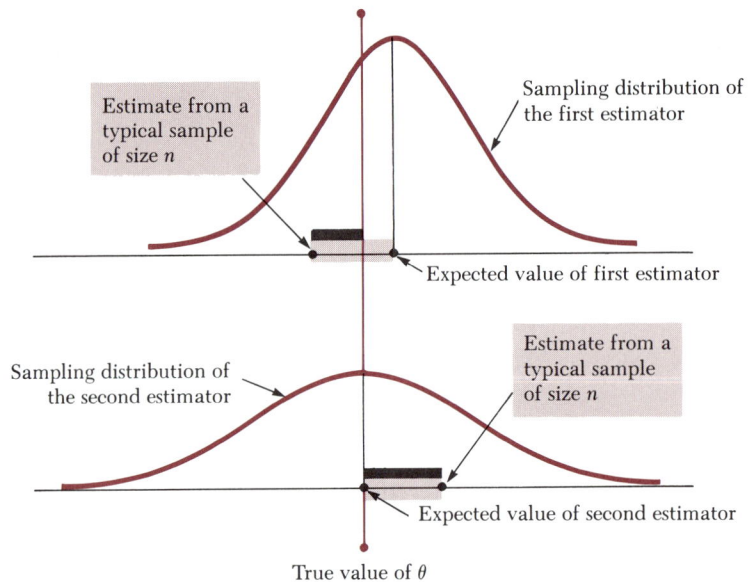

Estimate from a typical sample of size n

Sampling distribution of the first estimator

Expected value of first estimator

Sampling distribution of the second estimator

Estimate from a typical sample of size n

Expected value of second estimator

True value of θ

We will be concerned about estimating three population parameters in this chapter: the population mean (μ), the population proportion (π), and the population standard deviation (σ). The point estimators used in this text are considered "best" for most applications in terms of least or no bias, minimum mean squared error, and convenient functional form. For a discussion of alternate estimators and other criteria for selecting an estimator, see the Wonnacott and Wonnacott reference at the end of the chapter.

9.2.2 | Interval estimation

The idea of interval estimation is to use two functions of the sample random variables $X_1, X_2, \ldots, X_n$—denoted by $f_1(X_1, X_2, \ldots, X_n)$ and $f_2(X_1, X_2, \ldots, X_n)$—which will produce two points on the real line that define a *random interval* that will contain the value of the population parameter with a specified probability. Figure 9.8 illustrates this process when

FIGURE 9.8 | Interval estimator of the population mean μ

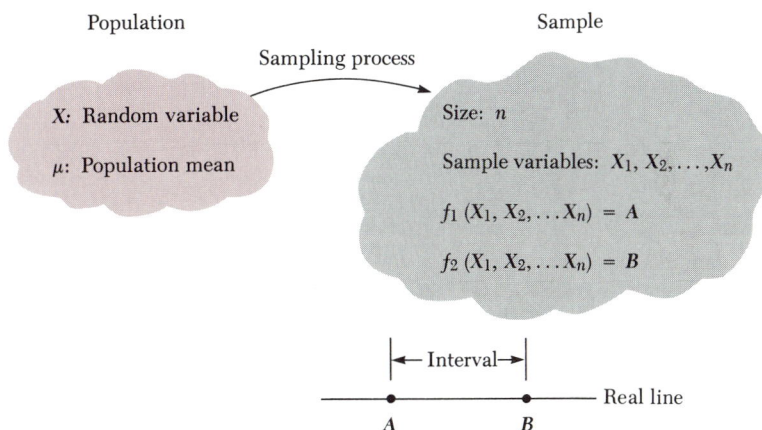

Population — Sampling process — Sample

X: Random variable

μ: Population mean

Size: n

Sample variables: $X_1, X_2, \ldots, X_n$

$f_1(X_1, X_2, \ldots X_n) = A$

$f_2(X_1, X_2, \ldots X_n) = B$

$\longleftarrow$ Interval $\longrightarrow$

Real line — A — B

constructing an interval estimator of the population mean μ. Two statistics (random variables) **A** and **B** are used to construct an interval on the real line that will contain the value of μ with the specified probability (usually, 0.90, 0.95, or 0.99). The interval estimator is often called a *confidence interval*—the use of the word *confidence* in this context will be explained in Section 9.4.

As with point estimators, there are "good" interval estimators. A good interval estimator is one in which the functions $f_1(X_1, X_2, \ldots, X_n)$ and $f_2(X_1, X_2, \ldots, X_n)$ produce the shortest interval **(A, B)** for a specified probability and specified sample size.

■ 9.3
Point estimation

9.3.1 | Population mean μ

The statistic $\overline{X}$, the sample mean, is used as the point estimator of μ. It is an unbiased estimator ($E(\overline{X}) = \mu$); and among all unbiased estimators of μ, it has the minimum variance for a specified sample size. Chapter 8 (Theorem

Standard error of an estimator is another term for the standard deviation of an estimator

8.1) said the standard deviation of $\overline{X}$ is given by $\sigma_{\overline{X}} = \sigma/\sqrt{n}$ (assuming that the finite population correction factor is not needed). In the context of point estimation, the standard deviation of $\overline{X}$ is referred to as the *standard error* of the estimator. Use of the standard error of the estimator is discussed in Section 9.4.

Point estimator of the population mean

Definition 9.5
Point estimator of the population mean μ and its standard error

The *point estimator of the population mean μ* is given by

$$\overline{X} = \sum_{i=1}^{n} X_i/n$$

where the random variables $X_1, X_2, \ldots, X_n$ represent a random sample of size n taken from the population.

The *standard error of the estimator $\overline{X}$* is given by

$$\sigma_{\overline{X}} = \sigma/\sqrt{n}$$

If the value of the population standard deviation (σ) is unknown, then the estimated standard error of the estimator $\overline{X}$ is given by

$$S_{\overline{X}} = S/\sqrt{n}$$

where S is the sample standard deviation whose value is given by

$$s = \sqrt{\frac{\sum_{i=1}^{n}(x_i - \overline{x})^2}{n-1}}$$

Note that when the standard deviation σ is unknown, we substitute the sample standard deviation S into the formula for the standard error of $\overline{X}$. (The reason for using S to substitute for σ is given in Section 9.3.3.)

Example 9.2 The NAC Chip Corporation produces an electronic component used in microcomputers. To fulfill government regulations, NAC must estimate the average lifetime in hours before failure of this component. The research director takes a random sample of 25 of these components, operates each until failure, and records the number of hours until each fails. These data are:

2,106	1,456	3,200	3,016	1,498
2,050	2,125	1,658	2,832	1,716
1,857	2,219	2,547	2,647	2,108
2,417	1,811	2,716	2,438	2,450
2,788	1,985	2,284	1,912	2,619

Determine a point estimate of the population mean lifetime of all produced components μ and its estimated standard error of the estimate.

Solution The value (estimate) of the estimator ($\overline{X}$) of μ for the selected sample of size $n = 25$ is given by

$$\bar{x} = \frac{\sum_{i=1}^{25} x_i}{n} = \frac{(2106 + 2050 + 1857 + \cdots + 2619)}{25} = 2258.20 \text{ hours}$$

Since the population standard deviation σ is unknown, we must calculate s, the sample standard deviation estimate:

$$s = \sqrt{\frac{\sum_{i=1}^{25} (x_i - \bar{x})^2}{n-1}}$$

$$= \sqrt{\frac{(2106 - 2258.2)^2 + (2050 - 2258.2)^2 + \cdots + (2619 - 2258.2)^2}{25-1}}$$

$$= 469.04 \text{ hours}$$

The estimated standard error of the estimate is given by

$$s_{\overline{X}} = s/\sqrt{n} = 469.04/\sqrt{25} = 93.81 \text{ hours}$$

Therefore, the estimate of μ is $\bar{x} = 2258.20$ hours, and the estimated standard error is 93.81 hours. Notice that the maximum number of hours that a component lasted in the sample was 3,200 hours, which is equal to 133.33 24-hour days! Thus it has taken the manufacturer more than four months to conduct the experiment and to gather the data. Getting the answers to questions by statistical sampling can take time!

Example 9.3 The administrator of the emergency room at All Saints' Hospital is concerned by the large number of admissions to the emergency room on Saturday evenings between the hours of 8 P.M. and 12 midnight. Typically, there are more admissions than the staff can accommodate during these hours. In an effort to request additional staff coverage during these hours, the administrator decides to take a random sample of 20 Saturday evenings and estimate the average number of admissions during the four-hour period from 8 P.M. to 12 midnight. Her hope is that she may use this estimate as evidence for the need to hire additional staff. The data (number of admissions) are:

| 46 | 22 | 34 | 41 | 52 | 19 | 28 | 40 | 36 | 22 |
| 31 | 35 | 42 | 36 | 44 | 27 | 34 | 36 | 41 | 48 |

Determine the point estimate of μ, the population mean number of admissions between 8 P.M. and midnight on Saturdays, and compute the estimated standard error of the estimate.

Solution   The estimate of μ is given by,

$$\bar{x} = \frac{\sum_{i=1}^{20} x_i}{n} = \frac{46 + 22 + 34 + \cdots + 48}{20} = 35.70 \text{ admissions}$$

Since the population standard deviation σ is unknown, we must find s, the sample standard deviation:

$$s = \sqrt{\frac{\sum_{i=1}^{20} (x_i - \bar{x})^2}{n - 1}}$$

$$= \sqrt{\frac{(46 - 35.70)^2 + (22 - 35.70)^2 + \cdots + (48 - 35.70)^2}{20 - 1}}$$

$$= 8.96 \text{ admissions}$$

Thus the estimated standard error is:

$$s_{\bar{x}} = s/\sqrt{n} = 8.96/\sqrt{20} = 2.00 \text{ admissions}$$

The estimate of μ, the average number of admissions on Saturday evenings between 8 P.M. and 12 midnight is $\bar{x} = 35.70$ admissions with an estimated standard error of 2.00 admissions.

9.3.2 | Population proportion π

The estimator of the population proportion π is the sample proportion $\mathbf{P}$, where $\mathbf{P} = \mathbf{X}/n$, $\mathbf{X} = $ Number of successes in n trials, where each trial produces either a success or a failure (Bernoulli experiment). From Theorem 8.5 in Chapter 8, we know that $E(\mathbf{P}) = \pi$. Thus this estimator is unbiased. Also, among all unbiased estimators of π, it has the minimum variance for a specified sample size. Further, from Theorem 8.5, we know that the standard deviation of $\mathbf{P}$ is given by $\sqrt{\pi(1 - \pi)/n}$. Thus the *standard error* of the estimator is $\sqrt{\pi(1 - \pi)/n}$. Since π is unknown, we must estimate the standard error of $\mathbf{P}$ by using the expression $\sqrt{\mathbf{P}(1 - \mathbf{P})/n}$.[1]

Example 9.4   An advertising firm is interested in the effectiveness of a 30-second advertisement designed for television. A random sample is selected of 100 people who agree to watch a two-hour television tape of TV shows with advertisements shown at the usual times during each hour and at the usual rate. The particular advertisement of interest is randomly imbedded among all advertisements for each showing of the tape to each of the 100 people. After the showing of the tape, the representative of the firm questions the people individually to determine whether or not they can recall a specified minimum number of details about the advertisement. Of the 100 people surveyed, only 18 can recall the minimum number of details (26

---

[1] An unbiased estimator of $\sqrt{\pi(1 - \pi)/n}$ is given by $\sqrt{\mathbf{P}(1 - \mathbf{P})/(n - 1)}$. Since n will usually be 25 or more for problems dealing with the population proportion, we will use a divisor of n rather than $n - 1$. The amount of error introduced by using n rather than $n - 1$ is negligible.

Definition 9.6

Point estimator of the population proportion π and its standard error

The *point estimator of the population proportion* π is given by

$$\mathbf{P} = \mathbf{X}/n$$

where $\mathbf{X}$ = Number of successes in n Bernoulli trials producing successes or failures.

The *standard error of the estimator* $\mathbf{P}$ is given by

$$\sigma_\mathbf{P} = \sqrt{\frac{\pi(1 - \pi)}{n}}$$

The *estimated standard error of the estimator* $\mathbf{P}$ is given by

$$S_\mathbf{P} = \sqrt{\frac{\mathbf{P}(1 - \mathbf{P})}{n}}.$$

---

people cannot recall the advertisement at all!). Estimate the population proportion π of all people seeing this ad who recall the minimum number of specific details, and find the estimate of the estimated standard error.

Solution   The point estimate of π based on this sample of 100 persons is:

$$p = x/n = 18/100 = 0.18$$

The estimate of the estimated standard error of $\mathbf{P}$ is:

$$s_\mathbf{P} = \sqrt{\frac{p(1 - p)}{n}} = \sqrt{\frac{(.18)(.82)}{100}} = 0.0384$$

Thus the point estimate of π is $p = 0.18$ (or 18 percent), and the estimated standard error is 0.0384.

Example 9.5   In Example 9.2, estimate the proportion of the components that will last more than 3,000 hours. Also find the estimated standard error.

Solution   Twenty-five components were sampled by the research director. The number of hours for those components that lasted more than 3,000 hours are 3,200 and 3,016. Thus only 2 of the 25 components lasted more than 3,000 hours. Calling the outcome "Lasted more than 3,000 hours" a success, we have

$$p = \frac{x}{n} = \frac{2}{25} = 0.08$$

The estimated standard error is:

$$s_\mathbf{P} = \sqrt{\frac{p(1 - p)}{n}} = \sqrt{\frac{(0.08)(0.92)}{25}} = 0.0543$$

9.3.3 | Population variance σ^2 and standard deviation σ

The estimator of the population variance σ^2 is the sample variance statistic $\mathbf{S}^2$, with a value in a specific sample of n values $x_1, x_2, \ldots, x_n$ given by:

$$s^2 = \frac{\sum_{i=1}^{n} (x_i - \bar{x})^2}{n - 1}$$

The sample variance $\mathbf{S}^2$ is an unbiased estimator of σ^2 $[E(\mathbf{S}^2) = \sigma^2]$, and the sampling distribution of $\mathbf{S}^2$ exhibits less variability than does any other unbiased estimator of σ^2.

The fact that $\mathbf{S}^2$ is unbiased means that if we took many repeated samples of size n from the population, the values of $\mathbf{S}^2$ computed in each sample will average out to σ^2. This is the reason the sum of the squared deviations in the formula for s^2 is divided by $(n - 1)$ instead of n. If the divisor were n instead, the expected value of the statistic,

$$\frac{\sum_{i=1}^{n} (\mathbf{X}_i - \bar{\mathbf{X}})^2}{n}$$

would not be σ^2—this estimator would be biased.

The estimator of the population standard deviation σ most frequently used is $\mathbf{S}$, the sample standard deviation statistic that is the square root of the variance statistic $\mathbf{S}^2$. The statistic $\mathbf{S}$ is a slightly biased estimator of $\sigma[E(\mathbf{S}) \neq \sigma]$. The amount of the bias decreases, however, as the sample size increases, which is comforting to know whenever we use a biased estimator. The reason the biased estimator $\mathbf{S}$ is used to estimate the population standard deviation σ is its convenient form; it is simply the square root of the statistic $\mathbf{S}^2$. Due to the complex forms of the standard errors of the estimators of σ^2 and σ, we will not give these formulas here.

Example 9.6 In Example 9.2, we needed to estimate σ in order to determine the estimated standard error of $\bar{\mathbf{X}}$. Using the data in Example 9.2, estimate the population variance σ^2 and the population standard deviation σ.

Solution For this sample of 25 lifetimes, these estimates are given by:

$$
\begin{aligned}
s^2 &= \frac{\sum_{i=1}^{n} (x_i - \bar{x})^2}{n - 1} \\
&= \frac{(2,106 - 2,258.2)^2 + (2,050 - 2,258.2)^2 + \cdots + (2,619 - 2,258.2)^2}{24} \\
&= \boxed{219,998.52 \text{ hours}^2}
\end{aligned}
$$

$$s = \sqrt{s^2} = \sqrt{219,998.52} = \boxed{469.04 \text{ hours}}$$

Table 9.1 summarizes the estimates of the population mean μ, the population proportion π, the population variance σ^2, and the population standard deviation σ.

Definition 9.7

Point estimators of the population variance σ^2 and the population standard deviation σ

The *point estimator of the population variance σ^2 is given by*

$$\mathbf{S}^2 = \frac{\sum\limits_{i=1}^{n} (\mathbf{X}_i - \overline{\mathbf{X}})^2}{n - 1}$$

where the random variables $\mathbf{X}_1, \mathbf{X}_2, \ldots, \mathbf{X}_n$ represent a random sample of size n with a sample mean of $\overline{\mathbf{X}}$.

The *point estimator of the population standard deviation σ is given by*

$$\mathbf{S} = \sqrt{\frac{\sum\limits_{i=1}^{n} (\mathbf{X}_i - \overline{\mathbf{X}})^2}{n - 1}}$$

TABLE 9.1 | Point estimates of μ, π, σ^2, and σ

Population parameter	Point estimate	Estimate of the estimated standard error
μ (mean)	$\bar{x} = \dfrac{\sum\limits_{i=1}^{n} x_i}{n}$ (sample mean)	$s/\sqrt{n}$
π (proportion)	$p = \dfrac{x}{n} = \dfrac{\text{Number of successes}}{\text{Number of trials}}$ (sample proportion)	$\sqrt{\dfrac{p(1 - p)}{n}}$
σ^2 (variance)	$s^2 = \dfrac{\sum\limits_{i=1}^{n} (x_i - \bar{x})^2}{n - 1}$ (sample variance)	*
σ (standard deviation)	$s = \sqrt{\dfrac{\sum\limits_{i=1}^{n} (x_i - \bar{x})^2}{n - 1}}$ (sample standard deviation)	*

* Not given due to the complexity of the formulas.

9.4.1 | Population mean μ

An interval estimator of μ may be developed from the sampling distribution of the statistic $\overline{X}$. If we assume that $\overline{X}$ is normally distributed, then 95 percent of the values of $\overline{X}$ will be contained within two standard deviations—more correctly, 1.96 standard deviations—of the mean of the sampling distribution of the statistic $\overline{X}$, as illustrated in Figure 9.9. In terms of a

FIGURE 9.9 | Sampling distribution of $\overline{X}$

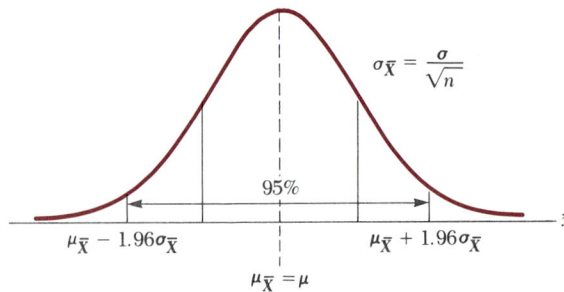

probability statement, the probability that a value of $\overline{X}$ will fall in this interval may be expressed as

$$P(\mu - 1.96\sigma_{\overline{x}} \le \overline{X} \le \mu + 1.96\sigma_{\overline{x}}) = 0.95$$

This probability statement may be "solved" for μ, producing the probability statement:

$$P(\overline{X} - 1.96\sigma_{\overline{x}} \le \mu \le \overline{X} + 1.96\sigma_{\overline{x}}) = 0.95$$

That is, the interval $\overline{X} \pm 1.96\sigma_{\overline{x}}$ will contain the value of μ with a probability of 0.95. The derivation of this probability statement is given in the appendix at the end of this chapter. Note that since $\sigma_{\overline{x}} = \sigma/\sqrt{n}$, this statement assumes that the population standard deviation σ is known. If this is not the case, then the statement does not apply. We will deal with the case when σ is unknown in Section 9.5.

If we compute the lower bound $\overline{X} - 1.96\sigma_{\overline{x}}$ and the upper bound $\overline{X} + 1.96\sigma_{\overline{x}}$ on μ in repeated samples of size n as illustrated in Figure 9.10, then on average in 95 percent of the samples the computed interval would contain the value of μ, whereas on average the intervals computed in 5 percent of the samples would not contain μ.

The interval $\bar{x} \pm 1.96\sigma_{\overline{x}}$ is called a *confidence interval* with "confidence" 95 percent. The confidence of 95 percent means that 95 percent of all samples of size n drawn from the population will produce intervals $\bar{x} \pm 1.96\sigma_{\overline{x}}$

FIGURE 9.10 | Interval estimates computed in repeated samples of size n

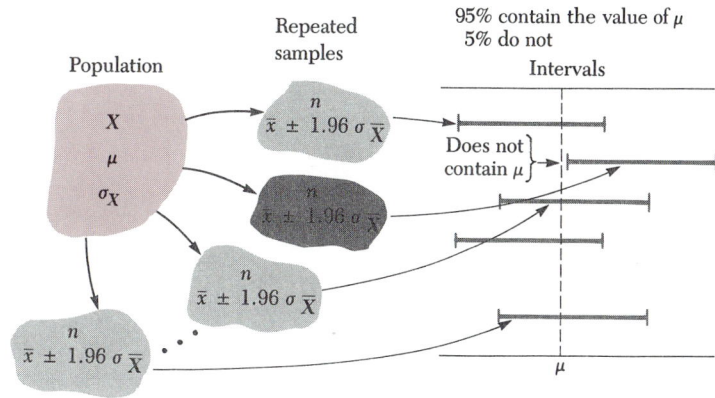

that contain the value of μ, and 5 percent of the samples will not. We refer to it as "confidence" because in practice we will be drawing only one sample, and our computed interval $\bar{x} \pm 1.96\sigma_{\bar{X}}$ either will contain the value of μ or will not contain the value of μ with probability one (with certainty). The 95 percent confidence measures the risk involved in the sampling process—we have a chance of 0.95 of securing one of the samples that will produce an interval containing the value of μ.

It should be apparent that we can specify whatever degree of confidence is desired for the confidence interval of μ. If we wish the confidence to be 90 percent, then we must move 1.64 standard deviations above and below the mean μ in the sampling distribution of $\bar{X}$. The 90 percent confidence interval is, therefore, $\bar{x} \pm 1.64\sigma_{\bar{X}}$.

We will now describe a method for constructing a confidence interval for μ with any specified degree of confidence. Define the *confidence coefficient* $(1 - \alpha)$ to be the desired confidence of the interval, measured as a proportion (0.95, 0.90, etc.), and define $z_{\alpha/2}$ to be the value in the standard normal table (Table B.3 in Appendix B) such that the area to the right of $z_{\alpha/2}$ is $\alpha/2$, as shown in Figure 9.11. Then the constructed confidence interval with confidence coefficient $(1 - \alpha)$ is given by

$$\bar{x} \pm z_{\alpha/2}\sigma_{\bar{X}}$$

Example 9.7 A manufacturer of fishing lines produces 10-pound lines—lines that on the average will not break if the weight of the fish on the line is 10 pounds or less. A consumer testing group randomly samples 25 of these lines and determines the number of pounds for each that the line will hold before breaking. The manufacturer claims that the standard deviation of the breaking weights is 0.5 pound. Assuming that the testing group accepts this

FIGURE 9.11 | Calculation of $z_{\alpha/2}$

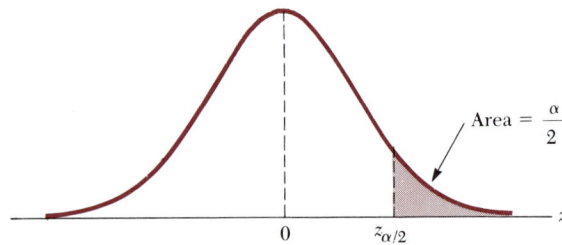

standard deviation as the population standard deviation σ of all 10-pound lines produced by this manufacturer, use the sample data below to construct a 99 percent confidence interval on the mean breaking weight μ for these lines.

12.0	10.1	10.2	11.6	9.9
9.8	10.0	10.5	10.9	10.2
10.2	11.4	11.0	10.7	10.2
10.4	10.6	10.5	10.8	10.6
11.1	10.6	10.2	10.1	10.4

Solution The point estimate of μ is:

$$\bar{x} = \frac{\sum_{i=1}^{25} x_i}{25} = \frac{12.0 + 9.8 + 10.2 + \cdots + 10.4}{25} = 10.56 \text{ pounds}$$

The confidence interval formula for μ is $\bar{x} \pm z_{\alpha/2}\, \sigma_{\bar{x}}$, where $\sigma_{\bar{x}} = \sigma/\sqrt{n}$. We must first determine the value of $z_{\alpha/2}$ for a 99 percent confidence statement. The confidence coefficient is $0.99 = (1 - \alpha)$, so that $\alpha = 0.01$ and $\alpha/2 = 0.005$. Thus we wish to find the value of Z on the standard normal distribution such that the area to the right of this value is 0.005 as shown in Figure 9.12.

From Table B.3 in Appendix B, the value of $z_{\alpha/2} = z_{0.005}$ is 2.58. Thus the 99 percent confidence interval is

$$\bar{x} \pm z_{0.005} \frac{\sigma}{\sqrt{n}} = 10.56 \pm 2.58 \frac{0.50}{\sqrt{25}} = 10.56 \pm (2.58)(0.10)$$
$$= 10.56 \pm 0.258$$

The 99 percent confidence interval is $(10.56 - 0.258)$ to $(10.56 + 0.258)$. The resulting interval is 10.302 pounds to 10.818 pounds.

The most frequently chosen levels of confidence for interval estimates are 0.90, 0.95, and 0.99. In Table 9.2, $z_{\alpha/2}$ and intervals for these levels of confidence are given.

FIGURE   9.12 | Computation of $z_{\alpha/2}$ for a 99 percent confidence interval

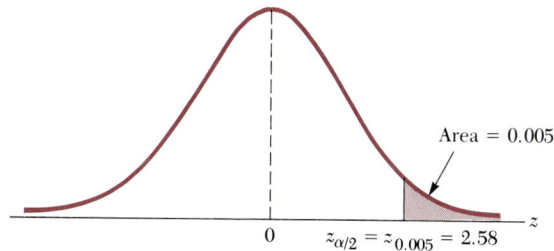

Area = 0.005

$$z_{\alpha/2} = z_{0.005} = 2.58$$

TABLE   9.2 | Confidence intervals for μ

Confidence coefficient	$z_{\alpha/2}$	Lower bound	Upper bound
0.90	1.64	$\bar{x} - 1.64\sigma/\sqrt{n}$	$\bar{x} + 1.64\sigma/\sqrt{n}$
0.95	1.96	$\bar{x} - 1.96\sigma/\sqrt{n}$	$\bar{x} + 1.96\sigma/\sqrt{n}$
0.99	2.58	$\bar{x} - 2.58\sigma/\sqrt{n}$	$\bar{x} + 2.58\sigma/\sqrt{n}$

The width of the confidence interval for estimating μ is $2z_{\alpha/2}\sigma/\sqrt{n}$, as shown in Figure 9.13. There are two options for decreasing the width of this interval: (1) by decreasing the confidence coefficient, which decreases $z_{\alpha/2}$—see Table 9.2—or (2) by increasing the sample size.

FIGURE   9.13 | Width of the interval estimate of μ

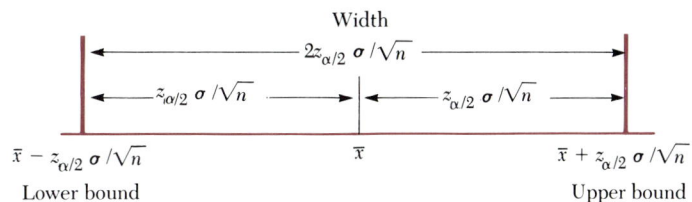

Width
$2z_{\alpha/2}\,\sigma/\sqrt{n}$

$z_{\alpha/2}\,\sigma/\sqrt{n}$ $z_{\alpha/2}\,\sigma/\sqrt{n}$

$\bar{x} - z_{\alpha/2}\,\sigma/\sqrt{n}$ $\bar{x}$ $\bar{x} + z_{\alpha/2}\,\sigma/\sqrt{n}$

Lower bound   Upper bound

That the width of the interval is a function of the sample size has intuitive appeal. As the sample size increases, the width of the confidence interval decreases and, taken to the limit as $n \to \infty$, the width of the interval approaches zero. But the center of the interval is the sample mean $\bar{x}$, which approaches μ as n becomes larger. Therefore, as n becomes large, the confidence interval for μ converges on the value of μ.

From Table 9.2 it should also be clear that a 100 percent confidence interval is not very meaningful. As the confidence coefficient increases, so does the value of $z_{\alpha/2}$ and thus so does the width of the interval. In fact, if we request a 100 percent interval, the answer is the entire real line from $-\infty$ to $+\infty$, which will contain the value of μ with certainty. In practice, of course, the interval would only have to extend over the entire range of values of the population variable.

Notice from Table 9.2 that as the confidence coefficient increases toward 1.00, the width of the confidence interval increases ($z_{\alpha/2}$ becomes larger). As the confidence level decreases, the width of the confidence interval decreases ($z_{\alpha/2}$ becomes smaller).

The formula for the confidence interval for the population mean μ is now summarized.

Interval estimate of the population mean, *population variance σ^2 known*

Formula 9.1

Interval estimate of the population mean μ

If a sample of n values $x_1, x_2, \ldots, x_n$ has been randomly drawn from a population whose random variable **X** is normally distributed, a confidence interval with confidence coefficient $(1 - \alpha)$ for the population mean μ is given by

$$\bar{x} \pm z_{\alpha/2} \frac{\sigma}{\sqrt{n}}$$

where $\bar{x}$ is the sample mean, σ is the population standard deviation, and $z_{\alpha/2}$ is the value on the standard normal distribution such that the area under the normal curve to the right of it is $\alpha/2$.

If the population random variable **X** is not normally distributed, then this formula will provide an approximate confidence interval when n is large ($n \geq$ 25 in most cases).

If the population standard deviation σ is not known, then a different formula for determining the confidence interval for μ must be used. It is described in Section 9.5.

A confidence interval is often used to give an indication of the *accuracy* or *reliability* of a point estimate. For example, in Example 9.7, the point estimate of the mean breaking weight of the 10-pound fishing lines was $\bar{x} = 10.56$ with a 99 percent confidence interval estimate of 10.302 to 10.818. Since the width of this interval is reasonably narrow ($10.818 - 10.302 = 0.516$), we can be reasonably assured that our point estimate of μ is a good estimate.

Notice that in the confidence interval formula for μ,

$$\bar{x} \pm z_{\alpha/2}\sigma/\sqrt{n}$$

the form of the interval estimate is the point estimate plus or minus the z-value times the *standard error* of the point estimate (see Definition 9.5). Often, the point estimate is given with its standard error so that the standard error can be used to give some idea of the reliability of the estimate. We might say, for example, that the estimate of μ in Example 9.7 is $\bar{x} = 10.56$ pounds *with a standard error of 0.10 pound* ($\sigma/\sqrt{n} = 0.50/\sqrt{25} = 0.10$). It is usually more informative to give the point estimate *and* a confidence interval estimate, however. So we would prefer to say that the point estimate of μ in Example 9.7 is $\bar{x} = 10.56$ pounds, and a 99 percent interval estimate is 10.302 pounds to 10.818 pounds.

9.4.2 | Population proportion π

An approximate confidence interval estimate of the population proportion π can be constructed by appealing to the central limit theorem. Recall from Chapter 8 that the sample proportion **P** will be approximately normally distributed for a sufficiently large sample size n. The rule given in Chapter 8 for the sample size n to be "sufficiently large" was based on the satisfaction of the two inequalities:

$$n\pi \geq 5 \quad \text{and} \quad n\pi(1 - \pi) \geq 5$$

If these two inequalities are satisfied for the sample size n, then the central limit theorem provides a theoretical basis for the following $100(1 - \alpha)$ percent confidence interval on π:

$$\mathbf{P} \pm z_{\alpha/2}\sigma_{\mathbf{P}}$$

From Section 9.3.2, the point estimator of π is the statistic **P**, and the standard error of the estimator **P** is given by

$$\sigma_{\mathbf{P}} = \sqrt{\frac{\pi(1 - \pi)}{n}}$$

Since the standard error of the estimator **P** depends on the unknown value of π, it is often estimated by using p in place of π, which produces this formula:

$$s_{\mathbf{P}} = \sqrt{\frac{p(1 - p)}{n}}$$

The approximate interval estimate is given by

$$p \pm z_{\alpha/2}\sqrt{\frac{p(1 - p)}{n}}$$

Interval estimate of
the population
proportion

Formula 9.2
Interval estimate of the population
proportion π for large samples ($n \geq 50$)

An approximate interval estimate of the population proportion π is given by

$$p \pm z_{\alpha/2} \sqrt{\frac{p(1-p)}{n}}$$

where p is the sample proportion of successes in n trials of a binomial experiment, and $z_{\alpha/2}$ is the value on the standard normal distribution such that $P(\mathbf{Z} \geq z_{\alpha/2}) = \alpha/2$.

From the inequalities $n\pi \geq 5$ and $n(1 - \pi) \geq 5$, assuming that $n \geq 50$ is a "sufficiently large" sample size requires that $0.10 \leq \pi \leq 0.90$. Since π is unknown, we can use approximate inequalities

$$np \geq 5 \quad \text{and} \quad n(1 - p) \geq 5$$

to check the appropriateness of using the above confidence interval formula. If a given sample size n and sample proportion p produce values np and $n(1 - p)$ that are close to the limit of 5, then caution should be exercised in making statements based on the confidence interval computed from the above formula. If either inequality is not satisfied, then other methods should be used (see the Ostle & Mensing reference at the end of this chapter, for example).

Example 9.8 The Harris Survey Company is interested in the proportion of residents in a certain city who favor a bond to pay for the construction of a new freeway in the city. A random sample of 100 residents is taken, and it is found that 60 of the 100 surveyed individuals favor the construction of the freeway financed by the bond. Construct a 95 percent confidence interval on the proportion π of all residents in the city who favor a bond to pay for the freeway construction.

Solution Since $p = 60/100 = 0.60$ and $n = 100$, $np = (100)(0.60) = 60$ and $n(1 - p) = (100)(0.40) = 40$. Therefore, it is appropriate to use the formula,

$$p \pm z_{\alpha/2} \sqrt{\frac{p(1-p)}{n}}$$

to set the confidence interval estimate on π; where $p = 60/100 = 0.60$ (the sample proportion) and $n = 100$. From either Table 9.2 or Table B.3 in Appendix B, the value of $z_{\alpha/2}$ for a 95 percent confidence interval is 1.96. Thus the interval estimate is

$$0.60 \pm 1.96 \sqrt{\frac{(0.60)(0.40)}{100}} = 0.60 \pm 1.96(0.049)$$

$$= 0.60 \pm 0.096$$

The 95 percent interval estimate is (0.60 − 0.096) to (0.60 + 0.096), or 0.504 to 0.696. Notice that since the interval is above 0.50, we can be reasonably confident that a majority of residents (greater than 50 percent) favor a bond to pay for the construction of the new freeway.

■ 9.5

Interval estimation of μ with an unknown population variance

In section 9.4.1, the formula for the interval estimate of μ is $\bar{x} \pm z_{\alpha/2}\sigma/\sqrt{n}$. To use this formula, we must know the value of the population standard deviation, σ. Typically, when we must estimate the population mean μ, the value of the population standard deviation will also be unknown. In this case, we must use an alternate formula for the interval estimator of μ.

The standardized statistic

$$\mathbf{Z} = \frac{\overline{\mathbf{X}} - \mu_{\bar{\mathbf{x}}}}{\sigma/\sqrt{n}}$$

is a standard normal variable provided that $\overline{\mathbf{X}}$ is normally distributed. If the value of the population standard deviation σ is not known, then we know from Definition 9.7 that we can estimate σ by using $\mathbf{S}$, the sample standard deviation statistic. If we substitute $\mathbf{S}$ for σ in the above standardized statistic, then the statistic is no longer a standard normal variable. Rather, the statistic

$$t = \frac{\overline{\mathbf{X}} - \mu_{\bar{\mathbf{x}}}}{\mathbf{S}/\sqrt{n}}$$

has a distribution known as the student's t-distribution if $\overline{\mathbf{X}}$ is normally distributed. The t-distribution was first reported by Gossett in 1908, who published the result under the pseudonym "Student."

The t-distribution is more "spread out" than the normal distribution, because by using $\mathbf{S}$ in place of σ, more uncertainty is introduced. The form of the distribution is a function of the sample size n. The fewer the sample values, the more spread out the distribution becomes, as indicated in Figure 9.14.

The distribution of t is tabulated according to its "degrees of freedom" rather than the sample size n. The degrees of freedom of the t-distribution are determined by the divisor of the estimated standard deviation in the t-statistic. In the t-statistic,

$$t = \frac{\overline{\mathbf{X}} - \mu_{\bar{\mathbf{x}}}}{\mathbf{S}/\sqrt{n}}$$

the sample standard deviation value $s = \sqrt{\sum_{i=1}^{n}(x_i - \bar{x})^2/(n - 1)}$ has $(n - 1)$ in its denominator under the radical. Hence, this t-statistic has $(n - 1)$ degrees of freedom (df).

The t-distribution is tabulated in Table B.5 in Appendix B. It is a symmetric distribution about the value $t = 0$, and Table B.5 gives the values of t corresponding to upper-tail areas of 0.10, 0.05, 0.025, 0.01, and 0.005. Notice that as the sample size increases, the t-values converge on the standard

FIGURE 9.14 | Comparison of the normal distribution with the
t-distribution for $n = 2$, 5, and 10

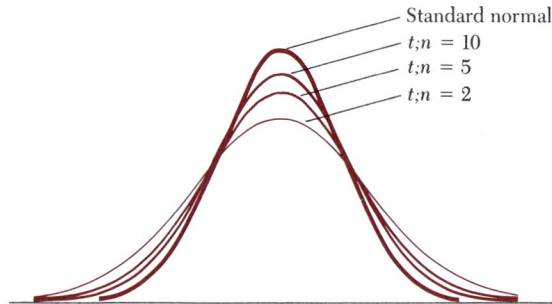

Standard normal
$t;n = 10$
$t;n = 5$
$t;n = 2$

TABLE 9.3 | Value of **t** corresponding to an upper-tail probability
of 0.025

df	Value of **t**
1	12.706
10	2.228
20	2.086
30	2.042
120	1.980
$+\infty$	1.960

normal values corresponding to the same tail areas. Table 9.3 shows the
convergence of the **t**-values for an upper-tail probability of 0.025. When the
sample size reaches $n = 31$ ($df = n - 1 = 30$), the **t**-value is quite close to
$z = 1.96$.

   If the population standard deviation σ is unknown and must be estimated,
the interval estimator of μ is changed by using a **t**-value instead of the
standard normal value $z_{\alpha/2}$. The interval estimate is:

$$\bar{x} \pm t_{\alpha/2;n-1} \frac{s}{\sqrt{n}}$$

where $t_{\alpha/2;n-1}$ is defined in Formula 9.3.

   Example 9.9   The length of time required for persons taking the civil
service PACE test is assumed to be normally distributed. A random sample
of 16 persons taking the test is conducted and their test times are recorded,
yielding an average test time of 60 minutes with a standard deviation of 12

Interval estimate of the population mean, *population variance σ^2 unknown*

Formula 9.3

Interval estimate of the population mean μ with unknown standard deviation σ

The confidence interval estimate of the population mean μ with a confidence coefficient of $(1 - \alpha)$, based on a random sample of n values $x_1, x_2, \ldots, x_n$ when the population standard deviation σ is unknown, is given by:

$$\bar{x} \pm t_{\alpha/2;n-1} \frac{s}{\sqrt{n}}$$

where $\bar{x}$ is the sample mean, s is the sample standard deviation given by

$$s = \sqrt{\frac{\sum_{i=1}^{n} (x_i - \bar{x})^2}{n - 1}}$$

n is the sample size, and $t_{\alpha/2;n-1}$ is the value on the t-distribution with $n - 1$ degrees of freedom (df) such that the area to the right of it is $\alpha/2$. An assumption for using this interval estimate is that the population random variable **X** is normally distributed, or very nearly so, if n is small (less than 25).

minutes. Find a 95 percent confidence interval for the population mean test time μ.

Solution   The population standard deviation is unknown and the population is normally distributed. Therefore the confidence interval formula using the t-distribution should be used:

$$\bar{x} \pm t_{\alpha/2;n-1} \frac{s}{\sqrt{n}}$$

For a 95 percent confidence interval, the confidence coefficient is $(1 - \alpha) = 0.95$, so that $\alpha = 0.05$ and $\alpha/2 = 0.025$. The degrees of freedom are $df = n - 1 = 16 - 1 = 15$. Thus the value of $t_{0.025;15}$ is 2.131. The required interval is

$$60 \pm 2.131 \frac{12}{\sqrt{16}} \quad \text{or} \quad 60 \pm 6.393$$

Thus we are 95 percent confident that the population mean test time μ is between 53.61 and 66.39 minutes.

If the standard normal distribution is used in place of the t-distribution, the value of $z_{\alpha/2}$ for a 95 percent interval is 1.96 and the resulting interval (from Section 9.4.1) is

$$60 \pm 1.96 \frac{12}{\sqrt{16}} \quad \text{or} \quad 60 \pm 5.88$$

This interval extends from 54.12 to 65.88 minutes, a shorter interval than the correct one determined above. The fact that the interval based on the *t*-distribution is wider than the one based on the standard normal distribution reflects that additional uncertainty has been introduced due to the estimation of σ by the sample standard deviation *s*.

An interesting difference between Formula 9.1 ($\bar{x} \pm z_{\alpha/2}\sigma/\sqrt{n}$) and Formula 9.3 ($\bar{x} \pm t_{\alpha/2;n-1}s/\sqrt{n}$) concerns the *width* of the confidence interval. Since Formula 9.3 uses the *estimated* standard error of $\bar{X}$, the width of the confidence interval will vary from sample to sample because the sample standard deviation *s* is used in place of the population standard deviation σ, and each sample will have a different sample standard deviation.

For instance, if we computed the 95 percent confidence interval estimate using Formula 9.3 in repeated samples of size *n*, then on the average in 95 percent of the samples the computed interval would contain the value of μ as illustrated in Figure 9.15. Compare Figures 9.15 and 9.10, where the 95

FIGURE 9.15 | Ninety-five percent interval estimates computed by Formula 9.3 in repeated samples of size *n*

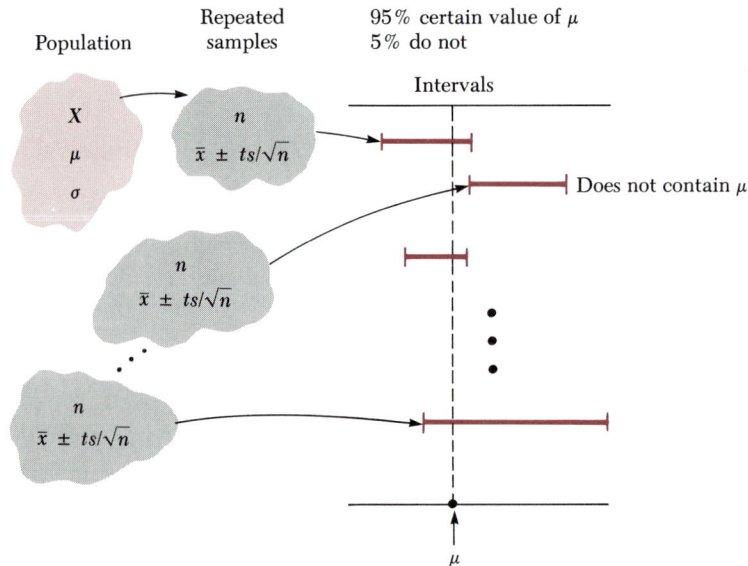

percent confidence intervals were computed using Formula 9.1. In both cases, on the average, Formula 9.1 and 9.3 will produce confidence intervals that contain μ in 95 percent of the repeated samples; but Formula 9.3 will produce intervals of different widths, since the estimated standard error of $\bar{X}(s/\sqrt{n})$ differs from sample to sample due to the sample standard deviation *s* value being different for each sample.

There are two confidence interval estimates for the population mean μ:

$$\text{Formula 9.1:} \quad \bar{x} \pm z_{\alpha/2} \frac{\sigma}{\sqrt{n}}$$

$$\text{Formula 9.3:} \quad \bar{x} \pm t_{\alpha/2;n-1} \frac{s}{\sqrt{n}}$$

Both formulas assume that $\overline{\mathbf{X}}$ is normally distributed. If this is not the case, then *neither* formula should be used unless $n \geq 25$. The central limit theorem guarantees that $\overline{\mathbf{X}}$ will be approximately normally distributed for sufficiently large n. Usually, when n is 25 or greater, the approximation is adequate to use Formula 9.1 or Formula 9.3 (whichever formula applies). It should be understood that when Formula 9.1 or Formula 9.3 is used under the assumption that $\overline{\mathbf{X}}$ is not normally distributed, *approximate* interval estimates of the population means μ result. The approximation improves as the sample size n increases.

The difference between Formula 9.1 and Formula 9.3 is based on whether the population standard deviation σ is known (Formula 9.1) or is not known (Formula 9.3). When Formula 9.3 is used, we must use Table B.5 in Appendix B to find $t_{\alpha/2;n-1}$. Notice that this table provides the values of **t** for degrees of freedom (df) from 1 to 100 [an abridged t-table is given on the inside back cover for $df = 1, 2, \ldots , 30, 40, 60,$ and 120]. Since in Formula 9.3, $df = n - 1$, what happens when the sample size n is greater than 100? Technically speaking, Formula 9.3 is the correct formula when the population standard deviation σ is unknown *for any sample size n*. But, what do we do for the **t**-value? When $n > 100$, the value of **t** will be very close to the value of **Z**. It is common, therefore, to substitute z for t in Formula 9.3 when $n > 100$. This produces the following formula:

Formula 9.4

Interval estimate of the population mean μ
with unknown standard deviation σ and sample size $n > 100$

$$\bar{x} \pm z_{\alpha/2} \frac{s}{\sqrt{n}}$$

where $\bar{x}$ is the sample mean, s is the sample standard deviation, $z_{\alpha/2}$ is the value on the z-distribution so that the area to the right of it is $\alpha/2$, and $n > 100$.

■ **9.6**

Confidence bounds on μ and π (advanced section)

In some instances, it is desirable to state either an upper or a lower bound on a population parameter rather than an interval. For example, we may wish to state that we are 95 percent confident that μ *does not exceed* a specified value. This is an example of setting an *upper confidence bound* on μ. The upper and lower confidence bounds are determined in a way similar to how the confidence interval is set. The difference is that we are giving only one end of the confidence interval, with the appropriate change in the determination of the **Z** or **t**-value in the formula for the bound. Since only one "end" of the confidence interval is given, we do not split α into $\alpha/2$ for the determination of **Z** or **t**. The error α is placed entirely in one end of the "interval" for a confidence bound. The formulas for the upper and lower confidence bounds on μ and π are given in Table 9.4.

TABLE 9.4 | Confidence bounds for μ and π

Parameter	Lower $100(1 - \alpha)$ percent confidence bound	Upper $100(1 - \alpha)$ percent confidence bound	Conditions
μ	$\bar{x} - z_\alpha \dfrac{\sigma}{\sqrt{n}}$	$\bar{x} + z_\alpha \dfrac{\sigma}{\sqrt{n}}$	σ is known, $n \geq 25$ unless **X** is normally distributed*
	$\bar{x} - t_{\alpha;n-1} \dfrac{s}{\sqrt{n}}$	$\bar{x} + t_{\alpha;n-1} \dfrac{s}{\sqrt{n}}$	σ is unknown, $n \geq 25$ unless **X** is normally distributed†
π	$p - z_\alpha \sqrt{\dfrac{p(1-p)}{n}}$	$p + z_\alpha \sqrt{\dfrac{p(1-p)}{n}}$	n is large (usually 100 or more)

Note: z_α is given by $P(\mathbf{Z} > z_\alpha) = \alpha$; $t_{\alpha;n-1}$ is given by $P(\mathbf{t} > t_{\alpha;n-1}) = \alpha$.
* If **X** is normally distributed, the formula applies to any sample size ($n \geq 1$).
† If **X** is normally distributed, the formula applies when $n > 1$.

Example 9.10 A small car rental agency is interested in purchasing a fleet of cars of a new diesel model sold by the WV automobile company. One condition for purchasing the fleet of cars of this model is that they produce *no less than* an average of 30 miles per gallon in city driving. To test whether this condition is met, the WV company provides the following data. In a recent test of 30 cars of this model in city driving, the cars produced an average of 32 miles per gallon with a standard deviation of 4 miles per gallon. Should the rental agency purchase the cars of this model?

Solution It is decided to determine a 99 percent lower confidence bound on μ, the population average miles per gallon for this car model. By doing so, it will be possible to make the statement: "We are 99 percent confident that the population average miles per gallon for this model is *not less than* the lower confidence bound." Since σ is unknown, the *t*-distribution should be used in setting the bound. The appropriate **t**-value with $(n - 1) = 29$ degrees of freedom is 2.462 from Table B.5 in Appendix B. The lower confidence bound is:

$$\bar{x} - t \cdot \frac{s}{\sqrt{n}} = 32 - 2.462 \cdot \frac{4}{\sqrt{30}} = 30.2$$

Therefore we are 99 percent confident that μ, the population average miles per gallon, is not less than 30.2. Accordingly, the rental agency should purchase the fleet of cars of this model.

■ 9.7
Confidence interval estimators of the population variance σ^2 and the population standard deviation σ (advanced section)

The confidence interval estimator of the population variance σ^2 is based on the sampling distribution of the statistic $\mathbf{S}^2$. The sampling distribution of the statistic $\mathbf{S}^2$ has been worked out by standardizing this random variable in a manner similar to the $\mathbf{Z}$ standardization. The standardized random variable

$$\chi^2 = \frac{(n-1)\mathbf{S}^2}{\sigma^2}$$

is called a *chi-square* random variable if the population random variable $\mathbf{X}$ is normally distributed. The sampling distribution of χ^2 is called the chi-square distribution. It is an asymmetric distribution skewed to the right and is defined for positive real numbers. The distribution of χ^2, like the t-distribution, depends on the sample size. Three members of the chi-square family of distributions are illustrated in Figure 9.16. As the sample size increases, the

FIGURE 9.16 | Three members of the chi-square family of distributions

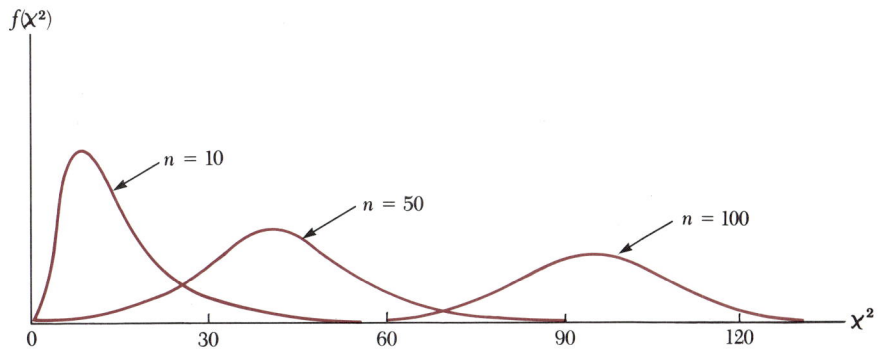

form of the chi-square distribution approaches the normal distribution. The $n = 100$ chi-square distribution is distinctly more like the normal distribution than is the $n = 10$ chi-square distribution.

The tabulated values of the chi-square distribution are based on the degrees of freedom, $df = n - 1$, as are the tabulated t-distribution values. Table B.6 in Appendix B gives the values of χ^2. The table is split into two parts—the left page gives the lower-tail values of χ^2, and the right page gives the upper-tail values of χ^2. Thus Table B.6 gives:

$$\text{Lower tail:} \quad P(\chi^2 \leq \chi^2) = \alpha$$
$$\text{Upper tail:} \quad P(\chi^2 \geq \chi^2) = \alpha$$

The chi-square distribution may be used to construct an interval estimator for σ^2 by solving the probability statement

$$P\left(\chi^2_{L;n-1} \leq \frac{(n-1)S^2}{\sigma^2} \leq \chi^2_{U;n-1}\right) = 1 - \alpha$$

for σ^2, where $\chi^2_{L;n-1}$ and $\chi^2_{U;n-1}$ are the lower and upper values on the χ^2 distribution with $(n-1)$ degrees of freedom which locate half of α in each tail. The resulting interval estimate is:

$$\frac{(n-1)s^2}{\chi^2_{U;n-1}} \leq \sigma^2 \leq \frac{(n-1)s^2}{\chi^2_{L;n-1}}$$

Interval estimate of the population variance

Formula 9.5

Interval estimate of the population variance σ^2

An interval estimate of the population variance σ^2 with confidence coefficient $(1 - \alpha)$ is given by

$$\frac{(n-1)s^2}{\chi^2_{U;n-1}} \leq \sigma^2 \leq \frac{(n-1)s^2}{\chi^2_{L;n-1}}$$

where s^2 is the sample variance computed on the basis of n randomly drawn values from a population that is normally distributed, and where $\chi^2_{L;n-1}$ and $\chi^2_{U;n-1}$ are the lower and upper χ^2 values which locate half of α in each tail of the chi-square distribution with $(n-1)$ degrees of freedom.

Example 9.11 A machine designed to fill soap boxes with 16 ounces of soap is supposed to fill in such a way that the standard deviation of the fills is no more than 0.1 ounce. If the machine operates properly, the variance of the fills must be less than 0.01 ounce^2. A sample of 20 soap boxes is selected from a lot produced by this machine; the sample variance s^2 is 0.015. Find a 90 percent confidence interval for the population variance σ^2.

Solution The confidence interval formula for σ^2 is

$$\frac{(n-1)s^2}{\chi^2_{U;n-1}} \leq \sigma^2 \leq \frac{(n-1)s^2}{\chi^2_{L;n-1}}$$

For a 90 percent confidence interval, $(1 - \alpha) = 0.90$ so that $\alpha = 0.10$ and $\alpha/2 = 0.05$. Thus we wish to find values on the χ^2 distribution with $n - 1 = 19$ degrees of freedom so that 5 percent of the area is in each tail. From Table

B.6, the values of $\chi^2_{L;19}$ and $\chi^2_{U;19}$ are 10.1 and 30.1, respectively. The resulting interval is therefore

$$\frac{(19)(0.015)}{30.1} \leq \sigma^2 \leq \frac{(19)(0.015)}{10.1} \qquad \text{or} \qquad 0.009 \leq \sigma^2 \leq 0.028$$

Thus we are 90 percent confident that the true variance of the fills lies between 0.009 and 0.028 squared ounces. Since the machine is operating properly when the variance is less than 0.01 squared ounces and the interval on the variance is 0.009 to 0.028 ounces squared, we must conclude that the machine is not operating properly—there is a good chance that the variance exceeds 0.01 since the upper limit of the confidence interval is 0.028.

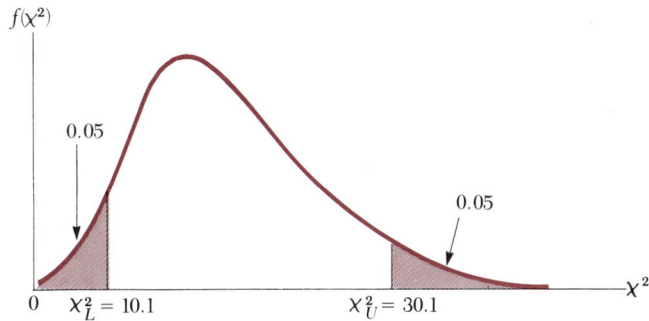

A formula for the shortest interval for the standard deviation σ for small samples will not be given because both the development and use of its form are quite complicated. An acceptable interval estimate of σ is given by taking the square roots of the upper and lower bounds of the population variance interval estimate. This interval has intuitive appeal because we are simply taking square roots of the limits given for the interval for σ^2, and it is sufficiently accurate for our purposes. For example, in Example 9.11 an approximate 90 percent confidence interval estimate of σ is given by

$$0.095 = \sqrt{0.009} \leq \sigma \leq \sqrt{0.028} = 0.167$$

9.8

Determination of the sample size

Up to this point in this chapter, we have assumed that the sample size n is known in forming point and interval estimates of the population mean μ and the population proportion π. The determination of the sample size n to be drawn from the population is a very important step in the statistical inference process.

Mathematical statisticians have spent a great deal of time developing ways to determine the sample size that achieves a desired level of "goodness" of the inferences. We shall look at two of these results that apply to the estimation of the population mean μ and the population proportion π.

9.8.1 | Determination of sample size; μ estimation

In determining the sample size in a statistical experiment, we must know two things:

1. How close we wish our estimate to be to the true value of the population parameter.

2. How certain we wish to be that our estimate will be within the selected number of units of the value of the parameter.

For example, we may request that our point estimate of the population mean μ not be more than 10 units away from the true value of μ with a confidence of 95 percent. By this, we mean that we are willing to take a 5 percent chance that the sample we draw of size n will produce an estimate $\bar{x}$ of μ that will be more than 10 units away from the actual value of μ.

We can use the sampling distribution of the statistic $\overline{X}$ to determine what size the sample must be to meet these requirements. We know from the sampling distribution of $\overline{X}$ shown in Figure 9.17 that the interval $\mu \pm 2\sigma_{\overline{X}}$ contains approximately 95 percent of the values of the statistic $\overline{X}$.

FIGURE 9.17 | Sampling distribution of the statistic $\overline{X}$

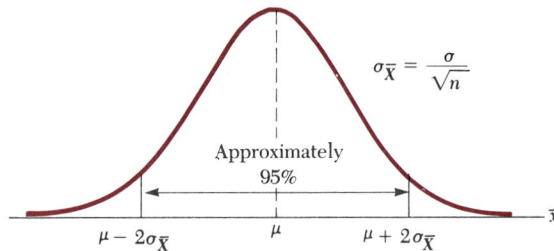

If we wish to be no more than E units from μ with our estimate $\bar{x}$, then we must set $2\sigma_{\overline{X}} = E$, or

$$2 \frac{\sigma}{\sqrt{n}} = E$$

Solving for n, we have

$$n = \frac{4\sigma^2}{E^2}$$

This formula presumes a confidence coefficient of $(1 - \alpha) = 0.9544$. If we want a confidence coefficient of $(1 - \alpha)$, then we may set

$$z_{\alpha/2}\sigma_{\overline{X}} = E \qquad \text{or} \qquad z_{\alpha/2}\frac{\sigma}{\sqrt{n}} = E$$

which results in the formula

$$n = \frac{z_{\alpha/2}^2 \sigma^2}{E^2} = \left[\frac{z_{\alpha/2}\sigma}{E}\right]^2$$

Perhaps it is easier to understand this formula by its derivation from the z-transformation formula,

$$z = \frac{\bar{x} - \mu}{\sigma_{\bar{X}}}$$

Solving this formula for $(\bar{x} - \mu)$ produces

$$(\bar{x} - \mu) = z_{\alpha/2}\sigma_{\bar{X}} = z_{\alpha/2}\frac{\sigma}{\sqrt{n}}$$

The quantity $(\bar{x} - \mu)$ represents the difference between the estimate $\bar{x}$ and the population mean μ. We want this difference to be no more than E units, the specified acceptable sampling error. Therefore we have

$$E = z_{\alpha/2}\frac{\sigma}{\sqrt{n}}$$

Solving this equation for n produces the formula

$$n = \left[\frac{z_{\alpha/2}\sigma}{E}\right]^2$$

Since σ is usually not known, an estimate of σ must be used in this formula. Notice that the sample standard deviation s cannot be used because no sample has yet been taken—we are trying to determine the sample size. Many times a rough estimate of σ will be available, usually based on past experience with the population random variable **X**.

If no estimate of σ is available, then it is common to take a small sample, called a *pilot sample,* for the sole purpose of estimating σ by calculating the sample standard deviation s from the pilot sample.

Pilot sample

Formula 9.6

Sample size for estimating the population mean μ

Sample size for estimating the population mean

If it is desired to estimate the population mean μ so that the estimate $\bar{x}$ will be no more than E units from the true value of μ with confidence coefficient $(1 - \alpha)$, and $\bar{\mathbf{X}}$ is approximately normally distributed, then n should be chosen such that

$$n = \left[\frac{z_{\alpha/2}\sigma}{E}\right]^2$$

where $z_{\alpha/2}$ is the value on the standard normal distribution such that $P(\mathbf{Z} \geq z_{\alpha/2}) = \alpha/2$ and σ is the population standard deviation.

Example 9.12 Past experience has indicated that the salaries of factory workers in a certain industry are approximately normally distributed with a standard deviation of $500. How large a sample of factory workers would be required if we wish to estimate the population mean salary μ to within $60 with a confidence of 99 percent?

Solution In the sample size formula,

$$n = \left[\frac{z_{\alpha/2}\sigma}{E}\right]^2$$

$\sigma = \$500$, $E = \$60$, and $z_{\alpha/2} = 2.58$ (from Table B.3). Thus

$$n = \left[\frac{(2.58)(500)}{60}\right]^2 = 462.25$$

We would require a sample size of at least $n = 463$ factory workers to produce a point estimate of μ with the required degree of precision. When the answer is not a whole number as above, the usual convention is to round up to the next larger integer. Thus we round 462.25 to 463.

9.8.2 | Determination of sample size; π estimation

The same argument may be applied to the population proportion π. The point estimator of π is the statistic **P**, which has a standard deviation of $\sqrt{\pi(1 - \pi)/n}$. If we wish to be within E units of π with our estimate p with confidence coefficient $(1 - \alpha)$, then we must set

$$z_{\alpha/2}\sqrt{\frac{\pi(1 - \pi)}{n}} = E \quad \text{or for } n, \quad n = \frac{z_{\alpha/2}^2[\pi(1 - \pi)]}{E^2}$$

Since the value of π is unknown, this formula appears to result in a dead end. However, the formula can be used in three ways. First, as with Formula 9.6, it may be possible to take a "pilot" sample, a small random sample taken to estimate parameters needed in sample size formulas. We could, therefore, take a pilot sample and estimate π by the sample proportion p in this sample in the formula for n. Second, the formula can be used to provide an approximate upper bound on n by noting that the product $\pi(1 - \pi)$ can be at most 0.25—this occurs when $\pi = 0.5$ noting that $0 \leq \pi \leq 1$. Thus we can replace the product $\pi(1 - \pi)$ in the formula by 0.25. Third, we may have prior information about the process that gives us an upper or lower bound for π. For example, if we have a production process and are interested in the proportion of defectives produced by the process, we may know that this proportion will be at most 5 percent. The upper or lower bound then can be used in the formula to obtain the sample size n.

Example 9.13 The Suny Corporation is interested in the proportion of defective units produced of its Super Betamex II VCR model. The quality control director wishes to estimate π, the proportion of defective units, to within 0.05 unit with 95 percent confidence coefficient $[(1 - \alpha) = (1 - 0.05) = 0.95]$. From the past history on similar VCRs, she knows that the maxi-

Sample size deter-
mination for a
population propor-
tion

Formulas 9.7 and 9.8

Sample size determination for estimating the population proportion π

If it is desired to estimate the population proportion π so that the estimate p will be no more than E units from the true value of π with confidence coefficient $(1 - \alpha)$, and it is assumed that **P** is approximately normally distributed, then n may be chosen by using either:

(1) $n = \dfrac{z_{\alpha/2}^2 p(1 - p)}{E^2}$ where p is an estimate of π based on a pilot   (9.7)
sample or a prior estimate

or

(2) $n = \dfrac{z_{\alpha/2}^2 (0.25)}{E^2}$ if no estimate of π is available   (9.8)

where $z_{\alpha/2}$ is the value of the standard normal distribution such that $P[\mathbf{Z} \geq z_{\alpha/2}] = \alpha/2$.

---

mum proportion of defective units is 0.05. What sample size is required to meet her criteria for estimating π?

Solution   Since we have a prior estimate of π, we can use Formula 9.7. In Formula 9.7, $E = 0.05$ unit, and $z_{\alpha/2} = z_{0.05/2} = z_{0.025} = 1.96$ from Table B.3 in Appendix B. Using $p = 0.05$ as the prior estimate in Formula 9.7, we have

$$n = \frac{[1.96]^2(0.05)(1 - 0.05)}{[0.05]^2} = 72.99$$

Rounding to the next larger integer, she would require a sample of 73 VCRs to estimate π while satisfying her estimation criteria.

Example 9.14   Suppose in Example 9.13 the director decides to first take a small "pilot" sample of VCRs to provide a preliminary estimate of π. In the pilot sample of 25 VCRs, she finds one defective unit. Using the pilot sample information, what sample size is now required to estimate π to within 0.05 unit with 95 percent confidence?

Solution   We now have a pilot sample to use for the value of p in Formula 9.7. In the pilot sample of size 25, we have one defective. So, $x = 1$ and $p = x/n = 1/25 = 0.04$. Using $p = 0.04$ in Formula 9.7, and for 95 percent confidence $z_{\alpha/2} = 1.96$ as in Example 9.13, we have:

$$n = \frac{[1.96]^2(0.04)(0.96)}{(0.05)^2} = 59.007$$

Rounding to the next larger integer produces a required sample size of $n = 60$. Since she has already sampled $n = 25$ units (her pilot sample), she need only sample $60 - 25 = 35$ additional units to meet her required sample size of 60.

In comparing Examples 9.13 and 9.14, the answers are very similar. In Example 9.13 she is depending on prior experience with similar VCRs suggesting that π is no more than 0.05. In Example 9.14, she has taken a pilot sample and found that the sample proportion of defectives is 0.04 (1 out of 25). She may place a little more confidence in the use of the pilot sample, since it is a sample of the VCRs of interest. Notice also that the pilot sample can be used as part of the sample size determined from the formula. Thus she has lost little in taking the pilot sample and in doing so has gained useful information about the proportion of defectives.

It is interesting to calculate the sample size n necessary to meet her requirements assuming nothing is known about π. With $E = 0.05$ unit and $z_{0.025} = 1.96$, from Formula 9.8, we have:

$$n = \frac{[1.96]^2(0.25)}{(0.05)^2} = 384.16$$

Rounded to the next larger integer, this is a sample size of 385. Clearly, it is very useful to have some information about π, either through prior knowledge or from a pilot sample!

A final word of caution in using either Formula 9.7 or Formula 9.8. These formulas assume that the sample proportion **P** is at least approximately normally distributed. From Section 8.7 in Chapter 8, usually the sample size must be 50 or more for this to be true by the central limit theorem. Therefore if either formula produces a sample size (say less than 50), then the formula and the resulting sample size should not be used. In this instance other methods should be used. Among these methods are the nonparametric procedures presented in Chapter 20.

■ 9.9

Use of statistical software packages

There are now many statistical software packages available that will calculate confidence interval estimates. Shown in Figure 9.18 is the output of an interactive session using MINITAB. The Example 9.7 problem is used to illustrate the use and interpretation of the MINITAB output. In Example 9.7 a consumer group is concerned about the average breaking weight of a 10-pound fishing line. A sample of 25 lines is taken and a 99 percent confidence interval estimate is desired. Further, it is assumed that σ, the population standard deviation is equal to 0.5 pound.

The data are entered into MINITAB by the SET command. The 25 sample observations are placed in column C1 by the statement SET INTO C1. The next three lines are used to enter the 25 numbers. The END command on line 5 tells MINITAB that there are no more numbers.

The DESCRIBE command produces useful information about the sample of 25 numbers. The mean ($\bar{x}$) is 10.560; the median is 10.50; and the trimmed mean (TMEAN) is 10.530. The trimmed mean is calculated by ordering the 25 numbers from the smallest to the largest, eliminating the lower and upper 5 percent of the numbers, and calculating the mean of the remaining numbers in the array. The standard deviation (s) is 0.54, and the standard error of the

FIGURE 9.18 | MINITAB output for the Example 9.7 problem

```
MTB  > SET INTO C1
DATA > 12.0 9.8 10.2 10.4 11.1 10.1 10.0 11.4 10.6
DATA > 10.6 10.2 10.5 11.0 10.5 10.2 11.6 10.9 10.7
DATA > 10.8 10.1 9.9 10.2 10.2 10.6 10.4
DATA > END
MTB  > DESCRIBE C1


            C1
N           25
MEAN     10.560
MEDIAN   10.500
TMEAN    10.530
STDEV     0.540
SEMEAN    0.108
MAX      12.000
MIN       9.800
Q3       10.850
Q1       10.200
MTB > ZINT CONF = 99 SIGMA = 0.5 ON C1

THE ASSUMED SIGMA = 0.500

                   N     MEAN   STDEV   SE MEAN    99.0 PERCENT C.I.
C1                25   10.560   0.540     0.10   ( 10.30,    10.82)


MTB > TINT CONF = 99 ON C1

                   N     MEAN   STDEV   SE MEAN    99.0 PERCENT C.I.
C1                25   10.560   0.540     0.11   ( 10.26,    10.86)
```

mean ($s/\sqrt{n}$) is 0.108. The maximum value is 12.000 and the minimum value is 9.800. The first and third quartiles are 10.200 and 10.8550, respectively.

The command to produce a 99 percent confidence interval estimate assuming that $\sigma = 0.5$ pound using Formula 9.1 is ZINT for z interval. The command for Example 9.7 is

```
ZINT CONF = 99 SIGMA = 0.05 ON C1
```

The 99 indicates that we want a 99 percent confidence interval estimate, SIGMA = 0.5 specifies the assumed value of σ, and the data are in column C1. The resulting confidence interval estimate is (10.30, 10.82), the same answer as in Example 9.7.

The next command, TINT, stands for the *t*-interval estimate when the value of σ is not known. Thus

```
TINT CONF = 99 on C1
```

asks for a 99 percent confidence interval estimate using Formula 9.3. The resulting confidence interval estimate is (10.26, 10.86). Notice that there is not much difference in the two confidence interval estimates, since the assumed value of σ in ZINT (0.50) is very close to the sample standard deviation value ($s = 0.54$).

Such software packages as MINITAB are readily available and make the computation of point and confidence interval estimates quite easy.

TABLE 9.5 | Confidence interval formulas for μ and π

Parameter	Formula	Conditions	Nature of interval
μ	$\bar{x} \pm z_{\alpha/2} \dfrac{\sigma}{\sqrt{n}}$	σ is known, **X** is normally distributed.	Exact for any sample size
	$\bar{x} \pm z_{\alpha/2} \dfrac{\sigma}{\sqrt{n}}$	σ is known, **X** is not normally distributed, but $n \geq 25$.	Approximate
	$\bar{x} \pm t_{\alpha/2;n-1} \dfrac{s}{\sqrt{n}}$	σ is unknown, **X** is normally distributed.	Exact for $n > 1$
	$\bar{x} \pm t_{\alpha/2;n-1} \dfrac{s}{\sqrt{n}}$	σ is unknown, **X** is not normally distributed, but $n \geq 25$.	Approximate
	$\bar{x} \pm z_{\alpha/2} \dfrac{s}{\sqrt{n}}$	σ is unknown, **X** is not normally distributed, but $n > 100$.	Approximate
π	$p \pm z_{\alpha/2} \sqrt{\dfrac{p(1-p)}{n}}$	σ_P is unknown, $n \geq 50$.	Approximate

Note: For other cases, such as $n < 25$, **X** not normally distributed and population variance unknown in estimating μ, other methods should be used. Some of these are given in Chapter 20.

■ **9.10**

Summary

In this chapter, we have developed the statistical inference method of estimating the value of a population parameter.

There are two types of statistical estimation: point estimation and interval estimation. A point estimate yields one value or point, the estimate, for a particular sample set $x_1, x_2, \ldots, x_n$. An interval estimate produces an interval on the real line that will contain the value of the population parameter with a prescribed level of confidence, usually taken to be 90, 95, or 99 percent.

The interval estimator is almost always based on the sampling distribution of the best point estimator of a given population parameter.

The appropriate confidence intervals for μ and π depend on the size of the sample, the distribution of the population random variable **X**, and the knowledge of the population variance (when estimating μ). Table 9.5 summarizes the confidence interval formulas for μ and π.

An important step in the inference process is determining the sample size—the number of units to be drawn from the population. If it is possible to state the maximum difference between the value of the population parameter and the estimate that will be tolerated and the level of confidence for which this condition will be satisfied, then a determination of the sample size is possible. Formulas for determining the sample size when estimating the population mean μ and the population proportion π are given in this chapter.

■ **References**

Introductory

Anderson, D. R.; D. J. Sweeney; and T. A. Williams. *Statistics for Business and Economics.* 2nd ed. St. Paul Minn.: West Publishing, 1984, Chapters 7 and 8.

Mendelhall, W., and J. E. Reinmuth. *Statistics for Management and Economics.* 4th, ed. Boston: Duxbury Press, 1982, Chapters 8 and 9.

Wonnacott, T., and N. Wonnacott. *Introductory Statistics for Business and Economics.* 3rd ed. New York: John Wiley & Sons, 1984, Chapters 7 and 8.

Advanced

Ostle, B., and N. Mensing. *Statistics in Research.* 3rd ed. Ames: Iowa State University Press, 1975, Chapter 5.

Wonnacott, T., and R. Wonnacott. *Introductory Statistics.* 4th ed. New York: John Wiley & Sons, 1985, Chapters 7 and 8.

■ **Appendix**

Derivation of confidence interval estimate formula for the population mean μ; σ assumed known.

$$P[\mu - z_{\alpha/2}\sigma_{\overline{\mathbf{X}}} \leq \overline{\mathbf{X}} \leq \mu + z_{\alpha/2}\sigma_{\overline{\mathbf{X}}}] = (1 - \alpha); 0 \leq \alpha \leq 1 \qquad \text{[from sampling distribution of } \overline{\mathbf{X}}]$$

$$P[-z_{\alpha/2}\sigma_{\overline{X}} \leq \overline{X} - \mu \leq z_{\alpha/2}\sigma_{\overline{X}}] = (1 - \alpha)$$

[subtracting μ from each term in the inequalities]

$$P[-\overline{X} - z_{\alpha/2}\sigma_{\overline{X}} \leq -\mu \leq -\overline{X} + z_{\alpha/2}\sigma_{\overline{X}}] = (1 - \alpha)$$

[subtracting $\overline{X}$ from each term in the inequalities]

$$P[\overline{X} + z_{\alpha/2}\sigma_{\overline{X}} \geq \mu \geq \overline{X} - z_{\alpha/2}\sigma_{\overline{X}}] = (1 - \alpha)$$

[multiplying each term of the inequalities by (-1); note change in direction of inequalities]

$$P[\overline{X} - z_{\alpha/2}\sigma_{\overline{X}} \leq \mu \leq \overline{X} + z_{\alpha/2}\sigma_{\overline{X}}] = (1 - \alpha)$$

[reordering terms, putting smallest term to the left of the statement]

$$P[\overline{X} - z_{\alpha/2}\sigma_{\overline{X}} \leq \mu \leq \overline{X} + z_{\alpha/2}\sigma_{\overline{X}}] = (1 - \alpha)$$

■ Problems

Section 9.1 Problems

9.1 Distinguish between a *parameter* and a *statistic*.

9.2 What are the two basic ways to draw statistical inferences about a population parameter? How do the two approaches differ?

9.3 What is the objective of statistical inference?

Section 9.2 Problems

9.4 Define *point estimation*.

9.5 Define *interval estimation*.

9.6 What is the difference between the terms, *estimator* and *estimate* of a population parameter? Give an example to illustrate the difference.

9.7 What are the two types of statistical estimation? How do they differ?

9.8 Discuss the criteria for selecting a point estimator of a population parameter.

9.9 Why might we not always want to restrict ourselves to using an *unbiased estimator* of a population parameter?

9.10 Define a *biased estimator*.

Section 9.3 Problems

9.11 Define the *standard error of an estimator*.

9.12 A company has 100 salespeople who regularly travel each month. In a random sample of 20 of these salespeople, the company finds that the average amount of travel-related expenses in a specific month is $1,452, with a standard deviation of $440.

  a. What is the point estimate of the mean travel-related expenses of all 100 salespeople for this month?

  b. What is the estimated standard error of this estimate?

9.13 Northern airlines is interested in determining the proportion of business travelers who use

Northern on flights between Boston and New York. In a random sample of 500 travelers who have taken recent trips between Boston and New York, Northern finds that the proportion who flew on Northern is 0.22.

a. What is the point estimate of the proportion of all business travelers who regularly fly between Boston and New York on Northern?

b. What is the estimated standard error of this estimate?

9.14 There are approximately 325 faculty members at Texas Christian University. Recently, the faculty has been asked to consider a proposal to revise the undergraduate curriculum. The revision substantially increases the "core" of the program by adding math, writing, speaking, and foreign language required courses. The faculty is split on the support of this proposal. In a random sample of 100 faculty members, it is found that 40 support the proposal, 50 are opposed to it, and 10 are undecided.

a. Find the point estimate of the proportion of the faculty who support the proposal, and find the estimated standard error of this estimate.

b. Find the point estimate of the proportion of the faculty who are opposed to the proposal, and find the estimated standard error of this estimate.

9.15 A large portfolio contains several hundred investments. An audit is taken to determine the average rate of return on investments. A sample of 25 investments in this portfolio produces the following rates of return on investment (in percentages):

5.4	6.9	10.2	4.4	6.7
8.9	10.0	9.6	12.2	6.6
3.4	9.5	3.4	9.9	13.2
11.6	10.4	7.5	7.8	4.4
2.0	8.8	11.4	8.0	6.8

a. Find a point estimate of the mean rate of return for all investments in this portfolio.

b. What is the estimated standard error of this estimate?

9.16 A random sample of 100 starting salaries of MBA graduates employed by financial institutions in New York City produced an average starting salary of $35,000 with a standard deviation of $4,000. Find point estimates of the average starting salary and standard deviation of all MBA graduates working in financial institutions in New York City.

9.17 At a large university, the issue of faculty unionization has become critical. One hundred faculty members are randomly sampled and asked whether or not they favor unionization. Forty say they do. Find a point estimate of the proportion of all faculty members at the university who favor unionization.

9.18 In a large federal agency, the number of employees with 10 years or more of service is 3,000. In a sample of 25 of these employees, the numbers of years of service are:

10, 22, 11, 10, 14, 16, 12, 25, 15,
12, 12, 30, 14, 10, 15, 14, 18,
26, 21, 11, 10, 15, 14, 20, 13

a. Find an estimate of the average number of years of service for these 3,000 employees.

b. Find an estimate of the standard deviation of the number of years of service for these 3,000 employees.

c. Estimate the proportion of employees among these 3,000 who have served 20 years or more.

9.19 A soft drink machine is designed to place an average of 9.5 ounces of soft drink into a 10-ounce container. In a sample of 18 fills of the 10-ounce containers, the following number of ounces of soft drink were found to be placed into the containers by the machine:

9.6 9.2 9.0 9.4 9.7 9.7 9.5 9.2 9.5
9.7 9.6 9.6 9.4 9.5 9.5 9.5 9.7 9.3

a. Determine a point estimate of the average fill, and determine the estimated standard error of this estimate.

b. Determine a point estimate of the standard deviation of the fill amounts by this machine.

Section 9.4 Problems

9.20 Why is a confidence interval estimate usually given with a point estimate of a population parameter?

9.21 When we form a 95 percent confidence interval on the population mean, what is meant by "95 percent confidence"?

9.22 A machine produces a miniature amplifier capable of producing a maximum of 1.5 watts of power. The amplifier is designed to be used in the space packs of astronauts exploring the moon and other bodies in our solar system. It powers certain sensing devices necessary for the astronauts' survival. Based on prior experience in making this amplifier, it is known that the standard deviation of the maximum power of the amplifiers is 0.1 watt. A random sample of 16 amplifiers is selected from the production line, and it is found that the average maximum power for the 16 is 1.45 watts. Assume that the maximum output power of the amplifier is normally distributed.

   a. Construct a 95 percent confidence interval for the average maximum power output among all amplifiers produced by this machine.

   b. Is the rated maximum output of 1.5 watts contained in your interval estimate?

   c. Based on parts *a* and *b*, what would you recommend regarding the production process if the machine is supposed to be producing parts with an average maximum output of 1.5 watts?

9.23 The APT Corporation produces photocopying machines. With the purchase of the machine, APT offers a service contract covering one year during which labor and parts required for repairs are fully covered by the contract. APT is interested in the proportion of customers who purchase the service contract with the photocopying machines. In a random sample of 100 recent sales, APT finds that 60 customers purchased the service contract.

   a. What is the point estimate of the proportion π of all purchases of photocopying machines who also purchase a service contract?

   b. Find a 95 percent confidence interval estimate of the proportion π of all purchases who also purchase a service contract.

9.24 The proportion of bad widgets produced in a factory is unknown. A random sample of 100 widgets is selected, and 10 bad widgets are found. Determine a 95 percent confidence interval estimate of the proportion of bad widgets.

9.25 The Dogwood City Council is concerned about the proportion of residents in its community who oppose the construction of a nuclear power plant within the city limits. In a random sample of 1,000 residents contacted in a telephone survey, the council finds that 760 respondents oppose the construction of the nuclear power plant.

   a. Find a point estimate of the proportion of all residents in the city who oppose the construction of the nuclear power plant. What is the standard error of this estimate?

   b. Find a 95 percent confidence interval for the proportion of residents who oppose the construction of the nuclear power plant.

9.26 A machine process is known to produce miniamplifiers that produce an average of μ watts of power with a standard deviation of 2.6 watts. In a sample test of 100 amplifiers, it is found that their average output is 50 watts. Determine a 92 percent confidence interval on the population mean, μ.

Section 9.5 Problems

9.27 A car rental agency is concerned about the time required to complete a reservation process by telephone by its customers. In a random sample of 36 calls, the average time required to complete the transaction is four minutes with a standard deviation of one minute.

   a. Find a 99 percent confidence interval estimate of the population mean telephone transaction time, using the formula $\bar{x} \pm z_{\alpha/2} s/\sqrt{n}$.

   b. Find a 99 percent confidence interval estimate of the population mean, using the formula $\bar{x} \pm t_{\alpha/2;n-1} s/\sqrt{n}$.

   c. Compare these two interval estimates. Which is more appropriate? Why?

9.28 Tests on a popular brand of paint gave the following results on the square feet of coverage per gallon: 150, 159, 162, 144, 150, 162,

140, 155, 164, 157. Assuming that the coverages are normally distributed, determine a 95 percent confidence interval on the population average square feet coverage.

9.29 An educational association is concerned about the average number of hours that 10 year olds watch TV each day. In a random sample of 100 10 year olds in a survey conducted for a particular day, it is found that the sample average number of hours of TV viewing is 10 hours, with a sample standard deviation of 2 hours.

a. Find a point estimate of the average number of hours that 10 year olds view TV. What is the standard error of this estimate?

b. Find a 95 percent confidence interval for the average number of hours daily that 10 year olds view TV, using this sample information.

9.30 The MGM Great Hotel in Las Vegas is interested in the average time taken by its telephone reservationists to make reservations. In a random sample of 100 reservation telephone calls to the reservation center, the average time taken to complete a reservation is six minutes with a standard deviation of two minutes. Find a 95 percent confidence interval on the population average time taken to complete a reservation by telephone by the reservationists at the MGM Great Hotel.

9.31 A production process is designed to produce bolts with a mean length of $\mu = 7$ inches and a standard deviation of $\sigma = 0.01$ inch. A random sample of 25 bolts produced an average length of 5.99 inches and a standard deviation of 0.03 inches. Assume that bolt lengths are normally distributed.

a. Determine an estimate of μ, and find the estimated standard error of the estimate.

b. Determine a point estimate of the population standard deviation.

c. Set a 95 percent confidence interval on μ.

d. Set a 99 percent confidence interval on μ.

e. Based on the sample, is the machine operating at specifications? Explain.

Section 9.6 Problems

9.32 A manufacturer claims that the average lifetime of a filter used in a coolant system for nuclear power plants will be at least three months in continuous use before it fails and needs to be replaced. To test this claim, a federal agency places 36 filters in simulated coolant systems and finds that the average time to failure of the filters is 2.8 months with a standard deviation of 0.6 month.

a. Find a 95 percent upper confidence bound on μ, the average lifetime of the filters.

b. Based on the upper confidence bound determined in part a, is the manufacturer's claim plausible? Explain.

9.33 Two Democratic candidates, Harry and David, are involved in a runoff election to determine who will run against the Republican candidate in the general election. The campaign manager for Harry takes a random sample of 100 voters who claim they will vote in the runoff election and finds that 42 claim they will vote for Harry.

a. Find a 90 percent upper confidence bound for π, the proportion of voters who will vote for Harry in the runoff election.

b. Based on the upper confidence bound determined in part a, does Harry stand a reasonable chance to win the runoff election? Explain.

9.34 The General Cereal Company believes that at least 20 percent of the consumers prefer its cereal to other cereals for breakfast. In a sample of 200 breakfast cereal eaters, it is found that 25 prefer the General Cereal brand to the other brands.

a. Find a 90 percent upper confidence bound for π, the proportion of cereal eaters who prefer the General Cereal brand to other cereals.

b. Based on the upper confidence bound determined in part a, what do you think about the General Cereal Company's contention that at least 20 percent of the consumers prefer its cereal to other brands?

9.35 The Baseball Card Company states that in its lots of 1,000 mixed cards by Tapps in a given year, "You will find on the average at least 500 different cards." A consumer takes a random sample of 25 lots and finds that the average number of different cards in each lot is 523, with a standard deviation of 42.

a. Determine a 95 percent lower confidence bound on the average number of different cards in all lots produced by the Baseball Card Company.

b. Comment on the statement by the Baseball Card Company. Does this sample evidence support their claim?

Section 9.7 Problems

9.36 A random variable is known to be normally distributed with mean $\mu = 50$, but unknown variance. A random sample of 72 values of the variable is taken, and it is found that the sample variance is 80. Construct a 95 percent confidence interval on the population variance.

9.37 A random sample of 100 teletype operators indicates that their salaries fluctuate quite a bit. The sample standard deviation of their daily salaries is $10. Construct a 90 percent confidence interval on the population standard deviation of the daily salaries.

9.38 A packaging process is supposed to place 16 ounces of material in a cylindrical container. The tolerable standard deviation is 0.5 ounce. A random sample of 25 containers produces the following amount of material placed in each:

15.92	15.90	16.05	16.04	15.88
15.89	15.92	15.81	16.09	15.92
16.01	15.99	15.75	16.13	16.09
15.94	15.90	15.87	15.98	16.04
16.20	15.78	15.90	15.88	15.89

a. Find a point estimate of the standard deviation of amounts placed in the containers.

b. Find a 95 percent confidence interval on the standard deviation of the amounts placed in the containers. Assume that the amounts placed in the containers are normally distributed.

9.39 The standard deviation of a sample of 80 times to complete an assembly task is 6.62 minutes.

a. What is the point estimate of the population standard deviation of all assembly times?

b. Find a 90 percent confidence interval on the population standard deviation of all assembly times.

Section 9.8 Problems

9.40 A statistician is asked to conduct a survey to determine an estimate of the proportion of people who favor the recall of a local politician. He is told that his estimate should not differ from the true proportion by more than 2 percent, with 95 percent confidence. How large should his random sample be to produce an estimate of the proportion satisfying this condition?

9.41 A consumer group is concerned about the truthfulness of a certain producer in stating the net weight of a box of SNAP cereal. It is decided that n boxes of SNAP will be purchased and the sample mean weight will be used to decide whether or not to press charges against the producer of SNAP. The boxes claim to contain a net weight of 12 ounces. It is decided that the sample size n should be made large enough to ensure that the sample average weight differs from the population average weight by no more than 0.1 ounce with 99 percent confidence. A very conservative guess at the population standard deviation, based on a few boxes of SNAP, is 0.8 ounce. How large should n be?

9.42 Past experience shows that the standard deviation of the yearly income of garment workers in a certain state is $400. How large a sample of garment workers would one need to take to estimate the population mean income to within $50 with a probability of 0.95 of being correct?

9.43 In the next presidential election, it is desired to estimate π, the proportion voting for the Democratic candidate, to within 0.01 unit with a probability of 0.99 of being correct. The estimate is to be made five days prior to the national election. How large a sample size of registered voters who will vote is required?

9.44 The EPA wishes to conduct a mileage test on a specific foreign import model of automobile. The research statistician at EPA wishes to estimate the mean miles per gallon μ for this model to within 1 mile per gallon with 95 percent confidence. Assuming that the population standard deviation σ is 2.5 miles per gallon, what sample size (number of automobiles of this model) is required to conduct the test?

9.45 The Lone Star Gas Company is concerned about the number of days required for its customers to pay their monthly bills. The company wishes to estimate μ, the average number of days required for a customer to pay his or her bill. They wish the estimate of μ to be within one day of the actual value of μ with 95 percent confidence. In a pilot sample of 25 randomly selected accounts for the past January billing month, they found the sample standard deviation s for these 25 accounts to be eight days. Using s to estimate σ in the sample size determination formula for μ, determine the sample size required to estimate μ with the specified precision.

9.46 The bumpers on the New Bimmer Sports Sedan are designed to withstand a 10 MPH crash with no damage. The quality control director at Bimmer wishes to estimate the proportion π of bumpers that will sustain no damage in 10 MPH crashes in laboratory testing. She wishes to estimate π to within 0.05 unit with 95 percent confidence.
  a. Assuming that she has no prior knowledge concerning the value of π, what sample size is required to meet her precision criteria?
  b. Suppose that she conducts a pilot study of 20 Bimmers and finds that 18 sustain no damage in a 10 MPH simulated crash. Using the pilot sample proportion estimate of π of $18/20 = 0.90$, now what sample size is required to meet her precision criteria?

9.47 The supervisor of a very large secretarial pool in the corporate offices of a national insurance company wishes to estimate the proportion of secretaries who arrive late to work each day. She wishes to estimate the proportion π to within 0.10 unit with 90 percent confidence. Based on preliminary testing, she believes that the proportion is about 0.10 (10 percent). Using the preliminary estimate, what sample size (number of secretaries) is

required to estimate π so that her precision criteria are met?

Additional Problems

9.48 A sample of 16 observations has been taken from a population in which the random variable is normally distributed. The sample mean is 50, and the sample standard deviation is 10.
  a. Determine a 95 percent confidence interval on the population mean.
  b. Interpret this interval.

9.49 Under what circumstances is it necessary to use the t-distribution instead of the z-distribution in constructing a confidence interval on the population mean?

9.50 Why is the determination of the sample size important in designing statistical experiments?

9.51 A report stated that in a study of a school district's attendance record of its students, a 90 percent confidence interval on the average number of school days missed by each student per year is three to nine days. Two improper interpretations of this statement follow. Explain why each is incorrect.
  a. Ninety percent of the students in the population miss from three to nine days of school each year.
  b. If many random samples are taken, 90 percent of them would produce sample means from three to nine days.

9.52 If, in setting a confidence interval on a population mean, the sample size is doubled, holding the confidence coefficient and the population standard deviation constant, what is the effect on the width of the confidence interval?

9.53 The Arthur Old Accounting Firm hires clerical help with a 90-day initial provisional period. If the individual's performance is not satisfactory during this 90-day period, then he or she is not retained. The personnel director is concerned about the proportion of provisional employees who are not retained following this 90-day period. He decides to take a random sample of the files of provisional employees for whom the 90-day period has elapsed. He wishes to estimate π, the proportion who are not retained to within 0.075 unit with 95 percent confidence.

a. Assuming that he has no prior estimate of π, what sample size is required?

b. From the study conducted last year, he is willing to assume that π is about 0.20. Using this information, what sample size is required?

9.54 A labor dispute involves the issue of starting salaries for a specific professional position in a large national corporation. The union claims that women hired for this job receive a smaller average starting salary than do men. The union wishes to estimate the average starting salary for men and women in this professional position to within $500 with 95 percent confidence. From a previous study, the union is willing to assume that the standard deviations of the starting salaries for men and women in this position are $4,000 and $2,500, respectively.

a. What sample size of men's starting salaries is required to estimate the average male starting salary with the stated precision requirements?

b. What sample size of women's starting salaries is required to estimate the average female starting salary with the stated precision requirements?

9.55 The Cears Company is interested in the proportion of purchases that are made using its own credit cards as opposed to cash payments (including personal checks). Also, they are interested in determining if the average purchase amount in dollars is different for credit card and cash purchases. In a random sample of 50 recent purchases, they find that 30 were made by Cears credit card, and 20 were made by paying cash. The amounts of the purchases, rounded to the nearest dollar, for the two types of purchases are shown below:

Credit card purchases			Cash purchases	
100	154	504	12	112
54	66	222	26	74
27	47	49	47	18
276	36	57	28	28
112	52	44	66	10
94	88	56	242	42
76	496	28	82	61
102	212	108	15	27
54	48	146	24	42
22	166	24	44	36

a. Find a point estimate and a 95 percent confidence interval estimate of the proportion π of all Cears customers who make purchases using the Cears credit card.

b. Find a point estimate and an 95 percent confidence interval estimate of the mean purchase amount (in dollars) of Cears customers who make credit card purchases.

c. Find a point estimate and a 95 percent confidence interval estimate of the mean purchase amount (in dollars) of Cears customers who make cash purchases.

d. Based on the confidence interval estimates determined in parts (b) and (c), do you believe that the average purchase amount (in dollars) differs for credit card and cash purchases? Explain.

10

Statistical inference: Hypothesis testing

■ **Introduction to hypothesis testing**

In Chapter 9 we discussed the two approaches to statistical inference: estimation and hypothesis testing. Although both approaches are extensively used in business and economic applications, hypothesis testing has taken a more important position than has estimation. Hypothesis testing has a language of its own. Although we will see that hypothesis testing and confidence interval estimation are very closely related methods of statistical inference, this would not be readily apparent to the casual reader due to, among other things, terminology differences associated with the two approaches.

In Part I of this chapter, the "language" of hypothesis testing is presented. Following the discussion of basic terms and concepts, the so-called "classical" method of hypothesis testing is developed. The procedure described has been referred to as the classical method due to its origin from developmental work in mathematical statistics during the early and mid-1900s. Two variants of the classical hypothesis-testing method are discussed next: standardized testing and the probability value ("p-value") approaches. The relationship between hypothesis testing and confidence interval estimation is presented, followed by an example illustrating the four approaches to hypothesis testing applied to the same problem. Finally, MINITAB is used to illustrate hypothesis testing with a computer statistical software package.

In Part II of this chapter, advanced topics are introduced. These include the concept of "power" in statistical hypothesis testing, the determination of sample size in conducting hypothesis tests, and hypothesis testing concerning the population variance.

This is a very long chapter, due to the diverse topics, concepts, and terms that have come to be known as hypothesis testing. Part I contains the basics of hypothesis testing, and Part II contains advanced topics that can be skipped without loss of continuity in covering the remaining chapters of the text.

PART I: | THE BASICS OF HYPOTHESIS TESTING

■ **10.1**

Introduction

The sampling process we shall use to develop the method of hypothesis testing is illustrated in Figure 10.1 for inferences concerning the population mean μ. The method of hypothesis testing requires us to hypothesize a value of the population mean μ, denoted by μ_0 in Figure 10.1. A random sample is

FIGURE 10.1 | Sampling process in statistical hypothesis testing

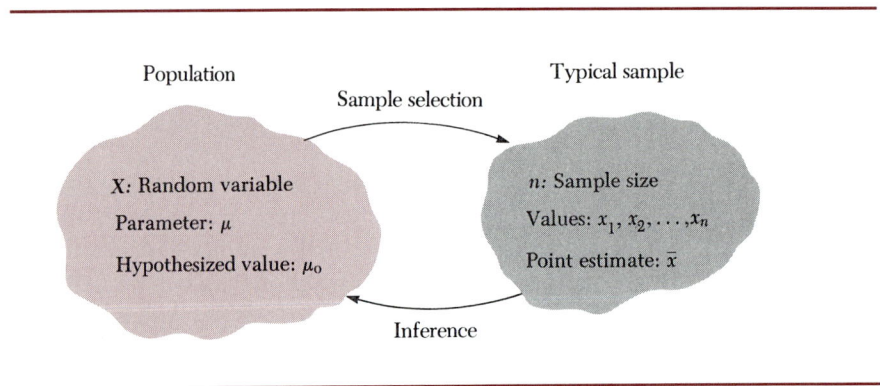

then taken from the population, and the value of the point estimator $\overline{X}$ in the sample is used to determine whether or not the hypothesized value of μ is reasonable.

There are three forms of the hypothesis statement about a population parameter that we can use. Let us suppose that the population parameter of interest is μ, the population mean. The three forms are:

$$\text{Form I:} \quad \begin{aligned} H_0 &: \quad \mu = \mu_0 \\ H_A &: \quad \mu \neq \mu_0 \end{aligned}$$

$$\text{Form II:} \quad \begin{aligned} H_0 &: \quad \mu \geq \mu_0 \\ H_A &: \quad \mu < \mu_0 \end{aligned}$$

$$\text{Form III:} \quad \begin{aligned} H_0 &: \quad \mu \leq \mu_0 \\ H_A &: \quad \mu > \mu_0 \end{aligned}$$

In these three forms, the symbol H_0 is called the *null hypothesis,* the symbol H_A is called the *alternate hypothesis,* and μ_0 is the hypothesized value of the population mean μ. It is the null hypothesis, H_0, that we will either reject or not reject on the basis of the sample information in our hypothesis-testing experiment. In Form I we are hypothesizing that μ takes on a specific value, μ_0. In Form II we are hypothesizing that μ takes on a *range* of values (greater than or equal to μ_0). Also, in Form III we are hypothesizing that μ takes on a *range* of values (less than or equal to μ_0).

The null hypothesis statement is evaluated for reasonableness based on the sample evidence. If H_0 is not reasonable based on the sample evidence, then we *reject* H_0. If H_0 is reasonable based on the sample evidence, then we *do not reject* H_0. As you might suspect, we will be spending quite a bit of time in Part I explaining what we mean by "reasonable."

Notice that the null and alternate hypothesis in Forms I, II, and III are complementary. In Form III, for example, H_A: $\mu > \mu_0$ contains all the values of μ not contained in H_0: $\mu \leq \mu_0$.

The forms of the statistical hypotheses are important. Each form addresses a different issue. We can illustrate these differences by way of an example. The Scholastic Aptitude Test (SAT) is designed to have a mean score (μ) of 1,000. The possible test scores range from 400 to 1,600. Concerning the population mean score μ, we might formulate the following questions:

1. Is the mean (μ) of the test scores equal to 1,000 as designed, or is there strong evidence based on a sample of test scores that μ is not 1,000?

    H_0: $\quad \mu = 1,000$

    H_A: $\quad \mu \neq 1,000$

2. Is the mean (μ) of the test scores greater than or equal to 1,000, or is there strong evidence that μ is less than 1,000?

    H_0: $\quad \mu \geq 1,000$

    H_A: $\quad \mu < 1,000$

3. Is the mean (μ) of the test scores less than or equal to 1,000, or is there strong evidence that μ is more than 1,000?

    H_0: $\quad \mu \leq 1,000$

    H_A: $\quad \mu > 1,000$

Notice the use in these three questions of the expression *strong evidence*. The approach to hypothesis testing that we will develop requires selecting one of the three forms (I, II, or III), which depends on which question among the three types above needs to be answered. Then sample data is evaluated to decide whether the null hypothesis (H_0) should be *rejected* or should *not be rejected*. If we decide to reject H_0, then we will be looking for strong evidence in the sample to arrive at this decision; that is, evidence that clearly makes the proposition stated in the null hypothesis unreasonable. If the evidence exists, then we say that we have made a *strong decision*. If the evidence in the sample is not strong or compelling (that H_0 should be rejected), then we decide *not to reject* the null hypothesis. This is often referred to as the *weak decision*. (We will return to the notion of strong and weak decisions later.)

In addition to hypotheses concerning the population mean μ, we may also be interested in hypotheses concerning other population parameters—such as the population proportion π and the population standard deviation σ (or the population variance σ^2). The three forms of the statistical hypotheses above apply to these population parameters, too.

We will now consider examples of formulating a statistical hypothesis, including the decision as to which form (Form I, II, or III) is appropriate for the particular hypothesis-testing problem.

Example 10.1 The Crunchy Cereals Company has been approached concerning a new packaging machine designed and built by the Ace Packaging Corporation. According to the Ace sales representative, the new machine will package 12-ounce cereal boxes twice as fast as the existing equipment used by the Crunchy people. In addition, the new machine is purportedly less expensive to operate than the old one. The Ace sales representative agrees to allow Crunchy to use the new machine without cost on a one-month trial basis. Of primary concern to the manager of Crunchy is the ability of the new machine to produce an average fill of 12 ounces in the cereal boxes packaged by the machine. During the trial period, he wishes to test the hypothesis that the average fill will indeed be 12 ounces. What is the form of the statistical hypothesis?

Solution The population parameter of interest is the population mean μ, the average fill for the population of cereal boxes. The question being asked is, "Is μ, the population average fill, equal to 12 ounces, or is there strong evidence that μ is not 12 ounces (that it is either less than 12 or greater than 12)? This is Form I:

$$H_0: \quad \mu = 12 \text{ ounces}$$
$$H_A: \quad \mu \neq 12 \text{ ounces}$$

In Example 10.1 the population may be conceptualized as all boxes filled by the machine when it is set for the 12-ounce filling weight during the operating cycle of the machine prior to wear out or a required maintenance period. The sample used to test the null hypothesis would be composed of a random selection of cereal boxes filled during the one-month trial period.

In the filling process, the variability of the cereal box fill weight would normally be of equal importance to the mean fill weight. We may wish to answer the question, "Is the standard deviation (σ) of the fill weights less than or equal to 0.4 ounces, or is there strong evidence (based on a sample of cereal box fill weights) that σ is greater than 0.4 ounces?" In this case, we have Form III:

$$H_0: \quad \sigma \leq 0.4 \text{ ounces}$$
$$H_A: \quad \sigma > 0.4 \text{ ounces}$$

The manager for Crunchy would use the sample of selected cereal boxes' fill weights to test both the hypothesis concerning the mean fill weight μ and the hypothesis concerning the standard deviation of the fill weights σ.

Example 10.2 The Commander Computer Company has recently released its new microcomputer, the Amica. The marketing director is very interested in the proportion of these machines sold for home use rather than for business or office use. Before the release of the Amica, the director hypothesized that the proportion of sales for home use would be 0.50 (50 percent of all Amicas sold would be used in the home). To test this hypothesis, what is the form of the null and alternate hypotheses?

Solution The population parameter of interest is the population proportion π, where π represents the proportion of sales for home use in the population of all sales of the Amica microcomputer. The director wishes to answer the question, "Is the proportion (π) of home sales of the Amica microcomputer equal to 0.50, or is there strong evidence (based on the sample) that π is not 0.50?" The correct form of the statistical hypotheses are:

$$H_0: \quad \pi = 0.50$$
$$H_A: \quad \pi \neq 0.50$$

Example 10.3 The quality control director of the new Novus Automobile Division of a major U.S. automotive manufacturer is concerned about the proportion of defective ignition microprocessors in the new Novus automobile. The ignition microprocessor is in fact a small computer that controls starting and running the car at maximum efficiency. The quality control director wishes to demonstrate that the proportion of defective ignition microprocessors is less than 0.001 (less than 1 in 1,000). What form of the null and alternate hypotheses should be used in this case?

Solution We are now interested in a *range* of values of the population proportion π, where π represents the proportion of defective ignition microprocessors. The range of interest is a proportion less than 0.001, or from 0.001 to 0.

The quality control director is posing the following question: "Is π, the proportion of defective ignition microprocessors, greater than or equal to 0.001, or is there strong evidence (based on the sample) that π is less than 0.001?" The correct form of the hypotheses is Form III:

$$H_0: \quad \pi \geq 0.001$$
$$H_A: \quad \pi < 0.001$$

Of course the director hopes that the sample will provide strong evidence in favor of $H_A: \pi < 0.001$ so that the null hypothesis can be rejected.

The general rules for choosing one of the three forms of the statistical hypothesis test are given in the following table.

The rules in Table 10.1 apply also to stating hypotheses concerning other population parameters, such as the population proportion π and the population standard deviation σ (or the population variance σ^2).

TABLE 10.1 | General rules for choosing the correct form of the null and alternate statistical hypotheses for the population mean μ

Form	*Hypothesis statement condition*
I H_0: $\mu = \mu_0$ H_A: $\mu \neq \mu_0$	A specific (single) value is being hypothesized about μ. The question to be answered is: "Is μ equal to μ_0, or is there strong evidence that μ is not equal to μ_0?"
II H_0: $\mu \geq \mu_0$ H_A: $\mu < \mu_0$	A range of values is being hypothesized. The question to be answered is: "Is μ greater than or equal to μ_0, or is there strong evidence that μ is less than μ_0?"
III H_0: $\mu \leq \mu_0$ H_A: $\mu > \mu_0$	A range of values is being hypothesized. The question to be answered is: "Is μ less than or equal to μ_0, or is there strong evidence that μ is greater than μ_0?"

As we shall see, the approach to statistical hypothesis testing *assumes* that the null hypothesis is correct and looks for strong evidence in the sample to suggest the contrary. Because of this, the alternate hypothesis is very important. It states the values of the population parameter over which the null hypothesis is not true. Consider the SAT mean (μ) score example mentioned earlier. If we conjecture that μ is equal to 1,000, then the null and alternate hypotheses are:

$$H_0: \quad \mu = 1,000$$
$$H_A: \quad \mu \neq 1,000$$

From the alternate hypothesis, we know that if there is evidence in the sample to suggest that μ is either much less than 1,000 or much more than 1,000, then we have a basis for rejecting H_0: $\mu = 1,000$. [Again, what constitutes "much less" and "much more" will be dealt with in Section 10.2.]

If we conjecture that μ is less than or equal to 1,000 against the alternative that μ is more than 1,000, then the hypotheses are:

$$H_0: \quad \mu \leq 1,000$$
$$H_A: \quad \mu > 1,000$$

Now, evidence in the sample suggesting that μ is less than or equal to 1,000 *supports* the null hypothesis, and evidence in the sample that μ is much more than 1,000 *does not support* the null hypothesis. In this sense, the alternate hypothesis states the *direction* in which the null hypothesis is not reasonable.

Since the alternative establishes the direction of the sample evidence in which the null hypothesis is not reasonable, it is an important consideration in selecting one of the three forms (I, II, or III) of the null and alternate hypothesis pairs. We will use the guidelines in Table 10.1 for selecting the correct form. The selection process of the correct form is being somewhat simplified at this point. In Part II, Section 10.11, we shall return to the issue of selecting the hypothesis form in more detail.

Sample information is used to decide whether or not the null hypothesis (H_0) is rejected. The decision is *always* made about the null hypothesis. Before explaining how the decision rule to reject or not to reject the null hypothesis based on sample information is determined, we must understand that there are four possible outcomes resulting from our decision regarding the null hypothesis—two are favorable outcomes and two are not.

These four outcomes are illustrated in Table 10.2. If the null hypothesis is true (column 1), then we may decide not to reject H_0 or to reject H_0 (and

TABLE 10.2 | States of nature (reality) and the conclusions reached in statistical hypothesis testing

Conclusion reached			State of nature (reality)	
			H_0 *is true.*	H_A *is true.*
		H_0 (do not reject H_0)	Correct decision.	Incorrect decision, Type II error
		H_A (reject H_0)	Incorrect decision, Type I error.	Correct decision.

conclude that H_A is correct). If we do not reject H_0 when it is true, then we have made a correct decision. If we reject H_0 when it is true, then we have committed an error—it is called a *Type I error*.

If the null hypothesis is false, then H_A is true, and we may conclude either H_0 or H_A based on the sample evidence. If we conclude H_0, then we have committed an error, it is called a *Type II error*. If we conclude H_A (by rejecting H_0), then we have made a correct decision.

Understandably, we will want to know what our chances are of making either type of error in a given hypothesis-testing experiment. The measurements of the chance of committing these two types of error are denoted by α and β where $\alpha = P(\text{Type I error})$ and $\beta = P(\text{Type II error})$.

Type I error

Definition 10.1
Type I error

A *Type I error* in a statistical hypothesis-testing experiment is committed by rejecting the null hypothesis when it is true. The probability of committing a Type I error is denoted by α where

$$\alpha = P(\text{Type I error})$$
$$\alpha = P(\text{rejecting } H_0/H_0 \text{ is true})$$

Type II error

Definition 10.2

Type II error

A *Type II error* in a statistical hypothesis-testing experiment is committed by not rejecting the null hypothesis when it is false. The probability of committing a Type II error is denoted by β where

$$\beta = P(\text{Type II error})$$
$$\beta = P(\text{not rejecting } H_0/H_0 \text{ is false})$$

The relationship of α and β to various choices concerning whether to reject or not to reject the null hypothesis is indicated in Table 10.3.

TABLE 10.3 | States of nature (reality) and the conclusions reached; the probabilities of making correct and incorrect decisions

Conclusion reached		*State of nature* (reality)	
		H_0 *is true*	H_A *is true*
	H_0	$1 - \alpha$	β
	H_A	α	$1 - \beta$
	Total probability	1	1

We note two important properties in Table 10.3. First, since one of the two conclusions must be reached in testing a statistical hypothesis (H_0 or H_A), by complementary probability arguments, *each column* in Table 10.3 must sum to 1. That is,

$$P(\text{do not reject } H_0 | H_0 \text{ is true}) + P(\text{reject } H_0 | H_0 \text{ is true})$$
$$= (1 - \alpha) + \alpha = 1$$
$$P(\text{do not reject } H_0 | H_A \text{ is true}) + P(\text{reject } H_0 | H_A \text{ is true})$$
$$= \beta + (1 - \beta) = 1$$

Second, in examining each *row* in the table, we see that a Type I error can be made *only if* we reject the null hypothesis and that a Type II error can be made *only if* we do not reject the null hypothesis.

■ 10.2

The classical statistical hypothesis-testing method

In this section we will describe the classical hypothesis-testing procedure. In addition the computation of α, the probability of committing a Type I error, and the computation of β, the probability of committing a Type II error, will be demonstrated.

Suppose we are interested in testing a hypothesis about the population mean μ. Specifically, let us assume that the null and alternate hypotheses are:

$$H_0: \quad \mu = 100$$
$$H_A: \quad \mu \neq 100$$

Further, we will assume that the value of the population standard deviation σ is known and is equal to 25. Let us suppose that a sample size of $n = 25$ will be used to decide if the null hypothesis $H_0: \mu = 100$ is true or not. If the null hypothesis H_0 is *assumed* to be true, then it is possible to describe the sampling distribution of $\overline{X}$, the estimator of μ. Recalling that the standard deviation (also referred to as the *standard error*) of $\overline{X}$ is $\sigma_{\overline{x}} = \sigma/\sqrt{n}$, the

Standard error of $\overline{X}$

sampling distribution of $\overline{X}$ is shown in Figure 10.2.

FIGURE 10.2 | The sampling distribution of $\overline{X}$, assuming $H_0: \mu = 100$ is true and $\sigma_{\overline{x}} = \sigma/\sqrt{n} = 25/\sqrt{25} = 5$

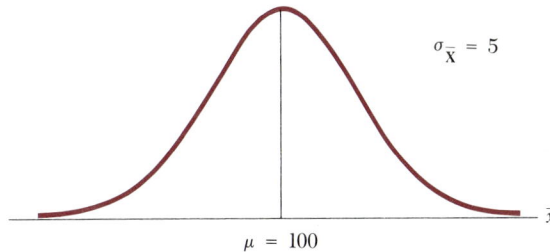

$\sigma_{\overline{x}} = 5$

$\mu = 100$

$\overline{x}$

In Figure 10.2 the sampling distribution of $\overline{X}$ is assumed to be normally distributed with a mean of 100 and a standard deviation of 5. From the normal distribution, we know that approximately 68 percent of the values will be between the mean $\pm$ one standard deviation, or from $100 - 5 = 95$ to $100 + 5 = 105$ in this case. Further, approximately 95 percent of the values will be between the mean $\pm$ two standard deviations, or from $100 - 2(5) = 90$ to $100 + 2(5) = 110$ in this case. Therefore, if μ *is really equal to 100*, it would not be very likely to take a sample of size 25 and find that the sample mean $\overline{X}$ has a value less than 90 or more than 110. In fact, the probability that $\overline{X}$ will be less than 90 or more than 110 is approximately 0.05, since the probability that $\overline{X}$ will be between 90 and 110 is approximately 0.95. It seems *reasonable* that if we take a sample of size $n = 25$ and find that the value of $\overline{X}$

is less than 90 or more than 110, then we would reject the null hypothesis H_0: $\mu = 100$ and conclude that the alternate hypothesis is true, $H_A: \mu \neq 100$.

In fact this discussion sets the basis for the classical method of hypothesis testing. The two key concepts are:

Key concepts of the classical method

1. *Assume that the null hypothesis is true.* In the above illustration, this means that we center the sampling distribution of $\overline{X}$ over $\mu = 100$.

2. *Reject the null hypothesis, H_0, if the value of the estimator of the population parameter is too far removed from the hypothesized value of the population parameter.* In the above illustration, we may decide to reject $H_0: \mu = 100$ if the sample mean $\bar{x}$ is less than 90 or more than 110, for example.

An obvious question at this point considering concept 2 is: How do we define "too far removed" from the hypothesized value of the population parameter? Deciding how much $\bar{x}$ computed from the sample can differ from the hypothesized value of μ before we decide to reject the null hypothesis as being unreasonable—depends on the values of α and β (the probabilities of committing a Type I error and Type II error, respectively). To illustrate this, we first need to define the action limits for the classical hypothesis testing method.

Action limits

Decision rule

Definition 10.3
The action limits of a hypothesis testing experiment

The action limits are points on the sampling distribution of the point estimator of the population parameter of interest that determine a *decision rule*. The decision rule determines when the null hypothesis should be rejected or not be rejected on the basis of the value of the point estimator.

Example 10.4 Suppose we have the following null and alternate hypotheses concerning the population mean μ, as above:

$$H_0: \quad \mu = 100$$
$$H_A: \quad \mu \neq 100$$

We will assume that the value of the population standard deviation σ is 25 and that we are planning to take a sample from the population of size $n = 25$. Let us suppose that we set the *action limits* at $\bar{x}_L = 90$ and $\bar{x}_U = 110$; that is, if the sample mean $\bar{x}$ from our sample of size $n = 25$ is less than $\bar{x}_L = 90$ or more than $\bar{x}_U = 110$, then we will *reject* the null hypothesis. Determine the decision rule based on these action limits, and show the decision rule on the sampling distribution of $\overline{X}$.

Solution The decision rule based on the action limits $\bar{x}_L = 90$ and $\bar{x}_U = 110$ is:

> Reject H_0: $\mu = 100$ if $\bar{x} < \bar{x}_L = 90$ or if $\bar{x} > \bar{x}_U = 110$.
> Do not reject H_0: $\mu = 100$ if $\bar{x}_L = 90 \leq \bar{x} \leq \bar{x}_U = 110$.

This decision rule is shown on the sampling distribution of $\overline{X}$ in Figure 10.3. Since we are hypothesizing that H_0: $\mu = 100$, $\mu_0 = 100$ is the conjectured value in the null hypothesis. Notice that the sampling distribution in Figure 10.3 is centered at $\mu_0 = 100$. This tells us that the *conjectured* value of μ is 100, not that the actual value of μ is known to be 100.

FIGURE 10.3 | The decision rule for the Example 10.4 hypothesis-testing problem

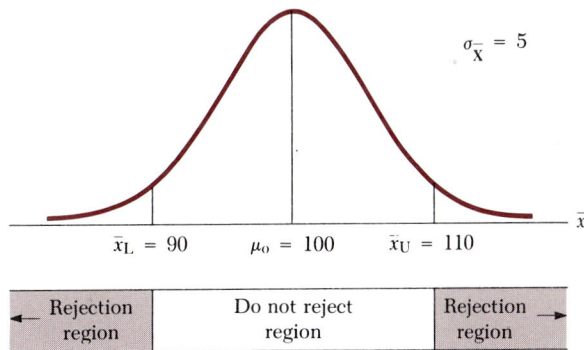

Notice in Figure 10.3 that we refer to the set of values $\bar{x}$ when $\bar{x}$ is less than 90 or more than 110 as the *rejection region* and the set of values $\bar{x}$ when $\bar{x}$ falls between 90 and 110 as the *do not reject region*. For example, if $\bar{x}$ equals 120 in our sample of size $n = 25$ taken from the population, then $\bar{x}$ falls in the upper rejection region, and we would reject the null hypothesis H_0: $\mu = 100$.

In Example 10.4, the action limits $\bar{x}_L = 90$ and $\bar{x}_U = 110$ were chosen arbitrarily. We still must determine a method of setting these limits. The next two examples help explain the method for setting the action limits by the classical hypothesis-testing method.

Rejection region

Do not reject region

Example 10.5 Assume for the null and alternate hypotheses in Example 10.4 that the action limits $\bar{x}_L = 90$ and $\bar{x}_U = 110$ are used, resulting in the decision rule illustrated in Figure 10.3. What is the probability of committing a Type I error for this decision rule?

Solution Recall that the probability of committing a Type I error is:

$$\alpha = P(\text{Type I error}) = P(\text{reject } H_0/H_0 \text{ is true}).$$

If the null hypothesis H_0: $\mu = 100$ is true, then for the decision rule in Example 10.4 we will reject H_0 if $\bar{x} < \bar{x}_L = 90$ or if $\bar{x} > \bar{x}_U = 110$, *thereby committing a Type I error*. Thus α, the probability of committing a Type I error, is the area under the sampling distribution of $\overline{\mathbf{X}}$ to the left of $\bar{x}_L = 90$ and to the right of $\bar{x}_U = 110$, as shown in Figure 10.4.

FIGURE 10.4 | The probability of committing a Type I error for the decision rule: Reject H_0 if $\bar{x} < 90$ or if $\bar{x} > 110$; do not reject H_0 if $90 \le \bar{x} \le 110$

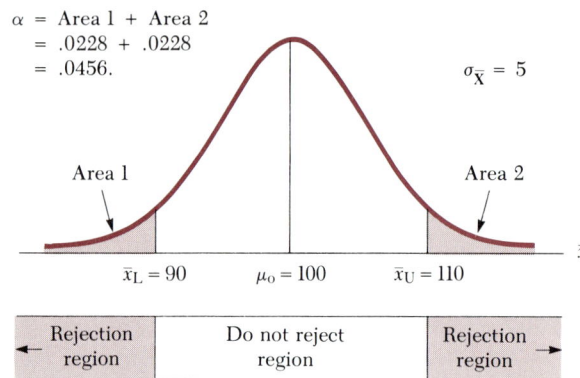

Computation of the area representing the probability of committing a Type I error follows the methods described in Section 7.3, Chapter 7, for finding areas under the normal distribution. We wish to find $P(\overline{\mathbf{X}} < 90)$ and $P(\overline{\mathbf{X}} > 110)$. We can use the standard normal distribution and Table B.3 in Appendix B by standardizing the values 90 and 110 using this transformation:

$$\mathbf{Z} = \frac{\overline{\mathbf{X}} - \mu_0}{\sigma_{\overline{\mathbf{x}}}}$$

For $\bar{x} = 90$, we have:

$$z = \frac{90 - 100}{5} = \frac{-10}{5} = -2$$

and for $\bar{x} = 110$:

$$z = \frac{110 - 100}{5} = \frac{10}{5} = 2$$

Now we have $P(\overline{X} < 90) = P(Z < -2)$ and $P(X > 110) = P(Z > 2)$. From the standard normal table (Table B.3), $P(Z < -2) = 0.0228$ and $P(Z > 2) = 0.0228$. Therefore,

$$\alpha = P(\text{Type I error}) = 0.0228 + 0.0228 = 0.0456$$

With this decision rule, before the sample is taken and the value of $\overline{X}$ is known, we stand a 4.56 percent chance of rejecting H_0: $\mu = 100$ when in fact H_0 is true.

Example 10.6   Suppose that we now change the decision rule in Example 10.4. The full description of the hypothesis test with the new decision rule is given below:

H_0: $\mu = 100$
H_A: $\mu \neq 100$
Sample size $n = 25$.
Population standard deviation $\sigma = 25$.
Standard error of $\overline{X}$: $\sigma_{\overline{x}} = 5$.
Action limits: $\bar{x}_L = 85$ and $\bar{x}_U = 115$.
Decision rule: Reject H_0 if $\bar{x} < 85$ or if $\bar{x} > 115$.
Do not reject H_0 if $85 \leq \bar{x} \leq 115$.

For this decision rule, find α, the probability of committing a Type I error.

Solution   Notice that the action limits $\bar{x}_L = 85$ and $\bar{x}_U = 115$ have been moved farther away from the hypothesized value of $\mu(\mu_0 = 100)$ when compared with the action limits in the Example 10.4 decision rule [$\bar{x}_L = 90$ and $\bar{x}_U = 110$]. The area representing α is shown in Figure 10.5. Standardizing $\bar{x} = 85$ and $\bar{x} = 115$, we have:

$$z = \frac{85 - 100}{5} = -3 \quad \text{and} \quad z = \frac{115 - 100}{5} = 3.$$

Then

$$P(\overline{X} < 85) = P(Z < -3) = 0.00135 \quad \text{(from Table B.3)}$$
$$P(\overline{X} > 115) = P(Z > 3) = 0.00135 \quad \text{(from Table B.3)}$$

Therefore

$$\alpha = P(\text{Type I error}) = 0.00135 + 0.00135 = 0.0027$$

Notice that the probability of committing a Type I error with this new decision rule is less than the probability of committing a Type I error with the old decision rule in Example 10.4. The relationship between the size of the

FIGURE 10.5 | The probability of committing a Type I error for the decision rule: Reject H_0 if $\bar{x} < 85$ or if $\bar{x} > 115$; do not reject H_0 if $85 \le \bar{x} \le 115$

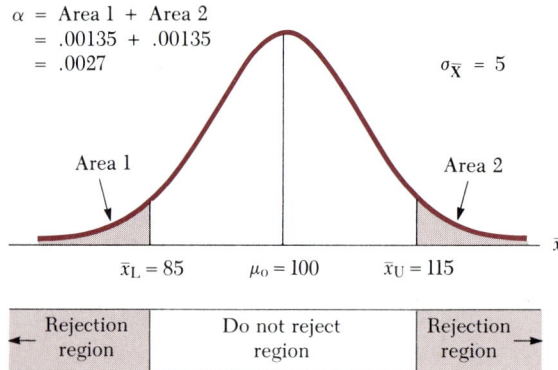

$$\alpha = \text{Area 1} + \text{Area 2}$$
$$= .00135 + .00135$$
$$= .0027$$

$\sigma_{\bar{x}} = 5$

Area 1

Area 2

$\bar{x}_L = 85$ $\qquad$ $\mu_0 = 100$ $\qquad$ $\bar{x}_U = 115$ $\qquad$ $\bar{x}$

| Rejection region | Do not reject region | Rejection region |

rejection region and the probability of committing a Type I error is important:

> As the size of the rejection region decreases (and the size of the do not reject region increases), the probability of committing a Type I error *decreases*.

In Examples 10.4 and 10.5, we saw how changing the action limits and therefore the decision rule changed the value of α, the probability of committing a Type I error. By reversing the process of finding α, given the action limits of the decision rule, we can stipulate the value of α that we desire and then find the corresponding decision rule with action limits as the following example illustrates.

Example 10.7 Suppose in the Example 10.4 problem that we wish to *set* α, the probability of committing a Type I error, equal to 0.05. Find the action limits $\bar{x}_L$ and $\bar{x}_U$, and therefore the decision rule for which α, the probability of committing a Type I error, will be 0.05.

Solution We can determine the required action limits by reversing the process used in Examples 10.5 and 10.6. From these examples, notice that α is the sum of two symmetric areas in the left and right tail of the distribution. The probability $\alpha/2$ is in the left tail, and the probability $\alpha/2$ is in the right tail of the sampling distribution for $\overline{\mathbf{X}}$. Also we used the transformation equation:

$$\mathbf{Z} = \frac{\overline{\mathbf{X}} - \mu_0}{\sigma_{\overline{\mathbf{X}}}}$$

where μ_0 is the conjectured value of μ in Form I in the null hypothesis ($H_0: \mu = \mu_0$).

If we can find the values of $\mathbf{Z}$ on the standard normal distribution so that the areas in the left and right tails are $\alpha/2 = 0.05/2 = 0.025$, then we can solve the transformation equation above for $\overline{\mathbf{X}}$ to find the action limits on the sampling distribution of $\overline{\mathbf{X}}$. This process is illustrated in Figure 10.6.

From Table B.3, Appendix B the z-value so that the area in the upper tail is 0.025 (and therefore the area between 0 and the z-value is 0.475) is 1.96. By symmetry the z-value in the lower tail is -1.96. Solving the transformation equation for $\overline{\mathbf{X}}$ gives

$$\overline{\mathbf{X}} = \mu_0 + z\sigma_{\overline{\mathbf{X}}}, \text{ where } \sigma_{\overline{\mathbf{X}}} = \sigma/\sqrt{n}$$

FIGURE 10.6 | The determination of the action limits with α, the probability of committing a Type I error, set at 0.05

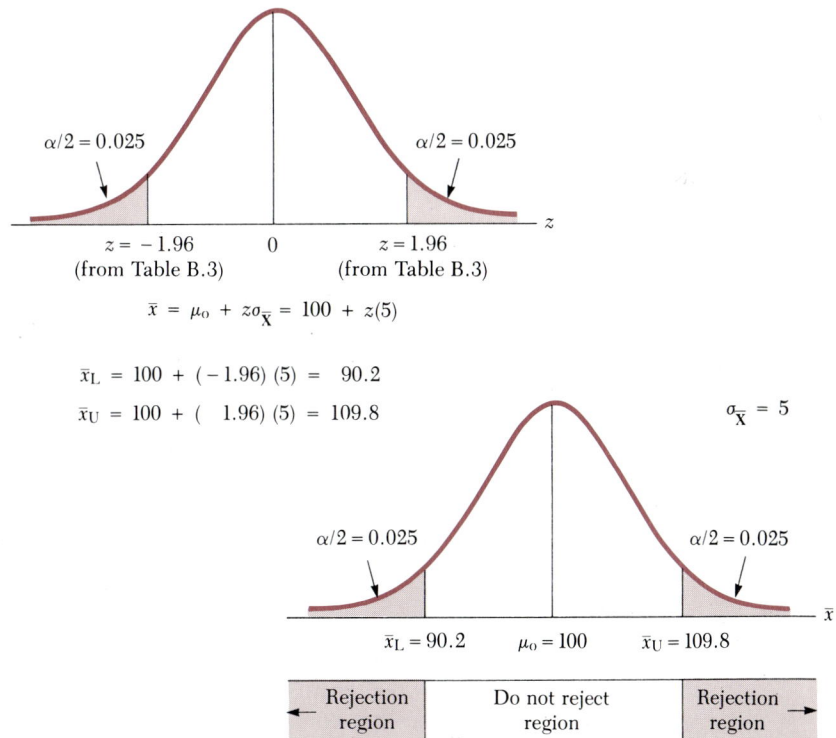

$\alpha/2 = 0.025$ $\alpha/2 = 0.025$

$z = -1.96$ 0 $z = 1.96$
(from Table B.3) (from Table B.3)

$\overline{x} = \mu_0 + z\sigma_{\overline{x}} = 100 + z(5)$

$\overline{x}_L = 100 + (-1.96)(5) = 90.2$

$\overline{x}_U = 100 + (1.96)(5) = 109.8$

$\sigma_{\overline{x}} = 5$

$\alpha/2 = 0.025$ $\alpha/2 = 0.025$

$\overline{x}_L = 90.2$ $\mu_0 = 100$ $\overline{x}_U = 109.8$

| Rejection region | Do not reject region | Rejection region |

The upper action point $\bar{x}_U$ and the lower action point $\bar{x}_L$ are then given by:

$$\bar{x}_U = 100 + (1.96)(5) = 109.8 \text{ and } \bar{x}_L = 100 + (-1.96)(5) = 90.2$$

The decision rule that will have $\alpha = P(\text{Type I error})$ set at 0.05 is therefore:

Reject H_0: $\mu = 100$ if $\bar{x} < 90.2$ or if $\bar{x} > 109.8$
Do not reject H_0: $\mu = 100$ if $90.2 \le \bar{x} \le 109.8$

The above action limits are very close to those in Example 10.4, for which we determined α to be 0.0456 of course.

Example 10.7 illustrates a final key concept in the classical method of hypothesis testing:

Final key concept of classical method

Set the value of $\alpha = P(\text{Type I error}) = P(\text{reject } H_0/H_0 \text{ is true})$. Use α to establish the action limits that define the decision rule.

For this reason, the classical method of hypothesis testing is said *to control the probability of making a Type I error*. We are now in a position to summarize the steps in the classical hypothesis-testing method.

We will apply these six steps in testing hypotheses concerning the population mean μ and the population proportion π in the next two sections. However, before we move on to specific examples of hypothesis testing for μ and π, we must discuss the other type of error that can be made in hypothesis testing: the Type II error, accepting the null hypothesis when it is false. Recall that the probability measure of this error is denoted by β and is given by:

$$\beta = P(\text{Type II error}) = P(\text{do not reject } H_0/H_0 \text{ is false})$$

In the hypothesis-testing method we are describing, the value of β is not set (the value of α is) and is unknown. However, if we *assume* a value for the population mean μ that *makes the null hypothesis false*, symbolized by μ_A, then we can calculate the value of β as the next two example problems illustrate.

STEPS IN THE CLASSICAL METHOD OF HYPOTHESIS TESTING

1. *State the null and alternate hypotheses.* We must choose one of the three forms of the hypotheses, Form I, Form II, or Form III.

2. *Select the statistic in the sample upon which to base the decision rule. This statistic is called the test statistic.* For example, in testing a hypothesis about a population mean μ, the test statistic we use is $\overline{X}$, the sample mean.

3. *Set the value of α, the probability of committing a Type I error (rejecting the null hypothesis when it is true).* The usual settings for α are 0.01, 0.05, and 0.10.

4. *Construct the decision rule with its action limits using the chosen value of α.*

5. *Draw a random sample of size n and compare the value of the test statistic in the sample to the action limits.* Testing a hypothesis about a population mean μ involves calculating the sample mean $\bar{x}$ from the sample observations.

6. *Make the decision to either reject H_0 or to not reject H_0 based on the decision rule—the comparison of the test statistic value to the action limits describing the rejection and do not reject regions.*

Example 10.8 Consider the hypothesis-testing problem described in Example 10.4:

H_0: $\mu = 100$
H_A: $\mu \neq 100$
Sample size: $n = 25$
Population standard deviation $\sigma = 25$
Standard error of $\overline{X}$: $\sigma_{\overline{x}} = 5$
Decision rule: Reject H_0 if $\bar{x} < \bar{x}_L = 90$ or if $\bar{x} > \bar{x}_U = 110$.
       Do not reject H_0 if $90 \leq \bar{x} \leq 110$.

Assume that μ is equal to $\mu_A = 105$, not the hypothesized value, $\mu_0 = 100$. If the value of μ is $\mu_A = 105$, then the null hypothesis H_0: $\mu = 100$ is false. For this decision rule, what is the probability of committing a Type II error, β?

Solution   If the population mean μ is assumed to be equal to $\mu_A = 105$, then the correct sampling distribution of $\overline{X}$ is shown in Figure 10.7. Notice that the sampling distribution is no longer centered over the hypothesized value of μ (100) but over the assumed true value of μ ($\mu_A = 105$). Now if the sample mean $\overline{X}$ has a value between 90 and 110 when it is calculated in the sample of size $n = 25$, the value falls in the do not reject region. We will therefore be led to the decision to not reject H_0: $\mu = 100$, when in fact this hypothesis is false (thereby we commit a Type II error). The probability of committing the Type II error is the area over the do not reject region under

FIGURE 10.7 | The computation of β, the probability of committing a Type II error for the decision rule: Reject H_0: $\mu = 100$ if $\bar{x} < 90$ or if $\bar{x} > 110$; do not reject H_0 if $90 \leq \bar{x} \leq 110$ [assuming $\mu_A = 105$]

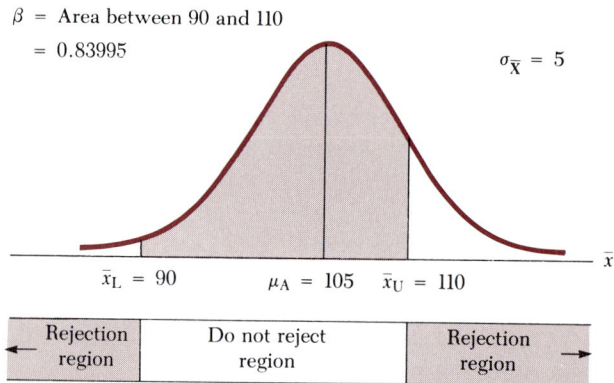

β = Area between 90 and 110
= 0.83995

$\sigma_{\bar{X}} = 5$

$\bar{x}_L = 90$ $\mu_A = 105$ $\bar{x}_U = 110$

| Rejection region | Do not reject region | Rejection region |

the sampling distribution of $\bar{X}$ as illustrated in Figure 10.7. We can find this area by using the standard normal distribution as before. First we must standardize the action limits whose values are 90 and 110:

$$z = \frac{90 - 105}{5} = -3 \quad \text{and} \quad z = \frac{110 - 105}{5} = 1$$

Now $P(90 < \bar{X} < 110) = P(-3 < Z < 1) = 0.49865 + 0.3414$ (from Table B.3). Thus $\beta = 0.49865 + 0.3413 = 0.83995$ (or rounded, 0.84). Thus there is perhaps a surprisingly high chance of committing a Type II error with this decision rule *assuming that $\mu = 105$*. Since $\mu_A = 105$ is not much different from hypothesized value ($\mu_0 = 100$), it is really not too surprising that this rule will result in not rejecting H_0: $\mu = 100$, when in fact $\mu = 105$ with such a high probability. If the true value of μ were farther from the hypothesized value, say $\mu = 120$, then much less of the area under the sampling distribution would fall over the do not reject region, resulting in a small value of β as illustrated in Figure 10.8. The computation of β is left as an exercise at the end of the chapter.

Example 10.9   We will now use the decision rule given in Example 10.6 and recompute the value of β. The revised decision rule is:

Reject H_0 if $\bar{x} < \bar{x}_L = 85$ or if $\bar{x} > \bar{x}_U = 115$.
Do not reject H_0 if $85 \leq \bar{x} \leq 115$

FIGURE 10.8 | The probability of committing a Type II error when we assume that the true value of μ is $\mu_A = 120$, using the decision rule in Example 10.8

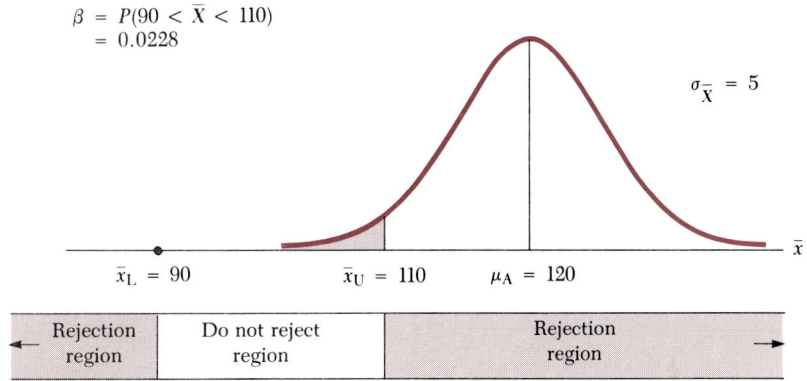

$\beta = P(90 < \bar{X} < 110)$
$\quad = 0.0228$

$\sigma_{\bar{X}} = 5$

$\bar{x}_L = 90$ $\qquad$ $\bar{x}_U = 110$ $\qquad$ $\mu_A = 120$

Rejection region	Do not reject region	Rejection region

What is the value of β, the probability of committing a Type II error, with this decision rule, assuming that the true value of $\mu_A = 105$ as in Example 10.8?

Solution The computation of β is illustrated in Figure 10.9.

FIGURE 10.9 | The computation of β, the probability of committing a Type II error, for the decision rule: Reject H_0: $\mu = 100$ if $\bar{x} < 85$ or if $\bar{x} > 115$; do not reject H_0 if $85 \leq \bar{x} \leq 115$ [assuming $\mu_A = 105$]

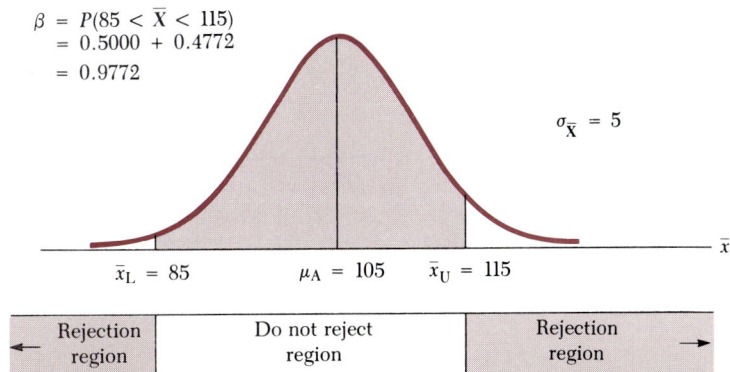

$\beta = P(85 < \bar{X} < 115)$
$\quad = 0.5000 + 0.4772$
$\quad = 0.9772$

$\sigma_{\bar{X}} = 5$

$\bar{x}_L = 85$ $\qquad$ $\mu_A = 105$ $\qquad$ $\bar{x}_U = 115$

Rejection region	Do not reject region	Rejection region

The computation of β proceeds as before. We first standardize the values of the action limits:

$$z = \frac{85 - 105}{5} = -4 \quad \text{and} \quad z = \frac{115 - 105}{5} = 2$$

Thus $\beta = P(85 < \overline{X} < 115) = P(-4 < \mathbf{Z} < 2) = 0.5000 + 0.4772$ from Table B.3 (notice that the area from -4 to 0 is for all practical purposes 0.5). Thus the probability of committing a Type II error for this decision rule, assuming that $\mu_A = 105$, is $0.5000 + 0.4772 = 0.9772$.

The comparisons of α, the probability of committing a Type I error, and β, the probability of committing a Type II error, for the two decision rules in Examples 10.5 and 10.6 (and repeated in Examples 10.8 and 10.9) are given in Table 10.4.

TABLE 10.4 | Comparison of the values of α and β for the decision rules in Examples 10.5 and 10.6 (repeated in Examples 10.8 and 10.9) $H_0 : \mu = 100$

Decision rule	$\alpha = P(\textit{Type I error})$	$\beta = P(\textit{Type II error})$ (with $\mu_A = 105$)
Reject H_0 if $\bar{x} < 90$ or if $\bar{x} > 110$; do not reject H_0 if $90 \leq \bar{x} \leq 110$	0.0456	0.83995
Reject H_0 if $\bar{x} < 85$ or if $\bar{x} > 115$; do not reject H_0 if $85 \leq \bar{x} \leq 115$	0.0027	0.9772

Notice in Table 10.4 that for the first decision rule, α is larger than for the second decision rule, but the reverse is true about β. The value of β is smaller for the first decision rule than it is for the second. This is an example of an interesting relationship between the values of α and β.

Inverse relationship of α and β

For a fixed sample size n, the values of α and β are inversely related; that is, as one becomes smaller, the other becomes larger.

This relationship can be seen by placing the sampling distribution for the decision rule with $\alpha = 0.0456$ (Figure 10.7) and for the decision rule with $\alpha =$

0.0027 (Figure 10.9) together. Notice that as α is decreased from 0.0456 to 0.0027, β increases from 0.83995 to 0.9772. (See Figure 10.10.)

We should comment that these values for β are very large and are not typically encountered in practice. They are large in this case due to the small

FIGURE 10.10 | Change in β as α is decreased from 0.0456 to 0.0027

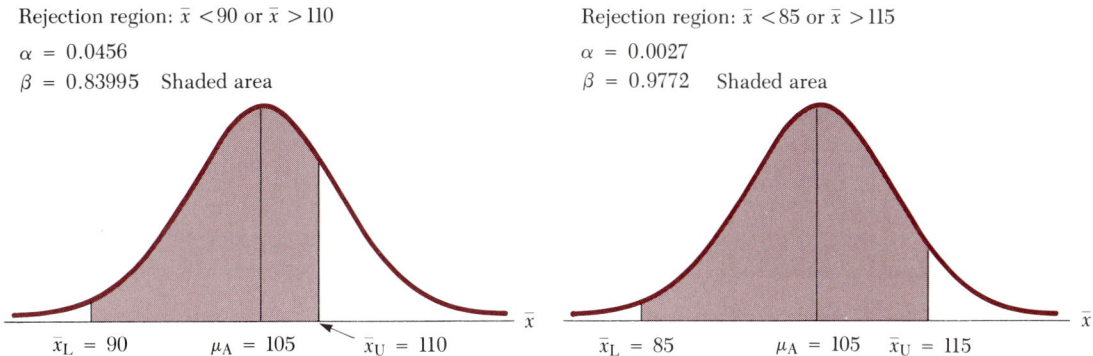

Rejection region: $\bar{x} < 90$ or $\bar{x} > 110$

$\alpha = 0.0456$

$\beta = 0.83995$ Shaded area

$\bar{x}_L = 90$ $\mu_A = 105$ $\bar{x}_U = 110$

Rejection region: $\bar{x} < 85$ or $\bar{x} > 115$

$\alpha = 0.0027$

$\beta = 0.9772$ Shaded area

$\bar{x}_L = 85$ $\mu_A = 105$ $\bar{x}_U = 115$

sample size ($n = 25$) and the closeness of $\mu_A = 105$ to the hypothesized value $\mu_0 = 100$. Therefore if we change the decision rule to decrease the value of α, then the value of β will increase. This is why α is not set too small in classical hypothesis testing—usually α is not set smaller than 0.01. The value of β depends on other factors, such as sample size and how close μ_A is to μ_0. However, on the sampling distribution of $\overline{X}$, we know that as α is decreased, β will increase.

Choosing the form of the null and alternate hypotheses is an important decision in the classical method of hypothesis testing. Since $\alpha = P[\text{Type I error}] = P[\text{reject } H_0/H_0 \text{ is true}]$ is set in classical hypothesis testing, its value is under our control.

If we decide to reject H_0, then either we have made a correct decision (H_A is true) or we have made a Type I error—rejecting H_0 when it is true. Thus by rejecting H_0 we know the probability that we may have made an error is α, the set value. If we decide not to reject H_0, then we have either made a correct decision (H_0 is true) or we have made a Type II error—not rejecting H_0 when it is false and H_A is true. Thus by not rejecting H_0, we know that the probability we may have made an error is β, but this value is not set.

To calculate β, we need to know the *true value* of μ (assuming it isn't equal to μ_0). We can calculate β *assuming* that $\mu = \mu_A$, a specific value of μ satisfying the alternate hypothesis. But since the actual value of μ is unknown to us in the population, so will the value of β be unknown. This is the reason the decision to reject H_0 is often called the *strong decision* (we know our chance of having made an error), and the decision not to reject H_0 is often called the *weak decision* (we do not know our chance of having made an error). Though the *actual* value of β is unknown, the calculation of β for *assumed values of μ* is important, as we shall see when we consider "the power of the test."

Because of the above discussion, it is usually desirable to choose a form of the null and alternate hypotheses so that the alternate hypothesis contains the proposition that we wish to demonstrate is correct. By rejecting the null hypothesis, we conclude that the alternate hypothesis is correct. By doing so, we know exactly the probability that we may have made an error—it is α.

Through the remainder of Part I, we will state the null and alternate hypotheses in examples used to illustrate the processes of hypothesis testing. In Part II, Section 10.12, we will return to the issue of formulating the null and alternate hypotheses in the light of our discussion here concerning α and β. We will now apply the classical hypothesis-testing method to testing hypotheses about the population mean μ and the population proportion π.

■ 10.3

Hypothesis testing concerning the population mean μ

The hypothesis-testing procedure for a population mean μ follows the six steps outlined in the previous section. The test statistic is the sample mean $\overline{X}$. Two cases must be considered when testing a hypothesis concerning the population mean μ: Case 1—the population standard deviation σ is known. Case 2—the population standard deviation σ is not known. These cases were used in Chapter 9 to differentiate between the two confidence interval estimate formulas for the population mean μ (one formula used the z-statistic, and the other formula used the t-statistic). For each of the two cases, the decision rules for each of the three possible forms (I, II, and III) of the null and alternate hypotheses will be given.

10.3.1 | Case 1: The population standard deviation σ is known

We will begin the testing procedures for a population mean μ when the population standard deviation σ is known with the first of the three forms for the null and alternate hypotheses:

$$\text{Form I:} \quad \begin{aligned} H_0: & \quad \mu = \mu_0 \\ H_A: & \quad \mu \neq \mu_0 \end{aligned}$$

The value μ_0 in the null and alternate hypotheses is the hypothesized value of the population mean μ. In Section 10.2 we discussed the classical hypothesis-testing procedure for the Form I hypotheses. The decision rule is summarized below.

Notice the requirement that $\overline{X}$, the sample mean, must be approximately normally distributed. The action limits $\bar{x}_L$ and $\bar{x}_U$ depend on a z-value ($z_{\alpha/2}$), and

$$Z = \frac{\overline{X} - \mu_0}{\sigma_{\overline{X}}}$$

Z-test

H_0: $\mu = \mu_0$
H_A: $\mu \neq \mu_0$

Hypothesis test about a population mean μ

Form I: H_0: $\mu = \mu_0$
$\qquad\qquad$ H_A: $\mu \neq \mu_0$

Test statistic: $\overline{X}$

Assumptions: 1. Population standard deviation σ is known.
$\qquad\qquad\quad$ 2. $\overline{X}$ is approximately normally distributed (sample size $n \geq$
$\qquad\qquad\qquad$ 25), *or* X is normally distributed (regardless of sample size).

Decision rule: Reject H_0 if $\bar{x} < \bar{x}_L$ or if $\bar{x} > \bar{x}_U$;
$\qquad\qquad\quad$ do not reject H_0 if $\bar{x}_L \leq \bar{x} \leq x_U$.

Action limits:

$$\bar{x}_L = \mu_0 - z_{\alpha/2}\sigma_{\overline{X}}$$
$$\bar{x}_U = \mu_0 + z_{\alpha/2}\sigma_{\overline{X}}$$

where $\sigma_{\overline{X}} = \sigma/\sqrt{n}$ and $z_{\alpha/2}$ is the value on the upper tail of the standard normal distribution so that $P(Z > z_{\alpha/2}) = \alpha/2$.

H_0: $\mu = \mu_0$
H_A: $\mu \neq \mu_0$

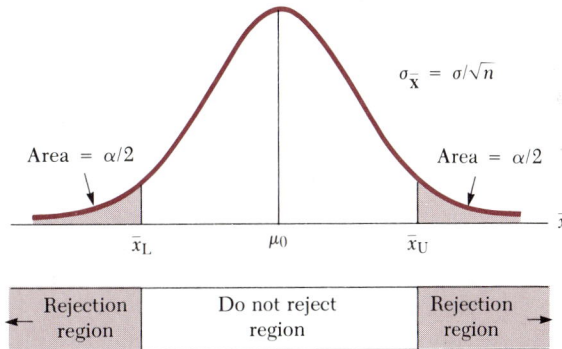

is a standard normal variable *if* $\overline{X}$ is normally distributed. The statistic $\overline{X}$ will be normally distributed if the population random variable X is normally distributed. If this is the case, then the testing procedure described above is exact. If the population random variable X is *not* normally distributed, then $\overline{X}$ is approximately normally distributed, provided that the sample size is large (via the central limit theorem). Usually a sample of size 25 or more will suffice. If this is the case (X is not normally distributed), then the testing procedure described above is approximate. The consequence in this event is that the probability of committing a Type I error (α) will not exactly equal the set value.

This test is commonly referred to as the Z-test, since its action limits are determined from the standard normal (Z) distribution.

The set value of the probability of committing a Type I error is referred to as the *significance level* of the test.

<table>
<tr><td>Significance level of test</td><td>

Definition 10.4

Significance level of the test

The *significance level* of the test is the selected value of α, where $\alpha = P(\text{Type I error}) = P(\text{reject } H_0/H_0 \text{ is true})$. The usual settings of α are 0.01, 0.05, and 0.10.

</td></tr>
</table>

Setting the value of the significance level (α) is the third step in the six-step hypothesis-testing procedure described in Section 10.2. This procedure will now be applied to an example problem.

Example 10.10 It is hypothesized that the mean starting salary of accounting graduates from accredited BBA degree programs with no prior work experience who are hired by one of the "Big 8" accounting firms is $24,000. A random sample of $n = 100$ starting salaries of recent graduates with no work experience hired by these firms is taken, and it is found that the sample average starting salary is $23,775. Assume that it is known from comprehensive studies of these starting salaries that the population standard deviation $\sigma = \$1,000$. Formulate the null and alternate hypotheses, and conduct the test of hypothesis using an $\alpha = 0.05$ significance level.

Solution It is being hypothesized that the population mean starting salary of accounting BBA graduates is $24,000. Thus the question is, "Is the mean starting salary (μ) equal to $24,000, or is there strong evidence in the sample suggesting that μ is either less than $24,000 or more than $24,000?" This is the Form I case:

$$H_0: \quad \mu = \$24,000$$
$$H_A: \quad \mu \neq \$24,000$$

The significance level is set at $\alpha = 0.05$. The value of the test statistic $\overline{X}$ is $\bar{x} = \$23,775$. To find the values of the action limits $\bar{x}_L$ and $\bar{x}_U$, we must first find the value of $z_{\alpha/2}$. Since $\alpha = 0.05$, then $\alpha/2 = 0.025$; and we want $z_{0.025}$ such that $P(\mathbf{Z} > z_{0.025}) = 0.025$. This uppertail area is illustrated under the standard normal distribution in Figure 10.11. With an area of 0.025 in the upper tail, the area between 0 and $z_{0.025}$ must be $0.50 - 0.025 = 0.475$. From

FIGURE 10.11 | The determination of $z_{\alpha/2} = z_{0.025}$ for the Example 10.10 problem

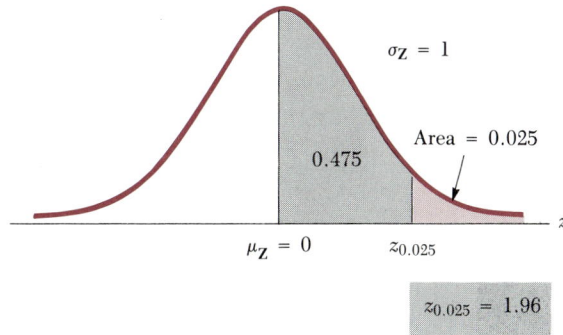

Table B.3, the z-value with an area of 0.475 between 0 and this value is 1.96. Thus $z_{0.025} = 1.96$. The value of the standard error of the mean $\overline{X}$ is:

$$\sigma_{\overline{X}} = \sigma/\sqrt{n} = \$1,000/\sqrt{100} = \$1,000/10 = \$100$$

The action limits for the decision rule are:

$$\overline{x}_L = \mu_0 - z_{\alpha/2}\sigma_{\overline{X}} = \$24,000 - (1.96)(\$100) = \$23,804$$
$$\overline{x}_U = \mu_0 - z_{\alpha/2}\sigma_{\overline{X}} = \$24,000 + (1.96)(\$100) = \$24,196$$

Therefore the decision rule is:

Reject H_0: $\mu = \$24,000$ if $\overline{x} < \$23,804$ or if $\overline{x} > \$24,196$.
Do not reject H_0: $\mu = \$24,000$ if $\$23,804 \leq \overline{x} \leq \$24,196$.

Since $\overline{x} = \$23,775 < \$23,804$, we reject the null hypothesis and conclude that the alternate hypothesis is true, H_A: $\mu \neq \$24,000$. The decision rule is shown in Figure 10.12.

In Forms II and III of the null and alternate hypotheses, we are testing a *range* of values in the null hypothesis. The nature of the decision rule changes from the Form I rule as we shall see. We will begin with the Form II case.

Z-test
H_0: $\mu \geq \mu_0$
H_A: $\mu < \mu_0$

Hypothesis test about a population mean μ

Form II: H_0: $\mu \geq \mu_0$

$\qquad\quad$ H_A: $\mu < \mu_0$

Test statistic: $\overline{X}$

Assumptions: 1. Population standard deviation σ is known.

$\qquad\qquad\quad$ 2. $\overline{X}$ is approximately normally distributed (sample size $n \geq$ 25) *or* X is normally distributed (regardless of sample size).

Decision rule: Reject H_0 if $\bar{x} < \bar{x}_L$; do not reject H_0 if $\bar{x} \geq \bar{x}_L$.

Action limit:

$$\bar{x}_L = \mu_0 - z_\alpha \sigma_{\overline{X}}$$

where $\sigma_{\overline{X}} = \sigma/\sqrt{n}$, and z_α is the value of the upper tail of the standard normal distribution so that $P(Z > z_\alpha) = \alpha$.

H_0: $\mu \geq \mu_0$

H_A: $\mu < \mu_0$

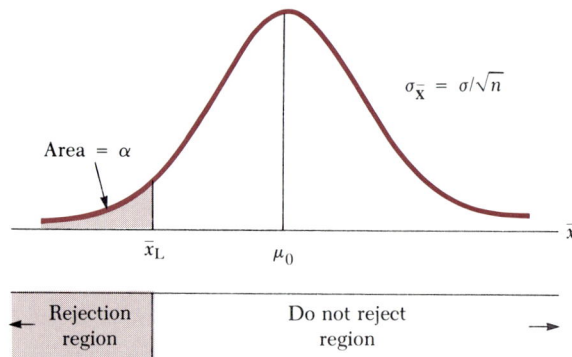

$\sigma_{\overline{X}} = \sigma/\sqrt{n}$

Area $= \alpha$

$\bar{x}_L$ $\qquad$ μ_0 $\qquad\qquad$ $\bar{x}$

| Rejection region | Do not reject region |

Notice in the figure that there is only one rejection region. Since we are hypothesizing that the population mean μ is *greater than or equal to* a specified value μ_0, any value of the sample mean $\overline{X}$ that falls to the right of μ_0 is consistent with the null hypothesis. Only when the sample mean value $\bar{x}$ becomes too much less than μ_0 (less than the lower action limit $\bar{x}_L$) are we

FIGURE 10.12 | The decision rule for the Example 10.10 problem

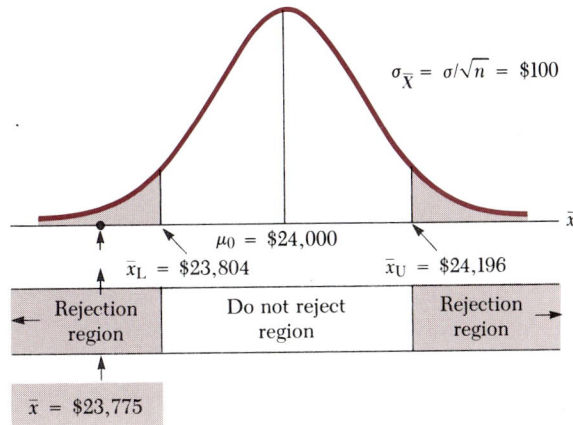

$\sigma_{\bar{X}} = \sigma/\sqrt{n} = \$100$

$\mu_0 = \$24,000$

$\bar{x}_L = \$23,804$ $\bar{x}_U = \$24,196$

| Rejection region | Do not reject region | Rejection region |

$\bar{x} = \$23,775$

persuaded to reject the null hypothesis. Since there is only one rejection region, the entire value of $\alpha = P(\text{Type I error})$ is placed as the area in the lower tail of the sampling distribution of $\overline{X}$. Because there is only one rejection region, this test is commonly referred to as a *one-tailed test*. The Form I test with $H_0: \mu = \mu_0$ and $H_A: \mu \neq \mu_0$ is referred to as a *two-tailed test*, since there are rejection regions in both tails of the sampling distribution of the test statistic.

Two-tailed test

Definition 10.5
A two-tailed hypothesis test

A *two-tailed hypothesis test* is a Form I test where the null hypothesis specifies a single value of the population parameter. For example, for the population mean μ, Form I is:

$H_0: \quad \mu = \mu_0$
$H_A: \quad \mu \neq \mu_0$

The decision rule for the Form I hypotheses will have a rejection region in both the upper and the lower tails of the sampling distribution of the test statistic ($\overline{X}$ for the population mean μ).

Definition 10.6

A one-tailed hypothesis test

A *one-tailed hypothesis test* is a Form II or III test where the null hypothesis specifies a range of values for a population parameter. For example, Forms II and III for the population mean μ are:

Form II	Form III
$H_0: \quad \mu \geq \mu_0$	$H_0: \quad \mu \leq \mu_0$
$H_A: \quad \mu < \mu_0$	$H_A: \quad \mu > \mu_0$

The decision rule for the hypothesis test will have a rejection region in either the lower tail only (Form II) or in the upper tail only (Form III) of the sampling distribution of the test statistic ($\overline{X}$ for the population mean μ).

We will now illustrate the use of the decision rule for testing a Form II hypothesis about a population mean μ.

Example 10.11   The Stay Bright Company claims that the mean lifetime in hours of a certain type of fluorescent light bulb it produces is "at least 1,600 hours." A consumer group takes a sample of 100 randomly selected fluorescent light bulbs of this type and finds that the mean lifetime in hours for these sampled bulbs is 1,560 hours. State the null and alternate hypotheses for this problem. From past studies of these and similar bulbs, it is assumed that the population standard deviation σ of the bulb lifetimes is 120 hours. Conduct the test of hypothesis at the $\alpha = 0.01$ level of significance.

Solution   The consumer group wishes to demonstrate that μ, the mean lifetime of the population of these bulbs is *not* "at least 1,600 hours." They are looking in the sample for strong evidence that this statement is false. The question to be answered is, "Is μ at least (greater than or equal to) 1,600 hours, or is there strong evidence to suggest that μ is less than 1,600 hours?" This is the Form II hypothesis statement:

$$H_0: \quad \mu \geq 1,600 \text{ hours}$$
$$H_A: \quad \mu < 1,600 \text{ hours}$$

The consumer group then hopes that it can reject H_0 based on the sample information. If it can reject H_0, then the probability that it has made an error cannot be more than $\alpha = P(\text{Type I error}) = P(\text{reject } H_0/H_0 \text{ is true}) = 0.01$. The value of the test statistic $\overline{X}$ is $\bar{x} = 1,560$ hours with a sample size $n = 100$. To find the action limit $\bar{x}_L$, we must first find z_α. Since $\alpha = 0.01$, we want $z_{0.01}$ so that $P(\mathbf{Z} > z_{0.01}) = 0.01$. From Table B.3, $z_{0.01} = 2.33$. The value of the standard error of $\overline{X}$ is: $\sigma_{\overline{X}} = \sigma/\sqrt{n} = 120/\sqrt{100} = 120/10 = 12$. Therefore, the action limit is:

$$\bar{x}_L = \mu_0 - z_{0.01}\sigma_{\overline{X}} = 1600 - (2.33)(12) = 1572.04$$

The decision rule is:

> Reject H_0: $\mu \geq 1,600$ hours if $\bar{x} < 1572.04$ hours.
> Do not reject H_0: $\mu \geq 1,600$ hours if $\bar{x} \geq 1572.04$ hours.

Since $\bar{x} = 1,560$ hours $< \bar{x}_L = 1,572.04$ hours, the consumer group should reject the null hypothesis and conclude that the population mean lifetime μ of these bulbs is less than 1,600 hours. This decision rule is illustrated in Figure 10.13.

FIGURE 10.13 | The decision rule for the Example 10.11 problem

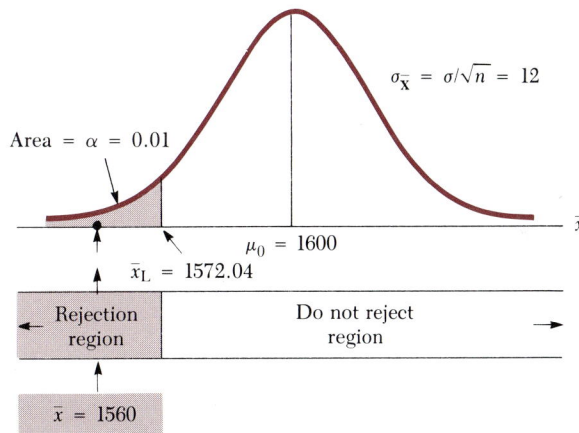

$\sigma_{\bar{x}} = \sigma/\sqrt{n} = 12$

Area $= \alpha = 0.01$

$\mu_0 = 1600$

$\bar{x}_L = 1572.04$

Rejection region

Do not reject region

$\bar{x} = 1560$

The description of the decision rule for the third of the three forms of the hypothesis test, Form III, is given on the next page.

We will now consider an example of the Form III hypothesis test concerning the population mean μ.

Example 10.12 The Texas Steel Company has recently changed its workweek from 40 hours to 36 hours per week for its workers. The workers' union has agreed to these changes. The company controller is concerned that the average number of hours worked per week may now exceed 40 hours due to the very favorable overtime provisions under the new 36-hour workweek union contract. She hypothesizes that μ, the average hours worked per week by all workers in the plant, is less than or equal to 40 hours.

Hypothesis test about a population mean μ

Form III: H_0: $\mu \leq \mu_0$
H_A: $\mu > \mu_0$

Test statistic: $\overline{X}$

Assumptions: 1. Population standard deviation σ is known.
2. $\overline{X}$ is approximately normally distributed (sample size $n \geq$ 25) *or* X is normally distributed (regardless of sample size).

Decision rule: Reject H_0 if $\bar{x} > \bar{x}_U$; do not reject H_0 if $\bar{x} \leq \bar{x}_U$.

Action limit:

$$\bar{x}_U = \mu_0 + z_\alpha \sigma_{\overline{X}}$$

where $\sigma_{\overline{X}} = \sigma/\sqrt{n}$, and z_α is the value of the upper tail of the standard normal distribution so that $P(\mathbf{Z} > z_\alpha) = \alpha$.

H_0: $\mu \leq \mu_0$
H_A: $\mu > \mu_0$

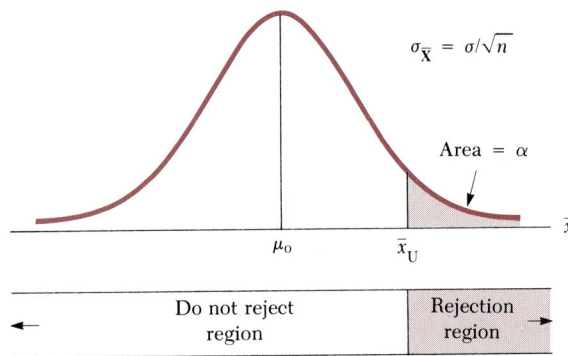

She is looking for strong evidence in the sample to reject this proposition so that she may conclude that μ is greater than 40 hours. A sample of 25 workers is taken for a recent week, and the number of hours each worked during that week is recorded. Those 25 observations are:

44	39	42	38	40	39	43	37	47	36	42	43	38
37	41	39	47	46	42	35	39	38	41	46	40	

Based on previous studies of the number of hours worked per week by the workers, she is willing to assume that the population standard deviation σ is 3.5 hours per week. State the null and alternate hypotheses, and test the null hypothesis at the $\alpha = 0.10$ level of significance.

Solution   The question the controller wishes to answer is, "Is μ, the average number of hours worked per week, less than or equal to 40, or is there strong evidence in the sample to indicate that μ is more than 40?" This is the Form III statement:

H_0: $\mu \leq 40$ hours
H_A: $\mu > 40$ hours

The test statistic value $\bar{x}$ must be computed for the 25 sample values:

$$\bar{x} = \frac{\sum_{i=1}^{n} x_i}{n} = \frac{(44 + 39 + \cdots + 46)}{25} = 40.76 \text{ hours}$$

Since $\alpha = 0.10$, we must find $z_\alpha = z_{0.10}$ so that $P(\mathbf{Z} > z_{0.10}) = 0.10$. From Table B.3, $z_{0.10} = 1.28$. The standard error of $\overline{\mathbf{X}}$ is $\sigma_{\bar{x}} = \sigma/\sqrt{n} = 3.50/\sqrt{25} = 3.50/5 = 0.70$. Thus the action limit is:

$$\bar{x}_U = \mu_0 + z_{0.10}\sigma_{\bar{x}} = 40 + (1.28)(0.70) = 40.896 \text{ hours.}$$

The decision rule is:

Reject H_0: $\mu \leq 40$ hours if $\bar{x} > 40.896$ hours.
Do not reject H_0: $\mu \leq 40$ hours if $\bar{x} \leq 40.896$ hours.

Since $\bar{x} = 40.76 < \bar{x}_U = 40.896$, the controller cannot reject the null hypothesis. Thus she makes the *weak decision* to not reject H_0: $\mu \leq 40$. Although the sample mean $\bar{x}$ is equal to 40.76 (and $40.76 > \mu_0 = 40$), there is not *strong* evidence to refute the null hypothesis proposition. By not rejecting H_0, she does not know her chance that an error may have been made (Type II error: Not rejecting H_0, given H_0 is false).

The decision rule is illustrated in Figure 10.14.

10.3.2 | Case 2: The population standard deviation σ is unknown

When the population random variable $\mathbf{X}$ is normally distributed, then $\overline{\mathbf{X}}$ is also normally distributed. If the population standard deviation σ is known, then the standardized variable

$$\mathbf{Z} = \frac{\overline{\mathbf{X}} - \mu}{\sigma/\sqrt{n}}$$

FIGURE 10.14 | The decision rule for the Example 10.12 problem

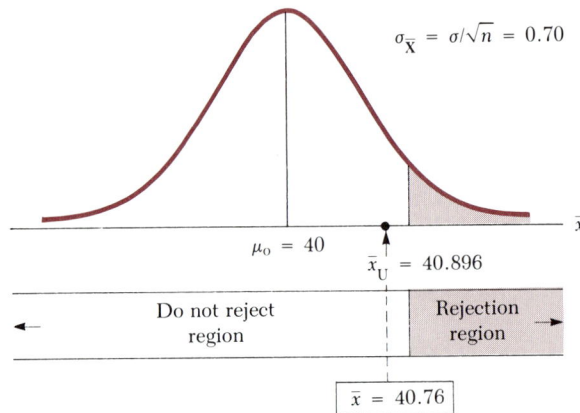

$\sigma_{\bar{X}} = \sigma/\sqrt{n} = 0.70$

$\mu_0 = 40$

$\bar{x}_U = 40.896$

Do not reject region

Rejection region

$\bar{x} = 40.76$

is a standard normal (mean = 0 and standard deviation = 1) random variable. If σ is not known and must be estimated using the sample standard deviation statistic **S**, then the standardized random variable

$$t = \frac{\bar{X} - \mu}{S/\sqrt{n}}$$

is t-distributed with $(n - 1)$ degrees of freedom, as discussed in Chapter 9. Thus if σ is not known, and s is used as an estimate of σ from sample data, then the t-distribution should be used rather than the z-distribution in the test statistics for the hypotheses about the population mean μ. Since the t-distribution is used in the test statistics, these tests have been generally referred to as t-tests.

Following the format for Case 1, we will give the testing procedures for a population mean μ when the population standard deviation σ is not known for the three forms for the null and alternate hypotheses, beginning with Form I:

$H_0: \quad \mu = \mu_0$
$H_A: \quad \mu \neq \mu_0$

The use of the Case 2 decision rule and action limits is similar to case 1—the t-value is substituted for the z-value in the action limits, and the t-table (Table B.5, Appendix B) is used instead of the z-table (Table B.3, Appendix B). The degrees of freedom for the t-value is $df = n - 1$, as is the case with the confidence interval estimate formula in Chapter 9. We will now show the use of the t-test for the population mean μ for the Form I null and alternate hypotheses statements.

Hypothesis test about a population mean μ

Form I: H_0: $\mu = \mu_0$
          H_A: $\mu \neq \mu_0$

Test statistic: $\overline{X}$

Assumption: 1. $\overline{X}$ is approximately normally distributed (sample size $n \geq$ 25) *or* X is normally distributed (regardless of sample size).

Decision rule: Reject H_0 if $\bar{x} < \bar{x}_L$ or if $\bar{x} > \bar{x}_U$. Do not reject H_0 if $\bar{x}_L \leq x \leq \bar{x}_U$.

Action limits:

$$\bar{x}_L = \mu_0 - t_{\alpha/2;df}\, s_{\overline{X}}$$
$$\bar{x}_U = \mu_0 + t_{\alpha/2;df}\, s_{\overline{X}}$$

where $s_{\overline{X}} = s/\sqrt{n}$, and s = sample standard deviation, and $t_{\alpha/2;df}$ is the value on the upper tail of the t-distribution so that $P(t > t_{\alpha/2;df}) = \alpha/2$ with the degrees of freedom $df = n - 1$.

---

t-test

H_0: $\mu = \mu_0$
H_A: $\mu \neq \mu_0$

H_0: $\mu = \mu_0$
H_A: $\mu \neq \mu_0$

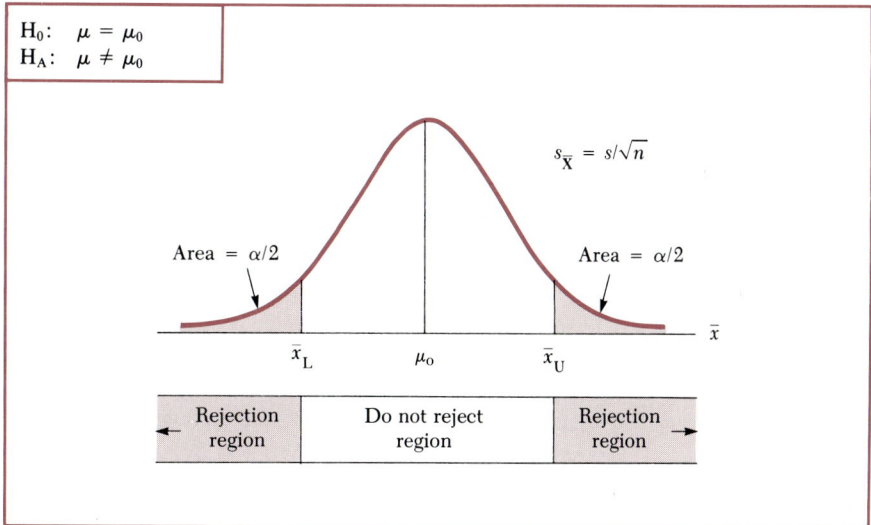

Example 10.13   A coffee machine is designed to deliver an average of 7.5 ounces of coffee into an 8-ounce cup. Before each machine is delivered, it is tested to see if the design target of an average 7.5 ounce fill is met by the machine. The company does not want to deliver a machine whose average fill is more than 7.5 ounces (the company will lose money by giving away "free" coffee) or whose average fill is less than 7.5 ounces (thereby suffering customer alienation from being "shorted" on the coffee cup fill amount).

The test for each machine is a random sample of 25 fills of 8-ounce coffee cups. The sample observations for a particular machine are:

$$
\begin{array}{ccccccccc}
7.2 & 7.4 & 7.2 & 7.6 & 7.5 & 7.5 & 7.7 & 7.0 & 7.3 \\
7.1 & 7.4 & 7.5 & 7.2 & 7.0 & 7.2 & 7.2 & 7.4 & 7.6 \\
7.3 & 7.2 & 7.2 & 7.4 & 7.5 & 7.1 & 7.5 & &
\end{array}
$$

Based on these data, does it appear that the machine is placing an average of 7.5 ounces of coffee in the cups? Use an $\alpha = 0.05$ significance level.

Solution The company is interested in whether or not this machine is producing a mean fill amount equal to 7.5 ounces. Thus the question to be answered is, "Is the mean fill amount μ for this machine equal to 7.5 ounces, or is there strong evidence that μ is either less than 7.5 ounces or more than 7.5 ounces?" This is the Form I hypothesis case:

H_0: $\mu = 7.5$ ounces
H_A: $\mu \neq 7.5$ ounces

Since the sample values are given, we must first calculate the sample statistics: $\bar{x}$, the sample mean, and s, the sample standard deviation:

$$
\bar{x} = \frac{\Sigma x}{n} = \frac{(7.2 + 7.4 + \cdots + 7.5)}{25} = 7.328 \text{ ounces}
$$

$$
s = \sqrt{\frac{\Sigma(x - \bar{x})^2}{n-1}} = \sqrt{\frac{\Sigma x^2 - \dfrac{(\Sigma x)^2}{n}}{n-1}}
$$

$$
= \sqrt{\frac{(7.2)^2 + (7.4)^2 + \cdots + (7.5)^2 - \dfrac{(7.2 + 7.4 + \cdots + 7.5)^2}{25}}{25 - 1}}
$$

$$
= 0.193 \text{ ounce}
$$

Since the significance level is $\alpha = 0.05$, $\alpha/2 = 0.025$, and the degrees of freedom are $df = n - 1 = 25 - 1 = 24$. Therefore we want $t_{\alpha/2;df} = t_{0.025;24}$ so that $P(t > t_{0.025;24}) = 0.025$. From Table B.5, Appendix B, $t_{0.025;24} = 2.064$. The determination of this t-value is illustrated in Figure 10.15. The value of $s_{\bar{x}}$, the estimated standard error of the mean $\bar{X}$ is:

$$
s_{\bar{x}} = s/\sqrt{n} = 0.193/\sqrt{25} = 0.193/5 = 0.039 \text{ ounce}
$$

The action limits for the decision rule are:

$$
\bar{x}_L = \mu_0 - t_{\alpha/2;df}\, s_{\bar{x}} = 7.5 - (2.064)(0.039) = 7.420 \text{ ounces}
$$
$$
\bar{x}_U = \mu_0 + t_{\alpha/2;df}\, s_{\bar{x}} = 7.5 + (2.064)(0.039) = 7.580 \text{ ounces}
$$

FIGURE 10.15 | The determination of $t_{\alpha/2;df} = t_{0.025;24}$ for the Example 10.13 problem

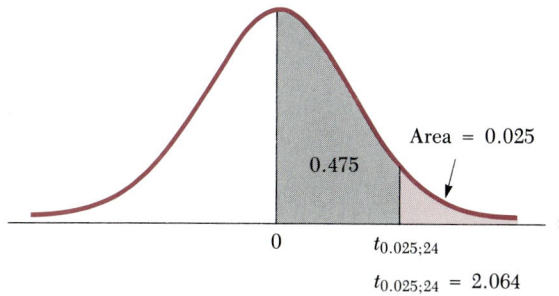

Area = 0.025

0.475

0

$t_{0.025;24}$

$t_{0.025;24} = 2.064$

t

Therefore the decision rule is:

Reject H_0: $\mu = 7.5$ ounces if $\bar{x} < 7.420$ or if $\bar{x} > 7.580$.
Do not reject H_0: $\mu = 7.5$ ounces if $7.420 \leq \bar{x} \leq 7.580$.

Since $\bar{x} = 7.328 < \bar{x}_L = 7.420$, we reject the null hypothesis and conclude that the alternate hypothesis H_A: $\mu \neq 7.5$ is true. The decision rule is shown in Figure 10.16.

FIGURE 10.16 | The decision rule for the Example 10.13 problem

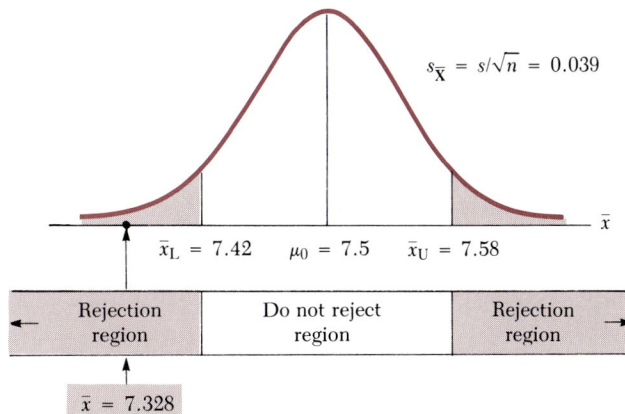

$s_{\bar{x}} = s/\sqrt{n} = 0.039$

$\bar{x}_L = 7.42$ $\mu_0 = 7.5$ $\bar{x}_U = 7.58$ $\bar{x}$

| Rejection region | Do not reject region | Rejection region |

$\bar{x} = 7.328$

Since $H_0: \mu = 7.5$ is rejected, there is strong evidence that this machine is not meeting the design specification on the average fill amount. The chance that we may have made an error [Type I error: rejecting H_0 given it is true] is 0.05 (α) or less. We are reasonably confident that this machine is underfilling the coffee cups on the average. (Since $\bar{x} = 7.328$ ounces is significantly below the target of $\mu = 7.5$ ounces.)

As with Case 1, we will now give the decision rules for Forms II and III of the null and alternate hypotheses, beginning with Form II.

Hypothesis test about a population mean μ

Form II: H_0: $\mu \geq \mu_0$
           H_A: $\mu < \mu_0$

<table>
<tr><td>

t-test

H_0: $\mu \geq \mu_0$
H_A: $\mu \leq \mu_0$

</td></tr>
</table>

Test statistic: $\overline{\mathbf{X}}$

Assumption: 1. $\overline{\mathbf{X}}$ is approximately normally distributed (sample size $n \geq 25$) or $\mathbf{X}$ is normally distributed (regardless of sample size).

Decision rule: Reject H_0 if $\bar{x} < \bar{x}_L$. Do not reject H_0 if $\bar{x} \geq \bar{x}_L$.

Action limit:

$$\bar{x}_L = \mu_0 - t_{\alpha;df}\, s_{\overline{\mathbf{X}}}$$

where $s_{\overline{\mathbf{X}}} = s/\sqrt{n}$, s = sample standard deviation, and $t_{\alpha;df}$ is the value on the upper tail of the t-distribution so that $P(\mathbf{t} > t_{\alpha;df}) = \alpha$ with the degrees of freedom $df = n - 1$.

H_0: $\mu \geq \mu_0$
H_A: $\mu < \mu_0$

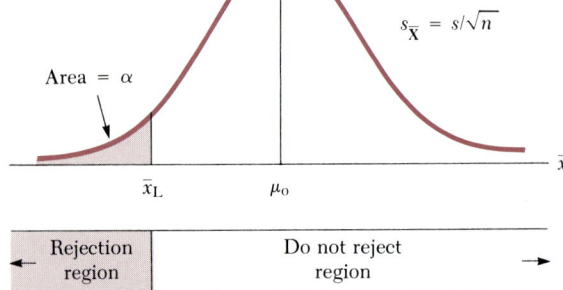

Example 10.14 The California State University—Yosemite is involved in a labor dispute concerning the average number of years it takes for tenured faculty to become full professors. The union claims that the average is at least 15 years, and the University administrators claim that the average is less than 15 years. The academic vice chancellor of the University randomly samples 36 full professor files and finds that the average number of years it took for these individuals to become full professors was 13.6 years with a standard deviation of three years. Do these data support the academic vice chancellor's claim that the average is less than 15 years? Use an $\alpha = 0.01$ significance level.

Solution The academic vice chancellor wants to make the strong decision that the mean years to tenure μ is less than 15 by rejecting the null hypothesis that μ is at least 15 years (as claimed by the union). Thus she wishes to answer the question, "Is μ, the mean years to tenure, at least (greater than or equal to) 15 years or is there strong evidence to suggest that μ is less than 15 years?" This is the Form II statement:

$$H_0: \quad \mu \geq 15 \text{ years}$$
$$H_A: \quad \mu < 15 \text{ years}$$

Since $\alpha = 0.01$, we must find $t_{\alpha;df} = t_{0.01;35}$, noting that $df = n - 1 = 36 - 1 = 35$. From Table B.5, Appendix B, $t_{0.01;35} = 2.438$. Now, the action limit is:

$$\bar{x}_L = \mu_0 - t_{0.01;35} s_{\bar{x}} = 15 - (2.438)(0.50) = 13.781$$

where $s_{\bar{x}} = s/\sqrt{n} = 3/\sqrt{36} = 3/6 = 0.50$. The decision rule is:

> Reject $H_0: \mu \geq 15$ years if $\bar{x} < 13.781$ years.
> Do not reject $H_0: \mu \geq 15$ years if $\bar{x} > 13.781$ years.

Since $\bar{x} = 13.6$ years $< \bar{x}_L = 13.781$ years, the null hypothesis $H_0: \mu \geq 15$ years should be rejected, and the academic vice chancellor should conclude that the alternate hypothesis, $H_A: \mu < 15$ years is true, thereby supporting her claim. The decision rule is shown in Figure 10.17.

We will now consider the third of the three forms of the hypothesis test, Form III, as given on the next page.

Example 10.15 The Candy Corporation claims that the average lifetime of its alkaline batteries used in portable computers exceeds 10 hours of continuous use. In order to substantiate this claim, the research director randomly samples 100 of these batteries from the production lot and records the time to failure of each in continuous use. The average time to failure for the 100 batteries is 10.05 hours, with a standard deviation of 0.75 hours. Do these sample data support the claim made by the Candy Corporation? Use an $\alpha = 0.05$ level of significance.

Hypothesis test about a population mean μ

Form III: H_0: $\mu \le \mu_0$
$\quad\quad\quad\quad$ H_A: $\mu > \mu_0$

t-test

H_0: $\mu \le \mu_0$
H_A: $\mu > \mu_0$

Test statistic: $\overline{X}$

Assumption: 1. $\overline{X}$ is approximately normally distributed (sample size $n \ge$ 25) *or* X is normally distributed (regardless of sample size).

Decision rule: Reject H_0 if $\bar{x} > \bar{x}_U$. Do not reject H_0 if $\bar{x} \le x_U$.

Action limit:

$$\bar{x}_U = \mu_0 + t_{\alpha;df}s_{\overline{X}}$$

where $s_{\overline{X}} = s/\sqrt{n}$, s = sample standard deviation, and $t_{\alpha;df}$ is the value on the upper tail of the t-distribution so that $P(t > t_{\alpha;df}) = \alpha$ with the degrees of freedom $df = n - 1$.

H_0: $\mu \le \mu_0$
H_A: $\mu > \mu_0$

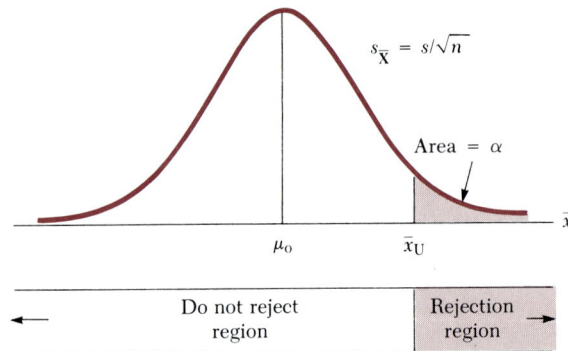

Solution   The research director desires to demonstrate the claim (μ = Mean lifetime of batteries > 10 hours) is true. He is interested in making the strong decision to reject H_0 and to conclude H_A, where the alternate hypothesis H_A contains the proposition $\mu > 10$. The question to be answered is, "Is μ, the mean lifetime of the batteries, less than or equal to 10 hours, or is μ greater than 10 hours? This is Form III:

H_0: $\mu \le 10$ hours
H_A: $\mu > 10$ hours

FIGURE 10.17 | The decision rule for the Example 10.14 problem

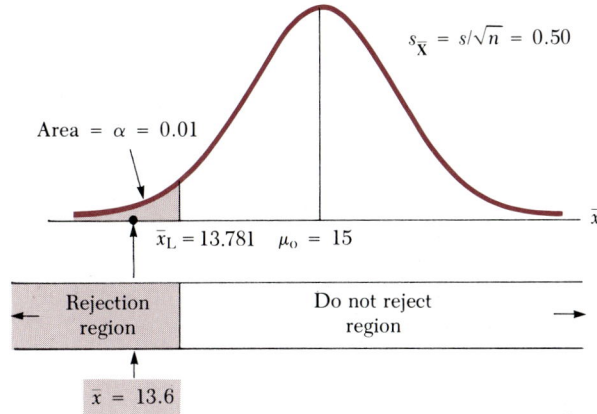

The research director is hoping that H_0 can be rejected based on the sample evidence so that the probability an error may have been made (Type I error: Rejecting H_0 given H_0 is true) is known.

The significance level is $\alpha = 0.05$, so we want $t_{\alpha;df} = t_{0.05;99}$, where $df = n - 1 = 100 - 1 = 99$. From Table B.5, Appendix B, $t_{0.05;99} = 1.660$. Thus the action limit is:

$$\bar{x}_U = \mu_0 + t_{0.05;99} s_{\bar{x}} = 10 + (1.660)(0.075) = 10.125$$

The decision rule is:

> Reject H_0: $\mu \leq 10$ hours if $\bar{x} > 10.125$ hours.
> Do not reject H_0: $\mu \leq 10$ hours if $\bar{x} \leq 10.125$ hours.

Since $\bar{x} = 10.05$ hours $< \bar{x}_U = 10.125$ hours, the research director cannot reject the null hypothesis, H_0: $\mu \leq 10$ hours. This is the weak decision (not rejecting H_0). Though $\bar{x}$ is greater than 10 hours, it is not *sufficiently* greater to provide *strong* evidence that $\mu > 10$ hours. Further, the director does not know the probability that a Type II error (not rejecting H_0 given H_0 is true) may have been made.

We have now completed the classical method of hypothesis testing for tests concerning the population mean μ. There are two cases: Case 1 when the population standard deviation σ is known (the z-test) and Case 2 when the population standard deviation σ is not known (the t-test). Under both cases, the approach is similar:

1. Choose one of three forms of the null and alternate hypotheses (Form I, II, or III).

2. Assume that the null hypothesis is true, so that the sampling distribution of $\overline{X}$ is centered at μ_0.

3. Establish the decision rule, using either the z-value (if σ is known) or the t-value (if σ is unknown).

4. Compare the sample statistic value $\bar{x}$ with the action limit (one-tailed test) or action limits (two-tailed test) and decide whether or not H_0, the null hypothesis, should be rejected.

There are two notes that we must add to the methods described in this section. First, we have ignored the finite population correction (FPC) factor in calculating the action limits. If the population is finite (size N) and $n/N > 0.05$, then the FPC should be used in calculating the standard error of $\overline{X}$:

$$\sigma_{\overline{x}} = \sqrt{\frac{N-n}{N-1}} \frac{\sigma}{\sqrt{n}} \qquad \text{(if } \sigma \text{ known)}$$

$$s_{\overline{x}} = \sqrt{\frac{N-n}{N-1}} \frac{s}{\sqrt{n}} \qquad \text{(if } \sigma \text{ unknown)}$$

Second, for Case 2 when σ is unknown, the use of the t-value from Table B.5, Appendix B, is correct. However, many introductory statistics texts recommend substituting the z-value for the t-value if $n > 30$ in Case 2. This is technically incorrect. The t-distribution for case 2 does converge to the z-distribution as n becomes large—in our view, a sample size of 30 or slightly greater is not sufficiently large. The t-table (Table B.5, Appendix B) goes up to $df = 100$ ($n = 101$ for this t-test). We recommend using the t-table for degrees of freedom up to and including $df = 100$, then switching to the z-distribution (Table B.3, Appendix B) when $df > 100$ for Case 2.

■ **10.4**

Hypothesis testing concerning the population proportion π

Chapter 8 presented and discussed the sampling distribution of $\mathbf{P}$, the sample proportion and the point estimator of π, the population proportion. The sampling distribution of $\mathbf{P}$ is shown in Figure 10.18, assuming that the normal distribution approximation to the binomial distribution applies (see Chapter 8, Section 8.7).

The form of the sampling distribution of $\mathbf{P}$ in Figure 10.17 is approximately normal provided that the conditions $n\pi \geq 5$ and $n(1 - \pi) \geq 5$ are met. Therefore, the sample size required to use the normal approximation varies, depending on the value of π. For example, if $\pi = 0.5$, then a sample size of $n = 10$ is sufficient [$10(0.5) = 5$ and $10(1 - 0.5) = 5$]. However, if $\pi = 0.1$, then a sample size of $n = 50$ or more is required [$50(0.1) = 5$ and $50(0.9) = 45$]. For the use of the sampling distribution in hypothesis testing, we will require that the sample size n be at least 25, with the understanding that if

---

Normal approximation conditions

FIGURE   10.18 | The sampling distribution of the sample proportion π

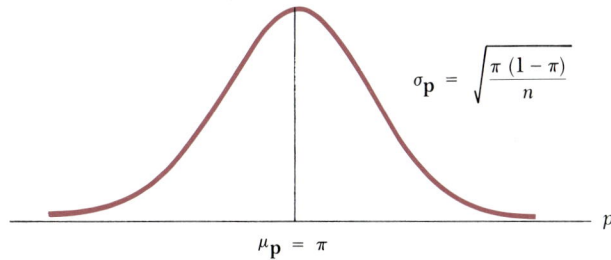

$$\sigma_p = \sqrt{\frac{\pi(1-\pi)}{n}}$$

$$\mu_p = \pi$$

there is evidence that π is either close to 0 or close to 1, then the sample size must be larger than 25—on the order of 50 to 100 or more, depending on how close π is to 0 or to 1.

The standard error of the point estimator **P** is given by:

$$\sigma_P = \sqrt{\frac{\pi(1-\pi)}{n}}$$

The estimated standard error is given by:

$$S_P = \sqrt{\frac{P(1-P)}{n}}$$

The estimated standard error is the formula for the standard error with the point estimator **P** substituted for π. Assuming that **P** is approximately normally distributed (do a normal approximation check to see if the sample size n is sufficiently large), then the standardized variable

$$Z = \frac{P - \pi}{\sqrt{\dfrac{P(1-P)}{n}}}$$

is approximately a standard normal random variable with a mean of 0 and a standard deviation of 1. Using the approximation sampling distribution of **P** and the estimated standard error of **P**, the decision rules and action limits for hypothesis tests concerning π follow in an analogous form to tests concerning μ, the population mean. For example, if we are testing the null hypothesis, $H_0: \pi = \pi_0$, where π_0 is the hypothesized value of π, then we will assume that the null hypothesis is true and center the sampling distribution of **P** over π_0. Further, since we assume that $H_0: \pi = \pi_0$ is true, we will set the value of the standard error of **P** by inserting π_0 into the formula for σ_P rather than inserting the sample proportion **P**. We will denote this formula by σ_{P_0}:

$$\sigma_{P_0} = \sqrt{\frac{\pi_0(1-\pi_0)}{n}}$$

Following the convention in the previous section, we will present the decision rule for each of the three forms of the null and alternate hypotheses, beginning with Form I.

Z-test

H_0: $\pi = \pi_0$
H_A: $\pi \neq \pi_0$

Hypothesis test about a population proportion π

Form I: H_0: $\pi = \pi_0$
$\quad\quad\quad$ H_A: $\pi \neq \pi_0$

Test statistic: **P**
Assumption: 1. **P** is approximately normally distributed (sample size $n \geq 25$; larger if there is an indication that π is close to 0 or to 1).
Decision rule: Reject H_0 if $p < p_L$ or if $p > p_U$.
$\quad\quad\quad\quad\quad$ Do not reject H_0 if $p_L \leq p \leq p_U$.
Action limits:

$$p_L = \pi_0 - z_{\alpha/2}\sigma_{P_0}$$
$$p_U = \pi_0 + z_{\alpha/2}\sigma_{P_0}$$

where $z_{\alpha/2}$ is the value on the upper tail of the z-distribution so that $P(\mathbf{Z} > z_{\alpha/2}) = \alpha/2$ and σ_{P_0} is given by:

$$\sigma_{P_0} = \sqrt{\frac{\pi_0(1 - \pi_0)}{n}}$$

H_0: $\pi = \pi_0$
H_A: $\pi \neq \pi_0$

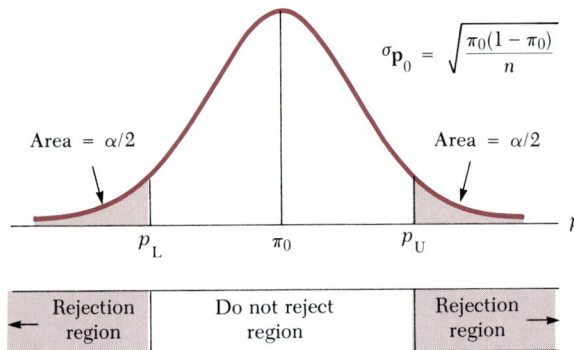

Example 10.16 The marketing survey department of a large food producer believes that consumers will be indifferent to two different proposed packaging designs for a new cereal. That is, it is believed that 50 percent of the consumers would choose package A and 50 percent would choose package B. In a random sample of 225 potential consumers of the cereal, 130 preferred the package A design. Test the hypothesis that the consumers are indifferent to the two package designs. Let α, the significance level of the test, be equal to 0.05.

Solution Let π = Proportion of consumers who prefer the package A design. Then the marketing survey department is hypothesizing that $\pi = 0.5$. That is, if the consumers are indifferent to the two package designs, then the proportion preferring package A should be 0.50 (and the proportion preferring package B should also be 0.50). This is the Form I hypothesis test:

$$H_0: \quad \pi = 0.50$$
$$H_A: \quad \pi \neq 0.50$$

Since $\alpha = 0.05$, $\alpha/2 = 0.025$ and $z_{\alpha/2} = z_{0.025} = 1.96$ from Table B.3, Appendix B, the standard error σ_{P_0} is given by:

$$\sigma_{P_0} = \sqrt{\frac{\pi_0(1 - \pi_0)}{n}} = \sqrt{\frac{(0.50)(1 - 0.50)}{225}} = 0.033$$

The action limits are:

$$p_L = \pi_0 - z_{0.025}\sigma_{P_0} = 0.50 - (1.96)(0.033) = 0.435$$
$$p_U = \pi_0 + z_{0.025}\sigma_{P_0} = 0.50 + (1.96)(0.033) = 0.565$$

The decision rule is:

> Reject H_0: $\pi = 0.50$ if $p < 0.435$ or if $p > 0.565$.
> Do not reject H_0: $\pi = 0.50$ if $0.435 \leq p \leq 0.565$.

Since the sample proportion $p = 130/225 = 0.578 > p_U = 0.565$, we reject the null hypothesis and conclude that the alternate hypothesis, H_A: $\pi \neq 0.50$ is true. More specifically, it appears that there is a preference for the package A design, since $p = 0.578$ is significantly greater than 0.50. The decision rule for this problem is illustrated in Figure 10.19.

FIGURE   10.19 | The decision rule for the Example 10.16 problem

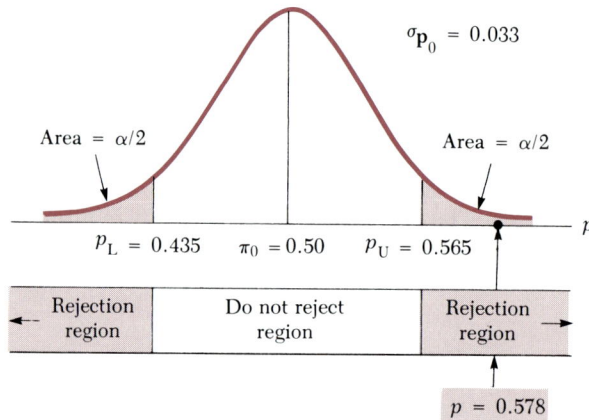

We now turn our attention to Forms II and III of the null and alternate hypotheses, beginning with Form II.

Hypothesis test about a population proportion π

Form II: H_0: $\pi \geq \pi_0$
           H_A: $\pi < \pi_0$

Test statistic: **P**
Assumption: 1. **P** is approximately normally distributed (sample size $n \geq$ 25; or larger if there is an indication that π is close to 0 or to 1).

Decision rule: Reject H_0 if $p < p_L$. Do not reject H_0 if $p \geq p_L$.
Action limit:

$$p_L = \pi_0 - z_\alpha \sigma_{P_0}$$

where z_α is the value on the upper tail of the z-distribution so that $P(Z > z_\alpha) = \alpha$ and σ_{P_0} is given by:

$$\sigma_{P_0} = \sqrt{\frac{\pi_0(1 - \pi_0)}{n}}$$

$H_0:$ $\pi \geq \pi_0$
$H_A:$ $\pi < \pi_0$

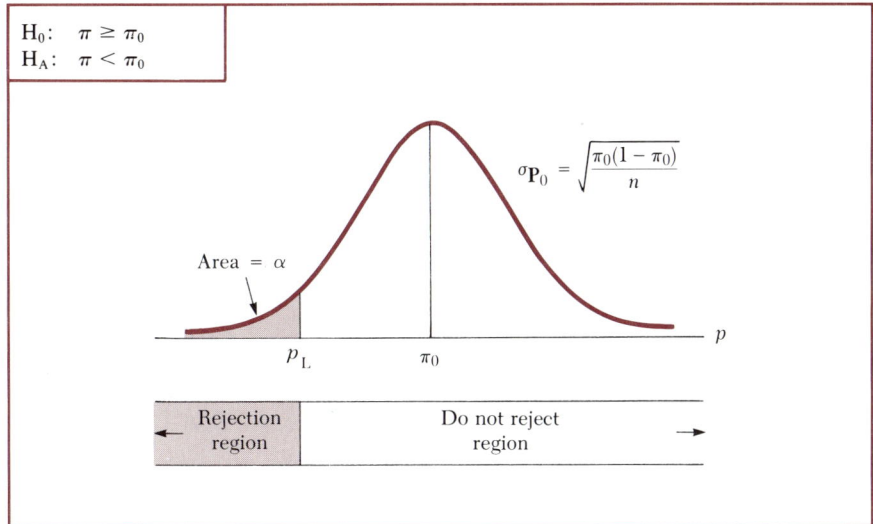

$\sigma_{P_0} = \sqrt{\dfrac{\pi_0(1 - \pi_0)}{n}}$

Area = α

p_L π_0 p

Rejection region   Do not reject region

Example 10.17   An investor has decided to sell all shares held in a stock if the proportion of analysts who believe that the stock will either hold its price or increase in price is *less than 0.20*. In a random sample of 100 analysts, he finds that only 15 believe that the stock will hold its price or increase in price. Can the condition that the population proportion of analysts who believe that the price will hold or rise is less than 0.20 be supported by this sample outcome? Use an $\alpha = 0.01$ level of significance.

Solution   Let π = Population proportion of analysts who believe that the stock in question will hold its current price or increase in price. The investor wishes to demonstrate that the proposition, $\pi < 0.20$, can be defended based on the sample evidence. The question to be answered is, "Is π greater than or equal to 0.20, or is there strong evidence to suggest that π is less than 0.20?" This is the Form II statement:

$H_0:$ $\pi \geq 0.20$
$H_A:$ $\pi < 0.20$

The investor hopes that the sample data will result in the strong decision to reject $H_0: \pi \geq 0.20$, thus allowing him to conclude that $H_A: \pi < 0.20$ is the likely state of affairs. Since $\alpha = 0.01$, $z_\alpha = z_{0.01} = 2.33$ from Table B.3, Appendix B, the standard error of the estimate is given by:

$$\sigma_{P_0} = \sqrt{\frac{\pi_0(1 - \pi_0)}{n}} = \sqrt{\frac{(0.20)(0.80)}{100}} = 0.040$$

The action limit is:

$$p_L = \pi_0 - z_{0.01}\sigma_{P_0} = 0.20 - (2.33)(0.040) = 0.107$$

The decision rule is:

> Reject H_0: $\pi \geq 0.20$ if $p < 0.107$.
> Do not reject H_0: $\pi \geq 0.20$ if $p \geq 0.107$.

Since the sample proportion $p = 15/100 = 0.15 > p_L = 0.107$, the investor must not reject the null hypothesis, H_0: $\pi \geq 0.20$. Even though the sample proportion $p = 0.15$ is less than 0.20, it is not sufficiently less to result in the rejection of the null hypothesis. That is, the sample outcome ($p = 0.15$) could happen with a reasonable chance if $\pi = 0.20$ (or more). Based on the sample evidence, the investor should retain the shares of the stock. The decision rule for this problem is illustrated in Figure 10.20.

FIGURE 10.20 | The decision rule for the Example 10.17 problem

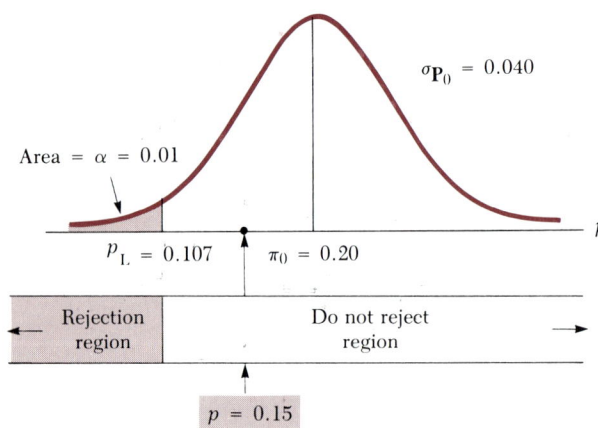

We will now consider the third of the three forms of the hypothesis test, Form III, as shown on page 449.

Example 10.18 ICM corporation receives hard disk drives from a Japanese supplier for its BT series microcomputer. Based on complaints by buyers of the BT microcomputer, ICM believes that the proportion of defective hard disk drives exceeds the tolerance of 5 percent. To test this belief, ICM contacts a random sample of 250 recent purchasers of the BT microcomputer and finds that 19 have had defective hard disks. Based on this

Z-test

$H_0: \quad \pi \leq \pi_0$
$H_A: \quad \pi > \pi_0$

Hypothesis test about a population proportion π

Form III: $H_0: \quad \pi \leq \pi_0$
$\ H_A: \quad \pi > \pi_0$

Test statistic: **P**

Assumption: 1. **P** is approximately normally distributed (sample size $n \geq 25$; or larger if there is an indication that π is close to 0 or to 1).

Decision rule: Reject H_0 if $p > p_U$. Do not reject H_0 if $p \leq p_U$.

Action limit:

$$p_U = \pi_0 + z_\alpha \sigma_{\mathbf{P}_0}$$

where z_α is the value on the upper tail of the z-distribution so that $P(\mathbf{Z} > z_\alpha) = \alpha$, and $\sigma_{\mathbf{P}_0}$ is given by:

$$\sigma_{\mathbf{P}_0} = \sqrt{\frac{\pi_0(1 - \pi_0)}{n}}$$

$H_0: \quad \pi \leq \pi_0$
$H_A: \quad \pi > \pi_0$

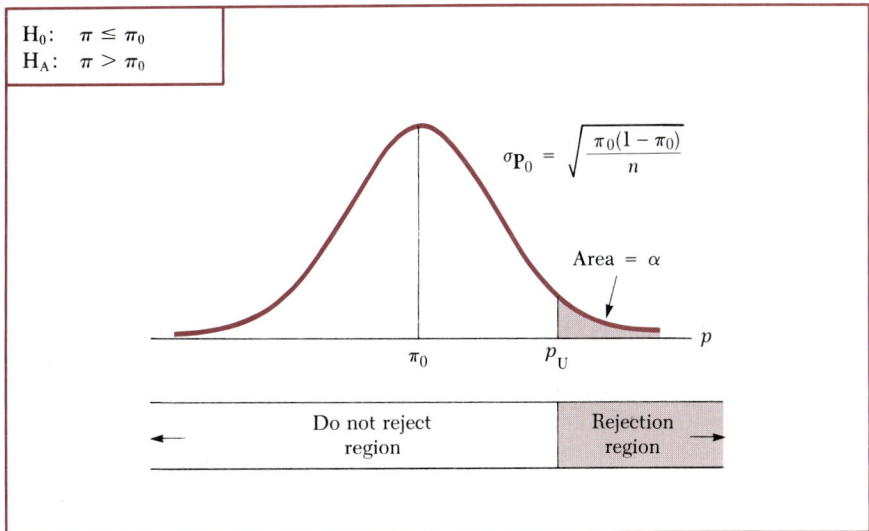

evidence, is ICM's belief supported or not? Use an $\alpha = 0.05$ level of significance.

Solution Let π = The proportion of defective hard disk drives in the population of BT microcomputers that have been sold. Since ICM is concerned about substantiating the proposition, $\pi > 0.05$, the question to be answered is, "Is π less than or equal to 0.05, or is there strong evidence based on the sample to support the proposition that π is greater than 0.05?" This is the Form III statement:

$$H_0: \quad \pi \leq 0.05$$
$$H_A: \quad \pi > 0.05$$

Since $\alpha = 0.05$, $z_\alpha = z_{0.05} = 1.64$ from Table B.3, Appendix B. The value of σ_{P0} is given by:

$$\sigma_{P0} = \sqrt{\frac{\pi_0(1 - \pi_0)}{n}} = \sqrt{\frac{(0.05)(0.95)}{250}} = 0.014$$

The action limit is therefore:

$$p_U = \pi_0 + z_\alpha \sigma_{P0} = 0.05 + (1.64)(0.014) = 0.073$$

The decision rule is:

> Reject H_0: $\pi \leq 0.05$ if $p > 0.073$.
> Do not reject H_0: $\pi \leq 0.05$ if $p < 0.073$.

Since the sample proportion $p = 19/250 = 0.076 > p_U = 0.073$, the null hypothesis, H_0: $\pi \leq 0.05$, should be rejected and the alternate hypothesis, H_A: $\pi > 0.05$, is supported by the sample data. ICM does have a problem with its supplier that it must straighten out. Since the strong decision to reject H_0 is made, the probability that ICM may have made a mistake in reaching this conclusion (the proportion of defective hard disk drives exceeds 5 percent) is $\alpha = 0.05$, the probability of a Type I error.

The decision rule is shown in Figure 10.21 for this problem.

This completes the section on hypothesis testing concerning the population proportion π. Two points deserve emphasis before moving on. First, all test statistics involving tests concerning π are based on the z-distribution. *The t-distribution is never used.* This separates hypothesis testing on π from hypothesis testing on the mean μ, where both the z-distribution (σ known) and the t-distribution (σ unknown) are used. Second, we have suggested a sample size of at least $n = 25$ when using these tests concerning π, noting that it must be larger if π is close to 0 or close to 1. It is usually worthwhile to check the normal approximation conditions $n\pi \geq 5$ and $n(1 - \pi) \geq 5$ to assure that the sample size is sufficiently large. This can be done by substituting p for π in these conditions. [Were these conditions met in Examples 10.16, 10.17, and 10.18?] If these conditions are not met, then the methods developed in this section for hypothesis tests concerning π are inappropriate. In this instance, the binomial distribution must be used for hypothesis testing. We will not discuss this approach. It can be found in several of the advanced references at the end of this chapter.

> Do not use t-distribution when testing hypotheses concerning π

FIGURE 10.21 | The decision rule for the Example 10.18 problem

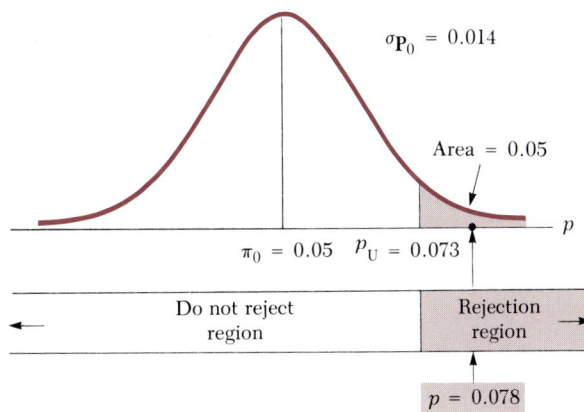

10.5

**Standardized
test statistics
and testing**

In Sections 10.2, 10.3, and 10.4, we described the classical hypothesis-testing method that establishes the decision rule on the sampling distribution of the estimator of the population parameter of interest. It is also possible to establish the decision rule on the z-distribution or on the t-distribution, depending on which distribution applies to the given hypothesis test. Establishing the decision rule directly on either the z-distribution or the t-distribution produces another version of the classical hypothesis-testing method. The motivation for describing this approach is related to the growth in use of computer statistical software packages, such as MINITAB. These packages almost always use the standardized form of hypothesis testing. We will use MINITAB in Section 10.11 to solve some of the example problems in this chapter.

To describe the standardized test statistic and testing procedure, let us return to testing a hypothesis about a population mean μ. Specifically, we will consider Case 1 (the population standard deviation σ is known) and Form I of the null and alternate hypotheses:

$$H_0: \quad \mu = \mu_0$$
$$H_A: \quad \mu \neq \mu_0$$

The decision rule for this test is: Reject H_0 if $\bar{x} < \bar{x}_L = \mu_0 - z_{\alpha/2}\sigma_{\bar{x}}$ or if $\bar{x} > \bar{x}_U = \mu_0 + z_{\alpha/2}\sigma_{\bar{x}}$; do not reject H_0 if $\bar{x}_L \leq \bar{x} \leq \bar{x}_U$.

Consider for the moment the lower rejection region defined by:

$$\bar{x} < \mu_0 - z_{\alpha/2}\sigma_{\bar{x}}$$

With a little bit of algebra, we can write this statement as

$$\frac{\bar{x} - \mu_0}{\sigma_{\bar{x}}} < -z_{\alpha/2}$$

The same manipulation can be made to the upper rejection region so that the decision rule for the hypothesis can be rewritten as:

Reject H_0 if $\dfrac{\bar{x} - \mu_0}{\sigma_{\bar{x}}} < -z_{\alpha/2}$ or if $\dfrac{\bar{x} - \mu_0}{\sigma_{\bar{x}}} > z_{\alpha/2}$.

Do not reject H_0 if $-z_{\alpha/2} \le \dfrac{\bar{x} - \mu_0}{\sigma_{\bar{x}}} \le z_{\alpha/2}$.

Notice that the value

$$z = \frac{\bar{x} - \mu_0}{\sigma_{\bar{x}}}$$

in the decision rule is the standardized statistic with a mean of 0 and a standard deviation of 1, *assuming that the null hypothesis is true so that* $\mu = \mu_0$. If we let this value be represented by the standardized statistic value z as above, then we can write the decision rule more simply as:

Reject H_0 if $z < -z_{\alpha/2}$ or if $z > z_{\alpha/2}$.
Do not reject H_0 if $-z_{\alpha/2} \le z \le z_{\alpha/2}$.

Therefore to use the standardized testing procedure for the null and alternate hypotheses,

H_0: $\mu = \mu_0$
H_A: $\mu \ne \mu_0$

we need to do the following:

1. Calculate the value of the standardized statistic:

$$z = \frac{\bar{x} - \mu_0}{\sigma_{\bar{x}}}$$

Standardized test statistic

where $\sigma_{\bar{x}} = \sigma/\sqrt{n}$. This value is called the value of the *standardized test statistic*.

2. Apply the decision rule above. That is, if $z < -z_{\alpha/2}$ or if $z > z_{\alpha/2}$, then *reject* H_0; if $-z_{\alpha/2} \le z \le z_{\alpha/2}$, then *do not reject* H_0.

Standardized test
statistic procedure

These steps summarize the *standardized test statistic procedure*. We will now apply this procedure to the Example 10.10 problem.

Example 10.19   (Example 10.10 revisited.) In Example 10.10 it was hypothesized that the mean starting salary of accounting graduates from accredited BBA degree programs with no prior work experience who are hired by one of the "Big 8" accounting firms is $24,000. Letting μ = The mean starting salary for the population of these graduates, the null and alternate hypotheses are:

$$H_0: \mu = \$24,000$$
$$H_A: \mu \neq \$24,000$$

A sample of 100 of these graduates is taken, and it is found that the sample average starting salary is $23,775. It is assumed that the population standard deviation σ = $1,000. Using the standardized test statistic method, test this null hypothesis at the α = 0.05 significance level.

Solution   We first calculate the value of the standardized test statistic z:

$$z = \frac{\bar{x} - \mu_0}{\sigma_{\bar{x}}} = \frac{\$23,775 - \$24,000}{\$100} = -2.25$$

where $\sigma_{\bar{x}} = \sigma/\sqrt{n} = \$1,000/\sqrt{100} = \$1,000/10 = \$100$. Now, with $\alpha = 0.05$, $\alpha/2 = 0.025$ and $z_{\alpha/2} = z_{0.025} = 1.96$ from Table B.3, Appendix B. The decision rule is therefore:

> Reject H_0 if $z < -1.96$ or if $z > 1.96$.
> Do not reject H_0 if $-1.96 \leq z \leq 1.96$.

Since $z = -2.25 < -1.96$, we reject $H_0: \mu = \$24,000$ and conclude that μ is less than $24,000 ($H_A: \mu \neq \$24,000$ is true). Notice that this is the same decision we arrived at by stating the decision rule on the sampling distribution of $\bar{X}$ in Example 10.10. The standardized test decision rule is shown in Figure 10.22.

The standardized testing procedure decision rules are summarized for the three forms of hypothesis tests about the population mean μ on page 455. Notice that the decision rule uses either the z-value (when the population standard deviation is known) or the t-value (when the population standard deviation is not known).

Example 10.20   It is hypothesized that it takes more than an average of 10 years to become a senior accountant partner in one of the major accounting firms. A random sample of 36 partner files is taken, and it is found that the average time taken to become a partner for these 36 individuals is 12 years, with a standard deviation of 3 years. Using the standardized testing

FIGURE   10.22 | The standardized test decision rule for the Example
10.19 problem

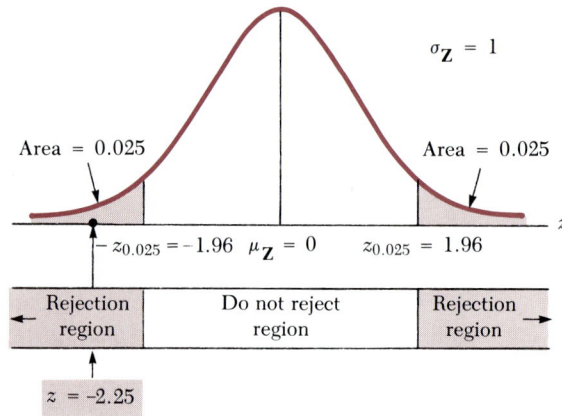

procedure, is the claim that it takes more than 10 years supported by these
sample data? Use an $\alpha = 0.05$ level of significance.

Solution   Let μ = Average number of years required to become a senior
partner in this firm. It is hoped that the sample data will provide strong
evidence that the proposition $\mu > 10$ years is true. The question to be
answered is therefore, "Is μ less than or equal to 10 years, or is there strong
evidence to suggest that μ is more than 10 years?" This is the Form III
statement:

H_0: $\mu \leq 10$ years
H_A: $\mu > 10$ years

The estimated standard error value is $s_{\bar{x}} = s/\sqrt{n} = 3/\sqrt{36} = 3/6 = 0.50$ year.
The value of the standardized test statistic is:

$$t = \frac{\bar{x} - \mu_0}{s_{\bar{x}}} = \frac{12 - 10}{0.50} = 4.00$$

Since $\alpha = 0.05$, $t_{\alpha;df} = t_{0.05;35} = 1.690$ from Table B.5, Appendix B, with $df = n - 1 = 36 - 1 = 35$. Now, the decision rule is:

Reject H_0 if $t > 1.690$.
Do not reject H_0 if $t \leq 1.690$.

Hypothesis testing about the population mean μ
Use of Standardized Test Statistics

Standardized test statistic value:	$z = \dfrac{\bar{x} - \mu_0}{\sigma_{\bar{x}}}$; $\sigma_{\bar{x}} = \sigma/\sqrt{n}$ if σ is known.
Standardized test statistic value:	$t = \dfrac{\bar{x} - \mu_0}{s_{\bar{x}}}$; $s_{\bar{x}} = s/\sqrt{n}$ if σ is not known.

Form I: H_0: $\mu = \mu_0$ $\qquad\qquad$ *Decision rule*
$\qquad\quad$ H_A: $\mu \neq \mu_0$ $\qquad$ Reject H_0 if $z < -z_{\alpha/2}$ or if $z > z_{\alpha/2}$.
$\qquad\qquad\qquad\qquad\qquad$ Do not reject H_0 if $-z_{\alpha/2} \leq z \leq z_{\alpha/2}$.
$\qquad$ Note: If σ is unknown and t is used, replace $z_{\alpha/2}$ in the
$\qquad\qquad\qquad$ decision rule with $t_{\alpha/2;df}$ where $df = n - 1$.

Form II: H_0: $\mu \geq \mu_0$ $\qquad\qquad$ *Decision rule*
$\qquad\quad$ H_A: $\mu < \mu_0$ $\qquad$ Reject H_0 if $z < -z_{\alpha}$.
$\qquad\qquad\qquad\qquad\qquad$ Do not reject H_0 if $z \geq -z_{\alpha}$.
$\qquad$ Note: If σ is unknown and t is used, replace z_{α} in the
$\qquad\qquad\qquad$ decision rule with $t_{\alpha;df}$ where $df = n - 1$.

Form III: H_0: $\mu \leq \mu_0$ $\qquad\qquad$ *Decision rule*
$\qquad\quad$ H_A: $\mu > \mu_0$ $\qquad$ Reject H_0 if $z > z_{\alpha}$.
$\qquad\qquad\qquad\qquad\qquad$ Do not reject H_0 if $z \leq z_{\alpha}$.
$\qquad$ Note: If σ is unknown and t is used, replace z_{α} in the
$\qquad\qquad\qquad$ decision rule with $t_{\alpha;df}$ where $df = n - 1$.

Since $t = 4.00 > 1.690$, we reject H_0: $\mu \leq 10$ years and conclude that H_A: $\mu > 10$ years is true. The decision rule on the t-distribution is shown in Figure 10.23.

The standardized testing procedure can also be applied to the population proportion π. The value of the standardized test statistic is given by:

$$z = \frac{p - \pi_0}{\sqrt{\dfrac{\pi_0(1 - \pi_0)}{n}}}$$

To use the standardized test statistic, we should check the normal approximation conditions to ensure that the sample size is sufficiently large for the use of the standard normal distribution. These conditions are: $np \geq 5$ and $n(1 - p) \geq 5$.

FIGURE 10.23 | The decision rule for the standardized test in Example 10.20

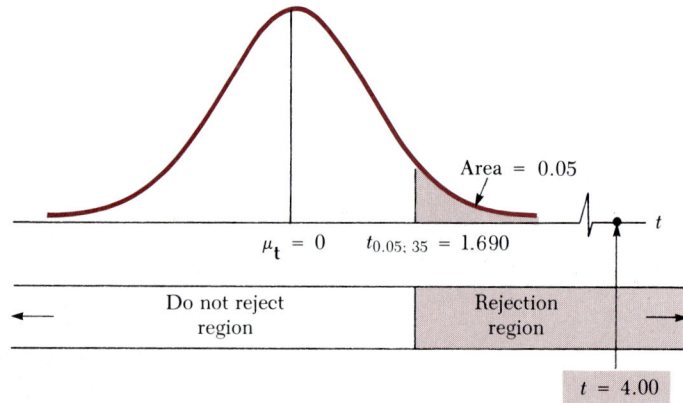

The following table summarizes the use of the standardized testing method for the hypotheses concerning the population proportion π.

Hypothesis testing about a population proportion π
Use of Standardized Test Statistics

Standardized test statistic value:	$z = \dfrac{p - \pi_0}{\sqrt{\dfrac{\pi_0(1 - \pi_0)}{n}}}$	Conditions: $np \geq 5$ and $n(1 - p) \geq 5$

Form I: H_0: $\pi = \pi_0$ *Decision Rule*
           H_A: $\pi \neq \pi_0$ Reject H_0 if $z < -z_{\alpha/2}$ or if $z > z_{\alpha/2}$.
                                     Do not reject H_0 if $-z_{\alpha/2} \leq z \leq z_{\alpha/2}$.

Form II: H_0: $\pi \geq \pi_0$ *Decision Rule*
           H_A: $\pi < \pi_0$ Reject H_0 if $z < -z_\alpha$.
                                  Do not reject H_0 if $z \geq z_\alpha$.

Form III: H_0: $\pi \leq \pi_0$ *Decision Rule*
           H_A: $\pi > \pi_0$ Reject H_0 if $z > z_\alpha$.
                                  Do not reject H_0 if $z \leq z_\alpha$.

Example 10.21 The Anthony Furniture Company believes that more than 50 percent of its sales are on credit through its own card or one of the major credit cards. To test this belief, a random sample of 100 sales receipts are taken, and it is found that 42 were credit card sales. Test this belief using an $\alpha = 0.05$ level of significance and the standardized testing procedure.

Solution Let π = The proportion of all sales that are made by credit card. The proposition to be demonstrated is, $\pi > 0.50$; this becomes the statement in the alternate hypothesis. The question to be answered is, "Is π less than or equal to 0.50, or is there strong evidence to support the proposition $\pi > 0.50$?" This is the Form III statement. Thus the null and alternate hypotheses are:

$$H_0: \quad \pi \le 0.50$$
$$H_A: \quad \pi > 0.50$$

The sample size is $n = 100$. This sample size is sufficiently large to support the use of the normal approximation: $np = 100(0.42) = 42 > 5$ and $n(1 - p) = 100(1 - 0.42) = 58 > 5$. Since $\alpha = 0.05$, $z_\alpha = z_{0.05} = 1.64$. The value of the test statistic is:

$$z = \frac{p - \pi_0}{\sqrt{\dfrac{\pi_0(1 - \pi_0)}{n}}} = \frac{0.42 - 0.50}{\sqrt{\dfrac{0.50(1 - 0.50)}{100}}} = \frac{-0.08}{0.05} = -1.60$$

The decision rule is:

> Reject H_0 if $z > 1.64$.
> Do not reject H_0 if $z \le 1.64$.

FIGURE 10.24 | The decision rule for the standardized test in Example 10.21

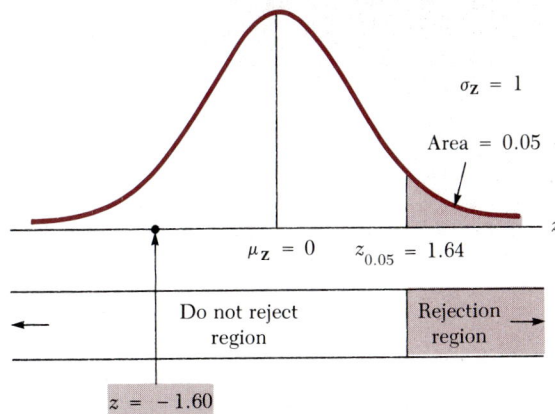

$\sigma_z = 1$

Area $= 0.05$

$\mu_z = 0$ $z_{0.05} = 1.64$ z

Do not reject region | Rejection region

$z = -1.60$

Since $z = -1.60 < 1.64$, the null hypothesis, $H_0: \pi \leq 0.50$ cannot be rejected. The sample does not support the belief that more than 50 percent of the sales are by credit card. The decision rule for this problem is shown in Figure 10.24.

■ **10.6**

The *p*-value method of hypothesis testing

In recent years an alternative to the classical method of hypothesis testing described in Section 10.2 has been developed. It is called the *p*-value method. The *p*-value method can provide additional information beyond that provided by the classical method of hypothesis testing. Most statistical computer software packages now use the *p*-value method alone or together with the classical method for hypothesis testing. Noteworthy examples of this are MINITAB, SPSSX (Statistical Package for the Social Sciences), and SAS (Statistical Analysis System), the most commonly used computer statistical programs. All three are now available on microcomputers as well as on mainframe and minicomputers. These packages are referenced at the end of this chapter. Due to the extra information provided by the *p*-value method when compared with the classical method of testing and the use of the *p*-value method in the statistical software packages, it is extensively used in hypothesis testing. The following example describes the *p*-value method.

Example 10.22   A cigarette manufacturer claims that the average tar content of Puff brand cigarettes is less than 5 milligrams per cigarette. To test this conjecture, 100 Puff cigarettes are sampled, and it is found that the average tar content is 4.94 milligrams. From previous experiments, it is known that the population standard deviation of the tar content is $\sigma = 0.2$ milligram. Test this conjecture at the $\alpha = 0.05$ significance level.

Solution   The proposition is $\mu < 5$ milligrams, where μ is the average tar content of the cigarettes. This becomes the alternate hypothesis, H_A. The question to be answered is, "Is μ greater than or equal to 5 milligrams, or is there strong evidence to suggest that the proposition, $\mu < 5$, is true?" This is the Form II statement. Thus the null and alternate hypotheses are:

H_0: $\mu \geq 5$ milligrams
H_A: $\mu < 5$ milligrams

Since $\alpha = 0.05$, $z_\alpha = z_{0.05} = 1.645$ (with interpolation) from Table B.3, Appendix B. The standard error is:

$$\sigma_{\bar{x}} = \sigma/\sqrt{n} = 0.02/\sqrt{100} = 0.002 \text{ milligram}$$

The action limit is:

$$\bar{x} = \mu_0 - z_{0.05}\sigma_{\bar{x}} = 5 - (1.645)(1.002) = 4.967$$

The decision rule is shown in Figure 10.25. Notice that the figure shows the rejection region and the region not to reject *above* the bell-shaped curve, whereas before we have shown these regions below the curve. This is simply

FIGURE 10.25 | Decision rule for the hypothesis test in Example 10.22

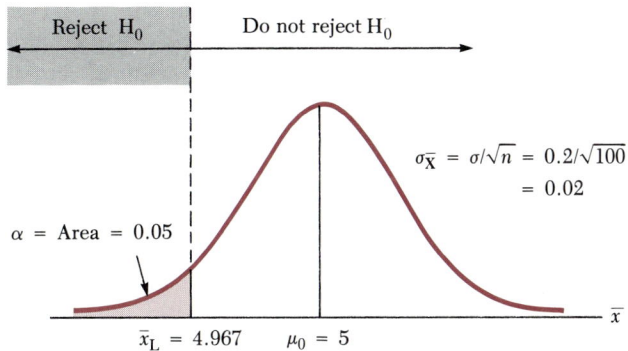

an alternate way of showing the decision rule graphically. The decision rule is:

Reject H_0 if $\bar{x} < \bar{x}_L = 4.967$.
Do not reject H_0 if $\bar{x} \geq \bar{x}_L = 4.967$.

Since $\bar{x} = 4.94 < \bar{x}_L = 4.967$, reject H_0: $\mu \geq 5$ and conclude that $\mu < 5$.

Thus, by the classical method of hypothesis testing, we would reject the null hypothesis at the $\alpha = 0.05$ level of significance. But the method does not give us an idea of the *strength* of the *conviction* that the decision is, in fact, correct. The *p*-value method determines how likely it is to sample a value of the statistic $\overline{\mathbf{X}}$ that is less than or equal to $\bar{x} = 4.94$, the sample average, when in fact $\mu = 5$. The probability of this event is called the *p-value* of the test. The computation of the *p*-value is illustrated in Figure 10.26.

The probability that $\overline{\mathbf{X}}$ is less than or equal to the sample average $\bar{x} = 4.94$ is 0.0013. This probability conveys important information about the *strength* of the decision to reject H_0: $\mu \geq 5$. We know from Figure 10.25 that $\bar{x} = 4.94$ falls in the rejection region so that with $\alpha = 0.05$, the probability that we have made a Type I error (rejected a true null hypothesis) is no greater than 0.05. But given that the *p*-value is 0.0013, we know that it is extremely unlikely that we could take a sample of size 100 and find that the average tar content of Puff cigarettes is 4.94 milligrams or less, *given that* $\mu = \mu_0 = 5$ milligrams. That is, if $\mu = \mu_0 = 5$, in only 13 out of 10,000 samples of size $n = 100$ would the value of the test statistic $\overline{\mathbf{X}}$ be equal to or less than 4.94. We

p-value

FIGURE 10.26 | Computation of the *p*-value in Example 10.22

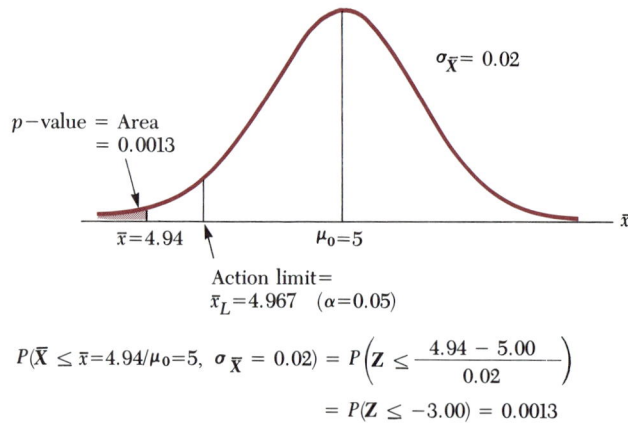

$$P(\overline{X} \le \bar{x}=4.94/\mu_0=5, \ \sigma_{\overline{X}} = 0.02) = P\left(Z \le \frac{4.94 - 5.00}{0.02}\right)$$

$$= P(Z \le -3.00) = 0.0013$$

have therefore very strong evidence to reject $H_0: \mu \ge 5$ and know that we have made the right decision almost with certainty.

The null hypothesis in this example is $H_0: \mu \ge 5$. Notice that we center the sampling distribution of $\overline{X}$ over $\mu_0 = 5$, the left-most value of μ that satisfies the proposition, $\mu \ge 5$. If H_0 is true and μ is greater than 5 ($\mu > 5$), then the *p*-value is even smaller than 0.0013. Say, for example, that μ is actually 10. Then the true sampling distribution of $\overline{X}$ is centered over 10 (moved to the right in Figure 10.26). The area to the *left* of $\bar{x} = 4.94$ would now be *much less* than 0.0013.

There is a direct connection between the *p*-value and the alternate hypothesis, H_A. In example 10.22, the null and alternate hypotheses are:

$H_0: \quad \mu \ge 5$
$H_A: \quad \mu < 5$

As we discussed earlier, the alternate hypothesis H_A proposition points in the direction of the sample evidence that *does not* support the null hypothesis, H_0. In this example, since $H_A: \mu < 5$, we look for evidence in the sample indicating that $\mu < 5$ to arrive at the strong decision to reject H_0. Notice the computation of the *p*-value in Example 10.22 requires evaluating the probability, $P[\overline{X} < \bar{x} = 4.94]$. The *direction* of the inequality in this probability statement is the *same* as the *direction* of the inequality in the statement of the alternate hypothesis, H_A. The *p*-value is calculating the probability that $\overline{X}$, the sample mean, will take on a value as extreme or more extreme as the sampled value $\bar{x}$ *in the direction of the alternate hypothesis*, given the null hypothesis is true (setting $\mu = \mu_0$).

Let's relate this concept to the numbers in Example 10.22. For the null hypothesis, $H_0: \mu > 5$, we set the mean of the sampling distribution at the *minimum* value in the range of values ($\mu \geq 5$) that makes H_0 true—$\mu_0 = 5$. In the sample we have observed the sample mean to be $\bar{x} = 4.94$. This value is in the direction of the alternate hypothesis, H_A: $\mu < 5$. Thus there is evidence to doubt the reasonableness of the null hypothesis, H_0: $\mu \geq 5$. The *p*-value answers the following question:

> How likely is it to determine that $\bar{x} = 4.94$ or less in the sample, given that H_0: $\mu \geq 5$ is true with the sampling distribution centered over $\mu_0 = 5$?

Another way of stating this question is:

> What is the probability of $\overline{X}$, the sample mean, taking on an extreme value of $\bar{x} = 4.94$ or a value that is more extreme in the direction of H_A, given that H_0: $\mu \geq 5$ is true with the sampling distribution centered at $\mu_0 = 5$?

We know from the solution to Example 10.22 that the answer to these two questions is: *p*-value $= 0.0013$. Therefore, it is *very unlikely* to determine that $\overline{X}$ is 4.94 or less, given that H_0: $\mu \geq 5$ is true with the sampling distribution centered at $\mu_0 = 5$.

As you may have noted from Example 10.22, the calculation of the *p*-value is made by determining an area under the sampling distribution of $\overline{X}$, the sample mean, where the sampling distribution is centered at μ_0. We will now illustrate the areas representing the *p*-value under the sampling distribution of $\overline{X}$ for the three forms of the null and alternate hypotheses. We will begin with Form II, since it corresponds to Example 10.22.

Form II

H_0: $\mu \geq \mu_0$
H_A: $\mu < \mu_0$

> *p*-value $= P[\overline{X} < \bar{x}]$

FIGURE 10.27 | The areas representing the *p*-value under the sampling distribution of $\overline{X}$ for the Form II hypotheses

(A)

(B)

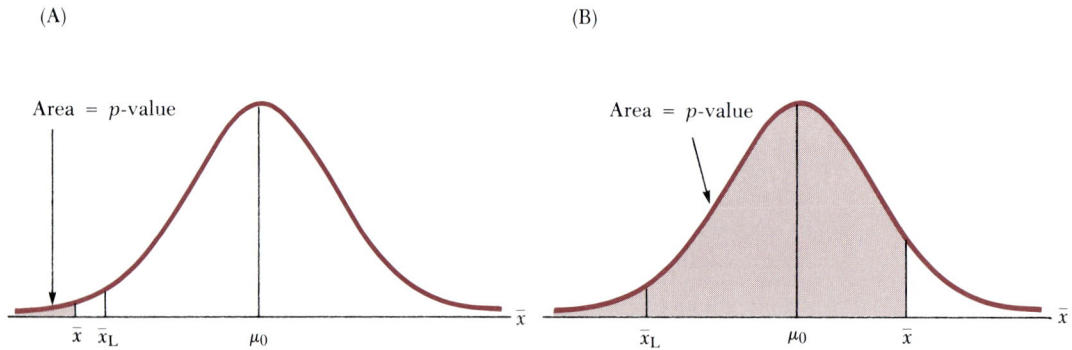

Figure 10.27 shows two examples of the area representing the *p*-value under the sampling distribution of $\overline{X}$. In Figures 10.27A and 10.27B, the action limit $\overline{x}_L$ is shown as if $\alpha = 0.05$. In Figure 10.27A the sample mean $\overline{x}$ is at the extreme end of the direction of $H_A: \mu < \mu_0$. It also falls in the rejection region ($\overline{x} < \overline{x}_L$). The *p*-value is the area to the *left* of $\overline{x}$, clearly a very small area strongly suggesting that $H_0: \mu > \mu_0$ should be rejected. In Figure 10.27B $\overline{x}$ is to the right of μ_0. In this instance it is falling in the "do not reject" region ($\overline{x} \geq \overline{x}_L$). The *p*-value is the area to the *left* of $\overline{x}$ (in the direction of $H_A: \mu < 5$). Clearly, this area is very large (bigger than 0.5). If $H_0: \mu \geq 5$ is true and the sampling distribution of $\overline{X}$ is centered at $\mu_0 = 5$, then there is a very high probability that we will encounter a value of $\overline{X}$ equal to or less than $\overline{x}$ when the sample is taken. We must therefore conclude that $H_0: \mu \geq 5$ should not be rejected.

Form III

$$H_0: \quad \mu \leq \mu_0$$
$$H_A: \quad \mu > \mu_0$$

$$\text{p-value} = P[\overline{X} > \overline{x}]$$

Figure 10.28 shows two examples of the area representing the *p*-value under the sampling distribution of $\overline{X}$ for the Form III hypotheses.

FIGURE 10.28 | The areas representing the *p*-value under the sampling distribution of $\overline{X}$ for the Form III hypotheses

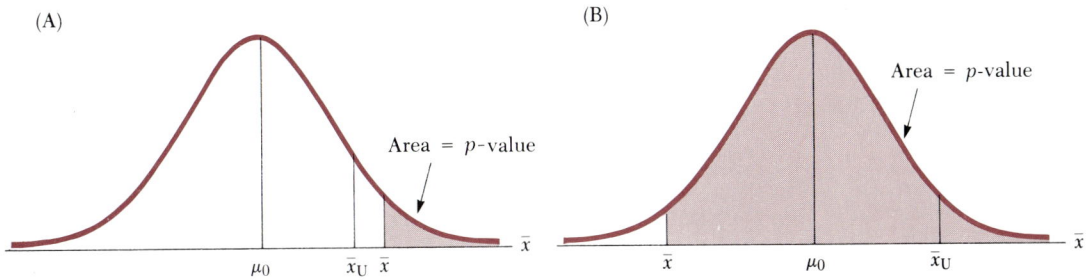

(A)

(B)

Area = *p*-value

Area = *p*-value

μ_0 $\overline{x}_U$ $\overline{x}$

$\overline{x}$ μ_0 $\overline{x}_U$

Notice in Figure 10.28A and B that the action limit $\overline{x}_U$ is on the upper tail of the sampling distribution. In Figure 10.28A, since $\overline{x} > \overline{x}_U$, we would reject the null hypothesis H_0: $\mu \leq \mu_0$. The *p*-value is the area to the *right* of $\overline{x}$ in the direction of the alternate hypothesis, H_A: $\mu > \mu_0$. In Figure 10.28B, since $\overline{x} \leq \overline{x}_U$, we would conclude the weak decision not to reject H_0: $\mu \leq \mu_0$. The *p*-value again is the area to the *right* of $\overline{x}$, a very large area in this case (greater than 0.50).

Form I

H_0: $\mu = \mu_0$
H_A: $\mu \neq \mu_0$

$$p\text{-value} = 2P[\overline{X} > \overline{x}] \qquad \text{if } \overline{x} > \mu_0$$
$$p\text{-value} = 2P[\overline{X} < \overline{x}] \qquad \text{if } \overline{x} < \mu_0$$

The computation of the *p*-value in the Form I case is different from the computation in Forms II and III. The reason for this can be understood by considering the classical method decision rule for the Form I hypotheses:

Reject H_0: $\mu = \mu_0$ if $\overline{x} < \overline{x}_L$ or $\overline{x} > \overline{x}_U$.
Do not reject H_0: $\mu = \mu_0$ if $\overline{x}_L \leq \overline{x} \leq \overline{x}_U$.

FIGURE 10.29 | The decision rule for the Form I hypotheses

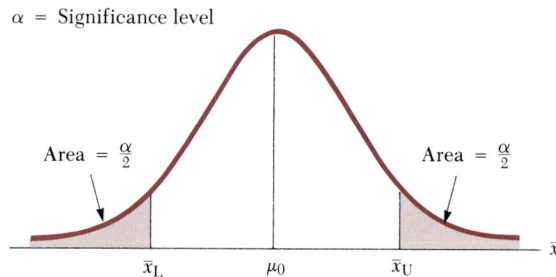

α = Significance level

Area = $\frac{\alpha}{2}$

Area = $\frac{\alpha}{2}$

$\bar{x}_L$ μ_0 $\bar{x}_U$ $\bar{x}$

This decision rule is illustrated in Figure 10.29. Notice in this figure that the action limits $\bar{x}_L$ and $\bar{x}_U$ are set by *halving* α, resulting in an area of $\alpha/2$ in each tail of the sampling distribution of $\overline{\mathbf{X}}$. The alternate hypothesis, $H_A: \mu \neq \mu_0$, establishes the *direction* in the sample mean $\bar{x}$ for which $H_0: \mu = \mu_0$ is unreasonable. Now there are *two* directions: *less* than μ_0 and *greater* than μ_0. Thus if $\bar{x}$ falls in either of the extreme tails, there is evidence to reject $H_0: \mu = \mu_0$.

The area representing the *p*-value is illustrated in Figure 10.30 for two examples. The action limits $\bar{x}_L$ and $\bar{x}_U$ are shown in Figures 10.30A and 10.30B. In Figure 10.30A the sample mean $\bar{x}$ falls in the lower rejection region ($\bar{x} < \bar{x}_L$). The *p*-value is calculated by finding the area to the *left* of $\bar{x}$ (since $\bar{x} < \mu_0$) and *doubling* this area. The reason the area is doubled is as

FIGURE 10.30 | The areas representing the *p*-value under the sampling distribution of $\overline{\mathbf{X}}$ for the Form I hypothesis

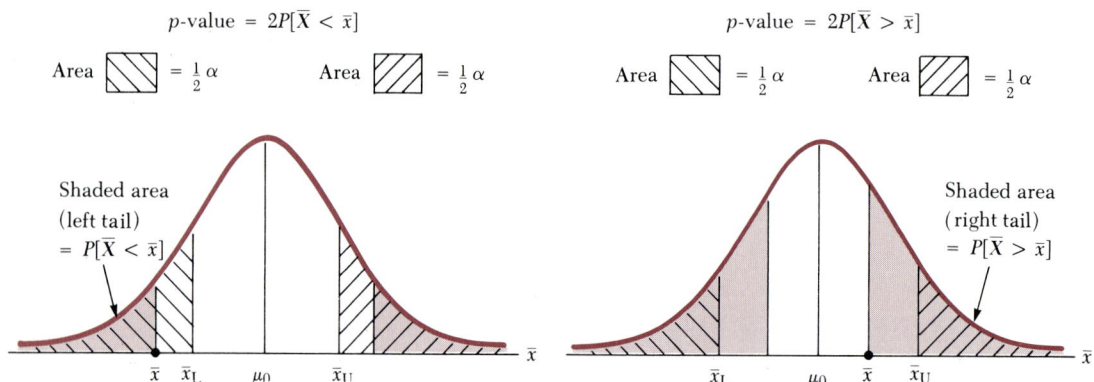

p-value = $2P[\overline{X} < \bar{x}]$

Area $\boxed{\diagdown}$ = $\frac{1}{2}\alpha$ Area $\boxed{\diagup}$ = $\frac{1}{2}\alpha$

Shaded area
(left tail)
= $P[\overline{X} < \bar{x}]$

$\bar{x}$ $\bar{x}_L$ μ_0 $\bar{x}_U$ $\bar{x}$

p-value = $2P[\overline{X} > \bar{x}]$

Area $\boxed{\diagdown}$ = $\frac{1}{2}\alpha$ Area $\boxed{\diagup}$ = $\frac{1}{2}\alpha$

Shaded area
(right tail)
= $P[\overline{X} > \bar{x}]$

$\bar{x}_L$ μ_0 $\bar{x}$ $\bar{x}_U$ $\bar{x}$

follows. In establishing the action limits $\bar{x}_L$ and $\bar{x}_U$ by the classical method, the significance level α was *halved*, placing an area of $\alpha/2$ in the upper and lower tails of the sampling distribution of $\overline{\mathbf{X}}$. To make the *p*-value comparable to α, we therefore *double* the left-tail area ($P[\overline{\mathbf{X}} < \bar{x}]$ so that the *p*-value area is equivalent to the sum of the two-shaded tail areas in Figure 10.30A. The reason for desiring the *p*-value to be comparable to α will be discussed shortly. In Figure 10.30B, the sample mean $\bar{x}$ falls to the right of μ_0 but not in the rejection region (to the right of $\bar{x}_U$). We find the area to the right of $\bar{x}$ [$P(\overline{\mathbf{X}} \geq \bar{x}]$ and double this area to make it comparable to α.

The rules for determining the *p*-value have been given above for the population parameter μ. They apply equally well to other population parameters. In Table 10.5 the rules are given for the general population parameter θ.

TABLE 10.5 | The rules for determining the *p*-value for the three forms of the null and alternate hypotheses for a population parameter θ^*

Form of hypotheses	Null and alternate hypotheses	Test statistic	p-value probability
I	H_0: $\theta = \theta_0$ H_A: $\theta \neq \theta_0$	θ	p-value = $2P[\theta < \theta]$ if $\theta < \theta_0$ p-value = $2P[\theta > \theta]$ if $\theta > \theta_0$
II	H_0: $\theta \geq \theta_0$ H_A: $\theta < \theta_0$	θ	p-value = $P[\theta < \theta_0]$
III	H_0: $\theta \leq \theta_0$ H_A: $\theta > \theta_0$	θ	p-value = $P[\theta > \theta_0]$

* If $\theta = \mu$, then $\theta = \overline{\mathbf{X}}$ and $\theta = \bar{x}$ (the population mean).
If $\theta = \pi$, then $\theta = \mathbf{P}$ and $\theta = p$ (the population proportion).
If $\theta = \sigma$, then $\theta = \mathbf{S}$ and $\theta = s$ (the population standard deviation).

Example 10.23 Suppose that we have the following null and alternate hypotheses:

H_0: $\mu = 10$
H_A: $\mu \neq 10$

Furthermore, suppose that we know that the population standard deviation σ is equal to 20 so that we have the Case 1 (σ known) situation. A sample of size $n = 25$ is taken, and the sample average $\bar{x}$ is 19. Test this hypothesis at the $\alpha = 0.10$ level of significance, and find the *p*-value of the test.

Solution Using the classical method of testing for a Form I and Case 1 test about the population mean μ, the action limits are:

$$\bar{x}_L = \mu_0 - z_{\alpha/2}\sigma_{\bar{x}}$$
$$\bar{x}_U = \mu_0 + z_{\alpha/2}\sigma_{\bar{x}}$$

With $\alpha = 0.10$, $\alpha/2 = 0.05$ and $z_{\alpha/2} = z_{0.05} = 1.64$. Also, $\sigma_{\bar{x}} = \sigma/\sqrt{n} = 20/\sqrt{25} = 20/5 = 4$. The action limits are then:

$$\bar{x}_L = 10 - (1.64)(4) = 3.44 \quad \text{and} \quad \bar{x}_U = 10 + (1.64)(4) = 16.56$$

The decision rule is

Reject H_0: $\mu = 10$ if $\bar{x} < 3.44$ or if $\bar{x} > 16.56$.
Do not reject H_0: $\mu = 10$ if $3.44 \leq \bar{x} \leq 16.56$.

Since $\bar{x} = 19 > \bar{x}_L = 16.56$, we reject H_0: $\mu = 10$ at the $\alpha = 0.10$ level of significance. The decision rule is shown in Figure 10.31.

FIGURE 10.31 | The decision rule for the Example 10.23 problem

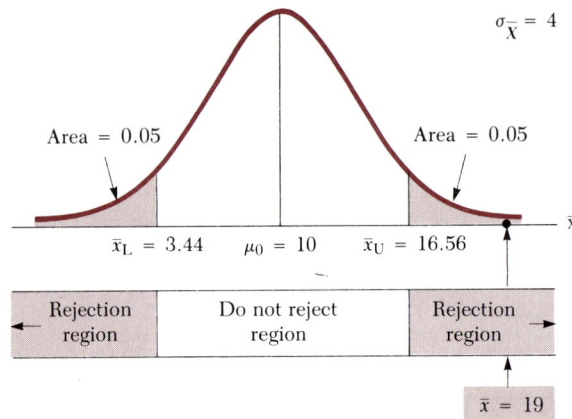

To find the p-value for this Form I hypotheses test, we first note that $\bar{x} = 19 > \mu_0 = 10$. Thus

$$p\text{-value} = 2P[\overline{\mathbf{X}} > \bar{x} = 19]$$

The area corresponding to the probability $P[\overline{X} > 19]$ is illustrated in Figure 10.32. This area under the sampling distribution of $\overline{X}$ can be computed as in Chapter 7, assuming that the sampling distribution is the normal distribution:

$$z = \frac{\bar{x} - \mu_0}{\sigma_{\overline{X}}} = \frac{19 - 10}{4} = 2.25$$

$$P[Z > 2.25] = 0.0122$$

where $P[Z > 2.25]$ has been determined from the standard normal table, Table B.3, Appendix B.

FIGURE 10.32 | The area to the right of $\bar{x} = 19$ $[P(\overline{X} > 19)]$

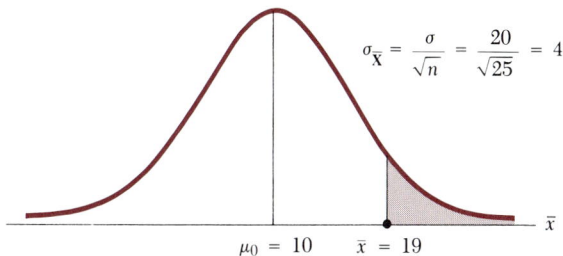

$$\sigma_{\overline{X}} = \frac{\sigma}{\sqrt{n}} = \frac{20}{\sqrt{25}} = 4$$

$\mu_0 = 10$ $\bar{x} = 19$

The *p*-value is now given by:

$$p\text{-value} = 2P[\overline{X} > 19] = 2(0.0122) = 0.0244$$

The *p*-value tells us that it is very unlikely to get a value of $\overline{X}$ that is *removed by more than* $19 - 10 = 9$ *units* from the hypothesized value $\mu_0 = 10$ in *either* direction from $\mu_0 = 10$. This provides strong evidence to reject $H_0: \mu = 10$.

The relationship between α, the level of significance of the test, and the *p*-value is important. To see this relationship, let's consider a simple example. Suppose we have the following Form II set of hypotheses:

$H_0:$ $\mu \geq 100$
$H_A:$ $\mu < 100$

Based on the setting of $\alpha = 0.05$, suppose that the action limit is $\bar{x}_L = 80$. Now consider the *p*-value for three sample values of $\overline{X}$: $\bar{x} = 70$, $\bar{x} = 80$, and $\bar{x} = 90$. The areas corresponding to the *p*-values are shown in Figure 10.33. The *p*-value is the area to the *left* of $\bar{x}$ for the Form II hypotheses.

FIGURE 10.33 | The relationship between α, the level of significance, and the *p*-value

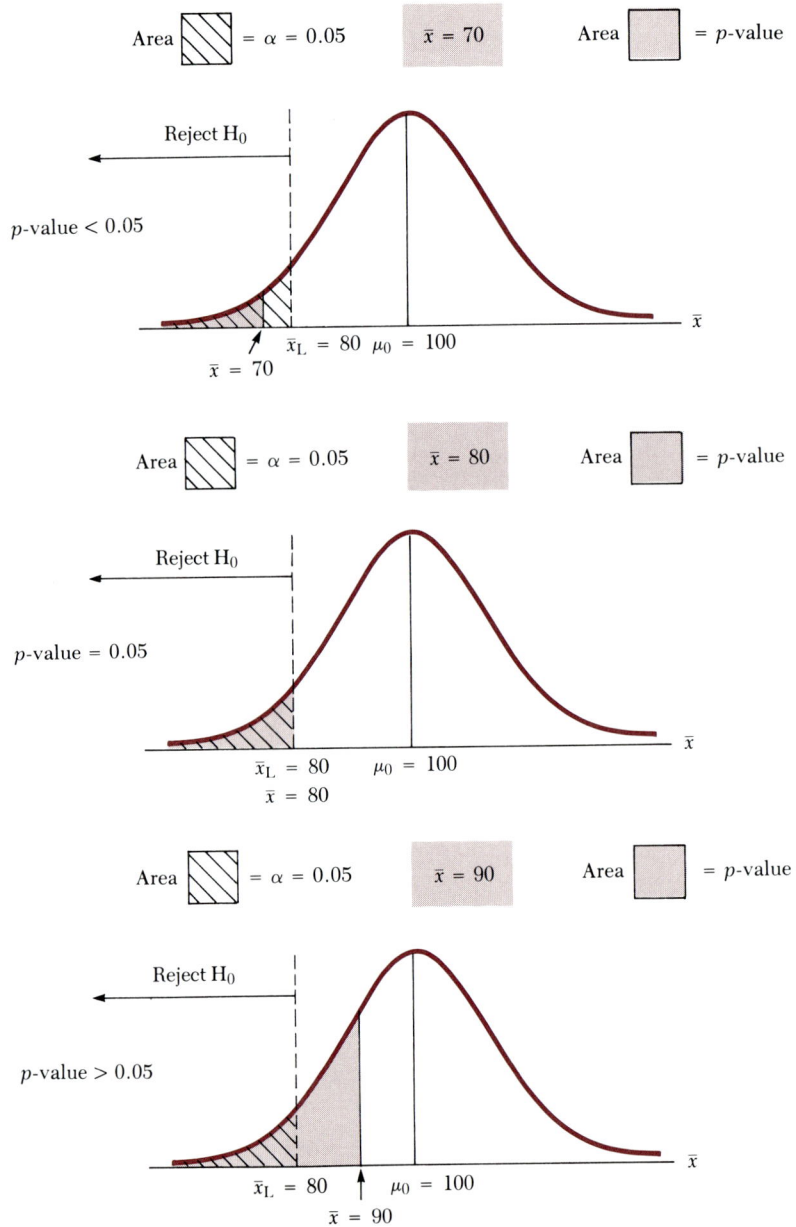

Area [hatched] $= \alpha = 0.05$ $\bar{x} = 70$ Area [white] $= p\text{-value}$

Reject H_0

p-value < 0.05

$\bar{x}_L = 80 \quad \mu_0 = 100$
$\bar{x} = 70$

Area [hatched] $= \alpha = 0.05$ $\bar{x} = 80$ Area [gray] $= p\text{-value}$

Reject H_0

p-value = 0.05

$\bar{x}_L = 80 \quad \mu_0 = 100$
$\bar{x} = 80$

Area [hatched] $= \alpha = 0.05$ $\bar{x} = 90$ Area [gray] $= p\text{-value}$

Reject H_0

p-value > 0.05

$\bar{x}_L = 80 \quad \mu_0 = 100$
$\bar{x} = 90$

An observation that you should be able to make from studying Figure 10.33 is the following:

Relationship between the *p*-value and the significance level α	If the *p*-value $< \alpha$, then reject H$_0$. If the *p*-value $\geq \alpha$, then do not reject H$_0$.

At the top of Figure 10.33, since $\bar{x} = 70 < \bar{x}_{\mathrm{L}} = 80$, we would reject H$_0$: $\mu \geq 100$. Note that *p*-value $< \alpha = 0.05$ in this case. In Figure 10.33, the lower portion, since $\bar{x} = 90 > \bar{x}_{\mathrm{L}} = 80$, we do not reject H$_0$: $\mu > 100$. Note that *p*-value $> \alpha = 0.05$ in this case. In Figure 10.33, the middle portion, $\bar{x} = \bar{x}_{\mathrm{L}} = 80$ so that *p*-value $= \alpha = 0.05$. We do not reject H$_0$: $\mu \geq 100$ in this case.

This relationship holds for Forms I and III of the null and alternate hypotheses as well. An important advantage of the *p*-value method is as follows: *Once the p-value has been computed, it can be compared with the current value of α (or a new value of α) to determine whether or not the null hypothesis H$_0$ should be rejected.*

This is the reason for doubling the tail area in the Form I hypotheses, H$_0$: $\mu = \mu_0$ versus H$_A$: $\mu \neq \mu_0$ to determine the *p*-value. By doing so the *p*-value becomes comparable to α and can be compared with α in deciding whether or not to reject H$_0$.

Example 10.24   A researcher has stated the null and alternate hypotheses for an experiment as follows:

$$H_0: \quad \pi = 0.50$$
$$H_A: \quad \pi \neq 0.50$$

where π is the population proportion. She decides to conduct this test at the $\alpha = 0.05$ level of significance. A sample is taken and, based on the sample proportion value (p), the *p*-value of the test is 0.024. Should the null hypothesis be rejected?

Solution   Since *p*-value $= 0.024 < \alpha = 0.05$, the null hypothesis should be rejected. The sample proportion p *must* fall in either the lowertail or the uppertail rejection region established from the classical method of hypothesis testing. Why?

Most computer software packages (such as MINITAB) print out the *p*-values for hypothesis tests. This makes it *very easy* to decide whether or not to reject the null hypothesis H$_0$. We merely need to compare the *p*-value with the set value of α: If *p*-value $< \alpha$, reject H$_0$, and if *p*-value $\geq \alpha$, do not reject H$_0$.

Another interesting way to consider the *p*-value in relation with the α, is the level of significance of the test. Consider the following Form III set of hypotheses:

$$H_0: \quad \mu \leq 500$$
$$H_A: \quad \mu > 500$$

Suppose a sample has been taken, and it is found that the area to the *right* of $\bar{x}$ is 0.05; that is, *p*-value = 0.05. Now compare this *p*-value to three possible settings of α (α = 0.10, 0.05, 0.01) and the corresponding action limits $\bar{x}_U$. These comparisons are shown in Figure 10.34.

In Figure 10.34A, since *p*-value < α, we reject H_0: $\mu \leq 500$. In Figure

FIGURE 10.34 | Comparing three values of α with a *p*-value equal to 0.05

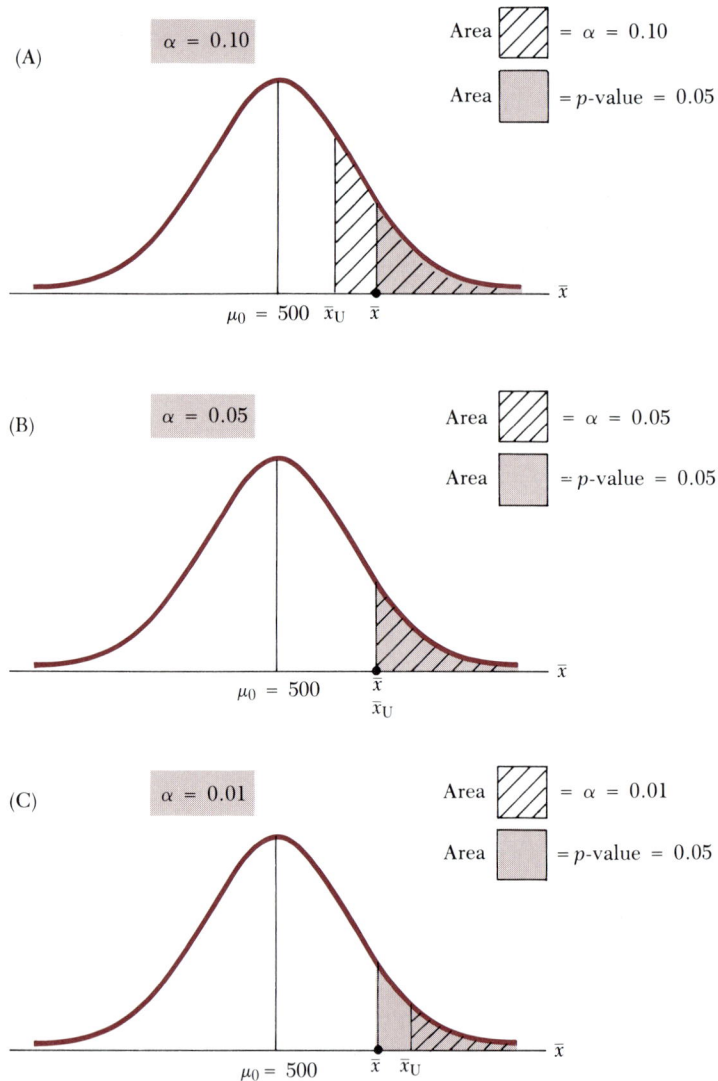

10.34B, since p-value $= \alpha$, we do not reject H_0: $\mu \le 500$. In Figure 10.34C, since p-value $> \alpha$, we do not reject H_0: $\mu \le 500$. Keeping these results in mind, one way of viewing the p-value is:

> The p-value is the *largest* value of α that leads to the weak decision not to reject the null hypothesis.

The decision not to reject H_0 will occur when $\bar{x} \ge \bar{x}_U$. The condition is not satisfied in Figure 10.34A [we reject H_0, since $\bar{x} < \bar{x}_U$], but it is satisfied in Figures 10.34B and C. The *largest* value of α for which the condition $\bar{x} \ge \bar{x}_U$ is satisfied is $\alpha = 0.05$, the situation illustrated in Figure 10.34B. When viewed in this way, the p-value of a test is often referred to as the *nominal* or *descriptive* significance level of the test. For a very good discussion of the p-value viewed in this way, see the Sachs advanced reference at the end of this chapter.

We now summarize the steps in the p-value method of hypothesis testing.

> **Steps in the p-value method of hypothesis testing**
>
> 1. State the null and alternate hypothesis.
>
> 2. Select the sample statistic, and specify its sampling distribution assuming that the null hypothesis is true.
>
> 3. Take a random sample of size n, and calculate the value of the test statistic.
>
> 4. Calculate the p-value for the test statistic by determining the appropriate probabilities in Table 10.5.
>
> 5. Decide to reject or not to reject the null hypothesis based on a comparison of the p-value determined in step 4 to your selected level of significance α.

In the remainder of the text, we will calculate or report p-values when appropriate in hypothesis-testing applications. As we shall see from our examples using statistical software packages, most programs calculate the p-value as a matter of routine in hypothesis testing.

■ 10.7

**Relationship
between
hypothesis
testing and
confidence
interval
estimation**

Hypothesis testing and confidence interval estimation are closely related inference methods. In fact, hypothesis testing can be conducted by using confidence interval estimation. We will first show the connection between hypothesis testing and confidence interval estimation with the Form I hypothesis test,

$$H_0: \quad \mu = \mu_0$$
$$H_A: \quad \mu \neq \mu_0$$

Then we will describe the connection between the hypothesis tests of Forms II and III, where a range of values is specified in H_0 and confidence bounds as discussed in Chapter 9.

10.7.1 | The Form I hypothesis and confidence intervals

The following example is used to show the connection between the Form I hypothesis test and confidence interval estimates.

Example 10.25 An air-conditioning/heating system is designed to maintain a 72°F temperature in a room that houses a large mainframe computer system. In 30 randomly sampled temperatures taken over a two-month period, the sample average temperature was 73.0°F with a standard deviation of 2.8°F. Is the heating/cooling system meeting its design specifications? Use an $\alpha = 0.05$ significance level.

Solution This is the Form I hypothesis test. Let μ = Average temperature in the room. Then the null and alternate hypotheses are:

$$H_0: \quad \mu = 72°F$$
$$H_A: \quad \mu \neq 72°F$$

Since the value of the population standard deviation σ is not given, this is the Case 2 situation requiring the use of the t-distribution. The value of the estimated standard error is:

$$s_{\bar{x}} = s/\sqrt{n} = 2.8/\sqrt{30} = 0.5112$$

The degrees of freedom for the t-value are $df = n - 1 = 29$. Since $\alpha = 0.05$, $\alpha/2 = 0.025$ and $t_{\alpha/2;df} = t_{0.025;29} = 2.045$ from Table B.5 in Appendix B. The action limits for the test are:

$$\bar{x}_L = \mu_0 - t_{\alpha/2;df}\,s_{\bar{x}} = 72 - (2.045)(0.5112) = 72 - 1.05 = 70.95$$
$$\bar{x}_U = \mu_0 + t_{\alpha/2;df}\,s_{\bar{x}} = 72 + (2.045)(0.5112) = 72 + 1.05 = 73.05$$

Thus the decision rule is:

Reject H_0: $\mu = 72°F$ if $\bar{x} < 70.95$ or if $\bar{x} > 73.05$.
Do not reject H_0: $\mu = 72°F$ if $70.95 \leq \bar{x} \leq 73.05$.

Since $\bar{x} = 73$ is between $\bar{x}_L = 70.95$ and $\bar{x}_U = 73.05$, the null hypothesis H_0: $\mu = 72°F$ should not be rejected.

Now, consider the $100(1 - \alpha)$ percent confidence interval estimate for the population mean μ for Case 2 with α set at 0.05:

$$\bar{x} \pm t_{\alpha/2;df} s_{\bar{x}} = 73 \pm t_{0.025;29} s_{\bar{x}} = 73 \pm (2.045)(0.5112)$$
$$= 73 \pm 1.05 \ (71.95 \ \text{to} \ 74.05)$$

Thus we are 95 percent confident that the population mean μ is between 71.95°F and 74.05°F.

Two facts are important to note about this confidence interval estimate:

1. The interval contains the hypothesized value of μ [H_0: $\mu = 72°F$]. Thus, based on the confidence interval estimate, the null hypothesis proposition seems reasonable, since the interval contains the hypothesized value.

2. The quantity $t_{\alpha/2;df} s_{\bar{x}}$ is used both in the determination of the action limits for the hypothesis test and the ends of the confidence interval estimate. In hypothesis testing, this quantity is added to and subtracted from μ_0, the hypothesized value; and in confidence interval estimation, this quantity is added to and subtracted from $\bar{x}$, the sample mean.

The two properties above that apply to this example problem can be shown to apply generally. This leads to an important and useful result for the Form I hypothesis on the population mean:

Relationship between confidence interval estimation and hypothesis testing

If the Form I hypothesis about μ is tested at an α level of significance, then the following equivalences will hold:

> The decision to not reject H_0 is equivalent to the $100(1 - \alpha)$ percent confidence interval containing the hypothesized value μ_0.

> The decision to reject H_0 is equivalent to the $100(1 - \alpha)$ percent confidence interval not containing the hypothesized value μ_0.

Therefore to test a Form I hypothesis about the population mean μ at the α level of significance, we could determine a $(100 - \alpha)$ percent confidence interval estimate and see if the interval contains the hypothesized value of μ or not. This method will also work for hypothesis tests about the population proportion π (with a cautionary note to be discussed shortly). The rules for using a confidence interval estimate to test hypotheses about μ or about π for Form I are given in Table 10.6.

TABLE 10.6 | Decision rule for rejecting the null hypothesis by the use of a confidence interval estimate

Null hypothesis	Estimation	Decision
$H_0: \mu = \mu_0$ or $H_0: \pi = \pi_0$	$100(1 - \alpha)$ percent confidence interval.	Reject H_0 if hypothesized value (μ_0 or π_0) *does not* lie in the confidence interval.

Example 10.26 (Example 10.16 revisited.) In Example 10.16 we considered the following Form I hypothesis about a population proportion π:

$$H_0: \quad \pi = 0.50$$
$$H_A: \quad \pi \neq 0.50$$

where $\pi =$ Proportion in population favoring the package A design. In a random sample of size $n = 225$ persons, 130 favored the package A design. At the $\alpha = 0.05$ significance level, the null hypothesis was rejected in Example 10.16. Determine a $100(1 - \alpha)$ percent $= 100(1 - 0.05)$ percent $= 95$ percent confidence interval estimate for π and show that based on the estimate the null hypothesis $H_0: \pi = 0.50$ is rejected.

Solution The formula for a confidence interval estimate on π is:

$$p \pm z_{\alpha/2} \sqrt{\frac{p(1 - p)}{n}}$$

where p is the sample proportion. Since $\alpha = 0.05$, $\alpha/2 = 0.025$ and $z_{\alpha/2} = z_{0.025} = 1.96$ from Table B.3, Appendix B. Since $p = 130/225 = 0.578$, the interval estimate is:

$$0.578 \pm (1.96) \sqrt{\frac{(0.578)(0.422)}{225}} = 0.578 \pm (1.96)(0.033)$$

$$= 0.578 \pm 0.065$$

Thus we are 95 percent confident that π is between $(0.578 - 0.065) = 0.513$ and $(0.578 + 0.065) = 0.643$. The hypothesized value of π is 0.50. Since this hypothesized value of π is *not* in the confidence interval, we reject $H_0: \pi = 0.50$ at the $\alpha = 0.05$ level of significance.

The confidence interval formula does require the normality distribution assumption regarding the statistic **P**. The normality assumption check conditions are $np \geq 5$ and $n(1 - p) \geq 5$. With $n = 225$ and $p = 0.578$, we have $(225)(0.578) = 103 > 5$ and $(225)(0.422) = 95 > 5$. Thus both conditions are satisfied.

There is one warning about using confidence interval estimates to do hypothesis testing about the population proportion π. The formula for the standard error of the point estimator **P** is:

$$\sigma_P = \sqrt{\frac{\pi(1 - \pi)}{n}}$$

The confidence interval formula for π estimates this standard error by replacing π with the sample proportion p. However, in hypothesis testing, π in the standard error formula is replaced by π_0, the hypothesized value of π. Therefore it may be possible to reach inconsistent decisions for the same hypothesis and sample information by using hypothesis testing and a confidence interval. However, this is unlikely to happen. Notice in Example 10.16 that the estimated standard error has a value of 0.033, the same value to three decimal places as the standard error in the confidence interval above *even though different values were inserted into the standard error formula for π.*

10.7.2 | The Forms II and III hypotheses and confidence bounds (advanced subsection)

A similar relationship exists between one-sided hypotheses and confidence bounds. For example, if we are testing the null hypothesis $H_0: \mu \geq \mu_0$ and decide to reject it at the 0.05 significance level based on the sample information, this decision would be consistent with determining a 95 percent *upper* confidence bound on μ and finding that μ_0 lies *above* this bound. Example 10.27 illustrates the relationship between the one-sided test and the confidence bound.

Example 10.27 A hamburger franchise chain sells a special hamburger called the "bacon and guacamole special." Since this hamburger requires special ingredients, the chain is concerned about the mean number μ of specials sold per week in each of its outlets. In order to justify the costs of the special ingredients, management has determined that it needs to sell an average of at least 750 "specials" per week. If the average number of specials sold is significantly less than 750, then they will discontinue the item. In a random sample of 61 outlets in the chain taken during a given week, it is found that an average of 722 specials are sold per week with a standard deviation of 80. By constructing a 95 percent upper confidence bound on μ, should this null hypothesis be rejected?

Solution The question being asked by management concerning the special is, "Is the average number (μ) of "specials" sold per week at least (greater than or equal to) 750, or is there strong evidence that μ is less than 750?" This is Form II of the null and alternate hypotheses:

$H_0: \quad \mu \geq 750$
$H_A: \quad \mu < 750$

The 95 percent upper confidence bound is given by:

$$\bar{x} + t_{df=60;\ \alpha=0.05}\left(\frac{s}{\sqrt{n}}\right) = 722 + 1.671\left(\frac{80}{\sqrt{61}}\right) = 739.12$$

Thus we are 95 percent confident that μ is not greater than 739.12 units. Since $\mu_0 = 750$ lies above this bound, we would reject H_0 at the 0.05 significance level.

Table 10.7 summarizes the rules for testing the Form II or the Form III hypotheses using confidence bounds.

TABLE 10.7 | Decision rules for rejecting a null hypothesis in Form II or in Form III using confidence bounds

Null hypothesis	Estimation	Decision
H_0: $\mu \geq \mu_0$ or H_0: $\pi \geq \pi_0$	$100(1 - \alpha)$ percent *upper* confidence bound	Reject H_0 if hypothesized value (μ_0 or π_0) is *greater than* upper confidence bound
H_0: $\mu \leq \mu_0$ or H_0: $\pi \leq \pi_0$	$100(1 - \alpha)$ percent *lower* confidence bound	Reject H_0 if hypothesized value (μ_0 or π_0) is *less than* lower confidence bound

■ 10.8

A summary example

In the previous sections, we have described four approaches to hypothesis testing:

1. The classical method.

2. Standardized testing.

3. The *p*-value method.

4. The use of confidence intervals.

We will now apply these four approaches to the same problem to summarize the approaches and to show the relationships among them.

Example 10.28 Suppose the null and alternate hypotheses are:

H_0: $\mu = 100$
H_A: $\mu \neq 100$

A random sample of size $n = 25$ is taken, and the sample mean turns out to be $\bar{x} = 120$. We will assume that $\sigma = 50$, so that we have the Case 1 (σ known) situation. Therefore the standard error is $\sigma_{\bar{x}} = \sigma/\sqrt{n} = 50/\sqrt{25} = 10$. Apply the four approaches to this hypothesis-testing problem. Use an $\alpha = 0.05$ level of significance.

Solution Figure 10.35 summarizes the four approaches to the hypothesis-testing problem.

In the classical method, since $\alpha = 0.05$, $\alpha/2 = 0.025$ and z is found from Table B.3, Appendix B, so that $P[\mathbf{Z} > z] = 0.025$; $z = 1.96$. Therefore the action limits are $\bar{x}_L = 80.4$ and $\bar{x}_U = 119.6$. The null hypothesis, H_0: $\mu = 100$, will be rejected if $\bar{x} < \bar{x}_L$ or if $\bar{x} > \bar{x}_U$. Since $\bar{x} = 120 > \bar{x}_U = 119.6$, the null hypothesis is rejected.

The *p*-value is calculated on the same curve—the sampling distribution of $\overline{\mathbf{X}}$. First we find $P[\overline{\mathbf{X}} > \bar{x}]$ since $\bar{x} > \mu_0$, and it is the Form I (two-tailed) hypothesis-testing statement. Then *p*-value $= 2P[\overline{\mathbf{X}} > \bar{x}] = 2P[\overline{\mathbf{X}} > 120] = 2(0.0228) = 0.0456$. Since *p*-value $= 0.0456 < \alpha = 0.05$, the null hypothesis is rejected.

FIGURE 10.35 The four approaches to hypothesis testing applied to the Example 10.28 problem

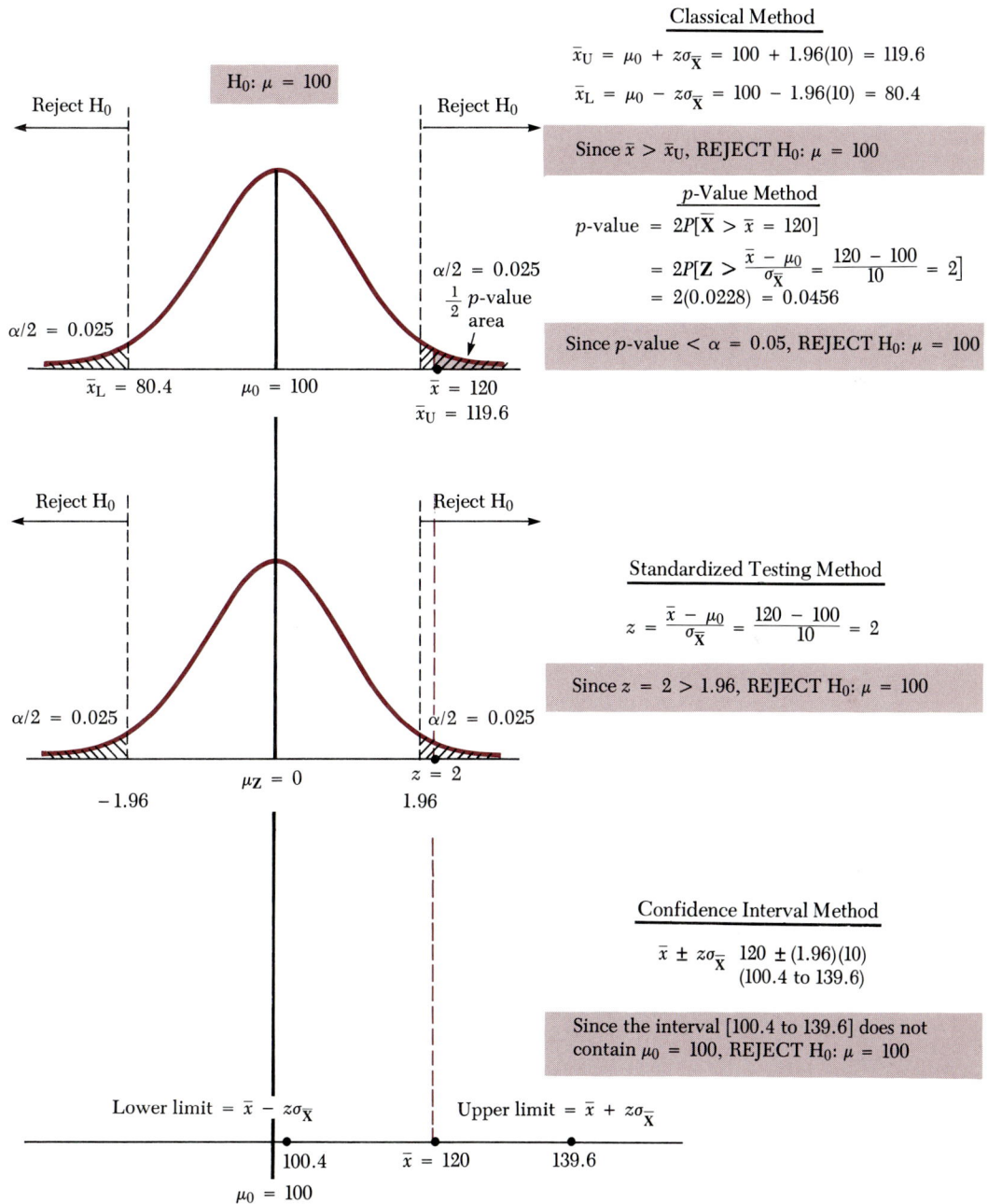

Classical Method

$$\bar{x}_U = \mu_0 + z\sigma_{\bar{X}} = 100 + 1.96(10) = 119.6$$

$$\bar{x}_L = \mu_0 - z\sigma_{\bar{X}} = 100 - 1.96(10) = 80.4$$

Since $\bar{x} > \bar{x}_U$, REJECT H_0: $\mu = 100$

p-Value Method

$$p\text{-value} = 2P[\bar{X} > \bar{x} = 120]$$
$$= 2P[Z > \frac{\bar{x} - \mu_0}{\sigma_{\bar{X}}} = \frac{120 - 100}{10} = 2]$$
$$= 2(0.0228) = 0.0456$$

Since p-value $< \alpha = 0.05$, REJECT H_0: $\mu = 100$

$H_0: \mu = 100$

Reject H_0 Reject H_0

$\alpha/2 = 0.025$

$\frac{1}{2}$ p-value area

$\alpha/2 = 0.025$

$\bar{x}_L = 80.4$ $\mu_0 = 100$ $\bar{x} = 120$
$\bar{x}_U = 119.6$

Reject H_0 Reject H_0

Standardized Testing Method

$$z = \frac{\bar{x} - \mu_0}{\sigma_{\bar{X}}} = \frac{120 - 100}{10} = 2$$

Since $z = 2 > 1.96$, REJECT H_0: $\mu = 100$

$\alpha/2 = 0.025$ $\alpha/2 = 0.025$

$\mu_Z = 0$ $z = 2$

-1.96 1.96

Confidence Interval Method

$$\bar{x} \pm z\sigma_{\bar{X}} \quad 120 \pm (1.96)(10)$$
$$(100.4 \text{ to } 139.6)$$

Since the interval [100.4 to 139.6] does not contain $\mu_0 = 100$, REJECT H_0: $\mu = 100$

Lower limit $= \bar{x} - z\sigma_{\bar{X}}$ Upper limit $= \bar{x} + z\sigma_{\bar{X}}$

100.4 $\bar{x} = 120$ 139.6

$\mu_0 = 100$

For the standardized test, we set up the action limits on the z-distribution. With $\alpha/2 = 0.025$, the action limits are -1.96 to $+1.96$. Since the standardized value of $\bar{x}$, $z = 2$ is greater than 1.96, the null hypothesis is rejected.

The confidence interval is centered at $\bar{x} = 120$. For a 95 percent confidence interval, $z = 1.96$. The resulting interval is 100.4 to 139.6. Since this interval does not contain $\mu_0 = 100$, the null hypothesis is rejected.

Take heart! You need not use *all* four approaches to test a hypothesis. Any one of the four approaches will do fine. Increasingly, the p-value approach is becoming the one favored in solving hypothesis-testing problems.

■ 10.9

The use of MINITAB in hypothesis testing

In this section we will use MINITAB to illustrate the use of statistical software packages for hypothesis testing.

In Example 10.12 the Texas Steel Company is concerned about the average number of hours worked per week (including overtime) by its union workers under the new union labor contract calling for a 36-hour workweek. The null and alternate hypotheses are:

H_0: $\mu \leq 40$ hours
H_A: $\mu > 40$ hours

where μ is the average number of hours worked per week (including overtime) under the new contract. The sample size is $n = 25$, and it is assumed that the population standard deviation σ is 3.5 hours per week (thus Case 1 prevails—σ is known). The significance level is set at $\alpha = 0.10$.

An interactive session on a computer terminal using MINITAB is illustrated below:

```
MTB> SET INTO COLUMN C1
DATA> 44 39 42 38 40 39 43 37 47 36
DATA> 42 43 38 40 37 41 39 47 46 42
DATA> 35 39 38 41 46
DATA> END
MTB> ZTEST MU=40 SIGMA=3.50 DATA IN C1;
SUBC> ALTERNATE = +1.
TEST OF MU = 40.0 VS MU G.T. 40.0
THE ASSUMED SIGMA = 3.50
          N     MEAN    STDEV   SE MEAN    Z     P-VALUE
C1       25    40.76    3.41     0.70    1.09     0.14
MTB>TTEST MU = 40 DATA IN C1;
SUBC>ALTERNATE = +1
TEST OF MU = 40.0 VS MU G.T. 40.0
          N     MEAN    STDEV   SE MEAN    T     P-VALUE
C1       25    40.76    3.41     0.68    1.12     0.14
```

The data are first set into column C1 using the SET command. We indicate that no more data is to be set into column C1 by typing "END" in the last DATA prompt from MINITAB. The test is performed by typing the command,

```
ZTEST MU=40 SIGMA=3.50 DATA IN C1;
```

with the subcommand,

```
ALTERNATE = +1.
```

The subcommand indicates that the alternative hypothesis H_A has the "greater than" ($>$) condition. In the ZTEST command, the statement MU $=$ 40 indicates that $\mu_0 = 40$ in the null hypothesis. Notice that the value of σ must be stated in the ZTEST command.

The results tell us that $n = 25$, $\bar{x} = 40.76$, $s = 3.41$, and $\sigma_{\bar{x}}$ (SE MEAN) is 0.70. The expression SE MEAN stands for the standard error of the mean. Next MINITAB prints the standardized test statistic value,

$$z = \frac{\bar{x} - \mu_0}{\sigma_{\bar{x}}} = \frac{40.76 - 40.0}{0.70} = 1.09$$

Finally the p-value is printed: p-value $P[\mathbf{Z} > z] = 0.14$. The p-value is all that is needed to decide whether or not to reject the null hypothesis, $H_0: \mu \leq 40$, at the $\alpha = 0.10$ level of significance. Since p-value $= 0.14 > \alpha = 0.10$, we cannot reject the null hypothesis.

If we wish to test the Form II hypotheses

$H_0: \quad \mu \geq \mu_0$
$H_A: \quad \mu < \mu_0$

then the subcommand is ALTERNATE $= -1$. If we wish to test the Form I hypotheses

$H_0: \quad \mu = \mu_0$
$H_A: \quad \mu \neq \mu_0$

then the subcommand is not needed.

The next command is used to conduct the Case 2 t-test when σ, the population standard deviation, is not known. We use the command TTEST to illustrate the Case 2 application, although in this case we know that $\sigma = 3.50$ hours. The t-test is performed by typing the command,

```
TTEST MU=40 DATA IN C1;
```

with the subcommand,

```
ALTERNATE = +1,
```

The results tell us that $n = 25$, $\bar{x} = 40.76$, $s = 3.41$, and the estimated standard error of the mean is 0.68. That is, $s_{\bar{x}} = s/\sqrt{n} = 3.41/\sqrt{25} = 0.68$. The standardized **t**-statistic value is

$$t = \frac{\bar{x} - \mu_0}{s_{\bar{x}}} = \frac{40.76 - 40.0}{0.68} = 1.12$$

The p-value is $P[\mathbf{t} > t] = 0.14$. Since p-value $= 0.14 > \alpha = 0.10$, the null hypothesis, $H_A: \mu \le 40$, cannot be rejected. Notice that the p-values for the Z-test and the t-test are the same (to two decimal places). This is because the sample standard deviation s $(s \doteq 3.41)$ in the t-test is so close to the assumed true value of the population standard deviation σ $(\sigma = 3.50)$ in the Z-test.

Computer software packages, such as MINITAB, make hypothesis testing very easy, since the p-value is calculated. All we have to do is compare the p-value to the level of significance α to decide whether or not the null hypothesis should be rejected.

10.10
Part I summary

In statistical hypothesis testing, we hypothesize a value for θ, a population parameter, and use the sample information to determine whether or not the hypothesis is reasonable. In this chapter we have considered three population parameters: μ, the population mean, π, the population proportion, and σ, the population standard deviation. The hypotheses statements may assume one of three forms:

Form I	Form II	Form III
$H_0: \theta = \theta_0$	$H_0: \theta \ge \theta_0$	$H_0: \theta \le \theta_0$
$H_A: \theta \ne \theta_0$	$H_A: \theta < \theta_0$	$H_A: \theta > \theta_0$

The form chosen for a particular problem is often clearly indicated by the nature of the problem. If one value of the parameter ($\theta = \mu$, π, or σ) is being hypothesized, the two-sided test (Form I) should be used. If a range of values of the parameter is being hypothesized, one of the one-sided tests (Form II or III) should be used. The determination of which one-sided form to use (Form II or III) is determined by the question to be answered. Is θ greater than or equal to θ_0, or is there strong evidence that θ is less than θ_0 (Form II)? Is θ less than or equal to θ_0, or is there strong evidence that θ is greater than θ_0 (Form III)?

The probability of committing a Type I error, rejecting H_0 when it is true, is denoted by α. The probability of committing a Type II error, not rejecting H_0 when it is false, is denoted by β.

There are four approaches to hypothesis testing: the classical method, the standardized testing method, the p-value method, and the use of confidence intervals (or bounds).

There are six steps when conducting a hypothesis test by the classical method:

1. State the null and alternate hypotheses.

2. Select the test statistic.

3. Set the value of $\alpha = P$ [Type I error].

4. Construct the decision rule.

5. Draw a random sample, and calculate the value of the test statistic.

6. Make the decision either not to reject or to reject the null hypothesis H_0.

The decision rule can either be established on the sampling distribution of the test statistic or on the standardized distribution (the applicable z- or t-distribution for testing hypotheses about the population mean μ or population proportion π). In the latter case, we refer to the testing method as the standardized testing method.

The p-value method is an alternative to the classical method when testing hypotheses. Most computer statistical software programs calculate the p-value. Once the p-value is calculated, we can decide to reject or not reject the null hypothesis by comparing the p-value to the significance level α:

Reject H_0 if p-value $< \alpha$.
Do not reject H_0 if p-value $\geq \alpha$.

Confidence intervals and bounds can also be used to test hypotheses of the three forms. For the Form I hypotheses,

$$H_0: \quad \mu = \mu_0$$
$$H_A: \quad \mu \neq \mu_0$$

the null hypothesis is rejected if the confidence interval does not contain μ_0.

Computer statistical software packages are becoming more widely used for hypothesis-testing experiments, particularly due to their recent availability on microcomputers. Three of the more popular packages are MINITAB, SAS, and SPSSX.

PART II: | ADVANCED TOPICS IN HYPOTHESIS TESTING

■ 10.11
Introduction to Part II

In Part II a selection of advanced topics are considered. These include more discussion on the formulation of hypotheses, the power of the statistical test, setting sample size for hypothesis testing, and hypothesis tests concerning the population variance and the population standard deviation.

■ 10.12
More on the forms of statistical hypothesis tests

In Section 10.2 of Part I, we considered the three most commonly used forms of the null and alternate hypotheses. Stated in terms of the population mean μ, these three forms are:

Form I: $H_0: \quad \mu = \mu_0$
               $H_A: \quad \mu \neq \mu_0$
Form II: $H_0: \quad \mu \geq \mu_0$
               $H_A: \quad \mu < \mu_0$
Form III: $H_0: \quad \mu \leq \mu_0$
               $H_A: \quad \mu > \mu_0$

In applying statistical hypothesis testing to problems, the first step is identifying the form of the null and alternate hypotheses. Recall from Section 10.2 that each form of the null and alternate hypotheses can be stated as questions:

Form I: Is μ equal to μ_0, or is there strong evidence in the sample that μ is either less than μ_0 or more than μ_0?

Form II: Is μ greater or equal to μ_0, or is there strong evidence in the sample that μ is less than μ_0:

Form III: Is μ less than or equal to μ_0, or is there strong evidence in the sample that μ is greater than μ_0?

The alternate hypothesis H_A is the determining factor in selecting one of the three forms. There is very good reason for this. If the null hypothesis H_0 is rejected, then we conclude that the alternate hypothesis is true. By rejecting the null hypothesis, there are two possible outcomes: (1) We reject H_0, and it is true; (2) we reject H_0, and it is false (H_A is true). The first outcome defines the Type I error, measured by α, the significance level of the test.

The second outcome results in a correct decision. Since $\alpha = P[\text{reject } H_0 / H_0$ is true] is *controlled* in the four methods of hypothesis testing described in this chapter (we set the value of α), by rejecting H_0 we know the probability that we may have made an error.

If on the other hand the null hypothesis H_0 is not rejected, then we conclude that there is insufficient evidence to reject H_0. By not rejecting the null hypothesis, there are two possible outcomes: (1) We do not reject H_0, and it is true; (2) we do not reject H_0, and it is false (H_A is true).

The first outcome results in a correct decision. The second outcome defines the Type II error, measured by β. The probability of committing a Type II error; $\beta = P[\text{do not reject } H_0 / H_0$ is false] *is not controlled* in the four methods of hypothesis testing described in this chapter. Thus by not rejecting H_0, we *do not know* the probability that we may have made an error.

For this reason, the decision to reject H_0 is called the *strong decision,* and the decision not to reject H_0 is called the *weak decision.*

As it turns out, it is possible to design hypothesis tests that control β rather than α. However, this is not typically done and requires advanced techniques beyond the scope of this introductory text.

The fact that rejecting H_0 is the strong decision in hypothesis testing establishes a strategy in selecting the forms of the null and alternate hypotheses. Generally, we wish to place the proposition to be supported by the sample data in the *alternate hypothesis* (and hope that we can reject the null hypothesis, thereby providing strong evidence that the alternate hypothesis is true). It isn't always possible to do this, though. For example, we may wish to demonstrate that the proposition $\mu = \mu_0$ is true. This is the Form I null hypothesis statement. However, we cannot place this statement in the alternate hypothesis. If we did so, then the null hypothesis proposition would become $\mu \neq \mu_0$, and we would have no idea where to center the sampling distribution of $\overline{X}$. This is best understood by defining *simple* and *composite* hypotheses, but this is a very advanced topic (see the Hogg and Craig citation in the reference section at the end of this chapter). Suffice it to say that the null hypothesis *must always have an equal sign in it* [$\mu = \mu_0$, $\mu \geq \mu_0$, or $\mu \leq \mu_0$].

This is why our three questions can be used as a guide in selecting one of the three forms of the null and alternate hypotheses:

1. If we wish to provide strong evidence that μ is different from μ_0 in either *direction* (greater than or less than), then choose Form I: $H_0: \mu = \mu_0$; $H_A: \mu \neq \mu_0$.

2. If we wish to provide strong evidence that μ is less than μ_0, then choose Form II: $H_0: \mu \geq \mu_0$; $H_A: \mu < \mu_0$.

3. If we wish to provide strong evidence that μ is more than μ_0, then choose Form III: $H_0: \mu \leq \mu_0$; $H_A: \mu > \mu_0$.

In this context, we can think of the null hypothesis as a straw man. We stand the straw man upright (state the null hypothesis), then try to knock him down (reject H_0 based on the sample evidence). If he fails to fall, then we feel varying degrees of frustration. Our recourses are to change the level of α, to increase the sample size, or to draw entirely new sample information. The frustration is in part due to not knowing our chance (β) that we may have not rejected a false null hypothesis (committed a Type II error).

In the next section we will discuss the concept of the *power* of the statistical hypothesis test. The power of the test does give us some idea of how high the probability of committing a Type II error may be for different settings of α, different *assumed* values of μ, and different sample sizes. As we shall see, the so-called *power curve* does provide useful information when the weak decision to not reject H_0 must be made.

Concerning the Forms II and III statements of the null and alternate hypotheses, the following rule may be handy in separating these two forms: Since the null hypothesis must contain the equality sign, look for it in the statement of the proposition in a problem. If it is contained in the proposition, then the proposition defines the *null hypothesis*. If it is not contained in the proposition, then the proposition defines the *alternate hypothesis*. For example,

1. "This car will get *more than* 40 miles per gallon on the average."
    H_0: $\mu \le 40$; H_A: $\mu > 40$.

2. "The mean time taken to complete the assembly task is *less than or equal to* 20 minutes." H_0: $\mu \le 20$; H_A: $\mu > 20$.

3. "The mean weight of the coffee in the can is *at least* one pound."
    H_0: $\mu \ge 1$; H_A: $\mu < 1$.

10.13

The power of the statistical test (advanced section)

The classical method of hypothesis testing (and the *p*-value method of testing) control only the Type I error—rejecting a true null hypothesis. As described in Section 10.2, the Type II error—not rejecting a false null hypothesis—cannot be controlled in the usual classical method of hypothesis testing. However, we can compute $\beta = P(\text{Type II error})$ if we *assume* a value of the population parameter that makes the null hypothesis false. We will review this computation and then define the *power of a statistical test*, a quantity that helps us decide if the decision rule for the test may lead to an unacceptable risk of committing a Type II error, not rejecting a false null hypothesis.

Example 10.29 Consider the following null and alternate hypotheses:

H_0: $\mu = 10$
H_A: $\mu \ne 10$

Assuming that the population random variable **X** is normally distributed and that the population standard deviation σ is equal to 20, a random sample of size $n = 25$ is taken, and the sample mean $\bar{x}$ is found to be equal to 19.

Let us assume that the decision maker has set a significance level of 0.05 for the test. Her selection of $\alpha = 0.05$ leads to the decision rule illustrated in Figure 10.36.

The rejection regions for the decision rule with $\alpha = 0.05$ in Figure 10.36 are $\bar{x} < 2.16$ and $\bar{x} > 17.84$, and the do not reject region is $2.16 \le \bar{x} \le 17.84$.

FIGURE   10.36 | Decision rule with $\alpha = 0.05$

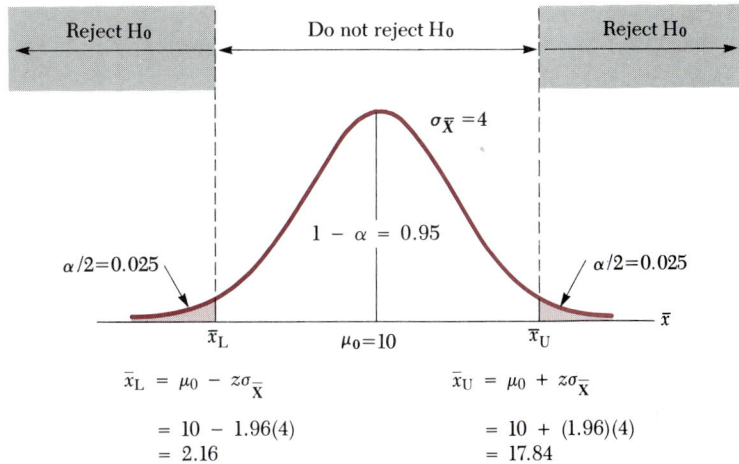

$$\bar{x}_L = \mu_0 - z\sigma_{\bar{X}}$$

$$= 10 - 1.96(4)$$
$$= 2.16$$

$$\bar{x}_U = \mu_0 + z\sigma_{\bar{X}}$$

$$= 10 + (1.96)(4)$$
$$= 17.84$$

Suppose that α is changed from 0.05 to 0.10. The decision rule with $\alpha = 0.10$ is shown in Figure 10.37.

The rejection regions for the decision rule with $\alpha = 0.10$ in Figure 10.37 are $\bar{x} < 3.42$ and $\bar{x} > 16.58$, and the do not reject region is $3.42 \leq \bar{x} \leq 16.58$. Notice the effects on the decision rules in Figures 10.36 and 10.37 by chang-

FIGURE   10.37 | Decision rule with $\alpha = 0.10$

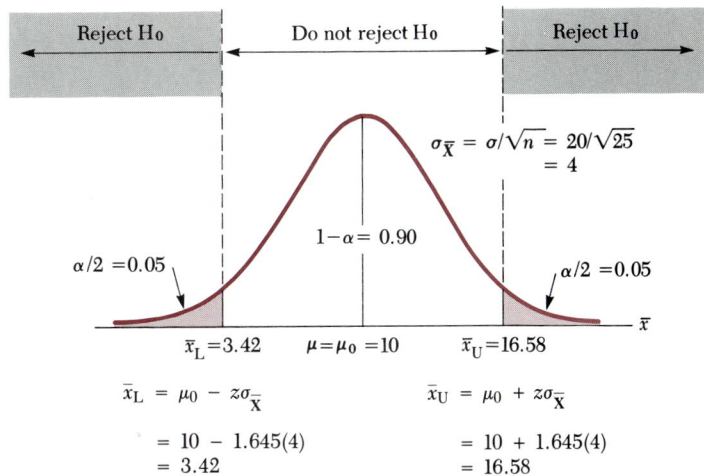

$$\bar{x}_L = \mu_0 - z\sigma_{\bar{X}}$$

$$= 10 - 1.645(4)$$
$$= 3.42$$

$$\bar{x}_U = \mu_0 + z\sigma_{\bar{X}}$$

$$= 10 + 1.645(4)$$
$$= 16.58$$

ing α. By *increasing* α from 0.05 to 0.10, the size of the rejection region (upper and lower tail) has been *increased* (the action limits have been brought closer to the center of the distribution at μ_0), and the size of the do not reject region has been *decreased*. Since the size of the rejection region has been *increased* by increasing α from 0.05 to 0.10, we would reject the null hypothesis more often. By doing so we have obviously increased our chance of rejecting a true null hypothesis ($\alpha = 0.05$ to $\alpha = 0.10$). But the probability of not rejecting a false null hypothesis (β) has been *decreased*, since the do not reject region decreased in size when α increased from 0.05 to 0.10.

Concomitantly, by *decreasing* α, say, from 0.10 to 0.05, we will *increase* the size of the do not reject region and thereby increase the probability of not rejecting a false null hypothesis (measured by β). Thus for a fixed sample size ($n = 25$ in our example), α and β are *inversely related;* that is, as one increases, the other decreases.

But how do we determine the value of β in a particular hypothesis-testing experiment? Our decision, based on the decision rule illustrated in Figure 10.36 using $\alpha = 0.05$, would be to reject H_0: $\mu = 10$, since $\bar{x} = 19 > \bar{x}_U = 17.84$.

Let us now suppose that $\bar{x} = 16$ rather than $\bar{x} = 19$ so that the decision would be not to reject H_0: $\mu = 10$ based on the decision rule illustrated in Figure 10.36. By not rejecting the null hypothesis, it is now possible that we may have committed a Type II error, whose probability is given by β. Can we now compute the probability of committing a Type II error? The answer is no! The reason is apparent by recalling the definition of β: $\beta = P$(do not reject H_0/H_0 is false). The null hypothesis is H_0: $\mu = 10$. In computing β the "given" event is that H_0 is false. But the complement of the event, $\mu = 10$, is the entire real number line, excluding the point $\mu = 10$! Thus we do not know where to center the sampling distribution of $\bar{X}$ to compute β. *We can, however, find β in a hypothetical sense.* Suppose, for example, we *assume* that μ is really equal to 12. In this instance, H_0: $\mu = 10$ is false. Now what is the probability that H_0 is *not* rejected? This probability is illustrated in Figure 10.38. Now,

$$\begin{aligned}
\beta &= P(\text{do not reject } H_0/H_0 \text{ is false}) \\
&= P(2.16 \leq \bar{X} \leq 17.84/\mu = 12, \sigma_{\bar{x}} = 4) \\
&= P\left(\frac{2.16 - 12}{4} \leq Z \leq \frac{17.84 - 12}{4}\right) \\
&= P(-2.46 \leq Z \leq 1.46) = 0.9210
\end{aligned}$$

Thus if the decision maker decides to reject H_0 if $\bar{x} < 2.16$ or if $\bar{x} > 17.84$, based on $\alpha = 0.05$, then if $\mu = 12$, there is a probability of 0.9210 that H_0: $\mu = 10$ will be incorrectly not rejected, thereby committing a Type II error.

By selecting a set of values for μ and computing β as illustrated above, the decision maker can get an idea of the *likelihood* that a Type II error will be made if the null hypothesis is not rejected. Rather than using β for this

FIGURE 10.38 | Illustration of β, the probability of committing a Type II error for the hypothesis test in Example 10.29

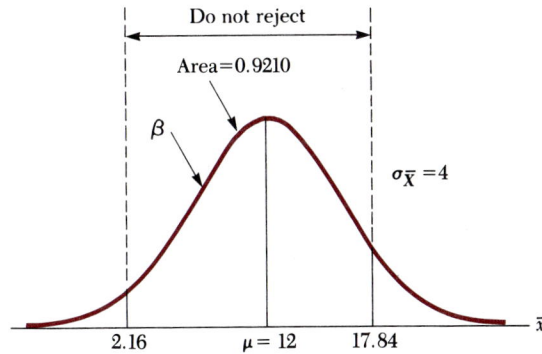

purpose, however, most statisticians determine the *power function* for the test.

Power and the power function

Definition 10.7
The power function

The *power function* for a statistical hypothesis test gives the probability of rejecting the null hypothesis H_0 for the set of assumed values of the population parameter θ:

$$\text{Power} = P(H_0 \text{ is rejected}/\theta)$$

10.13.1 | The determination of the power function in a two-sided test (Form I)

Consider the null and alternate hypotheses in Example 10.29:

$$H_0: \quad \mu = 10$$
$$H_A: \quad \mu \neq 10$$

Based on the decision rule to reject H_0 with the significance level of the test set at $\alpha = 0.05$, the region in which we would not reject H_0 was determined to be $2.16 \leq \bar{x} \leq 17.84$. Figure 10.39 illustrates the computation of values of the power function for this decision rule and the selected values of μ.

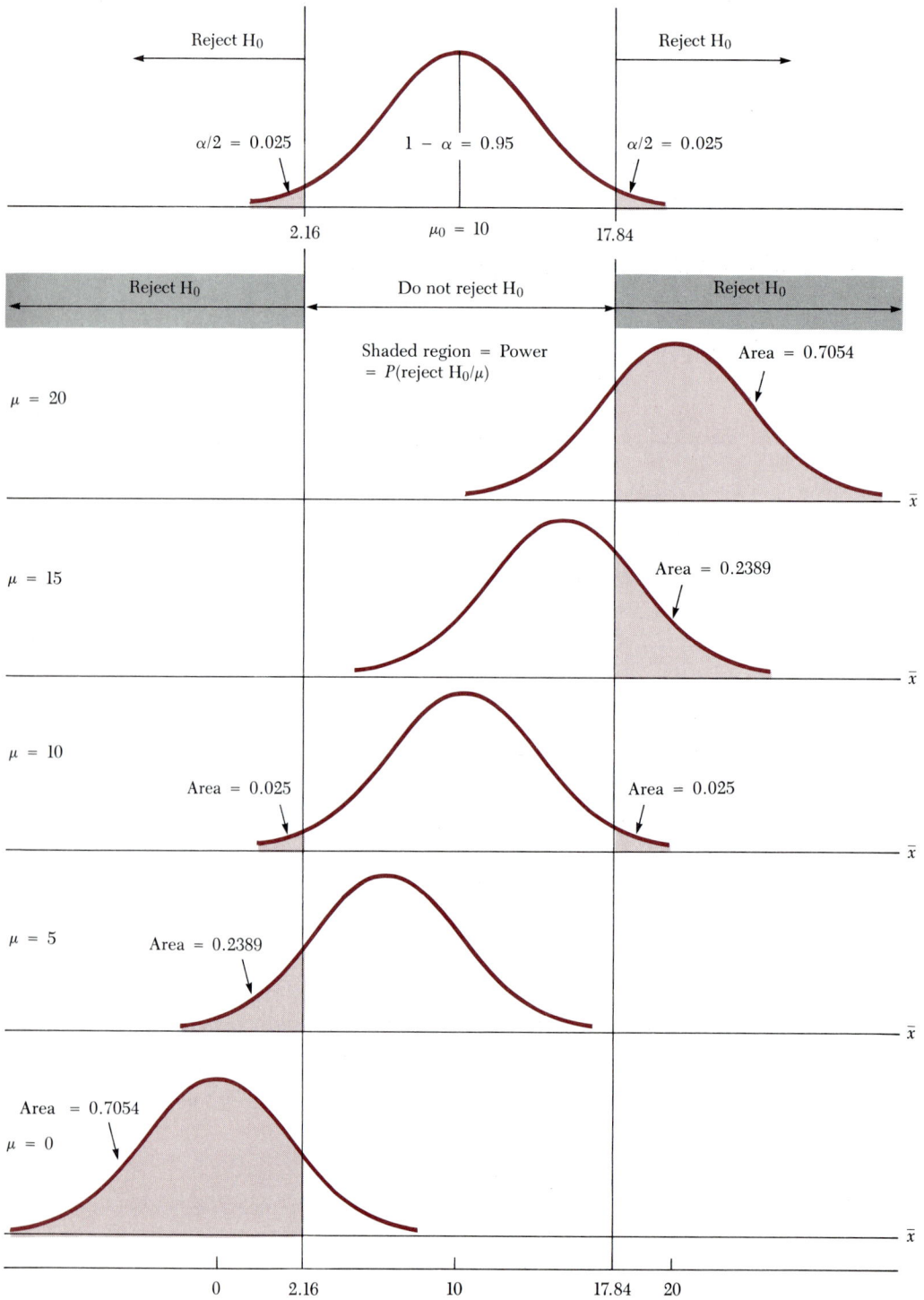

FIGURE 10.39 | Illustration of the probability representing the power of the test for the Example 10.29 hypotheses

By definition, Power = $P(H_0$ is rejected/μ). For example, in referring to Figure 10.39, the power of the test when $\mu = 20$ is computed as follows:

$$\text{Power} = P(\overline{X} < 2.16 \text{ or } \overline{X} > 17.84)$$
$$= P\left(Z < \frac{2.16 - 20}{4} \text{ or } Z > \frac{17.84 - 20}{4}\right)$$
$$= P(Z < -4.46 \text{ or } Z > -0.54)$$
$$= P(Z < -4.46) + P(Z > -0.54) = 0.0000 + 0.7054 = 0.7054$$

Thus if $\mu = 20$, the probability that H_0: $\mu = 10$ is rejected is equal to 0.7054 based on the decision rule to reject H_0 if the $\alpha = 0.05$ significance level is used. In a similar fashion, the power can be determined for the other values of μ given in Table 10.8.

TABLE 10.8 | Power of the test for selected values of μ and the rejection region

μ	Power = $P(reject\ H_0/\mu)$	β
0	0.7054	0.2946
5	0.2389	0.7611
10	0.0500	0.9500*
15	0.2389	0.7611
20	0.7054	0.2946

Rejection region: Reject H_0 if $\bar{x} > 17.84$ or if $\bar{x} < 2.16$. Sample size: $n = 25$, $\alpha = 0.05$.

* Actually, when $\mu = 10$, a Type II error cannot be made. Since H_0: $\mu = 10$ is true when $\mu = 10$, we cannot commit a Type II error—not rejecting H_0 when *it is false*. Thus $\beta = 0$ at μ equal *exactly* 10. Just on either side of $\mu = 10$, β will be very close to 0.95. For the continuity of drawing the power curves, we will therefore disregard this discontinuity.

Using the points in Table 10.8, it is possible to plot the power function by drawing a "smooth" curve through the points so determined, as shown in Figure 10.40. Notice that the power is equal to the significance level of the test ($\alpha = 0.05$) at the hypothesized value of μ, $\mu = 10$. Furthermore $1 - \text{Power} = \beta$; that is,

$$P(\text{reject } H_0/H_0 \text{ is false}) + P(\text{do not reject } H_0/H_0 \text{ if false}) = 1$$
$$\text{Power} \qquad + \qquad \beta \qquad = 1$$

Several values of β are given in Table 10.8 for the given set of μ values.

When H_0 is not true, the power represents the probability that H_0 is rejected given H_0 is false. Thus the power curve gives us some idea of the ability of the test to detect when the null hypothesis is not true as a function of μ. It seems intuitive that the power should increase as μ is farther away from μ_0, the hypothesized value in the null hypothesis. This phenomenon is reflected well in Figure 10.40—as the chosen values of μ depart from $\mu = 10$, the power or the probability of rejecting the null hypothesis H_0 increases.

FIGURE 10.40 | Power function for the test in Example 10.29

Power = $P(\text{rejecting } H_0/\mu)$

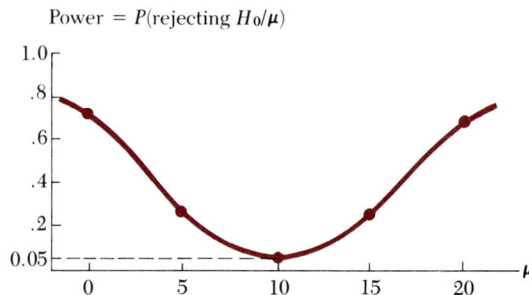

Ideally, we would like the power function to rise very rapidly as the chosen values of μ depart from μ_0. This would imply that the probability of rejecting H_0: $\mu = \mu_0$ is high when μ deviates only slightly from μ_0.

Factors affecting power function

Two factors affect the form of the power function: (1) the sample size and (2) the decision rule (significance level). Let us first consider the effect on the power function of a change in the sample size. Suppose in Example 10.29 the sample size is increased from $n = 25$ to $n = 100$. Now, $\sigma_{\bar{x}} = \sigma/\sqrt{n} = 20/\sqrt{100} = 2$.

The decision rule in terms of the test statistic $\bar{X}$ now changes due to the change in the sample size. The decision rule based on a sample of size $n = 100$ is illustrated in Figure 10.41. Now the power of the test for selected values of μ is given in Table 10.9.

TABLE 10.9 | Power of the test for selected values of μ and the specified rejection region

μ	$Power = P(reject\ H_0/\mu)$	β
0	0.9988	0.0012
5	0.7054	0.2946
10	0.0500	0.9500
15	0.7054	0.2946
20	0.9988	0.0012

Rejection region: Reject H_0 if $\bar{x} < 6.08$ or if $\bar{x} > 13.92$. Sample size: $n = 100$, $\alpha = 0.05$.

The power function when $n = 100$ is graphed in Figure 10.42 together with the power function when $n = 25$.

As the sample size increases, the power function will approach the ideal power function illustrated in Figure 10.43. In the ideal power function, the power becomes 1 if μ deviates *at all* from the hypothesized value of $\mu(\mu_0 =$

FIGURE   10.41   Decision rule in terms of the statistic $\bar{X}$ with a 0.05 significance level and $n = 100$

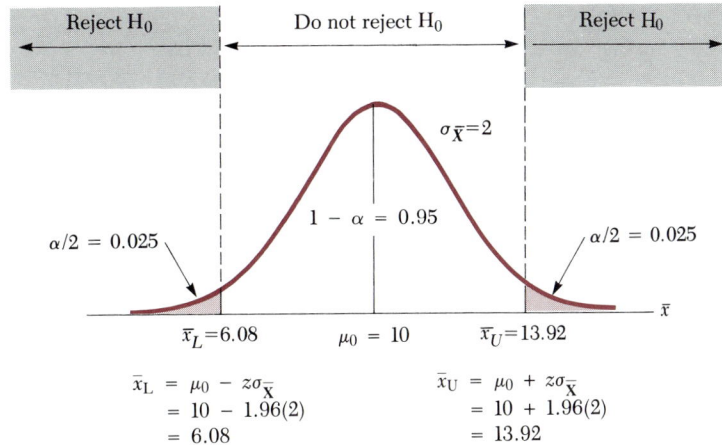

Reject H_0 Do not reject H_0 Reject H_0

$\sigma_{\bar{x}} = 2$

$1 - \alpha = 0.95$

$\alpha/2 = 0.025$ $\alpha/2 = 0.025$

$\bar{x}_L = 6.08$ $\mu_0 = 10$ $\bar{x}_U = 13.92$

$$\begin{aligned} \bar{x}_L &= \mu_0 - z\sigma_{\bar{x}} \\ &= 10 - 1.96(2) \\ &= 6.08 \end{aligned} \qquad \begin{aligned} \bar{x}_U &= \mu_0 + z\sigma_{\bar{x}} \\ &= 10 + 1.96(2) \\ &= 13.92 \end{aligned}$$

FIGURE   10.42   Power function for samples of size $n = 25$ and $n = 100$

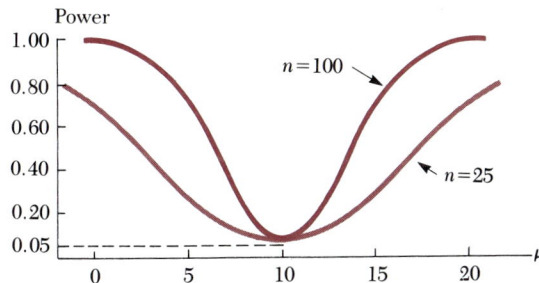

Power

$n = 100$

$n = 25$

10); that is, H_0: $\mu = 10$ will *always* be rejected if μ is not equal to 10. Unfortunately, the number of units that would have to be sampled is *extremely* large if it is desired to have a nearly ideal power curve. The statistician usually balances the cost of obtaining a large sample against the increased power by deciding on an acceptable form for the power curve (and hence an acceptable decision rule).

Let us now consider the effect on the power curve of changing the decision rule. Suppose the decision maker decides to reject H_0: $\mu = 10$ with a 0.10 significance level; that is, she is willing to allow a greater risk of committing a Type I error. Assuming we use a sample size $n = 25$, the decision

FIGURE 10.43 | Ideal power function

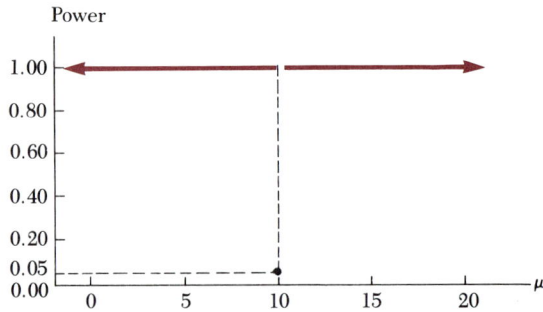

rule in terms of the statistic $\overline{\mathbf{X}}$ is illustrated in Figure 10.44. Table 10.10 gives the power and β for this new decision rule.

This power function is graphed in Figure 10.45 together with the power function for the previous decision rule, where $n = 25$ and $\alpha = 0.05$. Notice in Figure 10.45 that by allowing an increased risk of committing a Type I error (from 0.05 to 0.10), the power of the test is increased and, concomitantly, the probability of committing a Type II error is decreased (compare the columns for β in Tables 10.8 and 10.10.

FIGURE 10.44 | Decision rule with 0.10 significance level

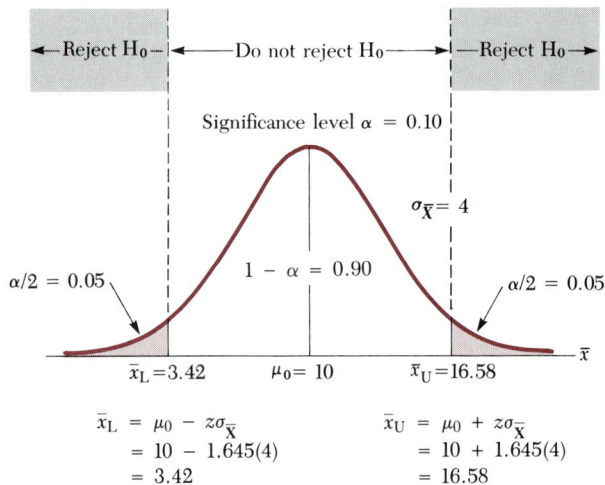

$$\overline{x}_L = \mu_0 - z\sigma_{\overline{X}} \qquad \overline{x}_U = \mu_0 + z\sigma_{\overline{X}}$$
$$= 10 - 1.645(4) \qquad = 10 + 1.645(4)$$
$$= 3.42 \qquad = 16.58$$

Rejection region: Reject H_0 if $\overline{x} < 3.42$ or if $\overline{x} > 16.58$
Region in which H_0 is not rejected: Do not reject H_0 if $3.42 \leq \overline{x} \leq 16.58$

TABLE 10.10 | Power of the test for selected values of μ and for the specified decision rule

μ	$Power = P(reject\ H_0/\mu)$	β
0	0.8051	0.1949
5	0.3465	0.6535
10	0.1000	0.9000
15	0.3465	0.6535
20	0.8051	0.1949

Decision rule: Reject H_0 if $\bar{x} < 3.42$ or if $\bar{x} > 16.58$. Sample size: $n = 25$, $\alpha = 0.10$.

FIGURE 10.45 | Power function graphs for decision rules A and B

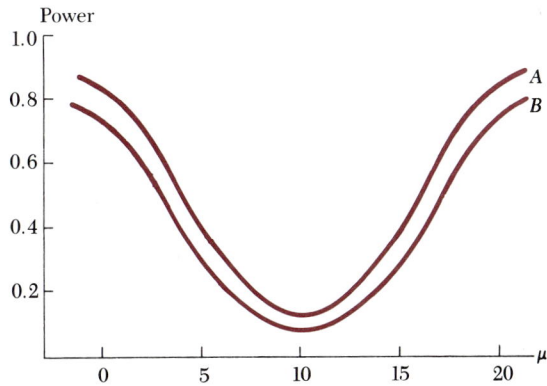

A: Decision rule: Reject H_0 if $\bar{x} < 3.42$ or if $\bar{x} > 16.58$
B: Decision rule: Reject H_0 if $\bar{x} < 2.16$ or if $\bar{x} > 17.84$

The power functions in Figure 10.42 and Figure 10.45 illustrate the following facts:

1. For a fixed sample size n, if the probability of committing a Type I error is *increased,* the probability of committing a Type II error is *decreased;* that is, α and β are inversely related.
2. For a fixed sample size n, if the probability of committing a Type I error is *increased,* the power of the test is *increased.*
3. As the sample size n is *increased,* the power of the test is *increased* and the probability of committing a Type II error is *decreased* for a fixed level of the probability of committing a Type I error.

Operating characteristic curve

An alternative to the power curve or function is the *operating characteristic (OC) curve*. The OC curve is the "complement" of the power curve and gives the probability of not rejecting the null hypothesis for various values of the population parameter. The OC curve corresponding to the power curve illustrated in Figure 10.40 can be generated by taking the complement of the power curve probabilities given in Table 10.8; that is, $P(\text{do not reject } H_0/\mu) = 1 - P(\text{reject } H_0/\mu)$. The OC curve is shown in Figure 10.46.

FIGURE 10.46 | OC curve corresponding to the power curve in Figure 10.40

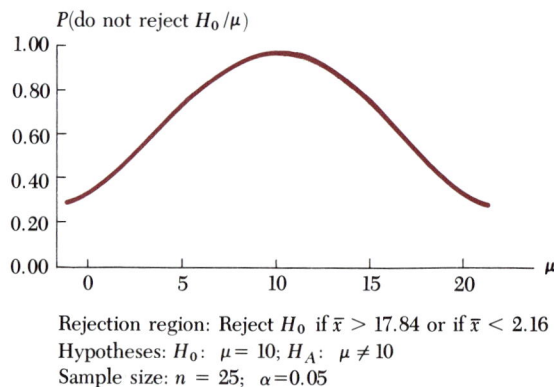

Rejection region: Reject H_0 if $\bar{x} > 17.84$ or if $\bar{x} < 2.16$
Hypotheses: H_0: $\mu = 10$; H_A: $\mu \neq 10$
Sample size: $n = 25$; $\alpha = 0.05$

10.3.2 | The determination of the power function in a one-sided test (Forms II and III)

Suppose that we wish to test the following hypotheses about a population proportion π:

H_0: $\pi \geq 0.96$
H_A: $\pi < 0.96$

A sample of size $n = 100$ is planned, and the significance level α is set at 0.05. The decision rule is shown in Figure 10.47. We are assuming that the sampling distribution of the test statistic **P** is approximately normally distributed (via the central limit theorem).

The power of the test for selected values of π is given in Table 10.11. The computation of the power probabilities given in Table 10.11 is illustrated in Figure 10.48. In Figure 10.48, the assumed true values of π are denoted by π_A. Notice that the standard error $\sigma_\mathbf{P}$ will change based on the assumed value of π. The formula for $\sigma_\mathbf{P}$ is:

$$\sigma_\mathbf{P} = \frac{\pi(1 - \pi)}{n}$$

FIGURE 10.47 | Rejection region and the region not to reject for the null hypothesis H_0: $\pi \geq 0.96$ with $n = 100$ and 0.05 significance level

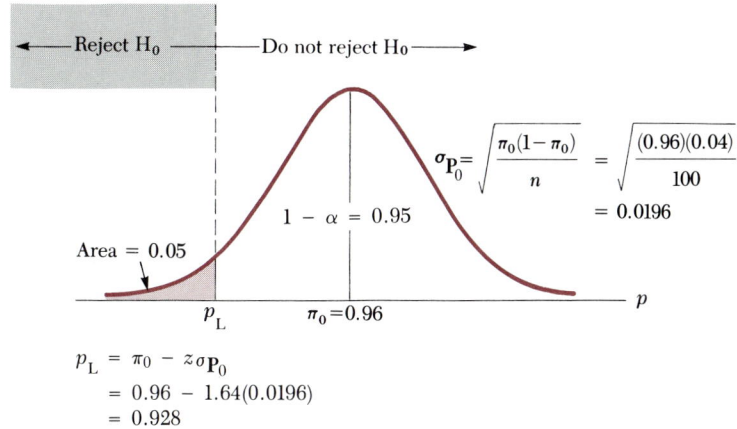

$$p_L = \pi_0 - z\sigma_{P_0}$$
$$= 0.96 - 1.64(0.0196)$$
$$= 0.928$$

TABLE 10.11 | Power probabilities and probability of a Type II error (β) for the rejection region in Figure 10.47

π	Power = $P(reject\ H_0/\pi)$	β
0.92	0.6179	0.3821
0.94	0.3050	0.6950
0.96	0.0516	0
0.98	0.0000	0

Since we are substituting the assumed value of π, denoted by π_A, into σ_P, we will denote it by:

$$\sigma_{P_A} = \frac{\pi_A(1 - \pi_A)}{n}$$

The power function is illustrated in Figure 10.49.

Notice in Table 10.11 that β is 0 when $\pi = 0.96$ *and* when $\pi = 0.98$. For these values of π, the null hypothesis is true, so it is *impossible* to commit a Type II error—failing to reject a false null hypothesis. When $\pi = 0.92$ or $\pi = 0.94$, $\beta = (1 - \text{Power})$ in Table 10.11. In fact, $\beta = 0$ when $\pi \geq 0.96$ and $\beta = (1 - \text{Power})$ when $\pi < 0.96$.

The plot of the probability of committing a Type II error (β) is illustrated in Figure 10.50. The error probability β is 0 from 0.96 and greater. It approaches 0.95 as π approaches 0.96 from the left.

FIGURE 10.48 | Computation of the power probabilities in Table 10.11

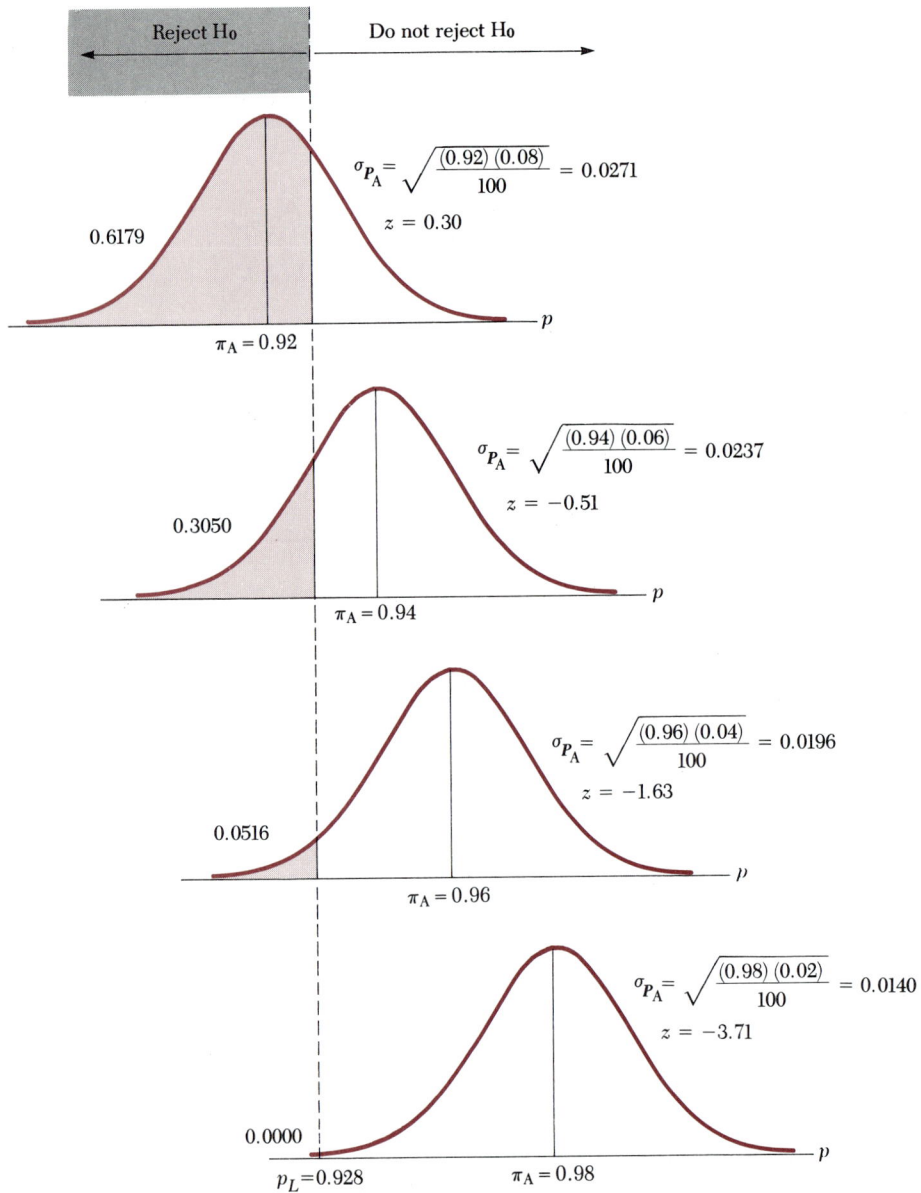

FIGURE   10.49 | Power function for the null hypothesis H_0: $\pi \geq 0.96$ with $n = 100$ and 0.05 significance level

Power = $P(\text{reject } H_0/\pi)$

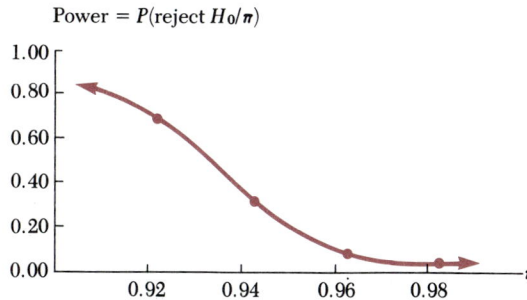

FIGURE   10.50 | The plot of β, the probability of a Type II error, from Table 10.11

β error

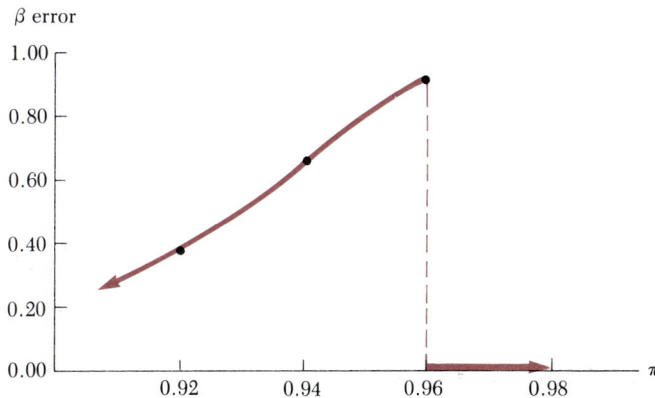

The determination of the power curve or function for various values of the significance level of the test (α) is very important in designing a hypothesis-testing experiment. To construct the power curve (the graph of the power function), a set of assumed values of the population parameter is chosen, and the power [$P(\text{reject } H_0/\text{parameter value})$] is calculated. By studying the power curve, the decision maker can assess whether or not the trade-off between the risk of committing a Type II error and the choice of the sample size is acceptable (as the sample size increases, the chance of a Type II error decreases, and the power increases). Furthermore, she can assess whether or not the trade-off between the risk of committing a Type II error and the risk of committing a Type I error (for a fixed sample size) is acceptable from the power curve. The detailed use of the power function as an aid to the decision maker in the selection of the sample size and in the determination of the significance level of the test (α) is a topic advanced beyond the scope of our introductory coverage of hypothesis testing.

■ 10.14

Setting the sample size for hypothesis testing (advanced section)

When we make the decision, "Do not reject H_0," we are saying in effect that based on the sample data set, we have insufficient information to reject H_0. Had we taken a larger sample, however, we may have been able to reject H_0.

If we carefully plan our hypothesis testing experiment, we can usually spare ourselves doubt when the decision is not to reject H_0. We can do this by specifying the *maximum* values of α and β we wish to tolerate in a particular experiment, where $\alpha = P(\text{reject } H_0/H_0 \text{ is true})$ and $\beta = P(\text{do not reject } H_0/H_0 \text{ is false})$. Selecting the sample size for the experiment is made possible by setting the maximum tolerable values of α and β, as illustrated in Example 10.30.

Example 10.30 Suppose we wish to test the null hypothesis H_0: $\mu = \mu_0$ against the alternative H_A: $\mu \neq \mu_0$ so that α is not greater than 0.05 ($\alpha \leq 0.05$) and β is not greater than 0.05 ($\beta \leq 0.05$). Having set the *maximum* allowable error probabilities, we now set α and β at these maxima: $\alpha = 0.05$ and $\beta = 0.05$. Notice that by setting β we have also set the power of the test. Recall that if H_0 is false, then Power $= 1 - \beta$. Therefore, Power $= 1 - 0.05 = .95$. Suppose $\mu = \mu_0 + d$, where d is the difference between the actual value of μ and the hypothesized value of μ (μ_0). We now illustrate the two probabilities α and β in Figure 10.51. In the top sampling distribution of $\overline{X}$ in Figure 10.51, A represents the upper-tail *action limit*. The area in the upper-tail rejection region is $\alpha/2 = 0.025$. In the bottom sampling distribution of $\overline{X}$ in Figure 10.51, the mean is set at the actual value $\mu = \mu_0 + d$, where $d = \mu - \mu_0$. If the sample mean $\overline{X}$ falls to the *left* of the action limit A, then H_0 will not be rejected. If $\mu = \mu_0 + d$, by not rejecting H_0: $\mu = \mu_0$, we have committed a Type II error (not rejecting H_0 when it is false).

Now in terms of sampling distribution of $\overline{X}$ under H_0 (the top sampling distribution in Figure 10.51), the action limit A can be found from:

$$A = \mu_0 + 1.96(\sigma/\sqrt{n})$$

In terms of the sampling distribution of $\overline{X}$ when $\mu = \mu_0 + d$,

$$A = (\mu_0 + d) - 1.64(\sigma/\sqrt{n})$$

In the first statement, $A = \mu_0 + z(\sigma/\sqrt{n})$ and $z = 1.96$, since the tail area is 0.025. In the second statement, $A = (\mu_0 + d) - z(\sigma/\sqrt{n})$ and $z = 1.64$, since the tail area is 0.05. Setting the two expressions for A equal to one another and solving for n, we have:

$$n = \frac{\sigma^2(1.96 + 1.64)^2}{d^2}$$

FIGURE 10.51 | Illustration of α and β in Example 10.30

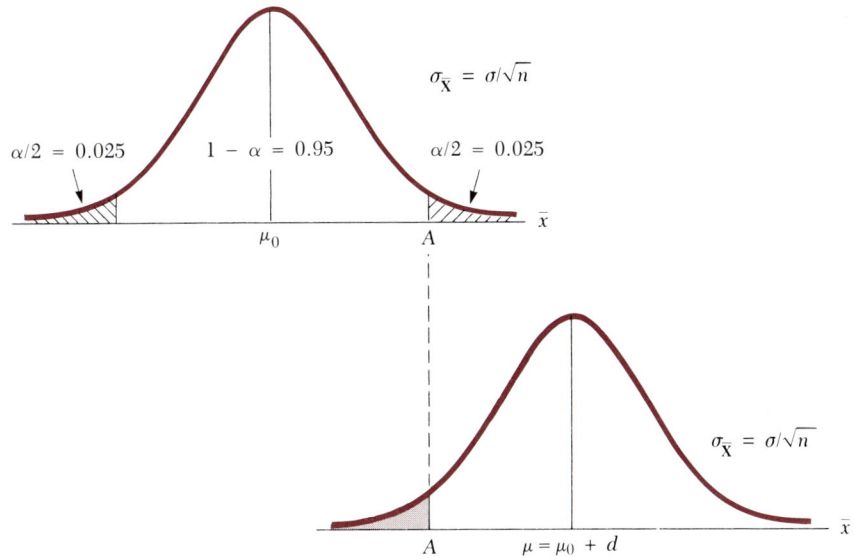

For example, if d is 2 (that is, μ is actually 2 units greater than the hypothesized value of μ) and $\sigma = 10$, say from a pilot study, then:

$$n = \frac{(10)^2(1.96 + 1.64)^2}{(2)^2} = 324$$

Thus to ensure that $\alpha \leq 0.05$ and $\beta \leq 0.05$, we must sample $n = 324$ units to test the null hypothesis H_0: $\mu = \mu_0$ against the alternative H_A: $\mu \neq \mu_0$ when $\mu = \mu_0 + 2$.

The statement regarding α, β, and power that is needed to determine the sample size n must take one of two forms:

1. The statement of the (maximum) value of α, the (maximum) value of β, and the difference to be detected, d.

2. The statement of the (maximum) value of α, a statement regarding the power of the test, and the stated difference to be detected. The power statement usually takes the form, "We wish to be 95 percent confident that if μ is more than d units removed from μ_0, then our decision rule and sample size will be able to detect this difference." The degree of confidence can of course be changed (90 percent, 99 percent, etc.). Since Power $= 1 - \beta$, $\beta = 1 -$ Power so that β can be determined from the statement of power (here expressed as a percent). Thus if Power $= 0.95$, then $\beta = 1 - 0.95 = 0.05$.

The general formula for the sample size to test a Form I hypothesis about a population mean μ is summarized below.

Formula 10.1

Sample size formula for testing H_0: $\mu = \mu_0$

$$n = \frac{\sigma^2(z_{\alpha/2} + z_\beta)^2}{d^2}$$

where

$\sigma^2 = $ Population variance
$z_{\alpha/2}: P(\mathbf{Z} \geq z_{\alpha/2}) = \alpha/2$
$z_\beta: P(\mathbf{Z} \geq z_\beta) = \beta$
$d = $ Difference between μ_0 and μ, which we wish to detect with Power = $1 - \beta$

In Formula 10.1 we must of course have a value to insert for σ^2. This parallels the sample size formula for estimating μ presented in Chapter 9. Normally the value of σ^2 is set by taking a small "pilot" sample to produce the sample variance s^2, which is used in place of σ^2.

Example 10.31 Suppose we are interested in testing the following null hypothesis:

H_0: $\mu = 300$
H_A: $\mu \neq 300$

and wish to have α no larger than 0.05. Further, if μ is really 500, we wish to have no more than a 10 percent chance that we will not reject H_0, thereby not rejecting a false null hypothesis. That is, when μ is 500, we wish β to be no more than 0.10. Further, suppose that we have a preliminary estimate of σ^2 based on a small sample. The sample estimate of the variance is $(700)^2$. What sample size is required to meet these conditions on α and on β?

Solution In Formula 10.1 $\alpha = 0.05$ and $\beta = 0.10$. Therefore $z_{\alpha/2} = z_{0.025} = 1.96$ and $z_\beta = z_{0.10} = 1.28$ from Table B.3, Appendix B. The difference we wish to detect is $d = 500 - 300 = 200$. Substituting into Formula 10.1 yields

$$n = \frac{(700)^2(1.96 + 1.28)^2}{(500 - 300)^2} = 128.59 \Rightarrow 129$$

Therefore, a sample size of 129 is required to ensure that $\alpha \leq 0.05$ and $\beta \leq 0.10$ when μ is actually 500.

Rounding noninteger "n" to next larger integer

Notice in Example 10.31 that we rounded n upward to the next larger integer ($n = 128.59 \Rightarrow 129$). This is the standard convention: If n turns out not to equal an integer value, round to the next larger integer.

Formula 10.1 can be modified to determine the sample size required to test hypotheses about the population mean μ for Forms II and III by replacing $z_{\alpha/2}$ with z_α.

Sample size formula for testing H_0: $\mu \geq \mu_0$ or H_0: $\mu \leq \mu_0$

Formula 10.2

Sample size formula for testing H_0: $\mu \geq \mu_0$ or H_0: $\mu \leq \mu_0$

$$n = \frac{\sigma^2(z_\alpha + z_\beta)^2}{d^2}$$

where

$\sigma^2 = $ Population variance

z_α: $P(Z \geq z_\alpha) = \alpha$

z_β: $P(Z \geq z_\beta) = \beta$

$d = $ Difference between μ_0 and μ, which we wish to detect with Power $= 1 - \beta$.

Example 10.32 Suppose we have the following null and alternate hypotheses:

H_0: $\mu \leq 200$
H_A: $\mu > 200$

The hypothesis is to be tested at the $\alpha = 0.01$ level of significance. Further, the risk of not rejecting H_0 when μ is actually 300 is set at no more than 0.05. That is, when $\mu = 300$, the null hypothesis is false (μ is not ≤ 200), and we want no more than a 0.05 probability of not rejecting H_0: $\mu \leq 200$. Equivalently, we want the power of the test to be at least $1 - 0.05 = 0.95$ when $\mu = 300$. Suppose we are willing to use 250 as an estimate of the population standard deviation σ based on a small sample from the population of interest. What sample size is required to meet these conditions on α and on β?

Solution In Formula 10.2 $\alpha = 0.01$ and $\beta = 0.05$. Thus $z_\alpha = z_{0.01} = 2.33$ and $z_\beta = z_{0.05} = 1.64$ from Table B.3, Appendix B. The difference we wish to detect when $\mu = 300$ is $d = 300 - 200 = 100$. Substituting into Formula 10.2 gives:

$$n = \frac{(250)^2(2.33 + 1.64)^2}{(100)^2} = 98.51 \Rightarrow 99$$

Therefore a sample of size at least 99 is required to test the null hypothesis with $\alpha = 0.01$ and β no more than 0.05 when μ is actually 300.

Sample size formulas can also be developed for testing hypotheses concerning the population proportion π. Consider the following null and alternate hypotheses:

$H_0:\quad \pi = \pi_0$
$H_A:\quad \pi \neq \pi_0$

Assume that we want to control α and β at specified values. Specifically, suppose we wish to detect a difference d between the hypothesized value π_0 and the assumed true value of π, denoted by π_A with Power $= 1 - \beta$. Figure 10.52 illustrates the error probabilities α and β. The symbol A denotes the

FIGURE 10.52 | Illustration of the error probabilities α and β

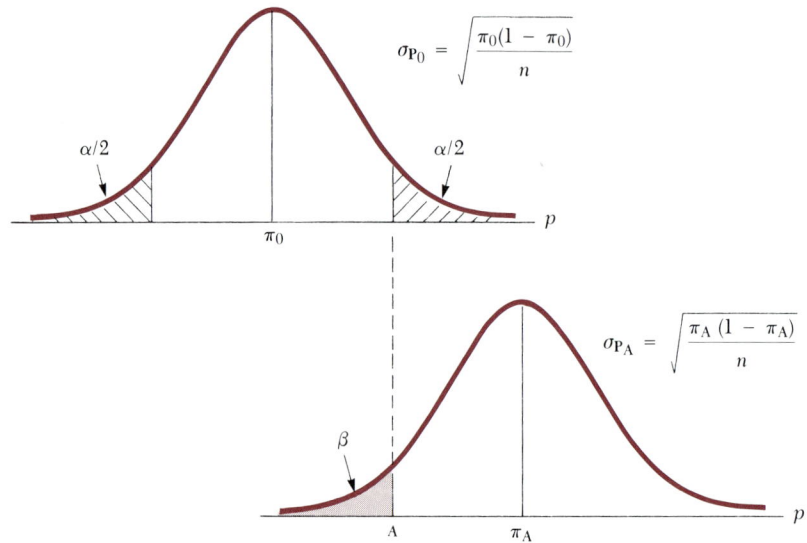

action limit for the hypothesis test on the sampling distribution of **P**. As before, we can write the formulas for the action limit based on the two sampling distributions of **P**; the one under H_0 centered at π_0 and the one centered at π_A:

$$A = \pi_0 + z_{\alpha/2}\sqrt{\frac{\pi_0(1 - \pi_0)}{n}}$$

and

$$A = \pi_A - z_{\beta}\sqrt{\frac{\pi_A(1 - \pi_A)}{n}}$$

Solving for n, we have:

$$n = \left[\frac{z_{\alpha/2} \sqrt{\pi_0(1 - \pi_0)} + z_\beta \sqrt{\pi_A(1 - \pi_A)}}{\pi_A - \pi_0} \right]^2$$

Letting $d = \pi_A - \pi_0$, the difference between the hypothesized value of π and the assumed value of π, we have:

$$n = \frac{[z_{\alpha/2} \sqrt{\pi_0(1 - \pi_0)} + z_\beta \sqrt{\pi_A(1 - \pi_A)}]^2}{d^2}$$

This formula is summarized below.

Sample size formula for testing
$H_0: \quad \pi = \pi_0$
$H_A: \quad \pi \neq \pi_0$

Formula 10.3

Sample size formula for testing $H_0: \pi = \pi_0$ against $H_A: \pi \neq \pi_0$

$$n = \frac{[z_{\alpha/2} \sqrt{\pi_0(1 - \pi_0)} + z_\beta \sqrt{\pi_A(1 - \pi_A)}]^2}{d^2}$$

where

π_A is the assumed value of π
$z_{\alpha/2}: P[Z \geq z_{\alpha/2}] = \alpha/2$
$z_\beta: P[Z \geq z_\beta] = \beta$
$d = $ Difference between π_A, the assumed value of π, and π_0 which we wish to detect with Power $= 1 - \beta$. $[d = \pi_A - \pi_0]$.

Example 10.33 In a market share problem for a specific product, it is hypothesized that the proportion of the market captured by the product is $\pi = 0.50$ (50 percent). This hypothesis is to be tested against the alternative, $\pi \neq 0.50$. If π is really 0.60 ($\pi_A = 0.60$), we wish to detect this difference ($d = \pi_A - \pi_0 = 0.60 - 0.50 = 0.10$) with a power equal to 0.99. The maximum probability of committing a Type I error is set at 0.05 ($\alpha = 0.05$). What sample size is required for this hypothesis test?

Solution Since Power $= 0.99 = 1 - \beta$, $\beta = 0.01$. Now, $\alpha = 0.05$ so that $\alpha/2 = 0.025$. Thus $z_{\alpha/2} = z_{0.025} = 1.96$. Since $\beta = 0.01$, $z_\beta = z_{0.01} = 2.33$. The difference to be detected is 0.10 ($d = 0.10$). Substituting into Formula 10.3, we have:

$$n = \frac{[1.96 \sqrt{0.5(1 - 0.5)} + 2.33 \sqrt{0.6(1 - 0.6)}]^2}{(0.10)^2}$$

$$= \frac{[2.1215]^2}{(0.10)^2} = 450.06 \Rightarrow 451$$

Thus a sample size of 451 is required to meet the stated conditions concerning α and β. The formula for either the Form II or Form III hypothesis test is given below.

Formula 10.4

Sample size formula for testing either $H_0: \pi \geq \pi_0$ or $H_0: \pi \leq \pi_0$

$$n = \frac{[z_\alpha \sqrt{\pi_0(1 - \pi_0)} + z_\beta \sqrt{\pi_A(1 - \pi_A)}]^2}{d^2}$$

where

π_A is the assumed value of π

$z_\alpha: P[\mathbf{Z} \geq z_\alpha] = \alpha$

$z_\beta: P[\mathbf{Z} \geq z_\beta] = \beta$

d = Difference between π_A, the assumed value of π, and π_0, which we wish to detect with Power = $1 - \beta$. $[d = \pi_A - \pi_0]$

Example 10.34 Suppose it is desired to test the null hypothesis, $H_0: \pi \leq 0.50$ against the alternative, $H_A: \pi > 0.50$. If π is really 0.60 ($\pi_A = 0.60$), we wish to detect this difference ($d = \pi_A - \pi_0 = 0.60 - 0.50 = 0.10$) with a power equal to 0.95. The maximum value of α is set at 0.50 ($\alpha = 0.05$). What sample size is required to meet these conditions?

Solution   From the information given, we have $\pi_A = 0.60$, $d = \pi_A - \pi_0 = 0.60 - 0.50 = 0.10$, $\alpha = 0.05$ and $\beta = 1 - \text{Power} = 1 - 0.95 = 0.05$. Thus $z_\alpha = z_{0.05} = 1.645$ and $z_\beta = z_{0.05} = 1.645$. Substituting into Formula 10.4

$$n = \frac{[1.645 \sqrt{0.5(1 - 0.5)} + 1.645 \sqrt{0.6(1 - 0.6)}]^2}{(0.10)^2}$$

$$= \frac{[1.6284]^2}{(0.10)^2} = 265.16 \Rightarrow 266$$

Thus a sample of size $n = 266$ is required to meet these conditions.

■ **10.15**

Testing hypotheses concerning the population variance σ^2 and the population standard deviation σ (advanced section)

There are occasions when we wish to test hypotheses concerning either the population variance σ^2 or the population standard deviation σ. One frequent application occurs in manufacturing, where we may be concerned about the variability of parts produced by machines. For instance, a machine may be designed to produce a steel rod of ½-inch diameter and length 2 feet. The diameter and length are target measurements that the machine is set up to meet. Also, it is usually very important that the *variability* in the diameter and length of the rods is controlled and is small. This is often set by establishing the acceptable *tolerance* of the machine process. For the diameter, we may specify the steel rod diameters should be 0.5 inch ± 0.001 inch; that is, we want the diameters to fall in the interval, [0.499 inch to 0.501 inch].

Hypothesis tests concerning the population variance or the population standard deviation are based on the χ^2 distribution first introduced in Chapter 9. These tests assume that the population random variable X is normally distributed. If X is not normally distributed, but is approximately so, then these tests should be used only for "large" samples ($n = 100$ or more is often required). The degrees of freedom for tests concerning the population variance and the population standard deviation are $df = n - 1$.

Figure 10.53 gives the statements of the three forms of the null and alternate hypotheses, the test statistic, the decision rules, and the areas representing the p-values.

FIGURE 10.53 | The p-value computation for the three forms of the null hypothesis concerning σ^2 or σ

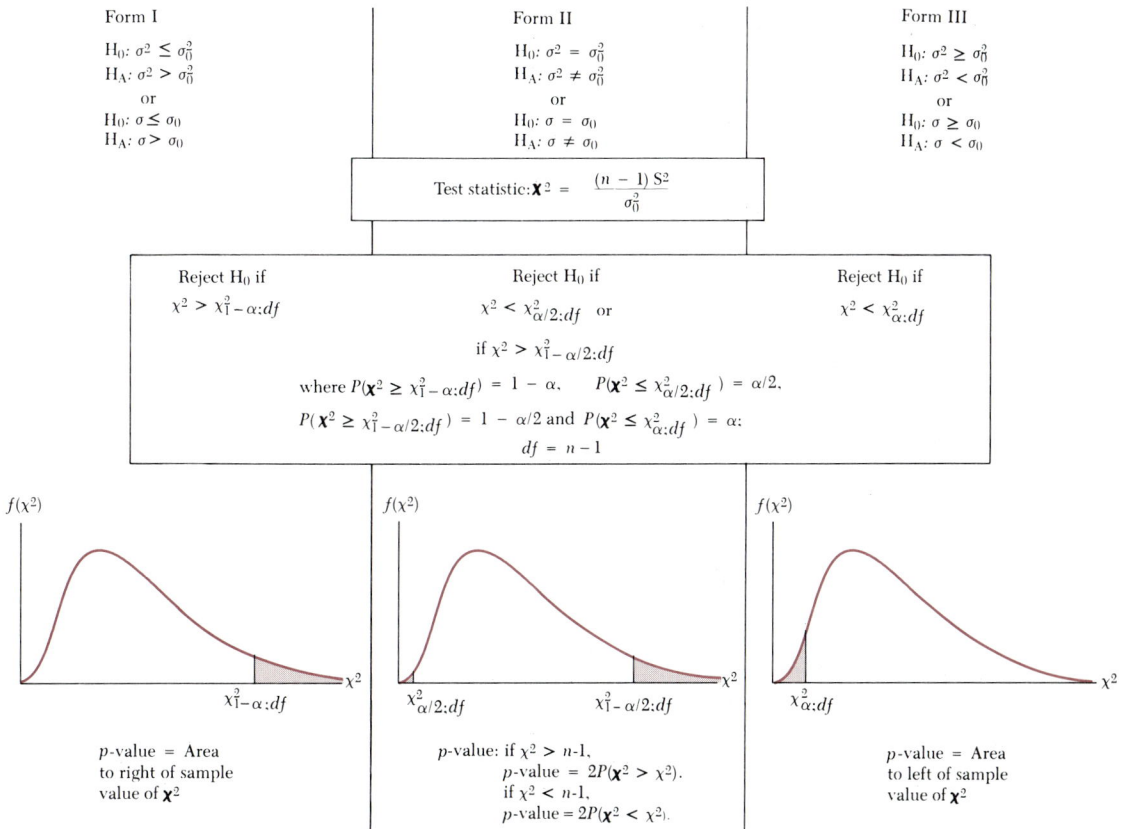

Form I

$H_0: \sigma^2 \leq \sigma_0^2$
$H_A: \sigma^2 > \sigma_0^2$

or

$H_0: \sigma \leq \sigma_0$
$H_A: \sigma > \sigma_0$

Form II

$H_0: \sigma^2 = \sigma_0^2$
$H_A: \sigma^2 \neq \sigma_0^2$

or

$H_0: \sigma = \sigma_0$
$H_A: \sigma \neq \sigma_0$

Form III

$H_0: \sigma^2 \geq \sigma_0^2$
$H_A: \sigma^2 < \sigma_0^2$

or

$H_0: \sigma \geq \sigma_0$
$H_A: \sigma < \sigma_0$

Test statistic: $X^2 = \dfrac{(n - 1) S^2}{\sigma_0^2}$

Reject H_0 if
$\chi^2 > \chi^2_{1 - \alpha:df}$

Reject H_0 if
$\chi^2 < \chi^2_{\alpha/2:df}$ or

if $\chi^2 > \chi^2_{1 - \alpha/2:df}$

where $P(\chi^2 \geq \chi^2_{1 - \alpha:df}) = 1 - \alpha$, $P(\chi^2 \leq \chi^2_{\alpha/2:df}) = \alpha/2$,

$P(\chi^2 \geq \chi^2_{1 - \alpha/2:df}) = 1 - \alpha/2$ and $P(\chi^2 \leq \chi^2_{\alpha:df}) = \alpha$:

$df = n - 1$

Reject H_0 if
$\chi^2 < \chi^2_{\alpha:df}$

$f(\chi^2)$

$\chi^2_{1 - \alpha:df}$

p-value = Area to right of sample value of X^2

$f(\chi^2)$

$\chi^2_{\alpha/2:df}$ $\chi^2_{1 - \alpha/2:df}$

p-value: if $\chi^2 > n-1$,
p-value = $2P(\chi^2 > \chi^2)$.
if $\chi^2 < n-1$,
p-value = $2P(\chi^2 < \chi^2)$.

$f(\chi^2)$

$\chi^2_{\alpha:df}$

p-value = Area to left of sample value of X^2

For example, if we wish to test the following null hypothesis,

$H_0: \quad \sigma^2 = \sigma_0^2$
$H_A: \quad \sigma^2 \neq \sigma_0^2$

then the decision rule is

Reject H_0 if $\chi^2 < \chi^2_{\alpha/2;df}$ or if $\chi^2 > \chi^2_{1-\alpha/2;df}$.

Do not reject H_0 if $\chi^2_{\alpha/2;df} \leq \chi^2 \leq \chi^2_{1-\alpha/2;df}$.

---

where $\chi^2_{\alpha/2;df}$ is the point on the left tail of the χ^2 distribution so that the area in the lower tail is $\alpha/2$ and $\chi^2_{1-\alpha/2;df}$ is the point on the right tail of the χ^2 distribution so that the area in the right tail is $\alpha/2$. These action limits may be looked up in Table B.6, Appendix B, as the following example illustrates.

Example 10.35 Consider the following null and alternate hypotheses:

H_0: $\sigma^2 = 75$
H_A: $\sigma^2 \neq 75$

when, in a sample of size 25, the sample variance $s^2 = 30$. We will assume that the population random variable $\mathbf{X}$ is normally distributed so that a sample of size 25 is suitable. Test the null hypothesis at the $\alpha = 0.05$ significance level.

Solution The value of the test statistic is:

$$\chi^2 = \frac{(n-1)s^2}{\sigma_0^2} = \frac{(24)(30)}{75} = 9.6$$

Since $\alpha = 0.05$, $\alpha/2 = 0.025$ and $1 - \alpha/2 = 0.975$. The degrees of freedom are $df = n - 1 = 25 - 1 = 24$. Therefore the action limits are $\chi^2_{\alpha/2;df} = \chi^2_{0.025;24} = 12.4$ and $\chi^2_{1-\alpha/2;df} = \chi^2_{0.975;24} = 39.4$. The decision rule is:

Reject H_0 if $\chi^2 < 12.4$ or if $\chi^2 > 39.4$.
Do not reject H_0 if $12.4 \leq \chi^2 \leq 39.4$.

Since $\chi^2 = 9.6 < 12.4$, the null hypothesis should be rejected at the $\alpha = 0.05$ significance level. Thus there is an indication that the population variance σ^2 is less than the hypothesized value of 75.

The p-value computation for this test is illustrated in Figure 10.54. Since $\chi^2 = 9.6$ is less than the mean of the χ^2-distribution $[E(\mathbf{X}) = n - 1 = 24]$, we find the area in the *lower tail* of the distribution. This area must be doubled to produce the p-value, since we are conducting a two-tailed test:

$$p\text{-value} = 2P(\chi^2 < \chi^2 = 9.6)$$

From Table B.6 with 24 degrees of freedom, $\chi^2_{0.001;24} = 8.08$ and $\chi^2_{0.005;24} = 9.89$. Therefore the p-value is between $2(0.001) = 0.002$ and $2(0.005) = 0.01$.

FIGURE 10.54 | The *p*-value area

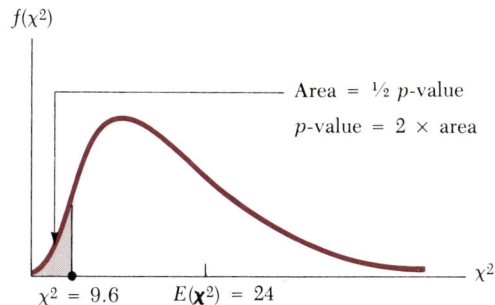

We have a very small chance that a Type I error has been committed by rejecting the null hypothesis $H_0: \sigma^2 = 75$.

Example 10.36 A process that produces ball bearings is considered to be out of control if the standard deviation of the bearing diameters *exceeds* 0.0007 inch. If the production process is not stopped in this case, a considerable cost is incurred to eventually replace the bearings. In a sample of $n = 100$ of these ball bearings, it is found that the sample standard deviation is $s = 0.00083$. Should the production process be stopped and considered to be out of control? Use $\alpha = 0.01$ for this test.

Solution Since it is desired that the data provide strong evidence if σ exceeds 0.0007, this proposition becomes the alternate hypothesis ($\sigma > 0.0007$). The null and alternate hypotheses are:

$H_0:$ $\sigma \leq 0.0007$ ($\sigma_0 = 0.0007$)
$H_A:$ $\sigma > 0.0007$

The sample χ^2 test statistic value is:

$$\chi^2 = \frac{(n-1)s^2}{\sigma_0^2} = \frac{(99)(0.00083)^2}{(0.0007)^2} = 139.2$$

Now, since $\alpha = 0.01$, $\chi^2_{1-\alpha;df} = \chi^2_{0.99;99}$. That is, we want the value of the upper tail of the χ^2-distribution so that the area in the upper (right) tail is 0.01, with $df = n - 1 = 100 - 1 = 99$. Using $df = 100$ (the table entry) as an approximation, the action limit is 135.8 from Table B.6, Appendix B. Thus the decision rule is:

> Reject H_0 if $\chi^2 > 135.8$.
> Do not reject H_0 if $\chi^2 \leq 135.8$.

Since $\chi^2 = 139.2$, the null hypothesis $H_0: \sigma \leq 0.0007$ is rejected. The production process appears to be out of control and should be stopped. The p-value for this test is given by:

$$p\text{-value} = P(\chi^2 > \chi^2 = 139.2)$$

The value 139.2 falls between 135.8 (an upper tail area of 0.01) and 140.2 (an upper tail area of 0.005) in Table B.6, Appendix B. Therefore the p-value is between 0.005 and 0.01. We should be confident, therefore, that we have made the right decision in rejecting $H_0: \sigma \leq 0.0007$. The p-value area is illustrated in Figure 10.55.

FIGURE 10.55 | The p-value area in Example 10.36

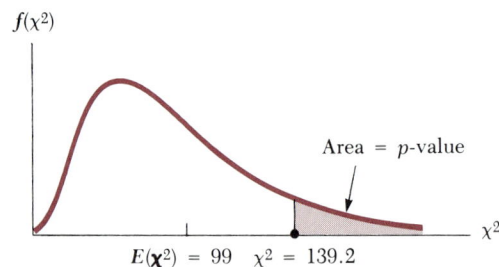

$$E(\chi^2) = 99 \quad \chi^2 = 139.2$$

■ 10.16

Part II summary

There are three forms usually used in statistical hypothesis testing (expressed in terms of the population mean μ):

Form I:	$H_0:$	$\mu = \mu_0$
	$H_A:$	$\mu \neq \mu_0$
Form II:	$H_0:$	$\mu \geq \mu_0$
	$H_A:$	$\mu < \mu_0$
Form III:	$H_0:$	$\mu \leq \mu_0$
	$H_A:$	$\mu > \mu_0$

These forms apply to the population proportion π, the population variance σ^2 and the population standard deviation as well. Form I should be used when strong evidence is sought in the data indicating that μ is not equal to μ_0 in either direction (less than or greater than). A simple rule for selecting between Forms II and III is: If the proposition of interest contains equality in the statement (greater than or equal to, at least, less than or equal to, etc.), then the proposition defines the *null hypothesis*. If the proposition does not contain equality in its statement (greater than, less than, etc.), then the proposition defines the alternate hypothesis.

The power of the test is the probability that the null hypothesis is rejected for a given value of the population parameter. The power curve is a device to aid a statistician in understanding the ability of a test to discriminate between the hypothesized value (or values) of a population parameter and the (assumed) value of the population parameter.

The determination of sample size is an important step in hypothesis testing. To set the sample size, the (maximum) tolerable values of $\alpha = P[\text{reject } H_0/H_0 \text{ is true}]$ and of $\beta = P[\text{do not reject } H_0/H_0 \text{ is false}]$ must be set. Alternatively, the power of the test for a specified alternate value of the

TABLE 10.12 | Test statistic values and sample size requirements for hypothesis testing concerning the population mean μ, the population proportion π, and the population variance σ^2

The population mean μ

Distribution of the population random variable **X**	Knowledge of the population standard deviation σ	Sample size	Test statistic value
Normal	Known	All	$z = \dfrac{\bar{x} - \mu_0}{\sigma/\sqrt{n}}$
	Unknown	All	$t = \dfrac{\bar{x} - \mu_0}{s/\sqrt{n}}$
Unknown	Known	$n \geq 25$	$z = \dfrac{\bar{x} - \mu_0}{\sigma/\sqrt{n}}$
	Unknown	$n \geq 25$	$t = \dfrac{\bar{x} - \mu_0}{s/\sqrt{n}}$
	Known or unknown	$n < 25$	Special small sample methods must be used (see, for example, Chapter 20 on nonparametric statistics)

The population proportion π

Distribution of the population random variable **X**		Sample size	
Binomial	Assumption: Normal approximation to binomial distribution applies.	Depends on normal approximation check, but usually $n \geq 25$ will do.	$z = \dfrac{p - \pi_0}{\sqrt{\dfrac{\pi_0(1 - \pi_0)}{n}}}$

The population variance σ^2

Distribution of the population random variable **X**	Sample size	
Normal	All	$\chi^2 = \dfrac{(n - 1)s^2}{\sigma_0^2}$
Unknown	$n \geq 100$	$\chi^2 = \dfrac{(n - 1)s^2}{\sigma_0^2}$

parameter may be stated in place of stating the (maximum) value of β, since Power $= 1 - \beta$ when the null hypothesis is false.

Hypothesis tests for the population variance σ^2 and the population standard deviation σ are given. The test statistic is chi-square distributed if the population random variable **X** is normally distributed.

Table 10.12 summarizes the test statistics for hypotheses concerning the population mean, the population proportion, and the population variance.

■ References

Introductory

Anderson, D. R.; D. J. Sweeney; and T. A. Williams. *Statistics for Business and Economics.* 2nd ed. St. Paul, Minn.: West Publishing, 1984, Chapter 9.

Mendenhall, W., and J. E. Reinmuth. *Statistics for Management and Economics.* 4th ed. North Scituate, Mass.: Duxbury Press, 1982.

Neter, J.; W. Wasserman; and G. A. Whitmore. *Applied Statistics.* 2nd ed. Boston: Allyn & Bacon, 1982, Chapters 11 and 12.

Wonnacott, T., and R. Wonnacott. *Introductory Statistics for Business and Economics.* 3rd ed. New York: John Wiley & Sons, 1984.

Advanced

Hogg, R. V., and A. T. Craig. *Introduction to Mathematical Statistics.* 4th ed. New York: MacMillan, 1978.

Ostle, B., and N. Mensing. *Statistics in Research.* 3rd ed. Ames: Iowa State University Press, 1975.

Sachs, Lothar. *Applied Statistics: A Handbook of Techniques.* 2nd ed. New York: Springer-Verlag, 1984.

Snedecor, G. W., and W. G. Cochran. *Statistical Methods.* 7th ed. Ames: Iowa State University Press, 1980.

Wonnacott, T., and R. Wonnacott. *Introductory Statistics.* 4th ed. New York: John Wiley & Sons, 1985.

Computer Statistical Software Packages

Nie, N.: C. Hull; Jean Jenkins; Kavin Steinbrenner; Dale Bent. *SPSSX, Statistical Package for the Social Sciences.* New York: McGraw-Hill Book Company, 1985.

Ryan, T.; B. Joiner; and B. Ryan. *MINITAB—A Student Handbook.* 2nd ed. North Scituate, Mass.: Duxbury Press, 1985.

SAS (Statistical Analysis System) SAS Institute, P.O. Box 8000, Cary, NC 27511.

■ Problems

Part I Problems

Section 10.2 Problems

10.1 How does hypothesis testing differ from statistical estimation?

10.2 What are the three basic forms of statistical hypotheses?

10.3 What is meant by a *one-tailed test?* A *two-tailed test?*

10.4 Carefully define a Type I error and a Type II error.

10.5 Explain the following terms used in statistical hypothesis testing:
a. Decision rule.
b. Do not reject region.
c. Rejection region.
d. Significance level of the test.

10.6 Explain what is meant by *action limits* in hypothesis testing.

10.7 How is the sampling distribution used in hypothesis testing?

10.8 What is meant by a *test statistic?*

10.9 What is the relationship between α and β? Explain. How can both be jointly decreased in a hypothesis-testing problem?

10.10 What role does the central limit theorem play in hypothesis testing?

10.11 If the null hypothesis is rejected in a hypothesis-testing experiment, then is it possible that a Type II error has been made? Explain.

10.12 If the null hypothesis is accepted in a hypothesis-testing experiment, then is it possible that a Type I error has been made? Explain.

10.13 Consider the simple hypothesis:

$$H_0: \quad \mu = 50$$
$$H_A: \quad \mu = 60$$

The decision rule, based on a sample of 100, is reject H_0 if $\bar{x} > 55$; otherwise, do not reject H_0. Assume $\sigma = 25$, and $\overline{X}$ is normally distributed.
a. Determine the probability of a Type I error.
b. Determine the probability of a Type II error.

10.14 Repeat parts *a* and *b* in Problem 10.13 for a sample of $n = 225$. Compare your answers here with those in Problem 10.13.

10.15 Consider the null and alternate hypotheses:

$$H_0: \quad \mu = 200$$
$$H_A: \quad \mu \neq 200$$

A sample of size $n = 25$ is to be taken, and the sample mean $\bar{x}$ is to be computed. It is assumed that the population standard deviation is $\sigma = 40$. The action limits for the test have been set at $\bar{x}_L = 184$ and $\bar{x}_U = 216$. Thus H_0 will be rejected if $\bar{x} < \bar{x}_L = 184$ or if $\bar{x} > \bar{x}_U = 216$.
a. What is the significance level α for this decision rule?

b. Suppose that μ is really 220. What is the probability that a Type II error, not rejecting H_0 when it is false, will be made for this decision rule?

10.16 Suppose that the decision rule is changed in Problem 10.15 to $\bar{x}_L = 180$ and $\bar{x}_U = 220$ so that H_0 is rejected if $\bar{x} < \bar{x}_L = 180$ or if $\bar{x} > \bar{x}_U = 220$.
a. What is the significance level α for this decision rule?
b. Suppose that μ is really 220, as in part b of Problem 10.15. What is the probability of committing a Type II error for this decision rule?

10.17 Consider the hypotheses:

$$H_0: \quad \pi = 0.30$$
$$H_A: \quad \pi = 0.40$$

The decision rule, based on a sample of size $n = 100$, is to reject H_0 if p, the sample proportion, is greater than 0.35. Assume that p is approximately normally distributed.
a. What is the probability of a Type I error?
b. What is the probability of a Type II error?

10.18 Consider the following null and alternate hypotheses:

$$H_0: \quad \pi \geq 0.50$$
$$H_A: \quad \pi < 0.50$$

A sample of size $n = 100$ is planned for testing the null hypothesis. The decision rule for this test is: Reject H_0 if $p < 0.40$. Do not reject H_0 if $p \geq 0.40$.
a. What is the probability of committing a Type I error for this decision rule?
b. If π is really 0.60, what is the probability of committing a Type II error for this decision rule?

10.19 Verify that $\beta = P[90 < \overline{X} < 110] = 0.0228$ in Figure 10.8 when the sampling distribution is centered at $\mu_A = 120$.

Section 10.3 Problems

10.20 In testing a hypothesis concerning a population mean, when is it appropriate to use the t-distribution rather than the z-distribution?

10.21 Describe the classical method of hypothesis testing.

10.22 It is known that the daily wages in a particular industry are approximately normally distributed with a mean of $54 and a standard deviation of $6. If a company in this industry employing 36 workers pays these workers an average daily salary of $51, can it be accused of paying inferior wages? Explain. Use an $\alpha = 0.05$ level of significance.

10.23 The mean lifetime of a sample of 100 fluorescent bulbs produced by a company is computed to be 1,570 hours with a standard deviation of 120 hours. If μ is the mean lifetime of all bulbs produced by the company, test the hypothesis $H_0: \mu \geq 1,600$ hours ($H_A: \mu < 1,600$ hours). Use an $\alpha = 0.10$ level of significance.

10.24 A manufacturer of CRT (cathode-ray tube) computer terminals purchases the tubes from one of a few large suppliers. The firm will not purchase the tubes from a particular supplier, however, unless it can be demonstrated that the average lifetime of the tubes exceeds 5,000 hours prior to initial failure. A random sample of nine tubes is tested, and the following sample values are obtained: $\bar{x} = 5,060$, $s^2 = 2,500$. Assume that the lifetimes are normally distributed. Test the null hypothesis $H_0: \mu \leq 5,000$ ($H_A: \mu > 5,000$) by using an $\alpha = 0.05$ level of significance.

10.25 A cigarette manufacturer claims the average nicotine content of one of its brand of cigarettes is "less than or equal to 10 grams." In a sample of 36 cigarettes of this brand, the average nicotine content was 12 grams with a standard deviation of 2 grams. Test the Form III null and alternate hypotheses:

$H_0: \quad \mu \leq 10$ grams
$H_A: \quad \mu > 10$ grams

at the $\alpha = 0.05$ level of significance.

10.26 A company issuing credit cards believes that the holders of its credit cards average at least $20,000 annual salary. A random sample of 145 of its 120,000 cardholders shows an average salary of $19,953 and a standard deviation of $156. The null and alternate hypotheses are stated as:

$H_0: \quad \mu \geq \$20,000$
$H_A: \quad \mu < \$20,000$

Test the null hypothesis at the $\alpha = 0.05$ level of significance.

10.27 In a production process, the mean weight of a steel rod is designed to be 10 pounds. The process is stopped if the mean weight is significantly below or above 10 pounds. A random sample of 25 rods is taken, and it is found that the sample mean is $\bar{x} = 10.058$ pounds. From previous quality control samples, it is assumed that the population standard deviation σ is 0.1 pound. Test the null and alternate hypothesis

$H_0: \quad \mu = 10$ pounds
$H_A: \quad \mu \neq 10$ pounds

at the $\alpha = 0.05$ level of significance, where μ is the mean weight of the population of these rods.

10.28 It is believed that the average test score on a new high school graduation certification test is 70 points on a 100-point exam. A random sample of 50 exam scores is selected. It is found that the average score is 68 and the standard deviation of the 50 scores is 15. Test the following null and alternate hypotheses at the $\alpha = 0.05$ level of significance:

$H_0: \quad \mu = 70$ points
$H_A: \quad \mu \neq 70$ points

where μ is the population average test score.

10.29 The seal in a joint of a solid fuel booster rocket is designed to seal the joint in less than 0.4 second upon ignition of the rocket engine. Therefore the null and alternate hypotheses to be tested are:

$H_0: \quad \mu \geq 0.4$ second
$H_A: \quad \mu < 0.4$ second

Twenty-five tests are run on the joint. In these tests, it is found that the average time required for the joint to be sealed is 0.371 seconds. The sample standard deviation of the 25 times is 0.05 second. Is there strong evidence that the seal is performing its job? Explain. Use $\alpha = 0.01$.

10.30 In a finite population of size $N = 500$, the following null and alternate hypotheses are tested concerning the population mean μ:

$H_0: \quad \mu = 100$
$H_A: \quad \mu \neq 100$

A sample of size $n = 25$ is taken. It is found that the sample mean is $\bar{x} = 90$ with a sample standard deviation $s = 20$. The null hypothesis is to be tested at the $\alpha = 0.05$ level of significance.

a. Test the hypothesis *ignoring* the fact that the population size is finite.
b. Test the hypothesis using the fact that the population size is finite. That is, use the adjusted standard error:

$$s_{\bar{x}} = \sqrt{\frac{N - n}{N - 1}} \frac{s}{\sqrt{n}}$$

c. Compare your results. Does ignoring the fact that the population is finite in part *a* have an effect on the decision?

Section 10.4 Problems

10.31 From past experience, it has been determined that a qualified operator on a certain machine turning out 400 items per day produces *20 or fewer* defectives per day (an error proportion of $20/400 = 0.05$ or less). A new operator is hired to run the same machine. The company is concerned that the proportion of the number of defective units produced in each daily run will now *increase*. Thus the hypotheses to be tested relative to the new operator are:

$$H_0: \quad \pi \leq 0.05$$
$$H_A: \quad \pi > 0.05$$

where π = Proportion of defectives produced by the new operator. A randomly sampled daily production lot from this operator produces 32 defective units in the 400 produced. Is there strong evidence that the defective production rate for this operator is greater than 0.05? Use $\alpha = 0.10$.

10.32 A production process is believed to produce *no more than* 2 percent defectives. In a random sample of 225 items taken from this process, it is found that there are 6 defective items.

a. State the null and alternate hypotheses.
b. Test the null hypothesis at the $\alpha = 0.01$ level of significance.

10.33 A company claims that at least 20 percent of the public prefers its product to that of its competitor (a proportion of 0.20 or more). The null and alternate hypotheses of interest are:

$$H_0: \quad \pi \geq 0.20$$
$$H_A: \quad \pi < 0.20$$

where π is the proportion of customers who prefer the company's product to that of its competitor. How small would the sample proportion p have to be in a sample of $n = 100$ persons in this target population to conclude that H_0 should be rejected? Use $\alpha = 0.05$.

10.34 Skuppy Peanut Butter, Inc. claims that the proportion of jars of its peanut butter that contain more than the federally regulated amount of "nonpeanut butter material" in the jar is less than or equal to 0.02 (as a percentage, 2 percent). A consumer advocate group takes a simple random sample of 64 jars of Skuppy peanut butter and finds that 4 jars contain more than the allowable amount of "nonpeanut butter material." Since the consumer group wishes to demonstrate that the proportion of nonpeanut butter material exceeds 0.02, the null and alternate hypotheses are:

$$H_0: \quad \pi \leq 0.02$$
$$H_A: \quad \pi > 0.02$$

Test the null hypothesis at the $\alpha = 0.05$ level of significance.

10.35 The City of Fort Worth believes that 60 percent of the people living in the city municipal area favor the construction of a new "Southwest" freeway. Various lobbying groups claim that the percentage is either less than 60 percent or is more than 60 percent (depending on their vested interests!). In a sample of 400 citizens in the city, 228 state that they favor the construction of the new freeway.

a. State the null and alternate hypotheses.
b. Test the null hypothesis using an $\alpha = 0.05$ level of significance.

10.36 In an Internal Revenue Service (IRS) region, the director hypothesizes that 8 percent of returns contain at least one arithmetic error. She is looking for strong evidence based on sample information to suggest that the percentage of returns containing at least one arithmetic error is either less than 8 percent or greater than 8 percent. The null and alternate hypotheses are:

$$H_0: \quad \pi = 0.08$$
$$H_A: \quad \pi \neq 0.08$$

A random sample of 400 is taken, and it is found that 48 contain at least one arithmetic error. Test the null hypothesis at the $\alpha = 0.05$ level of significance.

Section 10.5 Problems

10.37 Consider the following null and alternate hypotheses:

$$H_0: \quad \mu = 100$$
$$H_A: \quad \mu \neq 100$$

A random sample of size $n = 100$ is taken, and it is found that the sample average $\bar{x}$ is 88. Assume that the population standard deviation σ is 50. Test the null hypothesis at the $\alpha = 0.10$ level of significance using the standardized testing procedure.

10.38 The amount of impurities in each liter of a California table wine of a certain type and blend is specified by state regulations. A distributor must meet the following test: No more than 2 percent of the one-liter bottles produced can exceed the established amount of impurities set for each liter bottle. The state inspector is concerned about a particular distributor and believes that the distributor exceeds the 2 percent limit. She is looking for strong evidence in the sample to confirm her belief. The null and alternate hypotheses are therefore:

$$H_0: \quad \pi \leq 0.02$$
$$H_A: \quad \pi > 0.02$$

where π is the proportion of bottles exceeding the limit. In a sample of 400 bottles, she finds that 16 bottles exceed the limit for the amount of impurities.

a. Are the conditions met for the normal approximation to be applied for hypothesis testing for this problem? That is, are $np \geq 5$ and $n(1 - p) \geq 5$?
b. Using the standardized testing proce-

dure, test the null hypothesis at the $\alpha = 0.05$ level of significance.

10.39 The average time taken to complete a standardized examination is designed to be 60 minutes. The designers are looking for evidence that the exam takes either more than 60 minutes or less than 60 minutes on the average. A sample of 100 test times is taken, and the sample average $\bar{x}$ and the sample standard deviation s are computed (the population standard deviation σ is unknown). The standardized test statistic value is computed and found to be 2.67.

a. State the null and alternate hypotheses.
b. Test the null hypothesis using the standardized test procedure (is the value of the test statistic given a z-value or a t-value?). Use $\alpha = 0.05$.

10.40 A researcher is planning to test the following hypotheses at the $\alpha = 0.05$ level of significance:

$$H_0: \quad \mu \geq \mu_0$$
$$H_A: \quad \mu < \mu_0$$

She decides to use the standardized testing procedure. State the range of test statistic values for which the null hypothesis would be rejected (assume that the test statistic is the **t**-statistic).

10.41 Apply the standardized testing procedure to Problem 10.22. Summarize your decision.

10.42 Apply the standardized testing procedure to Problem 10.26. Summarize your decision.

10.43 Apply the standardized testing procedure to Problem 10.28. Summarize your decision.

10.44 Apply the standardized testing procedure to Problem 10.32. Summarize your decision.

Section 10.6 Problems

10.45 Describe the p-value method of hypothesis testing.

10.46 Explain the similarities between the classical method of testing and its p-value variant.

10.47 Suppose that the significance level of a test has been set at $\alpha = 0.05$ and the p-value of the test has been calculated to be 0.035. Should the null hypothesis for this test be rejected or not? Explain.

10.48 The proportion of returned 2-inch micro-television sets concerns a manufacturer of these sets. The sets are designed with cost-cutting in mind, so some customer dissatisfaction is expected. However, the manufacturer now believes that more than the acceptable 4 percent of the TVs are being returned. The null and alternate hypotheses of interest are:

$$H_0: \quad \pi \le 0.04$$
$$H_A: \quad \pi > 0.04$$

where π is the proportion of returned sets. In a sample 400 recent purchases of this set, it is found that 28 customers returned their sets.

   a. Find the p-value for this test.
   b. Suppose that the level of significance α is set at 0.005 for this test. Based on the p-value computed in part *a*, should the null hypothesis be rejected or not? Explain.

10.49 The average height of "six foot tall" bookcases manufactured by a furniture company is designed to be 70 inches (2 inches short of 6 feet). The manufacturer is concerned that the average height of these bookcases is less than the designed 70 inches. The boards for the bookcases are supplied by a wood distributor who has a history of "shorting" panel lengths to save money. The manufacturer is looking for strong evidence that the average height is less than 70 inches. A sample of 49 bookcases produces an average height of 69.7 inches. The manufacturer is willing to assume that the population standard deviation of the heights is $\sigma = 0.70$ inch.

   a. State the null and alternate hypotheses for this test.
   b. Determine the p-value for this test.
   c. Suppose that α, the significance level of the test, has been set at 0.01. Should the null hypothesis be rejected or not? Explain.

10.50 A random number generator on the computer is designed to produce an equal number of 0s and 1s in a random order. Let π represent the proportion of 1s in a random sequence of 0s and 1s produced by this generator. A researcher is suspicious about this generator. She suspects that π may be either significantly larger or smaller than 0.50. She sam-

ples 1,000 0s and 1s from this generator and finds that the number of 1s is 465.

   a. State the null and alternate hypotheses for this problem.
   b. Determine the p-value for this test.
   c. Suppose that the level of significance α for this test has been set at 0.05. Based on the p-value determined in part *b*, should the null hypothesis be rejected or not? Explain.

10.51 A firm is concerned about the average number of years its executives have vested in a retirement program. The controller of the firm believes that the average number of years vested exceeds 20. He is looking for strong evidence in a sample to support his proposition. The null and alternate hypotheses are:

$$H_0: \quad \mu \le 20$$
$$H_A: \quad \mu > 20$$

where μ is the average number of years vested by the population of executives in the firm. A sample of 25 files are drawn, and it is found that the average number of years vested is 23 with a standard deviation of 5.6 years.

   a. Find the p-value for the test. [Note: The p-value will have to be approximated from the t-tables.]
   b. With $\alpha = 0.05$ does the sample provide strong evidence to reject the null hypothesis based on the p-value computed in part *a*? Explain.

10.52 The average mortgage interest rate currently offered by lending institutions located in Fort Worth, Texas, on June 2, 1986, is hypothesized to be 10.50 percent. A random sample of 25 lending institutions are called on this date, and the mortgage interest rate is recorded [the interest rate observed is a 30-year conventional loan]. The null and alternate hypotheses of interest are:

$$H_0: \quad \mu = 10.50 \text{ percent}$$
$$H_A: \quad \mu \ne 10.50 \text{ percent}$$

where μ is the average mortgage interest rate. The data are analyzed by MINITAB, which prints a p-value of 0.015.

   a. If α is set at 0.05, should the null hypothesis be rejected or not based on the p-value? Explain.

b. If α is set at 0.01, should the null hypothesis be rejected or not based on the p-value? Explain.

c. If α is set at 0.005, should the null hypothesis be rejected or not based on the p-value? Explain.

10.53 Consider Problem 10.23.

a. What is the p-value for this test? [Note: the p-value will have to be approximated from the t-table.]

b. Based on the p-value determined in part a and the significance level stated in Problem 10.23, should the null hypothesis be rejected or not? Explain.

10.54 Consider Problem 10.27.

a. What is the p-value for this test?

b. Based on the p-value determined in part a and the significance level stated in Problem 10.27, should the null hypothesis be rejected or not? Explain.

10.55 Consider Problem 10.34.

a. What is the p-value for this test?

b. Based on the p-value determined in part a and the significance level stated in Problem 10.34, should the null hypothesis be rejected or not? Explain.

10.56 Consider Problem 10.35.

a. What is the p-value for this test?

b. Based on the p-value determined in part a and the significance level stated in Problem 10.35, should the null hypothesis be rejected or not? Explain.

Section 10.7 Problems

10.57 It is hypothesized that the mean μ in a problem is equal to 460 (H_0: $\mu = 460$ versus H_A: $\mu \neq 460$). The test is to be conducted with an $\alpha = 0.01$ level of significance. A 99 percent confidence interval is constructed, and the resulting interval estimate is 470 to 525. Based on this confidence interval, should the null hypothesis be rejected or not? Explain.

10.58 When testing a hypothesis about a population proportion π, why is it possible to arrive at conflicting results when using the classical

hypothesis-testing procedure and the confidence interval estimation approach to hypothesis testing?

10.59 Consider Problem 10.27.

a. Construct a 95 percent confidence interval estimate for the population mean.

b. Based on the confidence interval determined in part a, should the null hypothesis in Problem 10.27 be rejected at the $\alpha = 0.05$ level of significance? Explain.

10.60 Consider Problem 10.28.

a. Construct a 95 percent confidence interval estimate for the population mean.

b. Based on the confidence interval determined in part a, should the null hypothesis in Problem 10.28 be rejected at the $\alpha = 0.05$ level of significance? Explain.

10.61 Consider Problem 10.35.

a. Construct a 95 percent confidence interval estimate for the population proportion.

b. Based on the confidence interval determined in part a, should the null hypothesis in Problem 10.35 be rejected at the $\alpha = 0.05$ level of significance? Explain.

10.62 Consider Problem 10.36.

a. Construct a 95 percent confidence interval estimate for the population proportion.

b. Based on the confidence interval determined in part a, should the null hypothesis in Problem 10.36 be rejected at the $\alpha = 0.05$ level of significance? Explain.

10.63 The proportion of persons in a certain community who support "Star Wars" technology for missile defense is believed to be 0.75. The stated null and alternate hypotheses are:

$$H_0: \quad \pi = 0.75$$
$$H_A: \quad \pi \neq 0.75$$

A sample of 100 persons who live in this community is taken, and a telephone survey is conducted. This results in the following 95 percent confidence interval on π: 0.66 to 0.76. Based on this confidence interval, should the null hypothesis be rejected or not at the $\alpha = 0.05$ level of significance? Explain.

10.64 It is believed that the average time it takes to get to work by people living in a suburb of Dallas who work in downtown is 50 minutes.

The specified null and alternate hypotheses are:

$$H_0: \quad \mu = 50 \text{ minutes}$$
$$H_A: \quad \mu \neq 50 \text{ minutes}$$

where μ is the average travel time from their homes in the suburb to their jobs in downtown Dallas. A sample of 100 of these commuters is taken, and a 99 percent confidence interval estimate of μ is constructed. The resulting confidence interval estimate is: 51.6 minutes to 66.4 minutes. Should the null hypothesis be rejected or not at the $\alpha = 0.01$ level of significance based on this confidence interval estimate? Explain.

Part II Problems

Section 10.12 Problems

10.65 An auditor wishes to determine whether or not the proportion of invoices containing at least one error has changed since last year. She wants strong evidence from a sample that this proportion has either increased or decreased since last year. State the null and alternate hypotheses.

10.66 A government agency is conducting a study to determine whether or not the proportion of women who smoke in a certain age group has increased since the last study. The agency hopes that the study will provide strong evidence that the proportion has indeed increased since the last study. State the null and alternate hypotheses.

10.67 A business must determine whether or not its average order size for units of a product has changed from last year. The owner of the business is looking for strong evidence in the sample data to suggest that the average is either less than or greater than the average last year. State the null and alternate hypotheses.

10.68 An executive of a major credit card corporation claims that the average amount owed by all those using the card is $100. A sample will be taken to determine if there is strong evidence that the average amount owed is either greater than $100 or less than $100. State the null and alternate hypotheses.

10.69 A diet pill producer claims that its pills will result in at least a 10 pound weight loss if used according to the directions during the first week. A consumer's advocacy group challenges this claim. State the null and alternate hypotheses.

10.70 A marketing research firm is planning a study to demonstrate that the market share that a new product will capture during its first year on the market is "at least" 20 percent. State the null and alternate hypotheses.

10.71 The National Motors Corporation claims that due to its new manufacturing techniques adopted from Japanese methods the proportion of defective cars produced (cars with five or more critical defects) is no more than 1 percent. State the null and alternate hypotheses.

10.72 An MBA program director claims that the average number of years required to complete the MBA program for students enrolled in the part-time programs is no more than four years. State the null and alternate hypotheses that will test this claim.

Section 10.13 Problems

10.73 What is meant by *the power of the test* in hypothesis testing?

10.74 Why can't the exact (or true) probability of committing a Type II error (β) be computed when using the classical method of hypothesis testing described in the text? How is the problem of not knowing the true probability of committing a Type II error dealt with in hypothesis testing?

10.75 A packing process is designed to fill steel drums with 400 pounds of a chemical. To determine whether or not the process is working properly, a random sample of 36 drums will be selected for testing. The null and alternate hypotheses are:

$$H_0: \quad \mu = 400$$
$$H_A: \quad \mu \neq 400$$

where μ is the average amount of chemical filled in the drums. It is decided to reject H_0 if $\bar{x}$ is less than 390 or if $\bar{x}$ is greater than 410. Assume that the population standard deviation σ is equal to 42, and that the drum fills are normally distributed.

a. What is the probability of committing a Type I error for this decision rule?

b. Sketch the power curve by determining the points on the curve for the following possible values of μ: 380, 390, 400, 410, and 420.

c. Suppose that the decision rule is changed to the following: Reject H_0 if $\bar{x}$ is less than 395 or $\bar{x}$ is greater than 405. Repeat parts a and b for this new decision rule.

10.76 The shipping weight for a product is designed to be 100 pounds when the product is placed in its shipping crate. A shipper believes that the average weight, μ, is more than 100 pounds. He is therefore interested in testing the following null and alternate hypotheses:

$$H_0: \quad \mu \leq 100 \text{ pounds}$$
$$H_A: \quad \mu > 100 \text{ pounds}$$

From past records it is assumed that the population variance σ is equal to 10 pounds. To test the null hypothesis, a sample of 25 units will be selected and weighed in its shipping carton. Assume that the weights are normally distributed. The decision rule selected for the hypothesis test is: Reject H_0 if $\bar{x}$ is greater than 103.29 pounds, and do not reject H_0 if $\bar{x}$ is less than or equal to 103.29 pounds.

a. What is the probability of committing a Type I error for this decision rule?

b. Sketch the power curve by determining the points on the curve for the following values of μ: 98, 100, 102, 104, 106, and 108.

10.77 A survey firm believes that people in a metropolitan area are equally divided on a certain issue. Let π be the proportion who support the issue. The null and alternate hypotheses are:

$$H_0: \quad \pi = 0.50$$
$$H_A: \quad \pi \neq 0.50$$

A sample of size $n = 100$ will be taken to test the null hypothesis. It is decided to use the following decision rule: Reject H_0 if p, the sample proportion, is either less than 0.40 or greater than 0.60.

a. What is the probability of committing a Type I error for this decision rule?

b. Sketch the power curve by determining the points on the curve for the following values of π: 0.30, 0.40, 0.50, 0.60, and 0.70.

Section 10.14 Problems

10.78 It is desired to test a hypothesis concerning a population proportion π of the following form:

$$H_0: \quad \pi = \pi_0$$
$$H_A: \quad \pi \neq \pi_0$$

The sample size is to be determined for testing the null hypothesis so that two conditions are met:

a. α is no more than 0.05.

b. The power of the test to detect a difference between the hypothesized value of π and the true value of π that is 0.10 unit removed from π_0 is 0.99.

If the hypothesized value is $\pi_0 = 0.50$, what sample size is required to meet these two conditions? [Hint: Use either 0.40 or 0.50 for π_A in Formula 10.3.]

10.79 A manufacturer believes that the amount of weight required to break a nylon cable is more than 1,000 pounds. The null and alternate hypotheses established to test this claim are:

$$H_0: \quad \mu \leq 1,000 \text{ pounds}$$
$$H_A: \quad \mu > 1,000 \text{ pounds}$$

where μ is the average weight required to break the cable. The sample size is to be determined so that two conditions are satisfied:

a. Maximum value of α is 0.01.

b. Maximum value of β is 0.05 when μ is actually 1,010 pounds. What sample size is required to meet these two conditions? Assume that the population standard deviation of the breaking weights is $\sigma = 25$ pounds.

10.80 It is hypothesized that the proportion of defective units of a certain product that is produced in a machining process is not greater than 0.05. The stated null and alternate hypotheses are:

$$H_0: \quad \pi \leq 0.05$$
$$H_A: \quad \pi > 0.05$$

The sample size to be determined to test the null hypothesis must meet two conditions:

a. Maximum value of α is 0.05.
b. If π is actually equal to 0.10, the power of the test to detect the difference between π and its hypothesized value is 0.90.

How many units from the production process must be sampled to test the null hypothesis while meeting these two conditions?

10.81 It is desired to test the following hypotheses:

$$H_0: \quad \mu = 20$$
$$H_A: \quad \mu \ne 20$$

A sample is to be drawn to test the null hypothesis so that the following conditions are satisfied:

a. Maximum value of α is 0.01.
b. If μ differs by 10 units from the hypothesized value, the power of the test to detect this difference is 0.95.

Assuming that $\sigma = 5$, what sample size is required to meet these conditions?

10.82 The Ajax Auditing Company has a contract to audit the books of the Star Office Supply Corporation. In the past year, the average gross profit per sales invoice was $20. It is hypothesized that this average increased during the current year. Thus the null and alternate hypotheses are:

$$H_0: \quad \mu \le \$20$$
$$H_A: \quad \mu > \$20$$

The null hypothesis will be tested at the $\alpha = 0.05$ level of significance. If μ really is $21, it is decided to accept a probability of 0.10 of committing a Type II error (accepting a false null hypothesis). From the audit conducted last year, it is determined that the standard deviation of the gross profit per sales invoice is $8. What sample size is necessary to test the null hypothesis?

10.83 A foods manufacturer claims that the average amount of applesauce contained in its applesauce cans is 10 ounces. A consumer group decides to test the hypothesis that the average amount of applesauce in the cans is 10 ounces at the $\alpha = 0.05$ level of significance. If the mean fill μ is really 9.9 ounces, then the consumer group is willing to take a 0.05 risk of committing a Type II error. From past

studies, it is known that the standard deviation of the fill amounts is 0.2 ounces. What sample size is required to meet these specifications on α and β?

Section 10.15 Problems

10.84 Consider the following null and alternate hypotheses:

$$H_0: \quad \sigma = 20$$
$$H_A: \quad \sigma \ne 20$$

A sample of size $n = 36$ is taken, and it is found that the sample standard deviation is $s = 35$. Should the null hypothesis be rejected at the $\alpha = 0.05$ level of significance?

10.85 If it is desired to test hypotheses concerning the population standard deviation σ, how can the methods described in Section 10.10 on testing hypotheses about the population variance σ^2 be used?

10.86 With routine equipment like light bulbs, which wear out after a time, the standard deviation of the lifetime is an important factor in determining whether or not it is cheaper to replace all pieces at a fixed interval of time or to replace each piece individually when it breaks down. For a certain gadget, an industrial statistician has calculated that it will pay to replace at fixed time intervals if $\sigma < 6$ days. A sample of 71 pieces gives $s = 4.2$ days. Test the hypothesis $H_0: \quad \sigma \ge 6$ ($H_A: \quad \sigma < 6$) by using an $\alpha = 0.01$ level of significance.

10.87 The Graduate Management Admissions Test (GMAT) is designed to have a mean score of 500 and a standard deviation of 100 with scores ranging from 200 to 800 points. There is evidence that the standard deviation is much less than 100 points for current test takers. To test the claim that σ is now less than 100, the following null and alternate hypotheses are established:

$$H_0: \quad \sigma \ge 100$$
$$H_A: \quad \sigma < 100$$

A sample of 100 recent test scores is taken, and it is found that the sample standard deviation is $s = 70$. Test the null hypotheses at the $\alpha = 0.05$ level of significance.

10.88 The standard deviation of the diameters of 0.5 inch steel bearings is designed to be 0.001 inch. If there is strong evidence based on quality control samples that σ is greater than 0.001 inch, then the production process is shut down. The appropriate null and alternate hypotheses are:

$$H_0: \quad \sigma \le 0.001 \text{ inch}$$
$$H_A: \quad \sigma > 0.001 \text{ inch}$$

A sample of 100 bearings is taken, and it is found that the sample standard deviation is 0.0012 inch. At the $\alpha = 0.05$ level of significance, should the process be shut down? Explain.

10.89 The following null and alternate hypotheses concerning the population variance σ^2 are to be tested at the $\alpha = 0.05$ level of significance:

$$H_0: \quad \sigma^2 = 100$$
$$H_A: \quad \sigma^2 \ne 100$$

A sample of size $n = 50$ is taken, and it is found that the sample variance $s^2 = 102$. Should the null hypothesis be rejected or not? Explain. Assume that the population random variable **X** is normally distributed.

Additional Problems

10.90 The Hondo Corporation claims that the average miles per gallon for its new model, called the Mizer, is 50 miles per gallon, where the gas mileage is determined by the EPA formula that mixes city driving and highway driving. The Dotsun Corporation challenges this claim. Dotsun wishes to provide strong evidence that μ, the average miles per gallon, is *less* than 50 mpg. The appropriate null and alternate hypotheses are:

$$H_0: \quad \mu \ge 50 \text{ mpg}$$
$$H_A: \quad \mu < 50 \text{ mpg}$$

The Dotsun Corporation randomly contacts recent buyers of the Mizer and succeeds in getting 50 of these buyers to agree to participate in a gas mileage experiment with their cars that matches the EPA formula. The sample of 50 cars produced the following results: The average gas mileage was 47.8 miles per gallon, and the sample standard deviation was 8 miles per gallon. Test the null hypothesis at the $\alpha = 0.10$ level of significance.

10.91 The Department of Transportation (DOT) is concerned that Texas drivers are not driving within the 55 mph speed limit on highways. Formally, the null and alternate hypotheses are:

$$H_0: \quad \mu \le 55 \text{ mph}$$
$$H_A: \quad \mu > 55 \text{ mph}$$

In a random sample of $n = 324$ cars, it is found that the average speed is 57 mph. DOT is willing to assume, on the basis of previous studies, that σ, the population standard deviation, is 12 mph. Test the null hypothesis at the $\alpha = 0.01$ level of significance.

10.92 Refer to Problem 10.91. DOT further hypothesizes that 50 percent of the cars traveling between Ft. Worth and Dallas, Texas, on Interstate 30 violate the speed limit of 55 mph. In a random sample of $n = 300$ cars, it is found that 174 violate the 55 mph speed limit. Let π = Proportion of all cars traveling on Interstate 30 between Dallas and Ft. Worth.
   a. Formulate the null and alternate hypotheses.
   b. Test the null hypothesis at the $\alpha = 0.05$ level of significance.

10.93 The manager of a suburban shopping mall hypothesizes that cars parked in the parking lot remain there for an average of more than two hours on the weekend. A sample of 25 arriving cars on the weekend is taken, and the time each spends in the parking lot is recorded. It is found that the average time spent in the parking lot for the 25 cars is 2.18 hours with a standard deviation of 0.35 hours.
   a. Formulate the null and alternate hypotheses.
   b. Test the null hypothesis at the $\alpha = 0.05$ level of significance.

10.94 A study is conducted on job turnover. A random sample of $n = 200$ senior managers who have changed jobs during the previous year are interviewed. Thirty state that they changed jobs because of dissatisfaction with their salary. Does this result provide statistical evidence that the proportion of senior managers who change jobs for this reason is less than 0.20? Use an $\alpha = 0.05$ level of significance for the test.

a. Formulate the null and alternate hypotheses (Hint: We wish to demonstrate that the proportion who change jobs due to salary dissatisfaction is less than 0.20).

b. Test the null hypothesis using the specified level of significance.

10.95 It is now a law in many states that the driver and front seat passengers must wear seat belts. In the state of Texas, the highway patrol is interested in the proportion of drivers who are complying with the law. It is hypothesized that less than 50 percent of the drivers in Texas comply with the law. In a random sample of size $n = 500$ drivers, it is found that 225 drivers comply with the law.

a. Formulate the null and alternate hypotheses. Note that we wish to demonstrate that less than a proportion of 0.50 of Texas drivers comply with the law.

b. Test the null hypothesis at the $\alpha = 0.05$ level of significance.

10.96 Is it possible to commit both a Type I and a Type II error in making a decision to reject or not to reject a null hypothesis? Explain.

10.97 A study is to be conducted of MBA achievement scores on a graduate entrance exam. It is hypothesized that the average score of MBAs taking this test is 500:

$$H_0: \quad \mu = 500$$
$$H_A: \quad \mu \neq 500$$

A random sample of $n = 100$ MBA students is selected. The decision rule is: Reject H_0 if $\bar{x} > 525$ or if $\bar{x} < 475$. Assume that the standard deviation is 100, and $\overline{X}$ is normally distributed.

a. Find α for this test.

b. Find β for this test if μ is really 520.

10.98 When a machine is in perfect adjustment, it produces bolts with a mean diameter of 0.0600 inch and a standard deviation of 0.0150 inch. In order to ascertain whether or not the machine is still in adjustment, a sample of 36 bolts is selected. The sample mean is found to be 0.0575?

a. Test the null hypothesis $H_0: \mu = 0.0600$ at the $\alpha = 0.05$ level of significance.

b. Use a 95 percent confidence interval to test the hypothesis in part a.

10.99 A firm located in the downtown business district of a large city is concerned about the average time it takes for its workers to commute to and from work. The firm hypothesizes that the average time to commute to work each day for its employees is 60 minutes (60 minutes to get to work in the morning). In a sample of 36 employees, it is found that the average commute time in the morning is 54 minutes, with a standard deviation of 18 minutes.

a. Formulate the null and alternate hypotheses.

b. Calculate the p-value for this test.

c. If the significance level of the test is set at $\alpha = 0.01$, should the null hypothesis be rejected based on the p-value determined in part b?

10.100 Consider the following null and alternate hypotheses:

$$H_0: \quad \pi = 0.70$$
$$H_A: \quad \pi \neq 0.70$$

A sample of size $n = 100$ is to be taken and the sample proportion p calculated from the sample observations. The decision rule for this test has been established as:

Reject H_0 if $p < 0.60$ or if $p > 0.80$.
Do not reject H_0 if $0.60 \leq p \leq 0.80$.

a. What is the significance level α for this test?

b. Suppose that π is really 0.85. What is the probability of committing a Type II error, not rejecting a false null hypothesis, for this decision rule?

10.101 Suppose that the decision rule is changed in Problem 10.100 to:

Reject H_0 if $p < 0.58$ or if $p > 0.82$.
Do not reject H_0 if $0.58 \leq p \leq 0.82$.

a. What is the significance level α for this decision rule?

b. If π is really 0.85, what is the probability of committing a Type II error for this decision rule?

10.102 Consider the following null and alternate hypotheses:

$H_0: \quad \mu \leq 100$
$H_A: \quad \mu > 100$

It is planned to take a random sample of size $n = 25$ to test this null hypothesis. The decision rule for the test is:

Reject H_0 if $\bar{x} > 110$.
Do not reject H_0 if $\bar{x} \leq 110$.

It is assumed that the population standard deviation is $\sigma = 25$.

a. What is the probability of committing a Type I error for this decision rule?
b. If μ is really 120, what is the probability of committing a Type II error for this decision rule?

10.103 Consider Problem 10.25. Use the appropriate confidence bound to test the hypotheses in this problem.

10.104 Consider Problem 10.26. Use the appropriate confidence bound to test the hypotheses in this problem.

10.105 Consider Problem 10.31. Use the appropriate confidence bound to test the hypotheses in this problem.

10.106 Consider Problem 10.32. Use the appropriate confidence bound to test the hypotheses in this problem.

11

Statistical inference: Two populations

■ 11.1

Introduction

In many practical problems, we are concerned with comparing two populations with regard to some quantitative characteristic. For example, we may wish to compare the average time required to assemble a unit of a product by one production process with the average time required for assembly by an alternate process. In this instance the quantitative characteristic is the assembly time per unit, and the two populations are the sets of assembly times corresponding to the two production processes. The comparison of average assembly times for the two processes can be made by extending the methods of statistical estimation and hypothesis testing of Chapters 9 and 10. We may wish to construct a confidence interval on the difference between the two population mean assembly times or test a hypothesis that one mean is different from the other, for example.

To set the notation used in this chapter, consider the example of comparing the mean assembly times of a manufactured part by the current assembly process and an alternate assembly process. Since each assembly process produces a population of assembly times, we must distinguish between the two population parameters by the use of subscripts. For the current assembly process (population 1), we use the following notation for the population random variable, population mean, and the population standard deviation.

Notation for the two population inferences

X_1: Assembly time (in minutes) of the part by the current process.

μ_1: Mean assembly time (in minutes) of the part by the current process.

σ_1: Standard deviation of the assembly times of the part by the current process.

523

For the alternate process (population 2), we would have:

$\mathbf{X}_2$: Assembly time (in minutes) of the part by the alternate process.

μ_2: Mean assembly time (in minutes) of the part by the alternate process.

σ_2: Standard deviation of the assembly times of the part by the alternate process.

Values of the random variables $\mathbf{X}_1$ and $\mathbf{X}_2$ must be doubly subscripted:

Random variable $\mathbf{X}_1$: $x_{11}, x_{12}, x_{13}, \ldots$

Random variable $\mathbf{X}_2$: $x_{21}, x_{22}, x_{23}, \ldots$

With the notation for each value, the first subscript tells us whether it is a value of the first random variable (1) or of the second random variable (2), and the second subscript tells us which value it is for that random variable (1, 2, 3, . . .). The notation is illustrated in Figure 11.1. Notice that the sample means $\bar{x}_1$ and $\bar{x}_2$ must be subscripted, too.

The methods developed in this chapter will assume that the population random variables $\mathbf{X}_1$ and $\mathbf{X}_2$ are normally distributed. If this is not the case, then the procedures will be approximate and will be satisfactorily so only if n_1 and n_2 are "large." How large n_1 and n_2 must be for the methods to provide a good approximation depends on the specific cases considered. But as a general rule, n_1 and n_2 should each be larger than 25.

In this chapter we will focus our attention on two population parameters: the population mean μ and the population proportion π. When comparing two population means or two population proportions, we will consider the difference between the parameters in the two populations. For example, if we are interested in comparing two population means μ_1 and μ_2, then we will consider the difference $D = \mu_1 - \mu_2$. Point estimators and confidence interval estimators of D will be given. For hypothesis testing, we can also use the difference between the parameters in the two populations. For example, suppose that we wish to test the hypothesis that two population means are equal. We could write the null and alternate hypotheses as:

H_0: $\mu_1 = \mu_2$
H_A: $\mu_1 \neq \mu_2$

But these hypotheses can also be written as:

H_0: $\mu_1 - \mu_2 = 0$
H_A: $\mu_1 - \mu_2 \neq 0$

by subtracting μ_2 from both sides of the equation. Letting $D = \mu_1 - \mu_2$, we can now write these hypotheses as:

H_0: $D = 0$
H_A: $D \neq 0$

We can generalize this situation further by letting D_0 be the hypothesized difference between μ_1 and μ_2 by writing:

FIGURE 11.1 | Notation for the two-population case; mean assembly times of two assembly processes

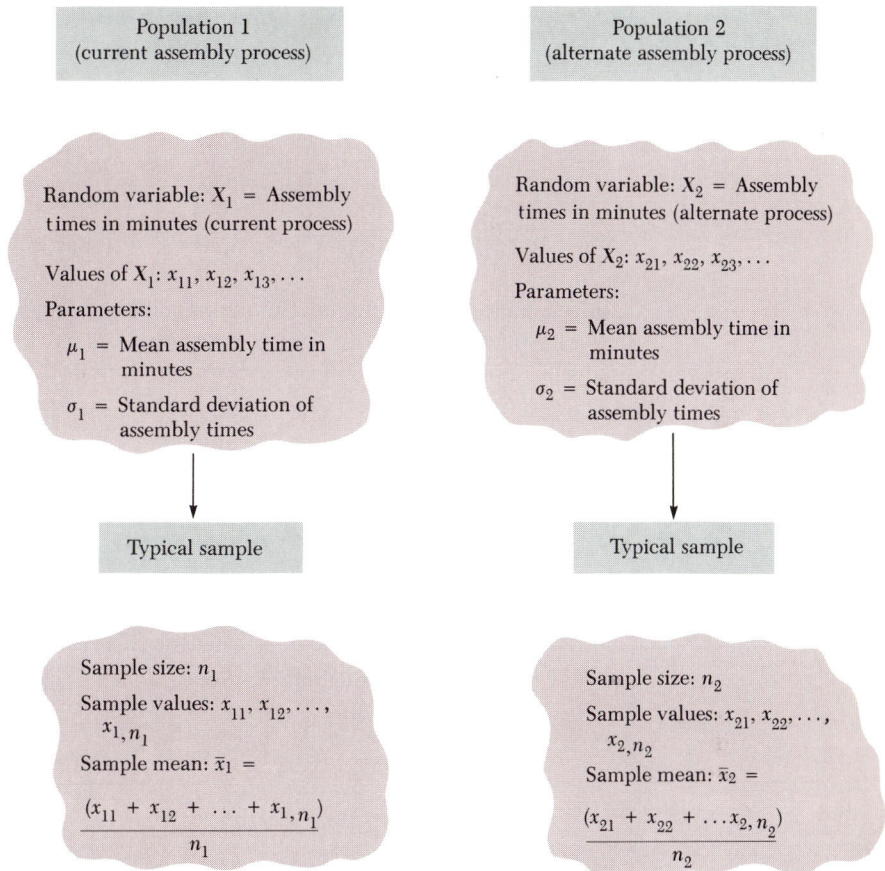

Population 1
(current assembly process)

Population 2
(alternate assembly process)

Random variable: X_1 = Assembly times in minutes (current process)

Values of X_1: $x_{11}, x_{12}, x_{13}, \ldots$

Parameters:

μ_1 = Mean assembly time in minutes

σ_1 = Standard deviation of assembly times

Random variable: X_2 = Assembly times in minutes (alternate process)

Values of X_2: $x_{21}, x_{22}, x_{23}, \ldots$

Parameters:

μ_2 = Mean assembly time in minutes

σ_2 = Standard deviation of assembly times

Typical sample

Typical sample

Sample size: n_1

Sample values: $x_{11}, x_{12}, \ldots,$
x_{1,n_1}

Sample mean: $\bar{x}_1$ =

$$\frac{(x_{11} + x_{12} + \ldots + x_{1,n_1})}{n_1}$$

Sample size: n_2

Sample values: $x_{21}, x_{22}, \ldots,$
x_{2,n_2}

Sample mean: $\bar{x}_2$ =

$$\frac{(x_{21} + x_{22} + \ldots x_{2,n_2})}{n_2}$$

$$H_0: \quad D = D_0$$
$$H_A: \quad D \neq D_0$$

where D_0 = hypothesized difference between μ_1 and μ_2. If $D_0 = 0$, then we are testing the equality of the two means. By using D_0, we can test a hypothesis about whether the difference D between the two means μ_1 and μ_2 is equal to a specific value D_0 (not necessarily 0).

Table 11.1 gives the hypothesis forms for two population means and two population proportions using the difference D.

In an advanced section, we will consider the comparison of two population variances σ_1^2 and σ_2^2. The comparison of two population variances will

TABLE 11.1 | Form of the hypotheses for tests about two population means and two population proportions

Population means: $D = \mu_1 - \mu_2$	
	$D_0 = $ Hypothesized difference
Population proportions: $D = \pi_1 - \pi_2$	

Form I	Form II	Form III
$H_0: \quad D = D_0$	$H_0: \quad D \geq D_0$	$H_0: \quad D \leq D_0$
$H_A: \quad D \neq D_0$	$H_A: \quad D < D_0$	$H_A: \quad D > D_0$

not use the *difference* between them ($\sigma_1^2 - \sigma_2^2$), but rather will use their *ratio* (σ_1^2/σ_2^2).

Most methods developed for making inferences concerning differences between parameters from two populations will be based on the following theorem.

<table>
<tr><td>

Variance of the difference of two random variables

</td></tr>
</table>

Theorem 11.1

Variance of the difference of two random variables

If a random variable X_1 has a variance of $\sigma_{X_1}^2$ and random variable X_2 has a variance of $\sigma_{X_2}^2$ and X_1 and X_2 are independent random variables, then the variance of the difference between X_1 and X_2 is the sum of the variances of X_1 and X_2. That is,

$$V(X_1 - X_2) = V(X_1) + V(X_2) = \sigma_{X_1}^2 + \sigma_{X_2}^2$$

This theorem was first introduced in Chapter 5. Theorem 11.1 applies to the random variables $\overline{X}_1$ and $\overline{X}_2$, the sample averages in sample 1 and sample 2. Recalling that $V(\overline{X}_1) = \sigma_{\overline{X}_1}^2 = \sigma_1^2/n_1$ and $V(\overline{X}_2) = \sigma_{\overline{X}_2}^2 = \sigma_2^2/n_2$, we have from Theorem 11.1,

$$V(\overline{X}_1 - \overline{X}_2) = V(\overline{X}_1) + V(\overline{X}_2) = \sigma_{\overline{X}_1}^2 + \sigma_{\overline{X}_2}^2 = \sigma_1^2/n_1 + \sigma_2^2/n_2$$

■ 11.2

Inferences concerning two population means, μ_1 and μ_2

The inference framework for comparing two population means is illustrated in Figure 11.2.

The sample means $\overline{X}_1$ and $\overline{X}_2$ (values in a typical sample are $\bar{x}_1$ and $\bar{x}_2$, respectively) and the sample standard deviations S_1^2 and S_2^2 (values in a typical sample are s_1^2 and s_2^2, respectively) will be used to make inferences concerning the difference between the two population means, $D = \mu_1 - \mu_2$. We will first consider point estimation, then interval estimation, and finally hypothesis testing for the difference $D = \mu_1 - \mu_2$.

FIGURE 11.2 | Two-population framework for comparing μ_1 and μ_2

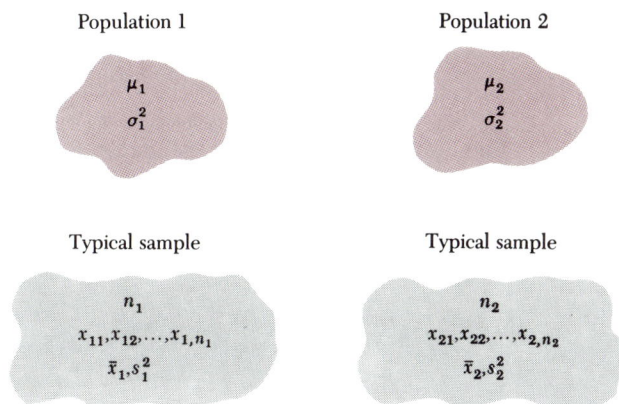

Population 1

μ_1

σ_1^2

Population 2

μ_2

σ_2^2

Typical sample

n_1

$x_{11}, x_{12}, \ldots, x_{1,n_1}$

$\bar{x}_1, s_1^2$

Typical sample

n_2

$x_{21}, x_{22}, \ldots, x_{2,n_2}$

$\bar{x}_2, s_2^2$

11.2.1 | Point estimation of $D = \mu_1 - \mu_2$

Point estimator of $\mu_1 - \mu_2$

Recall from Chapter 9 that the best estimator of a population mean μ is the sample mean, $\bar{X}$. Thus it seems reasonable that the best estimator of the difference, $D = \mu_1 - \mu_2$, is the difference between the sample means $\bar{X}_1$ and $\bar{X}_2$. In fact, the point estimator $\bar{X}_1 - \bar{X}_2$ is unbiased and has the minimum variance among all unbiased estimators of D.

11.2.2 | Confidence interval estimation of $D = \mu_1 - \mu_2$

In constructing confidence intervals for $D = \mu_1 - \mu_2$, it is necessary to consider three cases:

Case 1: The values of σ_1^2 and σ_2^2 are unknown.

Case 2: The values of σ_1^2 and σ_2^2 are unknown, but it is presumed that $\sigma_1^2 = \sigma_2^2$.

Case 3: The values of σ_1^2 and σ_2^2 are unknown, and it is presumed that $\sigma_1^2 \neq \sigma_2^2$.

The distinction among these cases at least in part can be drawn from the confidence interval formulas concerning one population mean μ. Recall that if σ^2 is known, then the confidence interval for μ developed in Chapter 9 is based on the standard normal variable $\mathbf{Z}$. However, if σ^2 is unknown and is estimated by the sample variance s^2, then the confidence interval for μ is based on the t-distribution.

As might be expected, the case 1 confidence interval formula is based on the $\mathbf{Z}$-statistic, and the interval formulas for cases 2 and 3 are based on the **t**-statistic.

Case 1: σ_1^2 and σ_2^2 known   When the values of σ_1^2 and σ_2^2 are known, a confidence interval for the difference between μ_1 and μ_2 can be constructed by using the general form of the standard normal statistic

$$\mathbf{Z} = \frac{(\text{Estimator}) - (\text{Parameter})}{(\text{Standard error of estimator})}$$

In this instance the estimator is $\overline{\mathbf{X}}_1 - \overline{\mathbf{X}}_2$, and the parameter is $D = \mu_1 - \mu_2$. The standard error of the estimator is another expression for the standard deviation of the estimator; and the standard deviation of the estimator, $\overline{\mathbf{X}}_1 - \overline{\mathbf{X}}_2$, is the square root of the variance of the estimator. Denoting the variance of the estimator by $\sigma_{\overline{\mathbf{X}}_1 - \overline{\mathbf{X}}_2}^2$ or simply by $V(\overline{\mathbf{X}}_1 - \overline{\mathbf{X}}_2)$, we have from Theorem 11.1,

$$\sigma_{\overline{\mathbf{X}}_1 - \overline{\mathbf{X}}_2}^2 = V(\overline{\mathbf{X}}_1 - \overline{\mathbf{X}}_2) = V(\overline{\mathbf{X}}_1) + V(\overline{\mathbf{X}}_2) = \frac{\sigma_1^2}{n_1} + \frac{\sigma_2^2}{n_2}$$

Therefore

$$\sigma_{\overline{\mathbf{X}}_1 - \overline{\mathbf{X}}_2} = \sqrt{\sigma_{\overline{\mathbf{X}}_1 - \overline{\mathbf{X}}_2}^2} = \sqrt{\frac{\sigma_1^2}{n_1} + \frac{\sigma_2^2}{n_2}}$$

Thus the form of the **Z**-statistic for constructing a confidence interval estimator for $\mu_1 - \mu_2$ is

$$\mathbf{Z} = \frac{(\overline{\mathbf{X}}_1 - \overline{\mathbf{X}}_2) - (\mu_1 - \mu_2)}{\sqrt{\dfrac{\sigma_1^2}{n_1} + \dfrac{\sigma_2^2}{n_2}}}$$

By using the fact that $P(-z_{\alpha/2} \le \mathbf{Z} \le z_{\alpha/2}) = 1 - \alpha$, it can be shown by solving the inequality statements for $(\mu_1 - \mu_2)$ that a $100(1 - \alpha)$ percent confidence interval on $(\mu_1 - \mu_2)$ is given by

$$(\overline{\mathbf{X}}_1 - \overline{\mathbf{X}}_2) \pm z_{\alpha/2} \sqrt{\frac{\sigma_1^2}{n_1} + \frac{\sigma_2^2}{n_2}}$$

The sampling distribution of $\overline{\mathbf{X}}_1 - \overline{\mathbf{X}}_2$ is shown in Figure 11.3. If the population random variables $\mathbf{X}_1$ and $\mathbf{X}_2$ are normally distributed, then $\overline{\mathbf{X}}_1 - \overline{\mathbf{X}}_2$ is normally distributed. If $\mathbf{X}_1$ and $\mathbf{X}_2$ are not normally distributed, then we must appeal to the central limit theorem to use a normal distribution to approximate the sampling distribution of $\overline{\mathbf{X}}_1 - \overline{\mathbf{X}}_2$. Usually, the sample sizes n_1 and n_2 must be at least 25 each for the approximation to be satisfactory.

Example 11.1   A company is in the process of deciding whether or not to produce a new electronic component. In the plant, there are two machines that could be adapted to produce the component. As a part of the overall decision, the company must select one of the machines to use if they decide to produce the component. In a test conducted on machine 1, the average production time per component was 5.23 minutes for a sample of 100 components. In a sample of 64 components, machine 2 averaged 5.37 minutes per

FIGURE 11.3 The sampling distribution of the statistic $\bar{X}_1 - \bar{X}_2$ for case 1 (σ_1^2 and σ_2^2 known)

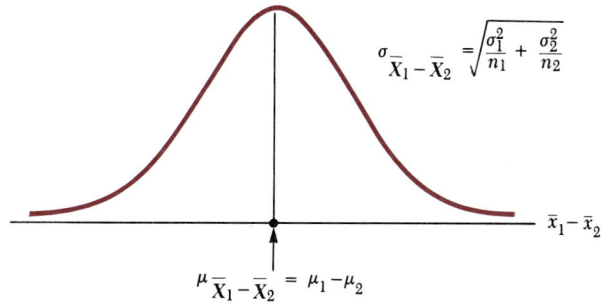

$$\sigma_{\bar{X}_1 - \bar{X}_2} = \sqrt{\frac{\sigma_1^2}{n_1} + \frac{\sigma_2^2}{n_2}}$$

$$\mu_{\bar{X}_1 - \bar{X}_2} = \mu_1 - \mu_2$$

$\bar{x}_1 - \bar{x}_2$

component. From past experience in using the machines to produce similar items, it is known that the standard deviations of the production times on machines 1 and 2 are 0.15 minute and 0.10 minute, respectively.

Determine a point estimate and a 95 percent confidence interval estimate of the difference between the population mean production times for machines 1 and 2.

Solution The first population is composed, conceptually, of the production times associated with the components that could be produced on machine 1. In a sample of 100 production times, the sample average production

Case 1 confidence interval estimate of $\mu_1 - \mu_2$

Formula 11.1

Confidence interval estimate of $\mu_1 - \mu_2$

Case 1: σ_1^2 and σ_2^2 known

A $100(1 - \alpha)$ percent confidence interval estimate of the difference between two population means, $\mu_1 - \mu_2$, is given by

$$(\bar{x}_1 - \bar{x}_2) \pm z_{\alpha/2} \sqrt{\frac{\sigma_1^2}{n_1} + \frac{\sigma_2^2}{n_2}}$$

where $\bar{x}_i$, σ_i^2, and n_i are the sample mean, population variance, and sample size associated with the ith population and sample ($i = 1, 2$). Additionally $z_{\alpha/2}$ is the point on the standard normal distribution such that $P(\mathbf{Z} \geq z_{\alpha/2}) = \alpha/2$.

The confidence interval is exact if the population random variables $\mathbf{X}_1$ and $\mathbf{X}_2$ are normally distributed. If this is not the case, then n_1 and n_2 should each be greater than or equal to 25, in which case the resulting confidence interval is approximate.

time is $\bar{x}_1 = 5.23$ minutes, and the population standard deviation of the production times is given as $\sigma_1 = 0.15$ minute. In a sample of 64 production times drawn from the second population, the sample average production time is $\bar{x}_2 = 5.37$ minutes, and the population standard deviation of the production times is given as $\sigma_2 = 0.10$ minute.

The point estimate of the difference $\mu_1 - \mu_2$ is:

$$\bar{x}_1 - \bar{x}_2 = 5.23 - 5.37 = -0.14 \text{ minute}$$

For a 95 percent confidence interval, $100(1 - \alpha) = 95$, so that $\alpha = 0.05$ and $\alpha/2 = 0.025$. From Table B.3 in Appendix B, $z_{\alpha/2} = z_{0.025} = 1.96$. Thus the 95 percent confidence interval estimate of $\mu_1 - \mu_2$ is given by

$$(\bar{x}_1 - \bar{x}_2) \pm z_{\alpha/2} \sqrt{\frac{\sigma_1^2}{n_1} + \frac{\sigma_2^2}{n_2}} = (5.23 - 5.37) \pm 1.96 \sqrt{\frac{(0.15)^2}{100} + \frac{(0.10)^2}{64}}$$

$$= -0.14 \pm 1.96(0.0195) = -0.14 \pm 0.038$$

and we are 95 percent confident that the true difference between μ_1 and μ_2 is between -0.102 and -0.178 minute. There is good evidence, therefore, to conclude that μ_1 is less than μ_2, since the difference, $\mu_1 - \mu_2$, appears to be negative. Therefore machine 1 appears to produce a smaller average production time than does machine 2 for the production of the electronic component. All things being equal (such as costs of the two machines), we would therefore prefer machine 1 to machine 2.

We should at this point remind ourselves of the meaning of "95 percent confidence." If many repeated pairs of samples of sizes n_1 and n_2 are drawn and a confidence interval for the difference between the two population means is constructed in each sample pair, on average, 95 percent of the confidence intervals will contain the true difference and on average, 5 percent will not.

In most problems where inferences about μ_1 and μ_2 are drawn, the values of the population variances σ_1^2 and σ_2^2 will not be known, of course. This leads us to cases 2 and 3.

Case 2: σ_1^2 and σ_2^2 unknown; $\sigma_1^2 = \sigma_2^2$ In case 2 the two population variances are unknown, but we assume that they are equal in value. Let σ^2 denote the common value of σ_1^2 and σ_2^2. The two-population schematic is illustrated in Figure 11.4.

In this case the standard deviation of the point estimator of $\mu_1 - \mu_2$ can be simplified:

$$\sigma_{\bar{X}_1 - \bar{X}_2} = \sqrt{\frac{\sigma_1^2}{n_1} + \frac{\sigma_2^2}{n_2}} = \sqrt{\frac{\sigma^2}{n_1} + \frac{\sigma^2}{n_2}} = \sigma \sqrt{\frac{1}{n_1} + \frac{1}{n_2}}$$

Pooled sample variance s_p^2

Since the populations have the same variance σ^2, the sample variances s_1^2 and s_2^2 are both point estimates of σ^2. It seems reasonable that s_1^2 and s_2^2 should be "pooled" in some way to form a combined point estimate of σ^2 based on the $n_1 + n_2$ sample units comprising both samples. The pooled variance, denoted by s_p^2, is given by

FIGURE 11.4 | Two-population framework for case 2

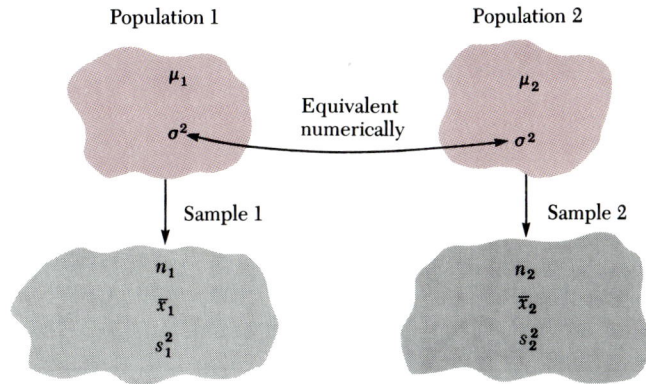

$$s_p^2 = \frac{(n_1 - 1)s_1^2 + (n_2 - 1)s_2^2}{(n_1 - 1) + (n_2 - 1)}$$

It is the weighted mean of s_1^2 and s_2^2.

The pooled variance $\mathbf{S}_p^2$ is the best point estimator of σ^2. Thus, the estimate of $\sigma_{\bar{\mathbf{x}}_1 - \bar{\mathbf{x}}_2}$ is

$$\hat{\sigma}_{\bar{\mathbf{x}}_1 - \bar{\mathbf{x}}_2} = s_p \sqrt{\frac{1}{n_1} + \frac{1}{n_2}}$$

Recall from Chapters 9 and 10 that when we had one population with mean μ and sample mean $\bar{\mathbf{X}}$, the standardized statistic

$$\mathbf{Z} = \frac{\bar{\mathbf{X}} - \mu}{\sigma_{\bar{\mathbf{x}}}}$$

is no longer a standard normal random variable when $\sigma_{\bar{\mathbf{x}}}$ has to be estimated. In this case, when the estimator $\sigma_{\bar{\mathbf{x}}} = \mathbf{S}/\sqrt{n}$ is used, the standardized statistic

$$\mathbf{t} = \frac{\bar{\mathbf{X}} - \mu}{\mathbf{S}/\sqrt{n}}$$

is t-distributed. Now, the estimator is $\bar{\mathbf{X}}_1 - \bar{\mathbf{X}}_2$, the parameter is $D = \mu_1 - \mu_2$, and the estimator of $\sigma_{\bar{\mathbf{x}}_1 - \bar{\mathbf{x}}_2}$ is $\mathbf{S}_p \sqrt{\frac{1}{n_1} + \frac{1}{n_2}}$:

$$\mathbf{t} = \frac{(\bar{\mathbf{X}}_1 - \bar{\mathbf{X}}_2) - (\mu_1 - \mu_2)}{\mathbf{S}_p \sqrt{\frac{1}{n_1} + \frac{1}{n_2}}}$$

where $\mathbf{S}_p$ is the pooled standard deviation statistic,

$$S_p = \sqrt{\frac{(n_1 - 1)S_1^2 + (n_2 - 1)S_2^2}{(n_1 - 1) + (n_2 - 1)}}$$

The degrees of freedom associated with this **t**-statistic are $n_1 + n_2 - 2$, since this number represents the divisor in the pooled sample standard deviation formula $[(n_1 - 1) + (n_2 - 1)]$.

Case 2 confidence interval estimate of $\mu_1 - \mu_2$

Formula 11.2

Confidence interval estimate of $\mu_1 - \mu_2$,

Case 2: σ_1^2 and σ_2^2 unknown; $\sigma_1^2 = \sigma_2^2$

A $100(1 - \alpha)$ percent confidence interval estimate of the difference between two population means, $\mu_1 - \mu_2$, is given by

$$(\bar{x}_1 - \bar{x}_2) \pm t_{\alpha/2;n_1+n_2-2}s_p \sqrt{\frac{1}{n_1} + \frac{1}{n_2}}$$

where $\bar{x}_i$ and n_i are the sample mean and sample size, respectively, of the ith sample $(i = 1, 2)$, s_p is the pooled standard deviation given by

$$s_p = \sqrt{\frac{(n_1 - 1)s_1^2 + (n_2 - 1)s_2^2}{(n_1 - 1) + (n_2 - 1)}}$$

and $t_{\alpha/2;n_1+n_2-2}$ is the point on the t-distribution with $n_1 + n_2 - 2$ degrees of freedom such that $P[t > t_{\alpha/2;n_1+n_2-2}] = \alpha/2$.

The confidence interval is exact if the population random variables X_1 and X_2 are normally distributed. If this is not the case, then n_1 and n_2 should each be greater than or equal to 25, in which case the resulting confidence interval is approximate.

The t-table [Table B.5, Appendix B] gives the t-value for degrees of freedom less than or equal to 100. If the degrees of freedom $[n_1 + n_2 - 2]$ exceed 100 in the application of Formula 11.2, then replace t in the formula with $z_{\alpha/2}$ from the standard normal distribution, where $z_{\alpha/2}$ is determined so that $P(Z > z_{\alpha/2}) = \alpha/2$. When the degrees of freedom exceed 100, the z-value may be used as an approximation of the t-value.

When using Formula 11.2, there must be a reasonable justification for assuming that the two population variances, σ_1^2 and σ_2^2, are equal. In manufacturing processes it is common when comparing two production processes (in mean time to produce a product) to find that the variation in production times by the two methods are very similar, though the means may differ. However, in many cases, as a mean μ increases in value, so does the variance σ^2. Therefore, if two means μ_1 and μ_2 differ, then so may the corresponding population variances σ_1^2 and σ_2^2. If the assumption of equal variances is in doubt, then case 3 should be used, or the hypothesis $H_0: \sigma_1^2 = \sigma_2^2$ should be tested (see Section 11.5).

Example 11.2 The SKM Drug Company has developed two antibiotics, drug A and drug B, for the treatment of a specific disease. In a planned experiment, 50 patients who have the disease at a similar stage of its development are randomly selected and are randomly assigned to two groups of 25 patients each. The first group will receive drug A, and the second group will receive drug B. For each group the number of days to recovery from the disease is recorded. At the completion of the experiment, we have the following sample results:

Drug A	*Drug B*
$n_1 = 25$ patients	$n_2 = 25$ patients
$\bar{x}_1 = 23.32$ days	$\bar{x}_2 = 16.12$ days
$s_1 = 4.35$ days	$s_2 = 3.60$ days

Find a 99 percent confidence interval estimate for the difference between the average number of days to recover from the disease from using drug A and drug B.

Solution We will assume that $\sigma_1^2 = \sigma_2^2$. (Does this assumption seem reasonable? More about this later.) For a 99 percent confidence interval, $100(1 - \alpha)$ percent $= 99$ percent, so that $\alpha = 0.01$ and $\alpha/2 = 0.005$. The degrees of freedom are $n_1 + n_2 - 2 = 25 + 25 - 2 = 48$. From Table B.5 in Appendix B, $t_{0.005;48} = 2.682$. The pooled standard deviation is:

$$s_p = \sqrt{\frac{(n_1 - 1)s_1^2 + (n_2 - 1)s_2^2}{(n_1 - 1) + (n_2 - 1)}} = \sqrt{\frac{(24)(4.35)^2 + (24)(3.60)^2}{48}}$$
$$= 3.99$$

Also,

$$\sqrt{\frac{1}{n_1} + \frac{1}{n_2}} = \sqrt{\frac{1}{25} + \frac{1}{25}} = \sqrt{0.04 + 0.04} = \sqrt{0.08} = 0.2828$$

The 99 percent confidence interval estimate is:

$$(\bar{x}_1 - \bar{x}_2) \pm t_{0.005;48} \, s_p \sqrt{\frac{1}{n_1} + \frac{1}{n_2}} = (23.32 - 16.12) \pm (2.682)(3.99)(0.2828)$$
$$= 7.20 \pm 3.03$$

Thus we are 99 percent confident that the true difference between the average number of days to recovery for drug A and drug B is between $(7.20 - 3.03) = 4.17$ days and $(7.20 + 3.03) = 10.23$ days. The point estimate of the difference $D = \mu_1 - \mu_2$ is $\bar{x}_1 - \bar{x}_2 = 7.20$ days. It would therefore appear that the average number of days to recovery for drug A is greater than the average number of days to recovery for drug B.

In this solution we assumed that the two population variances σ_1^2 and σ_2^2 are equal. A basis for this might be previous studies that have shown that the variation of number of days to recovery is similar for comparisons of drugs like drug A and drug B. If this is not the case, then we should use case 3 or test the null hypothesis, H_0: $\sigma_1^2 = \sigma_2^2$ by using the methods in Section 11.5. If it is assumed that σ_1^2 and σ_2^2 are not equal, then we have case 3.

Case 3: σ_1^2 and σ_2^2 unknown; $\sigma_1^2 \neq \sigma_2^2$ In this case we must estimate both σ_1^2 and σ_2^2. A logical sequence of steps would seem to be:

1. Since $V(\overline{\mathbf{X}}_1 - \overline{\mathbf{X}}_2) = \sigma_{\overline{\mathbf{X}}_1 - \overline{\mathbf{X}}_2}^2 = \sigma_1^2/n_1 + \sigma_2^2/n_2$, if $\overline{\mathbf{X}}_1$ and $\overline{\mathbf{X}}_2$ are independent, we could estimate $\sigma_{\overline{\mathbf{X}}_1 - \overline{\mathbf{X}}_2}^2$ by estimating σ_1^2 and σ_2^2 by s_1^2 and s_2^2, respectively. Thus the estimate of $V(\overline{\mathbf{X}}_1 - \overline{\mathbf{X}}_2)$ is:

$$\frac{s_1^2}{n_1} + \frac{s_2^2}{n_2}$$

2. Form the standardized statistic

$$\mathbf{t}' = \frac{(\overline{\mathbf{X}}_1 - \overline{\mathbf{X}}_2) - (\mu_1 - \mu_2)}{\sqrt{\dfrac{s_1^2}{n_1} + \dfrac{s_2^2}{n_2}}}$$

However, due to the unequal population variances, $\mathbf{t}'$ will not be t-distributed. There are two approximate approaches to setting a confidence interval on $\mu_1 - \mu_2$ in this case:

1. For moderately large sample sizes n_1 and n_2, use the $\mathbf{t}'$ statistic above. The sample size n_1 and n_2 must be at least 25 to use the $\mathbf{t}'$ statistic. As n_1 and n_2 become larger, $\mathbf{S}_1^2$ and $\mathbf{S}_2^2$ will converge in values to σ_1^2 and σ_2^2, respectively, and the sampling distribution of $\overline{\mathbf{X}}_1 - \overline{\mathbf{X}}_2$ will approach a normal distribution. Then we have, for all practical purposes, case 1. Thus if n_1 and n_2 are very large (usually 50 or more), then case 1 can be used. If the $\mathbf{t}'$ statistic is used, the degrees of freedom, denoted by Δ, must be computed by using the following formula:

$$\Delta = \frac{[(s_1^2/n_1) + (s_2^2/n_2)]^2}{\dfrac{(s_1^2/n_1)^2}{(n_1 - 1)} + \dfrac{(s_2^2/n_2)^2}{(n_2 - 1)}}$$

If Δ turns out to be fractional in value, round to the last smaller integer. Several statistical computer software packages calculate Δ. Among these are MINITAB and SPSS. Therefore, when these packages are used, it is not necessary to calculate Δ. We will use MINITAB in Section 11.6 to illustrate the application of this approximate test.

2. Use a method that does not require that $\overline{\mathbf{X}}_1 - \overline{\mathbf{X}}_2$ is normally distributed. One such method is called the Mann-Whitney method and is discussed in Chapter 20.

Note the distinction between Formula 11.3 and 11.4. If $\Delta \leq 100$, then Formula 11.3 should be used. When $\Delta > 100$, Formula 11.4 can be used. In

Case 3 confidence
interval estimate of
$\mu_1 - \mu_2$

Formulas 11.3 and 11.4
Confidence interval estimate of $\mu_1 - \mu_2$,
Case 3: σ_1^2 and σ_2^2 unknown; $\sigma_1^2 \neq \sigma_2^2$

Using the t' statistic, an approximate $100(1 - \alpha)$ percent confidence interval estimate of the difference between two population means, $\mu_1 - \mu_2$, is given by

$$(\bar{x}_1 - \bar{x}_2) \pm t_{\alpha/2;\Delta} \sqrt{\frac{s_1^2}{n_1} + \frac{s_2^2}{n_2}} \qquad (11.3)$$

where n_i, $\bar{x}_i$, and s_i^2 are the sample size, sample mean, and sample variance, respectively, of the ith sample ($i = 1, 2$), and $t_{\alpha/2;\Delta}$ is the point on the t-distribution with Δ degrees of freedom such that $P[t > t_{\alpha/2;\Delta}] = \alpha/2$, where:

$$\Delta = \frac{[(s_1^2/n_1) + (s_2^2/n_2)]^2}{\dfrac{(s_1^2/n_1)^2}{(n_1 - 1)} + \dfrac{(s_2^2/n_2)^2}{(n_2 - 1)}}$$

If Δ is fractional in value, then round to the last smaller integer. This is an approximate confidence interval estimate and should not be used unless n_1 and n_2 are both greater than or equal to 25.

If $\Delta > 100$, then the case 1 formula may be used with s_1^2 and s_2^2 substituted for σ_1^2 and σ_2^2, respectively, to produce an approximate $100(1 - \alpha)$ percent confidence interval estimate:

$$(\bar{x}_1 - \bar{x}_2) \pm z_{\alpha/2} \sqrt{\frac{s_1^2}{n_1} + \frac{s_2^2}{n_2}} \qquad (11.4)$$

the latter case, $t_{\alpha/2;\Delta} \doteq z_{\alpha/2}$; that is, when the degrees of freedom exceed 100, the t-value is very close to the z-value, thereby supporting the use of Formula 11.4.

The case 3 situation (σ_1^2, σ_2^2 unknown and $\sigma_1^2 \neq \sigma_2^2$) is a very interesting one in statistics. In this case $\bar{X}_1 - \bar{X}_2$ does not have an exact sampling distribution. This problem was first identified and studied in 1929 and became known as the Behrens-Fisher problem. (See reference for Fisher at the end of this chapter.) The problem was to find the exact sampling distribution of $\bar{X}_1 - \bar{X}_2$ when $\sigma_1^2 \neq \sigma_2^2$ and σ_1^2, σ_2^2 are unknown in value. As it turns out, there is no exact sampling distribution in this case. The approximation given in Formula 11.3 is due to Satterthwaite (reference at end of this chapter) and corrects the degrees of freedom so that a t-distribution with Δ degrees of freedom better fits the sampling distribution of t' than does a t-distribution with $n_1 + n_2 - 2$ degrees of freedom or the z-distribution. Most elementary statistics texts suggest using Formula 11.4 for case 3. Although the use of Formula 11.4 is easier than using Formula 11.3, this is unfortunate because

Formula 11.4 can produce some very misleading results when $\Delta < 100$ and particularly when the sample sizes n_1 and n_2 are small (10 to 30 or so). The use of Formula 11.4 when $\Delta < 100$ alters the value of $\alpha = P[\text{Type I error}]$ from the set value. Therefore a 95 percent confidence interval estimate using Formula 11.4 with small samples will not have 95 percent confidence.

For an interesting discussion of this case, see the Snedecor and Cochran reference (pp. 96–98) and the Ryan, Joiner, and Ryan reference for MINITAB (pp. 185–87) at the end of the chapter. MINITAB correctly handles case 3, using the Satterthwaite correction for degrees of freedom. We will use MINITAB to solve problems involving confidence interval estimates in this case in Section 11.6.

Example 11.3   The National Steel Company produces truck frames at its Memphis, Tennessee, plant. The plant operates on two shifts, a day shift and a night shift. In reviewing the records, the production manager is concerned about the difference between the average daily production of truck frames during the day shift and the average daily production of truck frames on the night shift. The manager randomly samples 30 working days during the day shift during the past four months and records the number of truck frames produced each day. The manager also samples 30 working nights on the night shift during the past four months and records the number of truck frames produced each night. The sample statistics are:

Day shift	Night shift
$n_1 = 30$ days	$n_2 = 30$ days
$\bar{x}_1 = 12.4$ truck frames	$\bar{x}_2 = 11.2$ truck frames
$s_1 = 1.6$ truck frames	$s_2 = 2.7$ truck frames

The production manager is not willing to assume that $\sigma_1^2 = \sigma_2^2$, since he knows that the variation of the number of truck frames produced daily on the night shift is larger than on the day shift. Find a 95 percent confidence interval estimate on the difference $D = \mu_1 - \mu_2$, where

$\mu_1 = $ Average number of truck frames produced daily on the day shift.
$\mu_2 = $ Average number of truck frames produced daily on the night shift.

Solution   For a 95 percent confidence interval estimate, $100(1 - \alpha)$ percent $= 95$ percent, so that $\alpha = 0.05$ and $\alpha/2 = 0.025$. The degrees of freedom are:

$$\Delta = \frac{[(s_1^2/n_1) + (s_2^2/n_2)]^2}{\dfrac{(s_1^2/n_1)^2}{(n_1 - 1)} + \dfrac{(s_2^2/n_2)^2}{(n_2 - 1)}} = \frac{[(1.6)^2/30 + (2.7)^2/30]^2}{\dfrac{[(1.6)^2/30]^2}{29} + \dfrac{[(2.7)^2/30]^2}{29}} = \frac{0.1078}{0.00229} = 47.13$$

Since Δ is fractional, we round down to the integer 47. Thus $\Delta = 47$; and since $\Delta \leq 100$, the t-value should be used. Therefore Formula 11.3 is appropriate. Now, $t_{\alpha/2;\Delta} = t_{0.025;47} = 2.012$ from Table B.5 in Appendix B.

The 95 percent confidence interval estimate is:

$$(\bar{x}_1 - \bar{x}_2) \pm t_{0.025;47} \sqrt{\frac{s_1^2}{n_1} + \frac{s_2^2}{n_2}} = (12.4 - 11.2) \pm (2.012) \sqrt{\frac{(1.6)^2}{30} + \frac{(2.7)^2}{30}}$$

$$= 1.2 \pm (2.012)(0.573)$$

$$= 1.2 \pm 1.153$$

Thus we are 95 percent confident that the true difference between the average number of truck frames produced daily on the day shift and the night shift is $(1.2 - 1.153) = 0.047$ truck frames to $(1.2 - 1.153) = 2.353$ truck frames. Since the confidence interval endpoints for the difference $\mu_1 - \mu_2$ are positive, this implies that $\mu_1 - \mu_2 > 0$. Thus the production manager may conclude that the day shift produces a greater average number of truck frames daily than does the night shift.

Example 11.4 The National Assurance Corporation hires insurance claim adjuster trainees monthly. During the past six months, two different training programs have been used. The first method is the traditional four-week program using mostly a lecture/classroom format. The second method is a newly proposed approach involving mostly computer-aided instruction and self-study materials. At the end of the most recent training period, 100 trainees are sampled who have been trained by the traditional program, and 100 trainees are sampled who have been trained by the proposed method. Each trainee is given a written and oral examination to determine how much information has been retained. The score on the exam ranges from 0 to 100. The sample statistics for each group are reported below.

Traditional program	Proposed program
$n_1 = 100$ trainees	$n_2 = 100$ trainees
$\bar{x}_1 = 82.41$	$\bar{x}_2 = 85.72$
$s_1 = 3.86$	$s_2 = 1.92$
$s_1^2 = 14.90$	$s_2^2 = 3.69$

Find a 95 percent confidence interval estimate on the difference $D = \mu_1 - \mu_2$, where μ_1 is the population mean score of all trainees who have completed the traditional program, and μ_2 is the population mean score of all trainees who have completed the newly proposed program.

Solution For a 95 percent interval estimate, $100(1 - \alpha)$ percent $= 95$ percent, so that $\alpha = 0.05$, and $\alpha/2 = 0.025$. The degrees of freedom are:

$$\Delta = \frac{[(s_1^2/n_1) + (s_2^2/n_2)]^2}{\dfrac{(s_1^2/n_1)^2}{(n_1 - 1)} + \dfrac{(s_2^2/n_2)^2}{(n_2 - 1)}} = \frac{[(14.90/100) + (3.69/100)]^2}{\dfrac{(14.90/100)^2}{99} + \dfrac{(3.69/100)^2}{99}}$$

$$= \frac{(0.1859)^2}{0.000238} = \frac{0.03456}{0.000238} = 145.21$$

Since Δ is fractional, round to 145. Now, $\Delta = 145 > 100$, so that Formula 11.4 may be used to find the confidence interval estimate:

$$(\bar{x}_1 - \bar{x}_2) \pm z_{\alpha/2} \sqrt{\frac{s_1^2}{n_1} + \frac{s_2^2}{n_2}} = (82.41 - 85.72) \pm 1.96 \sqrt{\frac{14.90}{100} + \frac{3.69}{100}}$$

$$= -3.31 \pm (1.96)(0.431)$$

$$= -3.31 \pm 0.845$$

where $z_{\alpha/2} = z_{0.025} = 1.96$ from Table B.3, Appendix B. Therefore, we are 95 percent confident that the difference between μ_1 and μ_2 is between $-3.31 - 0.845 = -4.155$ points and $-3.31 + 0.845 = -2.465$ points. Since it appears that $\mu_1 - \mu_2 < 0$ or $\mu_1 < \mu_2$, the newly proposed training program seems to produce higher average test scores than does the traditional program.

11.2.3 | Hypothesis testing concerning $D = \mu_1 - \mu_2$

Since hypothesis testing is directly related to confidence interval estimation, we must differentiate among the three cases discussed in Section 11.2.2. For cases 1, 2, and 3, we will state the null hypothesis as a difference in one of three forms:

$$H_0: \quad \mu_1 - \mu_2 = D \geq D_0$$
$$H_0: \quad \mu_1 - \mu_2 = D \leq D_0$$
$$H_0: \quad \mu_1 - \mu_2 = D = D_0$$

In these hypothesis statements, D_0 is the hypothesized value of the difference, usually taken to be zero.

For each of the cases and each form of the null hypothesis, H_0, we will give the decision rule using the *standardized testing method* described in Chapter 10. That is, instead of setting up the decision rule on the sampling distribution of $\bar{X}_1 - \bar{X}_2$, we will establish the decision rule on either the standard normal z-distribution or the standardized t-distribution. Also, the p-value formula for each hypothesis test is given and illustrated in each example problem. If the p-value method of hypothesis testing in Chapter 10 was not discussed, then this material may be ignored.

Case 1: σ_1^2 and σ_2^2 known   The test statistic in this case is

Standardized test-ing method

$$Z = \frac{(\bar{X}_1 - \bar{X}_2) - D_0}{\sqrt{\dfrac{\sigma_1^2}{n_1} + \dfrac{\sigma_2^2}{n_2}}}$$

Decision rules for the case 1 test of hypothesis: H_0: $\mu_1 - \mu_2 = D_0$

Hypothesis test

$$\mu_1 - \mu_2 = D$$

Case 1: σ_1^2 and σ_2^2 known

Hypothesis:

H_0: $D \leq D_0$	H_0: $D = D_0$	H_0: $D \geq D_0$
H_A: $D > D_0$	H_A: $D \neq D_0$	H_A: $D < D_0$

Test statistic value:

$$z = \frac{(\bar{x}_1 - \bar{x}_2) - D_0}{\sqrt{\dfrac{\sigma_1^2}{n_1} + \dfrac{\sigma_2^2}{n_2}}}$$

Rejection region:

Reject H_0 if $z > z_\alpha$	Reject H_0 if $z < -z_{\alpha/2}$ or $z > z_{\alpha/2}$	Reject H_0 if $z < -z_\alpha$

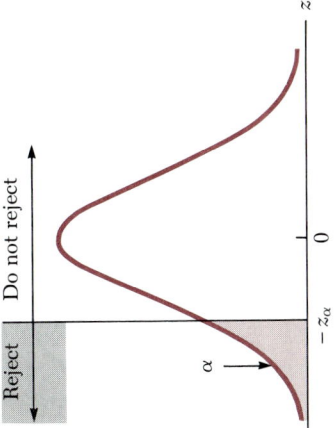

Reject | Do not reject

z_α 0

α

p-value = $P(Z > z)$

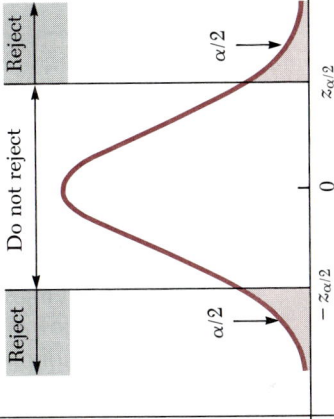

Reject | Do not reject | Reject

$-z_{\alpha/2}$ 0 $z_{\alpha/2}$

$\alpha/2$ $\alpha/2$

p-value = $2P(Z > |z|)$

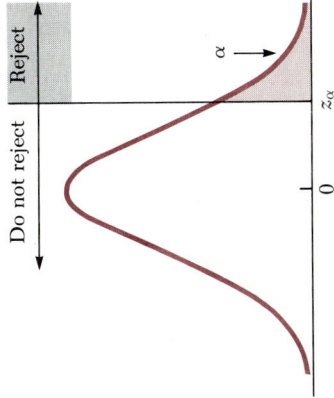

Reject | Do not reject

$-z_\alpha$ 0

α

p-value = $P(Z < z)$

z

This test statistic follows directly from the **Z**-statistic

$$Z = \frac{(\overline{X}_1 - \overline{X}_2) - (\mu_1 - \mu_2)}{\sqrt{\dfrac{\sigma_1^2}{n_1} + \dfrac{\sigma_2^2}{n_2}}}$$

used in Section 11.2.1 to develop the confidence interval formula for $\mu_1 - \mu_2$.

Example 11.5 The Flamerock Tire Company has decided to enter the radial tire market. The company decides to copy one of the European radial tire designs. The company has managed to procure the designs of two tires, type A and type B. Both tires are designed to last 40,000 miles. The company decides to produce the one that has the greater average tire mileage. From considerable previous testing with many similar radial tires, the company has found that the standard deviation of each type of radial tire is 3,000 miles. The company places 25 tires of type A on its mileage machine and finds that the average mileage is 39,780 miles. A sample of 25 type B tires produces an average mileage of 40,650 miles. The company wishes to determine whether or not there is a significant difference between the average mileage of all tires of types A and B. Test this hypothesis at the $\alpha = 0.05$ significance level.

Solution The hypotheses are:

$$H_0 : \quad D = \mu_1 - \mu_2 = 0$$
$$H_A : \quad D = \mu_1 - \mu_2 \neq 0$$

Hypothesizing that the difference is zero is of course equivalent to hypothesizing that the two population means are equivalent.

The sample data are:

$$n_1 = 25 \qquad n_2 = 25$$
$$\overline{x}_1 = 39{,}780 \qquad \overline{x}_2 = 40{,}650$$

Based on the information given, we will assume $\sigma_1^2 = \sigma_2^2 = (3{,}000)^2$; that is, it is assumed that the standard deviation of both tire types is 3,000 miles.

With $\alpha = 0.05$, the decision rule on the standard normal distribution is:

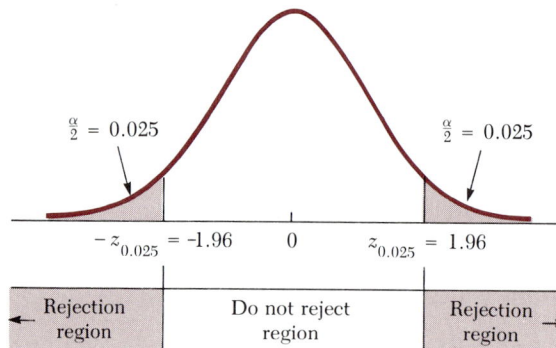

$\frac{\alpha}{2} = 0.025$		$\frac{\alpha}{2} = 0.025$
$-z_{0.025} = -1.96$	0	$z_{0.025} = 1.96$
Rejection region	Do not reject region	Rejection region

The value of the test statistic is:

$$z = \frac{(\bar{x}_1 - \bar{x}_2) - D_0}{\sqrt{\dfrac{\sigma_1^2}{n_1} + \dfrac{\sigma_2^2}{n_2}}} = \frac{(39,780 - 40,650) - 0}{\sqrt{\dfrac{(3,000)^2}{25} + \dfrac{(3,000)^2}{25}}} = -1.03$$

Since the value of the test statistic does not fall into either rejection region, we cannot reject the null hypothesis that $\mu_1 = \mu_2$, or equivalently, that $\mu_1 - \mu_2 = 0$.

Since this is a two-sided test, the p-value may be determined by:

$$\begin{aligned} p\text{-value} &= 2P(\mathbf{Z} > |z|) = 2P(\mathbf{Z} > |-1.03|) \\ &= 2P(\mathbf{Z} > 1.03) = 2(0.1515) = 0.303 \end{aligned}$$

Thus if we had decided to reject H_0: $\mu_1 - \mu_2 = 0$, we would have taken a risk equal to 0.303 of committing a Type I error (rejecting H_0/H_0 is true). (Recall that in the two-sided test, the p-value is determined by doubling the tail probability.)

As we discussed in Chapter 10, the p-value provides useful information in making a decision about the null hypothesis. At the $\alpha = 0.05$ significance level, it is evident that we should not reject H_0. Knowing that the p-value is relatively high (0.303) gives us some comfort in this decision. If we were to reject H_0, the probability of making an error is relatively high (0.303).

Case 2: σ_1^2 and σ_2^2 unknown; $\sigma_1^2 = \sigma_2^2$ In case 2 the **t**-statistic based on the pooled sample standard deviation is used. The test statistic and decision rules for this case are given on the next page. If the degrees of freedom $n_1 + n_2 - 2$ exceed 100, then the standard normal distribution value $z_{\alpha/2}$ may be substituted for $t_{\alpha/2;n_1+n_2-2}$. This follows the same convention used with setting confidence intervals on $\mu_1 - \mu_2$.

Example 11.6   A manufacturer claims an automobile battery it produces will last at least 100 hours longer, on the average, than competing batteries in the same price range when the batteries are subjected to continuous use in a controlled laboratory experiment. To test this claim, a competitor places samples of its battery and the supposedly superior one in a testing machine in which both batteries are subjected to identical conditions, and measures the lifetimes of the batteries. In a sample of 25 of its own batteries, the competitor finds that the average lifetime is 2,050 hours with a standard deviation of 80 hours; and 25 units of the supposedly superior battery last an average of 2,100 hours with a standard deviation of 100 hours. Test the manufacturer's claim at the $\alpha = 0.05$ significance level. Assume that the lifetimes of both batteries are normally distributed.

Solution   Let the supposedly superior batteries produced by the manufacturer making the claim compose the first population and the competitor's batteries compose the second population.

We wish to test the hypothesis that the mean lifetime of population 1 is at least 100 hours longer than the mean lifetime of population 2; that is, $\mu_1 - \mu_2 \geq 100$. Thus the hypotheses are:

Decision rules for the case 2 test of hypothesis: H_0: $\mu_1 - \mu_2 = D_0$

Hypothesis test

$$\mu_1 - \mu_2 = D$$

Case 2: σ_1^2, σ_2^2 unknown; $\sigma_1^2 = \sigma_2^2$

Hypothesis:

H_0: $D \le D_0$	H_0: $D = D_0$	H_0: $D \ge D_0$
H_A: $D > D_0$	H_A: $D \ne D_0$	H_A: $D < D_0$

Test statistic value:

$$t = \frac{(\bar{x}_1 - \bar{x}_2) - D_0}{s_p \cdot \sqrt{\dfrac{1}{n_1} + \dfrac{1}{n_2}}}$$

Rejection region:

Reject H_0 if
$t > t_{\alpha; n_1 + n_2 - 2}$

Reject H_0 if
$t < -t_{\alpha/2; n_1 + n_2 - 2}$
or
$t > t_{\alpha/2; n_1 + n_2 - 2}$

Reject H_0 if
$t < -t_{\alpha; n_1 + n_2 - 2}$

p-value $= P(t > t)$

p-value $= 2P(t > |t|)$

p-value $= P(t < t)$

$H_0 : \quad D = \mu_1 - \mu_2 \geq 100$
$H_A : \quad D = \mu_1 - \mu_2 < 100$

The sample data are:

$$
\begin{array}{ll}
n_1 = 25 & n_2 = 25 \\
\bar{x}_1 = 2{,}100 & \bar{x}_2 = 2{,}050 \\
s_1 = 100 & s_2 = 80
\end{array}
$$

If we assume that $\sigma_1^2 = \sigma_2^2$, then the case 2 test statistic based on the pooled sample standard deviation is appropriate:

$$
s_p = \sqrt{\frac{(n_1 - 1)s_1^2 + (n_2 - 1)s_2^2}{(n_1 - 1) + (n_2 - 1)}} = \sqrt{\frac{(24)(100)^2 + (24)(80)^2}{24 + 24}} = 90.55
$$

The decision rule is:

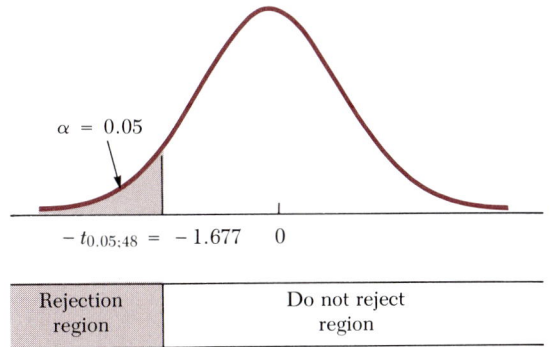

$\alpha = 0.05$

$-t_{0.05;48} = -1.677 \quad 0$

Rejection region	Do not reject region

The test statistic value is:

$$
t = \frac{(\bar{x}_1 - \bar{x}_2) - D_0}{s_p \sqrt{\dfrac{1}{n_1} + \dfrac{1}{n_2}}} = \frac{(2{,}100 - 2{,}050) - 100}{90.55 \sqrt{\dfrac{1}{25} + \dfrac{1}{25}}} = -1.95
$$

The **t**-statistic has degrees of freedom $n_1 + n_2 - 2 = 25 + 25 - 2 = 48$. With $\alpha = 0.05$, the standardized action limit is $-t_{0.05;48} = -1.677$ from Table B.5, Appendix B. Therefore, we will reject H_0: $\mu_1 - \mu_2 \geq 100$ if $t < -1.677$. Since $t = -1.95 < -1.677$, we reject H_0 and accept that the alternate hypothesis, H_A: $\mu_1 - \mu_2 < 100$ is true. Thus the competitor has been able to demonstrate statistically that the manufacturer's claim of a superior battery, one that lasts 100 hours or more than the competitors' batteries, is not warranted.

The p-value for this test is:

$$
p\text{-value} = P(t < t) = P(t < -1.95)
$$

From Table B.5, Appendix B, with 48 degrees of freedom, $P(t < -2.011) = 0.025$, and $P(t < -1.677) = 0.05$. Therefore the p-value for this test is between 0.025 and 0.05. By interpolation, the p-value equals 0.029.

Decision rule for the case 3 test of hypothesis: H_0: $\mu_1 - \mu_2 = D_0$

Hypothesis test
$\mu_1 - \mu_2 = D$
Case 3: σ_1^2, σ_2^2 unknown; $\sigma_1^2 \neq \sigma_2^2$

Hypothesis:	H_0: $D \leq D_0$ $\quad$ H_A: $D > D_0$	H_0: $D = D_0$ $\quad$ H_A: $D \neq D_0$	H_0: $D \geq D_0$ $\quad$ H_A: $D < D_0$

Test statistic value:
$$t' = \frac{(\bar{x}_1 - \bar{x}_2) - D_0}{\sqrt{\dfrac{s_1^2}{n_1} + \dfrac{s_2^2}{n_2}}}$$

Rejection region:	Reject H_0 if $t' > t_{\alpha;\,\triangle}$	Reject H_0 if $t' < -t_{\alpha/2;\,\triangle}$ or $t > t_{\alpha/2;\,\triangle}$	Reject H_0 if $t' < -t_{\alpha;\,\triangle}$

p-value $= P(t > t')$

p-value $= 2P(t > |t'|)$

p-value $= P(t < t')$

$$\triangle = \frac{[(s_1^2/n_1) + (s_2^2/n_2)]^2}{\dfrac{(s_1^2/n_1)^2}{n_1 - 1} + \dfrac{(s_2^2/n_2)^2}{n_2 - 1}}$$

Case 3: σ_1^2 and σ_2^2 unknown; $\sigma_1^2 \neq \sigma_2^2$ In case 3 the approximate t-statistic is used paralleling case 3 for the construction of the confidence interval estimator.

Following the convention used in constructing confidence interval estimates in case 3, if $\Delta > 100$, then the **Z**-statistic may be substituted for the **t**-statistic. That is, the **t'**-statistic would be computed and compared with z_α when H_0: $D \leq D_0$, with $-z_{\alpha/2}$ and $+z_{\alpha/2}$ when H_0: $D = D_0$, and with $-z_\alpha$ when H_0: $D \geq D_0$. The p-values would be computed by using the standard normal table (Table B.3, Appendix B) rather than the t-table (Table B.5, Appendix B).

Example 11.7 The Suzuka Automotive Company claims that its Accent model gets the best gasoline mileage (in terms of miles per gallon, MPG) of any car on the market. More specifically, it contends that the Accent gets better gasoline mileage than Hondo Slavic CPX model. To test this claim, Suzuka retains an independent testing agency to compare randomly selected Accent and Slavic automobiles that are equivalently equipped. The testing agency takes a random sample of 30 automobiles of each type and tests the cars on a 100-mile test drive. The average gasoline mileage and the standard deviation of the mileages are recorded for each make. The sample data are:

	Accent model	Slavic CPX model
	$n_1 = 30$ cars	$n_1 = 30$ cars
	$\bar{x}_1 = 44.24$ mpg	$\bar{x}_2 = 41.60$ mpg
	$s_1 = 3.16$ mpg	$s_2 = 1.74$ mpg

Is Suzuka's claim warranted at the $\alpha = 0.05$ significance level?

Solution Let μ_1 = Average miles per gallon for the Accent model and μ_2 = Average miles per gallon for the Slavic CPX model. Suzuka wishes to demonstrate that $\mu_1 > \mu_2$ or equivalently that $\mu_1 - \mu_2 > 0$. Following the convention in Chapter 10 for the one-sided tests, we will place this condition in the alternate hypothesis. Then the null and alternate hypotheses are:

H_0 : $\mu_1 - \mu_2 \leq 0$
H_A : $\mu_1 - \mu_2 > 0$

We first calculate the degrees of freedom Δ for the **t'** statistic:

$$\Delta = \frac{[(s_1^2/n_1) + (s_2^2/n_2)]^2}{\dfrac{(s_1^2/n_1)^2}{(n_1 - 1)} + \dfrac{(s_2^2/n_2)^2}{(n_2 - 1)}} = \frac{[(3.16)^2/30 + (1.74)^2/30]^2}{\dfrac{[(3.16)^2/30]^2}{29} + \dfrac{[(1.74)^2/30]^2}{29}}$$

$$= \frac{0.1882}{0.00417} = 45.115$$

Since Δ is fractional, we round to 45. Also, since $\Delta \leq 100$, we must use the t-distribution to establish the decision rule. The action limit is $t_{\alpha;\Delta} = t_{0.05;45} = 1.679$ from Table B.5, Appendix B. The value of the test statistic t' is:

$$t' = \frac{(\bar{x}_1 - \bar{x}_2) - D_0}{\sqrt{\dfrac{s_1^2}{n_1} + \dfrac{s_2^2}{n_2}}} = \frac{(44.24 - 41.60) - 0}{\sqrt{\dfrac{(3.16)^2}{30} + \dfrac{(1.74)^2}{30}}} = \frac{2.64}{0.6586} = 4.01$$

Since $t' = 4.01 > t_{0.05;45} = 1.679$, we reject $H_0: \mu_1 - \mu_2 \leq 0$ and conclude that the alternate hypothesis, $H_A: \mu_1 - \mu_2 > 0$, is true. Thus Suzuka's claim that $\mu_1 > \mu_2$ is supported by the sample evidence.

The p-value for this test is:

$$p\text{-value} = P(\mathbf{t} > t') = P(\mathbf{t} > 4.01)$$

From Table B.5, Appendix B, with 45 degrees of freedom, $P(\mathbf{t} > 3.281) < 0.001$. Therefore the p-value is less than 0.001. Since the p-value is so small, Suzuka can feel confident that it has demonstrated its claim beyond all reasonable doubt.

Finally, we observe that the use of the case 3 procedure is certainly warranted in this problem, since $s_1^2 = (3.16)^2 = 9.9856$ is more than three times $s_2^2 = (1.74)^2 = 3.0276$.

Although constructing the power curve for cases with two population parameters is a straightforward extension of the power curve construction in the one-parameter case (see Chapter 10), we will not develop power curves in this chapter. But it is important to remember the purpose and value of a power curve for a statistical test of a hypothesis. It is an important factor to consider in conjunction with the p-value (or the setting of the significance level) in the decision to reject or not to reject the null hypothesis.

The tests described in this section for comparing two population means assume that the population random variables $\mathbf{X}_1$ and $\mathbf{X}_2$ are normally distributed. If this is not the case, then n_1 and n_2 should both be 25 or more for the application of these procedures. Furthermore, case 2 assumes that the population variances, σ_1^2 and σ_2^2, are equal. Experimental results have demonstrated that mild departures from the assumptions of normality and equivalent variances do not seriously affect the conclusions drawn from using these tests, provided the sample sizes are equal or nearly equal and the population distributions are not more than moderately skewed. If the departures from the assumptions are mild, it will usually turn out that the probability of committing a Type I error, α, will in actuality be somewhat larger than the value chosen to establish the decision rule for the test.

In Chapter 19 we shall learn how to test the hypothesis that a set of data came from a normal distribution. If the departure from the assumption that the data are normal is not mild, then these procedures for comparing two population means should not be used. Testing procedures in this case will be presented in Chapter 20.

If the data do support the assumption of normality, we must still check the assumption of equal variances, for this dictates whether case 2 or case 3 applies.

■ **11.3**

Paired *t*-test for comparing two population means

In Section 11.2, when comparing two population means, we assumed that the two random samples drawn from the two populations were independent. In many instances, this will not be true. In fact, many times we purposefully design the experiment so that the samples drawn from the two populations will be dependent.

Let us first consider a two-population experiment in which the samples are dependent by the nature of the experiment. We will then discuss an experiment that is *designed* so that the two samples are dependent.

Example 11.8 A doctor wishes to test the efficacy of a new dietary plan. He decides to select 10 people randomly from a large group who are interested in losing weight. To test the efficacy of this plan, he will weigh the persons in the sample at the beginning of the dietary program and again at the end of six weeks. He wishes to demonstrate, of course, that the population mean weight at the end of the six-week program is less than the population mean weight at the beginning of the program.

It is clear that the 10 weights composing the first sample and the 10 weights composing the second sample are not independent. In fact, the weights will be "paired"; a weight in the first sample is taken from one individual who will produce a weight (one hopes it will be less) in the second sample.

Whenever measurements are taken from the same units (usually people) at two different times and the means of the measurements at the two times are compared, the two sets of sample measurements are dependent.

In many experiments, we may design the experiment so that units in the two samples are paired to eliminate effects in which we have no interest.

Example 11.9 A professor in charge of an introductory statistics course at a large university is interested in developing a new teaching method. She decides to form two classes, each containing 25 students, for the purpose of testing the new teaching method. The old teaching method will be used in the first class, and the new teaching method will be used in the second class. In the term during which she will compare the two methods, more than 500 students will take this course. The professor decides to select the 50 students she will need for her study in pairs where the two students composing each pair are as similar as possible—same sex, roughly the same SAT quantitative scores, about the same set of previous courses, same grade point average, and so on. She will flip a coin to see which member of each pair is assigned to the first class—the "loser" will be assigned to the second class.

The two samples (classes) will be dependent because the assignments of students to the two groups were made in pairs—the formation of the two samples was not completely random. Yet, by pairing the sample observations, many factors such as grade point average and IQ that may cloud the determination of which teaching method is "better" have been eliminated. If the samples had been formed completely at random, we might by chance find that the better students among the 50 selected were placed in the first class (old teaching method) while the less prepared students were placed in the second class. Certainly, this circumstance would not bode well for the new teaching method, even if it was really more effective than the old method.

The analysis of the paired observation experiment is different than if the two samples were chosen at random. Suppose the experiment yields n pairs of values denoted by (x_{11}, x_{21}), (x_{12}, x_{22}), (x_{13}, x_{23}), . . . , (x_{1n}, x_{2n}). That is, x_{ij} is the member of the jth pair drawn from the ith population. For the jth pair, define

$$d_j = x_{1j} - x_{2j}$$

That is, d_j is the difference between the two values forming the jth pair. It seems reasonable that a hypothesis test to determine whether the two population means are equal is based on the differences d_j, $j = 1, 2, \ldots, n$, where n is the number of observations in each sample. These differences are used in the following way. We first form the average of the differences, $\bar{d}$, and the variance of the differences, s_d^2, where

$$\bar{d} = \frac{\sum_{j=1}^{n} d_j}{n} \quad \text{and} \quad s_d^2 = \frac{\sum_{j=1}^{n} (d_j - \bar{d})^2}{n - 1}$$

If we assume that the collection of differences, d_j, where $j = 1, 2, \ldots,$ n, forms a random sample from a normal population with mean $\mu_d = \mu_1 - \mu_2$ and variance σ_d^2, then the population random variable **d** whose values are the paired differences in the population has as its sampling distribution a normal distribution as illustrated in Figure 11.5.

FIGURE 11.5 | The sampling distribution of **d**

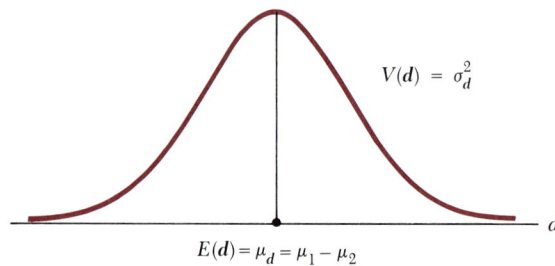

$$V(d) = \sigma_d^2$$

$$E(d) = \mu_d = \mu_1 - \mu_2$$

Then, the statistic

$$t = \frac{\bar{d} - \mu_d}{S_d / \sqrt{n}}$$

is a value from the t-distribution with $(n - 1)$ degrees of freedom.

Example 11.10 Suppose the weights of the 10 participating people in the dietary plan described in Example 11.8 are those listed in Table 11.2. Test the hypothesis that $\mu_d = 0$ at the $\alpha = 0.01$ significance level.

Decision rule for the paired samples test of hypothesis:
$H_0: \mu_d = \mu_1 - \mu_2 = 0$

Hypothesis test
$\mu_d = \mu_1 - \mu_2$
Paired observations

Hypothesis:

$H_0: \mu_d = \mu_1 - \mu_2 \geq 0$
$H_A: \mu_d = \mu_1 - \mu_2 < 0$

$H_0: \mu_d = \mu_1 - \mu_2 = 0$
$H_A: \mu_d = \mu_1 - \mu_2 \neq 0$

$H_0: \mu_d = \mu_1 - \mu_2 \leq 0$
$H_A: \mu_d = \mu_1 - \mu_2 > 0$

Test statistic value:

$$t = \frac{\bar{d} - \mu_{d_0}}{s_d / \sqrt{n}}$$

Rejection region:

Reject H_0 if
$t < -t_{\alpha;\, n-1}$

Reject H_0 if
$t < -t_{\alpha/2;\, n-1}$ or
$t > t_{\alpha/2;\, n-1}$

Reject H_0 if
$t > t_{\alpha;\, n-1}$

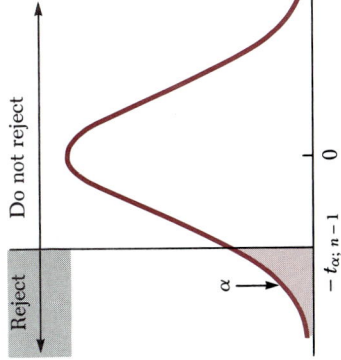

p-value = $P(t < t)$

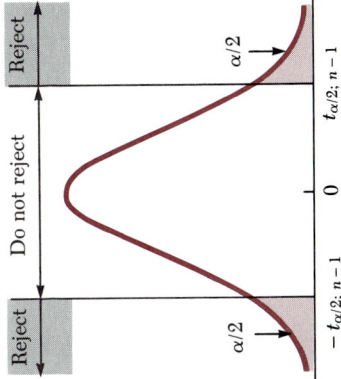

p-value = $2P(t > |t|)$

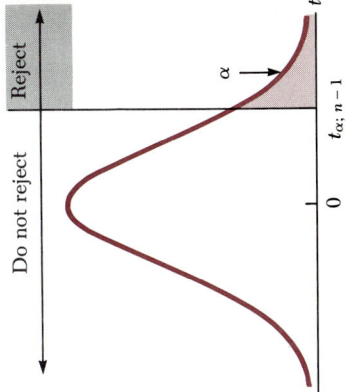

p-value = $P(t > t)$

TABLE 11.2

Sample data			
Beginning x_{1j}	Six weeks x_{2j}	d_j	d_j^2
155	154	1	1
228	207	21	441
172	165	7	49
141	147	−6	36
162	157	5	25
211	196	15	225
185	180	5	25
122	121	1	1
164	150	14	196
199	204	−5	25
		58	1,024

Solution   The hypothesis is:

$$H_0: \quad \mu_1 - \mu_2 = \mu_d = 0$$
$$H_A: \quad \mu_d \neq 0$$

For the $\alpha = 0.01$ significance level, the rejection region is

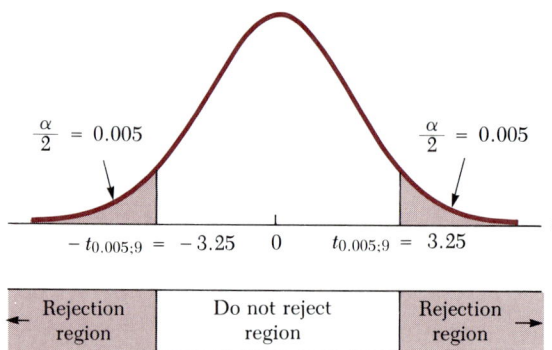

Rejection region	Do not reject region	Rejection region

In forming the value of the test statistic, we must compute $\bar{d}$ and s_d:

$$\bar{d} = \frac{\sum\limits_{j=1}^{10} d_j}{10} = \frac{58}{10} = 5.8$$

$$s_d^2 = \frac{\sum\limits_{j=1}^{10} (d_j - \bar{d})^2}{10 - 1} = \frac{\sum\limits_{j=1}^{10} d_j^2 - \dfrac{\left(\sum\limits_{j=1}^{10} d_j\right)^2}{10}}{9} = \frac{1,024 - \dfrac{3,364}{10}}{9} = 76.4$$

$$s_d = \sqrt{76.4} = 8.74$$

The test statistic value is:

$$t = \frac{\bar{d} - \mu_{d_0}}{s_d/\sqrt{n}} = \frac{5.8 - 0}{8.74/\sqrt{10}} = 2.10$$

Since $t = 2.10$ falls in the region where H_0 cannot be rejected, we cannot reject the null hypothesis that $\mu_d = 0$. We therefore conclude that the doctor's dietary plan was not very effective.

The p-value for this test is given by:

$$p\text{-value} = 2P(\mathbf{t} > |t|) = 2P(\mathbf{t} > 2.10)$$

Since $P(\mathbf{t} > 1.833) = 0.05$ and $P(\mathbf{t} > 2.262) = 0.025$ from Table B.5 in Appendix B, the p-value lies between twice 0.025 and twice 0.05, or between 0.05 and 0.10. Note that the significance level α was set very low for this test ($\alpha = 0.01$). We would not reject the null hypothesis if $\alpha = 0.01$ or 0.05, but we would reject the null hypothesis if $\alpha = 0.10$.

It should be apparent that we can use the t-statistic given above to construct a confidence interval estimator of $\mu_d = \mu_1 - \mu_2$.

Confidence interval estimate of $\mu_d = \mu_1 - \mu_2$ in paired samples

Formula 11.5

Confidence interval estimate of $\mu_d = \mu_1 - \mu_2$, paired samples

A $100(1 - \alpha)$ percent confidence interval estimate for $\mu_1 - \mu_2$ when the sample observations (x_{11}, x_{21}), (x_{12}, x_{22}), (x_{13}, x_{23}), . . . , (x_{1n}, x_{2n}) are paired is given by

$$\bar{d} \pm t_{\alpha/2;n-1} \frac{s_d}{\sqrt{n}}$$

where

$$\bar{d} = \frac{\sum\limits_{j=1}^{n} d_j}{n}, \quad s_d = \sqrt{\frac{\sum\limits_{j=1}^{n} (d_j - \bar{d})^2}{n - 1}}, \quad d_j = x_{1j} - x_{2j}, \quad j = 1, 2, \ldots, n$$

Example 11.11 For the dietary data in Example 11.9, determine a 99 percent confidence interval estimate on the mean difference, $\mu_d = \mu_1 - \mu_2$.

Solution From the solution to Example 11.9, we have: $\bar{d} = 5.8$, $s_d = 8.74$, and $t_{\alpha/2;n-1} = t_{0.005;9} = 3.25$. Then,

$$\bar{d} \pm t_{\alpha/2;n-1} \frac{s_d}{\sqrt{n}} = 5.8 \pm 3.25 \frac{8.74}{\sqrt{10}} = 5.8 \pm 8.98$$

The 99 percent confidence interval estimate is $(5.8 - 8.98) = -3.18$ pounds to $5.8 + 8.98 = 14.78$ pounds. Notice that since this interval contains 0, we would not reject H_0: $\mu_d = 0$. Thus the 99 percent confidence interval estimate outcome is consistent with the outcome of the $\alpha = 0.01$ significance test in Example 11.9.

A few comments regarding the paired experiment are in order. First, the paired experiment is usually more "efficient" than the independent sample experiment. By this, we mean that fewer sample observations are generally required to detect a true difference between μ_1 and μ_2 when the paired sample experiment is used instead of the two-independent sample experiment. This is because the pairing usually eliminates effects that may make it difficult for the two-independent sample experiments to detect a true difference. Second, the paired experiment does not require that the population variances, σ_1^2 and σ_2^2, be equal. When it is known that σ_1^2 is not equal to σ_2^2, if the sample values can be paired in some meaningful way, then the paired t-test can be used in place of the approximate case 3 method described in Section 11.2.3.

■ 11.4

Inferences concerning two population proportions, π_1 and π_2

The framework for comparing two population proportions is illustrated in Figure 11.6. Inferences concerning the population proportions π_1 and π_2 are based on the sample proportions p_1 and p_2, since **P** is the best point estimator of π from Chapter 9.

FIGURE   11.6 | Comparing two population proportions

Notation for comparing two population proportions

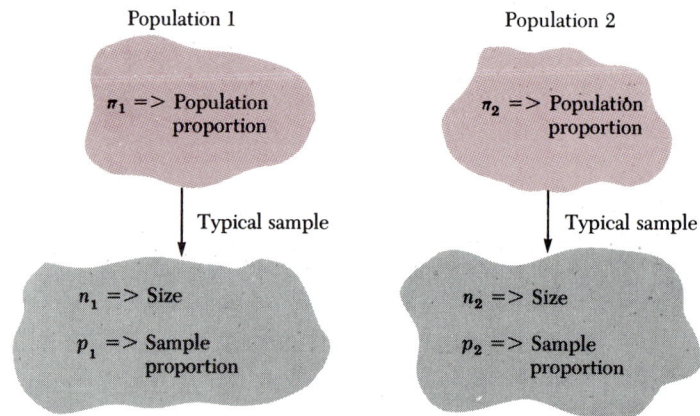

Population 1

$\pi_1 \Rightarrow$ Population proportion

Typical sample

$n_1 \Rightarrow$ Size

$p_1 \Rightarrow$ Sample proportion

Population 2

$\pi_2 \Rightarrow$ Population proportion

Typical sample

$n_2 \Rightarrow$ Size

$p_2 \Rightarrow$ Sample proportion

In this section, we shall develop a large sample method for constructing confidence intervals and testing hypotheses on the difference between two population proportions. If the sample sizes are small (n_1 or n_2 or both less than 25), then this method should not be used. A small sample test for comparing population proportions is given in Chapter 19.

11.4.1 | Point estimation of $D = \pi_1 - \pi_2$

Recall from Chapter 9 that our best point estimator of the population proportion π is the sample proportion $\mathbf{P}$, where $\mathbf{P} = \mathbf{X}/n$ with $\mathbf{X} =$ Number of successes in n Bernoulli trials of successes and failures. Thus it seems reasonable that the best point estimator of $D = \pi_1 - \pi_2$ is the difference between the sample proportions $\mathbf{P}_1$ and $\mathbf{P}_2$ in the two samples. In fact the point estimator $\mathbf{P}_1 - \mathbf{P}_2$ is unbiased and among all unbiased estimators has the minimum variance in its sampling distribution. Therefore, $\mathbf{P}_1 - \mathbf{P}_2$ is used as the point estimator of $D = \pi_1 - \pi_2$.

11.4.2 | Confidence interval estimation of $\pi_1 - \pi_2$

The confidence interval estimator of $\pi_1 - \pi_2$ can be constructed by using the general form of the standard normal statistic

$$\mathbf{Z} = \frac{\text{(Estimator)} - \text{(Parameter)}}{\text{(Standard error of estimator)}}$$

In this instance, the estimator is $\mathbf{P}_1 - \mathbf{P}_2$, and the parameter is $D = \pi_1 - \pi_2$. The standard error of the estimator (equivalently, the standard deviation of the estimator) is the square root of the variance of the estimator, $\mathbf{P}_1 - \mathbf{P}_2$. Denoting the variance of the estimator by $\sigma^2_{\mathbf{P}_1 - \mathbf{P}_2}$ or simply by $V(\mathbf{P}_1 - \mathbf{P}_2)$, we have from Theorem 11.1,

$$\sigma^2_{\mathbf{P}_1 - \mathbf{P}_2} = V(\mathbf{P}_1 - \mathbf{P}_2) = V(\mathbf{P}_1) + V(\mathbf{P}_2) = \sigma^2_{\mathbf{P}_1} + \sigma^2_{\mathbf{P}_2}$$

where

$$\sigma^2_{\mathbf{P}_1} = \frac{\pi_1(1 - \pi_1)}{n_1} \quad \text{and} \quad \sigma^2_{\mathbf{P}_2} = \frac{\pi_2(1 - \pi_2)}{n_2}$$

The sampling distribution of $\mathbf{P}_1 - \mathbf{P}_2$, assuming that $\mathbf{P}_1 - \mathbf{P}_2$ is approximately normally distributed, is shown in Figure 11.7.

FIGURE 11.7 | The sampling distribution of $\mathbf{P}_1 - \mathbf{P}_2$

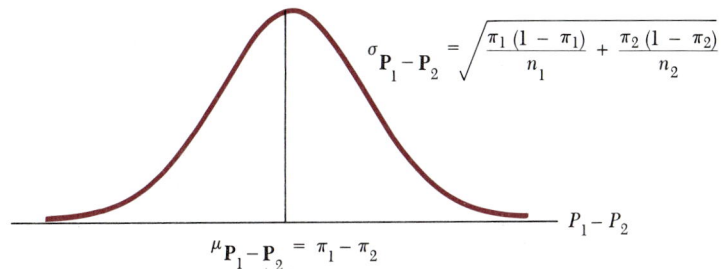

$$\sigma_{\mathbf{P}_1 - \mathbf{P}_2} = \sqrt{\frac{\pi_1(1 - \pi_1)}{n_1} + \frac{\pi_2(1 - \pi_2)}{n_2}}$$

$$\mu_{\mathbf{P}_1 - \mathbf{P}_2} = \pi_1 - \pi_2$$

$$P_1 - P_2$$

The appropriate **Z**-statistic is

$$Z = \frac{(P_1 - P_2) - (\pi_1 - \pi_2)}{\sqrt{\dfrac{\pi_1(1 - \pi_1)}{n_1} + \dfrac{\pi_2(1 - \pi_2)}{n_2}}}$$

This statistic can be used to construct a confidence interval estimator of $\pi_1 - \pi_2$ by estimating the denominator

$$\sigma_{P_1 - P_2} = \sqrt{\frac{\pi_1(1 - \pi_1)}{n_1} + \frac{\pi_2(1 - \pi_2)}{n_2}}$$

with

$$S_{P_1 - P_2} = \sqrt{\frac{P_1(1 - P_1)}{n_1} + \frac{P_2(1 - P_2)}{n_2}}$$

The substitution of $S_{P_1-P_2}$ for $\sigma_{P_1-P_2}$ produces a statistic that is approximately normally distributed for "large" samples. We denote this statistic by **Z′**, where

$$Z' = \frac{(P_1 - P_2) - (\pi_1 - \pi_2)}{\sqrt{\dfrac{P_1(1 - P_1)}{n_1} + \dfrac{P_2(1 - P_2)}{n_2}}}$$

Usually $n_1 \geq 25$ and $n_2 \geq 25$ will suffice for **Z′** to be approximately normally distributed. The conditions $n_1\pi_1 \geq 5$, $n_1(1 - \pi_1) \geq 5$, $n_2\pi_2 \geq 5$, and $n_2(1 - \pi_2) \geq 5$ can be used as well. Since π_1 and π_2 are unknown in value, we could use the conditions $n_1 p_1 \geq 5$, $n_1(1 - p_1) \geq 5$, $n_2 p_2 \geq 5$, and $n_2(1 - p_2) \geq 5$ to check if n_1 and n_2 are sufficiently large to assume that **Z′** is approximately normally distributed, where p_1 and p_2 are the point estimates from the two samples of π_1 and π_2, respectively.

The **Z′** statistic can be used to construct the following confidence interval estimate of $\pi_1 - \pi_2$.

| Confidence interval estimate of $\pi_1 - \pi_2$ |

Formula 11.6

Confidence interval estimate of $\pi_1 - \pi_2$

A $100(1 - \alpha)$ percent confidence interval estimate of $\pi_1 - \pi_2$ is given by

$$(p_1 - p_2) \pm z_{\alpha/2} \sqrt{\frac{p_1(1 - p_1)}{n_1} + \frac{p_2(1 - p_2)}{n_2}}$$

where n_i and p_i are the sample size and sample proportion, respectively, associated with the ith sample ($i = 1, 2$) and $z_{\alpha/2}$ is the point on the standard normal curve such that $P(Z \geq z_{\alpha/2}) = \alpha/2$.

Example 11.12 During a recent flu epidemic, a study was made at a large university to determine whether or not flu shots help people resist contracting the flu. A sample of 200 students who had received the series of flu shots at the campus medical center produced 40 victims of the flu. A sample of 100 students who did not receive the shots produced 35 students who got the flu. Find a 90 percent confidence interval for the difference between the proportion of flu victims among those students receiving the shots (π_1) and the proportion of flu victims among those not receiving the shots (π_2).

Solution For a 90 percent confidence interval, $z_{\alpha/2} = z_{0.05} = 1.64$ from Table B.3 in Appendix B. The 90 percent confidence interval for $\pi_1 - \pi_2$ is

$$
(p_1 - p_2) \pm z_{\alpha/2} \sqrt{\frac{p_1(1 - p_1)}{n_1} + \frac{p_2(1 - p_2)}{n_2}}
$$

$$
= \left(\frac{40}{200} - \frac{35}{100}\right) \pm 1.64 \sqrt{\frac{(0.2)(0.8)}{200} + \frac{(0.35)(0.65)}{100}}
$$

$$
= -0.15 \pm 0.09
$$

Thus we are 90 percent confident that the true difference between π_1 and π_2 is between -0.24 and -0.06. Since 0 is not included in this interval, we would conclude that the flu shots probably were helpful in resisting the flu bug.

11.4.3 | Hypothesis testing concerning $\pi_1 - \pi_2$

The hypothesis-testing procedure based on large samples for comparing π_1 and π_2 is developed from the **Z′**-statistic used in the preceding section to construct an interval estimator of $\pi_1 - \pi_2$. The test statistic for testing the null hypothesis H_0: $\pi_1 - \pi_2 = D_0$ is

$$
\mathbf{Z'} = \frac{(\mathbf{P}_1 - \mathbf{P}_2) - D_0}{\sqrt{\dfrac{\mathbf{P}_1(1 - \mathbf{P}_1)}{n_1} + \dfrac{\mathbf{P}_2(1 - \mathbf{P}_2)}{n_2}}}
$$

unless $D_0 = 0$. In the instance where $D_0 = 0$, we are hypothesizing that $\pi_1 - \pi_2 = 0$ or, equivalently, that $\pi_1 = \pi_2$. If the two population proportions are equal, then $\pi_1 = \pi_2 = \pi$, where π is the population proportion in both populations. In this instance $\mathbf{P}_1$ and $\mathbf{P}_2$, the two sample proportions, are estimators of π. Hence we can "pool" these estimators to produce an estimator of π based on both sample proportions. The pooled estimator of π is given by:

Pooled estimator of
$\pi = \pi_1 = \pi_2$

$$
\frac{n_1\mathbf{P}_1 + n_2\mathbf{P}_2}{n_1 + n_2}
$$

The variance of $\mathbf{P}_1 - \mathbf{P}_2$ in this instance becomes

$$
\sigma^2_{\mathbf{P}_1 - \mathbf{P}_2} = V(\mathbf{P}_1 - \mathbf{P}_2) = \frac{\pi_1(1 - \pi_1)}{n_1} + \frac{\pi_2(1 - \pi_2)}{n_2} = \frac{\pi(1 - \pi)}{n_1} + \frac{\pi(1 - \pi)}{n_2}
$$

since $\pi_1 = \pi_2 = \pi$. Factoring $\pi(1 - \pi)$ gives

$$\sigma_{P_1-P_2}^2 = \pi(1 - \pi)\left(\frac{1}{n_1} + \frac{1}{n_2}\right)$$

Estimator of $\sigma_{P_1-P_2}^2$ assuming $\pi_1 = \pi_2$ using the pooled estimator of $\pi = \pi_1 = \pi_2$

To estimate π in $\sigma_{P_1-P_2}^2$, we use the pooled estimator of π so that

$$S_{P_1-P_2}^2 = \left[\frac{n_1 P_1 + n_2 P_2}{n_1 + n_2}\right]\left[1 - \frac{n_1 P_1 + n_2 P_2}{n_1 + n_2}\right]\left[\frac{1}{n_1} + \frac{1}{n_2}\right]$$

The test statistic then becomes

Test statistic assuming $\pi_1 = \pi_2$

$$Z' = \frac{(P_1 - P_2) - D_0}{\sqrt{\left[\frac{n_1 P_1 + n_2 P_2}{n_1 + n_2}\right]\left[1 - \frac{n_1 P_1 + n_2 P_2}{n_1 + n_2}\right]\left[\frac{1}{n_1} + \frac{1}{n_2}\right]}}$$

where $D_0 = 0$. These results are summarized on page 557.

Example 11.13 The quality control director at the Turbo Equipment Company believes that the proportion of defective turbo chargers produced by his company for automobiles is the same by two production processes. Since the first process is less expensive than the second production process, he would like to demonstrate that his belief is true so that the first process can be used. As an experiment, he has produced 100 units by the first process and 100 units by the second process and counts the number of defectives in each sampled batch. The sample statistics are:

First process	Second process
First process	*Second process*
$n_1 = 100$ units	$n_2 = 100$ units
$x_1 = $ Number of defectives	$x_2 = $ Number of defectives
$= 6$	$= 3$
$p_1 = $ Proportion of defectives	$p_2 = $ Proportion of defectives
$= 6/100 = 0.06$	$= 3/100 = 0.03$

Test the hypothesis that $\pi_1 = \pi_2$, where $\pi_1 = $ Population proportion of defectives produced by the first process and $\pi_2 = $ population proportion of defectives produced by the second process. Use an $\alpha = 0.05$ significance level.

Solution The null and alternate hypotheses are:

$$\begin{array}{l} H_0: \quad \pi_1 = \pi_2 \\ H_A: \quad \pi_1 \neq \pi_2 \end{array} \left\{\begin{array}{l}\text{or, equivalently} \\ \text{with } D = \pi_1 - \pi_2\end{array}\right\} \quad \begin{array}{l} H_0: \quad D = 0 \\ H_A: \quad D \neq 0 \end{array}$$

Since $D_0 = 0$ in the null hypothesis, we must use the following formula to estimate $\sigma_{P_1-P_2}$:

Hypothesis test
$$D = \pi_1 - \pi_2$$

Hypothesis:

$H_0: D \leq D_0$	$H_0: D = D_0$	$H_0: D \geq D_0$
$H_A: D > D_0$	$H_A: D \neq D_0$	$H_A: D < D_0$

Test statistic value:
$$z' = \frac{(p_1 - p_2) - D_0}{s_{p_1 - p_2}}$$
where $s_{p_1 - p_2}$ is given below.

Rejection region:

Reject H_0 if $z > z_\alpha$

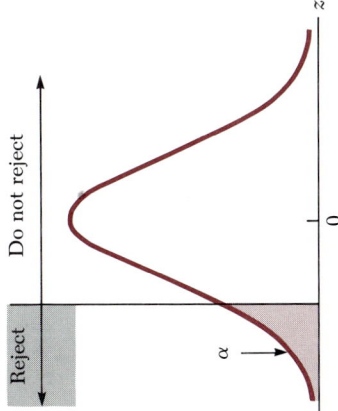

Reject H_0 if $z < -z_{\alpha/2}$ or $z > z_{\alpha/2}$

Reject H_0 if $z < -z_\alpha$

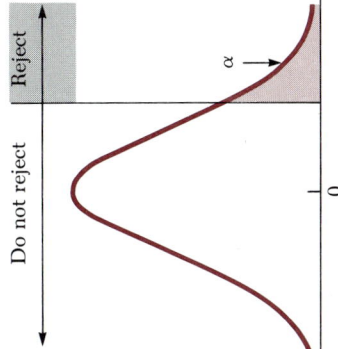

$p\text{-value} = P(Z > z)$

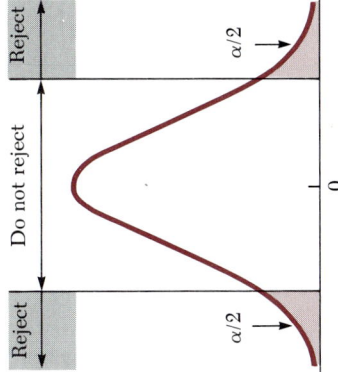

$p\text{-value} = 2P(Z > |z|)$

$p\text{-value} = P(Z < z)$

$$s_{p_1 - p_2} = \sqrt{\frac{n_1 p_1 + n_2 p_2}{n_1 + n_2}\left[1 - \frac{n_1 p_1 + n_2 p_2}{n_1 + n_2}\right]\left[\frac{1}{n_1} + \frac{1}{n_2}\right]} \quad \text{if } D_0 = 0$$

$$s_{p_1 - p_2} = \sqrt{\frac{p_1(1 - p_1)}{n_1} + \frac{p_2(1 - p_2)}{n_2}} \quad \text{if } D_0 \neq 0$$

$$s_{P_1-P_2} = \sqrt{\left[\frac{n_1 p_1 + n_2 p_2}{n_1 + n_2}\right]\left[1 - \frac{n_1 p_1 + n_2 p_2}{n_1 + n_2}\right]\left[\frac{1}{n_1} + \frac{1}{n_2}\right]}$$

$$= \sqrt{\left[\frac{(100)(0.06) + (100)(0.03)}{100 + 100}\right]\left[1 - \frac{(100)(0.06) + (100)(0.03)}{100 + 100}\right]\left[\frac{1}{100} + \frac{1}{100}\right]}$$

$$= \sqrt{(0.045)(0.955)(0.02)} = 0.0293$$

The value of the test statistic is:

$$z' = \frac{(p_1 - p_2) - D_0}{s_{P_1-P_2}} = \frac{(0.06 - 0.03) - 0}{0.0293} = 1.024$$

With an $\alpha = 0.05$ level of significance, $\alpha/2 = 0.025$ and $z_{\alpha/2} = z_{0.025} = 1.96$ from Table B.3, Appendix B. The decision rule is

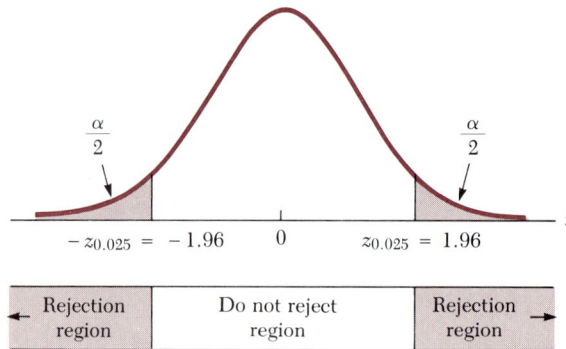

| Rejection region | Do not reject region | Rejection region |

Since $z = 1.024$ falls in the do not reject region, we do not reject the null hypothesis, H_0: $\pi_1 = \pi_2$. Although $p_1 = 0.06$ and $p_2 = 0.03$, the difference $p_1 - p_2 = 0.06 - 0.03 = 0.03$ is not large enough to conclude that $\pi_1 - \pi_2 \neq 0$ in the populations. Thus the quality control director must conclude that there is not a statistically significant difference in the proportion of defectives produced by these two processes based on this sample evidence.

The p-value for this test is:

$$p\text{-value} = 2P(\mathbf{Z} > |z|) = 2P(\mathbf{Z} > 1.024)$$
$$= 2(0.15294) = 0.3059 \text{ (by interpolation)}$$

Thus, there is a 0.3059 probability of making a Type I error (rejecting H_0/H_0 is true) if the decision is made to reject H_0. Since the p-value (0.3059) is so high, we would certainly decide not to reject H_0.

Example 11.14 The marketing director of Poor's beer knows that the preference for Poor's beer varies regionally across the country. She is convinced that the southwest region may require a stronger advertising effort to increase sales, bringing the sales up to a more comparable level to the sales in the western region. In an effort to increase the marketing budget for the southwest, she wishes to demonstrate that the percentage of beer drinkers who express a preference for Poor's beer is more than 20 percent higher in the western region when compared with the southwestern region. To test this belief, she contracts an independent survey company to take samples of beer drinkers in the two regions regarding their preference for Poor's beer over other beers. The sample data are:

Western region	Southwestern region
n_1 = 400 beer drinkers	n_2 = 500 beer drinkers
x_1 = Number who prefer Poor's beer	x_2 = Number who prefer Poor's beer
$\quad$ = 216	$\quad$ = 125
p_1 = Proportion who prefer Poor's beer	p_2 = Proportion who prefer Poor's beer
$\quad$ = 216/400 = 0.54	$\quad$ = 125/500 = 0.25

Is the marketing director's claim justified based on this sample information? Use an $\alpha = 0.01$ level of significance.

Solution Letting π_1 = Proportion of all beer drinkers in the western region who prefer Poor's beer and π_2 = Proportion of all beer drinkers in the southwestern region who prefer Poor's beer, she wishes to demonstrate that $\pi_1 - \pi_2 > 0.20$; that is, the proportion preferring Poor's in the West exceeds the proportion preferring Poor's beer in the southwest by 0.20 (20 percent). Placing the statement she wishes to demonstrate is true in this one-sided test in the alternate hypothesis, the null and alternate hypotheses are:

$$H_0 : \quad \pi_1 - \pi_2 \leq 0.20$$
$$H_A : \quad \pi_1 - \pi_2 > 0.20.$$

Since $D_0 = 0.20 \neq 0$, we use the following formula to estimate $\sigma_{P_1 - P_2}$:

$$s_{P_1 - P_2} = \sqrt{\frac{p_1(1 - p_1)}{n_1} + \frac{p_2(1 - p_2)}{n_2}} = \sqrt{\frac{(0.54)(0.46)}{400} + \frac{(0.25)(0.75)}{500}}$$
$$= \sqrt{0.001} = 0.0316$$

The value of the test statistic is:

$$z' = \frac{(p_1 - p_2) - D_0}{s_{P_1 - P_2}} = \frac{(0.54 - 0.25) - 0.20}{0.0316} = 2.85$$

with $\alpha = 0.01$, $z_\alpha = z_{0.01} = 2.33$ from Table B.3, Appendix B. The decision rule is:

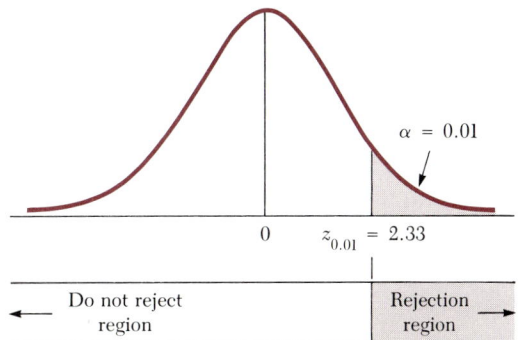

$$\alpha = 0.01$$

$$0 \qquad z_{0.01} = 2.33$$

Do not reject region	Rejection region

Since $z = 2.85 > z_{0.01} = 2.33$, we would reject H_0: $\pi_1 - \pi_2 \leq 0.20$ and conclude H_A: $\pi_1 - \pi_2 > 0.20$. Therefore the marketing director's claim is justified by these sample data. Thus the proportion of beer drinkers who express a preference for Poor's beer in the west region is more than 0.20 higher than the proportion who express preference for Poor's beer in the southwestern region.

The p-value for this test is:

$$p\text{-value} = P[\mathbf{Z} > z'] = P[\mathbf{Z} > 2.85] = 0.0022$$

Therefore the level of significance for the test, α, would have to be changed from $\alpha = 0.01$ to $\alpha = 0.0022$ (the p-value) before the null hypothesis cannot be rejected. By rejecting H_0, there is only a 0.0022 probability of getting a test statistic value of $z' = 2.85$ or larger given H_0 is true. Therefore, her conclusion that $\pi_1 - \pi_2 > 0.20$ is strongly supported by the sample data.

■ 11.5

Hypothesis testing concerning two population variances σ_1^2 and σ_2^2 (advanced section)

In this section, hypothesis testing concerning comparisons between two population variances is presented. No confidence interval estimates are given for these comparisons. For confidence interval estimate formulas, see the Ostle and Mensing reference at the end of this chapter.

We will consider hypothesis tests about σ_1^2 and σ_2^2 in three forms:

$$H_0: \quad \sigma_1^2 = \sigma_2^2 \qquad H_0: \quad \sigma_1^2 \geq \sigma_2^2 \qquad H_0: \quad \sigma_1^2 \leq \sigma_2^2$$
$$H_A: \quad \sigma_1^2 \neq \sigma_2^2 \qquad H_A: \quad \sigma_1^2 < \sigma_2^2 \qquad H_A: \quad \sigma_1^2 > \sigma_2^2$$

In each hypothesis statement, if we divide both sides of the statement by σ_2^2, we can rewrite these hypotheses as:

$$H_0: \quad \sigma_1^2/\sigma_2^2 = 1 \qquad H_0: \quad \sigma_1^2/\sigma_2^2 \geq 1 \qquad H_0: \quad \sigma_1^2/\sigma_2^2 \leq 1$$
$$H_A: \quad \sigma_1^2/\sigma_2^2 \neq 1 \qquad H_A: \quad \sigma_1^2/\sigma_2^2 < 1 \qquad H_A: \quad \sigma_1^2/\sigma_2^2 > 1$$

These statements of the hypotheses suggest the test statistic form:

$$\mathbf{F} = \mathbf{S}_1^2/\mathbf{S}_2^2$$

That is, the test statistic is the ratio of the sample variances $\mathbf{S}_1^2$ and $\mathbf{S}_2^2$. The test statistic has a sampling distribution known as the F-distribution when the following two assumptions are satisfied:

1. The two population random variables $\mathbf{X}_1$ and $\mathbf{X}_2$ are normally distributed.

2. The two samples are random samples that have been drawn independently of one another.

If these assumptions hold, then $\mathbf{F} = \mathbf{S}_1^2/\mathbf{S}_2^2$ follows the F-distribution shown in Figure 11.8. The F-distribution is skewed to the right with values that range from 0 to $+\infty$. There are two sets of degrees of freedom associated with F; the numerator degrees of freedom, denoted by r_1, and the denominator degrees of freedom, denoted by r_2. For the statistic

$$\mathbf{F} = \mathbf{S}_1^2/\mathbf{S}_2^2$$

FIGURE 11.8 | F-distribution

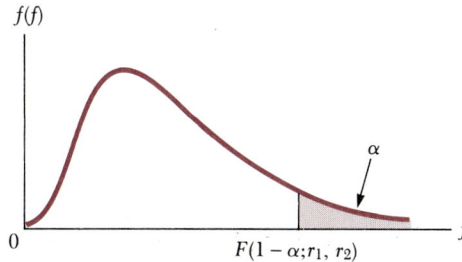

$r_1 = (n_1 - 1)$, the degrees of freedom associated with S_1^2, and $r_2 = (n_2 - 1)$, the degrees of freedom associated with S_2^2. A point on the F-distribution is denoted by $F(1 - \alpha; r_1, r_2)$ as shown in Figure 11.8. An area of $1 - \alpha$ is to the left of $F(1 - \alpha; r_1, r_2)$ and an area of α is to the right of $F(1 - \alpha; r_1, r_2)$.

Table B.7 in Appendix B gives the cumulative density function for the F-distribution, denoted by $F(1 - \alpha; r_1, r_2)$, where $F(1 - \alpha; r_1, r_2)$ is given by

$$P[\mathbf{F} \leq F(1 - \alpha; r_1, r_2)] = 1 - \alpha$$

For example, if $n_1 = 6$ and $n_2 = 5$ with $\alpha = 0.05$, the value of $F(1 - \alpha; r_1, r_2)$ $= F(1 - 0.05; r_1 = n_1 - 1, r_2 = n_2 - 1) = F(0.95; 5, 4) = 6.26$ is illustrated in Figure 11.9.

FIGURE 11.9 | $F[1 - \alpha; r_1 = (n_1 - 1), r_2 = (n_2 - 1)]$ on the F-distribution with $n_1 - 1 = 5$ and $n_2 - 1 = 4$; $\alpha = 0.05$.

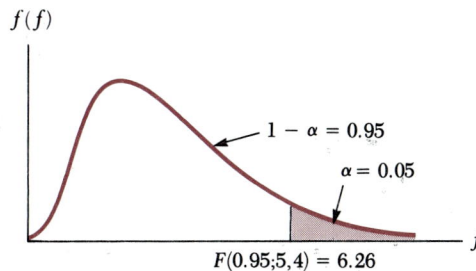

Table B.7 gives F-values for $r_1, r_2 = 1$ to 10, 12, 15, 20, 24, 30, 60, 120, and ∞, and for $1 - \alpha = 0.5, 0.9, 0.95, 0.975, 0.99, 0.995$, and 0.999.

For values of α greater than 0.50, the required probability can be determined by using the relationship,

$$F(1 - \alpha; r_1, r_2) = \frac{1}{F(\alpha; r_2, r_1)}$$

For example, suppose we wish to find $F(1 - 0.95; 5, 4)$—the lower-tail probability such that $P[\mathbf{F} < F(0.05; 5, 4)] = 0.05$. From the above relationship,

$$F(0.05; 5, 4) = \frac{1}{F(0.95; 4, 5)} = \frac{1}{5.19} = 0.1927$$

Notice in this computation that the degrees of freedom are switched when using the relationship to find lower-tail probabilities.

Having to use the lower-tail critical point on the F-distribution can be avoided in the two-sided test by placing the larger sample variance in the numerator and the smaller sample variance in the denominator of the test statistic:

$H_0: \quad \sigma_1^2 = \sigma_2^2$
$H_A: \quad \sigma_1^2 \neq \sigma_2^2$

$$\text{Test statistic value: } f = \frac{\text{Larger sample variance}}{\text{Smaller sample variance}}$$

The null hypothesis is rejected if f exceeds the upper-tail action point on the F-distribution given by $F(1 - \alpha/2; r_1, r_2)$, where r_1 is one less than the sample size of the sample corresponding to the larger sample variance, and r_2 is one less than the sample size of the sample corresponding to the smaller sample variance. The test statistic values and the decision rules for the three forms of the null hypothesis are given on page 563. Notice in the one-sided tests that the test statistics are different.

Example 11.15 A company is considering two machines that produce ½-inch ball bearings used in wheel assemblies. It is extremely important that the diameter of these bearings does not vary much from the designed diameter of 0.5 inch. A random sample of 25 bearings is taken from the production lot of machine 1, a sample of 21 bearings is taken from the production lot of machine 2, and the standard deviation of the diameters is calculated in each sample. The sample data are:

Machine 1	Machine 2
$n_1 = 25$ bearings	$n_2 = 21$ bearings
$s_1 = 0.001$ inch	$s_2 = 0.0018$ inch

Test the hypothesis that the two population variances, σ_1^2 and σ_2^2, are equal at the $\alpha = 0.05$ level of significance.

Tests for two population variances

Comparison of two population variances σ_1^2/σ_2^2

Hypothesis test

Hypothesis:	$H_0: \sigma_1^2 \le \sigma_2^2$ $H_A: \sigma_1^2 > \sigma_2^2$	$H_0: \sigma_1^2 = \sigma_2^2$ $H_A: \sigma_1^2 \ne \sigma_2^2$	$H_0: \sigma_1^2 \ge \sigma_2^2$ $H_A: \sigma_1^2 < \sigma_2^2$
Test statistic value:	$f = s_1^2/s_2^2$	$f = \dfrac{\text{Larger sample variance}}{\text{Smaller sample variance}}$	$f = s_2^2/s_1^2$
	$F(1 - \alpha; r_1, r_2)$ $r_1 = n_1 - 1$ $r_2 = n_2 - 1$	$F(1 - \alpha/2; r_1, r_2)$ $r_1 = (n - 1)$ for sample with larger variance $r_2 = (n - 1)$ for sample with smaller variance	$F(1 - \alpha; r_1, r_2)$ $r_1 = n_2 - 1$ $r_2 = n_1 - 1$
	$p\text{-value} = P[F > f]$	$p\text{-value} = 2P[F > f]$	$p\text{-value} = P(F > f)$

Solution   The null and alternate hypotheses are:

$$H_0: \quad \sigma_1^2 = \sigma_2^2$$
$$H_A: \quad \sigma_1^2 \neq \sigma_2^2$$

Since $s_2^2 = (0.0018)^2 = 0.00000324$ is larger than $s_1^2 = (0.001)^2 = 0.000001$, the test statistic is $\mathbf{F} = \mathbf{S}_2^2/\mathbf{S}_1^2$, whose value is $f = (0.0018)^2/(0.001)^2 = 3.24$. Since $\alpha = 0.05$, $\alpha/2 = 0.025$, and the degrees of freedom are:

$$r_1 = (n_2 - 1) = (21 - 1) = 20 \qquad r_2 = (n_1 - 1) = (25 - 1) = 24$$

The action limit is $F(1 - \alpha/2; r_1, r_2) = F(0.975; 20, 24) = 2.33$ from Table B.7, Appendix B. The decision rule is shown below.

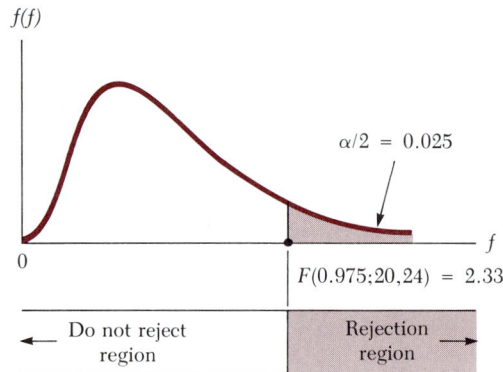

Since $f = 3.24 > F(0.975; 20, 24) = 2.33$, we reject the null hypothesis H_0: $\sigma_1^2 = \sigma_2^2$ and conclude that the two population variances are not equal.

As with the cases involving the t-distribution, the p-value can only be approximated when using the F-distribution and Table B.7. Noting from Table B.7 that $F(0.995; 20, 24) = 3.06$ and $F(0.999; 20, 24) = 3.87$, and that $f = 3.24$, we can say that the p-value is between $2(0.005) = 0.01$ and $2(0.001) = 0.002$.

If we are concerned about comparisons of population standard deviations, then we can use these tests for variances. For example, if the null hypothesis is H_0: $\sigma_1 = \sigma_2$, then this is equivalent to H_0: $\sigma_1^2 = \sigma_2^2$.

■ 11.6

Use of computer software packages

Consider the Example 11.2 problem wherein two drugs, drug A and drug B, are tested for the treatment of a specific disease. The number of days to recovery from the disease is recorded for each drug for 25 patients receiving drug A and for 25 patients receiving drug B. In Example 11.2 a 99 percent confidence interval is requested for the difference between the population means μ_1 and μ_2, where μ_1 = Average number of days to recovery using drug A, and μ_2 = Average number of days to recovery using drug B.

An interactive MINITAB session illustrates the computer analysis of this problem. The data are first set into columns. Column C1 contains the recovery days for drug A, and column C2 contains the recovery days for drug B.

```
MTB  > SET INTO C1
DATA > 16 22 31 25 20 22 18 27 32 24 24 21 22 20 29
DATA > 18 30 22 21 26 24 17 27 20 25
DATA > END
MTB  > SET INTO C2
DATA > 10 8 16 21 15 16 16 18 19 12 10 14 20 20 19
DATA > 16 18 16 20 14 17 13 18 22 15
DATA > END
```

The TWOSAMPLE command sets a confidence interval on $\mu_1 - \mu_2$ and also tests the null hypothesis that $\mu_1 - \mu_2 = D_0$. The POOLED subcommand indicates that the two sample standard deviations are to be pooled [case (2)].

```
MTB  > TWOSAMPLE 99% FOR DATA IN C1,C2;
SUBC > POOLED.

TWOSAMPLE T FOR C1 VS C2
          N      MEAN     STDEV    SE MEAN
C1        25     23.32    4.35     0.87
C2        25     16.12    3.60     0.72


99 PCT CI FOR MU C1 - MU C2: (4.17, 10.23)
TTEST MU C1 = MU C2 (VS NE): T=6.38 P=0.0000
DF=48.0
```

MINITAB prints out the mean, standard deviation (STDEV), and the standard error of the mean (SE MEAN) for each column. It then gives the 99 percent confidence interval estimate (4.17 to 10.23). It also conducts the case 2 hypothesis test of the equality of μ_1 and μ_2. The standardized t-statistic is $t = 6.38$ with 48 degrees of freedom. The p-value ($P = 0.0000$) is zero to four places. Thus we would certainly reject the null hypothesis and conclude that the mean recovery days for the two drugs differ—drug B has a statistically significant less mean recovery days than does drug A.

Example 11.16 It is possible to use MINITAB to do the paired t-test when the sampled observations are matched. Suppose that we consider a weight reduction program as in Example 11.10. We will change the data from Example 11.10, but the experiment is the same. The first observation of the pair is the weight before the program, and the second observation of the pair is the weight after the program is completed. The MINITAB READ command is used to read the 10 pairs into columns C1 and C2.

```
MTB  > READ INTO C1,C2
DATA > 124 121
DATA > 235 221
DATA > 267 256
DATA > 143 151
DATA > 132 126
DATA > 110 102
DATA > 233 216
DATA > 221 209
DATA > 165 147
DATA > 144 122
DATA > END
```

The 10 pairs are read into columns C1 and C2. Thus the first person began the program weighing 124 pounds and completed the program weighing 121 pounds.

```
MTB > SUBTRACT C2 FROM C1, PUT DIFFERENCE INTO C3
```

We then subtract column C2 from C1 to form the differences.

```
MTB > PRINT C1,C2,C3
  ROW    C1    C2    C3

    1   124   121     3
    2   235   221    14
    3   267   256    11
    4   143   151    -8
    5   132   126     6
    6   110   102     8
    7   233   216    17
    8   221   209    12
    9   165   147    18
   10   144   122    22
```

Columns C1, C2, C3 are printed to confirm that column C3 does contain the differences. Notice that the differences in column C3 represent *weight losses*. Thus the −8 for individual 4 is a negative loss, or a *gain* of 8 pounds for this individual!

```
MTB > TTEST ON DATA IN C3

TEST OF MU = 0 VS MU N.E. 0

          N      MEAN    STDEV  SE MEAN       T   P VALUE
C3       10     10.30     8.63      2.7    3.77    0.0044
```

We now ask MINITAB to conduct a *t*-test on the differences in column C3.

The null hypothesis in this case is H_0: $\mu_d = 0$; that is, the average difference is zero. MINITAB tells us that the mean difference (weight loss) is 10.30 pounds, the standard deviation of the differences is 8.63 pounds, and the standard error of the mean differences is 2.7 pounds. The standardized **t**-statistic value is $t = 3.77$. The *p*-value for the test is 0.0044. Thus α, the level of significance for the test, would have to be set at 0.0044 to arrive at the decision not to reject the null hypothesis, H_0: $\mu_d = 0$. For example, if α were set at 0.05, since *p*-value = 0.0044 < α = 0.05, we would reject the null hypothesis, H_0: $\mu_d = 0$.

This problem should be viewed as a one-sided test of course. The null and alternate hypotheses are:

H_0 : $\mu_d \leq 0$
H_A: $\mu_d > 0$

The program director wishes to reject H_0: $\mu_d \leq 0$ and conclude that H_A: $\mu_d > 0$ is correct; that is, μ_d = Average weight loss is greater than zero. MINITAB will not perform the one-sided test. However, we can use the two-sided test results for the null hypothesis, H_0: $\mu_d = 0$. In the two-sided test, the *p*-value is *twice* the probability in either the upper or the lower tail of the sampling distribution. Therefore, for a one-sided test, the *p*-value is *one half* the *p*-value for a two-sided test. Therefore, for the one-sided null hypothesis, H_0: $\mu_d \geq 0$, *p*-value = 0.0044/2 = 0.0022. Since the *p*-value is so small, we would certainly reject H_0 and conclude H_A: $\mu_d > 0$; that is, the program does produce a mean weight loss greater than zero.

■ **11.7**

Summary

Statistical hypothesis testing procedures are developed for testing the difference between population means (μ_1, μ_2) and population proportions (π_1, π_2), and for testing the equivalence of two population variances. In

SUMMARY TABLE | Hypothesis testing and confidence intervals for two populations

Population parameter	Hypothesis	Confidence interval formula	Test statistic value
μ_1, μ_2 (means)	$H_0: \ \mu_1 - \mu_2 = D_0$		
	Case 1: σ_1^2, σ_2^2 known	$(\bar{x}_1 - \bar{x}_2) \pm z_{\alpha/2} \sqrt{\dfrac{\sigma_1^2}{n_1} + \dfrac{\sigma_2^2}{n_2}}$	$z = \dfrac{(\bar{x}_1 - \bar{x}_2) - D_0}{\sqrt{(\sigma_1^2/n_1) + (\sigma_2^2/n_2)}}$
	Case 2: σ_1^2, σ_2^2 unknown $\sigma_1^2 = \sigma_2^2$	$(\bar{x}_1 - \bar{x}_2) \pm t_{\alpha/2;n_1+n_2-2}\, s_p \sqrt{\dfrac{1}{n_1} + \dfrac{1}{n_2}}$ $s_p = \sqrt{\dfrac{(n_1 - 1)s_1^2 + (n_2 - 1)s_2^2}{(n_1 - 1) + (n_2 - 1)}}$	$t = \dfrac{(\bar{x}_1 - \bar{x}_2) - D_0}{s_p \sqrt{\dfrac{1}{n_1} + \dfrac{1}{n_2}}}$ Degrees of freedom $= n_1 + n_2 - 2$
	Case 3: σ_1^2, σ_2^2 unknown $\sigma_1^2 \neq \sigma_2^2$	(approximate) $(\bar{x}_1 - \bar{x}_2) \pm t_{\alpha/2;\Delta} \sqrt{(s_1^2/n_1) + (s_2^2/n_2)}$ $\Delta = \dfrac{[(s_1^2/n_1) + (s_2^2/n_2)]^2}{\dfrac{(s_1^2/n_1)^2}{(n_1 - 1)} + \dfrac{(s_2^2/n_2)^2}{(n_2 - 1)}}$	$t' = \dfrac{(\bar{x}_1 - \bar{x}_2) - D_0}{\sqrt{(s_1^2/n_1) + (s_2^2/n_2)}}$ Degrees of freedom $= \Delta$
π_1, π_2 (proportions)	$H_0: \ \pi_1 - \pi_2 = D_0$	$(p_1 - p_2) \pm z_{\alpha/2} \sqrt{\dfrac{p_1(1 - p_1)}{n_1} + \dfrac{p_2(1 - p_2)}{n_2}}$	$z' = \dfrac{(p_1 - p_2) - D_0}{\sqrt{\dfrac{p_1(1 - p_1)}{n_1} + \dfrac{p_2(1 - p_2)}{n_2}}}$
σ_1^2, σ_2^2 (variances)	$H_0: \ \sigma_1^2 = \sigma_2^2$	(See Ostle & Mensing reference)	$f = \dfrac{\text{Larger sample variance}}{\text{Smaller sample variance}}$ Degrees of freedom: see Section 11.5

n_i = Size of ith sample; σ_i^2 = Variance of ith population; μ_i = Mean of ith population; $\bar{x}_i$ = Mean of the ith sample; s_i^2 = Variance of the ith sample; π_i = Proportion in the ith population; p_i = Proportion in the ith sample; z = Standard normal statistic value; t = **t** statistic value; f = **F** statistic value.

testing hypotheses concerning the difference between two population means, two different experimental designs may be used. The two samples drawn randomly from the two populations may be independent (the independent samples test) or dependent (the paired samples test). It is important to know the distinction between these two experimental designs and to know which design is appropriate in a given situation.

■ References

Elementary

Anderson, D. R.; D. J. Sweeney; and T. A. Williams. *Statistics for Business and Economics.* 2nd ed. St. Paul, Minn.: West Publishing, 1984, Chapter 10.

Conover, W. J., and R. L. Iman. *Introduction to Modern Business Statistics.* New York: John Wiley & Sons, 1983, Chapter 9.

Ryan, T.; B. Joiner; and B. Ryan. *MINITAB Student Handbook.* 2nd ed. Boston: Duxbury Press, 1985.

Wonnacott, T. H., and R. J. Wonnacott. *Introductory Statistics for Business and Economics.* 3rd ed. New York: John Wiley & Sons, 1984, Chapters 8 and 9.

Advanced

Fisher, R. A. "The Comparison of Samples with Possibly Unequal Variances," *Annals of Eugenics* 9 (1939), pp. 174–80.

Fisher, R. A., and F. Yates. *Statistical Tables.* 5th ed. (Tables VI, VI_1, and VI_2). Edinburgh, England: Oliver and Boyd, Publishers, 1957.

Ostle, B., and R. W. Mensing. *Statistics in Research.* 3rd ed., Ames: The Iowa State University Press, 1975, Chapter 6.

Satterthwaite, F. E. "On the Behrens-Fisher Problem," *Biometrics Bulletin* 2 (1946), p. 110.

Snedecor, G. W., and W. G. Cochran. *Statistical Methods.* 7th ed. Ames: The Iowa University Press, 1980, Chapter 6.

■ Problems

Section 11.1 Problems

11.1 Let $D = \mu_1 - \mu_2$.
  a. Are the hypotheses H_0: $\mu_1 \geq \mu_2$ and H_0: $D \geq 0$ equivalent?
  b. In terms of μ_1 and μ_2, restate the hypothesis H_0: $D \leq 2$.

11.2 If the random variables $\mathbf{X}$ and $\mathbf{Y}$ are independent, and $\sigma_\mathbf{X}^2 = 10$, $\sigma_\mathbf{Y}^2 = 25$, what is the variance of the difference $\mathbf{X} - \mathbf{Y}$? What do we mean in words when we say "$\mathbf{X}$ and $\mathbf{Y}$ are independent random variables"?

Section 11.2 Problems

11.3 In constructing confidence intervals on the difference $\mu_1 - \mu_2$, distinguish among the three cases concerning the variances, σ_1^2 and σ_2^2. Why is it necessary to define three cases for setting a confidence interval on $\mu_1 - \mu_2$?

11.4 Describe the sampling distribution of the statistic $\overline{\mathbf{X}}_1 - \overline{\mathbf{X}}_2$. What is its mean and variance? What is the approximate form of the distribution?

11.5 Two machines, A and B, are used to fill one-pound coffee cans with ground coffee. From lengthy experience with the machines, it is known that the standard deviation of the fills for both machines is 0.20 ounce ($\sigma_A = \sigma_B = 0.20$). But it is suspected that the mean fills for the two machines differ. A random sample of 25 cans taken from each machine produced the sample average fills, $\bar{x}_A = 15.96$ ounces and $\bar{x}_B = 15.67$ ounces. Assume that the fills are normally distributed.

a. Determine a 95 percent confidence interval for $\mu_A - \mu_B$.
b. At the $\alpha = 0.05$ level, test the null hypothesis that $\mu_A = \mu_B$.
c. What is the p-value for this test?

11.6 A market research analyst believes that items placed on shelves at or near eye level will sell more rapidly than those placed on the bottom shelf in a store. An experiment is set up where 10 weeks are randomly selected (during the first half of the year), and the product is placed at eye level during five of these weeks and on the bottom shelf during the remaining five weeks. The sales data in units sold per week are:

Bottom shelf	Eye-level shelf
33	41
35	40
32	45
38	42
44	48

Assuming that $\sigma_1^2 = \sigma_2^2$,
a. Set a 90 percent confidence interval on the difference between the average number of units sold per week at eye level and on the bottom shelf.
b. At $\alpha = 0.10$, are the means significantly different?
c. What is the p-value for this test?
d. What assumptions must you make in parts a and b to use the methods in this chapter? Are the assumptions valid in this case?

11.7 Random samples of two brands of whole milk are checked for the amount of fat content in grams. Twenty-five half-gallon containers of each brand are selected, and the fat content in grams is weighed. The data are given here.

Brand A			Brand B		
30	26	36	24	33	17
26	33	35	27	20	21
31	20	32	22	18	18
27	28	27	31	26	25
37	27	29	30	25	27
28	31	33	25	20	24
31	35	27	22	22	24
29	30	30	26	24	20
25			29		

a. Is this a paired experiment or an independent sample experiment? Explain.
b. Assuming $\sigma_A^2 = \sigma_B^2$, do the two brands have different average gram weight fat content? Use an $\alpha = 0.05$ level of significance.
c. What is the p-value for this test?

11.8 A car manufacturer is interested in testing the ability of two bumper designs to withstand low-speed crashes. Twenty-five samples of each type of bumper are installed on test cars, which are run into a brick wall at 10 m.p.h. An estimate of the cost to repair the damage is then made based on studying the impact in slow motion and carefully examining the car. Based on the experiment, the following sample statistics are computed:

	Mean	Standard deviation
Bumper type 1	$\bar{x}_1 = \$55$	$s_1 = \$10$
Bumper type 2	$\bar{x}_2 = \$70$	$s_2 = \$ 5$

a. Assuming that $\sigma_1^2 = \sigma_2^2$, find a 95 percent confidence interval on the difference $\mu_1 - \mu_2$. Is one bumper design significantly more effective than the other? Explain.

b. Is the assumption of equal population variances reasonable in this case? Comment without testing any hypotheses.

11.9 The braking distances at 60 MPH for two types of automobiles are compared. For the first make (A), 49 randomly selected cars were tested, and for the second make (B), 36 randomly selected cars were tested. The sample statistics are:

Make	Mean	Standard deviation
A	$\bar{x}_1 = 167$ ft.	$s_1 = 10$ ft.
B	$\bar{x}_2 = 180$ ft.	$s_2 = 20$ ft.

a. Is the difference between the mean stopping distances for the two cars significantly different at the $\alpha = 0.05$ significance level?

b. What is the *p*-value for this test?

11.10 A random sample of 25 engineers in company A produces a mean salary of $18,000 with a standard deviation of $3,000; and a random sample of 36 engineers in company B produces a mean salary of $22,000 with a standard deviation of $4,000.

a. Can we conclude that company B pays its engineers more than company A? Use an $\alpha = 0.05$ level of significance.

b. What is the *p*-value for this test?

11.11 The buying habits in two suburban communities located around a large city were surveyed. Of the 200 homemakers in the first suburb who were randomly sampled, it was found that they spent an average of $150 per week on food. In the second suburb, it was found that an average of $180 per week was spent on food among the 225 sampled homemakers. The standard deviation in the first suburb was found to be $25 and in the second suburb, $40. Determine a 99 percent confidence interval on the difference between the two population means.

11.12 A student is interested in graduate school education upon her graduation from an undergraduate business administration program. A consideration in her decision is the average

salary of the graduates of the particular degree program she has in mind. Among two candidate schools, she finds that the average salary among 10 sampled graduates from the first school is $20,000, with a standard deviation of $2,000; and the average salary among 10 sampled graduates from the second school is $24,000, with a standard deviation of $3,000. Assume that the salaries are normally distributed.

a. Formulate the null and alternate hypotheses.

b. Test the null hypothesis at the $\alpha = 0.05$ level of significance.

c. What is the *p*-value for this test?

11.13 Consider the following null and alternate hypotheses:

$$H_0: \quad \mu_1 - \mu_2 = 0$$
$$H_A: \quad \mu_1 - \mu_2 \neq 0$$

Samples of size $n_1 = 25$ and $n_2 = 25$ will be taken from each population. It is known that $\sigma_1 = 20$ and $\sigma_2 = 30$. Further, suppose it is known that the two population random variables X_1 and X_2 are normally distributed.

a. If the significance level is set at $\alpha = 0.05$, determine and describe the decision rule for this hypothesis test.

b. Suppose that when the samples are taken, it is found that $\bar{x}_1 - \bar{x}_2 = 15$. Based on the decision rule developed in part *a*, should the null hypothesis be rejected or not rejected? Explain.

c. What is the *p*-value for the test in part *b*?

Section 11.3 Problems

11.14 What are the consequences of analyzing a paired sample experiment as if it were an independent sample experiment?

11.15 The management wishes to make changes in the assembly process of a particular product. Workers are paid on the basis of their output, so it is necessary to demonstrate that the rate of assembling a unit will be increased under the new system. Ten workers are randomly selected to participate in an experiment in which each worker assembles one unit under the old process and one unit under the new process. The assembly times in minutes are:

Worker	Old process	New process
1	23	20
2	25	22
3	20	20
4	18	14
5	26	21
6	18	21
7	15	12
8	16	13
9	21	21
10	24	20

Assume that the assembly times are normally distributed.

a. Are the assembly times for the old and new processes independent samples? Explain.

b. Is the average assembly time for the new process faster than the average assembly time for the old process? Use an $\alpha = 0.05$ level of significance.

c. What is the p-value for this test?

11.16 In attempting to sell subscriptions to a local newspaper door to door, two different sales approaches are tried. Each of 20 sales representatives use the two approaches alternately for the same period and for the same number of household contacts. The sales data (number of subscriptions sold per period) are given here.

Salesman	Approach 1	Approach 2
1	10	8
2	15	12
3	26	24
4	13	15
5	18	10
6	10	10
7	16	20
8	21	16

Salesman	Approach 1	Approach 2
9	5	0
10	3	7
11	12	11
12	14	17
13	24	17
14	18	20
15	13	13
16	12	4
17	18	12
18	21	19
19	6	3
20	11	16

a. Is this a paired experiment or an independent sample experiment? Discuss.

b. Do the two sales approaches produce significantly different average sales? Use an $\alpha = 0.05$ level of significance.

c. What is the p-value for this test?

Section 11.4 Problems

11.17 Describe the sampling distribution of the statistic $P_1 - P_2$. What is its mean and variance? What is the approximate form of the distribution?

11.18 In one area, a random sample of 100 persons at grocery stores produces 60 who prefer Best soap to all others. In another area 75 out of 225 liked Best best.

a. Determine a 90 percent confidence interval on the difference between the proportions in the two areas who prefer Best.

b. What assumptions are required in part a?

11.19 It is suspected that the proportion of voters who prefer Henry Liberal in the Northeast is greater than the proportion who prefer him in the South. In two preelection polls, it is found that 100 of 150 prefer Henry to his competitor in the Northeast, and 75 of 150 prefer Henry in the South. Do these polls support the conjecture that Henry is more preferred in the Northeast than in the South?

a. Formulate the null and alternate hypotheses.

b. Test the null hypothesis at the $\alpha = 0.05$ level of significance.

c. What is the p-value for this test?

11.20 In a poll of workers in a large plant, 50 of 100 white-collar workers prefer the adoption of a new retirement plan, and 80 of 120 blue-collar workers prefer the new plan. Are the proportions favoring the new plan among the two groups significantly different?

  a. Formulate the null and alternate hypotheses.

  b. Test the null hypothesis at the $\alpha = 0.05$ level of significance.

  c. What is the p-value for this test?

11.21 The administration at a large state university has proposed to the faculty a new policy for promotion. It is believed that the proportion favoring the proposal among faculty who are not tenured is different from the proportion favoring the proposal among tenured faculty. In a sample of 64 faculty without tenure, it is found that only 20 percent favor the proposal; in a sample of 36 tenured faculty, 55 percent favor the proposal. Determine a 95 percent confidence interval on the difference between the two population proportions, assuming that the sample proportions are normally distributed. Interpret the resulting confidence interval.

11.22 Consider the following null and alternate hypotheses:

$$H_0: \quad \pi_1 - \pi_2 = 0$$
$$H_A: \quad \pi_1 - \pi_2 \neq 0$$

Samples of size $n_1 = 100$ and $n_2 = 100$ are planned.

  a. If the significance level is set at $\alpha = 0.05$, determine and describe the decision rule for the null hypothesis.

  b. Suppose that when the samples are taken it is found that $p_1 = 0.26$, and $p_2 = 0.31$. Based on the decision rule in part a, should the null hypothesis be accepted or rejected? Explain.

  c. What is the p-value for the test in part b?

Section 11.5 Problems

11.23 It is suspected that two machines that are set up to make ⅛-inch diameter ball bearings do so with different degrees of precision. A random sample of 50 bearings is taken from each machine, and the sample standard deviation is calculated. The sample statistics are:

$$n_1 = 50 \qquad n_2 = 50$$
$$s_1 = 0.001 \qquad s_2 = 0.004$$

Test the null hypothesis, $H_0: \quad \sigma_1^2 = \sigma_2^2$ at the $\alpha = 0.05$ level of significance.

11.24 Two machines fill bottles with catsup on an assembly line. It is suspected that the fills are more variable with machine A than with machine B. A random sample of 24 fills from A and a random sample of 18 fills from B produce the following data:

$$n_A = 24 \qquad n_B = 18$$
$$s_A = 0.5 \text{ oz.} \qquad s_B = 0.3 \text{ oz.}$$

Assume that the fills are normally distributed. Does machine A produce more variable fills than machine B? That is, test the null hypothesis,

$$H_0: \quad \sigma_A^2 \leq \sigma_B^2$$

against the alternative hypothesis

$$H_A: \quad \sigma_A^2 > \sigma_B^2$$

Use an $\alpha = 0.05$ level of significance.

11.25 In Problem 11.7 test the hypothesis that the two population variances are equal at the $\alpha = 0.05$ level of significance. In Problem 11.7 it was assumed that the two variances are equal. Was this assumption reasonable? What is the p-value for this test?

11.26 In Problem 11.8 it was assumed that the two population variances are equal. Test this assumption at the $\alpha = 0.05$ significance level. What is the p-value for this test?

Additional Problems

11.27 What is the Behrens-Fisher problem? How does this problem affect the construction of confidence interval estimates and hypothesis testing for the difference between two means?

11.28 When should the paired t-test of two means be used instead of the two-independent sample t-test of two means?

11.29 A 95 percent confidence interval is calculated for the difference between two means using the two-independent sample t-test. The confidence interval estimate of $\mu_1 - \mu_2$ is 8 to 20. If we were interested in testing the null hypothesis H_0: $\mu_1 = \mu_2$ at the $\alpha = 0.05$ level of significance, then what would be the decision (reject or not reject H_0) based on the confidence interval estimate? Explain.

11.30 It is hypothesized that the difference between two population proportions is 0.30. That is, the null hypothesis is H_0: $\pi_1 - \pi_2 = 0.30$. A 99 percent confidence interval estimate on the difference $\pi_1 - \pi_2$ is 0.26 to 0.31. Should the null hypothesis be rejected or not rejected based on the confidence interval estimate? Explain. What is the significance level of the test if the confidence interval estimate is used to accept or to reject H_0?

11.31 For each of the following inferences, carefully describe the assumptions required to use the methods developed in this chapter for hypothesis testing.

a. Comparison of two population means; independent samples.
b. Comparison of two population means; paired samples.
c. Comparison of two population proportions.
d. Comparison of two population variances.

11.32 If it is possible to use either a paired or independent sample experiment in a particular problem, why is a paired experiment preferred?

11.33 Consider the following null and alternate hypotheses:

$$H_0: \quad \mu_1 - \mu_2 \leq 0$$
$$H_A: \quad \mu_1 - \mu_2 > 0$$

Samples of size $n_1 = 25$ and $n_2 = 25$ are planned to test this null hypothesis. Suppose that it is known that $\sigma_1 = 10$ and $\sigma_2 = 20$. Further, suppose that the two population random variables X_1 and X_2 are normally distributed.

a. If the null hypothesis is to be tested at the $\alpha = 0.05$ level of significance, then determine and describe the decision rule for accepting or rejecting the null hypothesis.
b. Suppose when the samples are taken it is found that $\bar{x}_1 - \bar{x}_2 = 10$. Based on the decision rule developed in part a, should

the null hypothesis be rejected or not rejected? Explain.
c. What is the p-value for this test?

11.34 The Educational Testing Service (ETS) is concerned that the Graduate Management Admissions Test (GMAT) scores may differ for U.S. citizens and foreign students who take the test. In particular, they formulate the following null and alternate hypotheses:

$$H_0: \quad \mu_1 - \mu_2 \leq 0$$
$$H_A: \quad \mu_1 - \mu_2 > 0$$

where μ_1 is the population mean score of U.S. citizens taking the GMAT, and μ_2 is the population mean score of foreign students taking the exam. Independent random samples of size $n_1 = n_2 = 100$ are taken from both populations of test scores. The results are:

Sample	Size	Mean	Standard deviation
U.S. citizens	100	496	64
Foreign students	100	454	102

a. Using an $\alpha = 0.05$ level of significance, should the null hypothesis be rejected or not rejected?
b. What is the p-value for this test?

11.35 The Abson Test Preparation Company claims that students who take their Graduate Management Admissions Test (GMAT) preparation course can improve their test scores on the average by more than 50 points. To investigate this claim, the Educational Testing Service randomly samples $n = 20$ test takers who have (1) taken the test once, (2) attended the Abson course after taking the test, and (3) taken the test again. The table below gives the score results for the first time the test is taken and the repeat test after taking the Abson course for each of the 20 sampled individuals.

Individual	First test score	Second test score
1	386	420
2	298	350
3	454	510
4	323	361
5	410	400
6	482	536
7	505	585
8	433	531
9	601	644
10	333	432
11	475	500
12	422	496
13	348	376
14	415	510
15	380	444
16	404	410
17	560	589
18	310	350
19	466	515
20	355	423

a. Is this a paired t-test experiment or an independent samples experiment? Explain.

b. Using $\alpha = 0.05$, is the Abson claim supported by these data? Let μ_d = Mean increase in the test score. Then test the null hypothesis, H_0: $\mu_d \leq 50$ against the alternate hypothesis, H_A: $\mu_d > 50$ (the claim).

c. What is the p-value for this test?

d. Are the differences in test scores due only to the Abson course? How about the experience gained in taking the test the first time? Comment on Abson's use of the paired t-test to substantiate the company's claim.

11.36 Statistics are kept on the average number of robberies per month for the top 20 cities by population size.

a. Suppose city A claims that its average number of robberies per month is less than that of city B. State the null and alternate hypotheses for this claim.

b. Independent random samples of 25 months during the past five years are drawn from each city's records on robberies. It is found that $\bar{x}_A = 400$ per month, with a standard deviation of $s_A = 40$; and that $\bar{x}_B = 430$ per month, with a standard deviation of $s_B = 50$. Using an $\alpha = 0.05$ significance level, can city A's claim be supported by these data?

c. What is the p-value for the test in part b?

d. Is this a fair way to compare crime statistics between cities? What important factor has not been taken into consideration by using the average-number-of-robberies-per-month statistic?

11.37 The Rocker Company sells photocopiers for business office use. Recently a new sales technique has been proposed that promises to increase by 5 percent the proportion of people—among those contacted by a sales representative—who buy the copier. An experiment is devised wherein one set of sales representatives uses the old sales technique when contacting customers, and another set uses the new sales technique when contacting customers. A random sample of 100 sales contacts is taken from each set of sales representatives. The resulting proportions of sales for these contacts are:

$p_1 = 0.22$ (old sales technique)
$p_2 = 0.33$ (new sales technique)

a. Do these sample proportions support the claim made for the new sales technique? That is, test the null hypothesis,

H_0: $\pi_1 \geq \pi_2$, against the alternative hypothesis,
H_A: $\pi_1 < \pi_2$. Use $\alpha = 0.05$.

b. What is the p-value for this test?

11.38 The Poor's beer marketing manager in the Southwest region of the country suspects that the proportion of people who recognize the Poor's beer brand is different in the Texas cities of Dallas and Austin. She wishes to test the hypothesis that the two population proportions are equal, expecting that the sample data will lead to the rejection of the null hypothesis. Let π_D be the proportion of surveyed people in Dallas who recognize the

Poor's beer brand and π_A be the proportion of surveyed people in Austin who recognize the Poor's beer brand. One hundred people are surveyed in each city, and it is found that $p_D = 0.75$, and $p_A = 0.90$.

a. Formulate the null and alternate hypotheses.

b. Determine a 95 percent confidence interval estimate for the difference $\pi_D - \pi_A$.

c. Based on the confidence interval estimate in part a, should the null hypothesis be rejected or accepted? Explain.

11.39 The Best Advertising Agency claims that the proportion of people who recognize a brand name for a particular product will be raised by 0.25 when it runs its $1 million advertising campaign for the product. To check this claim, 100 people in the target market are randomly sampled by telephone before the campaign is run—20 recognize the brand name. After the campaign is run, another 100 people in the target market are sampled by telephone—61 people surveyed recognize the brand name.

a. Formulate the null and alternate hypotheses.

b. Test the null hypothesis at the $\alpha = 0.05$ level of significance.

c. What is the p-value for this test?

11.40 In a large corporation, the affirmative action officer believes that the proportion of men and the proportion of women who are promoted from manager trainees to assistant managers is the same. To investigate this claim, 25 files are randomly drawn from the personnel files for men and women who were manager trainees during the past year. It is found that 80 percent of the men were promoted and 60 percent of the women were promoted.

a. Determine a 99 percent confidence interval on the difference between the population proportions, $\pi_1 - \pi_2$.

b. Using the confidence interval estimate in part a, should the null hypothesis, H_0: $\pi_1 = \pi_2$ be accepted or rejected at the $\alpha = 0.01$ significance level?

11.41 The Alcon Lab, Inc. has developed a new medication for the treatment of a disease. To test the efficacy of the new medication, 100

patients are sampled who have the disease. Fifty of the 100 patients are randomly selected and are given the new medication. Of these 50 treated patients, 40 have improved conditions after taking the medication. The other group of 50, the control group, is given a placebo—a sugar pill. Twenty-five of these 50 have improved conditions after taking the placebo.

a. Is the medication more effective than the placebo in improving the condition of the patient after receiving it? Formulate the null and alternate hypotheses, and test the null hypothesis at the $\alpha = 0.05$ level of significance.

b. What is the p-value for this test?

11.42 The director of the loan division at the University Bank suspects that there is a difference between the proportion of loan defaults for two of his loan officers, Sue and Kathy. To investigate this claim, he takes a random sample of 120 loans made by Sue in the past two years and finds that 18 of them are classified as defaulted loans. In a random sample of 160 loans made by Kathy in the past two years, it is found that there are 15 loan defaults.

a. At the $\alpha = 0.05$ level of significance, is there a difference between the proportions of loan defaults for these two loan officers?

b. Determine a 95 percent confidence interval estimate on the difference in the proportions of defaulted loans for the two officers.

11.43 A new computerized checkout system has been proposed for use in the Roger's Grocery chain of supermarkets. To test the proposed system, the average checkout time is recorded for 100 customers checked out by the old system and 100 customers checked out by the new system. The results are:

Old system	New system
$n_1 = 100$	$n_2 = 100$
$\bar{x}_1 = 5.2$ minutes	$\bar{x}_2 = 4.4$ minutes
$s_1 = 2.1$ minutes	$s_2 = 1.5$ minutes

a. Is there a difference in the population mean checkout times for the two systems? Use an $\alpha = 0.05$ level of significance.

b. Determine a 95 percent confidence interval estimate for the difference in the two population means.

11.44 An expert in job satisfaction guarantees that his program for employees will reduce absenteeism significantly. The Beswick Corporation decides to try the expert's program. In a random sample of 100 employees taken before the two-day program is given, the average number of days of absence per month is found to be 1.24 days, with a standard deviation of 0.24 day. In a random sample of 100 employees after the program has been given, the average number of days of absence is found to be 0.87, with a standard deviation of 0.18 day.

a. The expert claims that the average number of days of absence will decrease after the employees have taken the program.

Formulate the null and the alternate hypotheses to test this claim.

b. Using an $\alpha = 0.05$ level of significance, can the claim be supported or rejected by these data?

11.45 The AllWorld Insurance Company claims that the proportion of individuals who file automobile insurance claims is much higher for the 16–20-year age bracket than it is for the 21–25-year age bracket. In a random sample of 100 individuals in the 16–20 age bracket who have insurance, it is found that 32 percent filed claims last year. In a random sample of 100 individuals in the 21–25 age bracket, it is found that 19 percent filed claims last year.

a. Formulate the null and alternate hypotheses to test the claim that the proportion of those filing for claims is higher in the 16–20 age bracket than it is in the 21–25 age bracket.

b. Test the null hypothesis at the $\alpha = 0.05$ level of significance.

12

Analysis of variance

■ **12.1**

Introduction

In this chapter we shall describe methods that may be used to compare two or more sample means and infer whether or not the corresponding population means are the same. To illustrate the kind of problem we will be addressing in this chapter, consider the following example.

Example 12.1 Farmco Investment Corporation has recently purchased a large chicken farm for the production of broilers intended for overseas markets. An animal breeding consultant engaged by Farmco has advised that three feeds are commonly used to ensure quality broilers and to produce rapid weight gains. He suggests conducting a statistical experiment to determine which feed produces the fastest weight gains, the factor considered most important by Farmco. Taking his advice, Farmco instructs its research staff to apply each feed to a randomly selected group of chicks and measure the weight gains over a specified interval of time. In a pilot experiment, it is decided to apply each feed to five randomly selected chicks from a large brood hatched at approximately the same time. The sample mean weight gains for each feed over the specified time interval will be compared to infer whether or not the corresponding population mean weight gains for the three feeds—Growbig, Max, and Topnotch—are equivalent.

The form of the sample data corresponding to this experiment is outlined in Table 12.1. For the ith feed, five responses (weight gains) x_{ij}, $j = 1, 2, \ldots, 5$, are recorded where $x_{ij} = $ weight gain in grams for the jth chick subjected to the ith feed. The mean weight gain corresponding to the ith feed is denoted by $\bar{x}_i$.

The three sample mean weight gains—$\bar{x}_{1.}$, $\bar{x}_{2.}$, and $\bar{x}_{3.}$—may be used to draw inferences about the equivalence of the three population means μ_1, μ_2,

TABLE 12.1 | Form of the sample data

	Feed			
i =	Growbig (1)	Max (2)	Topnotch (3)	j
	x_{11}	x_{21}	x_{31}	(1)
	x_{12}	x_{22}	x_{32}	(2)
	.	.	.	.
	.	.	.	.
	.	.	.	.
	x_{15}	x_{25}	x_{35}	(5)
Means	$\bar{x}_{1.}$	$\bar{x}_{2.}$	$\bar{x}_{3.}$	

x_{ij} = Weight gain for the jth chick subjected to the ith feed.
$\bar{x}_{i.}$ = Mean weight gain for the ith feed.

and μ_3. The ith population mean μ_i represents the average weight gain in grams for all chicks subjected to the ith feed under the selected experimental conditions. For example, the true relationship among the three population means *may* be as illustrated in Figure 12.1. In the figure, Max produces the greatest average weight gain, and Topnotch produces the least.

FIGURE 12.1 | Comparison of three population means

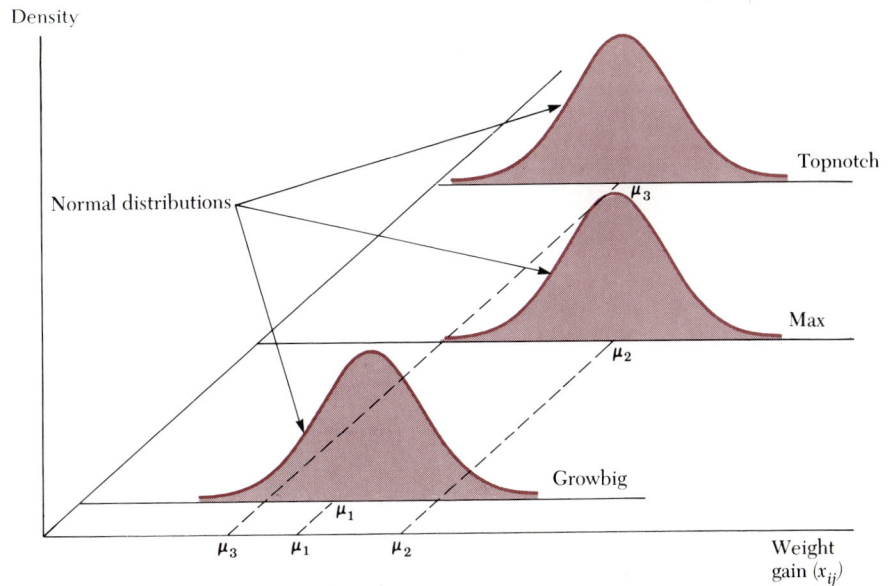

If the true relationship among the means is the one depicted in Figure 12.1, then the means are not equivalent, and we should reject the null hypothesis that $\mu_1 = \mu_2 = \mu_3$. The methods in this chapter provide a statistical procedure for determining whether or not the null hypothesis should be rejected based on the sample information—specifically the values of $\bar{x}_{1.}$, $\bar{x}_{2.}$, and $\bar{x}_3$.

The procedures developed in this chapter are called *analysis of variance* and are equivalent to the two-sided *t*-test in Chapter 11 when exactly two means are being compared. The name of the procedure, *analysis of variance,* may appear to be inappropriate because we wish to compare means, not variances. But, as we shall see, the name does accurately describe the procedure used to test the equivalence of two or more population means.

The analysis of variance procedure may be extended to study more than one factor in the same experiment. For example, Farmco may wish to study the effects of two or more kinds of vitamin supplements on chick weight gain in addition to the feed effects. We will discuss the methods of conducting a two-factor or multiple-factor experiment in this chapter. The application of analysis of variance to a two-factor (or multiple-factor) experiment will allow us to test the equivalence of the means associated with each factor simultaneously. In addition, we can determine whether the two factors significantly *interact*. For example, two vitamin supplements may produce approximately the same weight gain when used in conjunction with the feed Growbig, but they may produce quite dissimilar weight gains in conjunction with another feed, say Topnotch, as illustrated in Figure 12.2. In this instance, we say that the two factors, vitamin supplement and feed, *interact*.

Interaction of two factors

FIGURE 12.2 | Two-factor interaction effect between feeds and vitamin supplements

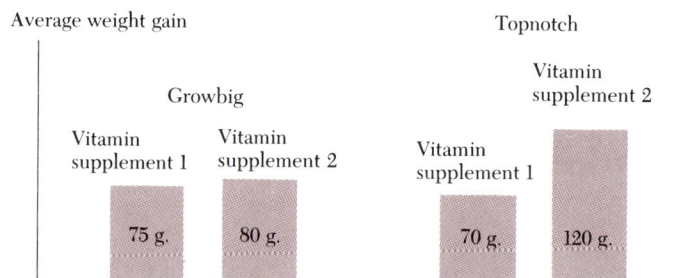

12.2

Assumptions of analysis of variance

The proper use of analysis of variance requires that certain assumptions about the nature of the data are satisfied.

The first assumption of analysis of variance is that the dependent (or response) variable is normally distributed in each of the populations being compared. To relate this assumption to Example 12.1, it requires that weight gain x_{ij} is a value of a normal distribution for each of the three feeds, $i = 1, 2,$ and 3, as illustrated in Figure 12.1.

Assumptions of
analysis of variance

The second assumption of analysis of variance is that the distributions of the dependent variable in each population have the same variance. This assumption is called *homogeneity of variance*.

If these two assumptions are satisfied, the distributions of the dependent variable for each feed in Example 12.1 must be normal and have a common variance. In Figure 12.1, notice that the distributions are bell-shaped and appear to have the same variability.

We will discuss further assumptions in the context of specific models used to compare, by analysis of variance, a set of population means. In Section 12.7, methods will be presented to check the assumptions of normality and homogeneity of variance. If the assumptions are not satisfied by the data, procedures to transform the dependent variable so that they conform to the assumptions will be discussed.

12.3

One-factor, completely randomized design

Let us now return to the discussion of Example 12.1. Suppose 15 chicks have been randomly selected from a large brood and have been randomly separated into three groups of five each. Each group is fed one of the three feeds, Growbig, Max, and Topnotch, over a fixed interval of time. At the end of the time, the weight gain of each chick is recorded. The gains are given in Table 12.2.

TABLE 12.2 | Weight gain of chicks in grams

	Feed		
	Growbig	Max	Topnotch
	42	112	70
	96	96	17
	81	88	49
	95	135	24
	76	119	40
Means:	78	110	40

These data conform to a *one-factor, completely randomized design*. The one factor is the feed given to the chicks. It is a completely randomized design because the experimental units—the 15 chicks—have been randomly assigned to the three kinds of feed.

The statistical hypothesis we wish to test is:

$H_0:$ $\mu_1 = \mu_2 = \mu_3$
$H_A:$ At least one mean is different from the others.

If the null hypothesis is true, then the three population mean weight gains are equivalent. Let μ denote the common population mean. In this case the three random samples of size five are drawn from one population with mean weight gain μ, and the model for the dependent random variable $\mathbf{X}_{ij}$ is specified by:

$$\mathbf{X}_{ij} = \mu + \mathbf{e}_{ij}; \qquad i = 1, 2, 3; \qquad j = 1, 2, 3, 4, 5$$

where $\mathbf{e}_{ij}$ is the error random variable component associated with the weight gain of the jth chick subjected to the ith feed.

From studying Table 12.2, it is apparent that the dependent random variable $\mathbf{X}_{ij}$ values are not randomly fluctuating about a common mean μ—the variation among the responses is too great to be entirely attributed to random sampling error of values drawn from one population. Rather, it *appears* more likely that each set of random samples of five per feed is drawn from a different population. This contention is strongly supported by the values of the three sample means, $\bar{x}_{1.} = 78$, $\bar{x}_{2.} = 110$, and $\bar{x}_{3.} = 40$; they *appear* to be too dissimilar to have come from one population.

FIGURE 12.3 | Relationship of the sample means to the null hypothesis

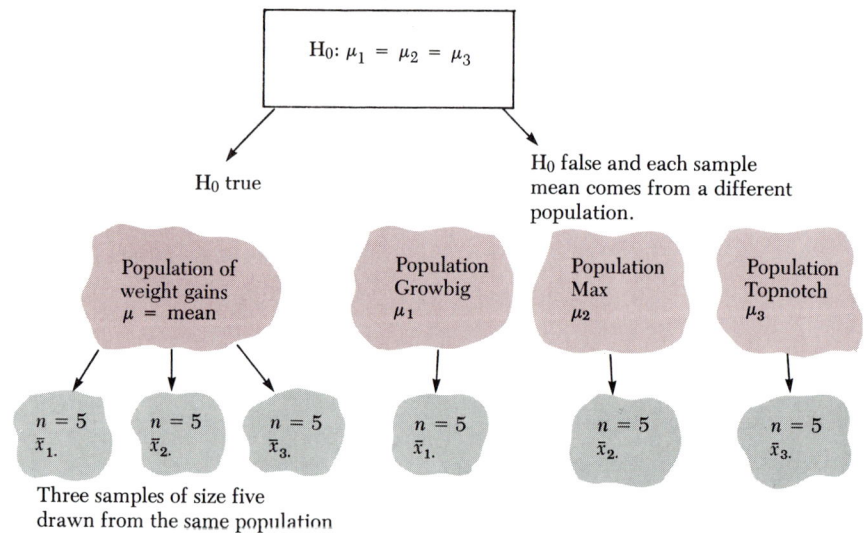

The relationship of the sample means to the null hypothesis is illustrated in Figure 12.3. If H_0 is true, then the three sample means have been drawn from the same population with mean weight gain μ. Any differences in weight gains are attributable to sampling variability. The null hypothesis may be false in a number of ways. In Figure 12.3 the sample means are each drawn from different populations; that is, the three population means μ_1, μ_2, and μ_3 are all different. It is of course possible that two of the population means are the same, and the third is different. For example, $\bar{x}_1$ and $\bar{x}_2$ may be drawn from the same population and $\bar{x}_3$ may be drawn from a second and different population. This would imply that the means μ_1 and μ_2 are equal, and the mean μ_3 differs from μ_1 and μ_2.

We will now summarize the one-factor, completely randomized design model.

One-factor, completely randomized design model

> **Definition 12.1**
> One-factor, completely randomized design model
>
> $$\mathbf{X}_{ij} = \mu_i + \varepsilon_{ij}; \quad i = 1, 2, \ldots, t; \quad j = 1, 2, \ldots, n$$
>
> where
>
> $\mathbf{X}_{ij}$ = Dependent response variable corresponding to the jth experimental unit subjected to the ith treatment.
> μ_i = Population mean response corresponding to the ith treatment (the ith population mean).
> ε_{ij} = Random error variable component associated with the response variable $\mathbf{X}_{ij}$.
>
> The (tn) experimental units are randomly selected and are randomly assigned in groups of size n to the t treatments. The null and alternate hypotheses are:
>
> H_0: $\mu_1 = \mu_2 = \cdots = \mu_t$
> H_A: At least one μ_i differs from the rest.

The error component ε_{ij} represents the difference between the response $\mathbf{X}_{ij}$ and the effect of the ith treatment mean μ_i:

$$\varepsilon_{ij} = \mathbf{X}_{ij} - \mu_i$$

The error component ε_{ij} thus explains whatever effects in the response variable $\mathbf{X}_{ij}$ that cannot be explained by the ith population treatment mean effect μ_i. For instance, in Example 12.1, consider the weight gain for the first chick ($j = 1$) subjected to the first treatment ($i = 1$), the Growbig feed. We can write this response variable as

$$\mathbf{X}_{11} = \mu_1 + \varepsilon_{11}$$

Whatever weight gain made by the first chick that is not attributable to the Growbig feed with mean effect μ_1 is measured by ε_{11}, the error component. Figure 12.4 shows the error component ε_{11} value, denoted by ε_{11}, and the response variable (weight gain) $\mathbf{X}_{11}$ value, denoted by x_{11}, on the distribution of all weight gains for chicks that are fed the Growbig feed.

FIGURE 12.4 | The response x_{11} and the error component value ε_{11} for the first chick that is fed the Growbig feed with mean weight gain μ_1.

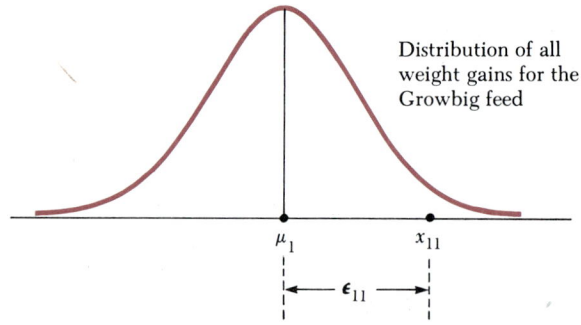

Distribution of all weight gains for the Growbig feed

μ_1 x_{11}

ε_{11}

12.3.1 | Assumptions: One-factor, completely randomized design

Assumptions of one-factor completely randomized design model

In Section 12.2 the requirements that the dependent random variable $\mathbf{X}_{ij}$ be normally distributed in the ith population and that the variance in each of the t populations be equal were presented as assumptions that must be met to use the analysis of variance procedure. These two assumptions plus the fact that $E(\mathbf{X}_{ij}) = \mu_i$, $i = 1, 2, \ldots, t$, can be stated more concisely in terms of the error component ε_{ij}. Specifically, we require that:

1. ε_{ij} is normally distributed for each value of i and j.

2. $E(\varepsilon_{ij}) = 0$ for each value of i and j.

3. $V(\varepsilon_{ij}) = \sigma^2$ for each value of i and j.

Homoscedasticity of variance

These three assumptions concerning the random variable ε_{ij} will ensure that $\mathbf{X}_{ij}$ is normally distributed and that the homoscedasticity (equal variances) assumption is satisfied.

To see that this is so, consider the model for $\mathbf{X}_{ij}$:

$$\mathbf{X}_{ij} = \mu_i + \varepsilon_{ij} = \text{Constant} + \varepsilon_{ij}$$

Since μ_i represents the ith population mean, it is a parameter and is therefore a constant. Assumptions placed on the response variable $\mathbf{X}_{ij}$ can be stated in terms of the error variable ε_{ij}, since $\mathbf{X}_{ij}$ and ε_{ij} are equal apart from a constant (μ_i). The mean and variance of $\mathbf{X}_{ij}$ can be expressed as follows:

$$E(\mathbf{X}_{ij}) = E(\mu_i + \varepsilon_{ij}) = E(\mu_i) + E(\varepsilon_{ij}) = \mu_i + 0 = \mu_i$$

since $E(\text{constant}) = \text{Constant}$, and $E(\varepsilon_{ij}) = 0$ by assumption; and

$$V(\mathbf{X}_{ij}) = V(\mu_i + \varepsilon_{ij}) = V(\varepsilon_{ij})$$

since $V(\text{constant}) = 0$. Also, if ε_{ij} is normally distributed, then so is X_{ij}, since X_{ij} and ε_{ij} are equal apart from a constant. And, if $V(\varepsilon_{ij}) = \sigma^2$, then $V(X_{ij}) = \sigma^2$ for each i and j.

Finally, we must assume that the error terms ε_{ij} are independently distributed random variables.

4. The error terms ε_{ij} are independent random variables.

This assumption requires that the error components in the model are independent from one another. The assumption is needed in constructing the hypothesis-testing methods that will follow. The process of randomly allocating the experimental units to the treatments is done to ensure that the errors are independent, so that no relationship exists among the errors that may favor one or more treatment responses.

12.3.2 | Analysis of variance procedure

The procedure for testing the null hypothesis,

$$H_0: \quad \mu_1 = \mu_2 = \cdot \cdot \cdot = \mu_t$$

is based on the following assumptions:

1. The response variable X_{ij} is normally distributed.

2. The variance of the response variable X_{ij} is σ^2 for $i = 1, 2, \ldots, t$ and $j = 1, 2, \ldots, n$.

The decision rule for rejecting or not rejecting the null hypothesis follows the ideas developed in Chapter 10. We will assume that the null hypothesis, $H_0: \quad \mu_1 = \mu_2 = \cdot \cdot \cdot = \mu_t$, is true. If the null hypothesis is true, then it is possible to develop two estimators of the population variance σ^2. By comparing the values of the two estimators of σ^2, we can decide if the null hypothesis seems reasonable (do not reject H_0) or not (reject H_0). We will use the Example 12.1 data on weight gains of chicks to illustrate the computation of the two estimates of σ^2.

Estimate number 1: The among treatments estimate of σ^2 If the null hypothesis is true, then each value of the response variable X_{ij} has been drawn from the *same* population with a mean value of μ ($\mu_1 = \mu_2 = \cdot \cdot \cdot = \mu_t = \mu$) and a variance of σ^2. The distribution of X_{ij} is illustrated in Figure 12.5. The t-treatment random samples each of size n are therefore drawn from the same population distribution. Figure 12.6 illustrates the sampling process for the Example 12.1 problem with $t = 3$ treatments (types of feeds) and $n = 5$ observations of weight gains for each of the treatments. In Figure 12.6 $\bar{x}_{1.}$, $\bar{x}_{2.}$, and $\bar{x}_{3.}$ represent the three sample means, each based on $n = 5$ sample observations, and s_1^2, s_2^2, and s_3^2 represent the three sample variances. Since each of the three samples has been drawn from the same population, we can combine the $tn = (3)(5)$ observations to produce an estimate of

FIGURE 12.5 | The population distribution of the random variable X_{ij}

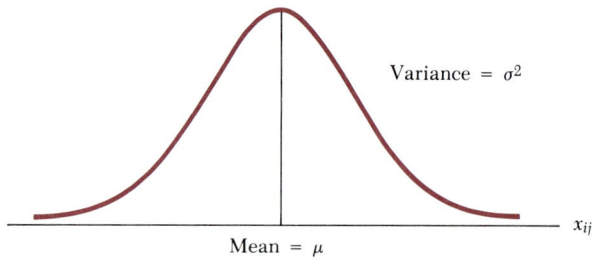

FIGURE 12.6 | Sampling from the same population assuming H_0 is true for the Example 12.1 problem

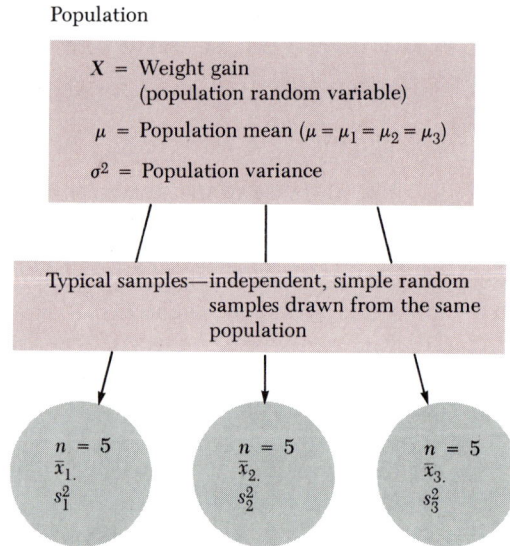

μ, the population mean. This overall sample mean is denoted by $\bar{x}_{..}$ and is given by

Overall sample mean

$$\bar{x}_{..} = \frac{\sum\limits_{i=1}^{t} \sum\limits_{j=1}^{n} x_{ij}}{tn}$$

Thus to calculate $\bar{x}_{..}$, we sum all the sample observations and divide by the total number of observations:

$$\bar{x}_{..} = \frac{(42 + 96 + 81 + 95 + \cdots + 49 + 24 + 40)}{(3)(5)} = \frac{1{,}140}{15} = 76$$

The average weight gain for all $tn = (3)(5) = 15$ chicks is 76 grams. Thus our estimate of the population mean weight gain μ is 76 grams.

Each sample of size $n = 5$ produces a sample mean $\bar{x}_{i.}$ that also estimates the population mean μ, assuming that the null hypothesis is true. The three sample means are:

Three sample
means

$$\bar{x}_{1.} = \frac{\sum\limits_{j=1}^{n} x_{1j}}{n} = \frac{(42 + 96 + 81 + 95 + 76)}{5} = 78$$

$$\bar{x}_{2.} = \frac{\sum\limits_{j=1}^{n} x_{2j}}{n} = \frac{(112 + 96 + 88 + 135 + 119)}{5} = 110$$

$$\bar{x}_{3.} = \frac{\sum\limits_{j=1}^{n} x_{3j}}{n} = \frac{(70 + 17 + 49 + 24 + 40)}{5} = 40$$

Each of these sample means is drawn from the same sampling distribution of the statistic $\overline{X}$, the sample mean, if the null hypothesis is true. From Chapter 8 the sampling distribution of the statistic $\overline{X}$ is shown in Figure 12.7.

FIGURE 12.7 | The sampling distribution of the statistic $\overline{X}$

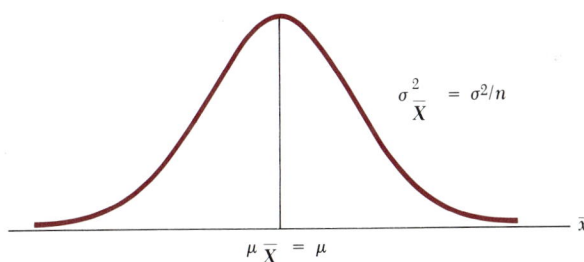

$$\sigma_{\overline{X}}^{2} = \sigma^{2}/n$$

$$\mu_{\overline{X}} = \mu$$

The variance of $\overline{X}$ is given by $\sigma_{\overline{X}}^{2} = \sigma^{2}/n$, where n is the sample size. To estimate the variance $\sigma_{\overline{X}}^{2}$, we can use the three sample means $\bar{x}_{1.}$, $\bar{x}_{2.}$, $\bar{x}_{3.}$ and the overall mean $\bar{x}_{..}$. The estimate, denoted by $s_{\overline{X}}^{2}$, is given by:

Estimate of the
variance $\sigma_{\overline{X}}^{2}$

$$s_{\overline{X}}^{2} = \frac{\sum\limits_{i=1}^{t} (\bar{x}_{i.} - \bar{x}_{..})^{2}}{t - 1}$$

From above, the three sample means and the overall mean are: $\bar{x}_{1.} = 78$, $\bar{x}_{2.} = 110$, $\bar{x}_{3.} = 40$, and $\bar{x}_{..} = 76$. Substituting these values into the equation for $s_{\overline{X}}^{2}$ produces:

$$s_{\bar{X}}^2 = \frac{(78 - 76)^2 + (110 - 76)^2 + (40 - 76)^2}{3 - 1} = \frac{2456}{2} = 1,228$$

From the sampling distribution of $\bar{X}$, we know that $\sigma_{\bar{X}}^2 = \sigma^2/n$. Solving this equation for σ^2 produces:

$$\sigma^2 = n\sigma_{\bar{X}}^2$$

where n is the sample size. Therefore, an estimate of σ^2 can be computed by multiplying n by the estimate of $\sigma_{\bar{X}}^2$, or

$$s^2 = ns_{\bar{X}}^2$$

Since $n = 5$, the estimate s^2 is

$$s^2 = (5)(1228) = 6,140$$

This represents our first estimate of σ^2.

The among treatment estimate of σ^2

$$s^2 = ns_{\bar{X}}^2 = \frac{n \sum_{i=1}^{t} (\bar{x}_{i.} - \bar{x}_{..})^2}{t - 1}$$

Estimate number 2: The within treatments estimate of σ^2 The second estimate of σ^2 does not assume that the null hypothesis, H_0, is true. It is based on the variation "within" each sample of size n. Referring to Figure 12.6, the sample variance s_i^2 can be computed for each sample of size n. The formula for the sample variance is:

$$s_i^2 = \frac{\sum_{j=1}^{n} (x_{ij} - \bar{x}_{i.})^2}{n - 1} = \frac{\sum_{j=1}^{n} x_{ij}^2 - \frac{\left(\sum_{j=1}^{n} x_{ij}\right)^2}{n}}{n - 1}$$

The second formula above is the computing formula for the sample variance presented in Chapter 2. Using the computing formula for the sample variance, the three sample variances are:

$$s_1^2 = \frac{(42)^2 + (96)^2 + \cdots + (76)^2 - \frac{(42 + 96 + \cdots + 76)^2}{5}}{5 - 1}$$

$$= \frac{32,342 - \frac{(390)^2}{5}}{4} = \frac{32,342 - 30,420}{4} = \frac{1,922}{4} = 480.5$$

$$s_2^2 = \frac{(112)^2 + (96)^2 + \cdots + (119)^2 - \dfrac{(112 + 96 + \cdots + 119)^2}{5}}{5 - 1}$$

$$= \frac{61,890 - \dfrac{(550)^2}{5}}{4} = \frac{61,890 - 60,500}{4} = \frac{1,390}{4} = 347.5$$

$$s_3^2 = \frac{(70)^2 + (17)^2 + \cdots + (40)^2 - \dfrac{(70 + 17 + \cdots + 40)^2}{5}}{5 - 1}$$

$$= \frac{9,766 - \dfrac{(200)^2}{5}}{4} = \frac{9,766 - 8,000}{4} = \frac{1,766}{4} = 441.5$$

The population variance σ^2 is assumed to be the same for all distributions of X_{ij}; therefore s_1^2, s_2^2, and s_3^2 represent three independent estimates of σ^2. Thus we can pool or combine the three estimates to produce an estimate of σ^2:

$$s^2 = \frac{\displaystyle\sum_{j=1}^{n} (x_{1j} - \bar{x}_{1.})^2 + \sum_{j=1}^{n} (x_{2j} - \bar{x}_{2.})^2 + \cdots + \sum_{j=1}^{n} (x_{tj} - \bar{x}_{t.})^2}{(n-1) + (n-1) + \cdots + (n-1)}$$

$$= \frac{\displaystyle\sum_{i=1}^{t} \sum_{j=1}^{n} (x_{ij} - \bar{x}_{i.})^2}{tn - n} = \frac{\displaystyle\sum_{i=1}^{t} \sum_{j=1}^{n} (x_{ij} - \bar{x}_{i.})^2}{t(n-1)}$$

The pooled estimate of σ^2 for the Example 12.1 data is:

$$s^2 = \frac{1,922 + 1,390 + 1,766}{3(5-1)} = \frac{5,078}{12} = 423.17$$

This represents our second estimate of σ^2.

The within treatments estimate of σ^2

$$s^2 = \frac{\displaystyle\sum_{i=1}^{t} \sum_{j=1}^{n} (x_{ij} - \bar{x}_{i.})^2}{t(n-1)}$$

We now have two estimates of σ^2: $s^2 = 6,140$, and $s^2 = 423.17$. The two estimates appear to be very different. What could explain this difference? To answer this question, we need to consider the expected (average) value of the two estimators of σ^2. It can be shown that

Expectations of the two estimators of σ^2

$$E\left[\frac{n\sum_{i=1}^{t}(\overline{\mathbf{X}}_{i.} - \overline{\mathbf{X}}_{..})^2}{t-1}\right] = \sigma^2 + \frac{n\sum_{i=1}^{t}(\mu_i - \mu)^2}{t-1}$$

Among treatments estimator

$$E\left[\frac{\sum_{i=1}^{t}\sum_{j=1}^{n}(\mathbf{X}_{ij} - \overline{\mathbf{X}}_{i.})^2}{t(n-1)}\right] = \sigma^2$$

Within treatments estimator

Notice that the expected value of the among treatments estimator of σ^2 is equal to σ^2 plus a function of the differences between the means μ_1, μ_2, ..., μ_t and the overall mean μ. If the null hypothesis, H_0: $\mu_1 = \mu_2 = \cdots = \mu_t$, is true, then $\mu_1 = \mu_2 = \cdots = \mu_t = \mu$, and this term drops out of the expectation. In this instance the among treatments estimator of σ^2 is an unbiased estimator. The within treatments estimator produces an unbiased estimator of σ^2, *whether or not the null hypothesis is true.*

The statistical test for the null hypothesis compares the *ratio* of the two estimators of σ^2. Let's define the two estimators of σ^2 as follows:

MSTR: Among treatments estimator of σ^2

$$\mathbf{MSTR} = \frac{n\sum_{i=1}^{t}(\overline{\mathbf{X}}_{i.} - \overline{\mathbf{X}}_{..})^2}{t-1}$$

Among treatments estimator

MSE: Within treatment estimator of σ^2

$$\mathbf{MSE} = \frac{\sum_{i=1}^{t}\sum_{j=1}^{n}(\mathbf{X}_{ij} - \overline{\mathbf{X}}_{i.})^2}{t(n-1)}$$

Within treatments estimator

MSTR stands for *Mean Square due to TReatments* and is commonly used to designate the among treatments estimator of σ^2. **MSE** stands for *Mean Square Error* and is used to designate the within treatments estimator of σ^2. The ratio of these two estimators, **MSTR/MSE,** is used to test the null hypothesis, H_0. If the null hypothesis is true, then the expected value of this ratio should be about 1:

$$E\left(\frac{\mathbf{MSTR}}{\mathbf{MSE}}\right) = \text{Approximately 1}$$

This ratio will not exactly equal one, because in general the expected value of a ratio is not equal to the ratio of expected values. If the expected value of the ratio **MSTR/MSE** is greater than one, then it must be due to the fact that not all treatment means μ_i are equal so that

$$E(\mathbf{MSTR}) = \sigma^2 + \frac{n\sum_{i=1}^{t}(\mu_i - \mu)^2}{t-1}$$

This is probably what is happening in the Example 12.1 case, since the value of **MSTR**, denoted by MSTR, is 6,140, and the value of **MSE**, denoted by

MSE, is 423.17. It is likely that MSTR is much bigger than MSE because the treatment means μ_1, μ_2, and μ_3 are not equal due to the contribution of the term

$$\frac{n \sum_{i=1}^{t} (\mu_i - \mu)^2}{t - 1}$$

to $E(\textbf{MSTR})$.

If the null hypothesis, H_0: $\mu_1 = \mu_2 = \cdots = \mu_t$, is true, then **MSTR** and **MSE** are two unbiased estimators of σ^2. The ratio of two independent sample variances follows an F-distribution as its sampling distribution (the F-distribution for the ratio of two independent sample variances was introduced in Chapter 11). If the assumptions of the analysis of variance procedure are satisfied by the data, then the statistic

| F-Statistic |

$$\textbf{F} = \textbf{MSTR/MSE}$$

follows an F-distribution with $r_1 = (t - 1)$ numerator degrees of freedom and $r_2 = t(n - 1)$ denominator degrees of freedom. The F-distribution is a skewed to the right distribution with a minimum value of 0 and a maximum value of $+\infty$. Since values of the **F**-statistic that are greater than 1 suggest that the null hypothesis should be rejected, the decision rule has a rejection region in the upper tail of the distribution. The critical value is denoted by $F(1 - \alpha; r_1, r_2)$ where α represents the probability of committing a Type I error (rejecting H_0 given it is true). The complete decision rule is:

| Decision rule for the null hypothesis: H_0: $\mu_1 = \mu_2 = \cdots = \mu_t$ |

Reject H_0 if $f = \text{MSTR/MSE} > F(1 - \alpha; r_1, r_2)$.
Do not reject H_0 if $f = \text{MSTR/MSE} \leq F(1 - \alpha; r_1, r_2)$.

An example of the F-distribution together with the decision rule is shown in Figure 12.8.

For the Example 12.1 data, we have:

$$f = \frac{\text{MSTR}}{\text{MSE}} = \frac{6,140}{423.17} = 14.51$$

Table B.7 in Appendix B gives the critical values of the **F**-statistic based on the numerator degrees of freedom r_1 and the denominator degrees of freedom r_2 for an area between 0 and $F(1 - \alpha; r_1, r_2)$ of $1 - \alpha$. Thus if we wish to set $\alpha = 0.05$, then $1 - \alpha = 0.95$ in looking up the critical **F**-value in Table B.7. The degrees of freedom are $r_1 = t - 1 = 3 - 1 = 2$ and $r_2 = t(n - 1) = 3(5 - 1) = 12$. From Table B.7, $F(0.95; 2, 12) = 3.89$. The decision rule is:

FIGURE 12.8 | The decision rule on the F-distribution for the null hypothesis, H_0: $\mu_1 = \mu_2 = \cdots = \mu_t$

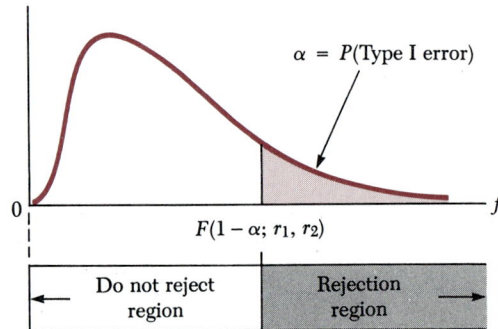

Reject H_0 if $f > 3.89$.
Do not reject H_0 if $f \leq 3.89$.

Since $f = 14.51 > 3.89$, we reject H_0: $\mu_1 = \mu_2 = \mu_3$ and conclude that at least one of these means is different than the others.

It is possible to calculate the p-value for this test by using the methods developed in Chapter 10. Thus we desire to determine:

$$p\text{-value} = P(\mathbf{F} > 14.51/r_1 = 2, r_2 = 12)$$
$$= 1 - P(\mathbf{F} < 14.51/r_1 = 2, r_2 = 12)$$

From Table B.7 in Appendix B, $P(\mathbf{F} < 13/r_1 = 2, r_2 = 12) = 0.999$. Thus we can say that the p-value for this test is less than 0.001 (from a computer-assisted computation, the p-value turns out to be 0.0006). Thus it is *very* likely that the three feeds do produce different average weight gains.

It should be apparent now why this procedure is called *analysis of variance*. The analysis of the two estimates of the variance σ^2 is used to decide whether or not to reject the null hypothesis, H_0: $\mu_1 = \mu_2 = \cdots \mu_t$. In the next subsection, we show how this procedure can be summarized in a tabular format, called the *analysis of variance table,* or ANOVA table for short.

| Calculation of the p-value |

12.3.3 | Analysis of variance table

A convenient computational format for calculating the statistics necessary to determine whether the null hypothesis should be rejected is provided by the *analysis of variance table*. Its form is presented in Table 12.3.

TABLE 12.3 | Analysis of variance table

Source of variation	Degrees of freedom	Sum of squares	Mean square	F-ratio
Among treatments	$t - 1$	SSTR	$\text{MSTR} = \dfrac{\text{SSTR}}{t - 1}$	$f = \dfrac{\text{MSTR}}{\text{MSE}}$
Experimental error (within treatments)	$t(n - 1)$	SSE	$\text{MSE} = \dfrac{\text{SSE}}{t(n - 1)}$	
Total	$tn - 1$	SST		

The first row of the table, "Among treatments," produces the first estimate of the variance σ^2—it is denoted by MSTR (mean square for treatments) in the table. The second row of the table, "Experimental error," produces the pooled estimate of the population variance σ^2—it is denoted by MSE (mean square for error) in the table. The **F**-statistic is the ratio of the mean square for treatments, MSTR, and the mean square for error, MSE.

The formula for MSTR is:

$$\text{MSTR} = \frac{n \sum_{i=1}^{t} (\bar{x}_{i.} - \bar{x}_{..})^2}{t - 1}$$

The numerator of MSTR is called the *sum of squares among treatments* and is denoted by SSTR in the table. The formula for MSE is:

$$\text{MSE} = \frac{(n - 1) \sum_{i=1}^{t} s_i^2}{t(n - 1)} = \frac{\sum_{i=1}^{t} \sum_{j=1}^{n} (x_{ij} - \bar{x}_{i.})^2}{t(n - 1)}$$

The numerator of MSE is called the *error sum of squares* and is denoted by SSE in the table. The "Degrees of freedom" column in the table gives the appropriate divisors of the sums of squares to produce the mean squares.

The last row in the table gives the "Total degrees of freedom"—$(tn - 1) = (t - 1) + t(n - 1)$—and the total sum of squares given by:

$$\text{SST} = \sum_{i=1}^{t} \sum_{j=1}^{n} (x_{ij} - \bar{x}_{..})^2$$

Notice that SST gives the sum of the squared deviations of each observation x_{ij} about the grand mean of the data $\bar{x}_{..}$. Thus this quantity is a measure of the total variability in the dependent variable.

An important relationship given in the analysis of variance table is:

$$SST = SSTR + SSE$$

Partitioning of the total sum of squares: SST = SSTR + SSE

$$\sum_{i=1}^{t} \sum_{j=1}^{n} (x_{ij} - \bar{x}_{..})^2 = n \sum_{i=1}^{t} (\bar{x}_{i.} - \bar{x}_{..})^2 + \sum_{i=1}^{t} \sum_{j=1}^{n} (x_{ij} - \bar{x}_{i.})^2$$

$$\begin{pmatrix} \text{Total sum} \\ \text{of squares} \end{pmatrix} = \begin{pmatrix} \text{Sum of squares} \\ \text{due to} \\ \text{treatments} \end{pmatrix} + \begin{pmatrix} \text{Error sum} \\ \text{of squares} \end{pmatrix}$$

That is, the total sum of squares can be partitioned into the sum of squares due to treatments plus the error sum of squares.

The deviations of the observations about the grand mean ($\bar{x}_{..} = 76$) and about their respective treatment means ($\bar{x}_{1.} = 78$, $\bar{x}_{2.} = 110$, $\bar{x}_{3.} = 40$) are illustrated in Figure 12.9. The light red lines connecting each observation to the treatment mean give the deviations of the observations about these means—the sum of the squares of these distances produces SSE. Thus SSE is a measure of the sampling variability within each treatment group of observations. The reddish gray lines give the deviations of the treatment means about the grand mean—n times the sum of the squares of these distances produces SSTR.

In these data, it is clear that the treatment sum of squares accounts for a considerable amount of the total sum of squares (remember the sum of the squares of the reddish gray line deviations is multiplied by $n = 5$ to produce SSTR). After SSTR has been partitioned from SST, what is left is the sum of squared deviations of each observation about its respective group mean—this is the "within treatment" sum of squares.

Imagine now that the observations corresponding to treatments 2 and 3 are more closely clustered about the grand mean $\bar{x}_{..} = 76$ so that the values $\bar{x}_{2.}$ and $\bar{x}_{3.}$ are both close to 76. In this case the treatment deviations (the reddish gray lines) will be small. In fact, when we "average" the treatment sum of squares to produce the treatment mean square MSTR = SSTR/$(t - 1)$ and when we "average" the error sum of squares to produce the error mean square MSE = SSE/$t(n - 1)$, we should find that MSTR and MSE are approximately equal. This will produce an F-ratio of approximately 1 and will not allow us to reject the null hypothesis of equal treatment effects.

Let us now verify that by using the analysis of variance table format we produce the same value of the **F**-statistic as computed earlier.

Treatment sum of squares, SSTR

$$SSTR = n \sum_{i=1}^{t} (\bar{x}_{i.} - \bar{x}_{..})^2$$

$$= 5[(78 - 76)^2 + (110 - 76)^2 + (40 - 76)^2]$$
$$= 5(4 + 1{,}156 + 1{,}296) = 12{,}280$$

FIGURE 12.9 | Deviations of observations about the grand mean and group means, and deviations of group means about the grand mean

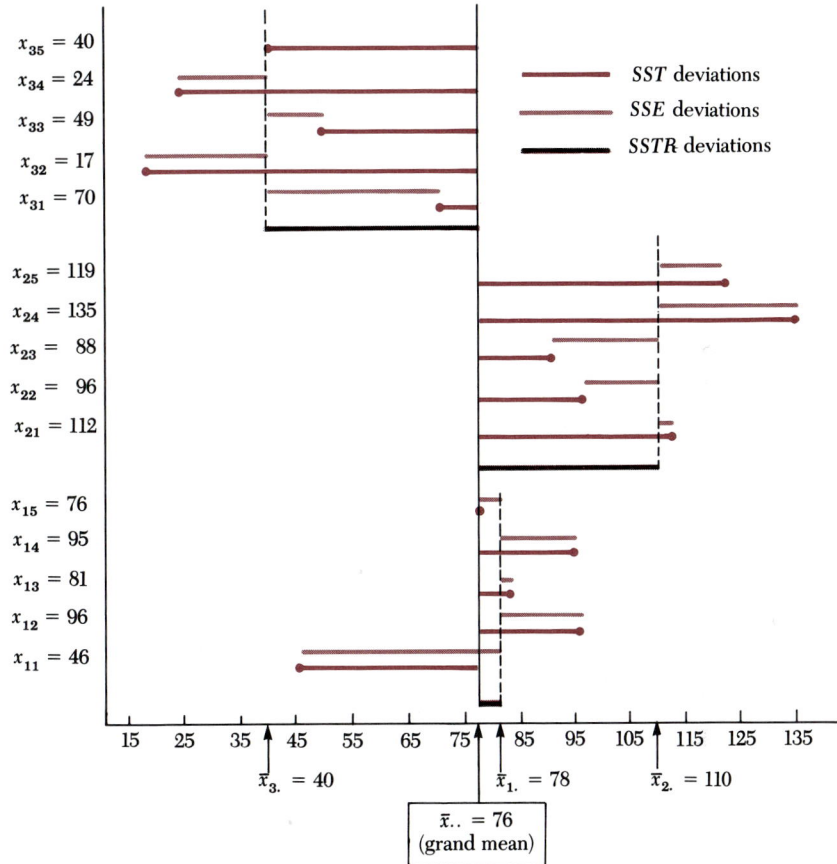

$x_{35} = 40$
$x_{34} = 24$
$x_{33} = 49$
$x_{32} = 17$
$x_{31} = 70$

$x_{25} = 119$
$x_{24} = 135$
$x_{23} = 88$
$x_{22} = 96$
$x_{21} = 112$

$x_{15} = 76$
$x_{14} = 95$
$x_{13} = 81$
$x_{12} = 96$
$x_{11} = 46$

SST deviations
SSE deviations
SSTR deviations

15 25 35 45 55 65 75 85 95 105 115 125 135

$\bar{x}_{3.} = 40$ $\bar{x}_{1.} = 78$ $\bar{x}_{2.} = 110$

$\bar{x}_{..} = 76$
(grand mean)

Total sum of squares, SST

$$SST = \sum_{i=1}^{t} \sum_{j=1}^{n} (x_{ij} - \bar{x}_{..})^2$$

$$= (42 - 76)^2 + (96 - 76)^2 + (81 - 76)^2 + \cdots + (40 - 76)^2$$

$$= 17{,}358$$

Error sum of squares, SSE

$$SST = SSTR + SSE; \qquad SSE = SST - SSTR$$

$$SSE = 17{,}358 - 12{,}280 = 5{,}078$$

Analysis of variance table

Source of variation	Degrees of freedom	Sum of squares	Mean square	F-ratio
Among treatments	$t - 1 = 2$	12,280	$\dfrac{12,280}{2} = 6,140$	$\dfrac{6,140}{423.17} = 14.51$
Experimental error	$t(n - 1) = 12$	5,078	$\dfrac{5,078}{12} = 423.17$	
Totals	$tn - 1 = 14$	17,358		

$F(0.95;\ 2,\ 12) = 3.89;\ 14.51 > 3.89;$ therefore reject H_0: $\mu_1 = \mu_2 = \mu_3$.

Table 12.4 was produced by using the SAS (Statistical Analysis System) analysis of variance program. The SAS software package contains many programs for performing statistical analyses. Notice that the SAS program prints out the p-value (PR > F); that is, $P(F > 14.51) = 0.0006$. Most computer programs written to perform statistical analyses now print the p-value so that interpolation in the statistical tables (e.g., the tables in Appendix B) is not required. We will use the SAS output for selected analysis of variance problems in the remainder of this chapter.

TABLE 12.4 | Computer analysis of Example 12.1 data

ANALYSIS OF VARIANCE PROCEDURE

DEPENDENT VARIABLE: YIELD

SOURCE	DF	SUM OF SQUARES	MEAN SQUARE	F VALUE	PR > F
MODEL	2	12280.000	6140.000	14.51	0.0006
ERROR	12	5078.000	423.167		
TOTAL	14	17358.000			

The analysis of variance table, along with the formulas for the sums of squares, is given in Table 12.5.

12.3.4 | Computational formulas for the sums of squares

The computation of SSTR and SST can be somewhat simplified by using the computational formulas:

Computational formulas for SSTR & SSE

$$\text{SSTR} = n \sum_{i=1}^{t} (\bar{x}_{i.} - \bar{x}_{..})^2 = \frac{1}{n} \sum_{i=1}^{t} x_{i.}^2 - C$$

$$\text{SST} = \sum_{i=1}^{t} \sum_{j=1}^{n} (x_{ij} - \bar{x}_{..})^2 = \sum_{i=1}^{t} \sum_{j=1}^{n} x_{ij}^2 - C$$

TABLE   12.5 | Analysis of variance table for a one-factor, completely randomized design with equal numbers of observations within each treatment group

Source of variation	Degrees of freedom	Sum of squares	Mean square	F-ratio
Among treatments	$t - 1$	$SSTR = n \sum\limits_{i=1}^{t} (\bar{x}_{i.} - \bar{x}_{..})^2$	$MSTR = \dfrac{SSTR}{t - 1}$	$f = \dfrac{MSTR}{MSE}$
Experimental error	$t(n - 1)$	$SSE = \sum\limits_{i=1}^{t} \sum\limits_{j=1}^{n} (x_{ij} - \bar{x}_{i.})^2$	$MSE = \dfrac{SSE}{t(n - 1)}$	
Total	$tn - 1$	$SST = \sum\limits_{i=1}^{t} \sum\limits_{j=1}^{n} (x_{ij} - \bar{x}_{..})^2$		

If $f > F[1 - \alpha; (t - 1), t(n - 1)]$, reject the null hypothesis H_0: $\mu_1 = \mu_2 = \cdots = \mu_t$.

$$p\text{-value} = P[\mathbf{F} > f/r_1 = t - 1, r_2 = t(n - 1)]$$

where

$$C = \frac{\left(\sum\limits_{i=1}^{t} \sum\limits_{j=1}^{n} x_{ij} \right)^2}{nt} \qquad \text{and} \qquad x_{i.} = \text{Total for the } i\text{th treatment}$$

Thus

$$C = \frac{(42 + 96 + \cdots + 40)^2}{(5)(3)} = \frac{(1,140)^2}{15} = 86,640$$

$$SSTR = \frac{1}{5} [(390)^2 + (550)^2 + (200)^2] - C$$

$$= \frac{494,600}{5} - 86,640 = 98,920 - 86,640 = 12,280$$

and,

$$SST = [(42)^2 + (96)^2 + \cdots + (40)^2] - C$$
$$= (1,764 + 9,216 + \cdots + 1,600) - C = 103,998 - 86,640 = 17,358$$

In this problem, the definitional forms of SSTR and SST given in Table 12.5 are easier to use than the computational forms because the sample means $\bar{x}_{1.}$, $\bar{x}_{2.}$, $\bar{x}_{3.}$, and $\bar{x}_{..}$ are whole numbers. If the sample means are fractional, the computational formulas for the sums of squares are usually easier to use.

In summary, the analysis of variance procedure tests the hypothesis of equal effects among t-treatments (population means) by analyzing the variation in the dependent (population) variable. If there are no differences among the treatment effects, then the treatment mean square (MSTR) and the error mean square (MSE) should independently provide unbiased estimates of the population variance. If the treatment effects differ, then MSE will still provide an unbiased estimate of the population variance, but MSTR will not. Its expected value will be larger than the population variance due to the differences among the treatment means.

12.3.5 | Unequal sample sizes

In the preceding example, the number of observations per treatment was five. Often we will not be so fortunate as to have a *balanced* experiment—one in which each treatment is applied to the same number of experimental units. In Example 12.1, one or more chicks may die before the time interval has elapsed, producing unequal sample sizes for the three treatments.

In the unequal sample size case, let n_i, where $i = 1, 2, \ldots, t$, denote the number of experimental units subjected to the ith treatment. The appropriate model for a one-factor, completely randomized design now becomes:

One-factor model with unequal sample sizes

$$\mathbf{X}_{ij} = \mu_i + \varepsilon_{ij}$$

$$i = 1, 2, \ldots, t; \quad j = 1, 2, \ldots, n_i$$

This model is identical to the one in the balanced case, except that it allows for unequal sample sizes ($j = 1, 2, \ldots, n_i$).

The formulas in the analysis of variance table do change from those given in Table 12.5 due to the unequal sample sizes. The appropriate analysis of variance table for this case is given in Table 12.6.

TABLE 12.6 | Analysis of variance table for a one-factor, completely randomized design with unequal numbers of observations within treatment groups

Source of variation	Degrees of freedom	Sum of squares	Mean square	F-ratio
Among treatments	$t - 1$	$\text{SSTR} = \sum_{i=1}^{t} n_i(\bar{x}_{i.} - \bar{x}_{..})^2$	$\text{MSTR} = \dfrac{\text{SSTR}}{t - 1}$	$f = \dfrac{\text{MSTR}}{\text{MSE}}$
Experimental error	$\sum_{i=1}^{t} (n_i - 1)$	$\text{SSE} = \sum_{i=1}^{t} \sum_{j=1}^{n_i} (x_{ij} - \bar{x}_{i.})^2$	$\text{MSE} = \dfrac{\text{SSE}}{\sum_{i=1}^{t} (n_i - 1)}$	
Total	$\left(\sum_{i=1}^{t} n_i\right) - 1$	$\text{SST} = \sum_{i=1}^{t} \sum_{j=1}^{n_i} (x_{ij} - \bar{x}_{..})^2$		

If $f > F[1 - \alpha; (t - 1), \sum_{i=1}^{t} (n_i - 1)]$, reject the null hypothesis H_0: $\mu_1 = \mu_2 = \cdots = \mu_t$

Example 12.2   A teacher is experimenting with a new way of teaching a statistical concept. To determine whether or not the new method produces better comprehension of the concept than does the standard procedure, he randomly selects five students from a large group who have similar abilities and mathematical backgrounds, subjects them to the new teaching method, and tests them for comprehension at the end of the teaching period. In another section, he subjects eight different students drawn from the same population of students to the standard teaching method and tests each for comprehension. The test score range is 0 to 100. The comprehension test scores for the two groups of students are given in the table.

	Treatment	
(1)	*(2)*	
New method	*Standard method*	
86	75	
91	90	
75	62	
80	65	
88	70	
	55	
	81	
	70	

$n_1 = 5 \quad n_2 = 8 \quad$ Grand mean $\bar{x}_{..} = 76$
$x_{1.} = 420 \quad x_{2.} = 568$
$\bar{x}_{1.} = 84 \quad \bar{x}_{2.} = 71$

Are the two teaching methods significantly different at the $\alpha = 0.05$ level?
   Solution   The null hypothesis is H_0: $\mu_1 = \mu_2$, where μ_1 is the population mean test score for the new method, and μ_2 is the population mean test score for the standard method. We now complete the analysis of variance table:

$$\text{SSTR} = \sum_{i=1}^{t} n_i(\bar{x}_{i.} - \bar{x}_{..})^2 = 5(84 - 76)^2 + 8(71 - 76)^2$$
$$= 320 + 200 = 520$$

$$\text{SST} = \sum_{i=1}^{t} \sum_{j=1}^{n_i} (x_{ij} - \bar{x}_{..})^2 = (86 - 76)^2 + (91 - 76)^2 + \cdots + (70 - 76)^2$$
$$= 1{,}538$$

$$\text{SSE} = \text{SST} - \text{SSTR} = 1{,}018$$

Source of variation	Degrees of freedom	Sum of squares	Mean square	F-ratio
Among treatments	$t - 1 = 1$	520	$\dfrac{520}{1} = 520$	$\dfrac{520}{92.54} = 5.62$
Experimental error	$\displaystyle\sum_{i=1}^{t} (n_i - 1) = 4 + 7 = 11$	1,018	$\dfrac{1{,}018}{11} = 92.54$	
Total	$\left(\displaystyle\sum_{i=1}^{t} n_i\right) - 1 = 5 + 8 - 1$ $= 12$	1,538		Critical F: $F(0.95;\ 1,\ 11) = 4.85$

Since $f = 5.62$ is greater than 4.85, we reject the null hypothesis H_0: $\mu_1 = \mu_2$; there is evidence to suggest that the new teaching method is significantly different from the standard method.

Note that the critical value $F(0.95;\ 1,\ 11)$ must be interpolated from Table B.7 in Appendix B, since the table gives only $F(0.95;\ 1,\ 10)$ and $F(0.95;\ 1, 12)$. Also, from Table B.7, it is apparent that the p-value lies between 0.05 and 0.025. Thus we would *not* reject the null hypothesis at the 0.01 significance level, for example.

Computational formulas for SSTR and SSE: Unequal sample sizes

Computational forms for the unbalanced case

$$\text{SSTR} = \sum_{i=1}^{t} n_i(\bar{x}_{i.} - \bar{x}_{..})^2 = \sum_{i=1}^{t} \frac{x_{i.}^2}{n_i} - C$$

$$\text{SST} = \sum_{i=1}^{t}\sum_{j=1}^{n_i} (x_{ij} - \bar{x}_{..})^2 = \sum_{i=1}^{t}\sum_{j=1}^{n_i} x_{ij}^2 - C$$

where

$$C = \frac{\left[\displaystyle\sum_{i=1}^{t}\sum_{j=1}^{n_i} x_{ij}\right]^2}{\displaystyle\sum_{i=1}^{t} n_i} \qquad \text{and} \qquad x_{i.} = \text{Total for } i\text{th treatment}$$

Applied to the data in Example 12.2, we obtain:

$$C = \frac{(86 + 91 + \cdots + 70)^2}{5 + 8} = 75{,}088$$

$$\text{SSTR} = \left[\frac{(420)^2}{5} + \frac{(568)^2}{8}\right] - C$$
$$= (35{,}280 + 40{,}328) - 75{,}088 = 520$$

$$\text{SST} = [(86)^2 + (91)^2 + \cdots + (70)^2] - C$$
$$= 76{,}626 - 75{,}088 = 1{,}538$$

12.3.6 | Relationship of the two-population *t*-test to analysis of variance

Since there are two treatments in Example 12.2, we should be able to perform the test of the null hypothesis, H_0: $\mu_1 = \mu_2$, where μ_i is the population average test score for the ith teaching method, by using the two-population independent sample t-test given in Section 11.2.

The t-test for the null hypothesis H_0: $\mu_1 = \mu_2$ is:

1. Compute

$$t = \frac{\bar{x}_1 - \bar{x}_2}{\sqrt{s_p^2 \left(\frac{1}{n_1} + \frac{1}{n_2}\right)}}$$

where $\bar{x}_i = i$th sample mean, $n_i = i$th sample size ($i = 1, 2$), and s_p^2 is the pooled estimate of the population variance given by:

$$s_p^2 = \frac{(n_1 - 1)s_1^2 + (n_2 - 1)s_2^2}{n_1 + n_2 - 2}$$

where $s_i^2 = i$th sample variance, $i = 1, 2$.

2. Reject H_0: $\mu_1 = \mu_2$ if $t > t_{\alpha/2; n_1 + n_2 - 2}$ or if $t < -t_{\alpha/2; n_1 + n_2 - 2}$. In this problem, $\bar{x}_1 = 84$, $\bar{x}_2 = 71$, $n_1 = 5$, and $n_2 = 8$.

$$s_1^2 = \frac{\sum_{i=1}^{n_1} (x_i - \bar{x}_1)^2}{n_1 - 1} = \frac{(86 - 84)^2 + (91 - 84)^2 + \cdots + (88 - 84)^2}{4}$$

$$= \frac{166}{4} = 41.5$$

$$s_2^2 = \frac{\sum_{i=1}^{n_2} (x_i - \bar{x}_2)^2}{n_2 - 1} = \frac{(75 - 71)^2 + (90 - 71)^2 + \cdots + (70 - 71)^2}{7}$$

$$= \frac{852}{7} = 121.714$$

$$s_p^2 = \frac{(n_1 - 1)s_1^2 + (n_2 - 1)s_2^2}{n_1 + n_2 - 2} = \frac{(4)(41.5) + (7)(121.714)}{5 + 8 - 2}$$

$$= \frac{166 + 852}{11} = \frac{1{,}018}{11} = 92.54$$

Now, we obtain

$$t = \frac{\bar{x}_1 - \bar{x}_2}{\sqrt{s_p^2 \left(\frac{1}{n_1} + \frac{1}{n_2}\right)}} = \frac{84 - 71}{\sqrt{92.54 \left(\frac{1}{5} + \frac{1}{8}\right)}} = \frac{13}{\sqrt{30.08}}$$

$$= \frac{13}{5.48} = 2.37$$

From Table B.5 in Appendix B, $t_{0.025;11} = 2.201$. Since $2.37 > 2.201$, we reject H_0: $\mu_1 = \mu_2$, the same conclusion we reached by analysis of variance.

Notice that $t^2 = f$; $(2.37)^2 = 5.62$. This certainly did not occur by chance. If we square the **t**-statistic with a value given by

$$t = \frac{\bar{x}_1 - \bar{x}_2}{\sqrt{s_p^2 \left(\dfrac{1}{n_1} + \dfrac{1}{n_2}\right)}}$$

it can be shown algebraically that $t^2 = \text{MSTR}/\text{MSE} = f$. Therefore, if we wish to test the equivalence of two population means, we can use either the t-test or analysis of variance. But if more than two population means are to be compared, analysis of variance must be used.

Table 12.7 contains a MINITAB interactive session for the Example 12.2 data. The SET commands are used to set the scores on the comprehensive test for the new teaching method and the standard teaching method in columns C1 and C2, respectively. The AOVONEWAY command is used to produce the analysis of variance table. This command also produces individual 95 percent confidence intervals based on the pooled standard deviation. The POOLED T command produces the pooled standard deviation t-test of the hypothesis, H_0: $\mu_1 = \mu_2$. Specifically, this command produces a confidence interval for the difference $\mu_1 - \mu_2$ between the two means (0.9 to 25.1), calculates the **t**-statistic value ($t = 2.37$), and calculates the p-value for the test ($p = 0.037$). Based on the p-value, $\alpha = P(\text{Type I error}) = P(\text{reject } H_0/H_0 \text{ is true})$ would have to be set at 0.037 to arrive at the decision not to reject H_0. Notice that the value of **t** ($t = 2.37$) when squared ($t^2 = 5.62$) equals the value of the **F**-statistic in the analysis of variance table ($f = 5.62$).

Although the null hypothesis, H_0: $\mu_1 = \mu_2$, is rejected at the $\alpha = 0.05$ level of significance, notice that the 95 percent confidence intervals printed by MINITAB overlap, suggesting that $\mu_1 - \mu_2$ may be zero. This apparent contradiction in the results is due to the fact that the 95 percent confidence intervals are calculated *individually* for the two population means, μ_1 and μ_2. Their *joint* confidence level is not 95 percent (it is higher), and the F-test correctly sets the joint confidence at 95 percent (confidence $= (1 - \alpha)100$ percent).

The one-factor, completely randomized design model can be written in alternate form to the one given in Definition 12.1. The null hypothesis for this model given in Definition 12.1 is:

$$H_0: \quad \mu_1 = \mu_2 = \cdots = \mu_t$$

where $t = $ Number of means to be compared. If the null hypothesis is true, then all t means are equal to the same mean value, say μ. If the t means differ, then we can specify the differences by introducing a new parameter, τ_i (Greek letter tau), where

$$\tau_i = \mu_i - \mu, \, i = 1, 2, \ldots, t$$

Thus τ_i measures the difference between the ith population mean μ_i and the overall mean μ. Since μ is the mean of the t population means $\mu_1, \mu_2, \ldots, \mu_t$, it must be true that

TABLE 12.7 | The MINITAB interactive session for the Example 12.2 problem

```
MTB > SET INTO C1
DATA> 86  91  75  80  88
DATA> END
MTB > SET INTO C2
DATA> 75  90  62  65  70  55  81  70
DATA> END
MTB > AOVONEWAY ON DATA IN C1,C2

ANALYSIS OF VARIANCE
SOURCE      DF          SS          MS          F
FACTOR       1        520.0       520.0       5.62
ERROR       11       1018.0        92.5
TOTAL       12       1538.0
                                INDIVIDUAL 95 PCT CI'S FOR MEAN
                                BASED ON POOLED STDEV
LEVEL    N      MEAN    STDEV   -------+---------+---------+---------
C1       5     84.00     6.44                    (--------*--------)
C2       8     71.00    11.03        (------*------)
                                -------+---------+---------+---------
POOLED STDEV =   9.62                 70        80        90
MTB > POOLED T FOR DATA IN C1 AND C2

TWOSAMPLE T FOR C1 VS C2
        N       MEAN     STDEV   SE MEAN
C1      5      84.00      6.44      2.9
C2      8      71.0      11.0       3.9

95 PCT CI FOR MU C1 - MU C2: (0.9, 25.1)
TTEST MU C1 = MU C2 (VS NE): T=2.37 P=0.037 DF=11.0

MTB > STOP
```

$$\sum_{i=1}^{t} \tau_i = \sum_{i=1}^{t} (\mu_i - \mu) = 0$$

Treatment effects

The τ_i parameters are often referred to as *treatment effects*, since each τ_i measures the deviation of the ith population mean μ_i from the overall mean μ. Solving the equation

$$\tau_i = \mu_i - \mu$$

for μ_i produces

$$\mu_i = \mu + \tau_i$$

Using this expression for μ_i, we can now define the one factor, completely randomized design using the treatment effects, τ_i. This alternate definition to Definition 12.1 is given below:

Definition 12.2
One factor, completely randomized design model

$$\mathbf{X}_{ij} = \mu + \tau_i + \boldsymbol{\varepsilon}_{ij}; \ i = 1, 2, \ldots, t; \ j = 1, 2, \ldots, n \ (\text{or } n_i),$$

where

$\mathbf{X}_{ij}$ = Dependent response random variable corresponding to the ith treatment and the jth experimental unit subjected to the ith treatment.

μ = The overall mean—the average of the t population means μ_1, μ_2, $\ldots, \mu_t$.

τ_i = ith treatment effect equal to the deviation of the ith population mean μ_i from the overall mean μ.

$\boldsymbol{\varepsilon}_{ij}$ = Random error component associated with the response random variable $\mathbf{X}_{ij}$.

In most applications of analysis of variance, Definition 12.2 is used. In the rest of this chapter, we will use the "treatment effects" approach for defining the remaining models that we will consider. However, it is important to remember that analysis of variance is essentially a procedure for comparing more than two population means as specified in Definition 12.1.

If we are testing the equivalence of t population means ($t \geq 2$) and, based on the analysis of variance, we reject the null hypothesis, we should be interested in the reason for this outcome. One or more population means must be different from the rest, but which one(s)? In Section 12.6 we shall discuss one method for analyzing the nature of the difference among a set of means. We now turn to a design that attempts to control sources of variation that may adversely affect our ability to determine by analysis of variance whether a set of treatment effects do or do not differ.

■ 12.4

The randomized block design

A randomized block design allocates treatments randomly to experimental units that have first been sorted into homogeneous groups called *blocks*. As an illustration of using this design, reconsider Example 12.2 in which two teaching methods are compared. The two teaching methods are randomly assigned to students selected from a large group with *similar abilities*. Ordinarily, students in a basic statistics course vary considerably in ability, and these differences, unless accounted for, may result in effects that are *con-*

Definition of blocks

founded with the treatment effects. For instance, by chance we may select students with less ability for the standard teaching method than those selected for the new teaching method. This may result in the new method "looking better" than the standard method when, in fact, it is not. If this were the case, we would say that student differences are *confounded* with the effect of the teaching method—on the basis of the students' performances on the comprehension test at the end of the term, it is not possible to determine whether any observed differences between the two groups are attributable to teaching methods alone.

To correct this situation, we could *block* on the abilities of the students. That is, we could sort the students into, say, three homogeneous ability groups (blocks) determined by IQ tests, grade point averages, a mathematical abilities test given the first day of class, or a combination of these ability measures. To each block we would randomly assign the two teaching methods to the students composing the block.

Thus an important reason for using a randomized block design is to remove a source of variation (by blocking) from the experimental error. The important consequence of blocking is therefore the reduction of the mean square error, MSE. In the randomized block design, the sum of squares due to the blocks (differences among student abilities) will be partitioned from the total sum of squares for the dependent variable:

$$SST = SSTR + SSB + SSE$$

If a completely randomized model were used in which the two teaching methods were randomly assigned to the students when differences in student abilities existed, then the sum of squares attributable to these differences (SSB) would be improperly ascribed to the experimental error (SSE).

The need for blocking in experimental designs occurs frequently; the following examples illustrate instances when blocking is required.

Example 12.3 To study the effectiveness of four types of automobile oil filters, an experiment is performed using a specific brand of oil. Twelve quarts of the oil are selected, and the same amount of impurities is added to each. Due to the nature of the experiment, only four tests can be performed per day. Since humidity and temperature differences may possibly affect the ability of a filter to trap impurities, it is decided to *block* on days. Designating the four filters by A, B, C, and D, the selected experimental design is:

Blocks (days)	Treatments (types of filters)			
1	A	B	D	C
2	C	A	B	D
3	D	B	A	C

Margin notes:

Confounded treatment effects

Blocking reduces MSE

Partitioning of total sum of squares for the randomized block design

The order in which the four filters are tested each day is randomly selected. For instance, on the first day, the filters are tested in the order A, B, D, and C.

In this design, any effects due to differences among days will be accounted for by blocking on days. The four treatments are randomly assigned to the experimental units (four cans of oil per day) within each block (day). Notice that each block represents a completely randomized design—each treatment is applied randomly to the experimental units in the block. Thus the blocks play the role of *replicates* of a completely randomized design.

Example 12.4 Frequently, subjects are treated as blocks in social science research. For example, a social scientist may be interested in the effects of five treatments on subjects. Let us suppose that four subjects are selected for the experiment. A possible experimental design is:

Blocks (subjects)	Treatments				
1	A	C	B	D	E
2	E	A	C	D	B
3	B	D	C	E	A
4	E	B	D	A	C

Designating the treatments by A, B, C, D, and E, the five treatments are assigned in random order to the four subjects. In this design the subjects serve as blocks, and the experimental units within each block are considered to be the different times that treatments may be applied to the subject.

Although this is a randomized block design, social scientists commonly refer to it as a *repeated measures design*—the treatment measures are repeated for each subject.

We now give the description of the randomized block design.

Repeated measures design

12.4.1 | Assumptions: Randomized block design

Assumptions of the randomized block design

1. The treatment and block effects are *fixed*.

2. The error components ε_{ij}, where $i = 1, 2, \ldots, t; j = 1, 2, \ldots, b$, are normally and independently distributed random variables with 0 means and a common variance σ^2.

3. There is no treatment by block interaction effect.

4. The design is balanced—each block contains the same number of experimental units subjected to the t treatments in equal numbers.

5. The treatment effects and the block effects sum to zero ($\Sigma \tau_i = 0$ and $\Sigma B_j = 0$).

Definition 12.3
One-factor, randomized block design

The model is specified by the linear equation:

$$\mathbf{X}_{ij} = \mu + \tau_i + B_j + \varepsilon_{ij}$$
$$i = 1, 2, \ldots, t; \qquad j = 1, 2, \ldots, b$$

where

$\mathbf{X}_{ij}$ = Response random variable corresponding to the ith treatment in the jth block.
μ = Overall population mean.
τ_i = Effect of the ith treatment.
B_j = Effect of the jth block.
ε_{ij} = Random error component associated with the response random variable $\mathbf{X}_{ij}$.

The t treatments are *randomly* assigned to the b experimental units in each block. The null and alternate hypotheses are:

$H_0:$ $\tau_1 = \tau_2 = \cdots = \tau_t$
$H_A:$ Not $\{\tau_1 = \tau_2 = \cdots = \tau_t\}$

12.4.2 | The analysis of variance table

The analysis of variance table for the one-factor, randomized block design is given in Table 12.8.

TABLE 12.8 | Analysis of variance table for one-factor, randomized block design with one experimental unit per treatment within a block

Source of variation	Degrees of freedom	Sum of squares	Mean square	F-ratio
Among treatments	$t - 1$	$\text{SSTR} = b \sum_{i=1}^{t} (\bar{x}_{i.} - \bar{x}_{..})^2$	$\text{MSTR} = \dfrac{\text{SSTR}}{t - 1}$	$f = \dfrac{\text{MSTR}}{\text{MSE}}$
Among blocks	$b - 1$	$\text{SSB} = t \sum_{j=1}^{b} (\bar{x}_{.j} - \bar{x}_{..})^2$		
Experimental error	$(t - 1)(b - 1)$	$\text{SSE} = \text{SST} - \text{SSTR} - \text{SSB}$	$\text{MSE} = \dfrac{\text{SSE}}{(t - 1)(b - 1)}$	
Total	$bt - 1$	$\text{SST} = \sum_{i=1}^{t} \sum_{j=1}^{b} (x_{ij} - \bar{x}_{..})^2$		

Critical F-value: $F[1 - \alpha; (t - 1), (t - 1)(b - 1)]$. If $f > F[1 - \alpha; (t - 1), (t - 1)(b - 1)]$, reject the null hypothesis $H_0:$ $\tau_1 = \tau_2 = \cdots = \tau_t$.

Example 12.5 Let us suppose that in the Example 12.3 problem, the following responses have been recorded:

Blocks (days)	Amounts of impurities trapped by filter			
1	17(A)	19(B)	14(D)	17(C)
2	18(C)	16(A)	21(B)	14(D)
3	17(D)	23(B)	18(A)	19(C)

Are the amounts of impurities trapped by the four filters significantly different at the $\alpha = 0.05$ level?

Solution The computation of the sums of squares for the analysis of variance table can be simplified somewhat by using the equivalent calculating forms to those given in Table 12.8.

Calculating formulas for sums of squares

Total sum of squares

$$\text{SST} = \sum_{i=1}^{t} \sum_{j=1}^{b} (x_{ij} - \bar{x}_{..})^2 = \sum_{i=1}^{t} \sum_{j=1}^{b} x_{ij}^2 - \frac{\left(\sum_{i=1}^{t} \sum_{j=1}^{b} x_{ij} \right)^2}{bt}$$

$$= (17)^2 + (19)^2 + \cdots + (19)^2 - \frac{[(17) + (19) + \cdots + (19)]^2}{(3)(4)}$$

$$= 3{,}855 - \frac{(213)^2}{12} = 3{,}855 - 3{,}780.75 = 74.25$$

Treatment sum of squares

$$\text{SSTR} = b \sum_{i=1}^{t} (\bar{x}_{i.} - \bar{x}_{..})^2 = \frac{\sum_{i=1}^{t} x_{i.}^2}{b} - \frac{\left(\sum_{i=1}^{t} \sum_{j=1}^{b} x_{ij} \right)^2}{bt}$$

Note that $x_{i.} = \sum\limits_{j=1}^{b} x_{ij}$; it is the *sum* of the ith treatment responses over the b blocks.

$$x_{1.} = 17 + 16 + 18 = 51 \qquad x_{2.} = 19 + 21 + 23 = 63$$
$$x_{3.} = 17 + 18 + 19 = 54 \qquad x_{4.} = 14 + 14 + 17 = 45$$

$$\text{SSTR} = \frac{(51)^2 + (63)^2 + (54)^2 + (45)^2}{3} - 3{,}780.75$$

$$= \frac{11{,}511}{3} - 3{,}780.75 = 3{,}837 - 3{,}780.75 = 56.25$$

Block sum of squares

$$\text{SSB} = t \sum_{j=1}^{b} (\bar{x}_{.j} - \bar{x}_{..})^2 = \frac{\sum_{j=1}^{b} x_{.j}^2}{t} - \frac{\left(\sum_{i=1}^{t} \sum_{j=1}^{b} x_{ij}\right)^2}{bt}$$

Note that $x_{.j} = \sum_{t=1}^{t} x_{ij}$; it is the *sum* of the jth block responses over the t treatments.

$$x_{.1} = 17 + 19 + 14 + 17 = 67 \qquad x_{.2} = 18 + 16 + 21 + 14 = 69$$
$$x_{.3} = 17 + 23 + 18 + 19 = 77$$

$$\text{SSB} = \frac{(67)^2 + (69)^2 + (77)^2}{4} - 3,780.75$$

$$= \frac{15,179}{4} - 3,780.75 = 3,794.75 - 3,780.75 = 14$$

Error sum of squares
$$\text{SSE} = \text{SST} - \text{SSTR} - \text{SSB} = 74.25 - 56.25 - 14 = 4$$

The analysis of variance results are summarized in Table 12.9.

TABLE 12.9 | Analysis of variance table for Example 12.5

Source of variation	Degrees of freedom	Sum of squares	Mean square	F-ratio
Among treatments	$t - 1 = 3$	56.25	$\frac{56.25}{3} = 18.75$	$f = \frac{18.75}{0.67} = 28.125$
Among blocks	$b - 1 = 2$	14.00	$\frac{14.00}{2} = 7$	
Experimental error	$(b - 1)(t - 1) = 6$	4.00	$\frac{4.00}{6} = 0.67$	
Total	$bt - 1 = 11$	74.25		Critical F: $F(0.95; 3, 6) = 4.76$

Since $f = 28.125 > 4.76$, we reject the null hypothesis H_0: $\tau_1 = \tau_2 = \tau_3$. Therefore, it appears from these data that the three oil filters do trap significantly different average amounts of impurities.

Table 12.10 presents the analysis of variance output from the SAS randomized block program. Notice that the sources of variation for the treatment and block effects are combined in the upper portion of the table and are denoted by the source of variation "MODEL." The breakdown of the "MODEL" sum of squares into the treatment sum of squares ("FILTER") and the block sum of squares ("DAYS") is given below the main portion of the analysis of variance table. From this table, the p-value for testing the

TABLE 12.10 | Computer analysis of the Example 12.5 data

```
                    REMOVING OF IMPURITIES DATA
                    ANALYSIS OF VARIANCE PROCEDURE
DEPENDENT VARIABLE: IMPURE
SOURCE     DF    SUM OF SQUARES    MEAN SQUARE      F-VALUE    PR > F
MODEL      5         70.25          14.05000        21.075    0.0010
ERROR      6          4.00           0.66666
TOTAL     11         74.25


SOURCE     DF     ANOVA SS      F-VALUE    PR > F
DAYS       2        14.00        10.50     0.0110
FILTER     3        56.25        28.13     0.0006
```

differences among the treatment effects is given by 0.0006; that is, Prob($\mathbf{F} >$ $f = 28.13$) = 0.0006.

Notice in Example 12.5 that there appears to be a significant blocking effect since SSB = 14; that is, the filters do appear to behave differently on the three days. We could test to see whether this effect is significant at the $\alpha = 0.05$ significance level by forming the value of the $\mathbf{F}$-statistic,

$$f = \frac{\text{MSB}}{\text{MSE}} = \frac{7.00}{0.67} = 10.5$$

and comparing the calculated f value to the critical F given by $F[1 - \alpha; (b - 1), (b - 1)(t - 1)] = F(0.95; 2, 6) = 5.14$. Since $10.5 > 5.14$, we can conclude that the blocking effect (days) does appear to be statistically significant.

Since the blocks were not chosen in a random fashion, it is really not appropriate to conduct the F-test for the blocking effect. However, by conducting the test, we can get *some* indication of whether or not the use of blocks has absorbed a significant amount of variability in the response random variable $\mathbf{X}$.

If the blocking effect had been ignored, then the design would be a completely randomized design with only two sources of variation: among treatments and experimental error. In this case, the sum of squares for blocks, SSB, would appear in the error sum of squares, SSE. The error mean square would then become

$$\text{MSE} = \frac{14.00 + 4.00}{2 + 6} = \frac{18.00}{8} = 2.25$$

instead of the proper value of 0.67. Although we still would reject the treatment effects null hypothesis, the decision to reject H_0 would be much closer.

By blocking on days, we have properly partitioned out an important source of variation that could have confounded the treatment effects.

In the previous section, the equivalence of the two-independent sample t-test and the one-factor, completely randomized design with $t = 2$ was demonstrated. As it turns out, the paired-sample t-test in Chapter 11 is equivalent to the randomized block design with $t = 2$ treatments. Blocks represent the matched pairs in the paired-sample t-test experiment.

It is possible to use a randomized block design in which more than one experimental unit is subjected to each treatment in each block. Within each block, each treatment is randomly assigned to n experimental units. This design is called a *randomized block design with subsampling* and requires that the tbn experimental units have been sorted into b homogeneous groups of tn units each. The interested reader is referred to the Ostle and Mensing reference at the end of this chapter for this and further extensions of the basic one-factor, randomized block design.

| Randomized block design with sub-sampling |

■ 12.5

Factorial experiments

In most studies we are interested in the changes in the dependent variable as a function of more than one independent variable (factor). To this point, we have been concerned with one-factor designs—the one-factor completely randomized design and the one-factor randomized block design in which the blocking variable is not thought of as a treatment factor, but rather as a variable introduced to control an unwanted source of variation in the dependent variable. We will now generalize the analysis of variance procedure for the analysis of experiments involving two or more factors.

Recall Example 12.1 where Farmco is concerned with the weight gain of chicks raised for overseas shipment. An experiment was conducted to determine whether three feeds—Growbig, Max, and Topnotch—produced different average weight gains among the chicks. As suggested in Section 12.1, Farmco may be interested in the effect on weight gain of a second factor—vitamin supplements. Let us suppose that two vitamin supplements, 1 and 2, are introduced into the experiment. The form of the sample data for this experiment is indicated in Table 12.11.

From Table 12.11 it is clear that the experiment is balanced; that is, n chicks are randomly assigned to each treatment—a combination of one level of each factor. The word *treatment* used in analysis of variance arose from experiments of this kind. The chick associated with the weight gain x_{111} has been "treated" with the Growbig feed and vitamin supplement 1. Treatments are, therefore, combinations of one level of each factor in the experiment. In this experiment, we have six ($i = 3$ times $j = 2$) treatments.

| Treatment |

To introduce the appropriate model and analysis of variance table for the two-factor, completely randomized design, we will generalize the preceding experiment to allow for any number of feeds, say a, and any number of vitamin supplements, say b.

The table of population means that correspond to the sample observations and the sample means shown in Table 12.11 is given in Table 12.12.

TABLE 12.11 | Form of the sample data for a two-factor experiment

Factor A (feed) / Factor B (vitamin supplement)	Growbig $i = 1$	Max $i = 2$	Topnotch $i = 3$	Totals	Means
Vitamin supplement 1, $j = 1$	x_{111} x_{112} $\vdots$ x_{11n}	x_{211} x_{212} $\vdots$ x_{21n}	x_{311} x_{312} $\vdots$ x_{31n}	$x_{.1.}$	$\bar{x}_{.1.}$
Vitamin supplement 2, $j = 2$	x_{121} x_{122} $\vdots$ x_{12n}	x_{221} x_{222} $\vdots$ x_{22n}	x_{321} x_{322} $\vdots$ x_{32n}	$x_{.2.}$	$\bar{x}_{.2.}$
Totals	$x_{1..}$	$x_{2..}$	$x_{3..}$	$x_{...}$	
Means	$\bar{x}_{1..}$	$\bar{x}_{2..}$	$\bar{x}_{3..}$		$\bar{x}_{...}$

TABLE 12.12 | Table of the population means for the two-factor experiment with a levels of Factor A and b levels of Factor B

Factor A / Factor B	$i = 1$	$i = 2 \;\cdots\; i = a$	Means
$j = 1$ $j = 2$	μ_{11} μ_{12}	μ_{21} μ_{22} $\qquad \mu_{a1}$ $\qquad \mu_{a2}$	$\mu_{.1}$ $\mu_{.2}$
$\vdots$ $j = b$	μ_{1b}	μ_{2b} $\qquad \mu_{ab}$	$\mu_{.b}$
Means	$\mu_{1.}$	$\mu_{2.}$ $\qquad \mu_{a.}$	$\mu_{..}$

Population means

The population means in Table 12.12 are defined as follows:

μ_{ij}: The population mean of the random variable $\mathbf{X}_{ijk}$ for the ith level of Factor A and the jth level of Factor B.

$\mu_{i.}$: The population mean of the random variable $\mathbf{X}_{ijk}$ for the ith level of Factor A, where

$$\mu_{i.} = \frac{\sum\limits_{j=1}^{b} \mu_{ij}}{b}$$

$\mu_{.j}$: The population mean of the random variable $\mathbf{X}_{ijk}$ for the jth level of Factor B, where

$$\mu_{.j} = \frac{\sum\limits_{i=1}^{a} \mu_{ij}}{a}$$

$\mu_{..}$: The overall population mean of the random variable $\mathbf{X}_{ijk}$ given by

$$\mu_{..} = \frac{\sum\limits_{i=1}^{a} \sum\limits_{j=1}^{b} \mu_{ij}}{ab}$$

From Table 12.11, $\bar{x}_{i..}$ estimates $\mu_{i.}$, $\bar{x}_{.j.}$ estimates $\mu_{.j}$, and $\bar{x}_{...}$ estimates $\mu_{..}$. For the response random variable $\mathbf{X}_{ijk}$, the two-factor, completely randomized design model may be written as:

| Two-factor, completely randomized design |

$$\mathbf{X}_{ijk} = \mu_{ij} + \varepsilon_{ijk}$$

where

μ_{ij} is the parameter representing the population mean response for the ith level of Factor A and the jth level of Factor B.

ε_{ijk} is the error term, assumed to be normally distributed, independent, and with a mean of 0 and a variance of σ^2.

$i = 1, 2, \ldots, a;\quad j = 1, 2, \ldots, b;\quad \text{and } k = 1, 2, \ldots, n.$

As with the one-factor, completely randomized design, it is possible to re-parameterize the model by defining factor effects and the interaction effect between the two factors. Define

| Reparameterizing the model |

$$\mu_{ij} = \mu_{..} + \alpha_i + \beta_j + (\alpha\beta)_{ij}$$

where

$$\alpha_i = \mu_{i.} - \mu_{..}$$

$$\beta_j = \mu_{.j} - \mu_{..}$$

$$(\alpha\beta)_{ij} = \mu_{ij} - \mu_{i.} - \mu_{.j} + \mu_{..}$$

The ith effect of Factor A, α_i, is the difference between the population mean for the ith level of Factor A, $\mu_{i.}$, and the overall mean, $\mu_{..}$. The jth effect of Factor B, β_j, is the difference between the population mean for the jth level of Factor B, $\mu_{.j}$, and the overall population mean, $\mu_{..}$. The term $(\alpha\beta)_{ij}$ represents the interaction effect between the ith level of Factor A and the jth level of Factor B. It is given by subtracting the ith level of the Factor A mean, $\mu_{i.}$, and the jth level of the Factor B mean, $\mu_{.j}$, from the cell mean, μ_{ij}, and adding the overall mean, $\mu_{..}$ to the difference.

The complete description of the reparameterized two-factor, completely randomized design model is given in Definition 12.4 below. For simplicity, the overall mean, $\mu_{..}$ is denoted by μ in this model.

Reparameterized two-factor completely randomized design

Definition 12.4

Two-factor, completely randomized design

The model is

$$\mathbf{X}_{ijk} = \mu + \alpha_i + \beta_j + (\alpha\beta)_{ij} + \varepsilon_{ijk}$$
$$i = 1, 2, \ldots, a; j = 1, 2, \ldots, b; \text{ and } k = 1, 2, \ldots, n$$

where

$\mathbf{X}_{ijk}$ = Response (dependent) random variable associated with the kth experimental unit subjected to the ith level of Factor A and the jth level of Factor B.

μ = Overall population mean (equivalent to $\mu_{..}$).

α_i = Effect of the ith level of Factor A.

β_j = Effect of the jth level of Factor B.

$(\alpha\beta)_{ij}$ = Interaction effect of the ith level of Factor A and the jth level of Factor B.

ε_{ijk} = Random error term associated with the response $\mathbf{X}_{ijk}$.

The abn experimental units are randomly selected and assigned to the ab treatments. The hypotheses of interest are:

1. H_0: $\alpha_1 = \alpha_2 = \cdots = \alpha_a$.
    H_A: Not all α_i are equal.
2. H_0: $\beta_1 = \beta_2 = \cdots = \beta_b$.
    H_A: Not all β_j are equal.
3. H_0: $(\alpha\beta)_{ij} = 0$, $i = 1, 2, \ldots, a$ and $j = 1, 2, \ldots, b$.
    H_A: Not all $(\alpha\beta)_{ij}$ are equal to zero.

Assumptions of the two-factor completely randomized design model

12.5.1 | Assumptions: Two-factor, completely randomized design

1. The factor effects, α_i and β_j, are fixed.

2. The error components, ε_{ijk} where $i = 1, 2, \ldots, a; j = 1, 2, \ldots, b;$ $k = 1, 2, \ldots, n$, are normally and independently distributed random variables with 0 means and a common variance of σ^2.

3. The design is balanced; that is, the same number of experimental units n ($n \geq 2$) are subjected to each of the ab treatments.

TABLE 12.13 | Analysis of variance table for two-factor, completely randomized design

Source of variation	Degrees of freedom	Sum of squares	Mean square	F-ratio
Factor A	$a - 1$	$\displaystyle SSA = bn \sum_{i=1}^{a} (\bar{x}_{i..} - \bar{x}_{...})^2$	$\displaystyle MSA = \frac{SSA}{a-1}$	$\displaystyle f = \frac{MSA}{MSE}$
Factor B	$b - 1$	$\displaystyle SSB = an \sum_{j=1}^{b} (\bar{x}_{.j.} - \bar{x}_{...})^2$	$\displaystyle MSB = \frac{SSB}{b-1}$	$\displaystyle f = \frac{MSB}{MSE}$
AB interaction	$(a-1)(b-1)$	$\displaystyle SSAB = n \sum_{i=1}^{a} \sum_{j=1}^{b} (\bar{x}_{ij.} - \bar{x}_{i..} - \bar{x}_{.j.} + \bar{x}_{...})^2$	$\displaystyle MSAB = \frac{SSAB}{(a-1)(b-1)}$	$\displaystyle f = \frac{MSAB}{MSE}$
Experimental error	$ab(n-1)$	$\displaystyle SSE = \sum_{i=1}^{a} \sum_{j=1}^{b} \sum_{k=1}^{n} (x_{ijk} - \bar{x}_{ij.})^2$	$\displaystyle MSE = \frac{SSE}{ab(n-1)}$	
Total	$abn - 1$	$\displaystyle SST = \sum_{i=1}^{a} \sum_{j=1}^{b} \sum_{k=1}^{n} (x_{ijk} - \bar{x}_{...})^2$		

4. $\sum_{i=1}^{a} \alpha_i = 0, \quad \sum_{j=1}^{b} \beta_j = 0, \quad \sum_{i=1}^{a} (\alpha\beta)_{ij} = 0, \quad \text{and} \quad \sum_{j=1}^{b} (\alpha\beta)_{ij} = 0$

$$j = 1, 2, \ldots, b \quad i = 1, 2, \ldots, a$$

12.5.2 | Analysis of variance table

The analysis of variance table for the two-factor, completely randomized design is given in Table 12.13.

Example 12.6 Let us suppose that 18 chicks are randomly selected from a large brood and are randomly separated into six treatment groups of three each. The resulting weight gains are given in Table 12.14.

TABLE 12.14 | Weight gain (in grams) in the two-factor experiment

Factor A / Factor B	Growbig	Max	Topnotch	Totals	Means
Vitamin supplement 1	46 75 59 $x_{11.} = 180$ $\bar{x}_{11.} = 60$	120 94 122 $x_{21.} = 336$ $\bar{x}_{21.} = 112$	36 49 50 $x_{31.} = 135$ $\bar{x}_{31.} = 45$	$x_{.1.} = 651$	$\bar{x}_{.1.} = 72.33$
Vitamin supplement 2	92 87 85 $x_{12.} = 264$ $\bar{x}_{12.} = 88$	116 96 88 $x_{22.} = 300$ $\bar{x}_{22.} = 100$	33 47 46 $x_{32.} = 126$ $\bar{x}_{32.} = 42$	$x_{.2.} = 690$	$\bar{x}_{.2.} = 76.67$
Totals	$x_{1..} = 444$	$x_{2..} = 636$	$x_{3..} = 261$	$x_{...} = 1{,}341$	
Means	$\bar{x}_{1..} = 74$	$\bar{x}_{2..} = 106$	$\bar{x}_{3..} = 43.5$		$\bar{x}_{...} = 74.5$

The notation used in Table 12.14 is defined as follows:

$x_{ijk} =$ Weight gain for the kth chick subjected to the ith feed and the jth vitamin supplement.

$x_{i..} = \displaystyle\sum_{j=1}^{b} \sum_{k=1}^{n} x_{ijk} =$ Total weight gain of the chicks subjected to the ith feed.

$\bar{x}_{i..} = x_{i..}/bn =$ Average weight gain of the chicks subjected to the ith feed.

$x_{.j.} = \displaystyle\sum_{i=1}^{a} \sum_{k=1}^{n} x_{ijk} =$ Total weight gain of the chicks subjected to the jth vitamin supplement.

$\bar{x}_{.j.} = x_{.j.}/an =$ Average weight gain of the chicks subjected to the jth vitamin supplement.

$$x_{...} = \sum_{i=1}^{a} \sum_{j=1}^{b} \sum_{k=1}^{n} x_{ijk} = \text{Total weight gain of all chicks } (abn) \text{ in the experiment.}$$

$$\bar{x}_{...} = x_{...}/(a)(b)(n) = \text{Average weight gain of all chicks in the experiment.}$$

$$x_{ij.} = \sum_{k=1}^{n} x_{ijk} = \text{Total weight gain of the } n \text{ chicks subjected to the } i\text{th feed and } j\text{th vitamin supplement.}$$

$$\bar{x}_{ij.} = x_{ij.}/n = \text{Average weight gain of the } n \text{ chicks subjected to the } i\text{th feed and } j\text{th vitamin supplement; } a = 3, b = 2, n = 3.$$

The calculations may be somewhat eased by using the calculating forms of the sums of squares:

$$C = \frac{\left(\sum\limits_{i=1}^{a=3} \sum\limits_{j=1}^{b=2} \sum\limits_{k=1}^{n=3} x_{ijk} \right)^2}{nab = (3)(3)(2)} = \frac{(1{,}341)^2}{18} = 99{,}904.5$$

$$\text{SSA} = \frac{\sum\limits_{i=1}^{a=3} x_{i..}^2}{nb = (3)(2)} - C = \frac{(444)^2 + (636)^2 + (261)^2}{6} - C$$
$$= 111{,}625.5 - 99{,}904.5 = 11{,}721$$

$$\text{SSB} = \frac{\sum\limits_{j=1}^{b=2} x_{.j.}^2}{na = (3)(3)} - C = \frac{(651)^2 + (690)^2}{9} - C$$
$$= 99{,}989 - 99{,}904.5 = 84.5$$

$$\text{SST} = \sum_{i=1}^{a=3} \sum_{j=1}^{b=3} \sum_{k=1}^{n=3} x_{ijk}^2 - C = [(46)^2 + (75)^2 + \cdots + (46)^2] - C$$
$$= 114{,}627 - 99{,}904.5 = 14{,}722.5$$

$$\text{SSAB} = \frac{\sum\limits_{i=1}^{a=3} \sum\limits_{j=1}^{b=2} x_{ij.}^2}{n = 3} - \text{SSA} - \text{SSB} - C$$

$$= \frac{(180)^2 + (336)^2 + \cdots + (126)^2}{3} - \text{SSA} - \text{SSB} - C$$
$$= 113{,}031 - 11{,}721 - 84.5 - 99{,}904.5 = 1{,}321$$
$$\text{SSE} = \text{SST} - \text{SSA} - \text{SSB} - \text{SSAB}$$
$$= 14{,}722.5 - 11{,}721 - 84.5 - 1{,}321 = 1{,}596$$

The analysis of variance table is summarized in Table 12.15 using the output from the SAS two-factor analysis of variance program.

Another sum of squares can be defined for computational purposes:

Treatment sum of squares

$$\text{SSTR} = n \sum_{i=1}^{a} \sum_{j=1}^{b} (\bar{x}_{ij.} - \bar{x}_{...})^2 = \frac{\sum\limits_{i=1}^{a} \sum\limits_{j=1}^{b} x_{ij.}^2}{n} - C$$

TABLE 12.15 | Computer analysis of the Example 12.6 data

```
                    FACTORIAL EXPERIMENT ON FEED DATA
                     ANALYSIS OF VARIANCE PROCEDURE
DEPENDENT VARIABLE: YIELD
SOURCE      DF    SUM OF SQUARES    MEAN SQUARE     F-VALUE    PR > F
MODEL        5       13126.50         2625.30        19.74    0.0001
ERROR       12        1596.00          133.00
TOTAL       17       14722.50


SOURCE      DF      ANOVA SS      F-VALUE    PR > F
FEED         2      11721.00       44.06     0.0001
SUPP         1         84.50        0.64     0.4409
FEED*SUPP    2       1321.00        4.97     0.0268
```

SSTR stands for the treatment sum of squares and is equal to:

$$SSA + SSB + SSAB$$

That is, the treatment in the two-factor, completely randomized design is the consequence of the factor A effect, the factor B effect, and the effect of the interaction between factors A and B. Therefore, the total sum of squares, SST, may be written as:

$$SST = SSTR + SSE$$
$$= SSA + SSB + SSAB + SSE$$

where

$$SSTR = SSA + SSB + SSAB$$

For the Example 12.6 data,

$$SSTR = \frac{(180)^2 + (336)^2 + \cdots + (126)^2}{3} - 99{,}904.5$$
$$= 113{,}031 - 99{,}904.5$$
$$= 13{,}126.5$$

| Testing the model or treatment effects |

Notice in Table 12.15 that the SAS analysis prints SSTR as the source of variation due to the "MODEL." This allows for a simple test to determine if any of the effects are significant (α_i effects, β_j effects, and $(\alpha\beta)_{ij}$ interaction effects). The F-value for this test is:

$$f = 19.74$$

with 5 numerator and 12 denominator degrees of freedom (information taken from the SAS output in Table 12.15). The critical F-value from Table B.7, Appendix B, is 3.11 with $\alpha = 0.05$. Since $f = 19.74 > F(0.95; 5, 12)$, we reject the null hypothesis that there are no treatment effects and conclude that one or more of the effects (due to factor A, factor B, or the interaction between factors A and B) are statistically significant.

Notice that since the p-value $= 0.0001$ from the SAS output, we can arrive at this conclusion by noting that p-value $= 0.0001 < \alpha = 0.05$. Given that there are significant treatment effects, each of the effects can be tested. We will start with the interaction effects, $(\alpha\beta)_{ij}$.

To test the hypotheses

Testing for interac-
tion effects

H_0 : $(\alpha\beta)_{ij} = 0$; $i = 1, 2, \ldots, a$; $j = 1, 2, \ldots, b$.
H_A : Not all $(\alpha\beta)_{ij}$ are equal to zero.

at the $\alpha = 0.05$ level, we compare the computed value of **F** in the table for the interaction effect, $f = 4.97$, with the appropriate tabular value of **F**, $F(0.95; 2, 12) = 3.89$. Since $4.97 > F(0.95; 2, 12) = 3.89$, we conclude that there is a significant interaction effect. From Table 12.15 the p-value for this test is equal to 0.0268.

The significant interaction effect is readily apparent from studying Table 12.15. The first vitamin supplement appears to produce *less* average weight gain in conjunction with Growbig than does the second vitamin supplement. Yet the first vitamin supplement appears to produce *more* average weight gain in conjunction with Max and Topnotch than does the second vitamin supplement.

Treatment mean
curve plots

This interaction effect can be illustrated by drawing the *treatment mean curve plots* as in Figure 12.10. The factor A (feeds) levels are placed on the horizontal axis, where A_1, A_2, and A_3 represent the feed levels Growbig,

FIGURE 12.10 | Feed and vitamin supplement interaction effects using treatment mean curve plots

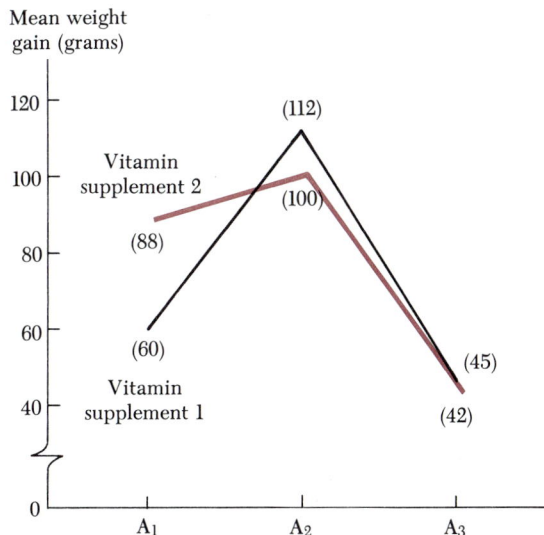

Max, and Topnotch, respectively. The treatment means are plotted for each of the levels of factor B (vitamin supplements). For example, the mean weight gains for the vitamin supplement 1 in combination with the levels of factor A are 60 grams, 112 grams, and 45 grams for levels A_1, A_2, and A_3, respectively. The crossing of the two plots indicates a dramatic interaction effect. Vitamin supplement 2 in combination with factor A levels A_2 and A_3 appears to be inferior to vitamin supplement 1, but clearly outperforms vitamin supplement 1 when both are given in combination with factor A level A_1 (the Growbig feed).

FIGURE 12.11 | Treatment mean curve plots showing no interaction between factors A and B (parallel lines)

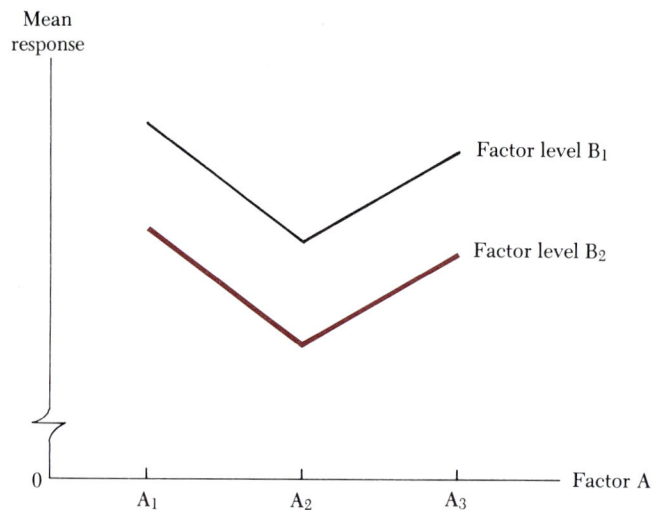

The treatment mean plots provide a very effective way to determine if interaction effects exist in two factor experiments. If there are no interaction effects, then the treatment mean plots are parallel, as illustrated in Figure 12.11. If there are interaction effects, then the treatment mean plots are not parallel, as is the case in the Example 12.6 problem and its treatment mean plots shown in Figure 12.10.

To test the hypotheses concerning feeds:

Testing for factor A effects

H_0: $\alpha_1 = \alpha_2 = \alpha_3$.
H_A: Not all α_i equal, $i = 1, 2, 3$

the appropriate tabular value of **F** at the $\alpha = 0.05$ level is $F(0.95; 2, 12) = 3.89$, and since $f = 44.06 > 3.89$, we conclude that the three feeds do produce different average weight gains among the chicks. From Table 12.14, the p-value is equal to 0.0001.

To test the hypotheses concerning vitamin supplements:

$$H_0: \quad \beta_1 = \beta_2$$
$$H_A: \quad \beta_1 \neq \beta_2$$

the appropriate tabular value of F at the $\alpha = 0.05$ level is $F(0.95; 1, 12) = 4.75$. Since $f = 0.64 < 4.75$, we conclude that there are no significant differences in average weight gains among the chicks for the two vitamin supplements. From Table 12.15, the p-value for this test is equal to 0.4409.

Caution must be exercised when drawing inferences about main effects (the primary factors) in the presence of significant interaction effects. For example, consider the means in Table 12.16 for a 3×3 experiment (two factors, each at three levels). In this experiment, there is a strong interaction effect—note the differences in responses for levels of factor A between the first level of factor B and the other two levels of B. The treatment mean plots are shown in Figure 12.12. These plots show rather dramatically the difference in the first level of B mean responses across the levels of factor A and the second and third levels of B mean responses across the levels of factor A.

TABLE 12.16 | Means of a 3×3 factorial experiment

Factor B \\ Factor A	1	2	3	
1	$\bar{x}_{11.} = 10$	$\bar{x}_{21.} = 90$	$\bar{x}_{31.} = 80$	$\bar{x}_{.1.} = 60$
2	$\bar{x}_{12.} = 90$	$\bar{x}_{22.} = 45$	$\bar{x}_{32.} = 55$	$\bar{x}_{.2.} = 63.33$
3	$\bar{x}_{13.} = 80$	$\bar{x}_{23.} = 45$	$\bar{x}_{33.} = 45$	$\bar{x}_{.3.} = 56.67$
	$\bar{x}_{1..} = 60$	$\bar{x}_{2..} = 60$	$\bar{x}_{3..} = 60$	$\bar{x}_{...} = 60$

From the margin means, it is clear that the analysis of variance procedure would not find significant differences among the levels of A and would most likely not find significant differences among the levels of B. But suppose that the first level of B were dropped from the experiment. Then the first level of A would appear to produce a significantly different response than would the other two levels of A. That is, excluding the first level of B, the first level of A appears to produce the greatest response in the dependent variable—one that is significantly different from the responses produced by levels 2 and 3 of factor A.

If the interaction is significant, it ordinarily does not make sense to test the hypothesis of significant main effects. As the preceding discussion suggests, the interaction effect may cover up (or uncover) significant main effects. But that is consistent with the nature and intent of multifactor exper-

FIGURE 12.12 | Strong interaction effects for the 3 × 3 factorial experiment with treatment means in Table 12.16

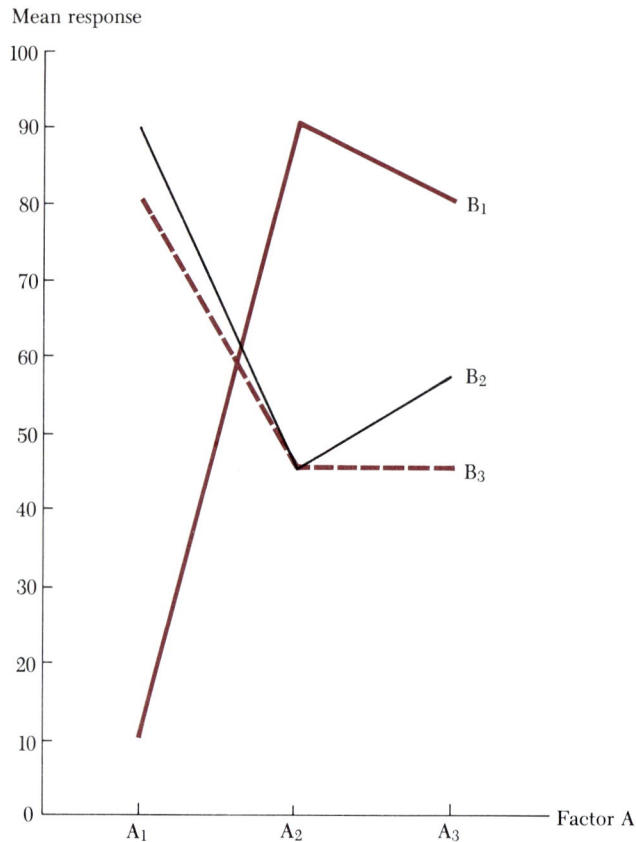

iments. In the Farmco example, it is understood that the intent is to provide *both* a feed and a vitamin supplement for the chicks. Although it may be possible to demonstrate that vitamin supplements 1 and 2 produce significantly different average weight gains *alone* by running a one-factor experiment excluding the three types of feed, it would not produce a useful inference since the supplements will be used in *conjunction* with the feed types. As it turned out, the two factors did interact with the result that Max in conjunction with the first vitamin supplement produced the greatest sample average weight gain among the chicks.

The analysis of variance procedure to analyze more than two factors is applied in a straightforward extension of the two-factor case. Although the computations become more tedious, multiple-factor experiments are desirable because of the efficiency they afford—many factors and their interactions can be studied by conducting one experiment. Additionally, a factorial experiment can be conducted in a randomized block design. Suppose we have two factors, A and B; a_1 and a_2 denote the two levels of factor A, and b_1 and b_2 denote the two levels of factor B. In Table 12.17 a randomized block design for the two-factor experiment is indicated.

Two-factor, randomized block design

TABLE 12.17 | Two-factor, randomized block design

	Block 1		Block 2		Block 3	
	a_1b_2	a_1b_2	a_2b_2	a_1b_1	a_2b_2	a_1b_1
	a_2b_2	a_2b_1	a_1b_1	a_1b_2	a_1b_2	a_2b_1
	a_1b_1	a_1b_1	a_1b_2	a_2b_1	a_2b_1	a_1b_2
	a_2b_1	a_2b_2	a_2b_1	a_2b_2	a_1b_1	a_2b_2

In this design, there are three blocks and four treatments (a_1b_1, a_1b_2, a_2b_1, and a_2b_2). Within each block, there are two replicates—each treatment is repeated twice.

The analysis of variance table for the two-factor, randomized block design as well as for more complicated designs can be found in most texts on analysis of variance or experimental design. A particularly good reference for basic designs is the Ostle and Mensing book listed at the end of the chapter.

■ **12.6**

Multiple comparisons (advanced section)

When we determine by analysis of variance that the levels of a factor differ significantly, we ordinarily are interested in which mean(s) differ from the rest. Consider, for example, the one-factor, completely randomized experiment to determine whether the three feeds—Growbig, Max, and Topnotch—produce significantly different weight gains among chicks. We found in Section 12.3 that the three feeds do produce significantly different weight gains. Now can we further determine whether all three feeds are different or whether two are similar but differ from the third? The answer is yes, and the procedures that may be used for comparisons among factor level means are called *multiple comparisons*. We will consider one of these procedures: the Bonferroni contrast method.

Multiple comparisons: Bonferroni method

The three sample average weight gains are given in Table 12.18.

TABLE 12.18 | Three sample means in Example 12.1

	Growbig	Max	Topnotch
Population mean	μ_1	μ_2	μ_3
Sample mean	$\bar{x}_1 = 78$	$\bar{x}_2 = 110$	$\bar{x}_3 = 40$

From the sample means, we may conjecture that $\mu_2 > \mu_1 > \mu_3$. Bonferroni's method will allow us to test this conjecture by using *contrasts*. If c_1, $c_2, \ldots, c_t$ are known constants such that

$$\sum_{t=1}^{t} c_i = c_1 + c_2 + \cdots + c_t = 0$$

then

$$L = c_1\mu_1 + c_2\mu_2 + \cdots + c_t\mu_t$$

is called a *contrast*. We have already used implicitly the concept of a contrast in testing the equivalence of two populations means in Chapter 11. The hypothesis $H_0: \mu_1 = \mu_2$ can be written as $H_0: \mu_1 - \mu_2 = 0$, and the difference $\mu_1 - \mu_2$ is clearly a contrast with $c_1 = 1$, $c_2 = -1$. With three means ($t = 3$), other possible contrasts are as shown in the table.

| | Contrast coefficients | | |
Contrast	c_1	c_2	c_3
$\mu_1 - \mu_3$	1	0	-1
$\mu_2 - \mu_3$	0	1	-1
$2\mu_1 - \mu_2 - \mu_3$	2	-1	-1
$5\mu_1 - 8\mu_2 + 3\mu_3$	5	-8	3

In fact, an infinite number of contrasts may be constructed with three means. The Bonferroni method applies to a *set* of contrasts. Suppose we have m contrasts ($m \geq 1$) of the form:

$$L_1 = c_{11}\mu_1 + c_{21}\mu_2 + \cdots + c_{t1}\mu_t = \sum_{i=1}^{t} c_{i1}\mu_i$$

$$L_2 = c_{12}\mu_1 + c_{22}\mu_2 + \cdots + c_{t2}\mu_t = \sum_{i=1}^{t} c_{i2}\mu_i$$

$$\vdots$$

$$L_m = c_{1m}\mu_1 + c_{2m}\mu_2 + \cdots + c_{tm}\mu_t = \sum_{i=1}^{t} c_{im}\mu_i$$

where $\sum_{i=1}^{t} c_{i1} = 0$, $\sum_{i=1}^{t} c_{i2} = 0, \ldots, \sum_{i=1}^{t} c_{im} = 0$.

The estimate of the jth contrast, L_j, is given by

$$\hat{L}_j = c_{1j}\bar{x}_{1.} + c_{2j}\bar{x}_{2.} + \cdots + c_{tj}\bar{x}_{t.} = \sum_{i=1}^{t} c_{ij}\bar{x}_{i.}$$

and the $100(1 - \alpha)$ percent confidence interval estimates of the m contrasts are given by

$$\hat{L}_j \pm Bs(\hat{L}_j), j = 1, 2, \ldots, m;$$

where

$$B = t_{\alpha/2m;df}$$

$$s^2(\hat{L}_j) = \frac{\text{MSE}}{n} \sum_{i=1}^{t} c_{ij}^2$$

n = Sample size for each of the t treatments in the one-factor, completely randomized design model.

MSE = Mean square error from the analysis of variance table.

$df = t(n - 1)$ for the one-factor, completely randomized design model.

The Bonferroni B coefficient in the interval estimate is a **t**-value with significance level α/m. The Bonferroni method guarantees that the confidence coefficient is at least $1 - \alpha$ for the *set* of m interval estimates.

Example 12.7 Apply the Bonferroni method to the Example 12.1 problem to investigate how the means μ_1, μ_2, and μ_3 may differ.

Solution From the analysis of variance for the Example 12.1 data, we know that the three mean responses (weight gains) μ_1, μ_2, and μ_3 differ at the $\alpha = 0.05$ level of significance. From Table 12.17, we might conjecture that $\mu_2 > \mu_1 > \mu_3$. To investigate how the means do differ with statistical significance, we consider all pairwise comparisons of the three means. Stated as contrasts, these are:

$$L_1 = (1)\mu_1 + (-1)\mu_2 + (0)\mu_3 = \mu_1 - \mu_2$$
$$L_2 = (1)\mu_1 + (0)\mu_2 + (-1)\mu_3 = \mu_1 - \mu_3$$
$$L_3 = (0)\mu_1 + (1)\mu_2 + (-1)\mu_3 = \mu_2 - \mu_3$$

The point estimates of these contrasts are:

$$\hat{L}_1 = (1)\bar{x}_{1.} + (-1)\bar{x}_{2.} + (0)\bar{x}_{3.} = 78 - 110 + 0 = -32$$
$$\hat{L}_2 = (1)\bar{x}_{1.} + (0)\bar{x}_{2.} + (-1)\bar{x}_{3.} = 78 + 0 - 40 = 38$$
$$\hat{L}_3 = (0)\bar{x}_{1.} + (1)\bar{x}_{2.} + (-1)\bar{x}_{3.} = 0 + 110 - 40 = 70$$

In applying the Bonferroni method to construct confidence interval estimates of L_1, L_2, and L_3, we must first set the confidence coefficient $1 - \alpha$ for the set of $m = 3$ contrasts. Suppose that we set $1 - \alpha = 0.95$ (95 percent confidence intervals) so that $\alpha = 0.05$. Now, from the analysis of variance table for the Example 12.1 data, MSE = 423.17 and the degrees of freedom are $t(n - 1) = 12$. The Bonferroni B coefficient is given by:

$$B = t_{\alpha/2m;df} = t_{0.05/(2)(3);12} = t_{0.008;12}$$

Thus we want the **t**-value from the t-table so that the area in the upper tail of the t-distribution is 0.008 with $df = 12$. From Table B.5, Appendix B, we have $t_{0.010;12} = 2.681$, and $t_{0.005;12} = 3.055$. Since the upper tail area of 0.008 falls between the upper tail area of 0.005 and 0.010, we must interpolate to find the desired **t**-value. By interpolation, $t_{0.008;12} = 2.831$. The confidence interval estimates are given as follows.

For the contrast $L_1 = \mu_1 - \mu_2$:

$$\hat{L}_1 = (1)\bar{x}_1 + (-1)\bar{x}_2 + (0)\bar{x}_3 = 78 - 110 + 0 = -32$$

$$s^2(\hat{L}_1) = \frac{MSE}{n} \sum_{i=1}^{t} c_{i1}^2 = \frac{423.17}{5} [(1)^2 + (-1)^2 + (0)^2] = 169.27$$

$$s(\hat{L}_1) = 13.01$$
$$\hat{L}_1 \pm Bs(\hat{L}_1) = -32 \pm 2.831(13.01) = -32 \pm 36.83$$

The interval estimate of $L_1 = \mu_1 - \mu_2$ is therefore -68.83 to 4.83.

For the contrast $L_2 = \mu_1 - \mu_3$:

$$\hat{L}_2 = (1)\bar{x}_1 + (0)\bar{x}_2 + (-1)\bar{x}_3 = 78 + 0 - 40 = 38$$

$$s^2(\hat{L}_2) = \frac{MSE}{n} \sum_{i=1}^{t} c_{i2}^2 = \frac{423.17}{5} [(1)^2 + (0)^2 + (-1)^2] = 169.27$$

$$s(\hat{L}_2) = 13.01$$
$$\hat{L}_2 \pm Bs(\hat{L}_2) = 38 \pm 2.831(13.01) = 38 \pm 36.83$$

The interval estimate of $L_2 = \mu_1 - \mu_3$ is therefore 1.17 to 74.83.

For the contrast $L_3 = \mu_2 - \mu_3$:

$$\hat{L}_3 = (0)\bar{x}_1 + (1)\bar{x}_2 + (-1)\bar{x}_3 = 0 + 110 - 40 = 70$$

$$s^2(\hat{L}_3) = \frac{MSE}{n} \sum_{i=1}^{t} c_{i3}^2 = \frac{423.17}{5} [(0)^2 + (1)^2 + (-1)^2] = 169.27$$

$$s(\hat{L}_3) = 13.01$$
$$\hat{L}_3 \pm Bs(\hat{L}_3) = 70 \pm 2.831(13.01) = 70 \pm 36.83$$

The interval estimate of $L_3 = \mu_2 - \mu_3$ is therefore 33.17 to 106.83.

The family of three contrast interval estimates can be summarized as follows. The contrast $L_1 = \mu_1 - \mu_2$ estimate is -68.83 to 4.83. Since this interval estimate contains 0, the hypothesis that $\mu_1 - \mu_2 = 0$ (or, alternatively, $\mu_1 = \mu_2$) cannot be rejected. The contrast $L_2 = \mu_1 - \mu_3$ estimate is 1.17 to 74.83. Since this interval does not contain zero, we can conclude that $\mu_1 - \mu_3 = 0$; rather, there is evidence that $\mu_1 - \mu_3 > 0$ implying that $\mu_1 > \mu_3$. Finally, the contrast $L_3 = \mu_2 - \mu_3$ estimate is 33.17 to 106.83. Since this interval does not contain zero, we can conclude that $\mu_2 - \mu_3 = 0$; rather, there is evidence that $\mu_2 - \mu_3 > 0$ implying that $\mu_2 > \mu_3$. Therefore we conclude that $\mu_1 > \mu_3$ and $\mu_2 > \mu_3$, but there is insufficient evidence to conclude that $\mu_1 \neq \mu_2$. Thus both the Growbig feed (mean response μ_1) and the Max feed (mean response μ_2) produce significantly larger mean weight gains than the Topnotch feed (mean response μ_3).

A multiple comparison method such as Bonferroni's allows one to compare treatment means after we have determined that the means are significantly different by performing an analysis of variance test. It must be remembered that the Bonferroni procedure can be applied only if significant differences in the means have been demonstrated by analysis of variance. If the null hypothesis of equal means cannot be rejected by analysis of variance and Bonferroni's method is applied to contrasts of the means, spurious significant contrasts may occur.

One may ask why the three pairwise tests of μ_1, μ_2, and μ_3 could not have been made by using the simple t-test presented in Chapter 11. The answer is that by using the method of contrasts, the $(1 - \alpha)$ 100 percent level of confidence will be maintained for the *set* of contrasts by *increasing* the level of confidence for each contrast in the set. Had we used three t-tests to compare the three means in pairs, the effective *system* confidence would be reduced to $[(1 - \alpha) \, 100 \text{ percent}]^3$. For example, if the tests were conducted at the $\alpha = 0.05$ level, then the t-tests would each have an effective confidence of $(0.95)^3 = 0.857$, not 0.95. The method of contrasts ensures that the *family* of confidence intervals will have a confidence of 95 percent or more.

The Bonferroni method is easily applied to experimental design models other than the one-factor, completely randomized design model. It is necessary to change only the Bonferroni B coefficient and the formula for $s^2(\hat{L}_j)$. The Bonferroni B coefficient is:

$$B = t_{\alpha/2m;df}$$

To apply the Bonferroni method to other models, the degrees of freedom df must be set to the degrees of freedom associated with the mean square error MSE for the particular model of interest. Thus, from the analysis of variance table, df is set to the degrees of freedom in the "Experimental error" row of the table. The formula for $s^2(\hat{L}_j)$ is given generally by:

$$s^2(\hat{L}) = \text{MSE} \sum_i \sum_j \cdots \frac{c_{ij\cdots}^2}{n^*}$$

where n^* is the number of observations used to calculate the sample mean estimates ($\bar{x}$'s) of the population means (μ's) involved in the contrast. In Example 12.7 each of the sample means $\bar{x}_1$, $\bar{x}_2$, and $\bar{x}_3$ is calculated from $n = 5$ observations. Thus in this application $n^* = 5$.

With the changes in B and $s^2(\hat{L})$, the Bonferroni method can be applied to the other design models presented in this chapter. We will now summarize the changes for these models.

For the one-factor, completely randomized design model with unequal sample sizes for each treatment (Subsection 12.3.5), the degrees of freedom in B are given by

$$df = \sum_{i=1}^{t} (n_i - 1)$$

and

$$s^2(\hat{L}_j) = \text{MSE} \sum_{i=1}^{t} \frac{c_{ij}^2}{n_i}$$

For the randomized block design model (Section 12.4), the degrees of freedom in B are given by

$$df = (t - 1)(b - 1)$$

and

$$s^2(\hat{L}_j) = \frac{\text{MSE}}{b} \sum_{i=1}^{t} c_{ij}^2$$

For the two-factor, completely randomized design model (Section 12.5), contrasts can be constructed for three sets of means: the factor A means $\mu_{i.}$, the factor B means $\mu_{.j}$, and the treatment means μ_{ij}, where $i = 1, 2, \ldots, a$ and $j = 1, 2, \ldots, b$. For each of the sets of means, the degrees of freedom in B are given by

$$df = ab(n - 1)$$

where a = Number of levels of factor A, b = Number of levels of factor B, and n = Sample size for each treatment (combination of a level of factor A and a level of factor B). The formula for $s^2(\hat{L})$ will change depending on which set of means the contrasts are to be applied.

Factor A means $\mu_{i.}$ For factor A, the population means are denoted by $\mu_{1.}, \mu_{2.}, \ldots, \mu_{a.}$. For instance, in Example 12.6 there are three levels of factor A with means $\mu_{1.}, \mu_{2.}$, and $\mu_{3.}$. The estimates of these means are $\bar{x}_{1..}$, $\bar{x}_{2..}$, and $\bar{x}_{3..}$, respectively. Each of the estimated means is calculated from $bn = (2)(3) = 6$ observations. The divisor of $s^2(\hat{L})$ is therefore six (in general, bn). If we had the contrasts,

$$L_1 = \mu_{1.} - \mu_{3.}$$
$$L_2 = \mu_{1.} - 2\mu_{2.} + \mu_{3.}$$

for the Example 12.6 problem, then

$$s^2(\hat{L}_1) = \text{MSE} \frac{[(1)^2 + (0)^2 + (-1)^2]}{bn} = 133 \frac{[2]}{6} = 44.33$$
$$s^2(\hat{L}_2) = \text{MSE} \frac{[(1)^2 + (-2)^2 + (1)^2]}{bn} = 133 \frac{[6]}{6} = 133.00$$

Factor B means $\mu_{.j}$ For factor B, the population means are denoted by $\mu_{.1}, \mu_{.2}, \ldots, \mu_{.b}$. In Example 12.6 there are two levels of factor B with means $\mu_{.1}$ and $\mu_{.2}$. The estimates of these means are $\bar{x}_{.1.}$ and $\bar{x}_{.2.}$, respectively. Each of the estimated means is calculated from $an = (3)(3) = 9$ observations. Thus the divisor of $s^2(\hat{L})$ is 9 (in general, an). If we had the contrast,

$$L_1 = \mu_{.1} - \mu_{.2}$$

for the Example 12.6 problem, then

$$s^2(\hat{L}_1) = \text{MSE} \frac{[(1)^2 + (-1)^2]}{an} = 133 \frac{[2]}{9} = 29.56$$

Treatment means μ_{ij} We can also apply the Bonferroni method to the treatment means, μ_{ij}, $i = 1, 2, \ldots, a$ and $j = 1, 2, \ldots, b$. In Example 12.6 there are six treatment means $\mu_{11}, \mu_{12}, \mu_{21}, \mu_{22}, \mu_{31}$, and μ_{32}. The estimates of these means are $\bar{x}_{11.}$, $\bar{x}_{12.}$, $\bar{x}_{21.}$, $\bar{x}_{22.}$, $\bar{x}_{31.}$, and $\bar{x}_{32.}$, respectively. Each of the estimated means is calculated for $n = 3$ observations. Thus the divisor of $s^2(\hat{L})$ is 3 (in general, n). If we had the contrasts,

$$L_1 = \mu_{11} + \mu_{21} + \mu_{31} - \mu_{12} - \mu_{22} - \mu_{32}$$
$$L_2 = \mu_{11} - \mu_{12}$$
$$L_3 = \mu_{11} - 2\mu_{21} + \mu_{31}$$

then

$$s^2(\hat{L}_1) = \text{MSE} \frac{[(1)^2 + (1)^2 + (1)^2 + (-1)^2 + (-1)^2 + (-1)^2]}{n}$$

$$= 133 \frac{[6]}{3} = 266.00$$

$$s^2(\hat{L}_2) = \text{MSE} \frac{[(1)^2 + (-1)^2]}{n} = 133 \frac{[2]}{3} = 88.67$$

$$s^2(\hat{L}_3) = \text{MSE} \frac{[(1)^2 + (-2)^2 + (1)^2]}{n} = 133 \frac{[6]}{3} = 266.00$$

The Bonferroni method is only one of several methods for constructing a set of confidence intervals using contrasts. Other commonly used methods are due to Scheffe and Tukey. (For a description of these methods and a comprehensive coverage of the use of contrasts for many experimental design models, see the Neter, Wasserman, and Kutner reference at the end of this chapter.)

■ 12.7

Violation of assumptions

There are four basic assumptions in applying the analysis of variance method to experimental data: (1) homogeneity of variances, (2) normality, (3) additivity, and (4) independence.

12.7.1 | Homogeneity of variance

This assumption requires that the population variances associated with each treatment in the experiment are equal. If the sample sizes corresponding to each treatment are equal, a very simple test for the hypothesis

$$H_0: \quad \sigma_1^2 = \sigma_2^2 = \cdots = \sigma_t^2$$
$$H_A: \quad \text{Not } \{\sigma_1^2 = \sigma_2^2 = \cdots = \sigma_t^2\}$$

is called the Hartley test, where there are t treatment populations, and it is assumed that each population is normally distributed. Hartley's test statistic (commonly called the F-max statistic) is:

> Hartley (F-max) statistic

$$\mathbf{H} = \frac{\text{Max } (s_j^2)}{\text{Min } (s_j^2)}$$

where max (s_j^2) is the largest among the t sample variances and min (s_j^2) is the smallest. Values of $\mathbf{H}$ near 1 support H_0, and large positive values support H_A. The distribution of $\mathbf{H}$ depends on the number of populations t and the sample size n. The table of critical values is given in Table B.9 in Appendix B.

Example 12.8   For the one-factor, completely randomized design in Example 12.1, test the hypothesis of equal variances.

Solution   he Farmco problem, there were three treatments (levels of the factor feed), and each sample size was five. Hence $t = 3$ and $n = 5$. The three sample variances were $s_1^2 = 480.5$, $s_2^2 = 347.5$, and $s_3^2 = 441.5$. Thus

$$h = \frac{s_1^2 = 480.5}{s_2^2 = 347.5} = 1.38$$

With $\alpha = 0.05$, the critical value of **H** is $H_{0.05;3,5} = 15.5$. Since $h < H_{0.05;3,5}$, we cannot reject the null hypothesis of equivalent population variances. Based on the experimental data, we would conclude, therefore, that the assumption of homogeneous variances (homoscedasticity) appears to be satisfied.

If the design is not balanced (the sample sizes are not equal), a more complicated test due to Bartlett (see the Neter, Wasserman, and Kutner reference) may be used. This test also assumes that the populations are normally distributed.

12.7.2 | Normality

The normality assumption requires that each treatment population is normally distributed. Two methods will be presented to test the "goodness of fit" of sample data to a normal distribution: the chi-square test (see Chapter 19) and the Kolmogorov-Smirnov test (see Chapter 20).

12.7.3 | Additivity

The additivity assumption requires that the effects in the model behave in an additive fashion. Nonadditivity may be caused by multiplicative true effects existing in the model, by the exclusion of significant interactions in the model, or by "outliers"—observations that are inconsistent in value with the majority of the responses in the experiment.

A test for this assumption that is commonly used is due to Tukey, but we shall not introduce it in this text. For the description of the Tukey test and an example of its application to experimental data, the interested reader is directed to the Neter, Wasserman, and Kutner reference at the end of this chapter.

12.7.4 | Independence

The independence assumption requires that the errors are statistically independent or, assuming that the treatment populations are normally distributed, that the errors are uncorrelated. This is a crucial assumption in analysis of variance. The use of randomization in experimental design (e.g., randomly assigning experimental units to treatments) is an "insurance policy" against correlated error terms. A test for randomness of errors is given in Chapter 20.

Regarding the assumptions of the analysis of variance, there are two important questions: (1) Are the assumptions satisfied in a particular experiment? (2) If not, what then?

The satisfaction of the assumptions can often be assured by properly designing the experiment and carefully using randomization throughout the experiment. For each assumption, a test is given in this chapter or elsewhere in the text, or a test is referenced to determine whether the data conform to the assumption. If the assumptions are not satisfied by the data, then we must be cautious in analyzing the data by analysis of variance.

| Robust procedure

In general, analysis of variance is very robust to modest departures from the assumptions. By *robust*, we mean that if significant differences exist among the factor means, analysis of variance will typically "sense" that the differences exist even if the assumptions are not entirely satisfied. That is, the power of the F-test used in analysis of variance is not seriously affected by small to moderate departures from the assumptions. This is particularly true relative to the normality assumption and, if the experiment is balanced, to the homogeneity of variance assumption.

If the independence assumption is not satisfied, the consequence is somewhat more serious. If the errors deviate slightly from an independent set of errors, analysis of variance will not be seriously affected. If the departure from independence is considerable, then the results of the analysis of variance may be meaningless.

| Transformation of
| data

If the departures from the assumptions in the experiment are serious, often the situation can be corrected by *transforming* the data. For example, we may take the logarithm of each response in the experiment and perform the analysis of variance on the logarithms if the original responses do not conform to the assumptions, but the log responses do. Various forms of transformations are given to correct for nonnormality, nonadditivity, nonhomogeneity of variance, and, to a lesser extent, nonindependence. If the errors are not independent, most often the corrective measure is a modification of the model—the inclusion of another factor or the use of blocking may be necessary to produce random errors. (For a more detailed discussion of the assumptions of analysis of variance, the interested reader is directed to the Neter, Wasserman, and Kutner reference.)

■ 12.8

Use of MINITAB

In previous sections of this chapter, we have used the SAS statistical software package to analyze the analysis of variance models presented. The SAS package is preferred to MINITAB because it calculates the values of the **F**-statistics used in hypothesis testing and prints out the *p*-value corresponding to each **F**-statistic value. MINITAB does not calculate the **F**-statistic values and does not print out the *p*-values (a persistent deficiency in the package that should be corrected).

MINITAB is applied to the two-factor, completely randomized design data in Example 12.6 for comparison to the SAS output in Table 12.15. To use MINITAB for these data, the responses (weight gains) are read into column C1 and the levels of factor A and the levels of factor B read into columns C2 and C3, respectively:

```
READ INTO COLUMNS C1, C2, and C3
    46,           1,           1,
    75,           1,           1,
    59,           1,           1,
   120,           2,           1,
    94,           2,           1,
   122,           2,           1,
    36,           3,           1,
    49,           3,           1,
    50,           3,           1,
    92,           1,           2,
    87,           1,           2,
    85,           1,           2,
   116,           2,           2,
    96,           2,           2,
    88,           2,           2,
    33,           3,           2,
    47,           3,           2,
    46,           3,           2,
```

To apply the two-factor, completely randomized design model to these data, we type:

```
TWOWAY ANALYSIS OF VARIANCE, DATA IN COLUMN C1,
LEVELS IN COLUMNS C2 and C3
```

The resulting output is:

```
-- TWOWAY ANALYSIS OF VARIANCE, DATA IN C1, LEVELS IN C2 AND C3


ANALYSIS OF VARIANCE

DUE TO            DF          SS        MS=SS/DF
FEED               2       11721.        5861.
VIT                1          85.          85.
FEED * VIT         2        1321.         661.
ERROR             12        1596.         133.
TOTAL             17       14723.

CELL MEANS
ROWS ARE LEVELS OF FEED    COLS ARE LEVELS OF VIT
                                      ROW
                 1          2        MEANS
     1          60.0       88.0      74.0
     2         112.0      100.0     106.0
     3          45.0       42.0      43.5
COL.
MEANS           72.3       76.7      74.5

POOLED ST. DEV. =        11.5

INDIVIDUAL 95 PERCENT C. I. FOR LEVEL MEANS OF FEED
(BASED ON POOLED STANDARD DEVIATION)
    +---------+---------+---------+---------+---------+---------+
1                            I******I******I
2                                                 I******I******I
3        I******I******I
    +---------+---------+---------+---------+---------+---------+
   30.       45.       60.       75.       90.      105.      120.


INDIVIDUAL 95 PERCENT C. I. FOR LEVEL MEANS OF VIT
(BASED ON POOLED STANDARD DEVIATION)
    +---------+---------+---------+---------+---------+---------+
1              I***************I***************I
2                 I***************I***************I
    +---------+---------+---------+---------+---------+---------+
   60.0      65.0      70.0      75.0      80.0      85.0      90.0
```

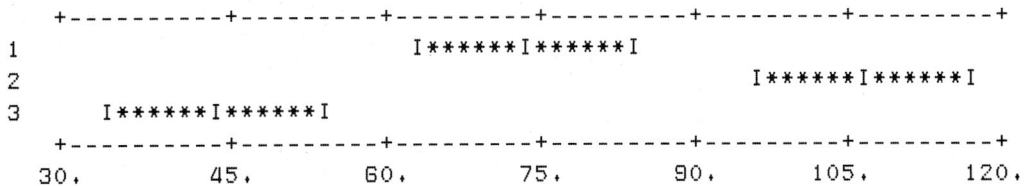

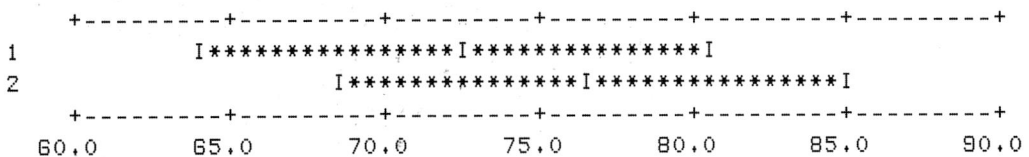

Compare the analysis of variance table above with the SAS analysis of variance table in Table 12.15. Notice that MINITAB does not compute the F-statistic values for the main effects FEED and VIT (for vitamin supplement), or for the interaction effect, FEED * VIT. MINITAB does print out a table of cell means, row means and column means—these means are given in Table 12.14. The confidence intervals for the main effects FEED and VIT are not based on the contrast method. Thus the stated confidence does not apply to the *set* of intervals, only to each individual interval. For example, based on the three confidence intervals for the FEED factor, we would conclude that $\mu_2 > \mu_1 > \mu_3$. But in Section 12.7, based on the Bonferroni multiple comparison method, we could not conclude that $\mu_2 > \mu_1$, only that $\mu_2 > \mu_3$ and $\mu_1 > \mu_3$.

MINITAB has no command for the randomized block design, although the TWOWAY command can be used to fit the randomized block design model. For analysis of variance problems, SAS is preferable to MINITAB. SAS can fit a variety of analysis of variance models and calculates the F-statistic values and the *p*-values; MINITAB does not.

■ 12.9
Summary

The analysis of variance procedure is appropriate for testing the equivalence of a set of two or more population means when certain assumptions are satisfied: (1) homogeneity of variances, (2) normality of the error terms, (3) additivity in the model, and (4) independence of the error terms. Two basic designs are given: the completely randomized design and the randomized block design. In the first design, experimental units are randomly assigned to treatments. In the second design, an extraneous factor that may be confounded with treatments is isolated by blocking. Within each block, the experimental units are randomly assigned to treatments.

More than one factor can be studied in either experimental design. Factorial experiments have the advantages of efficiently utilizing experimental units and allowing for the study of interaction effects.

If it is found that the factor level means differ significantly, Bonferroni's multiple comparison procedure may be used to identify how the means differ from one another.

Most analysis of variance problems are solved on the computer by using computer software codes available at most computer facilities. These codes typically perform tests for the assumptions mentioned above as well as other tests. Understanding how to perform these tests and how to compute the analysis of variance table is necessary for properly interpreting the results of a computer analysis and, more importantly, for checking the validity of the results.

References

Introductory

Harnett, D. L. *Statistical Methods*. 3rd ed. Reading, Mass.: Addison-Wesley, 1982, Chapter 14.

Neter, J.; W. Wasserman; and G. A. Whitmore. *Applied Statistics*. Boston: Allyn & Bacon, 1982, Chapter 21.

Wonnocott, R., and T. Wonnocott. *Introductory Statistics*. 4th ed. New York: John Wiley & Sons, 1985, Chapter 10.

Advanced

Neter, J.; W. Wasserman; and M. Kutner. *Applied Linear Statistical Models*. 2nd ed. Homewood, Ill.: Richard D. Irwin, 1985, Chapters 16 through 31.

Ostle, B., and R. Mensing. *Statistics in Research*. 3rd ed. Ames: Iowa State University Press, 1975, Chapters 9 through 12.

Snedecor, W. G., and W. Cochran. *Statistical Methods*. 7th ed. Ames: Iowa State University Press, 1980, Chapters 11 through 16.

Problems

Section 12.3 Problems

12.1 To determine whether or not three brands of light bulbs have equal mean lifetimes, a random sample of five light bulbs of each type is taken, and their lifetimes (in hours) are measured. The results are:

Brand A	Brand B	Brand C
75	85	82
63	80	71
67	82	75
70	90	79
72	78	72

At the $\alpha = 0.05$ level, test the hypothesis that the mean lifetimes of the light bulbs are equal. What, approximately, is the p-value for this test?

12.2 New employees of the Lax Company undergo a training program during their first six weeks on the job. Recently, the personnel department has suggested changing the nature of the program. To determine the effectiveness of the new program, 20 new employees are selected from the set of new employees and are randomly divided into two groups. The first group receives the "old program," and the second group receives the "new program." At the end of the training period, all 20 individuals are given an examination to determine how much information has been assimilated. The results are:

Old program	New program
65	80
60	65
45	92
50	78
48	74
62	58
66	72
42	83
45	90
56	78

a. Using analysis of variance, test the null hypothesis of equal population test scores at the $\alpha = 0.10$ level. What, approximately, is the p-value for this test?

b. Using the t-test of Chapter 11, repeat the test in part a.

c. Did you reach the same conclusion in both parts? Should you have? Discuss.

12.3 A car rental agency is in the process of deciding the brand of tire to purchase as standard equipment for its fleet. As a part of the decision process, they are interested in studying the treadlife of five competing brands. Based on testing, the research department determined that each of five tires of each brand lasted the following number of miles (in 1,000s):

		Tire brands		
A	B	C	D	E
40	45	30	35	28
42	40	32	40	32
45	40	31	42	34
38	44	35	36	28
40	42	28	38	32

Test the hypothesis at the $\alpha = 0.05$ level that the five tire brands will have identical average miles of wear. What, approximately, is the p-value for this test?

12.4 A meat producer is interested in the ability of various feeds to quickly fatten livestock. Fifteen heifers similar in age and weight are chosen to participate in an experiment to test the effectiveness of three feeds. The results in pounds gained over the duration of the experiment are:

	Feeds	
A	B	C
250	200	400
400	240	420
420	230	360
360	260	*
300	*	*

* Heifers became ill or had to be withdrawn from the experiment for other reasons.

Test the hypothesis of equal feed effectiveness in producing weight gain at the $\alpha = 0.05$ level. What, approximately, is the p-value for this test?

12.5 It is suspected that four machines in a canning operation fill cans to different levels on the average. Random samples of the cans produced from each machine were taken and the fill in ounces was measured. The results were:

	Machine		
1	2	3	4
12.10	12.18	12.20	12.01
12.08	12.10	12.25	12.00
12.22	12.10	12.25	12.03
12.12			12.00

Do the machines appear to be filling the cans at different average levels? (Calculate the approximate p-value and decide, based on its value, whether to not reject or to reject the null hypothesis.)

12.6 A marketing research firm conducted an experiment to compare the effect of five different colors on product recognition time. Thirty people were used in the experiment. Each color was used on the product package and was exposed to each person in one of five groups of six people each, formed from the 30 people. The recognition time (in seconds) was then measured. The experiment produced the following data:

$$\bar{x}_{1.} = 0.62 \qquad \bar{x}_{2.} = 0.90$$
$$\bar{x}_{3.} = 1.04 \qquad \bar{x}_{4.} = 0.48$$
$$\bar{x}_{5.} = 0.65$$
$$\text{SSTR} = 1.20 \qquad \text{SSE} = 0.60$$

(Color code: 1 = yellow, 2 = brown, 3 = blue, 4 = orange, 5 = green.)

a. State the null and alternate hypotheses.

b. Write the analysis of variance table.

c. Test the null hypothesis at the $\alpha = 0.05$ level.

d. Calculate the approximate p-value for this test.

12.7 A firm has a choice of four methods for storing potter's clay. To determine whether the storage methods differ in terms of the average number of months before the clay hardens, 20 50-pound sacks of clay were selected for experimentation. Five sacks were stored by each method. The experiment produced the following results:

$$\bar{x}_{1.} = 7.2 \text{ months} \qquad \bar{x}_{2.} = 8.0 \text{ months}$$
$$\bar{x}_{3.} = 6.6 \text{ months} \qquad \bar{x}_{4.} = 9.4 \text{ months}$$
$$\text{SSTR} = 10 \qquad \text{SSE} = 4.2$$

a. State the null and alternative hypotheses.
b. Write the analysis of variance table.
c. Test the null hypothesis at the $\alpha = 0.05$ level of significance.
d. Calculate the approximate p-value for this test.

Section 12.4 Problems

12.8 A company is interested in buying a fleet of fuel-efficient cars for use by its salespeople. Four makes and models are under consideration: DATMOON 210, TOYA Silica, Chev X-BOD, and VZ Bunny. It is decided to select the make and model that produces the highest average miles per gallon over a 200-mile course that includes both city and highway driving. Four salespeople are selected for an experiment to compare the average miles per gallon for the four cars over the selected course. Each person will drive each car once over the course. Although the salespeople are instructed to drive the cars in a similar fashion (for example, no "rabbit" starts from a standing stop), it is expected that some "people" differences will arise. Therefore, it is decided to treat the salespeople as blocks. The data are shown in the table below.

Treatments	Blocks, salespeople			
	1	2	3	4
DATMOON 210	32	30	34	32
TOYA Silica	27	27	25	26
Chev X-BOD	24	29	28	23
VZ Bunny	36	40	39	36

a. Conduct the randomized block design analysis of variance on these data.
b. What is the approximate p-value for this test?
c. Based on the p-value, do the cars appear to produce different average miles per gallon?

12.9 An advertising firm is studying the effects of four different kinds of displays of a product in a grocery store in three different sales areas in the city. Within each sales area, four stores are selected, and each receives one of the four displays. Over the duration of the experiment, the number of units of the product sold is recorded. The data are shown in the table.

Display	Sales area		
	1	2	3
A	100	56	75
B	94	40	82
C	120	65	102
D	82	60	65

a. Which model is appropriate for analyzing these data? Explain.
b. Do the four displays result in different average sales? (Use the approximate p-value to decide whether or not to reject the null hypothesis.)

12.10 In manufacturing a specific part for an automobile, four machines are used in a large plant. To determine whether the machines produce different average daily outputs, output data are collected on five different days. Since daily temperature, humidity, and so forth can affect machine output, the days are used as blocks. The data are shown in the table below.

Days (blocks)	Machines			
	1	2	3	4
1	273	286	303	313
2	240	323	320	310
3	260	303	330	348
4	278	333	323	343
5	268	338	345	325

a. State the null and alternate hypotheses.
b. Test the null hypothesis at the $\alpha = 0.05$ significance level.
c. Calculate the approximate p-value for this test.

12.11 The Alpha Insurance Company employs secretaries who use word processing software on microcomputers for typing letters, reports, and financial statements. The company wishes to choose one of three word processing software programs: Wordright, The Word, and Wordperfect. To test which of the three packages is best for the company, a report is constructed to test each package. Four secretaries are chosen to type the report using each of the three word processing software packages. Since there may be differences in the abilities of the secretaries, it is decided to use a randomized block design treating the secretaries as blocks. The times taken by the secretaries to complete the report are recorded in minutes. The table below (from MINITAB) gives the results. The C2 factor is the secretary-blocking factor with four levels representing the four secretaries. The C3 factor is the word processing treatment: Level 1 is Wordright; level 2 is The Word; and level 3 is Wordperfect. The margins of the table give the mean times for the secretaries (the rows) and the mean times for the treatments (the columns).

	1**	2	3	Row means
1*	45.00	54.00	67.00	55.33
2	35.00	47.00	50.00	44.00
3	48.00	60.00	66.00	58.00
4	38.00	48.00	55.00	47.00
Column means	41.50	52.25	59.50	51.08

* Rows are levels of C2.
** Columns are levels of C3.

The analysis of variance table is given below. The C2 factor represents the blocks (secretaries), and the C3 factor represents the treatments (word processing packages).

Analysis of variance

Due to:	DF	SS	MS = SS/DF
C2	3	398.25	132.75
C3	2	656.17	328.08
Error	6	28.50	4.75
Total	11	1082.92	

a. At the $\alpha = 0.05$ level of significance, test the null hypothesis that the three word processing packages have equal means for the completion of the typing of the report.
b. What is the approximate p-value for the hypothesis test in part a?
c. Using the Bonferroni method, investigate the differences among the three pairs of the three treatment means. Use $\alpha = 0.05$ for the contrasts.

Section 12.5 Problems

12.12 Three varieties of corn and four fertilizers were studied in a factorial experiment. The yields (in bushels per acre) were recorded for two subplots exposed to each of the $3 \times 4 = 12$ treatment combinations. The results were as shown in the table.

Fertilizer \\ Corn Variety	1	2	3
1	56	26	39
	47	35	42
2	67	45	57
	70	60	61
3	95	92	91
	90	88	82
4	92	96	98
	85	99	93

a. Write the analysis of variance model for this experiment.
b. Conduct the appropriate tests of the in-

teraction effects and the main effects (corn variety and fertilizer), if necessary.

12.13 The Startech Manufacturing Company is interested in selecting one of three production methods for the assembly of machine parts used in electric motors. The company recognizes that the number of parts produced per hour varies by the day of the week. Historically, fewer parts per hour are assembled by its employees on Mondays and on Fridays. To account for different production rates on the days of the week, the company statisti-

cian decides to block on days. On each day, from 10:00 A.M. to 11:00 A.M., the statistician records the number of parts made per hour for the set of employees assigned to this task. The table below gives the number of parts per hour that were assembled by one of three assembly methods (the rows of the table) and on each day of the five-day work week (the columns of the table). Thus the rows of the table represent the treatment factor "production method" and the columns of the table represent the blocking factor "days."

	1**	2	3	4	5	Row means
1*	62.00	68.00	74.00	72.00	66.00	68.40
2	66.00	69.00	78.00	80.00	72.00	73.00
3	58.00	54.00	66.00	70.00	62.00	62.00
Column means	62.00	63.67	72.67	74.00	66.67	67.80

* Rows are levels of C2.
** Columns are levels of C3.

The analysis of variance table is shown below. The factor C2 represents the treatment factor "production method," and the factor C3 represents the blocking factor "days."

Analysis of variance

Due to:	DF	SS	MS = SS/DF
C2	2	305.20	152.60
C3	4	342.40	85.60
Error	8	48.80	6.10
Total	14	696.40	

a. At the $\alpha = 0.05$ level of significance, is there a significant difference among the three production methods?

b. What is the approximate p-value for the hypothesis test in part *a*?

Section 12.6 Problems

12.14 In the text Example 12.5, use the Bonferroni multiple comparisons method to determine

which pairs of averages for the three oil filters differ. Use $\alpha = 0.05$ for the confidence intervals.

12.15 In Problem 12.1 use the Bonferroni multiple comparisons method to determine which mean lifetimes of the light bulbs of brands A, B, and C differ. Use $\alpha = 0.05$ for the confidence intervals.

12.16 In Problem 12.3 use the Bonferroni multiple comparisons method to determine which tire brands produce average miles of wear that differ. Use $\alpha = 0.05$ for the confidence intervals.

12.17 In Problem 12.8 apply the Bonferroni method of multiple comparisons to the averages to determine which cars differ in miles per gallon. Use $\alpha = 0.05$ for the set of contrasts.

12.18 In Problem 12.9 apply the Bonferroni method of multiple comparisons to determine which displays produce different average sales. Use $\alpha = 0.05$ for the set of contrasts.

12.19 In Problem 12.10 use the Bonferroni method of multiple comparisons to investigate the differences among the four mean outputs.

Use $\alpha = 0.05$ for the set of confidence interval estimates.

12.20   In Problem 12.11 use the Bonferroni method of contrasts to investigate how the three treatment means differ. With the three means, there are three possible pairwise comparisons. Form contrasts for each, and determine which pairs of means differ. Use $\alpha = 0.05$ for the set of three contrasts.

12.21   In Problem 12.13 use the Bonferroni method of contrasts to investigate how the three production method means differ. Use $\alpha = 0.05$ for the set of confidence interval estimates.

Section 12.7 Problems

12.22   State the three assumptions of the error term in the analysis of variance models. Which of the three assumptions is most critical in validating an analysis of variance model fitted to a data set?

12.23   Using Hartley's test, determine whether or not the two population variances are equal for the experiment in Problem 12.2.

Additional Problems

12.24   Why is the method to compare t ($t \geq 2$) population means called *analysis of variance*?

12.25   What are the objectives of analysis of variance? Discuss.

12.26   State the assumptions underlying the following analysis of variance models:
   a.   One-factor, completely randomized design.
   b.   Randomized block design.

12.27   What is the difference between a one-factor, randomized block design and a two-factor, completely randomized design?

12.28   How does the randomized block design minimize the chance that a factor under study is confounded with other factors that are related to the factor under study?

12.29   If the null hypothesis of the equality of population means is rejected by the analysis of variance method, what is the next step in the study of the data?

12.30   Explain the relationship between the two definitions of the one-factor, completely randomized design model: $\mathbf{X}_{ij} = \mu_i + \boldsymbol{\epsilon}_{ij}$, and $\mathbf{X}_{ij} = \mu + \tau_i + \boldsymbol{\epsilon}_{ij}$. What does τ_i represent?

12.31   What is a *treatment mean plot*? How is it used to identify an interaction between two factors?

12.32   What is meant by a *contrast*? What role do contrasts play in analysis of variance?

12.33   Suppose that there are $t = 4$ treatment means in a one-factor, completely randomized design. Why can't the two-sample t-test be used to identify differences among all possible pairs of means instead of using analysis of variance and the method of contrasts?

12.34   If the departures from the assumptions of analysis of variance are serious, then how can the situation be corrected?

12.35   To test the efficacy of four brands of fertilizer, the following experiment is designed. Five homogeneous plots of land are selected as blocks, and each is divided into five subplots. One subplot receives no fertilizer, and the remaining four receive each of one of the brands of fertilizer. The experiment produced the following data (yields in bushels/acre) on treatment means.

No fertilizer	*A*	*B*	*C*	*D*
$\bar{x}_{1.} = 23$	$\bar{x}_{2.} = 28$	$\bar{x}_{3.} = 35$	$\bar{x}_{4.} = 38$	$\bar{x}_{5.} = 42$

$SSE = 160$ $SSTR = 1,000$ $SSB = 150$

   a.   Write the analysis of variance model for this experiment.
   b.   Test the null hypothesis that the mean effects are equal at the $\alpha = 0.05$ significance level. What is the approximate p-value for this test?
   c.   Use the Bonferroni method of multiple comparisons to investigate the differences among the five means and interpret the results. Use $\alpha = 0.05$ for the confidence intervals.
   d.   Using the Bonferroni method, test that the following contrast is 0

$$4\mu_1 - \mu_2 - \mu_3 - \mu_4 - \mu_5$$

Interpret each contrast. Use $\alpha = 0.05$ for constructing the confidence interval.

12.36 A one-factor, completely randomized design experiment produced the following results: $t = 5$, $n = 4$, SST = 17.5, SSTR = 10, $\bar{x}_{1.} = 2.5$, $\bar{x}_{2.} = 3$, $\bar{x}_{3.} = 4.5$, $\bar{x}_{4.} = 4$, $\bar{x}_{5.} = 3.5$.

a. Complete the analysis of variance table for these data.

b. Determine the approximate p-value for testing the hypothesis of equal treatment means.

c. Use the Bonferroni method of multiple comparisons to test the following contrasts (use $\alpha = 0.05$):

   i. $\mu_1 - \mu_2 = 0$
   ii. $\mu_1 + \mu_2 - \mu_3 - \mu_4 = 0$
   iii. $2\mu_5 - \mu_1 - \mu_2 = 0$

12.37 A manufacturer of automobiles is testing three new engine types: a four-cylinder conventional engine, a diesel engine, and a newly engineered four-cylinder "hemisphere" engine. These engines are being considered for the manufacturer's new commuter sub-sub-compact car. Three of these cars are tested with the standard four-cylinder engine, five are tested with the diesel engine, and four are tested with the new engine. The results in miles per gallon are:

Four-cylinder engine	Diesel engine	New engine
36	57	50
33	53	41
48	43	47
	54	42
	48	

At the $\alpha = 0.05$ level of significance, are the three population mean miles per gallon different? If so, then use the Bonferroni method to identify how the three means differ.

12.38 Randy Granola, a famous race car builder, is considering three types of engine oil for use in his new Indy racing car. The oils are tested and rated according to a scale from 0 (worthless oil) to 100 (superb oil). The ratings are based on such factors as the ability of the oil

not to break down under heat and pressure, the lubricating effectiveness of the oil, and the ability of the oil to dissolve carbon particles in the oil system. Five quarts of each oil are tested with the following results:

Brand A	Brand B	Brand C
97	92	79
96	88	77
85	95	89
94	93	87
88	82	78

a. Are there statistically significant differences in the mean ratings for the three brands of oil? Use the $\alpha = 0.05$ significance level.

b. If the means do differ significantly, then use the Bonferroni method of contrasts to investigate how the means differ.

c. What is the approximate p-value for the test in part a?

12.39 The Front Porch Ice Cream Shop sells wonderful waffle sugar ice cream cones. For a single scoop of ice cream, the shop is interested in the number of cones sold per day at three prices: $1.75, $1.50, and $1.25. To determine whether or not the average number of cones sold per day differs for the three price settings, 10 days are randomly sampled in each of three four-week periods during which the price is set at one of the three values. The table below (from MINITAB) gives the sample average and the sample standard deviation for the number of cones sold per day at each price. The definitions of the levels of price in this table are: C1—$1.75; C2—$1.50, and C3—$1.25.

Level	N	Mean	Standard deviation
C1	10	40.40	3.98
C2	10	43.90	4.31
C3	10	52.50	3.84

The resulting analysis of variance table is given below, where the source of variation listed as "factor" in the first row represents the price treatment.

Analysis of variance

Due to:	DF	SS	MS = SS/DF	F-ratio
Factor	2	775.4	387.7	23.69
Error	27	441.8	16.4	
Total	29	1217.2		

a. Do the three prices result in different mean sales of cones per day? Test the null hypothesis at the $\alpha = 0.05$ significance level.
b. What is the approximate p-value for the test in part a?
c. Using the Bonferroni method, form the three contrasts for all possible pairs of the three means. Which pairs of means appear to differ? (Use $\alpha = 0.05$).
d. Using the Hartley method, do the population variances for the three treatments appear to be equal? (Use $\alpha = 0.05$).

12.40 The Suny Corporation is investigating four different methods for the assembly of a new microcomputer. For each of the methods, five microcomputers are assembled. Most of the assembly process is done by robotics, and very little human labor is involved. For each method the number of minutes taken to complete the assembly process is recorded. The table below (from MINITAB) gives for each method the mean assembly time in minutes and the standard deviation of the assembly times in minutes. In the table C1, C2, C3, and C4 designate the four assembly methods.

Level	N	Mean	Standard deviation
C1	5	101.00	4.36
C2	5	120.00	3.39
C3	5	140.20	3.77
C4	5	101.60	6.19

The analysis of variance table is given below. The source of variation listed as "factor" in the table is the treatment factor—the four assembly methods.

Analysis of variance

Due to:	DF	SS	MS = SS/DF	F-ratio
Factor	3	5168.2	1722.7	83.02
Error	16	332.0	20.7	
Total	19	5500.2		

a. At the $\alpha = 0.05$ level of significance, do the four assembly methods produce different mean assembly times?
b. What is the approximate p-value for the hypothesis test in part a?

12.41 The Clean Vending Company is considering three machines designed to place coffee in 8-ounce cups. Each machine is set to place 7.60 ounces of coffee in the cups. As a part of the company's evaluation of these machines, the average amount of coffee placed in the cups is measured. Twelve cups are randomly sampled from the first machine, 10 cups are randomly sampled from the second machine, and 15 cups are randomly sampled from the third machine. The table below (from MINITAB) gives the mean fill amount in ounces and the standard deviation of the fill amounts for each of the three machines. Machines 1, 2, and 3 are designated as C1, C2, and C3 in this table.

Level	N	Mean	Standard deviation
C1	12	7.508	0.144
C2	10	7.630	0.106
C3	15	7.700	0.141

The analysis of variance table is shown below. The source of variation listed in the first row as "factor" represents the machine treatment.

Analysis of variance

Due to:	DF	SS	MS = SS/DF	F-ratio
Factor	2	0.2466	0.1233	6.87
Error	34	0.6102	0.0179	
Total	36	0.8568		

a. At the $\alpha = 0.05$ level of significance, do the three machines place the same or different average amounts of coffee in the cups?

b. What is the approximate p-value for the hypothesis test in part *a*?

c. If the null hypothesis of equal treatment means is rejected in part *a*, then use the Bonferroni method to investigate all pairwise differences among the three treatment means. Use $\alpha = 0.05$.

d. Using Hartley's test, is the assumption of equal treatment variances reasonable for these data? Use $\alpha = 0.05$. (Use a sample size of 11 per treatment in Table B.9, Appendix B.)

e. Based on the results of analysis of variance and the table of the fill amount means and standard deviations for the three machines, which machine is preferable? Why?

12.42 The Bedford police department is concerned about the speeding on the airport freeway between downtown Fort Worth and the Dallas–Fort Worth airport. In an attempt to slow drivers down, they wish to experiment with three difference patrolling policies. The first method places two patrol cars on the five-mile stretch of the airport freeway that runs through Bedford. These cars drive continuously back and forth on the stretch, stopping only to give tickets. The second method places one car in a stopped position along the stretch. The car moves only to chase down speeding cars to give tickets. The third method randomly places one car on the stretch of freeway approximately 25 percent of the time during which the stretch is to be patrolled. The mean speeds in miles per hour and the standard deviations for the miles per hour for 10 cars randomly sampled along the stretch of the freeway during each patrol method are given (from MINITAB) in the table below.

Level	N	Mean	Standard deviation
C1	10	57.10	3.70
C2	10	62.70	3.06
C3	10	67.00	4.99

The analysis of variance table is shown below. The term *factor* in the table represents the three patrolling method treatments.

Analysis of variance

Due to:	DF	SS	MS = SS/DF	F-ratio
Factor	2	492.9	246.4	15.44
Error	27	431.0	16.0	
Total	29	923.9		

a. At the $\alpha = 0.05$ level of significance, do the three patrol methods produce different mean speeds?

b. What is the approximate p-value for the hypothesis test in part *a*?

c. Use the Bonferroni method of contrasts to investigate how the three treatment means differ. Form contrasts for each of the three possible pairwise comparisons of the three means. Use $\alpha = 0.05$.

d. Using the Hartley method, does the assumption of equal treatment variances appear to be reasonable for these data? Use $\alpha = 0.05$.

12.43 The Verity Accounting Firm is concerned that the average number of hours worked per week varies for its secretarial staff among its five departments. Of particular concern is the overtime recorded by the secretaries in these departments. Any hours worked more than 40 hours per week is considered to be overtime. A secretary is paid at the rate of 1.5

times his or her normal hourly rate for each hour of overtime worked. A random sample of 10 secretaries is taken from each of the five departments, and the number of hours worked last week is recorded. The table below gives the mean number of hours worked and the standard deviation of the number of hours worked for each department.

Department	Mean	Standard deviation
A	41.70	2.00
B	44.30	2.45
C	41.20	1.69
D	41.40	1.65
E	44.40	2.22

The analysis of variance table is given below.

Due to	DF	SS	MS = SS/DF	F-ratio
Department	4	103.40	25.85	6.30
Error	45	184.60	4.10	
Total	49	288.00		

a. At the $\alpha = 0.05$ level of significance, do the mean number of hours worked per week differ among the five departments?

b. What is the approximate p-value for the hypothesis test in part a?

c. Using the Bonferroni method of contrasts, construct contrasts for the 10 possible pairs of means to investigate how the five population mean number of hours may differ by department. Use $\alpha = 0.05$ for determining whether or not the means differ for each of the 10 pairs.

d. Use the Hartley method to test the hypothesis that the five treatment variances are equal. Use $\alpha = 0.05$.

12.44 The Toka Automobile Company is experimenting with ways to increase the gas mileage of its Comet automobile. Two factors are under study: Factor 1—the gear ratio in the differential, and Factor 2—the mixture of air and gasoline in the injection system. Three different levels of gear ratios are considered: 1, 2, and 3. Three levels of air/gasoline mixture are considered: 1, 2, and 3. For each treatment combination, three trials are used to record the gas mileage in miles per gallon (mpg) on a test track for the same car. The results are listed in the table below.

	Mixture		
	1	2	3
Gear 1	24	22	20
	26	25	19
	25	21	24
Gear 2	27	24	22
	25	20	26
	27	19	20
Gear 3	30	28	24
	26	24	20
	27	27	21

The treatment and factor level means are given in the following table.

	Mixture			Gear level means
	1	2	3	
Gear 1	25.00	22.67	21.00	22.89
Gear 2	26.33	21.00	22.67	23.33
Gear 3	27.67	26.33	21.67	25.22
Mixture level means	26.33	23.33	21.78	23.81

In the following table, the standard deviations for the samples of size 3 for each treatment are given:

	Mixture		
	1	*2*	*3*
Gear 1	1.00	2.08	2.65
Gear 2	1.15	2.65	3.06
Gear 3	2.08	2.08	2.08

The analysis of variance table from MINITAB is given below.

Due to:	*DF*	*SS*	*MS = SS/DF*
Gear	2	27.63	13.81
Mixture	2	96.52	48.26
Gear × mixture	4	31.93	7.98
Error	18	86.00	4.78
Total	26	242.07	

a. Form the treatment mean plots to investigate the possible interaction between the gear and the mixture factor.
b. At the $\alpha = 0.05$ level of significance, is the gear by mixture interaction significant?
c. At the $\alpha = 0.05$ level of significance, is the gear factor significant?
d. At the $\alpha = 0.05$ level of significance, is the mixture factor significant?
e. Using the Bonferroni method, investigate the differences among the levels of the mixture factor by considering all pairwise comparisons of the three means for the mixture factor.
f. Use the Hartley method to test the null hypothesis that the nine treatment variances are equal. Use $\alpha = 0.05$.

12.45 A mail-order firm is considering six possible advertisements to run nationally for several weeks. Three different designs for the advertisements are considered. The three designs are designated as Type 1, Type 2, and Type 3. Another factor being considered is whether or not the advertisement is printed in color. The firm decides to try all six combinations of the three design types and the color or black-and-white option. Each of the six advertisements is run in a test market for two, three-day periods, and the number of calls inquiring about the advertised product are recorded. The table below summarizes the number of calls received in the two replications for the six treatments.

Design types	*Black-and-white*	*Color*
1	64	72
	79	82
2	65	94
	68	86
3	40	76
	28	85

The following table gives the treatment and factor level means.

	Cell means		
	1**	*2*	*Row means*
1*	71.50	77.00	74.25
2	66.50	90.00	78.25
3	34.00	80.50	57.25
Column means	57.33	82.50	69.92

* Rows are levels of type.
** Columns are levels of color.

The standard deviations for the samples of size $n = 2$ for each treatment combination are given in the following table.

	Cell standard deviations	
	1	*2*
1	10.61	7.07
2	2.12	5.66
3	8.49	6.36

The analysis of variance table is given below.

Analysis of variance

Due to:	DF	SS	MS = SS/DF
Type	2	994.7	497.3
Color	1	1900.1	1900.1
Type * color	2	844.7	422.3
Error	6	311.5	51.9
Total	11	4050.9	

a. Form the treatment means plot to determine whether or not the design type factor and the color factor interact.

b. Test the null hypothesis that there is no design type by color factor interaction at the $\alpha = 0.05$ level of significance.

c. Use the Bonferroni method to construct a set of confidence intervals for the following contrasts (use $\alpha = 0.05$ for the familywise α):

$$\mu_{11} - 2\mu_{12} + \mu_{13}$$
$$\mu_{1.} + \mu_{2.} - 2\mu_{3.}$$
$$\mu_{22} - \mu_{32}$$

Interpret each of the estimated contrasts. What is each contrast measuring?

d. Use the Hartley method to test the hypothesis that the nine treatment variances are equal at the $\alpha = 0.05$ level of significance.

12.46 The Monster Chemical Company is interested in maximizing the yield of a chemical substance that is the combination of several ingredients. The substance is produced by applying pressure and heat to the mixture of ingredients. An experiment is developed for which three levels of pressure and four levels of temperature are chosen for testing the yield in pounds of the substance produced. For each pressure-temperature treatment combination, four trials are taken—resulting in a sample of size four of yields in pounds for each treatment. The table below gives the treatment and factor level means.

Cell means

	1**	2	3	4	Row means
1*	77.50	84.25	76.75	63.25	75.44
2	82.25	83.00	81.50	75.50	80.56
3	87.00	81.25	79.25	71.25	79.69
Column means	82.25	82.83	79.17	70.00	78.56

* Rows are levels of pressure.
** Columns are levels of temperature.

The standard deviations for the samples of size $n = 4$ for each of the 12 treatments are given in the following table.

Cell standard deviations

	1	2	3	4
1	2.65	2.87	5.38	3.59
2	2.06	2.94	2.65	3.42
3	2.58	2.99	2.50	2.99

The analysis of variance table (from MINITAB) is given below.

Analysis of variance

Due to:	DF	SS	MS = SS/DF
Pressure	2	240.50	120.25
Temperature	3	1266.23	422.08
Pressure * temperature	6	312.83	52.14
Error	36	358.25	9.95
Total	47	2177.81	

a. Form the treatment mean plots to determine whether or not the pressure and temperature factors interact.

b. Test the null hypothesis that there is no pressure by temperature interaction at the $\alpha = 0.05$ level of significance.

c. Do the pressure and temperature factors produce different mean yields for the levels of each factor? Use $\alpha = 0.05$ for these tests. Is it appropriate to test the significance of the pressure and temperature factors independently? What caution must be exercised in testing each of the pressure and temperature factors independently?

d. Use the Hartley method to test the null hypothesis that the 12 treatment variances are equal. Use $\alpha = 0.05$.

e. Which treatment appears to give the largest mean yield? Is this yield significantly greater than the next largest treatment mean yield? Use $\alpha = 0.05$ for this comparison.

13

Simple linear regression and correlation

Linear regression analysis is a technique used to estimate the value of one quantitative variable by considering its relationship with one or more other quantitative variables. For example, if we know the relationship between height and weight in American adult males, we can use regression analysis to predict weight given a particular value for height.

The relationship between height and weight is familiar to each of us; generally, the taller a person is, the more he weighs. Another example of a familiar relationship is crop yield and the amount of fertilizer applied to the land; the more fertilizer applied to the land, the greater the yield—to a point. If too much fertilizer is applied, the crop will be killed by the fertilizer chemicals; the land will be "burned." An important relationship in business is between the allocation of dollars to advertising effort and the level of sales of a product; the more money expended in advertising, in general, the higher the level of sales. We will develop in this chapter a model for predicting expected sales levels given a fixed amount expended for advertising, for example.

Regression analysis began in the late 1800s, when Sir Francis Galton, an English expert on heredity, was studying the relationship between the heights of fathers and the heights of their sons. He found that there was a tendency for short fathers to have sons of about average height (sons taller than fathers), and fathers with above average height to have sons with heights toward the average (sons were shorter than their fathers). Galton wrote that the heights of sons "regressed" or reverted to the average, thus originating the term *regression*.

Regression

The basic idea behind relating the behavior of one quantitative variable to one or more other quantitative variables should not be new to us. Indeed, it

Correlation

is the rare individual who has not heard or used the term *correlation* in making such statements as: "There is definitely a correlation between how much I study and the grade I receive in that class!" While one would hope that effort is positively correlated with exam and other performances, the relationship is *rarely* a perfect one; that is, a host of other factors undoubtedly influence performance in a class and on an examination. We will study the nature of this relationship and will derive quantitative indices that express the strength of the relationship between variables. A description or study of the *nature* of the relationship between variables will be called *re-*

Regression analysis

gression analysis, while an investigation of the *relative strength* of such a relationship will be termed *correlation analysis.*

We will develop further the distinction between correlation and regres-

Correlation analysis

sion in the pages that follow, but first we turn our attention to the nature of a statistical relationship between variables.

■ 13.2

Relationships between variables

If a *functional relationship* exists between two variables, then it is possible to represent the relationship by a formula $y = f(x)$, where x is the independent or the predictor variable, and y is the dependent or the predicted variable.

Example 13.1   Suppose that for every unit of a product sold, a company makes a profit of \$3. Let $x =$ Number of units sold and $y =$ Total profit. Then $y = 3x$. Illustrate this linear relationship.

Functional relationship

Solution   This linear functional relationship is illustrated in Figure 13.1A. For example, if $x = 10$, $y = 3(10) = \$30$; if $x = 30$, $y = 3(30) = \$90$; and if $x = 50$, $y = 3(50) = \$150$. Notice that all three of the pairs (x, y) of points fall exactly on a straight line.

FIGURE   13.1

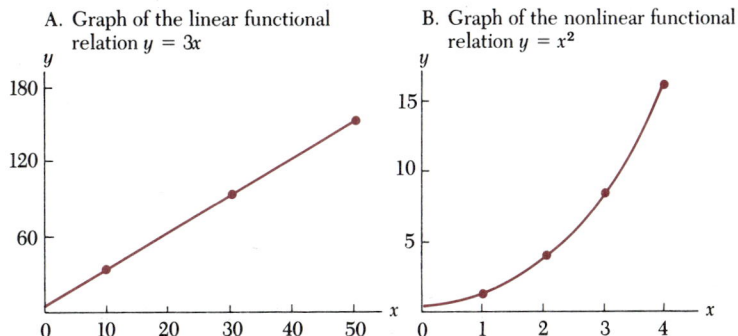

A. Graph of the linear functional relation $y = 3x$

B. Graph of the nonlinear functional relation $y = x^2$

In Example 13.1, the functional relationship is *linear*. An example of a *nonlinear* functional relationship is $y = x^2$. If, for example, $x = 2$, then $y = 4$. A graph of the functional relationship $y = x^2$ for $x > 0$ is illustrated in Figure 13.1B.

In a *statistical* relationship, the variables are not *perfectly* related as they are in a *functional* relationship. The pairs of points (x, y) will not all lie perfectly on the curve representing the relationship between the variables.

Statistical relationship

Example 13.2   An example of a *statistical* relationship is the relationship between dollars spent on advertising and sales of a product. Table 13.1 contains sales and advertising figures for a retail sales store in Niwot, Colorado, and several of its branches. Plot these data as a graph similar to Figure 13.1.

TABLE   13.1 | Sales and dollars expended on advertising for a retail store in Niwot, Colorado, and nine of its branches

Advertising expenditures x ($100)	Sales y ($1000)
18	55
7	17
14	36
31	85
21	62
5	18
11	33
16	41
26	63
29	87

Source: Company records.

Solution   These data are plotted in Figure 13.2. Clearly, the greater the number of dollars spent on advertising, the greater the general sales level. But the relationship is not a perfect one, as is evident in Figure 13.2. A straight line in Figure 13.2 has been drawn to fit reasonably well through the 10 points, and the points are scattered about this line. The scattering of points suggests that some of the variation in sales is not accounted for by advertising expenditures alone. The variation in sales not accounted for by advertising dollars may be considered to be random in nature, but may also be due to the failure to include other important independent predictor variables (price, general economic conditions, location, etc.). The randomness of the scattering of points about the fitted line is an important element in assessing the validity of a regression model. Before explaining how we fit a

FIGURE 13.2 | Plot of data in Table 13.1

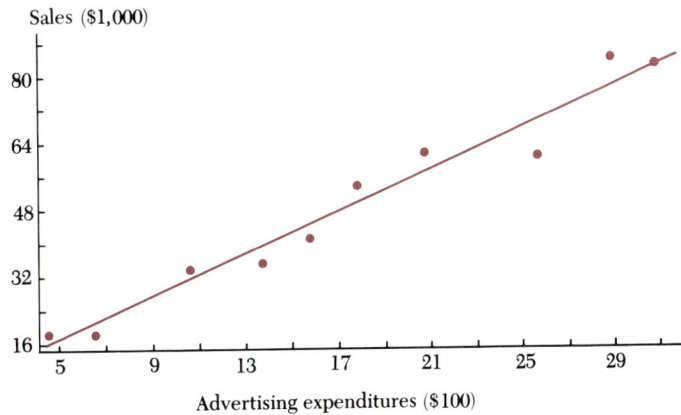

line such as displayed in Figure 13.2 to the data, we must first describe the regression model and the assumptions necessary for its correct application.

13.3

The simple linear regression model

The simple linear regression model is a mathematical way of stating the statistical relationship between two variables. The two principal elements of this statistical relationship are: (1) the tendency of the dependent random variable **Y** to vary in a systematic way with the independent variable x, and (2) the scattering of points about the "curve" that represents the relationship between x and **Y**. These two elements of a statistical relationship are represented in the simple linear regression model by assuming that: (1) there is a probability distribution of the random variable **Y** for *each value of* x, and (2) the means of these conditional probability distributions fall on a straight line. These two assumptions are illustrated in Figure 13.3 for the Example 13.2 data. The systematic way in which the random variable **Y** varies as a function of x is identified as a straight line, the *regression line of* **Y** *on* x. The regression line goes through the means of the conditional probability distributions of **Y**, given a value of x. The data are collected by taking random samples from the conditional probability distributions of **Y** for given values of x. For example, from Table 13.1, when $x = 16$, the random variable **Y** was observed to be 41. This particular value of the random variable **Y** represents a *random sample of size one* drawn from the conditional probability distribution of **Y** when $x = 16$. This value of **Y** is shown in Figure 13.3 as being somewhat below the mean of the conditional probability distribution.

There are two ways in which we can acquire the needed sample information as given in Table 13.1: by *experimentation* and by *survey*. To generate the sample data experimentally, we would *select* a set of values of x and for

Experiment

FIGURE 13.3 | Graphical form of the simple linear regression model

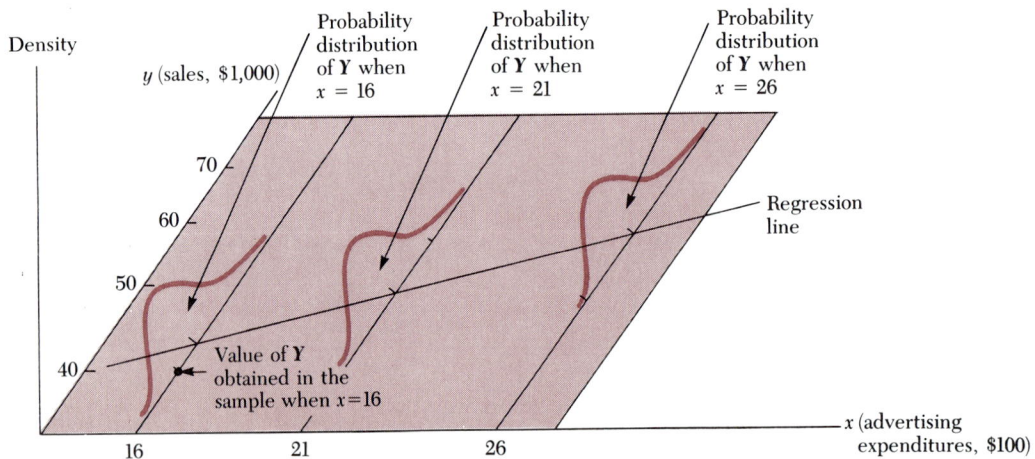

each, randomly sample one or more values of **Y**. Alternatively, we could generate the sample data by taking a survey. For example, we could randomly sample 10 branches of the firm in Example 13.2 to determine their advertising expenditures and corresponding sales. The survey method has the disadvantage that we must take whatever values of x occur in the survey; the selection of the set of values, the independent variable, is out of our control. We might be so unfortunate, for instance, to find that *all* 10 sales branches in our survey had advertising expenditures of $1,600—we ideally want a spread of x-values over the range of interest to us and over which the regression line will be built.

Survey

In regression analysis, it is always better to produce the sample data by experimentation, if possible, for then we can control the independent variable x—the experiment can be designed to suit our needs. When experimentation is not possible, surveys must be used to generate the data.

In discussing regression analysis, we will assume that the values of x, the independent variable, are fixed or are given in advance. This gives rise to the notation **Y** and x, where the dependent variable **Y** is a random variable, and the independent variable x is *not* a random variable, but is fixed. This relationship is shown in Figure 13.3—there is a conditional distribution of **Y**-values *for each* value of x. When discussing correlation in Section 13.10, it will be necessary for us to assume that both **Y** and **X** are random variables, and that their joint distribution is bivariate normal. This distinction is more fully developed in Section 13.10. For the present, we will assume that the independent variable x is not a random variable in our discussion of simple linear regression.

Y is a random variable

x is *not* a random variable in the regression model

The formal statement of the simple linear regression model is given here.

> ## Population regression model
>
> $$Y_i = \beta_0 + \beta_1 x_i + \epsilon_i, \qquad i = 1, 2, \ldots, n$$
>
> where:
>
> Y_i = ith dependent random variable.
> β_0, β_1 = Parameters in the regression model.
> x_i = ith level of the independent variable.
> ϵ_i = Random error term.
> n = Number of (x, y) pairs of observations.

The assumptions corresponding to the use of the population regression model are given in the following box.

> ## Assumptions underlying the use of the simple linear regression model
>
> 1. For the ith level of x, x_i, the expected value of the error component is 0 $[E(\epsilon_i) = 0]$, and the variance of the error component is σ^2 $[V(\epsilon_i) = \sigma^2]$ and is *constant* for all i, where $i = 1, 2, \ldots, n$.
> 2. The error components ϵ_i, ϵ_j between *any pair* of values of the dependent variable are uncorrelated.
> 3. β_0 and β_1 are unknown constants and must be estimated from the sample data.

The consequences of these assumptions are as follows:

1. The observed value of the random variable Y_i (i.e., y_i) when $x = x_i$ is the sum of two components, a constant and a *value of* a random variable:

$$y_i = \underbrace{\beta_0 \qquad + \qquad \beta_1 x_i}_{\text{constant}} + \underbrace{\epsilon_i}_{\substack{\text{value of random} \\ \text{variable } \epsilon_i}}$$

2. By using the expectation operator rules given in Chapter 5,

$$E(Y_i) = E(\beta_0 + \beta_1 x_i + \epsilon_i) = \beta_0 + \beta_1 x_i + E(\epsilon_i)$$

Since $E(\epsilon_i) = 0$,

$$E(\mathbf{Y}_i) = \beta_0 + \beta_1 x_i$$

Thus the mean of the conditional probability distribution of $\mathbf{Y}_i$ given a value of $x = x_i$ and denoted by $\mu_{\mathbf{Y}/x}$ (we omit the i subscript here) is equal to $\beta_0 + \beta_1 x$. Therefore, the regression function corresponding to the regression model is:

$$E(\mathbf{Y}) = \mu_{\mathbf{Y}/x} = \beta_0 + \beta_1 x$$

3.
$$\begin{aligned}
V(\mathbf{Y}_i) &= V(\beta_0 + \beta_1 x_i + \epsilon_i) \\
&= V(\beta_0 + \beta_1 x_i) + V(\epsilon_i) = 0 + V(\epsilon_i)
\end{aligned}$$

Since $V(\epsilon_i) = \sigma^2$,

$$V(\mathbf{Y}_i) = \sigma^2$$

Thus the variance of the conditional probability distribution of $\mathbf{Y}_i$ given $x = x_i$ and denoted by $\sigma^2_{\mathbf{Y}/x}$ is equal to σ^2, *and each conditional probability distribution has the same variance, σ^2.*

4. The observed value of $\mathbf{Y}_i$, y_i, when $x = x_i$, is larger or smaller than $\mu_{\mathbf{Y}/x}$ by the amount ε_i, the value of the error component, ϵ_i.

5. The second assumption requires that pairs ϵ_i, ϵ_j, $i \neq j$, of the error component are uncorrelated. Using the covariance operator given in Chapter 5, this assumption implies:

$$C(\epsilon_i, \epsilon_j) = 0 \qquad \text{for all } i \text{ and } j, \ i \neq j$$

Regression coefficients	

Regression coefficients—parameters β_0 and β_1
in the simple linear regression model

Two parameters in the simple linear regression model, β_0 and β_1, are called the *regression coefficients*. The coefficient β_0 is called the *Y-intercept*; it is the value of $\mathbf{Y}$ (according to the regression line) when $x = 0$. The coefficient β_1 is the *slope* of the regression line; its numerical value gives the change in the dependent random variable $\mathbf{Y}$ (either positive or negative) when there is a one-unit *increase* in the value of the independent variable, x.

Example 13.3 Suppose that in Example 13.2, the appropriate population regression model is:

$$\mathbf{Y}_i = 1.00 + 2.75x_i + \epsilon_i$$

Then, the regression function is given by $E(\mathbf{Y}_i) = 1.00 + 2.75\, x_i$. Graph this regression function.

Solution This regression function is shown in Figure 13.4. When $x = 11$, $\mu_{Y/x=11} = 1.00 + 2.75(11) = 31.25$; when $x = 21$, $\mu_{Y/x=21} = 58.75$; and when $x = 31$, $\mu_{Y/x=31} = 86.25$. The regression function passes through the conditional means $\mu_{Y/x} = 1.00 + 2.75x$. Furthermore, each conditional probability distribution has the same variance σ^2. And the error component is the deviation between the observed value of **Y** and $\mu_{Y/x}$. For example, when $x = 31$, from Table 13.1, the observed value of **Y** is 85; $\mu_{Y/x=31} = 86.25$. The error component is $\varepsilon_i = y_i - \mu_{Y/x=31} = 85 - 86.25 = -1.25$ as shown in Figure 13.4.

FIGURE   13.4 | Simple linear regression model

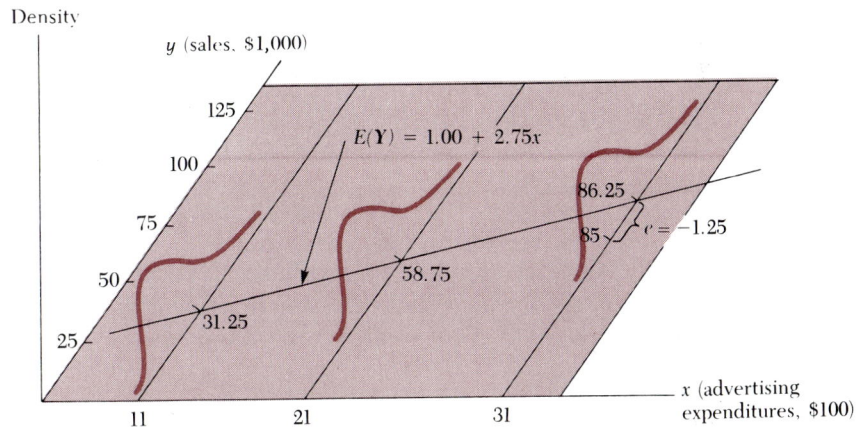

Example 13.4 In Example 13.3, we assumed that the correct regression function relating sales to advertising is $E(\mathbf{Y}) = 1.00 + 2.75x$. What are the values of the regression coefficients in this regression function?

Solution This function is graphed in Figure 13.5. The Y-intercept, $\beta_0 = 1.00$, is the value of the regression function when $x = 0$. The slope, β_1, tells us how much sales are *expected* to increase for each dollar increase in advertising expenditure. Since advertising expenditures x are expressed in $100s and sales **Y** are expressed in $1,000s, the regression coefficient $\beta_1 = 2.75$ indicates that sales are *expected* to increase by $2,750 for each additional $100 spent on advertising.

Example 13.5 The value of β_1, the slope, is particularly important in determining the nature of the regression function. If β_1 is positive, the regression line is *increasing* as the value of x increases. If β_1 is negative, the regression line is *decreasing* as x increases. If $\beta_1 = 0$, the regression line is horizontal (constant) for all values of x. Graph regression lines where $\beta_1 > 0$, where $\beta_1 < 0$, and where $\beta_1 = 0$.

Increasing line

Decreasing line

FIGURE 13.5 | Graph of the regression function $E(Y) = 1.00 + 2.75x$

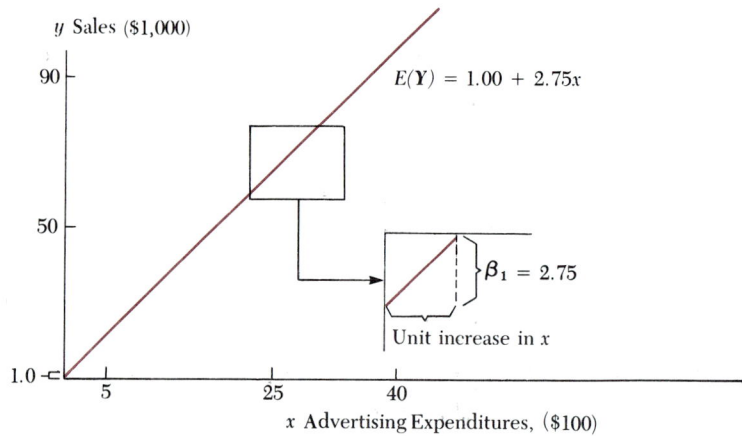

Solution These cases are illustrated in Figure 13.6. When the slope β_1 is 0, the regression function becomes $E(Y) = \beta_0$; that is, the mean of the conditional probability distribution is equal to β_0 *regardless* of the value of x. In this instance, the regression function is *not at all helpful* in predicting the value of **Y** given a value of x—the predicted value is *always* β_0.

FIGURE 13.6 | Nature of the regression line when β_1 is positive, negative, and 0

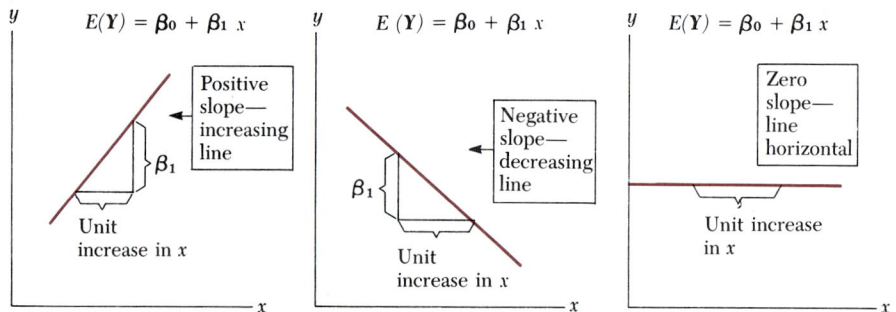

Since β_0 and β_1 will generally not be known in a particular problem, we now turn to a discussion of how a regression line is "fitted" to a set of (x, y) sample data points.

■ 13.4

The fitting of a simple linear regression model

Since β_0 and β_1 are generally not known in a regression problem, they must be *estimated* from sample data taken on the dependent random variable **Y** for a number of values of the independent variable x. These pairs of sample values are obtained either by experimentation or by survey. The data given in Table 13.1 were obtained by conducting a survey of 10 branch stores. The advertising expenditures (x) for the stores were not chosen experimentally—they were set by the store managers and recorded in the survey of the 10 stores.

Each sales branch in this example is called a *unit of association* because each branch included in the sample provides us with a value of the independent variable x (advertising expenditures) and a corresponding value of the dependent variable **Y** (sales). It is important to preserve this pairwise identity of (x, y) data pairs as we estimate the population regression line (function) from the sample data.

In Figure 13.7A, the scatterplot for the data in Table 13.1 is given. In Figures 13.7B and 13.7C, we have superimposed two "fitted" regression

Unit of association

FIGURE 13.7

A. Scatter diagram for data in Table 13.1

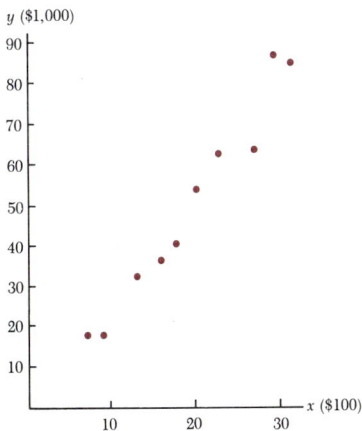

B. Scatter diagram and fitted line
$\hat{y} = 49.7 + 0x$ for data in Table 13.1

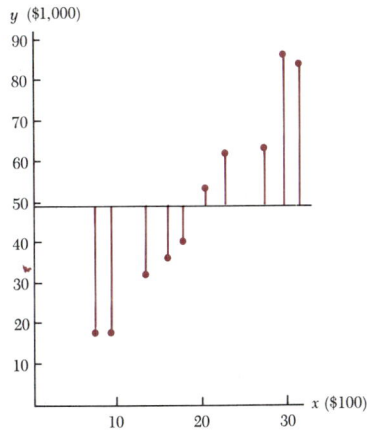

C. Scatter diagram and fitted line
$\hat{y} = 1.02 + 2.73x$ for data in Table 13.1

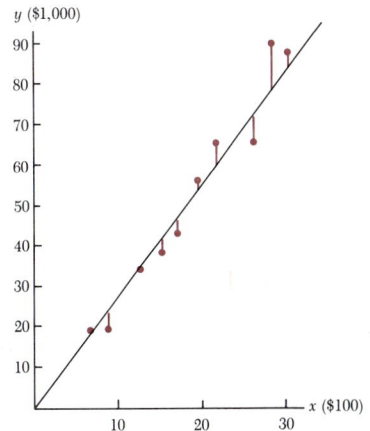

lines through the scatter of points, $\hat{y} = 49.7 + 0x$ and $\hat{y} = 1.02 + 2.73x$, respectively. It is apparent in Figure 13.7 that the line $\hat{y} = 1.02 + 2.73x$ fits the given data "better," but we must establish a criterion to evaluate when one line is "better" than another so that we can find the best-fitting line. The criterion we shall use is called *least squares*. For each sample observation pair (x_i, y_i), the least squares (LS) criterion considers the deviation of y_i from its expected value:

Least squares

$$y_i - E(\mathbf{Y}_i) = y_i - (\beta_0 + \beta_1 x_i) = \varepsilon_i$$

and requires that *values* of β_0 and β_1 be found that minimize:

$$\text{LS} = \sum_{i=1}^{n} (y_i - \beta_0 - \beta_1 x_i)^2 = \sum_{i=1}^{n} \varepsilon_i^2$$

This criterion is equivalent to minimizing the sum of the squared vertical "distances," shown by the vertical red lines, in Figure 13.7.

The specific *values* of β_0 and β_1 that minimize LS are the regression coefficient *estimates*, denoted by b_0 and b_1, respectively.[1] Thus the least squares criterion requires that we find a line, denoted by $\hat{\mathbf{Y}} = \mathbf{b}_0 + \mathbf{b}_1 x$, such that the sum of the squared *vertical* deviations between the line and the scatter of points is minimized. In Figure 13.7B, the vertical deviations corresponding to the line $\hat{y} = 49.7 + 0x$, where $b_0 = 49.7$ and $b_1 = 0$, are indicated. Obviously the line $\hat{y} = 1.02 + 2.73x$ in Figure 13.7C, where $b_0 = 1.02$ and $b_1 = 2.73$, does much better in the least squares sense because its vertical deviations from the scatter of points, when squared and summed, will be less than the sum of the squared deviations from the line $\hat{y} = 49.7 + 0x$.

It turns out that the *values* of $\mathbf{b}_0$ and $\mathbf{b}_1$ (b_0 and b_1, respectively) that minimize LS are solutions to the following two simultaneous equations, which are referred to as the *normal equations*:

Normal equations

Least squares normal equations

$$\Sigma y_i = n b_0 + b_1 \Sigma x_i$$
$$\Sigma x_i y_i = b_0 \Sigma x_i + b_1 \Sigma x_i^2$$

Solving the normal equations for b_0 and b_1 produces the point estimates of β_0 and β_1, respectively.[2] Computing formulas for deriving the estimates b_0 and b_1 are as follows:

---

[1] Following the notation of previous chapters, we will refer to the random variable estimators of the population regression coefficients β_0 and β_1 by $\mathbf{b}_0$ and $\mathbf{b}_1$, respectively. The sample values obtained will be denoted by nonboldface b_0 and b_1.

[2] Minimizing the least squares function LS with respect to β_0 and β_1 produces the point estimates b_0 and b_1. Partial differential calculus is required to perform this minimization. See the Draper and Smith reference at the end of the chapter for a derivation of the normal equations.

Estimates of the regression coefficients

> Computation formulas for least squares estimates b_0 and b_1

$$b_1 = \frac{\sum x_i y_i - \dfrac{(\sum x_i)(\sum y_i)}{n}}{\sum x_i^2 - \dfrac{(\sum x_i)^2}{n}} \tag{13.1}$$

$$b_0 = \bar{y} - b_1 \bar{x}, \quad \text{where } \bar{y} = \frac{\sum y_i}{n}, \quad \bar{x} = \frac{\sum x_i}{n} \tag{13.2}$$

Example 13.6 The sales-advertising data are given in Table 13.1. The scatterplot of these data in Figure 13.7A suggests the reasonableness of fitting the linear model (a line) to these sample data. Using the data provided, fit a simple linear regression model using sales as the dependent variable and advertising expenditures as the independent variable.

Solution The summations required for determining b_0 and b_1 are given in Table 13.2. The format in this table provides a convenient worksheet for finding the necessary components in the formulas for b_0 and b_1.

$$b_1 = \frac{\sum x_i y_i - \dfrac{(\sum x_i)(\sum y_i)}{n}}{\sum x_i^2 - \dfrac{(\sum x_i)^2}{n}} = \frac{10{,}820 - \dfrac{(178)(497)}{10}}{3{,}890 - \dfrac{(178)^2}{10}} = 2.7347 \doteq 2.73$$

$$b_0 = \bar{y} - b_1 \bar{x} = \frac{497}{10} - (2.7347)\left(\frac{178}{10}\right) = 1.02$$

Thus the fitted regression line is $\hat{y} = 1.02 + 2.73x$, and this is the best-fitting line based on the least squares criterion.

Notation for the dependent random variable **Y**

The notation for the dependent random variable **Y** is as follows. The ith level of the random variable **Y** is denoted by $\mathbf{Y}_i$, uppercase and boldfaced to indicate that it is a random variable whose value is unknown until the experiment or survey is conducted. Its value, once the experiment or survey has been conducted, is denoted by y_i, lowercase and not boldfaced. Similarly, y will denote a value of the dependent random variable **Y**. This notation follows the convention used in previous chapters.

The column headed by $\hat{y}$ in Table 13.2 gives the estimated value of $E(\mathbf{Y}) = \mu_{Y/x}$ for each value of x based on the fitted regression line. For example, when $x = 21$,

$$\hat{y} = 1.02 + (2.73)(21) = 58.45$$

that is, for an advertising expenditure of 21 (in \$100s), the estimate of $\mu_{Y/x=21}$, the average or expected sales amount, is \$58.45 (in \$1000s). The

TABLE 13.2 | Computation worksheet for determining b_0 and b_1 in Example 13.6*

Observation	x	y	x^2	y^2	xy	$\hat{y}$	$y - \hat{y}$	$(y - \hat{y})^2$
1	18	55	324.0	3,025.0	990.0	50.247	4.753	22.591
2	7	17	49.0	289.0	119.0	20.165	−3.165	10.015
3	14	36	196.0	1,296.0	504.0	39.308	−3.308	10.942
4	31	85	961.0	7,225.0	2,635.0	85.799	−0.799	0.638
5	21	62	441.0	3,844.0	1,302.0	58.451	3.549	12.594
6	5	18	25.0	324.0	90.0	14.695	3.305	10.922
7	11	33	121.0	1,089.0	363.0	31.104	1.896	3.596
8	16	41	256.0	1,681.0	656.0	44.777	−3.777	14.269
9	26	63	676.0	3,969.0	1,638.0	72.125	−9.125	83.266
10	29	87	841.0	7,569.0	2,523.0	80.329	6.671	44.498
Totals	178	497	3,890.0	30,311.0	10,820.0	497.000	0.000	213.332

* The quantities in the column labeled $\hat{y}$ were computer-generated and may differ *slightly* (because of roundoff error) from those determined by use of the equation $\hat{y} = 1.02 + 2.73x$.

| The fitted value $\hat{y}_i$ |

term, $\hat{y}_i$, denotes the estimated value of μ_{Y/x_i}. It is often referred to as the *fitted value,* since it represents the point on the fitted regression line for the value x_i.

The second to last column in Table 13.2, labeled $(y - \hat{y})$, gives the difference between the actual sales amount (y) and the estimated average sales amount ($\hat{y}$) for each of the 10 stores. This difference is called the *residual* associated with the ith observed value y_i.

| The ith residual |

The residual corresponding to the ith observed value y_i

The ith residual, denoted by e_i, is the difference between the observed value y_i and the corresponding fitted value, $\hat{y}_i$:

$$e_i = y_i - \hat{y}_i$$

| Distinction between value of error term and residual |

The distinction between the error term value in the regression model $\varepsilon_i = y_i - E(Y_i)$, and the residual, $e_i = y_i - \hat{y}_i$, is important. The error term value ε_i represents the vertical deviation of the ith observation y_i from the *unknown* population regression line, $E(Y) = \beta_0 + \beta_1 x$. The residual e_i represents the vertical deviation of the ith observation y_i from the *fitted* or *estimated regression line,* $\hat{y} = b_0 + b_1 x$.

The residuals have certain properties when the method of least squares is used to fit the regression line to a set of sample observations:

The residuals sum to zero

1. The sum of the residuals is zero:

$$\sum_{i=1}^{n} e_i = 0$$

Notice in Table 13.2 that the sum of the column $y - \hat{y}$ is zero to four decimal places.

The sum of squared residuals is a minimum

2. The sum of the squared residuals is a minimum:

$$\sum_{i=1}^{n} e_i^2 \quad \text{is a minimum.}$$

This is the criterion used by least squares in fitting the regression line to a sample set of observations. Since $e_i = y_i - \hat{y}_i$,

$$\sum_{i=1}^{n} e_i^2 = \sum_{i=1}^{n} (y_i - \hat{y}_i)^2$$

Notice that the last column in Table 13.2 sums the squared residuals (sum = 213.332).

There are other interesting properties that follow from the application of the least squares criterion. These are listed below without derivation (for the derivation of these properties, see the Neter, Wasserman, and Kutner reference at the end of this chapter).

Sum of observed values equals sum of fitted values

1. The sum of the observed values y_i equals the sum of the fitted values $\hat{y}_i$:

$$\sum_{i=1}^{n} y_i = \sum_{i=1}^{n} \hat{y}_i$$

Sum of residuals weighted by the x values is zero

2. The sum of the residuals weighted by the x values is zero:

$$\sum_{i=1}^{n} x_i e_i = 0$$

Sum of residuals weighted by fitted values is zero

3. The sum of the residuals weighted by the fitted values is zero:

$$\sum_{i=1}^{n} \hat{y}_i e_i = 0$$

Fitted regression line passes through the point $(\bar{x}, \bar{y})$

4. The fitted regression line *always* goes through the point $(\bar{x}, \bar{y})$.

These properties can be used to check computations in a regression analysis. Most computer programs check these properties and the property,

$$\sum_{i=1}^{n} e_i = 0$$

to determine the degree of numerical roundoff error in fitting the line to a set of sample observations. Due to computational roundoff, we cannot expect the residuals to sum to zero *exactly*. However, if the sum is not close to zero, then this indicates that a mistake has been made in determining the fitted regression line, $\hat{y} = b_0 + b_1 x$.

Other quantities in Table 13.2 (y^2, x^2 and xy) are used in other regression analysis computations to be discussed later in this chapter.

The following theorem establishes the properties of the least squares point estimators b_0 and b_1.

Best linear unbiased estimators of the regression coefficients

---

Theorem 13.1

Gauss-Markov Theorem (minimum variance, unbiased estimators of population regression coefficients β_0 and β_1)

When the assumptions of the simple linear regression model are satisfied, the least squares estimators b_0 and b_1 are unbiased and have minimum variance among *all* linear unbiased estimators of β_0 and β_1.

Thus we know from this theorem that $E(b_0) = \beta_0$ and $E(b_1) = \beta_1$—the estimators are unbiased. Furthermore, the theorem assures us that the estimators b_0 and b_1 are minimum variance estimators among the class of unbiased and linear estimators. Thus, over this class of estimators, the sampling distributions of b_0 and b_1 have less variability than do the sampling distributions of other point estimators of β_0 and β_1. The properties of LS estimators b_0 and b_1 are summarized in Figure 13.8. The sampling distribution of b_1, for instance, has a mean of β_1 [since $E(b_1) = \beta_1$], and its variability is less than that of the sampling distribution of any other linear and unbiased estimator of β_1.

■ 13.5

The estimator of the conditional probability distribution variance σ^2

We assume in the simple linear regression model that the conditional probability distributions of $\mathbf{Y}$, given a value of x, have the same variance, σ^2. When drawing inferences concerning the fitted regression line, it will be necessary to estimate σ^2 from sample data.

Recall that when we have a single population with an unknown variance σ^2, we use the sample variance

$$s^2 = \sum_{i=1}^{n} \frac{(x_i - \bar{x})^2}{n - 1}$$

as the point estimate of σ^2. The process of forming a point estimate s^2 of σ^2 is reviewed in Figure 13.9.

The divisor of $(n - 1)$ reflects that one degree of freedom is lost since the estimate $\bar{x}$ is used in place of the unknown population mean μ in the computation of s^2.

FIGURE 13.8 | Sampling distributions of b_0 and b_1

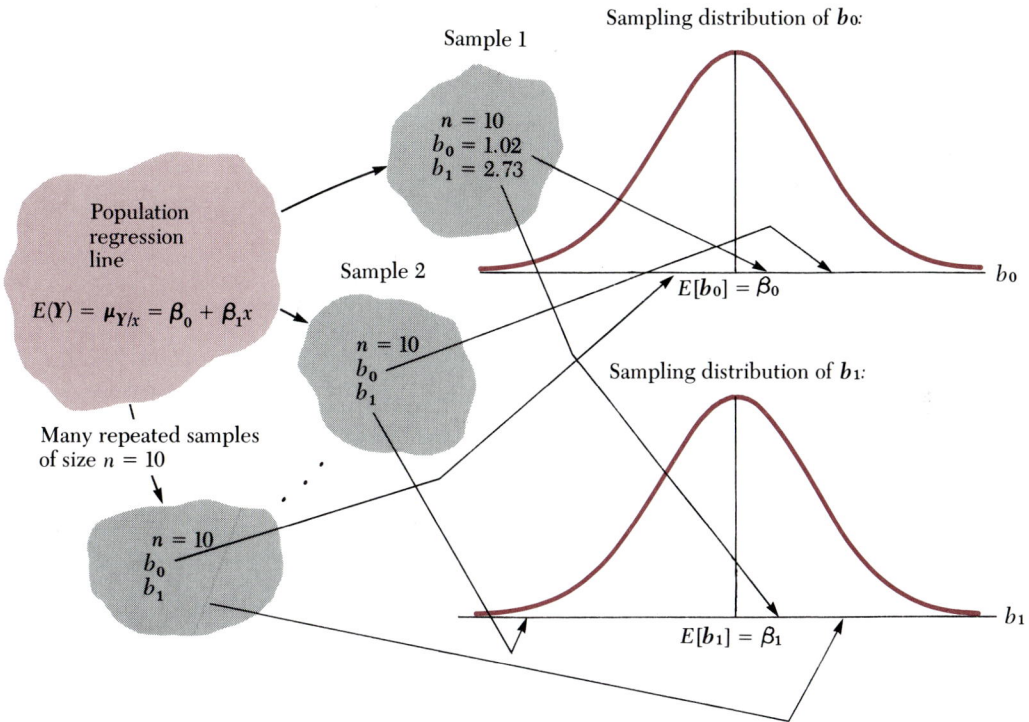

FIGURE 13.9 | Point estimate of σ^2

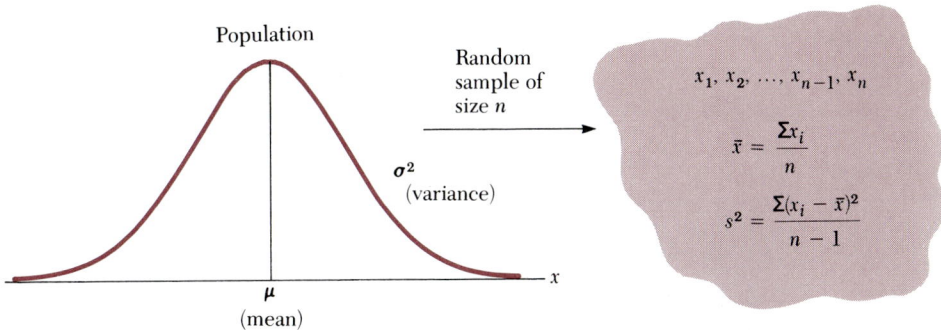

In regression analysis, we have random observations on **Y**, but they come from *two or more* population distributions as illustrated in Figure 13.10. The number of population distributions from which we sample depends on how many different levels of x we have in the experiment or survey; in Figure 13.10 it is assumed that we have n different levels of x.

FIGURE 13.10 | Conditional probability distributions of **Y**

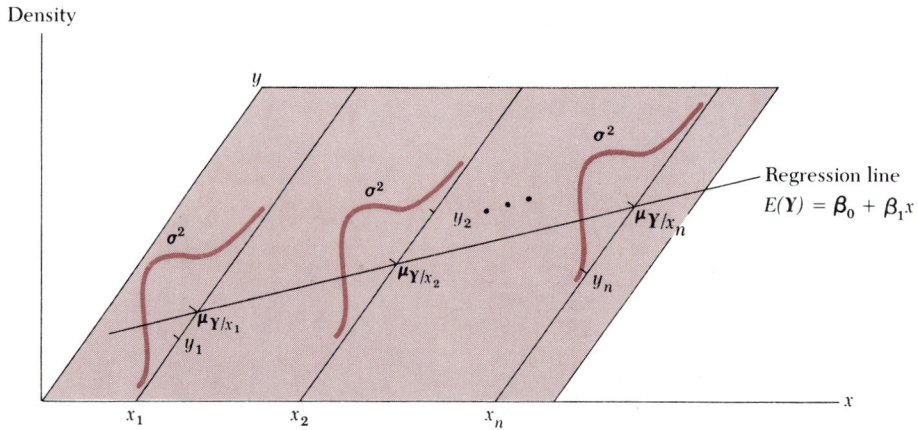

To form the point estimator of σ^2, we use the same basic form as the estimate of σ^2 based on a random sample taken from a single population. Since the conditional means $\mu_{\mathbf{Y}/x_i}$ are unknown (as was μ in the single-population case), we will estimate the population variance σ^2 using the predicted value of **Y**, given the value of x. The deviations we will use, therefore, are $y_i - \hat{y}_i = y_i - (b_0 + b_1 x_i)$; for each value of x, $\hat{y}_i = b_0 + b_1 x_i$ estimates the conditional mean $\mu_{\mathbf{Y}/x_i}$. Thus the estimate of σ^2 is given by the following formula (the square root of this quantity is called the *standard error of the estimate* and is given by $s_{\mathbf{Y}/x}$).

Standard error of the estimate

Estimate of σ^2

Estimate of the conditional distribution variance

$$s_{\mathbf{Y}/x}^2 = \sum_{i=1}^{n} \frac{(y_i - \hat{y}_i)^2}{n-2} = \sum_{i=1}^{n} \frac{[y_i - (b_0 + b_1 x_i)]^2}{n-2} \tag{13.3}$$

The reason we divide by $(n - 2)$ in equation 13.3 is that two parameters must be estimated in determining $s^2_{Y/x}$: namely, β_0 and β_1. Hence two degrees of freedom are lost in estimating σ^2 because b_0 and b_1 are used in place of the unknown population parameters β_0 and β_1.

Two commonly used alternate (computing) formulas for calculating $s^2_{Y/x}$ are given now.

Estimate of σ^2
computing formula
number 1

> Conditional distribution variance estimate—computing formula no. 1
>
> $$s^2_{Y/x} = \frac{1}{n - 2} \{\Sigma y_i^2 - b_0 \Sigma y_i - b_1 \Sigma x_i y_i\}$$ (13.4)

Estimate of σ^2
computing formula
number 2

> Conditional distribution variance estimate—computing formula no. 2
>
> $$s^2_{Y/x} = \frac{1}{n - 2} \left\{ \left[\Sigma y_i^2 - \frac{(\Sigma y_i)^2}{n} \right] - \frac{\left[\Sigma x_i y_i - \frac{(\Sigma x_i)(\Sigma y_i)}{n} \right]^2}{\Sigma x_i^2 - \frac{(\Sigma x_i)^2}{n}} \right\}$$ (13.5)

$S_{Y/x}$ is often called the *standard error of the estimator*. It is a measure of the *absolute* fit of the sample data points to the regression line. This is to be contrasted to the coefficient of determination described in the next section, which measures the *relative* goodness of fit of the sample regression line to the sample data points.

Although equation 13.5 looks most imposing for computing $s^2_{Y/x}$, it is usually the most accurate and fastest to use—remember, certain quantities in equation 13.5 have already been calculated in the determination of b_0 and b_1, as the following example illustrates.

Example 13.7   Let us return to the sales-advertising data given in Table 13.2. Using these data and equation 13.3, determine $s^2_{Y/x}$.

Solution   The last column in Table 13.2 gives the squared error deviations, $(y_i - \hat{y}_i)^2$. These deviations may be used to calculate $s^2_{Y/x}$ from equation 13.3. The quantity, $\Sigma_{i=1}^n (y_i - \hat{y})^2$, in equation 13.3 is called the *sum of squares due to the errors* and will be abbreviated by SSE. From Table 13.2, SSE = 213.332; therefore,

$$s_{\mathbf{Y}/x}^2 = \frac{\sum_{i=1}^{n} (y_i - \hat{y}_i)^2}{n - 2} = \frac{SSE}{n - 2} = \frac{213.332}{10 - 2} = 26.67$$

Thus, the point estimate of σ^2 is $s_{\mathbf{Y}/x}^2 = 26.67$.

This estimate can also be determined by using equation 13.5 and the data given in Table 13.2:

$$s_{\mathbf{Y}/x}^2 = \frac{1}{n - 2} \left\{ \left[\sum y_i^2 - \frac{(\sum y_i)^2}{n} \right] - \frac{\left[\sum x_i y_i - \frac{(\sum x_i)(\sum y_i)}{n} \right]^2}{\sum x_i^2 - \frac{(\sum x_i)^2}{n}} \right\}$$

$$= \frac{1}{10 - 2} \left\{ \left[30,311 - \frac{(497)^2}{10} \right] - \frac{\left[10,820 - \frac{(178)(497)}{10} \right]^2}{3,890 - \frac{(178)^2}{10}} \right\}$$

$$= \frac{213.332}{8} = 26.67$$

The terms in braces in equations 13.4 and 13.5 are simply alternate expressions for SSE. Equation 13.4 should be used with care—be certain to use plenty of decimal places for b_0 and b_1 in this formula, or else rather large roundoff errors may occur.

When regression data are generated by experimentation, it is often possible to test statistically the hypothesis that the variances of the n conditional probability distributions are all equal to σ^2 at a stated level of significance using analysis of variance techniques from Chapter 12. Often, plotting the sample data can serve as a "quick" guide in making judgments on the equality of variances. For example, a histogram plot of the data similar to that shown in Figure 13.10 for each level of x would indicate whether there was serious violation of the equality of variance assumption. In Section 13.7 we will describe the use of residual plots in visually testing whether the variances indeed do all appear to be equal.

13.6

The coefficient of determination r^2

Coefficient of determination $\mathbf{r}^2$

Population coefficient of determination ρ^2

To this point, we have not dealt with the *relative strength* of the linear relationship between the independent variable x and the dependent variable $\mathbf{Y}$. One measure of the relative strength of the linear relationship between x and $\mathbf{Y}$ is given by the *coefficient of determination*, $\mathbf{r}^2$.

The coefficient of determination $\mathbf{r}^2$ gives the proportion of variability in the sample dependent variable $\mathbf{Y}$ that is explained by the independent variable x through the fitting of the regression line. It is a point estimator of the *population coefficient of determination*, ρ^2. We will now describe how the sample statistic $\mathbf{r}^2$ measures the proportion of variability explained in the sample.

The variation in the observations on the dependent variable $\mathbf{Y}$ is measured in terms of the deviations $y_i - \bar{y}$: we square these deviations, where $i = 1, 2, \ldots, n$, and then sum them. If we divide this sum by $(n - 1)$, we have the sample variance of the random variable $\mathbf{Y}$:

$$s_Y^2 = \frac{\sum_{i=1}^{n}(y_i - \bar{y})^2}{n-1} = \frac{\sum y_i^2 - \frac{(\sum y_i)^2}{n}}{n-1}$$

Sum of squares total

Consider the *numerator* of s_Y^2: it is the total sum of squared deviations of the n observations about their mean, $\bar{y}$. We will denote the expression in the numerator of s_Y^2 by SST—*sum of squares total*. Although we typically use s_Y^2 to measure the variation in **Y**, **SST** certainly can measure the variation in **Y** as well. For the sales-advertising data, the value of **SST** is 5,610.10. The individual deviations from $\bar{y}$ before they are squared are shown in Figure 13.7B, where $\bar{y} = 49.7$.

$$SST = \sum_{i=1}^{n}(y_i - \bar{y})^2 = (5.3)^2 + (-32.7)^2 + (-13.7)^2 + (35.3)^2 + (12.3)^2$$
$$+ (-31.7)^2 + (-16.7)^2 + (-8.7)^2 + (13.3)^2$$
$$+ (37.3)^2 = 5,610.10$$

Sum of squares error

Figure 13.7C illustrates the regression line fitted to the sales-advertising data, and the computation of SSE, the error sum of squares, is given in Table 13.2 by $\sum_{i=1}^{10}(y_i - \hat{y}_i)^2 = 213.33$. The deviations in Figure 13.7C represent the unexplained deviations in the random variable **Y** *after* the sample regression line has been fitted. These deviations, in magnitude, are considerably less than those in Figure 13.7B *before* the regression line was fitted. As we have observed, the regression line does appear to account for considerable variability in **Y**, since there appears to be a rather strong linear relationship between x and **Y**.

The sample coefficient of determination, $\mathbf{r}^2$, is a simple ratio of the amount of variation explained by the regression line to the total variation in **Y**. Its value is given by:

> **Sample coefficient of determination**
>
> $$r^2 = \frac{SST - SSE}{SST} \qquad (13.6)$$

In the numerator of r^2, the amount of variation explained by the regression line is the difference between the total sum of squares (SST) and the error sum of squares (SSE), where SSE measures the variation unexplained by the fitted regression line.

For the sales-advertising example,

$$r^2 = \frac{5,610.10 - 213.33}{5,610.10} = 0.962$$

That is, 96.2 percent of the variability in **Y** (sales) has been accounted for by relating it to x (advertising) through the regression line. This means that we have fitted a very strong (relative) linear relationship between sales and advertising expenditures *for these sample data*.

The value of $\mathbf{r}^2$ is bounded between 0 and 1. This can be shown as follows. The total deviation $(\mathbf{Y}_i - \overline{\mathbf{Y}})$ can be partitioned as:

$$\underbrace{(\mathbf{Y}_i - \overline{\mathbf{Y}})}_{\substack{\text{total} \\ \text{deviation}}} = \underbrace{(\mathbf{Y}_i - \hat{\mathbf{Y}}_i)}_{\substack{\text{error} \\ \text{deviation}}} + \underbrace{(\hat{\mathbf{Y}}_i - \overline{\mathbf{Y}})}_{\substack{\text{deviation of estimator} \\ \text{about the mean}}}$$

The last term in this partitioning, $(\hat{\mathbf{Y}}_i - \overline{\mathbf{Y}})$, is required to make the statement an identity; that is, if we add $(\mathbf{Y}_i - \hat{\mathbf{Y}}_i)$ and $(\hat{\mathbf{Y}}_i - \overline{\mathbf{Y}})$, we get $(\mathbf{Y}_i - \overline{\mathbf{Y}})$. This last term represents the difference between the estimator of $E(\mathbf{Y}_i)$, given by $\hat{\mathbf{Y}}_i$, and the sample mean $\overline{\mathbf{Y}}$ of the values of **Y**. These deviations are shown in Figure 13.11 as broken lines. For example, for $y_1 = 55$,

$$(55 - 49.7) = (55 - 50.2) + (50.2 - 49.7)$$

FIGURE 13.11 | Fitted regression line $\hat{y} = 1.02 + 2.73x$, deviations $(y_i - \hat{y}_i)$, and computation of SSE

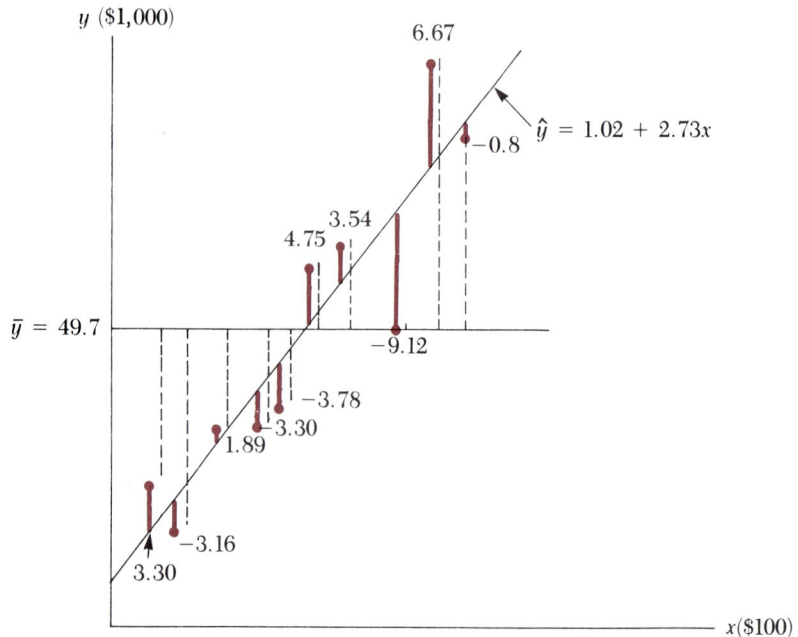

Remarkably, the sums of these squared deviations also bear this same relationship (we will not prove this):

$$\underbrace{\Sigma(\mathbf{Y}_i - \overline{\mathbf{Y}})^2}_{\text{SST}} = \underbrace{\Sigma(\mathbf{Y}_i - \hat{\mathbf{Y}}_i)^2}_{\text{SSE}} + \underbrace{\Sigma(\hat{\mathbf{Y}}_i - \overline{\mathbf{Y}})^2}_{\text{SSR}}$$

SST = SSE + SSR

Sum of squares due to regression

The last term in this expression is called the *sum of squares due to regression,* or **SSR** for short. Since they are sums of squares, SST $\geq$ 0 and SSE $\geq$ 0. Also, SST $\geq$ SSE (they could be equal if SSR = 0). These facts guarantee that the ratio (**SST** $-$ **SSE**)/**SST** can never be less than 0 or greater than 1; that is, $0 \leq \mathbf{r}^2 \leq 1$.

If $r^2 = 0$, then SSE = SST; the error sum of squares is equal to the total sum of squares. In this instance, the regression line has done nothing to reduce the variability in $\mathbf{Y}$. If $r^2 = 1$, then SSE = 0; the sample regression line has explained *all* the variability in the sample values. These two cases are illustrated in Figure 13.12. If $r^2 = 1$, the observations fall *perfectly* on the fitted regression line. If $r^2 = 0$, the fitted regression line has a slope of 0.

FIGURE 13.12 | Cases when $r^2 = 0$ and $r^2 = 1$

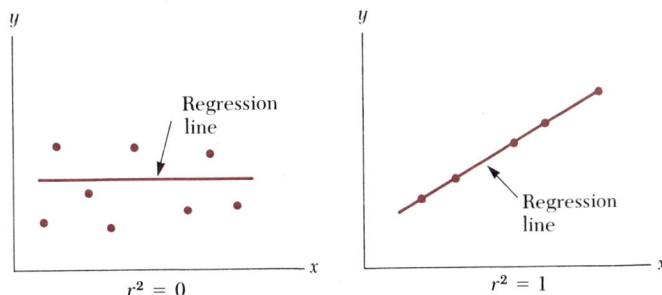

In our sales-advertising example, we determined that r^2 was 0.96—this indicates that we have fit a very strong linear relationship to the sample data points. It is unusual to experience a value of $\mathbf{r}^2$ that is this close to 1 in practice. In most business problems, for example, we are very fortunate to account for as much as 50 percent of the variability in a typical dependent random variable $\mathbf{Y}$ in our sample. The reason for this is that most often $\mathbf{Y}$ represents a very complex variable in business applications of regression analysis, and it is not easy to account for its variability through one or even more than one independent variable.

We complete this section by giving an alternate formula for r^2 and verifying that it does produce the same value as determined earlier for the sales-advertising example.

Computation formula for coefficient of determination

$$r^2 = \frac{\left[\sum x_i y_i - \frac{(\sum x_i)(\sum y_i)}{n}\right]^2}{\left[\sum x_i^2 - \frac{(\sum x_i)^2}{n}\right]\left[\sum y_i^2 - \frac{(\sum y_i)^2}{n}\right]} \qquad (13.7)$$

For the sales-advertising illustration, the value of r^2 is calculated with this formula by using the data in Table 13.2. Notice that using this formula for computing the value of r^2 justifies the inclusion of the y^2 column in Table 13.2, a column that we have not used up to this point.

$$r^2 = \frac{\left[10{,}820 - \frac{(178)(497)}{10}\right]^2}{\left[3{,}890 - \frac{(178)^2}{10}\right]\left[30{,}311 - \frac{(497)^2}{10}\right]} = 0.96$$

■ **13.7**

A discussion of the assumptions in regression analysis

In this section, we will discuss and review the assumptions underlying the use of the simple linear regression model. To summarize, the assumptions required in simple linear regression analysis are:

1. $E(\epsilon_i) = 0$ and $V(\epsilon_i) = \sigma^2$, for all i, where $1 \le i \le n$.

2. The error components ϵ_i, ϵ_j are uncorrelated; i.e., $C(\epsilon_i, \epsilon_j) = 0$, $i \ne j$.

3. β_0 and β_1 are parameters. They, and the values of x, are assumed to be *constants*.

4. The error components ϵ_i are normally distributed, where $1 \le i \le n$.

The last assumption is *not* required in calculating point estimates of β_0, β_1, σ^2, etc. However, this assumption *is* required if we wish to calculate confidence intervals or to construct hypothesis-testing decision rules for β_0, β_1, etc., as explained in the next chapter.

Table 13.3 repeats the values of the independent variable x, the values of the dependent variable **Y**, the values of the residuals $e = y - \hat{y}$, and the values of the *standardized residuals*. The standardized residuals are calculated by subtracting the mean of the residuals (which should be zero apart from numerical roundoff) and dividing by the standard error of the estimate $s_{Y/x}$; that is,

Standardized residuals

$$i\text{th standardized residual} = \frac{e_i - \bar{e}}{s_{Y/x}} = \frac{e_i - 0}{s_{Y/x}} = \frac{e_i}{s_{Y/x}}$$

TABLE 13.3 | The residuals and the standardized residuals

x (advertising)	y (sales)	$e = y - \hat{y}$ (residuals)	Standardized residuals
18	55	4.753	0.920
7	17	-3.165	-0.613
14	36	-3.308	-0.641
31	85	-0.799	-0.155
21	62	3.549	0.687
5	18	3.305	0.640
11	33	1.896	0.367
16	41	-3.777	-0.732
26	63	-9.125	-1.767
29	87	6.671	1.292

where

$$\bar{e} = \frac{\sum_{i=1}^{n} e_i}{n} = \frac{0}{n} = 0$$

and

$$s_{Y/x}^2 = \sum_{i=1}^{n} \frac{(y_i - \hat{y}_i)^2}{n - 2}$$

In Example 13.7 we found that $s_{Y/x}^2 = 26.67$. Therefore $s_{Y/x} = \sqrt{26.67} = 5.164$. The first standardized residual in Table 13.3 is determined from

$$\frac{e_1 - 0}{s_{Y/x}} = \frac{4.753 - 0}{5.164} = 0.920$$

The residuals or the standardized residuals may be used to assess the assumptions concerning the error term ϵ_i in the regression model. The error terms $\epsilon_i = Y_i - E(Y_i)$, $i = 1, 2, \ldots, n$ are assumed to be uncorrelated, normally distributed and homoscedastic (equal variances). The uncorrelated error term assumption may be replaced by assuming that the error terms are *independent*, since uncorrelated normally distributed random variables are statistically independent (this does not hold true if the errors are not normally distributed). Thus the commonly stated assumptions on the error terms ϵ_i, $i = 1, 2, \ldots, n$ are:

1. The ϵ_i's are normally distributed.

2. The ϵ_i's are independent.

3. The ϵ_i's have the same variance σ^2.

Assuming that the population regression line is an appropriate model to fit to the sample observations, the observed residuals e_i should reflect the above three assumptions on the error term ϵ_i. Thus by analyzing the residuals, e_i, $i = 1, 2, \ldots , n$, we can make judgments about whether or not the assumptions have been satisfied for a fitted regression line. This process is referred to as *residual analysis*.

Residual analysis

One method of residual analysis is the *residual plot*. In a residual plot, the residuals (either unstandardized or standardized) are plotted on the vertical axis and the values of the independent variable x are plotted on the horizontal axis.

Residual plot

Figure 13.13 gives the residual plot for the sales-advertising data in Table 13.3. The vertical axis represents the standardized residuals, and the horizontal axis represents the independent variable x.

FIGURE 13.13 | Residual plot for the sales-advertising example with the data given in Table 13.3

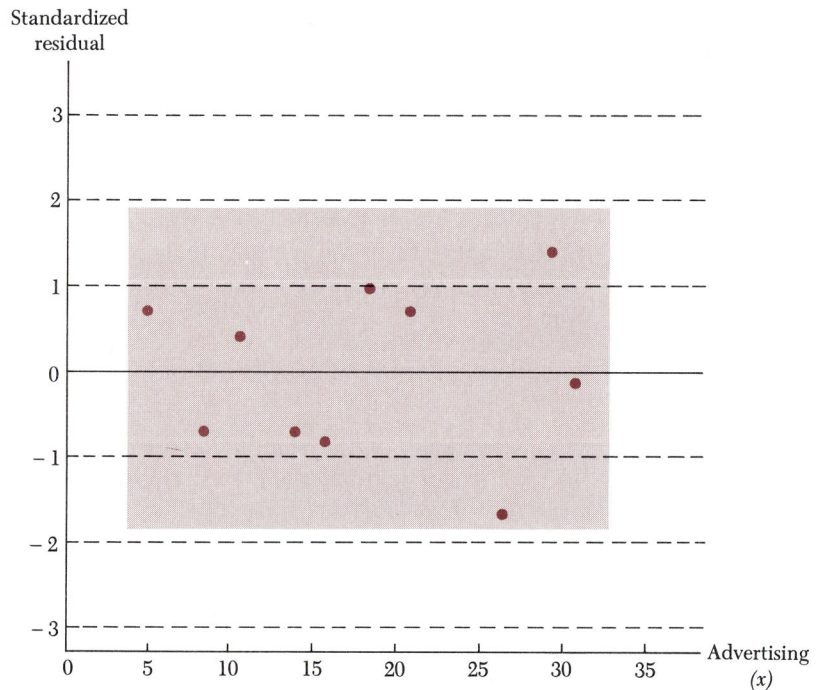

The three assumptions concerning the error terms, ϵ_i, $i = 1, 2, \ldots , n$ can be analyzed subjectively by using the residual plot. First, the error terms should have the same variance. Thus the residuals (and the standardized residuals) should reflect this assumption. Thus for any value of the independent variable x, the variability of the residuals should be approximately constant.

A residual plot satisfying this assumption is shown in Figure 13.14A. The width of the rectangle centered about the horizontal zero line should be constant for all values of x. Figure 13.14C shows a residual plot in which the variability of the residuals increases as the values of x increase. The residuals associated with larger values of x appear to have larger variances than the residuals associated with smaller values of x. In our plot of the sales-advertising residuals in Figure 13.13, the residuals appear to conform to a rectangle fairly well, although there is a slight suggestion of larger variances of the residuals for larger values of x.

FIGURE   13.14 | Four prototype residual plots

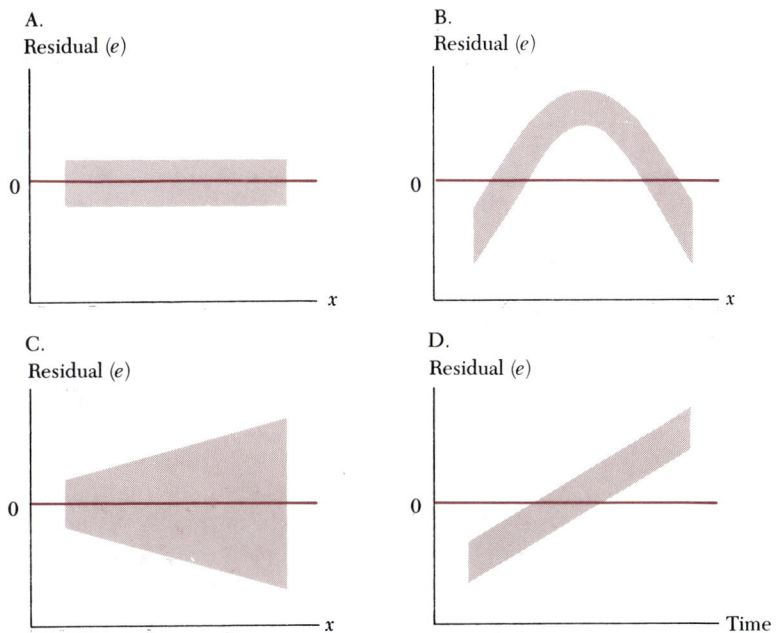

Second, the error terms must be uncorrelated (equivalently, independent if the error terms are normally distributed). This assumption requires that there are no patterns in the residual plot. Figure 13.14B is a residual plot that displays a definite pattern (a parabola) and would violate this assumption. If it is possible to plot the residuals as a function of time, then it may be possible to determine whether or not the residuals are correlated through time. This situation occurs when the values of the dependent variable **Y** are taken at different points in time, such as monthly sales of a product. By plotting the value of the residual corresponding to each month's sales against the time measure (months: 1, 2, . . .), the resulting residual plot will show the residuals as a function of time. When the residuals are correlated, it is often through time, and a plot such as Figure 13.14D can identify this situation. In Figure 13.14D, at the beginning of the time period, the residuals are negative; whereas at the end of the time period, the residuals are positive. This systematic relationship of the residuals with time would certainly lead to correlated error terms and the failure of the second regression analysis assumption to hold. The sales-advertising residual plot in Figure 13.13 appears to be a random scatter of points. Thus the assumption of uncorrelated error terms would appear to be satisfied for these data and the fitted regression model.

Third, the error terms must be normally distributed for the purposes of making inferences, the methods of which are described in Chapter 14. The sales-advertising example residual plot in Figure 13.13 is a residual plot of the *standardized* residuals. If the error terms are normally distributed, then the standardized error terms are distributed as standard normal (**Z**) random variables. Therefore we should expect about 68 percent of the standardized residuals to be between -1 and $+1$ (plus or minus one standard deviation), about 95 percent to be between -2 and $+2$ (plus or minus two standard deviations), and about 99 percent to be between -3 and $+3$ (plus or minus three standard deviations). Referring to either Table 13.3 or to Figure 13.13 (the dotted lines indicate the -1, $+1$, -2, $+2$, and -3, $+3$ boundaries), we find that 8/10, or 80 percent of the standardized residuals are between -1 and $+1$, 10/10, or 100 percent of the standardized residuals are between -2 and $+2$, and 10/10, or 100 percent of the standardized residuals are between -3 and $+3$. For a sample of size $n = 10$, these percentages (80 percent, 100 percent, and 100 percent) are fairly close to the expected percentages (68 percent, 95 percent, and 99 percent), although they do suggest that the residuals may be more closely grouped around the mean of 0 than should be expected if the error terms are in fact normally distributed. Another way of assessing the assumption of normality of the error terms is to form a histogram of the residuals to see if the resulting histogram looks like a symmetric normal distribution. The use of these two graphical methods (the residual plot and the histogram) can give us a reasonably good method of qualitatively assessing the normality assumption if the sample size is sufficiently large (25 or more). With a sample size of $n = 10$ in the sales-advertising example, the sample size is too small for a histogram to be of much use, for example.

Although the residual plot can give us a good idea of whether or not the assumptions of the regression model have been satisfied for a data set and a fitted regression model, there are statistical tests to determine whether or not the assumptions are indeed satisfied. We will defer the discussion of these statistical tests of the regression model assumptions until the development of the multiple regression model in Chapter 15.

What if there is an indication that one or more assumptions of the regression model are not satisfied for a data set and a fitted regression model to that data set? In general, regression analysis is not seriously affected by slight to moderate departures from the regression model assumptions. The regression model assumptions can be ranked in terms of the seriousness of the failure of the assumption to hold from the least serious to the most serious: normality of the error terms, equal variance of the error terms, and uncorrelated error terms. Regression analysis is *robust* to nonnormality of the error terms. By this we mean that modest to moderate departures from normality of the error terms will not seriously affect the analysis. Nonconstant error variance of the error terms is more serious, however. In this case it is usually necessary to take corrective measures depending on the degree to which the error variances differ. If the third assumption of uncorrelated errors is violated, then a "fatal" error has occurred. By this we mean that the fitted regression function should not be used for inference purposes. Either a different statistical technique must be used, or corrective measures must be taken to deal with the correlated residuals. A discussion of the corrective measures when one or more of the regression model assumptions is violated is beyond the scope of an introductory presentation of regression analysis. Generally these methods involve the transformation of the dependent variable $\mathbf{Y}$, such as log $\mathbf{Y}$ or $\sqrt{\mathbf{Y}}$. (For an excellent discussion of corrective measures when regression model assumptions are violated and a thorough description of residual plot analysis, see the Neter, Wasserman, and Kutner reference at the end of this chapter. Also, the Draper and Smith reference at the end of this chapter is an excellent source for further discussion of assessing the regression model assumptions by residual plots.)

The analysis of the residuals in a regression experiment is a fundamental step in conducting regression analysis. We will return to the use of residual plots and describe statistical tests for assessing the regression model assumptions in Chapter 15.

Robustness of regression analysis	

■ 13.8

The steps in building, analyzing, and validating a regression model

Now that we have been through a complete analysis of a data set (the sales-advertising example) by the sample regression model, it is appropriate to summarize the steps in the process. These steps will be used in the next section in a case study and will be used in Chapters 14 and 15 as we continue the development of regression analysis.

Step 1 | Identify the variables to be included in building the regression model

In Step 1 we identify the dependent variable and the independent variable(s) we will use in constructing an estimating equation. In this context the variable that is used to predict the variable of interest will be called the *independent variable*, and the variable of interest will be called the *dependent variable*. Often the independent variables to be used in predicting the dependent variable will be obvious. For example, when we wish to predict the sales of a product, we may examine the relationship between sales and advertising level, or between sales and the size of the population served, and so on, and select the model that we feel comes closest to predicting values of the dependent variable. At other times we may find that the regression analysis requires that a great deal of time be spent in identifying independent variables, or variables that can be used to predict values of the dependent variable. We generally will not know until Step 4 is completed whether it will be necessary to determine other independent variables to use in the estimating equation.

In simple linear regression there is only one independent variable x. As shall be seen in Chapter 15, we may use several independent variables in building a model to predict values of $\mathbf{Y}$ or to explain variability in $\mathbf{Y}$ in the multiple regression model.

Step 2 | Collect sample data

The data may be collected by a survey or by an experiment. In an experiment values of the independent variable(s) are selected and the corresponding values of $\mathbf{Y}$ (the dependent variable or output variable) are generated. In a survey, cases containing pairs of x and $\mathbf{Y}$ values are sampled so that we have no direct control over the values of x selected. Also the data may be collected at one point in time or over time. Frequently, data that have been collected over time violate one or more of the regression model assumptions (usually the uncorrelated errors assumption). Thus when data are collected over time, it is especially important to check the assumptions underlying regression analysis (Step 5) to assess whether or not the assumptions are satisfied by the data set.

Step 3 | Specify the relationship that exists between the dependent and the independent variable

In Step 3 we specify the nature of the relationship between the dependent variable and the independent variable. For example, is the relationship linear, or should nonlinear terms be included in the model? Frequently, a plot of the data points collected in Step 2 will provide us with clues to the form of the relationship between the variables of interest. We are concerned in this chapter with developing models in which the presumed relationship between the variables is linear, and in which *one* independent variable is used to estimate the value of the dependent variable. Such analysis is called *simple linear regression and correlation analysis—simple* because only one predictor or independent variable will be used in the estimating equation, and *linear* because of the linear manner in which the parameters of the regression model enter into the estimating equation.

Step 4 | Estimate parameters of the model specified

In this step, we estimate the parameters of the population regression model using sample data, since the population regression function relating the two variables will rarely be known in practice. In this chapter we obtain point estimates of the regression parameters, and in Chapter 14 we do hypothesis tests and construct confidence interval estimates of the parameters point estimated in this chapter. In this context, it will be necessary for us to specify the sampling distribution of the point estimators of the population regression parameters.

Step 5 | Determine whether the assumptions of the simple linear regression model have been met—validation

Although the determination of whether or not the underlying assumptions of the model have been met is a complex task, the construction of *residual plots* is often quite helpful in an initial determination of whether the assumptions have been met. This may lead us back to Step 1 and through the whole process once more if we discover that the model developed is not appropriate or that the underlying assumptions have been violated.

Step 6 | Statistically test the usefulness of the model developed

In this step we determine at a stated level of significance whether the model developed or the terms included in the model are statistically significant. This is a part of the hypothesis-testing and interval estimation process, so we will defer a discussion of this step until Chapter 14.

Step 7 | Use the model developed for prediction and estimation

Difference between estimation and prediction in the use of the fitted regression equation

Once we have developed the simple linear regression model, we will want to use the model for prediction and for estimation. The fitted regression equation, $\hat{y} = b_0 + b_1$, may be used for two purposes: (1) *estimating* the mean of $\mathbf{Y}$ for a given value of x (the mean is denoted by $\mu_{\mathbf{Y}/x}$) and (2) *predicting* an individual value of $\mathbf{Y}$ for a given value of x. To illustrate the difference between estimating $\mu_{\mathbf{Y}/x}$ and predicting an individual value of $\mathbf{Y}$, consider the sales/advertising data in Example 13.2. Suppose the amount of advertising expenditure (x) is fixed at 20 (in \$100s). On the one hand, we may be interested in *estimating* the mean sales, given $x = 20$, denoted by $\mu_{\mathbf{Y}/x=20}$. By doing so, we are estimating the mean sales of *all* stores that have an advertising budget of $x = 20$. On the other hand, we may be interested in *predicting* the sales for an *individual* store whose advertising budget is $x = 20$.

In most regression applications, we will be interested in estimating $\mu_{\mathbf{Y}/x}$ and predicting individual values of $\mathbf{Y}$ in the same experiment. The estimate of $\mu_{\mathbf{Y}/x}$ and the prediction of an individual value of $\mathbf{Y}$ for the *same* given value of x is the same. That is, the answer is the same in both cases and is given by $\hat{y} = b_0 + b_1 x$, where the given value of x is inserted in the fitted regression equation. Confidence intervals for estimating $\mu_{\mathbf{Y}/x}$ and for predicting individual values of $\mathbf{Y}$ do differ, as we shall see in Chapter 14.

In the next section we apply these steps in a case study of a regression analysis experiment.

13.9

A case study in simple linear regression model development

The analysis of a simple linear regression model is complex, and a great deal of care must be exercised to ensure that the analysis is properly done. From a computational standpoint, using a worksheet as given in Table 13.4 simplifies matters considerably. The totals from this worksheet provide the necessary numbers for calculating most of the estimates used in regression analysis. Furthermore, the worksheet provides a check to determine whether or not b_0 and b_1 have been calculated properly: $\Sigma(y - \hat{y}) = 0$.

TABLE 13.4 | Simple linear regression model worksheet

	x	y	x^2	y^2	xy	$\hat{y}$	$y - \hat{y}$	$(y - \hat{y})^2$
	x_1	y_1						
	x_2	y_2				Filled in *after* fitting the regression line $\hat{y} = b_0 + b_1 x$		
	.	.						
	.	.						
	.	.						
	x_n	y_n						
Totals	Σx	Σy	Σx^2	Σy^2	Σxy	$\Sigma \hat{y}$	$\Sigma(y - \hat{y})$	$\Sigma(y - \hat{y})^2$

In the following example, we will trace the development of a simple linear regression model through an examination of the underlying assumptions. This example should test your understanding of the steps involved in developing a simple linear regression model and provide you with experience in determining the regression model point estimates (Step 4).

Example 13.8 R. Griffin, a highly trained accountant, has been given the responsibility by his supervisor in the accounting firm, Ef Lundgren, to estimate federal and state income taxes for individuals for the coming year to serve as a guideline in testing the reasonableness of the computed taxes of its client customers. Ef would like some type of an estimating equation, as well as an indication of the coefficient of determination for the model developed, and an estimate of the conditional distribution standard deviation. He further would like R. to comment on the model developed and whether it appears to fit the underlying regression assumptions.

Solution *Step 1: Identify the variables to be included in building the regression model.* Although the income taxes paid by an individual are influenced by a host of factors that include adjusted gross income, unusual medical and other deductions, number of dependents, and other sources of income, R. decides to build a simple regression model as a first step to estimate this tax liability. The obvious choice for the independent variable in this example is the adjusted gross income of the individual, since the model is initially to include only one predictor variable.

Step 2: Collect sample data. R. Griffin randomly samples 10 of the accounting firm's previous clients and records each client's adjusted gross income and the corresponding federal and state tax liability. The client sampled in this instance is the unit of association, and the data were collected by survey rather than by experimentation. These data are reproduced in Table 13.5.

TABLE 13.5 | Survey of client adjusted gross income and tax liability

Client no.	Adjusted gross income	Tax liability
1	$18,945	$ 5,683
2	16,420	4,926
3	13,945	3,765
4	12,554	3,012
5	20,627	7,013
6	28,420	11,223
7	15,626	4,375
8	13,396	3,617
9	17,200	5,504
10	24,319	8,511

Step 3: Specify the relationship that exists between the dependent and the independent variable. R. Griffin obtains a scatter diagram of the data sampled and concludes that a linear relationship appears reasonable for these two variables. The scatter diagram constructed is given in Figure 13.15.

FIGURE 13.15 | Scatter diagram of data in Table 13.5

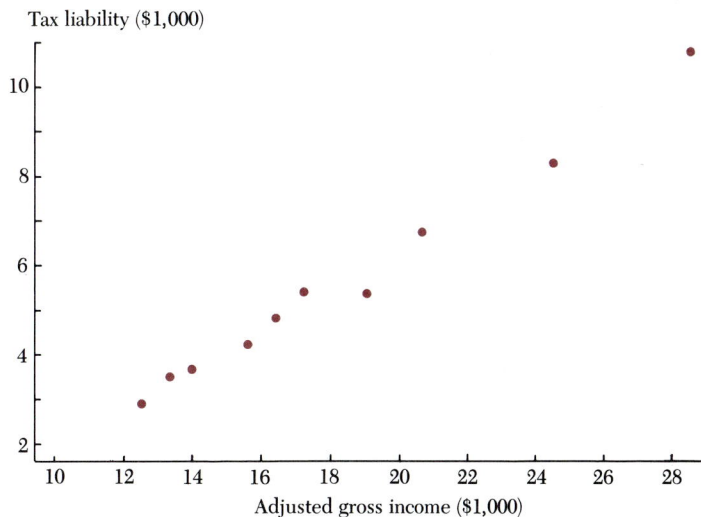

Step 4:   Estimate parameters of the model specified.   A worksheet similar to Table 13.4 is given in Table 13.6 for estimating the regression parameters. These estimates are given by:

$$b_1 = \frac{1,160,943,000 - \dfrac{(181,452)(57,629)}{10}}{3,524,642,000 - \dfrac{(181,452)^2}{10}} = 0.496$$

$$b_0 = \frac{57,629}{10} - (0.496)\left(\frac{181,452}{10}\right) = -3,245.15$$

$$\hat{y} = -3,245.15 + 0.496x$$

Furthermore, r^2 and $s_{Y/x}$ are determined to be (verify!):

$$r^2 = 0.990 \qquad s_{Y/x} = 267.82$$

Based on the high value of r^2, R. Griffin is quite pleased with the model
developed, since only one independent variable is used to predict the value
of the dependent variable (tax liability).

TABLE   13.6 | Simple linear regression model worksheet for data in Table 13.5

Trial	x	y	x^2	y^2	xy	$\hat{y}$	$y - \hat{y}$	$(y - \hat{y})^2$
1	18,945	5,683	358,913,000	32,296,480	107,664,400	6,159.95	−476.95	227,484.20
2	16,420	4,926	269,616,300	24,265,470	80,884,910	4,906.43	19.57	382.84
3	13,945	3,765	194,463,000	14,175,220	52,502,910	3,677.74	87.26	7,614.61
4	12,554	3,012	157,602,900	9,072,144	37,812,640	2,987.19	24.81	615.66
5	20,627	7,013	425,473,000	49,182,160	144,657,100	6,994.97	18.03	325.13
6	28,420	11,223	807,696,300	125,955,700	318,957,500	10,863.75	359.25	129,063.30
7	15,626	4,375	244,171,800	19,140,620	68,363,740	4,512.26	−137.26	18,839.71
8	13,396	3,617	179,452,800	13,082,680	48,453,320	3,405.19	211.81	44,862.88
9	17,200	5,504	295,839,900	30,294,010	94,668,800	5,293.66	210.34	44,242.85
10	24,319	8,511	591,413,700	72,437,120	206,979,000	8,827.84	−316.84	100,385.00
Totals	181,452	57,629	3,524,642,000	389,901,300	1,160,943,000		0.03	573,816.00

*Step 5:   Determine whether or not the assumptions of the simple linear
regression model have been met.*   A residual plot of the data is given in
Figure 13.16. This plot is patterned somewhat after plot C in Figure 13.14.
As the value of the independent variable x increases, it appears that the
variability of the conditional distributions increases also. We might suspect
this; it appears reasonable that as adjusted gross income increases, so too
would the variability in tax liabilities.

Outlier

The circled point in Figure 13.16 represents an *outlier*—a point in the
scatter plot that is set apart from the majority of the points. An outlier may

FIGURE 13.16 | Residual plot of regression model developed in Example 13.8

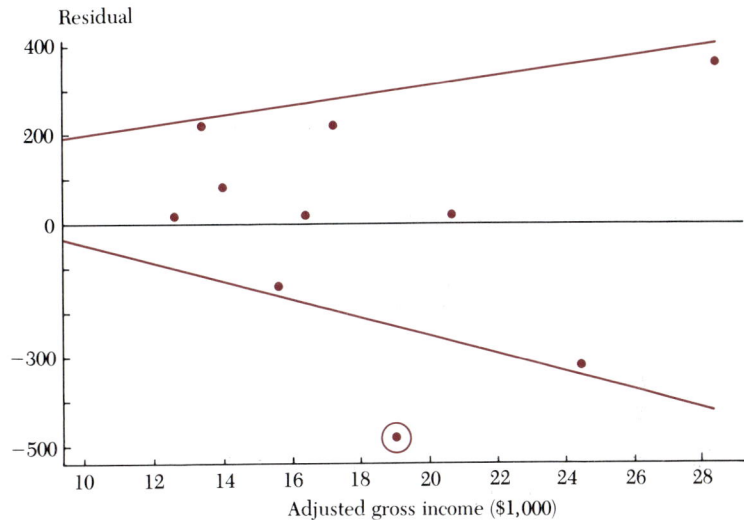

be a point corresponding to a case in which recording or other errors have been made. However, the outlier may be a legitimate case that is simply unusual and differs from the rest of the cases. In Example 13.8 the outlier in Figure 13.16 corresponds to case 1 with $x = \$18,945$ and $y = \$5,683$. In this case, we would expect the tax liability to be higher ($\hat{y} = \$6,159.95$) from the fitted regression equation than it actually is ($y = \$5,683$). This leads to a large negative residual ($y - \hat{y} = -\$476.95$). In general, outliers should be checked to ensure that they are legitimate points and that no recording or other errors have been made.

Step 5 can be one of the most important steps in developing a regression model since, as the data show, it may be possible to develop a simple linear regression model with a high value of r^2 and a very respectable value of $s_{Y/x}$ for estimation purposes. Yet the underlying assumptions of applying the model developed are not satisfied, thus decreasing the applicability of what otherwise would appear to be a very successful model. It may be possible that the model developed is still quite satisfactory for our purposes, but we should not be led into believing the model is any better than it really is. Before applying the model in any meaningful context, we should assure ourselves that the model developed is well suited to the uses for which it is intended.

Step 6: Statistically test the usefulness of the model developed. We will defer an analysis of the usefulness of the model developed until Chapter

14, when we learn hypothesis-testing and confidence interval estimation procedures for the simple linear regression parameters.

Step 7: Use the model developed for prediction and estimation. The end result of the simple linear regression analysis is a satisfactory regression model that will allow us to estimate or predict a value of the dependent random variable **Y** given a value of the independent variable x. If we have satisfied ourselves that the underlying assumptions of the simple linear regression model have been met, then we can use the model for estimation and prediction. For example, given a client of the accounting firm with an adjusted gross income of $19,500, we would predict her combined state and federal income tax liability to be:

$$\hat{y} = -3{,}245.15 + 0.496(19{,}500) = \$6{,}426.85$$

using the model developed. However, we must be cautious in using this predicted value in light of the residual plot in Figure 13.16. It appears that at least the equality of variance of the error terms assumption has been violated for this fitted regression equation.

Extrapolation

Often a fitted regression model is used for extrapolation—predicting value of **Y** for a value of x that is not within the range considered in the experiment. In general, this is a *very risky* use of regression and should be done with a great deal of caution. Just because a simple linear regression model does a good job of predicting values of the random variable **Y** when the values of x are within the range of values used in model development does not necessarily mean it will do a satisfactory job of predicting values of **Y** when values of x are outside the range of those values used for model development. It may be, for example, that the true relationship between x and **Y** is nonlinear; but within the range of x-values examined, a straight line does an excellent job of modeling the behavior of the random variable. As a case in point, we would predict the tax liability of a person with $0 adjusted gross income to be $-\$3{,}245.15$ using the model developed. Although the negative tax liability could be viewed as a transfer payment (the government subsidizes the individual with $0 income), this is well outside the meaning of the model considered. It is more likely that the true relationship between adjusted gross income and tax liability is nonlinear, but is well approximated between the levels of adjusted gross income investigated—$12,554 to $28,420—by a straight line. This situation is illustrated in Figure 13.17.

■ 13.10

Correlation

We turn now to a study of simple *linear* correlation, a second measure of the relative strength of the relationship between variables. When we speak of using correlation in the analysis of data sets, it suggests to most of us that we wish to study the degree of association between variables in the set. Most correlation methods are therefore closely related to regression analysis. In fact we will show that the most commonly used measure of correlation is directly related to the slope and to the coefficient of determination in simple linear regression. The coefficient of correlation we will develop in this chap-

FIGURE 13.17 | Curvilinear relationship and fitted linear regression function

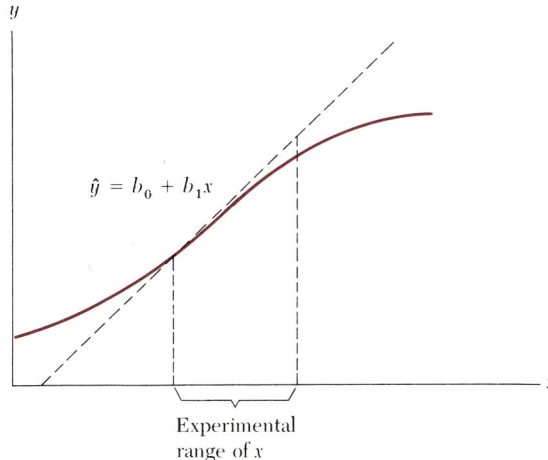

Experimental
range of x

ter is called *Pearson's product-moment correlation coefficient* and is denoted by ρ for the population and by r (the estimate of ρ) when using a sample data set to estimate ρ. Pearson's product-moment correlation coefficient is only one of several measures of the *relative strength* of the *linear association* between variables. In Chapter 20 we will consider another measure of correlation—*Spearman's rank correlation coefficient*.

The point estimator of ρ—the sample correlation coefficient, **r**—applies regardless of the joint distribution of **X** and **Y** in the population. But, if we wish to test hypotheses or set confidence intervals on ρ, we must assume that **X** and **Y** follow the *bivariate normal distribution* whose probability density function is given by:

$$f(x, y) = \frac{1}{2\pi\sigma_X\sigma_Y\sqrt{1 - \rho^2}} \exp\left\{ -\frac{1}{2(1 - \rho^2)} \left[\left(\frac{x - \mu_X}{\sigma_X}\right)^2 - 2\rho \left(\frac{x - \mu_X}{\sigma_X}\right)\left(\frac{y - \mu_Y}{\sigma_Y}\right) + \left(\frac{y - \mu_Y}{\sigma_Y}\right)^2 \right] \right\}$$

$$-\infty < x < \infty, \ -\infty < y < \infty$$

where

$\mu_X = E(X)$, the mean of the random variable **X**.
$\mu_Y = E(Y)$, the mean of the random variable **Y**.
$\sigma_X = \sqrt{V(X)}$, the standard deviation of the random variable **X**.
$\sigma_Y = \sqrt{V(Y)}$, the standard deviation of the random variable **Y**.
$\rho = $ Correlation between **X** and **Y**, given from Chapter 5 as

$$\rho = \frac{E\{[x - \mu_X][y - \mu_Y]\}}{\sigma_X \sigma_Y} = \frac{C(X, Y)}{\sigma_X \sigma_Y}$$

where $C(X, Y)$ is the covariance of X and Y.

The bivariate normal distribution is illustrated in Figure 13.18. It is a bell-shaped figure in three dimensions. In this section we will therefore assume that (x, y) pairs are from a bivariate normal distribution. This means of course that *both* X and Y are now random variables. In regression analysis, X is often treated as a random variable when data are obtained by survey rather than by experimentation. That is, if units of association are selected randomly yielding (x, y) data pairs, and we exercise no control over fixing the values of x in our study, then X is a random variable.

FIGURE 13.18 | "Slice" through the bivariate normal distribution—univariate normal distribution

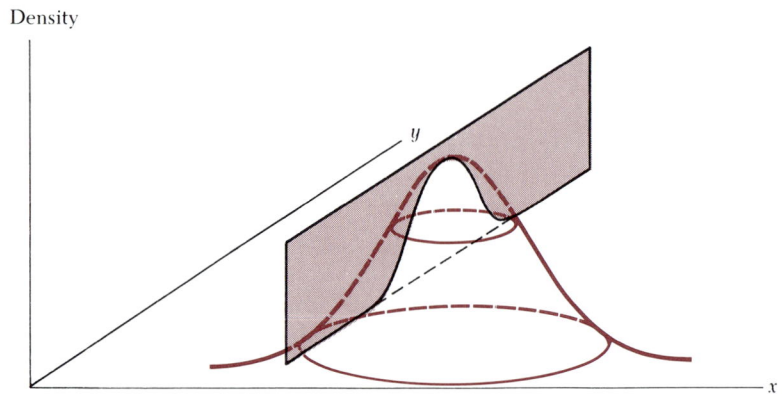

If we slice the bivariate normal distribution along either axis, the "slice" is a univariate normal distribution as illustrated in Figure 13.18. By forming slices that are parallel to the (x, y) plane, we form *contours*—slices as if we were looking down from above the "bell" as illustrated in Figure 13.19. On the left of the figure is shown one possible contour "slice" through the "bell." By repeating this process, we can form a set of contours that gives us in two dimensions a "feeling" for the nature of the three-dimensional surface. In Figure 13.20 are shown four representative contour sets for values of the population correlation coefficient, ρ. As the contours concentrate about their axis, ρ increases in magnitude.

In most practical situations, the value of the population correlation coefficient will not be known and will therefore have to be estimated from sample data. The process of forming a point estimate using sample data is illustrated

Contours

FIGURE 13.19 | Contours of the bivariate normal distribution

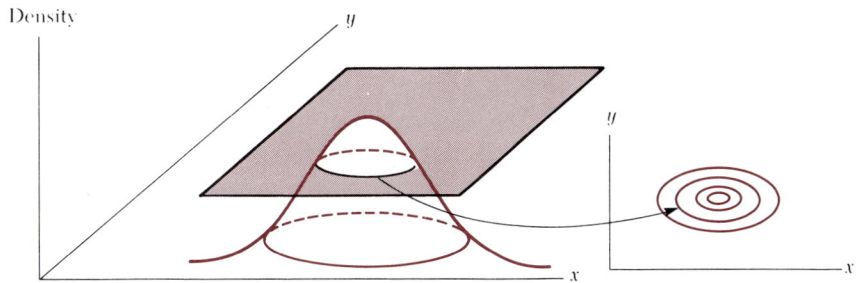

FIGURE 13.20 | Four representative contour sets for ρ values

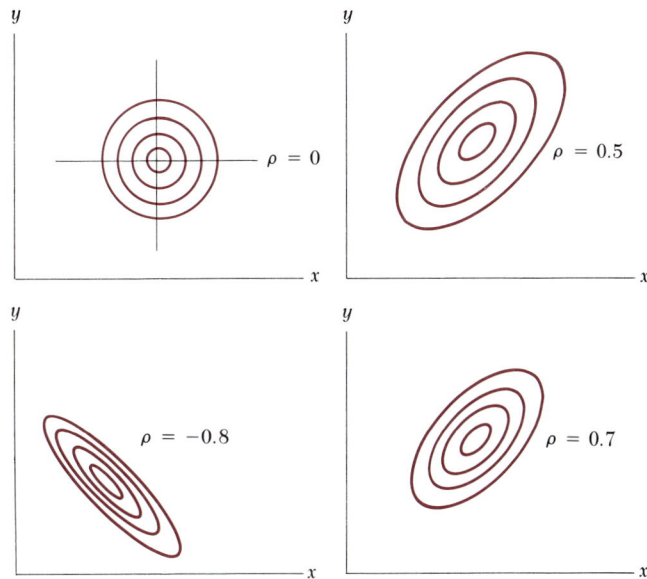

in Figure 13.21. A sample of n (x, y) data pairs are selected from the population, and these values are used to form a point estimate, r, of the population correlation coefficient.

One measure of the relationship between variables discussed in Chapter 5 is their covariance, a measure of how the variables "co-vary." Unfortunately, this measure of the linear association between variables is highly

Covariance of
X and **Y**

FIGURE   13.21 | Relationship between ρ and point estimate r

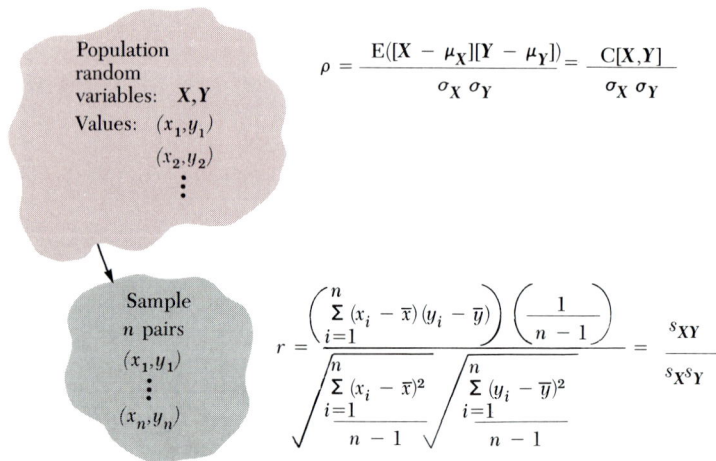

$$\rho = \frac{E([X - \mu_X][Y - \mu_Y])}{\sigma_X \, \sigma_Y} = \frac{C[X,Y]}{\sigma_X \, \sigma_Y}$$

Population random variables: X, Y

Values: (x_1, y_1)
(x_2, y_2)
$\vdots$

Sample n pairs
(x_1, y_1)
$\vdots$
(x_n, y_n)

$$r = \frac{\left(\sum\limits_{i=1}^{n} (x_i - \bar{x})(y_i - \bar{y}) \right) \left(\dfrac{1}{n-1} \right)}{\sqrt{\dfrac{\sum\limits_{i=1}^{n} (x_i - \bar{x})^2}{n-1}} \sqrt{\dfrac{\sum\limits_{i=1}^{n} (y_i - \bar{y})^2}{n-1}}} = \frac{s_{XY}}{s_X s_Y}$$

FIGURE   13.22 | Scatter diagrams with values of r

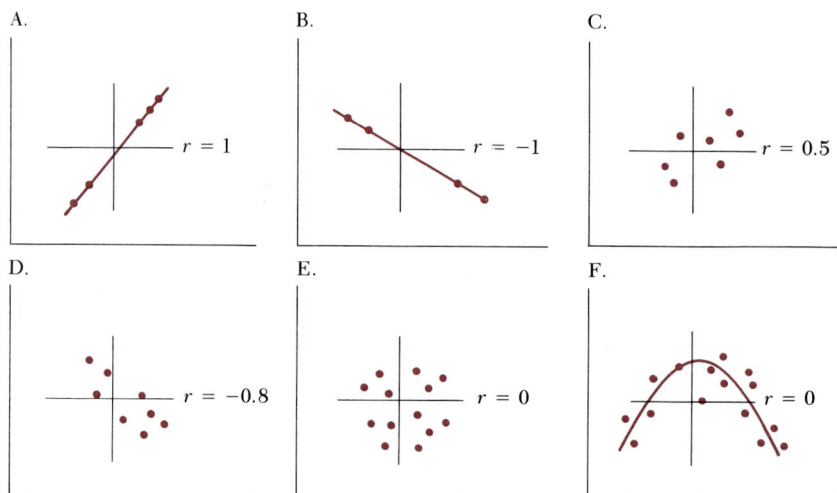

A. $r = 1$

B. $r = -1$

C. $r = 0.5$

D. $r = -0.8$

E. $r = 0$

F. $r = 0$

influenced by the units used to measure the variables. For this reason, we "standardize" the covariance of two random variables so that the standardized value is not influenced by the units of measurement. This standardization is accomplished by dividing the covariance of the random variables X and Y, $C(X, Y)$, by σ_X and σ_Y in the case of ρ, and dividing the point estimate of $C(X, Y)$, s_{XY}, by point estimates of σ_X and σ_Y, s_X and s_Y, respectively, in the case of a sample. This process yields values of the random variable $\mathbf{r}$, the sample correlation coefficient, bounded by -1 and $+1$, as illustrated in Figure 13.21.

As r approaches $+1$ or -1, the pairs of points must fall closer to a straight line; as r approaches 0, the points show a scatter that demonstrates no linear relationship. Examples of scatter plots with corresponding values of $\mathbf{r}$ are illustrated in Figure 13.22. Notice in particular plot F—$r = 0$, but there is a strong nonlinear (quadratic) relationship between X and Y. It is important to keep in mind that $\mathbf{r}$ measures only the *linear association* between the two random variables X and Y. If $r = 0$, it implies that there is no *linear relationship* in the sample, but some other relationship may exist between X and Y!

> Pearson product-moment correlation coefficient measures *linear* association

---

> Population correlation coefficient

Population correlation coefficient, ρ

$$\rho = \frac{E[(X - \mu_X)(Y - \mu_Y)]}{\sigma_X \sigma_Y} = \frac{C(X, Y)}{\sigma_X \sigma_Y}$$

> Sample correlation coefficient

Sample correlation coefficient, r (point estimate of ρ)

$$r = \frac{\left[\sum_{i=1}^{n}(x_i - \bar{x})(y_i - \bar{y})\right]\left(\dfrac{1}{n-1}\right)}{\sqrt{\dfrac{\sum_{i=1}^{n}(x_i - \bar{x})^2}{n-1}}\sqrt{\dfrac{\sum_{i=1}^{n}(y_i - \bar{y})^2}{n-1}}} = \frac{s_{XY}}{s_X s_Y}$$

$$= \frac{\left[\sum_{i=1}^{n}(x_i - \bar{x})(y_i - \bar{y})\right]}{\sqrt{\sum_{i=1}^{n}(x_i - \bar{x})^2}\sqrt{\sum_{i=1}^{n}(y_i - \bar{y})^2}}$$

<table>
<tr><td>Sample correlation coefficient: computing formula</td></tr>
</table>

Computing formula for r, the sample correlation coefficient

$$r = \frac{\sum\limits_{i=1}^{n} x_i y_i - \dfrac{\left(\sum\limits_{i=1}^{n} x_i\right)\left(\sum\limits_{i=1}^{n} y_i\right)}{n}}{\sqrt{\sum\limits_{i=1}^{n} x_i^2 - \dfrac{\left(\sum\limits_{i=1}^{n} x_i\right)^2}{n}} \sqrt{\sum\limits_{i=1}^{n} y_i^2 - \dfrac{\left(\sum\limits_{i=1}^{n} y_i\right)^2}{n}}} \qquad (13.8)$$

Experience in determining the sample correlation coefficient is provided in the following four examples.

Example 13.9 Consider the four points on the line $y = x + 2$ shown here: $(-1, 1)$, $(0, 2)$, $(1, 3)$, and $(2, 4)$. Determine the sample correlation coefficient for these data.

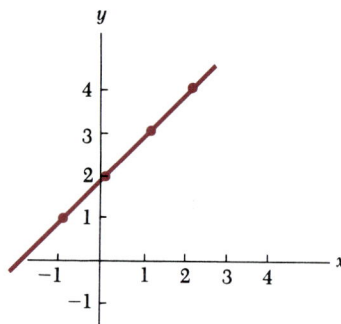

Solution We will work with the direct determination of the sample covariance in this example.

Point	x	$x - \bar{x}$	$(x-\bar{x})^2$	y	$y - \bar{y}$	$(y - \bar{y})^2$	$(x - \bar{x})(y - \bar{y})$
$(-1, 1)$	-1	-1.5	2.25	1	-1.5	2.25	2.25
$(2, 4)$	2	1.5	2.25	4	1.5	2.25	2.25
$(0, 2)$	0	-0.5	0.25	2	-0.5	0.25	0.25
$(1, 3)$	1	0.5	0.25	3	0.5	0.25	0.25
Totals	2	0	5	10	0	5	5.0

$$r = \frac{s_{XY}}{s_X s_Y} = \frac{\dfrac{5}{3}}{\sqrt{\dfrac{5}{3}} \sqrt{\dfrac{5}{3}}} = +1$$

Example 13.10   Consider the three points on the line $y = -2x + 1$: $(0, 1)$, $(-1, 3)$, and $(-2, 5)$. These points are illustrated here.

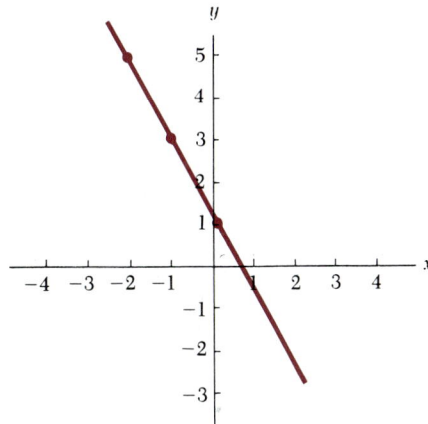

Compute r, the sample correlation coefficient for these data.
   Solution

$$r = \frac{-4}{\sqrt{2}\ \sqrt{8}} = -1 \qquad \text{(Verify!)}$$

Example 13.11   Consider the five data points shown in the graph: $(0, 0)$, $(4, 4)$, $(0, 4)$, $(4, 0)$, and $(2, 2)$.

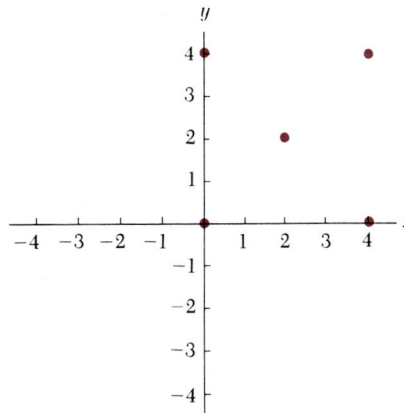

For these data, determine r.
   Solution

$$r = \frac{0}{\sqrt{16}\ \sqrt{16}} = 0 \qquad \text{(Verify!)}$$

These three examples suggest what can be proved:[3]

1. If all sample pairs fall on a line with a positive slope, then $r = 1$ (Example 13.9).

2. If all sample pairs fall on a line with a negative slope, then $r = -1$ (Example 13.10).

3. If the sample (x, y) pairs are scattered in the **X, Y** plane with no *linear* association whatever, then $r = 0$ (Example 13.11).

Example 13.12   Using the data in Table 13.2 for the sales-advertising example, compute r, the sample correlation coefficient, using the computational formula (13.8).

Solution   From Table 13.2,

$$r = \frac{10,820 - \dfrac{(178)(497)}{10}}{\sqrt{3,890 - \dfrac{(178)^2}{10}} \sqrt{30,311 - \dfrac{(497)^2}{10}}} = 0.9808$$

Thus, there is almost a perfect positive linear relationship between **X** and **Y** (note that we *must* now treat the variable x as a random variable **X** in order to use the correlation model) in these sample data.

The sample correlation coefficient, **r**, is directly related to the sample coefficient of determination, **r²**, in the simple linear regression model, **Y** $=$ $\beta_0 + \beta_1 x + \varepsilon$. If we write r in its *calculating form*:

$$r = \frac{\Sigma(x_i - \bar{x})(y_i - \bar{y})}{\sqrt{\Sigma(x_i - \bar{x})^2}\,\sqrt{\Sigma(y_i - \bar{y})^2}} = \frac{\Sigma xy - \dfrac{(\Sigma x)(\Sigma y)}{n}}{\sqrt{\Sigma x^2 - \dfrac{(\Sigma x)^2}{n}}\,\sqrt{\Sigma y^2 - \dfrac{(\Sigma y)^2}{n}}}$$

we see that $r = \pm\sqrt{r^2}$ from recalling the computing form of r^2 given in Section 13.6. Therefore the square of the correlation coefficient measures the amount of variability in the random variable **Y** when it is related linearly to the variable **X**. Notice that in Examples 13.9 and 13.10, $r^2 = 1$; that is, the scatter of points in each sample falls *perfectly* on a straight line, so that 100 percent of the variability in **Y** is accounted for by the linear relationship with **X** (**Y** $=$ **X** $+$ 2 in Example 13.9, and **Y** $= -2$**X** $+$ 1 in Example 13.10).

Furthermore, the sign of r, the sample correlation coefficient ($+$ or $-$), gives the direction ($+$ or $-$) of the *slope* of the regression line. Since b_1 in the fitted regression line, $\hat{y} = b_0 + b_1 x$, gives the slope, it is not surprising that r is related to b_1:

$$r = b_1 \left(\frac{s_X}{s_Y}\right)$$

Relationship between r^2, the coefficient of determination and r, the sample correlation coefficient

Relationship between r, the correlation coefficient and b_1, the slope of the fitted regression line

---

[3] We will not prove these results. The interested reader is referred to the Conover reference at the end of the chapter.

Thus there is a *direct relationship* between r and b_1. Since s_X and s_Y are both greater than or equal to 0, the signs $(+, -)$ of r and b_1 must be the same in a given sample. *In fact the difference in the values of r and b_1 obtained is simply due to the difference in the measurement scales used!*

Since simple linear regression and correlation are similar kinds of analyses, which one should be used when studying two variables, $X(x)$ and Y? The difference between the two models depends on the assumptions placed on each. In both models, the dependent variable Y is assumed to be a random variable. But, in simple regression, the independent variable x is considered to be "fixed"—it is a constant, whereas in correlation, the independent variable X is considered to be a random variable. Furthermore, for hypothesis testing, correlation analysis requires that the population random variable (X, Y) be distributed as a bivariate normal random variable, a much more restrictive requirement than the regression inference assumption that only the dependent variable Y be normally distributed.

Correlation attempts to determine the relative strength of the *linear relationship* between two variables, whereas regression analysis establishes an estimated linear functional relationship between them. The fitted regression line tells us not only *whether or not* the two variables are linearly related in the sample, but also *how* ($\hat{y} = b_0 + b_1x$).

In most cases we would therefore prefer regression analysis to correlation analysis; but in fact there is little difference between them. One difference worth keeping in mind is the sampling process for each method. In regression, we take a random sample of Y-values with the x-values being fixed or given, whereas in correlation analysis, we take a random sample of *pairs* of values, (X, Y). Thus we may not receive the "coverage" of the X random variable that is possible in correlation (by randomly sampling the x-values). This suggests that if we are *only* interested in establishing whether or not two variables are linearly related, correlation may be preferred to regression.

Relationship between correlation and causation

It is also *extremely* important to keep in mind that if two variables are highly correlated, it is not possible to claim an indication of cause and effect *without further study*. An interesting example of a "nonsense" correlation coefficient arose in a study in the Scandinavian countries around the turn of the century. Among the variables studied were X = Number of births per year and Y = Stork population per year. It was shown that X and Y were strongly correlated positively. Thus, as the stork population grew, so did the number of births. It may be passé nowadays, but a possible cause-effect explanation had it that births increased because there were more storks to fly in the babies! Upon further study of the data, it became apparent that there was a third variable that affected both X and Y; Z = The severity of the winter. In severe winters people had to keep their fireplaces going continuously to keep warm, and since storks build their nests in and about chimneys, this was definitely a plus for the stork population—fewer stork babies perished. And downstairs, being snowbound with a roaring fire and no TV, the human population flourished as well!

Although this may appear to be a casual example, it brings home the important point that strong correlation *does not imply* causation. There are some very serious examples of this today. What is the relationship between smoking and lung cancer or heart disease? The correlation is positive, but does this imply that smoking "causes," let us say, heart disease? Until we are able to rule out *any* and *all possible causative* variables, such as physiological factors that may make it more likely that a person will become a heavy smoker and will develop heart disease, we cannot make this connection. *Always* be cautious in interpreting results or statements of causality that are based on correlation analysis.

■ 13.11

Use of computer software packages

Virtually all regression analyses are performed on mainframe or microcomputers using a statistical software package. Among the most commonly used statistical software packages are MINITAB, SAS, BMD and SPSS (see the references at the end of this chapter). These software packages enable regression analysis to be performed without having to do the heavy computational work in evaluating the formulas in this chapter. As an illustration, consider the Example 13.2 problem in this chapter involving the sales-advertising data given in Table 13.1. To use MINITAB to fit the regression line, we could write the following program:

```
READ INTO COLUMNS C1 AND C2
    18    55
     7    17
    14    36
    31    85
    21    62
     5    18
    11    33
    16    41
    26    63
    29    87

PLOT C2 VS C1

REGRESSION OF Y IN C2 ON 1 PREDICTOR IN C1, STORE ST. RES. IN C3

CORRELATE C1 AND C2

PLOT C3 VS C1

STOP
```

The READ command is used to read the two columns of numbers into column numbers C1 and C2. The numbers are read into these two columns by the READ command in free-format, meaning that a space, spaces, or a comma may be used to separate the two columns. The advertising expenditures (x) are in C1, and the sales (y) are in C2. The PLOT command produces the scatterplot of the (x, y) pairs of observations. The REGRESSION command fits the model $\mathbf{Y} = \beta_0 + \beta_1 x + \epsilon$ and places the standardized residuals in column C3. There is 1 predictor—one x-variable or independent variable—in this case. In Chapter 15, in the multiple regression model, we will consider cases where there are more than one independent variable. The CORRELATE command calculates the Pearson product-moment correlation coefficient. The second use of the PLOT command produces the residual plot (the standardized residuals versus the values of the independent variable x). The STOP command indicates that no more commands follow. The output from this MINITAB program includes the following:

```
--  READ INTO COLUMNS C1 AND C2
```

MINITAB prints the column count for each column and prints out the first four rows for verification that the data have been read into the correct columns

```
    COLUMN          C1          C2
    COUNT           10          10
    ROW
     1             18,          55,
     2              7,          17,
     3             14,          36,
     4             31,          85,
          ,     ,     ,
```

The scatterplot of the (x, y) pairs

```
--  PLOT C2 VS C1
```

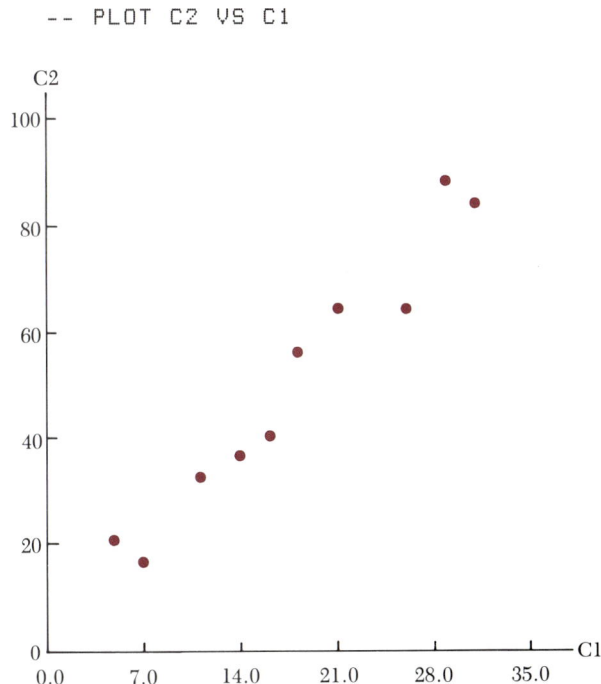

```
       -- REGRESSION OF Y IN C2 ON 1 PREDICTOR IN C1,
          STORE ST. RES. IN C3
```

The fitted regression line
```
THE REGRESSION EQUATION IS
Y =    1.02 +   2.73 X1
```

The estimate of the standard deviation σ.
```
THE ST. DEV. OF Y ABOUT REGRESSION LINE IS
S = 5.164
```

The coefficient of determination.
```
R-SQUARED = 96.2 PERCENT
```

A printout of the x-values, the y-values, the predicted y-values, the residuals, and the standardized residuals.[4]

ROW	X1 C1	Y C2	PRED. Y VALUE	RESIDUAL	ST.RES.
1	18.0	55.00	50.25	4.75	0.97
2	7.0	17.00	20.16	-3.16	-0.71
3	14.0	36.00	39.31	-3.31	-0.68
4	31.0	85.00	85.80	-0.80	-0.19
5	21.0	62.00	58.45	3.55	0.73
6	5.0	18.00	14.70	3.30	0.78
7	11.0	33.00	31.10	1.90	0.40
8	16.0	41.00	44.78	-3.78	-0.77
9	26.0	63.00	72.12	-9.12	-1.97
10	29.0	87.00	80.33	6.67	1.52

The Pearson product-moment correlation coefficient
```
-- CORRELATE C1 AND C2

     THE CORRELATION BETWEEN C1 AND C2 IS 0.981.

-- PLOT C3 VS C1
```

The residual plot

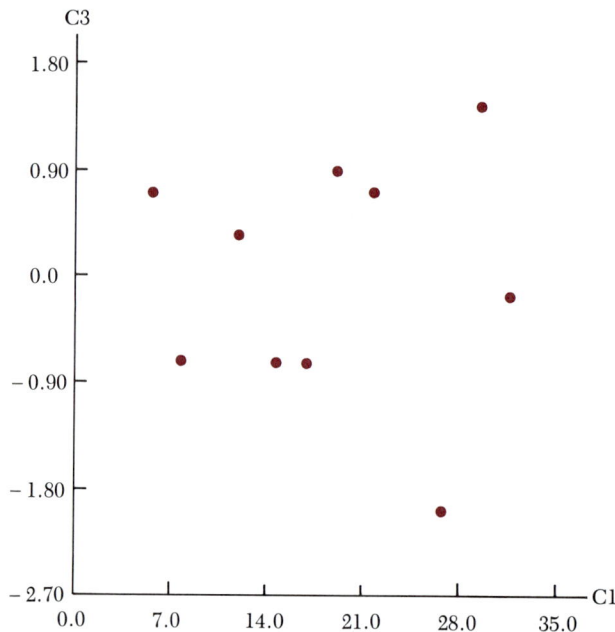

----

[4] MINITAB calculates the standardized residuals using the formula standardized residual = (residual)/$\sqrt{[s_{Y/x}^2 - (\text{standard deviation of the fit})^2]}$ rather than the formula in this chapter given by standardized residual = (residual)/$s_{Y/x}$. The term (standard deviation of the fit) used by MINITAB will be discussed in Chapter 14.

The scatterplot produced by MINITAB conforms well to the scatterplot produced in Figure 13.7A. Following the scatterplot, MINITAB prints the fitted regression line equation $\hat{y} = 1.02 + 2.73x$, the estimate of the standard deviation of the conditional probability distribution of $\mathbf{Y}$ given x ($s_{\mathbf{Y}/x} = 5.164$), and the coefficient of determination ($r^2 = 96.2$ percent). The standardized residuals are calculated in a different way than discussed in Section 13.7 as noted in the footnote. We will discuss the MINITAB method of calculating the standardized residuals in Chapter 14. Notice that although the standardized residuals are calculated using a different formula, the residual plot does not differ much in pattern from the one given in Figure 13.13—there is a slight indication of unequal variances in both plots due to the wider range of the residuals for larger values of x than for smaller values of x.

TABLE 13.7 | Summary formulas for computing point estimates of various regression and correlation statistics

Point estimate of regression slope:

$$b_1 = \frac{\Sigma x_i y_i - \dfrac{(\Sigma x_i)(\Sigma y_i)}{n}}{\Sigma x_i^2 - \dfrac{(\Sigma x_i)^2}{n}}$$

Point estimate of regression intercept:

$$b_0 = \bar{y} - b_1 \bar{x} = \frac{\Sigma y_i}{n} - b_1 \left(\frac{\Sigma x_i}{n}\right)$$

Point estimate of the variance of the conditional probability distributions:

$$s_{\mathbf{Y}/x}^2 = \frac{1}{n-2}\left\{\left[\Sigma y_i^2 - \frac{(\Sigma y_i)^2}{n}\right] - \frac{\left[\Sigma x_i y_i - \dfrac{(\Sigma x_i)(\Sigma y_i)}{n}\right]^2}{\Sigma x_i^2 - \dfrac{(\Sigma x_i)^2}{n}}\right\}$$

Sample coefficient of determination, r^2:

$$r^2 = \frac{\text{SST} - \text{SSE}}{\text{SST}} = \frac{\left[\Sigma x_i y_i - \dfrac{(\Sigma x_i)(\Sigma y_i)}{n}\right]^2}{\left[\Sigma x_i^2 - \dfrac{(\Sigma x_i)^2}{n}\right]\left[\Sigma y_i^2 - \dfrac{(\Sigma y_i)^2}{n}\right]}$$

Sample coefficient of correlation:

$$r = \frac{\Sigma x_i y_i - \dfrac{(\Sigma x_i)(\Sigma y_i)}{n}}{\sqrt{\Sigma x_i^2 - \dfrac{(\Sigma x_i)^2}{n}}\sqrt{\Sigma y_i^2 - \dfrac{(\Sigma y_i)^2}{n}}}$$

A statistical software package, such as MINITAB, eases the computational burden considerably when performing regression analyses. In the next two chapters, we will extensively use the MINITAB and the SAS statistical software packages for regression analyses.

■ **13.12**

Summary

This chapter presented the elements of simple linear regression and correlation analysis. Numerous assumptions were described for linear regression and correlation to be appropriate models for statistical analysis; and although statistical tests exist for assessing the reasonableness of these assumptions, residual plots presented in regression analysis are a good "first" step in assessing the reasonableness of the underlying assumptions of the approach. Numerous examples were provided to illustrate the use of the many regression and correlation statistics described.

Computation formulas presented in the chapter often served as shortcut methods for computing the quantities involved. Several of these formulas are summarized in Table 13.7. Table 13.4 gives a convenient format for computing many of the quantities involved in Table 13.7. The check that $\Sigma(y - \hat{y}) = 0$ or nearly equals 0 (allowing for computational roundoff) provides some assurance that the computations have been performed correctly.

The *relative* strength of a linear relationship as described in the chapter is measured by r^2, the coefficient of determination, and r, the coefficient of correlation, while a measure of the *absolute* strength of the linear relationship is given by $S_{Y/x}$. In the next chapter we will again address ourselves to measuring the strength of a linear relationship when we describe tests of hypotheses and confidence interval estimation of many of the parameters described in this chapter.

■ **References**

Brook, R. J., and G. C. Arnold. *Applied Regression Analysis and Experimental Design*. New York: Marcel Dekker, 1985.

Chatterjee, S., and B. Price. *Regression Analysis by Example*. New York: John Wiley & Sons, 1977.

Conover, W. J. *Practical Nonparametric Statistics*. New York: John Wiley & Sons, 1971.

Dixon, W. J., ed. *BMDP Statistical Software 1981*. Berkeley: University of California Press, 1981.

Draper, N. R., and H. Smith. *Applied Regression Analysis*. 2nd ed. New York: John Wiley & Sons, 1981.

Dutta, M. *Econometric Methods*. Cincinnati: South-Western Publishing, 1985.

Freedman, D.; R. Pisani; and R. Purves. *Statistics*. New York: W. W. Norton, 1978, Chapters 8–12.

Goldberger, A. S. *Econometric Theory*. New York: John Wiley & Sons, 1964.

Johnston, J. *Econometric Methods*. 2nd ed. New York: McGraw-Hill, 1972.

Mendenhall, W., and J. T. McClave. *A Second Course in Business Statistics: Regression Analysis*. San Francisco: Dellen Publishing, 1981.

Montgomery, D. C., and E. A. Peck. *Introduction to Linear Regression Analysis*. New York: John Wiley & Sons, 1982.

Neter, J.; W. Wasserman; and M. Kutner. *Applied Linear Statistical Models*. 2nd ed. Homewood, Ill.: Richard D. Irwin, 1985.

Ryan, T.; B. Joiner; and B. Ryan. *MINITAB Student*

Handbook. 2nd ed. Boston: PWS Publishers, 1985.

SAS User's Guide: Statistics. Cary, N.C.: SAS Institute, 1982.

SPSS-X User's Guide. New York: McGraw-Hill, 1983.

Younger, M. S. *A First Course in Linear Regression*. 2nd ed. Boston: Duxbury Press, 1985.

■ Problems

Section 13.2 Problems

13.1 Define a functional relationship between two variables.

13.2 Which of the following functional relationships are linear?
 a. $y = 2x - 1$
 b. $y = 4$
 c. $y = 4x^2 - 1$
 d. $y = e^x$
 e. $y = 2x + \log x$

13.3 Define a statistical relationship between two variables.

13.4 Given the following four sets of data, determine whether a functional or a statistical relationship exists between the variables, and whether or not the relationship is linear. Where the relationship is functional, determine the function that relates x and y.

a. x	y	c. x	y
3	109	3	109
4	112	4	116
6	118	5	125
9	127	7	149
12	136	9	181

b. x	y	d. x	y
3	107	3	4
5	117	5	16
7	119	7	36
9	129	8	49
11	131	10	81

Section 13.3 Problems

13.5 Distinguish between dependent and independent variables in a simple linear regression equation.

13.6 Discuss the assumptions made in using simple linear regression about the distributions of the conditional mean values.

13.7 Explain the requirement for the term ϵ_i in the population regression model.

13.8 Describe the two ways that the sample may be taken to produce data for fitting a regression model.

13.9 Graph the regression function $E(Y) = 10 - 2x$. What is the Y-intercept for this function? What is the slope of this function?

13.10 In the regression model, $Y = \beta_0 + \beta_1 x + \epsilon$, if $\beta_1 = 0$, then the model is said to be not useful in predicting values of Y. Explain why.

Section 13.4 Problems

13.11 Why is it important to plot a scatter diagram of the relationship between variables in a simple linear regression model?

13.12 What is meant by the term *least squares* in a simple linear regression model?

13.13 Describe the normal equations and how they are derived.

13.14 Describe the properties of the estimators of the regression coefficients, β_0 and β_1.

13.15 Plot each of the following sets of data as a scatter diagram. Which "curves" appear to best fit the given data?

a.

y Vehicle registrations	x Miles of primary highway
125	5,022
155	9,984
130	14,738
202	19,921
194	25,021
241	30,550
310	34,729
397	41,001
570	45,143
656	50,002

Source: Ten districts in states in Southeast Mexico.

b.

y Sales of premium unleaded gasoline (1000s of gallons)	x Price of premium unleaded gasoline (per gallon)
50.5	$1.37
57.3	1.36
60.0	1.35
60.1	1.31
67.8	1.29
70.1	1.27
68.2	1.26
74.8	1.25
75.1	1.23

Source: Gas station in Tumbleweed, Texas—nine months of premium gasoline sales and prices. (September, 1985).

c.

y Batting average	x Age
.304	21
.299	24
.293	27
.288	28
.280	30
.267	32
.260	33
.252	34
.264	35

Source: Nine individuals, selected at random, from the 1986 Metroplex Softball League.

d.

y	x
21	4
6	7
14	11
9	10
21	12
14	5
5	8
6	9
9	6

Source: "Random" pairs of observations.

13.16 A fuel-oil distribution company has collected data over a series of years to determine the statistical relationship between the average daily temperature and the consumption of fuel oil in single-family dwellings. Given the temperature, they desire to predict the consumption of fuel oil in order to better service customers on their delivery routes. The table shows a sample of the data that have been collected by the firm.

Average consumption of fuel oil in single- family dwellings (gallons)	Average daily temperature
7.0	10°
6.2	20°
5.1	30°
4.6	40°
3.5	50°
2.9	60°
1.2	70°
1.0	80°
0.7	90°

Source: Company records, district 3.

a. Plot the data as a scatter diagram, and construct a line through the data points that appears to minimize the sum of squares of the vertical deviations from the data points to the line.

b. Now estimate the linear regression equation using the method of least squares, and superimpose the line on the graph constructed in part a. How do the two lines compare?

c. What is the interpretation of b_1 in your regression equation?

d. What is the expected level of consumption of fuel oil given an average daily temperature of 55?

13.17 For the (x, y) pairs given below, calculate b_0 and b_1, the estimated intercept and slope coefficients for the fitted regression equation:

x	y
0	10
1	20
2	25
3	35
4	40
5	50

Section 13.5 Problems

13.18 Five pairs of (x, y) values are: (10, 100), (20, 90), (30, 90), (40, 75) and (50, 60). For these five pairs, calculate $s^2_{Y/x}$, by using either computing formula number 1 or computing formula number 2.

13.19 In a regression model fitted to $n = 25$ pairs of (x, y) points, the sum of squared residuals (SSE) is 100. What is the estimate of σ^2, the conditional distribution variance?

13.20 In a regression model fitted to $n = 100$ pairs of (x, y) points, it is found that the sum of squared residuals is

$$\sum_{i=1}^{n} e_i^2 = 200$$

What is the estimated conditional distribution variance in this regression experiment?

13.21 What assumption must be made about the conditional distribution variance σ^2 in a regression experiment?

Section 13.6 Problems

13.22 Consider the following five pairs of x and y values:

x	y
1	10
4	40
3	30
5	50
2	20

a. Form the scatterplot of the (x, y) pairs.
b. Calculate r^2 using formula 13.6.
c. Could r^2 have been determined from the scatterplot in part a without having to use formula 13.6? Explain.

13.23 In a given regression analysis of two random variables, the following quantities were determined: SSE = 129.45, SST = 792.67, and $\hat{y} = 1.2 - 4.9x$. What is the value of r?

13.24 State what r^2, the sample coefficient of determination, measures.

13.25 In a regression experiment with $n = 100$ observations, it is found that $\Sigma xy = 1,100$, $\Sigma x = 100$, $\Sigma y = 100$, $\Sigma x^2 = 400$, and $\Sigma y^2 = 200$. What is r^2 for the fitted regression line in this experiment?

13.26 What is meant by the coefficient of determination?

Section 13.7 Problems

13.27 What is meant by conducting a residual analysis in a regression experiment?

13.28 How is a residual plot constructed, and how is it used in a regression experiment?

13.29 Describe a residual plot that indicated the assumptions in a regression analysis have been satisfied for a fitted regression equation.

13.30 The values of the independent variable x and the residuals for a fitted regression equation are given below:

x	e
12	-1.5
6	0.5
15	0.1
4	-0.5
22	2.5
5	2.0
10	-0.5
18	-2.6

a. Form the residual plot.
b. Does the plot indicate that the variance of the error terms is constant? Explain.

Section 13.8 Problems

13.31   List the steps in building, analyzing, and validating a regression model and briefly describe each step.

Section 13.10 Problems

13.32   In comparing the degree of linear association between two variables, what advantages does correlation provide compared with simple linear regression?

13.33   How are the correlation coefficient and the coefficient of determination, r^2, in a simple linear regression, related?

13.34   What is meant by a "nonsense" correlation coefficient?

13.35   How does correlation differ from simple linear regression?

13.36   Explain and contrast the sampling requirements for simple linear regression and correlation?

13.37   On what scale is the correlation coefficient measured (nominal, ordinal, interval, ratio)? Does a correlation coefficient of 0.6 mean that the degree of linear association is twice as strong as when the coefficient is 0.3? Explain.

13.38   What is the proper interpretation of a correlation coefficient? What is the nature of the sample data when $r = 0$? When $r = +1$? When $r = -1$?

13.39   What is the relationship between the correlation coefficient and the slope in simple linear regression?

13.40   Explain what is meant by the *covariance* of two random variables.

Additional Problems

13.41   Given the following values for the two random variables **X** and **Y**:

x	y
0	2
1	4
1	3
2	5

a.   Determine their covariance.

b.   What percentage of the variability in **Y** can be accounted for or can be explained by relating it (linearly) to **X**?

13.42   The following data show federal individual income tax rates by taxable income (income after exclusions, deductions, and exemptions) for 15 income brackets in 1984:

Income bracket	Midpoint of bracket	Tax rate (percent)
$ 0 up to 2,300	$ 1,150	0%
2,300 up to 3,400	2,850	11
3,400 up to 4,400	3,900	12
4,400 up to 6,500	5,450	14
6,500 up to 8,500	7,500	15
8,500 up to 10,800	9,650	16
10,800 up to 12,900	11,850	18
12,900 up to 15,000	13,950	20
15,000 up to 18,200	16,600	23
18,200 up to 23,500	20,850	26
23,500 up to 28,800	26,150	30
28,800 up to 34,100	31,450	34
34,100 up to 41,500	37,800	38
41,500 up to 55,300	48,400	42
55,300 up to 81,800	68,550	48

a.   Plot the data as a scatter diagram, using the midpoint of each income level as the value of the independent variable.

b.   Describe the nature of the relationship that exists from your diagram between taxable income and income tax rate.

c.   Derive an estimating equation that predicts individual income tax rates as a function of the midpoint of each income level using the method of least squares.

d.   By examining a residual plot of the fitted linear regression equation, state whether the assumptions underlying the use of the simple linear regression model appear to be met in these sample data.

13.43   Joe Super-Jock, big-time star fullback of the local college football team, is deciding whether to accept an offer to play professional football (or else go to work for his father-in-law at $90,000 per year as a busboy in the family restaurant). Joe has collected the following data indicating yards gained in senior year and average annual salary from a sample of the pro-standouts with whom Joe is personally familiar.

Professional player (running back)	Yards gained in senior year in college	Annual salary
1	420	$ 48,000
2	480	42,000
3	550	52,000
4	680	66,000
5	740	60,000
6	880	75,000
7	910	90,000
8	960	90,000
9	1,020	100,000
10	1,450	250,000

a. Plot the data given as a scatter diagram, and estimate the parameters of the simple linear regression equation using the method of least squares.
b. Determine r^2 and $s_{Y/x}$ using the data provided.
c. Joe gained 999 yards his senior year, not counting the Gator Bowl yardage. Predict Joe's annual salary, given the yardage gained.
d. From a residual plot constructed using the given data and the fitted simple linear regression equation, do the assumptions of the simple linear regression model appear to have been met in this problem?

13.44 A department store, as part of its college recruiting program, regularly supplies college students with its latest survey of wage rates and years of postsecondary education of its current employees. This information is supplied to give *prospective* employees an indication of the wage rate (or equivalent salary) they might *expect* to receive before a formal job offer is made, and to provide them with an indication of the expected "monetary worth" in their firm of obtaining additional

education. The results of their latest survey are shown in the table.

Hourly wage rate and years of postsecondary education for 10 professional workers selected at random from a department store

Worker	Years of postsecondary education	Hourly wage rate
1	5.5	$18.20
2	0.0	6.50
3	1.0	3.90
4	4.0	11.70
5	3.5	10.40
6	2.0	7.80
7	6.0	15.60
8	2.5	9.10
9	4.5	13.00
10	5.0	14.30

Source: Sample of company employees.

a. Plot the data provided in the form of a scatter diagram.
b. Obtain an estimating equation of the form $\hat{y} = b_0 + b_1 x$ using the method of least squares.
c. Estimate the expected wage rate in this firm for an individual with four years of postsecondary education.
d. Calculate r^2 using the data provided.
e. Does the simple linear regression model appear to be an appropriate one for modeling the behavior of this random variable?
f. Comment on the "goodness" of the model developed.

13.45 Herman has not been doing very well on statistics examinations. He decides to sample eight of his friends to determine (a) the number of hours spent studying for the last statistics midterm examination and (b) the points (out of 100) received on that examination. Herman hopes he can surmise that time spent studying is *not* systematically related to exam grades! The results of Herman's sample appear in the table.

Student	Hours of exam preparation	Exam score
Lois	19	58
Don	12	42
Roger	34	70
Ann	42	98
John	9	37
Gail	18	71
Bucky	51	94
Patty	22	85

a. Plot a scatter diagram of the data given.
b. Determine the linear, unbiased, minimum variance estimates b_0 and b_1 of β_0 and β_1 assuming the assumptions of the simple linear regression model are satisfied.
c. Determine $s_{Y/x}$ and r^2.
d. Does a "strong" linear relationship appear to exist between exam scores and hours spent studying? Do the assumptions underlying the use of the simple linear regression model appear to be met?

13.46 A company wishes to determine whether or not the amount of money spent on advertising is linearly related to the sales of a particular product over a reasonable range of advertising dollars expended. To test this notion, a random sample of 10 sales regions produced the data in the table.

Region	Sales ($1000), y	Advertising ($1000), x
1	10	1.0
2	6	0.5
3	12	2.0
4	15	1.8
5	8	1.2
6	10	1.4
7	18	2.8
8	7	0.8
9	20	1.9
10	16	1.8

a. Form a scatter diagram for these data.
b. Determine the fitted linear regression equation.
c. Determine r^2.

d. Predict sales for the following advertising dollars expended (in $1,000s): (i) $1.8, (ii) $2.8, (iii) $2.5.
e. Determine $s_{Y/x}$.

13.47 The data in the following table show the number of weeks of experience in a job involving the production of a circuit board required for TV sets and the number of boards rejected during the past week for 25 workers.

Worker	No. of weeks of experience, x	No. of boards rejected, y
1	8	25
2	11	22
3	1	36
4	2	41
5	10	25
6	18	18
7	5	26
8	10	20
9	4	38
10	22	12
11	15	18
12	21	16
13	3	27
14	20	10
15	6	24
16	12	19
17	40	10
18	20	20
19	13	33
20	5	36
21	10	24
22	25	16
23	32	10
24	11	24
25	16	18

a. Determine the equation of the fitted linear regression line.
b. What is the value of r^2?
c. Predict the number of rejected boards for each of the following values of x: (i) 10, (ii) 35, (iii) 1, (iv) 0.
d. What is the value of $s_{Y/s}$?
e. From an examination of a residual plot, does it appear that the assumptions of the simple linear regression model have been met in this problem?

13.48 An admissions officer in an MBA (Master of Business Administration) program is interested in determining what relationship, if any, exists between the GMAT (Graduate Management Admissions Test) score and the grade point average (GPA) of graduating MBAs. A sample of six recent graduates produced the data in the table.

Student	GMAT score, x	GPA, y
1	610	3.80
2	440	3.33
3	525	3.40
4	555	3.10
5	480	3.65
6	505	3.75

a. Form a scatter diagram for these data. Do they appear to be linearly related?
b. Find the fitted regression line.
c. Determine the value of r^2.
d. Predict an MBA applicant's GPA, given that her GMAT score is 600.
e. Do you have much *faith* in the prediction determined in part *d*? Explain.

13.49 Ten employees were randomly selected in an employee benefits study in a plant. For each, the age and number of sick days used during the past six months were recorded.

Individual	Age, x	Sick days, y
1	32	2
2	34	0
3	27	1
4	25	2
5	21	0
6	44	4
7	48	3
8	22	0
9	36	4
10	52	6

a. Plot a scatter diagram for the pairs (x, y).
b. Calculate the coefficient of correlation, r.

13.50 In each of the following cases, state whether you would expect ρ to be positive, negative, or 0.

a. **X** = Years of education beyond high school. **Y** = Salary.
b. **X** = Family size. **Y** = Monthly food expenditure.
c. **X** = Sales in a retail store. **Y** = Inventory on hand.
d. **X** = Age of a production machine. **Y** = down time of the machine for repairs.
e. **X** = Income of family. **Y** = Family size.
f. **X** = Income of family. **Y** = Monthly food expenditure.

13.51 A firm is interested in determining whether or not a linear relationship exists between hourly pay and production output for a group of its employees. Ten employees in this group are randomly selected. The data are shown in the table below.

Individual	Pay per hour	Production in units/hour
1	4.10	20
2	5.50	25
3	4.00	28
4	5.60	21
5	6.20	22
6	7.00	32
7	4.25	27
8	5.10	30
9	5.50	24
10	5.50	30

a. Plot a scatter diagram of these data pairs.
b. Calculate r.
c. Would simple linear regression be preferred to correlation in this problem? Discuss.

13.52 What is the coefficient of correlation between two random variables **X** and **Y** under the following conditions?
a. One of the random variables is really a constant.
b. The value of the first random variable is always equal to 7.5 plus the value of the second random variable.
c. The value of the second random variable is always equal to 7.5 plus the value of the first random variable.
d. The random variables are constant multiples of one another.

13.53 A researcher has compiled data on the annual consumption of alcoholic beverages, **X**, and the number of automobile accidents, **Y**, in a certain locality. The data are shown in the table.

Year	X (1000 gallons)	Y (100 accidents)
1973	100	10.3
1974	96	9.9
1975	112	11.2
1976	123	11.6
1977	141	12.8
1978	149	13.6
1979	158	13.5
1980	174	14.7

a. Plot a scatter diagram of these pairs.
b. Calculate *r*.
c. Comment on the meaning (interpretation) of *r* in this problem.

13.54 The following characteristics: **X** = Years of schooling beyond high school; **Y** = Salary (in $1,000s); and **Z** = age, were measured for five randomly selected employees.

	Employee				
	1	*2*	*3*	*4*	*5*
x	4	4	6	2	4
y	12	20	25	14	30
z	22	30	32	26	35

a. Plot scatter diagrams for the pairs (x, y), (x, z), and (y, z).
b. Determine $r_{X,Y}$ and interpret.
c. Determine $r_{X,Z}$ and $r_{Y,Z}$, and interpret both.
d. Which variable, age (**Z**) or schooling (**X**), better explains salary (**Y**)? Discuss. (*Hint:* Determine the coefficient of determination in each case.)

13.55 A psychologist is interested in determining whether or not two IQ tests produce (linearly) related scores. A random sample of 10 subjects is taken, and each subject is administered both tests. A suitable period of time is left between tests to allow subjects to recover; five subjects take test A first, and the remaining five take test B first. The results are shown in the table that follows.

Subject	Test A	Test B
1	120	109
2	144	127
3	100	116
4	124	120
5	132	116
6	108	98
7	114	122
8	132	121
9	110	106
10	128	115

a. Plot a scatter diagram of these 10 pairs.
b. Calculate *r*.
c. Based on the value of **r**, how well do the two tests relate linearly? Explain.

13.56 The Pirax Corporation is interested in the relationship between the monthly sales in 100,000s of bottles of its liquid starch product per month and the retail price of the product. Sales in 100,000s of bottles, retail prices, and the average industrial prices for all brands of the product are collected for 10 randomly selected months:

Selected month	Sales	Price	Average industrial price
1	20	$2.47	$2.36
2	24	2.32	2.38
3	19	2.50	2.45
4	25	2.30	2.42
5	20	2.42	2.35
6	18	2.55	2.40
7	23	2.40	2.44
8	22	2.36	2.40
9	25	2.34	2.45
10	20	2.35	2.32

a. Form two scatterplots: sales in 100,000s of bottles (*y*) versus price (*x*) and sales in 100,000s of bottles (*y*) versus average industrial price (*x*). Which scatterplot appears to conform best to a line?
b. Determine the fitted linear regression equation for sales (*y*) versus price (*x*).

c. Determine the fitted linear regression equation for sales (y) versus average industrial price (x).

d. Form a new independent variable, Price difference = Price − Average industrial price. What is this new variable measuring? Determine the fitted regression equation for sales (y) versus this new independent variable, the difference between the price set by Pirax for its product and the average industrial price for all brands of the product.

e. Calculate the coefficients of determination for the two fitted regression equations in parts b, c, and d. Which of the three regression equations would you prefer to use in predicting sales of this product? Why?

f. For the next month, the company decides to set the retail price at $2.30. It estimates that the average industrial price for all brands of the product next month will be $2.35. Using the fitted regression equation you chose in part e, estimate the expected sales of this product in 100,000s of bottles for the next month.

13.57 Regression analysis is used extensively in finance. One use is to determine the relationship between the rate of return from investing in a stock and the rate of return of a broad-based index, such as the New York Stock Exchange Index. For a given stock, suppose that we have the following rates of return for the stock and the index for 12 months:

Month	Rate of return on stock	Rate of return of index
1	.060	.040
2	.075	.045
3	.080	.055
4	.065	.060
5	.055	.055
6	.050	.040
7	.035	.025
8	.020	.010
9	.025	.015
10	.020	.005
11	.010	−.010
12	−.005	−.015

a. Form the scatterplot of the rate of return of the stock (y) and the rate of return of the index (x). Does it appear from the scatterplot that the two variables are linearly related?

b. Determine the fitted regression equation for the rate of return on the stock (y) and the rate of return of the index (x). Interpret the intercept coefficient and the slope coefficient in the fitted regression equation.

c. Determine the coefficient of determination.

d. Suppose that in the next month it is estimated that the rate of return of the index will be 0.05. What is the predicted rate of return for the stock based on the fitted regression equation determined in part b?

e. Form the residual plot based on the *standardized residuals*. Does the plot indicate that one or more of the regression assumptions concerning the residuals has been violated? Explain.

13.58 Consider the following data set:

y	x
6	7
9	6
5	8
21	12
14	5
9	10
6	9
21	4
14	11

a. Determine the fitted linear regression equations.

b. Determine the coefficient of determination.

c. Based on parts a and b, does the relationship between x and y appear to be linear? Form a scatterplot of x and y. What is the relationship between x and y?

d. Form the residual plot using the *standardized residuals*. Based on the residual plot, do the assumptions of regression appear to be satisfied? Explain. What does the form of the residual plot appear to be suggesting?

13.59 Job satisfaction is a complex measure that depends on many factors, such as age of the employee, salary, job location, nature of the managerial organization of the company, prospects for advancement, and so on. In a particular company, a survey is conducted to investigate how satisfied the employees are with their jobs. An instrument is used that produces a score between 0 and 100—100 being completely satisfied and 0 being completely dissatisfied. Among the information collected is the percentage above or below the average salary level for individuals in the same job classification as the individual responding to the survey. For 10 randomly sampled employees involved in the survey, we have:

y Job satisfaction measure (0 to 100)	x Percent above or below average salary for job classification
80	45%
55	0
20	−50
50	−10
75	20
90	110
10	−85
60	5
40	−5
65	10

a. Form the scatterplot of the job satisfaction measure (y) and the percentage measure (x).

b. Determine the fitted regression equation for y and x.

c. Determine the coefficient of determination.

d. Based on parts a, b, and c, do you believe that the job satisfaction measure (y) and the percentage measure (x) are linearly related? Explain.

e. Form the residual plot using the *standardized residuals*. Interpret the plot.

14

Inferences in simple linear regression and correlation

■

■ 14.1

Introduction

In Chapter 13 we introduced the topic of simple linear regression and correlation analysis—*simple* because of the inclusion of one independent variable in the estimating equation or in the correlation analysis, and *linear* because of the manner in which the parameters of the regression model enter into the estimating equation. The steps in the regression model-building process were defined to be as follows:

Steps in regression analysis	

Steps in the simple linear regression model-building process

Step 1.   Identify the variables to be included in building the regression model.
Step 2.   Collect sample data.
Step 3.   Specify the relationship that exists between the dependent and the independent variable.
Step 4.   Estimate parameters of the model specified.
Step 5.   Determine whether or not the assumptions of the simple linear regression model have been met.
Step 6.   Statistically test the usefulness of the model developed.
Step 7.   Use the model developed for prediction and estimation.

Chapter 13 covered essentially Steps 1 through 5 and part of Step 7. In this chapter, we will learn how to test statistically for the usefulness of the model developed (Step 6), and we will learn how to construct a confidence interval for the mean and a prediction interval for an individual value of the random variable **Y** given a value of x (Step 7). We will also learn how to test for the significance of the correlation coefficient ρ based on the sample value obtained. Recall that the coefficient of correlation described—Pearson's product-moment correlation coefficient—measures the relative strength of the *linear* association between variables. Hence a hypothesis test for ρ will be a test of the *linear* association between **X** and **Y**.

In order to extend our inference techniques in simple linear regression analysis, we will have to add one more assumption to the regression model to those given in Chapter 13—that dealing with the distribution of the error terms, ϵ_i. By adding this fourth assumption, we will be able to expand greatly the statements that can be made concerning many regression parameters, and we will be able to complete the sixth step in the regression model-building process.

■ 14.2

Inferences in regression analysis

14.2.1 | The normality assumption

We have already discussed some aspects of inference making in the regression model in Chapter 13: namely, point estimation of the Y-intercept, β_0, and the slope, β_1, of the regression line, and point estimation of the conditional probability distribution mean value, $\mu_{Y/x}$ and variance, σ^2. We now wish to extend the inference techniques to include confidence intervals and hypothesis tests for these parameters; but to do so, we must add a fourth assumption to our regression model to the three given on page 653.

Normality assumption

Normality assumption

It is assumed that the error components, ϵ_i, that have a mean of 0 [$E(\epsilon_i) = 0$, a variance of σ^2 [$V(\epsilon_i) = \sigma^2$], and are uncorrelated are also *normally distributed*.

This assumption should not come as a surprise, since most of the confidence interval and hypothesis-testing techniques discussed in Chapters 9–12 are based on the normal distribution. As a consequence of this assumption, it is possible to say that the conditional probability distributions of **Y**, given x, are normally distributed, since $\mathbf{Y}_i$ and ϵ_i are related by an additive constant ($\mathbf{Y}_i = \beta_0 + \beta_1 x_i + \epsilon_i$, where $\beta_0 + \beta_1 x_i$ is a constant).

When the normality assumption is added to the list of regression model assumptions, the regression model is referred to as the *normal error regression model*. The full definition of this model is given below.

Normal error regression model

The population regression model $Y_i = \beta_0 + \beta_1 x_i + \epsilon_i$; $i = 1, 2, \ldots, n$, is called the normal error regression model when the following assumptions are satisfied:

1. For the ith level of x, x_i, the expected value of the error component is 0 [$E(\epsilon_i) = 0$], and the variance of the error component is σ^2 [$V(\epsilon_i) = \sigma^2$] and is constant for all i, where $i = 1, 2, \ldots, n$.
2. The error components ϵ_i, ϵ_j between any pair of values of the dependent variable are uncorrelated.
3. β_0, β_1, and σ^2 are unknown constants and must be estimated from the sample data.
4. The error component ϵ_i is normally distributed, $i = 1, 2, \ldots, n$.

The normality assumption is *not* required to obtain the point estimators of β_0, β_1, $\mu_{Y/x}$, and σ^2 (b_0, b_1, $\hat{Y}$, and $S^2_{Y/x}$, respectively). This assumption is required *only* when constructing confidence intervals and hypothesis-testing decision rules. Throughout the remainder of this chapter *unless otherwise specified*, we will consider the normality assumption to be satisfied.

14.2.2 | Inferences concerning the regression slope, β_1

The slope of a regression line is of interest to us usually in one of two ways. First, and most important, if we can infer from the sample that β_1 equals 0, then we know that the regression function is of *no use* to us as a predictor [recall that if $\beta_1 = 0$, then $E(Y) = \beta_0 + \beta_1 x = \beta_0$, so that for *every* value of x, $E(Y) = \beta_0$]. Second, β_1 gives the amount of increase or decrease in the dependent random variable (Y) per unit increase in the independent variable (x). If x represents advertising expenditures and Y represents sales, for example, then in the model $E(Y) = \beta_0 + \beta_1 x$, β_1 gives the increase (we hope) in sales per unit increase in advertising expenditure. Estimating the value of β_1 becomes quite important to a marketer in this instance!

To establish confidence interval and hypothesis-testing formulas for β_1, we must first describe the sampling distribution of its point estimator, b_1. The sampling distribution of b_1 arises conceptually by taking all possible repeated samples of n (x_i, y_i) pairs from the population regression function when the levels of x *are the same in each sample*, and from each sample, computing the point estimate of β_1 given by:

$$b_1 = \frac{\sum x_i y_i - \dfrac{(\sum x_i)(\sum y_i)}{n}}{\sum x_i^2 - \dfrac{(\sum x_i)^2}{n}}$$

Refer to Figure 13.8 for an illustration of the process used to generate the sampling distribution of b_1. We now give the characteristics of this sampling distribution without proof.

Theorem 14.1

Sampling distribution of b_1

The *sampling distribution* of b_1:

1. Is normal.
2. Has a mean value of β_1 [i.e., $E(b_1) = \beta_1$].
3. Has a variance, denoted by $\sigma_{b_1}^2$, equal to:

$$\sigma_{b_1}^2 = \frac{\sigma^2}{\Sigma x_i^2 - \dfrac{(\Sigma x_i)^2}{n}}$$

Since σ^2, the variance of the conditional probability distributions of Y, given a value of x, is typically unknown, it must be estimated from sample data. From Section 13.5, the point estimate of σ^2 is:

$$s_{Y/x}^2 = \frac{\sum_{i=1}^{n}(y_i - \hat{y}_i)^2}{n-2} = \frac{\text{SSE}}{n-2}$$

By inserting $s_{Y/x}^2$ for σ^2 in the formula for $\sigma_{b_1}^2$, we produce a point estimate of $\sigma_{b_1}^2$, denoted by $s_{b_1}^2$:

Point estimate of $\sigma_{b_1}^2$, the variance of the sampling distribution of b_1

$$s_{b_1}^2 = \frac{s_{Y/x}^2}{\Sigma x_i^2 - \dfrac{(\Sigma x_i)^2}{n}} \tag{14.1}$$

Since b_1 is normally distributed, the standardized random variable

$$\frac{b_1 - \beta_1}{\sigma_{b_1}}$$

is also normally distributed. If we now substitute S_{b_1} for σ_{b_1} in the above expression, the new standardized random variable

$$\frac{\mathbf{b}_1 - \beta_1}{\mathbf{S}_{\mathbf{b}1}}$$

is t-distributed with $(n - 2)$ degrees of freedom. This is analogous to the formation of the **t**-statistic in Chapter 10. This statistic is used to construct confidence intervals and hypothesis-testing decision rules for β_1, and thereby test the usefulness of the model developed.

<table>
<tr><td>

Confidence interval formula for β_1

</td><td>

Confidence interval estimate for β_1, the population slope

From the t-distributed standardized random variable $(\mathbf{b}_1 - \beta_1)/\mathbf{S}_{\mathbf{b}1}$ may be derived (we will not derive it) the $100(1 - \alpha)$ percent *confidence interval estimate* for β_1:

$$b_1 - t_{\alpha/2:n-2}s_{\mathbf{b}1} \leq \beta_1 \leq b_1 + t_{\alpha/2:n-2}s_{\mathbf{b}1}$$

</td></tr>
</table>

If this interval contains the number 0, we will conclude that the regression line is *not useful* for prediction purposes.

Example 14.1   Refer to the sales-advertising example first described in Example 13.2. Construct a 95 percent confidence interval estimate for β_1.

Solution   From Example 13.6, the point estimate of β_1 is $b_1 = 2.73$. For a 95 percent confidence interval, $(1 - \alpha) = 0.95$, so that $\alpha = 0.05$ and $\alpha/2 = 0.025$. From Table B.5 in Appendix B, $t_{0.025:10-2} = 2.306$. The value of $\mathbf{S}_{\mathbf{b}1}$ is given by:

$$s_{\mathbf{b}1} = \sqrt{\frac{s_{Y/x}^2}{\sum x_i^2 - \frac{(\sum x_i)^2}{n}}} = \sqrt{\frac{26.67}{3,890 - \frac{(178)^2}{10}}} = \sqrt{0.03696} = 0.1922$$

The confidence interval is:

$$2.73 - (2.306)(0.1922) \leq \beta_1 \leq 2.73 + (2.306)(0.1922)$$
$$2.29 \leq \beta_1 \leq 3.17$$

We would conclude, therefore, with 95 percent confidence, that the true slope β_1 lies between 2.29 and 3.17 and, since this interval does not contain 0, the sample regression line $\hat{y} = 1.02 + 2.73x$ is a useful predicting equation for estimating sales when advertising expenditures are between \$7 and \$31 (in \$00s).

The construction of a confidence interval for β_1 is equivalent to a test of a hypothesis, as was explained in Chapter 10. The null hypothesis is $H_0: \beta_1 = 0$, and the alternate hypothesis is $H_A: \beta_1 \neq 0$. The decision rule for rejection of the null hypothesis is given in Table 14.1.

The hypothesis testing procedure given in Table 14.1 is the standardized test for the null hypothesis, $H_0: \beta_1 = 0$. The method of standardized testing

TABLE 14.1 | Decision rule for testing whether the slope of a regression line is equal to 0

Standardized hypothesis test procedure for $H_0: \beta_1 = 0$

Hypothesis: $H_0: \quad \beta_1 = 0$
$H_A: \quad \beta_1 \neq 0$

Test statistic value: $t = \dfrac{b_1}{s_{b_1}}$

Decision rule: Reject if: $t > t_{\alpha/2; n-2}$
or
$t < -t_{\alpha/2; n-2}$

Do not reject if: $-t_{\alpha/2; n-2} \leq t \leq t_{\alpha/2; n-2}$

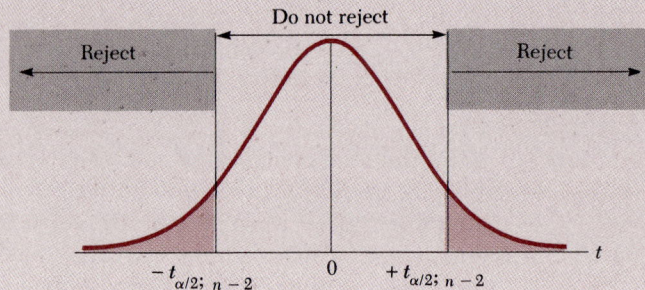

was presented in section 10.5 of Chapter 10. Equivalently, the decision rule for the null hypothesis could be established directly on the sampling distribution of $\mathbf{b}_1$. Most computer software packages test the null hypothesis using the standardized test statistic. In section 14.4 of this chapter, MINITAB is used to analyze the sales/advertising data in Example 13.2, and we will see that the standardized testing procedure is used to test the null hypothesis $H_0: \beta_1 = 0$. Therefore, we will not give the decision rule for the null hypothesis on the sampling distribution of $\mathbf{b}_1$.

Example 14.2 Refer to the sales-advertising example first described in Example 13.2 in Chapter 13. Test the null hypothesis, $H_0: \beta_1 = 0$ at the $\alpha = 0.05$ level of significance.

Solution From Example 13.6 the point estimate of β_1 is $b_1 = 2.73$. And the point estimate of σ_{b_1} is given by $s_{b_1} = 0.1922$ from Example 14.1. Therefore the value of the standardized test statistic is given by

$$t = \frac{b_1}{s_{b_1}} = \frac{2.73}{0.1922} = 14.20$$

With $\alpha = 0.05$ and $df = n - 2 = 8$, the action limits are:

$$-t_{0.05;8} = -2.306 \quad \text{and} \quad t_{0.05;8} = 2.306$$

Thus the decision rule is:

> Reject H_0: $\beta_1 = 0$ if $t < -2.306$ or if $t > 2.306$.
>
> Do not
> Reject H_0: $\beta_1 = 0$ if $-2.306 \le t \le 2.306$.

And the decision is:

> Reject H_0: $\beta_1 = 0$ since $t = 14.20 > 2.306$.

Therefore we can conclude that β_1 is not 0 and the independent variable x is useful in explaining variability in the dependent variable $\mathbf{Y}$ and in predicting its values.

It is important to remember that $b_1 = 2.73$ is an estimate of β_1 based on *one sample of size n* = 10. Another sample of size $n = 10$ would, of course, produce another estimate of β_1, which could be quite different in numerical value from the one produce by this sample (2.73). In the hypothesis test in Example 14.2, we are saying that the sample estimate $b_1 = 2.73$ is simply too far removed from the hypothesized value of 0 for β_1 (it is 14.20 standard deviations removed), and therefore the null hypothesis H_0: $\beta_1 = 0$ should be rejected.

The relationship expressed in Table 14.1 concerning the test of a hypothesis can also be shown by reference to the region in which we do not reject and the rejection region of the t-distribution as shown in the accompanying illustration. Thus if

$$-t_{\alpha/2;n-2} \le \frac{b_1}{s_{\mathbf{b}1}} \le t_{\alpha/2;n-2}$$

the hypothesis H_0: $\beta_1 = 0$ is *not rejected* at the specified level of significance. Note that the above interval is equivalent to the confidence interval

$$b_1 - s_{\mathbf{b}1} t_{\alpha/2;n-2} \le \beta_1 \le b_1 + s_{\mathbf{b}1} t_{\alpha/2;n-2}$$

If this interval includes 0, we do not reject H_0: $\beta_1 = 0$; if the interval does not contain 0, we reject H_0: $\beta_1 = 0$.

14.2.3 | The analysis of variance test for $H_0: \beta_1 = 0$

It is also possible to test the null hypothesis $H_0: \beta_1 = 0$ ($H_A: \beta \neq 0$) using the F-distribution discussed in Chapters 11 and 12. The reason for including the test here is that it can be generalized to include more than one independent variable in our estimating equation, a topic we will address in Chapter 15. As might be expected, the F-test for $H_0: \beta_1 = 0$ will produce results equivalent to those from the t-test for $H_0: \beta_1 = 0$. In fact, the *value* of the sample **F**-statistic is the square of the value of the sample **t**-statistic, as we will demonstrate in Example 14.3.

The form of the analysis of variance table for simple linear regression is given in Table 14.2. The degrees of freedom for the model (regression) are

TABLE 14.2 | Analysis of variance table for simple linear regression

Source of variation	Degrees of freedom	Sum of squares	Mean square	F-ratio
Regression (model)	1	SSR	MSR = SSR/1 = SSR	MSR/MSE
Error (residual)	$n - 2$	SSE	MSE = SSE/(n − 2)	
Total	$n - 1$	SST		

equal to the number of regression coefficients estimated, *excluding* β_0. Since in simple linear regression we are estimating the regression coefficients β_0 and β_1, the degrees of freedom for "regression" will always be 1. Similarly, the degrees of freedom for error (residual) is $n - 2$ in simple linear regression—the sample size n minus the number of regression coefficients estimated (2—β_0 and β_1). Notice that in the table, the degrees of freedom for "Total" are $n - 1$. The loss of the degree of freedom for estimating β_0 is taken from the sample size n in the total sum of squares row in the analysis of variance table.

The sums of squares in Table 14.2 are SST (total sum of squares), SSE (error sum of squares), and SSR (regression sum of squares). The formulas for SST, SSE, and SSR from Chapter 13 are:

$$\text{SST} = \sum_{i=1}^{n} (y_i - \bar{y})^2 = \Sigma y_i^2 - \frac{[\Sigma y_i]^2}{n}$$

$$\text{SSE} = \sum_{i=1}^{n} (y_i - \hat{y}_i)^2 = \left[\Sigma y_i^2 - \frac{(\Sigma y_i)^2}{n}\right] - \frac{\left[\Sigma x_i y_i - \frac{(\Sigma x_i)(\Sigma y_i)}{n}\right]^2}{\Sigma x_i^2 - \frac{(\Sigma x_i)^2}{n}}$$

$$\text{SSR} = \sum_{i=1}^{n} (\hat{y}_i - \bar{y})^2$$

$$= \text{SST} - \text{SSE}$$

The F-distribution for testing $H_0: \beta_1 = 0$ is shown in the accompanying figure. If the estimate of **MSR/MSE** exceeds $F(1 - \alpha; 1, n - 2)$ given in Table B.7 in Appendix B at the stated significance level α, then we reject $H_0: \beta_1 = 0$ and conclude that the independent variable is useful in predicting values of **Y** or in explaining variation in **Y**: if the estimate of **MSR/MSE** is less than $F(1 - \alpha; 1, n - 2)$, then we do not reject $H_0: \beta_1 = 0$ and conclude that the independent variable is of *no value* in predicting values of the dependent variable **Y** or in explaining variation in **Y** (i.e., the model is *not* useful).

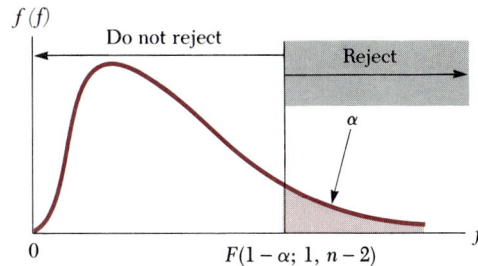

Example 14.3 Using the sales-advertising data given in Chapter 13, test the hypothesis $H_0: \beta_1 = 0$ at the 5 percent significance level using the F-distribution to determine whether to reject or not to reject $H_0: \beta_1 = 0$.
Solution From Chapter 13,

$$\text{SSE} = \sum_{i=1}^{n} (y_i - \hat{y}_i)^2 = 213.33$$

$$\text{SSR} = \sum_{i=1}^{n} (\hat{y}_i - \bar{y})^2 = 5{,}396.77$$

$$\text{SST} = \sum_{i=1}^{n} (y_i - \bar{y})^2 = 5{,}610.10; \text{ where } n = 10$$

The analysis of variance is given in Table 14.3 for the sales-advertising data.

TABLE 14.3 | Analysis of variance table for data in Example 14.3

Source of variation	Degrees of freedom	Sum of squares	Mean square	F-ratio
Regression	1	5,396.77	5396.77	202.38
Error	$n - 2 = 8$	213.33	26.67	
Total	$n - 1 = 9$	5,610.10		

From Table B.7 in Appendix B, $F(0.95; 1, 8) = 5.32$. Hence we reject $H_0: \beta_1 = 0$ and conclude that advertising expenditure is a useful predictor of sales for advertising expenditures between \$700 and \$3,100.

TABLE 14.4 | Computer analysis of data in Examples 14.1, 14.2, and 14.3

```
DEPENDENT VARIABLE: SALES
SOURCE                    DF  SUM OF SQUARES  MEAN SQUARE  F VALUE  PR > F
MODEL                      1       5396.768     5396.768   202.38  0.0001
ERROR                      8        213.332       26.667
CORRECTED TOTAL            9       5610.100
```

P-value

Table 14.4 shows a computer analysis of the data in Example 14.3. Note the correct interpretation of the p-value, PR > F = 0.0001, given in Table 14.4. Since the computer prints only four numbers to the right of the decimal point, either $\beta_1 \neq 0$, or else in sampling, we obtained a sample in which *fewer* than 1 in 10,000 yields an F-value of 202.38 or larger when in fact $\beta_1 = 0$. Hence, we conclude $\beta_1 \neq 0$ because of the odds against (*less than* 1 in 10,000) $\beta_1 = 0$.

Finally, note the relationship between the t-value determined in Example 14.2 and the F-value determined in Example 14.3

$$t^2 = \left(\frac{b_1}{s_{b_1}}\right)^2 = \left(\frac{2.7348}{0.1922}\right)^2 = 202.38 = f$$

(allowing for some discrepancy because of roundoff error).

Therefore it is possible to use either the t-test as in Example 14.2 or the F-test as in Example 14.3 to test the null hypothesis, $H_0: \beta_1 = 0$.

14.2.4 | Inferences concerning the population Y-intercept, β_0

Although confidence intervals and hypothesis tests on β_0 are not frequently used because the Y-intercept does not have important meaning in many practical applications, we include the appropriate formulas here for completeness. The sampling distribution of b_0, the point estimator of β_0, has the following characteristics:

Sampling distribution of b_0

Theorem 14.2

Sampling distribution of b_0

The *sampling distribution* of b_0:

1. Is normal.
2. Has a mean value of β_0 [i.e., $E(b_0) = \beta_0$].
3. Has a variance, denoted by $\sigma_{b_0}^2$, equal to:

$$\sigma_{b_0}^2 = \sigma^2 \left[\frac{1}{n} + \frac{\bar{x}^2}{\sum x_i^2 - \frac{(\sum x_i)^2}{n}} \right]$$

The point estimate of $\sigma^2_{b_0}$ is given by:

Point estimate of $\sigma^2_{b_0}$

Point estimate of $\sigma^2_{b_0}$, variance of sampling distribution of Y-intercept, β_0

$$s^2_{b_0} = s^2_{Y/x} \left[\frac{1}{n} + \frac{\bar{x}^2}{\Sigma x_i^2 - \dfrac{(\Sigma x_i)^2}{n}} \right] \tag{14.2}$$

The standardized statistic, $(b_0 - \beta_0)/S_{b_0}$ is t-distributed with $(n - 2)$ degrees of freedom. From this standardized t-statistic, the derived $100(1 - \alpha)$ percent confidence interval estimate is as follows:

Confidence interval formula for β_0

Confidence interval estimate for β_0, the population Y-intercept

$$b_0 - t_{\alpha/2;n-2} s_{b_0} \le \beta_0 \le b_0 + t_{\alpha/2:n-2} s_{b_0}$$

Caution must be exercised when making inferences concerning β_0. The population intercept β_0 is the value of the population regression function where it crosses the y-axis. In Figure 14.1 the simple linear regression function has been fitted to a population function that is clearly not linear. However, over the sampled range of x-values, a line fits quite well to the population function. But b_0 is obviously a very poor estimate of β_0 in this case. Using b_0 to estimate β_0 in this case represents *extrapolating beyond the range of the sampled x-values*. When the origin is far removed from the range of the sampled x-values, we must understand that b_0 may produce a very poor estimate of β_0.

In the next two sections, we consider the estimation of the *mean* of **Y**-values, given a specific input value x, and the prediction of an *individual* **Y**-value, given a specific input value x. To understand these two types of inferences and the difference between them, it will be helpful to consider a specific example. Suppose that we are interested in the relationship between the score on the Graduate Management Admissions Test (GMAT) of an applicant to a Master of Business Administration (MBA) program and the first-year grade-point average of the applicant, provided that the applicant is admitted to the MBA program and remains in the program for the duration of the first year. We will assume that the first-year grade-point average (GPA) is the dependent variable (**Y**) and is measured on a four-point scale, from 0.00 to 4.00. The GMAT score has a minimum of 200 and a maximum of 800.

FIGURE 14.1 | Consequences of using b_0 to estimate β_0 when extrapolating outside the range of the sampled x-values

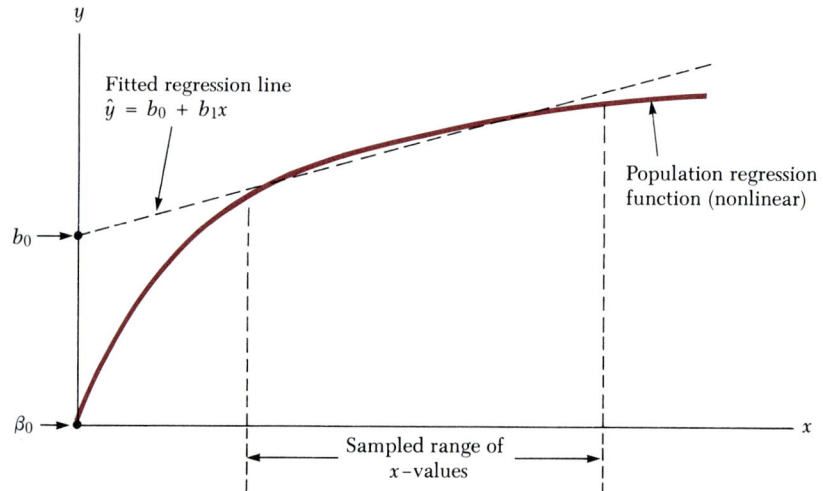

Assuming that the relationship between **Y** and x is linear, Figure 14.2 shows the population regression line $\mathbf{Y} = \beta_0 + \beta_1 x + \boldsymbol{\epsilon}$ and two conditional distributions of **Y**, when $x = 400$ and $x = 600$. Recall that the population regression line $\mathbf{Y} = \beta_0 + \beta_1 x + \boldsymbol{\epsilon}$ passes through the means of the conditional distributions of **Y**, given x. Thus, for example, there is a population of first-year GPAs in the MBA program for all admitted applicants who scored 400 on the GMAT. The mean of this distribution is denoted by $\mu_{\mathbf{Y}/x=400}$. In many applications of regression analysis, we are interested in making inferences about $\mu_{\mathbf{Y}/x}$, the mean of the conditional distribution of **Y**, given a value of x. In this illustration we might be interested in calculating a point estimate of $\mu_{\mathbf{Y}/x=400}$—that is, the mean first-year GPA of admitted applicants to an MBA program who scored 400 on the GMAT.

In addition to estimating $\mu_{\mathbf{Y}/x}$, we may be interested in predicting a new value of **Y**, given a value of x. Suppose, for example, that we have decided to admit an applicant whose GMAT score is 600 ($x = 600$). We would then like to predict his or her first-year GPA in the MBA program. The applicant's first year MBA GPA is *one individual value* on the conditional distribution of **Y**, given $x = 600$, as indicated in Figure 14.2. We will denote the predicted value as $\mathbf{Y}_{\text{new}}$, where the subscript "new" indicates that we are predicting a value of **Y** for a new value of x.

As we shall see in the following two sections, $\mu_{\mathbf{Y}/x}$, the *mean* of values of **Y**, can be estimated with greater precision (less error) than can an *individual* value of **Y**, whose estimate is denoted by $\hat{\mathbf{Y}}_{\text{new}}$. Both types of inferences are important and do occur frequently in regression analysis.

FIGURE 14.2 | Conditional distributions of **Y**, first-year MBA GPA, given x, score on GMAT illustrating a conditional mean $\mu_{Y/x}$ and an individual value Y_{new}

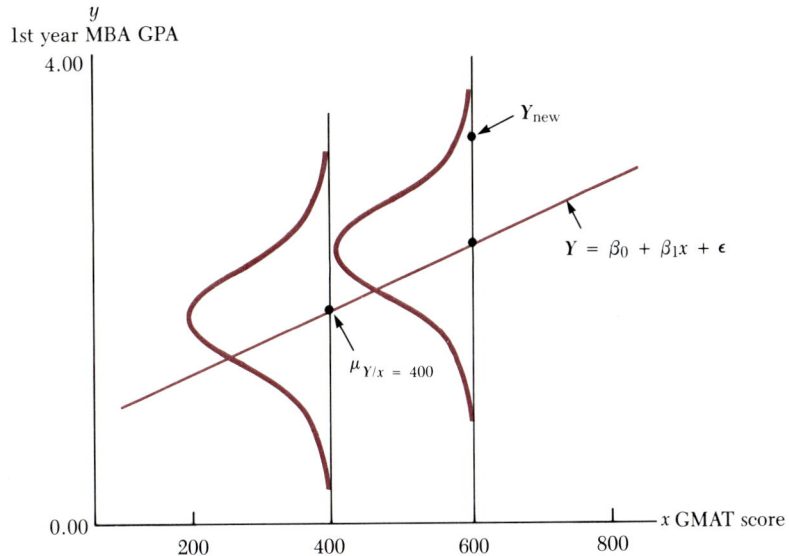

14.2.5 | Inferences concerning the mean of the conditional probability distribution of Y, given x: $\mu_{Y/x}$

The mean of the conditional distribution of **Y** given a specific value of x, say x_P, is:

$$\mu_{Y/x=x_P} = \beta_0 + \beta_1 x_P$$

One of the most important applications of regression analysis is the estimation of the conditional mean $\mu_{Y/x}$ for a specified value of x (Step 7). To determine a point estimate of $\mu_{Y/x=x_P}$, we would simply substitute the value of x_P into the fitted regression line $\hat{y} = b_0 + b_1 x$ for $x = x_P$; denote the resulting estimate by $\hat{y}_P = b_0 + b_1 x_P$. Such an estimate is a point estimate and represents our best estimate of the value of the mean (expected value). However, point estimates are often in error and often alone are not adequate. We usually desire some indication of the amount of error surrounding this estimate, for example. Such an estimate is provided by a confidence interval constructed for μ_{Y/x_P}.

We can set confidence intervals and establish hypothesis-testing decision rules for μ_{Y/x_P} by using the sampling distribution of $\hat{Y}_P$.

Sampling distribution of $\hat{Y}_P$, the estimator of $\mu_{Y/x}$, the conditional mean of Y, given x

Theorem 14.3

The sampling distribution of $\hat{Y}_P$

The *sampling distribution* of $\hat{Y}_P$:

1. Is normal.
2. Has a mean value of μ_{Y/x_P} [i.e., $E(\hat{Y}_P) = \mu_{Y/x_P} = \beta_0 + \beta_1 x_P$,
3. Has a variance, denoted by $\sigma^2_{\hat{Y}_P}$, given by

$$\sigma^2_{\hat{Y}_P} = \sigma^2 \left[\frac{1}{n} + \frac{(x_P - \bar{x})^2}{\Sigma x_i^2 - \frac{(\Sigma x_i)^2}{n}} \right]$$

Notice that the variance of $\hat{Y}_P$ depends on (1) the variance of the conditional probability distribution of Y given *any* value of x, σ^2; (2) the reciprocal of the sample size, $1/n$; (3) the distance between x_P, the selected value of x, and the mean, $\bar{x}$; and (4) the reciprocal of the sum of squares due to x, $\Sigma(x_i - \bar{x})^2 = \Sigma x_i^2 - [(\Sigma x_i)^2/n]$. The dependence of $\sigma^2_{\hat{Y}_P}$ on the first two factors is not surprising. We would expect the variability of $\hat{Y}_P$ to be related to σ^2, the variance of the conditional probability distributions, and to be a function of the reciprocal of the sample size, because $\hat{Y}_P$ is a population mean estimator, itself an expected value. The dependence of $\sigma^2_{\hat{Y}_P}$ on the third factor $(x_P - \bar{x})^2$ is best explained by examining Figure 14.3. In this figure are shown two fitted regression lines, both of which pass through the point $(\bar{x}, \bar{y})$. If we are estimating $\mu_{Y/x}$ for a value of x close to $\bar{x}$, say x_{P_1}, then the lines will give

FIGURE 14.3 | Effect of variability in b_1 on $\hat{Y}_P$

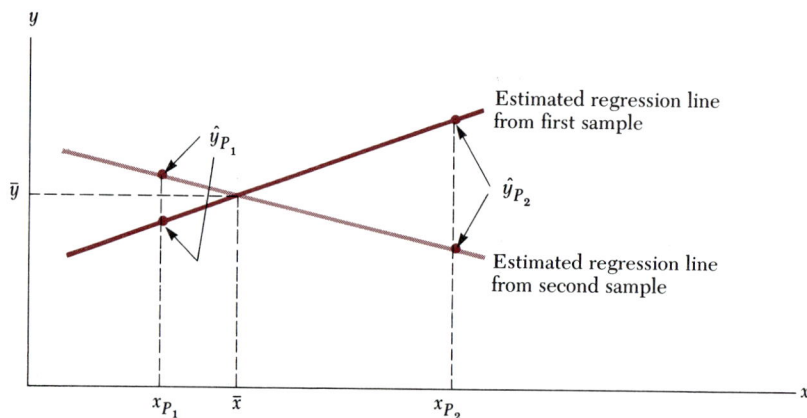

similar estimates. But, if we are estimating $\mu_{Y/x}$ for a value of x not close to $\bar{x}$, say x_{P2}, then the estimates may be quite dissimilar. Thus "the farther away" x_P is from $\bar{x}$, the greater the variability we should expect in the estimator $\hat{Y}_P$. The effect of the fourth factor, $\Sigma(x_i - \bar{x})^2$, on $\sigma^2_{\hat{Y}_P}$ is to moderate the term $(x_P - \bar{x})^2$. If the x-values are very spread out, then $\Sigma(x_i - \bar{x})^2$ will be large and will provide a small ratio, $(x_P - \bar{x})^2/\Sigma(x_i - \bar{x})^2$. This makes sense because if the x-values are spread out, then it should require a greater deviation $(x_P - \bar{x})$ to significantly increase the value of $\sigma^2_{\hat{Y}_P}$ than if the x-values are closely grouped. If the x-values are closely grouped, then $\Sigma(x_i - \bar{x})^2$ will be small and it will take a relatively small deviation $(x_P - \bar{x})$ to significantly increase the value of $\sigma^2_{\hat{Y}_P}$.

The point estimate of $\sigma^2_{\hat{Y}_P}$ is given by the following:

Point estimate of $\sigma^2_{\hat{Y}_P}$

> **Point estimate of $\sigma^2_{\hat{Y}_P}$, the variance of the sampling distribution of $\hat{Y}_P$**
>
> $$s^2_{\hat{Y}_P} = s^2_{Y/x}\left[\frac{1}{n} + \frac{(x_P - \bar{x})^2}{\Sigma x_i^2 - \dfrac{(\Sigma x_i)^2}{n}}\right]$$

The standardized random variable,

$$\frac{\hat{Y}_P - \mu_{Y/x_P}}{S_{\hat{Y}_P}}$$

Confidence interval formula for μ_{Y/x_P}

is t-distributed with $(n - 2)$ degrees of freedom. This **t**-statistic leads to the derived $100(1 - \alpha)$ percent confidence interval for μ_{Y/x_P}:

$$\hat{y}_P - t_{\alpha/2;n-2}s_{\hat{Y}_P} \le \mu_{Y/x_P} \le \hat{y}_P + t_{\alpha/2;n-2}s_{\hat{Y}_P}$$

Example 14.4 In the sales-advertising example from Chapter 13, find a point estimate of mean sales when the amount expended on advertising is $1,100 ($\mu_{Y/x=11}$), and determine a 90 percent confidence interval for $\mu_{Y/x=11}$.

Solution The estimated regression line was determined in Example 13.6—it is $\hat{y} = 1.02 + 2.73x$. Thus the point estimate of $\mu_{Y/x=11}$ is: $\hat{y}_{11} = 1.02 + 2.73(11) = 31.05$ (in $000). From Table B.5 in Appendix B, $t_{0.05;8} = 1.86$. Thus $s^2_{\hat{Y}_P}$ is:

$$s^2_{\hat{Y}_P} = s^2_{Y/x}\left[\frac{1}{n} + \frac{(x_P - \bar{x})^2}{\Sigma x_i^2 - \dfrac{(\Sigma x_i)^2}{n}}\right]$$

$$= (26.67)\left[\frac{1}{10} + \frac{(11 - 17.8)^2}{3,890 - \dfrac{(178)^2}{10}}\right] = 4.376$$

$$s_{\hat{Y}_P} = \sqrt{4.376} = 2.09$$

The 90 percent confidence interval estimate is:

$$31.05 - (1.86)(2.09) \leq \mu_{Y/x=11} \leq 31.05 + (1.86)(2.09)$$
$$27.16 \leq \mu_{Y/x=11} \leq 34.94$$

Thus we are 90 percent confident that mean sales lie between $27,160 and $34,940 for an advertising expenditure of $1,100.

In Table 14.5, 90 percent confidence interval estimates of $\mu_{Y/x}$ are given for a set of x values. (These estimates may differ slightly from other estimates because of the greater number of significant digits retained in calculating them.) These confidence intervals are also shown in Figure 14.4.

TABLE 14.5 | Ninety percent confidence intervals for $\mu_{Y/x}$ for a set of x-values in the sales-advertising example

Advertising expenditures, x_P ($00)	Estimated mean sales, $\hat{y}_P$ ($000)	Confidence interval estimate		
		Lower estimate	Upper estimate	Width
5	14.70	9.20	20.19	10.99
11	31.10	27.21	34.99	7.78
16	44.78	41.67	47.88	6.21
$\bar{x} = 17.8$				
21	58.45	55.21	61.70	6.49
26	72.13	67.90	76.35	8.45

First, we note in Table 14.5 that as x_P "moves away" from $\bar{x} = 17.8$, the width of the confidence intervals increases. This is impressively displayed in Figure 14.4—the five confidence intervals are shown as broken lines. If we calculated intervals for *all* values of x between 5 and 31, we would produce the "bands" shown in this figure by connecting all the upper limits and then all the lower limits by two smooth *curves*. The fact that the width of the confidence intervals increases as x_P departs from $\bar{x}$ substantiates our earlier discussion: $\hat{Y}_P$ is more variable as x_P departs from $\bar{x}$.

The confidence bands in Figure 14.4 do not take into account the *simultaneous estimation* of the *set* of means, $\mu_{Y/x}$. A correction for the estimation of a simultaneous set of means and the resulting confidence bands are given in the Neter, Wasserman, and Kutner (Chapter 5) reference at the end of this chapter.

We will not develop explicitly the hypothesis-testing decision rule for hypotheses concerning μ_{Y/x_P}, since confidence intervals may be used for this purpose. For example, suppose we hypothesize that average sales for this firm with $1,600 in advertising is $50,000, and we wish to test this hypothesis at the $\alpha = 0.10$ significance level. From Table 14.5 we would reject this

FIGURE 14.4 Confidence bands at the 90 percent confidence level for sales given advertising expenditures

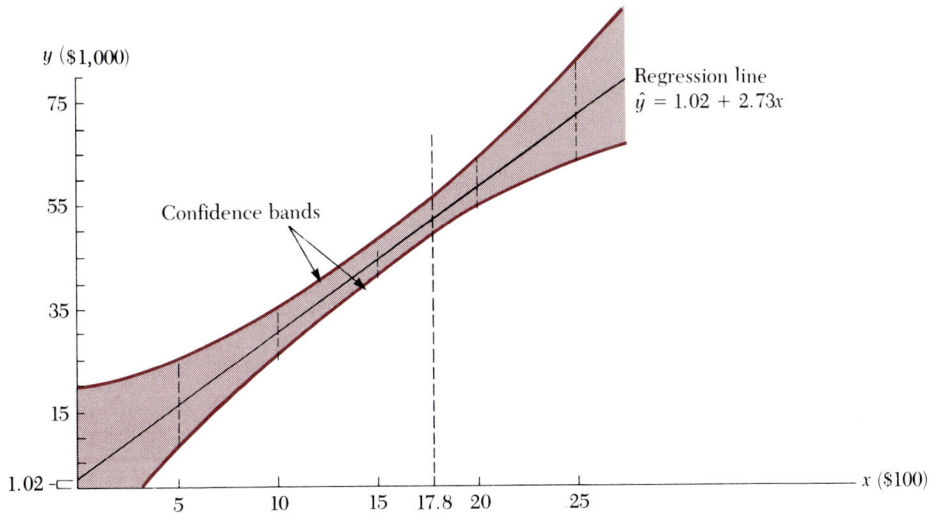

hypothesis since the 90 percent confidence interval does not contain 50 ($1,000) (the confidence interval estimate is $41,670–$47,880).

14.2.6 Prediction (forecast) interval for a "new" value, Y_{new}

In the previous section, we were concerned with drawing inferences about the mean $\mu_{Y/x}$ of the conditional probability distribution of **Y** for a given value of x, $x = x_P$. Often, however, we will be interested in inferences about a *single* or *new* observation (value of the random variable). For example, suppose in our sales-advertising illustration, we sample an additional sales branch, the new observation past the present 10 in our sample, and $1,900 is spent on advertising. We are interested in predicting sales for *this*, the next branch. This new observation y is viewed as the next trial in the experiment that produced the present 10 sample observations—this new observation is considered to be independent of the other sample observations. Denote the level of x, advertising expenditures, for this next observation as x_{new} and the corresponding **Y** observation as y_{new}.[1]

---

[1] Here, notationally, we are distinguishing the estimation of a mean value and the prediction of an individual observation. When we are estimating a mean value, we denote the level of x to be used in the estimation process (and the corresponding **Y**-statistics and estimates) by subscripting the variable with a P (e.g., x_P) and we denote the level of x to be used in the estimation process (and the corresponding **Y**-statistics and estimates) when we are predicting an *individual value* of the random variable by subscripting with the word new (e.g., x_{new}).

The prediction of $\mathbf{Y}_{new}$ is the same as the point estimate of the conditional mean $\mu_{\mathbf{Y}/x}$:

$$\hat{y}_{new} = b_0 + b_1 \cdot x_{new}$$

For instance, suppose the amount spent on advertising is \$1,900; that is, x_{new} = 19, and we are asked to predict (forecast) corresponding sales. Intuitively, we would give as our answer the estimated mean of the conditional probability distribution of $\mathbf{Y}$, given that x = 19. This prediction is: y_{new} = 1.02 + 2.73(19) = 52.89 (in \$1000).

But the estimation problem here is inherently different from the one presented in the previous section. Here, we are drawing inferences about a *single observation*, Y_{new}, whereas in Section 14.2.5, we were drawing inferences about a *mean* of a set of observations, the conditional mean $\mu_{\hat{\mathbf{Y}}/x_P}$. Although the two point estimates are the same, the confidence interval formulas are not—reflecting that it is much more difficult to predict a *single* value than to estimate a mean of a set of values.

Sampling distribution of $\hat{\mathbf{Y}}_{new}$

Theorem 14.4

The sampling distribution of $\hat{\mathbf{Y}}_{new}$

The *sampling distribution of* $\hat{\mathbf{Y}}_{new}$

1. Is normal.
2. Has a mean value of $\mu_{\mathbf{Y}/x_{new}}$ [i.e., $E(\hat{\mathbf{Y}}_{new}) = \mu_{\mathbf{Y}/x_{new}}$].
3. Has a variance, denoted by $\sigma^2_{\hat{\mathbf{Y}}_{new}}$, given by

$$\sigma^2_{\hat{\mathbf{Y}}_{new}} = \sigma^2 \left[1 + \frac{1}{n} + \frac{(x_{new} - \bar{x})^2}{\Sigma x_i^2 - \dfrac{(\Sigma x_i)^2}{n}} \right]$$

Now consider the variance formula $\sigma^2_{\hat{\mathbf{Y}}_{new}}$.

$$\sigma^2_{\hat{\mathbf{Y}}_{new}} = \sigma^2 \left[1 + \frac{1}{n} + \frac{(x_{new} - \bar{x})^2}{\Sigma x_i^2 - \dfrac{(\Sigma x_i)^2}{n}} \right] = \sigma^2 + \sigma^2 \left[\frac{1}{n} + \frac{(x_{new} - \bar{x})^2}{\Sigma x_i^2 - \dfrac{(\Sigma x_i)^2}{n}} \right] = \sigma^2 + \sigma^2_{\hat{\mathbf{Y}}_P}$$

That is, $\sigma^2_{\hat{\mathbf{Y}}_{new}}$ is equal to σ^2, the variance of the conditional probability distribution of $\mathbf{Y}$, given x, *plus* the variance of $\hat{\mathbf{Y}}_P$, the point estimator of the conditional mean $\mu_{\mathbf{Y}/x}$, given a specific value of x. Intuitively, this seems proper. In predicting a single observation y, we are confronted with the variability in estimating the conditional mean $\mu_{\mathbf{Y}/x}$ plus the *additional variability* of the specific conditional distribution of $\mathbf{Y}$, given a specific value of x. Thus, in predicting a single observation, we must take into account the variability in fixing the *location* of the conditional distribution (by estimating its mean $\mu_{Y/x}$) *and* take into account the variability *within* the conditional distribution (since we are attempting to predict a specific value that belongs to that conditional distribution).

The point estimate of $\sigma^2_{\hat{Y}_{new}}$ is given by the following:

Point estimate of
$\sigma^2_{\hat{Y}_{new}}$

> Point estimate of $\sigma^2_{\hat{Y}_{new}}$ the variance of the sampling distribution of $\hat{Y}_{new}$
>
> $$s^2_{\hat{Y}_{new}} = s^2_{Y/x} \left[1 + \frac{1}{n} + \frac{(x_{new} - \bar{x})^2}{\sum x_i^2 - \frac{(\sum x_i)^2}{n}} \right] \qquad (14.3)$$

The standardized random variable,

$$\frac{\hat{Y}_{new} - Y_{new}}{S_{\hat{Y}_{new}}}$$

is *t*-distributed within $(n - 2)$ degrees of freedom. This **t**-statistic leads to the derived $100(1 - \alpha)$ percent interval estimate for Y_{new}, called a *prediction or forecast interval*, since we are predicting a single value. The quantity $s_{\hat{Y}_{new}}$ is often called the *standard error of the forecast.*

Prediction interval
for $\hat{Y}_{new}$

> Prediction interval estimate for Y_{new}
>
> $$\hat{y}_{new} - t_{\alpha/2;n-2} s_{\hat{Y}_{new}} \leq Y_{new} \leq \hat{y}_{new} + t_{\alpha/2;n-2} s_{\hat{Y}_{new}}$$

Example 14.5 Determine a 90 percent prediction *(forecast)* interval for sales for an advertising expenditure of 19($100) (the "new" value) using the data provided.

Solution The point estimate of Y_{new} with $x_{new} = 19$ is:

$$\hat{y}_{new} = b_0 + b_1 x_{new} = 1.02 + 2.73(19) = 52.89(\$1,000)$$

The estimate of the variance of $\hat{Y}_{new}$ is given by:

$$s^2_{\hat{Y}_{new}} = s^2_{Y/x} \left[1 + \frac{1}{n} + \frac{(x_{new} - \bar{x})^2}{\sum x_i^2 - \frac{(\sum x_i)^2}{n}} \right] = 26.67 \left[1 + \frac{1}{10} + \frac{(19 - 17.8)^2}{3,890 - \frac{(178)^2}{10}} \right] = 29.39$$

$$s_{\hat{Y}_{new}} = \sqrt{29.39} = 5.42$$

From Table B.5 in Appendix B, $t_{0.05;8} = 1.86$. The 90 percent confidence interval is:

$$52.89 - (1.86)(5.42) \leq Y_{\text{new}} \leq 52.89 + (1.86)(5.42)$$
$$42.81 \leq Y_{\text{new}} \leq 62.97$$

Thus we are 90 percent confident that the next branch's sales will lie between $42,810 and $62,970.

The 90 percent confidence interval for the mean value, μ_{Y/x_P}, is given by the interval 49.82 to 55.96 (verify!) when $x_P = x_{\text{new}} = 19$. The point estimates of μ_{Y/x_P} and Y_{new} are, of course, the same, since the same value for the independent variable x is substituted into the regression equation. The 90 percent confidence interval for the mean and the 90 percent prediction interval for the next value are given in Figure 14.5. Note the greater width of the prediction interval, reflecting the added variability introduced by predicting a value of the random variable as opposed to estimating a mean value.

FIGURE 14.5 | Ninety percent prediction and confidence interval estimates for the data in Example 14.5

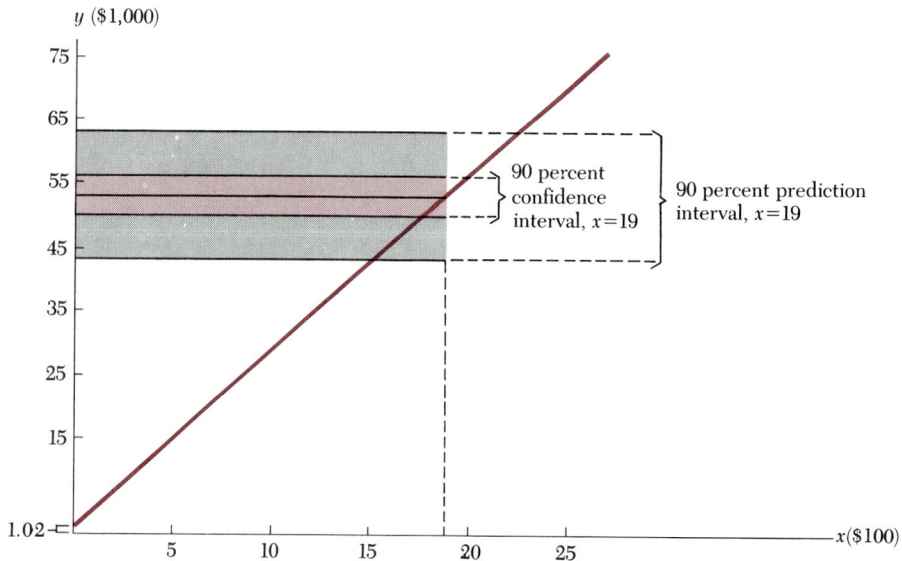

14.3

Inferences in correlation analysis

14.3.1 | Point estimation of the population correlation coefficient ρ

The sample correlation coefficient **r** is the point estimator of the population correlation coefficient, denoted by ρ. Figure 13.21 illustrates the relationship between r (the value of **r**) and ρ. The population correlation coefficient ρ is given by:

Population correlation coefficient ρ

$$\rho = \frac{E[(\mathbf{X} - \mu_{\mathbf{X}})(\mathbf{Y} - \mu_{\mathbf{Y}})]}{\sqrt{V(\mathbf{X})}\sqrt{V(\mathbf{Y})}} = \frac{C(\mathbf{X}, \mathbf{Y})}{\sigma_{\mathbf{X}} \cdot \sigma_{\mathbf{Y}}}$$

The numerator of ρ gives the covariance of the two random variables **X** and **Y** in the population. The denominator of ρ is the product of the population standard deviation of **X** and of **Y** ($\sigma_{\mathbf{X}}$ and $\sigma_{\mathbf{Y}}$, respectively).

The sample correlation coefficient r is the sample estimate of ρ and is given by:

Sample correlation coefficient **r**

$$r = \frac{\left(\dfrac{1}{n-1}\right) \sum\limits_{i=1}^{n} (x_i - \bar{x})(y_i - \bar{y})}{\sqrt{\dfrac{\sum\limits_{i=1}^{n} (x_i - \bar{x})^2}{n-1}} \sqrt{\dfrac{\sum\limits_{i=1}^{n} (y_i - \bar{y})}{n-1}}} = \frac{s_{\mathbf{XY}}}{s_{\mathbf{X}} s_{\mathbf{Y}}}$$

where $s_{\mathbf{XY}}$, $s_{\mathbf{X}}$, and $s_{\mathbf{Y}}$ are sample point estimates of $\sigma_{\mathbf{XY}} = C(\mathbf{X}, \mathbf{Y})$, $\sigma_{\mathbf{X}}$, and $\sigma_{\mathbf{Y}}$, respectively.

A difficulty arises in constructing confidence interval estimates and hypothesis tests for ρ that did not arise in constructing confidence interval estimates and hypothesis tests for the *regression* parameters discussed in Section 14.2. Namely, the sampling distribution of **r**, the sample correlation coefficient, is symmetric when $E(\mathbf{r}) = 0$ and is *not* symmetric when $E(\mathbf{r}) \neq 0$. This situation is depicted in Figure 14.6. Fortunately, Sir Ronald Fisher, a

FIGURE 14.6 | Sampling distributions of the sample correlation coefficient, **r**

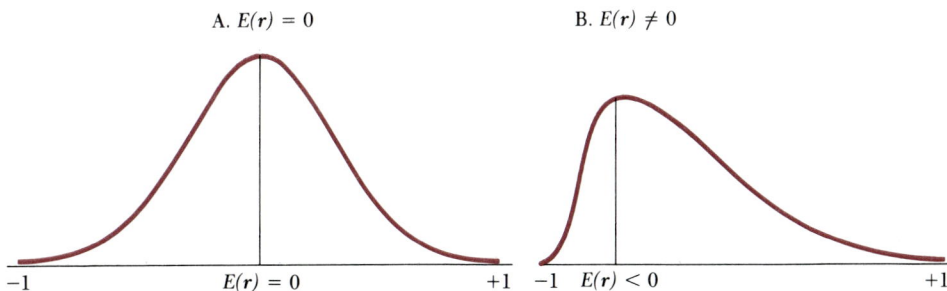

A. $E(r) = 0$ B. $E(r) \neq 0$

famous statistician, developed computational formulas for obtaining *approximate confidence interval estimates* for ρ when $E(\mathbf{r}) \neq 0$. We discuss confidence interval estimation of $\rho \neq 0$ in Section 14.3.3 and turn now to a test of the hypothesis $H_0: \rho = 0$.

14.3.2 | Hypothesis tests for the population correlation coefficient, $H_0: \rho = 0$

If we draw a random sample of n pairs, $(x_1, y_1), (x_2, y_2), \ldots, (x_n, y_n)$, from the bivariate normal distribution, then the standardized random variable $\mathbf{t} = \mathbf{r}/\mathbf{S_r}$ is t-distributed with $(n-2)$ degrees of freedom, where s_r is the point estimate of σ_r, the standard deviation of the sampling distribution of $\mathbf{r}$, and is given by the following:

Point estimate of σ_r, the standard deviation of the sampling distribution of $\mathbf{r}$

$$s_r = \sqrt{\frac{1 - r^2}{n - 2}} \tag{14.4}$$

Point estimate of σ_r

The statistic $\mathbf{S_r}$ is used when we wish to test the null hypothesis $H_0: \rho = 0$ when we have drawn a random sample of n pairs (x_i, y_i) from a bivariate normal distribution.

Example 14.6 Assume for the Example 13.2 sales-advertising data that the population joint distribution of $(\mathbf{X}, \mathbf{Y})$ is bivariate normal. Test the hypothesis, $H_0: \rho = 0$, using $\alpha = 0.05$.

Solution In Example 13.12, $r = 0.9808$; also,

$$s_r = \sqrt{\frac{1 - (0.9808)^2}{10 - 2}} = 0.069$$

$$t = \frac{r}{s_r} = \frac{0.9808}{0.069} = 14.23$$

Since $t > t_{0.025;8}$, we reject $H_0: \rho = 0$; ρ *appears* to be greater than 0.

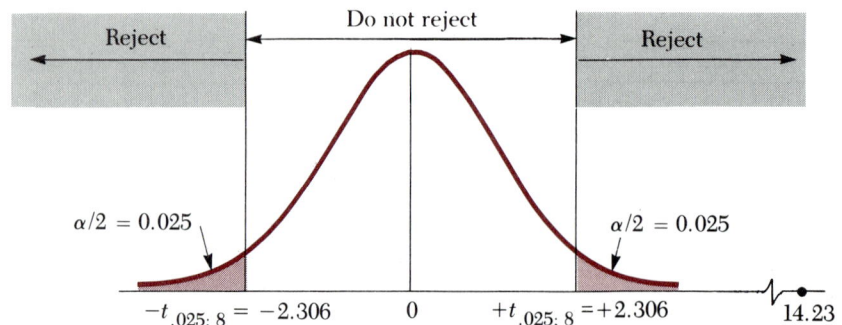

14.3.3 | Confidence interval estimation of ρ; $\rho \neq 0$

If we wish to test the null hypothesis that ρ is a value other than 0 *or* if we wish to set a confidence interval on ρ, the method becomes more complicated. Sir Ronald A. Fisher devised an approximate method for constructing confidence interval estimates for ρ, based on Fisher's Z-transformation:

Fisher's Z-transformation

$$\mathbf{Z} = \frac{1}{2} \log_e \left(\frac{1 + \mathbf{r}}{1 - \mathbf{r}} \right)$$

where $\log_e$ is the logarithm, base e. Fisher showed that $\mathbf{Z}$ is *approximately* normally distributed with mean and variance given by, respectively,

$$\mu_{\mathbf{Z}} = \frac{1}{2} \log_e \left(\frac{1 + \rho}{1 - \rho} \right) \quad \text{and} \quad \sigma_{\mathbf{Z}}^2 = \frac{1}{n - 3}$$

The approximation developed by Fisher is reasonably good as long as n is at least 25. By converting the sample correlation coefficient $\mathbf{r}$ to $\mathbf{Z}$ using this transformation, a confidence interval estimate for $\mu_{\mathbf{Z}}$ is given by:

$$l = z - z_{\alpha/2} \frac{1}{\sqrt{n - 3}} \leq \mu_{\mathbf{Z}} \leq u = z + z_{\alpha/2} \frac{1}{\sqrt{n - 3}}$$

The lower and upper limits on $\mu_{\mathbf{Z}}$, l and u, may then be converted to limits on ρ by using Table B.18 in Appendix B.

Example 14.7 Scores on an IQ/achievement test ($\mathbf{X}$) for $n = 100$ randomly selected workers were taken together with experimental measurements of the time ($\mathbf{Y}$) required to put together a complex optical device given the instructions. The sample correlation coefficient was computed to be $r = 0.75$. Determine an approximate 90 percent confidence interval for ρ.

Solution From Table B.18 in Appendix B, the value of $\mathbf{Z}$ that corresponds to $r = 0.75$ is 0.97. Then

$$l = 0.97 - (1.64) \frac{1}{\sqrt{97}} = 0.80 \quad \text{and} \quad u = 0.97 + (1.64) \frac{1}{\sqrt{97}} = 1.14$$

Now, referring to Table B.18 and entering the table with $z = 0.80$ and $z = 1.14$, the limits on ρ are:

$$0.66 \leq \rho \leq 0.81$$

Notice in Example 14.7 that Table B.18 does most of the work for us—there is no need to take any natural logarithms! Also, notice that the confidence interval is *not symmetric* about the point estimate, $r = 0.75$. This is because of the upper bound on ρ of $+1$; the sampling distribution of $\mathbf{r}$ must necessarily become *very skewed* if ρ is close to 1. This is why the $\mathbf{t}$-statistic cannot be used to set a confidence interval on ρ if $\rho \neq 0$; in this case, the statistic $\mathbf{r}/S_{\mathbf{r}}$ no longer has the symmetric t-distribution as its sampling distribution.

14.4

Use of the MINITAB software package

The computational burden of testing hypotheses concerning the regression coefficients or of developing confidence interval estimates of the conditional means of Y can be eased considerably by using a statistical software package, such as MINITAB. In this section, we will analyze the Example 13.2 sales-advertising example data set by the use of MINITAB. The first step is to read the sales-advertising data into columns of the MINITAB worksheet. The sales-advertising data are stored in the file named CHAPT13:

```
-- READ 'CHAPT13' INTO C1,C2
COLUMN          C1              C2
COUNT           10              10
ROW
  1             18,             55,
  2              7,             17,
  3             14,             36,
  4             31,             85,
     .     .     .
```

The READ command reads the data in the file CHAPT13 with C1 containing the 10 values of x (advertising expenditure) and C2 containing the 10 values of Y (sales). To ensure that all values have been correctly read into C1 and C2, the PRINT command is used to print out the contents of columns C1 and C2:

```
--  PRINT C1,C2
COLUMN          C1              C2
COUNT           10              10
ROW
  1             18,             55,
  2              7,             17,
  3             14,             36,
  4             31,             85,
  5             21,             62,
  6              5,             18,
  7             11,             33,
  8             16,             41,
  9             26,             63,
 10             29,             87,
```

Next, the NAME command is used to name the columns that will be used in the regression analysis:

```
-- NAME C1 'X'

-- NAME C2 'Y'

-- NAME C3 'STRES'
-- NAME C4 'YHAT'

-- NAME C5 'COEFFS'

-- NAME C6 'RESID'
```

Columns C1 and C2 are named X and Y, respectively, since C1 contains the values of the independent variable x, and C2 contains the values of the dependent variable $\mathbf{Y}$. Column C3 will contain the standardized residuals, so it is named STRES. Column C4 will contain the predicted values of $\mathbf{Y}$ for each of the 10 values of x. Therefore it is named YHAT. Column C5 will contain the regression coefficient estimates b_0 and b_1, so it is named COEFFS. Finally, column C6 will contain the residuals (not standardized). Hence it is named RESID. Next, the REGRESSION command is used (in MINITAB, only the first four letters of a command name need to be used):

```
-- REGR C2 1 C1 C3,C4,C5,M1
```

This command produces the simple linear fitted regression function of the dependent variable $\mathbf{Y}$ in C2 regressed on 1 predictor stored in C1 and stores the standardized residuals in C3, the estimated values of $\mathbf{Y}$ in C4, and the estimated regression coefficients in C5. The matrix M1 is used to store a matrix needed in calculating the estimated standard deviations for estimates and predictions of $\mathbf{Y}$. The output from the REGRESSION command is shown at the top of the next page. The boxed table contains the information needed to construct confidence interval estimates on β_0 and on β_1. The column entitled "COEFFICIENT" contains the values of b_0 and b_1; the column entitled "ST. DEV. OF COEF." contains the standard errors s_{b_0} and s_{b_1}. Finally, the column entitled "T-RATIO = COEF/S.D." contains the standardized t-statistic for testing the hypothesis H_0: $\beta_0 = 0$ and H_0: $\beta_1 = 0$. The formula used in Example 14.1 for the interval estimate of β_1 is:

$$b_1 \pm t_{\alpha/2;n-2}s_{b_1}$$

```
THE REGRESSION EQUATION IS
Y  =     1.02 +   2.73 X1
```

```
                                      ST. DEV.   T-RATIO =
            COLUMN    COEFFICIENT    OF COEF.    COEF/S.D.
              --             1.021       3.791        0.27
     X1    X             2.7348        0.1922        14.23
```

```
THE ST. DEV. OF Y ABOUT REGRESSION LINE IS
S = 5.164
WITH (   10- 2) =    8 DEGREES OF FREEDOM

R-SQUARED = 96.2 PERCENT
R-SQUARED = 95.7 PERCENT, ADJUSTED FOR D.F.
```

From the boxed MINITAB table, $b_1 = 2.7348$, and $s_{b_1} = 0.1922$. For a 95 percent confidence interval, $\alpha/2 = 0.025$, and $df = n - 2 = 10 - 2 = 8$. Thus, from Table B.5, Appendix B, $t_{0.025;8} = 2.306$. Therefore the 95 percent confidence interval on β_1 is:

$$2.7348 \pm (2.306)(0.1922) \Rightarrow 2.292 \text{ to } 3.178$$

From the same boxed table, a 95 percent confidence interval on β_0 is given by:

$$1.021 + (2.306)(3.791) \Rightarrow -7.721 \text{ to } 9.763$$

In Example 14.2 the null hypothesis H_0: $\beta_1 = 0$ was tested by using the t-test. From Example 14.2 the value of the standardized **t**-statistic was 14.20. Notice that this value is printed out in the boxed table under the column entitled "T-RATIO = COEF/S.D.." MINITAB calculates the value of the **t**-statistic with more accuracy than we did in Example 14.2. Hence the MINITAB value for the **t**-statistic is 14.23.

MINITAB also prints out the ANALYSIS OF VARIANCE table:

```
ANALYSIS OF VARIANCE

   DUE TO        DF             SS        MS=SS/DF
   REGRESSION    1         5396.75        5396.75
   RESIDUAL      8          213.33          26.67
   TOTAL         9         5610.08
```

Thus to test the null hypothesis H_0: $\beta_1 = 0$ by the F-test, the value of the **F**-statistic is:

$$f = \frac{5396.75}{26.67} = 202.35$$

In Example 14.3 the F-test was used to test the null hypothesis H_0: $\beta_1 = 0$. With an $\alpha = 0.05$ significance level, the action limit is $F(0.95; 1, 8) = 5.32$. Since $f = 202.35 > 5.32$, we reject the null hypothesis and conclude that the fitted regression function is useful in predicting values of **Y** and in explaining variation in **Y**.

By using a MACRO written in the MINITAB language (see the listing of the MACRO in the appendix at the end of this chapter), it is possible to

```
-- NOTE PRINT PREDICTED VALUE

-- NOTE

-- NOTE   *********

-- PRINT K5
    K5           31.1037

-- NOTE   *********

-- NOTE

-- NOTE   PRINT STANDARD DEVIATION OF THE ESTIMATE OF THE

-- NOTE   CONDITIONAL MEAN OF Y

-- NOTE

-- NOTE   *********

-- PRINT K6
    K6           2.09175

-- NOTE   *********

-- NOTE

-- NOTE   PRINT STANDARD DEVIATION OF THE PREDICTION OF AN

-- NOTE   INDIVIDUAL VALUE OF Y

-- NOTE

-- NOTE   *********

-- PRINT K7
    K7           5.57153

-- NOTE   *********
```

calculate the estimated value of **Y** together with the required estimated standard error for the estimation of $\mu_{\mathbf{Y}/x}$ and an individual value $\mathbf{Y}_{new}$. In Example 14.4 we determined a 90 percent confidence interval on $\mu_{\mathbf{Y}/x=11}$; that is, the mean of the conditional distribution of **Y** when the advertising expenditure is 11 (in thousands of \$'s). The MINITAB MACRO produces the results shown on the bottom of page 733. The predicted value is printed out together with the two standard errors, the first for $\mu_{\mathbf{Y}/x=11}$ and the second for $\mathbf{Y}_{new}$. In Example 14.4 the **t**-value for a 90 percent confidence interval was 1.860. Therefore the 90 percent confidence interval on $\mu_{\mathbf{Y}/x=11}$ is:

$$31.1037 + (1.860)(2.09175) \Rightarrow 27.213 \text{ to } 34.994$$

If we wished to determine a 90 percent prediction interval on $\mathbf{Y}_{new}$, an individual value of **Y** when $x = 11$, the interval would be:

$$31.1037 + (1.860)(5.57153) \qquad 20.741 \text{ to } 41.467$$

Notice that the 90 percent prediction interval for $\mathbf{Y}_{new}$ for $x = 11$ is much wider than the 90 percent confidence interval for $\mu_{\mathbf{Y}/x=11}$.

Finally, in Example 14.5 we wished to find a 90 percent prediction interval on an individual value $\mathbf{Y}_{new}$ when $x = 19$. From the MINITAB MACRO, we have:

```
  -- NOTE    PRINT PREDICTED VALUE
1-- NOTE

  -- NOTE    **********

  -- PRINT K5
     K5          52.9817

  -- NOTE    **********

  -- NOTE

  -- NOTE    PRINT STANDARD DEVIATION OF THE ESTIMATE OF THE

  -- NOTE    CONDITIONAL MEAN OF Y

  -- NOTE

  -- NOTE    **********

  -- PRINT K6
     K6           1.64920

  -- NOTE    **********

  -- NOTE

  --NOTE     PRINT STANDARD DEVIATION OF THE PREDICTION OF AN
```

```
-- NOTE   INDIVIDUAL VALUE OF Y

-- NOTE

-- NOTE   **********

-- PRINT K7
   K7          5.42092

-- NOTE   **********
```

Therefore the 90 percent prediction interval on $\mathbf{Y}_{new}$ when $x = 19$ is:

$$52.9817 + (1.860)(5.42092) \Rightarrow 42.899 \text{ to } 63.065$$

Notice that the answers from MINITAB are slightly different than the confidence intervals determined in Examples 14.4 and 14.5. This is due to the fact that we rounded numbers in the computations in Examples 14.4 and 14.5, although MINITAB carries more significant digits and hence produces more accurate answers.

TABLE 14.6 | Summary computational formulas for Chapter 14

Point estimate of the sampling distribution variance of $\mathbf{b}_1$

$$s_{b_1}^2 = \frac{s_{Y/x}^2}{\Sigma x_i^2 - \dfrac{(\Sigma x_i)^2}{n}}$$

Point estimate of the sampling distribution variance of $\mathbf{b}_0$

$$s_{b_0}^2 = s_{Y/x}^2 \left[\frac{1}{n} + \frac{\bar{x}^2}{\Sigma x_i^2 - \dfrac{(\Sigma x_i)^2}{n}} \right]$$

Point estimate of the sampling distribution variance of $\hat{\mathbf{Y}}_P$

$$s_{\hat{Y}_P}^2 = s_{Y/x}^2 \left[\frac{1}{n} + \frac{(x_P - \bar{x})^2}{\Sigma x_i^2 - \dfrac{(\Sigma x_i)^2}{n}} \right]$$

Point estimate of the sampling distribution variance of $\mathbf{r}$ (for testing $H_0: \rho = 0$)

$$s_r^2 = \frac{1 - r^2}{n - 2}$$

Point estimate of the sampling distribution variance of $\hat{\mathbf{Y}}_{next}$

$$s_{\hat{Y}_{next}}^2 = s_{Y/x}^2 \left[1 + \frac{1}{n} + \frac{(x_{next} - \bar{x})^2}{\Sigma x_i^2 - \dfrac{(\Sigma x_i)^2}{n}} \right]$$

14.5
Summary

In this chapter we expanded considerably upon the estimation techniques presented in Chapter 13. By adding a fourth assumption to the simple linear regression model, for example—that the error terms ε_i are normally distributed—we were able to develop tests of hypotheses and confidence interval estimation procedures for many of the regression parameters discussed in Chapter 13. The hypothesis test, whether β_1 equals 0 (i.e., $H_0: \beta_1 = 0$), leads to a direct statement (at a stated level of significance) about the usefulness of the regression model for estimation purposes (Step 6), for example. In Table 14.6 we summarize the variance estimates for the sampling distributions of the regression and correlation statistics presented in the chapter.

In Chapter 15 we shall consider the case in which more than one independent variable is used to predict the value of the dependent variable and where we wish to assess the strength of the linear relationship between a dependent variable and several independent variables. These types of analyses are called *multiple linear regression* and *multiple correlation analysis,* respectively.

References

Brook, R. J., and G. C. Arnold. *Applied Regression Analysis and Experimental Design*. New York: Marcel Dekker, 1985.

Chatterjee, S., and B. Price. *Regression Analysis by Example*. New York: John Wiley & Sons, 1977.

Conover, W. J. *Practical Nonparametric Statistics*. New York: John Wiley & Sons, 1971.

Dixon, W. J., ed. *BMDP Statistical Software 1981*. Berkeley: University of California Press, 1981.

Draper, N. R., and H. Smith. *Applied Regression Analysis*. 2nd ed. New York: John Wiley & Sons, 1981.

Dutta, M. *Econometric Methods*. Cincinnati: South-Western Publishing, 1985.

Freedman, D.; R. Pisani; and R. Purves. *Statistics*. New York: W. W. Norton, 1978, Chapters 8–12.

Goldberger, A. S. *Econometric Theory*. New York: John Wiley & Sons, 1964.

Johnston, J. *Econometric Methods*. 2nd ed. New York: McGraw-Hill, 1972.

Mendenhall, W., and J. T. McClave. *A Second Course in Business Statistics: Regression Analysis*. San Francisco: Dellen Publishing, 1981.

Montgomery, D. C., and E. A. Peck. *Introduction to Linear Regression Analysis*. New York: John Wiley & Sons, 1982.

Neter, J.; W. Wasserman; and M. Kutner. *Applied Linear Statistical Models*. 2nd ed. Homewood, Ill.: Richard D. Irwin, 1985.

Ryan, T.; B. Joiner; and B. Ryan. *MINITAB Student Handbook*. 2nd ed. Boston: PWS Publishers, 1985.

SAS User's Guide: Statistics. Cary, N.C.: SAS Institute, 1982.

SPSS-X User's Guide. New York: McGraw-Hill, 1983.

Younger, M. S. *A First Course in Linear Regression*. 2nd ed. Boston: Duxbury Press, 1985.

Appendix

Listing of the MINITAB MACRO to calculate the standard deviations used in confidence and prediction intervals

```
NOTE   MACRO NAME
NOTE   **********
NOTE   REGRESSION FORECAST STANDARD DEVIATION
NOTE
NOTE   MINITAB CODE
NOTE   ************
NOTE   COMPUTE UNSTANDARDIZED RESIDUALS
LET 'RESID' = 'Y' - 'YHAT'
NOTE   INITIALIZE  K48,K49,K50 TO N,P,AND S FROM THE OLS
       REGRESSION..
LET K48=COUNT('STRES')
LET K49=COUNT('COEFFS')
LET K50=SQRT(SSQ('RESID')/(K48-K49))
NOTE   COMPUTE PREDICTED VALUE,
NOTE   COMPUTE STANDARD DEVIATION OF THE ESTIMATE OF THE CONDITIONAL
NOTE   MEAN OF Y GIVEN X1, ... ,XP,    AND
NOTE   STANDARD DEVIATION OF THE PREDICTION OF AN INDIVIDUAL
NOTE   Y VALUE GIVEN X1, ... ,XP,
TRANS M2, PUT IN M3
COPY 'COEFFS' INTO M4
MULT M3 BY M4, PUT IN K5
MULT M3 BY M1, PUT IN M3
MULT M3 BY M2, PUT IN K1
LET K2=K1+1
LET K6=SQRT(K1)*K50
LET K7=SQRT(K2)*K50
NOTE
NOTE
NOTE   PRINT PREDICTED VALUE
NOTE
NOTE   **********
PRINT K5
NOTE   **********
NOTE
NOTE   PRINT STANDARD DEVIATION OF THE ESTIMATE OF THE
NOTE   CONDITIONAL MEAN OF Y
NOTE
NOTE   **********
PRINT K6
NOTE   **********
NOTE
NOTE   PRINT STANDARD DEVIATION OF THE PREDICTION OF AN
NOTE   INDIVIDUAL VALUE OF Y
NOTE
NOTE   **********
PRINT K7
NOTE   **********
END
```

To use this macro, store the macro code above in a file named, say, PRE-DICT. If we wish to estimate $\mu_{Y/x=11}$ or to predict $Y_{new=11}$, use the following commands following the MINITAB REGR command:

```
READ INTO A 2 BY 1 MATRIX M2
1
11
EXECUTE 'PREDICT'
```

The matrix M2 contains two rows and one column. The first row *always* contains a 1 and the second row contains the value of x for which estimates or predictions are required. Note that the MINITAB REGR command must list columns to contain the standardized residuals, the predicted values of $\mathbf{Y}$, the estimated coefficients, and the matrix M1. Further, the columns must be named exactly as in the MINITAB illustration since the MACRO uses the column names rather than column numbers.

Problems

Section 14.1 Problems

14.1 What is the interpretation of a confidence interval constructed to test whether or not $\beta_1 = 0$?

14.2 Explain the difference between a prediction or a forecast interval and a confidence interval constructed for $\mu_{Y/x}$.

14.3 Explain why it is nonsensical to construct a confidence interval for b_1 in a simple linear regression analysis of the linear relationship between variables $\mathbf{Y}$ and x.

14.4 There are three ways to test the null hypothesis $H_0: \beta_1 = 0$ against the alternative hypothesis $H_A: \beta_1 \neq 0$. There are the t-test, the F-test, and the use of a confidence interval on β_1.
   a. Explain how to use each of these approaches to test the null hypothesis $H_0: \beta_1 = 0$.

   b. Assuming that $\alpha = 0.05$ for both the t-test and the F-test and a 95 percent confidence interval is determined for β_1, will all three approaches reach the same decision regarding $H_0: \beta_1 = 0$? Explain.

14.5 Discuss what is meant by a prediction interval. When is a prediction interval used?

14.6 In a simple linear regression analysis experiment, it is found that b_1 is equal to 5.67 and that the standard error of b_1, s_{b_1}, is equal to 0.943 based on a random sample of 27 observations. Construct a 90 percent confidence interval for β_1.

14.7 Given the simple linear regression equation $\hat{y} = 23.0 + 6.21x$ determined by sampling 23 (x, y) data pairs where $s_{b_1} = 1.48$, test the null hypothesis $H_0: \beta_1 = 0$ using a significance level of $\alpha = 0.05$.

14.8 The following data show the dollar value of sales with corresponding advertising levels for a small retail chain in the Midwest.

Advertising	Sales
$584	$6,445
386	4,015
452	4,822
635	7,017
242	2,848
328	3,888
517	5,554

GPA	Starting salary
2.21	$16,400
2.23	16,900
2.45	24,450
2.61	20,800
2.63	22,420
2.87	21,990
3.19	23,250
3.26	23,800
3.28	24,400
3.77	25,500

a. Plot these data in the form of a scatter diagram. Does the linear model appear to be appropriate for estimating sales as a function of dollars spent on advertising?

b. Determine the least squares regression line.

c. Test the null hypothesis $H_0: \beta_1 = 0$ at the $\alpha = 0.05$ significance level using the t-distribution.

d. Test the null hypothesis $H_0: \beta_1 = 0$ at the $\alpha = 0.05$ significance level using the F-distribution.

e. What would be the level (value) of estimated sales if advertising expenditures were $622?

f. Construct a 95 percent confidence interval for mean sales when advertising expenditures are $555.

g. Construct a 95 percent prediction interval for the value of sales when advertising expenditures are $555.

14.9 In a simple linear regression analysis involving 19 observations, it is found that SST = 4.910.13 and SSE = 735.87. Use the F-distribution to test the hypothesis $H_0: \beta_1 = 0$ at the $\alpha = 0.01$ significance level.

14.10 An associate dean for undergraduate instruction at a large midwestern university is attempting to determine whether there is a strong linear relationship between undergraduate grade point average (GPA) and the starting salaries received. A sample of 10 recent graduates yielded the following results:

a. Determine the least squares regression equation using GPA as the independent variable.

b. Determine the coefficient of determination, r^2.

c. Find a 95 percent prediction interval for an individual whose grade point average is 3.00.

d. What is the estimated mean starting salary for graduates with a grade point average of 2.80, derived from your estimating equation?

e. What is the 95 percent confidence interval estimate of the mean starting salary for the salary determined in part d?

14.11 Given the regression equation $\hat{y} = 32.96 - 2.65x$ based on 26 observations with $s_{b_0} = 13.45$, test the null hypothesis $H_0: \beta_0 = 0$ at the $\alpha = 0.05$ significance level.

14.12 In a regression experiment involving 18 observations, it is found that SSE = 45.92 and SSR = 289.61. Use the F-distribution to test the hypothesis $H_0: \beta_1 = 0$ at the $\alpha = 0.05$ significance level.

14.13 A leading discount store sells six major brands of calculators. Unit prices along with corresponding sales amounts are given in the following table.

a. Plot the data in the form of a scatter diagram. Do units sold appear to be linearly related to the price of the unit?

b. Fit a linear function to the data using the method of least squares.

Calculator price	Units sold
$ 19.95	385
29.95	416
34.95	322
59.95	186
79.95	102
139.99	47

c. Test the hypothesis $H_0: \beta_1 = 0$ at the $\alpha = 0.05$ significance level. Is price a useful variable to include in a simple linear regression equation for estimating unit sales?

d. What is the p-value (approximate) associated with the test in part c?

e. What other variables might one include in a regression equation for estimating unit sales?

14.14 In a given regression experiment involving 11 observations, it is determined that SSE = 27.45 and SSR = 24.71. Using the F-distribution, determine the approximate p-value (by interpolation) for the hypothesis $H_0: \beta_1 = 0$.

14.15 As a part of the evaluation process of sales representatives, a consumer goods manufacturer estimates sales in the second year for each representative based on his or her sales during the first year. Data on nine randomly selected sales representatives are given in the table.

First-year sales ($000)	Second-year sales ($000)
85.3	109.6
65.2	88.5
73.9	92.6
109.6	134.3
123.4	143.9
98.7	89.3
50.5	63.1
77.8	95.6
101.1	111.9

a. Construct a linear regression equation to estimate second-year sales as a function of first-year sales.

b. What proportion of the variation in second-year sales is accounted for or explained by first-year sales?

c. Test the null hypothesis $H_0: \beta_1 = 0$ at the $\alpha = 0.10$ significance level.

d. Construct a 95 percent prediction interval for the estimated second-year sales of an individual who reported first-year sales of 98.7($1,000).

e. Construct a 95 percent confidence interval for mean second-year sales given first-year sales of 73.9($1,000).

14.16 *Referring to Problem 13.16 test the hypothesis $H_0: \beta_1 = 0$ at the $\alpha = 0.05$ significance level. What is the p-value for the test, and what is the correct interpretation of this p-value? Is average daily temperature significant in predicting fuel use? Why or why not?

14.17 Refer to Problem 13.43 and assume Joe gained 999 yards in his senior year.

a. Test the null hypothesis $H_0: \beta_1 = 0$ at the $\alpha = 0.01$ significance level. What is the p-value for the test, and how is this number interpreted? At which significance levels, 0.10, 0.05, or 0.01, would we reject the null hypothesis that $\beta_1 = 0$ based on the computed p-value?

b. Construct a 95 percent confidence interval and prediction interval for Joe's expected earnings. Which of these intervals is "wider," and why?

c. Which job should Joe select?

14.18 Refer to Problem 13.44

a. Is the number of years of postsecondary education significant in predicting the hourly wage rate in this firm? (Support your premise by interpreting the computed p-value of the test.)

b. Joe College is completing his fourth year at school and wants to determine what

---

* The problems with shaded boxes refer to problems in Chapter 13 at the end of the chapter.

wage he might expect to receive if he decides to accept an offer with the firm in question. Joe's situation is complicated somewhat by the fact that the firm will not be able to tell him the specific salary (or equivalent hourly wage rate) he will receive until next week when the personnel manager returns from a college recruiting trip. Another firm has extended an offer to Joe, and he must tell them by the end of the current week whether or not he intends to accept their offer. If he does not, the other firm wants to make the offer to someone else immediately so that they are assured of getting someone to fill the position. Based on Joe's four years of postsecondary education, obtain a 95 percent prediction interval to help estimate the wage Joe might expect to receive from this firm.

14.19 Consider the sales versus advertising data given in Problem 13.46.
- a. Test the hypothesis $H_0: \beta_1 = 0$ at the $\alpha = 0.05$ significance level.
- b. Obtain a 95 percent prediction interval for sales when advertising expenditures are equal to 1.2($1,000).

14.20 Consider the data in Problem 13.47.
- a. Test the hypothesis $H_0: \beta_1 = 0$ at the $\alpha = 0.05$ significance level. What is the approximate p-value for this test?
- b. Construct a 95 percent confidence interval for the mean number of boards rejected for an employee with 20 weeks of experience.

Section 14.3 Problems

14.21 Describe the differences in estimation and hypothesis-testing procedures when $E(\mathbf{r}) = 0$ as opposed to when $E(\mathbf{r}) \neq 0$. What is the nature of the sampling distribution of $\mathbf{r}$ in each instance?

14.22 Discuss the difficulties encountered when constructing a confidence interval on ρ when r turns out to be either $+1$ or -1 using the Fisher approximate method.

14.23 Assume that the data given in Problem 14.8 are distributed according to the bivariate normal distribution.
- a. Find the sample correlation coefficient.
- b. Test the null hypothesis $H_0: \rho = 0$ at the $\alpha = 0.05$ significance level.

14.24 In an analysis of the linear relationship between two random variables $\mathbf{X}$ and $\mathbf{Y}$, it is found that $\mathbf{r}^2 = 0.54$ based on a sample of 25 (x, y) data pairs. Does this indicate a significant (i.e., nonzero) correlation between $\mathbf{Y}$ and $\mathbf{X}$ at the $\alpha = 0.05$ significance level?

14.25 Assume that in analyzing the linear relationship between two random variables that are bivariate normally distributed, it is found that SST = 620.25 and SSR = 330.83 based on a sample of 23 observations. Further assume that the slope of a fitted sample regression line is negative.
- a. What is the coefficient of determination for this problem?
- b. Test the null hypothesis $H_0: \rho = 0$ at the $\alpha = 0.01$ significance level.

14.26 In a study involving 250 randomly selected individuals at a weight loss center, it is found that the correlation between the weight of an individual upon entering the program and the weight loss after six weeks in the program is $r = 0.52$. Construct a 99 percent confidence interval for the population correlation coefficient, ρ. Do weight loss after six weeks in the program and beginning weight appear to be "highly" correlated? Explain.

14.27 In a given analysis of two random variables, the following quantities were determined, based on a sample of 100 units:

$$SSE = 456.78$$
$$SST = 3,890.42$$
$$\hat{y} = 7.705 - 1.43x$$

- a. Determine a point estimate of ρ.
- b. Construct a 95 percent confidence interval for ρ.
- c. What is the interpretation of the interval constructed in part b?

14.28 In a correlation analysis involving 39 subjects, it is determined that $r = 0.36$. Construct a 90 percent confidence interval for ρ. Why do we have to be careful in interpreting the interval constructed?

14.29 Based on a random sample of 100 persons, a researcher finds that the correlation between number of pounds overweight and number of days ill during the past year is 0.42.
  a. Test the hypothesis at the $\alpha = 0.05$ significance level that $\rho = 0$.
  b. Determine a 95 percent confidence interval on ρ.

14.30 An MBA director is interested in determining whether or not a linear relationship exists between undergraduate grade-point average (GPA) and a student's MBA GPA upon graduation. She feels that the correlation between the two grade-point averages is positive and that the correlation coefficient ρ is probably about 0.50. She takes a random sample of 50 MBA students who have recently received their degrees and finds that $r = 0.37$. Using the confidence interval on ρ, test the hypothesis $H_0: \rho = 0.50$ at the $\alpha = 0.05$ significance level.

14.31 A high school counselor has surveyed 14 members of the senior class of 1970 and recorded (a) the individual's yearly income and (b) the number of years of higher education completed by the individual. The results of this survey are shown in the table.

Years of higher education	Annual income
0	$27,500
0	28,000
1	31,000
1	30,500
2	29,500
3	34,200
4	35,500
4	36,800
4	38,900
5	36,500
6	65,800
7	52,900
7	57,600
7	61,200

  a. What is the correlation between years of higher education and annual income in this sample of 14 individuals?

  b. Test the null hypothesis that $\rho = 0$ at the $\alpha = 0.01$ significance level.

14.32 An advertising firm believes that for its target market, the relationship between the ages of husbands and wives may be an important factor for consideration if the relationship is "strong": $\rho > 0.6$ or $\rho < -0.6$, where $\mathbf{X} =$ Age of husband and $\mathbf{Y} =$ Age of wife. In a random sample of 100 married couples, it is found that $r = 0.70$.
  a. Determine an approximate 95 percent confidence interval for ρ.
  b. By the firm's definition of a strong relationship, what would you conclude on the basis of part a? Explain.

14.33 For the data in Problem 13.51, test the null hypothesis $H_0: \rho = 0$ assuming that the paired (x, y) observations are bivariate normally distributed. Use the $\alpha = 0.05$ significance level for reporting your results.

14.34 Assuming the (x, y) pairs of values in Problem 13.53 are bivariate normally distributed, test the null hypothesis $H_0: \rho = 0$ at the $\alpha = 0.05$ significance level.

Section 14.4 Problems

14.35 In a certain industry, a study is conducted to determine the relationship between the total number of employees and the number of executives retained by the corporation. In general, it is known in this industry that as the total number of employees increases, the number of executives increases, too. Let $\mathbf{Y} =$ Number of executives in the corporation and $x =$ Number of employees in the corporation. The model, $\mathbf{Y} = \beta_0 + \beta_1 x + \epsilon$ is considered for relating x to $\mathbf{Y}$. Twenty corporations are involved in the study. The simple linear regression model is fitted by MINITAB, producing the following output:

C1 = Total number of employees.
C2 = Number of executives.

  a. What is the fitted regression model, $\hat{y} = b_0 + b_1 x$?
  b. What is the coefficient of determination, r^2?
  c. Test the hypothesis $H_0: \beta_1 = 0$ at the $\alpha = 0.05$ level of significance.

```
PRINT C1,C2
COLUMN          C1          C2
COUNT           20          20
ROW
  1         493,00         48,
  2        1024,00        122,
  3         846,00        136,
  4         596,00         58,
  5         332,00         44,
  6         295,00         36,
  7         692,00         78,
  8         714,00         88,
  9        1250,00        164,
 10        1812,00        206,
 11         342,00         41,
 12         456,00         64,
 13         621,00         80,
 14         964,00        141,
 15         920,00         96,
 16         743,00         90,
 17        1118,00        175,
 18         248,00         36,
 19         684,00         79,
 20         614,00         91,
```

PLOT C2 VS C1

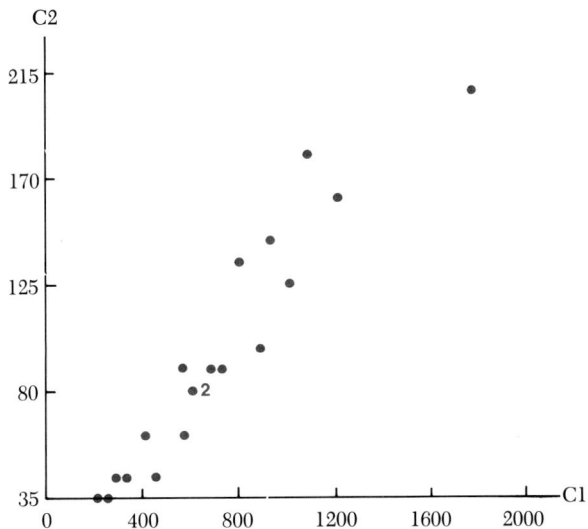

```
REGR C2 1 C1

THE REGRESSION EQUATION IS
Y =    1.90 + 0.124 X1

                                    ST. DEV.    T-RATIO =
        COLUMN      COEFFICIENT      OF COEF.   COEF/S.D.
        --              1.896          7.760       0.24
X1      C1          0.124294        0.009414      13.20

THE ST. DEV. OF Y ABOUT REGRESSION LINE IS
S = 15.44
WITH (  20- 2) =  18 DEGREES OF FREEDOM

R-SQUARED = 90.6 PERCENT
R-SQUARED = 90.1 PERCENT, ADJUSTED FOR D.F.

ANALYSIS OF VARIANCE

 DUE TO      DF          SS        MS=SS/DF
REGRESSION    1       41559.1      41559.1
RESIDUAL     18        4291.0        238.4
TOTAL        19       45850.1
```

```
-- NOTE   PRINT PREDICTED VALUE
-- NOTE
-- NOTE   **********
-- PRINT K5
   K5       103.320
-- NOTE   **********
-- NOTE
-- NOTE   PRINT STANDARD DEVIATION OF THE ESTIMATE OF THE
-- NOTE   CONDITIONAL MEAN OF Y
-- NOTE
-- NOTE   **********
-- PRINT K6
   K6       3.52931
-- NOTE   **********
-- NOTE
-- NOTE   PRINT STANDARD DEVIATION OF THE PREDICTION OF AN
-- NOTE   INDIVIDUAL VALUE OF Y
-- NOTE
-- NOTE   **********
-- PRINT K7
   K7       15.8381
-- NOTE   **********
-- END
```

d. Suppose that a firm adds 100 employees to its payroll. Predict the number of executives that would be added.

e. A firm not involved in the study has 816 employees. It is desired to predict the number of executives in this firm and to determine a 95 percent prediction interval on the number of executives for this *individual* firm. The appropriate MINITAB information from the prediction MACRO is shown on the bottom of page 744. Determine the point estimate and the prediction interval.

f. A firm not involved in the study has 2,000 employees. Using the MINITAB MACRO output below, predict the number of executives employed by this firm, and determine a 95 percent prediction interval for the number of employed executives.

g. Why should caution be exercised in using the prediction interval determined in part *f*?

h. The residual plot with the standardized residuals on the vertical axis and the predicted **Y** values on the horizontal axis is shown on page 746. Which regression assumption concerning the residuals appears to be violated in this plot?

14.36 An MBA program at a southwestern university admits students to its program for the fall semester 1985. The verbal, quantitative, and total GMAT (Graduate Management Admissions Test) scores, the undergraduate GPA (grade-point average), the junior-senior GPA, and the MBA GPA in the first year (academic year 1985–1986) in the program are recorded for each of the 48 admitted students who matriculated (entered the program) and completed the first year of studies in the two-year MBA program.

The admissions director is interested in relating the GMAT total score and the undergraduate GPA to the MBA GPA in the first year of the program. To explore these rela-

```
-- NOTE    PRINT PREDICTED VALUE
-- NOTE
-- NOTE    **********
-- PRINT K5
   K5          250.484
-- NOTE    **********
-- NOTE
-- NOTE    PRINT STANDARD DEVIATION OF THE ESTIMATE OF THE
-- NOTE    CONDITIONAL MEAN OF Y
-- NOTE
-- NOTE    **********
-- PRINT K6
   K6          12.3698
-- NOTE    **********
-- NOTE    PRINT STANDARD DEVIATION OF THE PREDICTION OF AN
-- NOTE    INDIVIDUAL VALUE OF Y
-- NOTE
-- NOTE    **********
-- PRINT K7
   K7          19.7839
-- NOTE    **********
```

PLOT C3,C4

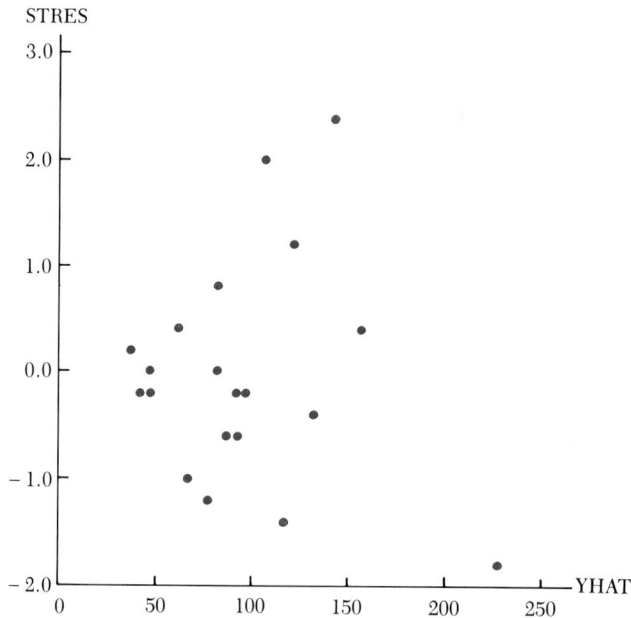

tionships, two simple linear regression models are fitted: $Y = \beta_0 + \beta_1 x_1 + \epsilon$, where Y = first-year MBA GPA and x_1 = GMAT score; and $Y = \beta_0 + \beta_1 x_2 + \epsilon$, where x_2 = undergraduate GPA. The MINITAB output for the two fitted regression equations is given below:

C1 = First-year MBA GPA.
C2 = GMAT score.
C3 = Undergraduate GPA.
C5 = Standardized residuals.
C6 = Predicted Y values.
C7 = Regression coefficients.

```
REGR C1 1 C2 C5 C6 C7 M1

THE REGRESSION EQUATION IS
Y =     1.72 +0.0028 X1

                                     ST. DEV.     T-RATIO =
          COLUMN      COEFFICIENT     OF COEF.     COEF/S.D.
          --            1.7218        0.5413        3.18
X1    GMAT            0.002850        0.001055      2.70

THE ST. DEV. OF Y ABOUT REGRESSION LINE IS
S = 0.5391
WITH (  48- 2) =   46 DEGREES OF FREEDOM

R-SQUARED = 13.7 PERCENT
R-SQUARED = 11.8 PERCENT, ADJUSTED FOR D.F.
```

```
ANALYSIS OF VARIANCE

  DUE TO      DF           SS        MS=SS/DF
 REGRESSION   1         2.1197       2.1197
 RESIDUAL    46        13.3669       0.2906
 TOTAL       47        15.4866
```

C1 = First-year MBA GPA.
C2 = GMAT score.
C3 = Undergraduate GPA.
C5 = Standardized residuals.
C6 = Predicted **Y** values.
C8 = Regression coefficients.

```
REGR C1 1 C3 C5 C6 C7 M1

THE REGRESSION EQUATION IS
Y =     2.15 + 0.331 X1

                                  ST. DEV.    T-RATIO =
          COLUMN     COEFFICIENT   OF COEF.   COEF/S.D.
          --           2.1543       0.4646       4.64
 X1       UDGGPA       0.3307       0.1493       2.22

THE ST. DEV. OF Y ABOUT REGRESSION LINE IS
S = 0.5515
WITH (  48- 2) =   46 DEGREES OF FREEDOM

R-SQUARED =   9.6 PERCENT
R-SQUARED =   7.7 PERCENT, ADJUSTED FOR D.F.

ANALYSIS OF VARIANCE

  DUE TO      DF           SS        MS=SS/DF
 REGRESSION   1         1.4934       1.4934
 RESIDUAL    46        13.9933       0.3042
 TOTAL       47        15.4866
```

```
CORRELATE C1-C3

                      Y       GMAT
         GMAT      0.370
         UDGGPA    0.311     0.106
```

PLOT C1 VS C2

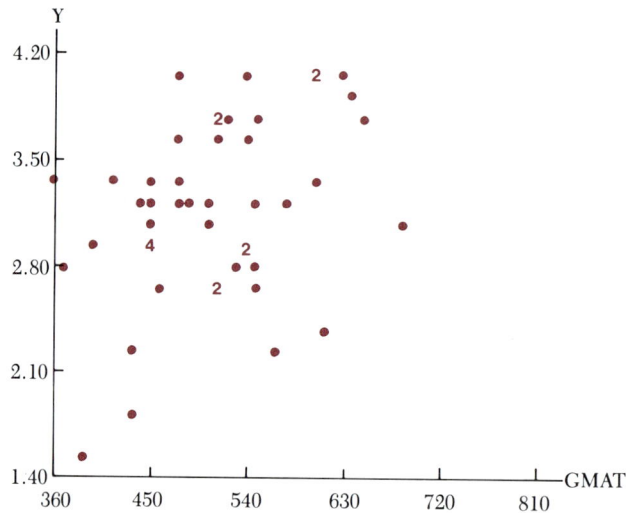

PLOT C1 VS C3

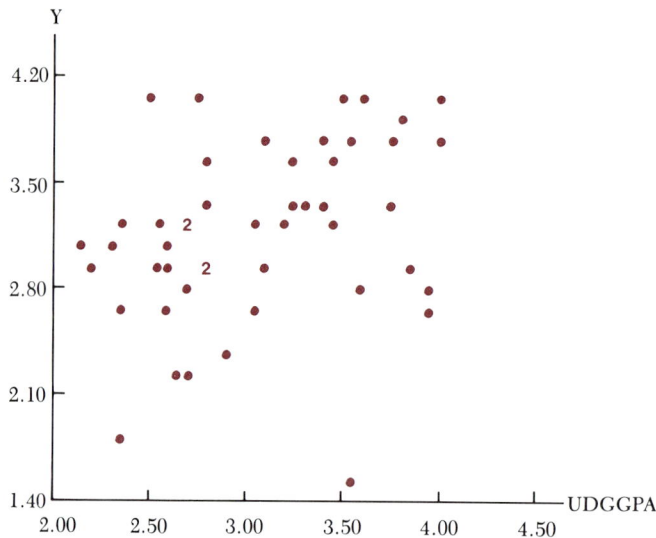

a. What are the two fitted regression equations?

b. Which one appears to do a better job of explaining variation in the dependent variable **Y**?

c. Test the null hypothesis, $H_0: \beta_1 = 0$, where β_1 is the slope coefficient on the GMAT score. Use an $\alpha = 0.05$ level of significance.

d. Test the null hypothesis, $H_0: \beta_1 = 0$, where β_1 is the slope coefficient on the undergraduate GPA. Use $\alpha = 0.05$ for the significance level.

e. The MBA director is interested in esti-

```
-- NOTE   PRINT PREDICTED VALUE
-- NOTE
-- NOTE   **********
-- PRINT K5
   K5          2.57671
-- NOTE   **********
-- NOTE
-- NOTE   PRINT STANDARD DEVIATION OF THE ESTIMATE OF THE
-- NOTE   CONDITIONAL MEAN OF Y
-- NOTE
-- NOTE   **********
-- PRINT K6
   K6          0.232521
-- NOTE   **********
-- NOTE
-- NOTE   PRINT STANDARD DEVIATION OF THE PREDICTION OF AN
-- NOTE   INDIVIDUAL VALUE OF Y
-- NOTE   **********
-- PRINT K7
   K7          0.587072
-- NOTE   **********
```

#1

```
-- NOTE   PRINT PREDICTED VALUE
-- NOTE
-- NOTE   **********
-- PRINT K5
   K5          2.98106
-- NOTE   **********
-- NOTE
-- NOTE   PRINT STANDARD DEVIATION OF THE ESTIMATE OF THE
-- NOTE   CONDITIONAL MEAN OF Y
-- NOTE
-- NOTE   **********
-- PRINT K6
   K6          0.116170
-- NOTE   **********
-- NOTE
-- NOTE   PRINT STANDARD DEVIATION OF THE PREDICTION OF AN
-- NOTE   INDIVIDUAL VALUE OF Y
-- NOTE
-- NOTE   **********
-- PRINT K7
   K7          0.563649
-- NOTE   **********
```

#2

mating the mean first-year MBA GPA for all students who have a 300 GMAT score. Using the MINITAB MACRO information on page 749, #1, determine the estimate of the mean MBA GPA, and determine a 95 percent confidence interval estimate of the mean MBA GPA.

f. The MBA director is interested in estimating the mean first-year MBA GPA for all students with a 2.50 undergraduate GPA. Using the MINITAB MACRO information on page 749, #2, determine a point estimate and a 95 percent confidence interval estimate of the mean MBA GPA.

g. The MBA director is evaluating an applicant with a 700 GMAT score. Using the MINITAB MACRO information below (#3), determine a point estimate and a 90 percent prediction interval for this applicant's first-year MBA GPA.

h. The MBA director is evaluating an applicant with a 3.75 undergraduate GPA. Using the MINITAB MACRO information on page 751, #4, determine a point estimate and a 90 percent prediction interval for this applicant's first-year MBA GPA.

i. The residual plot from the fitted regression relating the first-year MBA GPA (y) and the undergraduate GPA (x) is shown on page 751 from the MINITAB output. Does this residual plot suggest that the three regression assumptions have been satisfied: independent residuals, equal variance residuals, and normally distributed residuals?

14.37 Absenteeism of workers is a major concern for many corporations. High absenteeism is a complex phenomenon that results from a variety of factors, including the worker's age,

```
                                                              #3
-- NOTE    PRINT PREDICTED VALUE
-- NOTE
-- NOTE    **********
-- PRINT K5
   K5          3.71654
-- NOTE    **********
-- NOTE
-- NOTE    PRINT STANDARD DEVIATION OF THE ESTIMATE OF THE
-- NOTE    CONDITIONAL MEAN OF Y
-- NOTE
-- NOTE    **********
-- PRINT K6
   K6          0.217315
-- NOTE    **********
-- NOTE
-- NOTE    PRINT STANDARD DEVIATION OF THE PREDICTION OF AN
-- NOTE    INDIVIDUAL VALUE OF Y
-- NOTE
-- NOTE    **********
-- PRINT K7
   K7          0.581217
-- NOTE    **********
```

```
                                                                      #4
    -- NOTE
    -- NOTE   PRINT PREDICTED VALUE
    -- NOTE
    -- NOTE   **********
    -- PRINT K5
       K5           3.39443
    -- NOTE   **********
    -- NOTE
    -- NOTE   PRINT STANDARD DEVIATION OF THE ESTIMATE OF THE
    -- NOTE   CONDITIONAL MEAN OF Y
    -- NOTE
    -- NOTE   **********
    -- PRINT K6
       K6           0.129360
    -- NOTE   **********
    -- NOTE
    -- NOTE   PRINT STANDARD DEVIATION OF THE PREDICTION OF AN
    -- NOTE   INDIVIDUAL VALUE OF Y
    -- NOTE
    -- NOTE   **********
    -- PRINT K7
       K7           0.566514
    -- NOTE   **********
```

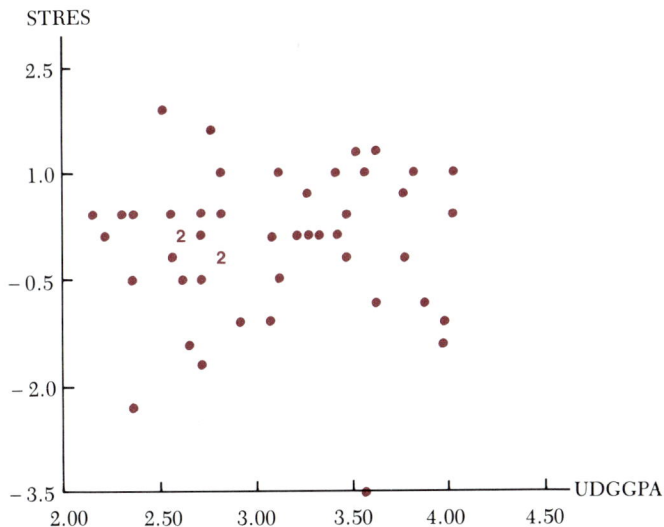

degree of seniority, salary, type of job, complexity of the job, and chance for advancement. In a six-month period, a firm records the number of absent days for a random sample of 30 employees. For each employee the firm has a job complexity measure, which indicates on a scale from 0 to 100 the degree of complexity in the job. The higher the rating, the more complex the job. It is decided to fit the model, $\mathbf{Y} = \beta_0 + \beta_1 x + \epsilon$, where $\mathbf{Y} =$ Number of absent days in the six-month period and $x =$ job complexity. The output for the fitted regression equation from MINITAB is given below:

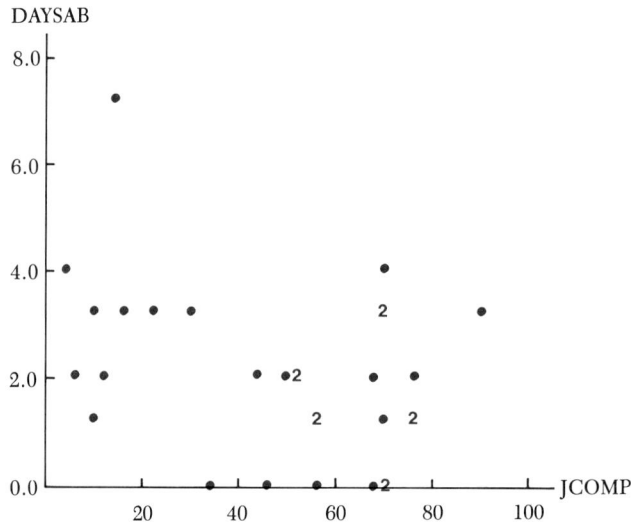

```
DESCRIBE C20,C21
    JCOMP    N =   30     MEAN =        47.767     ST.DEV. =      25.8
    DAYSAB   N =   30     MEAN =         1.9333    ST.DEV. =       1.55
--
?
CORRELATE C20,C21

    CORRELATION OF    JCOMP AND DAYSAB =-0.356

REGR C21 1 C20 C10 C11 C12 M1

THE REGRESSION EQUATION IS
Y =     2.96 -0.0215 X1

                                            ST. DEV.     T-RATIO =
             COLUMN        COEFFICIENT      OF COEF.     COEF/S.D.
               --             2.9587          0.5750         5.15
    X1       JCOMP          -0.02147          0.01063       -2.02

THE ST. DEV. OF Y ABOUT REGRESSION LINE IS
S = 1.476
WITH (   30- 2) =   28 DEGREES OF FREEDOM
```

```
R-SQUARED = 12.7 PERCENT
R-SQUARED =  9.6 PERCENT, ADJUSTED FOR D.F.

ANALYSIS OF VARIANCE

  DUE TO        DF          SS        MS=SS/DF
REGRESSION      1         8.875       8.875
RESIDUAL       28        60.991       2.178
TOTAL          29        69.867
```

a. Based on the scatterplot of the (x, y) pairs, does it appear that the number of days absent in the six-month period and the job complexity measure are linearly related?

b. What is the fitted regression equation?

c. What is the coefficient of determination, r^2?

d. Test the hypothesis, $H_0: \beta_1 = 0$, at the $\alpha = 0.05$ level of significance.

e. For a certain job, the job complexity measure has a value of 10. Determine a point estimate and a 95 percent confidence interval on the mean number of days absent in the six-month period for all employees with this job by using the MINITAB MACRO information below.

```
-- NOTE   PRINT PREDICTED VALUE
-- NOTE
-- NOTE   **********
-- PRINT K5
   K5          2.74402
-- NOTE   **********
-- NOTE
-- NOTE   PRINT STANDARD DEVIATION OF THE ESTIMATE OF THE
-- NOTE   CONDITIONAL MEAN OF Y
-- NOTE
-- NOTE   **********
-- PRINT K6
   K6          0.483640
-- NOTE   **********
-- NOTE
-- NOTE   PRINT STANDARD DEVIATION OF THE PREDICTION OF AN
-- NOTE   INDIVIDUAL VALUE OF Y
-- NOTE
-- NOTE   **********
-- PRINT K7
   K7          1.55311
-- NOTE   **********
```

```
-- NOTE   PRINT PREDICTED VALUE
-- NOTE
-- NOTE   **********
-- PRINT K5
   K5          1.02673
-- NOTE   **********
-- NOTE
-- NOTE   PRINT STANDARD DEVIATION OF THE ESTIMATE OF THE
-- NOTE   CONDITIONAL MEAN OF Y
-- NOTE
-- NOTE   **********
-- PRINT K6
   K6          0.523759
-- NOTE   **********
-- NOTE
-- NOTE   PRINT STANDARD DEVIATION OF THE PREDICTION OF AN
-- NOTE   INDIVIDUAL VALUE OF Y
-- NOTE
-- NOTE   **********
-- PRINT K7
   K7          1.56607
-- NOTE   **********
```

f. A specific individual has a job with a complexity measure of 90. Using the MINITAB MACRO information above, determine the prediction of the number of days absent for this employee in the six-month period, and determine a 95 percent prediction interval for the number of days absent in the six-month period.

Additional Problems

14.38 A firm is interested in determining whether or not its market share is linearly related to dollars expended on local television advertising in a small geographic area. Data collected on market share and corresponding local television advertising expenditures are shown in the table for a random sample of five months during the past year.

Television advertising expenditure ($000)	Market share (percent)
18	13
21	14
23	16
27	18
31	19
35	19

a. Assuming the data reported in the table are bivariate normally distributed, determine the sample correlation coefficient, r, beginning with the determination of the covariance of these two random variables.

b. Test the null hypothesis H_0: $\rho = 0$ at the $\alpha = 0.05$ significance level. Does there appear to be a high degree of linear relationship between these two random variables?

c. Suppose the firm decided in the next period to increase advertising expenditures to 65($1,000) to increase its market share. Why would it be dangerous in this instance to use the determined linear regression equation to estimate market share?

14.39 Consider the test score versus studying time data given in Problem 13.45.

a. Determine a 99 percent confidence interval for β_0. What is the correct interpretation of this interval?

b. Determine a 99 percent confidence interval for β_1. What is the correct interpretation of this interval?

c. Test the null hypothesis $H_0: \rho = 0$ at the $\alpha = 0.01$ significance level.

d. Why would it be inappropriate to construct a confidence interval for ρ in this problem?

e. What other factors might one include in a regression model for estimating exam scores?

14.40 The data reproduced below represents the heights and weights of 10 randomly selected U.S. adult males.

Height (inches)	Weight (pounds)
60	110
65	150
74	200
70	185
70	170
66	160
68	180
72	195
64	135
71	215

a. Predict, on the basis of the data provided, the weight of an individual who is 65 inches tall.

b. Predict, on the basis of the data provided, the height of an individual who weighs 148.79 pounds.

c. How do your estimates in parts a and b compare?

d. Compute r and r^2 in parts a and b above. Does the value of r depend on which variable is the dependent variable and which variable is the independent variable in the regression equation?

e. Form the residual plots for the two fitted regression equations determined in parts a and b. Do either or both of these residual plots appear to support the hypotheses that the assumptions underlying the use of the simple linear regression model have been met?

14.41 For the height-weight data given in problem 14.40 above,

a. Test the null hypothesis at the $\alpha = 0.05$ significance level that $\beta_1 = 0$ using both the t-test and the F-test in the regression model $\mathbf{Y} = \beta_0 + \beta_1 x + \boldsymbol{\epsilon}$, where $\mathbf{Y}$ is weight, and x is height.

b. Construct a 95 percent confidence interval on the mean weight of an individual who is 70 inches tall.

c. What other independent variables might be useful in a regression equation to predict an individual's weight?

14.42 In a random sample of 199 (x, y) pairs, it is determined that $r = -0.70$. Determine a 95 percent confidence interval for ρ.

14.43 An insurance company is interested in determining whether or not there is a relationship between age and the amount of whole life insurance carried. In a random sample of 250 persons, it is found that $r = 0.12$, where $\mathbf{X} =$ Age and $\mathbf{Y} =$ Amount of whole life insurance.

a. Test the hypothesis that $\rho = 0$ at the $\alpha = 0.05$ level.

b. Determine an approximate 95 percent confidence interval for ρ.

c. Interpret your results in parts a and b. What is your conclusion regarding the relationship between age and amount of whole life insurance carried?

14.44 Assuming the pairs of observations given in Problem 13.55 are bivariate normally distributed, test the null hypothesis at the $\alpha = 0.10$ level that the random variables $\mathbf{X}$ and $\mathbf{Y}$ are *not* correlated.

14.45 A researcher wishes to test the null hypothesis $H_0: \rho = 0$, where ρ is the Pearson product moment correlation coefficient. She finds that $r = 0.20$ and that $s_r = 0.14$, based on a sample of $n = 50$ (x, y) pairs. What should her decision be, assuming a significance level of $\alpha = 0.05$?

14.46 For the data given in Problem 13.53, test the null hypothesis at the $\alpha = 0.05$ significance level that the annual consumption of alcoholic beverages and the number of automobile accidents in a certain locality are not correlated.

14.47 A random sample of six families yielded the following data:

Family	Annual income (x)	Annual expenditures on clothing (y)
1	$40,000	$2,000
2	32,000	1,400
3	55,000	4,200
4	35,000	1,800
5	46,000	2,700
6	51,000	3,200

a. Form a scatterplot of the (x, y) pairs. Do the two variables appear to be linearly related?
b. Determine the slope and the intercept of the fitted regression line.
c. Determine the coefficient of determination, r^2.
d. Predict the annual expenditure on clothing for a family whose annual income is $38,000.
e. Determine a 95 percent confidence interval for β_1. Interpret the value of b_1, the estimate of β_1, and the confidence interval on β_1.

14.48 The capital asset pricing model (CAPM) is used extensively in finance to analyze potential investments. As a part of the model, the rate of return of a fund or security being considered as an investment is compared with the average market rate of return. The model used is $\mathbf{Y} = \beta_0 + \beta_1 x + \boldsymbol{\epsilon}$, where $\mathbf{Y} =$ Rate of return of the fund and $x =$ Market rate of return. The slope coefficient β_1 is called the beta coefficient in the CAPM model and mea-

sures the risk in investing in the fund or security. The greater the value of beta, the greater the fund's sensitivity to market fluctuations and the greater the risk in investing in the fund. If beta is less than 1, then the fund is considered to be less risky than the market average fund or security. If beta is greater than 1, the fund is considered to be more risky than the market average fund or security.

For a certain class of securities, the average annual market return together with the annual rate of return of a stock in the electronics industry are given in the table below:

Year	Rate of return of electronics security	Rate of return of average market security
1972	8.4	15.0
1973	−6.0	−18.2
1974	−2.0	−8.6
1975	5.0	2.1
1976	15.4	4.8
1977	12.6	3.6
1978	27.4	15.6
1979	22.2	9.7
1980	10.4	12.6
1981	16.8	22.4
1982	26.2	17.2
1983	6.4	8.6
1984	−4.6	0.2
1985	−9.6	−4.8

a. Determine the beta coefficient for the electronics stock.
b. Produce the scatterplot of the (x, y) pairs.
c. Calculate a 95 percent confidence interval for the beta coefficient.
d. On the basis of part c, does it seem reasonable that $\beta_1 = 1$? Explain.
e. Calculate and interpret r^2.

14.49 Consider Problem 13.56 (Pirax Corporation). For the regression model treating $\mathbf{Y}$ (monthly sales) as the dependent variable and x (difference between the company's price and the average industrial price) as the independent variable,

a. Test the null hypothesis H_0: $\beta_1 = 0$ at the $\alpha = 0.05$ level of significance.

b. Determine a 99 percent confidence interval on β_1.

14.50 Consider Problem 13.59 (job satisfaction). In this problem, Y = Job satisfaction measure, and x = Percent above or below the average salary for the job classification.

a. What does the value of b_1, the estimated slope, tell us in this problem?

b. Test the null hypothesis, H_0: $\beta_1 = 0$, at the $\alpha = 0.05$ level of significance. Interpret the result of this test.

c. Suppose that an employee's salary is 80 percent above the average salary for her job classification. Predict the job satisfaction measure value for this employee.

15

Multiple correlation and regression

■

Multiple linear regression and correlation analyses are extensions of the simple linear regression and correlation models studied in Chapter 13. In multiple linear regression and correlation, however, more than one independent variable is used to account for the variability in the dependent random variable **Y**.

One obvious advantage in using the multiple linear regression and correlation models over the simple linear regression and correlation models is in prediction and estimation—by having two or more independent variables to predict values of the random variable **Y**, we should be able to do a better job than by using just one independent variable. Another advantage is that in one statistical analysis, we can study three or more variables together rather than only two. A disadvantage of using multiple linear regression and correlation is the increased complexity of the analysis—in terms of both the computations required and the interpretation of the results. Computers have relieved much of the computational burden, however. The steps in the multiple linear regression model-building process are given on page 759.

The majority of these steps are much more involved in the multiple linear regression model-building process than they are in the simple linear regression case. For example, many more variables usually have to be identified to be included in the model, and it is important to specify not only the relationship that exists between the dependent and each of the independent variables, but also the relationship between the independent variables themselves (Step 3). Consider, for example, the estimation of sales where the independent variables to be used are the price of the unit and the dollars

Steps in the model-building process	Steps in the multiple linear regression model-building process

Step 1. Identify the variables to be included in building the multiple regression model.
Step 2. Collect sample data.
Step 3. Specify the relationship that exists between the dependent and the independent variables, and between the independent variables.
Step 4. Estimate the parameters of the model specified.
Step 5. Determine whether or not the assumptions of the multiple linear regression model have been met.
Step 6. Statistically test for the usefulness of the model developed and for the individual terms in the model.
Step 7. Use the model developed for prediction and estimation.

spent on advertising. The relationship between advertising and sales may not be independent of the price of the unit. That is, with a higher price, the effect of advertising (in terms of the slope of the regression equation) may not be so great as when the price is low and the influence of advertising is much greater. We show these types of relationships in Figure 15.1. In Figure

FIGURE 15.1 | Examples of interaction and no interaction regression models

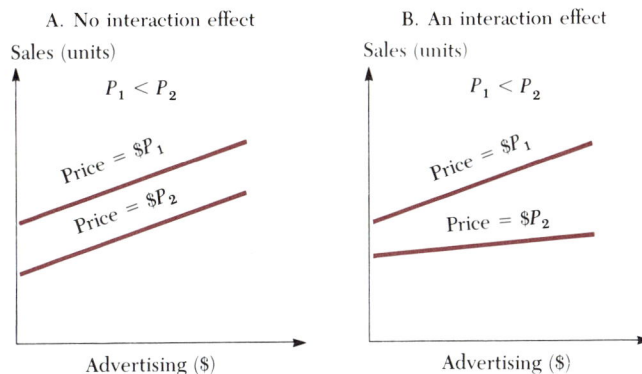

A. No interaction effect
Sales (units)
$P_1 < P_2$
Price = $\$P_1$
Price = $\$P_2$
Advertising ($)

B. An interaction effect
Sales (units)
$P_1 < P_2$
Price = $\$P_1$
Price = $\$P_2$
Advertising ($)

15.1A, the effect of advertising on sales is the same (as shown by the same slope for both regression lines) for each of the prices indicated (here, for explanatory purposes, we have shown only two different prices). In Figure 15.1B, on the other hand, the effect of advertising on sales differs depending on the sales price of the item. This is clearly shown by the different slopes of the regression lines. When the slope of the relationship between the dependent random variable $\mathbf{Y}$ and an independent variable x_1 depends on (or is not independent of) the value of one other independent variable, say x_2, the independent variables are said to *interact,* and it may be necessary to introduce a term reflecting the manner in which these variables interact into the regression equation to model the behavior of the dependent variable as a function of both independent variables. For this reason (as well as other reasons to be developed shortly), the development of a multiple linear regression model is much more complex than is the development of a simple linear regression model. Quite often, the statistician will find that he has to *repeat* many of the steps in the development of a multiple regression model to ensure that the model developed is useful for estimation and prediction purposes.

Because multiple linear regression and correlation analyses usually require the use of a computer to calculate estimates of the regression and correlation parameters, and because most colleges and universities have available for student use large-scale computers and appropriate statistical software to perform multiple regression and correlation analyses, our development of these two important techniques will concentrate on the interpretation of the output of two of the software computer routines, the general linear models procedure of the Statistical Analysis System (SAS) and MINITAB. Since many other available software packages (computer programs) have output similar to that of SAS and MINITAB, it should be relatively easy to interpret the output from these other software packages using the discussion in this chapter as a guide.

Multiple linear regression and correlation analyses are further best developed and understood by using matrix algebra. Since we do not presuppose a background for many students in matrix algebra, we will demonstrate the principal features of the analysis algebraically (without using matrix algebra) and show that they are in fact only extensions of simple linear regression and correlation. The student interested in exploring multiple linear regression and correlation further is advised to take a course in matrix or linear algebra, and *then* a course specifically devoted to multiple linear regression and correlation.

■ 15.2

The multiple linear regression model

The multiple linear regression model is specified in the box at the top of page 761. Note that this specification of the population linear regression model does not preclude us from including the interaction effects of different variables, nor does it preclude us from including higher-order terms in the model, such as x^2 and x^3. That is, $x_{i,3}$ could just as easily represent the combined effects of $x_{i,1}$ and $x_{i,2}$ in the form $x_{i,1} \cdot x_{i,2}$ as it could represent another independent variable. Similarly, $x_{i,j+1}$ could represent $x_{i,j}^2$, and so on. Some statisticians refer to a model with interaction and higher-order terms as a *complete model* and refer to a model lacking interaction and higher-order terms as a *reduced effects* or a *main effects model*. The model given

Complete model

The population multiple linear regression model

Population multiple linear regression model

$$\mathbf{Y}_i = \beta_0 + \beta_1 x_{i,1} + \beta_2 x_{i,2} + \cdots + \beta_k x_{i,k} + \boldsymbol{\epsilon}_i, \qquad i = 1, 2, \ldots, N$$

where:

$\mathbf{Y}_i$ = ith dependent random variable corresponding to $x_{i,1}, x_{i,2}, \ldots, x_{i,k}$
$\beta_0, \beta_1, \beta_2, \ldots, \beta_k$ are $(k + 1)$ parameters in the model.
$x_{i,j}$ = ith level of the jth independent variable, $j = 1, 2, \ldots, k$.
$\boldsymbol{\epsilon}_i$ = Random error term.

Main effects model

allows us to include both of these through a judicious choice of independent variable definitions.

The key assumptions of the multiple regression model are basically the same as those of the simple linear regression model.

Assumptions of the multiple linear regression model

Assumptions of the multiple linear regression model

1. The expected value of the error component $\boldsymbol{\epsilon}_i$ is 0 $[E(\boldsymbol{\epsilon}_i) = 0]$, and the variance of the error component $\boldsymbol{\epsilon}_i$ is σ^2 $[V(\boldsymbol{\epsilon}_i) = \sigma^2]$, for $i = 1, 2, \ldots, n$.
2. The error components are uncorrelated.
3. $\beta_0, \beta_1, \ldots, \beta_k$ are $(k + 1)$ parameters, and $x_{i,1}, x_{i,2}, \ldots, x_{i,k}$ are known constants.

And, for purposes of hypothesis testing and confidence interval estimation,

4. The error component $\boldsymbol{\epsilon}_i$ is normally distributed.

We shall illustrate the multiple linear regression model with two independent variables and no significant interaction terms by:

$$\mathbf{Y}_i = \beta_0 + \beta_1 x_{i,1} + \beta_2 x_{i,2} + \boldsymbol{\epsilon}_i, \qquad i = 1, 2, \ldots, n$$

By assumptions 1 and 3 given earlier, the regression function is given by $E(\mathbf{Y}_i) = \beta_0 + \beta_1 x_{i,1} + \beta_2 x_{i,2}$. The nature of this function is illustrated in Figure 15.2. Since there are now two independent variables, the dependent variable $\mathbf{Y}$ becomes the third axis in three-dimensional space; the regression function is a *plane*.

Regression plane

The interpretation of the regression coefficients β_0, β_1, and β_2 in this two independent variable model are analogous to the coefficients in the simple linear regression model: β_0 is the Y-intercept, the point on the y-axis where it

FIGURE 15.2 | Bivariate regression function

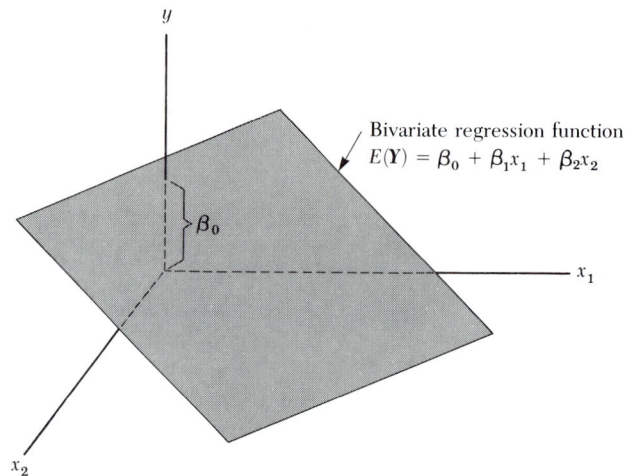

is intersected by the *plane*; β_1 gives the change in the value of the dependent variable **Y** when x_1 is incremented by one unit *and* x_2 is held constant; β_2 is similarly interpreted. The fitted regression plane $\hat{y} = b_0 + b_1x_1 + b_2x_2$ is determined by least squares: $\mathbf{b}_0$, $\mathbf{b}_1$, and $\mathbf{b}_2$ are least squares point estimators of β_0, β_1, and β_2, respectively. The least squares function is:

$$LS = \sum_{i=1}^{n} [y_i - (\beta_0 + \beta_1 x_{i,1} + \beta_2 x_{i,2})]^2 = \sum_{i=1}^{n} \varepsilon_i^2$$

and the values of β_0, β_1, and β_2 that minimize LS are the values of the least squares estimators, $\mathbf{b}_0$, $\mathbf{b}_1$, and $\mathbf{b}_2$, respectively.

We will continue our discussion of the bivariate regression model by explaining how one determines algebraically the point estimates of the regression coefficients β_0, β_1, and β_2; the point estimates of the variance of the error terms, σ^2, and the coefficient of multiple determination, ρ^2.

■ **15.3**

Point estimation of the population regression parameters— bivariate regression case

15.3.1 | **Point estimation of the regression coefficients, β_0, β_1, and β_2**

The values of $\mathbf{b}_0$, $\mathbf{b}_1$, and $\mathbf{b}_2$ that minimize LS are solutions to the following normal equations:

$$\Sigma y = nb_0 + b_1 \Sigma x_1 + b_2 \Sigma x_2$$
$$\Sigma x_1 y = b_0 \Sigma x_1 + b_1 \Sigma x_1^2 + b_2 \Sigma x_1 x_2$$
$$\Sigma x_2 y = b_0 \Sigma x_2 + b_1 \Sigma x_1 x_2 + b_2 \Sigma x_2^2$$

Normal equations

These equations can be solved algebraically for b_0, b_1, and b_2, but the resulting formulas are quite complicated. Usually, these equations are solved on a computer by using a multiple regression computer program. For illustrative purposes, we will use a "contrived" problem to develop point estimates of the multiple regression parameters. In this problem, the numbers are very "pleasant," since we will be using "hand" solution. In the next section, we will present a multiple regression, computer-analyzed case study to summarize the concepts discussed here.

Example 15.1 Consider the data given in Table 15.1. Fit the regression line $\hat{y} = b_0 + b_1 x_1 + b_2 x_2$ to these data by determining the necessary summations for use in the normal equations,[1] and then solve the resulting normal equations for b_0, b_1, and b_2.

TABLE 15.1 | Bivariate regression data

x_1	x_2	y
4	3	3
4	4	2
4	3	7
6	4	6
3	2	5
6	4	6
3	2	7
2	2	4

Solution The normal equations for the data are:

$$40 = 8b_0 + 32b_1 + 24b_2$$
$$164 = 32b_0 + 142b_1 + 104b_2$$
$$118 = 24b_0 + 104b_1 + 78b_2$$

Solution of these normal equations yields point estimates $b_0 = 6$, $b_1 = 2$, and $b_2 = -3$ (verify). Thus the fitted regression line is $\hat{y} = 6 + 2x_1 - 3x_2$.

Scatterplots of x_1 versus y and x_2 versus y are shown in Figure 15.3. The first independent variable, x_1, appears to have little, if any, linear relationship with the random variable $\mathbf{Y}$, and the second independent variable, x_2, appears to have a weak negative linear relationship with $\mathbf{Y}$.

15.3.2 | Point estimation of the variance, σ^2, of the error terms

The point estimate of the variance of the conditional probability distributions in multiple linear regression analysis is denoted by $s^2_{\mathbf{Y}/x_1, x_2, \ldots, x_k}$. Its com-

---

[1] Note that it is possible to drop the i-subscript from the x-values when we are not referring to specific values of x.

FIGURE 15.3 | Scatterplots of x_1 versus y and x_2 versus y

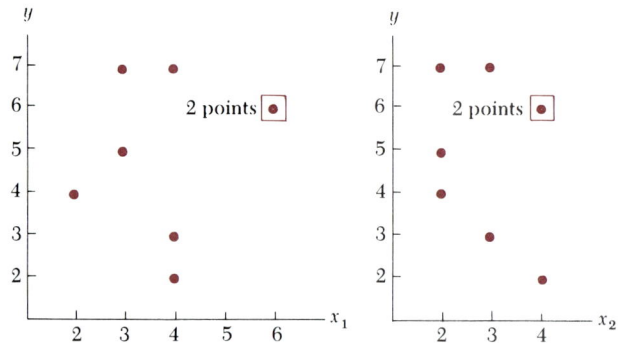

putation is shown in equation (15.1). Note in equation (15.1) that the numerator is the same as in the simple linear regression model—namely, we square the difference between the observed value of the dependent variable, y, and the estimate derived through the linear regression equation, $\hat{y}$; then sum these squares. The expression shown in the numerator is referred to as the error sum of squares, or SSE, just as in the simple linear regression model. Note in the denominator of equation (15.1), however, that the divisor is $n - (k + 1)$. This is because $(k + 1)$ degrees of freedom are lost in estimating the $(k + 1)$ parameters in the multiple regression model, $\beta_0, \beta_1, \beta_2, \ldots, \beta_k$. Experience in determining this estimate for the bivariate regression case is provided in the following example.

Conditional probability distribution variance point estimate

Point estimate of conditional probability distribution variance—multiple linear regression model

$$s^2_{Y/x_1, x_2, \ldots, x_k} = \frac{\sum_{i=1}^{n} (y_i - \hat{y}_i)^2}{n - (k + 1)} = \frac{SSE}{n - (k + 1)} \tag{15.1}$$

Example 15.2 Using the data given in Example 15.1, determine the point estimate of the conditional probability distribution variance.

Solution Relevant computations are given in Table 15.2. Using the data in this table,

$$s^2_{Y/x_1, x_2} = \frac{SSE}{n - (k + 1)} = \frac{10}{5} = 2$$

TABLE 15.2 | Determination of $s^2_{Y/x_1,x_2}$ using data from Example 15.1

	x_1	x_2	y	$\hat{y}$	$y - \hat{y}$	$(y - \hat{y})^2$
1	4	3	3	5	−2	4
2	4	4	2	2	0	0
3	4	3	7	5	2	4
4	6	4	6	6	0	0
5	3	2	5	6	−1	1
6	6	4	6	6	0	0
7	3	2	7	6	1	1
8	2	2	4	4	0	0
Totals	32	24	40	40	0	10

15.3.3 | Point estimation of the coefficient of multiple determination, ρ^2

The coefficient of multiple determination is defined similarly to the coefficient of determination in the simple linear regression case. That is, the coefficient of multiple determination is the proportion of variability in the random variable Y accounted for or explained by the independent variables $x_1, x_2, \ldots, x_k$. This sample statistic measures the *relative* strength of the linear relationship between the dependent random variable and the independent variables. The value of r^2 is computed using equation 15.2.

Coefficient of multiple determination	Sample coefficient of multiple determination, r^2
	$$r^2 = \frac{\text{SST} - \text{SSE}}{\text{SST}} = \frac{\text{SSR}}{\text{SST}} \qquad (15.2)$$

Example 15.3 Using the data given in Example 15.1, compute the sample coefficient of multiple determination.

Solution From Table 15.2, SSE = $\Sigma(y - \hat{y})^2$ = 10. And SST = $\Sigma(y - \bar{y})^2$ = 24 (verify!). Thus

$$r^2 = \frac{24 - 10}{24} = 0.5833$$

The bivariate regression accounts for 58.33 percent of the variability in the sample Y-values.

15.4

A computer-analyzed multiple regression example

The firm described in Section 13.2 desires to develop a multiple regression model to predict sales of each of its member branches. In addition to the independent variable "advertising expenditure" already included, the firm feels that data on the average income level in each of its sales districts can prove beneficial in estimating sales. Hence the firm has sampled 15 of its sales branches and recorded the sales amount, the amount spent on advertising, and the average income level in each sampled sales district. This information has been assembled and is presented in Table 15.3. Scatter plots of

TABLE 15.3 | Survey data for estimating sales

Sales district	Average income level ($000)	Advertising expenditures ($000)	Sales ($000)
1	15.7	1.814	55.420
2	13.1	1.112	45.819
3	6.2	0.718	16.940
4	10.7	1.421	35.818
5	16.3	3.085	85.090
6	17.5	2.119	62.025
7	8.4	0.525	17.918
8	11.1	1.108	32.845
9	13.2	1.621	41.180
10	14.8	2.645	62.910
11	18.9	2.927	87.013
12	7.6	0.847	22.150
13	7.8	0.621	16.660
14	10.0	0.981	27.050
15	12.1	1.394	39.121

sales and advertising expenditures and sales and income level are given in Figures 15.4 and 15.5, respectively. Both figures indicate the reasonableness of the straight-line model, although the firm may want to consider the introduction of nonlinear terms x_1^2 and x_2^2 in the model at some future time (we will return to this point later) to assess the impact of these terms on model development.

Table 15.4 contains the results of a computer analysis of the data in Table 15.3 using the main effects model. Much of the data shown in the table (printout) have not yet been discussed but will be discussed later in the chapter. We have shaded portions of Table 15.4 that correspond to the point estimates discussed in Section 15.3; we will describe the other quantities in Table 15.4 in the following section.

The least squares estimates of the β-parameters appear in the column labeled ESTIMATE, and a description of each of these parameters appears in the column labeled PARAMETER. From Table 15.4 one can see that $b_0 = -11.8888$ (INTERCEPT), $b_1 = 17.2336$ (ADDS), and $b_2 = 2.3500$ (INCOME). Hence the multiple regression estimating equation is:

FIGURE   15.4 | Scatter diagram of sales and advertising expenditures from data in Table 15.3

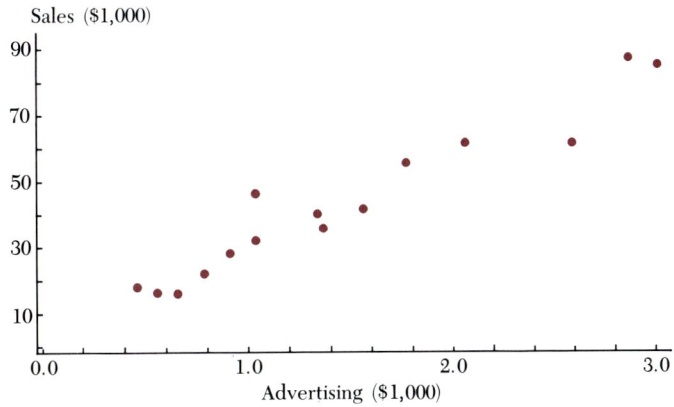

FIGURE   15.5 | Scatter diagram of sales and average income level from data in Table 15.3

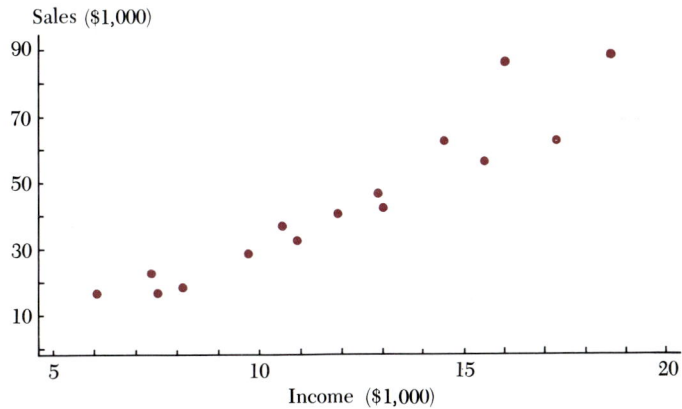

$$\hat{y} = -11.8888 + 17.2336x_1 + 2.3500x_2$$

where:

$\hat{y}$ = Estimated sales ($000)
x_1 = Advertising expenditures ($000)
x_2 = Average income level ($000)

TABLE 15.4 | Computer analysis of data in Table 15.3

DEPENDENT VARIABLE: SALES

SOURCE	DF	SUM OF SQUARES	MEAN SQUARE	F VALUE	PR > F
MODEL	2	7290.27547531	3645.13773766	215.87	0.0001
ERROR	12	202.63008162	16.88584013		
CORRECTED TOTAL	14	7492.90555693			

R-SQUARE	STD DEV
0.972957	4.10923839

PARAMETER	ESTIMATE	T FOR H0: PARAMETER = 0	PR > |T|	STD ERROR OF ESTIMATE
INTERCEPT	-11.88882579	-2.75	0.0175	4.31808626
ADDS	17.23357834	5.95	0.0001	2.89439663
INCOME	2.34998673	3.79	0.0026	0.62210994

To estimate average sales for a sales district with an average income of $16,500 and an advertising expenditure of $1,100, we would substitute the quantity 1.100 for x_1 and 16.5 for x_2 into the estimating equation to obtain:

$$\hat{y} = -11.8888 + 17.2336(1.100) + 2.3500(16.5) = 45.8432$$

Or we would estimate average sales to be $45,843.20 for the given levels of the independent variables.

The estimate of the population coefficient of multiple determination in this regression model is $r^2 = 0.97296$. Thus we would state that the two independent variables "average income level" and "advertising expenditure" account for approximately 97.30 percent of the variability in the random variable sales. Since r^2 is close to 1 in this example, we would state that there appears to be a strong (relative) linear relationship between sales and advertising expenditures and average income level in these sample data.

The estimate of the conditional distribution variance σ^2 is found in the column labeled MEAN SQUARE in the row labeled ERROR. That is,

$$s^2_{Y/x_1,x_2} = \text{MSE} = \frac{\text{SSE}}{df} = \frac{\text{SSE}}{n - (k + 1)} = \frac{202.6301}{15 - 3} = 16.886$$

where df represents the degrees of freedom used in calculating the estimate of the conditional probability distribution variance. Since we have 15 observations in this example and are using two independent variables to estimate sales, the degrees of freedom used in the computation of $s^2_{Y/x_1,x_2}$ are $15 - (2 + 1) = 12$. The quantity labeled STD DEV in Table 15.4 is the square root of the estimate of the population conditional distribution variance, or:

$$s_{Y/x_1,x_2} = \sqrt{16.886} = 4.1092$$

■ **15.5**

Hypothesis testing in multiple linear regression

15.5.1 | **Individual test of the regression coefficients, β_i**

To test the hypothesis that β_1 or β_2 is equal to 0 in the bivariate regression case, we need to estimate the standard deviation of the estimators $\mathbf{b}_1$ and $\mathbf{b}_2$. These formulas are quite complex algebraically but would be printed out in most multiple regression computer programs. For our "contrived" example, they are given in Table 15.5.

TABLE 15.5 | Point estimates, b_0 and b_1, and their standard errors, s_{b_1} and s_{b_2}

Parameter	Estimate	Standard error
β_1	$b_1 = 2$	$s_{b_1} = \sqrt{(2)(3/10)} = 0.775$
β_2	$b_2 = -3$	$s_{b_2} = \sqrt{(2)(7/10)} = 1.183$

Confidence interval
for β_i

A confidence interval placed on either β_1 or β_2 is calculated in the same way as the slope in the simple linear regression model, *except* that the degrees of freedom are now $n - (k + 1) = 8 - 3 = 5$. A 95 percent confidence interval for β_1, for example, is given by

$$b_1 - t_{\alpha/2;n-3}s_{b_1} \leq \beta_1 \leq b_1 + t_{\alpha/2;n-3}s_{b_1}$$
$$2 - (2.571)(0.775) \leq \beta_1 \leq 2 + (2.571)(0.775)$$
$$0.007 \leq \beta_1 \leq 3.993$$

Thus we are 95 percent confident that β_1 lies between 0.007 and 3.993. Since this interval does not contain 0, we would further conclude at the 0.05 significance level that β_1 is not 0. But, in the multiple regression model, this does not necessarily imply that this factor or this regression line is useful for prediction purposes. To determine whether the fitted regression line is a useful predictor, rather than testing $\beta_1 = 0$ and $\beta_2 = 0$ *separately*, a joint test should be performed. We will describe this joint test after testing whether $\beta_1 = 0$ or $\beta_2 = 0$ in our sales-advertising-income example.

Example 15.4   Using the data in Table 15.4, test the separate hypotheses that $\beta_1 = 0$ and $\beta_2 = 0$ at the 95 percent confidence level.

Solution   The form of the hypothesis test is given below. In Table 15.6 we have shaded the necessary quantities for performing these hypothesis tests. The STD ERROR OF ESTIMATE values are the estimates of the standard deviations of the sampling distributions of the point estimators of the regression coefficients, β_i. The quantities T FOR HO: PARAMETER = 0 are the ESTIMATES divided by the STD ERROR OF ESTIMATES. Thus, for example,

$$t = \frac{b_1}{s_{b_1}} = \frac{17.2336}{2.8944} = 5.95;$$

Hypothesis test for
β_i

Two-tailed hypothesis test of regression coefficients in multiple regression model

Model: $\mathbf{Y} = \beta_0 + \beta_1 x_1 + \beta_2 x_2 + \cdots + \beta_k x_k + \boldsymbol{\epsilon}$

Hypothesis: $H_0:\ \beta_i = 0$
$\qquad\qquad\quad H_A:\ \beta_i \neq 0$

Test statistic value: $t = \dfrac{b_i}{s_{b_i}}$

Decision rule: Reject if: $t > t_{\alpha/2;n-(k+1)}$ or $t < -t_{\alpha/2;n-(k+1)}$.
$\qquad\qquad\qquad$ Do not reject if: $-t_{\alpha/2;n-(k+1)} \leq t \leq t_{\alpha/2;n-(k+1)}$.

where:

n = Number of observations.
k = Number of independent variables.

TABLE 15.6 | Computer analysis of data in Table 15.3

DEPENDENT VARIABLE: SALES

SOURCE	DF	SUM OF SQUARES	MEAN SQUARE	F VALUE	PR > F
MODEL	2	7290.27547531	3645.13773766	215.87	0.0001
ERROR	12	202.63008162	16.88584013	R-SQUARE	STD DEV
CORRECTED TOTAL	14	7492.90555693		0.972957	4.10923839

PARAMETER	ESTIMATE	T FOR H0: PARAMETER = 0	PR > |T|	STD ERROR OF ESTIMATE
INTERCEPT	-11.88882579	-2.75	0.0175	4.31808626
ADDS	17.23357834	5.95	0.0001	2.89439663
INCOME	2.34998673	3.78	0.0026	0.62210994

$t_{0.025;12} = 2.179$. Since $t > t_{0.025;12}$, we reject the null hypothesis that $\beta_1 = 0$. For β_2, we have the following:

$$t = \frac{b_2}{s_{b_2}} = \frac{2.3500}{0.6221} = 3.78$$

Since $3.78 > 2.179$, we again reject the null hypothesis that the regression coefficient (β_2 in this case) is equal to 0.

One delightful aspect of using a canned computer program like the one shown (SAS) is that we can request the program to perform the hypothesis test for us. The quantities labeled PR $> |T|$ in Table 15.6 are the p-values for the regression coefficients as they were described in Chapter 10. The interpretation of these p-values is, for example, for β_2, that either β_2 is not equal to 0, or else we obtained one of the 26 samples in 10,000 (0.0026) that yields a t-statistic of 3.78 or larger when in fact $\beta_2 = 0$. Since this is highly unlikely, we reject the null hypothesis that $\beta_2 = 0$. The p-value for β_1 is likely much less than the estimate 0.0001 shown, but the computer program prints only four digits to the right of the decimal point.

We should emphasize that the tests do *not* indicate that both β_1 *and* β_2 are unequal to 0 at the stated significance levels. If, for example, we repeatedly perform hypothesis tests at the $\alpha = 0.05$ level, then inevitably there will be *some* rejections of the null hypothesis when in fact all β_i could be equal to 0. The F-test in the following section is a joint test for all β_i, where $i = 1, 2,$ . . . , k. The F-test will allow us to conclude either that all β_i are equal to 0 or that some coefficients in the model are significantly different from 0. The F-test solves what is referred to as the "family-wise" problem in determining the significance of the model (i.e., all terms in the model).

We can use the t-test to test for the significance of interaction terms in the model given in Table 15.4. That is, the multiple linear regression model that includes the interaction of advertising expenditures and average income level (denoted by $x_1 x_2$) can be examined, and a test of the hypothesis H_0: $\beta_3 = 0$ performed, where β_3 is the coefficient of the interaction term $x_1 x_2$ in the multiple linear regression model. Table 15.7 gives the multiple regression model with this term included. Note that the hypothesis H_0: $\beta_3 = 0$ would *not* be rejected at the $\alpha = 0.1$, 0.05, and 0.01 significance levels. This result can be read directly from the data given in Table 15.7, since the p-value for this hypothesis test is 0.1258.

15.5.2 | Tests of the significance of the model—the *F*-test

To test the significance of the model developed, we must perform a *joint* test of the regression coefficients. That is, we must test the following hypothesis:

H_0 : $\beta_1 = \beta_2 = \beta_3 = \cdots = \beta_k = 0.$
H_A : At *least* one of the coefficients is nonzero.

TABLE 15.7 | Multiple linear regression analysis of the model $Y = \beta_0 + \beta_1 x_1 + \beta_2 x_2 + \beta_3 x_1 x_2 + \epsilon_i$

DEPENDENT VARIABLE: SALES

SOURCE	DF	SUM OF SQUARES	MEAN SQUARE	F VALUE	PR > F
MODEL	3	7330.73788204	2443.57929401	165.75	0.0001
ERROR	11	162.16767489	14.74251590	R-SQUARE	STD DEV
CORRECTED TOTAL	14	7492.90555693		0.978357	3.83959840

PARAMETER	ESTIMATE	T FOR H0: PARAMETER = 0	PR > |T|	STD ERROR OF ESTIMATE
INTERCEPT	-0.66244604	-0.08	0.9346	7.88662644
ADDS(X1)	5.56061422	0.74	0.4767	7.54718047
INCOME(X2)	1.59886111	2.17	0.0529	0.73719684
X1X2	0.73959362	1.66	0.1258	0.44642972

The reason we must perform a joint test as outlined below is that if we, for example, perform a t-test at the $\alpha = 0.05$ significance level on both β_1 and β_2 in a bivariate regression model, the joint level of significance will not be $\alpha = 0.05$—it will be slightly greater than 0.05. For more than two independent variables in the regression model, the joint level of significance will be even greater! The correct test for the significance of the regression model is the F-test as depicted in Figure 15.6. The reader may recognize the procedure used to test the significance of the model as analysis of variance described in Chapter 12.

FIGURE 15.6 | Steps in the analysis of variance for testing the significance of the multiple linear regression model

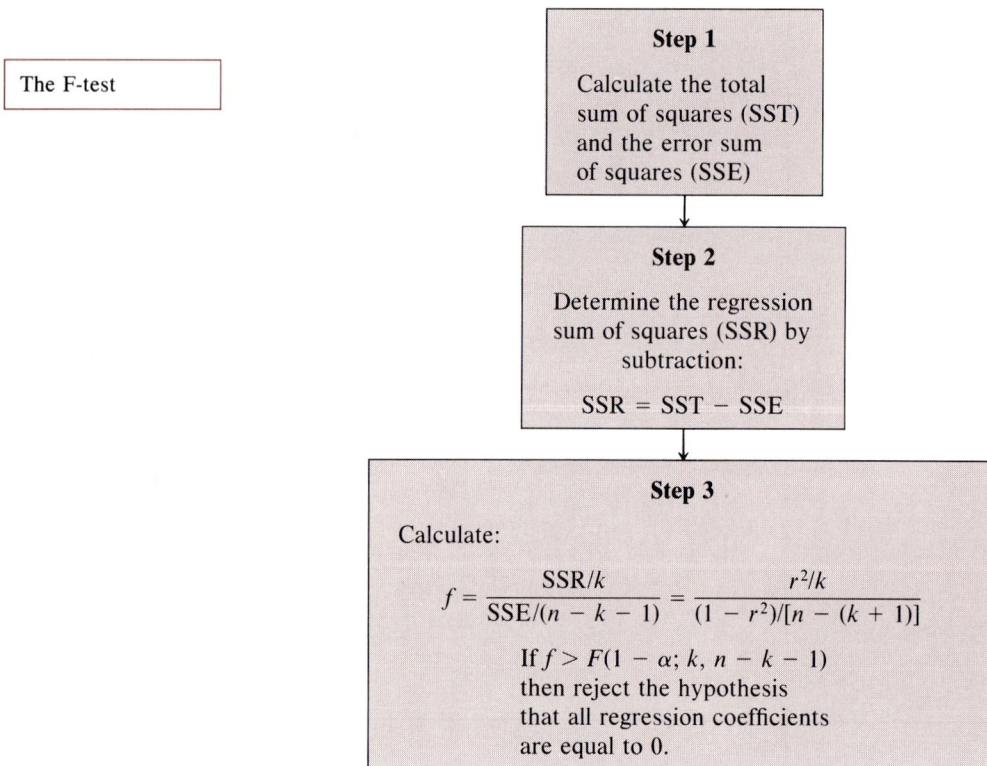

The F-test

Step 1

Calculate the total sum of squares (SST) and the error sum of squares (SSE)

Step 2

Determine the regression sum of squares (SSR) by subtraction:

$$SSR = SST - SSE$$

Step 3

Calculate:

$$f = \frac{SSR/k}{SSE/(n - k - 1)} = \frac{r^2/k}{(1 - r^2)/[n - (k + 1)]}$$

If $f > F(1 - \alpha; k, n - k - 1)$ then reject the hypothesis that all regression coefficients are equal to 0.

The format of the analysis of variance is given in Table 15.8, and Table 15.9 gives the analysis of variance for our "contrived" problem.

From Table B.7 in Appendix B, the critical F-value at the $\alpha = 0.05$ level is $F(0.95; 2, 5) = 5.79$. Since $f = 3.5 < 5.79$, we *cannot* reject the null hypothe-

TABLE 15.8 | Analysis of variance table for testing the significance of the regression line

Source of variation	Degrees of freedom	Sum of squares	Mean square	F-ratio
Regression (model)	k	SSR	MSR $=$ SSR$/k$	MSR/MSE
Error	$n - k - 1$	SSE	MSE $=$ SSE$/[n - (k + 1)]$	
Total	$n - 1$	SST		

TABLE 15.9 | Analysis of variance for example data

Source of variation	Degrees of freedom	Sum of squares	Mean square	F-ratio
Regression (model)	2	14	7	$7/2 = 3.5$
Error	5	10	2	
Total	7	24		

sis that *both* β_1 and β_2 are equal to 0. This implies that there is a high probability that the regression function is a *useless* predictor for this dependent variable.

Notice that our t-test (confidence interval) on β_1 leads us *not* to conclude that $\beta_1 = 0$ (although the interval almost contained 0). The problem in using the t-test is that when we perform a t-test, say at the $\alpha = 0.05$ level, on both β_1 and β_2, the *joint* level of significance for the two t-tests will *not* be $\alpha = 0.05$; it will be slightly greater than 0.05. The correct procedure in this case is the F-test as we have outlined it.

Example 15.5   Using the data given in Table 15.4, test the null hypothesis for the sales-advertising-income example that $\beta_1 = \beta_2 = 0$ using the analysis of variance F-test with $\alpha = 0.05$.

Solution   In the shaded portion of Table 15.10, we show that portion of Table 15.4 that corresponds to the analysis of variance $F(0.95; 2, 12) = 3.89$ (see Table B.7 in Appendix B). Since $f = 215.87 > 3.89$, we reject the null hypothesis that *both* $\beta_1 = 0$ *and* $\beta_2 = 0$. Note that the SAS output provides us with the p-value for the test under the column labeled PR > F. Either β_1 and β_2 are not both equal to 0, or else we obtained one sample out of 10,000 (0.0001) that would yield an F-value of 215.87 or larger when in fact β_1 and β_2 are both equal to 0. Since the probability of obtaining one of these samples is extremely small in this case, we reject the null hypothesis that both coeffi-

TABLE 15.10 | Analysis of variance table for data in Table 15.3

DEPENDENT VARIABLE: SALES

SOURCE	DF	SUM OF SQUARES	MEAN SQUARE	F-VALUE	PR > F
MODEL	2	7290.27547531	3645.13773766	215.87	0.0001
ERROR	12	202.63008162	16.88584013		
CORRECTED TOTAL	14	7492.90555693			

	R-SQUARE	STD DEV
	0.972957	4.10923839

PARAMETER	ESTIMATE	T FOR H0: PARAMETER = 0	PR > |T|	STD ERROR OF ESTIMATE
INTERCEPT	-11.88882579	-2.75	0.0175	4.31808626
ADDS	17.23357834	5.95	0.0001	2.89439663
INCOME	2.34998673	3.78	0.0026	0.62210994

cients are simultaneously equal to 0, and we conclude that the regression equation developed is useful for estimating average sales levels in the firm's branches for the range of advertising levels and the average income levels examined.

■ 15.6

Using the multiple linear regression model for estimation and prediction

In Section 14.2.5 we showed how to obtain a point estimate of $\mu_{Y/x}$ as well as a confidence interval estimate of this same quantity. Then in Section 14.2.6 we showed how to obtain a prediction of the "next" value of Y, Y_{next}, as well as a prediction interval for this quantity. You will recall that the point estimates in both cases were the same for the simple linear regression model—we simply substituted the appropriate value of x into the derived regression equation to obtain a point estimate of the mean of the Y-values or a prediction of the next Y-value. Constructing confidence and prediction intervals for these same quantities was quite another matter, however. The prediction interval for the next Y-value was much wider than the confidence interval for the mean, reflecting the added variability of predicting a specific conditional distribution of Y given a specific value of x.

The formulas for a prediction interval for a specific value (i.e., the next value) of a random variable Y and for a confidence interval for the mean of the conditional distribution of Y, given values of the independent variables are quite complicated algebraically in the multiple regression case. Fortunately, computer programs sometimes provide these estimates for us when we simply indicate that we desire such estimates by inputting a "key" parameter on our data input cards.

In Table 15.11 we show the SAS output of the predicted and observed values of the random variable Y sales for our sales-advertising-income example, as well as the residuals used for checking the assumptions of the multiple linear regression model. These estimates were determined by computer by substituting the given levels of the independent variables into the regression equation as was done in Section 15.4 to estimate (for example) expected sales when advertising expenditures were $1,100 and the average income was $16,000.

In Table 15.12 we show (1) the SAS-produced 95 percent confidence interval for the mean sales level when the amount spent on advertising is $621 and the average income is $7,800 and (2) the 95 percent prediction interval for the individual or the next sales level for an advertising expenditure of $621 and a corresponding average income level of $7,800. You will note from the estimates given in Table 15.12 that the interval for predicting the "next" Y-value is much wider, reflecting the added uncertainty in predicting an individual observation as opposed to estimating the mean of a distribution.

Not all computer software programs have the capacity to produce confidence interval estimates or prediction interval estimates of values of the dependent random variable Y in a multiple regression analysis. This is unfortunate, since the culmination of a regression analysis generally consists of estimating the mean of the conditional distribution and setting a confidence

TABLE 15.11 | Estimated values of the dependent random variable sales and corresponding residuals

Observation	Observed value	Predicted value	Residual
1	55.42	56.27	−0.85
2	45.82	38.06	7.76
3	16.94	15.05	1.89
4	35.82	37.74	−1.93
5	85.09	79.58	5.51
6	62.03	65.75	−3.73
7	17.92	16.90	1.02
8	32.85	33.29	−0.45
9	41.18	47.07	−5.89
10	62.91	68.47	−5.56
11	87.01	82.99	4.04
12	22.15	20.57	1.58
13	16.66	17.14	−0.48
14	27.05	28.52	−1.47
15	39.12	40.57	−1.45

Source: Computer analysis of data in Table 15.3.

TABLE 15.12 | Ninety-five percent confidence interval and prediction interval for sales when advertising expenditures = $621 and average income level = $7,800

```
                    Confidence  Interval
   OBSERVED         PREDICTED       LOWER 95% CL      UPPER 95% CL
    VALUE            VALUE          FOR MEAN          FOR MEAN

 16.66000000      17.14312288      13.54053760       20.74570815

                                                  Width of
                                              Confidence Interval

                                              |←——7.205——→|


                    Prediction  Interval
   OBSERVED         PREDICTED       LOWER 95% CL      UPPER 95% CL
    VALUE            VALUE          INDIVIDUAL        INDIVIDUAL

 16.66000000      17.14312288       7.49223318       26.79401258

                                                  Width of
                                              Prediction Interval

                                      |←————————19.302————————→|
```

Source: Computer analysis of data in Table 15.3.

interval on the mean, or predicting an individual value of Y and setting a prediction interval for the individual value of Y. Fortunately, many computer programs now produce both confidence intervals for the conditional distribution means and prediction intervals for individual values of Y. Many of the advanced references listed at the end of the chapter include the formulas necessary for obtaining these interval estimates, although their computation is often quite complex and requires the use of a computer or a *great deal of time* and a calculator to determine their values!

15.7
The treatment of qualitative variables in multiple linear regression

Our discussion of regression to this point has allowed only for the inclusion of such quantitative independent variables as advertising expenditures and income level in the estimating equation. Frequently, we would like to be able to include qualitative or indicator variables in the regression equation to estimate the value of a dependent random variable. Examples of indicator or qualitative variables (they are also called *dummy* or *binary* variables) are sex, geographic location, and season. For example, if we were to develop an estimating equation for the weight of an individual, we might attempt to relate (in a regression sense) weight to the individual's height—in general, the taller a person is, the more the person tends to weigh on the average. We might also include a variable that indicates the sex of the individual—in general, we would expect a male to weigh more than a female for a given height. The form of the estimating equation might be:

$$Y_i = \beta_0 + \beta_1 x_{i,1} + \beta_2 x_{i,2} + \epsilon_i$$

where

Y_i = Random variable weight.
$x_{i,1}$ = Height of a sampled unit in the population.
$x_{i,2}$ = Indicator variable for the sex of the sampled unit in the population, where, for example:

$$x_{i,2} = \begin{cases} 1 \text{ if the sampled unit is a male.} \\ 0 \text{ if the sampled unit is a female.} \end{cases}$$

ϵ_i = Random error term.

A graph of a population regression equation such as that given above might appear as in Figure 15.7. Note that if the sampled unit is a male, the population regression function becomes:

$$Y = (\beta_0 + \beta_2) + \beta_1 x_1 + \epsilon$$

and if the sampled unit is a female, the population regression function becomes:

$$Y = \beta_0 + \beta_1 x_1 + \epsilon$$

The slope in both estimating equations is β_1. The intercept is $(\beta_0 + \beta_2)$ for a male and β_0 for a female. Thus β_2 measures the differential effect of sex on average weight. If β_2 is positive, then expected weight is greater for a male

FIGURE 15.7 | Hypothetical population regression lines for weight as a function of height and sex

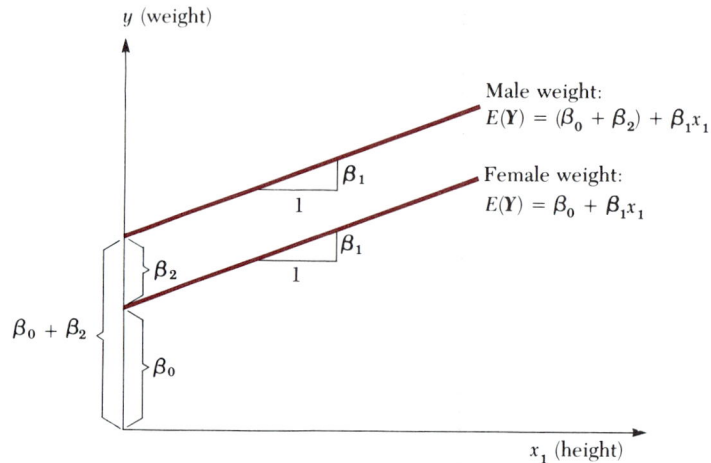

equal in height to a female; if β_2 is negative, expected weight is less for a male equal in height to a female. Since β_2 is positive in Figure 15.7, we would expect a male's weight to exceed that of his female counterpart who is equal in height.

We need not restrict ourselves to two classes in the use of the indicator or dummy variables. Indeed, if we desire to model the seasons of the year, for example (winter, spring, summer, fall), we would require the use of more than one indicator variable. In general, if there are m classes or distinguishable groups to be used as independent variables in a multiple regression, there will be $j = (m - 1)$ indicator or dummy variables required. To see how to model an indicator or a dummy variable where more than one class is involved, consider the following example.

Example 15.6 A ski and leisure resort in western Colorado has been attempting to predict accidents that require the use of plaster casts in its area as a function of the growth in the number of visitors (mostly skiers) in order to staff its needs for orthopedic medical facilities. Table 15.13 gives data for the past six years (by quarter) on the number of people visiting the facility and the corresponding number of accidents requiring plaster casts at the facility.

As a first step, the firm developed a simple linear regression model to estimate the number of accidents using visitors to the area as the independent variable. The results of this analysis are shown in Table 15.14. As shown in this table, the estimating equation is: $\hat{y} = 9.03 + 0.9589x$, where $\hat{y}$ = Predicted number of plaster casts, and x = Number of visitors to the facility (in thousands). The firm can fairly accurately predict the number of visitors to its facility, since demand far exceeds supply, and reservations

TABLE 15.13 | Visitors to a Colorado ski resort and number of accidents requiring the use of a plaster cast

Year and quarter	Visitors (000)	Casts	Year and quarter	Visitors (000)	Casts
1981			1984		
Winter	33.63	53	Winter	74.64	94
Spring	36.46	41	Spring	80.31	86
Summer	41.18	24	Summer	80.97	64
Fall	43.16	57	Fall	87.75	99
1982			1985		
Winter	46.45	70	Winter	88.07	109
Spring	50.63	60	Spring	94.00	101
Summer	54.41	41	Summer	96.16	77
Fall	58.66	77	Fall	96.98	110
1983			1986		
Winter	62.52	81	Winter	103.90	123
Spring	65.55	70	Spring	107.77	120
Summer	69.62	50	Summer	110.42	95
Fall	72.92	87	Fall	114.91	126

Source: Company records.

must be made well in advance (sometimes as much as 18 months) of the scheduled arrival of the tourists. The firm is disappointed with the proportion of sample variation explained by the simple linear regression equation ($r^2 = 0.7296$), and wonders whether taking seasonal factors into account (more accidents during snow!) might not help in predicting the need for plaster casts.

Construct a table corresponding to the input variables for the multiple linear regression equation where the demand for plaster casts is a function of the number of visitors to the facility and the seasons of the year. Be sure to define each variable you use in your estimating equation.

Solution The variables are defined as follows:

$x_{i,1}$ = Number (in thousands) of visitors to the facility in period i.

$$x_{i,2} = \begin{cases} 1 \text{ if period } i \text{ is the first quarter of the year (winter).} \\ 0 \text{ otherwise.} \end{cases}$$

$$x_{i,3} = \begin{cases} 1 \text{ if period } i \text{ is the second quarter of the year (spring).} \\ 0 \text{ otherwise.} \end{cases}$$

$$x_{i,4} = \begin{cases} 1 \text{ if period } i \text{ is the third quarter of the year (summer).} \\ 0 \text{ otherwise.} \end{cases}$$

$\hat{y}_i$ = Estimated demand for plaster casts in period i.

Hence the estimating equation becomes:

$$\hat{y}_i = b_0 + b_1 x_{i,1} + b_2 x_{i,2} + b_3 x_{i,3} + b_4 x_{i,4}$$

TABLE 15.14 | Simple linear regression of data in Table 15.13

DEPENDENT VARIABLE: CASTS

SOURCE	DF	SUM OF SQUARES	MEAN SQUARE	F-VALUE	PR > F
MODEL	1	13004.84571756	13004.84571756	59.37	0.0001
ERROR	22	4819.11261578	219.05057344	R-SQUARE	STD DEV
CORRECTED TOTAL	23	17823.95833333		0.729627	14.80035721

PARAMETER	ESTIMATE	T FOR H0: PARAMETER = 0	PR > |T|	STD ERROR OF ESTIMATE
INTERCEPT	9.02982441	0.93	0.3604	9.66787753
VISITORS	0.95890293	7.71	0.0001	0.12444981

Values of the dependent and independent variables are given in Table 15.15. Note that the regression equation for the fall quarter is $\hat{y}_i = b_0 + b_1 x_{i,1}$. In this context, the variables $x_{i,2}$, $x_{i,3}$, and $x_{i,4}$ denote, respectively, the expected increase (or decrease) from the fall quarter in the demand for plaster casts in the winter, spring, and summer.

TABLE 15.15 | Values of dependent and independent variables in Example 15.6

Period	Casts	Visitors (000)	Quarter 1	2	3	Period	Casts	Visitors (000)	Quarter 1	2	3
1	53	33.63	1	0	0	13	94	74.64	1	0	0
2	41	36.46	0	1	0	14	86	80.31	0	1	0
3	24	41.18	0	0	1	15	64	80.97	0	0	1
4	57	43.16	0	0	0	16	99	87.75	0	0	0
5	70	46.45	1	0	0	17	109	88.07	1	0	0
6	60	50.63	0	1	0	18	101	94.00	0	1	0
7	41	54.41	0	0	1	19	77	96.16	0	0	1
8	77	58.66	0	0	0	20	110	96.98	0	0	0
9	81	62.52	1	0	0	21	123	103.90	1	0	0
10	70	65.55	0	1	0	22	120	107.77	0	1	0
11	50	69.62	0	0	1	23	95	110.42	0	0	1
12	87	72.92	0	0	0	24	126	114.91	0	0	0

Source: Company records.

Season	Predicting equation
Winter	$\hat{y}_i = (b_0 + b_2) + b_1 x_{i,1}$
Spring	$\hat{y}_i = (b_0 + b_3) + b_1 x_{i,1}$
Summer	$\hat{y}_i = (b_0 + b_4) + b_1 x_{i,1}$
Fall	$\hat{y}_i = (b_0) + b_1 x_{i,1}$

The reasons we do not use a fourth indicator variable for the fall quarter are that (1) the additional variable is *not necessary* (the fourth season, the fall, is implied when $x_{i,2}$ and $x_{i,3}$ and $x_{i,4} = 0$), and (2) if the variable were included in the model, it would be impossible to solve the resulting normal equations for least squares estimates b_0, b_1, b_2, etc. This is because, if the additional variable were included in the model, it could be written as a linear combination of the variables already included. This situation is called perfect *multicollinearity* and is discussed in Section 15.9 and in more detail in the advanced references listed at the end of the chapter. For the present it is

Correct number of
dummy variables
for a qualitative
variable

sufficient to note that *when there are m categories of a qualitative indepen-*
dent variable, j = (m − 1) dummy or categorical variables **must** *be used* in
deriving an estimating equation for the dependent random variable. The
estimating equations for the four seasons of the year become as shown here.

The data in Table 15.15 were input to the SAS multiple linear regression
computer program, and the results of the analysis are shown in Table 15.16.
The derived equation is:

$$\hat{y}_i = 14.225 + 0.99x_{i,1} + 6.443x_{i,2} - 6.442x_{i,3} - 30.592x_{i,4}$$

where the variables are as previously defined. The coefficient of multiple
determination in Table 15.16 is 0.99297, indicating that a very strong linear
relationship exists in these sample data between the dependent variable
casts and the independent variables corresponding to the number of visitors
and the seasons of the year. From Table B.7 in Appendix B (and interpolation), $F(0.99; 4, 19) = 4.50$. Since the calculated F-value is 671.16, we
conclude that the regression model developed is useful for predicting the
demand for plaster casts. Furthermore, each of the regression coefficients is
significant as evidenced by the quantities in the column $PR > |T|$. Hence we
conclude that all regression coefficients are different from 0—all aid in estimating the demand for plaster casts. Note that the p-values given in Table
15.16 would cause us to reject the null hypothesis that $\beta_i = 0$ for all regression coefficients at the 0.10, 0.05, or 0.01 significance levels and would cause
us to reject the null hypothesis that all regression coefficients are *simulta-*
neously equal to 0 at the 0.10, 0.05, or 0.01 significance levels ($PR > F = 0.0001$).

Although the value of $\mathbf{r}^2$ increased from 0.73 to 0.99 when the categorical
variables corresponding to the seasons of the year were included in the
regression model, an interesting question that might be posed at this time is:
Can we statistically test for the **presence** of a *seasonal factor* in our developed model? The answer to this **question** is yes, and the analysis proceeds as
follows. First, rather than testing **whether** each β_i value is individually significantly different from 0, we will **test** the hypothesis $H_0: \beta_2 = \beta_3 = \beta_4 = 0$,
since these are the regression coefficients (population) corresponding to the
seasons of the year.

The test statistic that is useful for testing the above hypothesis is obtained
by first taking the difference between the amount of variation (SSE) left
unexplained by the regression equation before the indicator variables were
introduced and the amount of variation left unexplained (SSE) *after* the
categorical variables have been included in the model. Since we are taking
the difference between two sum of square error values, it is necessary for us
to subscript SSE to distinguish between the models. Let us designate the
sum of squares error *before* the categorical variables are introduced into the
estimating equation as SSE_B and the sum of squares error *after* the categorical variables have been included in the model as SSE_A. If we divide the
above difference by j, the number of additional variables in the revised
model, and divide the resulting quotient by $[SSE_A/(n - k - 1)]$, the resulting

TABLE 15.16 | Computer analysis of the data in Table 15.15

DEPENDENT VARIABLE: CASTS

SOURCE	DF	SUM OF SQUARES	MEAN SQUARE	F-VALUE	PR > F
MODEL	4	17698.69908874	4424.67477219	671.16	0.0001
ERROR	19	125.25924459	6.59259182	R-SQUARE	STD DEV
CORRECTED TOTAL	23	17823.95833333		0.992972	2.56760430

PARAMETER	ESTIMATE	T FOR H0: PARAMETER = 0	PR > |T|	STD ERROR OF ESTIMATE
INTERCEPT	14.22495659	7.03	0.0001	2.02319422
VISITORS	0.99213765	45.33	0.0001	0.02188722
QUARTER1	6.44293516	4.29	0.0004	1.50134838
QUARTER2	-6.44197011	-4.33	0.0004	1.48945001
QUARTER3	-30.59166399	-20.61	0.0001	1.48450349

statistic is F-distributed with j and $(n - k - 1)$ degrees of freedom when the null hypothesis is true. The form of the hypothesis test is shown here.

Test for a subset of independent variables

Hypothesis test for subsets of independent variables in a multiple linear regression model, $j = (m - 1)$ categorical variables

Notation: Let the sum of squares error before the subset of variables is included for which it is desired to test the null hypothesis that they are all simultaneously equal to 0 be denoted by SSE_B, and let the sum of squares error after the subset of variables is included in the model be denoted by SSE_A. Let the number of independent variables in the subset be denoted by j, and let the population regression coefficients for variables in the subset be denoted by β_i, β_{i+1}, $\ldots$, β_{i+j-1}.

Hypothesis: H_0: $\beta_i = \beta_{i+1} = \cdots = \beta_{i+j-1} = 0$.

H_A: At least one population regression coefficient β_h is not equal to 0, where $i \leq h \leq i + j - 1$.

Test statistic value: $f = \dfrac{(SSE_B - SSE_A)/j}{(SSE_A)/(n - k - 1)}$

Reject if: $f > F(1 - \alpha; j, n - k - 1)$

Do not reject if: $f \leq F(1 - \alpha; j, n - k - 1)$

where:

n = Number of observations.
α = Significance level of the test.
k = Number of independent variables in the model.

Example 15.7 Using the data provided in Tables 15.14 and 15.16, test for the presence of a seasonal effect in the demand for plaster casts at the $\alpha = 0.05$ significance level.

Solution From Table 15.14, $SSE_B = 4819.11$, and from Table 15.16, $SSE_A = 125.26$. Thus the computed value of the **F**-statistic is:

$$f = \frac{(4{,}819.11 - 125.26)/3}{125.26/(24 - 4 - 1)} = 237.33$$

$F(0.95; 3, 19) = 3.13$ (by interpolation). Since 237.33 far exceeds 3.13, we reject the null hypothesis and conclude that there is a seasonal effect in the demand for plaster casts.

An advantage of the test described above is that it can be modified for testing hypotheses concerning the significance of *any* subset of independent variables in a multiple linear regression model. For example, if it were felt that the regression coefficients for a certain subset of variables were all equal to 0, a multiple regression could be constructed first with the subset of variables not in the model, and then with the subset of variables in the model. The test just described could then be used to test the significance of the subset of variables. Note that if the subset of variables consists of *one* independent variable, then the test described is equivalent to the *t*-test for the significance of the simple linear regression model when we test $H_0: \beta_1 = 0$. In Chapter 14 (Section 14.2.3) we noted the similarity of the **t**-statistic and the **F**-statistic, indicating that for the test of a single variable, $\mathbf{t}^2 = \mathbf{F}$.

We should also note at this time that it is completely arbitrary which of the *m* levels of the categorical variable is not assigned a binary (0 or 1) independent variable. In the previous example, the fall quarter was not assigned a categorical variable because fall can be represented as the *absence* of winter, summer, and spring. That is, we represent the fall quarter of each year by assigning the value 0 to $x_{i,2}$, $x_{i,3}$, *and* $x_{i,4}$. We could have just as easily assigned independent indicator variables to the spring, summer, and fall quarters of the year, in which case the winter quarter would have been represented by each of the three binary categorical variables being set equal to 0. From this same line of reasoning, it should be apparent that in the example of relating the weight of an individual to the individual's sex, we could have defined our independent variable to be:

$$x_{i,2} = \begin{cases} 1 \text{ if the sampled unit is a female} \\ 0 \text{ if the sampled unit is a male} \end{cases}$$

Again, the choice is completely arbitrary, and this is why it is extremely important to define carefully each independent variable to aid the user of the developed regression equation.

■ 15.8

Model validation

Regression model assumptions

An important step in multiple regression analysis is Step 5—determine whether or not the assumptions of the multiple regression model have been met for a particular fitted regression model to a data set. The three assumptions of the multiple regression model can be stated in terms of the error components ϵ_i of the model:

1. The ϵ_i's are normally distributed.

2. The ϵ_i's have the same variance (σ^2).

3. The ϵ_i's are uncorrelated.

Methods of checking the assumptions

When a regression model has been fitted to *n* sampled values of the dependent variable, $y_1, y_2, \ldots, y_n$, we can use the sampled residuals $e_i = y_i - \hat{y}_i$ to check these assumptions, where $\hat{y}_i$ is the predicted value of the dependent variable **Y** from the fitted regression equation, $\hat{y} = b_0 + b_1 x_1 + b_2 x_2 + \cdots + b_k x_k$. There are two ways in which the sampled residuals $e_1, e_2, \ldots, e_n$ can be used to check the three assumptions listed above: (1) *residual plots* and (2) *statistical tests*. A residual plot is a graphical method of assessing whether or not the assumptions have been satisfied. As such it is *subjective* in that there are only guidelines to suggest whether or not a residual plot indicates that one or more of the assumptions have been violated. There are statistical tests designed to check the assumptions as well. These are *objective* methods in that formal hypothesis testing procedures are used to determine whether or not it is reasonable to assume that the assumptions of the regression model have been met by a specific fitted regression model fitted to a sample data set. Usually both approaches—residual plots and statistical tests—are used to *validate* the model; that is, to demonstrate that the assumptions have been satisfied for a specific fitted regression model. The validation of a fitted regression model is a very important step in the multiple regression process. It should *always* be done when using regression analysis, and the validation results should be reported together with the fitted regression model results.

> Model validation

15.8.1 | Residual plots

There are two basic types of residual plots: (1) plot of the residuals (e_i's) against the predicted values ($\hat{y}_i$'s) and (2) plot of the residuals (e_i's) against the values of each independent variable (x_i's). Usually the residuals are first *standardized* before the residual plots are made. There are several ways to standardize the residuals. To illustrate the idea of a standardized residual, we will consider the sales example whose data are given in Table 15.3. Table 15.17 gives the values of the dependent variable **Y**, the predicted y-values

> Standardized residuals

TABLE 15.17 | Sales data from Table 15.3; y, ŷ, e, and two sets of standardized residuals

Sales ($000) y	Predicted sales ($000) ŷ	Residual $e = y - \hat{y}$	Standardized residuals e_i'	e_i^* (MINITAB)
55.4200	56.2676	−0.84764	−0.22281	−0.23000
45.8190	38.0598	7.75923	2.03954	2.16547
16.9400	15.0549	1.88510	0.49550	0.54739
35.8180	37.7450	−1.92697	−0.50651	−0.49281
85.0900	79.5814	5.50861	1.44796	1.79934
62.0250	65.7538	−3.72881	−0.98013	−1.07408
17.9180	16.8988	1.01923	0.26791	0.27278
32.8450	33.2909	−0.44588	−0.11720	−0.11398
41.1800	47.0666	−5.88661	−1.54732	−1.48996
62.9100	68.4737	−5.56372	−1.46244	−1.60811
87.0130	82.9685	4.04451	1.06311	1.16494
22.1500	20.5680	1.58202	0.41584	0.42752
16.6600	17.1432	−0.48322	−0.12701	−0.12845
27.0500	28.5172	−1.46725	−0.38567	−0.37590
39.1210	40.5696	−1.44864	−0.38078	−0.36612

from the fitted regression equation, $\hat{y} = -11.8888 + 17.2336x_1 + 2.3500x_2$, the residuals $e_i = y_i - \hat{y}_i$, and two sets of standardized residuals. The first set of standardized residuals are determined by using the **Z**-statistic transformation, whose value is given by

$$z = \frac{x - \mu}{\sigma}$$

from Chapter 7. Applied to the residuals, we have

$$e_i' = \frac{e_i - \bar{e}}{s_e}$$

where e_i is the ith residual, e_i' is the ith standardized residual, $\bar{e}$ and s_e are the mean and standard deviation of the sampled residuals, e_i, $i = 1, 2, \ldots, n$, which are calculated from

$$\bar{e} = \frac{\sum\limits_{i=1}^{n} e_i}{n} \quad \text{and} \quad s_e = \sqrt{\frac{\sum\limits_{i=1}^{n} (e_i - \bar{e})^2}{n - 1}}$$

Since by least squares fitting, $\bar{e} = 0$, the standardized residuals may be calculated by using

$$e_i' = \frac{e_i}{s_e}$$

The second set of standardized residuals is produced from the computer software package MINITAB. The formula used is

$$e_i^* = \frac{e_i}{s_e^*}$$

MINITAB stan-
dardized residuals

where e_i is the ith residual, e_i^* is the standardized residual, and s_e^* is calculated from

$$s_{e,i}^* = \sqrt{\text{MSE} - [\text{Estimated standard deviation of estimator of } E(Y_i)]^2}$$

Thus each residual has a different standard deviation. The MINITAB method uses the standard deviation for each predicted value of **Y** to standardize the residuals (see p. 258, *MINITAB Handbook, 2nd ed.*). Notice in Table 15.17 that the standardized residuals do differ somewhat depending on the standardization method used. We will use the MINITAB standardized residuals in the remainder of this section.

The first residual plot used in validation is a plot of the standardized residuals (e_i^* from MINITAB) against the predicted values ($\hat{y}_i$). The plot for the sales example whose data are given in Table 15.3 is shown in Figure 15.8. The vertical axis labeled STRES contains the MINITAB standardized residuals, and the horizontal axis labeled C7 contains the y values.

As discussed in Chapter 13, if the error terms satisfy the three regression model assumptions (error terms are normally distributed, uncorrelated, and

FIGURE 15.8 | Residual plot for the sales data in Table 15.3

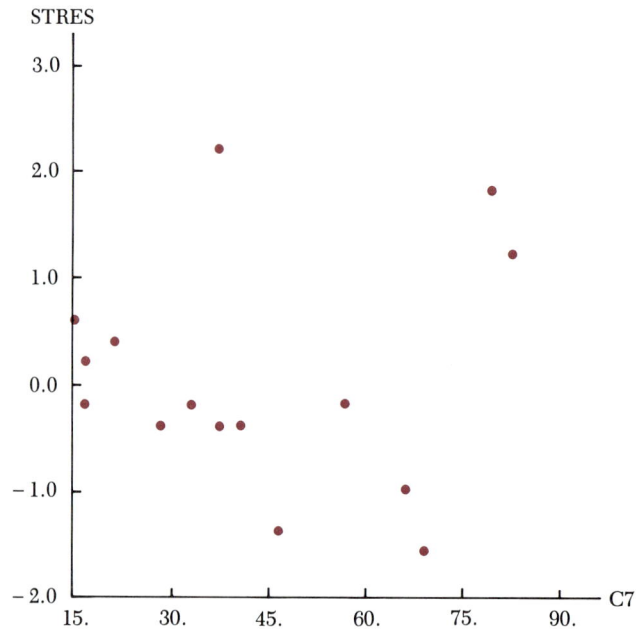

have the same variance), then we should expect the residuals to reflect the satisfaction of these assumptions provided that the model is an appropriate one to fit to the sample data. Thus we should expect the plotted pairs $(\hat{y}_i, e_i^*)$ to fall within a rectangular region with no patterns. An example of a "good" residual plot, a plot suggesting that the assumptions on the error terms have been met by the fitted regression equation, is shown in Figure 15.9. Comparing Figure 15.8 to Figure 15.9, the standardized residuals for the sales data appear to be conforming reasonably well. However, there are two possible problems. First, if the residuals are normally distributed, then we would expect an equal number of positive and negative residuals. From the last column in Table 15.17, there are nine negative residuals and six positive residuals. Thus there are somewhat more negative residuals than positive residuals, something unexpected if the error terms are normally distributed. Since the residuals are standardized, we should expect to find about 68 percent of the residuals between ± 1 (plus and minus one standard deviation), about 95 percent between ± 2, and about 99 percent between ± 3. From the last column in Table 15.17, we find that $9/15 = 0.60$, or 60 percent fall between ± 1; $14/15 = 0.933$, or 93.3 percent, fall between ± 2; and $15/15 = 100$ percent fall between ± 3. These percentages are not too different from the expected values (68 percent, 95 percent, and 99 percent). Given the small sample size ($n = 15$), the residuals appear not to deviate much from what we

FIGURE   15.9 | A residual plot for which the three regression model assumptions are satisfied

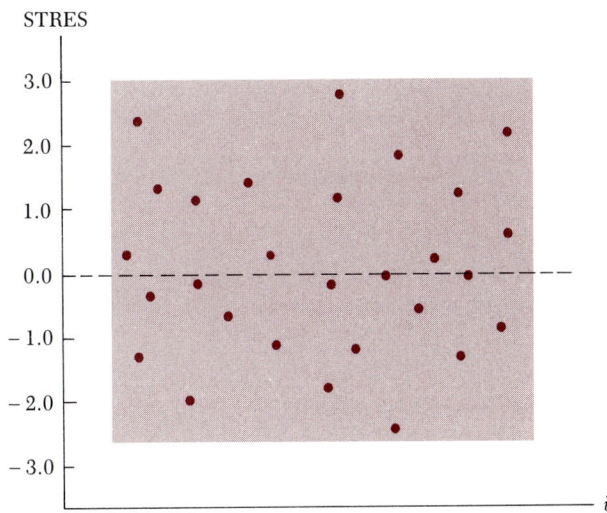

would expect if the error terms are normally distributed. Second, there appears to be more spread (variability) of the residuals plotted in Figure 15.8 as the values of $\hat{y}$ increase. If the error terms have the same variance, then the variability of the residuals should be approximately constant for all values of $\hat{y}$ (from small to large). This plot indicates that the variance of the error terms may not be constant.

Interpreting the residual plot of the standardized residuals (e_i^*s) and the predicted values ($\hat{y}$s) is subjective. What we are looking for are obvious indications that the assumptions have been violated. We will now consider some residual plots that clearly indicate one (or more) of the assumptions are violated. First, consider the equal variance assumption. Figure 15.10 illustrates two residual plots indicating unequal variances. In Figure 15.10A, the residual variation decreases as the values of $\hat{y}$ increase. In Figure 15.10B, the variation of the residuals is greater for the middle values of $\hat{y}$ and smaller for small and large values of $\hat{y}$. Second, consider the no correlation assumption. If the error terms are uncorrelated, then there should be no patterns in the residual plot. Figure 15.11 illustrates residual plots with correlated error terms. In Figure 15.11A, the residuals are conforming to a line. Clearly, the definite pattern of these residuals suggests that the error terms are not uncorrelated. The residuals are following one after another in a linear fashion, a very distinctive pattern. In Figure 15.11B, the residuals are oscillating about 0 in a very definite pattern. If the residual plot shows a distinctive pattern, then the error terms are almost certainly correlated. Third, consider the

FIGURE 15.10 | Residual plots indicating unequal variances

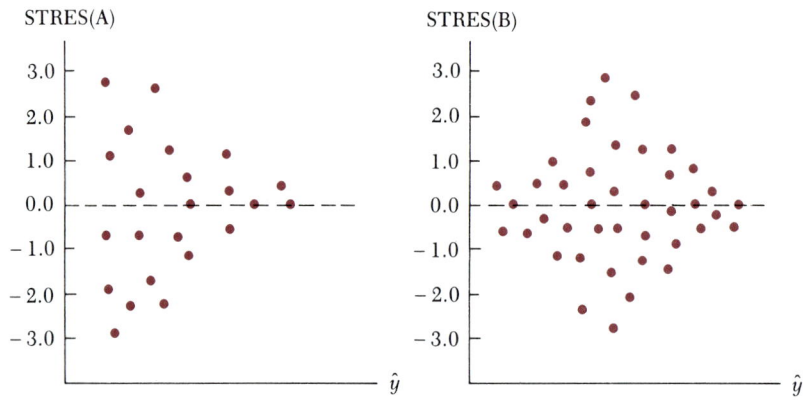

FIGURE 15.11 | Residual plots indicating correlated residuals

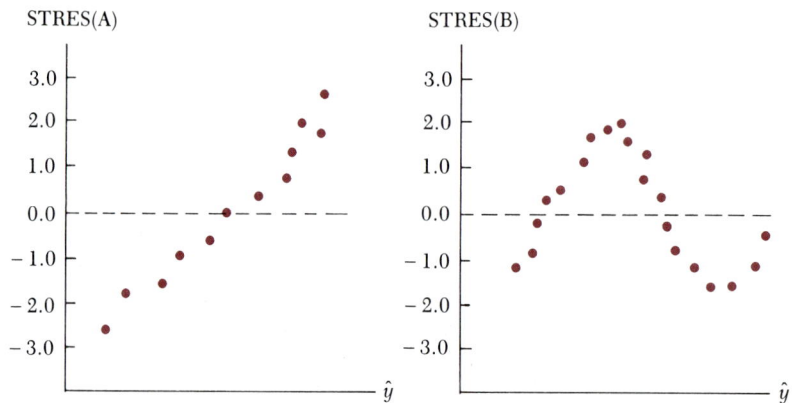

normality assumption. If the error terms are normally distributed, then the standardized residuals should conform approximately to the 68 percent, 95 percent, and 99 percent rules: 68 percent of the residuals should be between ± 1, 95 percent should be between ± 2, and 99 percent should be between ± 3. Figure 15.12 illustrates two residual plots where the residuals are not conforming to these rules, thus suggesting that the error terms are not normally distributed. In Figure 15.12A, there are clearly too many residuals between ± 1. In Figure 15.12B, the distribution of the residuals is rather seriously skewed toward the positive residuals—there are a very few large positive

FIGURE 15.12 | Residual plots indicating nonnormal residuals

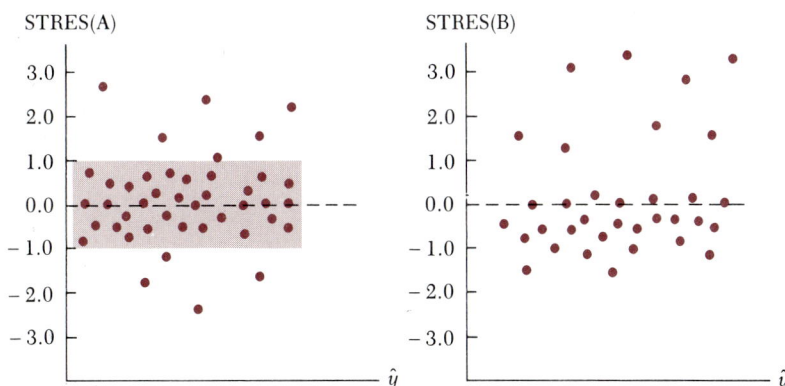

residuals and many not-so-large negative residuals. Since the normal distribution is symmetric, a skewed distribution of the residuals suggests that the error terms are not normally distributed.

Figure 15.13 indicates the presence of an outlier—a residual pair $(\hat{y}_i, e_i^*)$ that is far removed from the rest of the residuals. Usually, an outlier will cause one or more of the assumptions to be violated—equal variances or normality (due to skewing). Often, outliers are caused by miscoded data; an incorrect number was entered in coding the data for computer analysis, or the observation was initially incorrectly recorded.

FIGURE 15.13 | Residual plot with an outlier

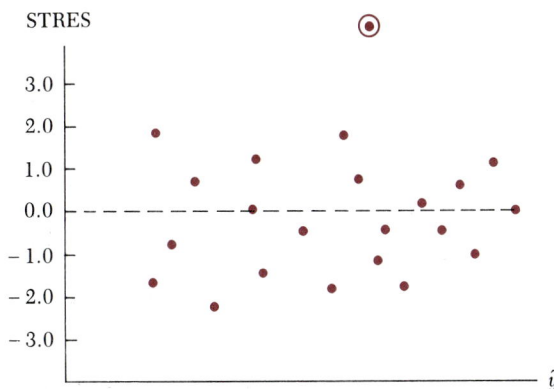

A second set of residual plots that are useful are the plots of the standardized residuals against the values of each of the independent variables (x_i's). In general these plots can be used in the same way that the residual plot of the standardized residuals against the predicted values ($\hat{y}$'s) is used. The plot should conform to a rectangular array of residuals—any deviations from this indicates one or more of the assumptions has been violated. In addition, the plot of the standardized residual against the value of an x-variable can provide useful information about the specification of the model. For example, suppose we observed the plot illustrated in Figure 15.14. This plot indicates

FIGURE 15.14 | A residual plot of the standardized residuals against values of an independent variable indicating a misspecified model

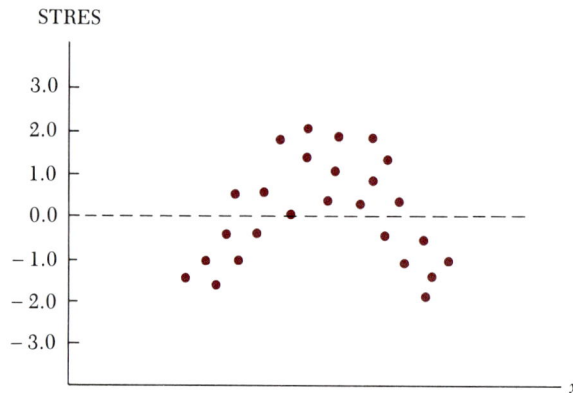

that the error terms are not uncorrelated. Almost certainly, the cause of this problem is a model that has not been correctly specified—the *square* of the independent variable x needs to be added to the model (e.g., $Y = \beta_0 + \beta_1 x + \beta_2 x^2 + \varepsilon$). A parabolic residual plot using an independent x-variable as the horizontal axis usually indicates that the independent variable should enter the model as a quadratic term. We will return to this point in Section 15.10—the case study.

Interpreting residual plots is more of an art than a science. Experience in viewing many plots is necessary to be able to "read" plots accurately. It is fairly easy to identify plots that indicate serious violations of one or more assumptions. But often, when one or more assumptions have been violated, the plot doesn't clearly indicate that this is the case. For this reason, there are statistical tests that may be used to check the regression model assumptions.

15.8.2 | Statistical tests

The most serious of the three assumptions to be violated is the uncorrelated residuals assumption. This assumption is the basis for all t-tests, F-tests, and confidence interval estimates used in regression analysis. Thus it is very important that this assumption be checked for a specific fitted regression equation. Returning to the sales data in Table 15.3, the MINITAB output for fitting the regression equation $Y = \beta_0 + \beta_1 x_1 + \beta_2 x_2 + \varepsilon$, where $Y = $ Sales, $x_1 = $ Advertising budget, and $x_2 = $ Income, is given below.

```
REGRESS Y IN C5 ON 2 PREDICTORS IN C3,C2, C6,C7,C8

THE REGRESSION EQUATION IS
Y = -  11.9 +  17.2 X1 +  2.35 X2
                                          ST, DEV,      T-RATIO =
              COLUMN        COEFFICIENT    OF COEF,      COEF/S.D.
              --            -11.889          4.318         -2.75
  X1   ADD                  17.233           2.894          5.95
  X2   INC                  2.3500           0.6221         3.78

THE ST, DEV, OF Y ABOUT REGRESSION LINE IS
S = 4.109
WITH (   15- 3) =   12 DEGREES OF FREEDOM

R-SQUARED = 97.3 PERCENT
R-SQUARED = 96.8 PERCENT, ADJUSTED FOR D.F.

ANALYSIS OF VARIANCE
  DUE TO        DF           SS        MS=SS/DF
REGRESSION      2         7290.19       3645.10
RESIDUAL       12          202.63         16.89
TOTAL          14         7492.82
```

	X1	Y	PRED. Y	ST.DEV.		
ROW	ADD	SALES	VALUE	PRED. Y	RESIDUAL	ST.RES.
2	1.11	45.82	38.06	2.01	7.76	2.17R
5	3.08	85.09	79.58	2.74	5.51	1.80 X

```
R DENOTES AN OBS. WITH A LARGE ST. RES.
X DENOTES AN OBS. WHOSE X VALUE GIVES IT LARGE INFLUENCE.

DURBIN-WATSON STATISTIC = 2.08
--
```

Notice that the MINITAB results conform well to those produced by SAS in Table 15.10—the differences are attributable to computer round-off errors. In the top shaded box, MINITAB prints out unusual residual values. The symbol R is used to denote a large standardized residual value, a potential sign of an outlier. The symbol X is used to denote an observation whose x-values (x_1, x_2) have a large influence on the regression equation. That is, if the data for sales district 5 were excluded from the sample of $n = 15$ districts, then the new regression equation would have potentially much different regression coefficient estimates. In the bottom shaded box, the Durbin-Watson statistic is given. The Durbin-Watson test is the most frequently used test for uncorrelated error terms. The test procedure is presented below.

| Durbin-Watson test |

The Durbin-Watson test for first-order autocorrelation of the error terms

Hypotheses: H_0: $\rho = 0$
$\qquad\qquad\;\; H_A$: $\rho > 0$
where ρ is the first-order autocorrelation of the error terms.
Test statistic value:

$$DW = \frac{\sum_{i=2}^{n} (e_t - e_{t-1})^2}{\sum_{i=1}^{n} e_i^2}$$

where e_i is the ith residual, and n is the sample size.

Decision rule:
$\quad$ If $DW > d_u$, do not reject H_0: $\rho = 0$.
$\quad$ If $DW < d_L$, reject H_0: $\rho = 0$.
$\quad$ If $d_L \leq DW \leq d_u$, the test is inconclusive.
The critical values d_L and d_u are given in Table B.20, Appendix B for $\alpha = 0.05$ and $\alpha = 0.01$.

The Durbin-Watson test is a test for the first order autocorrelation, ρ, of the error terms. The first-order autocorrelation ρ measures the correlation between consecutive pairs of error terms: (ε_1, ε_2), (ε_2, ε_3), (ε_3, ε_4), ... Although the error terms may be correlated in more complex ways (e.g., (ε_1, ε_3), (ε_2, ε_4), (ε_3, ε_5), ...), usually when they are correlated, it is through the first-order correlation of consecutive pairs. The **DW** test statistic has a minimum value of 0 and a maximum value of 4. If the null hypothesis is true ($\rho = 0$), then the expected value of the test statistic is 2. Hence a value of

DW close to 2 indicates that $\rho = 0$, and a small value toward 0 indicates that the error terms have first-order *positive* autocorrelation ($\rho > 0$). The decision rule is based on critical values found in Table B.20, Appendix B. Note that if the test statistic value DW falls between d_L and d_u, the test is inconclusive. In this case other methods must be used to assess the uncorrelated error terms assumption.

The test can also be used to test for *negative* first-order autocorrelation. In this case the null and alternate hypotheses are: H_0: $\rho = 0$ and H_A: $\rho < 0$. The test statistic value is DW* $= 4 -$ DW, and the decision rule is: Do not reject H_0: $\rho = 0$ if DW* $> d_u$, reject H_0: $\rho = 0$ if DW* $< d_L$, and the test is inconclusive if $d_L \leq$ DW* $\leq d_u$.

From the MINITAB output above, the value of the test statistic is DW $=$ 2.08 for the sales example data set in Table 15.3. From Table B.20, Appendix B, $d_L = 0.95$, and $d_u = 1.54$ for $n = 15$ and $\alpha = 0.05$ with two predictor variables ($k = 2$). Since DW $= 2.08 > 1.54$, do not reject H_0: $\rho = 0$. Thus we can conclude that there is insufficient evidence to suggest that the error terms exhibit first-order positive autocorrelation.

Normal plot

A common test for normality is based on *normal plots*. For the set of standardized residuals, an *expected set* of residuals is calculated from the standard normal z-distribution. When the standardized residuals are plotted against the expected standardized residuals, the pairs of values should conform to a line if the error terms are normally distributed. Using normal plots is very easy with the MINITAB computer software package. The NSCORES command is used to calculate the expected standardized residuals. Thus, if the standardized residuals were contained in column C6, then the command

```
NSCORES C6, C15
```

would calculate the expected residuals and place them in column C15 (see pp. 178–79, *MINITAB Handbook* for a description of the NSCORES command). The plot of the standardized residuals against the NSCORES for the sales data is shown in Figure 15.15. In Figure 15.15, the pairs of points conform fairly well to a line, indicating that the error terms do not appear to violate the normality assumption.

Since the correlation coefficient r presented in Chapter 13 measures the degree of linear relationship between the variables, we can calculate r for these 15 pairs of points and see how close it is to $+1.00$, indicating a perfect linear relationship with a positive slope. Using the MINITAB CORRELATE command with the standardized residuals stored in column C6 and the NSCORES stored in column C15, we have:

FIGURE 15.15 | The probability plot of the standardized residuals for the regression equation fitted to the sales data in Table 15.3

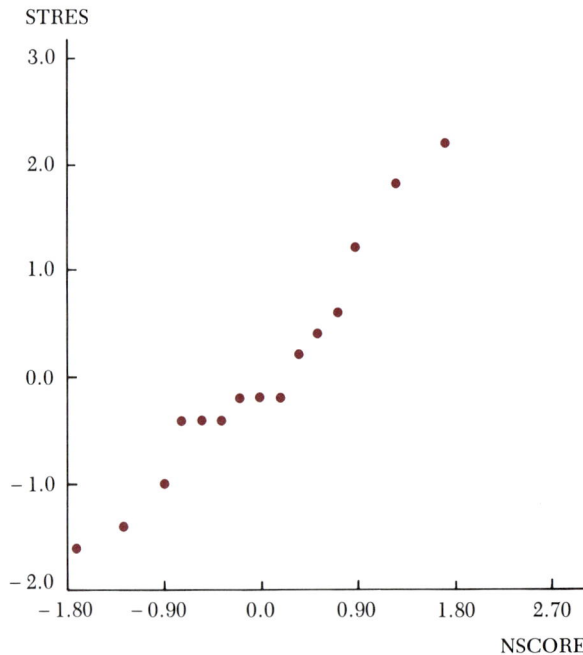

```
CORRELATE C6,C15
    CORRELATION OF    STRES AND NSCORE   = 0.979
```

Thus the correlation between the standardized residuals and the expected standardized residuals (NSCORES) is 0.979—very close to 1. Using the correlation coefficient, we can form a statistical test to determine whether or not r is "close enough" to 1 to satisfy the normality assumption. This test is given on the top of page 799. From Table B.21 with $n = 15$ and $\alpha = 0.05$, $r_c = 0.9383$. Since $r = 0.979 > r_c = 0.9383$, do not reject H_0: Error terms are normally distributed. Thus we can conclude that there is insufficient evidence to suggest that the error terms are nonnormal—the normality assumption appears to be satisfied.

Goldfeld-Quandt test

A test used for nonconstant variance is the Goldfeld-Quandt test. The test was originally designed for simple linear regression and assumed that if the variance of the error terms is not constant, then the variance is proportional

The normality plot test for the normality
assumption concerning the residuals

Hypotheses: H_0:   Error terms are normally distributed.
              H_A: Error terms are not normally distributed.

Test statistic value: r = Correlation between the standardized residuals and
                        the expected standardized residuals (NSCORES).

        Decision rule: Reject H_0 if $r < r_c$.
                         Do not reject H_0 if $r \geq r_c$.

where r_c is given in Table B.21, Appendix B with $\alpha = 0.01$, 0.05, and 0.10.

to the values of x, the independent variable. Due to the complexity of this test, particularly the extension to the multiple regression model, we will not present it here. The test is given in several of the references at the end of this chapter. Due to the absence of simple and good tests for the equal variance assumption, the residual plots are most often used to make assessments concerning this assumption.

15.8.3 | Corrections for violations of assumptions

When one or more of the three assumptions concerning the error terms in the multiple regression model is violated, what course of action should be taken? If only the normality assumption is violated, then we usually choose to ignore this violation. The reason for this is many studies have shown that the least squares regression technique is *robust* to nonnormality. That is, the technique will give usable results even if this assumption is not satisfied. It takes rather severe nonnormality before the regression results are adversely affected. When the assumption of equal variances is violated, the usual procedure is to *transform* the dependent variable **Y** using a shrinkage-type of transformation such as ln **Y** (natural log of **Y**) of $\sqrt{Y}$. These transformations shrink the variation in **Y**, given the x-values. When the assumption of uncorrelated errors is violated, a method that corrects this problem *must* be used. If the residuals suggest first-order autocorrelation of the error terms, for example, then methods have been devised to *transform* the dependent variable **Y** to correct for the correlation among the error terms. Two such methods are Prais-Winsten and Cochrane-Orcutt. The discussion of these techniques is well beyond our introductory treatment of regression analysis. For an excellent discussion of transformations to correct for nonnormality and unequal variance, see the Neter, Wasserman, and Kutner reference at the end of this chapter. For an excellent discussion of the econometric transformations to correct for correlated errors, see the Johnston reference at the end of this chapter.

Robust to nonnormality

Transformations

In the multiple regression model,

$$\mathbf{Y} = \beta_0 + \beta_1 x_1 + \beta_2 x_2 + \cdots + \beta_k x_k + \boldsymbol{\varepsilon}$$

the interpretation of the regression coefficients β_1, β_2, . . . , β_k as slope coefficients is very useful and important in many applications of the model. For example, β_1 is the increase (if positive) or decrease (if negative) in $\mathbf{Y}$ for a one-unit increase in x_1, holding the rest of the x's $(x_2, x_3, \ldots, x_k)$ fixed. This interpretation of a regression coefficient is not applicable when the independent variables x_1, x_2, . . . x_k are highly correlated among themselves. When the independent variables are highly intercorrelated, we say that severe *multicollinearity* exists among the independent variables.

| Definition of multi-collinearity |

Definition 15.1
Multicollinearity

Multicollinearity exists in a multiple regression model when the independent variables x_1, x_2, . . . , x_k are correlated among themselves (intercorrelated).

The consequence of multicollinearity is that the regression coefficient estimators $\mathbf{b}_0$, $\mathbf{b}_1$, $\mathbf{b}_2$, . . . $\mathbf{b}_k$ of the population regression coefficient parameters β_0, β_1, β_2, . . . , β_k are inefficient; that is, the sampling distributions of $\mathbf{b}_0$, $\mathbf{b}_1$, . . . , $\mathbf{b}_k$ have inflated variances. Therefore, the estimate b_i of β_i may be far removed from the value of β_i due to the high variability in the sampling distribution of $\mathbf{b}_i$. In this instance, we often refer to the estimate b_i as a nonsense regression coefficient estimate. Its value may not make any sense in that it is not what we would expect based upon the theory or our knowledge of the problem to which multiple regression analysis has been applied.

The ideal in multiple regression analysis is to have the independent variables x_1, x_2, . . . x_k be uncorrelated so that they each explain a separate percentage of the variation in the dependent variable as indicated in Figure 15.16. The pie represents the total variation in $\mathbf{Y}$. The x's are separate and distinct pieces of the pie, each explaining separate amounts of variability in $\mathbf{Y}$. In the more typical case, the pieces of pie attributable to the x's overlap, as indicated in Figure 15.17.

| Collinear x's |

An extreme case of multicollinearity is when the x's are *collinear*. This means that one of the x's can be written as a linear combination of the remaining x's. Suppose, for example, that $k = 2$ so that we have x_1 and x_2 in the model. Further, suppose that $x_1 = 10 + 2x_2$; that is, x_1 can be written as a linear function of x_2. Common sense suggests that if this is the case, then we do not need both x_1 and x_2 in the model. In fact, if x_1 and x_2 are *both* included,

FIGURE   15.16 | Explaining variability in **Y** by independent variables x_1, $x_2, \ldots, x_k$ that are uncorrelated.

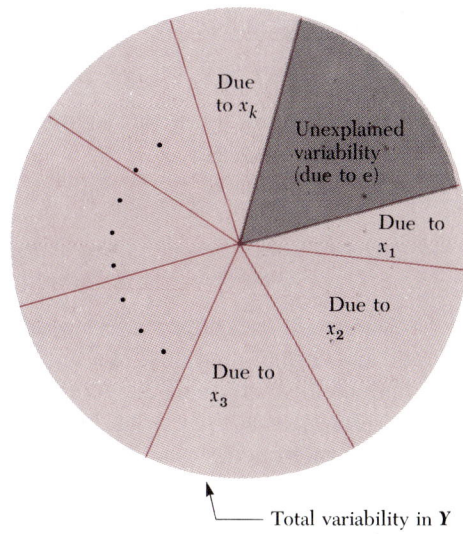

FIGURE   15.17 | Two correlated independent variables x_1 and x_2

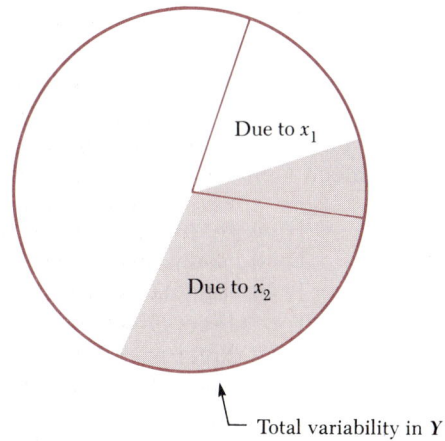

it is impossible to calculate the regression coefficient estimates by least squares. The variances of the sampling distributions of $\mathbf{b}_0$, $\mathbf{b}_1$, and $\mathbf{b}_2$ become infinite. Usually, the problem we are confronted with is not collinearity, but rather severe multicollinearity where the pairs of x's in the model have correlations that are very close to either -1 or $+1$ indicating a strong linear relationship.

There are two issues related to multicollinearity in the multiple regression model: (1) How is severe multicollinearity detected, and (2) how is the problem corrected?

15.9.1 | Detection of multicollinearity

There are two approaches to detecting multicollinearity: informal methods and formal methods. We will give examples of both approaches. For a complete discussion of the detection of multicollinearity, see one or more of the references at the end of the chapter, such as the Neter, Wasserman, and Kutner reference.

| Informal indications of severe multicollinearity

The most common informal methods for detecting severe multicollinearity include the following:

1. Estimated regression coefficients (b's) with algebraic signs opposite of that expected from theoretical considerations or prior experience with the problem.

2. Large changes in the estimated regression coefficients when a variable is added or deleted in the model.

3. Nonsignificant t-tests for individual regression coefficients for important independent variables.

4. Large correlations between pairs of independent variables in the model.

To illustrate these informal methods, consider the following example.

Example 15.8 A manufacturing plant manager is analyzing the production costs for one of the plant's products. Monthly data have been taken on the production costs for this product from January 1985 to June 1986— eighteen months of data. The variables recorded are:

$\mathbf{Y}$ = Total production cost per month in thousands of dollars.
x_1 = Total production of the product in thousands of units.
x_2 = Total variable overhead costs per month in thousands of dollars.
x_3 = Total direct labor hours used per month in hundreds of hours.
x_4 = Total direct machine hours used per month in hundreds of hours.

The data for the eighteen months are given in Table 15.18 from MINITAB output.

The MINITAB output for fitting the model $\mathbf{Y} = \beta_0 + \beta_1 x_1 + \beta_2 x_2 + \beta_3 x_3 + \beta_4 x_4 + \varepsilon$, is given on pages 803 and 804.

TABLE 15.18 Production cost data for the production of one specific product in the manufacturing plant

	Month/ year	Total cost	Total production (units)	Total overhead costs	Total direct labor costs	Total direct machine hours
COLUMN	C1	C2	C3	C4	C5	C6
COUNT	18	18	18	18	18	18
ROW						
1	185.00	108.250	53.6000	11.4000	33.2000	22.1000
2	285.00	101.650	49.9800	10.0000	30.2000	20.0000
3	385.00	124.220	62.1200	12.7000	37.3000	25.0000
4	485.00	140.050	70.3300	14.2000	42.0000	27.9000
5	585.00	170.780	83.7600	19.0000	68.4000	36.2000
6	685.00	187.940	92.0500	20.9000	75.0000	40.0000
7	785.00	192.550	93.9900	21.7000	81.6000	41.4000
8	885.00	132.180	64.8800	12.6000	38.0000	26.0000
9	985.00	130.090	65.4200	11.8000	37.9000	25.6000
10	1085.00	139.220	70.7800	14.3000	42.9000	28.2000
11	1185.00	144.860	72.9700	15.6000	43.2000	28.4000
12	1285.00	183.120	89.3300	20.9000	87.9000	39.0000
13	186.00	190.410	93.2500	21.7000	66.1000	40.5000
14	286.00	200.030	96.8800	23.5000	69.9000	43.4000
15	386.00	177.500	84.4200	20.0000	68.0000	41.6000
16	486.00	192.960	92.4300	22.9000	67.9000	41.7000
17	586.00	188.790	90.1100	21.6000	65.7000	40.3000
18	686.00	156.080	74.0400	18.1000	54.4000	33.5000

-- DESCRIBE C2-C6

C2	N = 18	MEAN =	158.93	ST.DEV. =	31.9
C3	N = 18	MEAN =	77.797	ST.DEV. =	14.8
C4	N = 18	MEAN =	17.383	ST.DEV. =	4.52
C5	N = 18	MEAN =	56.089	ST.DEV. =	18.1
C6	N = 18	MEAN =	33.378	ST.DEV. =	7.88

```
THE REGRESSION EQUATION IS
Y =    4.62 +  1.31 X1 +   1.07 X2
      -0.0240 X3 +  1.05 X4

                                    ST. DEV.    T-RATIO =
          COLUMN    COEFFICIENT    OF COEF.    COEF/S.D.
          --          4.622          2.933        1.58
   X1     UNITS      1.3100         0.1178       11.12
   X2     OVHD       1.0693         0.4968        2.15
   X3     LABOR     -0.02396        0.05131      -0.47
   X4     MACH       1.0530         0.2969        3.55
```

```
THE ST. DEV. OF Y ABOUT REGRESSION LINE IS
S = 1.377
WITH ( 18- 5) =  13 DEGREES OF FREEDOM

R-SQUARED = 99.9 PERCENT
R-SQUARED = 99.8 PERCENT, ADJUSTED FOR D.F.

ANALYSIS OF VARIANCE

 DUE TO        DF           SS      MS=SS/DF
REGRESSION     4       17300.01     4325.00
RESIDUAL      13          24.64        1.90
TOTAL         17       17324.64

FURTHER ANALYSIS OF VARIANCE
SS EXPLAINED BY EACH VARIABLE WHEN ENTERED IN THE ORDER GIVEN

 DUE TO        DF           SS
REGRESSION     4       17300.01
UNITS          1       17193.87
OVHD           1          81.93
LABOR          1           0.39
MACH           1          23.84

            X1         Y      PRED. Y    ST.DEV.
ROW       UNITS      COST      VALUE     PRED. Y     RESIDUAL    ST.RES.
  8        64.9    132.180   129.556      0.576        2.624      2.10R
 12        89.3    183.120   182.954      1.085        0.166      0.20 X
 15        84.4    177.500   178.774      1.228       -1.274     -2.04RX

R DENOTES AN OBS. WITH A LARGE ST. RES.
X DENOTES AN OBS. WHOSE X VALUE GIVES IT LARGE INFLUENCE.

DURBIN-WATSON STATISTIC = 1.44
--
```

The correlation matrix from MINITAB is given below.

	COST	UNITS	OVHD	LABOR
UNITS	0.996			
OVHD	0.988	0.978		
LABOR	0.928	0.923	0.923	
MACH	0.990	0.978	0.987	0.931

Notice that each of the independent variables is highly correlated with the dependent cost variable. The correlations between x_1 and y, x_2 and y, x_3 and y, and x_4 and y are 0.996, 0.988, 0.928, and 0.990, respectively. Clearly, every one of the four independent variables *individually* is important in explaining variability in **Y**. Now note the very high correlations among the independent variables. The correlations between x_1 and x_2, x_1 and x_3, x_1 and x_4 are 0.978, 0.923, and 0.978. The correlations between x_2 and x_3, x_2 and x_4, and x_3 and x_4 are 0.923, 0.987, and 0.931, respectively. These very high correlations among the x's are the first informal indication of severe multicollinearity in the model containing all four independent variables.

Another informal indication of severe multicollinearity is apparent by studying the algebraic signs of the regression coefficients. Each of the four independent variables should be contributing a positive amount to the total production cost **Y**. Thus we should expect all four coefficients to have positive signs. However, the estimated regression coefficient b_3 has a value of -0.024. It would tell us that as the number of labor units increased by 1 (100 hours), the total cost **Y** would *decrease* by -0.024 (thousands of dollars) holding x_1, x_2, and x_4 fixed. This of course does not make sense.

A further informal indication of severe multicollinearity is the **t**-statistic value for testing the hypothesis, $H_0: \beta_3 = 0$. From the MINITAB output, this value is -0.47, indicating that the null hypothesis cannot be rejected. However, we know that x_3 by itself is an important predictor variable of the total cost **Y**.

Finally, in Table 15.19 the results of fitting each independent variable separately using a simple linear regression model are given.

TABLE 15.19 | Separate simple linear regression models for the four independent variables in Example 15.8.

Variable	Definition	Slope coefficient	t-statistic value	r^2
x_1	Number of units produced	2.15	45.86	99.2%
x_2	Overhead costs	6.99	26.04	97.7
x_3	Labor hours used	1.64	9.99	86.2
x_4	Machine hours used	4.01	27.98	98.0

Notice from Table 15.19 that if x_1, x_2, and x_4 are dropped from the model resulting in the model $\mathbf{Y} = \beta_0 + \beta_1 x_3 + \boldsymbol{\varepsilon}$, the estimated regression coefficient for x_3 is 1.64. This is a considerable change from the estimate (-0.024) when x_1, x_2, and x_4 are included with x_3. This is the last informal indication of severe multicollinearity—as variables are added or deleted from the regression model, the regression coefficient estimates change in value considerably.

A formal method of determining whether or not multicollinearity is severe is based on *variance inflation factors* (VIFs).

Variance inflation factor

Definition 15.2
Variance inflation factor (VIF)

The variance inflation factor for an independent variable (x_i) is given by

$$VIF_i = \frac{1}{(1 - r_i^2)}$$

where r_i^2 is the coefficient of multiple determination (stated as a proportion) from regressing the remaining $k - 1$ independent variables on x_i using the model,

$$\mathbf{X}_i = \beta_0 + \beta_1 x_1 + \cdots + \beta_{i-1} x_{i-1} + \beta_{i+1} x_{i+1} + \cdots + \beta_k x_k + \epsilon$$

Multicollinearity is considered to be severe when the maximum VIF_i, $i = 1, 2, \ldots, k$ is greater than 10 or the average of the VIF_i, $i = 1, 2, \ldots, k$ is considerably larger than 1.

Table 15.20 gives the computations for the VIFs in the cost data example. Each set of three independent variables is regressed on the remaining independent variable (x_i). Thus, for the fitted regression equation $\hat{x}_1 = b_0 + b_1 x_2 + b_2 x_3 + b_3 x_4$, $r^2 = 0.963$ in Table 15.20.

TABLE 15.20 | The computation of the variance inflation factors for the cost data in Example 15.8

Variable	r^2	$(1 - r^2)$	$VIF = \dfrac{1}{(1 - r^2)}$
x_1	0.963	0.037	27.03
x_2	0.978	0.022	45.45
x_3	0.871	0.129	7.75
x_4	0.980	0.020	50.00

The maximum VIF is 50.0, and the average of the four VIFs is 32.56. Since the maximum VIF (50.0) is greater than 10, and the average VIF (32.56) is considerably larger than 1, multicollinearity is severe.

The Example 15.8 problem is an example of extreme multicollinearity if all four independent variables are included in the model. Clearly, if the model is used with all four independent variables, poor estimates of the regression coefficients result. The extreme case is the coefficient on x_3, which has the wrong sign (-0.024). If it is desirable to interpret the regression coefficients in the usual way, then this model should not be used.

15.9.2 | Correction for severe multicollinearity

If the model is to be used *only* for the prediction of **Y**-values, the predictions are made only over the region of the values of the independent variables, and the estimated regression coefficients will not be used for interpretation purposes concerning the relationships of the independent variables with the dependent variable **Y**, then multicollinearity, even when severe, does not present a problem. For instance, in Example 15.8, if we wished to use the model to predict total production costs for sets of x_1, x_2, x_3, and x_4 values that are in the sampled region of the x's, then the severe multicollinearity in the model is acceptable. Determining the sampled region of the x's can be difficult, however. If there is one independent variable, then the "region" is an interval on the real line between the minimum value of x and the maximum value of x in the sample. However, with four independent variables, the sampled region is in the four-dimensional space of the x's, and its boundaries are not obvious. Thus caution must be exercised so that the prediction does not represent an extrapolation beyond the sampled region of the x's when severe multicollinearity exists.

If we are interested in *interpretations* made from the estimated regression coefficients (b's), then severe multicollinearity cannot be tolerated. There are several ways to correct this problem. We will mention the two most commonly used methods.

Dropping indepen-
dent variables

First, an obvious solution is to not include all the highly intercorrelated independent variables in the model. In Example 15.8 from Table 15.19, it is clear that x_1 alone produces a model that will predict **Y**-values well, since $r^2 = 0.992$, or 99.2 percent. That is, 99.2 percent of the variability in **Y** is explained by x_1 alone. Adding x_2, x_3, and x_4 to the model increases r^2 to 99.9 percent, an increase of only 0.7 percent. This increase is hardly worth the introduction of severe multicollinearity and the resulting nonsense regression coefficients.

Ridge regression

Second, a method referred to as *ridge regression* is commonly used to correct for multicollinearity. Ridge regression produces *biased* estimators of the regression coefficients, but more efficient estimators than the least squares estimators when severe multicollinearity exists. The discussion of the ridge regression estimators is well beyond the scope of our introductory treatment of regression analysis. For a discussion of ridge regression, see the Neter, Wasserman, and Kutner reference at the end of this chapter.

Since the early 1970s, the price of gasoline has increased dramatically with wide fluctuations driven by the availability of oil. As a consequence, the U.S. government has stipulated minimum average miles per gallon gasoline usage figures for automobiles produced by U.S. manufacturers. In recent years, this has resulted in intensive studies of the factors that affect gasoline mileage in automobiles. In Appendix C, Table C2, are given data on 115 automobiles produced in the United States. The dependent variable is the Environmental Protection Agency (EPA) combined miles per gallon (MPG) estimate, which is based on a combination of city and highway driving. The independent variables are:

1. x_1 = Number of cylinders (4, 6, 8).

2. x_2 = Size of gasoline tank (in gallons).

3. x_3 = Engine size (in liters).

4. x_4 = Horsepower rating.

5. x_5 = Weight (curb weight in pounds).

Also, D is used to designate automobiles among the 115 cars that have diesel engines. This leads to a sixth independent dummy variable:

6. $x_6 = \begin{cases} 1 & \text{if engine is diesel} \\ 0 & \text{if engine is gasoline} \end{cases}$

Note that the values of the dependent variable **Y** are listed in column 2 in Table C2 in Appendix C. The independent variable values are given in columns 1 (x_1), 3 (x_2), 4 (x_3), 5 (x_4), 6 (x_5), and 7 (x_6). Since x_1 is an ordinal variable, it could also be treated as a dummy variable by introducing two new variables: $x_7 = 1$ if four-cylinder engine, 0 otherwise; and $x_8 = 1$ if six-cylinder engine, 0 otherwise. As it turns out, the number of cylinders is not as important as other variables and is not used in the final model. Therefore we will not bother constructing these dummy variables.

The objective is to develop a model that can be used to predict the miles per gallon for a specific automobile. The seven steps listed in Section 15.1 will be used in sequence to demonstrate how a regression equation is developed.

Step 1 | **Identify the variables to be included in the multiple regression model**

In this example, the independent x variables have been selected from a survey of automobile magazines (*Motor Trend, Road and Track,* etc.). Clearly, such variables as the weight of the automobile, the size of the engine, and the horsepower are important variables in predicting the miles per gallon that an automobile achieves. Other variables that may be useful are the type of transmission (automatic or stick shift) and time in seconds for the quarter mile from a standing stop. These variables were not available to the survey, however. Had we designed an *experiment* to collect the data, then we could have made a point to acquire data on these variables as well as any other variables that may contribute to MPG.

Step 2 | Collect the sample data

The data on the 115 automobiles were collected by a survey of available information in print on 1983 automobiles. Although the sample does not contain all U.S. production automobiles (and does include a few Japanese automobiles imported by Chrysler!) produced in 1983 in the United States, it is a very representative sample nevertheless. A very desirable way of collecting the data would have been through a designed experiment, where a sample of automobiles is taken and all relevant data measured for each automobile. As is often the case with business problems, it is often necessary due to time and cost limitations to use available survey information. Such is the case in this instance.

Step 3 | Specify the relationship that exists between the dependent and the independent variables, and between the independent variables

This is the primary modeling step in the process. We begin by plotting (using MINITAB) each of the independent variables against the dependent variable Y. These plots are shown in Figure 15.18. In these plots, note that MINITAB prints a number at a point (x, y) when two or more points fall at the same point. A striking feature of the first five plots is the relationship between y and x: The functional form is a "slide," with smaller values of Y being associated with larger values of each independent variable x. This form is the reciprocal relationship $y = 1/x$, whose graph is illustrated in Figure 15.19. In these scatterplots, it appears that horsepower (HP − x_4) and weight (WT − x_5) conform best to the reciprocal relationship $y = 1/x$. This makes sense. As weight and horsepower increase, we know that the miles per gallon falls at a rapid and increasing rate. The relationship between MPG and engine size (ENGS − x_3) appears to follow the same pattern, although a linear relationship would probably do just as well. The correlation matrix given in Table 15.21 gives further information about the relative importance of each independent variable. The correlation between MPG (y) and HP(x_4) is −0.794, for example. The weight variable—WT(x_5)—is next in importance; the correlation between MPG and WT is −0.737. The correlations between MPG(y) and NOCYL(x_1), TANK(x_2), and ENGS(x_3) are −0.674, −0.658, and −0.658. Recall that correlation measures the *linear* relationship between y and x. Since all of these correlations are negative, the best fitting line in each case is one with a negative slope, a "decreasing" line. Further, if these correlations are squared, then we have the coefficient of multiple determination, r^2. Thus if we wished to use the model, $Y = \beta_0 + \beta_1 x_4 + \epsilon$,

FIGURE 15.18 | The scatterplots of the independent variable (**Y**) with each independent variable (*x*)

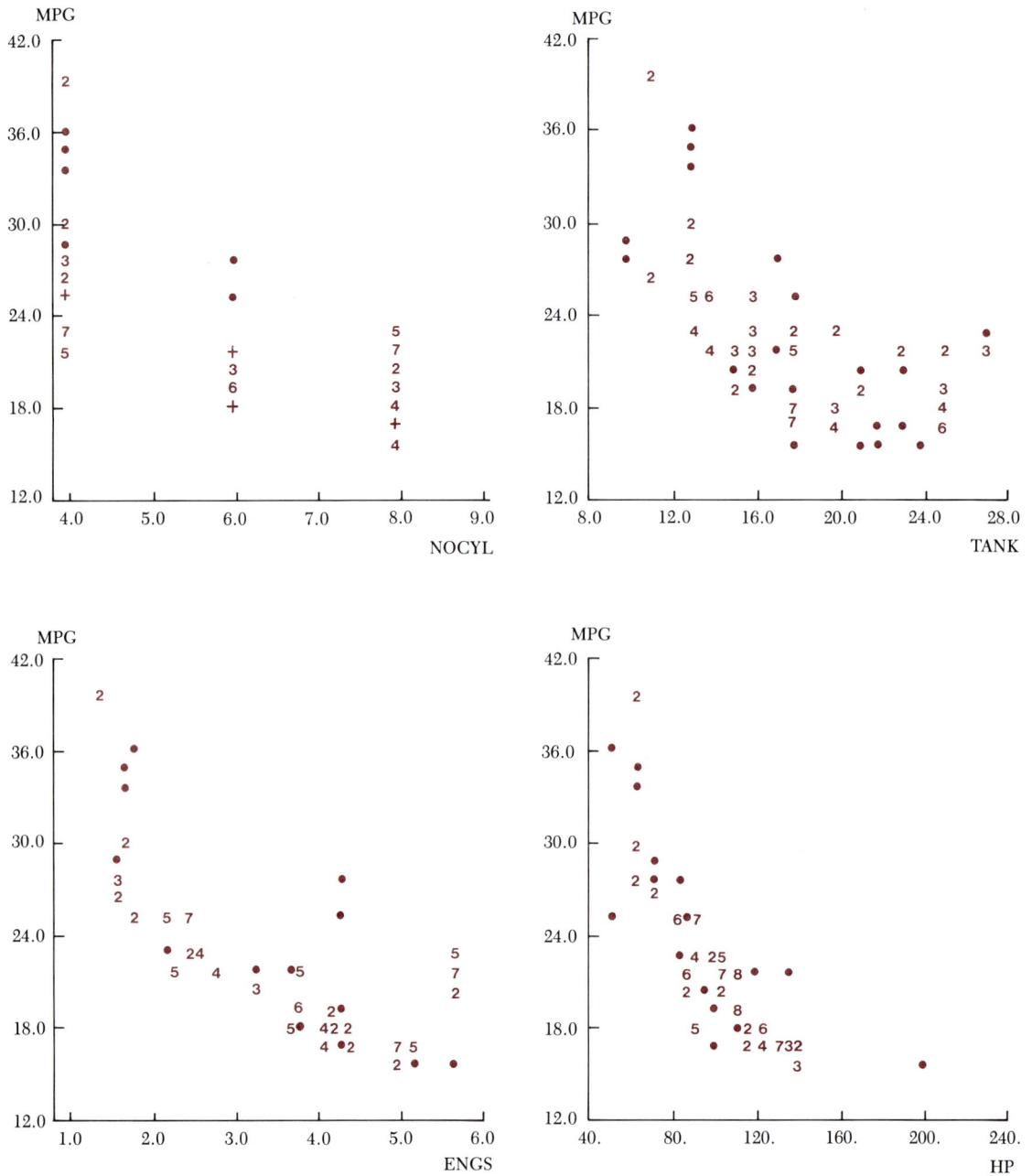

FIGURE   15.18 | *(concluded)*

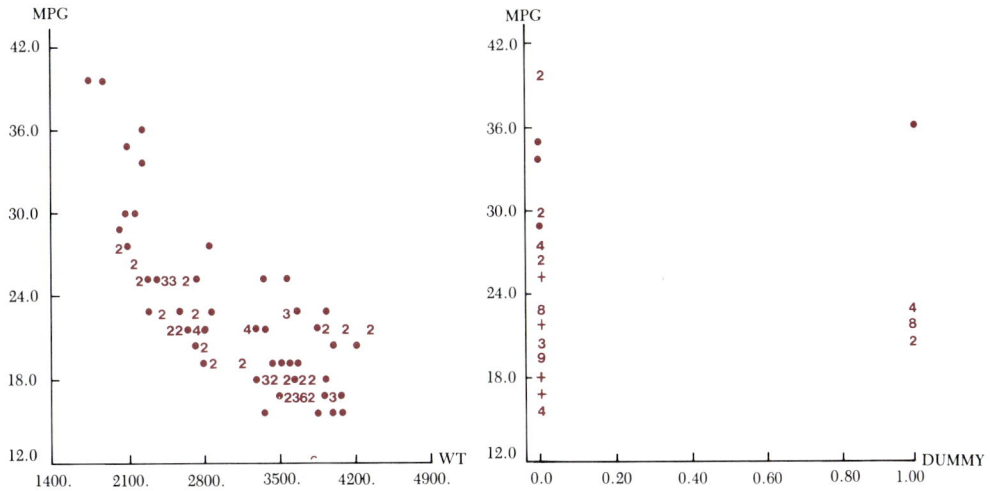

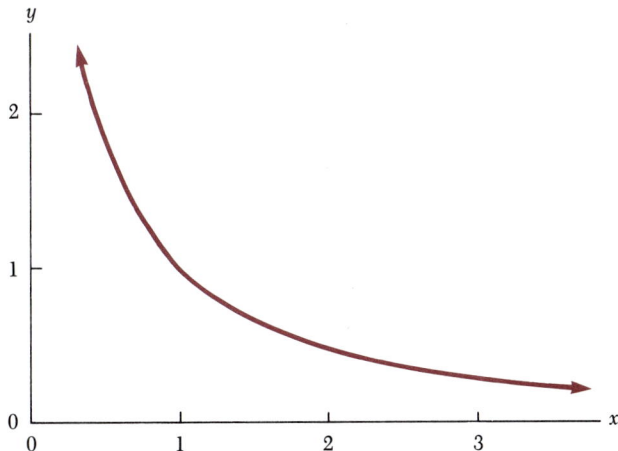

FIGURE   15.19 | The graph of the function *y* = 1/*x*

where HP is used alone and as a *linear* function, then $r^2 = (-0.794)^2 = 0.63$. Thus by regressing horsepower (x_4) on miles per gallon (y), we would explain 63 percent of the variation in the dependent variable $\mathbf{Y}$. However, from the scatterplots, we know that using the reciprocal function would produce a

TABLE 15.21 | The correlation matrix for the variables in Table C.2, Appendix C

```
-- CORRELATE C1-C7

            NOCYL      MPG     TANK     ENGS       HP       WT
MPG        -0.674
TANK        0.722   -0.614
ENGS        0.930   -0.658    0.786
HP          0.745   -0.794    0.626    0.732
WT          0.846   -0.737    0.879    0.897    0.716
DUMMY       0.345    0.095    0.390    0.483   -0.017    0.380
```

better predictor for the horsepower variable. This model is represented by $Y = \beta_0 + \beta_1(1/x_4) + \epsilon$, where $x_4 = $ Horsepower (HP). This model is easily fitted using MINITAB by first constructing the column of values $1/x_4$. Since the values of x_4 are stored in column C5, the MINITAB DIVIDE command can be used to form the $1/x_4$ values:

```
DIVIDE 1 BY C5, PUT ANSWER IN C15
```

The column C15 now contains the reciprocals of the values in column C5. The regression model $Y = \beta_0 + \beta_1(1/x_4) + \epsilon$ may now be fitted by using the MINITAB REGRESSION command

```
REGRESS Y IN C2 ON 1 PREDICTOR IN C15
```

The resulting fitted model is

$$\hat{y} = 5.974 + 1523.83x_4^*$$

where $x_4^* = 1/x_4$. The $\mathbf{r}^2$-value for this model is 0.707. Thus 70.7 percent of the variation in Y has been explained by using the reciprocal of horsepower (x_4^*). Clearly, using the reciprocal model is better than using the linear model based on horsepower (x_4).

Curvilinear regression

Using the reciprocal model is an illustration of *curvilinear regression*. We are creating a new independent variable, $1/x_4$, because the scatterplot suggested the relationship between y and x_4 follows the reciprocal function. The use of polynomial functions is also common in multiple regression analysis. For example, consider the scatterplot illustrated in Figure 15.20. We would

FIGURE 15.20 | Scatterplot of (x, y) values indicating a quadratic relationship

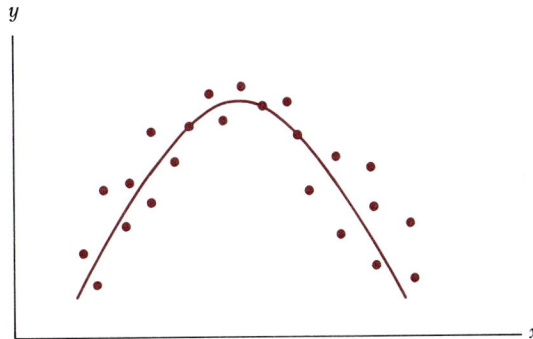

then consider the model $\mathbf{Y} = \beta_0 + \beta_1 x + \beta_2 x^2 + \boldsymbol{\epsilon}$. This model is a second-order (quadratic) polynomial in x. MINITAB can easily be used to fit such a model. Suppose that y is stored in column C1, and x is stored in column C2. Then the following commands will fit the quadratic model:

```
MULTIPLY C2 BY C2, PUT ANSWERS IN C3
REGRESS Y IN C1 ON 2 PREDICTORS IN C2 AND C3
```

There are now two predictors: x in C2 and x^2 in C3. Figure 15.21 illustrates some common functions that are useful in constructing multiple regression models based on the scatterplots of $\mathbf{Y}$ with each independent variable x.

At this point, let us review what we have concluded about the independent variables. It appears that horsepower (HP — x_4) and weight (WT — x_5) are the two most important variables. Although both of these variables are linearly related to $\mathbf{Y}$, it is clear from the scatterplots that a reciprocal relationship would produce better fitting models. For example, if HP is fitted

FIGURE 15.21 | Common functions of independent variables

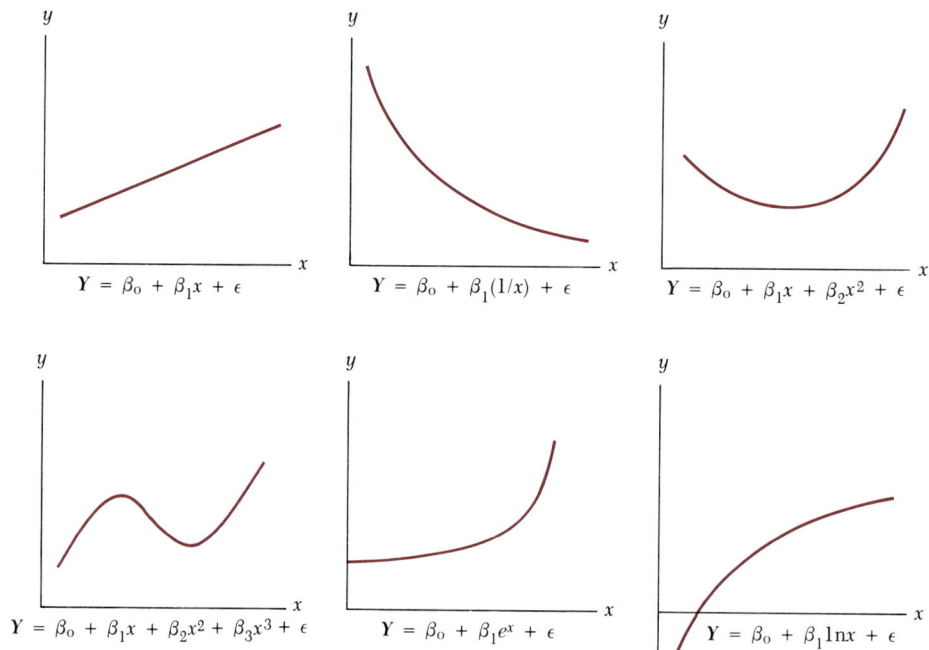

$Y = \beta_0 + \beta_1 x + \epsilon$

$Y = \beta_0 + \beta_1(1/x) + \epsilon$

$Y = \beta_0 + \beta_1 x + \beta_2 x^2 + \epsilon$

$Y = \beta_0 + \beta_1 x + \beta_2 x^2 + \beta_3 x^3 + \epsilon$

$Y = \beta_0 + \beta_1 e^x + \epsilon$

$Y = \beta_0 + \beta_1 \ln x + \epsilon$

linearly, then r^2 is 0.63, and if HP is fitted as a reciprocal relationship, then r^2 is 0.707.

Now the problem is to determine *how many* independent variables or functions of the independent variables to include in the model. From the correlation matrix in Table 15.21, it is clear that the multicollinearity will be a problem if too many predictor variables are used in the model. For example, the correlation between horsepower (HP) and weight (WT) is 0.716, indicating that these two variables are highly related linearly. This can also be determined from the scatterplot of x_4(HP) and x_5(WT). This scatterplot is shown in Figure 15.22.

The scatterplots of pairs of x variables can be used to suggest both the degree of multicollinearity introduced to the model by a specific pair of x variables and functional forms that we may wish to add to the model, such as x_1/x_2 or $x_1 x_2$. Once functional forms of the independent variables are considered, then the number of possible models becomes very large. Clearly, we cannot consider *all* possible models. The scatterplots must be used to suggest models worthy of investigation. In this case, there are six independent variables (x's). Plotting each possible pair of independent variables produces 15 scatterplots! Only one of these plots is shown here (Figure 15.22), but all should be constructed for modeling purposes.

FIGURE 15.22 | The scatterplot of x_4 (HP) with x_5 (WT)

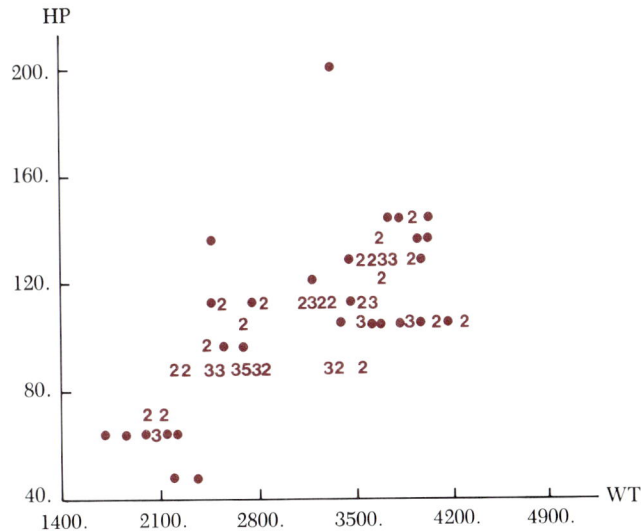

Table 15.22 contains several models that were fitted to the 115 sample observations. For each model, the coefficient of determination r^2 and the estimated standard deviation about the regression line s is listed.

TABLE 15.22 | Multiple regression models for the MPG case data

Number	Model	r^2	s
1	$Y = \beta_0 + \beta_1 NOCYL + \beta_2 TANK + \beta_3 ENGS + \beta_4 HP + \beta_5 WT + \beta_6 DUMMY + \epsilon$	0.759	2.380
2	$Y = \beta_0 + \beta_1(1/HP) + \beta_2(1/WT) + \epsilon$	0.761	2.330
3	$Y = \beta_0 + \beta_1(1/HP) + \epsilon$	0.707	2.568
4	$Y = \beta_0 + \beta_1 HP + \beta_2(1/HP) + \epsilon$	0.707	2.579
5	$Y = \beta_0 + \beta_1 ENGS + \beta_2 HP + \beta_3 WT + \epsilon$	0.703	2.606
6	$Y = \beta_0 + \beta_1 HP + \beta_2 WT + \beta_3(1/HP) + \beta_4(1/WT) + \beta_5(HP*WT) + \epsilon$	0.817	2.065
7	$Y = \beta_0 + \beta_1 NOCYL + \beta_2(1/TANK) + \beta_3(1/ENGS) + \beta_4(1/HP) + \beta_5(1/WT) + \epsilon$	0.768	2.324

The first model in Table 15.22 uses all six independent variables *linearly*. Notice that the second model produces a slightly higher r^2 and a slightly lower s based on functions (reciprocals) of only two independent variables.

Since s is used in calculating confidence intervals, the smaller the value of s, the shorter the confidence interval width is, a desirable goal. Thus both r^2 and s are reported for each model. Most statisticians prefer using s to r^2 when selecting models due to the fact that a smaller s produces shorter confidence and prediction intervals. There are selection measures that are functions of both r^2 and s that are often used by statisticians. We will not consider these selection measures. Several references at the end of this chapter discuss selection measures in considerable detail.

In general r^2 and s do not necessarily rank models under consideration in the same way. That is, it is possible when comparing two models that the first model has a higher r^2 than the second model, but the second model has a lower s than the first model. As it turns out, this did not occur with the models in Table 15.22. The ranking of the models in Table 15.22 based on highest r^2 and lowest s is: models 6, 7, 2, 1, 3, 4, and 5.

Model 6 gives the best results based on having the highest r^2 and the lowest s. The model uses (1) the reciprocal of HP, (2) the reciprocal of WT, (3) the product of HP and WT, and the original independent variables (4) HP and (5) WT. The problem in this model is the very high degree of multicollinearity resulting from having these five predictor variables in the model. Using the variance inflation factors (VIFs), there are five VIFs (regressing each of the five predictor variables on the remaining four). The maximum VIF is 333.33, and the average of the five VIFs is 172.78. Since the maximum VIF exceeds 10, and the average VIF considerably exceeds 1, this model contains a severe degree of multicollinearity. An additional consideration is *parsimony*—attempting to keep the model as simple as possible using a few, easily understandable terms. For example, models 2, 3, and 4 are parsimonious when compared with models 1, 5, 6, and 7.

| Parsimony |

At this point we will choose model 2 as our best candidate among the models in Table 15.22. It has the following desirable properties:

1. Model 2 is parsimonious—it has only two terms in the model (1/HP) and (1/WT), and both terms are understandable. We would expect gas mileage to be inversely related to horsepower (HP) and automobile weight (WT).

2. It has the third highest r^2 and the third lowest s. Models 6 and 7 have higher r^2's and lower s's. However, models 6 and 7 are more complex models with several terms and severe multicollinearity problems (multicollinearity of contending models is usually assessed in Step 5).

In selecting model 2 over models 6 and 7, model 2 is therefore preferable due to its being parsimonious, having relatively high r^2 and low s, and having no serious multicollinearity problems (as we shall see in Step 5).

Selecting a model is a subjective matter. Experience in using regression analysis is essential to performing this step well. Often we must return to Step 2 due to the failure to validate in Step 5 the model selected. Thus it is useful to carry a few of the promising models through Step 5 to determine which models can be validated (shown to satisfy the assumptions of regression analysis). We will use model 2 in the remaining steps of the regression model-building process.

Step 4 | Estimate the parameters of the model specified

Our model is $Y = \beta_0 + \beta_1(1/HP) + \beta_2(1/WT) + \epsilon$. Using MINITAB the reciprocals $(1/HP)$ and $(1/WT)$ have been stored in columns C15 and C16, respectively. The MINITAB output for fitting this model is shown below.

```
?
REGRESS Y IN C2 ON 2 PREDICTORS IN C15,C16,RESIDS C20,YHAT C21


THE REGRESSION EQUATION IS
Y =    3.93 +   975. X1 +23026. X2
                                     ST. DEV.     T-RATIO =
          COLUMN      COEFFICIENT    OF COEF.     COEF/S.D.
          --             3.9271        0.9771        4.02
X1        1/HP           975.2         137.6         7.09
X2        1/WT          23026          4582          5.02

THE ST. DEV. OF Y ABOUT REGRESSION LINE IS
S = 2.330
WITH ( 115- 3) = 112 DEGREES OF FREEDOM

R-SQUARED = 76.1 PERCENT
R-SQUARED = 75.7 PERCENT, ADJUSTED FOR D.F.

ANALYSIS OF VARIANCE

 DUE TO       DF            SS        MS=SS/DF
REGRESSION     2        1934.047      967.023
RESIDUAL     112         607.919        5.428
TOTAL        114        2541.967

FURTHER ANALYSIS OF VARIANCE
SS EXPLAINED BY EACH VARIABLE WHEN ENTERED IN THE ORDER GIVEN

 DUE TO       DF            SS
REGRESSION     2        1934.047
1/HP           1        1797.008
1/WT           1         137.040

              X1          Y    PRED. Y    ST.DEV.
ROW         1/HP        MPG     VALUE     PRED. Y    RESIDUAL     ST.RES.
  2        0.0196     36.000    33.396     0.947      2.604       1.22 X
  4        0.0156     39.000    32.481     0.732      6.519       2.95RX
```

```
 12    0.0156    39.000    31.510    0.605     7.490     3.33RX
 13    0.0159    35.000    30.277    0.517     4.723     2.08R
 21    0.0196    25.000    32.771    1.016    -7.771    -3.71RX
 25    0.0074    22.000    20.223    0.673     1.777     0.80 X
115    0.0050    15.000    15.693    0.646    -0.693    -0.31 X

R DENOTES AN OBS. WITH A LARGE ST. RES.
X DENOTES AN OBS. WHOSE X VALUE GIVES IT LARGE INFLUENCE.

DURBIN-WATSON STATISTIC = 1.73
--
```

Thus the fitted model is

$$\hat{y} = 3.93 + 975(1/HP) + 23026(1/WT)$$

Step 5 | Validate the model

The three assumptions of the regression model must be checked:

1. The error terms are normally distributed.

2. The error terms have the same variance.

3. The error terms are uncorrelated.

The residual plot of the standardized residuals (RESID on the vertical axis) and the predicted values (YHAT on the horizontal axis) is shown in Figure 15.23. In the residual plot in Figure 15.23, the standardized residuals appear to conform to a rectangle about 0.0 fairly well, although there is a slight indication of nonequal variances due to the larger spread of residuals at higher values of YHAT. The one residual point with coordinates (32.771, −3.71) appears to be an outlier. Notice in the MINITAB output for the REGRESS command, this point is listed as a residual that has both a large standardized residual value and one whose x-values give it a large influence on the fitted regression line. This residual corresponds to row 21—a Chevrolet Cavalier with a four-cylinder engine and a 25 MPG figure. Compare the MPG with row 2—a Chevrolet Chevette with the same size engine (1.8 liters) and a 36 MPG figure. Is there a coding error in these two rows? If so, then what is the most likely error that has been made? It is a good idea to carefully check the residual cases printed out in MINITAB indicated by an R or an X. Some of the cases corresponding to these residuals may include miscoded data. Can the remaining six residuals printed out by MINITAB as special cases (X's, R's, or both) be explained?

We will first test for normality of the error terms. The normal scores are calculated using the MINITAB NSCORES command. The NSCORES are

FIGURE 15.23 | The residual plot for the regression equation

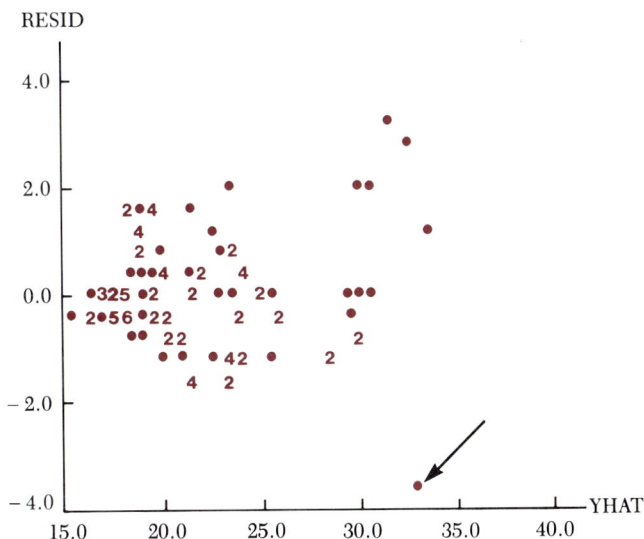

then correlated with the standardized residuals, and the pairs are plotted. The MINITAB results are given on the top of page 820.

The correlation between the NSCORES and the standardized residuals is 0.979. Referring to Table B.21, Appendix B, at the $\alpha = 0.05$ level, the critical value is 0.9835 with $n = 75$. Since $n = 115$ in this case, and the critical values are increasing with increasing sample size, we must reject the null hypothesis that the error terms are normally distributed. From the plot of the standardized residuals (STRES) and the NSCORES (NSCORE), the points conform well to a line with the exception of the outlier (a residual of -3.71, automobile number 21). Since regression analysis is robust to nonnormality of the error terms and the NSCORES plot looks very much like a line with the exclusion of one point, the model is viewed as being acceptable at this point.

The test for uncorrelated error terms is based on the Durbin-Watson test, which tests for first-order autocorrelation among the error terms only. From the MINITAB output, DW = 1.73. The null and alternate hypotheses are:

H_0: $\rho = 0$.
H_A: $\rho > 0$.

At the $\alpha = 0.05$ significance level, $d_L = 1.63$, and $d_u = 1.72$. (Note that there are two predictor variables, and we are using the row for $n = 100$ in Table B.20, Appendix B, since $n = 100$ is the largest value given.) Since DW = $1.73 > d_u = 1.72$, we cannot reject the null hypothesis, H_0: $\rho = 0$.

```
NSCORES C20, PUT NSCORES IN C25
--
?
CORRELATE C20 AND C25

    CORRELATION OF     STRES AND NSCORES = 0.979
--
?
PLOT C20 VS C25
```

STRES

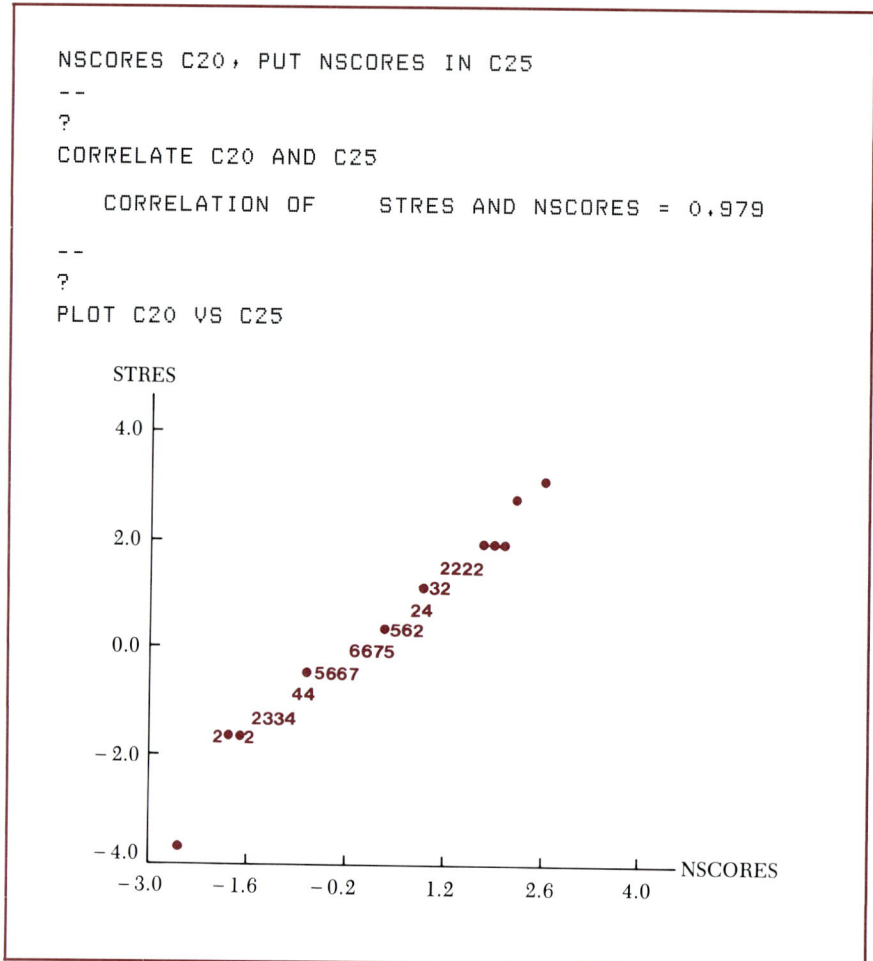

Finally, we need to consider the degree of multicollinearity between the two predictor variables (1/HP) and (1/WT). The correlation between (1/HP) and (1/WT) is 0.794, a relatively high correlation. In regressing (1/HP) on (1/WT), the resulting r^2-value is 0.630. Thus the variance inflation factor (VIF) is $1/(1 - 0.630) = 2.703$. Since there is only one VIF, 2.703 represents both the maximum VIF and the average VIF. Since the maximum VIF is less than 10, and the average VIF is not much greater than 1, we conclude that this model does not have a serious multicollinearity problem.

In conclusion, we find the model $Y = \beta_0 + \beta_1(1/HP) + \beta_2(1/WT) + \epsilon$ is acceptable. There appears to be modest nonnormality and unequal variances of the error terms, although these results may be due to the one (or more) outlier. The error terms appear to be uncorrelated based on the residual plot (no obvious pattern) and the Durbin-Watson test outcome. Also, the degree of multicollinearity is modest and is within acceptable limits.

Step 6 | Statistically test for the usefulness of the model developed

Since r^2 is high and s is low relative to the other models, it is evident that this model is useful. To further substantiate this fact, consider the null and alternate hypotheses,

H_0: $\beta_1 = \beta_2 = 0$.
H_A: At least one β is not zero.

for the model $Y = \beta_0 + \beta_1(1/HP) + \beta_2(1/WT) + \epsilon$. From the MINITAB output, the **F**-statistic value is

$$f = \frac{967.023}{5.428} = 178.155$$

The critical F-value from Table B.7, Appendix B, is determined as follows. The numerator degrees of freedom are $r_1 = k$ (k is the number of predictors in the model), and the denominator degrees of freedom are $r_2 = n - k - 1$ (n is the sample size). Since $k = 2$ and $n = 115$, we have $r_1 = 2$ numerator degrees of freedom and $r_2 = n - k - 1 = 115 - 2 - 1 = 112$ denominator degrees of freedom. From Table B.7, Appendix B, the critical value on the F-distribution with $\alpha = 0.05$ is 3.07 (using $r_2 = 120$ to avoid interpolation). Since $f = 178.155 > 3.07$, we reject H_0: $\beta_1 = \beta_2 = 0$ and conclude that the fitted regression model is useful. Notice also from the MINITAB output that the t-ratios for testing H_0: $\beta_1 = 0$ and H_0: $\beta_2 = 0$ are 7.09 and 5.02, respectively, indicating that both predictor variables are useful.

Step 7 | Use the model developed for prediction and estimation

The fitted regression model is $\hat{y} = 3.93 + 975x_1 + 23026x_2$, where y is MPG, x_1 is (1/HP) and x_2 is (1/WT). Thus to predict the EPA MPG for an automobile, the reciprocals of the automobile's HP and WT would have to be calculated and inserted into the regression function. For example, suppose that we wished to *predict* the miles per gallon of a specific automobile with a horsepower of 200 and a weight of 4,000. Then (1/HP) = 0.005 and (1/WT) = 0.0002. The predicted value is:

$$\hat{y} = 3.93 + 975(0.005) + 23026(0.0002) = 14.56$$

Using the MINITAB MACRO PREDICT (given in the Appendix in Chapter 14) produces the results on the top of page 822.

To four decimal places, the predicted value is 14.5592 MPG for a car with 200 horsepower and 4,000 pounds in weight. If we wish to set a 95 percent confidence interval on the *average* or *expected* MPG, the standard deviation is 0.529053. Note that the MINITAB MACRO PREDICT prints out two standard deviations—one for the conditional mean of **Y** and the other for an

```
1-- NOTE   PRINT PREDICTED VALUE
 -- NOTE
 -- NOTE   *********
 -- PRINT K5
    K5          14.5592
 -- NOTE   *********
 -- NOTE
 -- NOTE   PRINT STANDARD DEVIATION OF THE ESTIMATE OF THE
 -- NOTE   CONDITIONAL MEAN OF Y
 -- NOTE
 -- NOTE   *********
 -- PRINT K6
    K6          0.529053
 -- NOTE   *********
 -- NOTE
 -- NOTE   PRINT STANDARD DEVIATION OF THE PREDICTION OF AN
 -- NOTE   INDIVIDUAL VALUE OF Y
 -- NOTE
 -- NOTE   *********
 -- PRINT K7
    K7          2.38908
 -- NOTE   *********
 -- END
    END OF EXECUTION OF STORED INSTRUCTIONS
```

individual value of **Y**. Since we wish to set the prediction interval for one specific car and not the average MPG of *all* cars with 200 horsepower and a weight of 4,000 pounds, we must use the standard deviation for an individual value of **Y**. Using $z = 1.96$ to approximate the **t**-value with df = 112 and 95 percent confidence, the interval is:

$$14.5592 \pm 1.96(2.38908)$$
$$(9.88 \text{ to } 19.24 \text{ MPG})$$

This interval is rather wide, indicating that although this model explains 76.1 percent of the variation in MPG (**Y**), there remains uncertainty in predicting MPG, particularly when we are at the extremes of the sampled region of the x variables. Notice that a 200 horsepower rating and a weight of 4,000 pounds represents an *extrapolated* point, one outside the surveyed range of the horsepower and weight independent variables. This contributes to the large width of the confidence interval.

| Extrapolated point |

The seven steps of the multiple regression model-building process are quite demanding in terms of time required and the costs involved for computer time to perform the required regression analyses. It does take some experience to become good at implementing these steps. Finding the "best" model can be a lot of fun. It is a highly subjective process that requires analyzing many scatterplots and usually repeated attempts at validation until a suitable model is identified. Many colleges and universities offer one or two semester courses in regression analysis. If this widely applied statistical method interests you, you are strongly encouraged to take one or more of the courses available at your university or college in regression analysis.

■ **15.11**

Cautions in the use of multiple regression

Performing a competent multiple regression analysis requires a great deal of care. In executing the seven steps described in this chapter, there are numerous opportunities to make mistakes that lead to an unacceptable multiple regression model. In this section, a few of the more common mistakes are discussed.

Insufficient sample size   It should be remembered that multiple regression analysis is an inferential statistics technique. In a typical multiple regression model, $\mathbf{Y} = \beta_0 + \beta_1 x_1 + \beta_2 x_2 + \cdots + \beta_k x_k + \boldsymbol{\epsilon}$, there are $k + 2$ parameters to estimate: the $k + 1$ regression coefficients (including β_0) and σ^2, the variance of the error terms. A commonly used rule is to sample at least 10 observations for every parameter that must be estimated. Therefore, the sample size should be at least $10(k + 2)$. For example, in the model used in the case problem in Section 15.10, $k = 2$ so that $n \geq 10(k + 2) = 10(2 + 2) = 40$. In fact, $n = 115$ in the case problem, so that the sample size condition is satisfied. By having too small a sample, poor estimates of the parameters may result—leading to nonsense and unusable results.

Ignoring a high degree of multicollinearity of the model   A high degree of multicollinearity in a multiple linear regression model may result in nonsense estimates of the regression coefficient parameters, as illustrated in the cost data problem in Section 15.9. When multicollinearity is severe, it is possible to get an estimate of a regression coefficient (β) that has the wrong algebraic sign ($+$ or $-$). In addition, severe multicollinearity inflates the variance of the estimator $\mathbf{b}$ of a regression coefficient β. As a result, the estimated variance $s_{\mathbf{b}}^2$, may be very large. This in turn can produce a t-statistic value, $t = b/s_{\mathbf{b}}$, that is close to zero. As a consequence, it may not be possible to reject the null hypothesis, $H_0: \beta = 0$, when the regression coefficient β is associated with a very important independent variable x in explaining variation in $\mathbf{Y}$. In addition, when multicollinearity is present, one can get nonsense predictions when extrapolating beyond the region of the sampled values of the independent variables.

When multicollinearity is severe, one or more of the predictor variables in the model should be dropped or other methods such as ridge regression should be used.

Using the t-tests rather than the F-test   An important hypothesis to test in the multiple linear regression model, $\mathbf{Y} = \beta_0 + \beta_1 x_1 + \ldots + \beta_k x_k + \boldsymbol{\epsilon}$, is $H_0: \beta_1 = \beta_2 = \ldots = \beta_k = 0$. If this null hypothesis cannot be rejected, then we are wasting our time with the model. A common mistake in regression analysis is to use the t-tests to test the hypotheses, $H_0: \beta_i = 0$, $i = 1, 2, \ldots, k$. There are two problems in doing this. First, if the significance level α is set at 0.05 for each of the k t-tests, the overall or *family-wise* significance level will not be 0.05. In fact, if the null hypothesis $H_0: \beta_1 = \beta_2 = \ldots = \beta_k = 0$ is true, it is almost certain that one of the t-tests will reject the null hypothesis $H_0: \beta_i = 0$ if k is more than 5, indicating that at least one predictor variable is useful in explaining variation in $\mathbf{Y}$. The F-test controls for the family-wise significance level. The F-test should *always* be used to test the null hypothesis $H_0: \beta_1 = \beta_2 = \ldots = \beta_k = 0$. The t-tests then can be used to determine which among the k predictor variables are important, although this must be used with caution as well due to the problem that multicollinearity can create as discussed above.

Using only linear terms in the multiple linear regression model
Multiple *linear* regression refers to a statistical method whereby the regression coefficients enter linearly into the model. By this we mean that the regression coefficients form a linear combination in the regression model as in the general statement of the model, $\mathbf{Y} = \beta_0 + \beta_1 x_1 + \beta_2 x_2 + \ldots + \beta_k x_k + \boldsymbol{\varepsilon}$. The model, $\mathbf{Y} = \beta_0 e^{-\beta_1 x} + \boldsymbol{\varepsilon}$ is a *nonlinear model,* since the regression coefficients β_0 and β_1 are not represented in the model in an additive fashion. Ordinary least squares cannot be used to fit this model so that the estimation methods discussed in this chapter do not apply. Special methods referred to as nonlinear regression must be used to fit this model. Several references at the end of this chapter discuss the methodology of nonlinear regression estimation techniques. However, the independent variables *can* enter the model in a nonlinear fashion. This should be evident in the discussion of the case in Section 15.10. A frequent mistake is assuming that the x's also must enter the model only in linear terms. This is clearly not the case in the multiple linear regression model, where the term *linear* refers to the additive assumptions placed on the regression coefficients, not on the possible functional forms of the independent variables.

In the sales, advertising, and income example in Section 15.4, a better fitting model is $\mathbf{Y} = \beta_0 + \beta_1 x_1 + \beta_2 x_2 + \beta_3 x_1 x_2 + \beta_4 x_1^2 + \beta_5 x_2^2 + \boldsymbol{\varepsilon}$, a full quadratic in the two independent variables (x_1 = Average income level, and x_2 = Advertising expenditures). Although this model contains severe multicollinearity, a subset of it, $\mathbf{Y} = \beta_0 + \beta_1 x_1 + \beta_2 x_2^2 + \boldsymbol{\varepsilon}$, produces "better" results than the model used in Section 15.4. For this model, the coefficient of determination is $r^2 = 0.983$ compared with $r^2 = 0.973$ for the model $\mathbf{Y} = \beta_0 + \beta_1 x_1 + \beta_2 x_2 + \boldsymbol{\varepsilon}$. Further, the residual plots look better for the model $\mathbf{Y} = \beta_0 + \beta_1 x_1 + \beta_2 x_2^2 + \boldsymbol{\varepsilon}$. An exercise at the end of this chapter requires the fitting and validation of this model.

Failing to validate the fitted model Surprisingly, it is not uncommon in business applications of the multiple linear regression model to find that the model has not been validated. That is, it has not been shown that the error terms are normally distributed, have the same variance, and are uncorrelated. This step should *always* be performed in a multiple regression analysis.

Misuse of r^2, the coefficient of determination The coefficient of determination, r^2, is an estimator of a population coefficient of determination, ρ^2. In the multiple linear regression model, $Y = \beta_0 + \beta_1 x_1 + \ldots + \beta_k x_k + \varepsilon$, ρ^2 represents the proportion of variation in **Y** that is explained by the model for the *population* of $(Y, x_1, x_2, \ldots, x_k)$ values. The sample coefficient of determination, r^2, measures the proportion of the variation in **Y** that is explained by the model for the *sample* of $(Y, x_1, x_2, \ldots, x_k)$ values. As it turns out, r^2 is a *biased* estimator of ρ^2. If $\rho^2 = 0$ (that is, the model explains *none* of the variation in **Y** in the population), then

$$E[r^2 / \rho^2 = 0] = \frac{k + 1}{n}$$

For example, suppose $n = 2$ and $k = 1$. With $k = 1$, we have the simple linear regression model, $Y = \beta_0 + \beta_1 x + \varepsilon$. If this model explains no variation in **Y** in the population ($\rho^2 = 0$), then the bias in r^2 is $(k + 1)/n = (1 + 1)/2 = 1$. In this case, r^2 must equal 1, as can be seen from Figure 15.24. When there are

FIGURE 15.24 | r^2 when $n = 2$ and $k = 1$ in the simple linear regression model

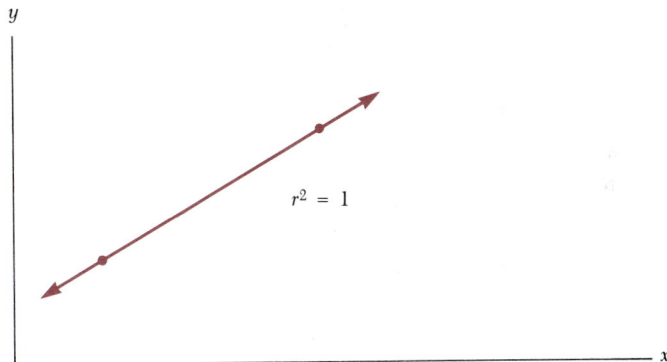

two points ($n = 2$), a line can be drawn that fits exactly through the two points, resulting in $r^2 = 1$. In the sample, all the variation has been explained in this case. The bias in r^2 is another reason the sample size n should be as large as possible. As n increases in size, the bias $(k + 1)/n$ diminishes.

Another property of r^2 causes it to be misused. If we take an existing model and add one additional variable, then the new value of r^2 must be at least as large as the one in the model before the variable is added. Therefore, we can make the value of r^2 arbitrarily large (close to 1) by regressing **Y** on a large number of independent variables or functions of the independent variables.

Because of these two points, to say that a specific regression model is "good" because the value of $\mathbf{r}^2$ is close to 1 may be misleading. A value of $\mathbf{r}^2$ close to 1 may be due to an insufficient sample size or to having too many predictors in the multiple regression equation. In selecting a good model, other factors such as s, the estimated standard deviation about the regression model, and parsimony, keeping the model simple and understandable, should be considered in addition to $\mathbf{r}^2$.

The bias in r^2 when $\rho^2 = 0$ is the reason that most computer programs print out a "corrected r^2" in addition to r^2. In the MINITAB output, the corrected r^2 is referred to as *R-SQUARED ADJUSTED FOR DEGREES OF FREEDOM*. The adjustment in r^2 attempts to take into account the bias in r^2 as a function of sample size. In the MPG case problem, notice in Section 15.10.4 that the r^2 and the adjusted r^2 printed out by MINITAB are very close ($r^2 = 0.761$ and adjusted $r^2 = 0.757$). This is due to the very large sample size ($n = 115$). As the sample size becomes large, the adjusted r^2 will converge to r^2. In small samples, there can be quite a difference between the r^2 and the adjusted r^2. In general it is advisable to give both r^2 and the adjusted r^2 when reporting results from a regression analysis. If there is a large difference between the two, then the adjusted r^2 is a better measure of the amount of variation in $\mathbf{Y}$ explained by the fitted regression model, since it makes an attempt to adjust for the bias in r^2 due to small or moderate sample sizes.

Misuse of the correlation coefficient in model building The Pearson product-moment correlation coefficient $\mathbf{r}$ measures the *linear* association between two variables. Thus, though the value of $\mathbf{r}$ may be close to zero between the dependent variable $\mathbf{Y}$ and one of the independent variables x, there may be a strong functional relationship—such as $1/x$ or x^2. To dismiss an independent variable x due to a correlation close to 0 is clearly a mistake. A scatterplot should be constructed, plotting values of $\mathbf{Y}$ against values of x, to ensure that there is also no nonlinear relationship.

Also, it must be remembered that a strong correlation between $\mathbf{Y}$ and an independent variable x *does not* imply causality. It only suggests that the two variables are linearly related.

There are other errors that can be made in applying multiple linear regression analysis, but the list above includes some of the more common ones. (For a complete discussion of the proper use of regression analysis, see the Neter, Wasserman, and Kutner reference at the end of this chapter.)

■ 15.12

Multiple correlation analysis

15.12.1 | The coefficient of multiple correlation

Multiple correlation bears the same relationship to simple linear correlation as multiple linear regression bears to simple linear regression. In multiple correlation, we are attempting to assess the relative strength of the relationship between the dependent random variable $\mathbf{Y}$ and the independent random variables $\mathbf{X}_1, \mathbf{X}_2, \ldots, \mathbf{X}_k$. The strength of the linear relationship is measured by the coefficient of multiple correlation, $\mathbf{r}$. Unlike the coefficient of correlation in the single independent variable case, however, the coefficient of multiple correlation is bounded by 0 and 1; that is, $0 \leq r \leq 1$. This is because r does not indicate the slope of the regression equation, since it is not possible to indicate the signs of all the regression coefficients that relate the dependent variable Y to the independent variables X_i. Similar to the simple correlation case, we must treat the independent variables X_i as random variables, since one assumption of the multiple correlation model is that the distribution of Y and the X values is multivariate normal. This assumption is necessary in constructing confidence intervals and performing hypothesis tests on the coefficient of multiple correlation.

As in the case of simple linear correlation, the measure r^2 is much simpler to interpret than is the coefficient of multiple correlation, r. That is, r^2 measures the proportion of variability in the dependent random variable Y explained by or accounted for by the independent random variables. The coefficient of multiple correlation is computed by simply taking the square root of the coefficient of multiple determination. Sometimes the coefficient of multiple correlation is subscripted to indicate the dependent random variable being explained by the independent random variables. Using such notation in the bivariate case, the coefficient of multiple correlation becomes $r_{Y/X_1.X_2}$. Experience in computing the coefficient of multiple correlation is provided in the following example.

Example 15.9 Using the data given in Table 15.6, compute the coefficient of multiple correlation.

Solution The coefficient of multiple correlation is given by:

$$r = \sqrt{r^2} = \sqrt{0.97296} = 0.9864$$

15.12.2 | Partial correlation

The coefficient of multiple correlation measures the strength of the linear relationship between the dependent variable Y and the independent variables X_1, X_2, . . . , X_k. We may also be interested in measuring the strength of the relationship (linear) between the dependent variable Y and *one* of the independent variables, where the effect of all the other variables has been removed or has been held constant. This is to be distinguished from the correlation between pairs of variables, as is the case in simple linear regression. In Table 15.23, for example, we show the correlation matrix giving the simple linear correlation between every pair of variables in the sales-advertising-income example introduced in Section 15.4. Table 15.23 was produced by the SAS statistical program CORR, which is similar to the regression program we have been discussing thus far. This matrix gives not only the simple correlation between every pair of variables but also the p-value for the significance of the correlation measure under the null hypothesis that the simple coefficient of correlation is equal to 0. As shown in Table 15.23, all pairs of linear correlations are unequal to 0 beyond the 0.10, 0.05, and 0.01 significance levels. From Table 15.23, the correlation of any variable with itself is 1.00, the correlation between income and sales is $r_{YX_2} =$

TABLE 15.23 | Correlation matrix of data in Table 15.3

CORRELATION COEFFICIENTS AND PROB > |R| UNDER HO:
RHO = O FOR N = 15

	SALES	INCOME	ADS
SALES	1.00000	0.94502	0.96995
	0.0000	0.0001	0.0001
INCOME	0.94502	1.00000	0.89000
	0.0001	0.0000	0.0001
ADS	0.96995	0.89000	1.00000
	0.0001	0.0001	0.0000

$r_{X_2Y} = 0.945$, and the simple linear correlation between income and advertising expenditures is $r_{X_2X_1} = r_{X_1X_2} = 0.89$. Thus these two independent random variables are themselves highly related! Note how we have subscripted the simple coefficient of correlation to indicate the random variables involved.

The partial correlation coefficient measures the strength of the linear relationship between the dependent random variable Y and *one* independent variable X_i where the linear effect of the remaining independent variables is held constant. As was the case with simple linear regression and correlation, it is much easier to interpret the *partial coefficient of determination,* which is the square of the partial correlation coefficient, and so we shall do so. The notation we will use for the partial coefficient of determination is $r^2_{Y/x_i \cdot x_1, x_2, ..., x_{i-1}, x_{i+1}, ..., x_k}$ where the independent variables to the right of the dot indicate the independent variables whose effect is being held constant. The coefficient of partial correlation is of course the square root of the above quantity.

In Table 15.24 are given the simple linear regression and correlation analyses of sales and advertising expenditures and of sales and income using the data given in Table 15.3. From Table 15.6, the amount of variation left unexplained by the multiple linear regression when both advertising and income are included as independent variables (SSE) is 202.63. In the bivariate case, the coefficient of partial determination is given by the following:

Coefficient of
partial determina-
tion

Coefficient of partial determination—bivariate case

$$r^2_{Y/x_i \cdot x_j} = \frac{\text{Additional variation in } Y \text{ explained by the addition of } X_i}{\text{Variation in } Y \text{ unexplained by } X_j \text{ alone}}$$

TABLE 15.24 | Simple linear regression and correlation analyses of data in Table 15.3

$$\hat{y} = b_0 + b_1 x_{\text{ADS}}$$

DEPENDENT VARIABLE: SALES

SOURCE	DF	SUM OF SQUARES	MEAN SQUARE	F VALUE	PR > F
MODEL	1	7049.32975962	7049.32975962	206.60	0.0001
ERROR	13	443.57579731	34.12121518	R-SQUARE	STD DEV
CORRECTED TOTAL	14	7492.90555693		0.940801	5.84133676

PARAMETER	ESTIMATE	T FOR H0: PARAMETER = 0	PR > |T|	STD ERROR OF ESTIMATE
INTERCEPT	1.96334786	0.61	0.5551	3.24106207
ADS	26.96437275	14.37	0.0001	1.87598194

$$\hat{y} = b_0 + b_1 x_{\text{INCOME}}$$

DEPENDENT VARIABLE: SALES

SOURCE	DF	SUM OF SQUARES	MEAN SQUARE	F VALUE	PR > F
MODEL	1	6691.64680572	6691.64680572	108.57	0.0001
ERROR	13	801.25875121	61.63528855	R-SQUARE	STD DEV
CORRECTED TOTAL	14	7492.90555693		0.893064	7.85081452

PARAMETER	ESTIMATE	T FOR H0: PARAMETER = 0	PR > |T|	STD ERROR OF ESTIMATE
INTERCEPT	-25.84262257	-3.73	0.0025	6.92908808
INCOME	5.64666488	10.42	0.0001	0.54192630

From Table 15.24, the total variation in **Y** to be explained is

$$\Sigma(y - \bar{y})^2 = 7,492.91$$

Based on the simple linear regression when advertising expenditures are included in the estimating equation, the amount of unexplained variation is SSE = 443.58. The amount of variation unexplained when both advertising expenditures *and* average income level are included in the estimating equation is SSE = 202.63. Thus the *extra* amount of variation explained when average income level is included in the model is 240.95 = 443.58 − 202.63. Thus the *proportion of previously unexplained variation* that is explained when average income level is included in the model is:

$$r^2_{\mathbf{Y}/x_2 \cdot x_1} = \frac{240.95}{443.58} = 0.543$$

TABLE 15.25 | Calculation of partial coefficient of determination for sales-advertising-income data

$r^2_{\mathbf{Y}/x_2 \cdot x_1}$:

	94.08%	3.22%
Total variation: 7,492.91	Variation explained by ads alone: 7,049.33	Extra variation explained by income: 240.95
		2.70% — Unexplained variation: 202.63

$$r^2_{\mathbf{Y}/x_2 \cdot x_1} = \frac{240.95}{443.58} = 0.543 \qquad \text{Variation unexplained by ads alone: 443.58}$$

$r^2_{\mathbf{Y}/x_1 \cdot x_2}$:

	89.31%	7.99%
Total variation: 7,492.91	Variation explained by income alone: 6,691.65	Extra variation explained by ads: 598.63
		2.70% — Unexplained variation: 202.63

$$r^2_{\mathbf{Y}/x_1 \cdot x_2} = \frac{598.63}{801.26} = 0.747 \qquad \text{Variation unexplained by income alone: 801.26}$$

The partial correlation coefficient for the average income level is thus:

$$r_{Y/x_2 \cdot x_1} = \sqrt{0.543} = 0.7369$$

The partial coefficient of determination for the variable advertising expenditures is shown in Table 15.25. Note that the "area" in Table 15.25 corresponding to the extra variation explained by ads is larger than the "area" for extra variation explained by income, reflecting the larger partial coefficient of determination for the independent variable advertising expenditures. The coefficient of partial determination measures the proportion of variation explained by the variable *of that remaining* after all other independent variables are included in the estimating equation.

■ **15.13**

Summary

Multiple linear regression analysis is one of the most widely used statistical procedures today. We cannot do justice to the analysis in this book, but it is hoped that this introduction suggests its wide applicability while discussing the rudiments of the procedure. (An excellent reference, and quite readable is the Neter, Wasserman, and Kutner reference cited at the end of this chapter, for those desiring more information on multiple linear regression and correlation analysis.)

■ **References**

Brock, R. J., and G. C. Arnold. *Applied Regression Analysis and Experimental Design*. New York: Marcel Dekker, 1985.

Chatterjee, S., and B. Price. *Regression Analysis by Example*. New York: John Wiley & Sons, 1977.

Conover, W. J. *Practical Nonparametric Statistics*. New York: John Wiley & Sons, 1971.

Dixon, W. J., ed. *BMDP Statistical Software 1981*. Berkeley: University of California Press, 1981.

Draper, N. R., and H. Smith. *Applied Regression Analysis*. 2nd ed. New York: John Wiley & Sons, 1981.

Dutta, M. *Econometric Methods*. Cincinnati: South-Western Publishing, 1985.

Freedman, D.; R. Pisani; and R. Purves. *Statistics*. New York: W. W. Norton, 1978, Chapters 8–12.

Goldberger, A. S. *Econometric Theory*. New York: John Wiley & Sons, 1964.

Johnston, J. *Econometric Methods*. 2nd ed. New York: McGraw-Hill, 1972.

Mendenhall, W., and J. T. McClave. *A Second Course in Business Statistics: Regression Analysis*. San Francisco: Dellen Publishing, 1981.

Montgomery, D. C., and E. A. Peck, *Introduction to Linear Regression Analysis*. New York: John Wiley & Sons, 1982.

Neter, J.; W. Wasserman; and M. Kutner. *Applied Linear Statistical Models*. 2nd ed. Homewood, Ill.: Richard D. Irwin, 1985.

Ryan, T.; B. Joiner; and B. Ryan. *MINITAB Student Handbook*. 2nd ed. Boston: PWS Publishers, 1985.

SAS User's Guide: Statistics. Cary, N.C.: SAS Institute, 1982.

SPSS-X User's Guide. New York: McGraw-Hill, 1983.

Younger, M. S. *A First Course in Linear Regression*. 2nd ed. Boston: Duxbury Press, 1985.

Problems

Note: Problems with an asterisk indicate that a computer analysis is required.

Section 15.2 Problems

15.1 What is meant by the *reduced effects* or the *main effects* model when referring to the multiple linear regression model?

15.2 State the assumptions of the multiple linear regression model.

15.3 Consider the multiple linear regression model $Y = \beta_0 + \beta_1 x_1 + \beta_2 x_2 + \epsilon$.
 a. What are the interpretations of the regression coefficients β_0, β_1, and β_2?
 b. What is the geometrical form of this model in three-dimensional space?

Section 15.3 Problems

15.4 What role do the normal equations play in estimating the regression coefficient parameters in the multiple linear regression model?

15.5 In a bivariate regression model fitted to $n = 10$ observations, the following normal equations result:

$$100 = 10b_0 + 40b_1 + 60b_2$$
$$50 = 40b_0 + 200b_1 + 100b_2$$
$$200 = 60b_0 + 100b_1 + 400b_2$$

Solve the normal equations to find the estimates b_0, b_1, and b_2.

15.6 In a multiple linear regression analysis involving four independent variables, it is decided to drop one of the independent variables from the model and use three variables for modeling purposes. Explain the effect on the point estimate of the conditional probability distribution variance (σ^2) from dropping this independent variable. Under what conditions will the estimate increase, and under what conditions will the estimate decrease?

15.7 In the bivariate regression model $Y = \beta_0 + \beta_1 x_1 + \beta_2 x_2 + \epsilon$, a sample 63 observations is used to fit the model. The resulting sum of squares due to error (SSE) is 1,200. What is the estimate of the conditional probability distribution variance?

15.8 In a regression model experiment, 120 observations are used to fit the model $Y = \beta_0 + \beta_1 x_1 + \beta_2 x_2 + \beta_3 x_3 + \epsilon$. The resulting sums of squares are: SST = 486, SSR = 396, and SSE = 90.
 a. What is the value of the sample coefficient of multiple determination?
 b. What does this value tell us about the fitted regression model $\hat{y} = b_0 + b_1 x_1 + b_2 x_2 + b_3 x_3$?

Section 15.4 Problems

15.9 A bivariate regression model of the form $Y = \beta_0 + \beta_1 x_1 + \beta_2 x_2 + \epsilon$ is fitted using $n = 115$ observations by using SAS with the results shown on the top of page 833.
 a. What is the estimated regression equation?
 b. What is the value of the sample coefficient of multiple determination? What does this number tell you about the fitted regression equation?
 c. What is the estimate of the conditional distribution variance σ^2?

15.10 The following model is fitted using $n = 52$ observations using SAS, $Y = \beta_0 + \beta_1 x_1 + \beta_2 x_2 + \beta_3 x_3 + \beta_4 x_4 + \beta_5 x_5 + \epsilon$. The partial SAS output is shown at the bottom of page 833. What are the missing values $a, b, c, d, e,$ and f?

Section 15.5 Problems

15.11 In a multiple regression analysis of yield as a function of three different nutrients, it is found that PR > F = 0.0876. Interpret this *p*-value statistic.

15.12 In a bivariate regression analysis involving 22 observations, it is found that $b_1 = 23.56$ and $b_2 = 0.0052$, with $s_{b_1} = 13.68$ and $s_{b_2} = 0.0013$. Test the null hypotheses $H_0: \beta_1 = 0$ and $H_0: \beta_2 = 0$ using the *t*-test with $\alpha = 0.05$ for each *t*-test.

```
DEPENDENT VARIABLE: MPG

SOURCE          DF      SUM OF SQUARES   MEAN SQUARE   F-VALUE   PR > F
MODEL           2          1934.047       967.023      178.155   0.0001
                                                                 STD DEV
ERROR           112         607.919         5.428               2.330
CORRECTED
 TOTAL          114        2541.967
PARAMETER  ESTIMATE
INTERCEPT      3.93
X1           975.21
X2         23026.05
```

15.13 Which of the variables, x_1 or x_2, is more statistically significant in Problem 15.12?

15.14 Explain why in a test of $H_0: \beta_1 = 0$ and $H_0: \beta_2 = 0$ using individual t-test where the significance level of each test is 0.05, we cannot claim that the significance of the model with both variables included is 0.05.

15.15 In a regression analysis involving 29 observations and three independent variables, the following estimating equation is determined, with standard errors for the coefficients shown in parentheses:

$$\hat{y} = 23 + 7x_1 - 4x_2 + 3x_3$$
$$\quad\;\; (2.6)\;\;\; (1.9)\;\; (1.8)$$

a. Test the null hypotheses $H_0: \beta_1 = 0$, $H_0: \beta_2 = 0$, and $H_0: \beta_3 = 0$ at the $\alpha = 0.05$ significance level using t-tests for the three hypotheses.

b. Given your results from part a, which variable would you say is the most statistically significant?

c. Assuming x_1 were to increase by 23, x_2 to decrease by 13, and x_3 to increase by 12, what would be the estimated change in y?

15.16 In a multiple linear regression analysis involving 33 observations and 7 independent variables, it is determined that SST = 5,672.67 and that SSE = 1,945.78. Test for the significance of the model developed at the $\alpha = 0.05$ significance level.

15.17* The state government in California is interested in the relationship between the number of automobile accidents per year and the number of licensed vehicles in several small communities in northern California. For each of 10 communities in northern California, the

```
DEPENDENT VARIABLE: Y

SOURCE          DF     SUM OF SQUARES   MEAN SQUARE   F-VALUE   PR > F
MODEL           a          4583.6          e           6.576
                                                                STD DEV
ERROR           b            d           139.4                    f
CORRECTED
 TOTAL          c         10994.2
```

number of accidents in 1985, the number of licensed vehicles, and the size of the police force is recorded. [It is pointed out to the investigator in charge of this study that the size of the police force in each community might be an important predictor (independent) variable.] The data are:

Community	Number of accidents	Number of licensed autos*	Size of police force
1	213	11	10
2	296	8	9
3	105	4	19
4	422	12	8
5	542	15	2
6	404	16	8
7	178	5	12
8	92	6	15
9	433	9	10
10	312	12	9

* In thousands

Let Y = Number of auto accidents in 1985, x_1 = Number of licensed autos (in thousands), and x_2 = Size of police force.

a. Form scatterplots of Y and x_1, and Y and x_2. Does there appear to be a relationship between Y and x_1? Between Y and x_2? Comment on these relationships.

b. Fit the regression model $\mathbf{Y} = \beta_0 + \beta_1 x_1 + \beta_2 x_2 + \epsilon$ to these observations.

c. Test the null hypothesis $H_0: \beta_1 = \beta_2 = 0$ at the $\alpha = 0.05$ level of significance.

d. Using the t-test, test each of the null hypotheses, $H_0: \beta_1 = 0$ and $H_0: \beta_2 = 0$. Use $\alpha = 0.05$ for each test. Based on the scatterplots in part a and the results of these t-tests, which of the two independent variables appears to be more important in explaining the number of accidents per year? Why?

15.18 A firm has undertaken an evaluation of the presence of a catalyst (amount) and the temperature of a chemical process in influencing the resulting yield. The data collected by the firm, as well as a multiple linear regression analysis of the results of the evaluation, are reported below.

Yield (000s)	Temperature (°F)	Catalyst
27	179	69
24	175	71
21	172	57
19	169	45
32	183	76
45	190	81
37	185	77
29	181	70
22	174	64
28	182	72
26	177	68
18	168	42
23	176	63
36	185	74

Dependent variable: Yield

SOURCE	DEGREES OF FREEDOM	SUM OF SQUARES	MEAN SQUARE	F-VALUE	PR > F
MODEL	2	717.9232	358.9616	91.21	0.0001
ERROR	11	43.2910	3.9355		
TOTAL	13	761.2142			

$r^2 = 0.943129$; standard deviation $s = 1.9838$.

PARAMETER	ESTIMATE	t-STATISTIC	PR > |t|	STANDARD ERROR OF ESTIMATE
INTERCEPT	-220.9176	-7.09	0.0001	31.1389
TEMPERATURE	1.4674	6.83	0.0001	0.2149
CATALYST	-0.1968	-1.63	0.1312	0.1207

The SAS results above are for fitting the model, $\mathbf{Y} = \beta_0 + \beta_1 x_1 + \beta_2 x_2 + \epsilon$, where $\mathbf{Y}$ is the yield, x_1 is the temperature, and x_2 is the amount of catalyst.

a. Obtain a point estimate of the mean yield when the temperature is 180° and the amount of catalyst is 71.

b. Test the null hypothesis, $H_0: \beta_1 = \beta_2 = 0$ at the $\alpha = 0.05$ level of significance using the F-test.

c. Test the null hypotheses $H_0: \beta_1 = 0$ and $H_0: \beta_2 = 0$ at the $\alpha = 0.05$ level of significance using t-tests.

d. What is the p-value for the F-test for the model, and what is the interpretation of this p-value?

15.19 In a multiple linear regression model with $k = 5$ independent variables, $n = 50$ observations are used to fit the model. The resulting value of the coefficient of multiple determination is $r^2 = 0.80$.

a. What is the $\mathbf{F}$-statistic value for testing the null hypothesis that all the regression slope coefficients are zero?

b. Using $\alpha = 0.05$ and the F-value determined in part a, should the null hypothesis be rejected? Explain.

15.20 Why should the F-test be used instead of the t-tests when testing the null hypothesis that all of the regression slope coefficients are zero with $k \geq 2$?

Section 15.6 Problems

15.21 What is the difference between *predicting* an individual value of $\mathbf{Y}$ and *estimating* the conditional distribution mean $\mu_{\mathbf{Y}/x}$?

15.22 How is a *prediction interval* different than a *confidence interval*?

Section 15.7 Problems

15.23 It is decided to use dummy variables to represent the months of the year in a regression experiment. Thus there are $m = 12$ categories (months) for the qualitative variable being represented by the dummy variables. Why must eleven dummy variables be used to represent this qualitative variable?

15.24 In a multiple linear regression analysis of income and marital status, it is determined that the number of "levels" of the independent variable "marital status" is five: single, married, separated, divorced, and widowed. Define the binary variables to model the qualitative characteristic marital status. How many binary variables are required to model this qualitative variable?

15.25* Reproduced below are data on sales and advertising expenditures for a firm marketing a semiluxury consumer good.

Sales and corresponding advertising levels

Year and quarter	Sales ($000)	Advertising expenditures ($000)
1981		
I	105	10.54
II	96	11.34
III	55	12.70
IV	117	13.25
1982		
I	132	14.18
II	125	15.39
III	82	16.47
IV	145	17.70

Sales and corresponding advertising levels *(concluded)*

Year and quarter	Sales ($000)	Advertising expenditures ($000)
1983		
I	151	18.81
II	143	19.66
III	100	20.84
IV	165	21.78
1984		
I	177	22.24
II	168	23.89
III	127	24.04
IV	186	26.02
1985		
I	202	26.07
II	194	27.80
III	149	28.40
IV	212	28.59
1986		
I	224	30.62
II	223	31.73
III	176	32.48
IV	234	33.77

a. Fit a simple linear regression line to these data, using advertising expenditures as the independent variable to estimate sales. Do advertising expenditures account for a significant amount of the variability in sales?

b. Using binary variables corresponding to the quarters of the year, fit a multiple linear regression plane to the data, estimating sales as a function of advertising expenditures and the quarter of the year. Has the "predictability" of the fitted regression improved by including the quarter of the year in the estimating equation?

c. Using the methods described in the text, test for the presence of a seasonal (quarter) factor at the $\alpha = 0.05$ significance level.

d. Given the results determined in part *c*, does the firm appear to be following a wise policy in spending its advertising dollars? Explain.

Section 15.8 Problems

15.26 State the three regression assumptions concerning the error terms. Which of the three assumptions is most critical in validating the multiple linear regression model? Why?

15.27 What are the two ways that the residuals may be used to check the assumptions concerning the error terms in the multiple linear regression model?

15.28 What are the two basic types of residual plots? How is each used to validate a fitted multiple linear regression equation?

15.29 The residual plots below are plots of standardized residuals (vertical axis) versus predicted y-values (horizontal axis). All four plots indicate the violation of one or more assumptions regarding the error terms. Indicate for each plot which assumption is most obviously violated.

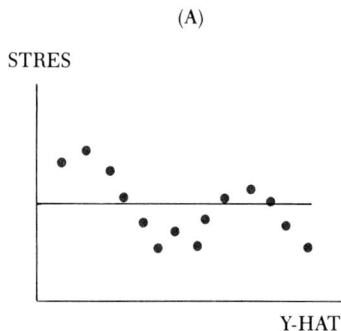

(A)

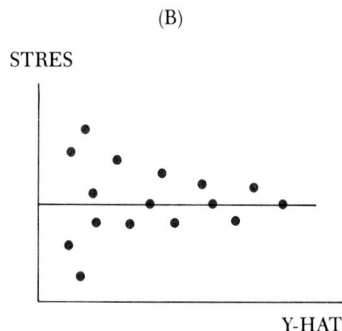

(B)

(C)

STRES

Y-HAT

(D)

STRES

Y-HAT

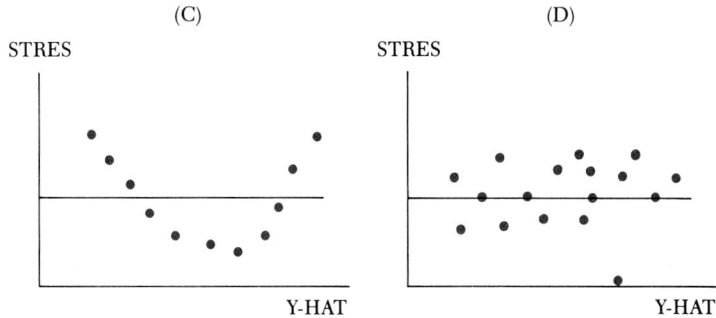

15.30 Over a period of 15 months, retail sales revenue is recorded together with the total
amount in dollars spent on advertising for a
chain of 20 electronics stores in the Southwest United States. Treating retail sales revenue as the dependent variable, a simple linear
regression model is fitted with the advertising
expenditures as the independent variable for
the $n = 15$ monthly periods. The resulting
standardized residuals are:

Month	Standardized residual	Month	Standardized residual
1	1.90	9	1.43
2	0.95	10	0.95
3	0.00	11	0.00
4	−0.48	12	−0.95
5	−1.43	13	−1.43
6	−0.95	14	−0.48
7	0.00	15	0.00
8	0.48		

Calculate the Durbin-Watson statistic value
DW for these residuals. Test the hypothesis
that the error terms have no first-order autocorrelation against the alternative hypothesis
that the error terms have positive first-order
autocorrelation at the 0.05 level of significance.

15.31 The standardized residuals in Problem 15.30
and the corresponding normal scores
(NSCORES) are given in the table at the bottom of the page.

a. Form the probability plot of the standardized residuals (vertical axis) and the
normal scores (horizontal axis) as in Figure 15.15 in the text. Does the probability plot suggest that the error terms are
normally distributed? Explain.

b. The correlation between the standardized residuals and the normal scores is
0.997. Test the null hypothesis that the
error terms are normally distributed at
the 0.05 level of significance.

Month	Standardized residual	Normal score	Month	Standardized residual	Normal score
1	1.90	1.74	9	1.43	1.24
2	0.95	0.82	10	0.95	0.82
3	0.00	0.08	11	0.00	0.08
4	−0.48	−0.42	12	−0.95	−0.82
5	−1.43	−1.45	13	−1.43	−1.45
6	−0.95	−0.82	14	−0.48	−0.42
7	0.00	0.08	15	0.00	0.08
8	0.48	0.51			

Section 15.9 Problems

15.32 Explain the meaning of multicollinearity and one of its effects on the model developed.

15.33 Discuss the informal indications of multicollinearity. What is the formal method of indicating severe multicollinearity? Describe the use of the formal method.

15.34 A multiple linear regression model with three independent variables is fitted to $n = 100$ observations. There is reason to believe that the three independent variables are highly intercorrelated so that the fitted regression equation may exhibit a high degree of multicollinearity. To investigate this belief, each possible pair of independent variables is regressed on the remaining variable resulting in the following coefficients of multiple determination:

Dependent variable	Independent variables	r^2
x_1	x_2, x_3	0.80
x_2	x_1, x_3	0.45
x_3	x_1, x_2	0.65

Using the variance inflation factors, is there reason to be concerned about the multicollinearity in the fitted regression equation? Explain.

15.35 Discuss one approach to correcting fitted models exhibiting moderate to high multicollinearity.

Section 15.10 Problems

Note: Problems 15.36–15.38 require the use of the car data set with a computer software package able to perform multiple regression analysis.

15.36* In the case study (Section 15.10), model 6 in Table 15.22 gave the highest r^2 and lowest s, but the degree of multicollinearity was very high as measured by the variance inflation factors (VIFs). By regressing each set of four of the five independent variables used in the model (HP, WT, 1/HP, 1/WT, HP*WT) on

the remaining independent variable, verify that the maximum VIF is 333.33 and the average of the five VIFs is 172.78.

15.37* In the case study (Section 15.10), model 1 in Table 15.22 includes all the original independent variables entered into the model as linear and additive terms. Fit this model to the data.

a. What is the estimated regression function?

b. Verify that r^2 is 0.759 and $s = 2.380$ for this model.

c. Are there informal indications of severe multicollinearity?

d. Calculate the variance inflation factors, and determine whether or not multicollinearity is severe for this model.

e. Form the residual plot using the standardized residuals plotted against the predicted values. Does this residual plot indicate that the three regression assumptions concerning the residuals are satisfied? Explain.

15.38* In the case study (Section 15.10), it is possible to construct other independent variables, such as (HP/ENGS), (HP/WT), and (HP/ENGS)/WT. These variables measure the efficiency of the engine in various ways. By forming scatterplots and by regressing each of these variables on MPG, miles per gallon, determine from the three regressions whether one or more of these variables are useful in predicting the miles per gallon for the sampled automobiles.

Section 15.11 Problems

15.39 Briefly discuss the problems created by using samples of insufficient size when fitting a multiple linear regression model to a set of observations.

15.40 What are the consequences of ignoring a high degree of multicollinearity in a fitted regression equation?

15.41 What does the term *linear* refer to in the multiple *linear* regression model?

15.42 A multiple linear regression model with $k = 15$ terms is fitted to a set of $n = 18$ observations. Assuming that the population coefficient of multiple determination is zero, what is the expected value of $\mathbf{r}^2$, the sample coefficient of multiple determination for the fitted regression equation?

Section 15.12 Problems

15.43 What does the coefficient of partial determination, $r^2_{Y/X_i \cdot X_j}$, measure?

15.44 Verify that the correlation between the independent variables advertising expenditures and average income level is 0.89 using the data given in Table 15.3.

15.45 In a simple linear regression analysis of the yield of a chemical process as a function of temperature and pressure, the following quantities are determined: when both independent variables (temperature and pressure) are included in the estimating equation, $SST = 3,645$ and $SSE = 734$; when temperature only (x_1) is included in the model, $SSE = 1,095$; and when pressure only (x_2) is included in the model, $SSE = 1,436$.
  a. Compute the coefficient of multiple determination for these data.
  b. Compute the coefficient of multiple correlation for these data.
  c. Compute $r^2_{Y/X_2 \cdot X_1}$.
  d. Compute $r^2_{Y/X_1 \cdot X_2}$.
  e. Construct a diagram similar to Table 15.25 showing the relationships between the partial coefficients of determination determined in parts c and d.

15.46 The total variation in a random variable $\mathbf{Y}$ is determined to be 4,798 units. The model $\hat{y} = b_0 + b_1x_1 + b_2x_2$ leaves 854 units unexplained based on 27 observations. When the model is extended to include x_3, 222 more units of the variation in $\mathbf{Y}$ are accounted for.
  a. Compute r^2.
  b. Compute $r^2_{Y/X_3 \cdot X_1, X_2}$.

15.47 Using 18 observations, a multiple linear regression equation involving four independent variables is estimated using the method of least squares. It is determined that $SSR = 700$ and $SST = 900$. Furthermore, the amount of variation explained jointly by x_1, x_2, and x_3 is 550.
  a. Compute r^2.
  b. Test for the significance of the model developed at the $\alpha = 0.05$ significance level.
  c. Compute $r^2_{Y/X_4 \cdot X_1, X_2, X_3}$ and interpret its meaning.

15.48* Consider Problem 15.18 with the yield of a chemical as the dependent variable and temperature and amount of catalyst as the independent variables.
  a. Compute the simple correlation between temperature and catalyst. What implications does this have in estimating yield by the use of multiple linear regression analysis using a model with the temperature and catalyst variables as independent variables?
  b. What is the increase in r^2 when temperature is added to the model that includes only catalyst as the independent variable?
  c. What is the increase in r^2 when catalyst is added to the model that includes only temperature as the independent variable?

Additional Problems

15.49 Explain the differences and the similarities between simple linear regression analysis and multiple linear regression analysis.

15.50 Distinguish among the coefficient of determination, the coefficient of multiple determination, and the coefficient of partial determination in a bivariate regression model. How are these measures related to the coefficient of determination between the independent variables?

15.51 Discuss why it might be useful to include more than one independent variable in a regression equation to model the behavior of a dependent variable.

15.52 An evaluation has been undertaken of the salary of 12 members of the graduating class of 1970 10 years after graduation. These data are given in the tables, along with each student's undergraduate grade-point average (GPA) and the number of hours spent per week (average) in extracurricular activities while enrolled as an undergraduate (hours). Also given are the results of three regression analyses to predict salary 10 years after graduation. The first model includes only GPA, the second only hours, and the third both GPA and hours. The SAS results are given on page 841.

Student no.	Annual salary ($000)	GPA	Hours
1	16.5	2.02	0
2	29.4	3.44	7
3	31.9	2.87	17
4	18.4	2.41	4
5	35.6	3.61	15
6	22.8	2.36	5
7	37.1	3.22	2
8	30.5	3.11	10
9	26.3	2.54	9
10	27.9	2.88	12
11	56.9	3.21	22
12	25.4	2.62	6

a. Give the estimating equation associated with each model, and comment on both the significance of the individual terms included in the model and the overall model. Support your comments with appropriate statistical data.

b. Using the third model, estimate an individual's annual salary 10 years after graduation if she had an undergraduate grade-point average of 3.12 and nine hours per week spent in extracurricular activities. How does this estimate compare with the estimate derived using each of the models with one independent variable? Which model do you have more faith in and why?

c. Compute the correlation between GPA and hours. What implications (if any) does this correlation have in estimating salary?

15.53 In a regression analysis involving 27 observations and three independent variables, it is determined that SSR = 35,023.98, and SSE = 2,219.56. Determine the coefficient of multiple determination and the coefficient of multiple correlation for these data.

15.54 For the data in Problem 15.52, test the null hypothesis H_0: $\beta_1 = \beta_2 = \beta_3 = 0$ at the $\alpha = 0.05$ significance level.

15.55 Assuming the data in Problem 15.52 are multivariate normally distributed, compute the coefficients of partial determination for each of the independent variables (GPA and hours). What interpretation should be given to these coefficients?

15.56 In a multiple linear regression analysis involving three independent variables and 21 observations, SSE is determined to be 440.41. The inclusion of a fourth independent variable explains 13.21 more units of variability in the random variable Y.
a. Determine $s_{Y/x_1.x_2.x_3}$.
b. Determine $s_{Y/x_1.x_2.x_3.x_4}$.

15.57* Referring to Problem 13.44 in Chapter 13 and Problem 14.18 in Chapter 14, Joe College was able to obtain data on the 10 employees as to the length of service in the firm. These data, along with their hourly wage rate and years of postsecondary education, are shown in the table.

Worker	Years of postsecondary education	Years of service	Hourly wage rate
1	5.5	12	$18.20
2	0.0	3	6.50
3	1.0	1	3.90
4	4.0	10	11.70
5	3.5	6	10.40
6	2.0	20	7.80
7	6.0	14	15.60
8	2.5	7	9.10
9	4.5	16	13.00
10	5.0	13	14.30

a. Using multiple linear regression, derive an estimating equation for hourly wage rate using years of postsecondary education and years of service in the firm as independent variables.

b. What is the expected wage rate for a person with four years of college and six years of service in the firm?

c. What is the expected wage rate for a person with six years of college and four years of service in the firm?

Dependent variable: Salary   Independent variable: GPA

Source	Degrees of freedom	Sum of squares	Mean square	F-value	PR > F
Model	1	581.44675312	581.44675312	9.14	0.0128
Error	10	635.92241355	63.49224135		
Corrected total	11	1,217.36916667			

| | | R-square | | Standard deviation | |
| | | 0.477626 | | 7.97447436 | |

Parameter	Estimate	t for H_0: Parameter = 0	PR > $\|t\|$	Standard error of estimate
Intercept	−13.41025990	−0.92	0.3770	14.50422069
GPA	15.15378008	3.02	0.0128	5.01150389

Dependent variable: Salary   Independent variable: Hours

Source	Degrees of freedom	Sum of squares	Mean square	F-value	PR > F
Model	1	693.03704785	693.03704785	13.22	0.0046
Error	10	524.33211881	52.43321188		
Corrected total	11	1,217.36916667			

| | | R-square | | Standard deviation | |
| | | 0.569291 | | 7.24107809 | |

Parameter	Estimate	t for H_0: Parameter = 0	PR > $\|t\|$	Standard error of estimate
Intercept	18.77762376	5.07	0.0005	3.70334359
Hours	1.22356436	3.64	0.0046	0.33655173

Dependent variable: Salary   Independent variables: GPA, hours

Source	Degrees of freedom	Sum of squares	Mean square	F-value	PR > F
Model	2	834.31988297	417.15994148	9.80	0.0055
Error	9	383.04928370	42.56103152		
Corrected total	11	1,217.36916667			

| | | R-square | | Standard deviation | |
| | | 0.685347 | | 6.52388163 | |

Parameter	Estimate	t for H_0: Parameter = 0	PR > $\|t\|$	Standard error of estimate
Intercept	−3.30212536	−0.26	0.7987	12.56960850
GPA	8.83697116	1.82	0.1018	4.85025765
Hours	0.87436480	2.44	0.0375	0.35871323

d. Test the null hypotheses $H_0: \beta_1 = 0$ and $H_0: \beta_2 = 0$ at the $\alpha = 0.01$ significance level.

e. What conclusions can be drawn regarding the efficacy of years of postsecondary education and length of service in the firm in predicting the expected wage rate?

f. Test the null hypothesis $H_0: \beta_1 = \beta_2 = 0$ at the $\alpha = 0.05$ significance level.

g. Do residual plots appear to support the hypothesis that the underlying assumptions of the multiple linear regression model have been met?

15.58 Assuming the data in Problem 15.57 are multivariate normally distributed, determine the following:

a. The sample coefficient of multiple determination.

b. The sample coefficient of multiple correlation.

c. The correlation between the dependent variable and each of the independent variables.

d. The correlation between the two independent variables.

e. The coefficient of partial determination for each of the independent variables.

15.59* Sophie Stoltz has recently become concerned about the level of pay of females versus males in the creosoting division of the company in which she works. From "informed sources," she has been told that the average pay of females in her division is less than $22,000 annually, and the average pay for males is nearly $26,000. She has informed her employer that she is considering sueing for sex discrimination in the firm. A sample of 20 individuals in the creosoting division has been taken, producing the data shown in the table.

a. Develop an estimating equation for salary as a function of length of service in the firm and sex.

b. Using the F-test described in the text, test for the presence of a sex factor (bias) in the data at the $\alpha = 0.10, 0.05$, and 0.01 levels.

c. Using the results obtained in part b, if you were the judge, and using the information provided, would you rule in her favor or not? Explain.

Employee number	Annual salary ($000)	Years of service	Sex
1	24.6	5	M
2	17.5	1	F
3	22.1	4	M
4	34.3	11	M
5	23.5	8	F
6	32.1	14	M
7	20.5	1	M
8	31.5	10	M
9	21.2	4	F
10	26.4	6	M
11	21.9	4	F
12	18.7	2	M
13	25.9	5	F
14	27.2	9	M
15	20.5	3	F
16	26.9	8	F
17	19.2	2	F
18	20.8	4	M
19	27.2	7	M
20	19.4	2	F

15.60 Assuming the data in Problem 15.59 are multivariate normally distributed, determine the following:

a. The sample coefficient of multiple determination.

b. The sample coefficient of multiple correlation.

c. The coefficient of partial determination for the independent variables sex and length of service.

d. What interpretation can be placed on the coefficient of partial determination for sex, and how would this influence the ruling on sex discrimination in your opinion?

15.61 Suppose the total variation in a random variable Y is 500 units, and a multiple linear regression analysis involving three independent variables is performed to help account for the variability in Y. Based on 27 observations, it is determined that SSE $= 100$. Further, assume that the variation explained by only the variables x_1 and x_2 is 320 units.

a. Test the null hypothesis, $H_0: \beta_1 = \beta_2 = \beta_3 = 0$ at the $\alpha = 0.05$ level of significance.

b. Determine r^2.

c. Compute $r^2_{Y/x_3 \cdot x_1, x_2}$.

15.62 For the data given in Problem 15.18, compute the coefficients of partial determination for each of the independent variables (tempera-

ture and catalyst). What is the interpretation of these coefficients, and how does each compare with the coefficient of determination?

15.63 Reproduced below are the heights and weights of 33 randomly selected individuals.

Student	Height (inches)	Weight (pounds)	Sex	Student	Height (inches)	Weight (pounds)	Sex
1	68	141	M	18	73	180	M
2	62	114	F	19	69	163	M
3	64	114	F	20	72	145	M
4	74	180	M	21	68	162	M
5	72	185	M	22	65	109	F
6	70	139	M	23	68	146	M
7	64	115	F	24	74	173	M
8	68	180	M	25	72	163	M
9	75	160	M	26	68	183	M
10	69	180	M	27	71	173	M
11	65	137	M	28	66	122	M
12	62	133	F	29	72	170	M
13	67	138	M	30	68	145	M
14	72	205	M	31	72	135	M
15	67	160	F	32	72	200	M
16	72	183	M	33	·66	132	M
17	67	140	M				

a. What is the population frame for these data?

b. Fit a multiple linear regression plane to the data to estimate weight, using a binary variable to indicate sex, where 1 corresponds to a male, and 0 corresponds to a female.

c. Test for the significance of the model developed at the $\alpha = 0.05$ level of significance.

d. Test for the significance of each of the independent variables at the $\alpha = 0.01$ level of significance.

e. What proportion of the variability in weights is accounted for by height and sex in this sample data?

f. Predict the weight of a male who is 66 inches tall.

g. Predict the weight of a female who is 66 inches tall.

15.64 In Problem 15.63, the model $Y = \beta_0 + \beta_1 x_1 + \beta_2 x_2 + \epsilon$, where $x_1 = $ height, $x_2 = 0$ or 1, and $Y = $ weight. This model assumes that each independent variable enters the model linearly and additively. Can you identify a better fitting model by considering functions of the independent variables? If so, what is this model?

15.65 A business school in the southwestern United States had 48 students admitted to its fall semester MBA program on a full-time basis who completed the first year of studies in a two-year program. For each student, the graduate grade-point average (GGPA) at the completion of the first year in the program, the total score on the Graduate Management Admissions Test (GMAT), and the undergraduate grade-point average (UGPA) is recorded. These data are given below.

COLUMN COUNT ROW	GGPA 48	GMAT 48	UGPA 48
1	3.00000	450.	2.80000
2	3.25000	550.	3.43000
3	2.93000	540.	2.55000
4	3.00000	400.	2.60000
5	3.68000	480.	3.25000
6	2.67000	510.	2.59000
7	2.93000	540.	3.10000
8	3.75000	650.	4.00000
9	4.00000	600.	3.59000
10	2.85000	370.	3.60000
11	3.20000	450.	3.05000
12	3.18000	580.	2.71000
13	3.08000	450.	2.30000
14	3.75000	520.	3.42000
15	3.13000	500.	2.14000
16	3.33000	600.	2.82000
17	3.00000	450.	2.19000
18	3.15000	490.	2.34000
19	3.33000	410.	3.75000
20	2.65000	550.	3.94000
21	3.33000	480.	3.42000
22	4.00000	600.	2.49000
23	4.00000	480.	3.50000
24	3.68000	510.	2.80000
25	3.83000	509.	3.56000
26	3.20000	440.	2.68000
27	1.77000	430.	2.35000
28	3.75000	510.	3.74000
29	3.93000	640.	3.78000
30	1.50000	390.	3.57000
31	3.10000	680.	2.59000
32	2.93000	450.	2.82000
33	3.25000	480.	3.22000
34	4.00000	540.	2.77000
35	3.58000	540.	3.44000
36	3.75000	550.	3.10000
37	2.18000	430.	2.70000
38	3.20000	500.	2.54000
39	2.25000	570.	2.65000
40	3.33000	360.	3.28000
41	3.00000	450.	3.84000
42	2.65000	460.	3.07000

```
        43              2.68000         510.        2.35000
        44              2.83000         530.        3.95000
        45              4.00000         630.        4.00000
        46              3.33000         450.        3.26000
        47              2.43000         610.        2.89000
        48              2.75000         550.        2.68000
  -- DESCRIBE C1-C3
     GGPA   N = 48   MEAN =   3.1685   ST.DEV. =   0.574
     GMAT   N = 48   MEAN =   507.69   ST.DEV. =    74.5
     UGPA   N = 48   MEAN =   3.0669   ST.DEV. =   0.539
```

The MBA director is interested in the relationship between performance in the first year in the MBA program and the selection criteria—GMAT score and undergraduate grade-point average.

a. The scatterplot of GGPA (vertical axis) and GMAT score (GMAT) is given below. What relationship is indicated by this plot? How strong is the relationship?

b. The scatterplot of GGPA (vertical axis)

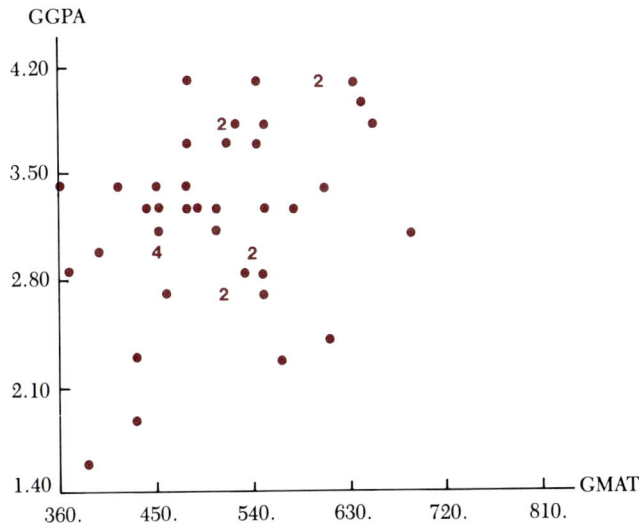

and UGPA is given on page 846. What relationship is indicated by this plot? How strong is the relationship?

c. The scatterplot of GMAT score (vertical axis) and UGPA is given on page 846. Based on this scatterplot, do you believe that multicollinearity may be a problem if both variables are entered into the re-

gression equation as independent variables in a linear fashion? Explain.

d. The correlation matrix for the three variables GGPA, GMAT score, and UGPA is given on page 846. Interpret the three correlation coefficients. What does each tell you about the relationships between pairs of variables?

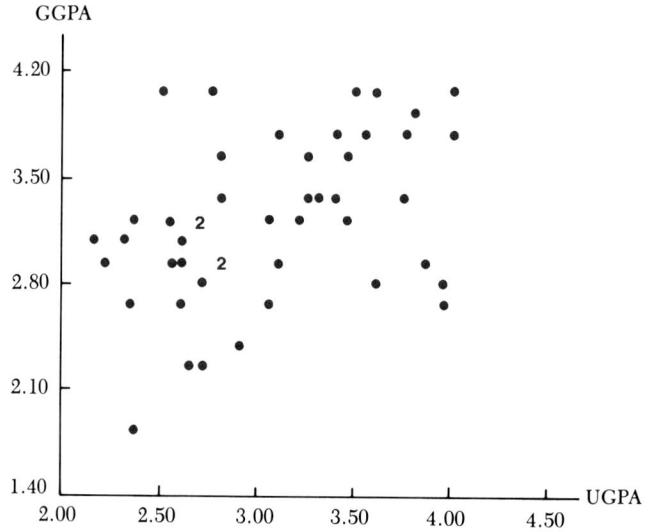

GGPA

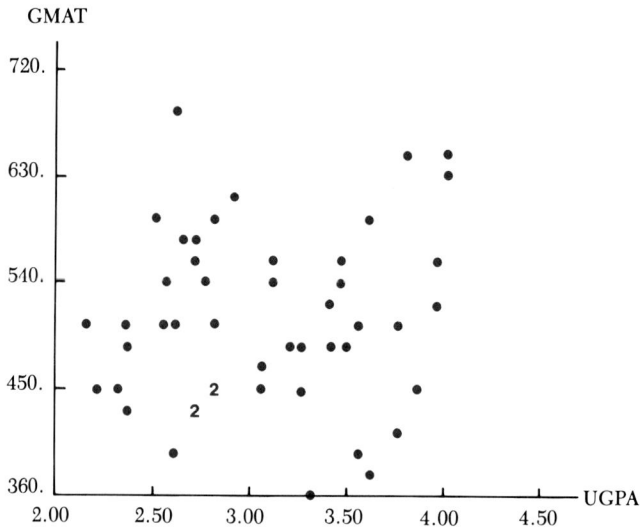

GMAT

```
1-- CORRELATE C1-C3

              GGPA        GMAT
GMAT          0.370
UGPA          0.311       0.106
```

15.66 Continuing Problem 15.65, the model $Y = \beta_0 + \beta_1 x_1 + \beta_2 x_2 + \epsilon$ is fitted where $x_1 =$ GMAT score, and $x_2 =$ UGPA. The MINITAB output for this model is given below.

```
-- REGRESS Y IN C1 ON 2 PREDICTORS IN C2,C3, STORE C4,C5,C6

THE REGRESSION EQUATION IS
Y =   0.940 +0.0026 X1 + 0.292 X2

                                  ST. DEV.    T-RATIO =
                                  OF COEF.    COEF/S.D.
          COLUMN    COEFFICIENT
          --          0.9402       0.6463        1.45
   X1     GMAT       0.002625     0.001026       2.56
   X2     UGPA        0.2921       0.1418        2.06


THE ST. DEV. OF Y ABOUT REGRESSION LINE IS
S = 0.5210
WITH (  48- 3) =   45 DEGREES OF FREEDOM


R-SQUARED = 21.1 PERCENT
R-SQUARED = 17.6 PERCENT, ADJUSTED FOR D.F.


ANALYSIS OF VARIANCE

   DUE TO         DF         SS      MS=SS/DF
REGRESSION       2        3.2715      1.6357
RESIDUAL        45       12.2151      0.2714
TOTAL           47       15.4866


FURTHER ANALYSIS OF VARIANCE
SS EXPLAINED BY EACH VARIABLE WHEN ENTERED IN THE ORDER GIVEN


   DUE TO         DF         SS
REGRESSION       2        3.2715
GMAT             1        2.1197
UGPA             1        1.1518


          X1          Y      PRED. Y    ST.DEV.
   ROW   GMAT       GGPA      VALUE     PRED. Y     RESIDUAL     ST.RES.
    8    650       3.7500     3.8146     0.2009      -0.0646      -0.13 X
   30    390       1.5000     3.0066     0.1648      -1.5066      -3.05R
   31    680       3.1000     3.4815     0.2098      -0.3815      -0.80 X
   45    630       4.0000     3.7621     0.1881       0.2379       0.49 X


R DENOTES AN OBS. WITH A LARGE ST. RES.
X DENOTES AN OBS. WHOSE X VALUE GIVES IT LARGE INFLUENCE.


DURBIN-WATSON STATISTIC = 1.85
```

a. What is the fitted regression equation?

b. For the fitted regression equation, test the hypothesis, $H_0: \beta_1 = \beta_2 = 0$ using an $\alpha = 0.05$ level of significance. Discuss the resulting hypothesis test decision.

c. What is the coefficient of multiple determination, r^2, for the fitted model. Interpret this number.

d. For the fitted regression equation, test the hypotheses $H_0: \beta_1 = 0$ and $H_0: \beta_2 = 0$ using the t-tests at the $\alpha = 0.05$ level of significance. What do these test results indicate?

15.67 Continuing Problem 15.66, answer the following questions related to validating the fitted regression equation in Problem 15.66.

a. Given below is the residual plot using the standardized residuals as the vertical axis and the predicted values as the horizontal axis. Based on the residual plot, do you believe that the three regression assumptions concerning the error terms (normality, equal variance, and uncorrelated error terms) are satisfied for this fitted model? Explain.

b. The Durbin-Watson statistic for the re-

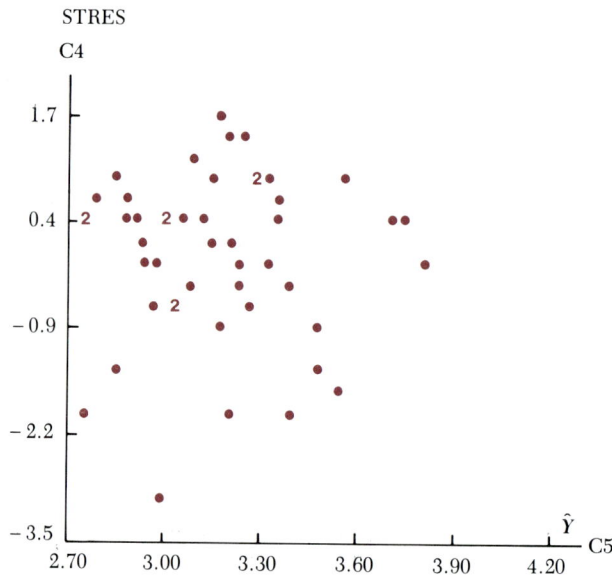

siduals is DW = 1.85. Using the $\alpha = 0.05$ level of significance, test the null hypothesis $H_0: \rho = 0$ against the alternative $H_A: \rho > 0$ to determine whether or not the first-order autocorrelation is significant.

c. The plot of the standardized residuals against the NSCORES is given on page 849 together with the correlation between the standardized residuals and the NSCORES.

Based on this plot and the correlation coefficient, are the error terms normally distributed (use $\alpha = 0.05$)? Explain.

15.68 Continuing Problem 15.67, answer the following regarding the degree of multicollinearity between $x_1 =$ GMAT score and $x_2 =$ UGPA.

a. From the MINITAB output, are there any informal indications of a high degree of multicollinearity between x_1 and x_2? Explain.

b. The regression model, $X_1 = \beta_0 + \beta_1 x_2 + \epsilon$ is fitted resulting in $r^2 = 0.011$. What does this tell you about the relationship between GMAT score and UGPA?

c. Using $r^2 = 0.011$ from part b, determine

```
-- NSCORES C4, PUT NSCORES IN C7
-- PLOT C4 VS C7
```

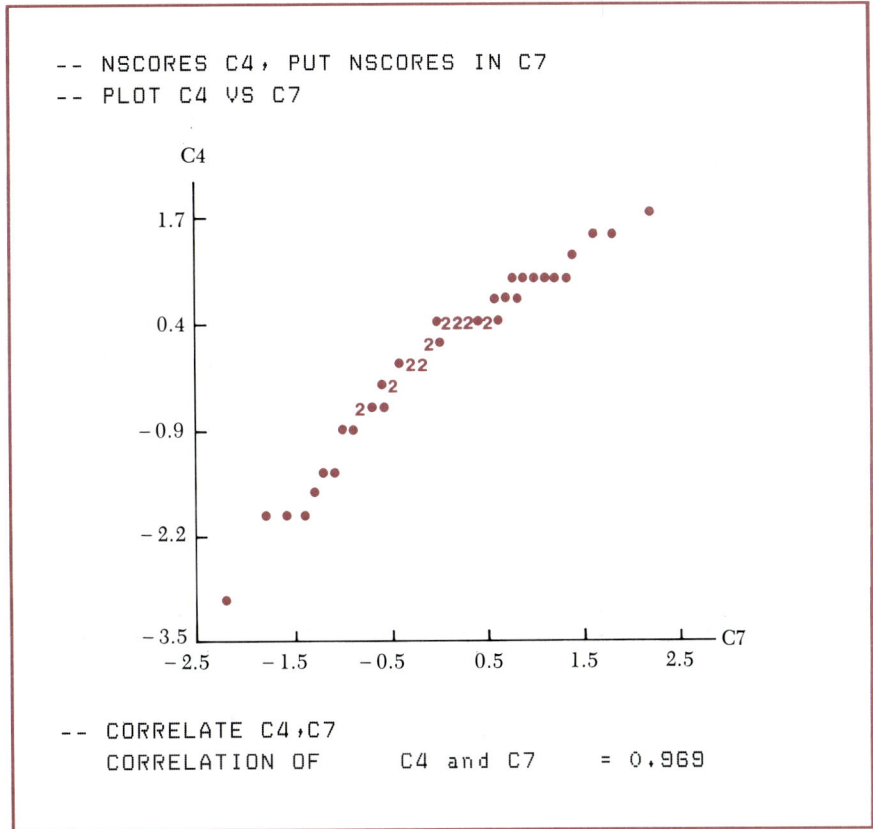

```
-- CORRELATE C4,C7
   CORRELATION OF     C4 and C7     = 0.969
```

the variance inflation factor (VIF). Based on the calculated VIF, is there a problem with multicollinearity? Explain.

15.69 Continuing Problem 15.68, the MBA director wishes to use the fitted regression equation to predict the first-year grade-point average (GGPA) for students applying to the program.

a. Predict GGPA for an applicant with a 580 GMAT score and a 3.75 UGPA; and determine a 95 percent prediction interval

for this person's GGPA. The MINITAB MACRO prediction information is given below.

b. Predict the GGPA for an applicant with a 300 GMAT score and a 2.00 UGPA. The MINITAB MACRO prediction information is given on page 850.

15.70 Suppose that there are $k = 10$ independent variables considered as potential predictors of the values of a dependent variable Y. Ex-

```
-- NOTE   PRINT PREDICTED VALUE
-- NOTE
-- NOTE   **********
-- PRINT K5
   K5          3.55787
-- NOTE   **********
-- NOTE
```

```
-- NOTE   PRINT STANDARD DEVIATION OF THE ESTIMATE OF THE
-- NOTE   CONDITIONAL MEAN OF Y
-- NOTE
-- NOTE   **********
-- PRINT K6
   K6         0.137876
-- NOTE   **********
-- NOTE
-- NOTE   PRINT STANDARD DEVIATION OF THE PREDICTION OF AN
-- NOTE   INDIVIDUAL VALUE OF Y
-- NOTE
-- NOTE   **********
-- PRINT K7
   K7         0.538941
-- NOTE   **********
```

```
-- NOTE   PRINT PREDICTED VALUE
-- NOTE
-- NOTE   **********
-- PRINT K5
   K5         2.31178
-- NOTE   **********
-- NOTE
-- NOTE   PRINT STANDARD DEVIATION OF THE ESTIMATE OF THE
-- NOTE   CONDITIONAL MEAN OF Y
-- NOTE
-- NOTE   **********
-- PRINT K6
   K6         0.258933
-- NOTE   **********
-- NOTE
-- NOTE   PRINT STANDARD DEVIATION OF THE PREDICTION OF AN
-- NOTE   INDIVIDUAL VALUE OF Y
-- NOTE
-- NOTE   **********
-- PRINT K7
   K7         0.581802
-- NOTE   **********
```

plain the process by which one would arrive at a "best" model containing one or more of these $k = 10$ independent variables in an appropriate functional form.

15.71 Why is it dangerous to use the coefficient of multiple determination r^2 alone for selecting the best regression model for a particular sample data set?

16

Time series

16.1

Introduction

Observations of data are frequently made over time. That is, we observe or record the value of a variable as of, or for, a specific instance and then again at some later time. For example, Figure 16.1 shows the growth in the gross national product (the total value of goods and services produced) in the United States recorded for the years 1929 to 1985. As indicated in the figure, there is a general pattern of increase in the level of GNP from 1929 to 1985, but with some variation in specific years or in specific groups of years (e.g., 1940–46). Figure 16.1 is an example of a time series.

Time series

> **Definition 16.1**
> Time series
>
> A *time series* consists of statistical data that are collected, recorded, or observed over successive increments of time.

Changes in a time series (such as shown in Figure 16.1) are the result of a variety of forces that make themselves felt in various ways. For example, the long-term growth in GNP can be associated with such factors as a general increase in the population, as well as a host of other factors. Deviations

FIGURE 16.1 | Gross National Product (GNP) of the United States, 1929–1985

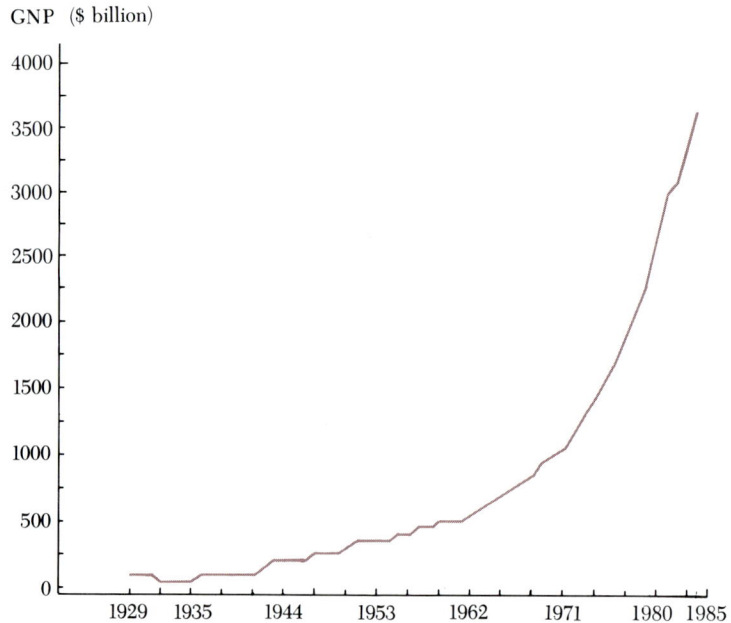

GNP ($ billion)

Source: U.S. Department of Commerce, *Survey of Current Business* (various issues).

from the general long-term growth pattern of GNP for specific years are usually attributed to a different set of factors, which would include the weather, wars, and political climate. Finally, if measurements of data are made within a year (or other period), seasonal factors often play a somewhat different role in influencing the direction or the magnitude of a given time series.

The objective of studying time series lies in the ability to estimate the value or the level of the series for a specified interval. Sociologists, economists, psychologists, environmentalists, and others ad infinitum often study patterns of growth of many or a few series to gain an understanding of certain phenomena in society and, of course, to plan or to estimate for the future. The estimation or prediction of the total demand for satellites for monitoring the discovery and depletion of specific natural resources, for example, was derived through the use of a logistic or a growth curve time series described in Section 16.3.

Whereas a general historical understanding of a time series enhances our ability to identify and analyze the underlying factors contributing to movements in the series, the greatest potential of a time series does in fact lie in the ability to predict an unknown value of the series. From this information, intelligent choices can be made concerning capital investment decisions or decisions about an appropriate level of production and inventory, etc. If one is willing to assume that there are *regular* and *repeating* components that interact in predictable ways to produce a given time series, he or she can then analyze these specific components to develop a reliable prediction of the series from which a decision can be made.

Regular and repeating components

In this chapter we will be concerned with describing "models" of time series as well as analyzing in detail the time series' component parts. Section 16.2 describes the factors that are usually attributed to or associated with movements in a time series, and specifies several common models that give the manner in which these factors interact. Section 16.3 then begins a discussion of the long-term growth component (secular trend) of a time series, and Section 16.4 presents statistical techniques necessary for isolating or measuring this component of a series. The effects of seasonal factors are addressed in Section 16.5, and methods are presented for "adjusting" statistical data for the effects of seasonal factors.

The statistical procedures to be employed in the chapter consist of linear regression discussed in the previous three chapters, as well as moving averages to isolate specific components. Because we will attempt to isolate specific components of a time series or to *decompose* it into its specific components, and because we will be employing classical techniques of linear regression and moving averages, the results of our analysis in this chapter will be termed *classical decomposition of time series*. In the following chapter, Chapter 17, we will study other methods of predicting the value of some time series, but will augment the techniques described in this chapter to predict the value of a time series at some future time, or with regard to another quantitative variable, etc. For example, we will use the multiple linear regression model for predicting accidents described in Section 15.7 to show how other than a specific "time" variable can be used to estimate the value of a time series at some future point. Because many of the assumptions underlying the use of the multiple linear regression model are violated when studying time series data (especially the assumption that the error terms are not correlated), we will have to be satisfied with obtaining point estimates of the time series only in this chapter and in the next.

Classical decomposition

■ 16.2

Components and models of a time series

16.2.1 | Models of a time series

A model of a time series is a specification of the forces that contribute to movements in the series as well as an analysis of the manner in which these forces interact in influencing the series' direction and magnitude. The most widely used model of a time series is the multiplicative model, in which the series is described as the *product* of four components—trend (T), seasonal (S) (if observations are recorded within a year), cyclical (C), and irregular (I)—although other models may be appropriate in analyzing specific series. This general model is described by the following equation.

Multiplicative model	**Multiplicative model of a time series** $$Y = T \times S \times C \times I$$

In the equation, **Y** is viewed as the result of the four elements acting in combination to produce the series. This particular model is well suited to situations in which *percentage* changes best represent the movement in the series. It turns out that a wide variety of economic as well as other data are best represented by the multiplicative model—hence its widespread acceptance as the "standard" or the "norm."

Any particular time series *may,* of course, be better represented by some model other than the multiplicative one. For example, if the components of trend, seasonal, cyclical, and irregular forces interact in an additive fashion to produce a given series, then the *additive model* may be appropriate.

Additive model	**Additive model of a time series** $$Y = T + S + C + I$$

If it is observed that certain components interact in an additive fashion in producing a given series, and that other factors interact in a multiplicative manner, then a variation of the two models may be used.

Mixed additive and multiplicative models	**Mixed additive and multiplicative models of a time series** $$Y = T \times C + S \times I$$ $$Y = T \times C \times I + S$$

Because of the widespread use of the multiplicative model and the large variety of different time series it represents, it will be the only model described in our treatment of time series analysis. One should keep in mind, however, that not *all* time series are best represented by this model; if an examination of the data reveals that certain components do not interact in the prescribed fashion, then a different model than the multiplicative one is appropriate.

16.2.2 | Components of a time series

A given time series consists of the following components:

1. Trend (T).

2. Seasonal (S).

3. Cyclical (C).

4. Irregular (I).

Secular trend

Definition 16.2
Secular trend

The *long-term trend* or *secular trend* of a time series is the smooth component of the series that represents the general long-run growth or decline in the time series over an extended period of time.

For example, Figure 16.1 depicts the growth of the gross national product of the United States from 1929 to 1985. In Figure 16.2 we have superimposed on this figure a representative trend curve, which reflects the general long-term growth in GNP over this interval. Although deviations from the long-term growth of GNP are apparent in Figure 16.2, one can recognize a long-term growth pattern. This long-term growth pattern is called the *trend* or *secular trend* of the series, and it is apparent that a curve as opposed to a straight line might better model the long-term growth in this series.

If observations in data are recorded within a year or shorter period, then *seasonal* variation, which reflects climate, customs (timing of vacations and holidays), and such other factors as the length of calendar months, may be present in the time series.

Seasonal variation

Definition 16.3
Seasonal variation

Seasonal variations are periodic patterns in a time series that complete themselves within a year and then are repeated according to the same periodic pattern in subsequent years.

FIGURE 16.2 | Gross national product of the United States (1929–1985) and representative long-term growth curve

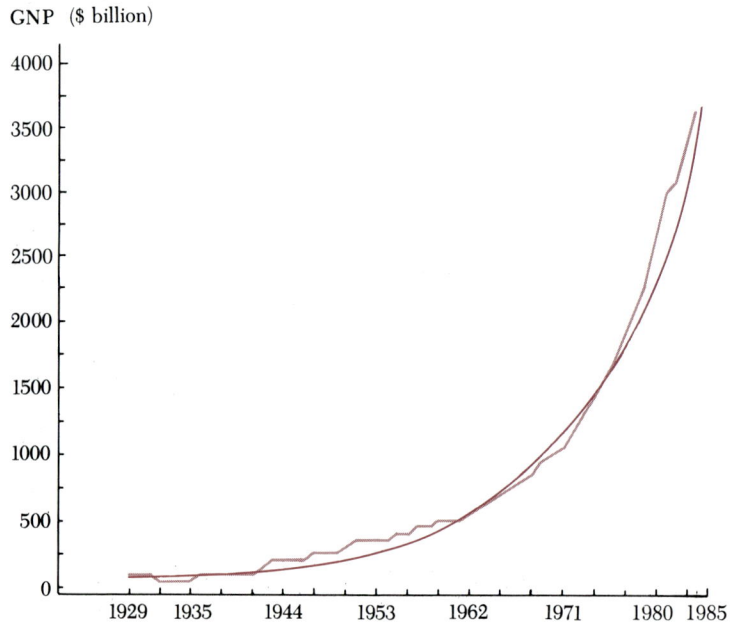

GNP ($ billion)

Source: U.S. Department of Commerce, *Survey of Current Business* (various issues), and computer analysis.

Figure 16.3 depicts the demand for plaster casts as a time series using the data given in Table 15.13 in Section 15.7. In addition to the apparent long-term (secular) growth in this series, there is a very definite seasonal pattern within each year as we discovered in Section 15.7. This seasonal pattern is more easily recognized by plotting the time series for selected years by quarter, as shown in Figure 16.3B for the years 1984–1986. Readily apparent from the data in Figure 16.3B is the tendency for the demand for casts to be higher (relative to other quarters) in the winter and fall quarters, and to be lower (relative to other quarters) in the summer quarter for each year. We can also isolate the seasonal component using residual plots described in Chapters 13 and 15, a topic we will defer until we attempt to assess the effects of seasonal factors in Section 16.5.

This wavelike, oscillating pattern is readily apparent in the data shown in Figure 16.2, especially for the years 1940–46. Different factors undoubtedly account for the cyclical movement in this time series and would include wars, uneven growth in the population, and others.

FIGURE 16.3

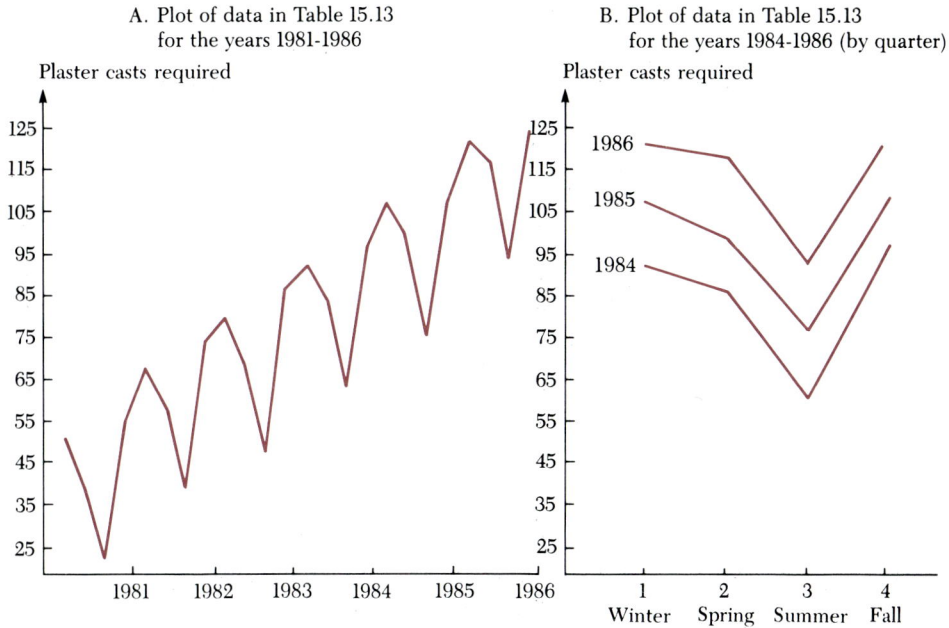

A. Plot of data in Table 15.13
for the years 1981-1986

Plaster casts required

B. Plot of data in Table 15.13
for the years 1984-1986 (by quarter)

Plaster casts required

Source: Company records.

Cyclical component	**Definition 16.4** Cyclical component of a time series The *cyclical component of a time series* refers to the recurring movements above and below the trend of the time series. These fluctuations last from two to 10 years (or even longer) when measured from peak to peak or from trough to trough. The duration of a cyclical component is more than one year.

Irregular variation	**Definition 16.5** Irregular variation The *irregular variation* in a time series is composed of nonrecurring, sporadic forces that are not described as or attributed to trend, cyclical, or seasonal factors.

We are rarely able to account for all the variation present in a time series, having broken the series down into its component parts. This residual variation, or the variation that remains after accounting for trend, cyclical, and possibly seasonal factors, is called *irregular variation*. It is desired that the irregular component will not comprise a large portion of the series and hence will be relatively unimportant in analyzing its movements.

The irregular component of a time series is analogous to the error terms ϵ in the linear regression models studied in previous chapters. In fact, when a residual plot forms a rectangle about 0, we are relatively assured that we have properly isolated the components of the time series under investigation.

16.3

Patterns of secular trend

In this section we will present several different mathematical functions for measuring or modeling the secular trend component of a time series. The curve used in modeling the trend component depends of course on the long-term growth in the series over the period for which trend is to be estimated. Although the selection of a particular mathematical model for assessing the trend component is as much an art as it is a science, two techniques—graphing and using the methods of first and second differences and constant rate increase—are described, which are often helpful in selecting an appropriate model to estimate trend. The use of these two approaches for assessing a model of trend is encouraged much in the same way as residual plots were encouraged in Chapter 13—to *suggest* the appropriateness of a certain technique. The student who wishes further information on the analysis of time series is encouraged to take an advanced course either in time series or in econometrics.

16.3.1 | Arithmetic straight-line trend

By far the most widely *applied* trend curve is the arithmetic straight-line trend, fitted by the method of least squares introduced in Chapter 13. This particular trend curve is applicable for series in which period-to-period changes are constant in absolute *amount*. (This is to be distinguished from the semilogarithmic trend curve, discussed shortly, which models time series that are changing at a constant *rate*.)

When the given time series is increasing at an approximately constant amount each period, then the straight-line or arithmetic trend represented by the equation

$$E(\mathbf{Y}) = \beta_0 + \beta_1 x$$

is used to model the long-term or secular growth of the series. If, for example, we want to estimate the secular trend in the demand for plaster casts shown in Figure 16.3A, then we would most likely use the arithmetic straight-line model, since the long-term growth over this period appears from the graph to be a constant amount each period (approximately).

Frequently, a given time series is modeled by a series of arithmetic straight-line trends, especially when a causative factor can be identified that contributes to a definite change in the amount of change per period.

Graphing the data is often the first step in assessing whether the trend component in the series is appropriately modeled by a straight line of the form $E(Y) = \beta_0 + \beta_1 x$. For example, it would appear that a straight line would be a poor "fit" of the growth in GNP in the United States between 1929 and 1985 from a visual inspection of the data plotted in Figure 16.1.

Method of first differences

Another "rapid" means for assessing the appropriateness of the straight-line model is the *method of first differences*. If the differences between successive observations of a series are constant (or nearly so), this suggests that an arithmetic straight line may be an appropriate representation of the trend component. The method of first differences is illustrated in Table 16.1 for the given series. Since the differences in successive observations are nearly constant, the arithmetic straight line is an appropriate model for assessing the trend component of the series.

TABLE 16.1 | Method of first differences

Year	Sales, y ($1000)	First difference
1977	$160	
		42
1978	202	
		40
1979	242	
		36
1980	278	
		37
1981	315	
		39
1982	354	
		41
1983	395	
		41
1984	436	
		42
1985	478	
		40
1986	518	

Source: Company records.

16.3.2 | Second-degree parabola trend

A polynomial of the form

$$E(Y) = \beta_0 + \beta_1 x + \beta_2 x^2$$

Second-degree parabola

is called a *second-degree parabola* and may be an appropriate model for the secular trend component of a time series when the data appear not to fall in a straight line. Figure 16.4 is a graph of a time series represented by the second-degree parabola $E(Y) = \beta_0 + \beta_1 x + \beta_2 x^2$, where $\beta_0 = 12$, $\beta_1 = 4$, and $\beta_2 = 3$. The method used to fit a second-degree parabola to a given time series is multiple linear regression described in Chapter 15, where

$$E(Y) = \beta_0 + \beta_1 x_1 + \beta_2 x_2 \quad \text{and} \quad x_1 = x, \, x_2 = x^2$$

FIGURE 16.4 | Plot of the second-degree parabola $y = 12 + 4x + 3x^2$ for values of x equal to 1, 2, . . . , 9

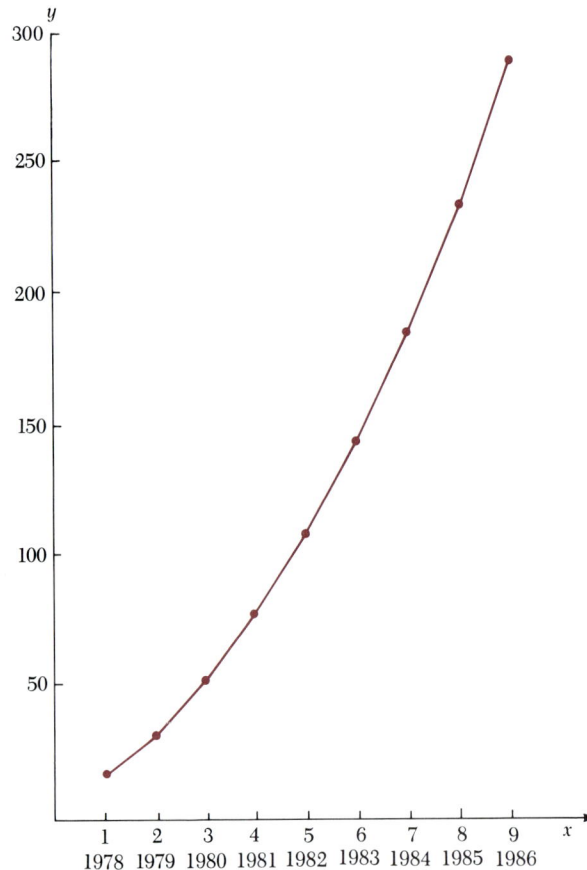

A test of the appropriateness of a second-degree parabola in modeling a given time series is based on the *method of second differences*. This method is illustrated in Table 16.2 for the parabola $y = 12 + 4x + 3x^2$. Note that when the differences are taken on the first differences, the result is a constant between successive observations of the given time series. Hence whenever the second differences of a given time series are constant (or are nearly so), a second-degree parabola fitted by the method of multiple linear regression is likely to be an appropriate model for measuring the secular trend of a given time series.

Method of second differences

TABLE 16.2 | Method of second differences

	$y = 12 + 4x + 3x^2$	First difference	Second difference
1	19		
2	32	13	6
3	51	19	6
4	76	25	6
5	107	31	6
6	144	37	6
7	187	43	6
8	236	49	6
9	291	55	

16.3.3 | Semilogarithmic trend

When a given time series is increasing at a constant *rate* and is *not* approaching some imputed upper limit, it is usually approximated best by an exponential curve, given by

$$E(\mathbf{Y}) = \beta_0 \beta_1^x$$

where the random variable $\mathbf{Y}$ measures the time series variable, and x is the period (weeks, months, years, etc.). By taking the log (base 10 is usually used) of both sides, we have

$$\log E(\mathbf{Y}) = \log\beta_0 + x\log\beta_1$$
$$= \beta_0' + \beta_1'x$$

where $\beta_0' = \log\beta_0$, and $\beta_1' = \log\beta_1$.

Table 16.3 gives information on the sales of electronic calculators at Midget Electronics, Inc. for the years 1979–1986, inclusive. A scatter dia-

TABLE 16.3 | Sales of electronic calculators—Midget Electronics, Inc.

Year	Sales ($000)
1979	$30.0
1980	32.1
1981	34.3
1982	36.8
1983	39.4
1984	42.3
1985	45.2
1986	48.5

Source: Company records.

FIGURE 16.5

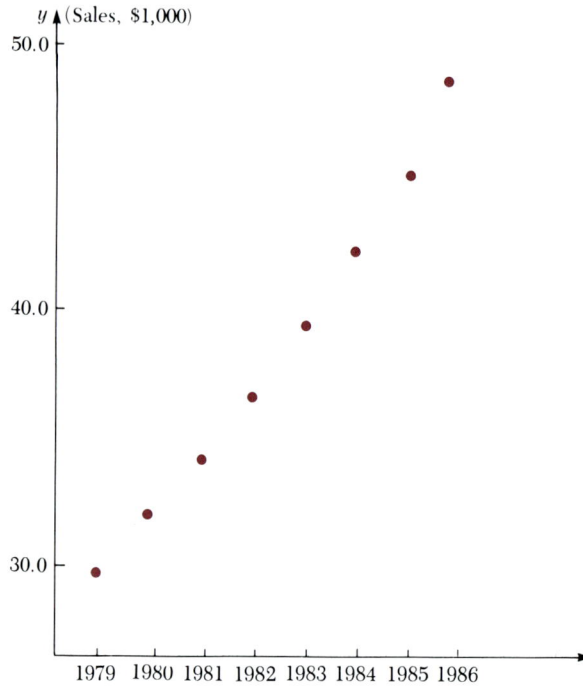

A. Scatter diagram of sales data in Table 16.3
(arithmetic scale)

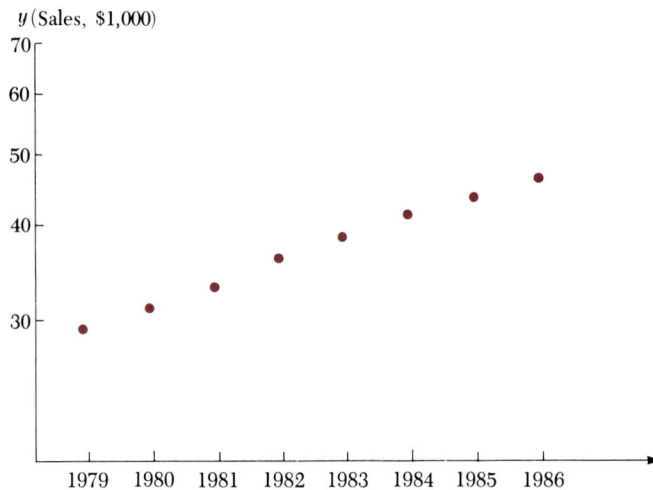

B. Semilogarithmic plot of sales data in Table 16.3

gram of the data in Table 16.3 is given in Figure 16.5A. From this figure, it appears as if a straight line may be an appropriate model of the secular trend of this series. However, when first and second differences of the data in Table 16.3 are taken (see Table 16.4), we find that a straight line or even a second-degree parabola may not be the best model for the secular trend of this series, as neither set of differences appears to be constant.

TABLE   16.4 | First and second differences for data in Table 16.3

x	y($000)	First difference	Second difference
1979	$30.0		
		2.1	
1980	32.1		0.1
		2.2	
1981	34.3		0.3
		2.5	
1982	36.8		0.1
		2.6	
1983	39.4		0.3
		2.9	
1984	42.3		0.0
		2.9	
1985	45.2		0.4
		3.3	
1986	48.5		

Rate of increase

Table 16.5 gives information on the *rate of increase* of calculator sales using the data in Table 16.3; the rate of increase appears to be fairly constant at around 7 percent each year. The data in Table 16.3 have also been plotted on semilogarithmic graph paper in Figure 16.5B. [With semilogarithmic graph paper, the units on the x-axis are evenly spaced, whereas the vertical scale (y-axis) is logarithmic, hence the term *semilogarithmic*.] Note from Figure 16.5B that when the data in Table 16.3 are plotted on semilogarithmic

Semilogarithmic graph paper

TABLE   16.5 | Assessing the rate of increase in a time series for semilogarithmic model

x	y($000)	Rate of increase (percent)
1979	$30.0	
		7.0
1980	32.1	
		6.8
1981	34.3	
		7.3
1982	36.8	
		7.1
1983	39.4	
		7.4
1984	42.3	
		6.8
1985	45.2	
		7.3
1986	48.5	

Source: Company records.

graph paper, the points appear to fall on a straight line. Hence whenever the percentage increase in a given time series is constant, or whenever a scatter diagram of the data points of the series on a semilogarithmic scale appears to cluster about a straight line, the semilogarithmic model for measuring the trend component is suggested.

Methods will be presented in the next section for fitting a semilogarithmic trend to a series of data using the method of least squares described in Chapter 13.

16.3.4 | Growth curves

Figure 16.6 depicts two growth curves that exhibit a pattern of growth in which there is an initial period of slow *absolute* growth, followed by a period of very rapid absolute expansion. Such growth patterns are apparent in the time series of sales in certain industries and in the population growth of various cities and states.

FIGURE 16.6 | Gompertz and Pearl-Reed growth curves

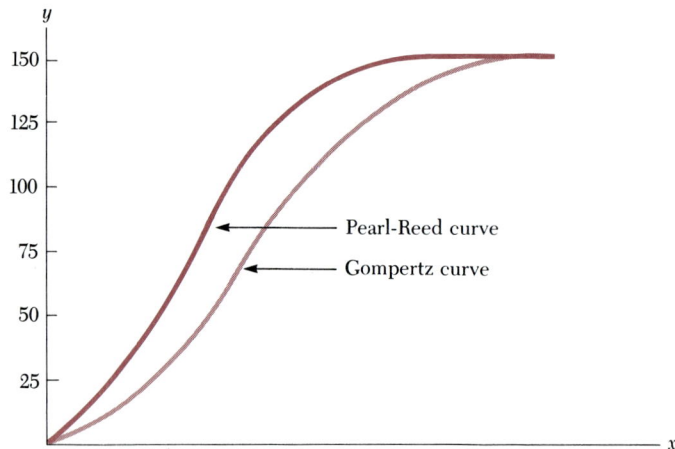

Growth curves

Gompertz curve	$y = ka^{b^x}$ (where k, a, and b are constants)
Pearl-Reed curve	$y = \dfrac{1}{k + ab^x}$ (where k, a, and b are constants)

Series such as those in Figure 16.6 are typically modeled by *growth curves;* two of the most widely used are the *Gompertz curve* and the *logistic* or *Pearl-Reed curve.* The formulas for these two families of growth curves are also given in Figure 16.6. Note that both curves depict trend patterns increasing at a decreasing *rate,* but the approach to maturity is much more rapid with the Pearl-Reed than with the Gompertz curve.

Both growth curves approach a finite limit (as shown), and this must be taken into account when fitting a given time series to one of the curves. Often a key resource is known to exist up to some finite amount, and this can be used to establish a limit on the growth of the time series in question. Increasingly, for example, "new cities" are being planned with an eye toward limiting growth, with the amount of planned land available having an upper limit. Time series analysis using growth curves analyzes or depicts the manner in which (or the rate at which) the series is approaching its limit. Thus whenever a given time series is increasing at a decreasing rate, but is understood to be approaching a finite limit in a predictable manner, the use of growth curves may be appropriate.

16.3.5 | Other patterns of secular trend

Other patterns of secular trend than those identified can occur, and special methods exist for fitting trend curves using these advanced techniques. For example, a polynomial of the form

Polynomial function

$$E(\mathbf{Y}) = \beta_0 + \beta_1 x + \beta_2 x^2 + \beta_3 x^3 + \cdots$$

may be used to represent a given time series. The various curve-fitting procedures for these advanced models can be found in the advanced references at the end of the chapter. We will now be concerned with fitting a given series using the arithmetic or the semilogarithmic trend curves using the method of least squares.

■ 16.4

Classical decomposition: Measuring the trend component of a time series

16.4.1 | Straight-line trend estimation

Whenever a scatter plot of the data or an examination of the first differences of the time series indicates that a straight-line trend of the series is appropriate, the trend component can be estimated using the method of linear least squares presented in Chapter 13. Unfortunately, most assumptions of the simple linear regression model—$E(\epsilon_i) = 0$, constant variance, uncorrelated error terms, normality, etc.—are not met when time series data are examined. Hence we cannot attach a measure of statistical confidence to the results obtained. We must be content merely to obtain *point estimates* (sometimes biased) of trend components, and we *may not* make statements as to the confidence that can be placed in these point estimates when the linear least squares method is used to assess the trend component of a time series. Example 16.1 introduces the notion of estimating linear trend by the method of least squares.

Example 16.1 The data in Table 15.13 represent the number of plaster casts placed on "patrons" to a ski and leisure resort in western Colorado between 1981 and 1986, by quarter. A graph of these data is given in Figure 16.3. From Figure 16.3 it appears as though a straight-line trend is the appropriate model for the *secular trend component* of this series. Using the method of least squares, compute a trend equation for the data provided. Use a straight-line to measure the trend component, and estimate the trend component for 1987.

Solution Since the demand for casts is being viewed as a function of time in the trend equation, time (in quarters) becomes the independent variable, x, and the demand for casts becomes the dependent variable, $\mathbf{Y}$, in the regression (trend) equation $E(\mathbf{Y}) = \beta_0 + \beta_1 x$. Because all time series models involve the common element of time, we shall denote the independent variable time by t. Here t is analogous to x in the simple linear regression model. Similarly, we will denote the values of the dependent random variable, $\mathbf{Y}$, by y_t to indicate that $\mathbf{Y}$ is a time series random variable. Thus our estimating equation will become:

$$\hat{y}_t = b_0 + b_1 t$$

By using the variable t for the independent variable in a trend equation, it will be easy to distinguish between time series variables and variables that are not necessarily time series. The computations, of course, will be the same whether we use x or t to denote the independent variable in our predicting equation.

Since β_0 and β_1 are rarely known in practice, they must be estimated by the least squares estimators b_0 and b_1, respectively. Equations for determining point estimates b_0 and b_1 are given as 16.1 and 16.2 to facilitate the computation of the estimates of the trend parameters.

Slope estimate in the straight-line trend equation	Least squares estimate of the slope of a straight-line trend equation

$$b_1 = \frac{\Sigma ty - \dfrac{(\Sigma t)(\Sigma y)}{n}}{\Sigma t^2 - \dfrac{(\Sigma t)^2}{n}} \qquad (16.1)$$

Intercept estimate in the straight-line trend equation	Least squares estimate of the intercept in a straight-line trend equation

$$b_0 = \bar{y} - b_1 \bar{t} = \frac{1}{n}(\Sigma y - b_1 \Sigma t) \qquad (16.2)$$

where $\bar{y} = (\Sigma y)/n$ and $\bar{t} = (\Sigma t)/n$.

TABLE 16.6 | Computation worksheet for fitting a straight-line trend to demand for plaster casts, 1981–1986

Year and period	Quarter	t	y	ty	t^2
1981					
1	Winter	0	53	0	0
2	Spring	1	41	41	1
3	Summer	2	24	48	4
4	Fall	3	57	171	9
1982					
5	Winter	4	70	280	16
6	Spring	5	60	300	25
7	Summer	6	41	246	36
8	Fall	7	77	539	49
1983					
9	Winter	8	81	648	64
10	Spring	9	70	630	81
11	Summer	10	50	500	100
12	Fall	11	87	957	121
1984					
13	Winter	12	94	1,128	144
14	Spring	13	86	1,118	169
15	Summer	14	64	896	196
16	Fall	15	99	1,485	225
1985					
17	Winter	16	109	1,744	256
18	Spring	17	101	1,717	289
19	Summer	18	77	1,386	324
20	Fall	19	110	2,090	361
1986					
21	Winter	20	123	2,460	400
22	Spring	21	120	2,520	441
23	Summer	22	95	2,090	484
24	Fall	23	126	2,898	529
		276	1,915	25,892	4,324

$$b_1 = \frac{25,892 - \dfrac{(276)(1,915)}{24}}{4,324 - \dfrac{76,176}{24}} = 3.3648 \qquad b_0 = \frac{1}{24}(1,915 - 3.3648 \cdot 276) = 41.0965$$

Before we begin to compute the estimates b_0 and b_1, note that the arithmetic is going to be quite burdensome because of the magnitude of the quantities involved. We can simplify the number of computations performed if we *code the data* and let the periods involved be represented by a sequence of nonnegative, increasing integers, where at the "base" (first) period, $t = 0$. This coding, along with relevant computational data for the trend equation, is given in Table 16.6. The trend equation for the data provided is:

Coding the data

$$\hat{y} = 41.0965 + 3.3648t, \qquad t = 0 \text{ at quarter I, 1981}$$

where t is in quarter years, and $\hat{y}$ is in single units.

Note that we are obligated (irrespective of the manner used to compute values of t) to state the origin of the trend equation ($t = 0$ at quarter I, 1981) so that trend values can be computed accordingly. Specific units for both t and y are often necessary also.

The trend (arithmetic straight-line) values for each quarter in 1981 to 1986 are given in Table 16.7. These values are derived by substituting the values $t = 0, 1, 2, \ldots$ into the trend equation for the quarters of the years 1981, 1982, 1983, and so on. A projection of the trend component for this time series for each quarter of 1987 is shown in the table.

	Quarter	*Trend projection*
I	Winter	$41.0965 + 3.3648(24) = 121.852$
II	Spring	$41.0965 + 3.3648(25) = 125.217$
III	Summer	$41.0965 + 3.3648(26) = 128.581$
IV	Fall	$41.0965 + 3.3648(27) = 131.946$

A graph of the derived trend equation and actual values is given in Figure

FIGURE 16.7 | Straight-line trend equation for the demand for plaster casts

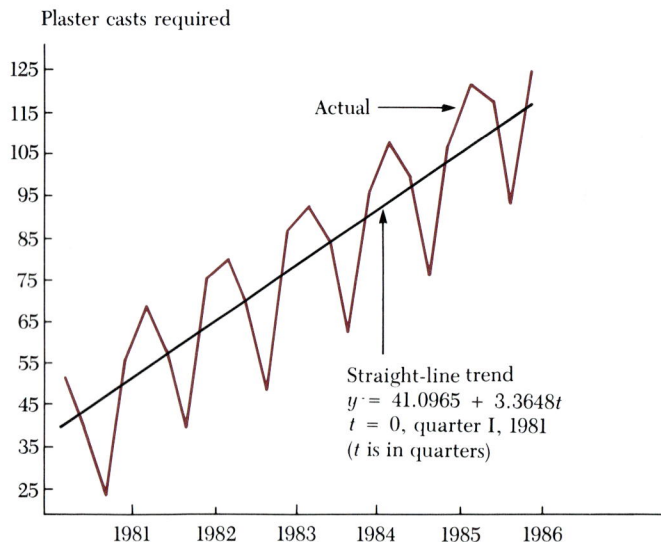

Plaster casts required

Actual

Straight-line trend
$y = 41.0965 + 3.3648t$
$t = 0$, quarter I, 1981
(t is in quarters)

1981 1982 1983 1984 1985 1986

TABLE 16.7 | Straight-line trend estimates for data in Table 16.6

Year and quarter	Observed value	Predicted value	Residual
1981			
Winter	53	41.097	11.903
Spring	41	44.461	−3.461
Summer	24	47.826	−23.826
Fall	57	51.191	5.809
1982			
Winter	70	54.556	15.444
Spring	60	57.920	2.079
Summer	41	61.285	−20.285
Fall	77	64.650	12.350
1983			
Winter	81	68.015	12.985
Spring	70	71.380	−1.380
Summer	50	74.744	−24.744
Fall	87	78.109	8.890
1984			
Winter	94	81.474	12.526
Spring	86	84.839	1.161
Summer	64	88.204	−24.204
Fall	99	91.568	7.431
1985			
Winter	109	94.933	14.067
Spring	101	98.298	2.702
Summer	77	101.663	−24.663
Fall	110	105.027	4.972
1986			
Winter	123	108.392	14.608
Spring	120	111.757	8.242
Summer	95	115.122	−20.122
Fall	126	118.487	7.513

Source: Tables 15.13 and 16.6.

16.7. The trend line computed appears to represent the long-term growth in this time series. As we learned in Section 15.7, however, seasonal factors play a role in determining the value of the time series *within each year*. This is also readily apparent in the graph in Figure 16.7. In Section 16.5 we will learn another method for determining the seasonal component of this time series, the method of ratio to moving average.

16.4.2 | Shifting the origin and the units of measurement in a straight-line trend equation

We frequently find that we would like to shift the origin or the units of measurement of a trend equation to get a forecast for a specific period in the future. Changing the origin when the units of time (years) do not change poses no special problems. The Y-intercept (b_0-value) simply becomes the trend value for that year, and the value b_1 (slope) remains the same.

Often, we would like a trend value for specific *months* within a year, and this necessitates our changing the units of the trend equation from years to months for an accurate trend extrapolation. The method for accomplishing this is presented in the following example.

Example 16.2 A trend equation for a manufacturer of a consumer durable good has been assessed by the method of least squares (straight-line trend), yielding the following trend equation:

$$\hat{y} = 19.88 + 2.86t, \quad t = 0 \text{ at } 1976$$

where t is in years, and $\hat{y}$ is in millions of dollars. Obtain an estimate of the trend component of this time series for July and August 1987 using the trend equation provided.

Solution Division of the given trend equation by 12 yields:

$$\hat{y} = 1.66 + 0.2383t$$

Note, however, that t is still expressed in years! That is, with our trend equation (centered at July 1 of each year), $t = 0$ implies July 1976, $t = 1$ implies July 1977, etc. To express the trend equation in consecutive *months*, it is necessary to divide the coefficient of t by 12 once more. The procedure for changing the units of measurement from years to months is

$$\begin{array}{cc} Years & Months \\ \hat{y} = b_0 + b_1 t & \hat{y} = \dfrac{b_0}{12} + \left(\dfrac{b_1}{144}\right) t \end{array}$$

Hence for our example,

$$\hat{y} = 1.66 + \left(\frac{0.2383}{12}\right) t = 1.66 + 0.02t,$$

$t = 0$ at July 1, 1976, t is in months, and $\hat{y}$ is in millions of dollars.

Quite often, one final adjustment is made to the preceding equation. Since the midpoint in each month is more representative of the trend value for that month, the monthly trend equation is centered. For the equation in question,

$$\hat{y} = 1.66 + 0.02\left(\frac{1}{2}\right) + 0.02t = 1.67 + 0.02t$$

where $t = 0$ at July 15, 1976, t is in months, and $\hat{y}$ is in millions of dollars.

Returning to the original problem, the projection of trend for July and August 1987 becomes:

$$\begin{aligned} \hat{y}_{\text{July 1987}} &= 1.67 + 0.02(132) = \$4.31 \text{ million} \\ \hat{y}_{\text{Aug 1987}} &= 1.67 + 0.02(133) = \$4.33 \text{ million} \end{aligned}$$

16.4.3 | Semilogarithmic trend estimation

It was noted in Section 16.3.3 that whenever a series is increasing at a constant *rate*, it is usually best represented by a semilogarithmic trend equation. Examples of time series exhibiting a constant rate of growth include U.S. gross national product from 1929 to 1985 (see Figure 16.1) and personal consumption expenditures for automobiles and parts from 1950 to 1985 (see Figure 16.8). The appropriateness of the semilogarithmic model is suggested whenever (1) a plot of the time series data on semilogarithmic graph paper tends to show points clustering about a straight line, and (2) the percentage increment between successive periods is constant or nearly so.

FIGURE 16.8 | Personal consumption expenditures for automobiles and parts, 1950–1985 (actual), and semilogarithmic trend

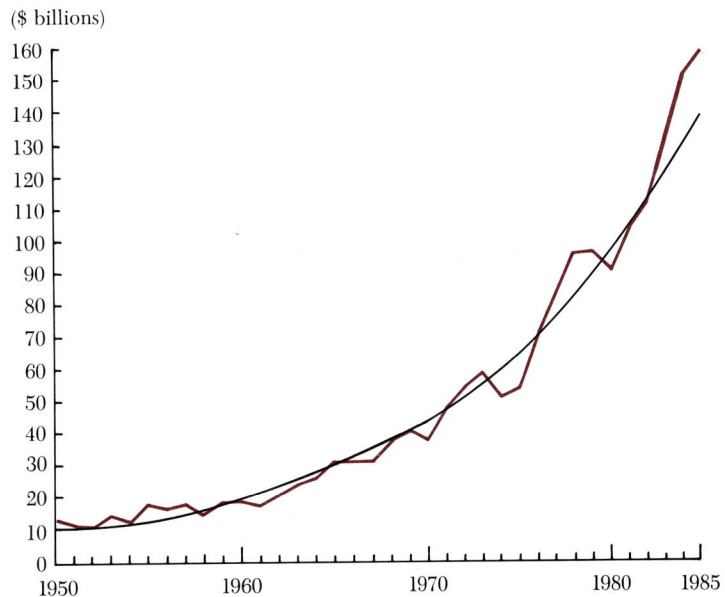

Source: U.S. Department of Commerce, *Survey of Current Business* (various issues), and computer analysis.

Rate of growth series

Estimates b_0 and b_1 of the parameters of the trend equation modeling a constant *rate of growth* series are given in equations (16.3) and (16.4). These formulas have been determined by the method of least squares. Note that they differ from the formulas for straight-line trend only in that log y replaces y in the computations.

Least squares estimate of the slope in a semilogarithmic trend equation

$$b_1 = \frac{\Sigma t \log y - \dfrac{(\Sigma t)(\Sigma \log y)}{n}}{\Sigma t^2 - \dfrac{(\Sigma t)^2}{n}} \qquad (16.3)$$

Least squares estimate of the intercept in a semilogarithmic trend equation

$$b_0 = \frac{1}{n}(\Sigma \log y - b_1 \Sigma t) \qquad (16.4)$$

An illustration of the use of these formulas for estimating trend is given in the following example.

Example 16.3   Personal consumption expenditures for automobiles and parts for the years 1950 to 1985 are given in the table below. Estimate the trend component in this series using the semilogarithmic model presented.

Year	Expenditures for automobiles and parts ($ billion)	Year	Expenditures for automobiles and parts ($ billion)
1950	12.9	1968	37.0
1951	11.1	1969	40.2
1952	10.4	1970	37.3
1953	13.2	1971	46.7
1954	12.6	1972	53.1
1955	17.2	1973	57.5
1956	15.8	1974	49.8
1957	17.1	1975	53.4
1958	13.9	1976	69.7
1959	18.1	1977	81.6
1960	18.8	1978	94.3
1961	17.1	1979	95.5
1962	20.4	1980	89.7
1963	24.3	1981	101.6
1964	25.8	1982	108.7
1965	29.8	1983	129.3
1966	30.3	1984	149.8
1967	30.5	1985	157.9

Source: U.S. Department of Commerce, *Survey of Current Business* (various issues).

Solution Relevant computations are shown in Table 16.8. Table B.8 in Appendix B gives the necessary data for computing the logarithms (to base 10) of the observed values. Note that it is again useful to code the data in some fashion. In this case we have simply let the first year in the series equal 0; the second year equal 1; etc. The relevant trend equation is:

$$\log \hat{y} = 0.970905 + 0.033455t, \quad t = 0 \text{ at } 1950$$

where t is in years and $\log \hat{y}$ is in log billions of dollars.

To determine the trend value from the semilogarithmic model, it is necessary to compute the antilogarithm of $\hat{y}$ for specific years. For example, to compute the trend value for 1955, we must find the antilogarithm of 1.13818 $= 0.970905 + 0.033455(5)$, which is (approximately) \$13.74 billion. Trend estimates for the remaining years are given in Table 16.12.

The rate of growth implicit in a semilogarithmic trend is often of interest. It is derived by solving the following equation

$$\log(1 + r) = b_1$$

where b_1 is the slope of the semilogarithmic trend equation, and r is the rate of growth (per year). For the data in question,

$$\log(1 + r) = 0.033445$$
$$1 + r = \text{antilogarithm } (0.033455)$$
$$r = 1.080 - 1 = 0.08$$

Hence we would state that the rate of growth in personal consumption expenditures for automobiles and parts between 1950 and 1985 (inclusive) has been about 8 percent.

There are other coding procedures for the variable "time" than the ones described thus far. For example, if there is an odd number of years (say, 7) and instead of letting the independent variable time assume the values 0, 1, 2, 3, 4, 5, and 6, we let the independent variable time assume the values -3, -2, -1, 0, 1, 2, and 3, the formulas for computing estimates b_0 and b_1 become somewhat simplified. Because computers are performing so many of the calculations in determining time series trend estimates, and because most pocket calculators have preprogrammed functions for measuring trend (straight-line) by the method of least squares, we will not go into detail in describing these other coding schemes for the time variable in a time series model. Several references at the end of the chapter present these other coding schemes for the time variable; the interested reader can find them there.

■ **16.5**

Classical decomposition: Measuring the seasonal component of a time series

It is difficult for anyone who regularly reads a newspaper or listens to the news not to have some notion of what are seasonally adjusted data. The government is continually bombarding us with seasonally adjusted data when it reports: "Unemployment for the month of June was up 5.2 percent *on a seasonally adjusted basis*" or "Gross national product rose last month by 3 percent on a seasonally adjusted basis." Naturally, what is meant is *not* that unemployment rose an *actual* 5.2 percent in June, but that when sea-

TABLE 16.8 | Computation worksheet for fitting a semilogarithmic trend to data on expenditures for automobiles and parts, 1950–1985

Year	Year in transformed units	Expenditures for automobiles and parts ($ billions)	$log_{10} y$	$t \, log_{10} y$	t^2
1950	0	12.9	1.11059	0.00000	0
1951	1	11.1	1.04532	1.04532	1
1952	2	10.4	1.01703	2.03407	4
1953	3	13.2	1.12057	3.36172	9
1954	4	12.6	1.10037	4.40148	16
1955	5	17.2	1.23553	6.17764	25
1956	6	15.8	1.19866	7.19194	36
1957	7	17.1	1.23300	8.63097	49
1958	8	13.9	1.14301	9.14412	64
1959	9	18.1	1.25768	11.31911	81
1960	10	18.8	1.27416	12.74158	100
1961	11	17.1	1.23300	13.56296	121
1962	12	20.4	1.30963	15.71556	144
1963	13	24.3	1.38561	18.01288	169
1964	14	25.8	1.41162	19.76268	196
1965	15	29.8	1.47422	22.11324	225
1966	16	30.3	1.48144	23.70308	256
1967	17	30.5	1.48430	25.23310	289
1968	18	37.0	1.56820	28.22763	324
1969	19	40.2	1.60423	30.48030	361
1970	20	37.3	1.57171	31.43418	400
1971	21	46.7	1.66932	35.05565	441
1972	22	53.1	1.72509	37.95208	484
1973	23	57.5	1.75967	40.47236	529
1974	24	49.8	1.69723	40.73350	576
1975	25	53.4	1.72754	43.18853	625
1976	26	69.7	1.84323	47.92405	676
1977	27	81.6	1.91169	51.61563	729
1978	28	94.3	1.97451	55.28633	784
1979	29	95.5	1.98000	57.42010	841
1980	30	89.7	1.95279	58.58377	900
1981	31	101.6	2.00689	62.21370	961
1982	32	108.7	2.03623	65.15935	1024
1983	33	129.3	2.11160	69.68275	1089
1984	34	149.8	2.17551	73.96740	1156
1985	35	157.9	2.19838	76.94337	1225
	630		56.02957	1110.49215	14910

$$b_1 = \frac{1,110.492 - \dfrac{630 \times 56.02957}{36}}{14,910 - \dfrac{(630)^2}{36}} = 0.033455$$

$$b_0 = \left(\frac{1}{36}\right)(56.02957 - 0.033455 \times 630) = 0.970905$$

$$\log \hat{y} = 0.970905 + 0.033445t$$

$$t = 0 \text{ at } 1950$$

t is in years; log $\hat{y}$ is in log $ billions.
Source: U.S. Department of Commerce, *Survey of Current Business* (various issues).

sonal factors are taken into account (e.g., recent graduates entering the job market), the rate of unemployment *still* was up 5.2 percent. The actual increase was most likely much larger than 5.2 percent, but because in general we expect it to be higher in June, we make adjustments in the data (deseasonalize) for this.

We first encountered the notion of a seasonal component of a time series in Section 15.7 when, using the analysis of variance procedure, we tested for the presence of a seasonal factor influencing the demand for plaster casts at a Colorado ski resort. A method for *identifying* the presence of a seasonal factor in addition to those already discussed is the use of residual plots described in Chapters 13 and 15. That is, when data are available on a monthly, quarterly, or other seasonal basis, we fit a trend line or curve to the data and examine a plot of the residuals. Such a plot is given in Figure 16.9

Seasonally adjusted basis

FIGURE   16.9 | Residual plot of the data in Table 16.7

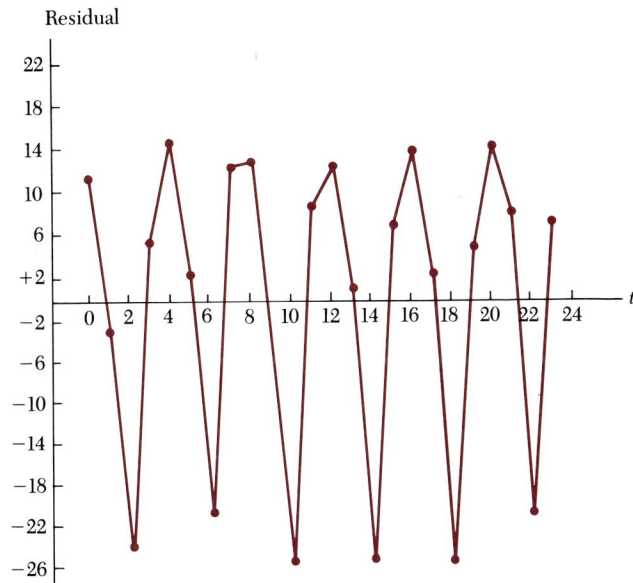

for the Table 16.7 data. Here we have "connected" the residuals between successive periods in order to emphasize their recurrent pattern between years. The analysis of the residual plot in Figure 16.9 would lead one to conclude not only that the residuals are not random, but also that there is systematic variability among them that should be able to be modeled. The modeling process consists of the identification and computation of seasonal indices.

Method of ratio to
moving average

We shall examine the data in Table 15.13 once more and assess the seasonal component by a different method than was discussed in Chapter 15—namely, by the method of *ratio to moving average*.

The first two columns of Table 16.9 give information on the demand for plaster casts repeated as a time series from Table 15.13. These data will be used to describe the ratio to moving average method, as we attempt to determine a quarterly seasonal index for the demand for plaster casts.

TABLE 16.9 | Computation of specific seasonal relatives using method of ratio to moving average

(1) Date	(2) Plaster casts	(3) Four-quarter moving total	(4) Eight-quarter centered moving total	(5) Centered moving average (4) ÷ 8	(6) Specific seasonal relative [(2) ÷ (5)] × 100
1981					
Winter	53				
Spring	41				
		175			
Summer	24		367	45.88	52.30
		192			
Fall	57		403	50.38	113.20
		211			
1982					
Winter	70		439	54.88	127.60
		228			
Spring	60		476	59.50	100.80
		248			
Summer	41		507	63.38	64.70
		259			
Fall	77		528	66.00	116.70
		269			
1983					
Winter	81		547	68.38	118.50
		278			
Spring	70		566	70.75	98.90
		288			
Summer	50		589	73.63	67.90
		301			
Fall	87		618	77.25	112.60
		317			
1984					
Winter	94		648	81.00	116.00
		331			
Spring	86		674	84.25	102.10
		343			
Summer	64		701	87.63	73.00
		358			
Fall	99		731	91.38	108.30
		373			
1985					
Winter	109		759	94.88	114.90
		386			
Spring	101		783	97.88	103.20
		397			
Summer	77		808	101.00	76.20
		411			
Fall	110		841	105.13	104.60
		430			
1986					
Winter	123		878	109.75	112.10
		448			
Spring	120		912	114.00	105.30
		464			
Summer	95				
Fall	126				

In the third column of Table 16.9, the total demand for casts is recorded for consecutive four-quarter periods. These totals are centered between the quarters, because the midpoint in any consecutive four-quarter period lies at the very end of the second quarter in the series. For example, for the four quarters of the year, the midpoint of the period is July 1, or the *start* of the third quarter. Hence, the four-quarter moving total for the four consecutive quarters of a year are placed between the second and the third quarters, or at the midpoint of the year.

To compute the second four-quarter moving total in Table 16.9, we subtract from the current total (175) the value for the first quarter of 1981 (53) and add to this value the time series value for the first quarter of 1982 (70). This moving total = $175 - 53 + 70 = 192$ is then centered between the third and fourth quarters of 1981. This procedure for obtaining the four-quarter moving totals is shown in detail in Table 16.10.

TABLE 16.10 | Computation of four-quarter moving total

Date	Actual data	Four-quarter moving total
1981		
Winter	53	
Spring	41	175
Summer	24	192
Fall	57	211
1982		
Winter	70	
Spring	60	
Summer	41	
Fall	77	

In order to center the moving totals so they will correspond to specific quarters, and using the midpoint of each quarter as representative of the value for that quarter, eight-quarter *centered* moving totals are computed by adding consecutive two-period, four-month moving totals, such as indicated in column (4) of Table 16.9. Division of this quantity by 8 then gives the centered moving average for the quarter involved. These centered moving averages are indicated in column (5) of Table 16.9. A graph of these centered moving averages as well as the time series values is given in Figure 16.10. Their computation is straightforward; for the third quarter of 1981, the centered moving average is $(175 + 192 = 367)/8 = 45.88$, for example. Note from Figure 16.10 the "smoothing" effect the moving average has on the time series data. This is a point we will return to in the next chapter, when we discuss moving averages as a method of forecasting the value of a time series.

FIGURE 16.10 | Time series values for the demand for plaster casts and their centered moving averages

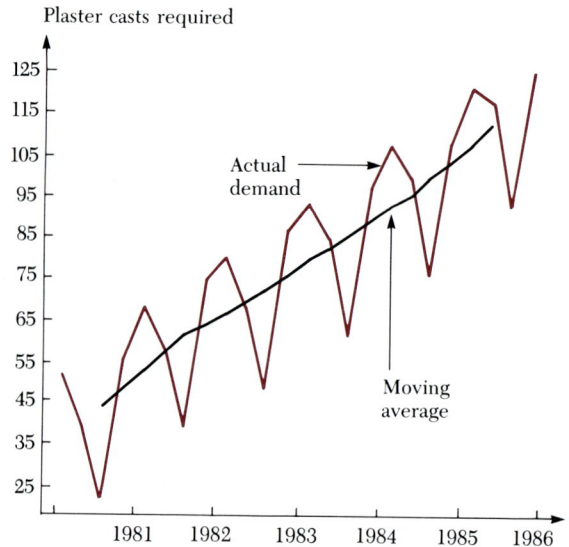

Source: Tables 15.13 and 16.9.

Specific seasonal relative

The ratio to moving average method derives its name from the division that takes place between the actual data and the corresponding centered moving average, yielding a *specific seasonal relative*. The term *specific* is attached to these seasonal relatives because each relative refers to a specific quarter and year. Note that by using quarterly data and the method of ratio to moving average, specific seasonal relatives do not exist for the first two quarters of the series and the last two quarters of the series, because it is impossible to construct a four-quarter moving average for these quarters. If instead we were computing monthly seasonal indices, then specific seasonal relatives would not exist for the first six months and the last six months of the series, and so on.

The data in column (6) of Table 16.9 have been rearranged in Table 16.11 listing the relatives by quarter for each of the years involved. Seasonal (quarterly) indices are then computed by either of two methods from the specific seasonal relatives listed.

Computation of seasonal indices: Use of the median and the mean

1. The *median* of the specific seasonal relatives for each quarter becomes the unadjusted seasonal index for that quarter, and these unadjusted seasonal indices are multiplied by the ratio of 400 (4 quarters, 100 base value per quarter) to their total to normalize their total to 400. The result is an *adjusted seasonal index* based on medians.

TABLE 16.11 | Computation of adjusted seasonal indices

	Specific seasonal relatives				
Year	Q1 Winter	Q2 Spring	Q3 Summer	Q4 Fall	
1981			52.30	113.20	
1982	127.60	100.80	67.40	116.70	
1983	118.50	98.90	67.90	112.60	
1984	116.00	102.10	73.00	108.30	
1985	114.90	103.20	76.20	104.60	
1986	112.10	105.30			
					Totals
Median of specific seasonal relatives	116.00	102.10	67.90	112.60	398.60
Adjusted seasonal index	116.40	102.46	68.14	113.00	400.00
Average of specific seasonal relatives	117.82	102.06	67.36	111.08	398.32
Adjusted seasonal index	118.31	102.49	67.64	111.56	400.00

2.   The *arithmetic mean* of the specific seasonal relatives for each quarter is the unadjusted seasonal index for that quarter, and all unadjusted seasonal indices are normalized (as above) so that they total 400. The result is an *adjusted seasonal index* based on averages.

Quarterly, seasonal indices based on both the median and the arithmetic mean are indicated in Table 16.11. Some economists would argue that a third method should be examined in addition to the two discussed above. Namely, any "outliers" from the data should be omitted, and the arithmetic mean should be taken of the remaining specific seasonal relatives. This method is

Modified mean

referred to as computing unadjusted seasonal indices using a *modified mean*. The difficulty in applying this third method lies in the determination of exactly what constitutes an outlier, and what does not. The data in Table 16.11 would appear to have no quarter with a specific seasonal relative that deviates significantly from the rest of the values.

Whichever method is used to compute the unadjusted seasonal indices, part of the irregular variation (I) in the model $\mathbf{Y} = \mathbf{T} \cdot \mathbf{S} \cdot \mathbf{C} \cdot \mathbf{I}$ is "averaged out" in the process of determining the adjusted seasonal indices. This smoothing effect of the averaging process eliminates many irregular changes from period to period and hence often dampens much of the irregular variation in the series.

If the data on the demand for plaster casts had been given monthly, then it would have been imperative to compute monthly seasonal indices as opposed to quarterly indices. In the third column of Table 16.9, we would have computed a 12-month moving total, and in the fourth column, we would have computed a 24-month centered moving total. The quantity in column (4) would have then been divided by 24, yielding the centered moving average, and this quantity would have then been divided into the actual demand for the month, yielding the specific seasonal relative for the month. The rest of the analysis would have proceeded similarly.

The use of seasonal indices in decision making is illustrated in the following examples.

Example 16.4 The firm desiring to estimate the demand for plaster casts has obtained an estimate of *yearly* demand for 1987 using a straight-line arithmetic trend of 521 casts. From this annual estimate, derive an estimate for each of the quarters of 1987 using the seasonal indices in Table 16.11 based on the arithmetic mean.

Solution Since annual demand has been estimated to be 521 units, the demand for each quarter if there were no seasonal influence in these data would be 130.25 units per quarter (521/4). An estimate of the demand for each quarter is obtained by multiplying the average quarterly estimate of 130.25 by the specific seasonal relative and dividing the result by 100 as shown in the table.

(1)	*(2)*	*(3)*	*(4)*
	Average quarterly	*Seasonal*	*Quarterly forecast*
Quarter	*demand (estimated)*	*index*	$[(2) \cdot (3)]/100$
Winter	130.25	118.31	154.1
Spring	130.25	102.49	133.5
Summer	130.25	67.64	88.1
Fall	130.25	111.56	145.3
	521.0		521.0

We will compare these estimates with those derived from a time series forecasting equation in the next chapter, where indicator or dummy variables will be used in a multiple regression model as described in Chapter 15 to assess the seasonal component in this time series.

Example 16.5 The seasonal index for a particular consumer good for January is 110.24. Actual demand for January 1986 was 1,085 units. Based on only this information, what would you estimate sales to be in 1986?

Solution Since January in general represents a month in which demand is 10.24 percent above an "average" month, average monthly demand based on 1,085 units being demanded in January is

$$\frac{1,085}{110.24} \times 100 = 984 \text{ units}$$

From the information provided, we would estimate total yearly sales to be

$$984 \times 12 = 11,808 \text{ units}$$

Example 16.6 Mayor Roger, after four clean years in office, has come under attack for his poor record on unemployment in the greater Glenwood Springs metropolitan area. "For the past two quarters," his opponent, council member Gene says, "unemployment has risen drastically!"

Gene presents the following evidence to support his contention:

Quarter	Number of unemployed in Glenwood Springs
I	285
II	327
III	340

Mayor Roger has been provided with the following quarterly indices of unemployment by the Glenwood Springs Chamber of Commerce. Not being a very good statistician, however, Mayor Roger does not know how to use this information to help his cause. How can the statistics quoted by Gene be recast in order to better judge the policies of Mayor Roger on unemployment?

Quarter	Index
I	0.90
II	1.10
III	1.15
IV	0.85

Solution In essence, the problem calls for the data presented by Gene to be deseasonalized. That is, we need to assess the net change in unemployment when factors affecting the seasonality of the job market are taken into account. To deseasonalize the data, we must divide the actual unemployment by the seasonal (quarterly) indices provided.

Quarter	Actual unemployment	Quarterly index	Deseasonalized unemployment
I	285	0.90	317
II	327	1.10	297
III	340	1.15	296
IV	—	0.85	—
		4.00	

Note that in this case we do not multiply the quotients by 100 because the quarterly indices are expressed on a basis totaling 4 for the four quarters of the year, rather than 400.

When the deseasonalized data are examined, it appears that Mayor Roger's policy on unemployment has been effective in actually reducing the seasonal level of unemployment in the Glenwood Springs area over the past two quarters.

This example should sharpen the distinction between seasonalized and deseasonalized data! Moreover, it should alert the user (voter) to the dangers in interpreting time-phased statistical data *of any type* when all the facts have not been reported. This is especially true with respect to "performance" data when they are measured between successive increments of time.

■ 16.6

Classical decomposition: Measuring the cyclical component of a time series

---
Model for annual data

The model of a time series when data are collected on an annual basis is given by:

$$\mathbf{Y} = \mathbf{TCI}$$

(no seasonal factor is involved since measurements are not made within a year). Previous sections have discussed the estimation of the trend component, **T**, where the trend is given either by a straight line or by a semilogarithmic curve. An estimate of the movement in a time series of the form given above due to cyclical (and irregular) forces can be derived by simply dividing the preceding equation by T, the trend value, because:

$$\frac{TCI}{T} = CI.$$

An estimate of the movements in the time series of personal consumption expenditures for automobiles and parts for 1950–1985 examined in Section 16.4.3 is given in Table 16.12. A graph of the fluctuating component (C and I) of this time series is given in Figure 16.11. The CI component of this time series is assessed by dividing the actual value of the time series, y, by the trend estimate, $\hat{y}$. From Figure 16.11 note how the fluctuating component of personal consumption expenditures for automobiles and parts appears to follow rather closely the fluctuations in general business activity, being down in periods of near or mild recession (1952, 1958, 1961, 1970, 1974–1975) and up in periods of high prosperity (1950, 1955, 1978, 1983–1984). Any projection or estimate (forecast) of expenditures for automobiles and parts should thus take into account the general level of business and automobile activity.

The process of dividing the actual value of the time series by the trend estimate to assess the CI component is valid no matter which of the curve-fitting procedures described in the chapter are used to measure the trend component. Thus, if a straight-line or a power curve would best assess the secular trend component of a given time series, we would still obtain an estimate of the cyclical and irregular component by dividing the actual value

TABLE 16.12 | Fluctuating component in expenditures for automobiles and parts as a percent of trend (semilogarithmic model), 1950–1985

Transformed year (t = 0 at 1950)	Actual ($billions)	Semilogarithmic trend	Actual as a percent of trend
0	12.9	9.4	137.938
1	11.1	10.1	109.891
2	10.4	10.9	95.327
3	13.2	11.8	112.022
4	12.6	12.7	99.002
5	17.2	13.7	125.125
6	15.8	14.8	106.419
7	17.1	16.0	106.636
8	13.9	17.3	80.254
9	18.1	18.7	96.755
10	18.8	20.2	93.046
11	17.1	21.8	78.357
12	20.4	23.6	86.548
13	24.3	25.5	95.451
14	25.8	27.5	93.829
15	29.8	29.7	100.341
16	30.3	32.1	94.460
17	30.5	34.6	88.034
18	37.0	37.4	98.877
19	40.2	40.4	99.464
20	37.3	43.7	85.446
21	46.7	47.1	99.048
22	53.1	50.9	104.272
23	57.5	55.0	104.541
24	49.8	59.4	83.828
25	53.4	64.2	83.224
26	69.7	69.3	100.574
27	81.6	74.9	109.015
28	94.3	80.8	116.641
29	95.5	87.3	109.367
30	89.7	94.3	95.109
31	101.6	101.9	99.740
32	108.7	110.0	98.798
33	129.3	118.8	108.808
34	149.8	128.3	116.713
35	157.9	138.6	113.903

of the time series by the trend estimate. Economists often study business and other ''cycles'' in order to warn of a sharp increase or decrease in general economic activity away from the secular trend component of the time series. Residual plots as described in the text are often an effective means for isolating and measuring the cyclical component after the trend of the time series has been suitably modeled.

FIGURE 16.11 | Personal consumption expenditures for automobiles and parts as a percent of trend (semilogarithmic), 1950–1985

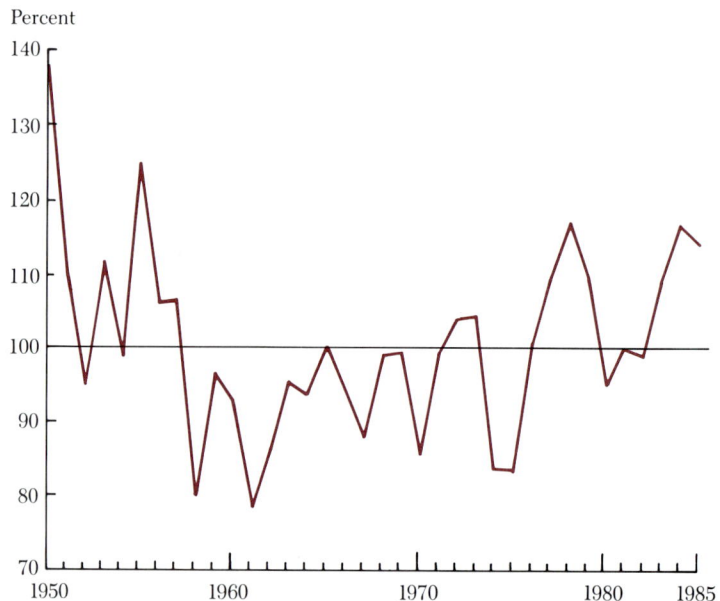

Source: U.S. Department of Commerce, *Survey of Current Business* (various issues), and computer analysis.

■ **16.7**

Classical decomposition: Measuring the irregular component of a time series

Method of residuals

After the quantities $C \cdot I$ have been isolated as described above, an attempt is made to measure the irregular component by computing a measure of the cycle. The following division is then performed to isolate the irregular component:

$$\frac{CI}{C} = I$$

Several methods exist for "measuring" or assessing the cyclical component, such as averaging across several cycles and using the averages to compute a "cyclical index" for specific years. One such method is called isolating the cyclical component by the *method of residuals*.

After the components of trend, cycle, and season (if measurements are made within a year) are measured, an analysis of the residuals of the difference between the actual and the estimated values should yield a rectangle about 0. This is an indication that all that remains after the modeling process is completed is the irregular or the error component. It is hoped that this will be a small part of the total!

■ 16.8
Summary

This chapter has presented methods for describing data that are collected, recorded, or observed over successive increments of time. Such data are referred to as *time series*.

The components of time series are: (1) trend, which represents the general long-term growth or decline of the data; (2) cyclical, which is the irregular, oscillating pattern of movement in the time series of more than one year's duration; (3) seasonal, which, if data are measured within a year, gives the repetitive, *recurrent* pattern of movement in the time series; and (4) irregular, which is the nonrecurrent, sporadic movement in the series that is not described by the other three factors. Several patterns of trend were reported and included growth curves and semilogarithmic and straight-line trends. Least squares regression was used to measure the trend component.

The presentation and computation of several indices alerted the reader to the dangers inherent in interpreting statistical data that are measured over time when these data have not been adjusted for seasonal (and possibly other) factors. This is particularly true regarding measures of performance (both good and bad!).

We hope the chapter stimulated sufficient interest that some of the more advanced methods for analyzing a time series (listed in the references at the end of the chapter) will be investigated at a later time. With the general availability of large-scale computers and software programs, the use of these more advanced techniques is quite simple and straightforward in the majority of computer installations.

One topic that is of special interest regarding the analysis of seasonal indices is analysis of variance presented in Chapter 12. With this technique, we can test the hypothesis that the seasonal indices are all equal, that is, $\mu_{Jan.} = \mu_{Feb.} = \cdots = \mu_{Dec.}$. If the hypothesis is accepted that they are all equal, then the indices we are reporting could be due to chance alone. Rejection of the hypothesis that $\mu_{Jan.} = \mu_{Feb.} = \cdots = \mu_{Dec.}$ adds credibility to the use of the seasonal indices in analyzing seasonal data. This is just one area where one might expand his or her knowledge of the analysis of time series data, and it is similar to the analysis of variance application in Chapter 15 to test for the presence of a seasonal factor in a series.

■ References

Introductory

Hamburg, M. *Basic Statistics*. 3rd ed. New York: Harcourt Brace Jovanovich, 1985, Chapter 13.

Neter, J.; W. Wasserman; and G. A. Whitmore. *Applied Statistics*. 2nd ed. Boston: Allyn & Bacon, 1982, Chapter 24.

Wonnacott, R., and T. Wonnacott. *Introductory Statistics for Business and Economics*. 3rd ed. New York: John Wiley & Sons, Chapter 24.

Advanced

Bowerman, B. L., and R. T. O'Connell. *Time Series and Forecasting*. North Scituate, Mass.: Duxbury Press, 1979.

Box, G. E. P., and G. M. Jenkins. *Time Series Analysis: Forecasting and Control*. Rev. ed. San Francisco: Holden-Day, 1976.

Makridakis, S.; S. Wheelwright; and V. McGee. *Forecasting: Methods and Applications*. 2nd ed. New York: John Wiley & Sons, 1983.

Montgomery, D. C., and L. A. Johnson. *Forecasting and Time Series Analysis.* New York: McGraw-Hill, 1976.

Pindyck, R. S., and D. L. Rubinfeld. *Econometric Models and Economic Forecasts.* New York: McGraw-Hill, 1978.

■ Problems

Section 16.1 Problems

16.1 Define a time series and give three examples of time series.

16.2 What assumption is required to develop a reliable prediction from a time series?

16.3 What are the objectives of studying time series?

16.4 Why are the methods developed in this chapter referred to as *classical decomposition of a time series*?

t	y	First differences	Second differences	Rate of increase (percent)
1	14.9			
2	20.0	5.1	0.0	34.2
3	25.1	5.1	0.4	25.5
4	29.8	4.7	−0.4	18.7
5	34.9	5.1	0.2	17.1
6	39.8	4.9	0.6	14.0
7	45.3	5.5	−0.6	13.8
8	50.2	4.9		10.8

a. Plot the time series with time (t) as the horizontal axis and y as the vertical axis.

b. What mathematical function—straight-line trend, parabola, semilogarithmic trend, or growth curve—would you use to fit to these data? Explain.

16.12 In the following table, time series data are given together with first differences, second differences, and rate of increase.

Section 16.2 Problems

16.5 Define and explain the *multiplicative model of a time series.*

16.6 Define and explain the *additive model of a time series.*

16.7 What is meant by mixed additive and multiplicative models of a time series?

16.8 Describe in your own words what is meant by trend, seasonal, cyclical, and irregular movements in a time series.

Section 16.3 Problems

16.9 What are the most commonly used mathematical functions for representing the secular trend of a time series?

16.10 What factor must be taken into account when using a growth curve such as the Pearl-Reed curve?

16.11 In the following table, time series data are given together with first differences, second differences, and rate of increase.

t	y	First differences	Second differences	Rate of increase (percent)
1	−3.0			
2	2.0	5.0	2.1	355.0
3	9.1	7.1	1.8	97.8
4	18.0	8.9	2.1	61.1
5	29.0	11.0	2.1	45.2
6	42.1	13.1	2.1	

a. Plot the time series with time (t) as the horizontal axis and y as the vertical axis.

b. What mathematical function—straight-line trend, parabola, semilogarithmic trend, or growth curve—would you use to fit to these data? Explain.

16.13 In the following table, time series data are given together with first differences, second differences, and rate of increase.

t	y	First differences	Second differences	Rate of increase (percent)
1	11.2			28.6
2	14.4	3.2	-0.3	20.1
3	17.3	2.9	0.5	19.6
4	20.7	3.4	0.8	20.3
5	24.9	4.2	0.8	20.1
6	29.9	5.0	0.9	19.7
7	35.8	5.9	1.3	20.1
8	43.0	7.2	1.4	20.0
9	51.6	8.6	1.7	20.0
10	61.9	10.3		

a. Plot the time series with time (t) as the horizontal axis and y as the vertical axis.

b. What mathematical function—straight-line trend, parabola, semilogarithmic trend, or growth curve—would you use to fit to these data? Explain.

16.14 In the following table, time series data are given together with first differences, second differences, and rate of increase.

t	y	First differences	Second differences	Rate of increase (percent)
0	0.01			
2	0.03	0.02	0.02	200.0
4	0.07	0.04	0.06	133.3
6	0.17	0.10	0.09	142.9
8	0.36	0.19	0.05	111.8
10	0.60	0.24	0.04	66.7
12	0.80	0.22	0.08	33.3
14	0.92	0.12	0.07	15.0
16	0.97	0.05	0.03	5.4
18	0.99	0.02	0.01	2.1
20	1.00	0.01		1.0

a. Plot the time series with time (t) as the horizontal axis and y as the vertical axis.

b. What mathematical function—straight-line trend, parabola, semilogarithmic trend, or growth curve—would you use to fit to these data? Explain.

Section 16.4 Problems

16.15 Given a linear trend equation of the form:

$$\hat{y} = 90.5 + 1.027t$$

where $t = 0$ at 1983, and t is in years, estimate the trend value for 1987.

16.16 Given the following trend equation:

$$\log \hat{y} = 1.00554 + 0.029769t$$

where $t = 0$ at 1982 and t is in years, compute the trend value for 1987.

16.17 Given the trend equation of Problem 16.16, estimate the rate of growth implied by this equation.

16.18 Provided in the table below are sales data for a firm selling a semiluxury consumer good for the years 1980 to 1986, inclusive. Estimate the trend component of this time series using a second-degree parabola, and comment on the appropriateness of this model for the long-term movements in this time series.

Year	Sales ($000)
1980	$2,582
1981	2,613
1982	2,629
1983	2,714
1984	2,775
1985	2,861
1986	2,945

16.19 Given the following linear trend equation:

$$\hat{y} = 180.43 + 745.00t$$

where $t = 0$ at 1983 and t is in years, determine the slope of the resulting trend equation assuming it is desired to shift the unit of measurement from years to months.

16.20 What would be the intercept component of the trend equation after the units of measurement were changed to months in Problem 16.19?

16.21 Given your results from Problems 16.19 and 16.20, what is the trend component for this time series for August 1986?

16.22 Given below are corporate profits in the United States (after inventory valuation adjustments) for the years 1975 to 1984, inclusive:

Corporate profits after tax
($ billions)

Year	Profits
1975	69.9
1976	87.8
1977	105.8
1978	121.9
1979	122.0
1980	106.9
1981	116.4
1982	95.3
1983	116.2
1984	140.2

Source: Economic Indicators, July 1985.

Estimate the long-term growth component in this time series using a straight-line trend equation. Does a straight-line trend appear to be a reasonable one for modeling the long-term growth in this time series?

16.23 Using the data given in Problem 16.22, estimate the long-term growth pattern in these data using a semilogarithmic trend equation. How does the trend equation compare with the one determined in Problem 16.22?

16.24 For the data in Problem 16.11, fit the trend component using the selected mathematical function.

16.25 For the data in Problem 16.12, fit the trend component using the selected mathematical function.

16.26 For the data in Problem 16.13, fit the trend component using the selected mathematical function.

Section 16.5 Problems

16.27 Unemployment statistics for a large metropolitan area for 1986 are given in the table below. Also shown are seasonal indices for unemployment based on the previous eight years' experience in this area. What is the percentage monthly *change* in unemployment on a seasonably adjusted basis? On a nonadjusted basis?

Month	Seasonal indices	Unemployment
Jan.	97.3	525
Feb.	100.8	536
March	98.9	522
April	100.1	501
May	110.2	599
June	119.5	622
July	116.2	688
Aug.	101.4	616
Sept.	91.5	592
Oct.	94.2	594
Nov.	88.1	590
Dec.	81.5	470

16.28 If a store has sales of 1,043 for October and a seasonal index of 112 for October, what is its adjusted sales level for October?

16.29 A firm estimates sales for the next year to be $12 million. Using the seasonal (quarterly) indices shown below, prepare a forecast for the year by quarter.

Quarter	Seasonal index
I	90
II	120
III	80
IV	110

16.30 A trend line for sales of a clothing manufacturer is given in the following equation. Given this information and knowledge that the seasonal index for April is 106.7, estimate sales for April 1987.

$$\hat{y} = 4,560 + 65t$$
$$t = 0 \text{ at } 1978; t \text{ is in years}$$

16.31 What is the estimate of the trend component (i.e., not including an adjustment for seasonal factors) for sales in April 1987 for the Problem 16.30 data?

Section 16.6 Problems

16.32 In Problem 16.18 is there a cyclical component to the time series? Explain.

16.33 In Problem 16.23 is there a cyclical component to the time series? Explain.

16.34 Provided below is a listing of sales of personal calculators from one of the leading manufacturers of these units. Estimate the trend equation (straight-line) for this time series. Does your analysis appear to indicate a cyclical pattern in sales of these units?

Calculator sales
(000s of units)

Year	Sales
1971	98.0
1972	92.1
1973	97.1
1974	110.0
1975	113.2
1976	118.9
1977	126.3
1978	122.1
1979	129.1
1980	136.2
1981	131.1
1982	143.1
1983	148.1
1984	151.9
1985	151.8
1986	159.1

Source: Company records.

Additional Problems

16.35 Suppose you were provided with a given time series of data and were asked to analyze its general pattern and fluctuations. Describe in detail the steps you would follow in determining the pattern of trend and whether a seasonal and/or a cyclical component contributed to movements in the series.

16.36 Given the following time series of data:

Year	Sales
1983	100.1
1984	112.9
1985	119.6
1986	132.1

estimate the linear trend present using the method of least squares.

16.37 Given the trend equation derived in Problem 16.36, estimate the trend component of this time series for 1987.

16.38 Provided in the table below are sales histories of lumber for two competing companies over a 12-year period.
a. Determine the pattern of trend for each company, and estimate the trend equation(s) by the method of least squares.
b. Is there a cyclical component to each of these series, and, if so, are they similar?
c. Estimate company sales for 1987. On what did you base your projection?

Sales of lumber ($000)

Year	Company 1	Company 2
1975	156.7	100.1
1976	161.2	103.1
1977	162.1	104.2
1978	171.5	109.8
1979	178.9	112.2
1980	181.4	116.3
1981	182.3	118.4
1982	187.2	122.2
1983	190.1	126.8
1984	200.1	129.1
1985	204.1	132.3
1986	212.1	139.8

16.39 Reproduced in the table below are data on total nonagricultural employment in the United States from 1979 to 1984 inclusive.

a. Estimate the trend component of this time series using a straight-line trend equation.

b. Does the straight-line trend equation appear to be an appropriate one for modeling the growth component of this time series?

c. Does there appear to be a cyclical component to this time series, and, if so, what can it be tied to?

d. Obtain a projection of the value of this time series for 1985.

e. Obtain the actual employment amount for 1985 from your university library. How does this number compare with the estimate derived in part d? Could this difference have been anticipated by considering factors other than the long-term growth in the series? If so, what other factors might have been considered?

Total nonagricultural employment in the United States (000)

Year	Employment
1979	89,823
1980	90,406
1981	91,156
1982	89,566
1983	90,196
1984	94,461

Source: Economic Indicators, July 1985.

16.40 Below are reproduced (by quarter) statistics on housing starts in the United States over a 10-year period. Compute quarterly, seasonal indices by the method of ratio to moving average using the data provided.

Housing starts in the continental United States (000s of units)

Year	I	II	III	IV
1	299.0	487.1	447.5	387.2
2	335.2	476.8	415.9	359.0
3	298.4	479.3	407.8	357.2
4	299.2	419.1	307.3	216.5
5	217.8	381.7	382.1	340.3
6	298.5	453.2	423.3	372.6
7	336.2	468.4	386.2	308.5
8	264.0	399.1	408.4	396.0
9	388.7	603.9	578.5	513.4
10	510.3	667.3	692.9	558.0

Source: U.S. Department of Commerce, *Survey of Current Business* (various issues).

16.41 What trend pattern is implied by the data reported in Problem 16.40. Also, isolate the cyclical and irregular components $(C \cdot I)$ of this time series, and comment on the apparent "pattern" in the cyclical movement.

16.42 Given in the table below are exports of goods and services for the United States for the years 1969–1984. Estimate the long-term growth component of this time series using a semilogarithmic trend equation. Obtain a trend estimate for 1985, and compare this estimate with the actual. Does the semilogarithmic model appear to be appropriate for modeling the growth component of this time series?

Exports of goods and services ($ billion)

Year	Exports
1969	54.7
1970	62.5
1971	65.6
1972	72.7
1973	101.6
1974	137.9
1975	154.9
1976	170.9
1977	182.7
1978	218.7
1979	281.4
1980	338.8
1981	369.9
1982	348.4
1983	336.2
1984	364.3

Source: *Survey of Current Business* (various issues).

16.43 Figure 16.1 gives data on gross national product in the United States from 1929 to 1985. Obtain the actual estimates of GNP for these years, and compute a trend equation using the semilogarithmic model. What is the growth rate implied by your model? How does the rate you computed compare with the one given in the text?

16.44 If a firm has sales in January of 3,424 units and a seasonal index for January of 89.9, what would be your estimate of sales for the firm for the entire year? What would be your estimate if you did not know the value of the seasonal index for the firm? Which estimate do you have the most faith in, and why?

16.45 For the data given in Problem 16.34, convert the straight-line trend equation to months, and estimate sales for the second quarter of 1987.

17 Forecasting

Forecasting is one of the most important functions performed by many managers today. Through forecasts of demand for end items, production schedules can be derived delineating the work force size, the rate of production, and the components to be manufactured. Moreover, each of us uses a forecast daily to decide such routine matters as whether or not to carry an umbrella (forecast of the weather) and when to leave for work (forecast of driving time).

Previous chapters have touched upon the idea of forecasting or prediction. In the Chapter 13 problems, for example, we were interested in forecasting or predicting a future event (exam score) given the number of hours spent studying for the exam. In studying time series in Chapter 16, we were often interested in predicting the value of a series at some future time. In this chapter, we will bring together many of the foundations laid in previous chapters in order to make intelligent choices regarding future events. Because the variable of interest will often be time-oriented in the applications described, we will be concerned primarily with estimating the value of a *time series* at some future instance or span of time. Hence many of the tools described in Chapter 16 will be relevant to the task at hand. We will not be concerned in this chapter with investigating techniques available for forecasting the weather!

Two basically different approaches exist for estimating the future value of a time series: *prediction* and *forecasting*. Prediction involves the incorporation of subjective factors into the estimate developed. For example, in attempting to predict or estimate a future value of GNP for the United States,

Prediction and forecasting

it would seem wise to incorporate planned government expenditures, anticipated federal reserve policy, anticipated tax legislation, and other factors into a projection. To estimate or predict future sales of a product, one *may want* to incorporate such qualitative factors as competitive action and the status of the national economy. In many instances, one simply *cannot turn one's back* (so to speak) on the qualitative factors that do influence the level of or the value of the time series at some future date.

Alternatively, one can *forecast* the value of a series by "*casting forward*" the past performance or the historical data composing the series. Forecasting time series then involves the analysis of historical data (usually through mathematical analysis) and the simple extrapolation of the component parts of these data to estimate future values of the series. Forecasting is thus applicable whenever there is reason *a priori* to suspect that past patterns will repeat themselves in the forecast period and whenever factors not in the model will not appreciably affect the value of the time series in the forecast period.

We will turn now to a discussion of barometric indicators, which, though qualitative, have an appreciable influence on government policy in directing this nation's economy. Subsequent sections will investigate various quantitative forecasting techniques.

Before we begin our discussion of barometric indicators, it will be helpful to introduce some notation that will be used in describing the forecasting models developed. First, we shall use the variable t to denote the time period of interest; t is analogous to the independent variable x in the regression models discussed in Chapters 13–15 and, in fact, many of the forecasting equations developed will be derived by the method of least squares. Since the parameters of the forecasting model $\beta_0, \beta_1, \beta_2, \ldots$ will rarely be known in practice, it will be necessary to obtain point estimates $b_0, b_1, b_2, \ldots$ of them. An observed value of the series at time t will be denoted by y_t, and an estimated or forecasted value of the series for time period t will be denoted by $\hat{y}_t$. (Recall that because many assumptions of the least squares linear regression model are not satisfied when time series data are examined, it will not be possible for us to construct confidence intervals for the point estimates obtained.) This notation is similar to that used in Chapter 16.

■ 17.2

Barometric indicators

NBER

17.2.1 | **Economic indicators**

The National Bureau of Economic Research (NBER) has identified groups of economic time series that *in general* lead, are coincident with, or lag behind aggregate economic activity. The usefulness of the leading series is in signaling possible downturns in aggregate economic activity so that remedial action can be taken before the downturn begins or has an opportunity to progress very far. It is these series that are quoted frequently in news broadcasts or in print:

> The government's index of leading economic indicators rose for the third straight month, signaling the healthy recovery of the economy is still in progress!

TABLE 17.1 | New composite index of leading economic indicators (by economic process)

Economic process / Index	I. Employment and unemployment	II. Production and income	III. Consumption, trade, orders, and deliveries
Leading indicators	1. Average weekly hours of production or nonsupervisory workers, manufacturing 5. Average weekly initial claims for unemployment insurance, State Programs		8. Manufacturers' new orders in 1972 dollars, consumer goods and materials industries 32. Vendor performance, percent of companies receiving slower deliveries

We will here be concerned with describing components of the NBER's most frequently quoted index, the composite index of leading economic indicators. This index forewarns of both peaks and troughs in aggregate economic activity.[1] This main aggregate index (which consists of 12 individual series) was revised in 1966 and then again in 1975 to aid in predicting changes in future activity. Further, adjustments in the standardizing factor and weights of each of the individual components of the series are revised periodically as additional information becomes known to the NBER.

The current components of NBER's composite index of leading economic indicators (categorized by economic process) is given in Table 17.1. In order for a series to be included in this composite index, it must have "scored well" on a set of criteria outlined by the NBER. That is, it should signal the upcoming upturn or downturn approximately the same number of months before the upturn or downturn begins, and it should not falsely predict an upturn or a downturn when one is not on the horizon. The performance of this latest composite index for the period 1948–1985 is shown in Figure 17.1. Note in general how the series tends to turn down well in advance of a downturn in aggregate economic activity (indicated by the shaded portion of the graph). Note also how the composite series incorrectly signaled a downturn in 1966! Still, barometric indicators such as the NBER's main leading series do provide useful information in alerting public officials to potential changes in the *direction* of aggregate economic activity.

---

[1] A narrative explanation of the index, a description of the methods used to construct the index, and a listing of the weights each of the 12 series receives in the construction of the index can be found in *Handbook of Cyclical Indicators* (Washington, D.C.: U.S. Department of Commerce, Bureau of Economic Analysis, 1984, pp. 65–70). A brief explanation of the construction of this index is also given monthly in *Business Conditions Digest*.

IV. *Fixed capital investment*	V. *Inventories and inventory investment*	VI. *Prices, costs, and profits*	VII. *Money and credit*
12. Index of net business formation 20. Contracts and orders for plant and equipment in 1972 dollars 29. Index of new private housing units authorized by local building permits	36. Change in manufacturing and trade inventories on hand and on order in 1972 dollars, smoothed	99. Change in sensitive materials prices, smoothed 19. Index of stock prices, 500 common stocks	106. Money supply M2 in 1972 dollars 111. Change in business and consumer credit outstanding

Source: *Handbook of Cyclical Indicators,* (Washington, D.C.: U.S. Department of Commerce, Bureau of Economic Analysis), p. 66.

FIGURE 17.1 | Composite index of leading economic indicators

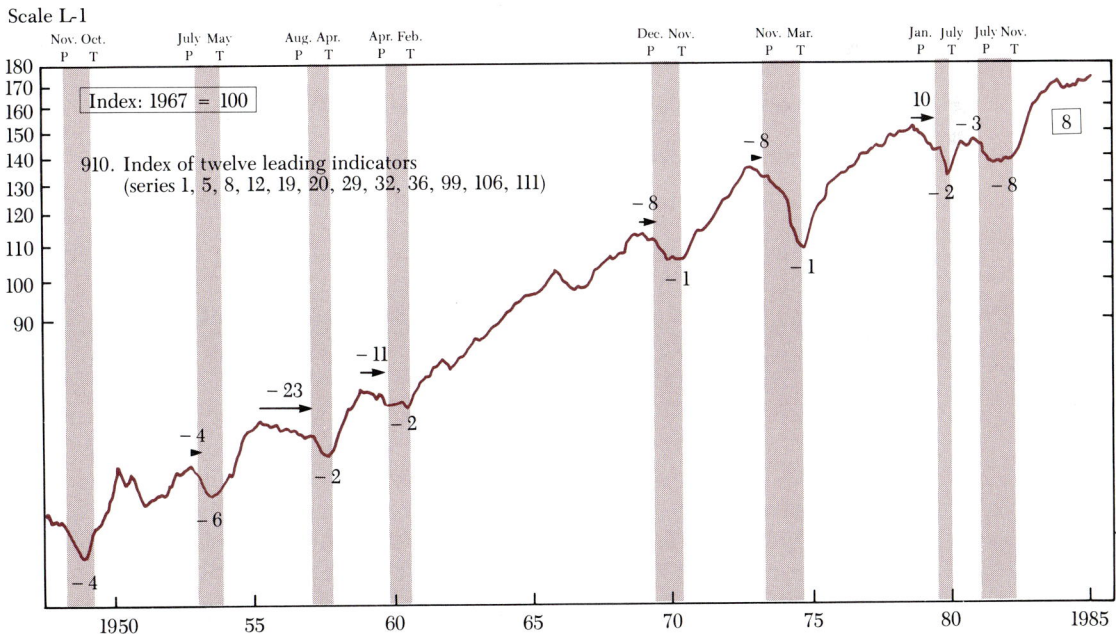

Note: Numbers entered on the chart indicate length of leads (−) and lags (+) in months from reference turning dates.

Source: U.S. Department of Commerce, *Business Conditions Digest*, September 1985.

17.2.2 | Diffusion indices

Diffusion index

The second barometric indicator we will describe is a *diffusion index,* which is designed to overcome one of the difficulties of leading indicator series—namely, what to do when some leading indicators signal a turn and others do not.

A diffusion index is designed to show the *percentage* of a collection of leading time series that are experiencing rises over some given time interval (e.g., three months). They are constructed using the following scheme. First, the series that will compose the diffusion index are selected. Then each series is differenced over succeeding periods of time. These differences are ignored with the exception of their *sign*. If the difference is positive, the observation is assigned a weight of 1; if the difference is negative, the observation receives a weight of 0; and if there is no difference between succeeding periods, the observation is assigned a weight of 0.5. The weights are then added cross-sectionally, and this total is divided by the total number of series in the index.

The Federal Reserve Board's (diffusion) index of industrial production consists of between 15 and 25 different time series, and the NBER has an all-inclusive index composed of between 600 and 700 different time series. Since the latter-series is dependent upon so many aspects of business, it is considered by many to be one of the best historical indices of the cyclical position of an economy yet devised.

Criteria for evaluating a diffusion index

In establishing the efficacy of a given diffusion index, the following criteria are generally assessed:

1. Average forecasting lead of the indicator.

2. Number of turning points accurately predicted.

3. Number of cyclical turns not predicted.

4. Number of turns that were predicted but never occurred.

Diffusion indices that have generally scored well on these criteria are published monthly in the U.S. Department of Commerce publication *Business Conditions Digest* (formerly *Business Cycle Developments*). Figure 17.2 is an example of two diffusion indices that are published monthly by the U.S. Department of Commerce.

It has been said of the diffusion indices that although they reduce irregularities and the diversity of movements in the leading economic indicator series, they do not eliminate these irregularities entirely. (Whenever the number of components in any given index is increased, the number of false leads is likely decreased.) Furthermore, those diffusion indices that use the greater number (3, 6, or 9) of months before peak to identify turning points are smoother than the one-month span indices, but the shorter-spanned indices are particularly important when big changes are taking place—hence the dilemma that exists when selecting the number of periods (time interval in constructing the index).

FIGURE 17.2 | Diffusion indices

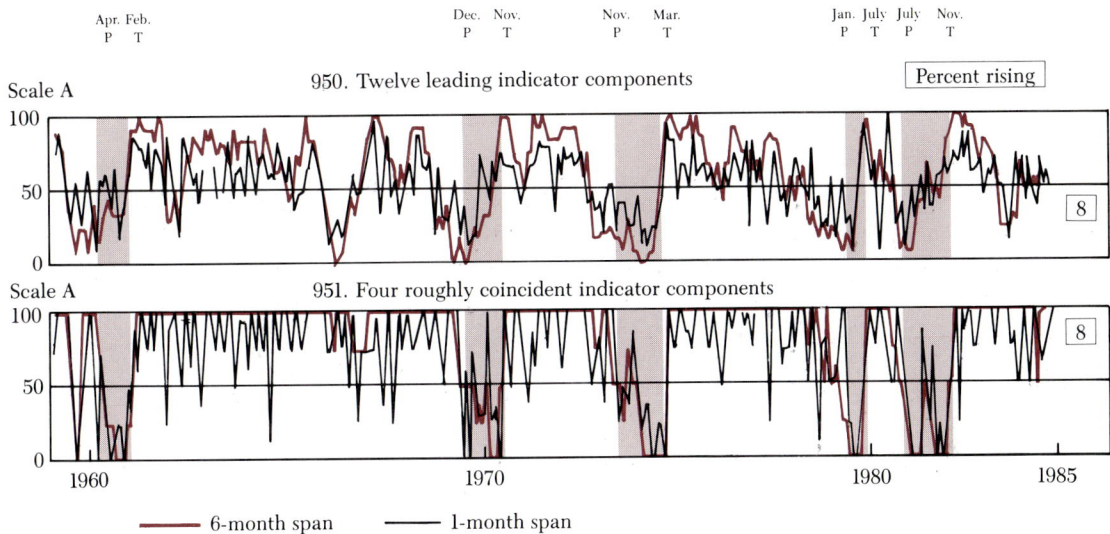

Source: U.S. Department of Commerce, *Business Conditions Digest,* September 1985.

The diffusion indices are also said to suffer one of the same, chief disadvantages of the leading indicators—that of predicting change but not the *magnitude* of the change. Hence many economic forecasters consider the use of barometric indicators a *wise supplement* to any "kit" of forecasting tools. They are by no means self-sufficient, yet they provide a healthy and systematic way of assessing the *direction of movement* in aggregate economic activity.

■ 17.3

Models for a stationary process

Stationary process

In this section we will examine two very similar models for forecasting a future value of a series where the process generating values of the random variable is *stationary*. By a stationary process, we mean that the average of the process is not changing over time. We will describe methods in Section 17.6 for reducing a nonstationary process to a stationary one (called *differencing*), but for the series examined in this section, it will be important to remember that the mean of the process is not changing over time, or else the models used are *incorrect* for the series being examined. An analysis of a residual plot would indicate such a misapplication, and we illustrate such a condition in Section 17.4 when we describe trend adjusted exponential smoothing and attempt to model a process with a mean changing over time with the methods described in this section. For the present, it is necessary for us to assume that the mean of the process is not changing, and we are attempting to model the movements of the series around this mean value. A scatter plot of the series being modeled is a first indication that the series is stationary, and it is always recommended that a scatter plot be constructed before any forecasting model is used.

17.3.1 | Moving average forecasts

The moving average time series forecasting model is perhaps one of the easiest to use and understand of the time series forecasting techniques. In it, we assume that the pattern followed by a sequence of observations can best be represented by an arithmetic average of past observations. Thus, in terms of a time series model, we assume that the slope component, β_1, is equal to 0 and that the underlying pattern fluctuates randomly around the constant term β_0. A time series exhibiting this type of behavior is depicted in Figure 17.3.

FIGURE 17.3 | Scatter diagram of data in Table 17.2

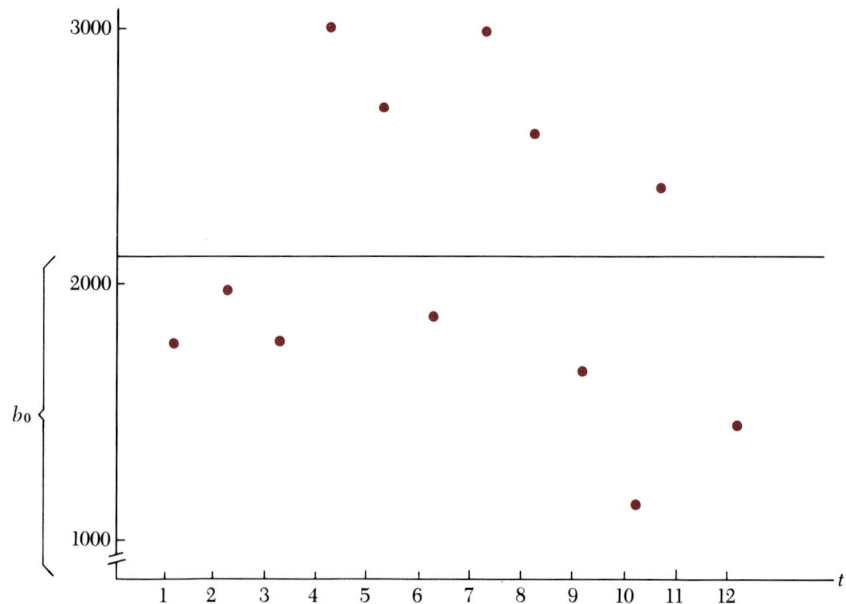

The simple moving average time series model is described below. Several properties of this model are worth noting. First, equal weights are given to each of the n most recent observations in determining the forecasted value $\hat{y}_t$, and zero weight is given to observations older than n periods. Second, each "new" prediction of $\mathbf{Y}_t$ is simply an updated adjustment of the previous estimate obtained by removing the oldest (nth) and adding the current observation. And third, the rate of response of the model depends greatly on n. Therefore, if slight changes in the underlying pattern are expected, then a large value of n should be selected to reduce the variance property of the forecasting model. If more than slight changes are expected in the underlying pattern, then a smaller value of n should be used to detect turning points and make an appropriate response. The use of the simple moving average time series forecasting model is illustrated in the following example.

| Moving average model |

Simple moving average forecasting model

Let an observation at time period t be given by

$$\mathbf{Y}_t = \beta_0 + \epsilon_t$$

where

β_0 = Constant, or level of the series.
ϵ_t = Random disturbance at time period t with mean 0 and variance σ_ϵ^2.

Then the simple moving average forecast of an observation for time period t is given by:

$$\hat{y}_t = \frac{y_{t-1} + y_{t-2} + y_{t-3} + \cdots y_{t-n}}{n}$$

That is, the simple moving average forecast for time period t is the arithmetic mean of the n most recent observations.

Example 17.1 Information on monthly sales of computer software from Daltons Software, Inc. in Fort Worth, Texas, for 1986 is given in Table 17.2. A scatter diagram of these data is given in Figure 17.3. A visual inspection of these data reveals an apparent absence of growth and seasonal movements in this series.

Using $n = 3$ and then $n = 6$ time periods, compute the three-month and six-month simple moving average forecasts for the remaining months of 1986 and January 1987. Plot the forecasts obtained on a scatter diagram similar to the one in Figure 17.3. How do these two forecasts compare?

Solution Relevant computations are given in Table 17.3, and a plot of the forecasted values and a scatter diagram of the actual data are given in Figure 17.4. Note how the moving total is "updated" each month by deleting the oldest observation and adding the most recent observation to the moving total in order to simplify the computations involved in computing the six-month moving average forecast. Note also the "smoothing" effect of the

TABLE 17.2 | Monthly sales of computer software from Daltons Software, Inc.

Month 1986	t	y_t	Month 1986	t	y_t
Jan.	1	$1,800	July	7	$3,000
Feb.	2	2,000	Aug.	8	2,600
March	3	1,800	Sept.	9	1,700
April	4	3,000	Oct.	10	1,200
May	5	2,700	Nov.	11	2,400
June	6	1,900	Dec.	12	1,500

Source: Daltons Software, Inc.

TABLE 17.3 | Three-month and six-month simple moving average forecasts for data in Table 17.2

Month	Most recent three-month sales	Forecast	Most recent six-month sales	Forecast
Jan.	—	—		—
Feb.	—	—		—
March	—	—		—
April	1,800 + 2,000 + 1,800	1,867		—
May	2,000 + 1,800 + 3,000	2,267		—
June	1,800 + 3,000 + 2,700	2,500		—
July	3,000 + 2,700 + 1,900	2,533	13,200*	2,200
Aug.	2,700 + 1,900 + 3,000	2,533	13,200 − 1,800 + 3,000 = 14,400	2,400
Sept.	1,900 + 3,000 + 2,600	2,500	14,400 − 2,000 + 2,600 = 15,000	2,500
Oct.	3,000 + 2,600 + 1,700	2,433	15,000 − 1,800 + 1,700 = 14,900	2,483
Nov.	2,600 + 1,700 + 1,200	1,833	14,900 − 3,000 + 1,200 = 13,100	2,183
Dec.	1,700 + 1,200 + 2,400	1,767	13,100 − 2,700 + 2,400 = 12,800	2,133
Jan.	1,200 + 2,400 + 1,500	1,700	12,800 − 1,900 + 1,500 = 12,400	2,067

* 13,200 = 1,800 + 2,000 + 1,800 + 3,000 + 2,700 + 1,900   ⌐Previous six-month moving total.

six-month moving average as compared with the three-month moving average as depicted in Figure 17.4. The three-month moving average forecast is much more responsive to movements in the observed values of sales over time.

17.3.2 | Single exponential smoothing

Moving average forecasts are a special case of single exponentially smoothed forecasts, and we will develop the correspondence between these

FIGURE 17.4 | Scatter diagram and three-month and six-month moving average forecasts for data in Table 17.2

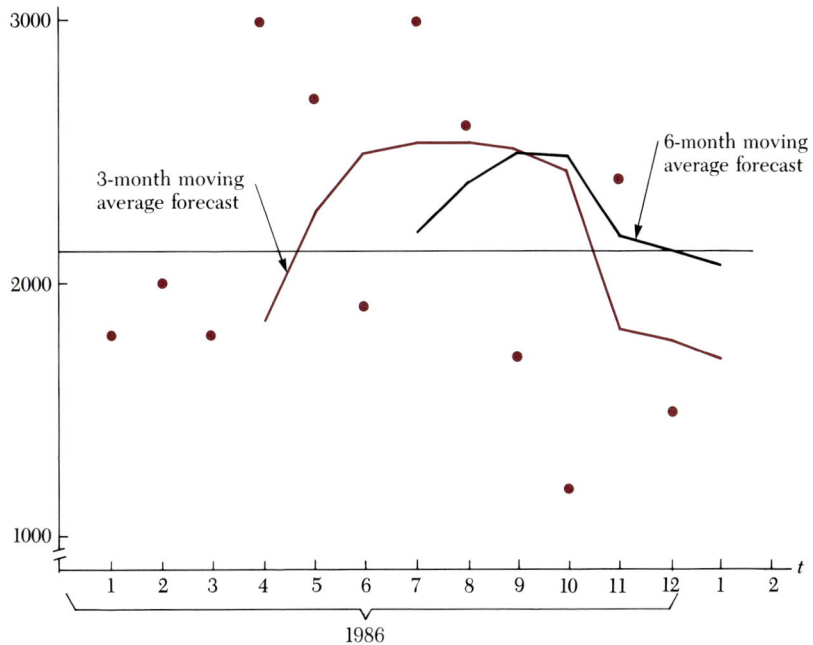

two very important forecasting techniques shortly. Single exponential smoothing is perhaps the most widely used time series forecasting technique today. It is *very easily* programmed for computer application and offers several computational advantages over the moving average time series forecasting model. Yet it differs from the simple moving average model only in the method of assigning weights to previous observations. In this section, we will describe the single exponential smoothing forecasting model under the assumption that the series being modeled is stationary. In Section 17.4 we will see how to generalize this technique for a process in which the mean *is* changing over time, called *trend adjusted exponential smoothing*.

A description of the single exponential smoothing model is given on the top of page 902. The forecasted value of the series for time t, $\hat{y}_t$, is equal to a fraction α of the forecast error of the previous period, $(y_{t-1} - \hat{y}_{t-1})$, plus the forecasted value of the previous period, $\hat{y}_{t-1}$. Thus to predict the value of the time series for time period t, we would use

$$\hat{y}_t = \alpha(y_{t-1} - \hat{y}_{t-1}) + \hat{y}_{t-1}$$

<table>
<tr><td>

Single exponential
smoothing model

</td><td>

Single exponential smoothing—constant process (stationary)

Let an observation at time period t be given by

$$\mathbf{Y}_t = \beta_0 + \epsilon_t$$

where

β_0 = Constant, or level of the series.
ϵ_t = Random disturbance at time period t with mean 0 and variance σ_ϵ^2.

Then the single exponential forecast of an observation for time is given by

$$\hat{y}_t = \alpha(y_{t-1} - \hat{y}_{t-1}) + \hat{y}_{t-1}$$

where α is a smoothing constant $(0 \le \alpha \le 1)$

</td></tr>
</table>

The formula for $\hat{y}_t$ may be rewritten as follows:

$$\hat{y}_t = \alpha(y_{t-1} - \hat{y}_{t-1}) + \hat{y}_{t-1} = (1 - \alpha)\hat{y}_{t-1} + \alpha y_{t-1}$$

Now, the formula for $\hat{y}_{t-1}$ is

$$\hat{y}_{t-1} = (1 - \alpha)\hat{y}_{t-2} + \alpha y_{t-2}$$

by substituting $\hat{y}_{t-1}$ for $\hat{y}_t$ in the formula for $\hat{y}_t$.

Then,

$$\hat{y}_t = (1 - \alpha)[(1 - \alpha)\hat{y}_{t-2} + \alpha y_{t-2}] + \alpha y_{t-1}$$
$$= \alpha y_{t-1} + \alpha(1 - \alpha)y_{t-2} + (1 - \alpha)^2\hat{y}_{t-2}$$

We can now write the formula for $\hat{y}_{t-2}$ and substitute the resulting expression for $\hat{y}_{t-2}$ in the above formula. If this process is continued, eventually we would have

$$\hat{y}_t = \alpha y_{t-1} + \alpha(1 - \alpha)y_{t-2} + \alpha(1 - \alpha)^2 y_{t-3} + \alpha(1 - \alpha)^3 y_{t-4} + \cdots$$

This expression clearly shows the nature of single exponential smoothing. The prediction in the current period, $\hat{y}_t$, is a function of the values of $\mathbf{Y}$ in all previous time periods. Since $0 \le \alpha \le 1$, the past observations are weighted in a way so that the most recent observations receive greater weights than older observations. For example, if $\alpha = 0.9$, then

$$\hat{y}_t = 0.9y_{t-1} + 0.9(0.1)y_{t-2} + 0.9(0.1)^2 y_{t-3} + 0.9(0.1)^3 y_{t-4} + \cdots$$
$$= 0.9y_{t-1} + 0.09y_{t-2} + 0.009y_{t-3} + 0.0009y_{t-4} + \cdots$$

If, on the other hand, $\alpha = 0.1$, then

$$y_t = 0.1y_{t-1} + 0.1(0.9)y_{t-2} + 0.1(0.9)^2 y_{t-3} + 0.1(0.9)^3 y_{t-4} + \cdots$$
$$= 0.1y_{t-1} + 0.09y_{t-2} + 0.081y_{t-3} + 0.0729y_{t-4} + \cdots$$

Thus if α is close to its upper bound of 1, then the most recent observations are weighted most heavily, and the weights on the older observations decrease rapidly. If α is close to its lower bound of 0, then the decay of the weights on the older observations is quite slow.

A difficulty in using exponential smoothing is the determination of α, where $0 \leq \alpha \leq 1$. We will consider this issue beginning with the following example.

Example 17.2   Using the data given in Table 17.2 and a forecast of sales for January 1986 of $2,100, forecast sales for February 1986 through January 1987 using α-values of 0.1 and 0.9. Plot the forecasts obtained on a scatter diagram similar to Figure 17.3. How do the forecasts compare?

Solution   Relevant computations are given in Table 17.4 and the forecasts obtained are depicted in Figure 17.5. Also shown in Figure 17.5 is a scatter diagram of the observed data. Note that in order to "start" the exponential smoothing process, it was necessary to obtain an initial forecast for the first month. This "seed" forecast can be based on a "best guess" or a moving average forecast determined from data from the n months preceding the start of the forecast period.

TABLE 17.4 | Single exponential forecast for data in Table 17.2

Time period t	Actual sales y_t	$\alpha = 0.1$		$\alpha = 0.9$	
		$\alpha(y_t - \hat{y}_t)$	$\hat{y}_t$	$\alpha(y_t - \hat{y}_t)$	$\hat{y}_t$
1	$1,800	-30*	2,100	-270	2,100
2	2,000	-7†	2,070	153	1,830
3	1,800	-26	2,063	-165	1,983
4	3,000	96	2,037	1,064	1,818
5	2,700	57	2,133	-164	2,882
6	1,900	-29	2,190	-736	2,718
7	3,000	84	2,161	916	1,982
8	2,600	36	2,245	-268	2,898
9	1,700	-58	2,280	-837	2,630
10	1,200	-102	2,222	-534	1,793
11	2,400	28	2,120	1,027	1,259
12	1,500	-65	2,148	-707	2,286
13			2,083		1,579

* $0.1(1,800 - 2,100) = -30$.
† $0.1(2,000 - 2,070) = -7$.

From a visual inspection of Figure 17.5, it is apparent that when α, the smoothing constant, is close to 0, the forecast behaves like the average of a large number of data. When α is close to 1, the forecast responds rapidly to changes in the pattern underlying the observations.

FIGURE 17.5 | Smoothed exponential forecasts for data in Table 17.2

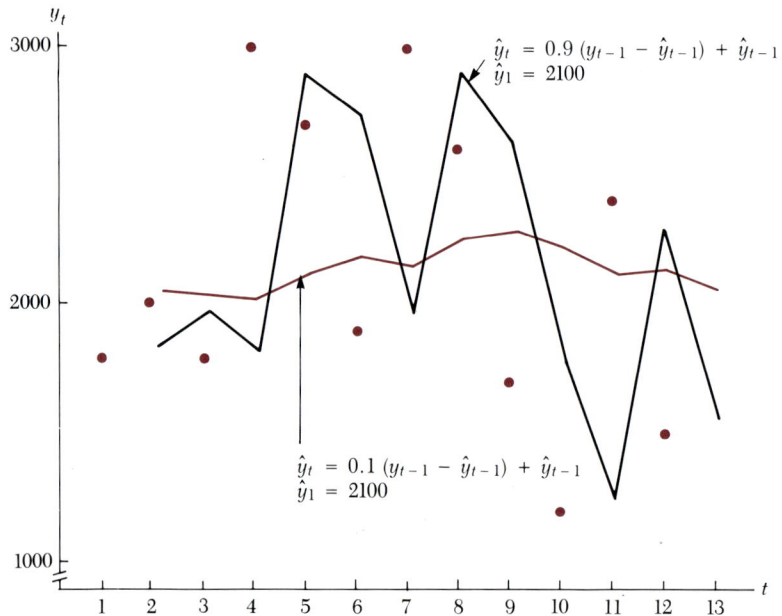

The term *exponential smoothing* is derived from the weight attached to preceding observations. Figure 17.6 is a graph of the weight attached to previous observations as a function of the age of the observation and the value of the smoothing constant, α. It should be apparent from this graph that *all* previous observations receive some weight in deriving the forecast, and not just the previous n as in the case of the simple moving average. This figure also shows how a plot of the weights yields an exponential curve, hence the term *exponential* smoothing.

Weights for the current period and the four previous periods for five values of α are also given in Figure 17.6. From these numbers it is apparent that as α increases, the weight attached to the more current observations increases. In general, weights attached to each of the previous n periods in the exponential smoothing model are as given in the following equation delineating the forecast for time period $(n + 1)$:

$$\hat{y}_{n+1} = \alpha \left[\sum_{t=0}^{n-1} (1 - \alpha)^t y_{n-t} \right] + (1 - \alpha)^n \hat{y}_0$$

The exponential smoothing model offers several computational advantages over the simple moving average. We need only retain, for example, the most recent forecast, the current observation, and the smoothing constant in order to obtain a forecast for the succeeding period. We need not retain the

FIGURE 17.6 | Weight attached to previous observations in the exponential smoothing model

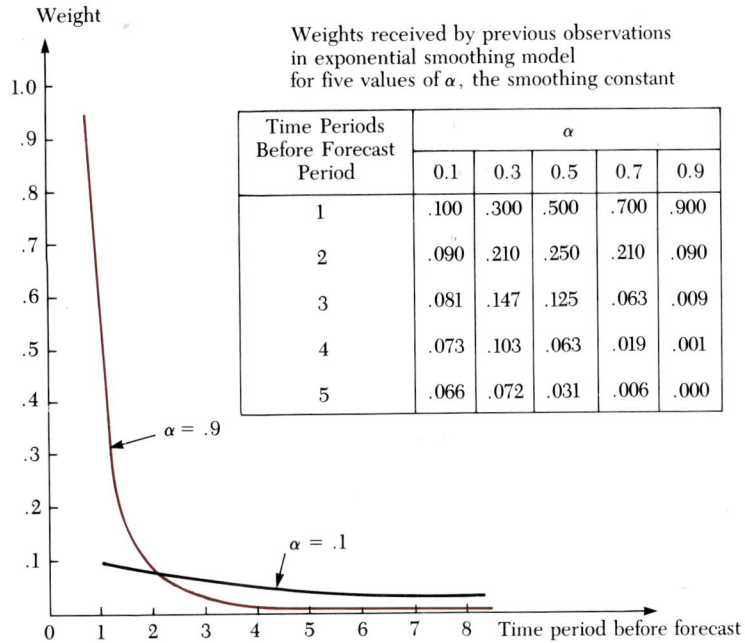

Weight

1.0
.9
.8
.7
.6
.5
.4

$\alpha = .9$

.3
.2
.1

$\alpha = .1$

0 1 2 3 4 5 6 7 8 Time period before forecast

Weights received by previous observations
in exponential smoothing model
for five values of α, the smoothing constant

Time Periods Before Forecast Period	α				
	0.1	0.3	0.5	0.7	0.9
1	.100	.300	.500	.700	.900
2	.090	.210	.250	.210	.090
3	.081	.147	.125	.063	.009
4	.073	.103	.063	.019	.001
5	.066	.072	.031	.006	.000

preceding n observations required in the simple moving average model. Both models, through the averaging process, tend to smooth irregularities in the observations. If the underlying pattern in the data is changing just slightly, then we would choose to use a large value for n in the simple moving average model and a small value of α in the exponential smoothing model. Similarly, if more than slight changes are anticipated in the underlying pattern, we might select a small value of n in the simple moving average model and a large value of α in the exponential smoothing model to detect turning points and make a more rapid adjustment to the change in the underlying pattern.

If we define an exponential smoothing model to be equivalent to a moving average model where equivalency is in the sense of a weighted average of the age of past observations, then

$$\alpha = \frac{2}{n + 1}$$

Table 17.5 gives the "equivalent" smoothing constant for several values of n in the moving average model. Note that with α set to some large value, little actual smoothing of the data takes place.

TABLE 17.5 | Equivalent values of n in a simple moving average forecasting model with α in exponential smoothing

α	No. of observations n in equivalent moving average model
0.1	19.00
0.3	5.67
0.5	3.00
0.7	1.86
0.9	1.22

In practice, the value of α to use in the exponential smoothing model is frequently determined in an iterative fashion as follows: α is incremented from 0.1 to 0.2 to 0.3, etc., and the sum of the squares of the observed minus the forecasted values of the series is retained. The value of α yielding the minimum sum of squared error deviations, $\Sigma(y_t - \hat{y}_t)^2$, is then used in the forecasting equation. Other criteria used to assess the efficacy of forecasting models are described in Section 17.9.

■ 17.4

Trend-adjusted exponential smoothing

When there is a growth pattern or a trend in the series under investigation, the single exponential smoothing forecasting model will *always* tend to lag movements in the actual observations. This is because the single exponential smoothing forecasting model is based on a stationary time series with only random disturbances about a mean value. When there is a definite trend in the given time series (either positive *or* negative), *trend-adjusted exponential smoothing* models the growth component of the series. The trend-adjusted exponential smoothing model for a linear trend is described below. Its use is illustrated in the following example.

Trend-adjusted exponential smoothing model

Trend-adjusted exponential smoothing—linear trend

Let an observation at period t be given by

$$\mathbf{Y}_t = \beta_0 + \beta_1 t + \epsilon_t$$

where

β_0 = Constant, or level of the series
β_1 = Growth, or trend component of the series
ϵ_t = Random disturbance at time period t with mean 0 and variance σ_ϵ^2

Then the double exponential smoothing forecast of an observation for time period t is given by

$$\hat{y}_t = \left[2 + \frac{\alpha}{(1 - \alpha)}\right] S_{t-1} - \left[1 + \frac{\alpha}{(1 - \alpha)}\right] S_{t-1}^{(2)}$$

where

$$S_{t-1} = \alpha y_{t-1} + (1 - \alpha) S_{t-2}$$

$$S_{t-1}^{(2)} = \alpha S_{t-1} + (1 - \alpha) S_{t-2}^{(2)}$$

$$S_0 = b_0(0) - \left[\frac{(1 - \alpha)}{\alpha}\right] b_1(0)$$

$$S_0^{(2)} = b_0(0) - 2\left[\frac{(1 - \alpha)}{\alpha}\right] b_1(0)$$

$$0 \le \alpha \le 1$$

and $b_0(0)$ and $b_1(0)$ are initial estimates of the intercept and slope, respectively, of the time series. (Estimates $b_0(0)$ *and* $b_1(0)$ are often the least squares estimates of β_0 and β_1 in the time series model.)

For the derivation of the formula for $\hat{y}_t$, see the Bowerman and O'Connell reference at the end of the chapter. The use of this formula is recursive in nature. We must start the process by finding S_1 and $S_1^{(2)}$ given by

$$S_1 = \alpha y_1 + (1 - \alpha) S_0$$

$$S_1^{(2)} = \alpha S_1 + (1 - \alpha) S_0^{(2)}$$

The initial estimates of S_0 and $S_0^{(2)}$ are found by the formulas above to initiate the process. Once S_1 and $S_1^{(2)}$ are determined, then we find S_2 and $S_2^{(2)}$, S_3 and $S_3^{(2)}$, and so forth until we reach S_{t-1} and $S_{t-1}^{(2)}$. Once these values are determined, then we can find the forecast of **Y** in period t from the formula for $\hat{y}_t$.

As with single exponential smoothing, we must set α to find $\hat{y}_t$, where $0 \le \alpha \le 1$. Typically, a set of values of α are used ($\alpha = 0.05, 0.10, 0.15$, etc.) and the value that minimizes the sum of squared errors or the sum of absolute errors is selected.

Example 17.3   Given in the table below are sales levels of a consumer product for nine periods (1978–1986). The least squares line through the

Period	Sales ($000)
1	68.27
2	76.06
3	79.06
4	89.55
5	109.25
6	110.93
7	114.10
8	135.72
9	162.63

scatter of pairs of periods and sales is $y_t = 51.073 + 10.798t$. Thus we will use $b_0(0) = 51.073$ and $b_1(0) = 10.798$ as initial values for calculating S_0 and $S_0^{(2)}$. Using a smoothing constant of $\alpha = 0.3$, develop a trend-adjusted exponential smoothing forecast for the periods in this series. Also, using an initial forecast of 55.29 and $\alpha = 0.3$, develop single exponential smoothing forecast for the periods in this series. Compare these two forecasts in average absolute value of the forecast error produced by each model, where the forecast error is taken to be the difference between the forecasted value of the time series and the actual value of the series in each period.

Solution   We first must calculate S_0 and $S_0^{(2)}$:

$$S_0 = b_0(0) - \left[\frac{(1-\alpha)}{\alpha}\right] b_1(0) = 51.073 - \left[\frac{1-0.3}{0.3}\right](10.798) = 25.878$$

$$S_0^{(2)} = b_0(0) - 2\left[\frac{(1-\alpha)}{\alpha}\right] b_1(0) = 51.073 - 2\left[\frac{1-0.3}{0.3}\right](10.798) = 0.682$$

With these initial values, we can then calculate S_1 and $S_1^{(2)}$, thus starting the recursive process:

$$S_1 = \alpha y_1 + (1-\alpha)S_0 = (0.3)(68.27) + (0.7)(25.878) = 38.596$$
$$S_1^{(2)} = \alpha S_1 + (1-\alpha)S_0^{(2)} = (0.3)(38.596) + (0.7)(0.682) = 12.056$$

Then

$$\hat{y}_2 = \left[2 + \frac{\alpha}{(1-\alpha)}\right] S_1 - \left[1 + \frac{\alpha}{(1-\alpha)}\right] S_1^{(2)}$$
$$= 2.4286 S_1 - 1.4286 S_1^{(2)} = 2.4286(38.596) - 1.4286(12.056)$$
$$= 76.511$$

The remaining forecasts are calculated in the same manner. For example, in time period $t = 2$,

$$S_2 = \alpha y_2 + (1-\alpha)S_1 = (0.3)(76.06) + (0.7)(38.596) = 49.835$$
$$S_2^{(2)} = \alpha S_2 + (1-\alpha)S_1^{(2)} = (0.3)(49.835) + (0.7)(12.056) = 23.390$$
$$\hat{y}_3 = 2.4286 S_2 - 1.4286 S_2^{(2)} = 2.4286(49.835) - 1.4286(23.390)$$
$$= 87.614$$

The values for S_t, $S_t^{(2)}$, and y_t are given in Table 17.6. Table 17.7 contains the computation of the forecast errors. Because there is a definite trend or growth pattern in these data, the trend-adjusted exponential smoothing model produces a lower average absolute deviation from the actual results than does the single exponential smoothing model. Note also that all the forecast errors using the single exponential smoothing model are negative indicating the lag present in the forecasted value.

In Figure 17.7 we give the residual plots of the data in Table 17.7. Notice in the trend-adjusted exponential smoothing model (Figure 17.7A) that the residuals fall much closer to the 0 residual line than do the residuals for the single exponential smoothing model (Figure 17.7B). And, the residuals fluc-

TABLE 17.6 | Computation worksheet for the trend-adjusted exponential smoothing forecasting model

t	y_t	S_t	$S_t^{(2)}$	$\hat{y}_t$
1	68.27	38.596	12.056	61.872
2	76.06	49.835	23.390	76.511
3	79.06	58.603	33.954	87.614
4	89.55	67.887	44.134	93.817
5	109.25	80.296	54.983	101.821
6	110.93	89.486	65.334	116.458
7	114.10	96.870	74.795	123.990
8	135.72	108.525	84.914	128.406
9	162.63	124.756	96.867	142.256

TABLE 17.7 | Computation of the forecast errors for Example 17.3

Period	Actual value ($000)	Single exponential forecast	Forecast error	Trend-adjusted exponential forecast	Forecast error
1	68.27	55.29	−12.98	61.87	6.40
2	76.06	59.18	−16.88	76.51	−0.45
3	79.06	64.25	−14.81	87.61	−8.55
4	89.55	68.69	−20.86	93.82	−4.27
5	109.25	74.95	−34.30	101.82	7.43
6	110.93	85.24	−25.69	116.46	−5.53
7	114.10	92.95	−21.15	123.99	−9.89
8	135.72	99.29	−36.43	128.41	7.31
9	162.63	110.22	−52.41	142.26	20.37

Average absolute error $\dfrac{235.51}{9} = 26.17$ Average absolute error $\dfrac{70.2}{9} = 7.80$

tuate above and below the 0 residual line. Notice that the residuals in Figure 17.7B for the single exponential smoothing model are all negative. An examination of a residual plot such as the one shown in Figure 17.7B would certainly alert the user or the developer of the forecasts to the fact that the model has been improperly specified.

It is also possible to define a three-parameter exponential smoothing forecasting model that is applicable when trend *and* seasonal components are present in a given time series. A discussion of this model as well as some more advanced applications of the exponential smoothing model is given in the advanced references at the end of the chapter.

FIGURE 17.7 | Residual plots of data in Table 17.7

A. Trend adjusted exponential forcasting model

B. Single exponential forecasting model

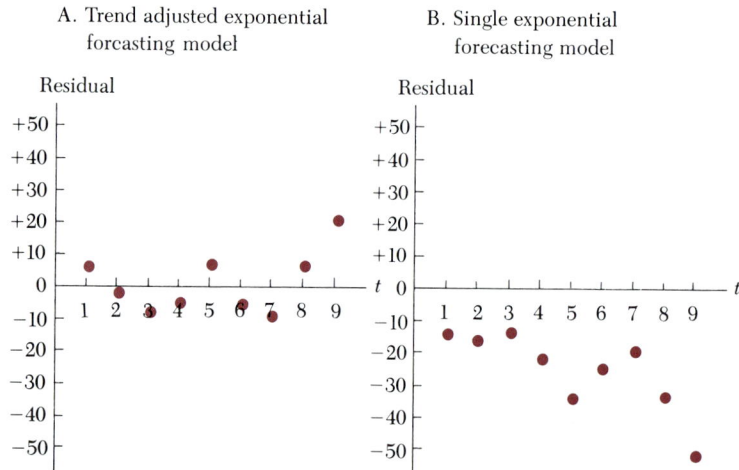

17.5

Autoregressive forecasting approaches

If observations in a given time series are *highly* correlated over time (as they would be, for example, in the presence of a trend component as described in the previous section), then it may be possible to forecast future values of the series using past observations as predictor variables. (We might use such a model in lieu of the trend-adjusted exponential smoothing forecasting model discussed in the previous section.)

If sales of a product bear a strong statistical relationship to (1) sales of the previous month, (2) sales of six months ago, and (3) sales of one year ago, then a forecasting equation of the form

$$\hat{y}_t = b_0 + b_1 y_{t-1} + b_2 y_{t-6} + b_3 y_{t-12}$$

may yield an accurate forecast of the time series for period t.

Correlogram

A *correlogram* is a useful statistical tool for assessing the time-lagged correlations, or the *autocorrelation* present in a time series of data. The correlogram is constructed using the techniques described in Chapter 13 for estimating the sample correlation coefficient where the "variables" are simply time-lagged observations of a given time series.

Autocorrelation

Suppose that we have a set of n sampled time series observations y_1, y_2, . . . , y_n. Then r_1, the first order autocorrelation, is the Pearson product-moment correlation coefficient (Chapter 13) for consecutive pairs of observations: (y_1, y_2), (y_2, y_3), . . . , (y_{n-1}, y_n). The second order autocorrelation, r_2, is the correlation coefficient with observations lagged by one in the pairs: (y_1, y_3), (y_2, y_4), . . . , (y_{n-2}, y_n). For the kth order autocorrelation, r_k, it is the correlation coefficient for the pairs (y_1, y_{k+1}), (y_2, y_{k+2}), . . . , (y_{n-k}, y_n). The correlogram is a graphical display of the kth order autocorrelations, for $k = 1, 2, 3, . . . $. Figure 17.8 gives two possible correlograms

FIGURE 17.8 | Possible correlograms for a given data series

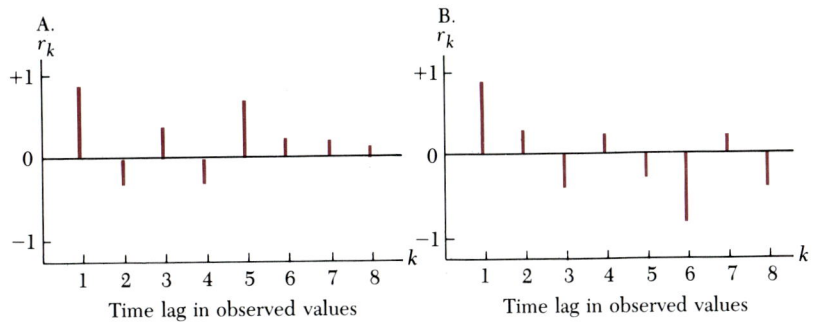

based on an analysis of a given time series. In Figure 17.8A, the observation for the current month appears to bear a strong statistical relationship (in the sense described in Chapters 13 and 14) to the observation of the previous month and the observation that is five periods old. In Figure 17.8B, the observation of the current period is highly correlated with the observation of the previous period, and it is highly negatively correlated with the observation that is six periods old (possible seasonal factor). These high autocorrelations can be useful in forecasting a future value of the series.

A kth-order autoregressive time series forecasting model is denoted by

$$\hat{y}_t = b_0 + b_1 y_{t-1} + b_2 y_{t-2} + \cdots + b_k y_{t-k}$$

A first-order ($k = 1$) autoregressive forecasting model is described here. Its use is illustrated in the following example.

First order autoregressive model

First-order autoregressive forecasting model

Let an observation at time t be given by

$$\mathbf{Y}_t = \beta_0 + \beta_1 y_{t-1} + \boldsymbol{\epsilon}_t$$

where

β_0 = Constant, or level of the series.
β_1 = Growth or trend component of the series.
ϵ_t = Random disturbance at time t with mean 0 and variance σ_ϵ^2.

Then the first-order autoregressive time series forecast of an observation for time t is given by

$$\hat{y}_t = b_0 + b_1 y_{t-1}$$

where b_0 and b_1 are point estimates of β_0 and β_1, respectively, determined by the method of least squares.

TABLE 17.8 | Sales of stereo equipment in Bloomington, Indiana, September 1984–December 1986

Date	Time period	Sales ($000)	Date	Time period	Sales ($000)
1984			1985		
Sept.	1	235.89	Nov.	15	292.80
Oct.	2	236.81	Dec.	16	301.85
Nov.	3	248.79	1986		
Dec.	4	254.16	Jan.	17	297.33
1985			Feb.	18	305.74
Jan.	5	247.47	March	19	307.20
Feb.	6	258.12	April	20	307.34
March	7	253.97	May	21	314.93
April	8	267.84	June	22	319.59
May	9	261.41	July	23	323.16
June	10	271.20	Aug.	24	322.94
July	11	266.47	Sept.	25	335.41
Aug.	12	286.69	Oct.	26	339.46
Sept.	13	299.34	Nov.	27	343.45
Oct.	14	292.34	Dec.	28	345.30

Example 17.4 Table 17.8 gives information on the sales of stereo equipment by a large retail stereo outlet in Bloomington, Indiana, over a 28-month period. Using the data provided, develop a first-order autoregressive forecasting model and estimate sales for January 1987. Does a high degree of first-order autocorrelation exist in these sample data?

Solution Relevant computations are given in Table 17.9. Using the data provided,

$$b_1 = \frac{\Sigma x_t y_t - \dfrac{(\Sigma x_t)(\Sigma y_t)}{n}}{\Sigma x_t^2 - \dfrac{(\Sigma x_t)^2}{n}} = \frac{2,306,666.45 - \dfrac{(7,791.69)(7,901.10)}{27}}{2,275,753.95 - \dfrac{(7,791.69)^2}{27}} = 0.975$$

$$b_0 = \bar{y} - b_1\bar{x} = \left(\frac{7,901.10}{27}\right) - 0.975\left(\frac{7,791.69}{27}\right) = 11.131$$

$$\hat{y}_t = 11.131 + 0.975 y_{t-1}$$

$$r = \sqrt{\frac{\text{SST} - \text{SSE}}{\text{SST}}} = \sqrt{\frac{25,901.75 - 1,316.96}{25,901.75}} = 0.974$$

Notice that the multiple correlation coefficient, r, gives the correlation between y_t and $x_t = y_{t-1}$ from the regression analysis. Thus, $r = r_1$, the first order autocorrelation coefficient. Of course, we could compute r_1 directly by finding the correlation coefficient for the pairs (235.89, 236.81), (236.81, 248.79), . . . , (343.45, 345.30). In Section 17.10, we will use the software package MINITAB to calculate the first and higher-order autocorrelations.

TABLE 17.9 | Computation worksheet for autoregressive forecasting model

$x_t = y_{t-1}$	y_t	x_t^2	y_t^2	$x_t y_t$	$\hat{y}_t$	$y_t - \hat{y}_t$	$(y_t - \hat{y}_t)^2$
235.89	236.81	55,654.04	56,077.56	55,860.88	241.24	−4.43	19.62
236.81	248.79	56,077.56	61,894.47	58,914.27	242.13	6.66	44.36
248.79	254.16	61,894.47	64,599.34	63,232.44	253.81	0.35	0.12
254.16	247.47	64,599.34	61,239.17	62,896.82	259.06	−11.59	134.33
247.47	258.12	61,239.17	66,624.64	63,875.18	252.53	5.59	31.25
258.12	253.97	66,624.64	64,502.28	65,554.88	262.92	−8.94	79.92
253.97	267.84	64,502.28	71,740.41	68,025.14	258.87	8.97	80.42
267.84	261.41	71,740.41	68,337.54	70,018.31	272.40	−10.99	120.78
261.41	271.20	68,337.54	73,547.81	70,894.83	266.13	5.06	25.60
271.20	266.47	73,547.81	71,008.13	72,266.81	275.68	−9.20	84.64
266.47	286.69	71,008.13	82,189.72	76,394.62	271.07	15.62	243.98
286.69	299.34	82,189.72	89,605.63	85,817.61	290.79	8.56	73.27
299.34	292.34	89,605.63	85,462.97	87,509.79	303.13	−10.79	116.42
292.34	292.80	85,462.97	85,729.20	85,595.98	296.30	−3.51	12.32
292.80	301.85	85,729.20	91,111.91	88,379.59	296.74	5.10	26.01
301.85	297.33	91,111.91	88,403.94	89,747.71	305.57	−8.25	68.06
297.33	305.74	88,403.94	93,475.72	90,904.46	301.16	4.57	20.88
305.74	307.20	93,475.72	94,369.69	93,921.64	309.37	−2.17	4.71
307.20	307.34	94,369.69	94,460.64	94,415.15	310.79	−3.45	11.90
307.34	314.93	94,460.64	99,178.07	96,790.62	310.94	3.99	15.92
314.93	319.59	99,178.07	102,134.57	100,645.47	318.33	1.25	1.56
319.59	323.16	102,134.57	104,432.06	103,276.93	322.88	0.28	0.08
323.16	322.94	104,432.06	104,293.15	104,362.58	326.36	−3.42	11.70
322.94	335.41	104,293.15	112,496.51	108,317.20	326.15	9.25	85.56
335.41	339.46	112,496.51	116,232.07	113,856.08	338.31	1.15	1.32
339.46	343.45	115,232.07	117,961.78	116,588.89	342.26	1.19	1.42
343.45	345.30	117,961.68	119,234.85	118,596.56	346.16	−0.86	0.74
7,791.69	7,901.10	2,275,753.95	2,339,343.77	2,306,660.45		−0.01	1,316.92

The forecast for January 1987 thus becomes

$$\hat{y}_{\text{Jan., 1987}} = 11.131 + 0.975(345.30) = \$347.799 \ (\$000)$$

The sample coefficient of correlation is 0.974 ($r^2 = 0.95$). Hence we would conclude that there is indeed a strong statistical relationship between sales in one period and sales in the previous period. The autoregressive model developed appears to be a good predictor of sales as judged by the high autocorrelation between one-period lagged observations.

Shown in the table below are the input data (Example 17.4) for a second-order autoregressive forecasting model. Note that in this case, $x_t = y_{t-2}$. That is, the dependent observations are lagged by two periods in computing the autoregressive model. Also note that for every increase in the autoregressive order (k), one additional data item is "lost" in computing the sample autocorrelation coefficient, or we have one less sample value to work with in computing the autoregressive model.

$x_t = y_{t-2}$	y_t	$x_t = y_{t-2}$	y_t
235.89	248.79	292.34	301.85
236.81	254.16	292.80	297.33
248.79	247.47	301.85	305.74
254.16	258.12	297.33	307.20
247.47	253.97	305.74	307.34
258.12	267.84	307.20	314.93
253.97	261.41	307.34	319.59
267.84	271.20	314.93	323.16
261.41	266.47	319.59	322.94
271.20	286.69	323.16	335.41
266.47	299.34	322.94	339.46
286.69	292.34	335.41	343.45
299.34	292.80	339.46	345.30

The second order autocorrelation coefficient, r_2, is computed in Section 17.10 by using MINITAB. It is the correlation coefficient for the pairs (235.89, 248.79), (236.81, 254.16), . . . , (339.46, 345.30).

■ 17.6

Autoregressive integrated moving average (ARIMA) models (advanced section)

An approach that uses both the autoregressive and moving average techniques for forecasting is due to George Box and Gwilym Jenkins and has become widely known as the Box-Jenkins methodology. The basic model is specified by:

$$y_t = \beta_0 + \beta_1 y_{t-1} + \beta_2 y_{t-2} + \cdots$$
$$+ \beta_p y_{t-p} - \theta_1 e_{t-1} - \theta_2 e_{t-2} - \cdots - \theta_q e_{t-q} + e_t$$

where $\beta_0, \beta_1, \ldots, \beta_p$ and $\theta_1, \theta_2, \ldots, \theta_q$ are parameters of the model, and e_t is the value of the random error component for the tth period ($e_t = y_t - \hat{y}_t$). Notice that the model contains two components. In the first, the current value y_t is "autoregressed" on the past p observations $y_{t-1}, y_{t-2}, \ldots, y_{t-p}$. This component represents the autoregressive part of the model (AR). In the second, the random error components are used to "regress" on the current value y_t. This component can be shown algebraically to be equivalent to a moving average process on the random error components, $e_{t-1}, e_{t-2}, \ldots, e_{t-q}$. Thus this component represents the moving average part of the model (MA). The model is generally referred to as the ARMA(p, q) model, meaning that we are autoregressing on the most recent p observations and the most recent q error component values. An ARMA(2, 0) model would be, for example,

$$y_t = \beta_0 + \beta_1 y_{t-1} + \beta_2 y_{t-2} + e_t$$

An ARMA(1, 1) model would be

$$y_t = \beta_0 + \beta_1 y_{t-1} - \theta_1 e_{t-1} + e_t$$

The Box-Jenkins model

The ARMA(p, q) model

The error component in the tth period, ϵ_t must be normally distributed, independent of all other error components, and have the same variance as all other error components.

Stationary series

To use the Box-Jenkins model, the time series y_t, $t = 1, 2, \ldots$ must be stationary. By stationary, we mean that its values must fluctuate about a constant mean, β_0. If the original series is not stationary, then a transformation must be used to produce a stationary series. Often, differencing will produce a stationary series. First differences are defined by

Differencing

$$z_t = y_t - y_{t-1}$$

Second differences are specified by

$$z_t = y_t - 2y_{t-1} + y_{t-2}$$

Once a stationary series has been produced, forecasts are made based on the stationary (z_t) series. Then forecasts for the original series are produced by solving the appropriate differencing equation for y_t. On occasion, $z_t = y_t$; that is, the original series is stationary, but this doesn't happen very often in practice. The degree of differencing is denoted by d in the Box-Jenkins model, which is more generally denoted by ARIMA(p, d, q).

The ARIMA(p, d, q) model

The Box-Jenkins model can be extended to represent seasonal effects. Due to the advanced nature of this extension, we will not consider it in this introductory treatment. For an excellent presentation of the Box-Jenkins methodology, including the extension for seasonal effects, see the Bowerman and O'Connell reference at the end of this chapter.

The Box-Jenkins methodology consists of three basic steps, plus the use of the model in actual forecasting.

Steps in the Box-Jenkins method

> ### Steps in the Box-Jenkins forecasting methodology
>
> 1. In step 1, the forecasting model is tentatively identified by analyzing sample autocorrelations (SAC) and sample partial autocorrelations (SPAC) to identify likely candidates (models) for predicting values of the series under investigation.
> 2. In step 2, the parameters of the ARIMA model are estimated using a nonlinear least squares procedure.
> 3. In step 3, diagnostic checking of the model takes place, in essence answering the question: Is the model adequate and correct? By analyzing the sample autocorrelations of the *residuals,* the model builder determines whether or not it is necessary to re-enter the model identification step and repeat the procedure.

In tentatively identifying the appropriate model, correlograms are constructed similar to the correlograms of the previous section for both the

Partial autocorrela-
tion function

autocorrelation functions and the *partial autocorrelation functions*. Recall from the discussion of partial correlation in Chapter 15 that partial correlation is a measure of the association (linear) between two variables when the effect of other variables is held constant. Partial *auto*correlations measure the degree of linear association between an observation at time t and at time $(t - p)$ where the effect of intervening time lags is held constant. The purpose of examining the partial autocorrelation functions is to help identify the appropriate ARIMA model.

In Figure 17.9, we indicated possible values of autocorrelation and of partial autocorrelation for three ARIMA modeled processes. In process A,

FIGURE 17.9 | Autocorrelation and partial autocorrelation functions for three ARIMA processes

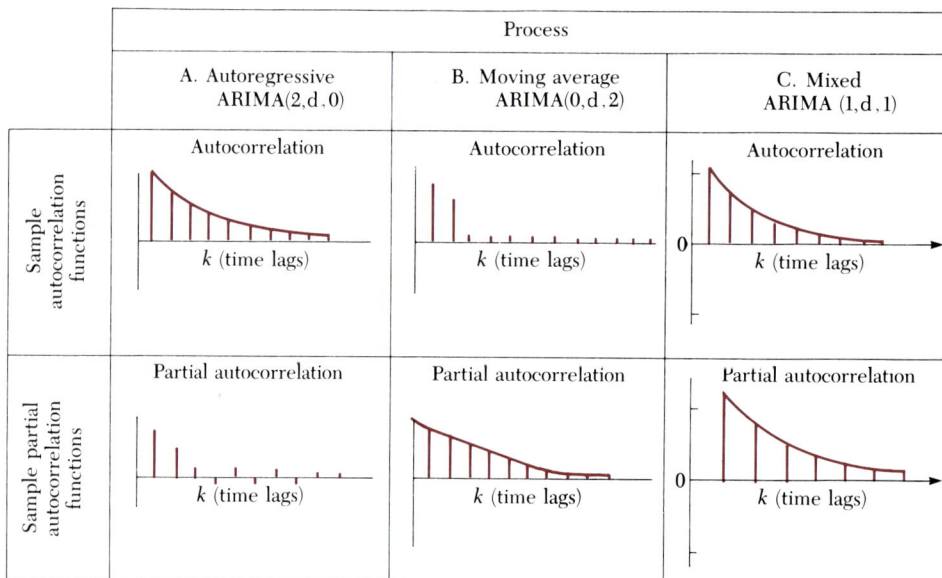

the autocorrelation functions decay exponentially as a function of the time lags of the observations, and the partial autocorrelation functions are initially large and then become instantly insignificant (it is possible to test statistically for the significance of these terms using the chi-square distribution discussed in Chapter 19). The order of the process is equal to the number of initial significant partial autocorrelations. For process A, the model is thus ARIMA(2, d, 0). In process B, the autocorrelations become 0 or insignificant after the second time lag, and the partial autocorrelation function decays exponentially as a function of the time lag of the observations. Such processes are modeled by moving averages, and because the autocorrelation function becomes insignificant after two time lags, the order of the process is 2. Hence the ARIMA model for process B is ARIMA(0, d, 2). In Figure 17.9C we have indicated a mixed ARIMA model; since both the partial autocorrelation and the autocorrelation functions decline exponentially, a model of the form ARIMA(1, d, 1) is suggested. Computers are usually used to provide the analyst with both values and graphs of the sample autocorrelation and sample partial autocorrelation functions, leaving the analyst free to concentrate on model development rather than on arithmetic calculations.

The graphs in Figure 17.9 are not intended to be all-inclusive representing all likely sample autocorrelation functions and sample partial autocorrelation functions. They are intended to provide the reader with examples of these functions that correspond to various processes. In Figure 17.10, for example, we show other sample autocorrelation functions for an autoregressive process of order one and of order two.

FIGURE 17.10 | Theoretical autocorrelation functions—autoregressive models

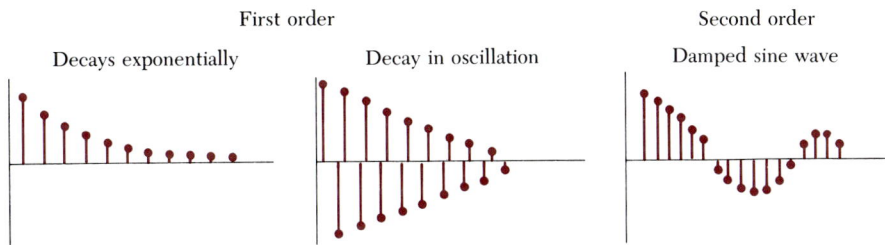

First order — Decays exponentially
First order — Decay in oscillation
Second order — Damped sine wave

| Stationary process |

To use the ARIMA models, it is imperative that the series being modeled is *stationary*. By *stationary,* we mean that the average of the process is not changing over time. A stationary series is shown in Figure 17.11C. Notice that the mean of this time series is not changing over time. The reason that the process must be stationary is that any type of trend that is present tends to introduce spurious autocorrelations into the data that dominate or mask the autocorrelation pattern. Fortunately, methods are available for reducing a nonstationary series to a stationary one for purposes of analysis, and the sample autocorrelation function can aid us in detecting the presence of a nonstationary series so that it can be reduced to a stationary one.

| Differencing |

The procedure for removing trend to render a process stationary is called *differencing*. In Chapter 16 a method for removing trend was indicated that divided each actual value of the time series by the trend estimate determined by the method of least squares. The method of differencing is an alternative to this method, which is more computationally efficient and better suited for

FIGURE 17.11 | Stationary and nonstationary time series processes

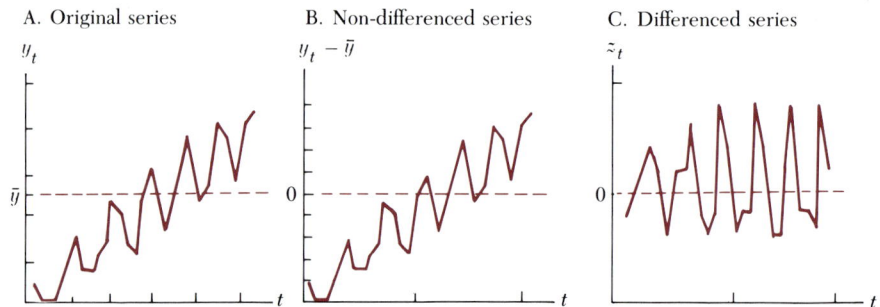

A. Original series   B. Non-differenced series   C. Differenced series

ARIMA models. We used differencing in Chapter 16 to determine the representativeness of a straight-line trend or a second-degree parabola in measuring the secular trend component in a time series, for example.

In Figure 17.11A we have indicated a time series that is definitely not stationary—an easily recognized trend is apparent. In Figure 17.11B we have indicated the same series minus the mean value for each observation in the series—this series too is nonstationary, as there is a definite trend in the data. Figure 17.11C is a scatter plot of the data $z_t = y_t - y_{t-1}$, where the Y-values are given in Figure 17.11A. Note how this figure is similar to Figure 17.3 in that the observations tend to scatter randomly around some nonchanging mean value. Such a series is *stationary,* and it is this *latter series* we use in developing our forecasting model. The autocorrelation function of the series z_t in Figure 17.11C would tend to have the first one or two autocorrelation values possibly significantly different from 0, but all remaining autocorrelation values would be very close if not equal to 0. In the event the first differenced series was not stationary, then differences of the first differences can be taken, and so on, until a stationary process is identified. Rarely does one have to proceed beyond second differencing in deriving a stationary series, or one that is approximately so.

Once a series has been identified as being stationary and the form of the ARIMA model has been identified through an examination of the sample autocorrelation and sample partial autocorrelation functions as described, a nonlinear least squares procedure is used to estimate the model coefficients. A computer is almost always required in estimating these quantities.

| Diagnostic checking |

After estimates are obtained for the parameters of the model, diagnostic checking of the *residuals* takes place to identify the appropriateness of the model constructed. Diagnostic checking of the residuals is accomplished through an examination of the sample autocorrelations of the residuals. If the model tentatively identified is adequate, the sample residuals will be normally distributed with mean 0 and variance σ^2. They will also be independent.

If the assumptions concerning the random error components ϵ_t, $t = 1, 2,$ . . . are not satisfied, then corrective measures must be taken. Notice that the assumptions (equal variances, independence, and normality) are the same assumptions required for the random error components in the regression models in Chapters 13–15.

The Box-Jenkins methodology is quite complex and usually requires the extensive use of computer analyses to perform the numerous computations required for identifying the model, estimating the parameters, and checking the model assumptions. In the identification stage, experience is required in matching the sample autocorrelations and the sample partial autocorrelations with the theoretical patterns of the population autocorrelation and partial autocorrelation functions. In Section 17.10, we will use MINITAB to illustrate the use of this methodology. However, it should not be used without more than an introductory coverage of the methodology. Several of the advanced references (Bowerman and O'Connell is excellent) go much further into the development of Box-Jenkins methodology than we can in these few pages. Rather than attempting to make the reader a sophisticated user of these techniques, we have attempted to indicate how two forecasting techniques can be integrated using the tools described in the text to derive a model that is much more sophisticated than the sum of the two individual components. These advanced references also describe the extensions of the Box-Jenkins methodology to cover the incorporation of independent variables other than strictly time-related ones by the use of *transfer functions*. In the next section, we consider the incorporation of other than time-phased observations of the time series in developing a forecasting model.

■ 17.7

Econometric model building: Time series and regression analysis revisited (advanced section)

The techniques of simple linear regression examined in Chapter 13 and multiple linear regression examined in Chapter 15 can be used to derive estimating equations for the value of a dependent variable we wish to predict as a function (linear) of one or more independent variables whose values are known. For example, in Section 15.7 estimating equations were developed for predicting the number of casts set on broken limbs at a ski and leisure resort in Colorado. The estimating equation developed for assessing the number of casts required using one independent variable was:

$$\hat{y} = 9.03 + 0.9589x$$

where $\hat{y}$ = Predicted number of plaster casts required, and x = Number of visitors to the facility (000). If we can accurately predict the number of visitors to the facility, then we can estimate the demand for plaster casts as a function of the number of visitors, as the following example illustrates.

Example 17.5 The firm in Example 15.6 is able to estimate fairly accurately the number of visitors to its facility because demand far exceeds supply, and reservations must be made well in advance of the scheduled arrival date. Using current information, the firm estimates the number of visitors to the facility in 1987 to be as shown in the table on page 920.

Quarter	Expected number of tourists (000)
Winter	118.42
Spring	121.54
Summer	125.84
Fall	128.75

Using the predicting equation given, estimate the expected number of casts that will be required as a function of the number of visitors to the facility.

Solution To estimate the number of casts required, we simply enter the expected number of visitors into the derived equation to estimate demand.

Quarter	Expected number of casts
Winter	$9.03 + 0.9589 \cdot 118.42 = 122.58$
Spring	$9.03 + 0.9589 \cdot 121.54 = 125.57$
Summer	$9.03 + 0.9589 \cdot 125.84 = 129.70$
Fall	$9.03 + 0.9589 \cdot 128.75 = 132.49$

We saw in Section 15.7 that there is a definite seasonal pattern in the necessity for setting broken limbs, and we were able to improve our estimates by taking seasonal factors into account. Using the four quarters of the year to predict the demand, the following estimating equation was developed:

$$\hat{y}_i = 14.225 + 0.99x_{i,1} + 6.443x_{i,2} - 6.442x_{i,3} - 30.592x_{i,4}$$

where

$\hat{y}_i$ = Demand for plaster casts (estimated).
$x_{i,1}$ = Number (000) of visitors to the facility in period i.
$x_{i,2}$ = {1 if period i is the first quarter of the year; 0 otherwise}.
$x_{i,3}$ = {1 if period i is the second quarter of the year; 0 otherwise}.
$x_{i,4}$ = {1 if period i is the third quarter of the year; 0 otherwise}.

The binary or dummy variables in the estimating equation allowed us to "capture" the effect of seasonal factors in influencing demand. In a similar fashion, the use of an independent variable allows us to better estimate the value of a dependent variable whenever there is a strong linear relationship between the dependent and the independent variable. If we know, for example, that the demand for automobiles and parts closely follows the pattern of overall economic activity as measured by, say, GNP, then it is possible to derive an estimating equation for automobiles and parts as a function of GNP. For such an estimating equation to be useful in predicting a future observation of the dependent variable, we must obtain an estimate of the value of the independent variable for the same period in the future. As we build simple linear and multiple linear regression models for estimating some future value of a dependent variable, we should be aware that to predict some future value of the dependent variable, we must have the value of the independent variable(s) if the predicting equation is to be useful.

In the same vein, an economist might develop an estimating equation for predicting sales of refrigerators, using as explanatory variables or independent variables a forecasted increase in disposable personal income, housing starts from two periods ago, and the price of the refrigerator units. The model developed may specify that sales of refrigerators in period t are estimated by disposable personal income in period t, housing starts in period $(t - 2)$, and price in period t. In estimating refrigerator sales, the economist must obtain an estimate of disposable personal income in period t, which is usually supplied by government agencies; also, the price charged in period t for the refrigerators must be known, as must be housing starts of two periods ago. Note the beauty in using a ''lagged'' estimate, such as housing starts of two periods previous—this quantity does itself not have to be estimated, since its value is known. The rationale for including this variable may be that the refrigerator is one of the last items added to the house, so its demand is not affected until two periods after the house begins construction. This is the essence of econometric model building—the identification and specification of causative factors to be used in a predicting equation. Recall, however, from the discussion in Chapter 13 that the use of statistical analysis—notably correlation and partial correlation analysis—does not allow us to claim cause and effect. In econometric model building, the economist is building his or her model from the construct that independent variables included should ''influence'' the behavior of the dependent variable in ways that can be explained on intuitive grounds.

When the values of the independent variables in a linear regression model are unknown, how can the regression model be used to estimate future values? First, we *may* be able to build an accurate model for predicting the value of the *independent* variables using forecasting model-building procedures described in the chapter. These forecasted values of the independent variables are then substituted into the regression equation to estimate the value of the dependent variable. Second, we can build our forecasting equation with independent variables in the model that are *either already known* or relatively simple to estimate. For example, we may be able to treat the dependent variable strictly in a time series fashion, as the next example illustrates.

Example 17.6   Assume that the firm in Example 17.5 is unable to estimate the demand for its facilities (tourists) in a reliable fashion. Construct a table of values of time series variables to be used to estimate the demand for plaster casts.

TABLE 17.10 | Values of dependent and independent variables in Example 17.6

		Quarter					Quarter		
Casts	Time	1	2	3	Casts	Time	1	2	3
53	1	1	0	0	94	13	1	0	0
41	2	0	1	0	86	14	0	1	0
24	3	0	0	1	64	15	0	0	1
57	4	0	0	0	99	16	0	0	0
70	5	1	0	0	109	17	1	0	0
60	6	0	1	0	101	18	0	1	0
41	7	0	0	1	77	19	0	0	1
77	8	0	0	0	110	20	0	0	0
81	9	1	0	0	123	21	1	0	0
70	10	0	1	0	120	22	0	1	0
50	11	0	0	1	95	23	0	0	1
87	12	0	0	0	126	24	0	0	0

Source: Company records.

Solution A table of independent variable values is given in Table 17.10. The time series, multiple regression equation is:

$$\hat{y}_t = b_0 + b_1 t + b_2 x_{t,1} + b_3 x_{t,2} + b_4 x_{t,3}$$

where

$\hat{y}_t$ = Expected demand for plaster casts in time period t.
t = 1 for first quarter 1981 (t is in quarters).
$x_{t,1}$ = {1 if quarter t is the first quarter of the year; 0 otherwise}.
$x_{t,2}$ = {1 if quarter t is the second quarter of the year; 0 otherwise}.
$x_{t,3}$ = {1 if quarter t is the third quarter of the year; 0 otherwise}.

The results of a computer analysis of the data in Table 17.10 are given in Table 17.11. In comparing this model (strictly time series) with the one given in Table 15.15, which uses visitors as one of its independent variables, the time series model developed appears quite favorable in relation to the multiple linear regression model developed in Chapter 15. This is due in part to the fact that the number of visitors to the facility is itself a time-related variable, increasing fairly steadily over the past six years. Note also that the seasonal estimates are quite similar in both models, and that the time series model has only a slightly larger SSE than the regression model developed in Chapter 15 (132.90 > 125.26).

We might now ask how the seasonal estimates given in Table 17.11 compare with the seasonal estimates computed by the ratio to moving average method in Chapter 16. The estimates given in Table 16.11 correspond to the *multiplicative time series model;* that is, to obtain an estimate by quarter, we would multiply the annual estimate divided by 4 (four quarters in a year) by the seasonal index, and divide by 100. In the time series model developed in

---

Multiplicative time series model

TABLE 17.11 | Computer analysis of time series data in Table 17.10

DEPENDENT VARIABLE: CASTS

SOURCE	DF	SUM OF SQUARES	MEAN SQUARE	F-VALUE	PR > F
MODEL	4	17691.06190476	4422.76547619	632.32	0.0001
ERROR	19	132.89642857	6.99454887	R-SQUARE	STD DEV
CORRECTED TOTAL	23	17823.95833333		0.992544	2.64472094

PARAMETER	ESTIMATE	T FOR HO: PARAMETER=0	PR > |T|	STD ERROR OF ESTIMATE
INTERCEPT	43.99166667	28.46	0.0001	1.54589916
TIME	3.47678571	44.00	0.0001	0.07902615
QUARTER 1	6.09702381	3.95	0.0009	1.54522571
QUARTER 2	−6.04642857	−3.94	0.0009	1.53508854
QUARTER 3	−30.68988095	−20.07	0.0001	1.52897398

this chapter, we would *add* an amount equal to the coefficient of the quarterly independent variable in the regression equation. In the former case, we are multiplying a trend estimate by a fixed *percentage;* greater than 1 if the seasonal effect indicates an increase in the given quarter, and less than 1 if the seasonal effect indicates a decrease over the average quarter. In the latter case, we are *adding* to our time series trend estimate a *constant amount* corresponding to the expected increase in the value of the dependent variable due to seasonal factors. Thus the former is in the scheme of a *multiplicative model,* and the latter is in the scheme of an *additive model.* Note, however, that if we momentarily consider the fall quarter to be 100, then the estimate for winter (quarter I) from Table 17.11 would be 106.10 (100 + 6.1); for spring it would be 93.95 (100 − 6.05); and for summer it would be 69.31 (100 − 30.69).

| Additive time series model

In both cases, the ascending sorted seasonal effect would be in the order summer, spring, fall, and winter. Both approaches give the same *relative ranking* of the effects of seasonal factors on the value of the time series. Both are measuring this component in a slightly different fashion, even though the comparative results are similar.

An advantage of the multiple regression methods over the "strict" time series methods frequently cited is that the forecasts obtained need not necessarily be time-dependent. This allows, of course, for the prediction of a horizon that may be considerably longer term than is appropriate for time series models. Furthermore, we can include any number of explanatory variables in our multiple linear regression model. We may find, for example, that sales are related to a host of independent variables, which may include GNP, advertising, prices, competition, and research and development expenditures. We can include several nontime-related variables in a multiple linear regression equation for estimation purposes. Still, if an estimate of the dependent variable is to be made for a definite period in the future, it may be necessary to obtain estimates of the independent variable values, putting us right back where we started: estimating the value of a variable in the future.

Today large-scale econometric models are being used to model the economy, selected industries within the economy, and even specific firms within the industry. Just as simple linear regression is a special case of multiple linear regression, multiple linear regression is a special case of econometrics. Econometric models can (and do) include any number of *simultaneous multiple linear regression equations*. Thus econometric models are systems of simultaneous equations involving several independent variables. We will not examine econometric modeling in depth in this text, as this is outside the scope of introductory statistics. We will continue in the following sections to augment our kit of forecasting techniques to include other regression and time series approaches for estimating the value of a dependent variable at some future time.

■ **17.8**

Sinusoidal models (advanced section)

Just as regression models may offer some advantages over strict time series approaches, sinusoidal forecasting models may offer certain advantages over other methods in forecasting the future value of some series. In the sinusoidal models, sine and cosine functions are used to model periodic or *seasonal* movements in the data. A computational advantage of the sinusoidal models over the method of ratio to moving averages for computing seasonal indices is that the amount of data storage and manipulation is greatly reduced with the sinusoidal methods. In this regard (minimal computer storage requirements), the sinusoidal models are similar to exponential smoothing and offer the same comparative advantages over other approaches. And, like the dummy or indicator variable techniques for handling or modeling seasonality, sinusoidal models most frequently use multiple linear regression techniques to estimate the parameters in the forecasting equation. Although the Box-Jenkins forecasting methodology is generalizable to include seasonal movements, sinusoidal models represent a quite different approach to estimating the future value of a series that possesses seasonal patterns that repeat themselves from year to year.

Figure 17.12 is a plot of the standard sine and cosine functions. Recall that

FIGURE 17.12 | Graph of the standard sine and cosine functions

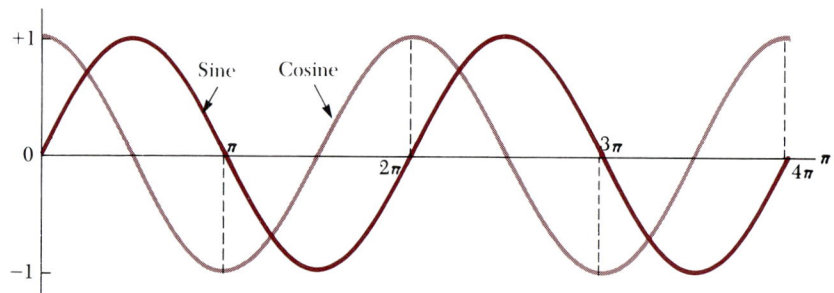

these transcendental functions repeat themselves every 2π radians or every 360° (one full cycle).

In Figure 17.13 is given a plot of the function $y_t = 80 + 10 \sin 30t°$. In terms of a forecasting model, this equation would represent a time series

FIGURE 17.13 | Graph of $y_t = 80 + 10 \sin 30t°$ and $y_t = 80 + 10 \sin 30t° + 10 \cos 30t°$

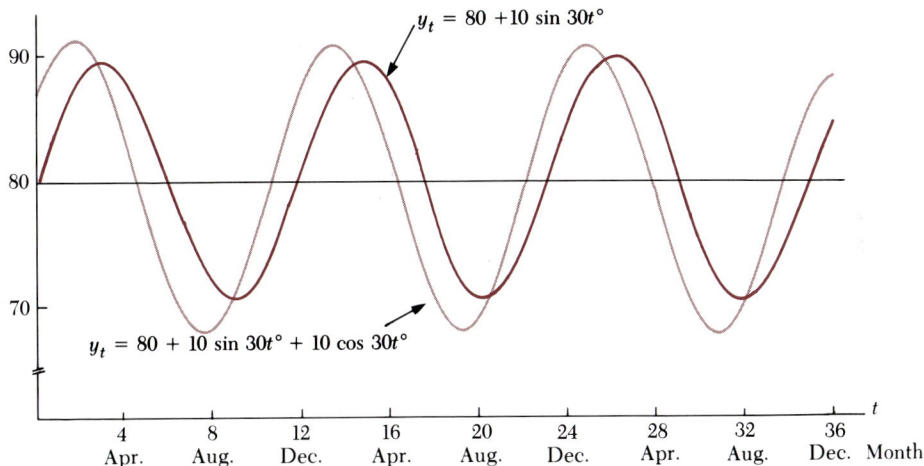

with a level or a constant value of 80, a seasonal movement corresponding to an observation dictated by 10 times the standard sine function every month (360/30 = 12 months), and a noted absence of trend, cyclical, and irregular factors. We have also superimposed months of the year on this figure to indicate how this sine wave can model movements in a given time series corresponding to months of the year. The *four-term* sinusoidal time series forecasting model corresponding to a time series with level, trend, *fundamental seasonal*, and irregular variation is described below.

Four-term sinusoidal model

> **Four-term sinusoidal forecasting model**
>
> Let an observation at time period t be given by
>
> $$Y_t = \beta_0 + \beta_1 t + \beta_2 \sin 30(t - t_0)° + \epsilon_t$$
>
> where
>
> β_0 = Constant, or level of the series.
> β_1 = Growth, or trend component of the series.
> β_2 = Amplitude of the seasonal peak (trough) in the series.

$\sin 30(t - t_0)° =$ Standard sine function measured at monthly (30°) intervals, with phase angle given by $(t - t_0)$.

$\epsilon_t =$ Random disturbance at time period t with mean 0 and variance σ_ϵ^2.

Then the *four-term sinusoidal time series forecast* of an observation for time period t is given by

$$\hat{y}_t = b_0 + b_1 t + b_2 \sin 30t° + b_3 \cos 30t°$$

where b_0, b_1, b_2, and b_3 are estimated from past data using the method of least squares (multiple linear regression).

In a sinusoidal time series forecasting model of the form

$$\hat{y}_t = b_0 + b_2 \sin 30t° + b_3 \cos 30°t$$

the coefficients b_2 *and* b_3 *together* determine the amplitude (height) and phase angle (start) of a standard sine wave with one observation every month (or every 30°).

From trigonometry

$$\sin(x - y) = \sin x \cos y - \cos x \sin y$$

Hence if we desire to shift the standard sine wave t_0 "months," we have

$$\sin 30(t - t_0)° = \sin 30t° \cos 30t_0° - \cos 30t° \sin 30t_0°$$
$$= b_2 \sin 30t° + b_3 \cos 30t°$$

where $b_2 = \cos 30t_0°$ and $b_3 = -\sin 30t_0°$. The use of this relationship is illustrated in the following example.

Example 17.7   A given time series is known to have an absence of trend, cyclical, and "generally" very minor irregular movements, a level of 200 units, and a seasonal pattern that follows the general sine function with an amplitude of 50 units, a peak in December of each year, and an absence of seasonal movements in September. Measurements of the time series are made each month. What is the general sinusoidal model for this time series?

Solution   In order to shift the peak of the standard sine function from March (90°) to December (360°) and have the seasonal movement equal to 0 in September (270°), t_0 must be set equal to 9 (verify by examining Figure 17.12).

$$\hat{y}_t = 200 + 50 \sin 30(t - 9)°$$
$$= 200 + 50(\cos 30 \cdot 9° \sin 30t° - \sin 30 \cdot 9° \cos 30t°)$$
$$= 200 + 50(\cos 270° \sin 30t° - \sin 270° \cos 30t°)$$
$$= 200 + 50(0 \sin 30t° - \cos 30t°)$$
$$= 200 - 50 \cos 30t°$$

It is possible, of course, to include more terms in the estimating equation than

$$\hat{y}_t = b_0 + b_1 t + b_2 \sin 30t° + b_3 \cos 30t°$$

FIGURE 17.14 | Graph of the equation $y_t = 80 + 10 \sin 30t° + 10 \cos 30t° + 10 \sin 60t° + 10 \cos 60t°$

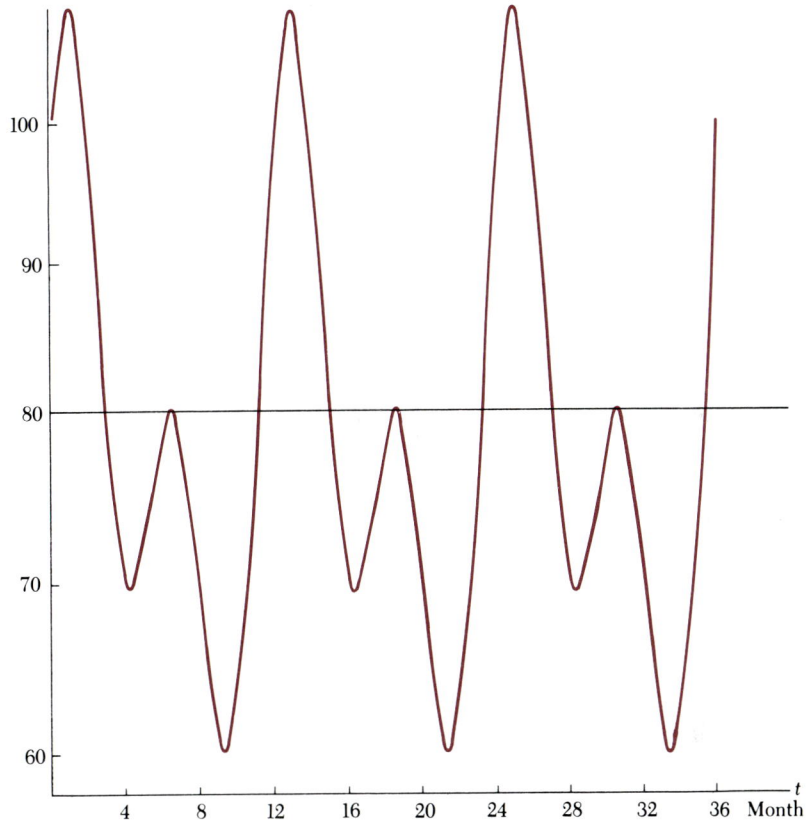

Figure 17.14, for example, represents a time series of the form

$$\hat{y}_t = 80 + 10 \sin 30t° + 10 \cos 30t° + 10 \sin 60t° + 10 \cos 60t°$$

where the latter two terms determine the amplitude and phase angle of a sine wave that completes a cycle in only six months. If trend and irregular variations are also present in the time series, the appropriate model becomes

$$\mathbf{Y}_t = \beta_0 + \beta_1 t + \beta_2 \sin 30t° + \beta_3 \cos 30t° + \beta_4 \sin 60t° + \beta_5 \cos 60t° + \epsilon_t$$

The general six-term sinusoidal forecasting model is defined below. The latter two terms in this model are sometimes referred to as *harmonics* and are included to match a variety of shapes and sizes of seasonal or cyclical patterns in a given time series that are not adequately modeled by the *fundamental* component alone.

It is possible, of course, to include more terms than even the six given in the above model, but this is rarely done in practice since the preceding equation will generally model most series satisfactorily. One exception might be the inclusion of a $\beta_6 t^2$ term if the trend component turned out to be a parabola. It can be shown that no more than six sine functions would ever be required to represent a seasonal pattern of the 12 months of the year.

<table>
<tr><td>Six-term sinusoidal model</td><td>

Six-term sinusoidal forecasting model

Let an observation at time period t be given by

$$Y_t = \beta_0 + \beta_1 t + \beta_2 \sin 30(t - t_0)° + \beta_3 \sin 60(t - t_1)° + \epsilon_t$$

where

β_0 = Constant, or level of the series.
β_1 = Growth, or trend component of the series.
$\beta_2 \sin 30(t - t_0)°$ = Amplitude and phase angle of a standard sine function measured at monthly (30°) intervals.
$\beta_3 \sin 60(t - t_1)°$ = Amplitude and phase angle of a standard sine function that repeats itself in six months.
ϵ_t = Random disturbance at time period t with mean 0 and variance σ_ϵ^2.

Then the *six-term sinusoidal time series forecast* of an observation for time period t is given by

$$\hat{y}_t = b_0 + b_1 t + b_2 \sin 30t° + b_3 \cos 30t° + b_4 \sin 60t° + b_5 \cos 60t°$$

where $b_0, b_1, b_2, b_3, b_4,$ and b_5 are estimated from past data using the method of least squares (multiple linear regression).

</td></tr>
</table>

Since the parameters $\beta_0, \beta_1, \beta_2, \ldots$ will rarely be known in practice, it is necessary to obtain point estimates $b_0, b_1, b_2, \ldots$ of them. The method used to obtain these point estimates is multiple linear regression as described in Chapter 15. The use of multiple linear regression in determining these point estimates is provided in the following examples.

Example 17.8 In Table 17.12 are given sales values of electronic components for small calculators for the 12 months of the years 1984, 1985, and 1986. A linear equation of the form

$$\hat{y}_t = b_0 + b_1 t + b_2 \sin 30t° + b_3 \cos 30t°$$

was fit to the given data (using the method of least squares), yielding the following results:

$$\hat{y}_t = 120.73 + 0.96t + 11.75 \sin 30t° + 6.82 \cos 30t°$$
$$s_{b_1} = 0.08 \qquad s_{b_2} = 1.17 \qquad s_{b_3} = 1.14$$

Plot a scatter diagram of the given data and the equation values. Does the equation determined appear to be a good fit of the data?

TABLE 17.12 | Sales of electronic components (in $00)

Month	Sales	Month	Sales
1984		1985	
Jan.	130.75	July	131.18
Feb.	136.01	Aug.	123.67
March	139.66	Sept.	132.74
April	127.55	Oct.	137.53
May	118.15	Nov.	142.00
June	120.79	Dec.	152.93
July	119.50	1986	
Aug.	109.80	Jan.	158.24
Sept.	121.12	Feb.	158.81
Oct.	120.31	March	151.74
Nov.	138.50	April	155.06
Dec.	135.41	May	148.79
1985		June	143.45
Jan.	145.86	July	138.64
Feb.	138.13	Aug.	132.29
March	151.69	Sept.	144.12
April	155.00	Oct.	136.83
May	137.68	Nov.	156.24
June	128.80	Dec.	165.25

Source: Company records.

Solution Relevant computations are given in Table 17.13, and a scatter plot of the actual data and the estimating equation is given in Figure 17.15. Also shown in this figure is an estimate of a simple linear trend equation, $\hat{y}_t = 123.9 + 0.79t$. The sinusoidal model derived appears to be a good fit of the data given.

Table 17.14 gives the SAS output for the multiple linear regression analysis of the data in Table 17.12, using both the simple linear regression model and the four-term sinusoidal model. Note the significance of the trend component (b_1) in both models. Note also how the sinusoidal model fits the data better as assessed by the lower mean square error (MSE) in this model and the higher value of r^2, the sample coefficient of determination.

In Figure 17.16 we give a residual plot for the linear trend model given in Table 17.14, and in Figure 17.17 the residual plot for these same data based on the four-term sinusoidal model given in Table 17.14. Note how the residual plot in Figure 17.16 denotes an incorrect model specification (or in this case an omission—the seasonal movements have not been appropriately accounted for). The pattern of the residuals in Figure 17.16 is too systematic for them to have occurred by chance! The residuals given in Figure 17.17 based on the four-term sinusoidal model, on the other hand, occur randomly within a rectangle about 0, indicating that the model has most likely been correctly specified in this instance.

TABLE 17.13 | Sinusoidal trend calculations for Example 17.8

t	$\sin 30t°$	$\cos 30t°$	$\hat{y}_t^*$	t	$\sin 30t°$	$\cos 30t°$	$\hat{y}_t^*$
1	0.5	0.866	133.47	19	−0.5	−0.866	127.15
2	0.866	0.5	136.24	20	−0.866	−0.5	126.30
3	1.00	0.00	135.37	21	−1.00	0.00	129.09
4	0.866	−0.5	131.34	22	−0.866	0.5	135.03
5	0.5	−0.866	125.50	23	−0.5	0.866	142.79
6	0.00	−1.00	119.66	24	0.00	1.00	150.54
7	−0.5	−0.866	115.65	25	0.5	0.866	156.47
8	−0.866	−0.5	114.80	26	0.866	0.5	159.24
9	−1.00	0.00	117.58	27	1.00	0.00	158.37
10	−0.866	0.5	123.53	28	0.866	−0.5	154.34
11	−0.5	0.866	131.29	29	0.5	−0.866	148.50
12	0.00	1.00	139.04	30	0.00	−1.00	142.67
13	0.5	0.866	144.97	31	−0.5	−0.866	138.65
14	0.866	0.5	147.74	32	−0.866	−0.5	137.80
15	1.00	0.00	146.87	33	−1.00	0.00	140.59
16	0.866	−0.5	142.84	34	−0.866	0.5	146.53
17	0.50	−0.866	137.00	35	−0.5	0.866	154.29
18	0.00	−1.00	131.16	36	0.00	1.00	162.05

* $\hat{y}_t = 120.73 + 0.96t + 11.75 \sin 30t° + 6.82 \cos 30t°$

FIGURE 17.15 | Scatter plot of data in Table 17.12 and sinusoidal model estimate

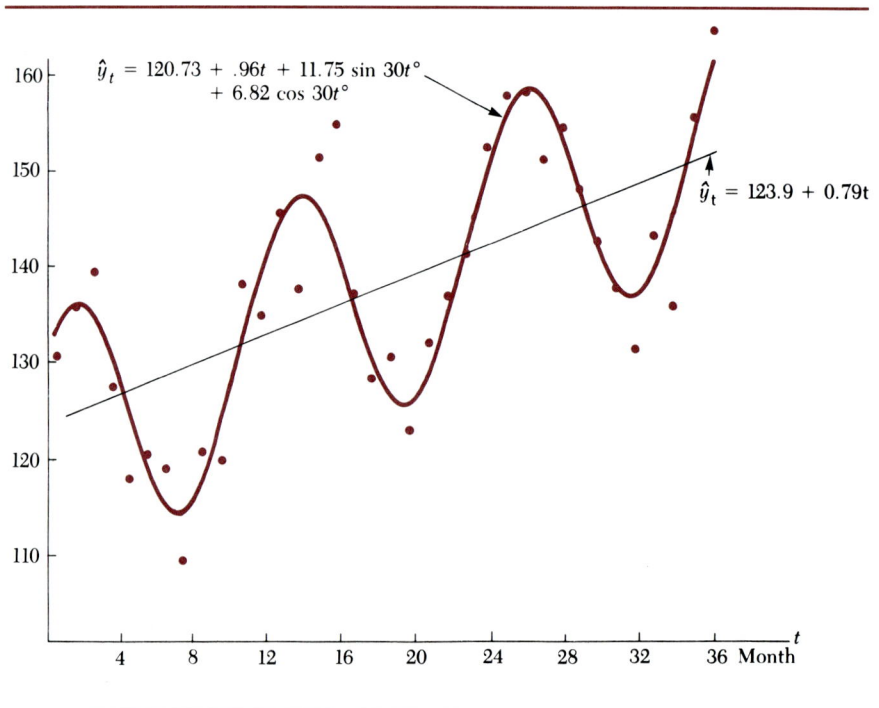

$\hat{y}_t = 120.73 + .96t + 11.75 \sin 30t° + 6.82 \cos 30t°$

$\hat{y}_t = 123.9 + 0.79t$

TABLE 17.14 | Least squares fit of data in Table 17.12

SIMPLE LINEAR TREND

DEPENDENT VARIABLE:
ELECTRIC

SOURCE	DF	SUM OF SQUARES	MEAN SQUARE	F-VALUE	PR > F
MODEL	1	2401.30088095	2401.30088095	20.67	0.0001
ERROR	34	3949.77530794	116.16986200	R-SQUARE	STD DEV
CORRECTED TOTAL	35	6351.07618889		0.378094	10.77821237

PARAMETER	ESTIMATE	T FOR HO: PARAMETER = 0	PR > |T|	STD ERROR OF ESTIMATE
INTERCEPT	123.90603175	33.77	0.0001	3.66891704
TIME	0.78619048	4.55	0.0001	0.17292239

FOUR-TERM SINUSOIDAL
       MODEL

DEPENDENT VARIABLE:
ELECTRIC

SOURCE	DF	SUM OF SQUARES	MEAN SQUARE	F-VALUE	PR > F
MODEL	3	5608.93062886	1869.64354295	80.62	0.0001
ERROR	32	742.14556003	23.19204875	R-SQUARE	STD DEV
CORRECTED TOTAL	35	6351.07618889		0.883146	4.81581237

PARAMETER	ESTIMATE	T FOR HO: PARAMETER = 0	PR > |T|	STD ERROR OF ESTIMATE
INTERCEPT	120.73095089	71.65	0.0001	1.68497709
T	0.95781926	11.96	0.0001	0.08008258
SIN 30T°	11.75149951	10.01	0.0001	1.17378401
COS 30T°	6.81627444	5.99	0.0001	1.13792099

FIGURE 17.16 | Residual plot based on simple linear trend of data in
                Table 17.12

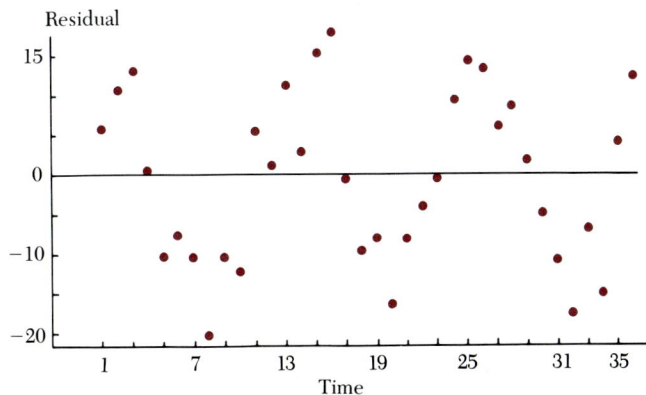

FIGURE 17.17 | Residual plot based on four-term sinusoidal model of data in Table 17.12

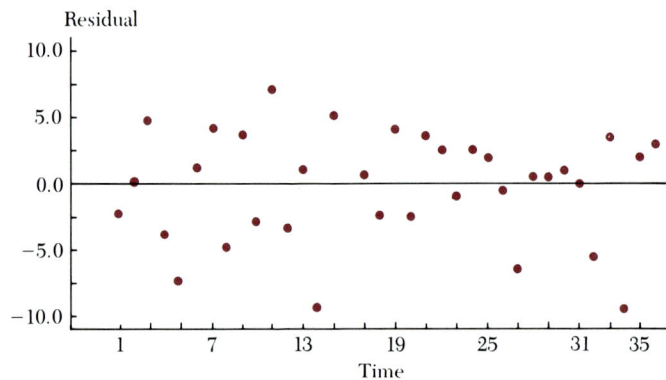

The use of the six-term sinusoidal forecasting model is illustrated using the data in Table 17.15, which gives sales of selected automobile parts from a NAPA vendor in Indianapolis, Indiana, for the years 1984, 1985, and 1986.

TABLE 17.15 | Sales of selected automobile parts (in $100)

Month	Sales	Month	Sales
1984		1985	
Jan.	174.41	July	174.85
Feb.	169.67	Aug.	157.33
March	159.66	Sept.	152.74
April	143.89	Oct.	153.88
May	144.49	Nov.	168.34
June	160.79	Dec.	192.93
July	163.16	1986	
Aug.	143.47	Jan.	201.91
Sept.	141.12	Feb.	192.48
Oct.	136.65	March	171.94
Nov.	164.84	April	171.41
Dec.	175.41	May	175.14
1985		June	183.45
Jan.	189.52	July	182.30
Feb.	171.79	Aug.	165.95
March	171.69	Sept.	164.12
April	171.34	Oct.	153.17
May	164.02	Nov.	182.58
June	168.80	Dec.	205.25

Source: Company records.

FIGURE 17.18 | Scatter diagram of the data in Table 17.15

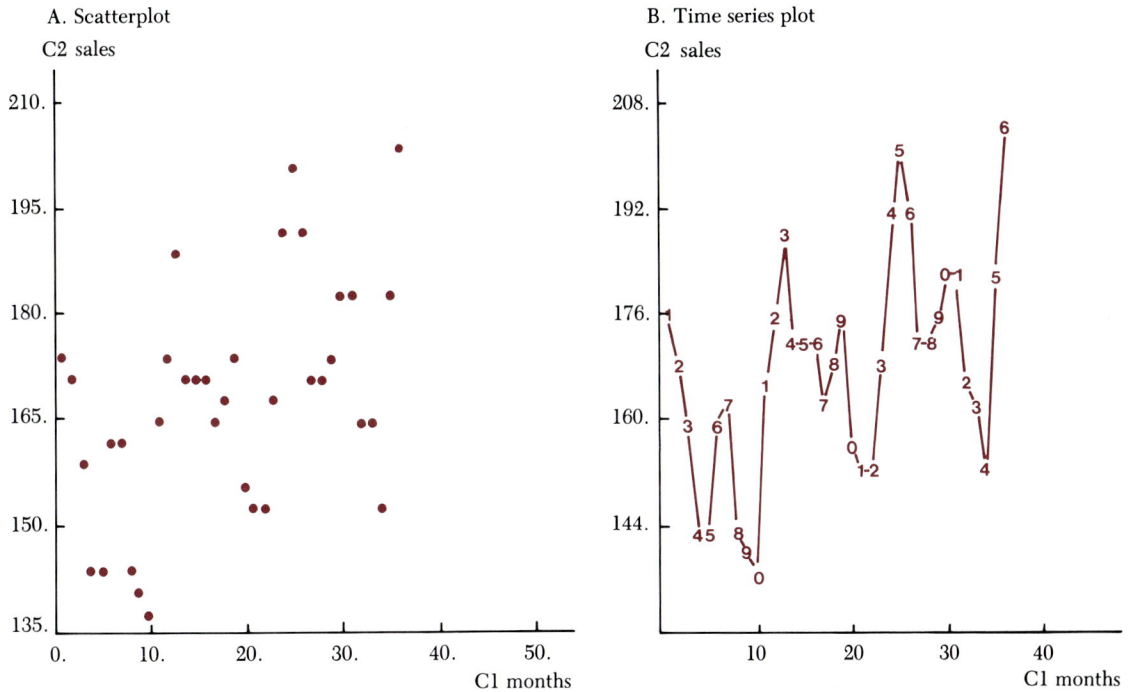

A. Scatterplot
C2 sales

B. Time series plot
C2 sales

A scatter plot of these data is given in Figure 17.18. Figure 17.18A illustrates the scatter plot from MINITAB for the 36 monthly observations on sales. Figure 17.18B gives the time series plot—the scatter plot, but with the scatter points connected by lines to show the cyclical nature of the observations. Table 17.16 gives a two-term (level and trend only), four-term, and a six-term time series forecasting fit, respectively, of the data in Table 17.15. Figures 17.19, 17.20, and 17.21 show the fits of the trend line, the four-term sinusoidal model, and the six-term sinusoidal model, respectively. The mean squared error in these models decreases as more terms are added to the forecasting equation. A visual inspection of the actual versus estimated values of this series given in Figure 17.21 verifies the general superiority of the six-term model for the data given in minimizing the sample mean squared error.

The sinusoidal models described offer several computational advantages over the seasonal indices computed in Chapter 16 by the method of ratio to moving average. First, recall that the most recent six months of data are not used in the computation of the seasonal indices for the latter months. This is not true for the sinusoidal models described. Furthermore, with the four-

TABLE 17.16 | Least squares fit of data in Table 17.15

SIMPLE LINEAR TREND

DEPENDENT VARIABLE: PLUMB

SOURCE	DF	SUM OF SQUARES	MEAN SQUARE	F-VALUE	PR > F
MODEL	1	2201.56510662	2201.56510662	10.21	0.0030
ERROR	34	7328.87505727	215.55514874	R-SQUARE	STD DEV
CORRECTED TOTAL	35	9530.44016389		0.231004	14.68179651

PARAMETER	ESTIMATE	T FOR HO: PARAMETER = 0	PR > |T|	STD ERROR OF ESTIMATE
INTERCEPT	154.53155556	30.92	0.0001	4.99770199
T	0.75278378	3.20	0.0030	0.23555031

FOUR-TERM SINUSOIDAL MODEL
DEPENDENT VARIABLE: PLUMB

SOURCE	DF	SUM OF SQUARES	MEAN SQUARE	F-VALUE	PR > F
MODEL	3	5369.63984064	1789.87994688	13.77	0.0001
ERROR	32	4160.80032325	130.02501010	R-SQUARE	STD DEV
CORRECTED TOTAL	35	9530.44016389		0.563420	11.40285096

PARAMETER	ESTIMATE	T FOR HO: PARAMETER = 0	PR > |T|	STD ERROR OF ESTIMATE
INTERCEPT	151.39842983	37.95	0.0001	3.98967841
T	0.92214472	4.86	0.0001	0.18961903
SIN 30T°	11.62978465	4.18	0.0002	2.77927857
COS 30T°	6.85150010	2.54	0.0160	2.69436233

SIX-TERM SINUSOIDAL MODEL
DEPENDENT VARIABLE: PLUMB

SOURCE	DF	SUM OF SQUARES	MEAN SQUARE	F-VALUE	PR > F
MODEL	5	8768.79360687	1753.75872137	69.08	0.0001
ERROR	30	761.64655702	25.38821857	R-SQUARE	STD DEV
CORRECTED TOTAL	35	9530.44016389		0.920083	5.03867230

PARAMETER	ESTIMATE	T FOR HO: PARAMETER = 0	PR > |T|	STD ERROR OF ESTIMATE
INTERCEPT	150.96044448	84.94	0.0001	1.77716294
T	0.94978136	11.22	0.0001	0.08466042
SIN 30T°	11.80079073	9.60	0.0001	1.22925263
COS 30T°	7.00656912	5.88	0.0001	1.19077941
SIN 60T°	8.04276101	6.73	0.0001	1.19548617
COS 60T°	11.20037703	9.39	0.0001	1.19230932

---

term or six-term sinusoidal model, many fewer coefficients have to be retained in order to develop a forecast by *months*. Where forecasts are made for large quantities of items and the coefficients are stored on a computer-readable medium, the savings in the amount of storage required can be significant.

FIGURE 17.19 | Simple linear trend estimate and scatter diagram of data in Table 17.15

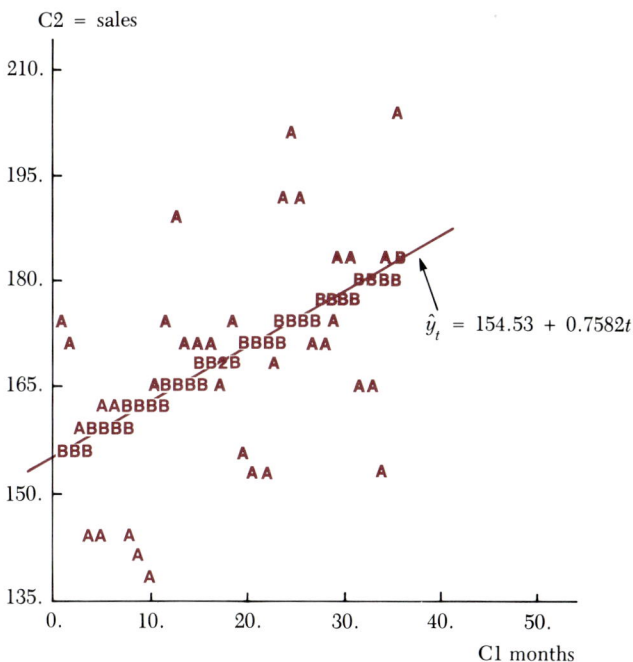

Legend: A represents a point for an observation y_t
B represents a point on the line $\hat{y}_t = 154.53 + 0.7528t$
2 represents a point with both a y_t and a $\hat{y}_t$ observation

Example 17.9   Using the six-term sinusoidal model given in Table 17.16, forecast sales of the selected automobile parts for January, February, and March 1987.

Solution   Relevant computations are given in Table 17.17. The given model forecasts a general decrease in sales over the period corresponding to anticipated seasonal movements in the given time series as indicated.

TABLE 17.17 | Estimated sales for data in Example 17.9

Month (1987)	t	$\sin 30t°$	$\cos 30t°$	$\sin 60t°$	$\cos 60t°$	$\hat{y}_t^*$
Jan.	37	0.5	0.866	0.866	0.50	210.92
Feb.	38	0.866	0.500	0.866	-0.50	203.36
Mar.	39	1.00	0.00	0.00	-1.0	189.35

* $\hat{y}_t = 150.96 + .95t + 11.80 \sin 30t° + 7.01 \cos 30t° + 8.04 \sin 60t° + 11.20 \cos 60t°$.

FIGURE 17.20 | Four-term sinusoidal time series model and scatter plot of data in Table 17.15

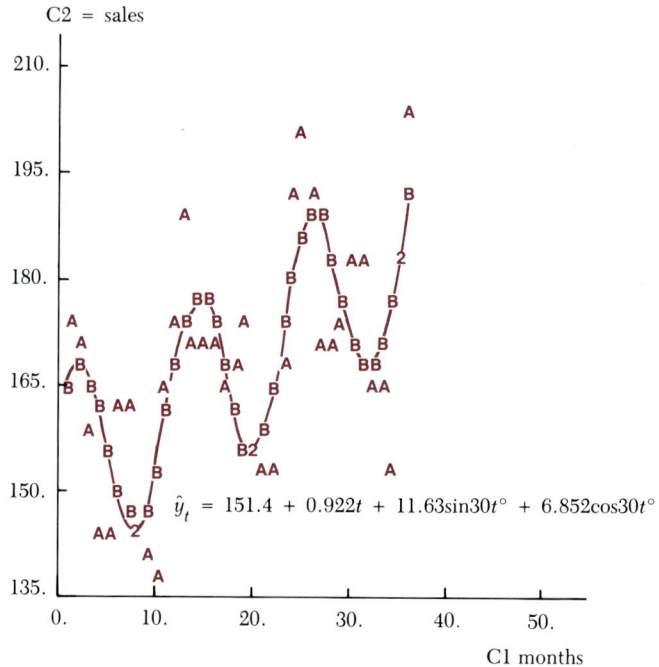

Legend: A represents a point for an observation y_t
B represents a point on the four term sinusoidal function $\hat{y}_t$
2 represents a point with both a y_t and a $\hat{y}_t$ observation

It is possible to test for the significance of the fundamental seasonal component in the four-term sinusoidal model (β_2) and to test for the significance of the fundamental component (β_2) and the harmonic component (β_3) in the six-term sinusoidal model, just as we tested for the significance of the presence of a seasonal factor in our dummy variable regression model using analysis of variance in Chapter 15. It turns out that the distribution of the fundamental and the harmonic components in the six-term sinusoidal model are chi-square distributed (Chapter 19) with two degrees of freedom. A description of the test for the significance of the seasonal components in these sinusoidal models is given in the Brown reference at the end of the chapter.

Sinusoidal models are a special case or subset of another statistical procedure called *spectral analysis*. Its name is derived from physics, where light is passed through a prism and broken down into components of its "spectrum." In time series analysis, we may wish to forecast a random variable

Spectral analysis

FIGURE 17.21 | Six-term sinusoidal time series model and scatter diagram of the data in Table 17.15

$$\hat{y}_t = 150.96 + 0.95t + 11.80\sin30t° + 7.01\cos30t° + 8.04\sin60t° + 11.20\cos60t°$$

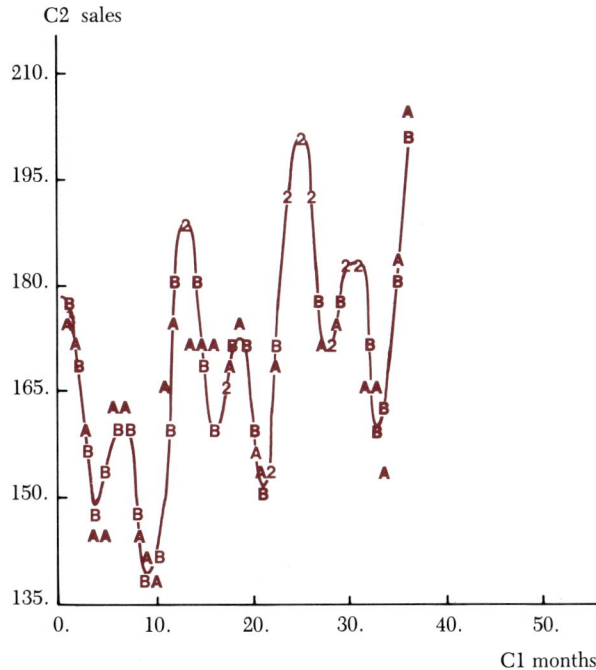

Legend: A represents a point for an observation y_t
B represents a point on the six term sinusoidal function $\hat{y}_t$
2 represents a point with both a y_t and a $\hat{y}_t$ observation.

whose mean is μ_t and whose variance is σ_t^2. If we can reduce σ_t^2, then a more accurate forecast can result. Using spectral analysis, σ_t^2 is broken down into components by "passing" the time series through "prisms, filters, or windows." The variance is broken down in such a way that we are able to associate the most important component with a particular period. An advantage of spectral analysis is that the variance is broken down into spectral components that are *statistically independent*. Thus one component can be manipulated without affecting the movement of other components.

In spectral analysis, the variance component is broken down into series of sine-cosine functions. Thus the concept itself is not too unlike the sinusoidal models discussed in this chapter. Both are particularly applicable in the analysis of seasonal or cyclical data. Several references at the end of the chapter include excellent discussions of this fast-emerging forecasting technique.

17.9

A brief comparison of forecasting models

Numerous forecasting techniques have been presented in the chapter. An evaluation or comparison of the various approaches for forecasting a future observation necessarily involves the specification of a criterion upon which the techniques are to be assessed or compared. Unfortunately, there is widespread difference of opinion as to the criterion to use in making a valid comparison.

One criterion that has been put forth for assessing the relative efficacy of alternate approaches is the minimization of the mean square error (MSE),

Mean square error

$$\text{Min MSE} = \frac{\sum\limits_{t=1}^{n} (y_t - \hat{y}_t)^2}{n}$$

Thus, in a least squares sense, we choose the technique that yields the minimum mean square error (MSE).

Another criterion that is used frequently to assess the merits of various time series forecasting techniques is the minimization of the mean absolute deviation (MAD),

Mean absolute deviation

$$\text{Min MAD} = \frac{\sum\limits_{t=1}^{n} |y_t - \hat{y}_t|}{n}$$

The relevance of this second criterion is that large deviations from actual results do not "penalize" the forecasting model as much as the squaring of these errors do in the computation of MSE.

Finally, a third criterion often used to assess various forecasting models is the minimization of the average absolute forecast error (AFE). That is, only part of the data is used in the computation of model coefficients. Then the model attempts to predict the latter observations that were not used in coefficient determination. Assuming the first m pieces of data are used in model specification and estimation, this latter criterion becomes

Average forecast error

$$\text{Min AFE} = \frac{\sum\limits_{t=m+1}^{n} |y_t - \hat{y}_t|}{n - m}$$

Many of the models described in the chapter simultaneously produce low values of MSE and MAD, but relatively large values of AFE. And those models that produce relatively low values of AFE simultaneously produce large values of MSE and/or MAD. The following general guidelines can be offered for selecting an appropriate forecasting technique.

1. Time series models in general are poor at detecting turning points and changes in the underlying pattern of data. Where underlying pattern changes can be anticipated, commonsense adjustments of forecasts are often very effective. The use of leading indicators can prove quite helpful in alerting the forecaster to a probable turning point, for which an examination of the underlying cause for the change can be assessed.

2. If no change in the underlying pattern of the data is expected and strong seasonal and cyclical movements in the data are present, use of the

sinusoidal models should be encouraged. For low-value, high-volume items, these models when fit by the method of least squares offer great computational advantages over the other models discussed.

3.   Where volatile changes in the series are anticipated, exponential smoothing may produce lower values of forecast errors than other techniques examined. Trend-adjusted exponential smoothing has been found to be among the best forecasting techniques in several simulation studies, for example. Exponential smoothing is even being used to predict general crime centers in large cities at different times of the day, and it is particularly helpful in routing surveillance vehicles in this regard.

4.   And finally, where a correlogram indicates a high correlation (either positive or negative) between lagged variables, an autoregressive forecasting model may be a correct choice of a forecasting technique, or, if resources (analyst time and computer time) permit, a mixed moving average autoregressive forecasting model (ARIMA) will likely be the best choice. Quite often the application of ARIMA models involves the greatest expense in terms of computer and analyst resources, and so is reserved for high-value inventory items or for making decisions where particularly large sums of money are involved in planning and the consequences of being right (or wrong!) are substantial.

■ 17.10

Use of computer software packages

Due to the extensive computations required, forecasting models require computers and computer software to identify models and to fit models to data. In this chapter we have illustrated computer output from the SAS statistical software package. In this section MINITAB is used, primarily to illustrate the application of the Box-Jenkins methodology.

We will consider the data set in Section 17.5, Table 17.8. These data represent sales of stereo equipment for 28 consecutive months beginning in September 1984 in Bloomington, Indiana. In Section 17.5 the first-order autoregressive forecasting model was fitted to these data, producing the forecasting equation,

$$\hat{y}_t = b_0 + b_1 y_{t-1} = 11.131 + 0.975 y_{t-1}$$

MINITAB can be used easily to produce this forecast equation. We first read the data—the month number and the sales amount—into columns C1 and C2:

```
READ INTO COLUMNS C1 AND C2
 1  235.89
 2  236.81
    .
    .
    .
28  345.30
```

The LAG command can be used to produce the column corresponding to $x_t = y_{t-1}$:

```
LAG BY 1 THE DATA IN COLUMN C2, PUT IN COLUMN C3
```

The results in columns C2 and C3 are:

```
COLUMN          C2              C3
COUNT           28              28
ROW
  1    235.890************
  2    236.810         235.890
  3    248.790         236.810
  4    254.160         248.790
  5    247.470         254.160
  6    258.120         247.470
  7    253.970         258.120
  8    267.840         253.970
  9    261.410         267.840
 10    271.200         261.410
 11    266.470         271.200
 12    286.690         266.470
 13    299.340         286.690
 14    292.340         299.340
 15    292.800         292.340
 16    301.850         292.800
 17    297.330         301.850
 18    305.740         297.330
 19    307.200         305.740
 20    307.340         307.200
 21    314.930         307.340
 22    319.590         314.930
 23    323.160         319.590
 24    322.940         323.160
 25    335.410         322.940
 26    339.460         335.410
 27    343.450         339.460
 28    345.300         343.450
```

Notice that the first row of column C3 contains asterisks. This indicates that there is no data in this position; it was lost due to the lagging of the column. The first-order autoregressive forecasting model may now be estimated by using the command:

```
REGRESS YT IN COLUMN C2 ON 1 PREDICTOR YT-1 IN COLUMN C3
```

The resulting output is:

```
        27 CASES USED
         1 CASES CONTAINED MISSING VALUES

    THE REGRESSION EQUATION IS
    Y =     11.1 + 0.975 X1

                                    ST. DEV.    T-RATIO =
            COLUMN     COEFFICIENT  OF COEF.    COEF/S.D.
              --          11.14       12.77        0.87
    X1    C3            0.97545      0.04400      22.17


    THE ST. DEV. OF Y ABOUT REGRESSION LINE IS
    S = 7.259
    WITH (  27- 2) =   25 DEGREES OF FREEDOM

    R-SQUARED = 95.2 PERCENT
    R-SQUARED = 95.0 PERCENT, ADJUSTED FOR D.F.

    ANALYSIS OF VARIANCE

     DUE TO         DF        SS      MS=SS/DF
    REGRESSION       1     25902.89   25902.89
    RESIDUAL        25      1317.43      52.70
    TOTAL           26     27220.31
```

Thus the forecasting equation is $\hat{y}_t = 11.14 + 0.975 y_{t-1}$.

To find the first-order sample autocorrelation coefficient, we must use the command,

```
CORRELATE YT IN COLUMN C2 AND YT-1 IN COLUMN C3
```

The output is:

```
CORRELATION OF     C2 AND C3    = 0.975
```

At the end of Section 17.5, a table gives the values (y_t, $x_t = y_{t-2}$) needed to fit the second-order autoregressive model. To find the second-order autocorrelation coefficient, we do the following:

```
LAG BY 2 THE DATA IN COLUMN C2, PUT IN COLUMN C4
```

The results in columns C2 and C4 are:

```
COLUMN          C2              C4
COUNT           28              28
ROW
   1      235.890************
   2      236.810************
   3      248.790         235.890
   4      254.160         236.810
   5      247.470         248.790
   6      258.120         254.160
   7      253.970         247.470
   8      267.840         258.120
   9      261.410         253.970
  10      271.200         267.840
  11      266.470         261.410
  12      286.690         271.200
  13      299.340         266.470
  14      292.340         286.690
  15      292.800         299.340
  16      301.850         292.340
  17      297.330         292.800
  18      305.740         301.850
  19      307.200         297.330
  20      307.340         305.740
  21      314.930         307.200
  22      319.590         307.340
```

```
23    323.160    314.930
24    322.940    319.590
25    335.410    323.160
26    339.460    322.940
27    343.450    335.410
28    345.300    339.460
```

Notice that there are two sets of asterisks in rows one and two of column C4, indicating that we have lost two observations by lagging the column by two. Further, these two columns match the y_t column (C2) and the $x_t = y_{t-2}$ column (C4) in the table at the end of Section 17.5. Then

```
CORRELATE YT IN COLUMN C2 AND YT-2 IN COLUMN C4
      CORRELATION OF      C2 AND C4      = 0.972
```

Thus the second-order autocorrelation coefficient is 0.972, indicating a high degree of correlation between the successive pairs (y_t, y_{t-2}).

These autocorrelation coefficients are estimates of the population autocorrelation coefficients and are computed by using the Pearson product-moment correlation measure. There are other estimators that can be used (e.g., see the Box and Jenkins reference at the end of this chapter for alternative estimators). These estimators will produce different estimates of the kth-order autocorrelation coefficient. MINITAB uses the estimator suggested by Box and Jenkins. The ACF (AutoCorrelation Function) command produces the estimated autocorrelations (and the correlogram) based on the Box-Jenkins estimator:

```
ACF FOR THE DATA IN COLUMN C2

          -1.0 -0.8 -0.6 -0.4 -0.2  0.0  0.2  0.4  0.6  0.8  1.0
          +----+----+----+----+----+----+----+----+----+----+
1    0.872                           XXXXXXXXXXXXXXXXXXXXXXXX
2    0.756                           XXXXXXXXXXXXXXXXXXXXX
3    0.652                           XXXXXXXXXXXXXXXXX
4    0.558                           XXXXXXXXXXXXXX
5    0.458                           XXXXXXXXXXXX
6    0.368                           XXXXXXXXXX
7    0.261                           XXXXXXXX
8    0.185                           XXXXXX
```

```
  9    0.104                          XXXX
 10    0.024                          XX
 11   -0.079                         XXX
 12   -0.126                        XXXX
 13   -0.168                       XXXXX
 14   -0.220                      XXXXXX
 15   -0.272                     XXXXXXXX
```

Notice that the correlogram is shown vertically instead of horizontally. The autocorrelations are shown in the left column, and the Xs are used to display them graphically. The MINITAB estimator produces an estimate of the first order autocorrelation coefficient equal to 0.872 while the Pearson product-moment measure estimate is 0.975. Also, the second-order autocorrelation coefficient estimates differ (MINITAB: 0.756, Pearson: 0.972). Most statisticians prefer the Box-Jenkins estimator used in MINITAB to the Pearson product-moment correlation estimator.

Suppose we wish to fit a Box-Jenkins model to the sales data in Table 17.8. The first step is to produce a time series plot of the 28 monthly sales values:

```
        TSPLOT THE DATA IN COLUMN C2
```

This produces the following output:

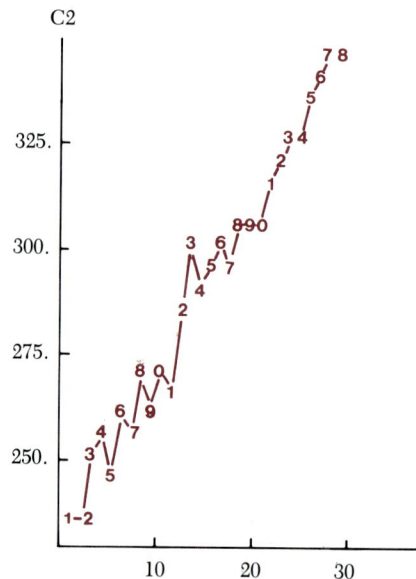

Notice that this plot indicates that the time series is not stationary. It is not fluctuating about a constant (horizontal line). Therefore, it is necessary to use differencing to produce a stationary series. We will try first differences: $z_t = y_t - y_{t-1}$.

```
DIFFERENCE BY 1 THE DATA IN COLUMN C2, PUT IN COLUMN C3
TSPLOT THE DATA IN COLUMN C3
```

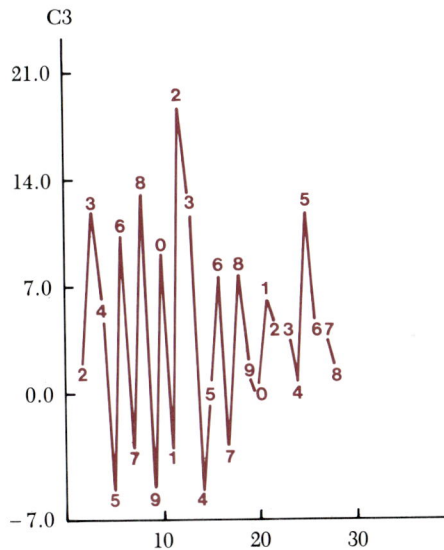

As the time series plot indicates, the resulting series appears to be randomly moving about a horizontal line, indicating that the $z_t = y_t - y_{t-1}$ series is stationary.

The next step is to identify the proper Box-Jenkins model. The sample autocorrelation function (ACF in MINITAB) and the sample partial autocorrelation function (PACF in MINITAB) are used to assist in the model identification:

```
ACF THE FIRST DIFFERENCED DATA  IN COLUMN C3
```

```
           -1.0 -0.8 -0.6 -0.4 -0.2  0.0  0.2  0.4  0.6  0.8  1.0
           +----+----+----+----+----+----+----+----+----+----+
  1  -0.488                        XXXXXXXXXXXXX
  2   0.072                                    XXX
  3  -0.138                                XXXX
  4   0.123                                    XXXX
  5  -0.072                                 XXX
  6   0.076                                    XXX
```

```
PACF THE FIRST DIFFERENCED DATA IN COLUMN C3
             -1.0 -0.8 -0.6 -0.4 -0.2  0.0  0.2  0.4  0.6  0.8  1.0
             +----+----+----+----+----+----+----+----+----+----+
  1  -0.488                        XXXXXXXXXXXXX
  2  -0.218                              XXXXXX
  3  -0.277                            XXXXXXXXX
  4  -0.111                               XXXX
  5  -0.112                               XXXX
  6  -0.020                                 X
```

These sample functions are matched with the theoretical autocorrelation and partial autocorrelation functions to suggest models to fit to the data. One model that this ACF and PACF pair suggests is the ARIMA (p, d, q) = ARIMA $(1, 1, 1)$ model with $p = 1$ (first-order autoregressive), $d = 1$ (first differencing to produce stationarity), and $q = 1$ (first-order moving average). The resulting model is:

$$z_t = \beta_0 + \beta_1 z_{t-1} - \theta_1 e_{t-1} + e_t$$

where

$$z_t = y_t - y_{t-1}$$

The MINITAB command to fit this model is:

```
ARIMA 1 1 1 FOR THE DATA IN COLUMN C2;
CONSTANT;
FORECAST 3.
```

The CONSTANT subcommand specifies that the constant β_0 be fitted in the model. The FORECAST subcommand is used to produce forecasts—we are asking for forecasts three periods beyond the last month in the time series data set. The resulting output is:

```
ESTIMATES AT EACH ITERATION
ITERATION          SSE       PARAMETERS
    0           1334.11     0.100      0.100     3.737
    1           1087.01    -0.050      0.197     4.305
    2           1035.97     0.034      0.347     3.963
    3            977.94     0.101      0.497     3.685
    4            913.82     0.149      0.647     3.477
    5            844.62     0.190      0.797     3.285
    6            765.47     0.237      0.947     3.066
    7            735.56     0.087      0.966     3.683
    8            734.29     0.047      0.959     3.860
    9            733.92     0.053      0.967     3.834
   10            733.69     0.049      0.964     3.852
   11            733.69     0.049      0.964     3.853
   12            733.69     0.049      0.964     3.853
   13            733.68     0.049      0.964     3.854
RELATIVE CHANGE IN EACH ESTIMATE LESS THAN   0.0010

FINAL ESTIMATES OF PARAMETERS
NUMBER       TYPE      ESTIMATE     ST. DEV.    T-RATIO
    1     AR   1        0.0491       0.2282        0.22
    2     MA   1        0.9637       0.1721        5.60
    3     CONSTANT      3.8543       0.1166       33.05

DIFFERENCING.  1 REGULAR
RESIDUALS.      SS =        732.915   (BACKFORECASTS EXCLUDED)
                DF =    24  MS =         30.538
NO. OF OBS.    ORIGINAL SERIES    28    AFTER DIFFERENCING    27

FORECASTS FROM PERIOD   28

                            95 PERCENT LIMITS
PERIOD      FORECAST      LOWER        UPPER       ACTUAL
   29        349.637      338.804      360.471
   30        353.704      342.831      364.577
   31        357.758      346.876      368.639
```

The estimate of β_0 is $b_0 = 3.8543$; the estimate of β_1 is $b_1 = 0.0491$; and the estimate of θ_1 is 0.9637. The mean squared error (MSE) for this model is 30.538. Notice that this is less than the mean squared error produced above for the first-order autoregressive model (MSE = 52.70). The FORECAST subcommand produces 95 percent confidence interval estimates in addition to the forecasts. Thus in month 29 (next month), the forecast is for sales of 349.637 (thousand) dollars with a confidence interval of 338.804 to 360.471.

The Box-Jenkins methodology has proven to be very effective in a variety of forecasting problems. The interested reader is encouraged to study the references at the end of this chapter and, if possible, take a course in forecasting with an emphasis on the Box-Jenkins methodology.

■ 17.11
Summary

The chapter presented several techniques for estimating the value of a dependent variable, usually at some future time, say the next month, the next year, or the next quarter. The techniques described included time series approaches often using multiple linear regression techniques, and multiple regression approaches using, in addition to time-related variables, other explanatory variables that are believed to contribute to movements in the series under investigation.

Point estimates only were provided for estimating the value of some series, since many of the assumptions underlying the use of the multiple linear regression model for inference purposes are violated when examining time series data. Indeed, when observations are highly correlated over time, autoregressive or mixed autoregressive moving average models may provide the analyst with powerful approaches for estimating the future value of some series.

When the data do not satisfy many assumptions of the multiple linear regression and correlation model, it may be possible to transform the data to conform to the assumptions. For example, when the error terms for given levels of the independent variable do not have the same variance component in the predicted value (conditional distribution), the variances are said to be *heteroscedastic*—unequal between different values of the independent variable(s). Taking logarithms of the data will often correct this situation. When autocorrelation is present in the data and it is desired to compensate for it, we can often do this by specifying a different model. For example, if the true fit of the series should be a second-degree parabola, then the residuals or the error terms will definitely be correlated over time if we fit a straight-line trend or other straight-line function to the data. Simply fitting the correct model will correct for autocorrelation of the error terms in this situation. Transformations of the data can often be used to remove the correlation between successive errors (autocorrelation) if the correct model has been specified. Multicollinearity is present in a model when the independent variables themselves are correlated. We described this situation briefly in Chapter 15. When independent variables are highly correlated, it may be possible to eliminate one of the explanatory variables, since the two variables describe the movement in the dependent variable in similar ways. The method of differencing was described in the chapter as one method of data transformation to create a stationary series from a nonstationary one (one with a mean that is changing over time). More transformations of data are possible other than taking differences of the data, taking logarithms of the data, etc. Several references at the end of the chapter describe the other approaches for transforming data values. The result of these transformations is generally a model that meets the assumptions of the multiple linear regression model, so that confidence limits can be set on estimated values of the dependent variable.

■ References

Introductory

Chambers, John C.; Satinder K. Mullick; and Donald D. Smith. "How to Choose the Right Forecasting Techniques." *Harvard Business Review,* July–August 1971, pp. 45–74.

Hamburg, M. *Basic Statistics.* 3rd ed. New York: Harcourt Brace Jovanovich, 1985, Chapter 13.

Neter, J.; W. Wasserman; and G. A. Whitmore. *Applied Statistics* 2nd ed. Boston: Allyn & Bacon, 1982, Chapter 24.

Wonnacott, R., and T. Wonnacott. *Introductory Statistics for Business and Economics.* 3rd ed. New York: John Wiley & Sons, Chapter 24.

Advanced

Bowerman, B. L., and R. T. O'Connell. *Time Series and Forecasting.* North Scituate, Mass.: Duxbury Press, 1979.

Box, G. E. P., and G. M. Jenkins. *Time Series Analysis: Forecasting and Control.* Rev. ed. San Francisco: Holden-Day, 1976.

Brown, R. G. *Smoothing, Forecasting, and Prediction of Discrete Time Series.* Englewood Cliffs, N.J.: Prentice-Hall, Inc., 1963.

Chow, C. M. "Adaptive Control of the Exponential Smoothing Constant." *Journal of Industrial Engineering* 16, no. 4 (1965), pp. 314–17.

Ferratt, T. W., and V. A. Mabert. "A Description and Application of the Box-Jenkins Methodology." *Decision Sciences* 3 (1972).

Groff, Gene K. "Empirical Comparison of Models for Short-Range Forecasting." *Management Science* 20, no. 1 (1973), pp. 22–31.

Kirbly, R. M. "A Comparison of Short- and Medium-Range Forecasting Methods." *Management Science* 13 no. 4 (1966), pp. 202–10.

Mabert, V. A., and R. C. Radcliffe. "A Forecasting Methodology as Applied to Financial Time Series." *The Accounting Review* XLIM (1974).

Makridakis, S.; S. Wheelwright; and V. McGee. *Forecasting: Methods and Applications.* 2nd ed. New York: John Wiley & Sons, 1983.

Montgomery, D. C., and L. A. Johnson. *Forecasting and Time Series Analysis.* New York: McGraw-Hill, 1976.

Nelson, C. R. *Applied Time Series Analysis for Managerial Forecasting.* San Francisco: Holden-Day, 1973.

Pack, D. J.; M. L. Goodman; and R. B. Miller. "Computer Programs for the Analysis of Univariate Time Series Using the Methods of Box-Jenkins." Technical Report Number 296, Department of Statistics, University of Wisconsin, Madison, April 1972.

Pindyck, R. S., and D. L. Rubinfeld. *Econometric Models and Economic Forecasts.* New York: McGraw-Hill, 1978.

Tiao, G. C., and H. E. Thompson. "Analysis of Telephone Data: A Case Study of Forecasting Seasonal Time Series." *The Bell Journal of Economics and Management Science* 2 (1971).

■ Problems

Section 17.2 Problems

17.1 In the computation of a diffusion index, if a given series composing the index is differenced, and the difference obtained is 0, what weighting does the observation receive?

17.2 How many series are in the NBER's composite index of leading economic indicators? When were the components of this series last revised?

Section 17.3 Problems

17.3 Using the data given in the table,

Period (months)	Actual sales
1	100.4
2	110.9
3	118.6
4	132.9
5	139.9
6	151.1
7	163.2
8	170.5
9	179.4
10	191.7

a. What would be the forecast for period 11 using a moving average forecast of six months' duration?

b. What would be the forecast for period 11 using a moving average forecast of three months' duration?

17.4 Using the data in Problem 17.3, what would be the forecast for period 11 using an exponentially smoothed forecast with an α-value of 0.3 and a forecast for period 10 of 185.55?

17.5 Using the data in Problem 17.3, what would be the forecast for period 11 using an exponentially smoothed forecast with an α-value of 0.1 and a forecast for period 9 of 183.4?

17.6 An exponential smoothing model with an α-value of 0.4 is related to a moving average forecasting model with how many terms, n?

17.7 Derive the relationship $\alpha = 2/(n + 1)$ in the exponential smoothing model.

17.8 Consider the sales data provided in the table below.

a. Using the method of least squares, obtain an estimate of the trend.

b. Form a residual plot from the least squares trend equation. Comment on the least squares fit for these data.

c. Using the single exponential smoothing model, estimate the trend for these data for $\alpha = 0.5$ and 0.9. Use 39.8 for forecasted sales in Nov., 1985.

d. Form a residual plot for the single exponential smoothing equations. Comment on the fit of this model for these data.

Period		t	Sales ($00)
1985	Nov.	1	45.2
	Dec.	2	47.1
1986	Jan.	3	45.3
	Feb.	4	51.3
	Mar.	5	51.9
	Apr.	6	55.5
	May	7	56.1
	June	8	60.1
	July	9	58.2
	Aug.	10	64.3
	Sept.	11	63.6
	Oct.	12	66.3
	Nov.	13	69.9
	Dec.	14	72.1

Section 17.4 Problems

17.9 For the data in Problem 17.8, fit the trend-adjusted exponential smoothing model using an α-value of 0.3. Form the residual plot. Comment on the fit of this model based on the residual plot.

17.10 The data below represent monthly sales of a product.

a. Fit the trend-adjusted exponential smoothing model to these data using an α-value of 0.1. Use $S_0 = 78$ and $S_0^{(2)} = 6$ for starting values.

b. Fit the trend-adjusted exponential smoothing model to these data using an α-value of 0.2. Use $S_0 = 118$ and $S_0^{(2)} = 86$ for starting values.

c. Form a residual plot for each of these fitted models. Which model ($\alpha = 0.1$ or $\alpha = 0.2$) would you use for forecasting purposes? Why?

Period		t	Sales ($000)
1986	Jan.	1	147
	Feb.	2	160
	Mar.	3	152
	Apr.	4	195
	May	5	190
	Jun.	6	220
	Jul.	7	256
	Aug.	8	212
	Sep.	9	210
	Oct.	10	205
	Nov.	11	210
	Dec.	12	240

17.11 For the data in Problem 17.10, fit the trend-adjusted exponential smoothing model to these data using an α-value of 0.1 and starting values of $b_0(0) = 0$ and $b_1(0) = 1$. Form the residual plot for the fitted model. Comparing your results with those in Problem 17.10, parts a and c (with $\alpha = 0.1$), comment on the effect produced by using different starting values for the S_t and $S_t^{(2)}$ recursive equations.

Section 17.5 Problems

17.12 For the data in Problem 17.8, fit the first-order autoregressive forecasting model. Does the autoregressive model developed appear to be a good predictor of the series? Why or why not?

17.13 Assess the second-order autocorrelation in the data given in Problem 17.8, and fit a sec-

ond-order autoregressive model to the data. How do the results compare with those determined with the first-order autoregressive model?

17.14 Given the following time series data of sales,

Period	Sales
1	2105
2	2000
3	1905
4	1995
5	2090
6	2005
7	1895

obtain the forecasts for periods 2 through 7, using the first-order autoregressive forecasting method.

Section 17.6 Problems

17.15 (Requires the use of MINITAB for Box-Jenkins.) The following data represent the number of telephone inquiries received by a new home and business security systems company:

Week	y_t	Week	y_t	Week	y_t	Week	y_t
1	135	11	136	21	356	31	450
2	140	12	245	22	366	32	460
3	145	13	256	23	378	33	472
4	153	14	267	24	392	34	470
5	165	15	285	25	408	35	480
6	178	16	300	26	418	36	492
7	186	17	312	27	424	37	504
8	196	18	322	28	434	38	517
9	210	19	334	29	440	39	524
10	225	20	345	30	445	40	534

a. Form the time series plot for these data.

b. Form the autocorrelation function (ACF) and the partial autocorrelation function (PACF) for these data.

c. Based on the time series plot in part a and the ACF in part b, is the time series stationary?

d. Form the first difference series, $z_t = y_t - y_{t-1}$. Form the time series plot and the ACF and the PACF for this series. Is this series stationary? Explain.

e. Fit the following three models to the z_t series:

ARIMA(1, 1, 0)   ARIMA(0, 1, 1)
ARIMA(1, 1, 1)

Which of these three models fits best to these data? Explain.

17.16 Describe the steps involved in the Box-Jenkins forecasting methodology. When does one usually conclude that the model developed is the correct one for the series under investigation?

Section 17.7 Problems

17.17 Explain how dummy variables are used in a regression model to represent seasonal effect.

17.18 Suppose a time series data set contains quarterly data. How many dummy variables are required to represent the four seasons in the year? What would happen if four dummy variables, one for each quarter of the year, were used in the regression model?

17.19 The following data represent quarterly inquiries for the rental for a period of one week of a condominium in Vail, Colorado:

Year	Quarter	y_t	Year	Quarter	y_t
1982	1	40	1984	1	46
	2	30		2	36
	3	15		3	19
	4	10		4	12
1983	1	43	1985	1	50
	2	34		2	39
	3	17		3	21
	4	11		4	14

a. Using a dummy regression model with a linear trend term, forecast the number of inquiries for the four quarters in 1986.

b. Form the residual plot for the 16 quarters of data from 1982 through 1985 for the fitted regression model. Does the model appear to fit well to these data? Explain.

Section 17.8 Problems

17.20 Assume in Example 17.4 in the chapter that the given time series peaks in June and attains its average value (level) in September. The appropriate sinusoidal model for predicting a given value of the time series is given by what equation?

17.21 Given on the top of page 953 is a computer analysis of a least squares fit of a six-term sinusoidal forecasting model for the data in Table 17.12. Using the results provided, obtain least squares estimates, $\hat{y}_t$, of the actual observations. Compare the six-term model with the four-term model described in Section 17.8 on the basis of mean square error and on the basis of mean absolute deviation. What do the results *suggest* regarding inference on the significance of the harmonic terms in the model?

17.22 Given the four-term sinusoidal forecasting model at the bottom of page 953, determine $\hat{y}_1$ and $\hat{y}_{14}$. If the assumptions underlying the use of the multiple linear regression model are met using these data, which of the terms included are significant?

Section 17.9 Problems

17.23 Explain the difference between the MAD and the MSE measures of the residuals produced by a forecasting model.

17.24 Two forecasting models, model A and model B, produce the following residuals for the same data set:

Forecast model A		Forecast model B	
t	e_t	t	e_t
1	−2	1	4
2	3	2	4
3	2	3	3
4	−1	4	2
5	−1	5	1
6	0	6	0
7	−4	7	−3
8	3	8	−3
9	0	9	−4
10	0	10	−4

```
     VAR,        COEFF,      ST, ERROR       T-VALUE
      t         0,9502       0,08267         11,49
  sin 30t°    11,73506      1,20037           9,78
  cos 30t°     6,82321      1,16298           5,87
  sin 60t°    -0,79673      1,16885          -0,68
  cos 60t°     0,3066       1,16298           0,26

  INTERCEPT    120,87896

  REGRESSION
    DEGREES OF FREEDOM      5
    SUM OF SQUARES          5630,61509
    MEAN SQUARE             1126,12302

  ERROR
    DEGREES OF FREEDOM      30
    SUM OF SQUARES          726,67698
    MEAN SQUARE             24,22257

  S,E, OF ESTIMATE          4,92164
  F-VALUE                   46,49
  MULTIPLE R-SQUARED        88,57
```

```
                      Sinusoidal Time Series Model
     VAR,        COEFF,      ST, ERROR       T-VALUE
      t         4,9586       0,07994         62,03
  sin 30t°    11,76639      1,17172         10,04
  cos 30t°     6,81481      1,13592          6

  INTERCEPT    180,72365

  REGRESSION
    DEGREES OF FREEDOM      3
    SUM OF SQUARES          92228,9088
    MEAN SQUARE             30742,9696

  ERROR
    DEGREES OF FREEDOM      32
    SUM OF SQUARES          739,54011
    MEAN SQUARE             23,11063

  S,E, OF ESTIMATE          4,80735
  F-VALUE                   1330,25
  MULTIPLE R-SQUARED        99,2
```

a. Which forecasting model, A or B, is preferable based on MAD?

b. Which forecasting model, A or B, is preferable based on MSE?

c. Do the residuals for model B indicate a problem? If so, what is it?

17.25 Two forecasting models, A and B, produce the following sets of residuals for the same data set:

Forecast model A		Forecast model B	
t	e_t	t	e_t
1	−1	1	5
2	0	2	0
3	2	3	0
4	−2	4	3
5	3	5	−5
6	0	6	−3
7	2	7	4
8	4	8	−4
9	−1	9	0
10	−7	10	0

a. Which forecasting model, A or B, is preferable based on MAD?

b. Which forecasting model, A or B, is preferable based on MSE?

Additional Problems

17.26 Given the following time series data of sales,

Period	Sales
1	2,105
2	2,000
3	1,905
4	1,995
5	2,090
6	2,005
7	1,895

obtain forecasts of sales for periods 2 through 7 using the single exponential smoothing model with an α-value of 0.3 and a forecast for the first period of 2,125.

17.27 Develop a first-order autoregressive forecasting model for the data given in Problem 17.26.

17.28 What type of model is suggested by the data given in Problem 17.26? What terms would you include in a model of this time series, and which of the methods described in the text would you use to estimate future values of this series?

17.29 Using the data in Problem 17.8, and a forecasted value of sales of 39.8 for November 1985, obtain forecasts for each of the remaining months using single exponential smoothing and α-values of 0.1, 0.2, 0.3, . . . , 0.9.

17.30 Obtain an estimate of trend for the data in Problem 17.8 using the method of least squares. Given your trend estimate and an analysis of the residuals using the single exponential smoothing model, comment on the appropriateness of the trend-adjusted exponential smoothing model for the data given.

17.31 Distinguish between prediction and forecasting. When should each be used in attempting to estimate the future value of a series?

17.32 Describe what is meant by a process being *stationary* and list one method available for reducing a nonstationary series to a stationary one.

18 Index numbers

18.1

Introduction

Rarely does a day pass without a newspaper having a headline such as the following:

August wholesale prices down 0.3%

Consumer prices up only .2% for fifth month!

These headlines are examples of the application of index numbers.

An index number is a statistical measure usually involving the simple ratio of two numbers. Index numbers are designed to show changes in one variable or in a group of related variables over time, or with respect to geographic location, or in terms of some other characteristic. For example, Table 18.1 provides information on the cost of living in selected U.S. cities in 1985. These indices are constructed by sampling living costs in the given cities (the cost of housing, food, transportation, medicine, etc.), then dividing the living costs sampled in each of the cities by the living costs in the base location (Gainsville, Florida), and multiplying the result of 100 (in order to quote them in percentages). The interpretation of the indices given in Table 18.1 might be, for example, that it costs 16.6 percent more to live in Los Angeles than it does in Gainsville in 1985 because of higher expenditures for those items sampled in determining living costs in each city.

It is difficult to imagine anyone who has not at some time either used or made reference to some type of an index number. The Consumer Price Index indicated on the front page of this chapter which we all hear mentioned so

955

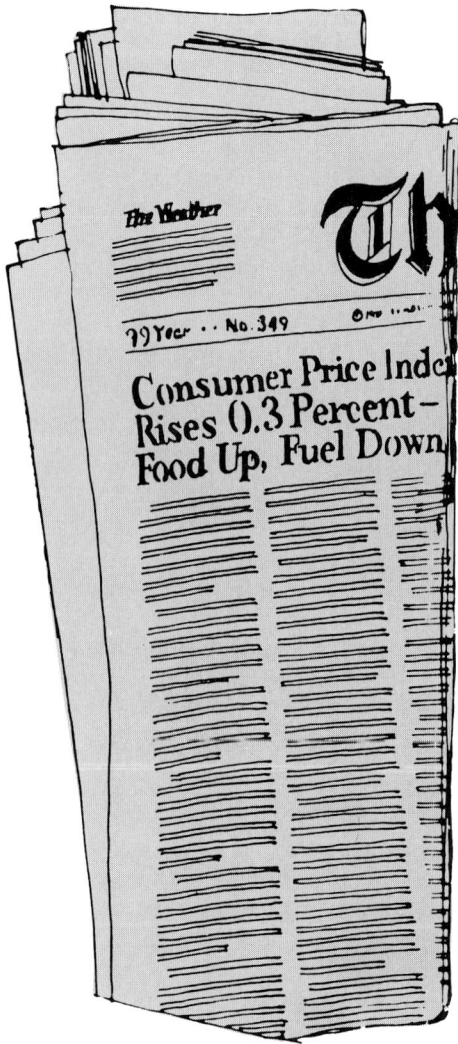

Social Security raise 3.1%

Low inflation rate keeps hike smallest in decade

United Press International

WASHINGTON—A modest 0.2 percent increase in consumer prices during September kept the annual inflation rate so far this year at its lowest level in nearly two decades and will give Social Security recipients their smallest increase in a decade, the government reported today.

The slow inflation means Social Security recipients in 1986 will see their monthly checks rise by only 3 percent—the lowest inc 1975, t' Soci

The Weather

99 Year · · No. 349

Consumer Price Index Rises 0.3 Percent— Food Up, Fuel Down

Dow Jones Surpasses 1800!

TABLE 18.1 | Index of living costs in selected U.S. cities, 1985

City	Cost of Living Index
Los Angeles, Calif.	116.6
Houston, Texas	109.0
Baltimore, Md.	105.0
Phoenix, Ariz.	104.7
Atlanta, Ga.	103.8
Gainesville, Fla.	100.0
St. Louis, Mo.	99.8
Louisville, Ky.	97.5
New Orleans, La.	97.4
Pueblo, Co.	88.4

Source: American Chamber of Commerce Research Assoc. "Intercity Cost of Living Index," 1st Quarter, 1985.

often and which attempts to assess the degree to which purchasing power has been eroded by price increases, looms central as the most talked about (and possibly least understood!) index number in use today. The incomes of more than 50 million people in the United States (including more than 8.5 million laborers, 32 million social security beneficiaries, 20 million food stamp recipients, and untold numbers of retired military personnel, postal workers, and alimony recipients) are affected by this little-understood index. Further, income tax schedules are now based upon rises in this index in order to end what politicians have come to refer to as "tax bracket creep." Other index numbers quoted frequently in the news are the Wholesale Price Index, the Dow-Jones Index of 30 Industrial Stocks, Standard & Poor's Index of 500 Stocks, and so on. In this chapter we shall build toward an understanding and interpretation of these as well as other index numbers. As we build toward this understanding, we should keep in mind the following questions and attempt to assess for ourselves whether the index number reported truly measures the changes in the series it is purporting to represent.

18.1.1 | What items should be included in the index?

The items included in the index should be representative of what the index is attempting to measure. For example, if we are attempting to assess the relative change in the price of rib roast over some period, the construction of the index number to accomplish this goal is fairly straightforward. But if we are attempting to assess the change in meat prices for a consumer over this same interval, then it is important for us to establish which meat products are to be included in the index and how they should be weighted in determining the value of the index number. If we are attempting to assess the change in prices of a food "basket" of goods for a consumer, then the determination of which food products and which weight each receives in the determination of the index becomes even more complex. The consumer price index discussed in Section 18.5 is based on approximately 400 items out of about 2,000 potential purchases, for example.

18.1.2 | What is the base period of the index?

The base period or the base location of an index number should reflect a fairly normal period or location in which the goods and their prices were not subject to unusual fluctuations that would have had a significant impact on the value of the index. For example, in the early 1970s the quantity of tuna consumed worldwide decreased substantially because of a mercury contamination scare. In sampling households for their consumption of tuna, one would likely find that the level of consumption would have been lower than normal because of the reported presence of mercury in these fish.

The base period of an index number should also be in the not-too-distant past, thus enabling the user of the index number to recall conditions existing in the reference period. Some policymakers regard a base period more than 15 years old as being too dated to allow users to have familiarity with the reference or the base period.

18.1.3 | What are the relative weights of the items to be included in the index?

In constructing a price index with quantity weights, for example, one should carefully assess whether or not the weights reflect approximate amounts of purchases to be made. Table 18.6 gives the relative weights of major groups of items in the consumer price index for four benchmark years. Note how the relative weights of the items in the index have changed as consumption patterns have changed. For this index to be a meaningful measure and reflect changes in the "purchasing power" of the dollar, the items purchased should closely parallel those that reflect current buying habits.

18.1.4 | How will the data be compiled, and how often should the index be revised?

The sampling plan used in collecting data is of central importance in obtaining an unbiased statistical measure of the change in the index. Many of the statistical sampling techniques described in Chapter 8 are used in obtaining estimates of changes in prices and quantities. Important questions in data collection reflect where the information is to be collected, when the information is to be collected, and how adjustments for sales are to be made. Whenever an appreciable change in buying patterns is detected, adjustment should be made in the construction of the index number. As a case in point, how should the purchase of a video recorder be reflected in the consumer price index, an item that was not even around for the mass market when the latest weights for the consumer price index were constructed in 1972-73? A well-designed sampling plan will detect changes in buying habits and alert the compositor of the index number to the change so that the index can be revised to reflect substantive changes in buying habits.

18.1.5 | How are quality changes to be handled?

This is undoubtedly one of the most difficult issues in the construction of an index number. How, for example, do we treat the increase in the price of an automobile when the price increases due to improvements such as catalytic converters and airbags? Many would argue that quality has decreased because of a catalytic converter, although one should keep in mind that *air* quality has likely increased by the presence of this item. Is the television set one purchases today of the same quality as the one purchased 10 years ago (which may have been black-and-white only or which may have had a round (as opposed to a rectangular) picture tube?

In the geographic example given in Table 18.1, how should one factor in differences in the "quality of living" among the various cities? The consumer price index, in its latest revision, attempts to account for some of the quality changes that have occurred by pricing quality changes separately, but admittedly, the CPI only approximates measuring some of the more important quality changes that have occurred over the years. As users of index numbers, it is important for us to keep in mind the changes in quality that have accompanied the macro changes in the index series and to interpret these changes accordingly.

We shall adopt the following notation in our discussion of index numbers. The index number itself will be denoted by a capital letter, and the components that determine the *value* of the index number will be denoted by lowercase letters. We will denote the attribute price by p, the attribute quantity by q, and the attribute value by $v = pq$. We will also refer to a base *period* of an index number because the greatest use of index numbers stems from their use in interpreting time series data. Note, though, that we could have just as easily referred to the base *location* of an index number, as we did in Table 18.1. The base period of an index number, or the period against which reference is made (given in Definition 18.1), shall be denoted by a 0 subscript, and the given period shall be denoted by an n subscript. The notation $P_{n/0}$ indicates a price index in period n measured against the base period 0. We alternately indicate the price index for 1987 with 1980 as a base year as $P_{1987/1980=100}$, for example.

> **Notation: Value of an index, attribute price, attribute quantity, base location**

Many index numbers require the summation of various prices, quantities, values, etc. We could, then, indicate the sum of M prices in a base year by

$$\sum_{i=1}^{M} p_0^{(i)} = p_0^{(1)} + p_0^{(2)} + \cdots + p_0^{(M)}$$

However, such notation is very cumbersome and confusing. We will thus indicate the sum of prices of a "basket of goods" in a base year, for example, by Σp_0, where it is understood by reference to the problem the items to be included in the summation.

> **Definition 18.1**
> Base period
>
> The *base period* of an index number (also called the *reference period*) is the period against which comparisons are made. Notationally, the base period of an index number I is indicated by an 0 subscript. The value of an index number in a base period is 100; i.e., $I_0 = 100$.

To express index numbers as percentages, we multiply their arithmetically-determined value by 100. This will ensure that the index number value in the base period will be equal to 100 (percent) rather than 1 (proportion). The majority of index numbers we hear reported each day are stated in this manner.

We will begin our discussion of index numbers with simple relative index numbers.

■ 18.2

Simple relative index numbers

18.2.1 | Simple price relatives

The simple relative index number is the simplest to compute of the index numbers. A price relative or a simple price relative, for example, is the ratio of the price of a *single commodity* in a given period to its price in the base period, multiplied by 100. The definition of a price relative is given in Definition 18.2, and some properties of this index number are given in Theorem 18.1.

> **Definition 18.2**
> Simple price relative index
>
> Let p_0 represent the price of a commodity in a base period 0, and p_n represent the price of the commodity during the given period n. Then the *simple price relative index* is defined to be:
>
> $$P_{n/0} = \left(\frac{p_n}{p_0}\right) \times 100$$

Example 18.1 Jack Hayya's wife has kept records on the prices paid for various meat products from 1983 to 1986 in order to budget Jack's paycheck from teaching at a rather large, prestigious business school. These records are given in Table 18.2. Using the data provided for rib roast, compute

Properties of a
simple price rela-
tive index

Theorem 18.1
Properties of a simple price relative index number

Let the simple price relative index, $P_{n/0}$, be defined as given in Definition 18.2, where the index is expressed as a proportion rather than as a percentage (i.e., before multiplying the index value by 100). Let $p_a, p_b, p_c, \ldots$ represent the price of a commodity in periods $a, b, c, \ldots$ Then $P_{n/0}$ possesses the following properties:

Identity property: $\qquad\qquad P_{a/a} = 1$

Time reversal property: $\qquad (P_{a/b})(P_{b/a}) = 1$ or

$$P_{a/b} = \frac{1}{P_{b/a}}$$

Circular property: $\qquad\qquad (P_{a/b})(P_{b/c})(P_{c/a}) = 1$

Modified circular property: $(P_{a/b})(P_{b/c}) = P_{a/c}$

TABLE 18.2 | Price of selected meat products consumed by a family of four for the years 1983–86*

Meat product	Year			
	1983	*1984*	*1985*	*1986*
Rib roast	$2.69	$2.89	$3.69	$3.87
Ham	2.89	2.98	3.18	3.29
Chicken	0.61	0.64	0.53	0.49
Tuna	0.79	0.81	0.82	0.89

* Price per pound, except for tuna, which is in 6.5-ounce-can units.

Source: Family records.

$P_{1986/1983=100}$ and verify that the four properties given in Theorem 18.1 are satisfied for these data where $a = 1983$, $b = 1984$, $c = 1985$, and the index number is expressed as a proportion.

Solution Using the data for rib roast in Table 18.2, and using 1983 as the base year and 1986 as the given year,

$$P_{n/0} = \left(\frac{3.87}{2.69}\right) \times 100 = 143.87$$

Thus the simple price relative index for the price Jack's wife pays for rib roast is 143.87, or prices have risen by 43.87 percent in this interval.

1. Identity property:

$$P_{a/a} = \frac{2.69}{2.69} = 1$$

This property simply states that the price relative for a commodity in a specific year a with respect to the same period is 1.

2. Time reversal property:

$$(P_{a/b})(P_{b/a}) = \left(\frac{2.69}{2.89}\right)\left(\frac{2.89}{2.69}\right) = 1$$

This property states that if the two periods are interchanged, the price relatives are reciprocals of one another.

3. Circular property:

$$(P_{a/b})(P_{b/c})(P_{c/a}) = \left(\frac{2.69}{2.89}\right)\left(\frac{2.89}{3.69}\right)\left(\frac{3.69}{2.69}\right) = 1$$

4. Modified circular property:

$$(P_{a/b})(P_{b/c}) = \left(\frac{2.69}{2.89}\right)\left(\frac{2.89}{3.69}\right) = \frac{2.69}{3.69} = P_{a/c}$$

Although simple price relatives are "simple" to compute, their usefulness is limited by the fact that they consider only one commodity. That is, Jack's wife would usually be more interested in how the prices of *all meat products* have changed, rather than in how the price of rib roast alone has changed. Certain industry groups nevertheless *are* interested in how the price of a single commodity is changing over time, as were many of the Latin American countries after the "freeze" of the mid-1970s influenced the price of coffee beans. Industry associations often accumulate price and quantity movements for a selected commodity for their member institutions as part of their routine functions.

18.2.2 | Simple quantity relative

Just as certain groups may be interested in the relative change in prices over an interval of time for a single commodity, they may also be interested in the relative change in the quantity of a commodity produced, consumed, purchased, etc., over an interval of time. For example, the tuna industry was *very interested* in the relative change in the quantities of tuna consumed after the contamination news severely depressed the market for tuna in 1970! The definition of a simple quantity relative is given in Definition 18.3, and properties of this index are provided in Theorem 18.2. In Table 18.3 are provided estimates of the weekly consumption of meat products listed in Table 18.2 for the years indicated. The computation of a simple quantity relative index number is left to the exercises at the end of the chapter.

18.2.3 | Simple value relative

Simple value relative

The simple value relative index is the ratio of the price times the quantity of a commodity in a given year, n, divided by the price times the quantity of the commodity in the base year, multiplied by 100. The simple value relative thus measures the relative value of a single commodity between two periods.

Definition 18.3
Simple quantity relative index

Let q_0 represent the quantity of a commodity produced, consumed, purchased, etc., in a base period 0, and let q_n represent the quantity produced, consumed, purchased, etc., during the given period n. Then the *simple quantity relative index* is defined to be:

$$Q_{n/0} = \frac{q_n}{q_0} \times 100$$

Theorem 18.2
Properties of a simple quantity relative index number

Let the simple quantity relative index, $Q_{n/0}$, be defined as given in Definition 18.3, where the index is expressed as a proportion rather than as a percentage (i.e., before multiplying the index value by 100). Let q_a, q_b, q_c, ... represent the quantity of a commodity produced, consumed, purchased, etc., in periods a, b, c, ... Then $Q_{n/0}$ possesses the following properties:

Identity property: $\quad\quad\quad\quad Q_{a/a} = 1$
Time reversal property: $\quad\quad (Q_{a/b})(Q_{b/a}) = 1$ or

$$Q_{a/b} = \frac{1}{Q_{b/a}}$$

Circular property: $\quad\quad\quad\quad (Q_{a/b})(Q_{b/c})(Q_{c/a}) = 1$
Modified circular property: $\quad (Q_{a/b})(Q_{b/c}) = Q_{a/c}$

TABLE 18.3 | Average weekly quantity of selected meat products consumed by a family of four for the years 1983–86*

Meat product	Year			
	1983	1984	1985	1986
Rib roast	5.00	5.00	4.00	3.00
Ham	2.00	3.00	2.00	3.00
Chicken	4.00	3.00	5.00	5.00
Tuna	1.00	0.50	0.50	0.25

* In pounds.
Source: Family records.

A definition of a simple value relative is given in Definition 18.4, and a list of the properties of the value relative are given in Theorem 18.3. In addition to the properties indicated for the simple relative price and simple relative quantity indices, the simple value relative satisfies the factor reversal test in which the product of the price and the quantity simple relatives is equal to the value relative, or,

$$V_{n/0} = \left(\frac{p_n q_n}{p_0 q_0}\right)(100) = \left(\frac{p_n}{p_0}\right)\left(\frac{q_n}{q_0}\right)(100)$$

Simple value relative index

Definition 18.4

Simple value relative index

Let q_0 represent the quantity of a commodity produced, consumed, purchased, etc., in a base period, and let p_0 represent its price in the same period. Let q_n represent the quantity produced, consumed, purchased, etc., in a given period n, and let p_n represent its price in the same period. Then the *simple value relative index* is defined to be:

$$V_{n/0} = \left(\frac{v_n}{v_0}\right)(100) = \left(\frac{p_n q_n}{p_0 q_0}\right)(100) = \left(\frac{p_n}{p_0}\right)\left(\frac{q_n}{q_0}\right)(100)$$

Properties of a simple value relative index

Theorem 18.3

Properties of a simple value relative index number

Let the simple value relative index, $V_{n/0}$, be defined as given in Definition 18.4, where the index is expressed as a proportion rather than as a percentage (i.e., before multiplying the index value by 100). Let v_a, v_b, v_c represent the values (i.e., $v_a = p_a q_a$, $v_b = p_b q_b$, etc.) of a commodity produced, consumed, purchased, etc., in periods a, b, c, ... Then $V_{n/0}$ possesses the following properties:

Identity property:	$V_{a/a} = 1$
Time reversal property:	$(V_{a/b})(V_{b/a}) = 1$ or
	$V_{a/b} = \dfrac{1}{V_{b/a}}$
Circular property:	$(V_{a/b})(V_{b/c})(V_{c/a}) = 1$
Modified circular property:	$(V_{a/b})(V_{b/c}) = V_{a/c}$
Factor reversal property:	$V_{a/b} = (P_{a/b})(Q_{a/b})$

The simple value relative attempts to measure the relative *change in value* of a commodity between periods. Thus, although the quantity of lobster caught may have declined, the price may have increased sufficiently so that the *value* of the catch may have actually increased!

The use of the simple value relative is illustrated in the following example.

Example 18.2 Using the data provided in Tables 18.2 and 18.3, compute a simple value relative for rib roast for 1986 with 1983 as a base.

Solution

$$V_{n/0} = \left(\frac{p_n q_n}{p_0 q_0}\right)(100) = \left(\frac{3.87 \cdot 3}{2.69 \cdot 5}\right)(100) = \left(\frac{11.61}{13.45}\right)(100) = 86.32$$

Thus the relative value of rib roast consumed by Jack's family each week has declined from 1983 to 1986. Although the price has increased from $2.69 to $3.87 per pound in this period, the average quantity consumed has declined from five to three pounds in this same period; hence the decline in the relative value of rib roast consumed.

18.3

Simple aggregate and simple average index numbers

18.3.1 | Simple aggregate price index

A simple aggregate price index is the ratio of total commodity prices in a given period to total commodity prices in a base period (multiplied by 100), where the summation of prices is over some well-understood set of commodities. The definition of a simple aggregate price index is given in Definition 18.5. This index possesses the same properties as the simple relative price index described in Section 18.2.

Simple aggregate price index

Definition 18.5

Simple aggregate price index

Let Σp_0 represent the sum of all commodity prices of a given set in a base period, and let Σp_n represent the sum of the same set of commodity prices in the given period n. Then the *simple aggregate price index* is defined to be:

$$\Sigma P_{n/0} = \left(\frac{\Sigma p_n}{\Sigma p_0}\right)(100)$$

Example 18.3 Compute a simple aggregate price index for meat products using the data in Table 18.2, with 1986 as the given year and 1983 as the base year.

Solution From Table 18.2, we have

$$\Sigma P_{n/0} = \left(\frac{3.87 + 3.29 + 0.49 + 0.89}{2.69 + 2.89 + 0.61 + 0.79}\right)(100) = \left(\frac{8.54}{6.98}\right)(100) = 122.35$$

18.3.2 | Simple aggregate quantity index

Simple aggregate
quantity index

A simple aggregate quantity index is defined similarly to a simple aggregate price index with the word *quantity* substituted for *price*. This index possesses the same properties as the simple aggregate price index described previously. The determination of a simple aggregate quantity index is provided in the following example.

Example 18.4 Compute a simple aggregate quantity index for the meat products given in Table 18.3, with 1983 as the base year and 1986 as the given year.

Solution From Table 18.3,

$$\Sigma Q_{n/0} = \Sigma Q_{1986/1983=100} = \left(\frac{3 + 3 + 5 + 0.25}{5 + 2 + 4 + 1}\right)(100) = \left(\frac{11.25}{12}\right)(100) = 93.75$$

Thus, we would conclude that the average weekly consumption of the four meat products in Jack's house was, in 1986, 93.75 percent of its level in 1983.

There are two basic disadvantages of simple aggregate price and quantity indices:

1. They do not take account of the relative importance of the various commodities.

2. The particular units used in the price or quantity quotations can exert a big influence on the value of the index.

Consider the following example.

Example 18.5 In Example 18.3, a price index was computed based on units of 1 pound of rib roast, 1 pound of ham, 1 pound of chicken, and 6.5 ounces of tuna. Recompute the index based on the price of tuna being expressed in units of a pound, and based on the price of tuna being expressed in units of cases (48 cans, at 6.5 ounces each).

Solution If 6.5 ounces of tuna cost $0.79 in 1983, then the price of one pound of tuna would have been $p_0 = (16/6.5)0.79 = \$1.94$. For 1986 the price would be $p_n = (16/6.5)0.89 = \$2.19$. Hence $\Sigma P_{n/0}$ becomes

$$\Sigma P_{n/0} = \left(\frac{3.87 + 3.29 + 0.49 + 2.19}{2.69 + 2.89 + 0.61 + 1.94}\right)(100) = \left(\frac{9.84}{8.13}\right)(100) = 120.97$$

which is 1.38 percentage points below the index computed in Example 18.3, simply by changing the units in which prices are assessed. If one 6.5 ounce can of tuna sold for $0.79 in 1983, one case sold for $p_0 = 48(0.79) = \$37.92$. For 1986, the price would be $p_n = 48(0.89) = \$42.72$. Hence $\Sigma P_{n/0}$ becomes

$$\Sigma P_{n/0} = \left(\frac{3.87 + 3.29 + 0.49 + 42.72}{2.69 + 2.89 + 0.61 + 37.92}\right)(100) = \left(\frac{50.37}{44.11}\right)(100) = 114.19$$

Because the "value" of simple aggregate indices can be significantly influenced by the units of measurement, simple aggregate indices are not used frequently in practice. One of the few simple aggregate indices that is still published is the *Dun & Bradstreet Wholesale Food Price Index*, which is based on the prices of one pound *each* of 31 basic commodities. This index appears monthly in *Dun's Statistical Review*.

18.3.3 | Simple average of price relatives index

The simple average of price relatives index number is delineated in Definition 18.6. (The simple average of quantity relatives index is similarly defined, substituting the word *quantity* for *price*.) An advantage of the simple average of price relatives index over the simple aggregate price index is that the former is independent of the units of measurement. That is, the same "value" would be obtained for the index no matter what the units of measurement (ounces, pounds, tons, etc.). As just seen, this is not true for the simple aggregate index.

Unfortunately, the simple average of price relatives index also suffers one of the chief disadvantages of the simple aggregate price index—both fail to account for the *relative importance* of the items composing the index. Chicken, for example, would receive the same weight as tuna in the determination of the index, even though much less tuna is consumed by the average family in a given period. *Weighted* index numbers attempt to remedy this by weighting commodities by their importance. Weighted price and quantity index numbers are addressed in the next section.

Simple average of price relatives index

Definition 18.6
Simple average of price relatives index

Let $\Sigma(p_n/p_0)$ represent the sum of the quotients for a given set of N commodities of the price in the given period divided by the price in the base period. Then the *simple average of price relatives index* is defined to be:

$$\overline{P}_{n/0} = \left(\frac{\sum \frac{p_n}{p_0}}{N} \right) (100)$$

■ 18.4

Weighted aggregate price index numbers

18.4.1 | Introduction

To overcome one of the difficulties encountered with simple aggregate price index numbers—namely, that the units of measurement can have a big influence on the value of the index—several weighted aggregate price indices have been developed, where the effect of the weights is to reflect (in some sense) the importance of the commodity. Thus, for example, in determining a price index for meat, we would weight the price of rib roast by the quantity of it consumed, the price of tuna by the quantity consumed, and so on.

A difficulty encountered in computing a weighted price index concerns the weights or the quantities that will be used. For example, should a weighted price index for food for 1986 with a base year of 1983 reflect consumption patterns in 1986 or in 1983? As we will see shortly, the determination of the weights to be used is chiefly influenced by the economies of data collection. And, although a weighted index presumably would assume more importance for a broader audience than would an unweighted one, the weighted indices possess few of the desirable properties of unweighted indices given in Theorems 18.1–18.3.

We will begin our discussion of weighted aggregate price indices by considering the Laspeyre price index—a price index with base-year weights.

18.4.2 | Laspeyre index—a price index with base-year weights

The Laspeyre price index is a weighted aggregate price index where the weights are determined by the quantities of the base period. A definition of the Laspeyre price index is given in Definition 18.7. Of those properties given in Theorem 18.1, the Laspeyre price index satisfies only the identity property.

| Laspeyre price index |

Definition 18.7

Laspeyre price index

Let $\Sigma p_n q_0$ represent the sum of the quantities of a given set of commodities produced, consumed, purchased, and so on, in the base period multiplied by their respective prices per unit in the given period, and let $\Sigma p_0 q_0$ represent the sum of the same quantities of the set of commodities multiplied by their respective unit prices in the base period. Then the weighted aggregate price index with base-year quantity weights or the *Laspeyre index* is defined to be:

$$L_{n/0}^* = \left(\frac{\Sigma p_n q_0}{\Sigma p_0 q_0}\right)(100)$$

Example 18.6 Using the data in Tables 18.2 and 18.3, compute the Laspeyre price index for 1986 with 1983 as a base.

Solution The computation of the Laspeyre price index for the data provided is given in Table 18.4. Note that the price of tuna has been converted to pounds since the quantities consumed are in pounds.

The Laspeyre price index is frequently criticized because, it is argued, it tends to overstate or have an upward bias. When prices increases, there is a tendency to reduce the consumption of the higher-priced items. Hence, by using base-year weights, too much weight is given to those items that have increased the most. Note, for example, how the quantity of rib roast consumed in Table 18.3 has declined as the price has increased.

TABLE 18.4 | Computation of Laspeyre price index for data in Tables 18.2 and 18.3 (1983 = 100)

(1) Product	(2) Weekly quantity consumed in base year, 1983	(3) Price in 1983	(4) $p_0 q_0$	(5) Price in 1986	(6) $p_n q_0$
Rib roast	5	$2.69	13.45	$3.87	19.35
Ham	2	2.89	5.78	3.29	6.58
Chicken	4	0.61	2.44	0.49	1.96
Tuna	1	1.94	1.94	2.19	2.19
			23.61		30.08

$$L^*_{1986/1980 = 100} = \left(\frac{30.08}{23.61}\right)(100) = 127.38$$

Similarly, when prices decline, consumers will shift their purchases to those items that have declined the most. By using base-period weights, sufficient weight will not be given to those items that have fallen the most in price, again overstating the index.

18.4.3 | Paasche index—a price index with given-year weights

The Paasche price index is a weighted aggregate price index in which the weights are determined by quantities in the *given* year. A definition of the Paasche price index is given in Definition 18.8. Again, the identity property is the only property exhibited by the Paasche index when the quantities between the base and the given year are not equal, as is generally the case. Use of the Paasche index is illustrated in the following example.

Paasche price index

Definition 18.8

Paasche price index

Let $\Sigma p_n q_n$ represent the sum of the quantities of a given set of commodities produced, consumed, purchased, etc., in the given period multiplied by their respective unit prices in the given period, and let $\Sigma p_0 q_n$ represent the sum of the same quantities of the set of commodities multiplied by their respective unit prices in the base period. Then the weighted aggregate price index with given period quantity weights or the *Paasche index* is defined to be:

$$P^*_{n/0} = \left(\frac{\Sigma p_n q_n}{\Sigma p_0 q_n}\right)(100)$$

Example 18.7 Using the data in Tables 18.2 and 18.3, compute the Paasche price index for 1986 with 1983 as a base. How does this compare with the Laspeyre price index determined in Example 18.6?

Solution A computation worksheet for determining the Paasche price index for the data in Tables 18.2 and 18.3 is given in Table 18.5. Using the

TABLE 18.5 | Computation of Paasche price index for data in Tables 18.2 and 18.3

(1) Product	(2) Weekly quantity consumed in given year, 1986	(3) Price in 1983	(4) $p_0 q_n$	(5) Price in 1986	(6) $p_n q_n$
Rib roast	3	$2.69	8.07	$3.87	11.61
Ham	3	2.89	8.67	3.29	9.87
Chicken	5	0.61	3.05	0.49	2.45
Tuna	0.25	1.94	0.49	2.19	0.55
			20.28		24.48

$$P^*_{1986/1983=100} = \left(\frac{24.48}{20.28}\right)(100) = 120.72$$

Paasche price index, we would conclude that prices for this particular group of commodities have increased 20.72 percent over the base year; using the Laspeyre price index, we would conclude that prices have increased 27.38 percent over the base year.

The difficulty in computing the Paasche index in practice is that revised weights or quantities have to be computed each year or each period, adding to the data-collection expense in preparation of the index. For this reason, the Paasche index is *not* used frequently when the number of commodities is large.

It is also possible to define a Laspeyre and a Paasche *quantity* index. These two are often referred to as the "symmetric" quantity indices. A discussion of these two symmetric quantity indices is left to the references at the end of the chapter.

18.4.4 | Typical-period price index—a price index with typical-period weights

The typical-period price index is a weighted aggregate price index in which the quantity weights are based on demand, consumption, and so on, in a "typical" period. Frequently, the typical period (year) is taken to be the average of some number of consecutive periods. The typical-period price index is applicable in those instances in which the quantity weights for the base or for the current period are known to be unrepresentative. For example, the inclusion of quantity weights for tuna in the base year of 1970 may not be representative because of the mercury contamination scare of 1970, which reduced the consumption of tuna considerably. The consumer price index described in the next section is basically a typical-period price index where the current weights are based on a consumer expenditure survey completed at a cost of $20 million in 1972–73, and the base period is 1967.

The typical-period price index is defined in Definition 18.9.

Typical-period price index

Definition 18.9

Typical-period price index

Let $\Sigma p_n q_t$ represent the sum of the quantities of a given set of commodities produced, consumed, purchased, and so on, in a "typical" period t multiplied by their respective unit prices in the given period n, and let $\Sigma p_0 q_t$ represent the sum of the same quantities of the set of commodities multiplied by their respective unit prices in the base period 0. Then the weighted aggregate price index with typical-period weights is defined to be:

$$T^*_{n/0} = \left(\frac{\Sigma p_n q_t}{\Sigma p_0 q_t}\right)(100)$$

18.4.5 | Fisher ideal price index

The Fisher ideal price index delineated in Definition 18.10 and Theorem 18.4 was designed to overcome several disadvantages of the Laspeyre and the Paasche price indices. Note, for example, from Theorem 18.4 that the Fisher ideal index possesses the time reversal and the factor reversal properties not possessed by the Laspeyre and the Paasche price indices, which gives it certain theoretical advantages over these two price indices. Note also that the Fisher ideal index requires the computation of quantity weights in the given year, giving it the added data-collection costs of the Paasche price index.

Use of the Fisher ideal price index is illustrated in the following example.

Example 18.8 Using the data in Tables 18.2 and 18.3, compute the Fisher ideal price index for 1986, with 1983 as a base.

Solution We can greatly simplify the computation of the Fisher ideal price index after the Laspeyre and the Paasche price indices are computed by realizing that the Fisher index is the geometric mean of the Laspeyre and the Paasche price indices. That is,

$$F^*_{n/0} = \sqrt{L^*_{n/0} P^*_{n/0}}$$

<table>
<tr><td>

Fisher ideal price
index

</td><td>

Definition 18.10
Fisher ideal price index

Let the quantities $\Sigma p_n q_0$, $\Sigma p_0 q_0$, $\Sigma p_n q_n$, and $\Sigma p_0 q_n$ be as defined in Definitions 18.7 and 18.8. Then the *Fisher ideal price index* is defined to be:

$$F_{n/0}^* = \left[\sqrt{\left(\frac{\Sigma p_n q_0}{\Sigma p_0 q_0}\right)\left(\frac{\Sigma p_n q_n}{\Sigma p_0 q_n}\right)} \right] (100)$$

Recalling that the geometric mean of two numbers a and b is $(ab)^{1/2}$, the Fisher ideal price index is the geometric mean of the Laspeyre and the Paasche price indices.

</td></tr>
</table>

<table>
<tr><td>

Properties of the
Fisher ideal price
index

</td><td>

Theorem 18.4
Properties of the Fisher ideal price index

Let the Fisher ideal price index, $F_{n/0}^*$, be defined as given in Definition 18.10, where the index is expressed as a proportion rather than as a percentage (i.e., before multiplying the index value by 100). Let the subscripts a, b, c, . . . refer to relevant prices or quantities in periods a, b, c, ... Then the Fisher ideal price index satisfies the following properties:

Identity property: $F_{a/a}^* = 1$

Time reversal property: $(F_{a/b}^*)(F_{b/a}^*) = 1$ or

$$F_{a/b}^* = \frac{1}{F_{b/a}^*}$$

Factor reversal property: $V(F_{a/b}^*) = P(F_{a/b}^*) Q(F_{a/b}^*)$

</td></tr>
</table>

From Tables 18.4 and 18.5,

$$F_{n/0}^* = \sqrt{(127.38)(120.72)} = 124.01$$

The Fisher ideal price index indicates that prices in 1986 for the commodities involved are 24.01 percent above base-year (1983) prices.

In general, we would expect the Fisher ideal price index to lie between the values of the Laspeyre and the Paasche indices, since it is the geometric mean of these two quantities. Recall from the discussion in Chapter 2 that the geometric mean is less affected by extremely large or small values than is the arithmetic mean. This adds to the desirability of the Fisher ideal price index, especially when one of the components of the index experiences an unusually large increase or decrease in the given or the base period.

■ **18.5**

The consumer price index

Perhaps the single index number that touches our lives most is the consumer price index (CPI) compiled by the Bureau of Labor Statistics (BLS). Indeed, provisions for "escalating" wages (escalator clauses) in the union contracts of more than 8.5 million workers are tied directly to the CPI. Politicians love to quote this index to show how well they are doing (or how badly the other party is doing!), and divorcees frequently receive a change in alimony payments with a change in the CPI! The CPI is further a major yardstick against which government economic policy is judged. A graph showing this index for the years 1949–84 is given in Figure 18.1. Note from

FIGURE 18.1 | Consumer price index, 1949–1984

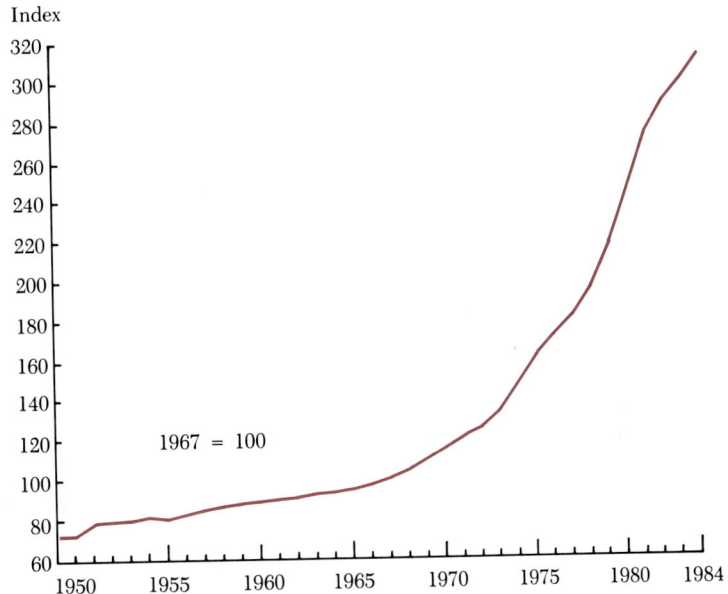

Source: U.S. Department of Commerce, *Survey of Current Business* (various issues).

this graph the substantial increase in inflation (as measured by the consumer price index) during the 1970s, and also note that the increases in this index number in the first half of the 1980s have been well below the increases witnessed in the latter half of the 1970s. An individual earning $10,000 yearly in 1967 would have to earn in excess of $31,000 in 1984 just to have the same buying power (that is, just to stay even)! Because the Federal income tax *rate* is substantially higher for individuals with $31,000 of yearly income as opposed to $10,000 of yearly income, certain federal income tax rates and deductions are now tied to increases in the consumer price index in order to slow tax bracket creep brought about by inflation.

Most of us view increases in the consumer price index as an erosion of our purchasing power. For this reason, increases in the index are most often viewed with distaste. When viewing the increases in the index over a number of years (as shown in Figure 18.1), we gain a better picture of what has happened to our purchasing power over the years. When viewed in this context, those 8–10 percent wage or salary increases over several years really haven't increased our purchasing power all that much. We suddenly feel worse off!

As bad as the inflation rate has been in the United States in recent years, it is indeed small when compared with the inflation rates in many of the Latin American Countries where it is routinely several hundred or a few thousand percent each year. Even this is a paltry sum when compared with inflation rates witnessed in other countries during select periods. In Figure 18.2, for

FIGURE 18.2 | A 500 million mark note issued during the inflationary period 1920–1923 in Germany

example, we show a 500-million German mark note issued during the 1920s. During 1920 wholesale prices in Germany increased about 80 percent. In 1921 the increase was 140 percent, and in 1922 the increase was a whopping 4,100 percent. Then it *really* started to grow! Between December 1922 and November 1923, it increased 100 million percent! By 1923 the government found that its printing presses couldn't keep up, even by printing notes as large as 500 million marks! One hears stories of workers being paid daily and then twice a day so that their wives could shop before their wages had been devalued so much that they could no longer afford to purchase a loaf of bread or other necessities of life. There is also the story frequently mentioned of a German lady going shopping and laying down her bushel basket filled with marks so that she could look closer at a few items. When she returned someone had stolen the basket, but left the money because it had no value! As amazing as all of this sounds, the inflation rate in Germany during the early 1920s is dwarfed by the inflation rate in Hungary from 1945 to 1946. At this time, the rate of inflation was 20,000 percent per month. During the final month before it started to subside (July 1946), inflation was 42 quadrillion percent! Needless to say, the printing presses in Hungary also had a difficult time keeping up with the demand for printed notes. Because the U.S. consumer price index is most often taken as the given measure for assessing inflation, it is important that we understand more about this index number and how it is computed. One hopes it will never have to measure the magnitude of increases indicated for both Germany and Hungary during these high inflationary periods!

The CPI has been published continually since 1913, although major revisions in the method of computation, in the data-collection methodology, and in the "market basket of goods" that compose the index have changed considerably since this time. Many commodities in the current CPI (there are about 400 of them at present) were not even around in 1913! Furthermore, with the most recent revision of the index (there have been five since its inception in 1913), more sophisticated statistical sampling procedures have been used both to measure the prices paid for the goods and services composing the index and to detect shifts in buying habits so the quantity weights used in compiling the index can be updated more systematically. For example, the selection of retail stores for obtaining prices of selected goods is based on a point-of-purchase survey of households, rather than on secondary sources. Food is now priced throughout the month, rather than during the first week of the month, as before. These and other refinements should lead to a more accurate picture of the change in the purchasing power of the dollar over time.

Computation of the CPI began during the inflationary conditions of World War I, when the Bureau of Labor Statistics investigated the cost of living in a number of shipbuilding centers and other industrial centers. Shortly thereafter, the index was given its official title of "Index of Changes in Prices of Goods and Services Purchased by City Wage Earner and Clerical Worker Families to Maintain Their Level of Living." The type of workers surveyed in the index has changed recently (1978) to include more than city wage earners and clerical workers. Today, for example, two consumer price indices are published instead of just the one. The CPI for City Wage Earners and Clerical Worker Families attempts to measure changes in the purchasing power of the dollar for about 40 percent of the noninstitutional civilian households in the United States. The more inclusive CPI for All Urban Consumers measures the change in the purchasing power of the dollar for *all* urban consumers, and covers in excess of 80 percent of the total noninstitutional civilian households in the United States. To obtain prices for commodities contained in these two indices, the Bureau of Labor Statistics has recently increased the number of urban areas sampled from 56 to 85. These areas are based on a probability sample using the 1970 census as its population data base.

TABLE 18.6 | Percent distribution of the consumer price index market basket by major expenditure groups, benchmark years

| Major group | Wage earners and clerical workers | | | | All urban consumers |
	1935–39*	1952**	1963†	1972–73§	1972–73§
Food and alcoholic beverages	35.4	32.2	25.2	20.4	18.8
Housing	33.7	33.5	34.9	39.8	42.9
Apparel	11.0	9.4	10.6	7.0	7.0
Transportation	8.1	11.3	14.0	19.8	17.7
Medical care	4.1	4.8	5.7	4.2	4.6
Entertainment	2.8	4.0	3.9	4.3	4.5
Personal care	2.5	2.1	2.8	1.8	1.7
Other goods and services	2.4	2.7	2.9	2.7	2.8

* Relative importance for the survey period 1934–36 (updated for price change).

** Relative importance for the survey period 1947–49 (updated for price change).

† Relative importance for the survey period 1960–61 (updated for price change).

§ Relative importance for the survey period 1972–73. Revised indexes which require expenditure weights updated for price change between the survey period and the link dates will differ from those shown.

Source: "The Consumer Price Index: Concepts and Content over the Years," U.S. Department of Labor, Bureau of Labor Statistics, Report 517 (May, 1978) [Revised].

Table 18.6 gives information on the relative importance of major components of the CPI for four benchmark years or periods. Note how the relative weighting of the proportions of income expended for housing for individuals contained in the wage earner and clerical worker index sample has increased from 33.7 percent in 1935–39 to 39.8 percent in 1972–73. This reflects the change in buying or purchasing habits of these consumers detected over this period. Note also that for the latest revision of the index, the proportion expended for housing differs depending upon whether one is in the wage earner, clerical worker sample or is in the all urban population sample. For the wage earner, clerical worker sample, the proportion of expenditures on food and alcoholic beverages has decreased from 35.4 percent to 20.4 percent from the period 1935–39 to 1972–73. Other quantities in Table 18.6 are interpreted similarly.

Three base periods for the wage earner, clerical worker index have existed since World War II: 1947–49, 1957–59, and 1967. The base period for the newer all urban population index, which first began publication in January 1978, is also 1967. The current weights for both of these indices, which are given in Table 18.6, are based on a Consumer Expenditure Survey conducted in 1972 and 1973.

Both consumer price indices just described are actually chain index numbers, where the index numbers for successive time periods are combined.

This facilitates the updating of the index when there is a detected change in purchasing habits. The CPI very closely approximates a typical-year price index described in the previous section whose base period of the index does not coincide with the dates of the items included in the index based on the 1972–73 Consumer Expenditure Survey. Since the weights used in computing the current index value are based on the Consumer Expenditure Survey of 1972–73, it is important to note that the index is *not* a cost-of-living index. To be a cost-of-living index, the index would have to reflect current purchasing habits, and purchasing habits have no doubt changed since 1972–73. We have already observed, for example, that as goods decline in price relative to others, there is a tendency to purchase more of the good that has decreased in price. Other things being equal, we tend to decrease our purchases of goods that are rising in price faster than others.

The use of the CPI in decision making is provided in the following example.

Example 18.9 The tables below give the CPI for the years 1980–84 (1967 = 100) and Jack Hayya's wage for the same period.

CPI	
Year	*CPI*
1980	246.8
1981	272.4
1982	289.1
1983	298.4
1984	311.1

Source: U.S. Bureau of Labor Statistics, *Monthly Labor Review.*

Jack's wages	
Year	*Wage rate*
1980	$10.27
1981	$11.32
1982	$12.19
1983	$12.42
1984	$12.59

Source: Jack's budget records.

On balance, has Jack's wage rate kept pace with the general increase in prices as measured by the CPI?

Solution   Relevant computations are shown in the table.

(1)	(2)	(3)	(4)
			Deflated wage rate
Year	*Wage rate*	*CPI**	$[(2) \div (3)](100)$
1980	$10.27	246.8	$4.16
1981	11.32	272.4	4.16
1982	12.19	289.1	4.22
1983	12.42	298.4	4.16
1984	12.59	311.1	4.05

* 1967 = 100.

On balance, Jack's wage increases have not kept pace with the inflation rate as measured by the CPI. In fact, real income (deflated wage rate) is actually lower in 1984 than it was in 1980! Jack has found himself in the same position as most other college teachers during this period!

■ **18.6**

Deflation of time series

Deflation of time series

Time series data are statistical data that are measured over time; an example would be the level of the gross national product in the United States for the most recent past. Most time series data, such as GNP, include the effect of price changes in their level or their value. To assess the amount of actual *real* growth in the series, it becomes necessary to adjust the data for increases in the price level. This adjusting process is called *deflation of time series*. Mathematically, deflation of time series is applicable only if the price index numbers satisfy the factor reversal test, as does the Fisher ideal price index.

The use of price index numbers for deflating time series data is illustrated in the following example.

Example 18.10 Given in the table below are unadjusted values of GNP for the United States for the years 1980–84.

	Year				
	1980	*1981*	*1982*	*1983*	*1984*
GNP ($ billions)	2,631.7	2,957.8	3,069.3	3,304.8	3,662.8

Source: "Economic Indicators," September 1985.

Table 18.7 gives information on gross national product implicit price deflators (aggregate price index numbers) for these same five years.

TABLE 18.7 | Gross national product implicit price deflators (1972 = 100)

	Year				
	1980	*1981*	*1982*	*1983*	*1984*
Price deflator	178.42	195.60	207.38	215.34	223.43

Source: "Economic Indicators," September 1985.

Using the information provided, estimate the percentage growth in *actual* GNP and in *real* GNP from 1980 to 1984 inclusive, and determine the level of GNP in "1972 dollars."

Solution We will do the second part first. In order to express GNP in 1972 dollars, we must divide the actual values of GNP by the implicit price deflators given and multiply the result by 100 as shown in the table.

Year	1980	1981	1982	1983	1984
GNP ($ billions)	2,631.7	2,957.8	3,069.3	3,304.8	3,662.8
Implicit price deflator (1972 = 100)	178.42	195.60	207.38	215.34	223.43
GNP in 1972 dollars ($ billions)	1,475.0	1,512.2	1,480.0	1,534.7	1,639.4

$$\text{Percentage growth in actual GNP, 1980–84} = \frac{3,662.8 - 2,631.7}{2,631.7} = 0.39$$

$$\text{Real growth in GNP (adjusted for price level changes), 1980–84} = \frac{1,639.4 - 1,475.0}{1,475.0} = 0.11$$

Thus, although actual GNP increased 39 percent in this interval, 72 percent of the increase is due solely to an increase in prices and does not represent *real* growth. Therefore 11 percent represents the real growth in GNP over this interval.

■ **18.7**

Index numbers as random variables

Throughout our discussion of index numbers, we have not emphasized that index numbers are random variables, and this may be unfortunate. To gain an appreciation of the inherent variability associated with the computation of index numbers, one has only to consider how compilers of index numbers arrive at the prices and quantities that determine the value of the index. For example, Jack's wife in Example 18.1 bases the prices of the amounts she spends for food items (meats) on trips to the grocery store. The prices she pays for meat products are bound to differ throughout the year, by the store where the purchase was made, whether or not there was a sale on the product the day she shopped, and other factors. Our discussion of the consumer price index in Section 18.5 included a brief discussion of some of the statistical sampling issues involved in determining both the quantity weights used in the construction of the index and in accumulating the prices of the goods priced. Thus the CPI reported each month is in reality one observation from a probability distribution of consumer price index values.

The danger in not treating an index number as a random variable is that policymakers may make important policy decisions based on the value of the index reported, when in fact the change reported may be due to nothing more than errors inherent in any sampling process.

Recently, statisticians have recognized the need not only for accumulating price index information per se, but also for reporting on whether the change in an index number is due to changes in the prices and goods composing the index or to inherent errors in statistical sampling. Noteworthy in this regard is the work of Hayya and of Hayya, Saniga, and Schaul (see the references at the end of this chapter). Under some not too restrictive assumptions regarding the distribution of the error terms in index number computation, these authors were able to derive a quadratic expression, which, when solved, yields confidence intervals and confidence bands for price index numbers. Table 18.8, for example, gives confidence intervals for the Bureau of the Census' price index for new one-family houses sold. The new houses sold must be similar to those sold in 1967 with respect to eight characteristics: floor area, number of stories, number of bathrooms, air conditioning, type of parking facility, type of foundation, geographic division within region, and location within a metropolitan area. (The value of ρ denotes the unknown correlation coefficient between the numerator and the denominator of the index.) Note how the width of the interval varies as a function of the unknown correlation coefficient between the numerator and the denominator of the index number expression (as given in the bottom row of Table 18.8).

TABLE 18.8 | Ninety percent probability intervals for the price index for new one-family houses sold, third quarter 1976 (1963 = 100)

Estimated index value: $P_{1976-73/1963-100} = 1.940$‡											
Confidence bounds: $a < P < b$											
	Correlation, ρ*										
	−1.0	−0.8	−0.6	−0.4	−0.2	0.0	0.2	0.4	0.6	0.8	1.0
a	1.914	1.915	1.917	1.918	1.920	1.922	1.923	1.926	1.928	1.931	1.937
b	1.967	1.965	1.964	1.962	1.961	1.959	1.957	1.955	1.952	1.949	1.943
Percent†	2.7	2.6	2.4	2.3	2.1	1.9	1.8	1.5	1.2	0.9	0.3

* The correlation ρ measures the degree of linear association between the numerator and the denominator of the index number expression.

† The percents give the width of the probability interval as a percentage of the point estimate of the index.

‡ Here, the index number is being quoted as a proportion, rather than as a percentage.

Source: Jack C. Hayya, Erwin M. Saniga, and Ronny A. Schaul, "Uncertainty in Price Indices," *R.A.I.R.O/Operations Research,* Vol. 16, No. 1 (February 1982), pp. 33–43.

Unfortunately, much more work needs to be done in this area before meaningful results can be used on a large scale. One may envision a newscaster of the future saying, "The Consumer Price Index rose by 0.3 percent last month, but this increase could be attributable to nothing more than errors inherent in the sampling process in gathering the data." For the time being, we, as users of index number data, should content ourselves with recognizing that index numbers are random variables and are but one real-

TABLE 18.9 | Summary formulas for computing index numbers

Index number type	Index number	Computation formula	Test satisfied				
			Identity test	Time reversal test	Circular test	Modified circular test	Factor reversal test
Simple relative (single commodity)	Simple price relative	$P_{n/0} = \left(\dfrac{p_n}{p_0}\right)(100)$	Yes	Yes	Yes	Yes	—
	Simple quantity relative	$Q_{n/0} = \left(\dfrac{q_n}{q_0}\right)(100)$	Yes	Yes	Yes	Yes	—
	Simple value relative	$V_{n/0} = \left(\dfrac{p_n q_n}{p_0 q_0}\right)(100)$	Yes	Yes	Yes	Yes	Yes
Simple aggregate	Simple aggregate price index	$\Sigma P_{n/0} = \left(\dfrac{\Sigma p_n}{\Sigma p_0}\right)(100)$	Yes	Yes	Yes	Yes	No
	Simple aggregate quantity index	$\Sigma Q_{n/0} = \left(\dfrac{\Sigma q_n}{\Sigma q_0}\right)(100)$	Yes	Yes	Yes	Yes	No
Weighted aggregate price index	Laspeyre index	$L_{n/0}^* = \left(\dfrac{\Sigma p_n q_0}{\Sigma p_0 q_0}\right)(100)$	Yes	No	No	No	No
	Paasche index	$P_{n/0}^* = \left(\dfrac{\Sigma p_n q_n}{\Sigma p_0 q_n}\right)(100)$	Yes	No	No	No	No
	Typical-year price index (CPI)	$T_{n/0}^* = \left(\dfrac{\Sigma p_n q_t}{\Sigma p_0 q_t}\right)(100)$	Yes	No	No	No	No
	Fisher index	$F_{n/0}^* = \left[\sqrt{\left(\dfrac{\Sigma p_n q_0}{\Sigma p_0 q_0}\right)\left(\dfrac{\Sigma p_n q_n}{\Sigma p_0 q_n}\right)}\right](100)$	Yes	Yes	No	No	Yes

ization from the distribution of index number values, and we should interpret them accordingly.

■ 18.8
Summary

This chapter has presented a discussion of the use, determination, and interpretation of index numbers—statistical measures designed to show relative changes in a variable or in a group of variables with respect to some base point or base period. The types of index numbers covered included simple price, quantity and value relatives, simple aggregate price and quantity indices, and four weighted aggregate price relatives—Laspeyre, Paasche, typical year, and the Fisher ideal index. The CPI, which touches so many of our lives, was also described; it is a special case of a weighted aggregate price index.

Five tests or properties of index numbers were described that contribute to their theoretical development and interpretation. These tests or properties include: the identity test, the time reversal test, the circular and modified circular tests, and the factor reversal tests. A summary table delineating computation formulas for the index numbers described and which tests are satisfied by each is given in Table 18.9. Unfortunately, those index numbers that should be the most useful to large groups of people are the ones that fail to pass many of the tests described.

■ References

Fisher, I. *The Making of Index Numbers.* Boston: Houghton Mifflin, 1923.

Hayya, Jack C. "Confidence Bands for Price Indices," American Statistical Association 1977 Proceedings of the Business and Economics Statistics Section, part I, pp. 362–67.

Hayya, Jack C.; Erwin M. Saniga; and Ronny A. Shaul. "Uncertainty in Price Indices," *R.A.I.R.O./Operations Research* 16, no. 1, (February 1982), pp. 33–43.

Layng, W. John. "The Revision of the Consumer Price Index," U.S. Department of Commerce, *The Statistical Reporter,* no. 78–5 (February 1978), pp. 140–48.

Mudgett, Bruce D. *Index Numbers.* New York: John Wiley & Sons, 1951.

Several documents are published periodically by the Department of Labor, Bureau of Labor Statistics, Washington, D.C., on descriptions of or revisions to the Consumer Price Index. These publications include the following:

The Consumer Price Index—January 1953

The Consumer Price Index as Related to Other Consumer Statistics—1959

An Abbreviated Description of the Revised Consumer Price Index—March 1964

Major Changes in the Consumer Price Index—March 1964

Conversion of the Consumer Price Index to 1967 Standard Reference Base Period—1971

Relative Importance of Components in the Consumer Price Index, 1970–1971—1972

Updating and Revising the Consumer Price Index—April 1974

Revising the CPI: A Brief Review of Methods—1976

The Consumer Price Index Revision—1978

Revising the Consumer Price Index—1978

The Consumer Price Index: Concepts and Content over the Years—1978

Problems

Section 18.1 Problems

18.1 What factors should be considered in the construction of an index? List the factors and briefly describe each.

18.2 Explain what is meant by the base period of an index. How is the base period determined?

18.3 When constructing an index, what is the most difficult factor for which to make adjustments? Why?

18.4 How are index numbers reported? What is the most common way of reporting an index value?

Section 18.2 Problems

18.5 Prove that $(P_{a/b})(P_{b/c}) = P_{a/c}$ and that $(P_{a/b})(P_{b/a}) = 1$.

18.6 The simple quantity relative for the year 1984 with 1980 as a base is 110, and the simple quantity relative for 1984 with 1982 as a base is 130. What is the simple quantity relative for 1982, with 1980 as a base?

18.7 Using the data provided in Table 18.3, compute a simple quantity relative for rib roast for 1986 using 1983 as a base, and for 1983 with 1986 as a base. Do these two indices satisfy the time reversal test (i.e., exhibit the time reversal property)?

18.8 A stock you have been observing over the past several years has exhibited the year-end closing prices shown in the table.

Year	Closing price
1981	$54.30
1982	62.35
1983	64.86
1984	79.94
1985	86.45
1986	75.32

Using 1981 as the base year, compute the simple price relative index number for this stock between 1981 and 1986.

Section 18.3 Problems

18.9 Which of the properties listed in Table 18.9 are satisfied by the simple average price relative index? Show by counter-example which properties do not hold.

18.10 Reproduced in the table below are data on unit sales prices of electronic calculators for three models for the years 1982–86, inclusive.

Calculator type*	1982	1983	1984	1985	1986
X	$110	$105	$ 90	$ 83	$ 45
Y	450	420	260	120	80
Z	750	650	525	460	320

* Usually relates to number of programmed functions available.

a. Compute the simple relative price index number for calculator type Y for 1986, with 1982 as the base year.

b. Compute the simple aggregate price index for calculator sales, using 1986 as the given year and 1982 as the base year.

c. Compute the simple average of price relatives index number, $\bar{P}_{n/0}$, for 1986, with 1982 as the base year.

18.11 Given in the table below are data on unit sales of electronic calculators, the prices of which are given in Problem 18.10.

Calculator type	1982	1983	1984	1985	1986
X	20	40	60	65	60
Y	5	10	8	15	60
Z	1	5	10	20	40

a. Compute the simple quantity relative, $Q_{n/0}$, for calculator type X for 1986, with 1982 as the base year.

b. Compute the simple aggregate quantity index for calculator sales, using 1986 as the given year and 1982 as the base year.

Section 18.4 Problems

18.12 Develop arguments similar to those developed in Section 18.4.2 to describe why the Paasche index tends to understate or have a downward bias.

18.13 Show that the Paasche price index does not in general satisfy the time reversal and the factor reversal tests using the data in Example 18.7.

18.14 Verify that the time reversal and the factor reversal tests are satisfied by the Fisher ideal price index using the data provided in Example 18.8.

18.15 Using the definition, prove that the Fisher ideal price index satisfies the time reversal test.

18.16 Use the data from Problems 18.10 and 18.11.

a. Compute the Laspeyre price index for 1986 with 1982 as the base year.

b. Compute the Paasche price index for 1986 with 1982 as the base year.

c. Compute the Fisher ideal price index for 1986 with 1982 as the base year.

18.17 Consider the information in the table.

Year	Price I	Quantity I	Price II	Quantity II
1983	25¢	10	50¢	40
1984	30	20	45	50
1985	35	30	48	55
1986	40	40	50	50

a. Construct a Laspeyre price index for 1986 using 1983 as the base year.

b. Construct Paasche's price index for each of the years indicated using 1983 as the base year.

c. Construct the Fischer ideal price index for 1986, with 1983 as the base year.

Section 18.5 Problems

Year	1977	1978	1979	1980	1981	1982	1983	1984
CPI (1967 = 100)	181.5	195.4	217.4	246.8	272.4	289.1	298.4	311.1
Salary (average per hour)	$5.93	$6.25	$6.48	$6.63	$7.00	$7.35	$7.79	$8.30

18.18 The table above shows the consumer price index (CPI) for 1977–84. Also shown is the hourly salary received by a statistics instructor at a large midwestern university.

a. Recast the salary amounts into constant 1967 dollars.

b. The *actual* percentage growth in salary over this interval is equal to what amount?

c. The *real* percentage growth in salary over this interval is equal to what amount?

18.19 For the Consumer Price Index for City Wage Earner and Clerical Worker Families and the Consumer Price Index for All Urban Consumers surveys, respectively, what percentage of the noninstitutional civilian households in the United States is represented at the present time?

18.20 The consumer price index for selected services increased from 156 to 167 over a certain period.

a. Given that Calvin Consumer budgeted $50 per month for these services at the beginning of this period, how much should he budget for these same services at the end of this period?

b. Suppose Calvin's monthly salary increased from $1,200 to $1,340 over this same period. What is the *real* change Calvin has experienced in the purchasing power for these services?

18.21 Suppose an executive's income increases from $56,000 to $73,000 over a period of three years, and the CPI increases from 167 to 197. What is the change in her real income over this period?

18.22 Given the wage rate and consumer price index shown in the table below for the years indicated, Jack's *real* wage rate (adjusted for inflation) has increased by what percentage from 1981 to 1984?

Year	Jack's wages (per hour)	CPI (1967 = 100)
1981	$5.40	272.4
1982	5.87	289.1
1983	6.30	298.4
1984	6.87	311.1

Section 18.6 Problems

18.23 The GNP of a developing nation has increased from $40 billion in 1983 to $44 billion in 1986. The price index used in calculating GNP has increased from 145 to 160 in this same period. What is the *real* growth rate of this country's GNP after accounting for the erosion of the purchasing power of their currency?

18.24 Using CPI estimates for the years 1981 to 1986, inclusive, (from your university library) compute the deflated year-end closing price for the stock given in Problem 18.8.

Additional Problems

18.25 Given the data in the table below, compute the following:

a. $P_{n/0}$ for $n = 1973$ and $0 = 1969$.

b. $\Sigma P_{n/0}$ for $n = 1973$ and $0 = 1969$, where the summation is over all cereals and bakery products listed.

c. For $n = 1973$, $0 = 1969$, $a = 1969$, $b = 1970$, and $c = 1971$, show that the properties listed in Theorem 18.1 hold for flour, wheat.

Average retail prices of selected foods,* 1969–1973

	1969	1970	1971	1972	1973
Cereals and bakery products					
Flour, wheat	11.6	11.8	12.0	11.9	15.1
Rice	18.8	19.1	19.6	19.6	26.0
Corn flakes	41.7	42.9	44.5	41.6	42.9
Bread, white	23.0	24.3	25.0	24.7	27.6

* Cents per pound.

Source: *Statistical Abstract of the United States, 1974.*

18.26 For the first three products given in Problem 18.25, find the average U.S. per capita consumption of each of the commodities for the years indicated. Then, for $n = 1973$ and $0 = 1969$, determine the following:

a. The Paasche index, $P_{n/0}^*$.

b. The Laspeyre index, $L_{n/0}^*$.

c. Fisher ideal index, $F_{n/0}^*$.

d. In general, prove that the Fisher ideal price index falls between $P_{n/0}^*$ and $L_{n/0}^*$.

19

Chi-square tests

■ **19.1**

Introduction

In Chapter 10 we used the chi-square distribution to construct decision rules for hypotheses concerning the population variance, σ^2. In this chapter we introduce several tests of hypotheses that are generally referred to as *chi-square tests*. Most of these testing procedures were first suggested by Karl Pearson in 1900 and were among the earliest methods of statistical inference. Although the chi-square tests are frequently misused, they are among the most popular statistical inference procedures today, and for good reason. They are typically easy to implement and can tell us a great deal about the characteristics of a random variable or sets of random variables.

■ **19.2**

Goodness-of-fit tests

The chi-square distribution may be used to test the hypothesis that a random variable has a specified theoretical statistical distribution. Perhaps the most important application of the chi-square distribution in this context is testing whether or not a random variable is normally distributed, because virtually all statistical inference procedures presented in the text to this point make the normality assumption about the distribution of the population random variable or the sampling distribution of a sample statistic.

The chi-square goodness-of-fit test is based on the difference between the frequencies in the classes of the observed sample and the class frequencies we would expect if the random variable conformed to the assumed theoretical distribution. The chi-square test statistic is:

$$\chi^2 = \sum_{i=1}^{k} \frac{(O_i - E_i)^2}{E_i}$$

986

where O_i represents the observed frequency in the ith class, E_i represents the expected frequency in the ith class, and k is the number of classes. The distribution of the χ^2 statistic, if the sampled observations do come from the assumed theoretical (population) distribution, is *approximately* the chi-square distribution with $(k - 1 - m)$ degrees of freedom, where m is the number of parameters that must be estimated to determine the expected frequencies.

k–l–m degrees of freedom

Since the test statistic is approximately distributed as a chi-square random variable, all hypothesis tests for which the decision rule is based on this test statistic are *approximate statistical tests*. We must keep this fact in mind when using the chi-square goodness-of-fit testing procedures.

Approximate statistical tests

If the observed frequencies O_i do not differ much from the expected frequencies E_i, the value of the test statistic χ^2 is small. Indeed, the *minimum* value of χ^2 is 0, and this value of the test statistic occurs when $O_i = E_i$ in each of the k classes. As the observed frequencies O_i begin to differ from the expected frequencies E_i, the value of χ^2 will increase because the statistic squares these differences, weights them by the reciprocal of the expected frequencies, and adds the resulting ratios. Thus a small value of χ^2 supports the null hypothesis that the random variable conforms to the specified theoretical statistical distribution; and a large value of χ^2 supports the alternate hypothesis that the random variable does not conform to the specified statistical distribution. Consequently, the decision rule for the chi-square goodness-of-fit test will have an upper-tail rejection region; its location is determined by the form of the chi-square distribution (determined by df) and the selected significance level α of the test. A typical decision rule is illustrated in Figure 19.1. The form of the chi-square distribution is solely a function of its degrees of freedom, df. Several chi-square distributions are illustrated in Figure 19.2 for different values of the degrees of freedom.

The form of the chi-square statistic,

$$\chi^2 = \sum_{i=1}^{k} \frac{(O_i - E_i)^2}{E_i}$$

FIGURE 19.1 | Decision rule for the chi-square goodness-of-fit test

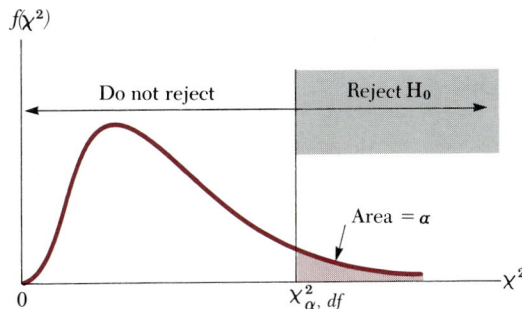

FIGURE 19.2 | Four χ^2 distributions

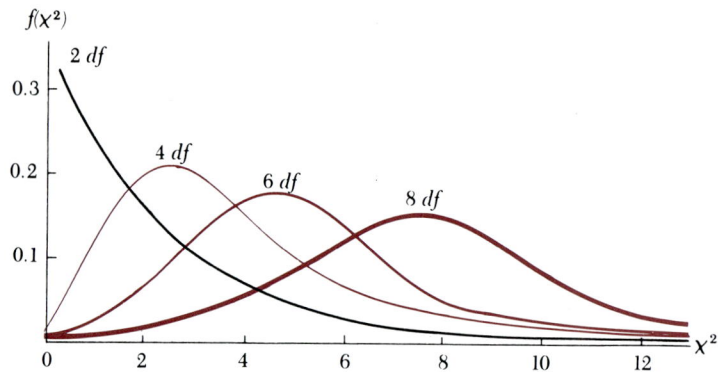

should not be unfamiliar to us. It adds weighted squared deviations and is similar in form to the population and sample variance formulas, which add weighted squared deviations of observations about their respective means. It is a very easy statistic to compute, which has no doubt contributed to its widespread use.

Example 19.1 A fair die is one in which each of the six faces has an equal probability of showing on the top when the die is tossed. Suppose we wish to test the "fairness" of a particular die experimentally. We decide to toss the die 120 times and record the number of dots showing on the top face each time. If the die is fair, we should expect to find that each number occurs with an equal frequency in the 120 tosses. Any deviation of the expected frequencies of $(1/6)(120) = 20$ for each number from the observed frequencies will give us cause to question the fairness of the die.

Table 19.1 gives the observed frequencies of the numbers in the 120 trials. The expected frequencies are easily computed in this problem. If the die is

TABLE 19.1 | Observed frequencies in Example 19.1

Number of dots showing on top face	Observed frequency O_i
1	12
2	14
3	31
4	29
5	20
6	14
Total	120

fair, then the probability that each number of dots will appear on the top face in any trial is $\frac{1}{6}$. Since there are 120 trials, the expected frequency of each number in the experiment is $(\frac{1}{6})(120) = 20$. The computation of the chi-square statistic value for this experiment is given in Table 19.2.

TABLE 19.2 | Computation of the chi-square statistic for Example 19.1

Class	i	Observed frequency, O_i	Expected frequency, E_i	$O_i - E_i$	$(O_i - E_i)^2$	$\dfrac{(O_i - E_i)^2}{E_i}$
1	1	12	20	-8	64	3.20
2	2	14	20	-6	36	1.80
3	3	31	20	11	121	6.05
4	4	29	20	9	81	4.05
5	5	20	20	0	0	0.00
6	6	14	20	-6	36	1.80
Totals		120	120	0		$\chi^2 = 16.90$

The degrees of freedom for this test are $df = k - 1 - m = 6 - 1 - 0 = 5$; that is, $m = 0$, since no parameters had to be estimated to determine the expected frequencies. Under the null hypothesis of a fair die, the expected frequencies are determined $[(\frac{1}{6})(120) = 20$ in each class] without additional computations.

If we select $\alpha = 0.05$ for the significance level of this test, then from Table B.6 in Appendix B, the critical tabular χ^2-value is $\chi^2_{0.05;5} = 11.1$. The decision rule is:

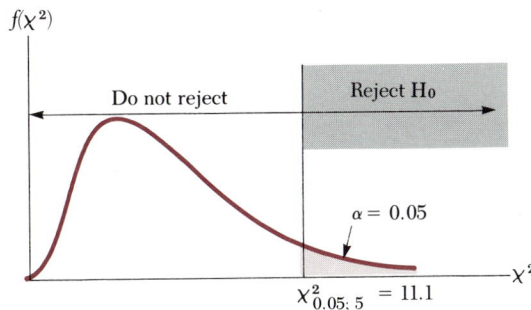

Since $\chi^2 = 16.9 > 11.1$, we reject the null hypothesis that the die is fair and conclude that the die is not fair—it appears to be "loaded" to produce more 3s and 4s than the other values.

The p-value for the chi-square tests in this chapter is given by $P(\chi^2 > \chi^2/df)$, where χ^2 is the calculated χ^2 value for the test, and df is the degrees of freedom. Frequently, it will be necessary to interpolate in Table B.6 in Appendix B to calculate the p-value. Alternatively, we can state approximately that the p-value is between, say, 0.05 and 0.025 to avoid interpolation.

For this test, the p-value is given by:

$$p\text{-value} = P(\chi^2 > 16.9/df = 5)$$

Since $P(\chi^2 > 16.7/df = 5) = 0.005$ and $P(\chi^2 > 20.5/df = 5) = 0.001$ from Table B.6, we may conclude that the p-value is between 0.005 and 0.001 (by interpolation, it is 0.0048). Thus if the null hypothesis is true, then we would expect to get a value of χ^2 greater than or equal to 16.9 with probability 0.0048. Consequently, the significance level α would have to be less than 0.0048 before the null hypothesis could not be rejected. Since α, the probability of committing a Type I error (rejecting H_0 given H_0 is true) would have to be set so low to not reject H_0, we can be confident that the decision to reject H_0 is well supported by the sample evidence.

Now let us make some observations concerning the use of the chi-square goodness-of-fit test in this example. First, the test is equivalent to testing the null hypothesis that the frequencies of the die numbers are uniformly distributed. In Chapter 6 we studied the discrete uniform distribution, the probability mass function of which is given by:

$$P[\mathbf{X} = x_i] = \frac{1}{k}, \qquad i = 1, 2, 3, \ldots, k$$

In this example $k = 6$, and the possible values of the random variable $\mathbf{X}$ are $x_1 = 1, x_2 = 2, \ldots, x_6 = 6$. A stick diagram for this discrete uniform distribution is illustrated in Figure 19.3. We can therefore conclude that the

FIGURE 19.3 | Discrete uniform distribution

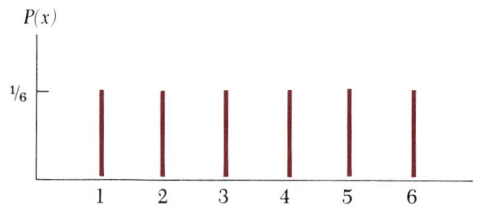

observed frequency distribution does not fit well to the discrete uniform distribution. In this sense we are testing the goodness-of-fit of the observed frequency distribution to a specified theoretical probability distribution in Example 19.1.

Goodness-of-fit

Our second observation is that this test is related to testing the hypothesis that the population proportions corresponding to the number of dots on the die faces are equal. Define π_i = population proportion of i's, or probability that the number of dots i appears face up. The hypotheses we have tested are equivalent to the hypotheses:

H_0: $\pi_1 = \pi_2 = \cdots = \pi_6$
H_A: At least one π_i is different

Population proportion test

When used in a situation like Example 19.1, the chi-square test can be considered as an *approximate* test of the equivalence of a set of population proportions. We would conclude in this example that the population proportions do not appear to be equal; the proportions of 3s and 4s appear to be greater than the other proportions.

A final note—the chi-square test must always be based on frequencies, not proportions. In Example 19.1 we could not compute the chi-square statistic as the squared difference between the observed and expected proportions divided by the expected proportion. If proportions were used, the resulting chi-square statistic would be too small (by a factor of $1/n$).

There is one requirement for the appropriate use of the chi-square tests— each class should have an expected frequency of at least five. This requirement is related to the fact that the chi-square statistic is approximately chi-square distributed. If the expected frequency becomes small (less than five) in one or more classes, the quality of the approximation suffers. Although the "at least five" minimum expected frequency is not a hard and fast rule (some statisticians use three, and a few use one as the minimum), we shall adopt it in this text. What if a class has less than five as an expected frequency? The solution, as we will see, is to group classes until the minimum of five is achieved.

Requirement for the use of the chi-square test

Each class should have an expected frequency of at least five. That is,

$$E_i \geq 5; i = 1, 2, \ldots, k$$

where k = number of classes.

Example 19.2   The number of service requests per hour in a plumber's shop is thought to be Poisson distributed with an average of three requests per hour. One hundred randomly selected hourly periods during a four-week interval produced the demands for service given in Table 19.3.

Let the random variable $\mathbf{X}$ = Number of service requests per hour. Since we are assuming that the average number of requests per hour is three, the Poisson parameter λ equals three, and the appropriate probability mass function is:

$$P(\mathbf{X} = x) = \frac{e^{-\lambda}\lambda^x}{x!} = \frac{e^{-3}(3)^x}{x!}, \qquad x = 0, 1, 2, \ldots$$

TABLE 19.3 | Number of requests for plumbing service per hour

Number of requests per hour	Number of hourly periods having the number of requests per hour
0	3
1	17
2	28
3	16
4	18
5	9
6	2
7	5
8	1
9	1
10	0
More than 10	0
Total	100

To determine the expected frequency in the first class, 0 requests per hour, we first compute the probability that $\mathbf{X} = 0$, using Table B.2 in Appendix B ($\lambda = 3$, $x = 0$):

$$P(\mathbf{X} = 0) = \frac{e^{-3}3^0}{0!} = e^{-3} = 0.0498$$

The expected number of hourly periods out of 100 sampled hours in which 0 service calls will be experienced is $E_1 = (100)(0.0498) = 4.98$ hourly periods.

Table 19.4 gives the computation of the value of the chi-square statistic

TABLE 19.4 | Computation of the chi-square statistic for Example 19.2

x	O_i	$P(\mathbf{X} = x)$	E_i	$O_i - E_i$	$\dfrac{(O_i - E_i)^2}{E_i}$
0	3 ⎫ 20	$e^{-3}3^0/0! = 0.0498$	4.98 ⎫ 19.92	0.08	0.000
1	17 ⎭	$e^{-3}3^1/1! = 0.1494$	14.94 ⎭		
2	28	$e^{-3}3^2/2! = 0.2241$	22.41	5.59	1.394
3	16	$e^{-3}3^3/3! = 0.2241$	22.41	−6.41	1.833
4	18	$e^{-3}3^4/4! = 0.1681$	16.81	1.19	0.084
5	9	$e^{-3}3^5/5! = 0.1009$	10.09	−1.09	0.118
6	2 ⎫	$e^{-3}3^6/6! = 0.0504$	5.04 ⎫		
7	5 ⎪	$e^{-3}3^7/7! = 0.0216$	2.16 ⎪		
8	1 ⎬ 9	$e^{-3}3^8/8! = 0.0081$	0.81 ⎬ 8.36	0.64	0.049
9	1 ⎪	$e^{-3}3^9/9! = 0.0027$	0.27 ⎪		
10	0 ⎪	$e^{-3}3^{10}/10! = 0.0007$	0.07 ⎪		
More than 10	0 ⎭	0.0001*	0.01 ⎭		
Total	100	1.000	100.00		$\chi^2 = 3.480$

* By subtraction, because the probabilities must total 1.

for these data. Notice that the first two classes and the last six classes must be combined to produce *expected* frequencies of five or more.

The degrees of freedom are $k - 1 - m = 6 - 1 - 0 = 5$; there are $k = 6$ classes actually used in the computation of the value of the chi-square statistic, and $m = 0$, since no parameters had to be estimated from the data (the value of the parameter λ needed to calculate the expected frequencies is given—$\lambda = 3$).

The hypotheses are

H_0 :   The data are Poisson-distributed with a mean of $\lambda = 3$ requests per hour.

H_A :   The data are not Poisson-distributed with a mean of $\lambda = 3$ requests per hour.

To test the null hypothesis at the $\alpha = 0.10$ level, the decision rule is:

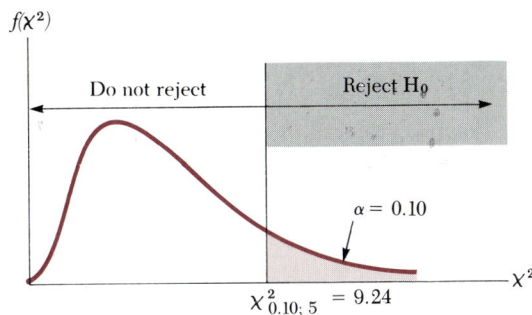

Since $\chi^2 = 3.48 < 9.24$, we cannot reject H_0 and must therefore conclude that the data conform reasonably well to a Poisson distribution with a mean of three.

From Table B.6 in Appendix B, since $P(\chi^2 < 1.61/df = 5) = 0.10$ and $P(\chi^2 > 9.24/df = 5) = 0.10$, and the table does not give values of χ^2 between 1.61 and 9.24, we can only say that the p-value is between 0.10 and 0.90. Since its value appears to be closer to 0.10, we can be quite confident that we have made a good decision to not reject H_0.

Note that the number of classes k used to determine the degrees of freedom is equal to the number of classes actually used to compute the chi-square statistic value (6), *not* the original number of classes (12).

In Example 19.2 the Poisson parameter λ is specified ($\lambda = 3$), but frequently this is not the case. We will wish to determine whether or not a set of data conforms to a theoretical probability distribution in which the parameters have not been specified. What if λ had not been specified in Example 19.2? How then could we compute the expected frequencies? One answer is to estimate the parameter from the data. For example, an estimate of the average number of service requests (λ) from the data in Table 19.3 is $302/100 = 3.02$. The estimate may be used in the computation of the probabilities as in Table 19.4 in place of the parameter, λ. The consequence of estimating λ is the loss of a degree of freedom—for every probability distribution parameter that must be estimated, one degree of freedom in the chi-square statistic is lost.

We now consider the goodness-of-fit test for normality, an extremely important and useful application of the chi-square statistic.

Example 19.3   A firm's reliability expert believes an electric component used in a switching device has a lifetime that is normally distributed. A sample of 500 component lifetimes provides the data in Table 19.5. The

TABLE 19.5 | Frequency distribution of 500 component lifetimes

Lifetime in hours	Number of components
40 but under 60	25
60 but under 80	50
80 but under 100	225
100 but under 120	125
120 but under 140	75
	500

reliability expert determines that the sample average and sample standard deviation of the 500 observations are $\bar{x} = 100$ and $s = 20$, respectively. Test the null hypothesis that these lifetimes are normally distributed.

Solution   The assumed form of the normal distribution is based on the estimates $\bar{x} - 100$ and $s - 20$, since the parameters μ and σ have not been specified. This normal distribution is illustrated in Figure 19.4. To find the expected frequency for, say, the class 40 but under 60, we would first find the probability that a component lifetime X lies between 40 and 60 hours. This probability is equal to the shaded area in Figure 19.4. By standardizing x, $z = (x - \bar{x})/s$, this area is equivalent to the shaded area in the standard normal distribution shown in Figure 19.4. From Table B.3 in Appendix B, this area is equal to 0.0215. Thus $P(40 \leq X \leq 60) \doteq 0.0215$ if the data are normally distributed (this probability is approximate because we used $\bar{x}$ for μ and s for σ). The expected frequency in this class is given by $(500)(0.0215) = 10.75$. The chi-square computations for these data are given in Table 19.6. Notice that two open-ended classes must be added to accommodate the tails of the normal distribution, though one is lost (the first) because its expected frequency is less than five. The hypotheses are:

H_0:   The data are normally distributed.
H_A: The data are not normally distributed.

The degrees of freedom are given by $k - 1 - m = 6 - 1 - 2 = 3$. Six classes are used to calculate the chi-square statistic value, and two degrees of freedom are lost due to the estimation of μ and σ by $\bar{x}$ and s, respectively. If we select the $\alpha = 0.01$ significance level, the decision rule is:

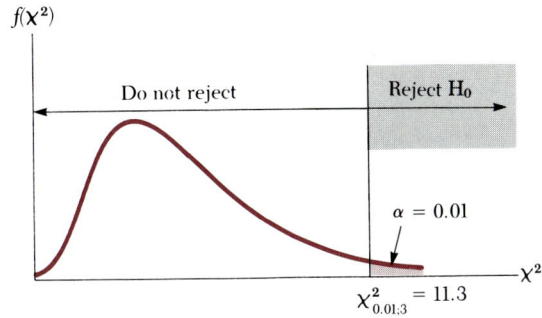

FIGURE 19.4 | Assumed theoretical distribution for component lifetimes

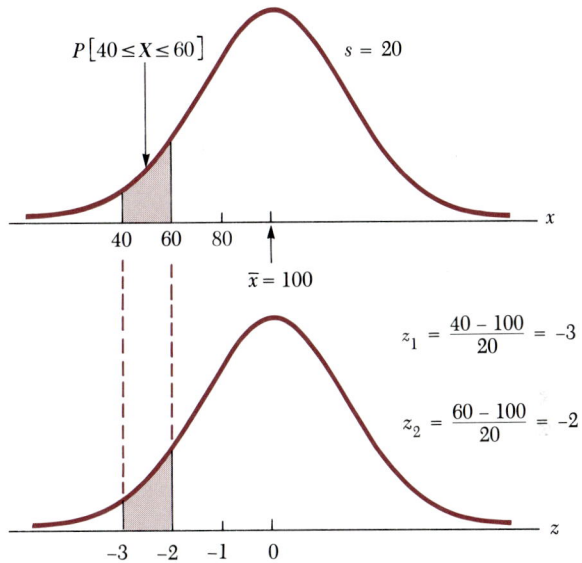

TABLE 19.6 | Chi-square computations for Example 19.3

Lifetime (in hours)	O_i	Probability	E_i	$O_i - E_i$	$\dfrac{(O_i - E_i)^2}{E_i}$
Less than 40	0 $\Big\}25$	$P(\mathbf{X} < 40) = 0.0013$	0.65 $\Big\}11.40$	13.60	16.225
40 but under 60	25	$P(40 \leq \mathbf{X} < 60) = 0.0215$	10.75		
60 but under 80	50	$P(60 \leq \mathbf{X} < 80) = 0.1359$	67.95	−17.95	4.742
80 but under 100	225	$P(80 \leq \mathbf{X} < 100) = 0.3413$	170.65	54.35	17.310
100 but under 120	125	$P(100 \leq \mathbf{X} < 120) = 0.3413$	170.65	−45.65	12.212
120 but under 140	75	$P(120 \leq \mathbf{X} < 140) = 0.1359$	67.95	7.05	0.731
Over 140	0	$P(\mathbf{X} \geq 140) = \underline{0.0228}$	$\underline{11.40}$	−11.40	$\underline{11.400}$
Total	500	1.0000	500.00		$\chi^2 = 62.620$

Since $\chi^2 = 62.62 > 11.3$, we would reject H_0 and conclude that the data are not normally distributed. This is evident from studying the frequency distribution of observed responses in Table 19.5—the data are definitely skewed toward the smaller lifetimes. Since $P(\chi^2 > 16.3/df = 3) = 0.001$ from Table B.6, we know that p-value $= P(\chi^2 > 62.62)$ is *much* less than 0.001. Therefore, we can be very confident that we have made the right decision in rejecting H_0.

The use of the chi-square statistic to test the goodness of fit of data to the normal distribution has one serious drawback—the data first must be grouped into a frequency distribution before the test can be conducted. Unfortunately, the chi-square statistic may be affected by how the frequency distribution is constructed (the number of classes, class widths, etc.) As a general rule, construct the frequency distribution with equal-width intervals and select a sufficient number of intervals to cover the range of the data (usually 5 to 10 will do), subject to the condition that each class has an expected frequency of five or more.

A further drawback with the chi-square statistic is its insensitivity to departures from normality in the tails of the distribution. Often, we must group classes in the left and right tails to meet the requirement that each class should have an expected frequency of five or more. By doing so, the chi-square loses power in its ability to detect departures from normality in the tails.

Chapter 20 describes alternative goodness-of-fit tests to the chi-square test. These tests do not require placing the data into frequency distributions, and typically they are more powerful than the chi-square test.

■ **19.3**

Tests for independence: Contingency tables

In Chapter 4 we introduced the concept of a *contingency table,* a table used to display the dependence of two or more variables on one another. We repeat the initial Chapter 4 contingency table example here.

Example 19.4   A survey of 100 patients at a local Veterans Administration hospital is conducted to determine whether or not there is a connection between smoking and lung cancer. Each patient is classified as a smoker or nonsmoker and as having lung cancer or not having lung cancer. The resulting frequencies are given in the contingency table.

Smoker \ Lung cancer	Yes	No	Totals
Yes	15	25	40
No	5	55	60
Totals	20	80	100

In Chapter 4 we showed that the events S (smoker) and C (lung cancer) are statistically dependent events; that is, $P(S \cap C) = 0.15 \neq P(S) \cdot P(C) =$

(0.4)(0.2) = 0.08. This result follows from viewing the 100 VA patients as a *population*. We now wish to consider these data as a *sample* from which we may draw inferences concerning the dependence of the two categories, smoking and cancer. Let π_{ij} be the population proportion in the (i, j)th cell. For example, π_{12} = proportion of persons who are smokers ($i = 1$; first row of the table) and who do not have lung cancer ($j = 2$; second column of the table). Let π_i^r and π_j^c represent the marginal probability distributions for the smoking category and the lung cancer category, respectively. The null hypothesis of independence is:

$$H_0: \quad \pi_{ij} = (\pi_i^r)(\pi_j^c); \quad i = 1, 2, \quad j = 1, 2$$

and the alternate hypothesis is:

$$H_A: \quad \pi_{ij} \neq (\pi_i^r)(\pi_j^c) \text{ for at least one pair } (i, j)$$

That is, if the two categories are independent in the population, then any cell probability (π_{ij}) should be the product of the ith row and jth column marginal probabilities ($\pi_i^r \pi_j^c$).

Let p_{ij} be the sample proportion in the (i, j)th cell, and p_i^r and p_j^c be the row and column sample proportions, respectively. Table 19.7 gives these sample proportions. Also shown in Table 19.7 are the sample expected cell

TABLE 19.7 | Sample proportion and the expected sample cell proportions (in parentheses) assuming independence

Smoker \\ Lung cancer	Yes	No	Totals
Yes	$p_{11} = 0.15$ ($p_1^r p_1^c = 0.08$)	$p_{12} = 0.25$ ($p_1^r p_2^c = 0.32$)	$p_1^r = 0.40$
No	$p_{21} = 0.05$ ($p_2^r p_1^c = 0.12$)	$p_{22} = 0.55$ ($p_2^r p_2^c = 0.48$)	$p_2^r = 0.60$
Totals	$p_1^c = 0.20$	$p_2^c = 0.80$	1.00

proportions if the two categories are independent in the population. The discrepancies between p_{ij} and $p_i^r p_j^c$ certainly suggest that the two categories are not statistically independent in the population.

The chi-square test for independence is based on the difference between the *observed frequencies* and the *expected frequencies* if the null hypothesis, $H_0: \pi_{ij} = (\pi_i^r)(\pi_j^c)$, is true. The fundamental idea of this test is therefore the same as the chi-square goodness-of-fit tests. Let O_{ij}, be the observed frequency and E_{ij} be the expected frequency in the (ij)th cell, respectively. The expected cell frequencies are given by

$$E_{ij} = np_i^r p_j^c$$

To find the expected frequency in the (ij)th cell, we multiply the marginal sample proportion in the ith row by the marginal sample proportion in the jth column—this gives the estimated cell proportion, assuming the categories are independent—and multiply this proportion by the sample size to produce a frequency. The observed and expected frequencies for these data are given in Table 19.8.

TABLE 19.8 | Observed and expected frequencies

Smoker \ Lung cancer	Yes	No	Totals
Yes	$O_{11} = 15$ $E_{11} = (0.4)(0.2)(100)$ $= 8$	$O_{12} = 25$ $E_{12} = (0.4)(0.8))100$ $= 32$	40
No	$O_{21} = 5$ $E_{21} = (0.6)(0.2)(100)$ $= 12$	$O_{22} = 55$ $E_{22} = (0.6)(0.8)(100)$ $= 48$	60
Totals	20	80	100

The value of the chi-square statistic is:

$$\chi^2 = \sum_{i=1}^{r} \sum_{j=1}^{c} \left[\frac{(O_{ij} - E_{ij})^2}{E_{ij}} \right]$$

where r and c are the numbers of rows and columns in the contingency table, respectively. The degrees of freedom for this chi-square statistic are $df = (r - 1)(c - 1)$. The degrees of freedom for the test are determined as follows:

$$df = (\text{Number of cells}) - 1 - (\text{Number of estimated parameters})$$

Degrees of freedom

This is similar to the way the degrees of freedom are determined for the chi-square goodness-of-fit test. One degree of freedom is lost in specifying the sample size. The number of estimated parameters is $(r + c - 2)$; we must estimate π_{ij} for each cell and its point estimate is $p_i^r p_j^c$ under the null hypothesis. Since $\sum_{i=1}^{r} p_i^r = 1$ for the sample proportions, only $(r - 1)$ must be determined—this follows because the sum of all row sample proportions must be 1. Similarly, only $(c - 1)$ column sample proportions must be determined. Therefore, the degrees of freedom are:

$$df = (rc) - 1 - [(r - 1) + (c - 1)] = (r - 1)(c - 1)$$

For these data

$$\chi^2 = \frac{(15 - 8)^2}{8} + \frac{(25 - 32)^2}{32} + \frac{(5 - 12)^2}{12} + \frac{(55 - 48)^2}{48}$$

$$= 6.125 + 1.531 + 4.083 + 1.021 = 12.76$$

From Table B.6 in Appendix B, $\chi^2_{\alpha=0.05:df=1} = 3.84$ with $\alpha = 0.05$, $df = (2 - 1)(2 - 1) = 1$. The decision rule is:

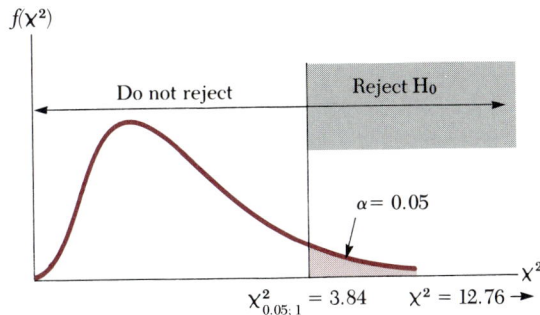

Since $\chi^2 = 12.76 > 3.84$, we reject the null hypothesis and conclude that in the population from which these data are drawn, there appears to be a significant statistical dependence between the two categories in the contingency table, smoking and lung cancer. Since $P(\chi^2 > 10.8/df = 1) = 0.001$ from Table B.6, we know that the p-value $= P(\chi^2 > 12.76/df = 1) < 0.001$. Therefore, we can be confident that we have made the right decision in rejecting H_0.

An important point to remember from our discussion of statistical dependence in Chapters 4 and 13 is that dependence does not imply cause and effect. These data suggest that in this population there appears to be a greater proportion of lung cancer victims among smokers than among non-smokers. This relationship may be caused, however, by some other factor, say a physiological one, that increases a person's chance of contracting lung cancer and of becoming a smoker.

An additional point—the expected cell frequencies can be determined directly from the row and column frequencies and the sample size. For example, the expected frequency in the (1, 1) cell may be determined by

> Statistical dependence does not imply cause and effect

$$(\cancel{100}) \left(\frac{40}{\cancel{100}} \right) \left(\frac{20}{100} \right) = \frac{(40)(20)}{100} = 8$$

That is, the expected frequency is determined by the product of the first-row and first-column marginal frequencies, divided by the sample size. We now summarize the chi-square test for independence for two categories.

Example 19.5 In a certain population, a psychologist wishes to determine whether or not salary is affected by level of education attained. Based on a random sample of 150 persons in this population, she forms the contingency table shown on the top of page 1001.

Summary of the chi-square test for independence, given a two-way contingency table

Category B

		1	2	$\cdots$	c	
	1	O_{11}	O_{12}	$\cdots$	O_{1c}	R_1
	2	O_{21}	O_{22}	$\cdots$	O_{2c}	R_2
Category A	$\cdot$	$\cdot$	$\cdot$		$\cdot$	$\cdot$
	$\cdot$	$\cdot$	$\cdot$		$\cdot$	$\cdot$
	$\cdot$	$\cdot$	$\cdot$		$\cdot$	$\cdot$
	r	O_{r1}	O_{r2}	$\cdots$	O_{rc}	R_r
		C_1	C_2	$\cdots$	C_c	T

where

T = total sample size
O_{ij} = observed frequency in (i, j)th cell
R_i = total of ith row
C_j = total of jth column
r = number of rows
c = number of columns

The value of the chi-square test statistic for the hypotheses

H_0: Category A is independent of category B.
H_A: Categories A and B are statistically dependent.

is

$$\chi^2 = \sum_{i=1}^{r} \sum_{j=1}^{c} \left[\frac{(O_{ij} - E_{ij})^2}{E_{ij}} \right]$$

where

$$E_{ij} = \frac{(R_i)(C_j)}{T}$$

$$df = (r - 1)(c - 1)$$

Educational level \ Salary (in $000s)	0–9.99	10–14.99	15–19.99	Above 20	Totals
High school education or less	10	10	14	16	50
High school education plus some college training	10	20	28	17	75
College degree or greater (some graduate work)	0	0	18	7	25
Totals	20	30	60	40	150

Are the two categories, salary and education level, statistically independent in this population?

Solution We first determine the expected frequencies.

Education category \ Salary category	1	2	3	4	Totals
1	$\frac{(50)(20)}{150} = 6.67$	$\frac{(50)(30)}{150} = 10$	$\frac{(50)(60)}{150} = 20$	$\frac{(50)(40)}{150} = 13.33$	$R_1 = 50$
2	$\frac{(75)(20)}{150} = 10$	$\frac{(75)(30)}{150} = 15$	$\frac{(75)(60)}{150} = 30$	$\frac{(75)(40)}{150} = 20$	$R_2 = 75$
3	$\frac{(25)(20)}{150} = 3.33$	$\frac{(25)(30)}{150} = 5$	$\frac{(25)(60)}{150} = 10$	$\frac{(25)(40)}{150} = 6.67$	$R_3 = 25$
Totals	$C_1 = 20$	$C_2 = 30$	$C_3 = 60$	$C_4 = 40$	$T = 150$

The value of the chi-square statistic is:

i	j	O_{ij}	E_{ij}	$O_{ij} - E_{ij}$	$(O_{ij} - E_{ij})^2/E_{ij}$
1	1	10	6.67	3.33	1.66
1	2	10	10.00	0.00	0.00
1	3	14	20.00	−6.00	1.80
1	4	16	13.33	2.67	0.53
2	1	10	10.00	0.00	0.00
2	2	20	15.00	5.00	1.67
2	3	28	30.00	−2.00	0.13
2	4	17	20.00	−3.00	0.45
3	1	0	3.33	−3.33	3.33
3	2	0	5.00	−5.00	5.00
3	3	18	10.00	8.00	6.40
3	4	7	6.67	0.33	0.02
					$\chi^2 = 20.99$

If we conduct the test at the $\alpha = 0.05$ level, the decision rule is:

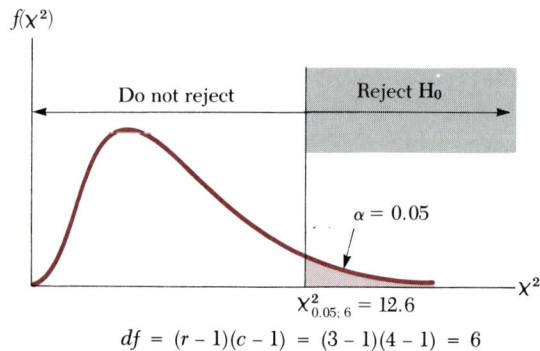

$$df = (r - 1)(c - 1) = (3 - 1)(4 - 1) = 6$$

Since $\chi^2 = 20.99 > 12.6$, we would reject H_0 and conclude that the salary of a person in this population is affected by his or her education level. Since $P(\chi^2 > 18.5/df = 6) = 0.005$ and $P(\chi^2 > 22.5/df = 6) = 0.001$, the p-value lies between 0.001 and 0.005 (by interpolation, p-value $= 0.002$). Thus, by rejecting H_0, we have incurred a small risk that we have made an error. It is interesting to note what is causing the high value of χ^2. In the computation of χ^2 above, notice that the largest contributions to the sum of 20.99 are due to the components 3.33 ($i = 3, j = 1$), 5.00 ($i = 3, j = 2$), and 6.40 ($i = 3, j = 3$). Thus individuals who have a college degree or greater and who fall in the lower three salary categories are contributing the most to the rejection of the null hypothesis. Note that those individuals with a high school education or less can make as much as college graduates, but the percentage who fall in the lower salary categories is higher. It is in this sense that a college education is often viewed as "insurance" (against falling in the lower salary ranges).

As with the chi-square goodness-of-fit test, there are two cautions that should be exercised in using the chi-square statistic for tests of independence. First, the expected cell frequencies should be greater than or equal to five. In Example 19.5, $E_{31} = 3.33$. To correct this problem, either we must increase the sample size with the hope that E_{31} will reach five or more, or we must change the classifications for the salary category. By extending the first class beyond 9.99 (thousand) and readjusting the subsequent classes as necessary, we can ensure that all $E_{ij} \geq 5$. The second caution is that this is an *approximate* test for independence. It is best suited when both factors (categories) are *qualitative*. When one or both factors are quantitative, the chi-square statistic does not exploit the numerical character of the factor. For instance, in Example 19.5 salary is quantitative, and its specific numerical values in the sample data have been lost by placing these values into four salary categories. Other techniques that directly use the numerical character of salary may be preferable to chi-square in this case. As an example, we could consider the average salary in the population μ_i for each education level $i = 1, 2, 3$, and test the hypothesis $H_0: \mu_1 = \mu_2 = \mu_3$ to determine whether the mean salary is affected by education level. The analysis of variance procedures in Chapter 12 would be well suited for this approach. To repeat the point, if one or both factors are quantitative, methods that exploit the numerical character of the factor(s) should be considered before using chi-square.

Finally, it is possible to develop exact tests for the 2×2 contingency table and to "correct" for the approximation in the general two-way table. The interested reader is directed to the Snedecor and Cochran reference at the end of this chapter for information on this technique and further uses of the chi-square statistic.

■ 19.4
Summary

The chi-square statistic lends itself to constructing *approximate* methods for two important statistical hypothesis tests: the test for goodness of fit of sample data to an assumed probability distribution and the test for independence of two factors. It is important to remember that these chi-square tests are approximate, and before using them it should be ascertained that "better" procedures, if they exist, are not feasible in a given problem.

These tests are commonly referred to as "nonparametric" or "distribution-free" statistical tests. Unlike the tests in Chapters 9–11, these do not assume that the population random variable is distributed according to any specified distribution, such as the normal distribution whose form is controlled by the parameters μ and σ. Rather, the only restriction is that the counts in each cell should be five or greater. Additional nonparametric tests are presented in Chapter 20.

References

Introductory

Anderson, D. R.; D. J. Sweeney; and T. A. Williams. *Statistics for Business and Economics.* 2nd ed. St. Paul, Minn.: West Publishing, 1984, Chapter 12.

Iman, R. L., and W. J. Conover. *Modern Business Statistics.* New York: John Wiley & Sons, 1983, Chapter 10.

Wonnacott, T. H., and R. J. Wonnacott. *Introductory Statistics for Business and Economics.* 3rd ed. New York: John Wiley & Sons, 1984, Chapter 17.

Advanced

Ostle, B., and R. W. Mensing. *Statistics in Research.* 3rd ed. Ames: The Iowa State University Press, 1975, Chapter 6.

Pearson, Karl. "On the Criterion that a Given System of Deviations from the Probable in the Case of a Correlated System of Variables Is Such that It Can Be Reasonably Supposed to Have Arisen from Random Sampling." *Philosophy Magazine,* series 5, vol. 50 (1900), p. 157.

Snedecor, G. W., and W. G. Cochran. *Statistical Methods.* 7th ed. Ames: Iowa State University Press, 1980, pp. 124–27; 194–210.

Problems

Section 19.2 Problems

19.1 Persons arriving at a service facility are believed to arrive randomly; that is, the number of arrivals in a fixed interval of time, say five minutes, should be uniformly distributed over a set of intervals. Five five-minute intervals are selected, and the following frequencies of the numbers of persons arriving in these intervals are recorded:

Five-minute interval	Number of arrivals
1	22
2	15
3	25
4	12
5	26
Total	100

Are the frequencies of arrivals uniformly distributed over the five intervals? Use $\alpha = 0.01$. What is the p-value for this test?

19.2 A set of four coins is tossed 1,000 times. The number of times that 0, 1, 2, 3, and 4 heads were obtained is given in the table.

Number of heads	Number of times
0	50
1	250
2	340
3	300
4	60
Total	1,000

Assume the random variable $\mathbf{X}$ = Number of heads when four coins are tossed follows the binomial distribution.

$$P(\mathbf{X} = x) = C_x^4 (1/2)^x (1 - 1/2)^{4-x},$$
$$x = 0, 1, 2, 3, 4$$

Do the data suggest that these four coins allow us to use this probability distribution for $\mathbf{X}$? Use $\alpha = 0.05$. What is the p-value for this test?

19.3 The marketing research firm for a large foods manufacturer is interested in determining whether there is any preference for the brand of vanilla ice cream purchased at the grocery store in a certain region. It is believed that among four brands, customers do not express a preference. That is, the four brands have approximately an equal proportion of this vanilla ice cream market. Samples of each vanilla ice cream are given to 100 randomly selected customers, and they are asked to identify the one they like best. The data are recorded in the table.

Brand of ice cream	Customers preferring
A	36
B	18
C	20
D	26
Total	100

Does it appear that customers in this vanilla ice cream market have no brand preference? Use $\alpha = 0.05$. What is the p-value for this test?

19.4 It is believed that the average number of service calls received at a service facility is two per every 15-minute interval. One hundred 15-minute intervals are sampled, and the frequencies of the numbers of calls are recorded in the table.

Number of calls	Number of intervals
0	22
1	50
2	15
3	10
4	2
5 or more	1
Total	100

Do these data fit to a Poisson distribution with a mean of two calls per each 15-minute interval? Calculate the p-value for this test. Should you reject or not reject H_0? Explain.

19.5 It is believed that the number of persons entering a store per hour is Poisson distributed. Two hundred hourly periods when the store is open are randomly chosen, and the number of persons entering the store in each period is recorded in the table.

Persons per hour	Hourly periods
0	10
1	30
2	60
3	40
4	20
5	18
6	10
7	2
8	5
9	4
10	1
11 or more	0
Total	200

Do these data support the contention that the random variable X = Number of persons entering the store per hour is Poisson distributed? (Note: Use the data to estimate the average number of persons entering the store per hour, λ.) Use $\alpha = 0.05$. What is the p-value for this test?

19.6 In the past year, a company had 850 persons employed on an hourly basis. A frequency distribution of their wages is given in the table.

Hourly wage	Frequency
$3.75–$3.99	62
4.00– 4.24	124
4.25– 4.49	267
4.50– 4.74	228
4.75– 4.99	106
5.00– 5.24	46
5.25– 5.49	12
5.50– 5.74	5
Total	850

The company statistician intends to use these data to draw inferences regarding the hourly wage scale for the company. The inference-making methods he plans to use assume that the variable is normally distributed. From the 850 values, he finds that the sample mean and standard deviation are \$4.50 and \$0.35, respectively. Do these data support the normality assumption, using the *estimated* mean and standard deviation? Calculate the *p*-value for this test. Should you reject or not reject H_0? Explain.

Score	Frequency
Below 300	1
300 but less than 350	5
350 but less than 400	8
400 but less than 450	20
450 but less than 500	30
500 but less than 550	25
550 but less than 600	8
Over 600	3
Total	100

19.7 The lifetimes in hours of 50 100-watt light bulbs in a quality control test are:

1,310	1,233	1,203	1,240	759
944	872	1,067	984	1,252
1,248	1,105	956	1,233	1,385
1,262	1,303	985	1,122	1,490
1,234	816	1,173	1,028	1,067
1,001	1,213	996	1,003	987
1,243	1,187	1,111	997	1,324
889	1,432	1,146	1,109	932
1,089	1,213	962	949	991
1,006	1,103	1,333	1,066	1,378

a. Group these data into five classes of equal width.

b. The lifetimes are supposed to have a mean $\mu = 1,100$ hours and a standard deviation $\sigma = 200$ hours. Using $\mu = 1,100$ and $\sigma = 200$, test the null hypothesis at the $\alpha = 0.10$ level that these data are normally distributed.

19.8 An aptitude test for graduate study in business is designed to have a mean test score of $\mu = 500$ with a standard deviation of $\sigma = 100$. It is assumed that test scores are normally distributed. A random sample of 100 recent test scores produced the frequency distribution shown in the table.

a. Test the hypothesis at the $\alpha = 0.05$ level that these data are normally distributed with a mean $\mu = 500$ and a standard deviation of $\sigma = 100$.

b. The mean and standard deviation of these data are $\bar{x} = 475$ and $s = 50$, respectively. Using the estimated mean and standard deviation, may we now conclude that these data are normally distributed? Use a 0.05 level of significance as in part *a*.

Section 19.3 Problems

19.9 Two shipment lots of manufactured items are randomly sampled to determine whether the proportion of defective items is different in the two lots. The data are shown in the matrix below. Using the chi-square contingency table statistic, test the hypothesis that the proportion of defectives in lot 1, denoted by π_1, is equal to π_2, the proportion of defectives in lot 2. Use $\alpha = 0.05$. What is the *p*-value for this test?

Lot \ Findings	Defectives	Nondefectives	Totals
1	11	64	75
2	14	36	50
Totals	25	100	125

19.10 In Problem 19.9, test the equivalence of π_1 and π_2 by constructing a 95 percent confidence interval on the difference $\pi_1 - \pi_2$, using the formula from Chapter 11:

$$p_1 - p_2 \pm z_{\alpha/2} \sqrt{\frac{p_1(1-p_1)}{n_1} + \frac{p_2(1-p_2)}{n_2}}$$

where $p_1 = {}^{11}/_{75}$, $p_2 = {}^{14}/_{50}$, $n_1 = 75$, and $n_2 = 50$. Compare your answer with the one determined in Problem 19.9.

19.11 Students selected at random from public and private high schools were given a standardized achievement test. The results were as shown in the matrix below.

a. Test the hypothesis at the $\alpha = 0.05$ level that an achievement test score is independent of the kind of school a student attends. What is the p-value for this test?

b. Can you suggest a better way to determine whether achievement scores differ between students at private and public schools?

School \ Test scores	0–300	301–400	401–500	501–600	601–1,000	Totals
Private	0	5	10	20	15	50
Public	5	15	30	35	15	100
Totals	5	20	40	55	30	150

19.12 Four kinds of automobiles (A, B, C, and D) were shown to 200 married couples. Husbands and wives were then asked separately to select their first choice among the four cars. The data are shown in the matrix.

Husband's choice \ Wife's choice	A	B	C	D	Totals
A	10	16	8	24	58
B	15	12	14	7	48
C	6	24	20	8	58
D	6	8	16	6	36
Totals	37	60	58	45	200

Are the two choices independent factors? Calculate the p-value for this test. Should you reject or not reject the null hypothesis? Explain.

19.13 A consulting firm is asked to determine whether the preference for one of three weekly news magazines (A, B, and C) is dependent on the region in the United States

(South, West, North, East). Two hundred and fifty persons randomly sampled from each of these four geographic regions were asked to state their first choice among the three magazines. The data are shown in the accompanying matrix. Do these data suggest that the preference is dependent on geographic region? Calculate the p-value for this test. Should you reject or not reject H_0? Explain.

Region \ Magazine	A	B	C	Totals
North	150	60	40	250
South	100	100	50	250
East	150	50	50	250
West	125	50	75	250
Totals	525	260	215	1,000

19.14 A company is concerned that the length of its training program has no influence on later job ratings of its production-line workers. One hundred fifty new workers are chosen and are randomly divided into three groups of 50 each. The first group receives one week of training, the second group receives two weeks, and the third group receives three weeks. Three months later, all 150 workers are rated on a three-point scale, "Excellent," "Average," and "Poor." The data are shown in the accompanying matrix. Do the ratings and amount of training appear to be statistically independent? Use $\alpha = 0.05$.

Training \ Ratings	Excellent	Average	Poor	Totals
One week	12	20	18	50
Two weeks	18	25	7	50
Three weeks	20	20	10	50
Totals	50	65	35	150

19.15 A manufacturer produces units of a product in three shifts: day, evening, and night. Quality control teams check the production lots for defects at the end of each shift. The teams have compiled the data in the matrix shown.

Do these data suggest that the number of defects produced is independent of the shift when the units are produced? Use $\alpha = 0.05$. What is the p-value for this test?

Shift \ Defects	Major	Minor	None	Totals
Day	25	45	330	400
Evening	30	30	240	300
Night	40	40	220	300
Totals	95	115	790	1,000

19.16 It is suspected that the number of accidents per year is dependent on the age of drivers. One thousand drivers who are insured by a major insurance company are randomly sampled. The contingency table at the top of page 1009 is formed. At the $\alpha = 0.05$ level of signifi-cance, is age independent of the number of annual accidents? (Although two cells have expected frequencies less than five, do not regroup the categories—use the expected frequencies that are less than five.) What is the p-value for this test?

Age Number of accidents	16–21	22–30	31–40	41–50	51–70	Totals
0	160	250	180	120	30	740
1	50	40	20	25	5	140
2 or more	40	30	10	25	15	120
Totals	250	320	210	170	50	1,000

Additional Problems

19.17 In a famous experiment by the geneticist Mendel, when two types of peas are crossed, the following types should occur in the proportion 9:3:3:1—round and yellow, round and green, wrinkled and yellow, and wrinkled and green. In a cross producing 1,000 peas, the following frequencies were observed:

Type	Frequency
Round and yellow	575
Round and green	193
Wrinkled and yellow	182
Wrinkled and green	50
Total	1,000

At the $\alpha = 0.05$ significance level, do these data appear to follow Mendel's proportions?

19.18 In an introductory statistics course, 40 students took the first exam. The scores are given below:

```
100  44  63  45  72  82  78  75
 72  88  82  91  81  78  76  70
 66  52  58  81  73  71  66  60
 67  59  48  92  84  78  58  66
 84  79  68  76  72  75  64  80
```

Are these scores normally distributed? Use a 0.05 level of significance for this test.

19.19 The Poor Man's Law Firm sends out its bills on the last day of each month. The number of days that pass until the payment is received is recorded. For the 50 bills sent out in a recent month, the number of days until payment are:

```
10  32  36  22  24  21  18  45  62  12
25  30  28  36  64  25  21  18  16  33
36  28  34  41  31  51  48  34  37  10
28  26  35  44  48  24  22  21  18  50
42  44  36  30  28  30  32  27  31  26
```

Place these 50 observations in five classes, and test the hypothesis that these data are normally distributed at the 0.05 level of significance.

19.20 Ralph gets involved in a dice game in his college dorm. After playing for one hour and losing money consistently, it dawns on Ralph that perhaps the dice are not fair. He quits playing, and after the end of the game, asks to see the dice. He flips each die 120 times and records the following frequencies:

Die #1		Die #2	
Number on top face	Frequency	Number on top face	Frequency
1	22	1	12
2	18	2	14
3	23	3	34
4	21	4	36
5	16	5	10
6	20	6	14
Total	120	Total	120

Using a level of significance of 0.05, use the chi-square test to determine if each of these die pass the test of fairness (a fair die is one in which each top face has an equal chance of occurring on each toss).

19.21 The Brilliant Soap Company is investigating the use of three colors for its bars of bath soap. The Company is interested in determining if consumers express any preference regarding the three selected colors. In a random sample of 100 possible consumers, the following frequencies are recorded (each individual in the test picks the color that he or she prefers):

Blue	Yellow	White
50	10	40

Test the hypothesis that the consumers express no color preference for the bars of bath soap. Use a 0.05 level of significance. [Hint: If the null hypothesis is true, then the expected frequencies should all be equal.]

19.22 The Rola-Cola Company is interested in determining if consumers prefer Rola-Cola to other brands of cola. In a random sample of 500 possible consumers of cola, the following frequencies were recorded indicating which cola was preferred:

Rola-Cola	Brand A	Brand B	Brand C
100	150	120	130

Use the chi-square goodness-of-fit test to determine whether or not consumers express a preference among these four colas. Use a 0.05 level of significance for this test. What about Rola-Cola's hope that consumers prefer their cola over the other three brands?

19.23 The Rola-Cola Company is contemplating a change in its original formula for its cola. The new formula will be sweeter and smoother. The company decides to survey potential consumers to see if men and women behave differently in their preference for the old cola or the new formula. One hundred potential consumers are surveyed, and the results are shown in the contingency table below.

Preference \ Gender	Male	Female	Total
Old cola	28	22	50
New cola	16	34	50
Total	44	56	100

Using the chi-square test of independence at the 0.05 level of significance, is the preference for the new cola or the old cola dependent on gender?

19.24 The Hercules Company receives from three suppliers components for a microcomputer board that the company produces. The company is very concerned about one particular component that appears to be defective more often than others. It decides to take a random sample of 100 components from each supplier and categorize each component as being "good," needing "minor repair," or needing "major repair." The contingency table at the top of page 1011 summarizes the findings:

Supplier \ Component condition	Good	Minor repair	Major repair	Total
Alpha	85	10	5	100
NAC	75	10	15	100
Nipon	70	10	20	100
Total	230	30	40	300

Using the chi-square test of independence at the 0.05 level of significance, do the two categories, "component condition" and "supplier," appear to be independent?

20

Nonparametric statistics

■ 20.1

Introduction

Parametric statistics

Nonparametric methods

In Chapters 9–15, the statistical inference methods described all have a common thread—a population parameter (μ, σ^2, or ρ, for instance) is identified about which we wish to draw inferences, a random sample is collected, a point estimator of the parameter ($\overline{\mathbf{X}}$, $\mathbf{S}^2$, or $\mathbf{r}$) is selected, and its sampling distribution is used to construct hypothesis-testing decision rules or confidence interval formulas. These methods commonly are called *parametric* statistics, for they require the identification of a population parameter, a point estimator, and its sampling distribution. Most of the parametric statistical methods we have described in this text depend on knowing the form of the sampling distribution of the point estimator of the parameter to be estimated. That is, the *parameters* of the sampling distribution must be specified, usually by producing estimates of the parameters from the sample. In fact, we usually assume that the sampling distribution is the normal distribution, or at least approximately so (via the central limit theorem).

In this chapter, we will describe a number of so-called *nonparametric* methods—methods that may be appropriate when one or more of the parametric method (see Chapters 9–15) assumptions are not satisfied by the data. Nonparametric methods can therefore be used when their parametric counterparts cannot, due to the failure of one or more of the parametric method assumptions to hold. However, when the assumptions do hold for a parametric test (such as the *t*-test), then the corresponding nonparametric method will not have as much power (ability to detect differences from hypothesized values) as the parametric method.

The determination of the p-value for tests of hypotheses using these methods can be extremely difficult due to the nature of the significance tables. Thus the classical method of testing will be used, in which the significance level α is preset, thus determining the decision rule for the test. In most instances, we will give only a bound on the p-value (e.g., p-value < 0.001) because of the difficulties associated with interpolation in the nonparametric test tables.

To properly characterize the circumstances for which nonparametric methods may be appropriate, we first describe the four basic scales of measurement.

■ 20.2

Scales of measurement

Nominal

Ordinal

Interval

Ratio

Values of a random variable may belong to any of four measurement scales: *nominal*, *ordinal*, *interval*, or *ratio*. We will describe these scales in turn, from the "weakest" (the nominal scale) to the "strongest" (the ratio scale).

The *nominal* scale uses numbers only to name categories to which the observations belong. For example, consider the qualitative variable, sex. We may use a 1 for male and a 0 for female, but clearly the number assignment is arbitrary, for the numbers serve only as category names. We could just as well have used 100 to name the female category and 0 to name the male category.

The *ordinal* scale uses numbers for measurements, where the ordering of the numbers is relevant. For example, we may have designed a survey question that asks the respondent to indicate a preference among three brands of ice cream, where 1 indicates most preferred, and 3 indicates least preferred. The ordering of the three numbers (1, 2, and 3) is now relevant, but their magnitude is not. We could have used any three numbers, say 1, 50, and 100, as long as the ordering of the numbers reflects relative preference for the brands of ice cream.

The *interval* scale takes into account the difference between measurements as well as their ordering. An interval scale requires fixing an arbitrary 0 point and a unit distance to measure the difference between measurements. Good examples of interval scales are the Fahrenheit and Celsius temperature scales. These two scales have different 0 points and different unit distances. In general, one interval scale may be transformed to another by changing its scale or the location of 0 or both.

The *ratio* scale applies when the order and distance between measurements are important as in the interval scale, but the scale further requires that the ratio between two measurements is important. For example, if the ratio of one measurement to another is two, then the first is twice as "big" as the second. This requires the ratio scale to have a fixed and *natural* 0 point. Height is a measurement that is properly made on the ratio scale, for example. If we measure height in inches, 0 inches is the natural 0 point, and a person who is 60 inches tall is three times as tall as a person who is 20 inches tall. The unit distance on the ratio scale is arbitrary. For heights, we could use one inch or one foot, for example.

To determine to which scale a set of measurements belongs, we must take into consideration the nature of the quantities being measured and how the measurements are taken. Are we simply naming categories of a qualitative variable with numbers, or do the numbers themselves have meaning, as in their order, distance among measurements, and/or their ratios?

Most parametric statistical methods require that the measurements belong to at least the interval scale. Based on the four scales of measurement, we now define nonparametric statistical methods as in Conover.[1]

Definition 20.1

Nonparametric statistical methods

A statistical method is called *nonparametric* if it satisfies at least one of the following conditions:

1. It is appropriate for nominally measured data.
2. It is appropriate for ordinally measured data.
3. It is appropriate for interval or ratio scale data, but the population distribution function of the random variable from which the data are taken is unspecified.

Definition of nonparametric methods

The set of statistical procedures used for condition 3 situations in Definition 20.1 have frequently been called "distribution free" statistical methods—methods that do not depend on the specifications of the probability distribution of the population random variable. Consistent with Definition 20.1, we shall call these *nonparametric methods*.

Nonparametric methods have a number of advantages over parametric methods. First, nonparametric statistics are generally easy to compute, although describing the statistics is often somewhat involved. Second, nonparametric statistics can be applied to data in situations where parametric statistics may not be applicable. This is usually the case when the data measurement scale is nominal or ordinal. Third, nonparametric methods do not assume that the population random variable has a specific probability distribution form. These methods are based on sampling distributions, but the form of the sampling distribution is not dictated by assuming a form for the population probability distribution. Fourth, if a nonparametric method applies to a "weak" scale of measurement, it will typically also apply to all "stronger" scales. For example, if a method is applicable for data measured on the ordinal scale, then it also applies to data measured on the interval and ratio scales.

Nonparametric methods are at a disadvantage compared with parametric procedures when a data set satisfies the assumption of a parametric method.

---

[1] W. J. Conover, *Practical Nonparametric Statistics*, 2nd ed. (New York: John Wiley & Sons, 1980).

In these cases the *power* of nonparametric test methods is almost always less than the power of the appropriate parametric test. Therefore, as a general rule, if a parametric test is appropriate for the required statistical analysis, it should be used over any nonparametric method that may also be appropriate.

We will now consider some of the more important and frequently used nonparametric methods.

■ **20.3**

One sample: The median test

The median test can be used to test the hypothesis that a set of *n* randomly drawn measurements came from a population with a specified median. If it is known that the distribution of the population random variable **X** is symmetric, the test on the median is equivalent to a test on the population mean because the mean is equal to the median in a symmetric distribution. The median test is the nonparametric analog of the *t*-test described in Chapter 10. However, the *t*-test requires that the sample average $\overline{\mathbf{X}}$ is *normally distributed*—the median test has no such requirement.

Median test

Definition 20.2

Median test

Assumptions:

The data consist of *n* measurements $\mathbf{X}_1, \mathbf{X}_2, \ldots, \mathbf{X}_n$. Let $\mathbf{D}_i$ denote the difference between $\mathbf{X}_i$ and the hypothesized median M. Then

1. Each $\mathbf{D}_i$ must be a continuous random variable.
2. The distribution of each $\mathbf{D}_i$ must be symmetric.
3. The measurements $\mathbf{X}_1, \mathbf{X}_2, \ldots, \mathbf{X}_n$ must represent a random sample from the population distribution.
4. The measurement scale for **X** must be at least interval.

Hypotheses:

H_0:   The population median is M.
H_A: The population median is not M.

Test statistic value:

Determine the differences $d_i = x_i - M$, $i = 1, 2, \ldots, n$. If any $d_i = 0$, drop it from the set and decrease *n* by one. Rank the absolute values, $|d_i|$. If ties occur among the ranks, average the ranks of the items involved in the tie, and use the average as the rank of each tied item. Each rank is suffixed with the sign of the difference corresponding to it. Let r^+ be the total of the positive ranks, and let r^- be the total of the negative ranks. The value of the test statistic is the smaller of r^+ and r^-; let this number be *r*.

Decision rule:

Reject H_0 at the α level of significance if *r* exceeds $W_{1-\alpha/2}$ or *r* is less than $W_{\alpha/2}$, where $W_{\alpha/2}$ is given in Table B.12 in Appendix B and $W_{1-\alpha/2} = [n(n + 1)/2] - W_{\alpha/2}$. Otherwise, do not reject H_0.

Example 20.1 Ten randomly selected cars of a specific year, make, and model and with similar equipment are subjected to an EPA gasoline mileage test. The resulting miles per gallon are: 24.6, 30.0, 28.2, 27.4, 26.8, 23.9, 22.2, 26.4, 32.6, and 28.8. Using the median test, test the hypothesis at the $\alpha = 0.10$ level that the population median is 30 miles per gallon.

Solution The measurements, d_i, $|d_i|$, and the ranks are given in the following table.

Measurement	d_i	$|d_i|$	Rank	r^+	r^-
24.6	−5.4	5.4	7		7
30.0	0.0	0.0			
28.2	−1.8	1.8	2		2
27.4	−2.6	2.6	3.5		3.5
26.8	−3.2	3.2	5		5
23.9	−6.1	6.1	8		8
22.2	−7.8	7.8	9		9
26.4	−3.6	3.6	6		6
32.6	2.6	2.6	3.5	3.5	
28.8	−1.2	1.2	1		1
			Total:	3.5	41.5

$$\text{Eliminate } d_2 = 0$$
$$r^+ = 3.5; \ r^- = 41.5$$

Since r^+ is smaller than r^-, $r = 3.5$ is the value of the test statistic. From Table B.12 $W_{\alpha/2} = W_{0.05} = 9$ with $n = 9$ (eliminating observation 2 because $d_2 = 0$) and $W_{1-\alpha/2} = W_{0.95} = 36$. Since $r < 9$, reject the null hypothesis that the population median is 30 miles per gallon. Since the value of the test statistic ($r = 3.5$) lies between $W_{0.005}$ and $W_{0.01}$, we can conclude that the p-value for the test is between $2(0.005) = 0.01$ and $2(0.01) = 0.02$. (Recall that we must double the lower-tail probabilities to get the p-value for a two-sided test.)

Notice the treatment of the tied ranks for $|d_4| = |d_9| = 2.6$. These two absolute differences tie for ranks 3 and 4; thus each is assigned the average rank of 3.5. Also note that the first assumption of the test does not allow for ties. Since the d_i values must be continuous, ties should not occur. If the d_i values are not continuous, such as in Example 20.1, we can use the test as an approximate test by dealing with the ties in the manner described above.

Two observations are important regarding Example 20.1. First, if we can assume that the distribution of miles per gallon is symmetric, then the median test is equivalent to testing the hypothesis that the mean miles per gallon is 30. Second, these data have been measured on the ratio scale, and the t-test given in Chapter 10 is appropriate if we assume that the distribution of miles per gallon is symmetric and normal. The first problem at the end of this chapter will ask you to test the hypothesis in Example 20.1 by using the t-test.

With ties, the test is approximate

■ 20.4

Two independent samples: The Mann-Whitney test

The Mann-Whitney test is the nonparametric analogue of the two independent sample t-test presented in Chapter 11. The two independent sample t-test in Chapter 11 requires that the difference between the two sample means is *normally distributed*—the Mann-Whitney test does not require this assumption.

Mann-Whitney test

Definition 20.3

Mann-Whitney test

Assumptions:
1. $\mathbf{X}_1, \mathbf{X}_2, \ldots, \mathbf{X}_n$ and $\mathbf{Y}_1, \mathbf{Y}_2, \ldots, \mathbf{Y}_m$ are two independent random samples of sizes n and m, respectively.
2. Both samples consist of continuous random variables.
3. The measurement scale is at least ordinal.

Hypotheses:
H_0: The two random samples have been drawn from the same population distribution.
H_A: The two random samples have been drawn from different population distributions.

Test statistic value:
Assign ranks 1 to $m + n$ to the combined sample. That is, order all $(m + n)$ observations from the smallest to the largest, and assign ranks 1, 2, . . . , $(m + n)$ to these $(m + n)$ ordered observations. Let $r(x_i)$ and $r(y_j)$ denote the ranks assigned to the $x_i, i = 1, 2, \ldots, n$, and to the $y_j, j = 1, 2, \ldots, m$. Let $s = \sum_{i=1}^{n} r(x_i)$. The test statistic value is:

$$t = s - \frac{n(n + 1)}{2}$$

Decision rule:
Reject H_0 with significance level α if t is less than $W_{\alpha/2}$ or t is greater than $W_{1-\alpha/2}$, where $W_{\alpha/2}$ is given in Table B.11 in Appendix B and $W_{1-\alpha/2} = nm - W_{\alpha/2}$. Otherwise, do not reject H_0.

The Mann-Whitney test is designed to determine whether two random samples have been drawn from the same or different populations. If we assume that any difference between the two population distributions is due only to the difference in location of the two distributions, then the Mann-Whitney test is equivalent to testing whether or not the two population means are equal. This is similar to the two independent sample t-test, where we must assume that the two population variances are equal.

The Mann-Whitney test is based on the notion that if the two independent random samples have been drawn from the same population, then the average of the sample ranks $r(x_i)$ and $r(y_j)$ should be approximately equal. If the average of the $r(x_i)$ is much greater or smaller than the average of the $r(y_j)$, then this indicates that the two samples likely came from different populations.

Example 20.2 New employees of the ABC Corporation are given a training program to acquaint them with business procedures and principles. Two groups of 10 each are selected randomly from a large set of new employees. The first group is trained using method A, and the second group is trained using method B. At the end of the training period, each group is given the same test to determine how much information has been assimilated. The data are given in the following table.

Method A		Method B	
55	81	50	88
70	72	91	84
70	58	90	78
65	67	62	82
62	50	75	80

At the $\alpha = 0.05$ level, do these samples come from the same population?

Solution The ordered combined sample with ranks is shown in the table.

Measurement	Rank	Measurement	Rank
50 50	1.5	72	11
		75	12
55	3	78	13
58	4	80	14
62 62	5.5	81	15
		82	16
65	7	84	17
67	8	88	18
70 70	9.5	90	19
		91	20

The shaded observations are from method A. Notice the treatment of ties. If two observations tie, the rank given to each is the average of the two relevant ranks. If three numbers tie, say for ranks 8, 9, and 10, then each number is assigned the rank of 9—the average of the three ranks. Now

$$s = \sum_{i=1}^{10} r(x_i) = 1.5 + 3 + 4 + 5.5 + 7 + 8 + 9.5 + 9.5 + 11 + 15 = 74$$

$$t = s - \frac{n(n+1)}{2} = 74 - \frac{(10)(11)}{2} = 19$$

From Table B.11 in Appendix B, $W_{0.025} = 24$ and $W_{0.975} = (10)(10) - 24 = 76$. Since $t = 19 < W_{0.025} = 24$, we reject the null hypothesis that the two samples come from the same population. Since $t = 19$ falls between $W_{0.005}$ and $W_{0.01}$, we can conclude that the p-value lies between $2(0.005) = 0.010$ and $2(0.01) = 0.02$.

If we can assume that the only difference between the two populations is due to location alone, then we may conclude based on the Mann-Whitney test that the two population means differ. Is that a reasonable assumption based on these sample data (see Problem 20.4 at the end of this chapter)?

■ 20.5

Two matched samples: The Wilcoxon signed rank test

The Wilcoxon signed rank test is the nonparametric analogue of the parametric paired t-test for matched samples. The parametric paired t-test described in Chapter 10 requires that the differences between the pairs of observations $(\mathbf{X}_i, \mathbf{Y}_i)$ are *normally distributed*—the Wilcoxon signed rank test does not have this requirement.

Wilcoxon signed rank test

Definition 20.4
Wilcoxon signed rank test

Assumptions:
The data consist of n matched pairs, $(\mathbf{X}_i, \mathbf{Y}_i)$. Let $\mathbf{D}_i$ denote the difference between $\mathbf{X}_i$ and $\mathbf{Y}_i$ for the ith pair. Then
1. Each $\mathbf{D}_i$ must be a continuous random variable.
2. The distribution of each $\mathbf{D}_i$ must be symmetric.
3. The pairs, $(\mathbf{X}_i, \mathbf{Y}_i)$, $i = 1, 2, \ldots , n$, represent a random sample from a bivariate distribution.
4. The measurement scale for the $\mathbf{X}$ and $\mathbf{Y}$ values is at least interval.

Hypotheses:
H_0: $\mu_x = \mu_y$
H_A: $\mu_x \neq \mu_y$

Test statistic value:
Determine the differences $d_i = x_i - y_i$, $i = 1, 2, \ldots , n$. If any $d_i = 0$, drop it from the set and decrease n by one. Rank the absolute values, $|d_i|$. If ties occur, average the ranks of the items involved in the tie, and use the average as the rank of each tied item. Each rank is suffixed with the sign of the difference d_i corresponding to it. Let r^+ be the total of the positive ranks and r^- be the total of the negative ranks. The test statistic value is the smaller of r^+ and r^-; call this number r.

Decision rule:
Reject H_0 at the α significance level if r exceeds $W_{1-\alpha/2}$ or r is less than $W_{\alpha/2}$, where $W_{\alpha/2}$ is given in Table B.12 in Appendix B and $W_{1-\alpha/2} = [n(n + 1)/2] - W_{\alpha/2}$. Otherwise, do not reject H_0.

Example 20.3   Ten employees of a company are randomly selected to determine whether or not a speed reading program can improve their reading rates. The individuals are given standardized speed reading exams before and after the program. The data in words per minute are shown in the table.

Individual	Before	After
1	200	225
2	150	375
3	300	650
4	400	380
5	120	100
6	250	250
7	320	410
8	175	180
9	100	130
10	500	600

At the $\alpha = 0.05$ significance level, test the null hypothesis that the two population means are equal, using the Wilcoxon signed rank test.

Solution   The differences, ranks of $|d_i|$, r^+, r^-, and r are given in the table.

Individual	d_i	Rank of $|d_i|$	r^+	r^-
1	-25	4		4
2	-225	8		8
3	-350	9		9
4	$+20$ }	2.5	2.5	
5	$+20$ }	2.5	2.5	
6	0			
7	-90	6		6
8	-5	1		1
9	-30	5		5
10	-100	7		7
		Totals:	$r = 5.0$	40

Eliminate: $d_6 = 0$, $r^+ = 2.5 + 2.5 = 5.0$, and $r^- = 40.0$; $r = 5.0$

From Table B.12 in Appendix B, $W_{0.025} = 6$ with $n = 9$. $W_{0.975} = [n(n + 1)/2] - W_{0.025} = [9(10)/2] - 6 = 39$. Since $r < W_{0.025}$, reject H_0; the speed reading program appears to be effective. Since $r = 5$ is between $W_{0.01}$ and $W_{0.025}$, we may conclude that the p-value lies between $2(0.01) = 0.02$ and $2(0.025) = 0.05$.

■ 20.6

Several independent samples: The Kruskal-Wallis test

The Kruskal-Wallis test is an extension of the Mann-Whitney test when there are more than two populations. It is the nonparametric analogue of the parametric single-factor, completely randomized analysis of variance design discussed in Chapter 12. The data for the Kruskal-Wallis test must be in the following form:

Sample 1	Sample 2	· · ·	Sample k
x_{11}	x_{21}		x_{k1}
x_{12}	x_{22}		x_{k2}
.	.		.
.	.		.
.	.		.
x_{1,n_1}	x_{2,n_2}		x_{k,n_k}

The total number of observations is given by $N = \Sigma_{i=1}^{k} n_i$. The test depends on ranks and is similar to the Mann-Whitney test. Assign ranks 1 to N to all N observations when they have been ordered from the smallest to the largest, disregarding from which of the k samples the observations came. Let r_i be the sum of the ranks assigned to the ith sample.

$$r_i = \sum_{j=1}^{n_i} r(x_{ij}); \qquad \bar{r}_i = \frac{r_i}{n_i}$$

where $r(x_{ij})$ = Rank assigned to x_{ij}. If the $\bar{r}_i$ values are approximately the same, it supports the null hypothesis that the k samples came from the same population. If the $\bar{r}_i$ values are not the same, then it indicates that one or more populations are likely composed of different values than the rest.

If they occur, ties are treated as in the Mann-Whitney test.

The Kruskal-Wallis test does not require that the sample means in each of the k samples are normally distributed as is the casé with the analysis of variance method described in Chapter 12. Therefore we can think of the Kruskal-Wallis test as relaxing one of the assumptions (normality) of the analysis of variance procedure.

We will work an example for which Table B.13 in Appendix B is appropriate. The chi-square approximation for the critical **T**-value appears to be good even if k and n_i are only slightly larger than 3 and 5, respectively. If, for example, $k = 6$ and we set $\alpha = 0.05$, the approximate critical **T**-value is $\chi^2_{\alpha=0.05:k-1=5} = 11.1$, from Table B.6 in Appendix B. In this case, if $t > 11.1$, we reject the null hypothesis.

If the population distributions differ, but only in location, then the null hypothesis in the Kruskal-Wallis test is equivalent to testing the equality of the k population means, $\mu_1, \mu_2, \ldots, \mu_k$.

Example 20.4 A manager wishes to study the production output of three machines, A, B, and C. The hourly output of each machine is measured for five randomly selected hours of operation.

Observation / Machine	A	B	C
1	25	18	26
2	22	23	28
3	31	21	24
4	26	*	25
5	20	24	32

* Observation lost due to machine failure.

Kruskal-Wallis test

Definition 20.5

Kruskal-Wallis test

Assumptions:
1. The k random samples are mutually independent.
2. All random variables X_{ij} are continuous.
3. The measurement scale is at least ordinal.

Hypotheses:
H_0: The k population distributions are equal.
H_A: At least one population tends to yield different observations than the rest.

Test statistic value:

$$t = \frac{12}{N(N + 1)} \sum_{i=1}^{k} \frac{[r_i - (\frac{1}{2})n_i(N + 1)]^2}{n_i}$$

where

$$n_i = i\text{th sample size}, \quad N = \sum_{i=1}^{k} n_i$$

$$r_i = \sum_{j=1}^{n_i} r(x_{ij}), \text{ where } i = 1, 2, \ldots, k$$

$$r(x_{ij}) = \text{Rank assigned to observation } x_{ij}$$

Decision rule:
Table B.13 in Appendix B gives critical **T**-values at exact significance levels α for $k = 3$ and samples up to and including a sample of size five. If $k > 3$ and/or $n_i > 5$ for at least one sample, the χ^2 distribution with $df = k - 1$ may be used to find the approximate critical **T**-value. Reject the null hypothesis if t is greater than the critical value in Table B.13 or from the χ^2 distribution. Otherwise, do not reject H_0. Note that this is a one-sided test.

Test the null hypothesis at the $\alpha = 0.05$ significance level that the three population distributions are equal by using the Kruskal-Wallis test.

Solution The ordered data are shown in the table.

Rank	A	B	C
1		18	
2	20		
3		21	
4	22		
5		23	
6.5		24	24
8.5	25		25
10.5	26		26
12			28
13	31		
14			32

$r_1 = 2 + 4 + 8.5 + 10.5 + 13 = 38$
$r_2 = 1 + 3 + 5 + 6.5 = 15.5$
$r_3 = 6.5 + 8.5 + 10.5 + 12 + 14 = 51.5$

Thus

$$t = \frac{12}{14(15)} \left\{ \frac{[38 - (1/2)(5)(15)]^2}{5} + \frac{[15.5 - (1/2)(4)(15)]^2}{4} + \frac{[51.5 - (1/2)(5)(15)]^2}{5} \right\}$$

$$= (0.057)[(0.05) + (52.5625) + (39.2)] = 5.233$$

From Table B.13 in Appendix B, the critical **T**-value is 5.6429. Notice that this value corresponds to $n_1 = 5$, $n_2 = 5$, and $n_3 = 4$ in the table. But the order of the sample sizes does not affect the critical value. Since $t = 5.233$ is not greater than 5.6429, we do not reject the null hypothesis that the three population distributions are equal. Since $t = 5.233$ lies between the 0.05 significance value (5.6429) and the 0.10 significance value (4.5229), we may conclude that the p-value for the test is between 0.05 and 0.10.

■ **20.7**

Rank correlation: Spearman's rho

Spearman's rho is the nonparametric equivalent of the Pearson product-moment correlation coefficient given in Chapter 13. The hypothesis test that **X** and **Y** are correlated, described in Chapter 14, based on the Pearson product-moment correlation coefficient requires that the random variables **X** and **Y** are distributed according to the bivariate normal distribution—Spearman's rho does not have this requirement.

If there are no ties, Spearman's rho can be calculated from Pearson's product-moment correlation coefficient,

$$r = \frac{\sum_{i-1}^{n} (x_i - \bar{x})(y_i - \bar{y})}{\left[\sum_{i-1}^{n} (x_i - \bar{x})^2 \sum_{i-1}^{n} (y_i - \bar{y})^2 \right]^{1/2}}$$

by replacing the x_i and y_i values by their ranks.

Definition 20.6

Spearman's rho coefficient

Assumptions:
  1. The *n* pairs, $(\mathbf{X}_i, \mathbf{Y}_i)$, represent a random sample drawn from a bivariate population distribution of continuous random variables $\mathbf{X}$ and $\mathbf{Y}$.
  2. The measurement scale is at least ordinal.

Hypotheses:
  H_0: The $\mathbf{X}_i$ and $\mathbf{Y}_i$ values are uncorrelated.
  H_A: Either there is a tendency for *larger* values of $\mathbf{X}$ to be paired with larger values of $\mathbf{Y}$, or there is a tendency for *smaller* values of $\mathbf{X}$ to be paired with larger values of $\mathbf{Y}$.

Test statistic value:
Let $r(x_i)$ be the ranks of the $\mathbf{X}$ values and $r(y_i)$ be the ranks of the $\mathbf{Y}$ values. Ties are handled as usual—assign to each tied value the average of the ranks that would have been assigned had there been no ties. Spearman's rho—the correlation measure and the test statistic value—is given by

$$\text{rho} = 1 - 6 \sum_{i=1}^{n} \frac{[r(x_i) - r(y_i)]^2}{n(n^2 - 1)}$$

Decision rule:
Reject H_0 if rho is greater than $\rho_{1-\alpha/2}$ or if rho is less than $\rho_{\alpha/2}$, where $\rho_{\alpha/2}$ and $\rho_{1-\alpha/2}$ are given in Table B.14 in Appendix B. Otherwise, do not reject H_0.

Example 20.5 In a target marketing population, 10 married couples are randomly selected to participate in an experiment to determine the relationship between the ratings husbands and wives express for a product. Each couple is shown the company's product, and each individual is then asked to rate the product on a scale from 0 (terrible product) to 100 (terrific product). The 10 pairs of ratings are shown in the table.

Couple	Husband x_i	Wife y_i
1	90	70
2	100	60
3	75	60
4	80	80
5	60	75
6	75	90
7	85	100
8	40	75
9	95	85
10	65	65

Determine Spearman's rho and test the null hypothesis at the $\alpha = 0.05$ level that **X** and **Y** are uncorrelated.

Solution The ranks of the x_i and y_i values are given in the table.

i	$r(x_i)$	$r(y_i)$
1	8	4
2	10	1.5
3	4.5	1.5
4	6	7
5	2	5.5
6	4.5	9
7	7	10
8	1	5.5
9	9	8
10	3	3

The value of rho is:

$$\text{rho} = 1 - \frac{6[(8-4)^2 + (10-1.5)^2 + (4.5-1.5)^2 + \cdots + (3-3)^2]}{10(10^2-1)}$$

$$= 1 - \frac{(6)(161)}{990} = 1 - 0.976 = 0.024$$

To test the null hypothesis, from Table B.14 in Appendix B, $\rho_{0.975} = 0.6364$ and $\rho_{0.025} = -\rho_{0.975} = -0.6364$. Since rho $= 0.024$ is between $\rho_{0.025}$ and $\rho_{0.975}$, we cannot reject the null hypothesis that **X** and **Y** are uncorrelated. Since the value of the test statistic (rho $= 0.024$) is less than 0.4424, the p-value for the test is greater than $2(0.100) = 0.200$.

20.8

Goodness-of-fit: The Kolmogorov-Smirnov and Lilliefors tests

In this section, we will study an exact test for goodness-of-fit when the population random variable is continuous and the population distribution has been completely specified—the Kolmogorov-Smirnov test. Another test of this type—the Lilliefors test—will be given for the special case of testing the goodness-of-fit of a set of sample observations to a normal distribution with unspecified population mean and variance.

Before describing these tests, we will review the definition of a cumulative distribution function and define a sample cumulative distribution.

Example 20.6 Find the cumulative distribution function for the uniform distribution specified by,

$$f(x) = \begin{cases} 1 & 0 \le x \le 1 \\ 0 & \text{elsewhere} \end{cases}$$

Solution The graph of the uniform density function is illustrated in Figure 20.1. To find its CDF, we must evaluate the function $F(x) = P(\mathbf{X} \le x)$.

Definition 20.7*
Cumulative distribution function (CDF)

Given a random variable **X**, its *cumulative distribution function* is specified by

$$F(x) = P(\mathbf{X} \leq x) \qquad -\infty < x < +\infty$$

* See also definitions 5.5 and 5.9.

FIGURE 20.1 | Uniform distribution

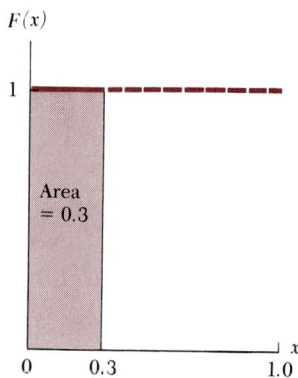

FIGURE 20.2 | CDF of the uniform distribution

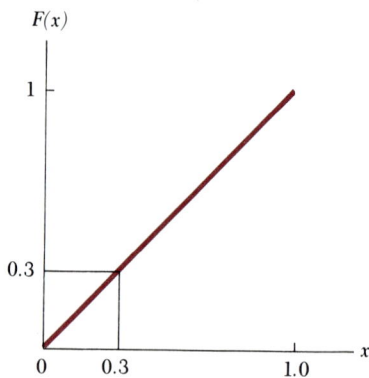

This function "accumulates" the probability to the left of x. For example, if $x = 0.3$, from Figure 20.1, $F(0.3) = P(\mathbf{X} \leq 0.3) = 0.3$. By selecting more points x, it becomes apparent that $F(x)$ takes the form illustrated in Figure 20.2.[2]

Sample cumulative distribution function (SCDF)

Definition 20.8

Sample cumulative distribution function (SCDF)

The *sample cumulative distribution function,* denoted by $S(x)$ for a specific value of $\mathbf{X}$, is formed by ordering a set of sample observations from the smallest to the largest and then plotting the cumulative relative frequencies as illustrated in Figure 20.3.

FIGURE 20.3 | Sample cumulative distribution function

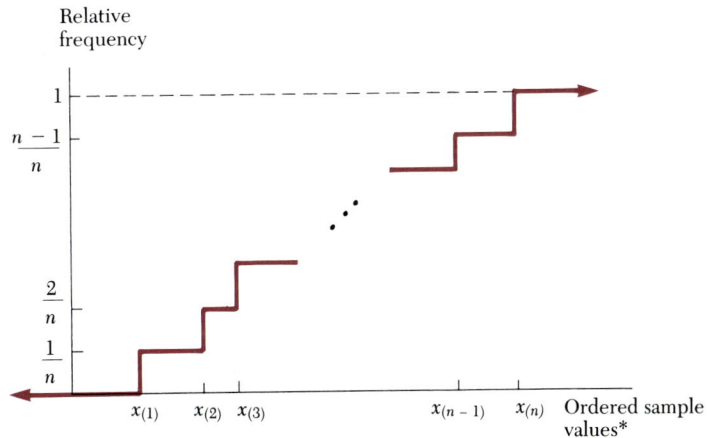

* The parentheses on the subscripts of the x_i values indicate that the $x_{(i)}$ values are the ordered observations, from the smallest to the largest.

The SCDF is often called a jump function—it jumps (in relative frequency) by $1/n$ each time an observation is encountered as we move along the x-axis from left to right. If the random variable is discrete, it is possible that two or more observations are equal in value. In this case, the SCDF jumps $1/n$ times the number of observations equal to the specific value.

---

[2] The CDF may alternatively be formed directly by integration: $F(x) = P(\mathbf{X} \leq x) = \int_{-\infty}^{x} f(x) \, dx = \int_{-\infty}^{0} (0) \, dx + \int_{0}^{x} (1) \, dx = 0 + x|_{0}^{x} = (x - 0) = x$, for $0 \leq x \leq 1$. If $x > 1$, $F(x) = 1.0$. If $x < 0$, $F(x) = 0.0$.

Example 20.7 A random sample, supposedly drawn from the standardized uniform distribution, produced the following 10 numbers:

0.36 0.14 0.60 0.20 0.43 0.81 0.12 0.02 0.55 0.25

Form the SCDF for these 10 numbers.

Solution The ordered numbers are:

0.02 0.12 0.14 0.20 0.25 0.36 0.43 0.55 0.60 0.81

The SCDF is illustrated in Figure 20.4.

FIGURE 20.4 | Sample CDF for Example 20.7

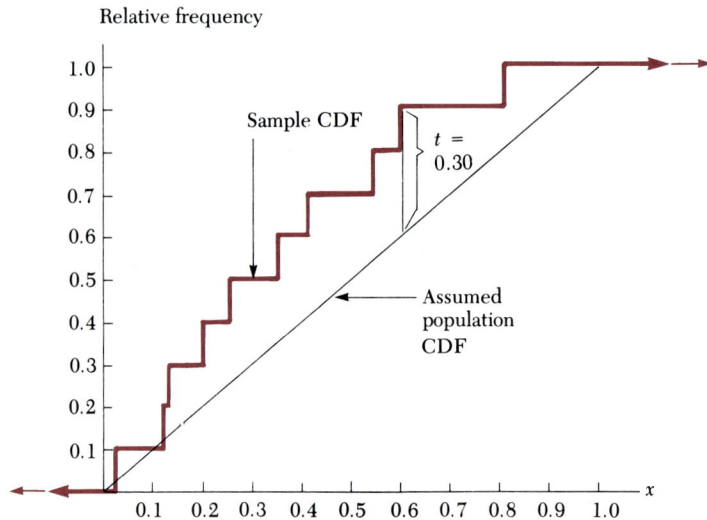

In Figure 20.4 the cumulative distribution function is also shown as the straight line from (0, 0) to (1, 1). Notice that the SCDF is above the CDF almost everywhere. This may suggest that the 10 numbers in Example 20.7 do not come from the standardized uniform distribution. In fact, the Kolmogorov-Smirnov goodness-of-fit test is based on the single greatest difference between the sample and assumed population CDFs.

Example 20.8 Using the data in Example 20.7, test the hypothesis at the $\alpha = 0.05$ significance level that these 10 data do represent a random sample from the standardized uniform distribution.

Solution From Figure 20.4, it is apparent that the greatest difference occurs when $x = 0.60$; $F_0(0.60) - S(0.60) = 0.60 - 0.90 = -0.30$. Therefore $t = |-0.30| = 0.30$. From Table B.15 in Appendix B, $W_{0.95} = 0.409$. Since $t < 0.409$, do not reject the null hypothesis that the data come from the standardized uniform distribution.

Kolmogorov-
Smirnov goodness-
of-fit test

Definition 20.9
Kolmogorov-Smirnov goodness-of-fit test

Assumptions:
1. The sample $X_1, X_2, \ldots, X_n$ is a random sample.
2. The hypothesized population CDF, denoted by $F_0(x)$, is continuous and completely known (if it has parameters, their values are known).

Hypotheses:
H_0: $F(x) = F_0(x)$
H_A: $F(x) \neq F_0(x)$ for at least some x, where $F(x)$ is the unknown CDF of $X_1, X_2, \ldots, X_n$.

Test statistic value:
Let $S(x)$ be the sample CDF based on the ordered sample observations, $x_{(1)}$, $x_{(2)}, \ldots, x_{(n)}$. The test statistic value is:

$$t = \sup_{X} |F_0(x) - S(x)|$$

The symbol sup means the greatest difference, and the test statistic value t is
     X
therefore defined as the greatest absolute difference between $F_0(x)$, the hypothesized CDF, and $S(x)$, the sample CDF.

Decision rule:
Reject the null hypothesis if t is greater than $W_{1-\alpha}$ given in Table B.15 in Appendix B. Otherwise, do not reject H_0.

The Kolmogorov-Smirnov test can be used for any hypothesized continuous CDF and is typically a very good approximate test if the population random variable is discrete. While this test is applicable in many situations, it is most widely used in goodness-of-fit for normality, since most parametric tests require the population random variable to be normally distributed, or approximately so.

Example 20.9 The Scholastic Aptitude Test scores are thought to be normally distributed with a mean of 500 and a standard deviation of 100. A random sample of 10 recent test results produced the following scores:

450 420 500 530 440 475 445 520 460 480

Use the Kolmogorov-Smirnov test to determine whether it is likely that these data have been drawn from the specified normal distribution (use $\alpha = 0.10$).

Solution To simplify the calculation of the value of the test statistic, we first standardize the 10 scores using

$$z_i = \frac{x_i - \mu}{\sigma}$$

These z-scores are:

$$-0.5 \quad -0.8 \quad 0.0 \quad 0.3 \quad -0.6 \quad -0.25 \quad -0.55 \quad 0.20 \quad -0.40 \quad -0.20$$

Denote by $z_{(i)}$ the ordered z-scores:

$$-0.80 \quad -0.60 \quad -0.55 \quad -0.50 \quad -0.40 \quad -0.25 \quad -0.20 \quad 0.00 \quad 0.20 \quad 0.30$$

The cumulative distribution function for the standardized normal distribution may be plotted by using Table B.3 in Appendix B. This CDF and the sample CDF $S(z)$ are illustrated in Figure 20.5. From Figure 20.5 it is appar-

FIGURE 20.5 | $F_0(z)$ and $S(z)$ functions for Example 20.9

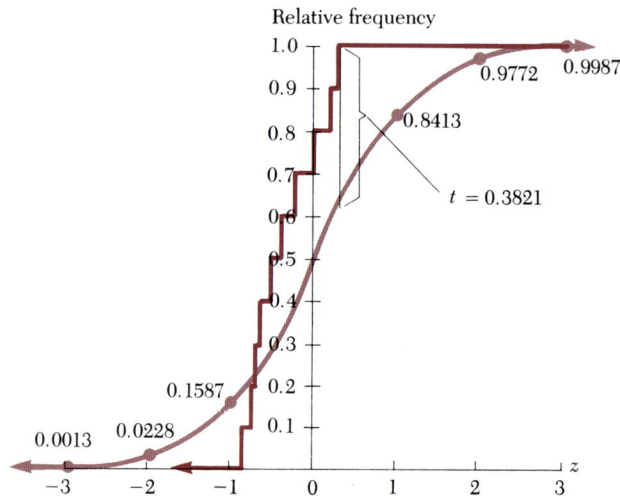

ent that the greatest difference between $F_0(z)$ and $S(z)$ occurs when $z = 0.30$; $F_0(0.30) = 0.50000 + 0.1179 = 0.6179$, and $S(0.30) = 1.00$. Thus $t = |0.6179 - 1.0000| = 0.3821$. From Table B.15, $W_{0.90} = 0.369$. Since $t > 0.369$, reject the null hypothesis. These data do not appear to come from a normal distribution with mean 500 and variance 100 at the $\alpha = 0.10$ significance level. Since $t = 0.3821$ lies between $W_{0.10}$ and $W_{0.05}$, we may conclude that the p-value for this test is between 0.10 and 0.05.

Often when we are interested in determining whether a random sample fits well to the normal distribution, we will not know the values of the population parameters μ and σ. But the Kolmogorov-Smirnov test requires that the hypothesized CDF, $F_0(x)$, be completely specified. Lilliefors modified the Kolmogorov-Smirnov critical tables so that the value of the test statistic **t** can be based on the sample mean $\bar{x}$ and the sample standard deviation s in place of μ and σ, respectively.

> **Definition 20.10**
> Lilliefors test for normality
>
> *Assumptions:*
>   1. The sample $X_1, X_2, \ldots, X_n$ is a random sample.
>   2. The hypothesized population CDF, denoted by $F_0(x)$, is continuous.
>
> *Hypotheses:*
> H_0: The random sample came from a normal distribution, with unspecified mean and standard deviation.
> H_A: The cumulative distribution function is not normal.
>
> *Test statistic value:*
> Standardize the x_i values by using
>
> $$z_i = \frac{x_i - \bar{x}}{s}, \text{ where } i = 1, 2, \ldots, n$$
>
> where $\bar{x}$ is the sample mean and s is the sample standard deviation. Let $S(z)$ denote the sample CDF of the z_i values. The test statistic value is:
>
> $$t = \sup_z |F_0(z) - S(z)|$$
>
> that is, it is the greatest difference between $F_0(z)$ and $S(z)$.
>
> *Decision rule:*
> Reject H_0 if t is greater than $W_{1-\alpha}$ given in Table B.16 in Appendix B. Otherwise, do not reject H_0.

Example 20.10 The lifetime of a certain electronics component is thought to be normally distributed. A random sample of eight components is selected, and the components are allowed to operate until failure. The failure times in hours are:

$$1{,}000 \quad 1{,}500 \quad 1{,}600 \quad 1{,}450 \quad 1{,}725 \quad 1{,}560 \quad 1{,}650 \quad 1{,}700$$

Using the Lilliefors test, test the null hypothesis that these data came from a normal distribution at the $\alpha = 0.05$ significance level.

Solution The sample mean and sample standard deviation are $\bar{x} = 1{,}523.125$ and $s = 231.439$, respectively. Using these values, the standardized variables are:

$$-2.26 \quad -0.10 \quad 0.33 \quad -0.32 \quad 0.87 \quad 0.16 \quad 0.55 \quad 0.76$$

The graphs of the standard normal CDF and the sample CDF are illustrated in Figure 20.6.

From Figure 20.6 the greatest difference between $F_0(z)$ and $S(z)$ occurs for a value of Z that is approaching $z_{(2)} = -0.32$ from the left, say $z^* = -0.31999$. Then $F(z^*) = 0.3745$, $S(z^*) = 0.125$, and $t = |0.3745 - 0.125| =$

FIGURE 20.6 | Population CDF and sample CDF for Example 20.10

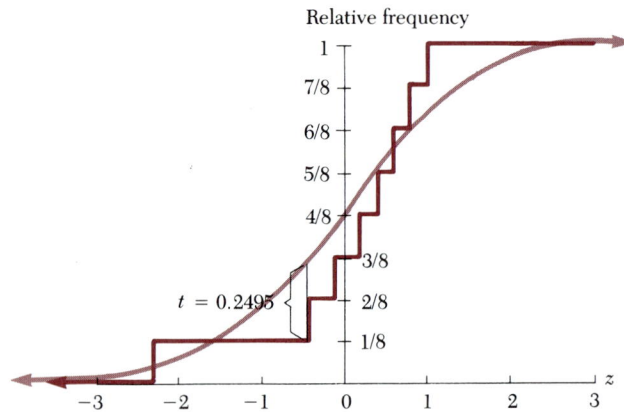

0.2495. From Table B.16 in Appendix B, $W_{0.95} = 0.285$. Since $t < W_{0.95}$, we fail to reject the null hypothesis at the $\alpha = 0.05$ significance level. There is insufficient evidence to reject the sample as not coming from a normal distribution. Since $t = 0.2495$ lies between $W_{0.15}$ and $W_{0.10}$, we may conclude that the p-value for this test is between 0.15 and 0.10.

■ **20.9**

Randomness test: The runs test

A very simple test has been devised to test for randomness of a set of observations. The test is based on the number of "runs" in a sequence of items, where a run is defined to be an unbroken row of like items that are preceded and followed by items of a different type and includes as many items as possible. Consider the sequence $(- - - + + +)$. This sequence has two runs, and the sequence $(- + - + - +)$ has six runs. The maximum number of runs among n items is n, and the minimum number of runs is one (all items are alike). We will now see how such a simple notion leads to a test for randomness among a set of observations.

Example 20.11 The following sequence is purported to be a set of random integers from 0 to 99. Use the runs test to test the hypothesis of randomness at the $\alpha = 0.05$ significance level. The sequence is:

$$28 \quad 4 \quad 23 \quad 98 \quad 44 \quad 10 \quad 6 \quad 25 \quad 54 \quad 81 \quad 12$$
$$6 \quad 4 \quad 33 \quad 67 \quad 55 \quad 71 \quad 66 \quad 22 \quad 18 \quad 49 \quad 85$$

Solution The median is $30.5[(33 + 28) \div 2]$. The sequence of $+$'s and $-$'s is, therefore,

$$-, -, -, +, +, -, -, -, +, +, -, -, -, +, +, +, +, +, -, -, +, +$$

The total number of runs is $t = 8$. From Table B.17 in Appendix B, let $N_1 =$ number of $-$'s and $N_2 =$ number of $+$'s. Then $N_1 = N_2 = 11$ and $W_{0.025} = 8$,

---

Definition 20.11
Runs Test

Assumption:
The observations are listed in the order obtained. Calculate the median of the set of observations. For all those observations *below* the median, assign a − sign, and for all those *above*, assign a + sign. This procedure generates a sequence of +'s and −'s.

Hypotheses:
H_0: The process that generates the sequence is a random process.
H_A: The process is not random—some items in the sequence are dependent on others in the sequence, or some items are distributed differently from others.

Test statistic value:

$$t = \text{Total number of runs of like items in the sequence}$$

Decision rule:
Reject H_0 if $t > W_{1-\alpha/2}$ or $t < W_{\alpha/2}$, where $W_{\alpha/2}$ and $W_{1-\alpha/2}$ are given in Table B.17 in Appendix B. Otherwise, do not reject H_0.

$W_{0.975} = 16$. Since $t = W_{0.025}$ we cannot reject the null hypothesis of randomness at the $\alpha = 0.05$ significance level. Since $W_{0.025} = 8$ and $W_{0.05} = 8$ and $t = 8$, we may conclude that the *p*-value is between $2(0.025) = 0.05$ and $2(0.05) = 0.10$.

The runs test may be used to check the assumption in regression analysis that the error terms are independent. See Problem 20.16 at the end of this chapter, for example.

■ **20.10**

Summary

In this chapter we have presented a number of methods that may be appropriate in analyzing data when the standard parametric procedures may not be appropriate. Nonparametric tests are typically easy to compute but are almost always less powerful than a parametric test if one is appropriate for the data set. Most nonparametric tests require that the data be at least ordinally measurable, and most parametric tests require interval or ratio scale measurements.

Nonparametric tests may be applied to nominally scaled measurements, although we did not consider any of these tests in this chapter. The most widely used nonparametric statistic for the analysis of nominal data is the chi-square statistic introduced in Chapter 19. The chi-square statistic is used to analyze contingency tables, where most often the categories in the table are ordinally measured. The chi-square statistic is therefore commonly thought of as a nonparametric statistic, although the chi-square distribution is also used for constructing decision rules for parametric tests (e.g., tests concerning the population variance σ^2—see Chapter 10).

There are of course many more nonparametric tests than we have described in this text. An excellent source of the most widely used tests is the Conover reference cited at the end of this chapter.

■ References

Introductory Level

Anderson, D. R.; D. J. Sweeney; and T. A. Williams. *Statistics for Business and Economics.* 2nd ed. St. Paul, Minn.: West Publishing, 1984, Chapter 18.

Iman, R. L., and W. J. Conover. *Modern Business Statistics.* New York: John Wiley & Sons, 1983, Chapters 4, 7, 8, 9, 16.

Kohler, H. *Statistics for Business and Economics.* Glenview, Ill.: Scott, Foresman, 1985, Chapter 11.

Wonnacott, T. H., and R. J. Wonnacott. *Introductory Statistics for Business and Economics.* 3rd ed. New York: John Wiley & Sons, 1984, Chapter 16.

Advanced Level

Conover, W. J. *Practical Nonparametric Statistics.* 2nd ed. New York: John Wiley & Sons, 1980.

Gibbons, J. D. *Nonparametric Statistical Inference.* New York: McGraw-Hill, 1971.

Ostle, B., and R. W. Mensing. *Statistics in Research.* 3rd ed. Ames: Iowa State University Press, 1976, Chapter 14.

Snedecor, G., and W. Cochran. *Statistical Methods.* 7th ed. Ames: Iowa State University Press, 1980, Chapter 5.

■ Problems

Section 20.3 Problems

20.1 Assuming the distribution of miles per gallon is symmetric so that the mean and median are equivalent, test the hypothesis that the population mean is 30 miles per gallon, using $\alpha = 0.10$ and the data in Example 20.1, by using the t-test in Chapter 10. What is the p-value for this test?

20.2 A marketing research group randomly samples 15 potential buyers of a product to determine their acceptance of the product. The measurement scale used is:

0	1	2	3	4
Strongly dislike	Dislike	Neutral	Like	Strongly like

The 15 responses were: 0, 2, 2, 1, 0, 1, 3, 2, 2, 0, 1, 4, 3, 2, 0. Using the median test, test the hypothesis that the population median is 2. (Use $\alpha = 0.1$.) What is the p-value for this test?

20.3 A brochure for a leading MBA program nationally claims that the median GMAT (Graduate Management Admissions Test) score for its admitted students is 600. A random sample of 25 scores is taken from the files of admitted students for the fall semester, 1986. The scores are:

666	545	592	498	604
575	605	675	725	790
646	609	555	529	544
598	614	717	580	592
522	645	571	560	525

a. Using the median test with a 0.05 level of significance, does the school's claim appear to be warranted?

b. What assumption would have to be made for these data to test the hypothesis that the *mean* score is 600 by the median test?

Section 20.4 Problems

20.4 Consider the data in Example 20.2.

a. Is it reasonable to assume that if the two populations differ, they differ only in location? Using the *F*-test in Chapter 11, test the hypothesis at the $\alpha = 0.05$ significance level that the two population variances σ_1^2 and σ_2^2 are equal.

b. If the null hypothesis in part *a*, $H_0: \sigma_1^2 = \sigma_2^2$, is not rejected, test the hypothesis that $\mu_1 = \mu_2$ at the $\alpha = 0.05$ level, pooling the two sample variances. Compare this result to one using the Mann-Whitney test in Example 20.2. What is the *p*-value for this test?

20.5 Two fertilizers, brands A and B, are used on two identical plots of land planted with corn. Each plot is divided into eight equal sections. At the end of the experiment, the yields per section for the two fertilizers are measured. The data are shown in the table.

Fertilizer A	Fertilizer B
80.2	95.2
76.8	84.7
93.2	88.9
90.1	98.6
85.7	100.8
81.5	89.8
79.0	99.6
82.0	101.4

Using the Mann-Whitney test and $\alpha = 0.05$, test the null hypothesis that the two samples came from the same population. What is the *p*-value for this test?

Section 20.5 Problems

20.6 Use the paired *t*-test given in Chapter 11 to test the equivalence of the two population means in Example 20.3 (use $\alpha = 0.05$). Compare your result with that in Example 20.3, which is based on the Wilcoxon signed rank test. What additional assumptions are required to use the paired *t*-test? What is the *p*-value for this test?

20.7 A drug company is interested in determining whether or not a chemical treatment for a specific form of cancer changes body temperature. Ten patients with the disease are selected at random from a set of patients under experimental control. Their temperatures are measured before and after taking the treatment. The data, given in degrees Fahrenheit, are listed in the table.

Patient	Before	After
1	98.4	99.6
2	98.2	100.9
3	98.0	97.6
4	99.0	99.9
5	98.6	98.2
6	97.0	98.4
7	98.4	98.4
8	100.0	102.6
9	99.8	102.2
10	98.6	98.8

Test the null hypothesis that the two population means are equal at the $\alpha = 0.01$ level by using the Wilcoxon signed rank test. What is the *p*-value for this test?

20.8 The FTC (Federal Trade Commission) has decided to investigate a firm that claims that it can raise a person's GMAT (Graduate Management Admissions Test) score over the person's most recently posted score. To test the truthfulness of this claim, the FTC asks to see the company's records and from the records randomly samples 10 individual cases. Each case contains an individual's score before enrolling in the six-week course and the person's score after enrolling in the six-week course. The data are:

Individual	1	2	3	4	5	6	7	8	9	10
Score before	453	310	520	560	440	610	570	480	500	460
Score after	510	450	490	560	460	600	555	492	525	454

Using the Wilcoxon signed rank test, is there an indication that the firm's claim is correct at the 0.05 level of significance? What is the p-value for this test?

Section 20.6 Problems

20.9 Use the analysis of variance method described in Chapter 12 to test the null hypothesis of the equality of the three population means in Example 20.4 ($\alpha = 0.05$). Compare your result with the Kruskal-Wallis test result in Example 20.4. What further assumptions must you make to use the analysis of variance F-test on these data? What is the p-value for this test?

20.10 The breaking strengths of three kinds of wire cord are compared by using four samples of each and measuring the force required to break the cord. The data are shown in this matrix.

Observation	Type A	Type B	Type C
1	200	310	275
2	250	265	210
3	243	262	255
4	227	288	242

Test the null hypothesis of equal population distributions at the $\alpha =$ approximately 0.05 level, using the Kruskal-Wallis test. What is the p-value for this test?

20.11 A large consulting firm hires a West Coast university to provide an MBA program for its employees. The basic statistics course is taught at four locations of the firm. After completion of the course, standardized tests are given to the participating employees at each location. The results are given in the table.

Observation	Location A	Location B	Location C	Location D
1	100	66	64	60
2	89	72	92	75
3	90	78	88	66
4	75	81	72	75
5	84	70	86	80
6	66	74	88	68
7	88	82	92	74
8	86	77	72	66
9	94		77	70
10	85		82	78
11	98		80	
12	84		86	
13	90			
14	76			
15	85			

Using the Kruskal-Wallis test, test the null hypothesis of equal population distributions. Calculate the p-value for this test. Should the null hypothesis be rejected based on the p-value? Explain.

Section 20.7 Problems

20.12 A production manager suspects that the level of production among a specific class of workers in the firm is related to their hourly pay. The following data are collected on eight randomly selected workers.

Worker	Hourly pay (x_i)	Production (y_i)*
1	$3.75	50
2	4.56	20
3	5.10	62
4	4.00	30
5	6.35	75
6	6.00	66
7	3.75	40
8	5.65	60

* In units per hour.

a. Calculate Spearman's rho for these data.
b. Calculate Pearson's product-moment correlation coefficient for these data. What assumptions are necessary to compute Pearson's r? What does it measure?

c. Using Spearman's rho, test the hypothesis at the $\alpha = 0.10$ level that **X** and **Y** are uncorrelated. What is the p-value for this test?

20.13 In a production process, it is suspected that the number of defectives produced per hour is related to a skills test score recorded by a worker. The units being produced require a high degree of manual dexterity, and the skills test is designed to test manual dexterity. Ten workers are selected, and each is asked to put together 25 units in one hour. Units that are not completed in the hour period are considered to be defective; those that are completed are inspected for defects. The data are given in the table below. Using Spearman's rho, determine whether the correlation between test score and number of defects is significantly different from 0 at the $\alpha = 0.05$ level of significance. What is the p-value for this test?

					Worker					
	1	2	3	4	5	6	7	8	9	10
Test score	90	72	62	66	50	84	78	96	80	75
Number of defectives	2	4	7	8	10	3	3	1	3	5

Section 20.8 Problems

20.14 A company believes that its color television tubes have an average lifetime of 18 continuous months of play and that these lifetimes are exponentially distributed. The company has its quality control department randomly sample 12 tubes, which are run to failure. The 12 lifetimes (in months) are:

5.6	22.4	16.8	12.2	19.4
23.5	19.0	16.4	21.0	23.7
14.8	24.9			

Use the Kolmogorov-Smirnov test to determine whether it is reasonable to assume that

these data came from the hypothesized exponential distribution. Use $\alpha = 0.05$. (Hint: Form the variable $\mathbf{Y} = \mathbf{X}/\lambda, \lambda = 18$. Use Table B.4 in Appendix B to plot the population CDF.) What is the p-value for this test?

20.15 The sales research division of a large corporation is conducting research on sales methods for selling one of the products of the corporation. The division has designed a completely randomized one-factor analysis of variance model to investigate the efficacy of three sales methods. The responses are measured in units of $100 in sales. The data are shown in the table.

Response	Sales method		
	A	B	C
1	22	12	30
2	21	16	27
3	22	15	46
4	20	28	30
5	36	14	34
6	44		27
7	28		36
8	24		26
9	31		
10	25		

a. Use the Lilliefors test to determine whether each treatment sample comes from a normal population. Use $\alpha = 0.05$.

b. What other assumptions must be checked before the parametric analysis of variance F-test may be used on these data?

Section 20.9 Problems

20.16 A regression analysis using the model $Y_i = \beta_0 + \beta_1 x_i + \epsilon_i$, $i = 1, 2, \ldots, 10$, $y_i =$ Amount of sales (in \$000); and $x_i =$ Amount of advertising expenditure (in \$000), was performed by a marketing research group to assess the effect on sales of the amount of advertising budget. After fitting the model, the group calculated the following 10 residuals to check the aptness of the regression model:

$$e_1 = 2.42 \qquad e_6 = -0.20$$
$$e_2 = 2.02 \qquad e_7 = -1.60$$
$$e_3 = -1.08 \qquad e_8 = -0.44$$
$$e_4 = -0.54 \qquad e_9 = -0.80$$
$$e_5 = -0.98 \qquad e_{10} = 1.20$$

a. Use the runs test to determine whether or not this is a random sequence of resid-uals. That is, are the residuals drawn from independent error terms in the regression model? Use $\alpha = 0.05$ for the statistical test.

b. For statistical hypothesis testing purposes, the error terms in regression analysis are supposed to be normally distributed with a mean of 0 and a standard deviation σ. Use the Lilliefors test to check this assumption for the sample of 10 residuals drawn from the regression model error terms. Use a $\alpha = 0.05$ level of significance for the test.

20.17 A computer software program is designed to produce sequences of random integers from 1 to 100. A random sample of 25 integers is produced from this program. The values, in the order from left to right, are:

45	98	76	55	88	66	72	03	12	22
36	08	29	90	71	62	01	21	55	68
88	10	18	28	75					

Use the runs test to determine if this sequence of integers is random. Use a 0.05 level of significance.

Additional Problems

20.18 A machine plant is interested in acquiring a new stamping machine and has narrowed its selection down to two machines. In order to rate the machines, each machine is set up to produce 10 units, and each unit is given a score based on an index that includes such factors as time required to produce the unit, number of flaws, and appearance of the unit. The higher the score, the better the outcome. The index scores are given in the table below. At the $\alpha = 0.05$ significance level, do these machines appear to be producing parts with the same index scores? What is the p-value for this test?

Machine A	50	45	60	60	75	40	46	58	62	72
Machine B	72	61	60	78	56	77	81	75	74	82

20.19 A major automobile subsidiary believes that its standard batteries have capacities that are normally distributed, with a mean of 100 ampere hours and a standard deviation of 20 ampere hours. Fifteen batteries are tested with the following results:

120	80	60	90	100
175	120	155	160	160
110	140	145	105	150

Test the null hypothesis that these data came from a normal distribution with mean 100 and standard deviation 20, using the Kolmogorov-Smirnov test with $\alpha = 0.05$. What is the p-value for this test?

20.20 Cears, Inc. claims that its super-road-hugger radial tire will produce the same distribution of miles of usage as its main competitor, the Mikilen XA5 radial tire. A consumer's union decides to test this claim by sampling 25 tires of each brand and placing each tire on a wear machine that measures the usable miles of wear for the tire. The data, given in 1,000s of miles of tread wear, are:

Cear's tires					Mikilen tires				
40	36	32	38	45	45	48	44	46	52
34	29	26	37	44	51	50	54	47	46
31	30	30	41	46	55	48	43	49	50
46	40	39	33	47	50	56	49	42	58
28	49	52	48	43	45	51	50	53	46

a. Using the Mann-Whitney test, does Cear's claim of equivalence of its tire to Mikilen's tire in tread wear seem reasonable? Use a 0.05 level of significance.

b. What assumption would have to be made on these data to use the Mann-Whitney test to test the equivalence of the population *mean* tread life for the two brands of tires?

c. What is the approximate p-value for the test in part a?

20.21 An automobile manufacturer is experimenting with four kinds of carburization methods to reduce pollutants. The manufacturer selects 20 cars of a similar make, model, horsepower rating, and so on. Each carburization method is tried on five of the 20 cars. The lower the score, the more effective the carburization system is. The data are:

Carburetor system			
A	B	C	D
20	33	18	36
24	26	12	12
27	18	14	21
22	12	10	27
21	24	13	33

Using the Kruskal-Wallis test, do the four carburization systems appear to produce the same scores? Use the 0.05 significance level. What is the p-value for this test?

20.22 A research laboratory believes that it has discovered a new medication to retard the effects of the disease AIDS on the immune system. Pairs of patients are matched on their current degree of ability to fight off infection, progression of the disease, age, weight, and other factors. One set of patients, the control group, is given the standard treatment, and the other set of patients is given the new medication. After 60 days the ability of the patient's immune system to fight off infection is measured. The measure is converted to a scale of from 0 to 100, 0 being no defense and 100 being normal and healthy immune reaction. The data for the 20 matched pairs of patients are:

Patient number	Control group	Test group	Patient number	Control group	Test group
1	20	32	11	45	54
2	35	36	12	66	75
3	10	15	13	31	35
4	78	89	14	76	84
5	33	28	15	90	92
6	87	81	16	44	45
7	78	50	17	28	39
8	14	27	18	72	66
9	22	33	19	72	88
10	50	60	20	61	91

a. Using the Wilcoxon signed rank test for matched pairs, does the hypothesis of equal mean scores for the control group and for the test group appear to be reasonable? Use the 0.05 level of significance.

b. What is the approximate p-value for this test?

20.23 The Marvin Advertising Firm has prepared four television advertisement spots for a leading manufacturer of compact disc audio devices. To assist the manufacturer in selecting the advertisement to use on national television, a representative of Marvin takes a random sample of 20 individuals and randomly assigns the 20 individuals to four groups of five each. Each group is shown one of the advertisements and is asked to rate it according to several measures. Each individual's rating results are converted to a score on a scale from 0 to 100, with higher scores indicating a more favorable reaction to the advertisement. The resulting data are shown in the table below:

Advertisement one	Advertisement two	Advertisement three	Advertisement four
60	25	46	90
48	32	52	86
55	28	44	60
40	48	39	78
50	36	48	80

a. Using the Kruskal-Wallis test, do these four sample sets appear to have come from the same population distribution? Use a 0.05 level of significance.

b. What assumption must be made for these data to use the Kruskal-Wallis test to test the equivalence of the four population means?

c. What is the approximate p-value for the test in part a?

20.24 A business school admissions officer is interested in the correlation between high school grade point average and the score on the SAT (Scholastic Aptitude Test). The officer takes a simple random sample of 20 student files and records each student's high school grade point average and SAT score. These data are:

Student number	HS GPA	SAT score	Student number	HS GPA	SAT score
1	3.20	1050	11	2.86	878
2	3.45	1205	12	2.22	705
3	2.50	1109	13	3.44	1080
4	2.10	690	14	3.12	1170
5	2.77	988	15	3.56	1345
6	3.76	1288	16	2.55	1020
7	3.62	1198	17	2.82	1000
8	2.85	967	18	2.30	780
9	3.60	1301	19	3.25	1133
10	3.28	1020	20	2.24	790

a. Use the Spearman's rho test to test the null hypothesis that the high school GPA is uncorrelated with the SAT score for each student. Use a 0.05 level of significance.

b. What is the approximate p-value for this test.

20.25 The completion times for a specific assembly task are thought to be normally distributed. To check this assumption, a random sample of 20 completion times are taken with the following results (in minutes):

```
25  34  33  42  36  34  33  37  28  32
29  30  36  32  30  30  45  27  30  34
```

a. Use the Lilliefors test to test the hypothesis that these completion times are normally distributed. Use a 0.05 level of significance.

b. What is the approximate p-value for this test?

21

Statistical decision theory: Decision making under uncertainty

■ **21.1**

Introduction

Bayesian statistics

In this chapter we begin a discussion of statistical decision theory. This discussion will continue for two more chapters, at which time we will have completed our introductory exposition of this topic.

The material discussed in Chapters 21–23 will mark a distinct deviation from what has been discussed previously. The preceding material is characterized as classical statistics or as classical statistical techniques or methods, whereas the material in these three chapters is characterized as Bayesian statistics, as Bayesian decision theory, or, in abbreviated form, as just decision theory. We will learn as we proceed through this material that there are differences between the two approaches or schools of thought. In Chapter 23, we attempt to summarize the major differences between these two approaches for using statistical data to arrive at a decision, so the reader can make an informed choice as to which method he or she chooses to follow.

In this chapter, we will learn how to characterize a statistical decision problem and will investigate four separate approaches for making decisions in situations in which it is impossible or extremely difficult to assess a probability distribution for a random variable about which we must make a decision. Then, in Chapter 22, we will see how Bayesian decision theory prescribes a course of action to follow when a probability distribution for the random variable of interest exists and is known, but there is no opportunity to obtain any additional information. Then, in Chapter 23, we will see how the Bayesian approach prescribes a decision rule to follow for selecting the sample size and the rejection or critical value of a test statistic in making a decision where the possibility for obtaining additional information exists (i.e., sampling).

Following the convention in previous chapters (notably Chapters 9 and 10), we will use the Greek symbol theta in boldface $\boldsymbol{\theta}$ to represent a random variable or population characteristic of interest (mean, proportion, etc.), and we will use nonboldfaced theta θ to represent possible values of the random variable $\boldsymbol{\theta}$. However, it will greatly simplify the discussion of decision theory if we also refer to the random variable $\boldsymbol{\theta}$ as being a particular state of nature. Thus, for our purposes, the following two statements will be equivalent: "The probability that the random variable $\boldsymbol{\theta}$ is equal to θ_j is denoted by $P(\boldsymbol{\theta} = \theta_j)$," and "The probability that we are in state θ_j or that state of nature θ_j occurred is denoted by $P(\boldsymbol{\theta} = \theta_j)$."

Notation: Random variable is bold-faced; its value is not

The use of the Greek letter $\boldsymbol{\theta}$ to denote a random variable underscores the fundamental difference between classical statistics and Bayesian statistics. In classical statistics (Chapters 1–20 in this text), Greek symbols (μ, σ, and ρ for example) are used to designate population parameters, assumed to be constants, usually unknown in value. In Bayesian statistics the population parameters are viewed as being *random variables*. When we speak of the "states of nature" in the Bayesian context, we are referring to the values that the population parameter can assume. In this and the following two chapters, the difference between these two viewpoints (classical statistics and Bayesian statistics) will be developed.

The fundamental difference between classical statistics and Bayesian statistics

We begin our discussion of statistical decision theory or Bayesian decision theory by an examination of alternate means available for characterizing a decision problem.

■ 21.2

Structuring a statistical decision problem

Several alternative methods exist for giving structure to a statistical decision problem. In this section, we will describe three methods: the payoff (or loss) matrix, the loss function, and decision trees. Each of these methods possesses a distinct format that is often helpful in analyzing certain decision problems.

21.2.1 | Payoff and regret tables

Payoff matrices are a convenient means of organizing and displaying the various ingredients of a decision problem when the number of possible states of nature and the number of possible actions are finite. Table 21.1 depicts the basic format of a payoff matrix, where Ω_{ij} represents the payoff (or loss, for Ω_{ij} negative) that results when action a_i is taken and the particular state of nature is θ_j. Table 21.1 is often referred to also as a *value matrix*—denoting the "value" of each act for each state of the world or state of nature that occurs.

Value matrix

It is customary to denote the various states of nature across the top of the table (columns) and the various possible acts (alternatives) on the left side of the matrix (rows) as is shown. The intersection of the act chosen and the state of nature existing thus denotes the payoff from the particular act–state

TABLE 21.1 | General form of a payoff matrix

Act \ State of nature	θ_0	θ_1	θ_2	$\cdots$	θ_j	$\cdots$	θ_n
a_0	Ω_{00}	Ω_{01} $\cdots$					$\cdots \Omega_{0n}$
a_1	Ω_{10}						
a_2							
$\vdots$							
a_i				$\cdots$	Ω_{ij}	$\cdots$	
$\vdots$							
a_m	Ω_{m0}	$\cdots$					$\cdots \Omega_{mn}$

of nature combination. In Table 21.1 we have indicated an act labeled a_0 and a state of nature denoted θ_0. These represent, respectively, the act "do nothing" (e.g., stock 0 items) and "nothing occurs," or the value of the random variable is equal to 0 (e.g., no demand, or demand equals 0). Not all payoff (or loss) tables have this 0 designation, but many do. From Table 21.1 we see that since we have allowed for the act–state of nature combination (a_0, θ_0), there are $(m + 1)$ possible actions that can be taken (which includes the act "do nothing"), and there are $N = (n + 1)$ values the random variable θ can assume.

Also, from Table 21.1, it can be seen that in a statistical decision problem, the payoff from selecting a particular act a_i and having a particular state of nature θ_j occur is known in advance of the decision. There is assumed to be no ambiguity regarding the payoff to be received when the particular act is selected and the state of nature occurs. Hence we assume that the results of a particular action can be determined without *error* when provided with the state of nature that has or is about to occur. That is, the results in terms of payoffs of our actions are certain; knowledge concerning the state of nature may or may not be certain. Thus the *stochastic* or the *probabilistic* elements involved in a statistical decision concern states of nature; the *deterministic* or *certain* elements involve the payoffs to be received from acts to be selected when the particular stochastic elements occur.

Use of the payoff matrix in characterizing the ingredients of a decision problem is illustrated in the following example.

Example 21.1 The George Thomas Travel Agency regularly schedules tours from Rochester, Minnesota, to a vacation resort in southern Florida (private resort area). These tours normally take place in the winter months, and George reserves seats on the regularly scheduled airlines by depositing $100 three months in advance of the scheduled departure date. George then sells the tour package to residents of Rochester for $150 (airfare only).

George has been given the right to purchase up to six seats in advance for each tour that he schedules.[1] If George has not sold the tour to as many people as he has reserved seats by one week before the scheduled departure date, he must notify the airline that he is relinquishing his right to the unsold seats, in which case the airline refunds him $80 of his $100 deposit. Thus George must reserve the seats well in advance of when he will know what the demand for the seats will be. For each seat reserved but unsold by one week before the scheduled departure date, he loses $20.

Characterize this decision problem both in terms of the various states of nature (*demands* for seats on the tour) and in terms of the various alternatives (*supplies* of reserved seats) open to George in the form of a payoff matrix.

Solution Table 21.2 is the payoff matrix for the problem. In this form, the payoff matrix might also be called a *cash-flow matrix,* since the quanti-

Cash-flow matrix

TABLE 21.2 | Payoff matrix for Example 21.1*

Act, a_i (supply) / State of nature, θ_j (demand)	0	1	2	3	4	5	6
0	0	0	0	0	0	0	0
1	−20	50	50	50	50	50	50
2	−40	30	100	100	100	100	100
3	−60	10	80	150	150	150	150
4	−80	−10	60	130	200	200	200
5	−100	−30	40	110	180	250	250
6	−120	−50	20	90	160	230	300

* Quantities are in dollars.
Source: Data in Example 21.1.

ties displayed represent the actual cash George will receive ($\Omega_{ij} \geq 0$) or payout ($\Omega_{ij} \leq 0$) if he books a_i seats and θ_j are demanded. Thus, for example, if George reserves five seats and sells only one, his net payoff is a negative $30 (loss).

---

[1] George can actually reserve many more seats than this and, in fact, usually does. However, it will greatly simplify the computational effort involved without our losing an understanding of the subject matter if we restrict the number of alternatives we have to examine in this problem to a more manageable number. This is why we have chosen to limit our discussion of the number of seats available to six.

The advantage of displaying the act–state of nature combinations in matrix or tabular form is that the consequence of every act–state of nature pair is readily available from the table. The payoff matrix itself, however, is often criticized because it gives the appearance that certain acts are equally "profitable" for various states of nature. For example, if a decision is made to reserve two seats and there is a demand for six, the cash flow or profit is $100, the same as it would be if the demand is for two. Obviously, the *potential for profit* is greater if the demand is six rather than if the demand is only two. Yet the payoff matrix does not reflect the differences in profit that would result *for a fixed act* when the number of units demanded is greater than the number of units stocked.

For this reason, many decision problems are resolved using an *opportunity loss* or a *regret* matrix to express the consequences of various act–state of nature combinations.

Opportunity loss

> **Definition 21.1**
> Opportunity loss
>
> The *opportunity loss* of selecting a particular act a_i is the difference between the cost or profit that was actually realized for that act and the cost or profit that would have resulted if the decision had been the best one for the state of nature that actually occurred.

The actual computation of the opportunity loss for a given act is very straightforward: Subtract each entry in the payoff or the value matrix from the largest entry in its column.

> Opportunity loss
>
> $$l(a_i, \theta_j) = (\underset{i}{\text{Max }} \Omega_{ij}) - \Omega_{ij} \qquad (21.1)$$

The opportunity loss, then, represents the difference between what was realized under a given act and what could have been realized if the state of nature that actually occurred had been known in advance. Since the opportunity loss is conditional upon the value of θ that occurs, it is also called a *conditional loss* or a *conditional opportunity loss*.

Example 21.2 Compute the opportunity loss or the regret matrix for the data given in Example 21.1.

Solution Table 21.3 gives the opportunity loss or the regret matrix for the data in Example 21.1. The opportunity losses for the entries in the matrix for $\theta = 3$ (three seats are demanded) are computed as follows:

$$\text{Max}_i(\Omega_{i3}) = \Omega_{33} = 150 \qquad i = 0, 1, 2, \ldots, 6$$

$$l(a_i, \theta_3) = 150 - \Omega_{i3}$$

Or, for $\theta = 3$, $l(a_i, \theta_3)$ is

$$
\begin{array}{ll}
150 - \quad 0 = 150 & 150 - 130 = 20 \\
150 - \quad 50 = 100 & 150 - 110 = 40 \\
150 - 100 = \quad 50 & 150 - \quad 90 = 60 \\
150 - 150 = \quad 0 &
\end{array}
$$

Note that with the opportunity loss matrix, the consequence or the opportunity loss suffered from a decision to reserve two seats when the demand is six is $200.

The opportunity loss table offers at least one other advantage over the payoff or the value matrix. Namely, the entries in the opportunity loss table are nonnegative (i.e., always either positive or equal to 0) quantities, and the best or the optimal act(s) under a given state of nature can always be found by simply locating those rows (acts) with a 0 entry for that column.

TABLE 21.3 | Opportunity loss matrix for Example 21.1*

Act, a_i (supply)	State of nature, θ_j (demand)						
	0	1	2	3	4	5	6
0	0	50	100	150	200	250	300
1	20	0	50	100	150	200	250
2	40	20	0	50	100	150	200
3	60	40	20	0	50	100	150
4	80	60	40	20	0	50	100
5	100	80	60	40	20	0	50
6	120	100	80	60	40	20	0

* Quantities are in dollars.
Source: Data in Example 21.1.

21.2.2 | Opportunity loss or payoff functions

Frequently, the most convenient means for expressing the consequences of various acts for various states of nature is the functional form. While it may or may not always be possible to indicate the consequences of act–state of nature combinations in this fashion, this form is necessary when the number of possible states of nature is infinite (e.g., the diameter of bearings produced on a certain machine) and is often quite helpful when the number of various acts or states of nature is quite large, so that the construction of a payoff or a loss table would be quite cumbersome. For the problem of Example 21.1, the loss function is:

$$l(a_i, \theta_j) = \begin{cases} 50(\theta_j - a_i) & \text{if } \theta_j > a_i \\ 0 & \text{if } \theta_j = a_i \\ 20(a_i - \theta_j) & \text{if } \theta_j < a_i \end{cases}$$

This loss function was derived by noting that no loss occurs (or the loss is 0) when the number of seats reserved is equal to the number of seats demanded ($\theta_j = a_i$), a \$20 loss occurs for every seat that is reserved and is not demanded ($\theta_j < a_i$), and a \$50 loss (opportunity loss) occurs for every seat that is demanded, but has not been reserved ($\theta_j > a_i$). The quantities ($\theta_j - a_i$) and ($a_i - \theta_j$) denote, respectively, the number of seats demanded but not available, and the number of seats reserved, but not demanded. These quantities are multiplied in the loss function by the per unit costs of being greater or less than the estimate of the quantity demanded.

Figure 21.1 is a graph of the conditional opportunity loss function for the acts reserve 3 and reserve 4. Several features of this graph are worth noting

FIGURE 21.1 | Conditional opportunity losses for acts reserve 3 and reserve 4 in Example 21.1

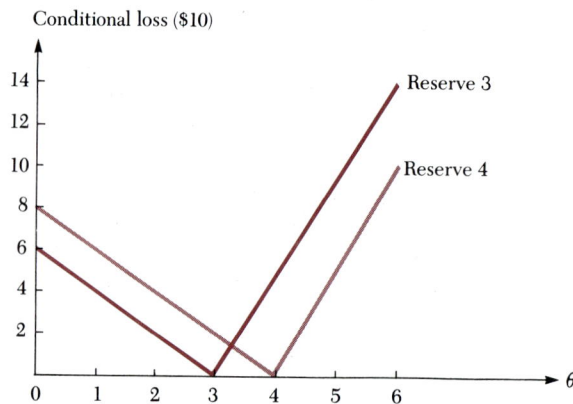

Source: Table 21.3.

at this time. First, the graph forms a characteristic asymmetric V, with a vertex at the point where conditional losses are 0. Note from the graph that conditional losses are 0 when the number of seats reserved is equal to the number demanded (or the greatest profit act is selected), and conditional losses are positive elsewhere. Second, the slopes of the sides of the V give the rate of decrease in potential profits resulting when the nonoptimal or the best act for a given demand is not selected. The slopes of the two sides are usually different, reflecting the difference in profit foregone (opportunity loss) from an overage as opposed to an underage of an item. *The majority of conditional loss functions exhibit graphs similar to the one shown in Figure 21.1.*

21.2.3 | Decision trees

A third alternative for giving the information for a decision problem when there are a finite number of decision alternatives and a finite number of states of nature is a *decision tree*. A decision tree gives the decision alternatives (the acts), the states of nature, and the consequences (usually in monetary terms), of each act–state of nature combination. We will construct a decision tree for the Example 21.1 problem with the payoff matrix given in Table 21.2.

Example 21.3 Construct a decision tree for the Example 21.1 problem whose payoff table is given in Table 21.2.

Solution The decision tree is shown in Figure 21.2. The tree begins from the left with a *decision branch,* indicated by a box. At the box, there are seven branches representing the decision alternatives (acts), one of which must be chosen by the decision maker. Following each decision branch is a *state of nature branch,* indicated by a circle. At each circle, there are seven branches representing the seven states of nature in the problem. At the end of each decision branch–state of nature branch combination, the payoff is listed.

A decision tree shows the same information as given in a payoff matrix table. It can be a very effective and easily understood way to show the important information in a decision problem, but may become quite large and confusing if the number of acts and/or the number of states of nature become large. We will return to the decision tree method of analyzing decision problems in Chapter 22.

■ 21.3

Decision making under uncertainty

Decisions under certainty

In any statistical decision problem, three assumptions can be made regarding knowledge of the true state of nature, θ. First, we can assume that θ is known *with certainty,* in which case the determination of the act to follow is straightforward. That is, suppose it is known that $\theta = 2$ in the Example 21.1 problem. Then, from the decision tree in Figure 21.2, reserve 2 is the best act to follow.

FIGURE 21.2 | Payoff decision tree for the tour problem in Example 21.1

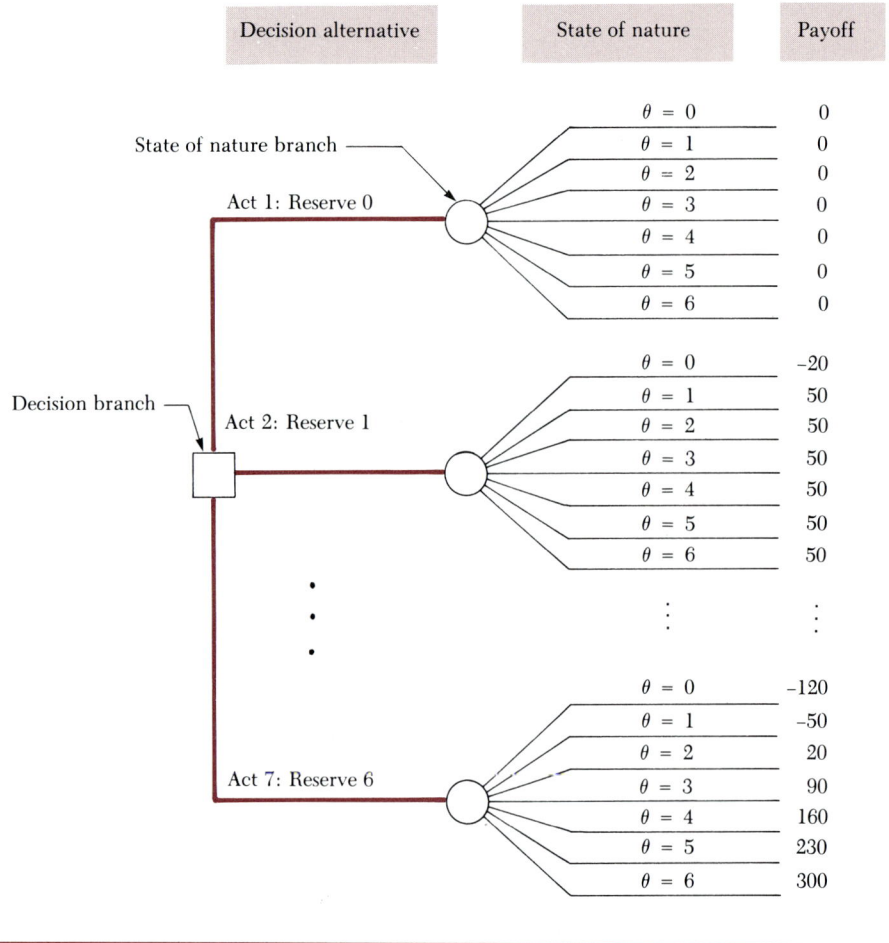

Decision alternative	State of nature	Payoff

State of nature branch

Act 1: Reserve 0

$\theta = 0$	0
$\theta = 1$	0
$\theta = 2$	0
$\theta = 3$	0
$\theta = 4$	0
$\theta = 5$	0
$\theta = 6$	0

Decision branch

Act 2: Reserve 1

$\theta = 0$	-20
$\theta = 1$	50
$\theta = 2$	50
$\theta = 3$	50
$\theta = 4$	50
$\theta = 5$	50
$\theta = 6$	50

Act 7: Reserve 6

$\theta = 0$	-120
$\theta = 1$	-50
$\theta = 2$	20
$\theta = 3$	90
$\theta = 4$	160
$\theta = 5$	230
$\theta = 6$	300

Decisions under uncertainty

Second, we can assume total ignorance of the value of θ, in which case we are said to be making a decision *under uncertainty*. Instances in which decisions are made under uncertainty, although few in number, do exist. These instances usually occur whenever something is being done for the first time, although even in these instances we may be able to say something about the probability of various values of θ based on a recent experience of a like occurrence.

Decisions under risk

And third, we can assess a probability distribution for various values of θ occurring, in which case we are making decisions *under risk*. The majority of both personal and business decisions are made with imperfect knowledge of the value of θ, and hence decision making under risk characterizes most decision situations. Other assumptions are often made concerning knowledge of θ, but a close examination of many situations reveals that decisions are made with somewhat less than perfect information, or are actually made under *risk*.

In this section, we will describe four strategies for resolving decisions under uncertainty: maximin profits, maximax profits, minimax regret, and Laplace, or the strategy of insufficient information. Chapters 22 and 23 will then discuss decision making under risk.

21.3.1 | Maximin profits

The maximin profits strategy dictates that we select the best of the worst—that is, that we maximize the worst that can happen to us. It is derived by selecting the minimum payoff for each act and maximizing the minimum of these payoffs.

Maximin

$$\text{Maximin profits}$$

$$\text{Max}_{i}[\text{Min}_{j}(\Omega_{ij})] \tag{21.2}$$

Example 21.4 Compute the maximin profits strategy for the data given in Example 21.1.

Solution For the tour problem of Example 21.1, the maximin profits strategy dictates that George schedule no seats in advance:

$$\text{Max}_{i}(0, -20, -40, -60, -80, -100, -120) = 0$$

With this strategy, George is *guaranteed* of losing no more than $0 irrespective of the number of seats or tours actually demanded. The maximin profits strategy is also depicted in Table 21.4.

The maximin strategy is *ultraconservative,* since it hedges against the *worst* that can possibly happen. Its use appears to be based on the premise that forces that control the state of nature are malevolent, that forces are "out to get" the decision maker. It frequently leads to inaction (e.g., reserve 0), since almost every decision *might* lead to losing money (a negative payoff), whereas doing nothing incurs no monetary losses. Still, the strategy does have applicability in certain one-on-one competitive situations as well as others.

21.3.2 | Maximax profits

Decision makers who use the maximax profits strategy have the *opposite* view of the world—as does the maximin strategist. They view the world as

TABLE 21.4 | Determination of maximin profits strategy for tour problem of Example 21.1*

Act, a_i	State of nature, θ_j 0	1	2	3	4	5	6	
0	0	0	0	0	0	0	0	←Maximum = 0 at a_0
1	−20	50	50	50	50	50	50	
2	−40	30	100	100	100	100	100	
3	−60	10	80	150	150	150	150	Minimum payoff of each act (shaded area)
4	−80	−10	60	130	200	200	200	
5	−100	−30	40	110	180	250	250	
6	−120	−50	20	90	160	230	300	

* Quantities are in dollars.
Source: Example 21.1.

being composed of "friendly forces" and hence choose to maximize the best consequence of all possible acts. They are termed *ultraoptimistic*.

Maximax

Maximax profits

$$\underset{i}{\text{Max}}[\underset{j}{\text{Max}}(\Omega_{ij})] \qquad (21.3)$$

Example 21.5 Using the data in Example 21.1, compute the maximax profits strategy.

Solution For the tour problem of Example 21.1, the maximax profits value is $300:

$$\underset{i}{\text{Max}}(0, 50, 100, 150, 200, 250, 300) = \$300$$

The act that corresponds to this maximax profit strategy is "reserve six seats."

This strategy poses a set of disadvantages similar to the maximin strategy. Just as the world is rarely judged to be malevolent, so too is it rarely judged as being composed entirely of friendly forces. This strategy can be appropriate when it is necessary to "go for broke," such as when a major discovery or innovation is required to prevent a firm from going under.

21.3.3 | Minimax loss (minimax opportunity loss, minimax regret)

The minimax loss strategy operates on the opportunity loss matrix in reverse of the way the maximin profits strategy operates on the payoff matrix. First, the maximum opportunity loss for each act is computed. Then the strategy corresponding to the minimum of these maximum losses is selected.

Minimax opportunity loss or minimax regret

Minimax opportunity loss

$$\underset{i}{\text{Min}}[\underset{j}{\text{Max}}\ l(a_i,\ \theta_j)] \qquad (21.4)$$

Example 21.6 Using the data in Example 21.2, compute the minimax loss strategy for George Thomas.

Solution From the opportunity loss matrix for the tour problem (Table 21.3), the minimax loss value is:

$$\underset{i}{\text{Min}}(300,\ 250,\ 200,\ 150,\ 100,\ 100,\ 120) = \$100$$

The strategies corresponding to the minimax opportunity losses of $100 are to reserve four *or* five seats for the tour.

Under the minimax loss strategy, George should reserve either four or five seats and minimize the maximum opportunity loss he would suffer. Note that in general the maximin profit and the minimax loss strategies are not equivalent, as shown in this example.

The minimax loss strategy has the desirable property that it focuses on the analysis of the opportunity costs that arise in decision problems. Thus, for example, the opportunity loss matrix shows more of a loss when two seats are reserved and five versus four are demanded, whereas the payoff matrix indicates the same profit (reward) for both demands for the one reserve level. Still, it is a pessimistic strategy, since it examines only the worst regrets for each act (alternative) and ignores all others.

21.3.4 | Principle of insufficient information or Laplace method

The principle of insufficient information strategy bridges the gap in a sense between decision making under uncertainty and decision making under risk. Under it, we assume that each state of nature is equally likely to occur. The Laplace value of a strategy then equals the arithmetic average of the values for that strategy in the payoff matrix.

Laplace

$$\text{Laplace value}$$

$$\underset{i}{\text{Max}} \left[\frac{1}{N} \sum_{j} \Omega_{ij} \right] \tag{21.5}$$

In the formula, N is equal to the number of possible states of nature ($N = 7$ for the tour problem of Example 21.1).

Example 21.7 Compute the Laplace values for each of the acts in Example 21.1, and determine the act to follow that maximizes the Laplace value for this problem.

Solution The Laplace values for the tour problem of Example 21.1 are:

Act (supply of seats reserved)	0	1	2	3	4	5	6
Laplace value	0	40	70	90	100	100	90

Hence George should reserve either four or five seats and maximize the Laplace value of 100.

The Laplace method does offer certain advantages over the other methods described thus far. First, it uses all the information available in the payoff matrix—it does not examine only the "extremes." Second, it forces the decision maker to think in probabilistic terms; if he has no reason to suspect that one event is more likely to occur than another, then he should be willing to weight them equally in arriving at a decision.

Still, caution should be used when applying the Laplace strategy or the principle of insufficient information whenever we suspect that events are not equally likely to occur. Whenever $P(\theta_0) \neq P(\theta_1) \neq \ . \ . \neq P(\theta_n)$, the Laplace method is not appropriate. Too, we should be cautious that the decision maker is not using the Laplace method as a means for avoiding the sometimes difficult task of assessing probabilities for the various states of nature.

■ **21.4**

Summary

This chapter has addressed statistical decision making under uncertainty. We use the theory provided whenever knowledge does not exist concerning probabilities of various states of nature occurring and the opportunity does not exist to obtain additional information.

Different alternatives or methodologies were described for structuring a decision problem: the payoff or loss matrix, the functional form, and decision trees. Payoff or loss matrices offer the advantage of portraying the payoff or loss of every act–state of nature combination when the number of possible states of nature and acts is finite and small. The functional form is a concise means of representing the effects of the interaction of act–state of nature pairs when the number of possible pairs is large or even infinite (in the case of a continuous probability distribution for θ). Decision trees are particularly helpful when it is useful to examine the results of various decisions (act selections) at different decision points.

Decision making, or act selection, was then categorized as being made under certainty, under uncertainty, or under risk, depending on knowledge concerning the state of nature θ. Four different strategies were then described for making decisions under uncertainty: maximin profits, maximax profits, minimax loss, and the principle of insufficient information. Maximin profits strategists believe the world to be malevolent and out to get them, whereas maximax profits strategists believe the world to be benevolent and composed exclusively of friendly forces. Minimax loss strategists are still pessimists in that they believe in examining only the worst consequences for each act. These strategists do, however, make an attempt at incorporating profit foregone (opportunity loss) into their decisions. And finally, the person using the principle of insufficient information would like to incorporate all consequences of each act into his or her decision (not just the best or the worst) and, lacking information on the various states of nature, is willing to assume that each state of nature is equally likely to occur. This person thus attempts to bridge the gap between decision making under uncertainty and decision making under risk. Table 21.5 is a summary of these four strategies for making decisions under uncertainty.

TABLE 21.5 | Strategies for making decisions under uncertainty

The set of acts: $a_i \in A$ Payoff or profit: $P(a_i, \theta_j) = \Omega_{ij}$	The set of possible states of nature: $\theta_j \in \theta$ Opportunity loss: $l(a_i, \theta_j) = (\underset{i}{\text{Max }} \Omega_{ij}) - \Omega_{ij} \geq 0$

Strategy	Derivation	Formula number	Assumptions about the world
Maximin profits	$\underset{i}{\text{Max}}[\underset{j}{\text{Min}}(\Omega_{ij})]$	(21.2)	Malevolent; forces are out to get the decision maker; strategy is ultraconservative
Maximax profits	$\underset{i}{\text{Max}}[\underset{j}{\text{Max}}(\Omega_{ij})]$	(21.3)	Benevolent; world is composed of friendly forces; strategy is ultraoptimistic
Minimax loss	$\underset{i}{\text{Min}}[\underset{j}{\text{Max}}\ l(a_i, \theta_j)]$	(21.4)	Malevolent; the worst will probably happen; strategy is conservative
Laplace (principle of insufficient information)	$\underset{i}{\text{Max}} \dfrac{\sum\limits_{j-1}^{N} \Omega_{ij}}{N}$	(21.5)	All states are equally likely to occur; strategy is neither optimistic nor pessimistic

Chapter 22 will describe decision making under uncertainty, where a probability distribution for the possible states of nature either exists or can be readily determined. Then Chapter 23 will again examine decision making under uncertainty, but for the special case in which there is an opportunity to obtain additional (sample) information regarding the various states of nature. We will see in Chapter 23 that statistical decision theory provides us with a model for determining how much information should be obtained and how much we should be willing to pay for it!

■ References

Introductory Level

Anderson, D. R.; D. J. Sweeney; and T. A. Williams. *An Introduction to Management Science.* 4th ed. St. Paul, Minn.: West Publishing, 1985, Chapter 16.

Cook, William D. *Quantitative Methods for Management Decisions.* New York: McGraw-Hill, 1985, Chapter 9.

Lapin, Lawrence L. *Quantitative Methods for Business Decisions.* 2nd ed. New York: Harcourt Brace Jovanovich, 1981, Chapters 5 and 6.

Advanced Level

Bierman, H.; C. P. Bonini; and W. H. Hausman. *Quantitative Analysis for Business Decisions.*

7th ed. Homewood, Ill.: Richard D. Irwin, 1986, Chapters 3–10.

Hillier, F. S., and G. J. Lieberman. *Introduction to Operations Research.* 3rd ed. San Francisco: Holden-Day, 1980, Chapter 15.

Raiffa, H. *Decision Analysis.* Reading, Mass.: Addison-Wesley Publishing, 1968.

Raiffa, H., and R. Schlaifer. *Applied Statistical Decision Theory.* Boston: Division of Research, Harvard University, 1961.

Schlaifer, R. *Introduction to Statistics for Business Decisions.* New York: McGraw-Hill, 1961.

Winkler, R. L. *An Introduction to Bayesian Inference and Decision.* New York: Holt, Rinehart & Winston, 1972.

■ Problems

Section 21.2 Problems

21.1 What is the payoff function (functional form) or the profit function for the tour problem of Example 21.1?

21.2 Given the following payoff table, construct the corresponding loss table (assume payoffs are profits):

Act \ State of nature	θ_1	θ_2	θ_3	θ_4	θ_5
a_1	−8	10	2	5	−1
a_2	12	6	4	−1	2
a_3	1	8	−10	6	4
a_4	0	0	1	4	10
a_5	6	2	8	−4	4

21.3 Paula Clark obtains long-stem roses from Fred's Flowers each Friday for sale to the public. Fred generally sells Paula flowers that, if left over the weekend, would likely be unsalable the next week. Hence he is willing to give Paula a good price on his remaining long-stem roses and to take back any roses that are not sold over the weekend and give Paula partial credit for them. Those are sold to Paula in quantities or units of one dozen each and at a price of $12 per dozen. Paula sells the roses for $15 per dozen, and any remaining unsold over the weekend can be returned to Fred for a credit of $10 per dozen. Although the number of roses remaining to be sold at the end of the week varies, Fred is willing to let Paula have as many as six dozen each weekend and supply her with fresh flowers at the same price if she desires more than the quantity of "old" roses that Fred has. Characterize this decision problem both in terms of the various states of nature (demands) and in terms of the various acts (supplies) open to Paula. Construct the payoff and loss matrices for this problem.

21.4 What is the functional form of the profit function for the inventory problem of Problem 21.3?

21.5 Freddie Flicker is the proprietor of Frieda Flicker's Wholesale Produce Shop. Each week Freddie must decide how many cases of lettuce to order for sale to retail outlets. In the past, the demand for cases has varied between zero and four per week, with no apparent "pattern" to the demand quantities. For each case sold, Freddie earns a nice profit of $30. For each case unsold at the end of the week, Freddie loses $10.
a. Construct a payoff table for this problem.
b. Construct a payoff decision tree for this problem.

21.6 Each month, the Expert Computer Systems Company (ECSC) can order up to and including five ATB model 6600 microcomputers from its supplier. Each unit costs ECSC $2,000, and ECSC can sell each of these microcomputers for $3,000. At the end of the month, ECSC *must* return any unsold ATB model 6600 microcomputer units to the sup-

plier for a credit of $1,800 (thus ECSC takes a loss of $200 on each unsold unit).
a. Construct a payoff matrix for this purchasing problem with the acts being the number of units ordered each month and the states of nature being the number of units ECSC is able to sell each month.
b. Construct a payoff decision tree for this problem.

Section 21.3 Problems

21.7 Explain the difference between decision making under risk and decision making under uncertainty. Why is the Laplace method considered to be a part of decision making under uncertainty as opposed to decision making under risk?

21.8 Describe a situation you encountered today in which you had to make a choice between alternatives without the benefit of obtaining additional information. Is it possible that you could have reduced the risk in the situation by obtaining additional information if circumstances had been different?

21.9 Given the payoff and loss table of Problem 21.2 what is the appropriate act to follow under each of the following strategies?
a. Maximin profits.
b. Maximax profits.
c. Minimax loss.
d. Laplace method.

21.10 Given the following payoff table (profits), construct the corresponding loss table:

Act \ State of nature	θ_1	θ_2	θ_3	θ_4	θ_5
a_1	100	-50	50	-100	0
a_2	80	20	20	60	50
a_3	200	-80	-20	50	100
a_4	50	40	60	-50	0

21.11 Given the payoff and loss table of Problem 21.10, which is the appropriate act to follow under each of the following strategies?

a. Maximin profits.
b. Maximax profits.
c. Minimax profits.
d. Laplace method.

21.12 Given the following payoff table, can any of the acts (with their associated payoffs) be removed from the table without affecting the optimal act to follow? If so, such acts are said to be *inadmissible* or *dominated* acts and can be dropped from further analysis.

State of nature					
Act	θ_1	θ_2	θ_3	θ_4	θ_5
a_1	7	−3	2	12	5
a_2	−2	−4	10	−4	−6
a_3	2	4	9	6	3
a_4	6	10	3	11	−2
a_5	−2	−2	11	4	−4
a_6	1	9	7	2	8
a_7	4	−4	0	−1	4

21.13 Using the data given in Problem 21.12, compute the appropriate act to follow under each of the following strategies.
a. Maximin profits.
b. Maximax profits.

21.14 Through inheritance, a trust has been established for a teenager to attend college. Conditions of the trust are such that the funds can be invested in one of four "portfolios" (alternatives), each of which matures in the next three years when the individual will graduate from high school and begin attending college. Depending on the economic conditions over the next three years, the net gain or loss in the amount of the trust is as shown in the payoff matrix below. Determine the optimal act to follow under each of the following strategies:
a. Maximin profits.
b. Maximax profits.
c. Laplace method.

State of nature (the economy)			
Act	θ_1	θ_2	θ_3
a_1	$4,000	$8,400	$−2,500
a_2	10,000	−5,000	4,000
a_3	8,000	8,000	8,000
a_4	−7,000	4,000	12,500

21.15 Construct an opportunity loss matrix for the data in Problem 21.14, and determine the optimal act to follow under the minimax loss strategy. What type of investment strategy would you guess is being followed under act 3?

21.16 Using the data given in Problem 21.3, determine the optimal act to follow under each of the following strategies.
a. Maximin profits.
b. Maximax profits.
c. Minimax loss.
d. Laplace method.

21.17 Using the payoff table constructed in Problem 21.5, determine the optimal act to follow under the following strategies.
a. Maximin profits.
b. Maximax profits.
c. Laplace method.

21.18 Using the payoff table constructed in Problem 21.5, construct the opportunity loss matrix, and determine the optimal act based on the minimax loss strategy.

21.19 Using the payoff matrix constructed in Problem 21.6, determine the optimal act based on the following strategies:
a. Maximin profits.
b. Maximax profits.
c. Laplace method.

21.20 Construct the opportunity loss matrix for the payoff matrix determined in part *a*, Problem 21.6, and determine the optimal act under the minimax loss strategy.

Additional Problems

21.21 The following matrix represents *costs* for the pairs of acts and states of nature:

States of nature Acts	θ_1	θ_2	θ_3	θ_4
a_1	$10,000	$6,000	$ 0	$-2,000
a_2	4,000	8,000	-1,000	1,000
a_3	12,000	5,000	2,000	500
a_4	8,000	8,000	8,000	8,000

a. Convert this cost matrix to a payoff matrix by taking the negative of all table entries (Note: A cost of $10,000 is equivalent to a negative payout of $-$10,000.

b. Using the payoff matrix constructed in part *a*, determine the optimal acts based on the following decision strategies.
  i. Maximax profits.
  ii. Minimax profits.
  iii. Laplace method.

c. Construct the loss table from the payoff table determined in part *a*, and find the optimal act based on the minimax loss strategy.

21.22 Whenever a decision matrix contains costs rather than payoffs (revenues or profits), either (1) the matrix must be converted to a payoff matrix by taking the negative of all table cell entries or (2) decision strategies must be used that apply to costs rather than payoffs (revenues or profits). The table below gives the equivalence of strategies for payoffs and costs:

Payoffs (revenues or profits)	Costs	Assumptions about the world
Maximin profits	Minimax costs	Malevolent
Maximax profits	Minimin costs	Benevolent
Minimax loss	Minimax loss	Malevolent
Laplace (max avg.)	Laplace (min. avg.)	Equally likely state of nature

For the cost matrix in Problem 21.21, apply the following strategies to the acts to determine which act is optimal under each strategy.
a. Minimax costs.
b. Minimin costs.
c. Laplace method.

21.23 Verify that the optimal acts determined for each of the strategies applied to the cost matrix in Problem 21.22 match the optimal acts determined for each of the strategies applied to the payoff matrix ("negative costs") in part *b* of Problem 21.21.

21.24 The opportunity loss matrix can be constructed for a decision matrix containing costs rather than payoffs (revenues or profits). The formula for the opportunity losses in this case is:

$$l(a_i, \theta_j) = (\underset{i}{\text{Min }} \Omega_{ij}) - \Omega_{ij} \geq 0.$$

That is, for each state of nature (θ_j), determine the *minimum* cost. Once the minimum cost for a particular state of nature (column) has been determined, the opportunity loss for an act is found by taking the difference between the minimum cost and the cost for that act.

a. For the decision matrix in Problem 21.21, where the dollar amounts in the table represent costs, construct the opportunity loss matrix.

b. Apply the minimax loss strategy to the opportunity loss matrix constructed in part a.

21.25 Verify that the optimal acts are the same whether they are (1) determined by applying the minimax loss strategy to costs to the opportunity loss table constructed in part a, Problem 21.24 or (2) determined by applying the minimax loss strategy to payoffs to the opportunity loss table constructed in part c, Problem 21.21.

21.26 Starion Film Production Company has recently completed a new full-length movie entitled "Super Rat." Starion must determine how to release this movie. Management is considering three alternatives:

A_1: Release as a major movie with a run of 12 weeks. After the run, release in the form of a videocassette for purchase and rentals for a period of 30 weeks. Finally, after the $12 + 30 = 42$ weeks, sell the rights to CINEMIN, a cable movie channel.

A_2: Release as a minor movie with a run of six weeks. After the run, release in the form of a videocassette for purchase and rentals for a period of 24 weeks. After the $6 + 24 = 30$ weeks, sell the rights to CINEMIN.

A_3: Sell the rights immediately to CINEMIN.

The financial consequences of each act depend on how well the movie is rated by movie critics. Starion considers three states of nature:

θ_1: "Super Rat" receives a high rating by the critics.

θ_2: "Super Rat" receives an average rating by the critics.

θ_3: "Super Rat" receives a poor rating by the critics.

Starion estimates the following payoffs (profits) for the pairs of acts and states of nature:

Acts \ States of nature	θ_1	θ_2	θ_3
A_1	$5,000,000	$1,000,000	−$8,000,000
A_2	$2,000,000	$ 200,000	−$3,000,000
A_3	$ 500,000	$ 500,000	$ 500,000

Determine the optimal act to follow under each of the following strategies:

a. Maximin profits.
b. Maximax profits.
c. Laplace method.

21.27 Construct the opportunity loss table for the payoff matrix in Problem 21.26, and determine the optimal act under the minimax loss strategy.

21.28 The Radix Computer Company receives microprocessor chips for its Radix model 1000 microcomputer from a supplier in Japan. The company has recently become concerned about the possibility of restrictions placed on the importation of the chips from outside the country. The management of Radix considers two alternate supply sources: Taiwan (which would also be subject to import restrictions, though not as restrictive as those applied to Japanese imports) and Motarola, a U.S. supplier. Radix thus considers two states of nature:

θ_1: Import restrictions do not occur.
θ_2: Import restrictions do occur.
The three decision alternatives are:
a_1: Buy the chips from the Japanese supplier.
a_2: Buy the chips from the Taiwanese supplier.
a_3: Buy the chips from Motarola, the U.S. supplier.
The chips are purchased in lots of 10,000. Based on the effects of the import restrictions

if they become law, Radix constructs the following decision *cost* matrix. (That is, the matrix entries are *costs* per 10,000 chips.)

States of nature Acts	θ_1	θ_2
a_1	$ 50,000	$120,000
a_2	80,000	105,000
a_3	100,000	100,000

a. Determine the optimal minimax loss (or equivalently, the maximin payoff) act.
b. Determine the optimal minimin loss (or equivalently, the maximax payoff) act.
c. Determine the optimal act by the Laplace method.
d. Construct the opportunity loss table for this cost matrix.
e. Determine the optimal minimax opportunity loss acts (or equivalently the minimax loss act).
f. Construct a decision tree for this problem.

22

Statistical decision theory: Decision making under risk

■ 22.1
Introduction

In this chapter, we will introduce the topic of statistical decision theory for the important case of decision making when there are not opportunities to obtain additional information, but information does exist on the probability distribution of θ, $P(\theta)$ [or $f(\theta)$]. These instances arise frequently in our daily lives as we are confronted with uncertain situations but must make an immediate decision. For example, we may look outside to decide whether or not to carry an umbrella on a specific day. We could, if time allowed, listen to the weather report to get additional information about the weather, but in so doing, we run the risk of missing our means of transportation, being late, or missing an appointment. Funds also can be a limiting factor in determining whether additional information can be obtained. For example, we may have spent all that we feel we can in searching out employment opportunities, and we must now arrive at a terminal decision on where we will work.

The case in point is that, for whatever reason, each of us daily faces uncertain situations in which a decision must be made *without benefit of additional information*. Statistical decision making under risk and with terminal acts serves as a medium for analyzing a decision situation and then prescribing a course of action to follow that satisfies a given measure of performance.

In this chapter we will often refer to the probability distribution of the random variable θ as the *prior distribution* or *prior probability distribution* of θ. For our purposes, the prior probability distribution of θ will simply consist of the probability distribution of θ that currently exists. In Chapter 23, we will learn how to revise this distribution in the light of new information, and in fact, will assess just how much effort (dollars) should be expended in

Prior distribution

obtaining additional information on the value of the random variable. For the moment, however, we will assume that the opportunity does *not* exist to obtain additional information and, hence, will work with the prior (existing) probability distribution of the random variable θ.

22.2.1 | Finite action decision problems with two or more alternatives (acts)

When making decisions under risk, we inherently assume that a probability distribution can be assessed for the various states of nature, θ_j. This distribution can be assessed using *any* of the methods described in Chapter 3, as well as can be based on sample results. In the majority of decision situations, it is possible to assess some probability distribution for the various states of nature (e.g., the probability of passing this course), even though this assessment may be a difficult task! In situations in which probabilities can be assessed for the various states of nature, the *equivalent* strategies of maximizing expected monetary value (EMV) (or payoff) and minimizing expected opportunity loss (EOL) may be appropriate.

Expected value decision making thus incorporates the likelihood or the probability of certain events occurring in the analysis of a decision. Unlike maximin profits, maximax profits, or minimax loss (or regret) from Chapter 21, expected value decision making takes into account *all* the values in the payoff or the loss matrix, as well as the relative likelihood of events occurring. And unlike the Laplace method, probabilities of events occurring are determined by objective and subjective estimates, rather than by the belief that all events are equally likely.

22.2.2 | Bayes criterion and expected monetary value

In Chapter 5 we defined an expected value of a discrete random variable to be the arithmetic mean or average of a random variable. That is, in computing an expected value, we weight each value of the random variable by its relative frequency or its probability of occurrence. Since an expectation determined in statistical decision theory usually involves monetary units, such an expectation is referred to as an *expected monetary value*

EMV, EOL

(EMV), or possibly as an *expected opportunity loss* (EOL).

The expected profit (payoff) maximization *criterion* in statistical decision theory states that the expected profit of each act should be computed and that act (course of action) selected that yields the greatest expected profit.

Expected monetary value: EMV

Maximize expected monetary value (EMV) (profit)

$$\text{Max}_i [E(\Omega_{ij})] = \text{Max}_i \sum_j \Omega_{ij} P(\theta_j) \qquad (22.1)$$

In the equation, Ω_{ij} = Profit if act a_i is selected and state of nature θ_j occurs, and $P(\theta_j)$ = Probability that state of nature θ_j occurs.

The Bayes criterion or the minimize expected opportunity loss criterion in statistical decision making states that the expected opportunity loss of each act should be computed and that act selected with the minimum expected opportunity loss.

Expected opportu-
nity loss: EOL

Bayes criterion (minimize EOL)

$$\operatorname*{Min}_{i} [E\{l(a_i, \theta_j)\}] = \operatorname*{Min}_{i} \left[\sum_{j} l(a_i, \theta_j) \cdot P(\theta_j) \right] \qquad (22.2)$$

We will illustrate the use of these two decision-making criteria in the following example.

Example 22.1   Assume that George in Example 21.1 has kept track of the number of seats sold on each tour and has assessed probabilities of various sales amounts occurring according to the long-run relative frequency interpretation of probability as given in Table 22.1.

TABLE 22.1 | Probabilities (long-run relative frequency) of various sales levels (demands)

State of nature, θ_j (sales)	0	1	2	3	4	5	6
Long-run relative frequency	0.01	0.08	0.32	0.31	0.19	0.06	0.03

Source: George's six-year travel agency records.

Compute the expected payoff (profit) and the expected opportunity loss of each act. Which act (number of seats to reserve) should George select?

Solution   For the tour problem, the act "reserve 2" yields the following expected profit:

$$-40(0.01) + 30(0.08) + 100(0.32) + 100(0.31) + 100(0.19)$$
$$+ 100(0.06) + 100(0.03) = \$93.00$$

Expected profits for each of the other possible acts are given in Table 22.2, where a stick diagram has been constructed depicting the probabilities of the various states of nature. Based on historical frequencies, George should reserve three seats on each tour and maximize his expected profit of that act

TABLE 22.2 | Calculation of expected monetary value of all acts for tour problem—empirical discrete distribution

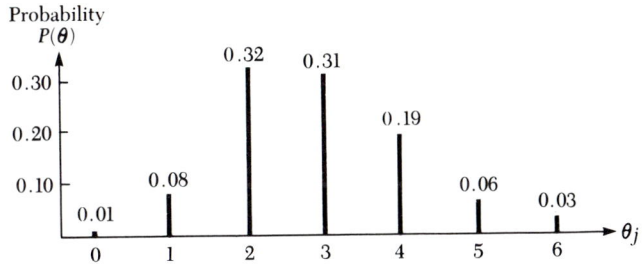

Act, a_i (supply)	State of nature, θ_j (demand)							Expected monetary value of act
	0	1	2	3	4	5	6	
0	0	0	0	0	0	0	0	$ 0.00
1	−20	50	50	50	50	50	50	49.30
2	−40	30	100	100	100	100	100	93.00
3	−60	10	80	150	150	150	150	114.30 ←
4	−80	−10	60	130	200	200	200	113.90
5	−100	−30	40	110	180	250	250	100.20
6	−120	−50	20	90	160	230	300	82.30

Source: Examples 21.1 and 21.2.

of $114.30 as indicated in Table 22.2. Any other act would reduce his expected profit below $114.30.

For the same problem, the act "reserve 2" yields an expected opportunity loss of $51.50 as follows:

$$40(0.01) + 20(0.08) + 0(0.32) + 50(0.31) + 100(0.19)$$
$$+ 150(0.06) + 200(0.03) = \$51.50$$

The opportunity losses for each of the other acts are given in Table 22.3. Using the Bayes criterion, George should reserve three seats on each tour, yielding a minimum expected opportunity loss of $30.20. Note that this is the same act selected using the profit (EMV) maximization criterion, and in fact the two methods are equivalent in the sense that they will always lead to the same act. (A proof of this relationship is left to Problem 22.1 at the end of the chapter.)

Equivalence of
EMV and EOL
acts

TABLE 22.3 | Calculation of expected opportunity loss of all acts for tour problem—empirical discrete distribution

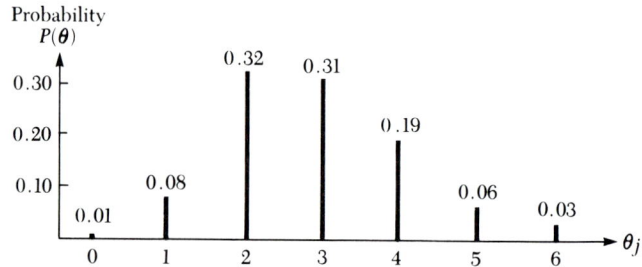

Act, a_i (supply) / State of nature, θ_j (demand)	0	1	2	3	4	5	6	Expected opportunity loss
0	0	50	100	150	200	250	300	$144.50
1	20	0	50	100	150	200	250	95.20
2	40	20	0	50	100	150	200	51.50
3	60	40	20	0	50	100	150	30.20 ←
4	80	60	40	20	0	50	100	30.60
5	100	80	60	40	20	0	50	44.30
6	120	100	80	60	40	20	0	62.20

Source: Examples 21.1 and 21.2.

22.2.3 | Expected profit under certainty

From Table 22.2, note that if demand is 1, profit is maximized if one seat is reserved; if demand is 2, profit is maximized if two seats are reserved; etc. Thus the *maximum* profits to be realized from alternate acts *conditional upon the value of θ* are given in the principal diagonal of Table 22.2. The sum of the products of the maximum payoff for each value of θ and the probability of θ occurring have a special meaning in statistical decision making called the *expected profit of a perfect predictor* (EPPP). It represents the expected profit that would be realized if one knew beforehand the value of θ that will occur—in other words, what expected profit would be if one could predict precisely the state of nature that might occur.

This quantity, the *expected profit of a perfect predictor,* is also called the *expected profit under certainty* and is given in equation (22.3).

Expected profit of perfect predictor

Expected profit under certainty

Expected profit of perfect predictor: EPPP	Expected profit of perfect predictor (expected profit under certainty) $$\text{EPPP} = \sum_{j} P(\theta_j)(\underset{i}{\text{Max }} \Omega_{ij}) \qquad (22.3)$$

Note how we should interpret the expected profit of a perfect predictor, EPPP. If we knew beforehand that there would be a demand for three seats, we would advise George to reserve three seats, and he would realize a profit of 3($50) = $150. On what percentage of the tours will three seats be demanded? Using the long-run relative frequency interpretation of probability, on 31 percent of the tours, three seats will be demanded. The product (0.31)($150) represents the profit that can be realized if we know in advance what demand will be, multiplied by the relative frequency or the probability with which this demand or profit potential will occur. When we continue to multiply these maximum profits by the probability that (or the frequency with which) they occur and then add the corresponding products, we obtain the *expected profit* from reserving the optimal number of seats *each time* a tour is selected. That is, we are simply weighting each profit figure by the frequency with which it will occur in the long run.

Example 22.2 Calculate the expected profit of the perfect predictor (EPPP) for George's payoff matrix given in Table 22.2

Solution The computation of EPPP is shown below:

(1) State of nature (θ_j) (demand: number of seats)	*(2)* Probability of state of nature $P(\theta_j)$	*(3)* Conditional (maximum) profit Max Ω_{ij}	*(4)* (2) × (3)
0	0.01	$ 0	0.00
1	0.08	50	4.00
2	0.32	100	32.00
3	0.31	150	46.50
4	0.19	200	38.00
5	0.06	250	15.00
6	0.03	300	9.00
	1.00		$144.50 = EPPP

Thus EPPP = $144.50. If George could determine the number of seats demanded each time he arranged a tour, then he could average $144.50 per tour. Notice how EPPP compares with EMV calculated in Table 22.2. By the EMV strategy, he will order three seats for *every* tour, thereby averaging $114.30 per tour. Clearly, he could do better if he could determine the demand on *each* tour and order exactly that number of seats.

22.2.4 | Expected value of perfect information

Using equation (22.1), we compute the expected profit to be realized by selecting the optimal act (alternative) under the prior probability distribution of θ. The difference between the profit to be realized with perfect knowledge of the various states of nature and the best we can do (maximize EMV criterion) *without this information* represents an upper bound for the amount we should be willing to pay for this information. That is, the difference between EPPP and the maximum EMV represents the *expected value of perfect information* whose formula is given in equation (22.4).

Expected value of perfect information: EVPI	

Expected value of perfect information

$$\text{EVPI} = \left[\sum_j P(\theta_j)(\underset{i}{\text{Max }} \Omega_{ij}) \right] - \left\{ \underset{i}{\text{Max}} \left[\sum_j P(\theta_j)\Omega_{ij} \right] \right\} \qquad (22.4)$$

$$= \quad \text{EPPP} \quad - \quad \text{Max[EMV]}$$

Example 22.3   Given the information provided in Examples 22.1 and 22.2, compute the expected value of perfect information (EVPI).

Solution   Formula (22.4) asks us to calculate EPPP $\left[\sum_j P(\theta_j)(\underset{i}{\text{Max }} \Omega_{ij}) \right]$ and subtract from this quantity the EMV for the optimal act based on the EMV strategy. From Example 22.2 EPPP = $144.50, and from Example 22.1 EMV = $114.30 for the optimal EMV act (order three seats). Thus

$$\text{EVPI} = \text{EPPP} - \text{EMV (act: order 3 seats)}$$
$$= \$144.50 - \$114.30 = \$30.20$$

Thus if George could somehow obtain *perfect* information regarding the number of seats demanded on each tour, he would over the long run expect to make $30.20 more on each tour than if he used the act that maximizes expected monetary value (EMV), which is reserve three seats for each and every tour. This quantity (EPVI) sets an upper limit on the amount George should be willing to pay for such information.

22.2.5 | Conditional value of perfect information

Recall that a conditional opportunity loss is defined to be the difference between the profit or payoff that could have been realized and what was realized from following a certain course of action. The expected value of perfect information can be expressed as the expectation (with respect to $\boldsymbol{\theta}$) of the conditional losses that result from selecting *that particular act that is optimal*. Thus the expected value of perfect information can be calculated by using either equation (22.2) or (22.4). Because the expected opportunity losses may have already been computed in determining the optimal act, equation (22.2) may be preferred in those cases. Note that whatever formula (22.2 or 22.4) is used to compute EVPI, EVPI can be computed only *after* the optimal act to follow has been selected, and it is computed using the *conditional losses for the optimal act*.

Expected value of perfect information equals expected opportunity loss for optimal act

Expected value of perfect information
(using opportunity loss matrix)

$$\text{EVPI} = \underset{i}{\text{Min}} \left[E\{l(a_i, \theta_j)\}\right] = \underset{i}{\text{Min}} \left[\sum_j l(a_i, \theta_j) \cdot P(\theta_j) \right] \qquad (22.5)$$

$$= \text{EOL (optimal act)}$$

Conditional value of perfect information (CVPI)

The conditional losses (conditional opportunity losses) for the optimal act are also referred to as the *conditional value of perfect information* (CVPI). That is, *after* the optimal act to follow has been determined, the conditional value of perfect information is equal to the opportunity loss for values of $\boldsymbol{\theta}$ that differ from the value that is optimal under the maximize EMV or the minimize EOL criterion. For example, referring to Table 22.3, the optimal act is to reserve three seats. The *conditional* value of perfect information that $\boldsymbol{\theta}$ is equal to 0 (no seats are demanded) is equal to $60. That is, if demand is 0, the decision maker will save $60 if he reserves no seats, as opposed to reserving the optimal number of seats. The decision maker should be willing to spend up to $60 for information that $\boldsymbol{\theta}$ will equal 0, because this is the amount that will be saved by reserving zero as opposed to three seats.

The "conditional" in the conditional value of perfect information thus refers to the value of perfect information being conditional on the value of $\boldsymbol{\theta}$ that occurs. For example, again referring to Table 22.3, what should the decision maker be willing to spend for information that the number of seats demanded on a tour is five? Since profit can be *increased* by $100 if five as opposed to three seats are reserved, the conditional value of perfect information that $\theta = 5$ in this instance is $100, and the decision maker should be willing to spend up to $100 for this information. Thus the conditional value of perfect information varies as the value of $\boldsymbol{\theta}$ deviates from its optimal value determined by the EMV or EOL criterion. The conditional value of perfect information is given in Definition 22.1.

Conditional value
of perfect informa-
tion: CVPI

Definition 22.1
Conditional value of perfect information (CVPI)

The *conditional value of perfect information* is equal to the conditional opportunity loss of the optimal act under the probability distribution of θ.

Given the definition of CVPI, the expected value of perfect information can alternately be computed using equation 22.6.

Formula for EVPI
in terms of CVPI

Expected value of perfect information (EVPI)

$$\text{EVPI} = \sum_j P(\theta_j)\text{CVPI}_\theta \qquad (22.6)$$

Note that formulas (22.5) and (22.6) are equivalent. In equation (22.6), the quantity CVPI represents the opportunity losses $l(a_i, \theta_j)$ for the optimal EOL act in formula (22.5).

If one examines carefully the sum of the expected opportunity loss and the expected payoff *for each act* in Examples 22.1 and 22.2, an interesting relationship appears. The sum of these two quantities is the same ($144.50) for each act and, in fact, is equal to the expected payoff of a perfect predictor or the expected profit under certainty. This identity, EPPP = EMV + EOL, for each act is often used in checking the calculations performed in solving for the expected payoff or the expected value and the expected opportunity loss of each act.

22.2.6 | Decision trees: The determination of the optimal act based on the EMV or EOL strategies

It is possible to determine the optimal act based on the EMV or EOL strategies by the use of decision trees as the following example illustrates.

Example 22.4 A major publisher of popular fiction books recently received a manuscript that is under consideration for publication. The publisher has three decision alternatives:

a_1: Do not publish the manuscript.
a_2: Publish the manuscript in paperback form.
a_3: Publish the manuscript in hardback cover form.

The publisher identifies three states of nature for this decision problem:

θ_1: Manuscript will be successful (many copies sold).

θ_2: Manuscript will experience average response (average number of copies sold).

θ_3: Manuscript will be unsuccessful (few copies sold).

The publisher arranges for the manuscript to be reviewed by critics of fiction work. Based upon those reviews, the publisher establishes the following probabilities that the states of nature will occur: $P(\theta_1) = 0.10$, $P(\theta_2) = 0.50$ and $P(\theta_3) = 0.40$. Based on financial considerations, the publisher estimates the profits (and losses) for each decision alternative—state of nature pair. These estimates (in dollars) are shown in the payoff matrix in Table 22.4.

TABLE 22.4 | Payoff matrix for Example 22.4

Act (a_i) / State of nature (θ_j)	θ_1 Successful	Response θ_2 Average	θ_3 Unsuccessful	EMV
a_1: Do not publish	$0	0	0	0
a_2: Publish (paperback)	1,000,000	100,000	−200,000	70,000
a_3: Publish (hardback)	5,000,000	1,000,000	−2,250,000	100,000
$P(\theta_j)$	0.10	0.50	0.40	

Table 22.4 also gives the probability that each state of nature occurs in the last row and the EMV for each act in the last column. For example,

$$\text{EMV}(a_3) = (\$5,000,000)(0.10) + (\$1,000,000)(0.50) + (-\$2,250,000)(0.40)$$
$$= \$100,000$$

For this problem, construct the decision tree and determine the optimal act using the EMV strategy.

Solution The decision tree is shown in Figure 22.1.

These are the usual conventions associated with a decision tree:

1. Decision branches are shown eminating from boxes. Thus the three decision alternatives $(a_1, a_2, \text{and } a_3)$ are shown in Figure 22.1 as branches from the box at the far left of the tree.

2. States of nature branches are shown eminating from circles. Following each decision alternative branch are circles from which the states of nature branches are drawn.

3. The decision alternatives $a_1, a_2,$ and a_3 are labeled on the tree's decision branches.

FIGURE 22.1 | The decision tree for the Example 22.4 problem

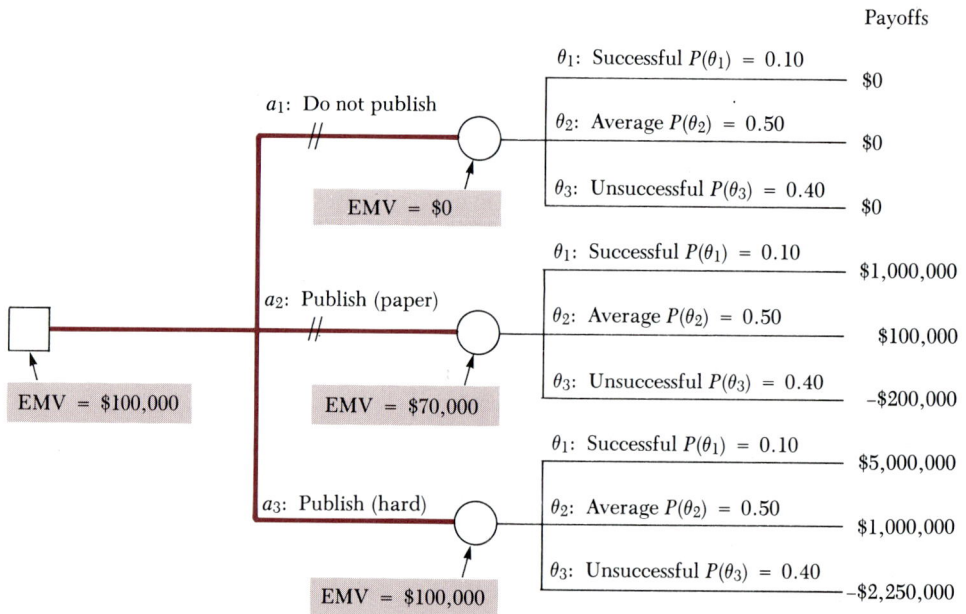

4. The states of nature θ_1, θ_2, and θ_3 and their probabilities are labeled on the tree's states of nature branches.

5. The payoff for each decision branch and state of nature combination is shown at the end branch (right end) of the tree.

6. The EMV (or EOL) is shown at each state of nature circle or decision alternative box in the tree.

7. The tree is evaluated from the *right* to the *left*. We first determine the EMV at the states of nature circles at the right of the tree ($0, $70,000, and $100,000 in Figure 22.1). Then working toward the left, we observe that the best decision branch to take is a_3. Thus we place the EMV for the optimal act (a_3) near the decision alternative box ($100,000 in Figure 22.1).

8. Less than optimal decision alternatives are indicated by placing the symbol // through their decision alternative branches (acts a_1 and a_2 in Figure 22.1).

The EMVs at each state of nature circle are determined by multiplying the payoffs at the end of its branches by the corresponding state of nature probabilities.

The decision tree captures graphically the essence of the decision problem. The publisher must first select a decision alternative, say a_2. Following that selection, one of three states of nature can prevail with payoffs $1 million, $100,000, and $-$200,000. These payoffs occur with probabilities 0.10, 0.50, and 0.40, respectively. Thus by always choosing decision a_2, the publisher will in the *long run*, average $70,000. Obviously, the publisher *should* select decision alternative a_3. By doing so, he will average $100,000, a much better EMV than for alternatives a_1 and a_2.

Example 22.5   For the publisher decision problem in Example 22.4:

a. Determine EPPP (expected profit with the perfect predictor).

b. Determine EVPI (expected value of perfect information) and interpret this quantity.

c. Construct the opportunity loss table.

d. Determine EVPI using the opportunity loss table.

Solution

a. To determine EPPP we first determine the maximum payoff for each state of nature. For θ_1 the maximum payoff is $5 million (act a_3); for θ_2 the maximum payoff is $1 million (act a_3); and for θ_3 the maximum payoff is $0 (act a_1). Then EPPP $=$ ($5,000,000)(0.10) $+$ ($1,000,000)(0.50) + ($0)(0.40) = $1,000,000.

b. EVPI is the difference between EPPP and the EMV for the optimal act (a_3). Thus

$$\text{EVPI} = \text{EPPP} - \text{EMV}(a_3) = \$1,000,000 - \$100,000 = \$900,000$$

This quantity ($900,000) sets an upper bound on the amount we should be willing to pay for perfect information (that is, to be told which state of nature will occur).

c. The opportunity loss table is shown in Table 22.5.

TABLE 22.5 | Opportunity loss matrix for Example 22.5

Act (a_i)	State of nature (θ_j) θ_1 Successful	θ_2 Average	θ_3 Unsuccessful	EOL
a_1: Do not publish	$5,000,000	1,000,000	0	$1,000,000
a_2: Publish (paperback)	4,000,000	900,000	200,000	930,000
a_3: Publish (hardback)	0	0	2,250,000	900,000
$P(\theta_j)$	0.10	0.50	0.40	

In Table 22.5 the opportunity losses are calculated as follows:

θ_1: Maximum payoff = $5,000,000 (act a_3). Thus opportunity loss for a_3 is $0.
  a_1: Opportunity loss = $5,000,000 − $0 = $5,000,000
  a_2: Opportunity loss = $5,000,000 − $1,000,000 = $4,000,000
θ_2: Maximum payoff = $1,000,000 (act a_3). Thus opportunity loss for a_3 is $0.
  a_1: Opportunity loss = $1,000,000 − $0 = $1,000,000
  a_2: Opportunity loss = $1,000,000 − $100,000 = $900,000
θ_3: Maximum payoff = $0 (act a_1). Thus opportunity loss for a_1 is $0.
  a_2: Opportunity loss = $0 − (−$200,000) = $200,000
  a_3: Opportunity loss = $0 − (−$2,250,000) = $2,250,000

d. In the opportunity loss matrix, EVPI is equal to the EOL for the optimal act (a_3). For example

$$EOL(a_2) = (\$4,000,000)(0.10) + (\$900,000)(0.50) + (\$200,000)(0.40)$$
$$= \$930,000$$

The minimum EOL occurs for act a_3 ($900,000). Therefore, EVPI = $900,000.

22.3

Utility—When maximizing expected monetary value may not be an appropriate guide to action

Previous sections have presented the maximization of EMV or the minimization of EOL (Bayes criterion) as the appropriate decision strategy for making decisions under risk. To see why either of these equivalent procedures *may not* be appropriate, consider the following two examples.

Example 22.6 Assume you are offered either of the following alternate courses of action:

A1: You receive $250,000.
A2: A fair coin is tossed in the air. If heads, you receive $750,000. If tails, you receive $0.

Which alternative, A1 or A2, would you select?

Solution The expected value (profit) of the second act (A2) is $375,000 [0.5($750,000 + 0.5($0)], and yet most of us would rather have $250,000 *for sure* than a 50–50 chance of winning $750,000. So, although the expected payoff is higher with A2, most of us would prefer to select A1 and guarantee ourselves a payoff of $250,000. The reason we would prefer A1 (if indeed you do) is that the *utility* of $250,000 for use is greater than the utility of a 50–50 chance of $750,000 (or else $0!). That is, the *relative value* to us of $250,000 is greater than the relative value of a 50–50 chance of $750,000. Consider also the following example.

Example 22.7 You have an opportunity to play in a game (or not) in which a fair coin is to be tossed in the air *once*. If it lands heads up, you receive $1; if it lands tails up, you lose 60 cents (−60 cents payoff). Would you play this game? Would you play the game if the payoffs are $10 and −$6, respectively; if the payoffs are $10,000 and −$6,000, respectively?

Utility

Solution   The expected payoff of not participating in any game is, of course, $0. The expected value of participating in the three games is (respectively) 20¢, $2, and $2,000.

Depending on one's attitude toward gambling, most of us would be willing to participate in the game when the expected payoff is 20 cents, since the loss of 60 cents will not materially affect our financial well-being. At an expected payoff of $2, a few of us would probably withdraw from the game since the potential loss of $6 is serious enough to make the bet undesirable. And with an expected payoff of $2,000, few of us would be left in the game since the loss of $6,000 may materially affect our financial well-being. We would prefer not to play in the game with the higher stakes even though the expected payoff is greater than the payoff of not participating.

The reason we might not choose to participate in the game with the larger stakes, or that we may in general not select the act with the highest expected payoff, is that there are factors involved *other than the monetary payoff* represented. For example, a loss of $6,000 may affect your ability to finance an automobile or a home, or you may simply face the additional harassment from your spouse for gambling with such high stakes! Similarly, most of us would prefer $250,000 for sure because it might give us an advantageous start in life, as opposed to the chance of winding up with nothing. These considerations and the St. Petersburg paradox (Problem 22.13 at the end of the chapter) led Daniel Bernoulli to investigate the maximization of expected *utility* as opposed to the maximization of expected monetary value (payoff) as a criterion for decision making under risk.

We will not describe in detail how one might assess utilities for various consequences—the actual assessment of a utility function or a utility curve is treated adequately in the references at the end of the chapter. However, whenever it is felt that factors other than those involved with the monetary payoff of an action will influence the choice of alternatives, the utility or the relative worth of each alternative should be assessed. Given the utilities of various actions, the appropriate strategy to follow is the maximization of expected utility (MEU). This is accomplished by substituting "utilities" for "payoffs" and then using the same procedures described in the text.

■ **22.4**

Decision making under risk—continuous probabilities (advanced section)

In this section, we will select from a set of facts when the number of states of nature is infinite! In the first section both the number of states of nature and the number of possible acts are (or are assumed to be) infinite. Then we will treat the situation where the number of states of nature is infinite, but the decision maker must choose between *two* alternatives (acts).

This latter situation is stressed because so many decision problems involve the choice between two alternatives in the final analysis.

22.4.1 | **Decision making with infinite states of nature and infinite alternatives (normal distribution)**

Since the random variable is now assumed to be continuous, its probability distribution is specified by its density function, $f(\theta)$. We will assume in this section that losses (opportunity losses) are linear in the random variable $\boldsymbol{\theta}$, and we will refer to losses of overstocking and understocking a demand item. These losses occur whenever we underestimate or overestimate the value of the random variable $\boldsymbol{\theta}$. Furthermore, we will assume that the probability distribution governing demand for the item is normal.

Let the cost of overstocking an item be represented by C_o and the cost of understocking an item be represented by C_u. Furthermore, assume a_i units are stocked to meet demand. Then a loss from overstocking occurs whenever $a_i > \theta$, the demand; and a loss from understocking occurs whenever $a_i < \theta$. No loss occurs whenever $a_i = \theta$. This type of loss function is given in equation (22.7). Figure 22.2 is a graph of the corresponding loss function superimposed on the probability distribution of demand.

FIGURE 22.2 | Loss function superimposed on the probability distribution of demand

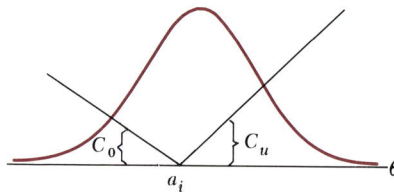

Linear loss of overstocking or understocking

Linear losses of overstocking and understocking a demand item (continuous distribution)

$$l(a_i, \theta) = \begin{cases} 0 & \text{if } a_i = \theta \\ C_o(a_i - \theta) & \text{if } a_i > \theta \\ C_u(\theta - a_i) & \text{if } a_i < \theta \end{cases} \qquad (22.7)$$

Shifting the quantity stocked from a_i to a_j, where $j \neq i$, shifts the vertex of the "loss function" from a_i to a_j, as shown in Figure 22.3. Thus, in a graphical sense, the solution of the statistical decision problem consists of determining where the vertex of the loss function should be placed on the θ-axis.

The Bayes criterion for solving this problem consists of minimizing the expected losses (opportunity losses) associated with overstocking or understocking the demand item. In the discrete case, this loss was given by the sum of the products of the conditional losses and the probability of these losses occurring. In the continuous case, the expected loss can be deter-

FIGURE 22.3 | Opportunity loss functions (conditional value of perfect information) for several stocking points a_i

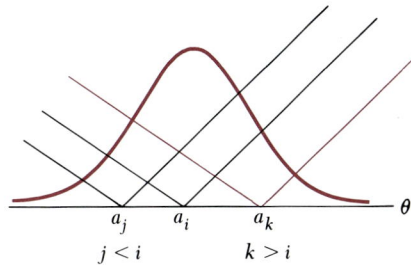

mined by multiplying each of the losses by the value of the probability density function, $f(\theta)$, where the loss occurs. Since we are dealing with a continuous distribution, this involves an infinite number of multiplications! To sum these multiplications, integral calculus must be used for the continuous distribution $f(\theta)$.

Fortunately, it can be shown that the optimal act is to stock a_i units, where the cumulative density function of θ, $F(a)$, is equal to the following:

Optimal fractile of demand distribution

Optimal fractile of demand distribution to minimize opportunity losses

$$F(a_i) = \frac{C_u}{C_o + C_u} \qquad (22.8)$$

Hence to minimize expected opportunity losses using the Bayes criterion, the decision maker should stock the $[C_u/(C_o + C_u)]$th fractile of his/her demand distribution. For example, if $C_u = C_o$, one should stock

$$\frac{C_u}{2C_u} = 0.50 \text{ fractile} = \text{Median}$$

If $3C_u = C_o$, one should stock

$$\frac{C_u}{4C_u} = 0.25 \text{ fractile}$$

If $C_u = 3C_o$, one should stock

$$\frac{C_u}{\frac{4}{3}C_u} = 0.75 \text{ fractile}$$

The use of equation (22.8) in determining the optimal number of units to stock is illustrated in the following example.

Example 22.8   A baker each morning must decide how many doughnuts to make to meet that day's demand. Each doughnut sold yields a net profit of 3 cents. Doughnuts left unsold at the end of the day are sold at a loss of 2 cents each. The baker feels that the demand distribution for doughnuts can be *approximated* by a normal distribution with an average demand of 1,200 doughnuts and a standard deviation of 100. How many doughnuts should the baker stock to minimize expected losses (expected opportunity losses)?

Solution   Since $C_u = 3¢$ and $C_o = 2¢$, the baker should stock the $3/(2 + 3) = \frac{3}{5} = 0.60$ fractile of his demand distribution. Since the baker approximates his actual demand distribution by a normal distribution with mean $\mu = 1,200$ and standard deviation $\sigma = 100$, he should stock doughnuts such that:

$$P(\theta \le 1,200 + z \cdot 100) = 0.60$$

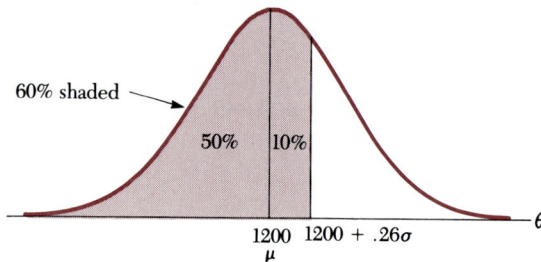

From Table B.3, Appendix B, the normal distribution table, $z = 0.26$. Hence the baker should stock $a_i = 1,200 + 0.26 \cdot 100 = 1,226$ doughnuts per day to minimize his expected losses (expected opportunity losses).

Example 22.9   Assume the same facts as in Example 22.8 except that $C_o = 3¢$ and $C_u = 2¢$. Superimpose the baker's linear loss function on his assumed distribution of demand for the optimal act.

Solution

$$F(a_i) = \frac{2}{2 + 3} = \frac{2}{5} = 0.40, \qquad a_i = 1,200 - z \cdot 100$$

where

$$P(\theta \le 1,200 - z \cdot 100) = 0.40$$

From the normal distribution table, $z \doteq 0.26$. Hence he should stock $a_i = 1,200 - 0.26 \cdot 100 = 1,174$ doughnuts.

The application of equation (22.8) is appropriate whenever the decision maker can choose from an infinite number of acts (or assumed infinite number), the number of states of nature is infinite, and the decision maker wishes to base the choice on minimizing expected losses. The reason for using the normal approximation whenever the number of possible alternatives is finite (but large!) is that a table of normal probabilities is frequently available, and

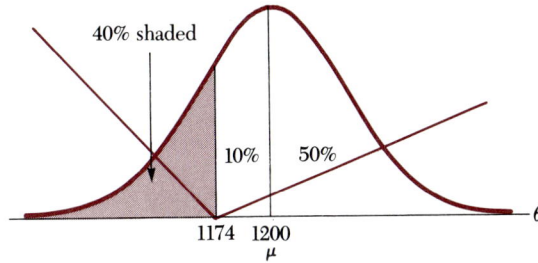

this greatly simplifies the computation of the decision quantity as described in Chapter 7.

22.4.2 | Decision making with infinite states of nature and the choice of two alternatives (normal distribution)

Frequently, the decision on whether to automate involves the choice of two alternatives. Should the machinery or whatever be purchased for the purpose of saving labor, or would the firm be just as well off because of uncertainty or risk in demand not to make a capital expenditure? Other problems also involve the choice between two alternatives—for example, when a firm is deciding between two particular pieces of equipment or two production processes and the number of states of nature is infinite.

When the number of states of nature is infinite and the number of alternatives is limited to two, the analysis proceeds similarly to the case in which the number of states of nature is finite and the number of actions is limited to two. That is, we compute a break-even value of the random variable θ and compare $E(\theta)$ with it. This comparison leads to one of the two actions being selected.

Break-even value

A difficulty encountered when the number of states of nature is infinite is shown in Figure 22.4. To compute the expected value of perfect information (EVPI), we need to multiply the conditional value of perfect information by

FIGURE 22.4 | Conditional value of perfect information—continuous example

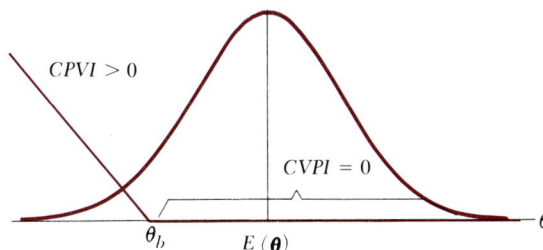

the probability density function of demand to the left or to the right of the break-even value an infinite number of times! These multiplications can be summed by using integral calculus. Fortunately, tables have been provided for accomplishing this multiplication (integration) in the case of a continuous distribution. Table B.10 in Appendix B is such a table giving the required summation of products for the case in which the demand distribution is normal. That is, $N(D)$ from Table B.10, when substituted into the expression for EVPI in equation 22.9, gives the expected value of perfect information.

> Normal loss table for $N(D)$, Table B.10

> Expected value of perfect information (EVPI) for the normal distribution

---

Expected value of perfect information—normal distribution

$$\text{EVPI} = C \cdot \sigma_\theta \cdot N(D) \qquad (22.9)$$

where

C = Absolute value of the slope of the conditional value or opportunity loss line

σ_θ = Standard deviation of the prior distribution

$$D = \frac{|\theta_b - E(\theta)|}{\sigma_\theta}$$

$N(D)$ is found in Table B.10 in Appendix B.

The use of Table B.10 and equation 22.9 in computing EVPI when the number of states of nature is infinite and the number of actions is limited to two is provided in the following example.

Example 22.10   Assume that a firm has the opportunity to invest in a certain piece of capital equipment that is marketed as having a rather large laborsaving potential. The cost to the firm of the particular piece of equipment is $10,000 per year (capital investment, depreciation, maintenance, etc.). The firm presently pays $5 an hour for labor. Hence, if the machine can save the company more than 2,000 labor hours per year, it will prove beneficial to purchase the equipment (i.e., $\theta_b = 2,000$ hours).

An in-house analysis is made of the potential number of labor hours saved based on previous experience with automated equipment, demand for the finished product, and other relevant information. The number of labor hours saved each year is estimated to be $E(\theta) = 2,250$, with a standard deviation of $\sigma_\theta = 150$. The firm also feels that the probability distribution of hours saved can be approximated by a normal distribution.

What is the optimal act under the prior distribution? What is the conditional value of perfect information? The expected value of perfect information?

Solution Let θ represent the number of labor hours saved. Then *profits* from the purchase of the piece of equipment can be represented by the linear function

$$\mathbf{P}_E = -10,000 + 5\theta$$

To maximize expected profits, we should compute the expectation of $\mathbf{P}_E$, $E(\mathbf{P}_E)$. Recall, however, that because the profit equation is a linear function of θ, the expected profit from use of the machine is given by

$$E(\mathbf{P}_E) = -10,000 + 5E(\theta) = -10,000 + 5 \cdot 2,250 = \$1,250$$

Since the expected profit with the machine alternative is greater than 0, the optimal act is to purchase the piece of equipment. Note that we could have derived the same conclusion by computing $\theta_b = 10,000/5 = 2,000$ hours and comparing it with $E(\theta)$. Since $E(\theta) > \theta_b$ (2,250 > 2,000), the optimal act is to purchase the piece of equipment. (Note that we are dealing with *profits* in this example and not costs. Hence the optimizing criteria is to maximize EMV, and the optimal strategy is therefore to purchase the equipment.)

Since the optimal act under the prior distribution is to purchase the piece of equipment, CVPI is positive only for values of θ less than θ_b. We lose (or profits are decreased) $5 an hour for each hour less than the $\theta_b = 2,000$ hours that are actually saved. Hence CVPI can be expressed as:

$$\text{CVPI} = \begin{cases} 0 & \text{if } \theta \geq \theta_b \\ 5(\theta_b - \theta) & \text{if } \theta < \theta_b \end{cases}$$

This functional relationship is superimposed on a graph of the probability density function of demand in Figure 22.5.

FIGURE 22.5 | Conditional value of perfect information— automation example

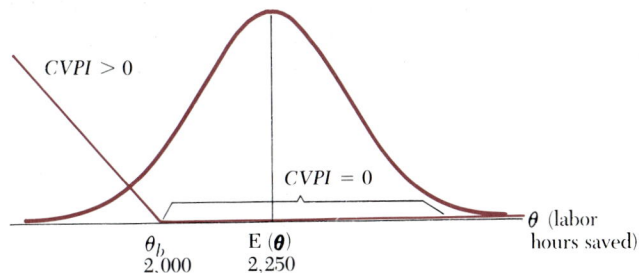

The expected value of perfect information is given by the sum of the products of the normal probability density function, $f(\theta)$, and the conditional value of perfect information to the "left" of θ_b, the break-even value.

For the current example,

$$\text{EVPI} = 5 \cdot 150 \cdot N(D) = 5 \cdot 150 \cdot N \left(\frac{|2,000 - 2,250|}{150} \right)$$

$$= 750 \cdot N(1.67) = 750 \cdot 0.01967 = \$14.75$$

Hence the expected value of perfect information is \$14.75. The reason, of course, for the relatively small value of EVPI is the distance of θ_b from $E(\boldsymbol{\theta})$ and the relatively small standard deviation, σ_θ, in relation to the mean, $E(\boldsymbol{\theta})$.

■ **22.5**

Summary

In this chapter, we examined decision making under risk where the probability distribution of the random variable either currently exists or can be readily assessed. The cases of decision making under risk included decision making with both a finite and an infinite number of states of nature and possible acts. When making decisions under risk, we inherently assume that a probability distribution can be assessed for the various states of nature, $\boldsymbol{\theta}$. By using the information provided by the probability mass function (discrete case) or the probability density function (continuous case), it becomes possible to maximize the expected monetary value or minimize the expected opportunity loss (Bayes criterion) of any decision. Hence the equivalent strategies of maximizing EMV or minimizing EOL were put forth as the desired objective to be reached in decision making under risk. Table 22.6 is a summary of the computations required to determine the optimal act for the payoff matrix.

TABLE 22.6 │ Expected value decision making

Criterion	Payoff or loss matrix representation	
Maximize expected payoff	$\underset{i}{\text{Max}} \left[\sum_j \Omega_{ij} P(\theta_j) \right]$	(22.1)
Minimize expected loss	$\underset{i}{\text{Min}} \left[\sum_j l(a_i, \theta_j) P(\theta_j) \right]$	(22.2)

The expected profit of a perfect predictor (EPPP) and the expected value of perfect information were then described. These measures denote, respectively, the expected profit to be realized if one could predict the state of nature perfectly, and the expected value of perfect information *if it could be obtained* delineating the true state of nature. The determination of the expected value of perfect information can be a difficult task depending on the probability distribution of $\boldsymbol{\theta}$, $P(\boldsymbol{\theta})$ or $f(\boldsymbol{\theta})$. Fortunately, tables of linear loss integrals, such as Table B.10 in Appendix B, have been derived to simplify

TABLE   22.7   Calculation of expected value of perfect information (EVPI)

Payoff matrix form, finite number of states of nature and acts	*Functional form, finite number of states of nature, two acts*	*Functional form, infinite number of states of nature, two acts, normal distribution*
$$\sum_{j} P(\theta_j)\left(\underset{i}{\text{Max }} \Omega_{ij}\right) - \underset{i}{\text{Max}}\left(\sum_{j} P(\theta_j)\Omega_{ij}\right)$$	$$\sum_{\theta=\theta_{b+1}}^{\theta_{\max}} P(\theta) \cdot \text{CVPI} \quad \text{if } E(\boldsymbol{\theta}) < \theta_b$$ $$\sum_{\theta=\theta_{\min}}^{\theta_{b-1}} P(\theta) \cdot \text{CVPI} \quad \text{if } E(\boldsymbol{\theta}) > \theta_b$$ CVPI is the slope of the opportunity loss line	$C \cdot \sigma_{\boldsymbol{\theta}} \cdot N(D)$ where: C is the absolute value of the slope of the opportunity loss line $\sigma_{\boldsymbol{\theta}} =$ standard deviation of the prior distribution $D = \dfrac{|\theta_b - E(\boldsymbol{\theta})|}{\sigma_{\boldsymbol{\theta}}}$ $N(D)$ is found in Table B.10

the computations in many instances. A summary table with the computation of EVPI under a variety of conditions is given in Table 22.7. Note that EVPI is the sum of the products of the conditional value of perfect information and the probability of $\boldsymbol{\theta}$.

The maximization of expected utility has been promoted as an alternative to the maximization of expected monetary value whenever the range of potential payoffs (losses) is so large that factors other than monetary ones need be considered in the evaluation of a decision.

The case of decision making with an infinite number of states of nature and possible acts was then described. To minimize losses or maximize profits, a decision maker should "stock" the $[C_u/(C_o + C_u)]$th fractile of $f(\theta)$, where C_u represents the cost of being understocked and C_o represents the cost of being overstocked a "unit" θ.

■   References

Introductory Level

Anderson, D. R.; D. J. Sweeney; and T. A. Williams. *An Introduction to Management Science.* 4th ed. St. Paul, Minn.: West Publishing, 1985, Chapter 16.

Cook, William D. *Quantitative Methods for Management Decisions.* New York: McGraw-Hill, 1985, Chapter 9.

Lapin, Lawrence L. *Quantitative Methods for Business Decisions.* 2nd ed. New York: Harcourt Brace Jovanovich, 1981, Chapters 5 and 6.

Advanced Level

Bierman, H.; C. P. Bonini; and W. H. Hausman. *Quantitative Analysis for Business Decisions.* 7th ed. Homewood, Ill.: Richard D. Irwin, 1986, Chapters 3–10.

Hillier, F. S., and G. J. Lieberman. *Introduction to Operations Research*. 3rd ed. San Francisco: Holden-Day, 1980, Chapter 15.

Raiffa, H. *Decision Analysis*. Reading, Mass.: Addison-Wesley, 1968.

Raiffa, H., and R. Schlaifer. *Applied Statistical De-cision Theory*. Boston: Division of Research, Harvard University, 1961.

Schlaifer, R. *Introduction to Statistics for Business Decisions*. New York: McGraw-Hill, 1961.

Winkler, R. L. *An Introduction to Bayesian Inference and Decision*. New York: Holt, Rinehart and Winston, 1972.

■ Problems

Section 22.2 Problems

22.1 Show that maximization of expected monetary value (expected payoff) is equivalent to minimization of expected opportunity loss for the case of a finite number of states of nature and possible acts. Equivalency should be established by showing that both will lead to the same action.

22.2 Every "football" Saturday, Joe Raisins wheels out his hot dog stand and sells hot dogs to attendees of the football game. Hot dogs cost Joe $1 each to sell, plus a 25 cent kickback per hot dog to the local syndicate for the "privilege" of operating the stand. Joe sells hot dogs for $2 each, and hot dogs unsold at the end of the game (plus buns, etc.) are sold back to his distributor for redistribution to college dormitory cafeterias for 50 cents each (and you always wondered why the "dorms" have hot dogs on football weekends!). For unsold hot dogs, Joe does *not* have to pay the required 25 cents to the syndicate. Joe has built up the following sales history over his past nine years in college.

Hot dogs sold (in dozens)	Frequency
10	4
11	5
12	7
13	9
14	8
15	8
16	4

Construct payoff and loss matrices similar to those in Tables 22.2 and 22.3 for this problem. What is the optimal act for Joe to follow? What is the expected value of perfect information?

22.3 Write the functional form of the payoff and loss quantities in Problem 22.2. Then graph the profit equations for each act, stock 10 dozen, stock 11 dozen, . . . , stock 16 dozen. What information does the graph provide?

22.4 A decision maker must select one of four acts. The payoff matrix and probabilities for the states of nature are shown below.

Act (a_i)	State of nature (θ_j) θ_1	θ_2	θ_3
a_1	50	−20	10
a_2	10	30	−20
a_3	10	10	10
a_4	−20	20	20
$P(\theta_j)$	0.20	0.40	0.40

a. Using the EMV strategy, construct a decision tree, and determine the act that maximizes EMV by using the tree.

b. Construct the opportunity loss table and determine the act that minimizes EOL.

c. Determine the expected profit with the perfect predictor (EPPP) for the payoff matrix.

d. What is EVPI for this problem?

22.5 For the inventory (flower) Problem 21.3 of Chapter 21, assume Paula is able to assess the following probability distribution of demand (θ) based on sales over past weekends:

θ	0	1	2	3	4	5	6
$P(\theta)$	0.05	0.05	0.10	0.10	0.30	0.30	0.10

Construct payoff matrices and opportunity loss matrices similar to Tables 22.2 and 22.3 for this problem. What is the conditional value of perfect information? The expected value of perfect information? How many dozens of roses should Paula stock to maximize EMV? To minimize EOL?

22.6 What is the functional form of the payoff and loss quantities in Problem 22.5?

22.7 Every Friday, a newstand operator in Dallas, Texas, must decide how many copies of the Sunday edition of the *Austin Statesman* newspaper to order from a distributor of newspapers. Based on past experience, the operator knows that the demand for the Austin Sunday newspaper fluctuates between 10 and 14 copies per week. Using the historical data on demand, the operator sets the following probabilities for the five values of demand:

Demand	Probability
10	0.10
11	0.20
12	0.30
13	0.30
14	0.10
	1.00

The operator will stock no less than 10 copies and no more than 14 copies of the newspaper. Thus the operator must choose one of five decision alternatives: Order 10 copies, order

11 copies, . . . , order 14 copies. Each copy of the Austin Sunday paper costs the operator 60 cents and sells in Dallas for $2. Unsold copies of the paper are thrown away.

a. Construct a payoff matrix for this decision problem.

b. What act should the operator choose if she wishes to maximize EMV?

c. What is EPVI for this problem? Interpret the value of EVPI.

22.8 Construct a decision tree for the decision problem in Problem 22.7. By maximizing EMV, show the optimal act on the tree.

22.9 A U.S. manufacturer of IBM-compatible microcomputers must decide which of three machines to market. The three machines will be built by a Japanese firm according to the specifications of the manufacturer. All three machines are essentially the same in visible physical appearance and in specifications but differ in manufacturing costs. The manufacturing cost of each machine is the sum of the fixed cost and the variable cost, and these costs do vary for each machine as indicated in the following table.

Machine	Fixed cost	Variable cost
PCA	$200,000	$500
PCB	$150,000	$550
RCC	$100,000	$580

For example, if the manufacturer decides on machine PCA and orders 2,000 units, the total cost is $200,000 + (2,000)($500) = $1,200,000. For the coming year, the manufacturer sets four possible sales levels based on market forecasts: 500 units, 1,000 units, 1,500 units, and 2,000 units. The probabilities assigned to these four states of nature are 0.40, 0.30, 0.20, and 0.10, respectively. The manufacturer plans to sell the machine selected to retailers for $2,000.

a. Construct the *profit* matrix for this decision problem.

b. Based on the profit matrix and the maximizing EMV strategy, which machine should the manufacturer select?

c. Determine EVPI for this problem. What does this value tell you about the decision problem?

22.10 In Problem 22.9, suppose that the demand probabilities are changed as follows:

Demand (θ_j)	Probability $P(\theta_j)$
500 units	0.20
1000 units	0.30
1500 units	0.30
2000 units	0.20
	1.00

Based on these new probabilities, now which machine should be selected? Calculate the EVPI with these new probabilities and interpret its value.

22.11 The PG Corporation must decide whether or not to package its toothpaste product in a new container that "serves one helping" of toothpaste by depressing a button on top of a cylindrical container. The research director in charge of the project to evaluate the new design estimates that if the container is successful (the product is well received in the market), the company stands to increase its toothpaste profits by $5 million annually. However, if the container is not successful (the product is rejected in the market), the *loss* annually is estimated to be $10 million. After much study and discussion, the research director sets the probability of success equal to 0.70. Using a decision tree and the maximizing EMV strategy, should the PG Corporation market the new container or not?

22.12 The BLP Corp. has $1 million to invest in the short term. The comptroller of BLP is considering three investments: treasury bills (A), a money market fund (B), and certificates of deposit (C). She knows that the returns from the three options are affected in different ways by the state of the economy. She describes three possible states: expanding economy with high inflation and ready access to money (I), a healthy economy with modest, noninflated growth (II), and a tight economy with recessionary effects including very tight access to money (III). Based on net present value considerations and the period over which funds can be invested in each option, she produces the following return ("profit") matrix:

Act \ State of nature	I expanding	II healthy	III recession
A. Treasury bills	60,000	50,000	40,000
B. Money market fund	75,000	60,000	50,000
C. Certificates of deposit	60,000	65,000	50,000

After a careful analysis, she establishes the following probabilities for the states of the economy: $P(\text{I}) = 0.50$, $P(\text{II}) = 0.40$, and $P(\text{III}) = 0.10$.

a. Are any of the decision alternatives dominated? Explain.

b. Using a decision tree and the EMV strategy, determine which investment act is best.

c. Calculate EVPI for this problem and interpret its value.

Section 22.3 Problems

22.13 St. Petersburg paradox: A coin is to be tossed into the air k times until a head appears (i.e., a head first appears on the kth trial, $k = 1, 2,$

. . .). Your reward is to be 2^k. Thus if a head appears on the first toss, you win $2^1 = 2; if a head does not appear until the second toss (i.e., the sequence TH results), you win $2^2 = 4; etc. What is the expected payoff of this game? Given this expected payoff, what would you be willing to pay to play the game once? What does this tell you about your utility for money?

22.14 Explain in your own words why it is "consistent" for an individual to purchase insurance and also make small bets. What roles does the variability or the dispersion of payoffs (losses) play in the determination of this type of behavior?

Section 22.4 Problems

22.15 Referring to the equipment investment problem (Example 22.10), assume that (a) the standard deviation of labor hours saved is 350, (b) the cost of labor is $10 (changing, of course, θ_b), and (c) both the standard deviation and labor costs are changed as in a and b. What is the revised expected value of perfect information in each of these three instances? What does this tell you about the relationship between the standard deviation and EVPI, the "distance" between θ_b and $E(\theta)$ and EVPI, and the interaction of the "distance" between $E(\theta)$ and θ_b and the standard deviation with EVPI?

22.16 Again referring to Example 22.10, what would be the optimal decision if $E(\theta) = 1,500$ and the other quantities remain unchanged? What is the EVPI in this instance? What is the functional form of CVPI?

22.17 A firm is considering making a capital improvement to one of its production processes that will increase the quality of the end product. If the improvement is made, the firm feels it may be able to charge $5 more for the end product, although there is a 50–50 chance the improvement will lead to less than a $4 or greater than a $6 increase in what the firm can charge for the product. The cost of the improvement is $20,000, and the number of units of expected sales over the life of the improvement (ignore present value considerations) is 3,600. Fit a normal distribution to

this situation, and calculate the expected profit of the best act and the expected value of perfect information. Should the investment be made?

22.18 Peter the paperboy sells on the average 100 newspapers per day, with a standard deviation of 21 newspapers. Papers cost Peter 15 cents each and are sold for 20 cents each. Papers remaining at the end of the day can be sold back to the publisher for 11 cents each. Assuming the demand for newspapers is normally distributed, determine the number of newspapers to stock each day to minimize the expected opportunity losses of the decision.

22.19 If papers can be returned to the publisher for 12 cents net return each, how does this affect the number of papers Peter should stock in Problem 22.18?

22.20 A firm has an opportunity to invest in a piece of equipment, the expected savings of which is 2,500 hours a year with a standard deviation of 500 hours (normally distributed). The cost of the equipment is $15,000 a year, and labor costs are $6.10 an hour. What course should the firm follow, and what is the expected value of perfect information?

Additional Problems

22.21 A decision maker is faced with selecting one of two alternatives. There are two states of nature, θ_1 and θ_2. For the two states of nature and the two decision alternatives, the payoff matrix is given below:

Act (a_i)	State of nature (θ_j)	
	θ_1	θ_2
a_1	100	200
a_2	−50	250

The decision maker sets the probabilities for the states of nature as follows: $P(\theta_1) = 0.40$ and $P(\theta_2) = 0.60$.

a. Based on the given probabilities, which decision alternative is optimal under the EMV strategy?

b. Suppose that the states of nature probabilities are changed as follows: $P(\theta_1) = 0.20$ and $P(\theta_2) = 0.80$. Now what decision alternative is optimal under the EMV strategy?

c. You should have found that different acts were optimal in parts a and b. Let $p = P(\theta_1)$ and $(1 - p) = P(\theta_2)$. What is the "break even" value of p? That is, at what value of p are you indifferent to acts a_1 and a_2? [Hint: Using p and $(1 - p)$, find the EMV for each act, and set $EMV(a_1) = EMV(a_2)$, then solve for p.]

22.22 For the data in Problem 21.14, Chapter 21, assume the trust officer is able to obtain a probability distribution for the various states of the economy over the next three years from his economics instructor in an evening course he is taking. The probabilities of the three states are as given in the table.

State of the economy	Probability
θ_1	0.2
θ_2	0.5
θ_3	0.3
	1.0

Determine the expected monetary value and the expected opportunity loss of each act, and determine the optimal act to follow under the maximize EMV or minimize EOL criterion. What is the expected value of perfect information regarding the state of the economy? What is the conditional value of perfect information?

22.23 Given the payoff table and probability distribution for the various states of nature shown below, determine the following.

a. The maximum expected payoff.

b. The minimum expected loss.

c. The expected profit with a perfect predictor.

d. The expected value of perfect information.

Act \ State of nature	I	II	III	IV	V
1	−10	12	4	7	−3
2	14	8	6	−4	−4
3	3	10	−10	8	4
4	7	0	1	4	9
5	8	4	11	−6	6

State of nature	I	II	III	IV	V
Probability	0.21	0.09	0.13	0.36	0.21

22.24 The M-Bank of Dallas, Texas, has acquired an oil lease for a parcel of land in West Texas. The lease belonged to the AJAX Wildcatting Firm, which defaulted on the repayment of a loan to the M-Bank. The M-Bank Investment Department Officer is faced with three decision alternatives:

a_1: Sell the lease outright.

a_2: Construct a well and drill at the leased site.

a_3: Find a partner who will assume 50 percent of the well-drilling costs and will take 50 percent of the profits; and drill the well at the leased site.

From geological studies acquired from AJAX for the site, there are three possible states of nature (outcomes from drilling):

θ_1: No oil or gas at the site.

θ_2: Gas at the site.

θ_3: Oil at the site.

From the geological studies, the officer at M-Bank sets the following probabilities for those outcomes: $P(\theta_1) = 0.75$, $P(\theta_2) = 0.15$, $P(\theta_3) = 0.10$. After considerable analysis with the help of the bank's finance and accounting departments, the officer constructs the following profit matrix:

State of nature (θ_j) Act (a_i)	θ_1 no gas or oil	θ_2 gas	θ_3 oil
a_1: Sell lease	−$50,000	−$50,000	−$50,000
a_2: Drill	−200,000	100,000	1,000,000
a_3: Drill with partner	−125,000	25,000	475,000

The profits in the payoff table include the remaining loan loss and take into account net present value considerations.

a. Use a decision tree *or* the payoff table to determine which act is optimal using the EMV strategy.

b. Determine EVPI for this problem, and interpret its value.

c. Suppose that the probabilities change to $P(\theta_1) = 0.60$, $P(\theta_2) = 0.40$, and $P(\theta_3) = 0.00$. Now, which act is optimal, based on EMV?

22.25 The M-Bank Investment Department Officer in Problem 22.24 is very concerned about the use of the EMV strategy to determine which act is optimal. She showed her analysis to her superior, the vice-president and comptroller of the bank. He pointed out that the use of the expected value concept is not applicable, since this "game" will be played only once and "long run" probabilities and expected values are meaningless in this context. If you were the M-Bank officer in this case, how would you respond to the vice-president's concern? Can you suggest ways to analyze this problem so that the criticism is no longer warranted?

23

23.1

Introduction

In Chapter 22 we introduced the topic of statistical decision making under risk, where the opportunity did not exist to obtain additional information—a terminal decision had to be made. The criterion for arriving at a decision involved the equivalent strategies of maximizing expected monetary value (EMV) (payoff) and minimizing expected opportunity loss (Bayes criterion). In this chapter we will continue the line of reasoning developed in the previous chapter in making decisions under risk, but we will include the provision for obtaining imperfect—or sample—information. In making our decision, we will continue to use the Bayes criterion and minimize expected opportunity losses (or, simply, total expected losses). Because we are in essence asking the question, "What should we do before we make a terminal decision?" the type of analysis described in the chapter is often called *preposterior analysis* or preposterior decision making.

The notion of obtaining additional information should not be new to us. Indeed, Chapter 10 and others were devoted entirely to obtaining additional information (through a sample) and evaluating various decisions based on this additional information. What will be new in this chapter will be the incorporation of the consequence of making an incorrect decision into the construction of a decision rule. We will learn that the decision to sample can be regarded much in the same light as the setting of a "critical value" or an "action limit" in a test of a hypothesis (see Chapter 10). Both these decisions have an important bearing on the choices to be made in arriving at a decision that maximizes some measure of performance (or minimizes losses). Because ours is an introductory exposition, we will of necessity have to limit our treatment of the subject. We will consider only two cases: (1) decision problems with a finite number of alternatives and of states of nature and (2) two alternative decision problems with a continuous state of nature variable, assumed to be normally distributed. More advanced and in-depth coverage of the decision to buy imperfect (sample) information can be found in the references at the end of this chapter.

■ 23.2

Decision making with sample information— Discrete state of nature variable case

Prior to making a final decision, it is often possible to obtain additional information through a sampling process or by other means. This information can be used to *revise* the probabilities assigned to the states of nature in the decision problem. When the state of nature variable is discrete, Bayes theorem in Chapter 4 may be used to produce the revised probabilities. We introduce the methods of decision making with sample information by using an oil-drilling example problem adapted from the Winkler test referenced at the end of this chapter.

Example 23.1 Henry Wildcat invests in oil land leases. These leases give Henry the right to drill on the land for oil. Further, the leases provide the legal right of ownership to any oil found at the site covered by the lease. Henry holds many such leases. At each leased site, he must decide to sell the lease or to drill for oil. He currently is trying to reach a decision about a site in West Texas, referred to hereafter as the Aztec site. At the Aztec site, he must choose one among four alternatives:

a_1: Drill with 100 percent interest (Henry assumes *all* costs of drilling).

a_2: Drill with 50 percent interest (find a partner to share equally the costs and the profits).

a_3: Farm out for $25,000 and a $1/10$ override (Henry sells the drilling rights for $25,000 *and* $1/10$ of the *net* profits, if positive, zero otherwise).

a_4: Do not drill—sell lease outright for $50,000.

The states of nature in this case represent the amount of oil found at the site. Although the state of nature variable could be modeled by a continuous random variable, Henry decides for simplicity and ease of coming to grips with the decision problem to define four states of nature, where each state represents the amount in gallons of crude oil at the site.

θ_1: 0 gallons.

θ_2: 500,000 gallons.

θ_3: 1,000,000 gallons.

θ_4: 2,000,000 gallons.

The cost of drilling at the site is $250,000. The net profit per gallon of crude oil discovered is $0.75. From this information, Henry constructs the decision payoff matrix for this problem, given in Table 23.1.

TABLE 23.1 | The payoff matrix for Henry's decision problem

Act (a_i)	State of nature (θ_j) gallons of crude oil	θ_1 0	θ_2 500,000	θ_3 1,000,000	θ_4 2,000,000	EMV
a_1: Drill with 100 percent interest		\$−250,000	\$125,000	\$500,000	\$1,250,000	\$31,250
a_2: Drill with 50 percent interest		−125,000	62,500	250,000	625,000	15,625
a_3: Farm out for \$25,000 and ¹/₁₀ override		25,000	37,500	75,000	150,000	45,625
a_4: Sell lease for \$50,000		50,000	50,000	50,000	50,000	50,000
$P(\theta_j)$		0.70	0.05	0.15	0.10	

Table 23.1 also contains the probabilities that Henry has assigned to each of the four states of nature [$P(\theta_1) = 0.70$, $P(\theta_2) = 0.05$, $P(\theta_3) = 0.15$, and $P(\theta_4) = 0.10$]. These probabilities are based on Henry's subjective determination of the likelihood of no oil or the three amounts of oil defining the states of nature *prior* to any additional information, such as geological testing at the site. We therefore refer to these probabilities as the *prior* probabilities of the states of nature occurring. Using the EMV maximization strategy, which act should Henry select? What is the expected value of perfect information (EVPI) for this problem?

| Prior probabilities

Solution The expected monetary values for the four acts are given also in Table 23.1. For example,

$$\text{EMV}(a_1) = (-\$250,000)(0.70) + (\$125,000)(0.05) + (\$500,000)(0.15)$$
$$+ (\$1,250,000)(0.10)$$
$$= \$31,250$$

The optimal act based on maximizing EMV is a_4, sell the lease outright for \$50,000. This solution is also shown by a decision tree in Figure 23.1. EVPI for this problem is calculated as follows:

$$\text{EVPI} = \text{EPPP} - \text{EMV(optimal act)}$$

The expected profit of the perfect predictor EPPP is given by:

$$\text{EPPP} = (\$50,000)(0.70) + (\$125,000)(0.05) + (\$500,000)(0.15)$$
$$+ (\$1,250,000)(0.10)$$
$$= \$241,250$$

FIGURE 23.1 | The decision tree for Henry's problem—prior probabilities

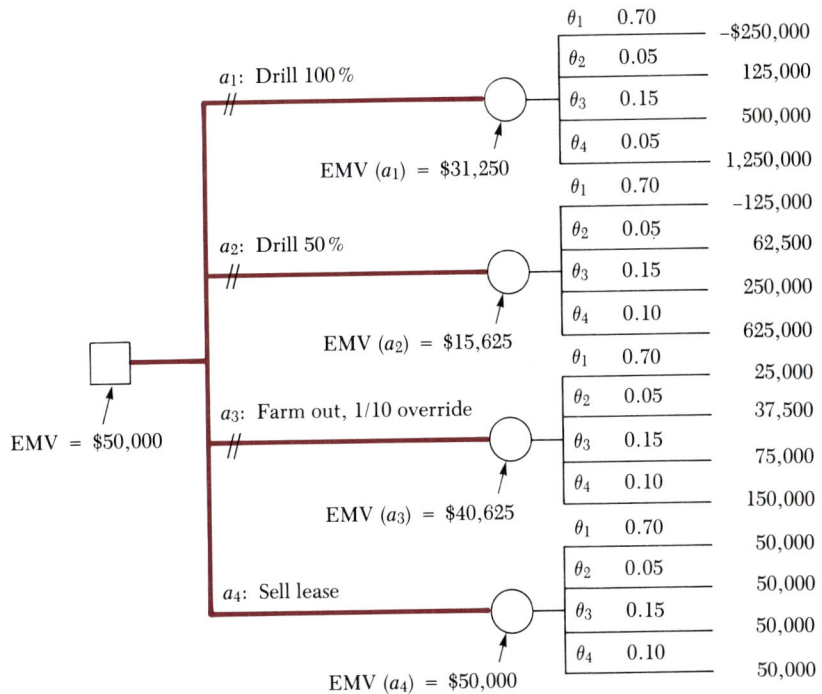

The shaded boxes in Table 23.1 represent the maximum profit for each state of nature. EPPP weights these values by the respective probabilities of the states of nature occurring. Therefore,

$$\text{EVPI} = \$241{,}250 - \$50{,}000 = \$191{,}250$$

Thus it would be worth up to $191,250 to Henry to know which state of nature prevails at the Aztec site.

In cases similar to the Example 23.1 problem, it is possible to acquire additional information about the problem that may affect the choice of the optimal act. In the Example 23.1 problem, a common procedure in wildcatting is to purchase a geological survey of the site. The outcome of the survey is used to *revise* the prior probabilities for the occurrence of the states of nature in the problem.

The method used to revise the prior probabilities $P(\theta_1)$, $P(\theta_2)$, . . . , $P(\theta_K)$ is Bayes theorem [Theorem 4.2 in Chapter 4]. We rewrite Bayes theorem now in context of the decision problem, where $\theta_1, \theta_2, \ldots, \theta_K$ represent the mutually exclusive and collectively exhaustive states of nature and B represents another event, such that $P(B) \neq 0$:

Bayes theorem

$$P(\theta_j/B) = \frac{P(B/\theta_j)P(\theta_j)}{\sum\limits_{j=1}^{K} P(B/\theta_j)P(\theta_j)}, \qquad j = 1, 2, \ldots, K \qquad (23.1)$$

We will now show how Bayes theorem is used to revise the prior probabilities $P(\theta_1)$, $P(\theta_2)$, . . . , $P(\theta_K)$ by continuing the Example 23.1 problem in Example 23.2.

Example 23.2 Henry Wildcat decides to contract with the Southern Company to conduct a geological survey at the site. The survey costs Henry \$25,000. The survey will produce one of three outcomes:

NS: There is no geological structure at the site (usually associated with the absence of oil).

OS: There is an open geological structure at the site (fair chance of oil at the site).

CS: There is a closed geological structure at the site (usually connected with the presence of oil).

From past geological studies of many sites, the following conditional probabilities have been estimated from the sample information:

$P(NS/\theta_1) = 0.80$	$P(NS/\theta_2) = 0.50$	$P(NS/\theta_3) = 0.20$	$P(NS/\theta_4) = 0.10$
$P(OS/\theta_1) = 0.10$	$P(OS/\theta_2) = 0.30$	$P(OS/\theta_3) = 0.20$	$P(OS/\theta_4) = 0.20$
$P(CS/\theta_1) = \underline{0.10}$	$P(CS/\theta_2) = \underline{0.20}$	$P(CS/\theta_3) = \underline{0.60}$	$P(CS/\theta_4) = \underline{0.70}$
1.00	1.00	1.00	1.00

For example, *given* that the state of nature θ_1 existed (no oil), then the probability of no structure is 0.80, the probability of an open structure is 0.10, and the probability of a closed structure is 0.10. Thus, *if* the site is dry, then it is very *likely* that there is no geological structure (NS). These conditional probabilities are commonly called *likelihoods,* for they tell us how likely each event (NS, OS, and CS) is, given the state of nature. Notice that in this problem it would be very easy to estimate these likelihoods from sample data. To get the last column of likelihoods above, we would simply have to go to our records, look up all sites that produced 2 million gallons of oil (or close to 2 million gallons), and find the proportion of those sites that had no structure (0.10), an open structure (0.20), or a closed structure (0.70).

Likelihoods

Let us now suppose that the Southern Company submits a geological survey to Henry Wildcat that indicates a *closed structure* (CS). Now, what are the *revised* probabilities associated with the states of nature θ_1, θ_2, θ_3, and θ_4?

Solution From Formula 23.1 (Bayes theorem), let the event B be CS (closed structure). Then we can write:

$$P(\theta_j/CS) = \frac{P(CS/\theta_j)P(\theta_j)}{\sum\limits_{j=1}^{4} P(CS/\theta_j)P(\theta_j)}, \qquad j = 1, 2, 3, 4$$

For example,

$$P(\theta_1/\text{CS}) = \frac{P(\text{CS}/\theta_1)P(\theta_1)}{\sum\limits_{j=1}^{4} P(\text{CS}/\theta_j)P(\theta_j)}$$

Recalling that $P(\theta_1) = 0.70$, $P(\theta_2) = 0.05$, $P(\theta_3) = 0.15$, and $P(\theta_4) = 0.10$, and using the likelihoods above, we have:

$$P(\theta_1/\text{CS}) = \frac{P(\text{CS}/\theta_1)P(\theta_1)}{P(\text{CS}/\theta_1)P(\theta_1) + P(\text{CS}/\theta_2)P(\theta_2) + P(\text{CS}/\theta_3)P(\theta_3) + P(\text{CS}/\theta_4)P(\theta_4)}$$

$$= \frac{(0.10)(0.70)}{(0.10)(0.70) + (0.20)(0.05) + (0.60)(0.15) + (0.70)(0.10)}$$

$$= \frac{0.07}{0.07 + 0.01 + 0.09 + 0.07} = \frac{0.07}{0.24} = 0.292$$

Thus, if a closed structure has been found, the probability of finding no oil (θ_1) has been *revised* downward from $P(\theta_1) = 0.70$ to $P(\theta_1/\text{CS}) = 0.292$. It is now much less likely that Henry will find no oil at the Aztec site.

Posterior probability

The revised (or conditional) probability $P(\theta_1/\text{CS})$ is often called a *posterior probability,* indicating that it has been calculated *after* the sample information has come to our attention. Rather than using Bayes theorem directly, the computation of the posterior probabilities can be simplified by using a tabular format as indicated in Table 23.2.

TABLE 23.2 | The tabular format for calculating posterior probabilities, Example 23.2 (closed structure—CS)

(1) State of nature (θ_j)	(2) Prior probability $P(\theta_j)$	(3) Likelihood $P(\text{CS}/\theta_j)$	(4) Product (2) × (3)	(5) Posterior probability $P(\theta_j/\text{CS})$
θ_1	0.70	0.10	0.07	0.07/0.24 = 0.292
θ_2	0.05	0.20	0.01	0.01/0.24 = 0.041
θ_3	0.15	0.60	0.09	0.09/0.24 = 0.375
θ_4	0.10	0.70	0.07	0.07/0.24 = 0.292
	1.00		$P(\text{CS}) = 0.24$	1.000

Notice in Table 23.2 that we are in fact calculating the posterior probabilities from Bayes theorem. Column 4 is the *denominator* of Bayes theorem: the sum of the products of the prior probabilities and the likelihoods for each state of nature. The denominator gives us the probability of finding a closed structure [$P(\text{CS}) = 0.24$]. Thus in 24 percent of the geological surveys executed by the Southern Company, a closed structure [event CS] was encountered. Given that the geological survey indicated a closed structure (CS), the probabilities associated with the states of nature have been revised rather substantially as indicated in Table 23.2. For example, the probability of finding no oil (θ_1) has been revised *downward* from 0.70 to 0.292, and the probability of finding 2 million gallons of crude oil (θ_4) has been revised *upward* from 0.10 to 0.292.

Example 23.2 Find the posterior probabilities *given* that no geological structure (NS) is found. Also, find the posterior probabilities *given* that an open geological structure (OS) is found.

Solution The two sets of posterior probabilities are given in Table 23.3 (no structure found—NS) and in Table 23.4 (an open structure found—OS). Notice that when no structure is found, the probability of finding no oil (θ_1) has been revised *upward* from 0.70 to 0.896. When the open structure is found (OS), the probability of finding no oil has been revised downward from 0.70 to 0.519. Also notice from Tables 23.2, 23.3, and 23.4,

$$P(CS) + P(NS) + P(OS) = 0.240 + 0.625 + 0.135 = 1.000$$

TABLE 23.3 | The tabular format for calculating posterior probabilities, Example 23.2 (no structure—NS)

(1) State of nature (θ_j)	*(2)* Prior probability $P(\theta_j)$	*(3)* Likelihood $P(NS/\theta_j)$	*(4)* Product $(2) \times (3)$	*(5)* Posterior probability $P(\theta_j/NS)$
θ_1	0.70	0.80	0.560	0.896
θ_2	0.05	0.50	0.025	0.040
θ_3	0.15	0.20	0.030	0.048
θ_4	0.10	0.10	0.010	0.016
	1.00		$P(NS) = 0.625$	1.000

TABLE 23.4 | The tabular format for calculating posterior probabilities, Example 23.2 (open structure—OS)

(1) State of nature (θ_j)	*(2)* Prior probability $P(\theta_j)$	*(3)* Likelihood $P(OS/\theta_j)$	*(4)* Product $(2) \times (3)$	*(5)* Posterior probability $P(\theta_j/OS)$
θ_1	0.70	0.10	0.070	0.519
θ_2	0.05	0.30	0.015	0.111
θ_3	0.15	0.20	0.030	0.222
θ_4	0.10	0.20	0.020	0.148
	1.00		$P(OS) = 0.135$	1.000

One of these three structures *must* be found by the Southern Company: Twenty-four percent of the time it is a closed structure; 62.5 percent of the time it is no structure; and 13.5 percent of the time it is an open structure.

The process of producing the posterior probabilities is illustrated in Figure 23.2—the application of the process is applied to Henry's problem with the closed structure (CS) geological survey report. To determine the posterior probabilities, two sets of probabilities must be known:

1. The prior probabilities associated with the states of nature: $P(\theta_1)$, $P(\theta_2)$, . . . , $P(\theta_K)$.

FIGURE 23.2 | The Bayesian revision process for the calculation of posterior probabilities, application to the Example 23.2 problem

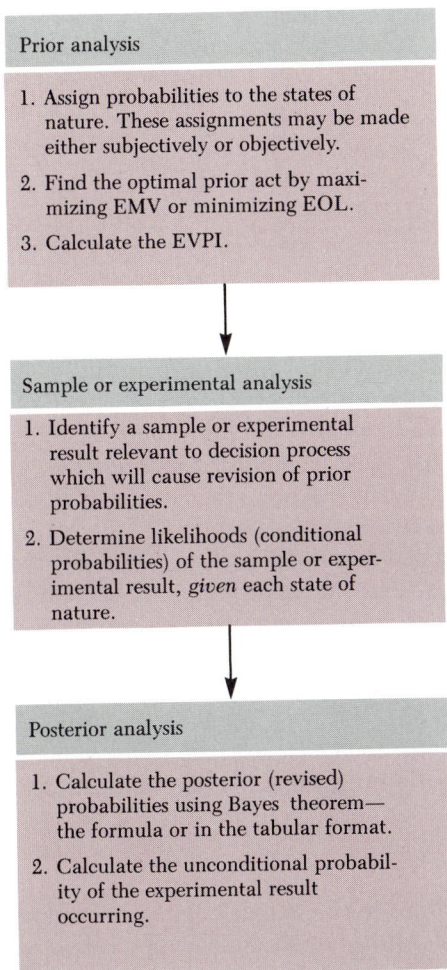

Prior analysis

1. Assign probabilities to the states of nature. These assignments may be made either subjectively or objectively.

2. Find the optimal prior act by maximizing EMV or minimizing EOL.

3. Calculate the EVPI.

Henry Wildcat's application with the closed structure (CS) survey outcome

1. $P(\theta_1) = 0.70$ $P(\theta_2) = 0.05$ $P(\theta_3) = 0.15$

   $P(\theta_4) = 0.10$ (determined *subjectively*)

2. EMV $(\theta_4) = \$50,000$ is the maximum EMV. Therefore, the *prior* decision is to sell the lease.

3. EVPI = \$191,250. Maximum limit for perfect or imperfect (sample) information.

Sample or experimental analysis

1. Identify a sample or experimental result relevant to decision process which will cause revision of prior probabilities.

2. Determine likelihoods (conditional probabilities) of the sample or experimental result, *given* each state of nature.

1. Experimental process is the geological survey—the results in this illustration are encountering a closed structure (CS).

2. Likelihoods—$P(CS/\theta_1) = 0.10$, $P(CS/\theta_2) = 0.20$, $P(CS/\theta_3) = 0.60$, $P(CS/\theta_4) = 0.70$; likelihoods determined from historical data.

Posterior analysis

1. Calculate the posterior (revised) probabilities using Bayes theorem—the formula or in the tabular format.

2. Calculate the unconditional probability of the experimental result occurring.

1. From Table 23.2, $P(\theta_1/CS) = 0.292$, $P(\theta_2/CS) = 0.041$, $P(\theta_3/CS) = 0.375$, and $P(\theta_4/CS) = 0.292$.

2. $P(CS) = 0.24$, also from Table 23.2.

2. The likelihoods (conditional probabilities) of the experimental or sample result, say event B, occurring, given each state of nature: $P(B/\theta_1)$, $P(B/\theta_2)$, . . . , $P(B/\theta_K)$.

With these two sets of probabilities, the posterior probabilities $P(\theta_1/B)$, $P(\theta_2/B)$, . . . , $P(\theta_K/B)$ can be calculated most easily in a tabular format, illustrated in Table 23.5.

TABLE 23.5 | The tabular form for calculating posterior probabilities using Bayes theorem with the experimental or sample result labeled event B

(1) *State of nature* (θ_j)	*(2)* *Prior probabilities* $P(\theta_j)$	*(3)* *Likelihood* $P(B/\theta_j)$	*(4)* *Product* $(2) \times (3)$	*(5)* *Posterior probabilities* $P(\theta_j/B)$
θ_1	$P(\theta_1)$	$P(B/\theta_1)$	$P(\theta_1)P(B/\theta_1)$	$P(\theta_1/B)$
θ_2	$P(\theta_2)$	$P(B/\theta_2)$	$P(\theta_2)P(B/\theta_2)$	$P(\theta_2/B)$
$\vdots$	$\vdots$	$\vdots$	$\vdots$	$\vdots$
θ_K	$+P(\theta_K)$	$P(B/\theta_K)$	$+P(\theta_K)P(B/\theta_K)$	$+P(\theta_K/B)$
	1.00		$P(B)$	1.00

Two final notes regarding the computation of posterior probabilities:

Note 1 In Table 23.5, *Column 3 does not have to sum to 1.* The likelihoods are conditional probabilities, each of which is conditioned on a *different* event (θ_1, θ_2, . . . , θ_K). Column 5—the posterior probabilities—must of course sum to 1. Notice that these conditional probabilities are conditioned on the *same* event (B). If event B occurs, then one and only one of the events θ_1, θ_2, . . . , θ_K can occur.

Note 2 The revision process described in Table 23.5 can be applied again if there is new evidence by experiment or sampling that may impact on the probabilities that the states of nature occur. To revise again, we treat the posterior probabilities in Table 23.5 as the new prior probabilities, entering their values in Column 1 and revise them according to the likelihoods associated with the new experimental result, say event C. In this way, Bayes theorem can be used to constantly update (revise) the states of nature probabilities as we gain new information regarding the decision process.

23.2.1 | The posterior analysis

Once the posterior probabilities have been calculated, it is then possible to reanalyze the problem, using the sample information contained in the revised or posterior probabilities. To illustrate the *posterior analysis* (the analysis of the problem using the posterior probabilities in place of the prior probabilities), we return to Henry Wildcat's problem in Example 23.4.

Example 23.4   The Southern Company's survey indicates that a closed structure (CS) exists at the Aztec site. Given this information, what is Henry's optimal act now?

Solution   To determine Henry's *posterior* optimal act, we substitute the posterior probabilities $P(\theta_1/CS)$, $P(\theta_2/CS)$, $P(\theta_3/CS)$, and $P(\theta_4/CS)$ for the prior probabilities $P(\theta_1)$, $P(\theta_2)$, $P(\theta_3)$, and $P(\theta_4)$ either in the payoff table (Table 23.1) or in the decision tree (Figure 23.1) and calculate the EMV for each act. The optimal posterior act is the one with the maximum EMV. We show these calculations in Table 23.6. The posterior analysis is also shown in the decision tree illustrated in Figure 23.3. Notice in the decision tree that the *posterior probabilities* are shown on the states of nature branches, and a slash(/) is used at the left end of the tree to show the reduction of EMV for the cost of the geological survey ($25,000).

TABLE 23.6   The determination of the optimal posterior act for Example 23.3 (closed structure—CS)

Act (a_i)	θ_1	θ_2	θ_3	θ_4	EMV
a_1: Drill (100 percent)	−$250,000	$125,000	$500,000	$1,250,000	$484,625
a_2: Drill (50 percent)	−125,000	62,500	250,000	625,000	242,312
a_3: Farm out ($25,000 and $1/10$ override)	25,000	37,500	75,000	150,000	80,762
a_4: Sell lease	50,000	50,000	50,000	50,000	50,000
posterior probabilities $P(\theta_j/CS)$	0.292	0.041	0.375	0.292	

From Table 23.6, it is obvious that Henry's optimal act is to now drill for himself (with 100 percent interest). By doing so, the EMV of this act is $484,625. However, the geological survey conducted by the Southern Company for Henry cost him $25,000. Thus his *net* expected monetary value is $484,625 − $25,000 = $459,625. We could also determine Henry's optimal acts given that the geological survey indicates no structure (NS) or an open structure (OS). Using the posterior probabilities associated with the survey result "no structure" (NS) and an "open structure" (OS), try to determine the posterior acts for each of the survey outcomes. The answers are given in subsection 23.2.2 (see Figure 23.4a for NS and Figure 23.4c OS outcomes, respectively).

FIGURE 23.3 | The determination of the optimal posterior act by use of a decision tree for Example 23.3 (closed structure—CS)

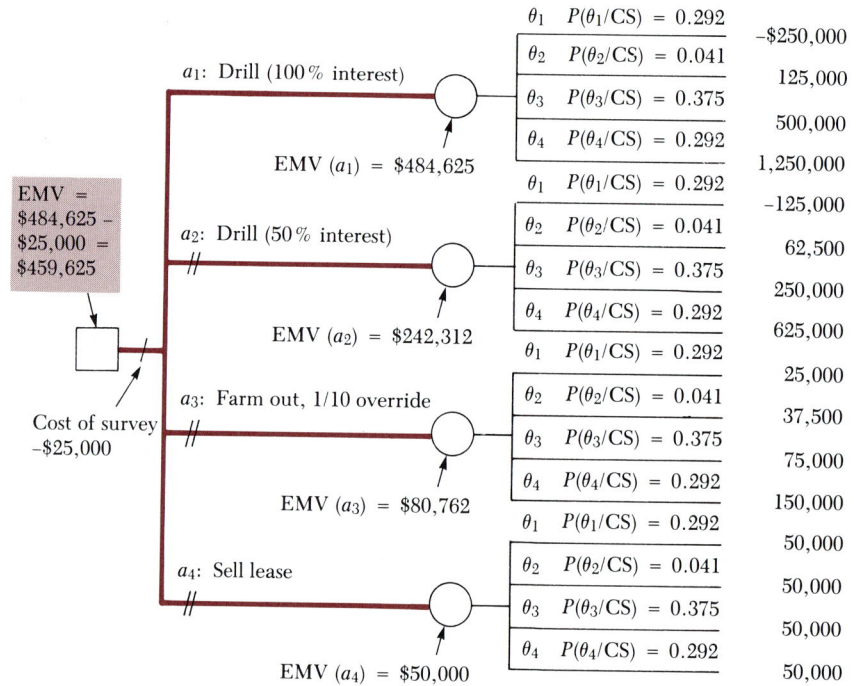

a_1: Drill (100% interest)	θ_1 $P(\theta_1/CS) = 0.292$	$-\$250,000$
	θ_2 $P(\theta_2/CS) = 0.041$	$125,000$
	θ_3 $P(\theta_3/CS) = 0.375$	$500,000$
EMV $(a_1) = \$484,625$	θ_4 $P(\theta_4/CS) = 0.292$	$1,250,000$
a_2: Drill (50% interest)	θ_1 $P(\theta_1/CS) = 0.292$	$-125,000$
	θ_2 $P(\theta_2/CS) = 0.041$	$62,500$
	θ_3 $P(\theta_3/CS) = 0.375$	$250,000$
EMV $(a_2) = \$242,312$	θ_4 $P(\theta_4/CS) = 0.292$	$625,000$
a_3: Farm out, 1/10 override	θ_1 $P(\theta_1/CS) = 0.292$	$25,000$
	θ_2 $P(\theta_2/CS) = 0.041$	$37,500$
	θ_3 $P(\theta_3/CS) = 0.375$	$75,000$
EMV $(a_3) = \$80,762$	θ_4 $P(\theta_4/CS) = 0.292$	$150,000$
a_4: Sell lease	θ_1 $P(\theta_1/CS) = 0.292$	$50,000$
	θ_2 $P(\theta_2/CS) = 0.041$	$50,000$
	θ_3 $P(\theta_3/CS) = 0.375$	$50,000$
EMV $(a_4) = \$50,000$	θ_4 $P(\theta_4/CS) = 0.292$	$50,000$

EMV = $484,625 − $25,000 = $459,625

Cost of survey −$25,000

23.2.2 | Preposterior analysis—the decision to sample or to experiment

When the possibility exists to revise the prior probabilities associated with the states of nature based on sample or experimental information, it is possible to determine whether or not the information is worth the cost *before the information is purchased*. Thus we can decide whether or not to purchase the sample or experimental information as a part of the overall decision process. This analysis is referred to as *preposterior analysis*—prior to the posterior analysis. To illustrate the application of preposterior analysis, let us return to Henry Wildcat's problem.

Preposterior analysis

Example 23.5 Recall that Henry Wildcat was offered the geological survey by the Southern Company for $25,000. In Examples 23.2, 23.3, and 23.4, we assumed that Henry bought the survey information (the outcome was a closed structure, event CS). Now, we will retrace our steps to the point in time when Henry must *decide whether or not to buy the geological survey information*. Should Henry buy the geological survey experiment from the Southern Company?

Solution The decision tree for the preposterior analysis is shown in Figure 23.4. First, consider the *structure* of the tree. The first decision (box) confronted by Henry at the left-most point of the tree is whether or not to purchase the geological survey. The decision to purchase is labeled event A_1, and the decision not to purchase is labeled event A_2. If action A_2 is taken, notice that the remaining part of the tree following the A_2 branch is the *prior analysis*. This portion of the tree is shown also in Figure 23.1, the prior analysis with no sample or experimental information. Thus, if action A_2 is taken, Henry knows that the EMV is $50,000.

If action A_2 is taken (Henry buys the survey), then the next set of branches represent the *outcome* of the geological survey—no structure, closed structure, or open structure. The remaining branches of the tree for each outcome represent the *posterior analysis, given* that outcome. For example, the tree for the closed structure (CS) branch, shown in detail in Figure 23.4, is a replication of Figure 23.3 shown earlier, the posterior analysis given the closed structure outcome. Figures 23.4(a) and 23.4(c) show the posterior analyses for the no structure (NS) and the open structure (OS) survey outcomes, respectively. Notice the use of the posterior probabilities along the branches of the states of nature in Figures 23.4(a), 23.4(b), and 23.4(c). Each set is conditioned on the appropriate survey outcome. Returning to the full tree in Figure 23.4, notice that the probability that the survey report will be no structure, closed structure, or open structure is shown on the outcome branches. These unconditional probabilities of the survey results were determined in Tables 23.3, 23.2, and 23.4, respectively.

The *analysis* of the tree is made from the *right* of the tree to the *left* of the tree. On the top portion of the tree, we know that if the survey outcome is no structure (NS), then the optimal act is a_4: Sell the lease with an EMV of $50,000. This is determined from the posterior analysis of the no structure outcome, shown as a decision tree in Figure 23.4(a). Similarly, if the survey outcome is closed structure (CS), then the optimal act is a_1: Drill with 100 percent interest with an EMV of $484,625; and if the survey outcome is open structure (OS), then the optimal act is also a_1: Drill with 100 percent interest with an EMV of $180,125. Since we know the probabilities of each outcome occurring, we can find the EMV of the act A_1, purchase the survey, by weighting these EMVs by the appropriate outcome result probabilities:

$$EMV(A_1) = (\$50,000)(0.625) + (\$484,625)(0.240) + (\$180,125)(0.135)$$
$$\text{(no structure)} \quad \text{(closed structure)} \quad \text{(open structure)}$$
$$= \$171,877$$

We must, however, remember to subtract the cost of the survey from this figure. Thus the *net* expected monetary value of act A_1 is:

$$\$171,877 - \$25,000 = \$146,877$$

Results of the preposterior analysis

The results of the preposterior analysis may now be fully summarized:

1. Henry should *purchase the geological survey*. The EMV of act A_1, decreased by the cost of the survey, is much more than the EMV of act A_2.

FIGURE 23.4 | The decision tree for the preposterior analysis of the Example 23.5 problem

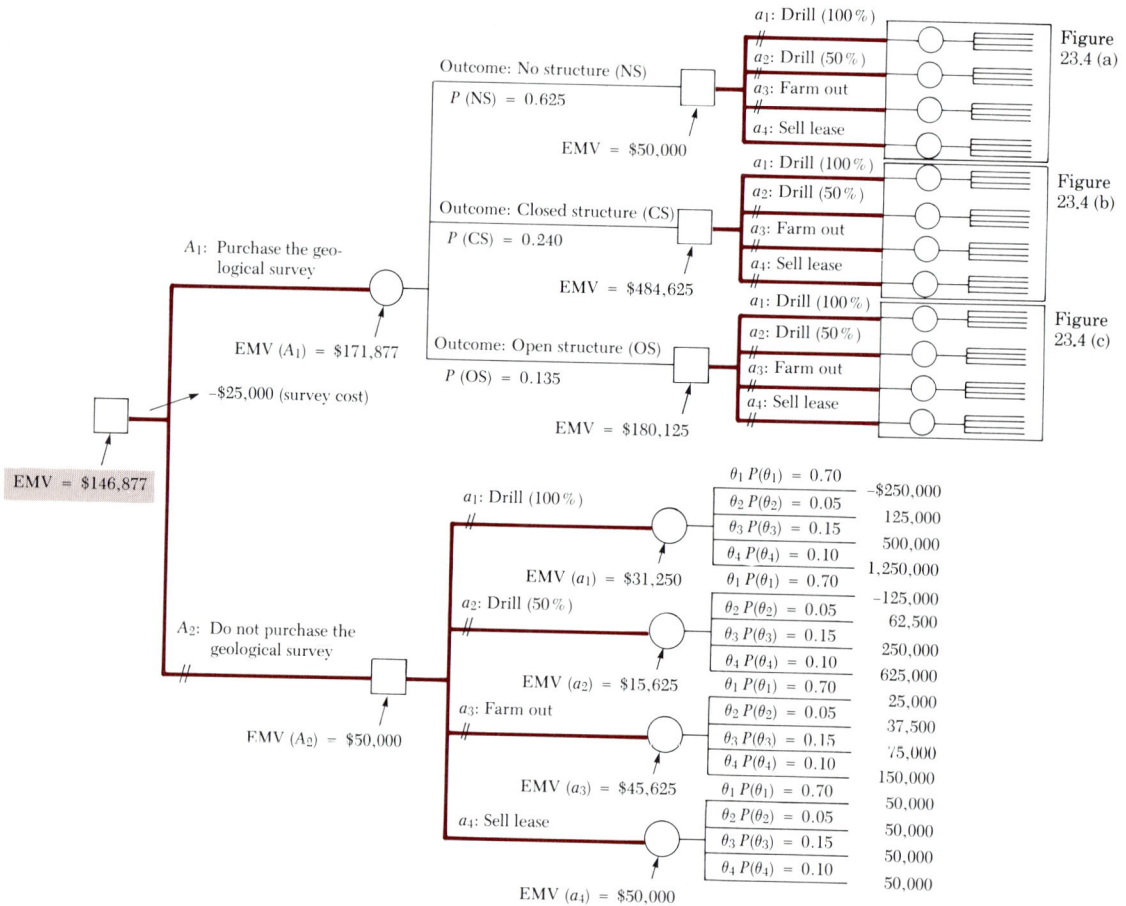

2. When the survey result is known, Henry should do the following:
   a. If the result is no structure (NS), he should *sell the lease.*
   b. If the result is either closed structure (CS) or open structure (OS), he should *drill with 100 percent interest.*

The preposterior analysis is very informative, though it requires a lot of calculations. It determines for the decision maker whether or not sample or experimental information is worthwhile. If it is, then the analysis also indicates which action is optimal, depending on the outcome of the sample or experiment. The required inputs are all profits associated with outcomes, the prior probabilities, and the likelihoods associated with each sample or exper-

FIGURE 23.4(a) | States of nature branches for the no structure outcome (NS)

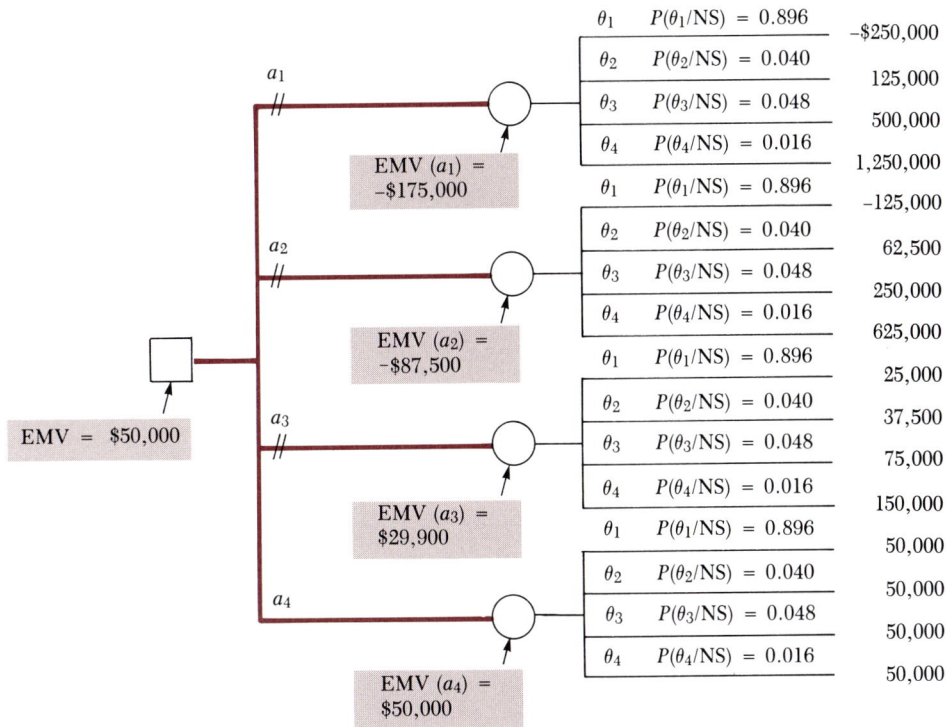

	θ_1 $P(\theta_1/\text{NS}) = 0.896$	$-\$250,000$
	θ_2 $P(\theta_2/\text{NS}) = 0.040$	$125,000$
a_1	θ_3 $P(\theta_3/\text{NS}) = 0.048$	$500,000$
	θ_4 $P(\theta_4/\text{NS}) = 0.016$	$1,250,000$

EMV (a_1) = $-\$175,000$

θ_1 $P(\theta_1/\text{NS}) = 0.896$ $\quad -125,000$
θ_2 $P(\theta_2/\text{NS}) = 0.040$ $\quad 62,500$
a_2
θ_3 $P(\theta_3/\text{NS}) = 0.048$ $\quad 250,000$
θ_4 $P(\theta_4/\text{NS}) = 0.016$ $\quad 625,000$

EMV (a_2) = $-\$87,500$

θ_1 $P(\theta_1/\text{NS}) = 0.896$ $\quad 25,000$
θ_2 $P(\theta_2/\text{NS}) = 0.040$ $\quad 37,500$
a_3
θ_3 $P(\theta_3/\text{NS}) = 0.048$ $\quad 75,000$
θ_4 $P(\theta_4/\text{NS}) = 0.016$ $\quad 150,000$

EMV (a_3) = $\$29,900$

θ_1 $P(\theta_1/\text{NS}) = 0.896$ $\quad 50,000$
θ_2 $P(\theta_2/\text{NS}) = 0.040$ $\quad 50,000$
a_4
θ_3 $P(\theta_3/\text{NS}) = 0.048$ $\quad 50,000$
θ_4 $P(\theta_4/\text{NS}) = 0.016$ $\quad 50,000$

EMV (a_4) = $\$50,000$

EMV = $\$50,000$

imental outcome. Notice that we do not necessarily need to incorporate directly the cost of the survey. Since EMV(A_1) = $\$171,877$ and EMV(A_2) = $\$50,000$, we know that Henry should be willing to pay up to $\$171,877 - \$50,000 = \$121,877$ for the survey information. It appears that he is getting a very good bargain at the survey price of $\$25,000$.

23.2.3 | The expected value of sample information (EVSI)

We have just noted that Henry is getting quite a bargain when he purchases the survey information for $\$25,000$. The measure of what he should be willing to pay up to for the survey information is called the Expected Value of Sample Information (EVSI). It's definition is given on page 1104.

In the following example, EVSI is calculated for Henry Wildcat's problem.

Example 23.6 Based on the preposterior analysis of Henry Wildcat's problem, what is the expected value of the sample information (EVSI)?

FIGURE 23.4(b) | States of nature branches for the closed structure outcome (CS)

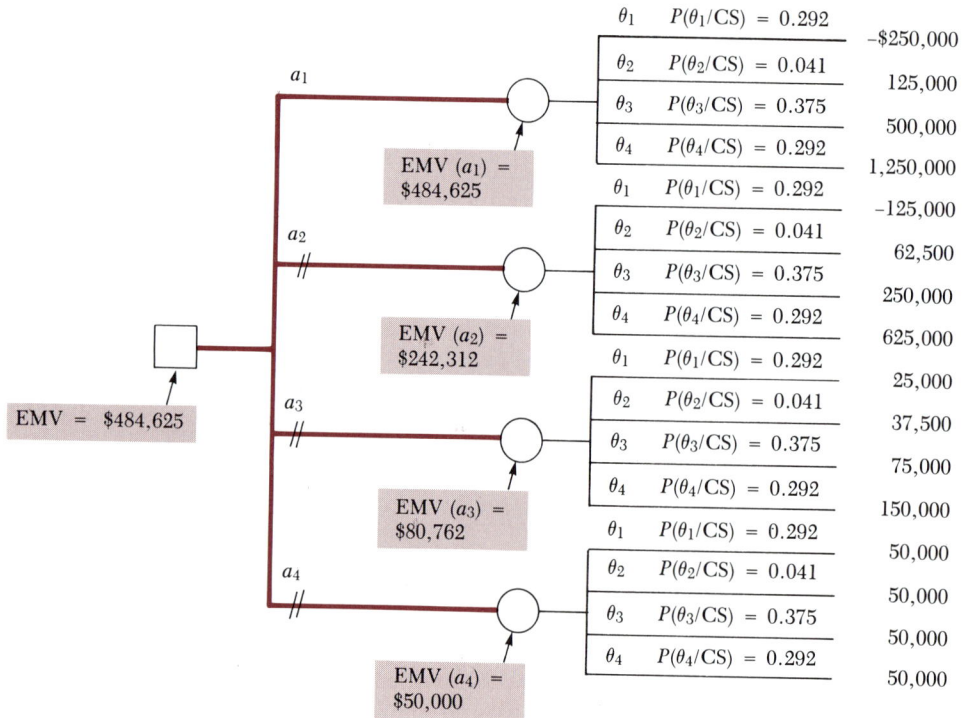

		θ_1	$P(\theta_1/CS) = 0.292$	$-\$250,000$
	a_1	θ_2	$P(\theta_2/CS) = 0.041$	$125,000$
		θ_3	$P(\theta_3/CS) = 0.375$	$500,000$
		θ_4	$P(\theta_4/CS) = 0.292$	$1,250,000$

EMV $(a_1) = \$484,625$

θ_1 $P(\theta_1/CS) = 0.292$ $-\$250,000$
θ_2 $P(\theta_2/CS) = 0.041$ $125,000$
θ_3 $P(\theta_3/CS) = 0.375$ $500,000$
θ_4 $P(\theta_4/CS) = 0.292$ $1,250,000$

a_2

θ_1 $P(\theta_1/CS) = 0.292$ $-125,000$
θ_2 $P(\theta_2/CS) = 0.041$ $62,500$
θ_3 $P(\theta_3/CS) = 0.375$ $250,000$
θ_4 $P(\theta_4/CS) = 0.292$ $625,000$

EMV $(a_2) = \$242,312$

a_3

θ_1 $P(\theta_1/CS) = 0.292$ $25,000$
θ_2 $P(\theta_2/CS) = 0.041$ $37,500$
θ_3 $P(\theta_3/CS) = 0.375$ $75,000$
θ_4 $P(\theta_4/CS) = 0.292$ $150,000$

EMV $(a_3) = \$80,762$

a_4

θ_1 $P(\theta_1/CS) = 0.292$ $50,000$
θ_2 $P(\theta_2/CS) = 0.041$ $50,000$
θ_3 $P(\theta_3/CS) = 0.375$ $50,000$
θ_4 $P(\theta_4/CS) = 0.292$ $50,000$

EMV $(a_4) = \$50,000$

EMV $= \$484,625$

Definition 23.1
The expected value of sample information (EVSI)

When maximizing EMV for profits or revenue:

$$\text{EVSI} = \begin{bmatrix} \text{EMV of optimal decision} \\ \text{with sample or} \\ \text{experimental information} \\ \textit{excluding} \text{ cost} \\ \text{of information} \end{bmatrix} - \begin{bmatrix} \text{EMV of optimal} \\ \text{decision without sample} \\ \text{or experimental} \\ \text{information} \end{bmatrix} \quad (23.2)$$

When minimizing expected costs or EOL:

$$\text{EVSI} = \begin{bmatrix} \text{Expected cost or EOL} \\ \text{of optimal act} \\ \text{without sample or} \\ \textit{experimental} \\ \text{information} \end{bmatrix} - \begin{bmatrix} \text{Expected cost or EOL of} \\ \text{optimal act with sample or} \\ \text{experimental information} \\ \textit{excluding} \text{ cost of} \\ \text{information} \end{bmatrix} \quad (23.3)$$

FIGURE 23.4(c) | States of nature branches for the open structure outcome (OS)

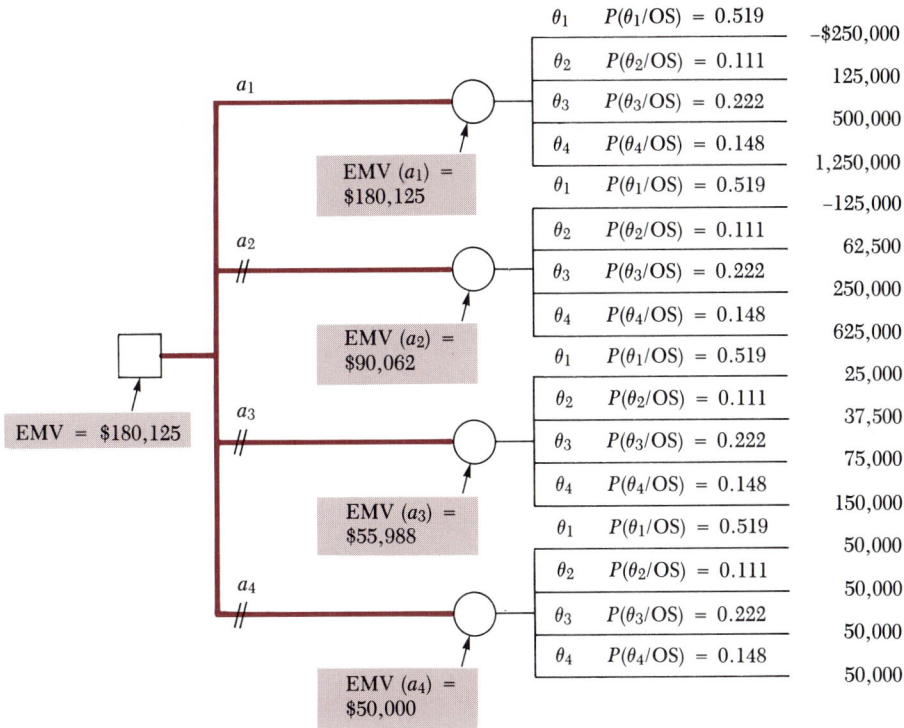

	θ_1 $P(\theta_1/OS) = 0.519$	$-\$250,000$
	θ_2 $P(\theta_2/OS) = 0.111$	$125,000$
a_1	θ_3 $P(\theta_3/OS) = 0.222$	$500,000$
	θ_4 $P(\theta_4/OS) = 0.148$	$1,250,000$
EMV (a_1) = $180,125	θ_1 $P(\theta_1/OS) = 0.519$	$-125,000$

Decision tree with EMV = $180,125:

EMV (a_1) = $180,125
EMV (a_2) = $90,062
EMV (a_3) = $55,988
EMV (a_4) = $50,000

States of nature values:
- a_1: $-\$250,000$; $125,000$; $500,000$; $1,250,000$
- a_2: $-125,000$; $62,500$; $250,000$; $625,000$
- a_3: $25,000$; $37,500$; $75,000$; $150,000$
- a_4: $50,000$; $50,000$; $50,000$; $50,000$

Probabilities for each: $P(\theta_1/OS) = 0.519$, $P(\theta_2/OS) = 0.111$, $P(\theta_3/OS) = 0.222$, $P(\theta_4/OS) = 0.148$

Solution Since the strategy is to maximize EMV for profits:

$$EVSI = \begin{bmatrix} \text{EMV for optimal decision} \\ \text{with sample information} \\ \text{excluding its cost} \end{bmatrix} - \begin{bmatrix} \text{EMV of optimal} \\ \text{decision without} \\ \text{sample information} \end{bmatrix}$$

From the decision tree for the preposterior analysis in Figure 23.4, the value of the first component above is $171,877 (notice that we have not subtracted the cost of the survey—$25,000). Either from the lower branch of the same decision tree or from the prior analysis (Figure 23.1), the value of the second component is $50,000. Thus

$$EVSI = \$171,877 - \$50,000 = \underline{\$121,877}$$

Thus Henry should be willing to pay up to $121,877 for the geological survey result.

Recall from Example 23.1 that EVPI = $191,250—this is the amount up to which Henry should be willing to pay for *perfect* information. A frequently used measure of the efficiency of gaining the sample or experimental information is now defined.

Definition 23.2

Efficiency of sample or experimental information

$$\text{ESI} = \frac{\text{EVSI}}{\text{EVPI}} \cdot 100 \text{ percent} \qquad (23.4)$$

Efficiency of sample information: ESI

For Henry Wildcat's problem,

$$\text{ESI} = \frac{\text{EVSI}}{\text{EVPI}} \cdot (100 \text{ percent}) = \frac{\$121,877}{\$191,250} \cdot (100 \text{ percent}) = 63.73 \text{ percent}$$

Therefore, the Southern Company, the supplier of the geological survey result, is 63.73 percent as efficient as perfect information. Naturally, we would like ESI to be as high as possible. If ESI is close to 100 percent, then we know that it is not worthwhile to search for more information—we have all that we need. If ESI is close to 0 percent, then we know that we must continue searching for new sources of information. In most business applications, ESI ratings of 50 percent or more are considered to be quite good. Henry has quite a good ESI rating, and he got the survey information for considerably less than he perhaps should have expected to pay for it.

■ 23.3

Preposterior decision making with a normal prior distribution and normal sampling (advanced section)

In Chapter 22 it was shown that the computation of EVPI is greatly simplified using Table B.10 in Appendix B if the probability distribution of the random variable is one of the theoretical probability distribution models, namely, the normal distribution. It turns out that the numerical calculations required in determining EVSI is *also* greatly simplified over the methods just discussed if the prior distribution is normal (as is sampling). EVSI, like EVPI, can be determined through the use of Table B.10. We will discuss the use of Table B.10 in determining EVSI and an optimal sampling plan in the following example.

Example 23.7 The Calcit Company is considering selling an electronic calculator with programming features in each of its 2,000 franchised outlets. The investment required to manufacture and promote the model under consideration is $400,000 over the two-year expected life of the unit (before a newer model must be introduced to stay competitive). The profit to the firm on sales of the calculator is $20 per unit.

The firm has been marketing similar (but nonprogrammable) calculators for several years and estimates demand to be 14 units per store, with a standard deviation of 4 units on the model under consideration. The standard deviation of the distribution of sales by each of the 2,000 franchised outlets is felt to be known and to be equal to 12. This quantity, σ_X, is an indication of the dispersion from the average sales of all stores.

The firm has the opportunity to sample its franchised outlets to estimate retail sales per outlet. The estimated cost of sampling a retail outlet is $130 per store because of the wide geographic dispersion of the outlets.

Given the information provided, should the firm undertake marketing the new product? What is the EVPI? The CVPI? What is EVSI?

Solution First, we must compute the break-even quantity of sales per store. Since the required investment is $400,000, each of 2,000 retail outlets must sell

$$\theta_b = \mu_b = \frac{\$400,000}{2,000 \cdot \$20} = 10 \text{ units}$$

for the firm to break even on its investment. Since the prior expected level of sales per store is $E(\theta) = \mu_0 = 14$ units, therefore $E(\theta) > \theta_b$ $(14 > 10)$.[1] Hence, the firm would make an expected profit if it marketed the programmable unit.

The CVPI for this example is shown in Figure 23.5. That is, since any additional information that sales per store will be greater than $\theta_b = \mu_b = 10$

FIGURE 23.5 | CVPI for Example 23.7

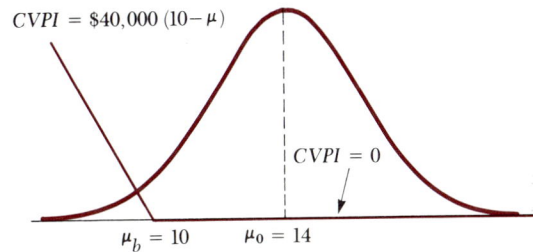

units leads to the same act (introduce the product), CVPI = 0 for values of $\mu \geq 10$. For every unit decrease in mean sales per store below 10 units, however, the firm will incur an opportunity loss of ($20)(2,000)(1), or $40,000. Hence the functional form of CVPI is:

$$\text{CVPI} = \begin{cases} \$40,000(10 - \mu) & \text{for } \mu < 10 \\ 0 & \text{for } \mu \geq 10 \end{cases}$$

The expected value of perfect information (EVPI) under the prior distribution is:

$$\text{EVPI} = C\sigma_0 N(D) = \$40,000 \cdot 4 \cdot N\left(\frac{|10 - 14|}{4}\right) \qquad (22.9)$$

---

[1] For the normal distribution, we will subscript the parameters of the prior distribution with 0 and subscript the parameters of the revised distribution with 1. Hence μ_0 and σ_0^2 will denote, respectively, the mean and the variance of the prior distribution.

From Table B.10, $N(1) = 0.08332$. Hence

$$\text{EVPI} = \$40,000 \cdot 4 \cdot 0.08332 = \$13,331.20$$

The EVPI sets an upper limit on the amount to spend for perfect information and, therefore, for sample (imperfect) information as well.

Since it costs \$130 to sample each outlet, we would never consider taking a sample larger than $n = (13,331.20/130) \doteq 102$ outlets. Let's assume for the moment that a decision is made to take a sample of 25 outlets. The cost of sampling is thus $\$130.00 \cdot 25 = \$3,250$.

We must now compute the variance of the sampling distribution. Since the producer estimates the standard deviation of the population to be 12,

$$\sigma_{\overline{X}} = \frac{\sigma_X}{\sqrt{n}} = \frac{12}{\sqrt{25}} = 2.4 \quad \text{and} \quad \sigma_{\overline{X}}^2 = (2.4)^2 = 5.76$$

Before the sample is actually selected, the estimate of the standard deviation used in the determination of EVSI is given by:

$$\sigma_* = \sqrt{\sigma_0^2 \left(\frac{\sigma_0^2}{\sigma_0^2 + \sigma_{\overline{X}}^2} \right)} \quad \text{or} \quad \sigma_* = \sqrt{4^2 \left[\frac{4^2}{4^2 + (2.4)^2} \right]} = 3.43$$

This quantity, σ_*, represents the "amount" of revision in the standard deviation from the prior to the posterior distribution and is derived through use of the relationship

$$\sigma_* = \sqrt{\sigma_0^2 - \sigma_1^2}$$

Given σ_*, EVSI for a normal prior and normal sampling is as defined here.

Expected value of sample information: Normal prior and normal sampling	**Definition 23.3** The expected value of sample information— normal prior distribution and normal sampling $$\text{EVSI} = C\sigma_* N(D_*) \qquad (23.5)$$ where $D_* = \|\mu_b - \mu_0\|/\sigma_*$ and $N(D_*)$ is found in Table B.10 in Appendix B.

For the data in Example 23.7,

$$\text{EVSI} = \$40,000 \cdot 3.43 \cdot N\left(\frac{|10 - 14|}{3.43} \right) = \$40,000 \cdot 3.43 \cdot N(1.17)$$
$$= \$40,000 \cdot 3.43 \cdot 0.05964 = \$8,182.61$$

For a sample of size 25, EVSI is $8,182.61. To take the sample of size $n = 25$ costs ($130)(25) = $3,250. Since EVSI is greater than the cost of sampling, we have experienced a monetary gain by taking the sample of size $n = 25$. The definition of the measure of gain from sampling is given below.

<table>
<tr><td>Expected net gain from sampling: ENG</td></tr>
</table>

Definition 23.4

Expected net gain from sampling for a sample of size $n - ENG(n)$

$$ENG(n) = EVSI(n) - C(n)$$

where $EVSI(n)$ is the expected value of sample information for a sample size n, and $C(n)$ is the cost of taking a sample of size n.

Since the cost is $130 to sample each store, $C(n) = \$130n$. Now, $ENG(25) = EVSI(25) - C(25) = \$8,182.61 - \$130(25) = \$8,182.61 - \$3,250 = \$4,932.61$. Hence total losses or total opportunity losses should be reduced by $4,932.61 by taking a sample of 25 outlets.

The ENGS for sample sizes of 20, 25, 30, and 40 are given in Table 23.7. From this table, it is apparent that the optimal sample size is between 25 and 40. If we vary n by units of one starting at $n = 30$ and calculate $ENGS(n)$, we will determine the optimal sample size. These results are also given in Table 23.7. From this table, we should take a sample of size $n = 30$ and compute the posterior mean, $E(\theta)$. If $E(\theta) < \theta_b = \mu_b$, we should not manufacture the product. If $E(\theta) > \theta_b$, we should manufacture the product. This decision rule will minimize the unconditional expected total losses of the decision.

TABLE 23.7 | Determination of optimal sample size for data in Example 23.7

Sample size, n	$\sigma_{\bar{x}} = \dfrac{12}{\sqrt{n}}$	σ_*	D_*	$N(D_*)$	EVSI	Cost of sampling	ENGS
20	2.68	3.32	1.20	0.05610	$7,450.08	$2,600	$4,850.08
25	2.40	3.43	1.17	0.05964	8,182.61	3,250	4,932.61
29	2.23	3.49	1.45	0.06273[a]	8,757.11	3,770	4,987.11
30	2.19	3.51	1.14	0.06336	8,895.75	3,900	4,995.75
31	2.15	3.52	1.14	0.06336	8,921.09	4,030	4,891.09
40	1.90	3.61	1.11	0.06727	9,713.79	5,200	4,513.79

[a] Determined by interpolation.

■ **23.4**

Revising probabilities: Bayes theorem for a normal prior distribution and normal sampling (advanced section)

In this section, we will describe the revision of probability for a random variable that is normally distributed. Only one important case of the normal distribution will be covered in this chapter; namely, the case in which the variance of the sampling distribution of the mean, $\sigma_{\bar{X}}^2$, is known and the mean (average) is unknown. The cases where either the variance of the sampling distribution is unknown and the mean is known, or where both the sample mean and the sampling distribution variance are unknown require the knowledge of probability distributions beyond the scope of this text. In most practical situations in which n is large ($n > 30$), we can treat the variance of the sampling distribution of the mean as being known through use of the relationship

$$\sigma_{\bar{X}}^2 = \frac{\sigma_{\bar{X}}^2}{n} \doteq \frac{S^2}{n}$$

as discussed in Chapter 8.

Before presenting the formulas necessary for revising normal probabilities as well as an example of their use, it will be useful to introduce the following notation. I represents the amount of information contained in a distribution, and it is equal to the reciprocal of the variance. For the prior distribution, for example, I_0 is:

$$I_0 = \frac{1}{\sigma_0^2}$$

The notion of setting the amount of information in a distribution equal to the reciprocal of the variance has great intuitive appeal, for the larger the value of σ^2, the less certain we are about the value of μ, and the smaller is the value of I, the amount of information available. This is illustrated in Figure 23.6. Obviously, the amount of information available in Figure 23.6A is less than that available in Figure 23.6B.

Once again, stressing that it is assumed that *the value of the variance of the sampling distribution of the mean is known,* the results for revising normal probabilities can now be given.

FIGURE 23.6 | Two normal distributions with equal means and unequal variances (unequal amounts of information)

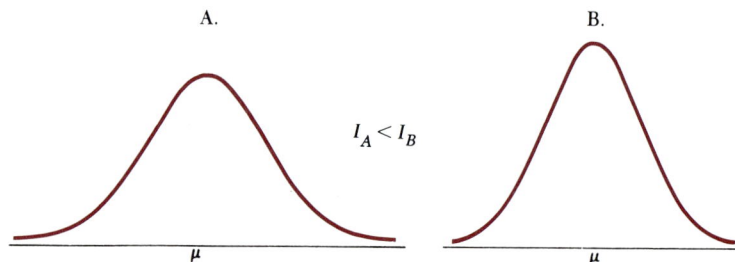

A.   B.

$I_A < I_B$

μ μ

The mean of the posterior distribution, μ_1, is a weighted average of the prior mean, μ_0, and sample mean, $\bar{x}$; the weights are determined by the amount of information contained in the prior distribution, I_0, and in the sampling distribution, $I_{\bar{x}}$. The standard deviation of the revised distribution, σ_1, reflects the pooling of information in the prior distribution and in the sample according to the relationship $I_1 = I_0 + I_{\bar{x}}$. The amount of information in the revised distribution is thus the sum of the information contained in the prior distribution and in the sample. The use of the formulas in Theorem 23.1 for revising normal probabilities is illustrated in the following example.

<table>
<tr><td>Revision of normal
prior probabilities</td></tr>
</table>

Theorem 23.1
Revision of normal prior probabilities with normal sampling
(variance of sampling distribution known)[2]

If the prior probability distribution of a population mean is normal with parameters μ_0 and σ_0, and if the sampling distribution is normal with known variance $\sigma_{\bar{x}}^2$ (or $\sigma_{\bar{x}}^2$ can be estimated from the sample variance for n large, $n \geq 30$), and a sample of size n is taken yielding an estimate $\bar{x}$ of μ, then the posterior distribution of $\theta = \mu$ is normal with parameters μ_1 and σ_1 given by:

$$I_0 = \frac{1}{\sigma_0^2}$$

$$I_{\bar{x}} = \frac{1}{\sigma_{\bar{x}}^2} \left(\text{or } I_{\bar{x}} = \frac{n}{s^2} \quad \begin{array}{l} \text{if } \sigma_{\bar{x}}^2 \text{ is unknown, } n \geq 30 \\ \text{and } s^2 \text{ is the sample variance} \end{array} \right)$$

$$\mu_1 = \frac{I_0 \mu_0 + I_{\bar{x}} \bar{x}}{I_0 + I_{\bar{x}}}$$

$$\sigma_1 = \sqrt{\frac{\sigma_0^2 \cdot \sigma_{\bar{x}}^2}{\sigma_0^2 + \sigma_{\bar{x}}^2}}$$

Example 23.8 A consumer products firm purchases bearings in lots of 500 units for installation on roller skates, which are subsequently marketed to the public. Each lot of bearings is sampled, and if the average diameter is within a "tolerable level," the whole lot of bearings is accepted. Because of the low value of these bearings in relation to the total product, a small sample ($n = 10$) is usually selected, and the information provided by the sample is combined with the prior information on the behavior of the process responsible for producing these bearings to arrive at a decision on whether or not to accept the shipment.

---

[2] The interested reader who desires more information on the development of these formulas as well as on the explicit form of the sampling and marginal distributions is referred to the Jedamus and Frame reference at the end of the chapter.

The prior distribution (derived from recent experience) on the average diameter of bearings produced is normal with parameters $\mu_0 = 1.50$ inches and standard deviation $\sigma_0 = 0.01$ inch. The standard deviation of bearings produced by the machine is known to be equal to $\sigma_X = 0.04472$ inch. A sample of 10 bearings yields a mean $\bar{x} = 1.48$.

If the revised probability of the average diameter of bearings in the lot being within ± 0.02 inch of 1.50 inches is not at least 0.90, the whole lot is rejected; otherwise, it is accepted.

Using the criteria outlined by the firm, should the current shipment of bearings be accepted or rejected?

Solution   First, we must compute the parameters of the revised, normal probability distribution. They are:

$$I_0 = \frac{1}{\sigma_0^2} = \frac{1}{0.0001} = 10,000$$

$$\sigma_{\bar{X}}^2 = \frac{\sigma_X^2}{n} = \frac{(0.04472)^2}{10} = 0.0002$$

$$I_{\bar{X}} = \frac{1}{\sigma_{\bar{X}}^2} = \frac{1}{0.0002} = 5,000$$

$$\mu_1 = \frac{10,000 \cdot 1.50 + 5,000 \cdot 1.48}{10,000 + 5,000} = 1.4933$$

$$\sigma_1 = \sqrt{\frac{(0.0001)(0.0002)}{0.0001 + 0.0002}} = \sqrt{0.000066} = 0.0082$$

The revised distribution is a normal distribution with parameters $\mu_1 = 1.4933$ and $\sigma_1 = 0.0082$. Calculation of the required probability is shown below. Since the revised probability that the average diameter of bearings in the lot is within 1.48–1.52 exceeds the firm's requirements ($0.9473 > 0.9000$), the lot of bearings should be accepted.

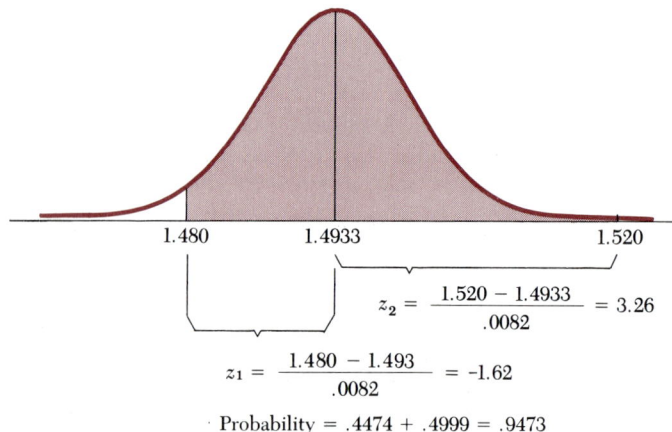

$$z_2 = \frac{1.520 - 1.4933}{.0082} = 3.26$$

$$z_1 = \frac{1.480 - 1.493}{.0082} = -1.62$$

Probability $= .4474 + .4999 = .9473$

Notice that the amount of information in the prior distribution, I_0, is twice as large as the amount of information in the sample, $I_{\bar{x}}$. This is reflected in the value of the posterior mean, which is closer in value to the prior mean than to the sample mean.

■ 23.5

Terminal decision making when a sample has already been selected— Normal process (advanced section)

Frequently a statistician is called in *after* a sample has already been selected and is requested to use the information that has been made available. If both the prior mean and the sample mean are less than or both are greater than μ_b (i.e., $\mu_0 > \mu_b$ and $\bar{x} > \mu_b$; or $\mu_0 < \mu_b$ and $\bar{x} < \mu_b$), then the act that was optimal under the prior is still optimal under the posterior, and we do not need to determine a posterior mean.

If, however, $\mu_0 < \mu_b$ and $\bar{x} > \mu_b$, or if $\mu_0 > \mu_b$ and $\bar{x} < \mu_b$, then the posterior mean *may not* bear the same inequality relationship to $\theta_b = \mu_b$. If this inequality relationship changes, then the optimal act may change also. This is illustrated in the following two examples.

Example 23.9 Assume that a sample of 10 stores selected in Example 23.7 yields the results given in the table. Does the optimal act to follow change under the posterior distribution?

Store	Calculators sold
1	8
2	6
3	18
4	22
5	10
6	11
7	12
8	10
9	14
10	9

Solution

$$\bar{x} = \frac{8 + 6 + 18 + 22 + 10 + 11 + 12 + 10 + 14 + 9}{10} = 12$$

Since $\bar{x} > \mu_b$ and $\mu_0 > \mu_b$, the optimal act to follow will still be to manufacture and promote the calculators. It is not necessary to compute μ_1.

Example 23.10 Assume now that a sample of 10 outlets is selected and yields the results given in the table. What is the optimal act under the revised distribution?

Store	Calculators sold
1	8
2	11
3	13
4	5
5	4
6	10
7	8
8	7
9	6
10	8

Solution

$$\bar{x} = \frac{8 + 11 + 13 + 5 + 4 + 10 + 8 + 7 + 6 + 8}{10} = 8$$

Since $\bar{x} < \mu_b$ and $\mu_0 > \mu_b$, it is necessary to calculate μ_1 using Theorem 23.1. Since σ_X is known to be 12, $\sigma_{\bar{X}} = 12/\sqrt{10} = 3.80$, and

$$I_0 = \frac{1}{4^2} = 0.0625$$

$$I_{\bar{x}} = \frac{1}{(3.80)^2} = 0.06925$$

$$\mu_1 = \frac{0.0625 \cdot 14 + 0.06925 \cdot 8}{0.0625 + 0.06925} = \frac{0.875 + 0.554}{0.13175} = 10.85$$

Because $\mu_1 > \mu_b$ (10.85 > 10.00), the optimal act is still to produce and market the programmable calculators.

■ 23.6

A comparison of Bayesian and classical approaches (advanced section)

In this section we characterize the major differences between the approaches taken by classical and Bayesian statisticians in arriving at a decision under risk. These differences include whether or not to admit a subjective probability distribution into the analysis of a problem, and they concern statements that can be made regarding the interpretation of results using the different approaches.

23.6.1 | Use of subjective probabilities

The major point of deviation between Bayesian and classical statisticians concerns the admittance of subjective probabilities (reflecting degrees of belief) into the analysis of a problem. The Bayesian, believing that *all* information pertinent to a decision should be incorporated into the decision process, is perfectly willing to use—and indeed, often encourages the use of—

subjective probability distributions in arriving at a decision. The classical statistician, on the other hand, will not use a subjective probability distribution to construct, say, a test of a hypothesis or a confidence interval. He bases his judgments solely on the sampling distribution or on the likelihood function described in Chapters 8–11. As such, the classical statistician is often called a "frequentist," basing all probabilities on the long-run frequency interpretation of probability.

23.6.2 | Estimating the value of a population parameter

The second point of departure between the two "camps" of statistics concerns the "value" of a population parameter. The classical statistician does not allow probability statements to be made regarding possible values of a population parameter. Either the population parameter is equal to some value or it is not, and it is "incorrect" to speak of the probability of the parameter being some value. The Bayesian, on the other hand, views an unknown population parameter as a random variable and thus is perfectly willing to make probability statements concerning it.

We show a graph of this distinction in Figure 23.7. Figure 23.7A represents the *posterior distribution* of the random variable μ, the population

FIGURE 23.7 | Posterior and sampling distribution

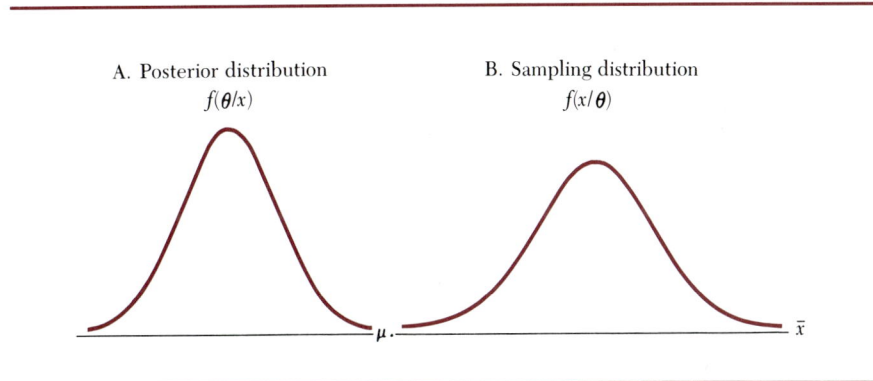

A. Posterior distribution
$f(\theta/x)$

B. Sampling distribution
$f(x/\theta)$

μ

$\bar{x}$

mean or expected value, and Figure 23.7B represents the *sampling distribution* of the random variable $\overline{X}$. The Bayesian statistician would use the posterior distribution of the *random variable* μ to make probability statements regarding the population mean, whereas the classical statistician would use the sampling distribution of the mean to base his statements on the population mean value. Note the difference in the two distributions: $P(\theta/x)$ $[f(\theta/x)]$ represents the probability distribution of the random variable θ given the sample result x, and $P(x/\theta)$ $[f(x/\theta)]$ represents a probability distribution of sample outcomes, conditional on values of the random variable $\boldsymbol{\theta}$. The use of these two distributions leads to some interesting interpretations of probability statements in estimation and hypothesis testing.

Let's take first the case of confidence interval estimation and consider a $(1 - \alpha)$ percent confidence interval on the mean of some process. The Bayesian statistician would state, "The probability is $(1 - \alpha)$ that the mean of the process lies in the stated interval," whereas the classical statistician would say, "$(1 - \alpha)$ percent of the confidence intervals so constructed contain the true mean of the process." The Bayesian is making statements about the interval at hand, while the classicist is making statements about the long run. To differentiate between the two interval estimates, the classical probability interval is called *confidence interval,* and the Bayesian interval is called a *credible interval.*

Credible interval

Now let's consider the case of hypothesis testing (for discussion purposes, we'll consider a one-tail test of the form:

$$H_0: \quad \mu \le \mu_0$$
$$H_A: \quad \mu > \mu_0$$

but the same distinction holds for either a one-tail or a two-tail test of a hypothesis). The classical statistician will either reject or not reject the null hypothesis H_0. The Bayesian statistician, on the other hand, will speak of the probability that H_0 is true *and* the probability that H_A is true. The classical statistician specifies the distribution under H_0, but not under H_A because of a concern with the probability of making a Type I error, α. The Bayesian statistician takes into account the distribution under both H_0 *and* H_A in making probability statements regarding both hypotheses.

23.6.3 | Formal versus informal consideration of losses of an incorrect decision

The classical statistician, in performing a test of a hypothesis, usually sets α (the probability of making a Type I error) equal to a predetermined value of 0.1, 0.05, or 0.01. The actual value of α (of the three) is selected to "reflect" the economic consequences of rejecting the null hypothesis when it is true. Thus we say that the determination of α is done somewhat informally in regard to the incorporation of losses into the decision. Given a value of α, a test is selected that gives the smallest value of β, the probability of making a Type II error.

The Bayesian statistician bases the determination of α *and* β on the economic consequences that might result if the decision is incorrect. Since the maximization of EMV or the minimization of EOL *formally* incorporates the costs or the losses of making an incorrect decision into a decision rule, we say that the Bayesian statistician formally incorporates the consequences of making both types of errors into his or her decision rule. It is safe to say that Bayesian statisticians can "tolerate" a much higher probability of making a Type I error (α) than can their classical counterparts!

There are other differences between Bayesian and classical statisticians than those mentioned above, to be sure. We have attempted in the preceding discussion to summarize a few of the major differences between the two approaches so that the reader can make an informed choice between them. References by Schlaifer and by Winkler at the end of the chapter are *excellent* treatments of advanced topics in statistical decision theory or in Bayesian decision theory, which the reader who is interested in more on this topical area may want to consider.

■ **References**

Introductory Level

Anderson, D. R.; D. J. Sweeney; and T. A. Williams. *Introduction to Management Science*. 4th ed. St. Paul, Minn.: West Publishing, 1985, Chapter 16.

Cook, William D. *Quantitative Methods for Management Decisions*. 2nd ed. New York: McGraw-Hill, 1985, Chapter 9.

Lapin, Lawrence L. *Quantitative Methods for Business Decisions*. 2nd ed. New York: Harcourt Brace Jovanovich, 1981, Chapters 5 and 6.

Advanced Level

Bierman, H.; C. P. Bonini; and W. H. Hausman. *Quantitative Analysis for Business Decisions*. 7th ed. Homewood, Ill.: Richard D. Irwin, 1986, Chapters 3–10.

Hillier, F. S., and G. J. Lieberman. *Introduction to Operations Research*. 3rd. ed. San Francisco: Holden-Day, 1980, Chapter 15.

Jedamus, Paul, and R. J. Frame. *Business Decision Theory*. New York: McGraw-Hill, 1969.

Luce, Duncan R., and Howard Raiffa. *Games and Decisions*. New York: John Wiley & Sons, 1957.

Pratt, J. W.; Howard Raiffa; and Robert Schlaifer. *Introduction to Statistical Decision Theory*. Preliminary edition. New York: McGraw-Hill, 1965.

Raiffa, Howard. *Decision Analysis: Introductory Lectures on Choices under Uncertainty*. Reading, Mass.: Addison-Wesley, 1968.

Raiffa, Howard, and Robert Schlaifer. *Applied Statistical Decision Theory*. Boston: Division of Research, Harvard University, 1961.

Schlaifer, Robert. *Analysis of Decisions under Uncertainty*. New York: McGraw-Hill, 1969.

————. *Computer Programs for Elementary Decision Analysis*. Boston: Division of Research, Harvard University, 1971.

————. *Introduction to Statistics for Business Decisions*. New York: McGraw-Hill, 1961.

————. *Probability and Statistics for Business Decisions*. New York: McGraw-Hill, 1959.

Winkler, Robert L. *An Introduction to Bayesian Inference and Decision*. New York: Holt, Rinehart and Winston, 1972.

■ **Problems**

Section 23.2 Problems

23.1 Suppose we have the following prior probability distribution for the states of nature (θ_j) in a decision problem:

θ_j	$P(\theta_j)$
θ_1	0.6
θ_2	0.4
	1.0

Further, suppose that for some event B, we have the following likelihoods: $P(B/\theta_1) = 0.9$ and $P(B/\theta_2) = 0.1$.

a. Determine $P(B)$.

b. Find the posterior probabilities $P(\theta_1/B)$ and $P(\theta_2/B)$.

23.2 Suppose we have the following prior probability distribution for the states of nature (θ_j) in a decision problem:

θ_j	$P(\theta_j)$
θ_1	0.2
θ_2	0.8
	1.0

Further, suppose that for some event B, we have the following likelihoods: $P(B/\theta_1) = 0.6$ and $P(B/\theta_2) = 0.2$.

a. Determine $P(B)$.

b. Find the posterior probabilities $P(\theta_1/B)$ and $P(\theta_2/B)$.

23.3 Suppose we have the following prior probability distribution for the states of nature (θ_j) in a decision problem:

θ_j	$P(\theta_j)$
θ_1	0.3
θ_2	0.7
	1.0

Further, suppose that for some event B, we have the following likelihoods: $P(B/\theta_1) = 0.1$ and $P(B/\theta_2) = 0.5$.

a. Determine $P(B)$.

b. Determine $P(\bar{B})$.

c. Find the posterior probabilities $P(\theta_1/B)$ and $P(\theta_2/B)$.

d. Find the posterior probabilities $P(\theta_1/\bar{B})$ and $P(\theta_2/\bar{B})$.

23.4 Suppose that we have the following prior probability distribution for the states of nature (θ_j) in a decision problem:

θ_j	$P(\theta_j)$
θ_1	0.10
θ_2	0.15
θ_3	0.25
θ_4	0.40
θ_5	0.10
	1.00

Further suppose that for some event B we have the following likelihoods: $P(B/\theta_1) = 0$, $P(B/\theta_2) = 0$, $P(B/\theta_3) = 0.5$, $P(B/\theta_4) = 0.9$ and $P(B/\theta_5) = 0.3$.

a. Determine $P(B)$.

b. Find the posterior probabilities $P(\theta_1/B)$, $P(\theta_2/B)$, $P(\theta_3/B)$, $P(\theta_4/B)$, and $P(\theta_5/B)$.

23.5 In Problem 22.4, Chapter 22, suppose that event B occurs and for the three states of nature θ_1, θ_2, and θ_3, it is known that the likelihoods are $P(B/\theta_1) = 0.90$, $P(B/\theta_2) = 0.20$, and $P(B/\theta_3) = 0.10$.

a. Find the posterior probabilities $P(\theta_1/B)$, $P(\theta_2/B)$, and $P(\theta_3/B)$.

b. What is the optimal posterior act using the EMV maximization strategy?

23.6 In Problem 22.23, Chapter 22, suppose that event B occurs and for the five states of nature I, II, III, IV and V, it is known that the likelihoods are $P(B/I) = 0$, $P(B/II) = 0$, $P(B/III) = 0.10$, $P(B/IV) = 0.80$, and $P(B/V) = 0.40$.

a. Determine $P(B)$.

b. Find the posterior probabilities $P(I/B)$, $P(II/B)$, $P(III/B)$, $P(IV/B)$, and $P(V/B)$.

c. What is the optimal posterior act using the EMV maximization strategy?

23.7 Refer to Problem 22.12, Chapter 22. Suppose that the national election for president is soon, and let the event B be A republican candidate wins the election. The comptroller of BLP has subjectively determined the following likelihoods:

$$P(B/I) = 0.20, \qquad P(B/II) = 0.50,$$
$$P(B/III) = 0.90$$

where the three states of nature I, II, and III are defined to be expanding economy, healthy economy, and recession, respectively. Given the prior probabilities $P(I) =$

0.50, $P(II) = 0.40$, and $P(III) = 0.10$, determine the following:

a. $P(B)$.

b. The posterior probabilities $P(I/B)$, $P(II/B)$, and $P(III/B)$.

c. The optimal posterior act using the EMV maximization strategy.

23.8 The Acme Corporation is contemplating the introduction of a new product. The new products director must choose from among two alternatives: Act 1, introduce new product, and Act 2, do not introduce new product. There are only two states of nature: θ_1, market is favorable for introduction of the new product, and θ_2, market is not favorable for introduction of the new product. The director, after consultation with the finance and accounting department personnel, constructs the following payoff matrix:

Act (a_i)	State of nature (θ_j) Favorable θ_1	Unfavorable θ_2
a_1: Introduce	$2,500,000	−$1,000,000
a_2: Do not introduce	$0	$0
$P(\theta_j)$	0.4	0.6

Notice that the table also contains the prior probabilities for the two states of nature. The director feels uncomfortable about the prior probabilities. As a result, he hires a consulting firm to conduct a market survey of the potential (target) market for the new product. The consulting firm will issue a report indicating that the new product is either going to be successful (event B) or not successful (event $\bar{B}$). Based on similar products in the recent past the consulting firm provides the director with the following likelihoods:

$$P(B/\theta_1) = 0.95 \quad P(\bar{B}/\theta_1) = 0.05$$
$$P(B/\theta_2) = 0.10 \quad P(\bar{B}/\theta_2) = 0.90$$

The director is very impressed with these likelihoods. The survey reliability values appear to be quite high [$P(B/\theta_1) = 0.95$, and $P(\bar{B}/\theta_2) = 0.90$].

a. What is the optimal prior act based on maximizing EMV?

b. What is the EVPI for this decision based on the prior probability distribution?

c. Suppose that the consulting firm's report indicates that the new product will be successful. Determine the posterior probability distribution, and find the optimal posterior act based on maximizing EMV.

d. Suppose that the consulting firm's report indicates that the new product will *not* be successful. Determine the posterior probability distribution, and find the optimal posterior act on maximizing EMV.

23.9 In Problem 23.8 above, use a decision tree to perform the preposterior analysis. In addition to the tree, specifically answer the following questions:

a. What is the EVSI for this problem?

b. What is the ESI for this problem?

c. Up to what dollar amount should the director be willing to pay for the consulting firm's report?

d. What is the probability that the firm will present the director with a report indicating a successful product if the firm is retained?

23.10 Refer to Problem 22.24 in Chapter 22. Suppose that the M-Bank officer must decide whether or not to pay the Western Company to conduct a geological survey of the parcel of land in West Texas. The geological survey will cost $25,000 and will produce one of two outcomes: Structure exists (event B), or no structure exists (event $\bar{B}$). The likelihoods for the survey outcomes are:

$$P(B/\theta_1) = 0.10 \quad P(\bar{B}/\theta_1) = 0.90$$
$$P(B/\theta_2) = 0.20 \quad P(\bar{B}/\theta_2) = 0.80$$
$$P(B/\theta_3) = 0.70 \quad P(\bar{B}/\theta_3) = 0.30$$

Using the preposterior analysis with a decision tree, and the prior probabilities $P(\theta_1) = 0.75$, $P(\theta_2) = 0.15$, and $P(\theta_3) = 0.10$, should the M-Bank officer pay the Western Company to conduct the geological survey? What is the EVSI for this problem? What is the ESI?

23.11 Each week the Steele Case Company receives a large shipment of components used in constructing a certain type of metal case. In each shipment there may be defective components. From past records on these weekly shipments, the percentage of defectives and the corresponding probabilities are given below.

Percentage of defectives	Probability
0%	0.20
1%	0.40
2%	0.20
3%	0.20
	1.00

The firm currently inspects each and every one of the components upon receipt of the shipment. Because of this, the company knows that it has never received more than 3 percent defectives and is very confident that the probabilities listed above are accurate. The current 100 percent inspection plan is costly. Currently, the cost is $500 per shipment to inspect each and every time. The production manager suggests an alternate plan. Do no inspection and rework or replace a defective item at the time it is used in construction. The production manager estimates the costs will be $0 if there is 0 percent defectives, $200 if there are 1 percent defectives, $500 if there are 2 percent defectives, and $750 if there are 3 percent defectives. Treating the percent defective as the state of nature, the payoff (*cost*) matrix is:

State of nature (θ_j) / Act (a_i)	0 θ_1	1 θ_2	2 θ_3	3 θ_4
a_1: 100 percent inspection	$500	$500	$500	$500
a_2: No inspection	0	200	500	750
$P(\theta_j)$	0.20	0.40	0.20	0.20

a. Based on the prior probability distribution, what is the optimal act?

b. What is the EVPI for this decision? From the current shipment, a random sample of four components is selected and inspected. One defective component is found.

c. Using the binomial distribution, find the likelihoods of this outcome, given each of the states of nature. Hint: Let $\mathbf{X}$ = Number of defectives in a sample of size $n = 4$ components. Then find $P(\mathbf{X} = 1/\theta_1)$, $P(\mathbf{X} = 1/\theta_2)$, $P(\mathbf{X} = 1/\theta_3)$, and $P(\mathbf{X} = 1/\theta_4)$.

d. With these likelihoods, now revise the prior probability distribution, forming the posterior distribution probabilities $P(\theta_1/\mathbf{X} = 1)$, $P(\theta_2/\mathbf{X} = 1)$, $P(\theta_3/\mathbf{X} = 1)$, and $P(\theta_4/\mathbf{X} = 1)$.

e. Determine the optimal posterior act, using the posterior distribution found in part d.

Section 23.3 Problems

23.12 In what way does the sample size n affect the amount of information present in a normal sampling distribution? Describe the exact relationship.

23.13 Assume the same facts as in Example 23.10 and that $\sigma_{\mathbf{X}} = 9$. What are EVSI and ENGS for the optimal decision rule (sample size)?

Section 23.4 Problems

23.14 Beginning with the relationship

$$I_1 = I_0 + I_{\bar{x}}$$

show that

$$\sigma_1 = \sqrt{\frac{\sigma_0^2 \cdot \sigma_{\bar{x}}^2}{\sigma_0^2 + \sigma_{\bar{x}}^2}}$$

23.15 A decision maker assesses a prior distribution of sales for a new product as being normal with mean $\mu_0 = 190$ units per store and standard deviation $\sigma_0 = 30$ units per store. A sample of 100 stores is selected at random for testing sales of the new product, yielding a sample mean $\bar{x} = 225$ units per store and a sample standard deviation $s = 25$ units. What is the posterior distribution of sales for the new product?

23.16 Assume in Problem 23.15 that the sample consists of only 20 stores, but that the statistics remain the same. What is the revised probability distribution?

Section 23.5 Problem

23.17 In Example 23.10, assume the following sample results were obtained.

Store	Calculators sold
1	8
2	5
3	7
4	5
5	2
6	8
7	8
8	7
9	6
10	8

What is the optimal act under the posterior distribution of sales?

Section 23.6 Problem

23.18 Herman, as his main source of income, inspects garbage disposals at the local plumbing fixtures factory. A lot of 10,000 disposals is awaiting shipment to retail outlets. Herman's boss, Fort Worth Annie, has instructed Herman to inspect 10 of the 10,000 garbage disposals, and at the 5 percent significance level (i.e., $\alpha = 0.05$) either conclude that 10 percent or fewer are defective and ship the batch of 10,000 disposals, or else conclude that more than 10 percent of the disposals are defective and inspect the entire lot of disposals before shipment. Construct a "classical" decision rule for Herman so that he can decide whether to ship the disposals as is or else inspect the entire lot and repair the defective ones. Comment on the choice of $\alpha = 0.05$, and the consequences that might result by setting α at this level (alternative losses).

Additional Problems

23.19 A corporation has undertaken a training program to introduce members of its accounting staff to new computer languages. After the program is completed, a test is administered to evaluate training efficiency. From previous experience, it is known that the probability of passing the test after training is 70 percent, whereas the probability of failing the test without the training is 50 percent. (Employees may, if they wish, take the examination, and if they pass, they are not expected to enroll in the training program.) Ten percent of the accounting staff has participated in the training program. An employee (accountant) selected at random is known to have passed the test. What is the probability that he participated in the training program? That he did not? What probability would you give to passing the test if you know nothing about whether the individual had the training or not?

23.20 A certain machine produces bearings in 500-unit lots. The frequency distribution for the number of defectives in a lot based on recent experience is given in the table.

Number of defectives	Frequency
5	0.10
10	0.20
15	0.40
20	0.20
25	0.10
	1.00

A sample of 20 bearings from one 500-unit lot is taken, and two defective units are found. Treating the number of defective units as a state of nature and using the binomial distribution to determine the likelihoods, what is the posterior distribution for defectives in the lot?

23.21 Assuming you are willing to let subjective judgments enter into a determination of a prior probability distribution, state formally the steps you would take and the factors you would consider in assessing a probability distribution for the outcome of your college's first football game.

23.22 Why is it difficult to determine the prior probabilities in a business decision process by objective rather than subjective methods? Give an example of a decision problem in which the prior probabilities *could* be determined as objective probabilities.

23.23 Explain how posterior probabilities are calculated, given a set of prior probabilities and a set of likelihoods for a given event B.

23.24 Suppose that we have the following prior probability distribution for the states of nature (θ_j) in a decision problem:

θ_j	$P(\theta_j)$
θ_1	0.2
θ_2	0.3
θ_3	0.4
θ_4	0.1
	1.0

Further, suppose that for some event B, we have the following likelihoods: $P(B/\theta_1) = 0.9$,

$P(B/\theta_2) = 0.7$, $P(B/\theta_3) = 0.5$, and $P(B/\theta_4) = 0.1$.

a. Determine $P(B)$.
b. Determine $P(\overline{B})$.
c. Find the posterior probabilities $P(\theta_1/B)$, $P(\theta_2/B)$, $P(\theta_3/B)$, and $P(\theta_4/B)$.

23.25 The Maypro Computer Company must decide whether to purchase the "motherboard" for its Maypro 16 computer or build the board itself. The decision depends to a great extent on the demand for the new Maypro 16 computer. The production manager defines three states of nature for the product demand: θ_1—demand is low, θ_2—demand is average, and θ_3—demand is high. Subjectively, he assigns the following *prior* probabilities to these states of nature: $P(\theta_1) = 0.40$, $P(\theta_2) = 0.30$, and $P(\theta_3) = 0.30$. The profit table for this problem is given below:

State of nature (θ_j) / Act (a_i)	Demand		
	Low θ_1	Average θ_2	High θ_3
a_1: Purchase board	$25,000	$50,000	$ 75,000
a_2: Build board	−20,000	60,000	100,000
$P(\theta_j)$	0.40	0.30	0.30

The profits are based on the revenue from sales of the Maypro 16 computer less production costs including the cost of either purchasing or building its motherboard. Further, the profits are calculated for the first production run of 100 units. A marketing research firm is contracted to *predict* the demand for this product. The firm will provide Maypro with one of three experimental outcomes:

E_1: Firm *predicts* low demand.
E_2: Firm *predicts* average demand.
E_3: Firm *predicts* high demand.

From the consulting firm, Maypro receives the following likelihoods:

$P(E_1/\theta_1) = 0.80$	$P(E_2/\theta_1) = 0.10$	$P(E_3/\theta_1) = 0.10$
$P(E_1/\theta_2) = 0.15$	$P(E_2/\theta_2) = 0.75$	$P(E_3/\theta_2) = 0.10$
$P(E_1/\theta_3) = 0.05$	$P(E_2/\theta_3) = 0.10$	$P(E_3/\theta_3) = 0.85$

a. What is the optimal prior act based on maximizing EMV?

b. What is the EVPI for this decision based on the prior probability distribution?

c. Suppose that the firm's report *predicts* low demand. Determine the posterior probability distribution, and find the optimal posterior act based on maximizing EMV.

d. Suppose that the firm's report *predicts* average demand. Determine the posterior probability distribution, and find the optimal posterior act based on maximizing EMV.

e. Suppose that the firm's report *predicts* high demand. Determine the posterior probability distribution and find the optimal posterior act based on maximizing EMV.

23.26 In Problem 23.25 above, use a decision tree to perform the preposterior analysis. In addition to the tree specifically answer the following questions:

a. What are the probabilities that the consulting firm will predict low, average, and high demands?

b. What is the EVSI for this problem?

c. What is the ESI for this problem?

d. Up to what dollar amount should the director be willing to pay for the consulting firm's report?

e. Suppose that the director purchases the report, and the report *predicts* high demand. What is the optimal posterior act?

23.27 The Doctor Gramble Corporation is considering a new container for one of its products. The marketing director is very concerned about switching to a new container, since the product has built up a large and devoted population of customers. She is afraid that some faithful customers may be turned off by the new container to the extent that they may switch brands. She decides to consider three possible states of nature appropriate for the use of the new container:

θ_1: Weak market response (considerable brand switching).

θ_2: Moderate market response (about the same market share).

θ_3: Strong market response (increased market share).

After considerable study and discussions in meetings, she decides to set the following prior probabilities for these states of nature: $P(\theta_1) = 0.25$, $P(\theta_2) = 0.50$, and $P(\theta_3) = 0.25$. The payoff (profit) matrix for the present net value of the two alternatives for the coming year is shown below:

Act (a_i) \ State of nature (θ_j)	Market response		
	θ_1 Weak	θ_2 Moderate	θ_3 Strong
a_1: Introduce new container	−$3,000,000	$0	$5,000,000
a_2: Retain old container	0	0	$0
$P(\theta_j)$	0.25	0.50	0.25

She must decide on whether or not to ask the director of the research and development (R&D) department to conduct a survey on the response to the new container. There would be two possible outcomes of the survey: Respondents in general like the new container (event B), and respondents in general do not like the new container (event $\bar{B}$). The likelihoods for the survey outcomes would be:

$P(B/\theta_1) = 0.10 \quad P(\overline{B}/\theta_1) = 0.90$
$P(B/\theta_2) = 0.40 \quad P(\overline{B}/\theta_2) = 0.60$
$P(B/\theta_3) = 0.50 \quad P(\overline{B}/\theta_3) = 0.50$

Using preposterior analysis, should the marketing director ask the R&D department to conduct the survey? If the survey should be conducted, explain the decision plan, depending on whether the survey outcome is positive (respondents like new container) or negative (respondents do not like the new container). What is the EVSI for the preposterior analysis?

23.28 The Oopsie Corporation is considering a change in its formula for Oopsie Cola. The formula used to make Oopsie Cola has been used for many years. But in the past few years, competitors have made a significant dent in Oopsie Cola's market share. There are two decision alternatives: a_1, change formula, and a_2, do not change formula. There are three states of nature: θ_1, response to new formula is poor, θ_2, response is average, and θ_3, response is good. The prior probabilities for these states of nature are set at $P(\theta_1) = 0.20$, $P(\theta_2) = 0.30$, and $P(\theta_3) = 0.50$. The profit matrix (in millions of dollars) for the planned quarterly production is shown below:

	State of nature (θ_j)	Response		
Act (a_i)		Poor θ_1	Average θ_2	Good θ_3
a_1: Change formula		−$50	−$5	$100
a_2: Do not change formula		0	0	0
$P(\theta_j)$		0.20	0.30	0.50

a. Based on the prior probability distribution, what is the optimal act?
b. What is the EVPI for this decision?

Suppose now that a market survey is planned. The survey would involve taste tests with consumers to determine which of the two formulas they preferred—the old formula or the new formula. The survey would

cost $5 million and would produce one of three results:

B_1: Survey *predicts* poor response.
B_2: Survey *predicts* average response.
B_3: Survey *predicts* good response.

The likelihoods are:

$P(B_1/\theta_1) = 0.70 \quad P(B_2/\theta_1) = 0.20 \quad P(B_3/\theta_1) = 0.10$
$P(B_1/\theta_2) = 0.30 \quad P(B_2/\theta_2) = 0.50 \quad P(B_3/\theta_2) = 0.20$
$P(B_1/\theta_3) = 0.20 \quad P(B_2/\theta_3) = 0.50 \quad P(B_3/\theta_3) = 0.30$

c. Given the survey outcome is B_1—predicts poor response, find $P(B_1)$ and the posterior probability distribution $P(\theta_1/B_1)$, $P(\theta_2/B_1)$, and $P(\theta_3/B_1)$.
d. Given the survey outcome is B_2—predicts average response, find $P(B_2)$ and the posterior probability distribution $P(\theta_1/B_2)$, $P(\theta_2/B_2)$, and $P(\theta_3/B_2)$.
e. Given the survey outcome is B_3—predicts good response, find $P(B_3)$ and the posterior probability distribution $P(\theta_1/B_3)$, $P(\theta_2/B_3)$, and $P(\theta_3/B_3)$.
f. Should the survey be taken? That is conduct the preposterior analysis of this decision problem using a decision tree.

23.29 A manufacturer of electronic calculators relies on mailed circulars to generate sales of its products. Before any new item is put into full production, a sample mailing is made and the proportion of sampled individuals purchasing the product is recorded. If the proportion of favorable responses (purchases) is greater than what the firm considers to be its break-even point, the item is put into full production in anticipation of forthcoming sales. If the proportion is less than the break-even point, the firm usually decides not to market the product or else to redesign the product so that it will be salable. In either case, the market is such that if the product is released late, competition will severely reduce sales, thus making the item unprofitable. If the item is put into production when there is an insufficient demand for it, large sums of money will be lost in additional circulars advertising the product at a reduced price. The firm has built up the sales experience shown in the table below in marketing calculators similar to the one now under consideration for production, the "High Performance IV." One hundred circulars advertising the new calculator are mailed, and they result in 5 purchases and 95 negative responses. What is the *revised probability distribution* of sales for the High Performance IV?

Proportion, π	Prior probability (long-run relative frequency from previous sales)
0.02	0.21
0.03	0.23
0.04	0.45
0.05	0.09
0.06	0.01
0.07	0.01
	1.00

Note: To be able to answer this question, you will have to work directly with the probability mass function of a binomial random variable, or you will have to obtain a table of binomial probabilities from your college or university library for $n = 100$.

23.30 Given the information in Problem 23.29 for sales of electronic calculators, assume that the firm in question decided to market the product based on the prior distribution and that 10,000 circulars were then mailed to prospective customers. How many calculators can the firm expect to sell? Justify your result, and comment on the probability of selling at least half again as many as you originally estimated.

23.31 Assume the same facts in Problem 23.30, except that the firm chooses to work with the posterior distribution after probabilities have been revised (see Problem 23.29). How many calculators can the firm now expect to sell? What is the probability of selling half again as many as originally estimated or more now that the probability distribution has been revised?

23.32 A competitor of the electronic calculator company of Problem 23.29 markets its calculators principally through retail department stores. In the past it has chosen to market only the larger, nonportable, non-LSI circuitry models. Recently, however, the firm has been "persuaded" that the market for calculators is shifting to the smaller units, so it has decided to test-market a new model conforming in size to a portable one. One hundred stores from the 10,000 retail outlets it currently has under contract are provided with the new units, resulting in an average sale per store of 200 units and a standard deviation of 25 units. The firm's prior distribution on sales of the new units (based on sales histories of the older models) is assumed normal with mean $\mu_0 = 120$ units and standard deviation $\sigma_0 = 40$ units. What is the firm's posterior distribution for sales of its new calculator?

Appendix A
Special topics: A calculus-based study
of continuous random variables

The student who has had calculus can appreciate the relationship that exists between the area under a curve [say $f(x)$—a probability density function] between two limits c and d, and the probability that a continuous random variable $\mathbf{X}$ assumes a value between c and d, i.e., $P(c \leq \mathbf{X} \leq d)$. That is, if $\mathbf{X}$ is a continuous random variable with probability density function $f(x)$ defined between the limits a and b, then the probability that the random variable $\mathbf{X}$ assumes a value between c and d, where $c \geq a$, $d \leq b$, and $d > c$, is given by:

$$P(c \leq \mathbf{X} \leq d) = \int_c^d f(x) \, dx$$

In what follows, we present the continuous analogue of many of the definitions for a discrete random variable given in Chapter 5.

The integration symbol $\int$ for a continuous random variable is analogous to the summation symbol for a discrete random variable. In the former case, we are "summing" areas, whereas in the latter case, we are summing over discrete values. Table A.1 summarizes the similarities and differences between the two types of random variables. This table is similar to Table 5.13, except that here we have indicated the integration required to determine the needed probabilities.

TABLE A.1 | Comparison of properties of discrete and continuous random variables

Property	Discrete random variable	Continuous random variable
Number of possible values of the random variable	Finite *or* countably infinite	Infinite (not countable)
$P(c \leq \mathbf{X} \leq d)$	$\sum_{x=c}^{x=d} P(x)$	Area under the curve $f(x)$ from c to $d = \int_c^d f(x) \, dx$
$P(\mathbf{X} = e)$	$P(e)$	Always zero, $P(\mathbf{X} = e) = \int_e^e f(x) \, dx \equiv 0$

In order to show that the probability a continuous random variable assumes a particular value, say $\mathbf{X} = e$, is *always equal to 0* using calculus, note the following. The notation $\int_a^b f(x)\, dx$ represents the area under the curve $f(x)$ from $x = a$ to $x = b$. By using the second property in Table A.1, we may write:

$$P(\mathbf{X} = e) = P(e \le \mathbf{X} \le e) = \int_e^e f(x)\, dx = 0$$

since the integral defined over the limits e to e is always equal to 0! Hence, using calculus, it is relatively simple to argue that the probability that a continuous random variable assumes a particular value indeed is always 0.

The student who has had calculus should begin to recognize some familiar notions at this point. First, the area under a continuous curve, say $f(x)$, can be determined by integration. Indeed, the second condition in Definition 5.8 [the area under the curve $f(x)$ from $x = a$ to $x = b$ must be 1] can be stated concisely as:

Definition 5.8
(Addendum)

$$\int_a^b f(x)\, dx = 1$$

For example, consider the following function:

$$f(x) = 1/\theta, \qquad 0 < x \le \theta;\, 0 < \theta < \infty$$

Is this function a density function? First, it is evident that $f(x) \ge 0$ for all values of x $(0 < x \le \theta)$. Second, noting that $a = 0$ and $b = \theta$ in Definition 5.8,

$$\int_a^b f(x)\, dx = \int_0^\theta \frac{1}{\theta}\, dx = \left.\frac{x}{\theta}\right|_0^\theta = \frac{\theta}{\theta} - \frac{0}{\theta} = 1.$$

Therefore, $f(x)$ is a density function.

Actually, we can extend the requirement that the area under $f(x)$ is 1 to

$$\int_{-\infty}^{\infty} f(x)\, dx = 1.$$

For example, a more complete definition of the density function above is:

$$f(x) = \begin{cases} 0 \text{ if } x \le 0 \\ 1/\theta \text{ if } 0 < x \le \theta \\ 0 \text{ if } x > \theta \end{cases}$$

Then,

$$\int_{-\infty}^{\infty} f(x)\, dx = \int_{-\infty}^{0} (0)\, dx + \int_0^\theta \frac{1}{\theta}\, dx + \int_\theta^\infty (0)\, dx = \int_0^\theta \frac{1}{\theta}\, dx = 1.$$

The expected value of a continuous random variable defined between the limits a and b involves the integration of the product of x and $f(x)$ over the interval (a, b); that is, it is the area under the curve of the function $xf(x)$ between a and b. We give the definition of the expected value of a continuous random variable in Definition A.1.

Definition A.1

Expectation of a random variable (continuous case)

Let **X** be a continuous random variable defined over the interval (a, b) with probability density function $f(x)$. The *mean* or the *expected value* of the random variable **X**, denoted by $E(\mathbf{X})$, is given by

$$E(\mathbf{X}) = \int_a^b xf(x)\, dx$$

Consider the continuous uniform random variable introduced in Example 5.8 in Chapter 5, whose density function is given by:

$$f(x) = 1, \quad 0 \le x \le 1 \quad [f(x) = 0 \text{ elsewhere}]$$

The expected value (mean) of **X** is given by:

$$E(\mathbf{X}) = \int_a^b xf(x)\, dx = \int_0^1 x(1)\, dx = \left.\frac{x^2}{2}\right|_0^1 = \frac{(1)^2}{2} - \frac{(0)^2}{2} = \frac{1}{2} - 0 = \frac{1}{2}.$$

Definition A.2 gives the formula for determining the variance of a continuous random variable. Integration is almost always required to determine the variance of a continuous random variable, as was the case for the expected value of a continuous random variable.

Definition A.2

Variance of a random variable (continuous case)

Let **X** be a continuous random variable defined over the interval (a, b) with probability density function $f(x)$. The *variance* of the random variable **X**, denoted by $V(\mathbf{X})$, is given by

$$V(\mathbf{X}) = \int_a^b [x - E(\mathbf{X})]^2 f(x)\, dx = \int_a^b x^2 f(x)\, dx - [E(\mathbf{X})]^2$$

For the continuous uniform random variable whose probability density function is given by $f(x) = 1$, $0 \le x \le 1$,

$$V(\mathbf{X}) = \int_a^b x^2 f(x) \, dx - [E(\mathbf{X})]^2 = \int_0^1 x^2 \,(1) \, dx - [1/2]^2$$

$$= \left.\frac{x^3}{3}\right|_0^1 - 1/4 = \left[\frac{(1)^3}{3} - \frac{(0)^3}{3}\right] - 1/4 = 1/3 - 1/4 = 1/12$$

The expected value and variance for the general continuous uniform distribution, whose probability density function is given by

$$f(x) = \frac{1}{b - a} \quad a \le x \le b \quad [f(x) = 0 \text{ elsewhere}],$$

are derived using calculus in the Appendix to Chapter 7. Notice if $b = 1$ and $a = 0$, then we have the specific member of the family of continuous uniform distributions used as an illustration above.

It is interesting to note the similarity in the computation of the expected value and the variance of a discrete and a continuous random variable as illustrated in Table A.2. The analogy to summing in the discrete case (Σ) is, of course, integrating in the continuous case ($\int$).

TABLE A.2 | Comparison of the expected value and variance formulas for discrete and continuous random variables

Property	Discrete random variable	Continuous random variable
Expected value, $E(\mathbf{X})$	$\Sigma x P(x)$	$\int x f(x) \, dx$
Variance, $V(\mathbf{X})$	$\Sigma [x - E(\mathbf{X})]^2 P(x)$	$\int [x - E(\mathbf{X})]^2 f(x) \, dx$

Finally, we present the definition of the expectation of a function of a continuous random variable. This definition is similar to Definition 5.10 for the expectation of a function of a discrete random variable.

Definition A.3

Expectation of a function of a random variable (continuous case)

Let **X** be a continuous random variable defined between the limits a and b (a < b), with probability density function $f(x)$ and consider a function of **X**, denoted by $h(\mathbf{X})$. The *average* or the *expected value* of $h(\mathbf{X})$ is given by:

$$E[h(\mathbf{X})] = \int_a^b h(x) f(x) \, dx$$

Example Consider the following probability density function,

$$f(x) = \frac{1}{\theta} \quad 0 < x \le \theta \quad [f(x) = 0 \text{ elsewhere}]$$

Find the expected value (average) of the following functions of the random variable $\mathbf{X}$ whose density function is given by $f(x)$:

a. $\qquad\qquad\qquad\qquad h(\mathbf{X}) = \mathbf{X}$

b. $\qquad\qquad\qquad\qquad h(\mathbf{X}) = 2\mathbf{X} - 1$

c. $\qquad\qquad\qquad\qquad h(\mathbf{X}) = \mathbf{X}^2$

Solution

a. $E[h(\mathbf{X})] = E[\mathbf{X}] = \int_0^\theta x \frac{1}{\theta} dx = \frac{x^2}{2\theta}\Big|_0^\theta = \frac{\theta^2}{2\theta} - \frac{(0)^2}{2\theta} = \frac{\theta^2}{2\theta} = \underline{\underline{\frac{\theta}{2}}}.$

Thus, the expected value of $\mathbf{X}$ is $\theta/2$.

b. $E[h(\mathbf{X})] = E[2\mathbf{X} - 1] = \int_0^\theta (2x - 1) \frac{1}{\theta} dx = \frac{1}{\theta} \left[\frac{2x^2}{2} - x \right]\Big|_0^\theta$

$= \frac{1}{\theta} (x^2 - x)\Big|_0^\theta = \frac{1}{\theta} [(\theta^2 - \theta) - (0 - 0)] = \frac{\theta^2 - \theta}{\theta} = \underline{\underline{\theta - 1}}$

Also, note that $E[2\mathbf{X} - 1] = 2E(\mathbf{X}) - 1 = 2\left(\frac{\theta}{2}\right) - 1 = \underline{\underline{\theta - 1}}$.

c. $E[h(\mathbf{X})] = E[\mathbf{X}^2] = \int_0^\theta x^2 \frac{1}{\theta} dx = \frac{x^3}{3\theta}\Big|_0^\theta = \frac{\theta^3}{3\theta} - \frac{(0)^3}{3\theta} = \underline{\underline{\frac{\theta^2}{3}}}.$

Now, $V(\mathbf{X}) = E[\mathbf{X}^2] - [E(\mathbf{X})]^2 = \frac{\theta^2}{3} - \left(\frac{\theta}{2}\right)^2 = \frac{\theta^2}{3} - \frac{\theta^2}{4}$

$= \frac{\theta}{12}$

The astute student may have observed that the density function $f(x) = 1/\theta$, $0 < x \le \theta$, is another special form of the general continuous uniform distribution $f(x) = 1/(b - a)$, $a \le x \le b$. Set $a = 0$ and $b = \theta$. This produces the "one parameter" (θ) continuous uniform distribution.

Appendix B
Statistical tables

Table interpolation illustration—standard normal distribution

1. $P(\mathbf{Z} \geq 1.364) = 0.50 - P(0 \leq \mathbf{Z} \leq 1.364)$:

z	$P(0 \leq \mathbf{Z} \leq z)$

$$0.010 \begin{bmatrix} 1.360 \\ 1.364 \\ 1.370 \end{bmatrix} 0.004 \qquad 0.0016 \begin{bmatrix} 0.4131 \\ * \\ 0.4147 \end{bmatrix} \frac{0.004}{0.010} (0.0016) = 0.00064$$

$$P(0 \leq \mathbf{Z} \leq 1.364) = 0.4131 + 0.00064 = 0.4137 = *$$
$$P(\mathbf{Z} \geq 1.364) = 0.5000 - 0.4137 = 0.0863$$

2. $P(0 \leq \mathbf{Z} \leq z) = 0.1977$, find z.

z	$P(0 \leq \mathbf{Z} \leq z)$

$$\frac{0.0027}{0.0035} (0.010) = 0.0077 \begin{bmatrix} 0.510 \\ z \\ 0.520 \end{bmatrix} 0.010 \qquad \begin{bmatrix} 0.1950 \\ 0.1977 \\ 0.1985 \end{bmatrix} 0.0027 \end{bmatrix} 0.0035$$

$$z = 0.5100 + 0.0077 = 0.5177; \quad P(0 \leq \mathbf{Z} \leq 0.5177) = 0.1977$$

TABLE B.1 | Binomial probability tables

Part I: Mass function probabilities

Tabulated values give the probability that the random variable $\mathbf{X}$ assumes the value $x = a$. These values are:

$$P[\mathbf{X} = x = a] = P(x = a), \quad P(x) = \frac{n!}{x!(n-x)!} \pi^x (1-\pi)^{n-x}, \quad x = 0, 1, 2, \ldots, n.$$

$n = 1$

a \ π	.010	.050	.100	.200	.300	.400	.500	.600	.700	.800	.900	.950	.990
0	0.990	0.950	0.900	0.800	0.700	0.600	0.500	0.400	0.300	0.200	0.100	0.050	0.010
1	0.010	0.050	0.100	0.200	0.300	0.400	0.500	0.600	0.700	0.800	0.900	0.950	0.990

$n = 2$

a \ π	.010	.050	.100	.200	.300	.400	.500	.600	.700	.800	.900	.950	.990
0	0.980	0.903	0.810	0.640	0.490	0.360	0.250	0.160	0.090	0.040	0.010	0.002	0.000
1	0.020	0.095	0.180	0.320	0.420	0.480	0.500	0.480	0.420	0.320	0.180	0.095	0.020
2	0.000	0.002	0.010	0.040	0.090	0.160	0.250	0.360	0.490	0.640	0.810	0.903	0.980

$n = 3$

a \ π	.010	.050	.100	.200	.300	.400	.500	.600	.700	.800	.900	.950	.990
0	0.970	0.857	0.729	0.512	0.343	0.216	0.125	0.064	0.027	0.008	0.001	0.000	0.000
1	0.030	0.136	0.243	0.384	0.441	0.432	0.375	0.288	0.189	0.096	0.027	0.007	0.000
2	0.000	0.007	0.027	0.096	0.189	0.288	0.375	0.432	0.441	0.384	0.243	0.136	0.030
3	0.000	0.000	0.001	0.008	0.027	0.064	0.125	0.216	0.343	0.512	0.729	0.857	0.970

TABLE B.1 | (continued)

Part I: Mass function probabilities

n = 4

a \ π	.010	.050	.100	.200	.300	.400	.500	.600	.700	.800	.900	.950	.990
0	0.961	0.815	0.656	0.410	0.240	0.130	0.063	0.026	0.008	0.002	0.000	0.000	0.000
1	0.038	0.171	0.292	0.409	0.412	0.345	0.250	0.153	0.076	0.025	0.004	0.000	0.000
2	0.001	0.014	0.048	0.154	0.264	0.346	0.375	0.346	0.264	0.154	0.048	0.014	0.001
3	0.000	0.000	0.004	0.025	0.076	0.153	0.250	0.345	0.412	0.409	0.292	0.171	0.038
4	0.000	0.000	0.000	0.002	0.008	0.026	0.062	0.130	0.240	0.410	0.656	0.815	0.961

n = 5

a \ π	.010	.050	.100	.200	.300	.400	.500	.600	.700	.800	.900	.950	.990
0	0.951	0.774	0.590	0.328	0.168	0.078	0.031	0.010	0.002	0.000	0.000	0.000	0.000
1	0.048	0.203	0.329	0.409	0.360	0.259	0.157	0.077	0.029	0.007	0.000	0.000	0.000
2	0.001	0.022	0.072	0.205	0.309	0.346	0.312	0.230	0.132	0.051	0.009	0.001	0.000
3	0.000	0.001	0.009	0.051	0.132	0.230	0.313	0.346	0.309	0.205	0.072	0.022	0.001
4	0.000	0.000	0.000	0.007	0.029	0.077	0.156	0.259	0.360	0.409	0.329	0.203	0.048
5	0.000	0.000	0.000	0.000	0.002	0.010	0.031	0.078	0.168	0.328	0.590	0.774	0.951

n = 6

a \ π	.010	.050	.100	.200	.300	.400	.500	.600	.700	.800	.900	.950	.990
0	0.941	0.735	0.531	0.262	0.118	0.047	0.016	0.004	0.001	0.000	0.000	0.000	0.000
1	0.058	0.232	0.355	0.393	0.302	0.186	0.093	0.037	0.010	0.002	0.000	0.000	0.000
2	0.001	0.031	0.098	0.246	0.324	0.311	0.235	0.138	0.059	0.015	0.001	0.000	0.000
3	0.000	0.002	0.015	0.082	0.186	0.277	0.312	0.277	0.186	0.082	0.015	0.002	0.000
4	0.000	0.000	0.001	0.015	0.059	0.138	0.235	0.311	0.324	0.246	0.098	0.031	0.001
5	0.000	0.000	0.000	0.002	0.010	0.037	0.093	0.186	0.302	0.393	0.355	0.232	0.058
6	0.000	0.000	0.000	0.000	0.001	0.004	0.016	0.047	0.118	0.262	0.531	0.735	0.941

n = 7

a \ π	.010	.050	.100	.200	.300	.400	.500	.600	.700	.800	.900	.950	.990
0	0.932	0.698	0.478	0.210	0.082	0.028	0.008	0.002	0.000	0.000	0.000	0.000	0.000
1	0.066	0.258	0.372	0.367	0.247	0.131	0.055	0.017	0.004	0.000	0.000	0.000	0.000
2	0.002	0.040	0.124	0.275	0.318	0.261	0.164	0.077	0.025	0.005	0.000	0.000	0.000
3	0.000	0.004	0.023	0.115	0.227	0.290	0.273	0.194	0.097	0.028	0.003	0.000	0.000
4	0.000	0.000	0.003	0.028	0.097	0.194	0.273	0.290	0.227	0.115	0.023	0.004	0.000
5	0.000	0.000	0.000	0.005	0.025	0.077	0.165	0.261	0.318	0.275	0.124	0.040	0.002
6	0.000	0.000	0.000	0.000	0.004	0.017	0.054	0.131	0.247	0.367	0.372	0.258	0.066
7	0.000	0.000	0.000	0.000	0.000	0.002	0.008	0.028	0.082	0.210	0.478	0.698	0.932

TABLE B.1 | *(continued)*

Part I: Mass function probabilities

$n = 8$

a \ π	.010	.050	.100	.200	.300	.400	.500	.600	.700	.800	.900	.950	.990
0	0.923	0.663	0.430	0.168	0.058	0.017	0.004	0.001	0.000	0.000	0.000	0.000	0.000
1	0.074	0.280	0.383	0.335	0.197	0.089	0.031	0.008	0.001	0.000	0.000	0.000	0.000
2	0.003	0.051	0.149	0.294	0.297	0.209	0.110	0.041	0.010	0.001	0.000	0.000	0.000
3	0.000	0.006	0.033	0.147	0.254	0.279	0.218	0.124	0.047	0.009	0.000	0.000	0.000
4	0.000	0.000	0.005	0.046	0.136	0.232	0.274	0.232	0.136	0.046	0.005	0.000	0.000
5	0.000	0.000	0.000	0.009	0.047	0.124	0.218	0.279	0.254	0.147	0.033	0.006	0.000
6	0.000	0.000	0.000	0.001	0.010	0.041	0.110	0.209	0.297	0.294	0.149	0.051	0.003
7	0.000	0.000	0.000	0.000	0.001	0.008	0.031	0.089	0.197	0.335	0.383	0.280	0.074
8	0.000	0.000	0.000	0.000	0.000	0.001	0.004	0.017	0.058	0.168	0.430	0.663	0.923

$n = 9$

a \ π	.010	.050	.100	.200	.300	.400	.500	.600	.700	.800	.900	.950	.990
0	0.914	0.630	0.387	0.134	0.040	0.010	0.002	0.000	0.000	0.000	0.000	0.000	0.000
1	0.083	0.299	0.388	0.302	0.156	0.061	0.018	0.004	0.000	0.000	0.000	0.000	0.000
2	0.003	0.063	0.172	0.302	0.267	0.161	0.070	0.021	0.004	0.000	0.000	0.000	0.000
3	0.000	0.007	0.045	0.176	0.267	0.251	0.164	0.074	0.021	0.003	0.000	0.000	0.000
4	0.000	0.001	0.007	0.066	0.171	0.250	0.246	0.168	0.074	0.017	0.001	0.000	0.000
5	0.000	0.000	0.001	0.017	0.074	0.168	0.246	0.250	0.171	0.066	0.007	0.001	0.000
6	0.000	0.000	0.000	0.003	0.021	0.074	0.164	0.251	0.267	0.176	0.045	0.007	0.000
7	0.000	0.000	0.000	0.000	0.004	0.021	0.070	0.161	0.267	0.302	0.172	0.063	0.003
8	0.000	0.000	0.000	0.000	0.000	0.004	0.018	0.061	0.156	0.302	0.388	0.299	0.083
9	0.000	0.000	0.000	0.000	0.000	0.000	0.002	0.010	0.040	0.134	0.387	0.630	0.914

$n = 10$

a \ π	.010	.050	.100	.200	.300	.400	.500	.600	.700	.800	.900	.950	.990
0	0.904	0.599	0.349	0.107	0.028	0.006	0.001	0.000	0.000	0.000	0.000	0.000	0.000
1	0.092	0.315	0.387	0.269	0.121	0.040	0.010	0.002	0.000	0.000	0.000	0.000	0.000
2	0.004	0.074	0.194	0.302	0.234	0.121	0.044	0.010	0.002	0.000	0.000	0.000	0.000
3	0.000	0.011	0.057	0.201	0.267	0.215	0.117	0.043	0.009	0.001	0.000	0.000	0.000
4	0.000	0.001	0.011	0.088	0.200	0.251	0.205	0.111	0.036	0.005	0.000	0.000	0.000
5	0.000	0.000	0.002	0.027	0.103	0.201	0.246	0.201	0.103	0.027	0.002	0.000	0.000
6	0.000	0.000	0.000	0.005	0.036	0.111	0.205	0.251	0.200	0.088	0.011	0.001	0.000
7	0.000	0.000	0.000	0.001	0.009	0.043	0.117	0.215	0.267	0.201	0.057	0.011	0.000
8	0.000	0.000	0.000	0.000	0.002	0.010	0.044	0.121	0.234	0.302	0.194	0.074	0.004
9	0.000	0.000	0.000	0.000	0.000	0.002	0.010	0.040	0.121	0.269	0.387	0.315	0.092
10	0.000	0.000	0.000	0.000	0.000	0.000	0.001	0.006	0.028	0.107	0.349	0.599	0.904

TABLE B.1 | (continued)

Part I: Mass function probabilities

n = 11

a \ π	.010	.050	.100	.200	.300	.400	.500	.600	.700	.800	.900	.950	.990
0	0.895	0.569	0.314	0.086	0.020	0.004	0.000	0.000	0.000	0.000	0.000	0.000	0.000
1	0.100	0.329	0.383	0.236	0.093	0.026	0.006	0.001	0.000	0.000	0.000	0.000	0.000
2	0.005	0.087	0.213	0.295	0.200	0.089	0.027	0.005	0.001	0.000	0.000	0.000	0.000
3	0.000	0.013	0.071	0.222	0.257	0.177	0.080	0.023	0.003	0.000	0.000	0.000	0.000
4	0.000	0.002	0.016	0.111	0.220	0.237	0.161	0.070	0.018	0.002	0.000	0.000	0.000
5	0.000	0.000	0.003	0.038	0.132	0.220	0.226	0.148	0.056	0.010	0.000	0.000	0.000
6	0.000	0.000	0.000	0.010	0.056	0.148	0.226	0.220	0.132	0.038	0.003	0.000	0.000
7	0.000	0.000	0.000	0.002	0.018	0.070	0.161	0.237	0.220	0.111	0.016	0.002	0.000
8	0.000	0.000	0.000	0.000	0.003	0.023	0.080	0.177	0.257	0.222	0.071	0.013	0.000
9	0.000	0.000	0.000	0.000	0.001	0.005	0.027	0.089	0.200	0.295	0.213	0.087	0.005
10	0.000	0.000	0.000	0.000	0.000	0.001	0.006	0.026	0.093	0.236	0.383	0.329	0.100
11	0.000	0.000	0.000	0.000	0.000	0.000	0.000	0.004	0.020	0.086	0.314	0.569	0.895

n = 12

a \ π	.010	.050	.100	.200	.300	.400	.500	.600	.700	.800	.900	.950	.990
0	0.886	0.540	0.282	0.069	0.014	0.002	0.000	0.000	0.000	0.000	0.000	0.000	0.000
1	0.108	0.342	0.377	0.206	0.071	0.018	0.003	0.000	0.000	0.000	0.000	0.000	0.000
2	0.006	0.098	0.230	0.283	0.168	0.063	0.016	0.003	0.000	0.000	0.000	0.000	0.000
3	0.000	0.018	0.085	0.237	0.240	0.142	0.054	0.012	0.002	0.000	0.000	0.000	0.000
4	0.000	0.002	0.022	0.132	0.231	0.213	0.121	0.042	0.007	0.001	0.000	0.000	0.000
5	0.000	0.000	0.003	0.054	0.158	0.227	0.193	0.101	0.030	0.003	0.000	0.000	0.000
6	0.000	0.000	0.001	0.015	0.079	0.177	0.226	0.177	0.079	0.015	0.001	0.000	0.000
7	0.000	0.000	0.000	0.003	0.030	0.101	0.193	0.227	0.158	0.054	0.003	0.000	0.000
8	0.000	0.000	0.000	0.001	0.007	0.042	0.121	0.213	0.231	0.132	0.022	0.002	0.000
9	0.000	0.000	0.000	0.000	0.002	0.012	0.054	0.142	0.240	0.237	0.085	0.018	0.000
10	0.000	0.000	0.000	0.000	0.000	0.003	0.016	0.063	0.168	0.283	0.230	0.098	0.006
11	0.000	0.000	0.000	0.000	0.000	0.000	0.003	0.018	0.071	0.206	0.377	0.342	0.108
12	0.000	0.000	0.000	0.000	0.000	0.000	0.000	0.002	0.014	0.069	0.282	0.540	0.886

n = 13

a \ π	.010	.050	.100	.200	.300	.400	.500	.600	.700	.800	.900	.950	.990
0	0.878	0.513	0.254	0.055	0.010	0.001	0.000	0.000	0.000	0.000	0.000	0.000	0.000
1	0.115	0.352	0.367	0.179	0.054	0.012	0.002	0.000	0.000	0.000	0.000	0.000	0.000
2	0.007	0.110	0.245	0.268	0.138	0.045	0.009	0.001	0.000	0.000	0.000	0.000	0.000
3	0.000	0.022	0.100	0.245	0.219	0.111	0.035	0.007	0.001	0.000	0.000	0.000	0.000
4	0.000	0.003	0.028	0.154	0.233	0.184	0.087	0.024	0.003	0.000	0.000	0.000	0.000
5	0.000	0.000	0.005	0.069	0.181	0.221	0.158	0.066	0.014	0.001	0.000	0.000	0.000
6	0.000	0.000	0.001	0.023	0.103	0.197	0.209	0.131	0.044	0.006	0.000	0.000	0.000
7	0.000	0.000	0.000	0.006	0.044	0.131	0.209	0.197	0.103	0.023	0.001	0.000	0.000
8	0.000	0.000	0.000	0.001	0.014	0.066	0.158	0.221	0.181	0.069	0.005	0.000	0.000
9	0.000	0.000	0.000	0.000	0.003	0.024	0.087	0.184	0.233	0.154	0.028	0.003	0.000
10	0.000	0.000	0.000	0.000	0.001	0.007	0.035	0.111	0.219	0.245	0.100	0.022	0.000
11	0.000	0.000	0.000	0.000	0.000	0.001	0.009	0.045	0.138	0.268	0.245	0.110	0.007
12	0.000	0.000	0.000	0.000	0.000	0.000	0.002	0.012	0.054	0.179	0.367	0.352	0.115
13	0.000	0.000	0.000	0.000	0.000	0.000	0.000	0.001	0.010	0.055	0.254	0.513	0.878

TABLE B.1 | *(continued)*

Part I: Mass function probabilities

$n = 14$

a \ π	.010	.050	.100	.200	.300	.400	.500	.600	.700	.800	.900	.950	.990
0	0.869	0.488	0.229	0.044	0.007	0.001	0.000	0.000	0.000	0.000	0.000	0.000	0.000
1	0.123	0.359	0.356	0.154	0.040	0.007	0.001	0.000	0.000	0.000	0.000	0.000	0.000
2	0.008	0.123	0.257	0.250	0.114	0.032	0.005	0.001	0.000	0.000	0.000	0.000	0.000
3	0.000	0.026	0.114	0.250	0.194	0.084	0.023	0.003	0.000	0.000	0.000	0.000	0.000
4	0.000	0.004	0.035	0.172	0.229	0.155	0.061	0.014	0.002	0.000	0.000	0.000	0.000
5	0.000	0.000	0.008	0.086	0.197	0.207	0.122	0.040	0.006	0.000	0.000	0.000	0.000
6	0.000	0.000	0.001	0.032	0.126	0.206	0.183	0.092	0.023	0.002	0.000	0.000	0.000
7	0.000	0.000	0.000	0.010	0.062	0.158	0.210	0.158	0.062	0.010	0.000	0.000	0.000
8	0.000	0.000	0.000	0.002	0.023	0.092	0.183	0.206	0.126	0.032	0.001	0.000	0.000
9	0.000	0.000	0.000	0.000	0.006	0.040	0.122	0.207	0.197	0.086	0.008	0.000	0.000
10	0.000	0.000	0.000	0.000	0.002	0.014	0.061	0.155	0.229	0.172	0.035	0.004	0.000
11	0.000	0.000	0.000	0.000	0.000	0.003	0.023	0.084	0.194	0.250	0.114	0.026	0.000
12	0.000	0.000	0.000	0.000	0.000	0.001	0.005	0.032	0.114	0.250	0.257	0.123	0.008
13	0.000	0.000	0.000	0.000	0.000	0.000	0.001	0.007	0.040	0.154	0.356	0.359	0.123
14	0.000	0.000	0.000	0.000	0.000	0.000	0.000	0.001	0.007	0.044	0.229	0.488	0.869

$n = 15$

a \ π	.010	.050	.100	.200	.300	.400	.500	.600	.700	.800	.900	.950	.990
0	0.860	0.463	0.206	0.035	0.005	0.000	0.000	0.000	0.000	0.000	0.000	0.000	0.000
1	0.130	0.366	0.343	0.132	0.030	0.005	0.000	0.000	0.000	0.000	0.000	0.000	0.000
2	0.010	0.135	0.267	0.231	0.092	0.022	0.004	0.000	0.000	0.000	0.000	0.000	0.000
3	0.000	0.031	0.128	0.250	0.170	0.064	0.014	0.002	0.000	0.000	0.000	0.000	0.000
4	0.000	0.004	0.043	0.188	0.218	0.126	0.041	0.007	0.001	0.000	0.000	0.000	0.000
5	0.000	0.001	0.011	0.103	0.207	0.186	0.092	0.025	0.003	0.000	0.000	0.000	0.000
6	0.000	0.000	0.002	0.043	0.147	0.207	0.153	0.061	0.011	0.001	0.000	0.000	0.000
7	0.000	0.000	0.000	0.014	0.081	0.177	0.196	0.118	0.035	0.003	0.000	0.000	0.000
8	0.000	0.000	0.000	0.003	0.035	0.118	0.196	0.177	0.081	0.014	0.000	0.000	0.000
9	0.000	0.000	0.000	0.001	0.011	0.061	0.153	0.207	0.147	0.043	0.002	0.000	0.000
10	0.000	0.000	0.000	0.000	0.003	0.025	0.092	0.186	0.207	0.103	0.011	0.001	0.000
11	0.000	0.000	0.000	0.000	0.001	0.007	0.041	0.126	0.218	0.188	0.043	0.004	0.000
12	0.000	0.000	0.000	0.000	0.000	0.002	0.014	0.064	0.170	0.250	0.128	0.031	0.000
13	0.000	0.000	0.000	0.000	0.000	0.000	0.004	0.022	0.092	0.231	0.267	0.135	0.010
14	0.000	0.000	0.000	0.000	0.000	0.000	0.000	0.005	0.030	0.132	0.343	0.366	0.130
15	0.000	0.000	0.000	0.000	0.000	0.000	0.000	0.000	0.005	0.035	0.206	0.463	0.860

TABLE B.1 | *(continued)*

Part I: Mass function probabilities

$n = 16$

a \ π	.010	.050	.100	.200	.300	.400	.500	.600	.700	.800	.900	.950	.990
0	0.851	0.440	0.185	0.028	0.003	0.000	0.000	0.000	0.000	0.000	0.000	0.000	0.000
1	0.138	0.371	0.330	0.113	0.023	0.003	0.000	0.000	0.000	0.000	0.000	0.000	0.000
2	0.010	0.146	0.274	0.211	0.073	0.015	0.002	0.000	0.000	0.000	0.000	0.000	0.000
3	0.001	0.036	0.143	0.246	0.147	0.047	0.009	0.001	0.000	0.000	0.000	0.000	0.000
4	0.000	0.006	0.051	0.200	0.204	0.102	0.027	0.004	0.000	0.000	0.000	0.000	0.000
5	0.000	0.001	0.014	0.120	0.210	0.162	0.067	0.014	0.002	0.000	0.000	0.000	0.000
6	0.000	0.000	0.002	0.055	0.165	0.198	0.122	0.039	0.005	0.000	0.000	0.000	0.000
7	0.000	0.000	0.001	0.020	0.101	0.189	0.175	0.084	0.019	0.001	0.000	0.000	0.000
8	0.000	0.000	0.000	0.006	0.048	0.142	0.196	0.142	0.048	0.006	0.000	0.000	0.000
9	0.000	0.000	0.000	0.001	0.019	0.084	0.175	0.189	0.101	0.020	0.001	0.000	0.000
10	0.000	0.000	0.000	0.000	0.005	0.039	0.122	0.198	0.165	0.055	0.002	0.000	0.000
11	0.000	0.000	0.000	0.000	0.002	0.014	0.067	0.162	0.210	0.120	0.014	0.001	0.000
12	0.000	0.000	0.000	0.000	0.000	0.004	0.027	0.102	0.204	0.200	0.051	0.006	0.000
13	0.000	0.000	0.000	0.000	0.000	0.001	0.009	0.047	0.147	0.246	0.143	0.036	0.001
14	0.000	0.000	0.000	0.000	0.000	0.000	0.002	0.015	0.073	0.211	0.274	0.146	0.010
15	0.000	0.000	0.000	0.000	0.000	0.000	0.000	0.003	0.023	0.113	0.330	0.371	0.138
16	0.000	0.000	0.000	0.000	0.000	0.000	0.000	0.000	0.003	0.028	0.185	0.440	0.851

$n = 17$

a \ π	.010	.050	.100	.200	.300	.400	.500	.600	.700	.800	.900	.950	.990
0	0.843	0.418	0.167	0.023	0.002	0.000	0.000	0.000	0.000	0.000	0.000	0.000	0.000
1	0.145	0.374	0.315	0.095	0.017	0.002	0.000	0.000	0.000	0.000	0.000	0.000	0.000
2	0.011	0.158	0.280	0.192	0.058	0.010	0.001	0.000	0.000	0.000	0.000	0.000	0.000
3	0.001	0.041	0.155	0.239	0.125	0.034	0.005	0.000	0.000	0.000	0.000	0.000	0.000
4	0.000	0.008	0.061	0.209	0.187	0.080	0.019	0.003	0.000	0.000	0.000	0.000	0.000
5	0.000	0.001	0.017	0.136	0.208	0.138	0.047	0.008	0.001	0.000	0.000	0.000	0.000
6	0.000	0.000	0.004	0.068	0.178	0.184	0.094	0.024	0.002	0.000	0.000	0.000	0.000
7	0.000	0.000	0.001	0.027	0.120	0.193	0.149	0.057	0.010	0.000	0.000	0.000	0.000
8	0.000	0.000	0.000	0.008	0.065	0.160	0.185	0.107	0.027	0.003	0.000	0.000	0.000
9	0.000	0.000	0.000	0.003	0.027	0.107	0.185	0.160	0.065	0.008	0.000	0.000	0.000
10	0.000	0.000	0.000	0.000	0.010	0.057	0.149	0.193	0.120	0.027	0.001	0.000	0.000
11	0.000	0.000	0.000	0.000	0.002	0.024	0.094	0.184	0.178	0.068	0.004	0.000	0.000
12	0.000	0.000	0.000	0.000	0.001	0.008	0.047	0.138	0.208	0.136	0.017	0.001	0.000
13	0.000	0.000	0.000	0.000	0.000	0.003	0.019	0.080	0.187	0.209	0.061	0.008	0.000
14	0.000	0.000	0.000	0.000	0.000	0.000	0.005	0.034	0.125	0.239	0.155	0.041	0.001
15	0.000	0.000	0.000	0.000	0.000	0.000	0.001	0.010	0.058	0.192	0.280	0.158	0.011
16	0.000	0.000	0.000	0.000	0.000	0.000	0.000	0.002	0.017	0.095	0.315	0.374	0.145
17	0.000	0.000	0.000	0.000	0.000	0.000	0.000	0.000	0.002	0.023	0.167	0.418	0.843

TABLE B.1 | (continued)

Part I: Mass function probabilities

n = 18

a \ π	.010	.050	.100	.200	.300	.400	.500	.600	.700	.800	.900	.950	.990
0	0.835	0.397	0.150	0.018	0.002	0.000	0.000	0.000	0.000	0.000	0.000	0.000	0.000
1	0.151	0.377	0.300	0.081	0.012	0.001	0.000	0.000	0.000	0.000	0.000	0.000	0.000
2	0.013	0.168	0.284	0.172	0.046	0.007	0.001	0.000	0.000	0.000	0.000	0.000	0.000
3	0.001	0.047	0.168	0.230	0.105	0.025	0.003	0.000	0.000	0.000	0.000	0.000	0.000
4	0.000	0.009	0.070	0.215	0.168	0.061	0.011	0.001	0.000	0.000	0.000	0.000	0.000
5	0.000	0.002	0.022	0.151	0.201	0.115	0.033	0.005	0.000	0.000	0.000	0.000	0.000
6	0.000	0.000	0.005	0.082	0.188	0.165	0.071	0.014	0.001	0.000	0.000	0.000	0.000
7	0.000	0.000	0.001	0.035	0.137	0.189	0.121	0.038	0.005	0.000	0.000	0.000	0.000
8	0.000	0.000	0.000	0.012	0.081	0.174	0.167	0.077	0.015	0.001	0.000	0.000	0.000
9	0.000	0.000	0.000	0.003	0.039	0.128	0.186	0.128	0.039	0.003	0.000	0.000	0.000
10	0.000	0.000	0.000	0.001	0.015	0.077	0.167	0.174	0.081	0.012	0.000	0.000	0.000
11	0.000	0.000	0.000	0.000	0.005	0.038	0.121	0.189	0.137	0.035	0.001	0.000	0.000
12	0.000	0.000	0.000	0.000	0.001	0.014	0.071	0.165	0.188	0.082	0.005	0.000	0.000
13	0.000	0.000	0.000	0.000	0.000	0.005	0.033	0.115	0.201	0.151	0.022	0.002	0.000
14	0.000	0.000	0.000	0.000	0.000	0.001	0.011	0.061	0.168	0.215	0.070	0.009	0.000
15	0.000	0.000	0.000	0.000	0.000	0.000	0.003	0.025	0.105	0.230	0.168	0.047	0.001
16	0.000	0.000	0.000	0.000	0.000	0.000	0.001	0.007	0.046	0.172	0.284	0.168	0.013
17	0.000	0.000	0.000	0.000	0.000	0.000	0.000	0.001	0.012	0.081	0.300	0.377	0.151
18	0.000	0.000	0.000	0.000	0.000	0.000	0.000	0.000	0.002	0.018	0.150	0.397	0.835

n = 19

a \ π	.010	.050	.100	.200	.300	.400	.500	.600	.700	.800	.900	.950	.990
0	0.826	0.377	0.135	0.014	0.001	0.000	0.000	0.000	0.000	0.000	0.000	0.000	0.000
1	0.159	0.378	0.285	0.069	0.009	0.001	0.000	0.000	0.000	0.000	0.000	0.000	0.000
2	0.014	0.178	0.285	0.154	0.036	0.004	0.000	0.000	0.000	0.000	0.000	0.000	0.000
3	0.001	0.054	0.180	0.218	0.087	0.018	0.002	0.000	0.000	0.000	0.000	0.000	0.000
4	0.000	0.011	0.080	0.218	0.149	0.047	0.008	0.001	0.000	0.000	0.000	0.000	0.000
5	0.000	0.002	0.026	0.164	0.192	0.093	0.022	0.002	0.000	0.000	0.000	0.000	0.000
6	0.000	0.000	0.007	0.095	0.192	0.145	0.052	0.009	0.001	0.000	0.000	0.000	0.000
7	0.000	0.000	0.002	0.045	0.152	0.180	0.096	0.023	0.002	0.000	0.000	0.000	0.000
8	0.000	0.000	0.000	0.016	0.098	0.179	0.144	0.053	0.008	0.000	0.000	0.000	0.000
9	0.000	0.000	0.000	0.005	0.051	0.147	0.176	0.098	0.022	0.002	0.000	0.000	0.000
10	0.000	0.000	0.000	0.002	0.022	0.098	0.176	0.147	0.051	0.005	0.000	0.000	0.000
11	0.000	0.000	0.000	0.000	0.008	0.053	0.144	0.179	0.098	0.016	0.000	0.000	0.000
12	0.000	0.000	0.000	0.000	0.002	0.023	0.096	0.180	0.152	0.045	0.002	0.000	0.000
13	0.000	0.000	0.000	0.000	0.001	0.009	0.052	0.145	0.192	0.095	0.007	0.000	0.000
14	0.000	0.000	0.000	0.000	0.000	0.002	0.022	0.093	0.192	0.164	0.026	0.002	0.000
15	0.000	0.000	0.000	0.000	0.000	0.001	0.008	0.047	0.149	0.218	0.080	0.011	0.000
16	0.000	0.000	0.000	0.000	0.000	0.000	0.002	0.018	0.087	0.218	0.180	0.054	0.001
17	0.000	0.000	0.000	0.000	0.000	0.000	0.000	0.004	0.036	0.154	0.285	0.178	0.014
18	0.000	0.000	0.000	0.000	0.000	0.000	0.000	0.001	0.009	0.069	0.285	0.378	0.159
19	0.000	0.000	0.000	0.000	0.000	0.000	0.000	0.000	0.001	0.014	0.135	0.377	0.826

TABLE B.1 | (continued)

Part I: Mass function probabilities

n = 20

a \ π	.010	.050	.100	.200	.300	.400	.500	.600	.700	.800	.900	.950	.990
0	0.818	0.358	0.122	0.012	0.001	0.000	0.000	0.000	0.000	0.000	0.000	0.000	0.000
1	0.165	0.378	0.270	0.057	0.007	0.001	0.000	0.000	0.000	0.000	0.000	0.000	0.000
2	0.016	0.189	0.285	0.137	0.027	0.003	0.000	0.000	0.000	0.000	0.000	0.000	0.000
3	0.001	0.059	0.190	0.205	0.072	0.012	0.001	0.000	0.000	0.000	0.000	0.000	0.000
4	0.000	0.013	0.090	0.219	0.131	0.035	0.005	0.000	0.000	0.000	0.000	0.000	0.000
5	0.000	0.003	0.032	0.174	0.178	0.075	0.015	0.002	0.000	0.000	0.000	0.000	0.000
6	0.000	0.000	0.009	0.109	0.192	0.124	0.037	0.004	0.000	0.000	0.000	0.000	0.000
7	0.000	0.000	0.002	0.055	0.164	0.166	0.074	0.015	0.001	0.000	0.000	0.000	0.000
8	0.000	0.000	0.000	0.022	0.115	0.180	0.120	0.036	0.004	0.000	0.000	0.000	0.000
9	0.000	0.000	0.000	0.007	0.065	0.159	0.160	0.071	0.012	0.001	0.000	0.000	0.000
10	0.000	0.000	0.000	0.002	0.031	0.117	0.176	0.117	0.031	0.002	0.000	0.000	0.000
11	0.000	0.000	0.000	0.001	0.012	0.071	0.160	0.159	0.065	0.007	0.000	0.000	0.000
12	0.000	0.000	0.000	0.000	0.004	0.036	0.120	0.180	0.115	0.022	0.000	0.000	0.000
13	0.000	0.000	0.000	0.000	0.001	0.015	0.074	0.166	0.164	0.055	0.002	0.000	0.000
14	0.000	0.000	0.000	0.000	0.000	0.004	0.037	0.124	0.192	0.109	0.009	0.000	0.000
15	0.000	0.000	0.000	0.000	0.000	0.002	0.015	0.075	0.178	0.174	0.032	0.003	0.000
16	0.000	0.000	0.000	0.000	0.000	0.000	0.005	0.035	0.131	0.219	0.090	0.013	0.000
17	0.000	0.000	0.000	0.000	0.000	0.000	0.001	0.012	0.072	0.205	0.190	0.059	0.001
18	0.000	0.000	0.000	0.000	0.000	0.000	0.000	0.003	0.027	0.137	0.285	0.189	0.016
19	0.000	0.000	0.000	0.000	0.000	0.000	0.000	0.001	0.007	0.057	0.270	0.373	0.165
20	0.000	0.000	0.000	0.000	0.000	0.000	0.000	0.000	0.001	0.012	0.122	0.358	0.818

n = 21

a \ π	.010	.050	.100	.200	.300	.400	.500	.600	.700	.800	.900	.950	.990
0	0.810	0.341	0.109	0.009	0.001	0.000	0.000	0.000	0.000	0.000	0.000	0.000	0.000
1	0.171	0.376	0.256	0.049	0.005	0.000	0.000	0.000	0.000	0.000	0.000	0.000	0.000
2	0.018	0.198	0.283	0.121	0.021	0.002	0.000	0.000	0.000	0.000	0.000	0.000	0.000
3	0.001	0.066	0.200	0.191	0.059	0.009	0.001	0.000	0.000	0.000	0.000	0.000	0.000
4	0.000	0.016	0.100	0.216	0.112	0.026	0.003	0.000	0.000	0.000	0.000	0.000	0.000
5	0.000	0.003	0.038	0.183	0.165	0.059	0.009	0.001	0.000	0.000	0.000	0.000	0.000
6 .	0.000	0.000	0.011	0.122	0.188	0.104	0.026	0.003	0.000	0.000	0.000	0.000	0.000
7	0.000	0.000	0.002	0.066	0.172	0.150	0.056	0.008	0.001	0.000	0.000	0.000	0.000
8	0.000	0.000	0.001	0.029	0.129	0.174	0.097	0.023	0.001	0.000	0.000	0.000	0.000
9	0.000	0.000	0.000	0.010	0.080	0.167	0.140	0.050	0.007	0.000	0.000	0.000	0.000
10	0.000	0.000	0.000	0.003	0.042	0.135	0.168	0.089	0.017	0.001	0.000	0.000	0.000
11	0.000	0.000	0.000	0.001	0.017	0.089	0.168	0.135	0.042	0.003	0.000	0.000	0.000
12	0.000	0.000	0.000	0.000	0.007	0.050	0.140	0.167	0.080	0.010	0.000	0.000	0.000
13	0.000	0.000	0.000	0.000	0.001	0.023	0.097	0.174	0.129	0.029	0.001	0.000	0.000
14	0.000	0.000	0.000	0.000	0.001	0.008	0.056	0.150	0.172	0.066	0.002	0.000	0.000
15	0.000	0.000	0.000	0.000	0.000	0.003	0.026	0.104	0.188	0.122	0.011	0.000	0.000
16	0.000	0.000	0.000	0.000	0.000	0.001	0.009	0.059	0.165	0.183	0.038	0.003	0.000
17	0.000	0.000	0.000	0.000	0.000	0.000	0.003	0.026	0.112	0.216	0.100	0.016	0.000
18	0.000	0.000	0.000	0.000	0.000	0.000	0.001	0.009	0.059	0.191	0.200	0.066	0.001
19	0.000	0.000	0.000	0.000	0.000	0.000	0.000	0.002	0.021	0.121	0.283	0.198	0.018
20	0.000	0.000	0.000	0.000	0.000	0.000	0.000	0.000	0.005	0.049	0.256	0.376	0.171
21	0.000	0.000	0.000	0.000	0.000	0.000	0.000	0.000	0.001	0.009	0.109	0.341	0.810

Part I: Mass function probabilities

n = 22

a \ π	.010	.050	.100	.200	.300	.400	.500	.600	.700	.800	.900	.950	.990
0	0.802	0.324	0.098	0.007	0.000	0.000	0.000	0.000	0.000	0.000	0.000	0.000	0.000
1	0.178	0.374	0.241	0.041	0.004	0.000	0.000	0.000	0.000	0.000	0.000	0.000	0.000
2	0.019	0.207	0.281	0.106	0.017	0.002	0.000	0.000	0.000	0.000	0.000	0.000	0.000
3	0.001	0.073	0.208	0.178	0.047	0.006	0.000	0.000	0.000	0.000	0.000	0.000	0.000
4	0.000	0.018	0.110	0.211	0.097	0.019	0.002	0.000	0.000	0.000	0.000	0.000	0.000
5	0.000	0.003	0.044	0.190	0.148	0.045	0.006	0.000	0.000	0.000	0.000	0.000	0.000
6	0.000	0.001	0.014	0.134	0.181	0.086	0.018	0.002	0.000	0.000	0.000	0.000	0.000
7	0.000	0.000	0.003	0.077	0.177	0.132	0.041	0.005	0.000	0.000	0.000	0.000	0.000
8	0.000	0.000	0.001	0.036	0.143	0.164	0.076	0.014	0.001	0.000	0.000	0.000	0.000
9	0.000	0.000	0.000	0.014	0.094	0.170	0.119	0.034	0.003	0.000	0.000	0.000	0.000
10	0.000	0.000	0.000	0.004	0.053	0.148	0.154	0.066	0.010	0.000	0.000	0.000	0.000
11	0.000	0.000	0.000	0.002	0.025	0.107	0.168	0.107	0.025	0.002	0.000	0.000	0.000
12	0.000	0.000	0.000	0.000	0.010	0.066	0.154	0.148	0.053	0.004	0.000	0.000	0.000
13	0.000	0.000	0.000	0.000	0.003	0.034	0.119	0.170	0.094	0.014	0.000	0.000	0.000
14	0.000	0.000	0.000	0.000	0.001	0.014	0.076	0.164	0.143	0.036	0.001	0.000	0.000
15	0.000	0.000	0.000	0.000	0.000	0.005	0.041	0.132	0.177	0.077	0.003	0.000	0.000
16	0.000	0.000	0.000	0.000	0.000	0.002	0.018	0.086	0.181	0.134	0.014	0.001	0.000
17	0.000	0.000	0.000	0.000	0.000	0.000	0.006	0.045	0.148	0.190	0.044	0.003	0.000
18	0.000	0.000	0.000	0.000	0.000	0.000	0.002	0.019	0.097	0.211	0.110	0.018	0.000
19	0.000	0.000	0.000	0.000	0.000	0.000	0.000	0.006	0.047	0.178	0.208	0.073	0.001
20	0.000	0.000	0.000	0.000	0.000	0.000	0.000	0.002	0.017	0.106	0.281	0.207	0.019
21	0.000	0.000	0.000	0.000	0.000	0.000	0.000	0.000	0.004	0.041	0.241	0.374	0.178
22	0.000	0.000	0.000	0.000	0.000	0.000	0.000	0.000	0.000	0.007	0.098	0.324	0.802

n = 23

a \ π	.010	.050	.100	.200	.300	.400	.500	.600	.700	.800	.900	.950	.990
0	0.794	0.307	0.089	0.006	0.000	0.000	0.000	0.000	0.000	0.000	0.000	0.000	0.000
1	0.184	0.372	0.226	0.034	0.003	0.000	0.000	0.000	0.000	0.000	0.000	0.000	0.000
2	0.020	0.216	0.277	0.093	0.013	0.001	0.000	0.000	0.000	0.000	0.000	0.000	0.000
3	0.002	0.079	0.215	0.164	0.038	0.004	0.000	0.000	0.000	0.000	0.000	0.000	0.000
4	0.000	0.021	0.120	0.204	0.082	0.014	0.001	0.000	0.000	0.000	0.000	0.000	0.000
5	0.000	0.004	0.050	0.194	0.133	0.035	0.004	0.000	0.000	0.000	0.000	0.000	0.000
6	0.000	0.001	0.017	0.145	0.171	0.070	0.012	0.001	0.000	0.000	0.000	0.000	0.000
7	0.000	0.000	0.005	0.088	0.178	0.113	0.030	0.003	0.000	0.000	0.000	0.000	0.000
8	0.000	0.000	0.001	0.045	0.153	0.151	0.058	0.009	0.001	0.000	0.000	0.000	0.000
9	0.000	0.000	0.000	0.018	0.109	0.168	0.097	0.022	0.001	0.000	0.000	0.000	0.000
10	0.000	0.000	0.000	0.006	0.065	0.157	0.137	0.046	0.005	0.000	0.000	0.000	0.000
11	0.000	0.000	0.000	0.002	0.034	0.123	0.161	0.083	0.014	0.001	0.000	0.000	0.000
12	0.000	0.000	0.000	0.001	0.014	0.083	0.161	0.123	0.034	0.002	0.000	0.000	0.000
13	0.000	0.000	0.000	0.000	0.005	0.046	0.137	0.157	0.065	0.006	0.000	0.000	0.000
14	0.000	0.000	0.000	0.000	0.001	0.022	0.097	0.168	0.109	0.018	0.000	0.000	0.000
15	0.000	0.000	0.000	0.000	0.001	0.009	0.058	0.151	0.153	0.045	0.001	0.000	0.000
16	0.000	0.000	0.000	0.000	0.000	0.003	0.030	0.113	0.178	0.088	0.005	0.000	0.000
17	0.000	0.000	0.000	0.000	0.000	0.001	0.012	0.070	0.171	0.145	0.017	0.001	0.000
18	0.000	0.000	0.000	0.000	0.000	0.000	0.004	0.035	0.133	0.194	0.050	0.004	0.000
19	0.000	0.000	0.000	0.000	0.000	0.000	0.001	0.014	0.082	0.204	0.120	0.021	0.000
20	0.000	0.000	0.000	0.000	0.000	0.000	0.000	0.004	0.038	0.164	0.215	0.079	0.002
21	0.000	0.000	0.000	0.000	0.000	0.000	0.000	0.001	0.013	0.093	0.277	0.216	0.020
22	0.000	0.000	0.000	0.000	0.000	0.000	0.000	0.000	0.003	0.034	0.226	0.372	0.184
23	0.000	0.000	0.000	0.000	0.000	0.000	0.000	0.000	0.000	0.006	0.089	0.307	0.794

n = 24

a \ π	.010	.050	.100	.200	.300	.400	.500	.600	.700	.800	.900	.950	.990
0	0.786	0.292	0.080	0.005	0.000	0.000	0.000	0.000	0.000	0.000	0.000	0.000	0.000
1	0.190	0.369	0.212	0.028	0.002	0.000	0.000	0.000	0.000	0.000	0.000	0.000	0.000
2	0.022	0.223	0.272	0.082	0.010	0.001	0.000	0.000	0.000	0.000	0.000	0.000	0.000
3	0.002	0.086	0.222	0.149	0.030	0.003	0.000	0.000	0.000	0.000	0.000	0.000	0.000
4	0.000	0.024	0.129	0.196	0.069	0.009	0.001	0.000	0.000	0.000	0.000	0.000	0.000
5	0.000	0.005	0.057	0.196	0.118	0.027	0.002	0.000	0.000	0.000	0.000	0.000	0.000
6	0.000	0.001	0.021	0.155	0.160	0.056	0.008	0.001	0.000	0.000	0.000	0.000	0.000
7	0.000	0.000	0.005	0.100	0.176	0.096	0.021	0.001	0.000	0.000	0.000	0.000	0.000
8	0.000	0.000	0.002	0.053	0.160	0.136	0.044	0.006	0.000	0.000	0.000	0.000	0.000
9	0.000	0.000	0.000	0.023	0.122	0.161	0.078	0.014	0.001	0.000	0.000	0.000	0.000
10	0.000	0.000	0.000	0.009	0.079	0.161	0.117	0.031	0.003	0.000	0.000	0.000	0.000
11	0.000	0.000	0.000	0.003	0.043	0.137	0.148	0.061	0.008	0.000	0.000	0.000	0.000
12	0.000	0.000	0.000	0.001	0.019	0.099	0.162	0.099	0.019	0.001	0.000	0.000	0.000
13	0.000	0.000	0.000	0.000	0.008	0.061	0.148	0.137	0.043	0.003	0.000	0.000	0.000
14	0.000	0.000	0.000	0.000	0.003	0.032	0.117	0.161	0.079	0.009	0.000	0.000	0.000
15	0.000	0.000	0.000	0.000	0.001	0.014	0.078	0.161	0.122	0.023	0.000	0.000	0.000
16	0.000	0.000	0.000	0.000	0.000	0.005	0.044	0.136	0.160	0.053	0.002	0.000	0.000
17	0.000	0.000	0.000	0.000	0.000	0.002	0.021	0.096	0.176	0.100	0.005	0.000	0.000
18	0.000	0.000	0.000	0.000	0.000	0.000	0.008	0.056	0.160	0.155	0.021	0.001	0.000
19	0.000	0.000	0.000	0.000	0.000	0.000	0.002	0.027	0.118	0.196	0.057	0.005	0.000
20	0.000	0.000	0.000	0.000	0.000	0.000	0.001	0.009	0.069	0.196	0.129	0.024	0.000
21	0.000	0.000	0.000	0.000	0.000	0.000	0.000	0.003	0.030	0.149	0.222	0.086	0.002
22	0.000	0.000	0.000	0.000	0.000	0.000	0.000	0.001	0.010	0.082	0.272	0.223	0.022
23	0.000	0.000	0.000	0.000	0.000	0.000	0.000	0.000	0.002	0.028	0.212	0.369	0.190
24	0.000	0.000	0.000	0.000	0.000	0.000	0.000	0.000	0.000	0.005	0.080	0.292	0.786

n = 25

a \ π	.010	.050	.100	.200	.300	.400	.500	.600	.700	.800	.900	.950	.990
0	0.778	0.277	0.072	0.004	0.000	0.000	0.000	0.000	0.000	0.000	0.000	0.000	0.000
1	0.196	0.365	0.199	0.023	0.002	0.000	0.000	0.000	0.000	0.000	0.000	0.000	0.000
2	0.024	0.231	0.266	0.071	0.007	0.000	0.000	0.000	0.000	0.000	0.000	0.000	0.000
3	0.002	0.093	0.227	0.136	0.024	0.002	0.000	0.000	0.000	0.000	0.000	0.000	0.000
4	0.000	0.027	0.138	0.187	0.057	0.007	0.000	0.000	0.000	0.000	0.000	0.000	0.000
5	0.000	0.006	0.065	0.196	0.103	0.020	0.002	0.000	0.000	0.000	0.000	0.000	0.000
6	0.000	0.001	0.024	0.163	0.148	0.045	0.005	0.000	0.000	0.000	0.000	0.000	0.000
7	0.000	0.000	0.007	0.111	0.171	0.080	0.015	0.001	0.000	0.000	0.000	0.000	0.000
8	0.000	0.000	0.002	0.062	0.165	0.120	0.032	0.003	0.000	0.000	0.000	0.000	0.000
9	0.000	0.000	0.000	0.030	0.134	0.151	0.061	0.009	0.000	0.000	0.000	0.000	0.000
10	0.000	0.000	0.000	0.011	0.091	0.161	0.097	0.021	0.002	0.000	0.000	0.000	0.000
11	0.000	0.000	0.000	0.004	0.054	0.146	0.133	0.044	0.004	0.000	0.000	0.000	0.000
12	0.000	0.000	0.000	0.002	0.027	0.114	0.155	0.076	0.011	0.000	0.000	0.000	0.000
13	0.000	0.000	0.000	0.000	0.011	0.076	0.155	0.114	0.027	0.002	0.000	0.000	0.000
14	0.000	0.000	0.000	0.000	0.004	0.044	0.133	0.146	0.054	0.004	0.000	0.000	0.000
15	0.000	0.000	0.000	0.000	0.002	0.021	0.097	0.161	0.091	0.011	0.000	0.000	0.000
16	0.000	0.000	0.000	0.000	0.000	0.009	0.061	0.151	0.134	0.030	0.000	0.000	0.000
17	0.000	0.000	0.000	0.000	0.000	0.003	0.032	0.120	0.165	0.062	0.002	0.000	0.000
18	0.000	0.000	0.000	0.000	0.000	0.001	0.015	0.080	0.171	0.111	0.007	0.000	0.000
19	0.000	0.000	0.000	0.000	0.000	0.000	0.005	0.045	0.148	0.163	0.024	0.001	0.000
20	0.000	0.000	0.000	0.000	0.000	0.000	0.002	0.020	0.103	0.196	0.065	0.006	0.000
21	0.000	0.000	0.000	0.000	0.000	0.000	0.000	0.007	0.057	0.187	0.138	0.027	0.000
22	0.000	0.000	0.000	0.000	0.000	0.000	0.000	0.002	0.024	0.136	0.227	0.093	0.002
23	0.000	0.000	0.000	0.000	0.000	0.000	0.000	0.000	0.007	0.071	0.266	0.231	0.024
24	0.000	0.000	0.000	0.000	0.000	0.000	0.000	0.000	0.002	0.023	0.199	0.365	0.196
25	0.000	0.000	0.000	0.000	0.000	0.000	0.000	0.000	0.000	0.004	0.072	0.277	0.778

TABLE B.1 | (concluded)

Part I: Mass function probabilities

$n = 50$

a \ π	.010	.050	.100	.200	.300	.400	.500	.600	.700	.800	.900	.950	.990
0	0.605	0.077	0.005	0.000	0.000	0.000	0.000	0.000	0.000	0.000	0.000	0.000	0.000
1	0.306	0.202	0.029	0.000	0.000	0.000	0.000	0.000	0.000	0.000	0.000	0.000	0.000
2	0.075	0.262	0.078	0.001	0.000	0.000	0.000	0.000	0.000	0.000	0.000	0.000	0.000
3	0.012	0.219	0.138	0.005	0.000	0.000	0.000	0.000	0.000	0.000	0.000	0.000	0.000
4	0.002	0.136	0.181	0.012	0.000	0.000	0.000	0.000	0.000	0.000	0.000	0.000	0.000
5	0.000	0.066	0.185	0.030	0.001	0.000	0.000	0.000	0.000	0.000	0.000	0.000	0.000
6	0.000	0.026	0.154	0.055	0.001	0.000	0.000	0.000	0.000	0.000	0.000	0.000	0.000
7	0.000	0.009	0.108	0.087	0.005	0.000	0.000	0.000	0.000	0.000	0.000	0.000	0.000
8	0.000	0.002	0.064	0.117	0.011	0.000	0.000	0.000	0.000	0.000	0.000	0.000	0.000
9	0.000	0.001	0.033	0.137	0.022	0.001	0.000	0.000	0.000	0.000	0.000	0.000	0.000
10	0.000	0.000	0.016	0.140	0.039	0.001	0.000	0.000	0.000	0.000	0.000	0.000	0.000
11	0.000	0.000	0.006	0.127	0.060	0.004	0.000	0.000	0.000	0.000	0.000	0.000	0.000
12	0.000	0.000	0.002	0.103	0.084	0.007	0.000	0.000	0.000	0.000	0.000	0.000	0.000
13	0.000	0.000	0.001	0.075	0.105	0.015	0.000	0.000	0.000	0.000	0.000	0.000	0.000
14	0.000	0.000	0.000	0.050	0.119	0.026	0.001	0.000	0.000	0.000	0.000	0.000	0.000
15	0.000	0.000	0.000	0.030	0.122	0.042	0.002	0.000	0.000	0.000	0.000	0.000	0.000
16	0.000	0.000	0.000	0.017	0.115	0.060	0.005	0.000	0.000	0.000	0.000	0.000	0.000
17	0.000	0.000	0.000	0.008	0.098	0.081	0.008	0.000	0.000	0.000	0.000	0.000	0.000
18	0.000	0.000	0.000	0.003	0.077	0.099	0.016	0.001	0.000	0.000	0.000	0.000	0.000
19	0.000	0.000	0.000	0.002	0.056	0.110	0.027	0.001	0.000	0.000	0.000	0.000	0.000
20	0.000	0.000	0.000	0.001	0.037	0.115	0.042	0.002	0.000	0.000	0.000	0.000	0.000
21	0.000	0.000	0.000	0.000	0.023	0.109	0.060	0.004	0.000	0.000	0.000	0.000	0.000
22	0.000	0.000	0.000	0.000	0.013	0.096	0.079	0.008	0.000	0.000	0.000	0.000	0.000
23	0.000	0.000	0.000	0.000	0.006	0.078	0.096	0.015	0.000	0.000	0.000	0.000	0.000
24	0.000	0.000	0.000	0.000	0.004	0.058	0.108	0.026	0.001	0.000	0.000	0.000	0.000
25	0.000	0.000	0.000	0.000	0.001	0.041	0.112	0.041	0.001	0.000	0.000	0.000	0.000
26	0.000	0.000	0.000	0.000	0.001	0.026	0.108	0.058	0.004	0.000	0.000	0.000	0.000
27	0.000	0.000	0.000	0.000	0.000	0.015	0.096	0.078	0.006	0.000	0.000	0.000	0.000
28	0.000	0.000	0.000	0.000	0.000	0.008	0.079	0.096	0.013	0.000	0.000	0.000	0.000
29	0.000	0.000	0.000	0.000	0.000	0.004	0.060	0.109	0.023	0.000	0.000	0.000	0.000
30	0.000	0.000	0.000	0.000	0.000	0.002	0.042	0.115	0.037	0.001	0.000	0.000	0.000
31	0.000	0.000	0.000	0.000	0.000	0.001	0.027	0.110	0.056	0.002	0.000	0.000	0.000
32	0.000	0.000	0.000	0.000	0.000	0.001	0.016	0.099	0.077	0.003	0.000	0.000	0.000
33	0.000	0.000	0.000	0.000	0.000	0.000	0.008	0.081	0.098	0.008	0.000	0.000	0.000
34	0.000	0.000	0.000	0.000	0.000	0.000	0.005	0.060	0.115	0.017	0.000	0.000	0.000
35	0.000	0.000	0.000	0.000	0.000	0.000	0.002	0.042	0.122	0.030	0.000	0.000	0.000
36	0.000	0.000	0.000	0.000	0.000	0.000	0.001	0.026	0.119	0.050	0.000	0.000	0.000
37	0.000	0.000	0.000	0.000	0.000	0.000	0.000	0.015	0.105	0.075	0.001	0.000	0.000
38	0.000	0.000	0.000	0.000	0.000	0.000	0.000	0.007	0.084	0.103	0.002	0.000	0.000
39	0.000	0.000	0.000	0.000	0.000	0.000	0.000	0.004	0.060	0.127	0.006	0.000	0.000
40	0.000	0.000	0.000	0.000	0.000	0.000	0.000	0.001	0.039	0.140	0.016	0.000	0.000
41	0.000	0.000	0.000	0.000	0.000	0.000	0.000	0.001	0.022	0.137	0.033	0.001	0.000
42	0.000	0.000	0.000	0.000	0.000	0.000	0.000	0.000	0.011	0.117	0.064	0.002	0.000
43	0.000	0.000	0.000	0.000	0.000	0.000	0.000	0.000	0.005	0.087	0.108	0.009	0.000
44	0.000	0.000	0.000	0.000	0.000	0.000	0.000	0.000	0.002	0.055	0.154	0.026	0.000
45	0.000	0.000	0.000	0.000	0.000	0.000	0.000	0.000	0.001	0.030	0.185	0.066	0.000
46	0.000	0.000	0.000	0.000	0.000	0.000	0.000	0.000	0.000	0.012	0.181	0.136	0.002
47	0.000	0.000	0.000	0.000	0.000	0.000	0.000	0.000	0.000	0.005	0.138	0.219	0.012
48	0.000	0.000	0.000	0.000	0.000	0.000	0.000	0.000	0.000	0.001	0.078	0.262	0.075
49	0.000	0.000	0.000	0.000	0.000	0.000	0.000	0.000	0.000	0.000	0.029	0.202	0.306
50	0.000	0.000	0.000	0.000	0.000	0.000	0.000	0.000	0.000	0.000	0.005	0.077	0.605

TABLE B.1

Part II: Cumulative probabilities

Tabulated values give the probability that the random variable **X** is less than or equal to $x = a$. These values are:

$$P[\mathbf{X} \le x = a/n, \pi] = \sum_{x=0}^{a} P(x), \quad P(x) = \frac{n!}{x!(n-x)!} \pi^x (1-\pi)^{n-x} \quad x = 0, 1, 2, \ldots, n$$

$n = 1$

a \ π	.010	.050	.100	.200	.300	.400	.500	.600	.700	.800	.900	.950	.990
0	.990	.950	.900	.800	.700	.600	.500	.400	.300	.200	.100	.050	.010

$n = 2$

a \ π	.010	.050	.100	.200	.300	.400	.500	.600	.700	.800	.900	.950	.990
0	.980	.903	.810	.640	.490	.360	.250	.160	.090	.040	.010	.002	.000
1	1.000	.998	.990	.960	.910	.840	.750	.640	.510	.360	.190	.098	.020

$n = 3$

a \ π	.010	.050	.100	.200	.300	.400	.500	.600	.700	.800	.900	.950	.990
0	.970	.857	.729	.512	.343	.216	.125	.064	.027	.008	.001	.000	.000
1	1.000	.993	.972	.896	.784	.648	.500	.352	.216	.104	.028	.007	.000
2	1.000	1.000	.999	.992	.973	.936	.875	.784	.657	.488	.271	.143	.030

TABLE B.1 | (continued)

Part II: Cumulative probabilities

$n = 4$

a \ π	.010	.050	.100	.200	.300	.400	.500	.600	.700	.800	.900	.950	.990
0	.961	.815	.656	.410	.240	.130	.063	.026	.008	.002	.000	.000	.000
1	.999	.986	.948	.819	.652	.475	.313	.179	.084	.027	.004	.000	.000
2	1.000	1.000	.996	.973	.916	.821	.688	.525	.348	.181	.052	.014	.001
3	1.000	1.000	1.000	.998	.992	.974	.938	.870	.760	.590	.344	.185	.039

$n = 5$

a \ π	.010	.050	.100	.200	.300	.400	.500	.600	.700	.800	.900	.950	.990
0	.951	.774	.590	.328	.168	.078	.031	.010	.002	.000	.000	.000	.000
1	.999	.977	.919	.737	.528	.337	.188	.087	.031	.007	.000	.000	.000
2	1.000	.999	.991	.942	.837	.683	.500	.317	.163	.058	.009	.001	.000
3	1.000	1.000	1.000	.993	.969	.913	.813	.663	.472	.263	.081	.023	.001
4	1.000	1.000	1.000	1.000	.998	.990	.969	.922	.832	.672	.410	.226	.049

$n = 6$

a \ π	.010	.050	.100	.200	.300	.400	.500	.600	.700	.800	.900	.950	.990
0	.941	.735	.531	.262	.118	.047	.016	.004	.001	.000	.000	.000	.000
1	.999	.967	.886	.655	.420	.233	.109	.041	.011	.002	.000	.000	.000
2	1.000	.998	.984	.901	.744	.544	.344	.179	.070	.017	.001	.000	.000
3	1.000	1.000	.999	.983	.930	.821	.656	.456	.256	.099	.016	.002	.000
4	1.000	1.000	1.000	.998	.989	.959	.891	.767	.580	.345	.114	.033	.001
5	1.000	1.000	1.000	1.000	.999	.996	.984	.953	.882	.738	.469	.265	.059

$n = 7$

a \ π	.010	.050	.100	.200	.300	.400	.500	.600	.700	.800	.900	.950	.990
0	.932	.698	.478	.210	.082	.028	.008	.002	.000	.000	.000	.000	.000
1	.998	.956	.850	.577	.329	.159	.063	.019	.004	.000	.000	.000	.000
2	1.000	.996	.974	.852	.647	.420	.227	.096	.029	.005	.000	.000	.000
3	1.000	1.000	.997	.967	.874	.710	.500	.290	.126	.033	.003	.000	.000
4	1.000	1.000	1.000	.995	.971	.904	.773	.580	.353	.148	.026	.004	.000
5	1.000	1.000	1.000	1.000	.996	.981	.938	.841	.671	.423	.150	.044	.002
6	1.000	1.000	1.000	1.000	1.000	.998	.992	.972	.918	.790	.522	.302	.068

$n = 8$

a \ π	.010	.050	.100	.200	.300	.400	.500	.600	.700	.800	.900	.950	.990
0	.923	.663	.430	.168	.058	.017	.004	.001	.000	.000	.000	.000	.000
1	.997	.943	.813	.503	.255	.106	.035	.009	.001	.000	.000	.000	.000
2	1.000	.994	.962	.797	.552	.315	.145	.050	.011	.001	.000	.000	.000
3	1.000	1.000	.995	.944	.806	.594	.363	.174	.058	.010	.000	.000	.000
4	1.000	1.000	1.000	.990	.942	.826	.637	.406	.194	.056	.005	.000	.000
5	1.000	1.000	1.000	.999	.989	.950	.855	.685	.448	.203	.038	.006	.000
6	1.000	1.000	1.000	1.000	.999	.991	.965	.894	.745	.497	.187	.057	.003
7	1.000	1.000	1.000	1.000	1.000	.999	.996	.983	.942	.832	.570	.337	.077

$n = 9$

a \ π	.010	.050	.100	.200	.300	.400	.500	.600	.700	.800	.900	.950	.990
0	.914	.630	.387	.134	.040	.010	.002	.000	.000	.000	.000	.000	.000
1	.997	.929	.775	.436	.196	.071	.020	.004	.000	.000	.000	.000	.000
2	1.000	.992	.947	.738	.463	.232	.090	.025	.004	.000	.000	.000	.000
3	1.000	.999	.992	.914	.730	.483	.254	.099	.025	.003	.000	.000	.000
4	1.000	1.000	.999	.980	.901	.733	.500	.267	.099	.020	.001	.000	.000
5	1.000	1.000	1.000	.997	.975	.901	.746	.517	.270	.086	.008	.001	.000
6	1.000	1.000	1.000	1.000	.996	.975	.910	.768	.537	.262	.053	.008	.000
7	1.000	1.000	1.000	1.000	1.000	.996	.980	.929	.804	.564	.225	.071	.003
8	1.000	1.000	1.000	1.000	1.000	1.000	.998	.990	.960	.866	.613	.370	.086

TABLE B.1 | *(continued)*

Part II: Cumulative probabilities

$n = 10$

a \ π	.010	.050	.100	.200	.300	.400	.500	.600	.700	.800	.900	.950	.990
0	.904	.599	.349	.107	.028	.006	.001	.000	.000	.000	.000	.000	.000
1	.996	.914	.736	.376	.149	.046	.011	.002	.000	.000	.000	.000	.000
2	1.000	.988	.930	.678	.383	.167	.055	.012	.002	.000	.000	.000	.000
3	1.000	.999	.987	.879	.650	.382	.172	.055	.011	.001	.000	.000	.000
4	1.000	1.000	.998	.967	.850	.633	.377	.166	.047	.006	.000	.000	.000
5	1.000	1.000	1.000	.994	.953	.834	.623	.367	.150	.033	.002	.000	.000
6	1.000	1.000	1.000	.999	.989	.945	.828	.618	.350	.121	.013	.001	.000
7	1.000	1.000	1.000	1.000	.998	.988	.945	.833	.617	.322	.070	.012	.000
8	1.000	1.000	1.000	1.000	1.000	.998	.989	.954	.851	.624	.264	.086	.004
9	1.000	1.000	1.000	1.000	1.000	1.000	.999	.994	.972	.893	.651	.401	.096

$n = 11$

a \ π	.010	.050	.100	.200	.300	.400	.500	.600	.700	.800	.900	.950	.990
0	.895	.569	.314	.086	.020	.004	.000	.000	.000	.000	.000	.000	.000
1	.995	.898	.697	.322	.113	.030	.006	.001	.000	.000	.000	.000	.000
2	1.000	.985	.910	.617	.313	.119	.033	.006	.001	.000	.000	.000	.000
3	1.000	.998	.981	.839	.570	.296	.113	.029	.004	.000	.000	.000	.000
4	1.000	1.000	.997	.950	.790	.533	.274	.099	.022	.002	.000	.000	.000
5	1.000	1.000	1.000	.988	.922	.753	.500	.247	.078	.012	.000	.000	.000
6	1.000	1.000	1.000	.998	.978	.901	.726	.467	.210	.050	.003	.000	.000
7	1.000	1.000	1.000	1.000	.996	.971	.887	.704	.430	.161	.019	.002	.000
8	1.000	1.000	1.000	1.000	.999	.994	.967	.881	.687	.383	.090	.015	.000
9	1.000	1.000	1.000	1.000	1.000	.999	.994	.970	.887	.678	.303	.102	.005
10	1.000	1.000	1.000	1.000	1.000	1.000	1.000	.996	.980	.914	.686	.431	.105

$n = 12$

a \ π	.010	.050	.100	.200	.300	.400	.500	.600	.700	.800	.900	.950	.990
0	.886	.540	.282	.069	.014	.002	.000	.000	.000	.000	.000	.000	.000
1	.994	.882	.659	.275	.085	.020	.003	.000	.000	.000	.000	.000	.000
2	1.000	.980	.889	.558	.253	.083	.019	.003	.000	.000	.000	.000	.000
3	1.000	.998	.974	.795	.493	.225	.073	.015	.002	.000	.000	.000	.000
4	1.000	1.000	.996	.927	.724	.438	.194	.057	.009	.001	.000	.000	.000
5	1.000	1.000	.999	.981	.882	.665	.387	.158	.039	.004	.000	.000	.000
6	1.000	1.000	1.000	.996	.961	.842	.613	.335	.118	.019	.001	.000	.000
7	1.000	1.000	1.000	.999	.991	.943	.806	.562	.276	.073	.004	.000	.000
8	1.000	1.000	1.000	1.000	.998	.985	.927	.775	.507	.205	.026	.002	.000
9	1.000	1.000	1.000	1.000	1.000	.997	.981	.917	.747	.442	.111	.020	.000
10	1.000	1.000	1.000	1.000	1.000	1.000	.997	.980	.915	.725	.341	.118	.006
11	1.000	1.000	1.000	1.000	1.000	1.000	1.000	.998	.986	.931	.718	.460	.114

$n = 13$

a \ π	.010	.050	.100	.200	.300	.400	.500	.600	.700	.800	.900	.950	.990
0	.878	.513	.254	.055	.010	.001	.000	.000	.000	.000	.000	.000	.000
1	.993	.865	.621	.234	.064	.013	.002	.000	.000	.000	.000	.000	.000
2	1.000	.975	.866	.502	.202	.058	.011	.001	.000	.000	.000	.000	.000
3	1.000	.997	.966	.747	.421	.169	.046	.008	.001	.000	.000	.000	.000
4	1.000	1.000	.994	.901	.654	.353	.133	.032	.004	.000	.000	.000	.000
5	1.000	1.000	.999	.970	.835	.574	.291	.098	.018	.001	.000	.000	.000
6	1.000	1.000	1.000	.993	.938	.771	.500	.229	.062	.007	.000	.000	.000
7	1.000	1.000	1.000	.999	.982	.902	.709	.426	.165	.030	.001	.000	.000
8	1.000	1.000	1.000	1.000	.996	.968	.867	.647	.346	.099	.006	.000	.000
9	1.000	1.000	1.000	1.000	.999	.992	.954	.831	.579	.253	.034	.003	.000
10	1.000	1.000	1.000	1.000	1.000	.999	.989	.942	.798	.498	.134	.025	.000
11	1.000	1.000	1.000	1.000	1.000	1.000	.998	.987	.936	.766	.379	.135	.007
12	1.000	1.000	1.000	1.000	1.000	1.000	1.000	.999	.990	.945	.746	.487	.122

TABLE B.1 | (continued)

Part II: Cumulative probabilities

n = 14

a \ *π*	.010	.050	.100	.200	.300	.400	.500	.600	.700	.800	.900	.950	.990
0	.869	.488	.229	.044	.007	.001	.000	.000	.000	.000	.000	.000	.000
1	.992	.847	.585	.198	.047	.008	.001	.000	.000	.000	.000	.000	.000
2	1.000	.970	.842	.448	.161	.040	.006	.001	.000	.000	.000	.000	.000
3	1.000	.996	.956	.698	.355	.124	.029	.004	.000	.000	.000	.000	.000
4	1.000	1.000	.991	.870	.584	.279	.090	.018	.002	.000	.000	.000	.000
5	1.000	1.000	.999	.956	.781	.486	.212	.058	.008	.000	.000	.000	.000
6	1.000	1.000	1.000	.988	.907	.692	.395	.150	.031	.002	.000	.000	.000
7	1.000	1.000	1.000	.998	.969	.850	.605	.308	.093	.012	.000	.000	.000
8	1.000	1.000	1.000	1.000	.992	.942	.788	.514	.219	.044	.001	.000	.000
9	1.000	1.000	1.000	1.000	.998	.982	.910	.721	.416	.130	.009	.000	.000
10	1.000	1.000	1.000	1.000	1.000	.996	.971	.876	.645	.302	.044	.004	.000
11	1.000	1.000	1.000	1.000	1.000	.999	.994	.960	.839	.552	.158	.030	.000
12	1.000	1.000	1.000	1.000	1.000	1.000	.999	.992	.953	.802	.415	.153	.008
13	1.000	1.000	1.000	1.000	1.000	1.000	1.000	.999	.993	.956	.771	.512	.131

n = 15

a \ *π*	.010	.050	.100	.200	.300	.400	.500	.600	.700	.800	.900	.950	.990
0	.860	.463	.206	.035	.005	.000	.000	.000	.000	.000	.000	.000	.000
1	.990	.829	.549	.167	.035	.005	.000	.000	.000	.000	.000	.000	.000
2	1.000	.964	.816	.398	.127	.027	.004	.000	.000	.000	.000	.000	.000
3	1.000	.995	.944	.648	.297	.091	.018	.002	.000	.000	.000	.000	.000
4	1.000	.999	.987	.836	.515	.217	.059	.009	.001	.000	.000	.000	.000
5	1.000	1.000	.998	.939	.722	.403	.151	.034	.004	.000	.000	.000	.000
6	1.000	1.000	1.000	.982	.869	.610	.304	.095	.015	.001	.000	.000	.000
7	1.000	1.000	1.000	.996	.950	.787	.500	.213	.050	.004	.000	.000	.000
8	1.000	1.000	1.000	.999	.985	.905	.696	.390	.131	.018	.000	.000	.000
9	1.000	1.000	1.000	1.000	.996	.966	.849	.597	.278	.061	.002	.000	.000
10	1.000	1.000	1.000	1.000	.999	.991	.941	.783	.485	.164	.013	.001	.000
11	1.000	1.000	1.000	1.000	1.000	.998	.982	.909	.703	.352	.056	.005	.000
12	1.000	1.000	1.000	1.000	1.000	1.000	.996	.973	.873	.602	.184	.036	.000
13	1.000	1.000	1.000	1.000	1.000	1.000	1.000	.995	.965	.833	.451	.171	.010
14	1.000	1.000	1.000	1.000	1.000	1.000	1.000	1.000	.995	.965	.794	.537	.140

n = 16

a \ *π*	.010	.050	.100	.200	.300	.400	.500	.600	.700	.800	.900	.950	.990
0	.851	.440	.185	.028	.003	.000	.000	.000	.000	.000	.000	.000	.000
1	.989	.811	.515	.141	.026	.003	.000	.000	.000	.000	.000	.000	.000
2	.999	.957	.789	.352	.099	.018	.002	.000	.000	.000	.000	.000	.000
3	1.000	.993	.932	.598	.246	.065	.011	.001	.000	.000	.000	.000	.000
4	1.000	.999	.983	.798	.450	.167	.038	.005	.000	.000	.000	.000	.000
5	1.000	1.000	.997	.918	.660	.329	.105	.019	.002	.000	.000	.000	.000
6	1.000	1.000	.999	.973	.825	.527	.227	.058	.007	.000	.000	.000	.000
7	1.000	1.000	1.000	.993	.926	.716	.402	.142	.026	.001	.000	.000	.000
8	1.000	1.000	1.000	.999	.974	.858	.598	.284	.074	.007	.000	.000	.000
9	1.000	1.000	1.000	1.000	.993	.942	.773	.473	.175	.027	.001	.000	.000
10	1.000	1.000	1.000	1.000	.998	.981	.895	.671	.340	.082	.003	.000	.000
11	1.000	1.000	1.000	1.000	1.000	.995	.962	.833	.550	.202	.017	.001	.000
12	1.000	1.000	1.000	1.000	1.000	.999	.989	.935	.754	.402	.068	.007	.000
13	1.000	1.000	1.000	1.000	1.000	1.000	.998	.982	.901	.648	.211	.043	.001
14	1.000	1.000	1.000	1.000	1.000	1.000	1.000	.997	.974	.859	.485	.189	.011
15	1.000	1.000	1.000	1.000	1.000	1.000	1.000	1.000	.997	.972	.815	.560	.149

TABLE B.1 | (continued)

Part II: Cumulative probabilities

n = 17

a \ π	.010	.050	.100	.200	.300	.400	.500	.600	.700	.800	.900	.950	.990
0	.843	.418	.167	.023	.002	.000	.000	.000	.000	.000	.000	.000	.000
1	.988	.792	.482	.118	.019	.002	.000	.000	.000	.000	.000	.000	.000
2	.999	.950	.762	.310	.077	.012	.001	.000	.000	.000	.000	.000	.000
3	1.000	.991	.917	.549	.202	.046	.006	.000	.000	.000	.000	.000	.000
4	1.000	.999	.978	.758	.389	.126	.025	.003	.000	.000	.000	.000	.000
5	1.000	1.000	.995	.894	.597	.264	.072	.011	.001	.000	.000	.000	.000
6	1.000	1.000	.999	.962	.775	.448	.166	.035	.003	.000	.000	.000	.000
7	1.000	1.000	1.000	.989	.895	.641	.315	.092	.013	.000	.000	.000	.000
8	1.000	1.000	1.000	.997	.960	.801	.500	.199	.040	.003	.000	.000	.000
9	1.000	1.000	1.000	1.000	.987	.908	.685	.359	.105	.011	.000	.000	.000
10	1.000	1.000	1.000	1.000	.997	.965	.834	.552	.225	.038	.001	.000	.000
11	1.000	1.000	1.000	1.000	.999	.989	.928	.736	.403	.106	.005	.000	.000
12	1.000	1.000	1.000	1.000	1.000	.997	.975	.874	.611	.242	.022	.001	.000
13	1.000	1.000	1.000	1.000	1.000	1.000	.994	.954	.798	.451	.083	.009	.000
14	1.000	1.000	1.000	1.000	1.000	1.000	.999	.988	.923	.690	.238	.050	.001
15	1.000	1.000	1.000	1.000	1.000	1.000	1.000	.998	.981	.882	.518	.208	.012
16	1.000	1.000	1.000	1.000	1.000	1.000	1.000	1.000	.998	.977	.833	.582	.157

n = 18

a \ π	.010	.050	.100	.200	.300	.400	.500	.600	.700	.800	.900	.950	.990
0	.835	.397	.150	.018	.002	.000	.000	.000	.000	.000	.000	.000	.000
1	.986	.774	.450	.099	.014	.001	.000	.000	.000	.000	.000	.000	.000
2	.999	.942	.734	.271	.060	.008	.001	.000	.000	.000	.000	.000	.000
3	1.000	.989	.902	.501	.165	.033	.004	.000	.000	.000	.000	.000	.000
4	1.000	.998	.972	.716	.333	.094	.015	.001	.000	.000	.000	.000	.000
5	1.000	1.000	.994	.867	.534	.209	.048	.006	.000	.000	.000	.000	.000
6	1.000	1.000	.999	.949	.722	.374	.119	.020	.001	.000	.000	.000	.000
7	1.000	1.000	1.000	.984	.859	.563	.240	.058	.006	.000	.000	.000	.000
8	1.000	1.000	1.000	.996	.940	.737	.407	.135	.021	.001	.000	.000	.000
9	1.000	1.000	1.000	.999	.979	.865	.593	.263	.060	.004	.000	.000	.000
10	1.000	1.000	1.000	1.000	.994	.942	.760	.437	.141	.016	.000	.000	.000
11	1.000	1.000	1.000	1.000	.999	.980	.881	.626	.278	.051	.001	.000	.000
12	1.000	1.000	1.000	1.000	1.000	.994	.952	.791	.466	.133	.006	.000	.000
13	1.000	1.000	1.000	1.000	1.000	.999	.985	.906	.667	.284	.028	.002	.000
14	1.000	1.000	1.000	1.000	1.000	1.000	.996	.967	.835	.499	.098	.011	.000
15	1.000	1.000	1.000	1.000	1.000	1.000	.999	.992	.940	.729	.266	.058	.001
16	1.000	1.000	1.000	1.000	1.000	1.000	1.000	.999	.986	.901	.550	.226	.014
17	1.000	1.000	1.000	1.000	1.000	1.000	1.000	1.000	.998	.982	.850	.603	.165

n = 19

a \ π	.010	.050	.100	.200	.300	.400	.500	.600	.700	.800	.900	.950	.990
0	.826	.377	.135	.014	.001	.000	.000	.000	.000	.000	.000	.000	.000
1	.985	.755	.420	.083	.010	.001	.000	.000	.000	.000	.000	.000	.000
2	.999	.933	.705	.237	.046	.005	.000	.000	.000	.000	.000	.000	.000
3	1.000	.987	.885	.455	.133	.023	.002	.000	.000	.000	.000	.000	.000
4	1.000	.998	.965	.673	.282	.070	.010	.001	.000	.000	.000	.000	.000
5	1.000	1.000	.991	.837	.474	.163	.032	.003	.000	.000	.000	.000	.000
6	1.000	1.000	.998	.932	.666	.308	.084	.012	.001	.000	.000	.000	.000
7	1.000	1.000	1.000	.977	.818	.488	.180	.035	.003	.000	.000	.000	.000
8	1.000	1.000	1.000	.993	.916	.667	.324	.088	.011	.000	.000	.000	.000
9	1.000	1.000	1.000	.998	.967	.814	.500	.186	.033	.002	.000	.000	.000
10	1.000	1.000	1.000	1.000	.989	.912	.676	.333	.084	.007	.000	.000	.000
11	1.000	1.000	1.000	1.000	.997	.965	.820	.512	.182	.023	.000	.000	.000
12	1.000	1.000	1.000	1.000	.999	.988	.916	.692	.334	.068	.002	.000	.000
13	1.000	1.000	1.000	1.000	1.000	.997	.968	.837	.526	.163	.009	.000	.000
14	1.000	1.000	1.000	1.000	1.000	.999	.990	.930	.718	.327	.035	.002	.000
15	1.000	1.000	1.000	1.000	1.000	1.000	.998	.977	.867	.545	.115	.013	.000
16	1.000	1.000	1.000	1.000	1.000	1.000	1.000	.995	.954	.763	.295	.067	.001
17	1.000	1.000	1.000	1.000	1.000	1.000	1.000	.999	.990	.917	.580	.245	.015
18	1.000	1.000	1.000	1.000	1.000	1.000	1.000	1.000	.999	.986	.865	.623	.174

TABLE B.1 | *(continued)*

Part II: Cumulative probabilities

n = 20

a \ π	.010	.050	.100	.200	.300	.400	.500	.600	.700	.800	.900	.950	.990
0	.818	.358	.122	.012	.001	.000	.000	.000	.000	.000	.000	.000	.000
1	.983	.736	.392	.069	.008	.001	.000	.000	.000	.000	.000	.000	.000
2	.999	.925	.677	.206	.035	.004	.000	.000	.000	.000	.000	.000	.000
3	1.000	.984	.867	.411	.107	.016	.001	.000	.000	.000	.000	.000	.000
4	1.000	.997	.957	.630	.238	.051	.006	.000	.000	.000	.000	.000	.000
5	1.000	1.000	.989	.804	.416	.126	.021	.002	.000	.000	.000	.000	.000
6	1.000	1.000	.998	.913	.608	.250	.058	.006	.000	.000	.000	.000	.000
7	1.000	1.000	1.000	.968	.772	.416	.132	.021	.001	.000	.000	.000	.000
8	1.000	1.000	1.000	.990	.887	.596	.252	.057	.005	.000	.000	.000	.000
9	1.000	1.000	1.000	.997	.952	.755	.412	.128	.017	.001	.000	.000	.000
10	1.000	1.000	1.000	.999	.983	.872	.588	.245	.048	.003	.000	.000	.000
11	1.000	1.000	1.000	1.000	.995	.943	.748	.404	.113	.010	.000	.000	.000
12	1.000	1.000	1.000	1.000	.999	.979	.868	.584	.228	.032	.000	.000	.000
13	1.000	1.000	1.000	1.000	1.000	.994	.942	.750	.392	.087	.002	.000	.000
14	1.000	1.000	1.000	1.000	1.000	.998	.979	.874	.584	.196	.011	.000	.000
15	1.000	1.000	1.000	1.000	1.000	1.000	.994	.949	.762	.370	.043	.003	.000
16	1.000	1.000	1.000	1.000	1.000	1.000	.999	.984	.893	.589	.133	.016	.000
17	1.000	1.000	1.000	1.000	1.000	1.000	1.000	.996	.965	.794	.323	.075	.001
18	1.000	1.000	1.000	1.000	1.000	1.000	1.000	.999	.992	.931	.608	.264	.017
19	1.000	1.000	1.000	1.000	1.000	1.000	1.000	1.000	.999	.988	.878	.642	.182

n = 21

a \ π	.010	.050	.100	.200	.300	.400	.500	.600	.700	.800	.900	.950	.990
0	.810	.341	.109	.009	.001	.000	.000	.000	.000	.000	.000	.000	.000
1	.981	.717	.365	.058	.006	.000	.000	.000	.000	.000	.000	.000	.000
2	.999	.915	.648	.179	.027	.002	.000	.000	.000	.000	.000	.000	.000
3	1.000	.981	.848	.370	.086	.011	.001	.000	.000	.000	.000	.000	.000
4	1.000	.997	.948	.586	.198	.037	.004	.000	.000	.000	.000	.000	.000
5	1.000	1.000	.986	.769	.363	.096	.013	.001	.000	.000	.000	.000	.000
6	1.000	1.000	.997	.891	.551	.200	.039	.004	.000	.000	.000	.000	.000
7	1.000	1.000	.999	.957	.723	.350	.095	.012	.001	.000	.000	.000	.000
8	1.000	1.000	1.000	.986	.852	.524	.192	.035	.002	.000	.000	.000	.000
9	1.000	1.000	1.000	.006	.032	.091	.332	.085	.009	.000	.000	.000	.000
10	1.000	1.000	1.000	.999	.974	.826	.500	.174	.026	.001	.000	.000	.000
11	1.000	1.000	1.000	1.000	.991	.915	.668	.309	.068	.004	.000	.000	.000
12	1.000	1.000	1.000	1.000	.998	.965	.808	.476	.148	.014	.000	.000	.000
13	1.000	1.000	1.000	1.000	.999	.988	.905	.650	.277	.043	.001	.000	.000
14	1.000	1.000	1.000	1.000	1.000	.996	.961	.800	.449	.109	.003	.000	.000
15	1.000	1.000	1.000	1.000	1.000	.999	.987	.904	.637	.231	.014	.000	.000
16	1.000	1.000	1.000	1.000	1.000	1.000	.996	.963	.802	.414	.052	.003	.000
17	1.000	1.000	1.000	1.000	1.000	1.000	.999	.989	.914	.630	.152	.019	.000
18	1.000	1.000	1.000	1.000	1.000	1.000	1.000	.998	.973	.821	.352	.085	.001
19	1.000	1.000	1.000	1.000	1.000	1.000	1.000	1.000	.994	.942	.635	.283	.019
20	1.000	1.000	1.000	1.000	1.000	1.000	1.000	1.000	.999	.991	.891	.659	.190

TABLE B.1 | *(continued)*

Part II: Cumulative probabilities

n = 22

a \ π	.010	.050	.100	.200	.300	.400	.500	.600	.700	.800	.900	.950	.990
0	.802	.324	.098	.007	.000	.000	.000	.000	.000	.000	.000	.000	.000
1	.980	.698	.339	.048	.004	.000	.000	.000	.000	.000	.000	.000	.000
2	.999	.905	.620	.154	.021	.002	.000	.000	.000	.000	.000	.000	.000
3	1.000	.978	.828	.332	.068	.008	.000	.000	.000	.000	.000	.000	.000
4	1.000	.996	.938	.543	.165	.027	.002	.000	.000	.000	.000	.000	.000
5	1.000	.999	.982	.733	.313	.072	.008	.000	.000	.000	.000	.000	.000
6	1.000	1.000	.996	.867	.494	.158	.026	.002	.000	.000	.000	.000	.000
7	1.000	1.000	.999	.944	.671	.290	.067	.007	.000	.000	.000	.000	.000
8	1.000	1.000	1.000	.980	.814	.454	.143	.021	.001	.000	.000	.000	.000
9	1.000	1.000	1.000	.994	.908	.624	.262	.055	.004	.000	.000	.000	.000
10	1.000	1.000	1.000	.998	.961	.772	.416	.121	.014	.000	.000	.000	.000
11	1.000	1.000	1.000	1.000	.986	.879	.584	.228	.039	.002	.000	.000	.000
12	1.000	1.000	1.000	1.000	.996	.945	.738	.376	.092	.006	.000	.000	.000
13	1.000	1.000	1.000	1.000	.999	.979	.857	.546	.186	.020	.000	.000	.000
14	1.000	1.000	1.000	1.000	1.000	.993	.933	.710	.329	.056	.001	.000	.000
15	1.000	1.000	1.000	1.000	1.000	.998	.974	.842	.506	.133	.004	.000	.000
16	1.000	1.000	1.000	1.000	1.000	1.000	.992	.928	.687	.267	.018	.001	.000
17	1.000	1.000	1.000	1.000	1.000	1.000	.998	.973	.835	.457	.062	.004	.000
18	1.000	1.000	1.000	1.000	1.000	1.000	1.000	.992	.932	.668	.172	.022	.000
19	1.000	1.000	1.000	1.000	1.000	1.000	1.000	.998	.979	.846	.380	.095	.001
20	1.000	1.000	1.000	1.000	1.000	1.000	1.000	1.000	.996	.952	.661	.302	.020
21	1.000	1.000	1.000	1.000	1.000	1.000	1.000	1.000	1.000	.993	.902	.676	.198

n = 23

a \ π	.010	.050	.100	.200	.300	.400	.500	.600	.700	.800	.900	.950	.990
0	.794	.307	.089	.006	.000	.000	.000	.000	.000	.000	.000	.000	.000
1	.978	.679	.315	.040	.003	.000	.000	.000	.000	.000	.000	.000	.000
2	.998	.895	.592	.133	.016	.001	.000	.000	.000	.000	.000	.000	.000
3	1.000	.974	.807	.297	.054	.005	.000	.000	.000	.000	.000	.000	.000
4	1.000	.995	.927	.501	.136	.019	.001	.000	.000	.000	.000	.000	.000
5	1.000	.999	.977	.695	.269	.054	.005	.000	.000	.000	.000	.000	.000
6	1.000	1.000	.994	.840	.440	.124	.017	.001	.000	.000	.000	.000	.000
7	1.000	1.000	.999	.928	.618	.237	.047	.004	.000	.000	.000	.000	.000
8	1.000	1.000	1.000	.973	.771	.388	.105	.013	.001	.000	.000	.000	.000
9	1.000	1.000	1.000	.991	.880	.556	.202	.035	.002	.000	.000	.000	.000
10	1.000	1.000	1.000	.997	.945	.713	.339	.081	.007	.000	.000	.000	.000
11	1.000	1.000	1.000	.999	.979	.836	.500	.164	.021	.001	.000	.000	.000
12	1.000	1.000	1.000	1.000	.993	.919	.661	.287	.055	.003	.000	.000	.000
13	1.000	1.000	1.000	1.000	.998	.965	.798	.444	.120	.009	.000	.000	.000
14	1.000	1.000	1.000	1.000	.999	.987	.895	.612	.229	.027	.000	.000	.000
15	1.000	1.000	1.000	1.000	1.000	.996	.953	.763	.382	.072	.001	.000	.000
16	1.000	1.000	1.000	1.000	1.000	.999	.983	.876	.560	.160	.006	.000	.000
17	1.000	1.000	1.000	1.000	1.000	1.000	.995	.946	.731	.305	.023	.001	.000
18	1.000	1.000	1.000	1.000	1.000	1.000	.999	.981	.864	.499	.073	.005	.000
19	1.000	1.000	1.000	1.000	1.000	1.000	1.000	.995	.946	.703	.193	.026	.000
20	1.000	1.000	1.000	1.000	1.000	1.000	1.000	.999	.984	.867	.408	.105	.002
21	1.000	1.000	1.000	1.000	1.000	1.000	1.000	1.000	.997	.960	.685	.321	.022
22	1.000	1.000	1.000	1.000	1.000	1.000	1.000	1.000	1.000	.994	.911	.693	.206

TABLE B.1 | (continued)

Part II: Cumulative probabilities

$n = 24$

a \ π	.010	.050	.100	.200	.300	.400	.500	.600	.700	.800	.900	.950	.990
0	.786	.292	.080	.005	.000	.000	.000	.000	.000	.000	.000	.000	.000
1	.976	.661	.292	.033	.002	.000	.000	.000	.000	.000	.000	.000	.000
2	.998	.884	.564	.115	.012	.001	.000	.000	.000	.000	.000	.000	.000
3	1.000	.970	.786	.264	.042	.004	.000	.000	.000	.000	.000	.000	.000
4	1.000	.994	.915	.460	.111	.013	.001	.000	.000	.000	.000	.000	.000
5	1.000	.999	.972	.656	.229	.040	.003	.000	.000	.000	.000	.000	.000
6	1.000	1.000	.993	.811	.389	.096	.011	.001	.000	.000	.000	.000	.000
7	1.000	1.000	.998	.911	.565	.192	.032	.002	.000	.000	.000	.000	.000
8	1.000	1.000	1.000	.964	.725	.328	.076	.008	.000	.000	.000	.000	.000
9	1.000	1.000	1.000	.987	.847	.489	.154	.022	.001	.000	.000	.000	.000
10	1.000	1.000	1.000	.996	.926	.650	.271	.053	.004	.000	.000	.000	.000
11	1.000	1.000	1.000	.999	.969	.787	.419	.114	.012	.000	.000	.000	.000
12	1.000	1.000	1.000	1.000	.988	.886	.581	.213	.031	.001	.000	.000	.000
13	1.000	1.000	1.000	1.000	.996	.947	.729	.350	.074	.004	.000	.000	.000
14	1.000	1.000	1.000	1.000	.999	.978	.846	.511	.153	.013	.000	.000	.000
15	1.000	1.000	1.000	1.000	1.000	.992	.924	.672	.275	.036	.000	.000	.000
16	1.000	1.000	1.000	1.000	1.000	.998	.968	.808	.435	.089	.002	.000	.000
17	1.000	1.000	1.000	1.000	1.000	.999	.989	.904	.611	.189	.007	.000	.000
18	1.000	1.000	1.000	1.000	1.000	1.000	.997	.960	.771	.344	.028	.001	.000
19	1.000	1.000	1.000	1.000	1.000	1.000	.999	.987	.889	.540	.085	.006	.000
20	1.000	1.000	1.000	1.000	1.000	1.000	1.000	.996	.958	.736	.214	.030	.000
21	1.000	1.000	1.000	1.000	1.000	1.000	1.000	.999	.988	.885	.436	.116	.002
22	1.000	1.000	1.000	1.000	1.000	1.000	1.000	1.000	.998	.967	.708	.339	.024
23	1.000	1.000	1.000	1.000	1.000	1.000	1.000	1.000	1.000	.995	.920	.708	.214

$n = 25$

a \ π	.010	.050	.100	.200	.300	.400	.500	.600	.700	.800	.900	.950	.990
0	.778	.277	.072	.004	.000	.000	.000	.000	.000	.000	.000	.000	.000
1	.974	.642	.271	.027	.002	.000	.000	.000	.000	.000	.000	.000	.000
2	.998	.873	.537	.098	.009	.000	.000	.000	.000	.000	.000	.000	.000
3	1.000	.966	.764	.234	.033	.002	.000	.000	.000	.000	.000	.000	.000
4	1.000	.993	.902	.421	.090	.009	.000	.000	.000	.000	.000	.000	.000
5	1.000	.999	.967	.617	.193	.029	.002	.000	.000	.000	.000	.000	.000
6	1.000	1.000	.991	.780	.341	.074	.007	.000	.000	.000	.000	.000	.000
7	1.000	1.000	.998	.891	.512	.154	.022	.001	.000	.000	.000	.000	.000
8	1.000	1.000	1.000	.953	.677	.274	.054	.004	.000	.000	.000	.000	.000
9	1.000	1.000	1.000	.983	.811	.425	.115	.013	.000	.000	.000	.000	.000
10	1.000	1.000	1.000	.994	.902	.586	.212	.034	.002	.000	.000	.000	.000
11	1.000	1.000	1.000	.998	.956	.732	.345	.078	.006	.000	.000	.000	.000
12	1.000	1.000	1.000	1.000	.983	.846	.500	.154	.017	.000	.000	.000	.000
13	1.000	1.000	1.000	1.000	.994	.922	.655	.268	.044	.002	.000	.000	.000
14	1.000	1.000	1.000	1.000	.998	.966	.788	.414	.098	.006	.000	.000	.000
15	1.000	1.000	1.000	1.000	1.000	.987	.885	.575	.189	.017	.000	.000	.000
16	1.000	1.000	1.000	1.000	1.000	.996	.946	.726	.323	.047	.000	.000	.000
17	1.000	1.000	1.000	1.000	1.000	.999	.978	.846	.488	.109	.002	.000	.000
18	1.000	1.000	1.000	1.000	1.000	1.000	.993	.926	.659	.220	.009	.000	.000
19	1.000	1.000	1.000	1.000	1.000	1.000	.998	.971	.807	.383	.033	.001	.000
20	1.000	1.000	1.000	1.000	1.000	1.000	1.000	.991	.910	.579	.098	.007	.000
21	1.000	1.000	1.000	1.000	1.000	1.000	1.000	.998	.967	.766	.236	.034	.000
22	1.000	1.000	1.000	1.000	1.000	1.000	1.000	1.000	.991	.902	.463	.127	.002
23	1.000	1.000	1.000	1.000	1.000	1.000	1.000	1.000	.998	.973	.729	.358	.026
24	1.000	1.000	1.000	1.000	1.000	1.000	1.000	1.000	1.000	.996	.928	.723	.222

TABLE B.1 | *(concluded)*

Part II: Cumulative probabilities

n = 50

a \ π	.010	.050	.100	.200	.300	.400	.500	.600	.700	.800	.900	.950	.990
0	.605	.077	.005	.000	.000	.000	.000	.000	.000	.000	.000	.000	.000
1	.911	.279	.034	.000	.000	.000	.000	.000	.000	.000	.000	.000	.000
2	.986	.541	.112	.001	.000	.000	.000	.000	.000	.000	.000	.000	.000
3	.998	.760	.250	.006	.000	.000	.000	.000	.000	.000	.000	.000	.000
4	1.000	.896	.431	.018	.000	.000	.000	.000	.000	.000	.000	.000	.000
5	1.000	.962	.616	.048	.001	.000	.000	.000	.000	.000	.000	.000	.000
6	1.000	.988	.770	.103	.002	.000	.000	.000	.000	.000	.000	.000	.000
7	1.000	.997	.878	.190	.007	.000	.000	.000	.000	.000	.000	.000	.000
8	1.000	.999	.942	.307	.018	.000	.000	.000	.000	.000	.000	.000	.000
9	1.000	1.000	.975	.444	.040	.001	.000	.000	.000	.000	.000	.000	.000
10	1.000	1.000	.991	.584	.079	.002	.000	.000	.000	.000	.000	.000	.000
11	1.000	1.000	.997	.711	.139	.006	.000	.000	.000	.000	.000	.000	.000
12	1.000	1.000	.999	.814	.223	.013	.000	.000	.000	.000	.000	.000	.000
13	1.000	1.000	1.000	.889	.328	.028	.000	.000	.000	.000	.000	.000	.000
14	1.000	1.000	1.000	.939	.447	.054	.001	.000	.000	.000	.000	.000	.000
15	1.000	1.000	1.000	.969	.569	.096	.003	.000	.000	.000	.000	.000	.000
16	1.000	1.000	1.000	.986	.684	.156	.008	.000	.000	.000	.000	.000	.000
17	1.000	1.000	1.000	.994	.782	.237	.016	.000	.000	.000	.000	.000	.000
18	1.000	1.000	1.000	.997	.859	.336	.032	.001	.000	.000	.000	.000	.000
19	1.000	1.000	1.000	.999	.915	.446	.059	.002	.000	.000	.000	.000	.000
20	1.000	1.000	1.000	1.000	.952	.561	.101	.004	.000	.000	.000	.000	.000
21	1.000	1.000	1.000	1.000	.975	.670	.161	.008	.000	.000	.000	.000	.000
22	1.000	1.000	1.000	1.000	.988	.766	.240	.016	.000	.000	.000	.000	.000
23	1.000	1.000	1.000	1.000	.994	.844	.336	.031	.000	.000	.000	.000	.000
24	1.000	1.000	1.000	1.000	.998	.902	.444	.057	.001	.000	.000	.000	.000

n = 50 (continued)

a \ π	.010	.050	.100	.200	.300	.400	.500	.600	.700	.800	.900	.950	.990
25	1.000	1.000	1.000	1.000	.999	.943	.556	.098	.002	.000	.000	.000	.000
26	1.000	1.000	1.000	1.000	1.000	.969	.664	.156	.006	.000	.000	.000	.000
27	1.000	1.000	1.000	1.000	1.000	.984	.760	.234	.012	.000	.000	.000	.000
28	1.000	1.000	1.000	1.000	1.000	.992	.839	.330	.025	.000	.000	.000	.000
29	1.000	1.000	1.000	1.000	1.000	.997	.899	.439	.048	.000	.000	.000	.000
30	1.000	1.000	1.000	1.000	1.000	.999	.941	.554	.085	.001	.000	.000	.000
31	1.000	1.000	1.000	1.000	1.000	.999	.968	.664	.141	.003	.000	.000	.000
32	1.000	1.000	1.000	1.000	1.000	1.000	.984	.763	.218	.006	.000	.000	.000
33	1.000	1.000	1.000	1.000	1.000	1.000	.992	.844	.316	.014	.000	.000	.000
34	1.000	1.000	1.000	1.000	1.000	1.000	.997	.904	.431	.031	.000	.000	.000
35	1.000	1.000	1.000	1.000	1.000	1.000	.999	.946	.553	.061	.000	.000	.000
36	1.000	1.000	1.000	1.000	1.000	1.000	1.000	.972	.672	.111	.000	.000	.000
37	1.000	1.000	1.000	1.000	1.000	1.000	1.000	.987	.777	.186	.001	.000	.000
38	1.000	1.000	1.000	1.000	1.000	1.000	1.000	.994	.861	.289	.003	.000	.000
39	1.000	1.000	1.000	1.000	1.000	1.000	1.000	.998	.921	.416	.009	.000	.000
40	1.000	1.000	1.000	1.000	1.000	1.000	1.000	.999	.960	.556	.025	.000	.000
41	1.000	1.000	1.000	1.000	1.000	1.000	1.000	1.000	.982	.693	.058	.001	.000
42	1.000	1.000	1.000	1.000	1.000	1.000	1.000	1.000	.993	.810	.122	.003	.000
43	1.000	1.000	1.000	1.000	1.000	1.000	1.000	1.000	.998	.897	.230	.012	.000
44	1.000	1.000	1.000	1.000	1.000	1.000	1.000	1.000	.999	.952	.384	.038	.000
45	1.000	1.000	1.000	1.000	1.000	1.000	1.000	1.000	1.000	.982	.569	.104	.000
46	1.000	1.000	1.000	1.000	1.000	1.000	1.000	1.000	1.000	.994	.750	.240	.002
47	1.000	1.000	1.000	1.000	1.000	1.000	1.000	1.000	1.000	.999	.888	.459	.014
48	1.000	1.000	1.000	1.000	1.000	1.000	1.000	1.000	1.000	1.000	.966	.721	.089
49	1.000	1.000	1.000	1.000	1.000	1.000	1.000	1.000	1.000	1.000	.995	.923	.395

TABLE B.2 | Poisson probabilities

X	λ									
	0.1	0.2	0.3	0.4	0.5	0.6	0.7	0.8	0.9	1.0
0	.9048	.8187	.7408	.6703	.6065	.5488	.4966	.4493	.4066	.3679
1	.0905	.1637	.2222	.2681	.3033	.3293	.3476	.3595	.3659	.3679
2	.0045	.0164	.0333	.0536	.0758	.0988	.1217	.1438	.1647	.1839
3	.0002	.0011	.0033	.0072	.0126	.0198	.0284	.0383	.0494	.0613
4		.0001	.0002	.0007	.0016	.0030	.0050	.0077	.0111	.0153
5				.0001	.0002	.0004	.0007	.0012	.0020	.0031
6							.0001	.0002	.0003	.0005
7										.0001

X	λ									
	1.5	2.0	2.5	3.0	3.5	4.0	4.5	5.0	6.0	7.0
0	.2231	.1353	.0821	.0498	.0302	.0183	.0111	.0067	.0025	.0009
1	.3347	.2707	.2052	.1494	.1057	.0733	.0500	.0337	.0149	.0064
2	.2510	.2707	.2565	.2240	.1850	.1465	.1125	.0842	.0446	.0223
3	.1255	.1804	.2138	.2240	.2158	.1954	.1687	.1404	.0892	.0521
4	.0471	.0902	.1336	.1680	.1888	.1954	.1898	.1755	.1339	.0912
5	.0141	.0361	.0668	.1008	.1322	.1563	.1708	.1755	.1606	.1277
6	.0035	.0120	.0278	.0504	.0771	.1042	.1281	.1462	.1606	.1490
7	.0008	.0034	.0099	.0216	.0385	.0595	.0824	.1044	.1377	.1490
8	.0001	.0009	.0031	.0081	.0169	.0298	.0463	.0653	.1033	.1304
9		.0002	.0009	.0027	.0066	.0132	.0232	.0363	.0688	.1014
10			.0002	.0008	.0023	.0053	.0104	.0181	.0413	.0710
11				.0002	.0007	.0019	.0043	.0082	.0225	.0452
12				.0001	.0002	.0006	.0016	.0034	.0113	.0264
13					.0001	.0002	.0006	.0013	.0052	.0142
14						.0001	.0002	.0005	.0022	.0071
15							.0001	.0002	.0009	.0033
16									.0003	.0014
17									.0001	.0006
18										.0002
19										.0001

X	λ				
	8.0	9.0	10.0	15.0	20.0
0	0.0003	0.0001	0.0000	0.0000	0.0000
1	0.0027	0.0011	0.0005	0.0000	0.0000
2	0.0107	0.0050	0.0023	0.0000	0.0000
3	0.0286	0.0150	0.0076	0.0002	0.0000
4	0.0573	0.0337	0.0189	0.0006	0.0000
5	0.0916	0.0607	0.0378	0.0019	0.0001
6	0.1221	0.0901	0.0631	0.0048	0.0002
7	0.1396	0.1171	0.0901	0.0104	0.0005
8	0.1396	0.1318	0.1126	0.0194	0.0013
9	0.1241	0.1318	0.1251	0.0324	0.0029
10	0.0993	0.1186	0.1251	0.0486	0.0058
11	0.0772	0.0970	0.1137	0.0663	0.0106
12	0.0481	0.0728	0.0948	0.0829	0.0176
13	0.0296	0.0504	0.0729	0.0956	0.0271
14	0.0169	0.0324	0.0521	0.1024	0.0387
15	0.0090	0.0194	0.0347	0.1024	0.0516
16	0.0045	0.0109	0.0217	0.0960	0.0646
17	0.0021	0.0058	0.0128	0.0847	0.0760
18	0.0009	0.0029	0.0071	0.0706	0.0844
19	0.0004	0.0014	0.0037	0.0557	0.0888
20	0.0002	0.0006	0.0019	0.0418	0.0888
21	0.0001	0.0003	0.0009	0.0299	0.0846
22		0.0001	0.0004	0.0204	0.0769
23			0.0002	0.0133	0.0669
24			0.0001	0.0833	0.0557
25				0.0050	0.0446
26				0.0029	0.0343
27				0.0016	0.0254
28				0.0009	0.0181
29				0.0004	0.0125
30				0.0002	0.0083
31				0.0001	0.0054
32				0.0001	0.0034
33					0.0020
34					0.0012
35					0.0007
36					0.0004
37					0.0002
38					0.0001
39					0.0001

Table entries are $P(\mathbf{X} = x/\lambda)$.

Source: John Neter, William Wasserman, and G. A. Whitmore, *Fundamental Statistics for Business and Economics,* 4th ed. (Boston: Allyn & Bacon, Inc., 1972 ©).

TABLE B.3 | Standard normal distribution areas

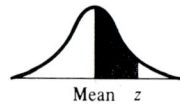

Mean z

z	.00	.01	.02	.03	.04	.05	.06	.07	.08	.09
0.0	.0000	.0040	.0080	.0120	.0160	.0199	.0239	.0279	.0319	.0359
0.1	.0398	.0438	.0478	.0517	.0557	.0596	.0636	.0675	.0714	.0753
0.2	.0793	.0832	.0871	.0910	.0948	.0987	.1026	.1064	.1103	.1141
0.3	.1179	.1217	.1255	.1293	.1331	.1368	.1406	.1443	.1480	.1517
0.4	.1554	.1591	.1628	.1664	.1700	.1736	.1772	.1808	.1844	.1879
0.5	.1915	.1950	.1985	.2019	.2054	.2088	.2123	.2157	.2190	.2224
0.6	.2257	.2291	.2324	.2357	.2389	.2422	.2454	.2486	.2518	.2549
0.7	.2580	.2612	.2642	.2673	.2704	.2734	.2764	.2794	.2823	.2852
0.8	.2881	.2910	.2939	.2967	.2995	.3023	.3051	.3078	.3106	.3133
0.9	.3159	.3186	.3212	.3238	.3264	.3289	.3315	.3340	.3365	.3389
1.0	.3413	.3438	.3461	.3485	.3508	.3531	.3554	.3577	.3599	.3621
1.1	.3643	.3665	.3686	.3708	.3729	.3749	.3770	.3790	.3810	.3830
1.2	.3849	.3869	.3888	.3907	3925	.3944	.3962	.3980	.3997	.4015
1.3	.4032	.4049	.4066	.4082	.4099	.4115	.4131	.4147	.4162	.4177
1.4	.4192	.4207	.4222	.4236	.4251	.4265	.4279	.4292	.4306	.4319
1.5	.4332	.4345	.4357	.4370	.4382	.4394	.4406	.4418	.4429	.4441
1.6	.4452	.4463	.4474	.4484	.4495	.4505	.4515	.4525	.4535	.4545
1.7	.4554	.4564	.4573	.4582	.4591	.4599	.4608	.4616	.4625	.4633
1.8	.4641	.4649	.4656	.4664	.4671	.4678	.4686	.4693	.4699	.4706
1.9	.4713	.4719	.4726	.4732	.4738	.4744	.4750	.4756	.4761	.4767
2.0	.4772	.4778	.4783	.4788	.4793	.4798	.4803	.4808	.4812	.4817
2.1	.4821	.4826	.4830	.4834	.4838	.4842	.4846	.4850	.4854	.4857
2.2	.4861	.4864	.4868	.4871	.4875	.4878	.4881	.4884	.4887	.4890
2.3	.4893	.4896	.4898	.1901	.4904	.4906	.4909	.4911	.4913	.4916
2.4	.4918	.4920	.4922	.4925	.4927	.4929	.4931	.4932	.4934	.4936
2.5	.4938	.4940	.4941	.4943	.4945	.4946	.4948	.4949	.4951	.4952
2.6	.4953	.4955	.4956	.4957	.4959	.4960	.4961	.4962	.4963	.4964
2.7	.4965	.4966	.4967	.4968	.4969	.4970	.4971	.4972	.4973	.4974
2.8	.4974	.4975	.4976	.4977	.4977	.4978	.4979	.4979	.4980	.4981
2.9	.4981	.4982	.4982	.4983	.4984	.4984	.4985	.4985	.4986	.4986
3.0	.49865	.4987	.4987	.4988	.4988	.4989	.4989	.4989	.4990	.4990
4.0	.4999683									

Source: John Neter, William Wasserman, and G. A. Whitmore, *Fundamental Statistics for Business and Economics,* 4th ed. (Boston: Allyn & Bacon, Inc., 1972 ©).

x	e^{-x}	x	e^{-x}	x	e^{-x}	x	e^{-x}
0.0	1.000000	2.35	0.095371	4.70	0.009096	7.35	0.000643
0.05	0.951229	2.40	0.090720	4.75	0.008652	7.40	0.000611
0.10	0.904837	2.45	0.086296	4.80	0.008230	7.45	0.000581
0.15	0.860708	2.50	0.082087	4.85	0.007829	7.50	0.000553
0.20	0.818731	2.55	0.078084	4.90	0.007447	7.55	0.000526
0.25	0.778801	2.60	0.074275	4.95	0.007084	7.60	0.000501
0.30	0.740818	2.65	0.070653	5.00	0.006738	7.65	0.000476
0.35	0.704688	2.70	0.067207	5.05	0.006410	7.70	0.000453
0.40	0.670320	2.75	0.063930	5.10	0.006097	7.75	0.000431
0.45	0.637628	2.80	0.060812	5.15	0.005800	7.80	0.000410
0.50	0.606531	2.85	0.057846	5.20	0.005517	7.85	0.000390
0.55	0.576950	2.90	0.055025	5.25	0.005248	7.90	0.000371
0.60	0.548812	2.95	0.052341	5.30	0.004992	7.95	0.000353
0.65	0.522046	3.00	0.049789	5.35	0.004748	8.00	0.000335
0.70	0.496586	3.05	0.047360	5.40	0.004517	8.05	0.000319
0.75	0.472367	3.10	0.045051	5.45	0.004297	8.10	0.000304
0.80	0.449329	3.15	0.042854	5.50	0.004087	8.15	0.000289
0.85	0.427415	3.20	0.040764	5.55	0.003888	8.20	0.000275
0.90	0.406570	3.25	0.038776	5.60	0.003698	8.25	0.000261
0.95	0.386741	3.30	0.036884	5.65	0.003518	8.30	0.000249
1.00	0.367880	3.35	0.035086	5.70	0.003346	8.35	0.000236
1.05	0.349938	3.40	0.033375	5.75	0.003183	8.40	0.000225
1.10	0.332872	3.45	0.031747	5.80	0.003028	8.45	0.000214
1.15	0.316638	3.50	0.030199	5.85	0.002880	8.50	0.000203
1.20	0.301195	3.55	0.028726	5.90	0.002740	8.55	0.000194
1.25	0.286506	3.60	0.027325	5.95	0.002606	8.60	0.000184
1.30	0.272533	3.65	0.025992	6.00	0.002479	8.65	0.000175
1.35	0.259242	3.70	0.024725	6.05	0.002358	8.70	0.000167
1.40	0.246599	3.75	0.023519	6.10	0.002243	8.75	0.000158
1.45	0.234572	3.80	0.022372	6.15	0.002134	8.80	0.000151
1.50	0.223132	3.85	0.021281	6.20	0.002030	8.85	0.000143
1.55	0.212250	3.90	0.020243	6.25	0.001931	8.90	0.000136
1.60	0.201899	3.95	0.019256	6.30	0.001836	8.95	0.000130
1.65	0.192052	4.00	0.018316	6.35	0.001747	9.00	0.000123
1.70	0.182686	4.05	0.017423	6.40	0.001662	9.05	0.000117
1.75	0.173776	4.10	0.016573	6.45	0.001581	9.10	0.000112
1.80	0.165301	4.15	0.015765	6.50	0.001504	9.15	0.000106
1.85	0.157239	4.20	0.014996	6.55	0.001430	9.20	0.000101
1.90	0.149571	4.25	0.014265	6.60	0.001360	9.25	0.000096
1.95	0.142276	4.30	0.013569	6.65	0.001294	9.30	0.000091
2.00	0.135337	4.35	0.012907	6.70	0.001231	9.35	0.000087
2.05	0.128737	4.40	0.012278	6.75	0.001171	9.40	0.000083
2.10	0.122459	4.45	0.011679	6.80	0.001114	9.45	0.000079
2.15	0.116486	4.50	0.011110	6.85	0.001060	9.50	0.000075
2.20	0.110805	4.55	0.010568	6.90	0.001008	9.55	0.000071
2.25	0.105401	4.60	0.010052	6.95	0.000959	9.60	0.000068
2.30	0.100261	4.65	0.009562	7.00	0.000912	9.65	0.000064
				7.05	0.000867	9.70	0.000061
				7.10	0.000825	9.75	0.000058
				7.15	0.000785	9.80	0.000055
				7.20	0.000747	9.85	0.000053
				7.25	0.000710	9.90	0.000050
				7.30	0.000676	9.95	0.000048
						10.00	0.000045

TABLE B.5 | The *t*-distribution

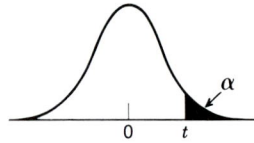

α d.f.	0.40	0.30	0.20	0.10	0.05	0.025	0.01	0.005	0.001	df
1	0.325	0.727	1.376	3.078	6.314	12.706	31.821	63.657	318.309	1
2	0.289	0.617	1.061	1.886	2.920	4.303	6.965	9.925	22.327	2
3	0.277	0.584	0.978	1.638	2.353	3.182	4.541	5.841	10.215	3
4	0.271	0.569	0.941	1.533	2.132	2.776	3.747	4.604	7.173	4
5	0.267	0.559	0.920	1.476	2.015	2.571	3.365	4.032	5.893	5
6	0.265	0.553	0.906	1.440	1.943	2.447	3.143	3.707	5.208	6
7	0.263	0.549	0.896	1.415	1.895	2.365	2.998	3.499	4.785	7
8	0.262	0.546	0.889	1.397	1.860	2.306	2.896	3.355	4.501	8
9	0.261	0.543	0.883	1.383	1.833	2.262	2.821	3.250	4.297	9
10	0.260	0.542	0.879	1.372	1.812	2.228	2.764	3.169	4.144	10
11	0.260	0.540	0.876	1.363	1.796	2.201	2.718	3.106	4.025	11
12	0.259	0.539	0.873	1.356	1.782	2.179	2.681	3.055	3.930	12
13	0.259	0.538	0.870	1.350	1.771	2.160	2.650	3.012	3.852	13
14	0.258	0.537	0.868	1.345	1.761	2.145	2.624	2.977	3.787	14
15	0.258	0.536	0.866	1.341	1.753	2.131	2.602	2.947	3.733	15
16	0.258	0.535	0.865	1.337	1.746	2.120	2.583	2.921	3.686	16
17	0.257	0.534	0.863	1.333	1.740	2.110	2.567	2.898	3.646	17
18	0.257	0.534	0.862	1.330	1.734	2.101	2.552	2.878	3.610	18
19	0.257	0.533	0.861	1.328	1.729	2.093	2.539	2.861	3.579	19
20	0.257	0.533	0.860	1.325	1.725	2.086	2.528	2.845	3.552	20
21	0.257	0.532	0.859	1.323	1.721	2.080	2.518	2.831	3.527	21
22	0.256	0.532	0.858	1.321	1.717	2.074	2.508	2.819	3.505	22
23	0.256	0.532	0.858	1.319	1.714	2.069	2.500	2.807	3.485	23
24	0.256	0.531	0.857	1.318	1.711	2.064	2.492	2.797	3.467	24
25	0.256	0.531	0.856	1.316	1.708	2.060	2.485	2.787	3.450	25
26	0.256	0.531	0.856	1.315	1.706	2.056	2.479	2.779	3.435	26
27	0.256	0.531	0.855	1.314	1.703	2.052	2.473	2.771	3.421	27
28	0.256	0.530	0.855	1.313	1.701	2.048	2.467	2.763	3.408	28
29	0.256	0.530	0.854	1.311	1.699	2.045	2.462	2.756	3.396	29
30	0.256	0.530	0.854	1.310	1.697	2.042	2.457	2.750	3.385	30
31	0.256	0.530	0.853	1.309	1.696	2.040	2.453	2.744	3.375	31
32	0.255	0.530	0.853	1.309	1.694	2.037	2.449	2.738	3.365	32
33	0.255	0.530	0.853	1.308	1.692	2.035	2.445	2.733	3.356	33
34	0.255	0.529	0.852	1.307	1.691	2.032	2.441	2.728	3.348	34
35	0.255	0.529	0.852	1.306	1.690	2.030	2.438	2.724	3.340	35

TABLE B.5 | *(continued)*

α d.f.	0.40	0.30	0.20	0.10	0.05	0.025	0.01	0.005	0.001	df
36	0.255	0.529	0.852	1.306	1.688	2.028	2.434	2.719	3.333	36
37	0.255	0.529	0.851	1.305	1.687	2.026	2.431	2.715	3.326	37
38	0.255	0.529	0.851	1.304	1.686	2.024	2.429	2.712	3.319	38
39	0.255	0.529	0.851	1.304	1.685	2.023	2.426	2.708	3.313	39
40	0.255	0.529	0.851	1.303	1.684	2.021	2.423	2.704	3.307	40
41	0.255	0.529	0.850	1.303	1.683	2.020	2.421	2.701	3.301	41
42	0.255	0.528	0.850	1.302	1.682	2.018	2.418	2.698	3.296	42
43	0.255	0.528	0.850	1.302	1.681	2.017	2.416	2.695	3.291	43
44	0.255	0.528	0.850	1.301	1.680	2.015	2.414	2.692	3.286	44
45	0.255	0.528	0.850	1.301	1.679	2.014	2.412	2.690	3.281	45
46	0.255	0.528	0.850	1.300	1.679	2.013	2.410	2.687	3.277	46
47	0.255	0.528	0.849	1.300	1.678	2.012	2.408	2.685	3.273	47
48	0.255	0.528	0.849	1.299	1.677	2.011	2.407	2.682	3.269	48
49	0.255	0.528	0.849	1.299	1.677	2.010	2.405	2.680	3.265	49
50	0.255	0.528	0.849	1.299	1.676	2.009	2.403	2.678	3.261	50
51	0.255	0.528	0.849	1.298	1.675	2.008	2.402	2.676	3.258	51
52	0.255	0.528	0.849	1.298	1.675	2.007	2.400	2.674	3.255	52
53	0.255	0.528	0.848	1.298	1.674	2.006	2.399	2.672	3.251	53
54	0.255	0.528	0.848	1.297	1.674	2.005	2.397	2.670	3.248	54
55	0.255	0.527	0.848	1.297	1.673	2.004	2.396	2.668	3.245	55
56	0.255	0.527	0.848	1.297	1.673	2.003	2.395	2.667	3.242	56
57	0.255	0.527	0.848	1.297	1.672	2.002	2.394	2.665	3.239	57
58	0.255	0.527	0.848	1.296	1.672	2.002	2.392	2.663	3.237	58
59	0.254	0.527	0.848	1.296	1.671	2.001	2.391	2.662	3.234	59
60	0.254	0.527	0.848	1.296	1.671	2.000	2.390	2.660	3.232	60
61	0.254	0.527	0.848	1.296	1.670	2.000	2.389	2.659	3.229	61
62	0.254	0.527	0.847	1.295	1.670	1.999	2.388	2.657	3.227	62
63	0.254	0.527	0.847	1.295	1.669	1.998	2.387	2.656	3.225	63
64	0.254	0.527	0.847	1.295	1.669	1.998	2.386	2.655	3.223	64
65	0.254	0.527	0.847	1.295	1.669	1.997	2.385	2.654	3.220	65
66	0.254	0.527	0.847	1.295	1.668	1.997	2.384	2.652	3.218	66
67	0.254	0.527	0.847	1.294	1.668	1.996	2.383	2.651	3.216	67
68	0.254	0.527	0.847	1.294	1.668	1.995	2.382	2.650	3.214	68
69	0.254	0.527	0.847	1.294	1.667	1.995	2.382	2.649	3.213	69
70	0.254	0.527	0.847	1.294	1.667	1.994	2.381	2.648	3.211	70
71	0.254	0.527	0.847	1.294	1.667	1.994	2.380	2.647	3.209	71
72	0.254	0.527	0.847	1.293	1.666	1.993	2.379	2.646	3.207	72
73	0.254	0.527	0.847	1.293	1.666	1.993	2.379	2.645	3.206	73
74	0.254	0.527	0.847	1.293	1.666	1.993	2.378	2.644	3.204	74
75	0.254	0.527	0.846	1.293	1.665	1.992	2.377	2.643	3.202	75

TABLE B.5 | *(concluded)*

d.f.	0.40	0.30	0.20	0.10	0.05	0.025	0.01	0.005	0.001	df
76	0.254	0.527	0.846	1.293	1.665	1.992	2.376	2.642	3.201	76
77	0.254	0.527	0.846	1.293	1.665	1.991	2.376	2.641	3.199	77
78	0.254	0.527	0.846	1.292	1.665	1.991	2.375	2.640	3.198	78
79	0.254	0.527	0.846	1.292	1.664	1.990	2.374	2.640	3.197	79
80	0.254	0.526	0.846	1.292	1.664	1.990	2.374	2.639	3.195	80
81	0.254	0.526	0.846	1.292	1.664	1.990	2.373	2.638	3.194	81
82	0.254	0.526	0.846	1.292	1.664	1.989	2.373	2.637	3.193	82
83	0.254	0.526	0.846	1.292	1.663	1.989	2.372	2.636	3.191	83
84	0.254	0.526	0.846	1.292	1.663	1.989	2.372	2.636	3.190	84
85	0.254	0.526	0.846	1.292	1.663	1.988	2.371	2.635	3.189	85
86	0.254	0.526	0.846	1.291	1.663	1.988	2.370	2.634	3.188	86
87	0.254	0.526	0.846	1.291	1.663	1.988	2.370	2.634	3.187	87
88	0.254	0.526	0.846	1.291	1.662	1.987	2.369	2.633	3.185	88
89	0.254	0.526	0.846	1.291	1.662	1.987	2.369	2.632	3.184	89
90	0.254	0.526	0.846	1.291	1.662	1.987	2.368	2.632	3.183	90
91	0.254	0.526	0.846	1.291	1.662	1.986	2.368	2.631	3.182	91
92	0.254	0.526	0.846	1.291	1.662	1.986	2.368	2.630	3.181	92
93	0.254	0.526	0.846	1.291	1.661	1.986	2.367	2.630	3.180	93
94	0.254	0.526	0.845	1.291	1.661	1.986	2.367	2.629	3.179	94
95	0.254	0.526	0.845	1.291	1.661	1.985	2.366	2.629	3.178	95
96	0.254	0.526	0.845	1.290	1.661	1.985	2.366	2.628	3.177	96
97	0.254	0.526	0.845	1.290	1.661	1.985	2.365	2.627	3.176	97
98	0.254	0.526	0.845	1.290	1.661	1.984	2.365	2.627	3.175	98
99	0.254	0.526	0.845	1.290	1.660	1.984	2.365	2.626	3.175	99
100	0.254	0.526	0.845	1.290	1.660	1.984	2.364	2.626	3.174	100
∞	0.253	0.524	0.842	1.282	1.645	1.960	2.326	2.576	3.090	∞

TABLE B.6 | The χ^2-distribution

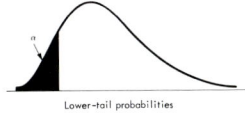

Lower-tail probabilities

df \ α	.001	.005	.010	.025	.050	.100
1	.000	.000	.000	.001	.004	.016
2	.002	.010	.020	.051	.103	.211
3	.024	.072	.115	.216	.352	.584
4	.091	.207	.297	.484	.711	1.06
5	.210	.412	.554	.831	1.15	1.61
6	.381	.676	.872	1.24	1.64	2.20
7	.598	.989	1.24	1.69	2.17	2.83
8	.857	1.34	1.65	2.18	2.73	3.49
9	1.15	1.73	2.09	2.70	3.33	4.17
10	1.48	2.16	2.56	3.25	3.94	4.87
11	1.83	2.60	3.05	3.82	4.57	5.58
12	2.21	3.07	3.57	4.40	5.23	6.30
13	2.62	3.57	4.11	5.01	5.89	7.04
14	3.04	4.07	4.66	5.63	6.57	7.79
15	3.48	4.60	5.23	6.26	7.26	8.55
16	3.94	5.14	5.81	6.91	7.96	9.31
17	4.42	5.70	6.41	7.56	8.67	10.1
18	4.90	6.26	7.01	8.23	9.39	10.9
19	5.41	6.84	7.63	8.91	10.1	11.7
20	5.92	7.43	8.26	9.59	10.9	12.4
21	6.45	8.03	8.90	10.3	11.6	13.2
22	6.98	8.64	9.54	11.0	12.3	14.0
23	7.53	9.26	10.2	11.7	13.1	14.8
24	8.08	9.89	10.9	12.4	13.8	15.7
25	8.65	10.5	11.5	13.1	14.6	16.5
26	9.22	11.2	12.2	13.8	15.4	17.3
27	9.80	11.8	12.9	14.6	16.2	18.1
28	10.4	12.5	13.6	15.3	16.9	18.9
29	11.0	13.1	14.3	16.0	17.7	19.8
30	11.6	13.8	15.0	16.8	18.5	20.6
35	14.7	17.2	18.5	20.6	22.5	24.8
40	17.9	20.7	22.2	24.4	26.5	29.1
45	21.3	24.3	25.9	28.4	30.6	33.4
50	24.7	28.0	29.7	32.4	34.8	37.7
55	28.2	31.7	33.6	36.4	39.0	42.1
60	31.7	35.5	37.5	40.5	43.2	46.5
65	35.4	39.4	41.4	44.6	47.4	50.9
70	39.0	43.3	45.4	48.8	51.7	55.3
75	42.8	47.2	49.5	52.9	56.1	59.8
80	46.5	51.2	53.5	57.2	60.4	64.3
85	50.3	55.2	57.6	61.4	64.7	68.8
90	54.2	59.2	61.8	65.6	69.1	73.3
95	58.0	63.2	65.9	69.9	73.5	77.8
100	61.9	67.3	70.1	74.2	77.9	82.4

TABLE B.6 | *(concluded)*

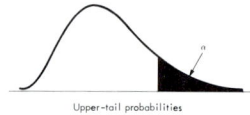

Upper-tail probabilities

df \ α	.100	.050	.025	.010	.005	.001
1	2.71	3.84	5.02	6.63	7.88	10.8
2	4.61	5.99	7.38	9.21	10.6	13.8
3	6.25	7.81	9.35	11.3	12.8	16.3
4	7.78	9.49	11.1	13.3	14.9	18.5
5	9.24	11.1	12.8	15.1	16.7	20.5
6	10.6	12.6	14.4	16.8	18.5	22.5
7	12.0	14.1	16.0	18.5	20.3	24.3
8	13.4	15.5	17.5	20.1	22.0	26.1
9	14.7	16.9	19.0	21.7	23.6	27.9
10	16.0	18.3	20.5	23.2	25.2	29.6
11	17.3	19.7	21.9	24.7	26.8	31.3
12	18.5	21.0	23.3	26.2	28.3	32.9
13	19.8	22.4	24.7	27.7	29.8	34.5
14	21.1	23.7	26.1	29.1	31.3	36.1
15	22.3	25.0	27.5	30.6	32.8	37.7
16	23.5	26.3	28.8	32.0	34.3	39.3
17	24.8	27.6	30.2	33.4	35.7	40.8
18	26.0	28.9	31.5	34.8	37.2	42.3
19	27.2	30.1	32.9	36.2	38.6	43.8
20	28.4	31.4	34.2	37.6	40.0	45.3
21	29.6	32.7	35.5	38.9	41.4	46.8
22	30.8	33.9	36.8	40.3	42.8	48.3
23	32.0	35.2	38.1	41.6	44.2	49.7
24	33.2	36.4	39.4	43.0	45.6	51.2
25	34.4	37.7	40.6	44.3	46.9	52.6
26	35.6	38.9	41.9	45.6	48.3	54.1
27	36.7	40.1	43.2	47.0	49.6	55.5
28	37.9	41.3	44.5	48.3	51.0	56.9
29	39.1	42.6	45.7	49.6	52.3	58.3
30	40.3	43.8	47.0	50.9	53.7	59.7
35	46.1	49.8	53.2	57.3	60.3	66.6
40	51.8	55.8	59.3	63.7	66.8	73.4
45	57.5	61.7	65.4	70.0	73.2	80.1
50	63.2	67.5	71.4	76.2	79.5	86.7
55	68.8	73.3	77.4	82.3	85.7	93.2
60	74.4	79.1	83.3	88.4	92.0	99.6
65	80.0	84.8	89.2	94.4	98.1	106.0
70	85.5	90.5	95.0	100.4	104.2	112.3
75	91.1	96.2	100.8	106.4	110.3	118.6
80	96.6	101.9	106.6	112.3	116.3	124.8
85	102.1	107.5	112.4	118.2	122.3	131.0
90	107.6	113.1	118.1	124.1	128.3	137.2
95	113.0	118.8	123.9	130.0	134.2	143.3
100	118.5	124.3	129.6	135.8	140.2	149.4

TABLE   B.7 | The *F*-distribution

Percentiles of the *F*-distribution: entry is $F(1 - \alpha; r_1, r_2)$ where $P[\mathbf{F} \leq F(1 - \alpha; r_1, r_2)] = 1 - \alpha$.

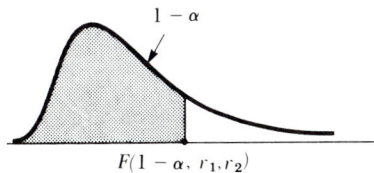

$F(1 - \alpha, r_1, r_2)$

TABLE B.7 | *(continued)*

r_2	$1-\alpha$	1	2	3	4	5	6	7	8	9
1	.50	1.00	1.50	1.71	1.82	1.89	1.94	1.98	2.00	2.03
	.90	39.9	49.5	53.6	55.8	57.2	58.2	58.9	59.4	59.9
	.95	161	200	216	225	230	234	237	239	241
	.975	648	800	864	900	922	937	948	957	963
	.99	4,052	5,000	5,403	5,625	5,764	5,859	5,928	5,981	6,022
	.995	16,211	20,000	21,615	22,500	23,056	23,437	23,715	23,925	24,091
	.999	405,280	500,000	540,380	562,500	576,400	585,940	592,870	598,140	602,280
2	.50	0.667	1.00	1.13	1.21	1.25	1.28	1.30	1.32	1.33
	.90	8.53	9.00	9.16	9.24	9.29	9.33	9.35	9.37	9.38
	.95	18.5	19.0	19.2	19.2	19.3	19.3	19.4	19.4	19.4
	.975	38.5	39.0	39.2	39.2	39.3	39.3	39.4	39.4	39.4
	.99	98.5	99.0	99.2	99.2	99.3	99.3	99.4	99.4	99.4
	.995	199	199	199	199	199	199	199	199	199
	.999	998.5	999.0	999.2	999.2	999.3	999.3	999.4	999.4	999.4
3	.50	0.585	0.881	1.00	1.06	1.10	1.13	1.15	1.16	1.17
	.90	5.54	5.46	5.39	5.34	5.31	5.28	5.27	5.25	5.24
	.95	10.1	9.55	9.28	9.12	9.01	8.94	8.89	8.85	8.81
	.975	17.4	16.0	15.4	15.1	14.9	14.7	14.6	14.5	14.5
	.99	34.1	30.8	29.5	28.7	28.2	27.9	27.7	27.5	27.3
	.995	55.6	49.8	47.5	46.2	45.4	44.8	44.4	44.1	43.9
	.999	167.0	148.5	141.1	137.1	134.6	132.8	131.6	130.6	129.9
4	.50	0.549	0.828	0.941	1.00	1.04	1.06	1.08	1.09	1.10
	.90	4.54	4.32	4.19	4.11	4.05	4.01	3.98	3.95	3.94
	.95	7.71	6.94	6.59	6.39	6.26	6.16	6.09	6.04	6.00
	.975	12.2	10.6	9.98	9.60	9.36	9.20	9.07	8.98	8.90
	.99	21.2	18.0	16.7	16.0	15.5	15.2	15.0	14.8	14.7
	.995	31.3	26.3	24.3	23.2	22.5	22.0	21.6	21.4	21.1
	.999	74.1	61.2	56.2	53.4	51.7	50.5	49.7	49.0	48.5
5	.50	0.528	0.799	0.907	0.965	1.00	1.02	1.04	1.05	1.06
	.90	4.06	3.78	3.62	3.52	3.45	3.40	3.37	3.34	3.32
	.95	6.61	5.79	5.41	5.19	5.05	4.95	4.88	4.82	4.77
	.975	10.0	8.43	7.76	7.39	7.15	6.98	6.85	6.76	6.68
	.99	16.3	13.3	12.1	11.4	11.0	10.7	10.5	10.3	10.2
	.995	22.8	18.3	16.5	15.6	14.9	14.5	14.2	14.0	13.8
	.999	47.2	37.1	33.2	31.1	29.8	28.8	28.2	27.6	27.2
6	.50	0.515	0.780	0.886	0.942	0.977	1.00	1.02	1.03	1.04
	.90	3.78	3.46	3.29	3.18	3.11	3.05	3.01	2.98	2.96
	.95	5.99	5.14	4.76	4.53	4.39	4.28	4.21	4.15	4.10
	.975	8.81	7.26	6.60	6.23	5.99	5.82	5.70	5.60	5.52
	.99	13.7	10.9	9.78	9.15	8.75	8.47	8.26	8.10	7.98
	.995	18.6	14.5	12.9	12.0	11.5	11.1	10.8	10.6	10.4
	.999	35.5	27.0	23.7	21.9	20.8	20.0	19.5	19.0	18.7
7	.50	0.506	0.767	0.871	0.926	0.960	0.983	1.00	1.01	1.02
	.90	3.59	3.26	3.07	2.96	2.88	2.83	2.78	2.75	2.72
	.95	5.59	4.74	4.35	4.12	3.97	3.87	3.79	3.73	3.68
	.975	8.07	6.54	5.89	5.52	5.29	5.12	4.99	4.90	4.82
	.99	12.2	9.55	8.45	7.85	7.46	7.19	6.99	6.84	6.72
	.995	16.2	12.4	10.9	10.1	9.52	9.16	8.89	8.68	8.51
	.999	29.2	21.7	18.8	17.2	16.2	15.5	15.0	14.6	14.3

TABLE B.7 | *(continued)*

r_2	$1-\alpha$	r_1								
		10	12	15	20	24	30	60	120	∞
1	.50	2.04	2.07	2.09	2.12	2.13	2.15	2.17	2.18	2.20
	.90	60.2	60.7	61.2	61.7	62.0	62.3	62.8	63.1	63.3
	.95	242	244	246	248	249	250	252	253	254
	.975	969	977	985	993	997	1,001	1,010	1,014	1,018
	.99	6,056	6,106	6,157	6,209	6,235	6,261	6,313	6,339	6,366
	.995	24,224	24,426	24,630	24,836	24,940	25,044	25,253	25,359	25,464
	.999	605,620	610,670	615,760	620,910	623,500	626,100	631,340	633,970	636,620
2	.50	1.34	1.36	1.38	1.39	1.40	1.41	1.43	1.43	1.44
	.90	9.39	9.41	9.42	9.44	9.45	9.46	9.47	9.48	9.49
	.95	19.4	19.4	19.4	19.4	19.5	19.5	19.5	19.5	19.5
	.975	39.4	39.4	39.4	39.4	39.5	39.5	39.5	39.5	39.5
	.99	99.4	99.4	99.4	99.4	99.5	99.5	99.5	99.5	99.5
	.995	199	199	199	199	199	199	199	199	200
	.999	999.4	999.4	999.4	999.4	999.5	999.5	999.5	999.5	999.5
3	.50	1.18	1.20	1.21	1.23	1.23	1.24	1.25	1.26	1.27
	.90	5.23	5.22	5.20	5.18	5.18	5.17	5.15	5.14	5.13
	.95	8.79	8.74	8.70	8.66	8.64	8.62	8.57	8.55	8.53
	.975	14.4	14.3	14.3	14.2	14.1	14.1	14.0	13.9	13.9
	.99	27.2	27.1	26.9	26.7	26.6	26.5	26.3	26.2	26.1
	.995	43.7	43.4	43.1	42.8	42.6	42.5	42.1	42.0	41.8
	.999	129.2	128.3	127.4	126.4	125.9	125.4	124.5	124.0	123.5
4	.50	1.11	1.13	1.14	1.15	1.16	1.16	1.18	1.18	1.19
	.90	3.92	3.90	3.87	3.84	3.83	3.82	3.79	3.78	3.76
	.95	5.96	5.91	5.86	5.80	5.77	5.75	5.69	5.66	5.63
	.975	8.84	8.75	8.66	8.56	8.51	8.46	8.36	8.31	8.26
	.99	14.5	14.4	14.2	14.0	13.9	13.8	13.7	13.6	13.5
	.995	21.0	20.7	20.4	20.2	20.0	19.9	19.6	19.5	19.3
	.999	48.1	47.4	46.8	46.1	45.8	45.4	44.7	44.4	44.1
5	.50	1.07	1.09	1.10	1.11	1.12	1.12	1.14	1.14	1.15
	.90	3.30	3.27	3.24	3.21	3.19	3.17	3.14	3.12	3.11
	.95	4.74	4.68	4.62	4.56	4.53	4.50	4.43	4.40	4.37
	.975	6.62	6.52	6.43	6.33	6.28	6.23	6.12	6.07	6.02
	.99	10.1	9.89	9.72	9.55	9.47	9.38	9.20	9.11	9.02
	.995	13.6	13.4	13.1	12.9	12.8	12.7	12.4	12.3	12.1
	.999	26.9	26.4	25.9	25.4	25.1	24.9	24.3	24.1	23.8
6	.50	1.05	1.06	1.07	1.08	1.09	1.10	1.11	1.12	1.12
	.90	2.94	2.90	2.87	2.84	2.82	2.80	2.76	2.74	2.72
	.95	4.06	4.00	3.94	3.87	3.84	3.81	3.74	3.70	3.67
	.975	5.46	5.37	5.27	5.17	5.12	5.07	4.96	4.90	4.85
	.99	7.87	7.72	7.56	7.40	7.31	7.23	7.06	6.97	6.88
	.995	10.2	10.0	9.81	9.59	9.47	9.36	9.12	9.00	8.88
	.999	18.4	18.0	17.6	17.1	16.9	16.7	16.2	16.0	15.7
7	.50	1.03	1.04	1.05	1.07	1.07	1.08	1.09	1.10	1.10
	.90	2.70	2.67	2.63	2.59	2.58	2.56	2.51	2.49	2.47
	.95	3.64	3.57	3.51	3.44	3.41	3.38	3.30	3.27	3.23
	.975	4.76	4.67	4.57	4.47	4.42	4.36	4.25	4.20	4.14
	.99	6.62	6.47	6.31	6.16	6.07	5.99	5.82	5.74	5.65
	.995	8.38	8.18	7.97	7.75	7.65	7.53	7.31	7.19	7.08
	.999	14.1	13.7	13.3	12.9	12.7	12.5	12.1	11.9	11.7

TABLE B.7 | *(continued)*

r_2	$1-\alpha$	1	2	3	4	5	6	7	8	9
8	.50	0.499	0.757	0.860	0.915	0.948	0.971	0.988	1.00	1.01
	.90	3.46	3.11	2.92	2.81	2.73	2.67	2.62	2.59	2.56
	.95	5.32	4.46	4.07	3.84	3.69	3.58	3.50	3.44	3.39
	.975	7.57	6.06	5.42	5.05	4.82	4.65	4.53	4.43	4.36
	.99	11.3	8.65	7.59	7.01	6.63	6.37	6.18	6.03	5.91
	.995	14.7	11.0	9.60	8.81	8.30	7.95	7.69	7.50	7.34
	.999	25.4	18.5	15.8	14.4	13.5	12.9	12.4	12.0	11.8
9	.50	0.494	0.749	0.852	0.906	0.939	0.962	0.978	0.990	1.00
	.90	3.36	3.01	2.81	2.69	2.61	2.55	2.51	2.47	2.44
	.95	5.12	4.26	3.86	3.63	3.48	3.37	3.29	3.23	3.18
	.975	7.21	5.71	5.08	4.72	4.48	4.32	4.20	4.10	4.03
	.99	10.6	8.02	6.99	6.42	6.06	5.80	5.61	5.47	5.35
	.995	13.6	10.1	8.72	7.96	7.47	7.13	6.88	6.69	6.54
	.999	22.9	16.4	13.9	12.6	11.7	11.1	10.7	10.4	10.1
10	.50	0.490	0.743	0.845	0.899	0.932	0.954	0.971	0.983	0.992
	.90	3.29	2.92	2.73	2.61	2.52	2.46	2.41	2.38	2.35
	.95	4.96	4.10	3.71	3.48	3.33	3.22	3.14	3.07	3.02
	.975	6.94	5.46	4.83	4.47	4.24	4.07	3.95	3.85	3.78
	.99	10.0	7.56	6.55	5.99	5.64	5.39	5.20	5.06	4.94
	.995	12.8	9.43	8.08	7.34	6.87	6.54	6.30	6.12	5.97
	.999	21.0	14.9	12.6	11.3	10.5	9.93	9.52	9.20	8.96
12	.50	0.484	0.735	0.835	0.888	0.921	0.943	0.959	0.972	0.981
	.90	3.18	2.81	2.61	2.48	2.39	2.33	2.28	2.24	2.21
	.95	4.75	3.89	3.49	3.26	3.11	3.00	2.91	2.85	2.80
	.975	6.55	5.10	4.47	4.12	3.89	3.73	3.61	3.51	3.44
	.99	9.33	6.93	5.95	5.41	5.06	4.82	4.64	4.50	4.39
	.995	11.8	8.51	7.23	6.52	6.07	5.76	5.52	5.35	5.20
	.999	18.6	13.0	10.8	9.63	8.89	8.38	8.00	7.71	7.48
15	.50	0.478	0.726	0.826	0.878	0.911	0.933	0.949	0.960	0.970
	.90	3.07	2.70	2.49	2.36	2.27	2.21	2.16	2.12	2.09
	.95	4.54	3.68	3.29	3.06	2.90	2.79	2.71	2.64	2.59
	.975	6.20	4.77	4.15	3.80	3.58	3.41	3.29	3.20	3.12
	.99	8.68	6.36	5.42	4.89	4.56	4.32	4.14	4.00	3.89
	.995	10.8	7.70	6.48	5.80	5.37	5.07	4.85	4.67	4.54
	.999	16.6	11.3	9.34	8.25	7.57	7.09	6.74	6.47	6.26
20	.50	0.472	0.718	0.816	0.868	0.900	0.922	0.938	0.950	0.959
	.90	2.97	2.59	2.38	2.25	2.16	2.09	2.04	2.00	1.96
	.95	4.35	3.49	3.10	2.87	2.71	2.60	2.51	2.45	2.39
	.975	5.87	4.46	3.86	3.51	3.29	3.13	3.01	2.91	2.84
	.99	8.10	5.85	4.94	4.43	4.10	3.87	3.70	3.56	3.46
	.995	9.94	6.99	5.82	5.17	4.76	4.47	4.26	4.09	3.96
	.999	14.8	9.95	8.10	7.10	6.46	6.02	5.69	5.44	5.24
24	.50	0.469	0.714	0.812	0.863	0.895	0.917	0.932	0.944	0.953
	.90	2.93	2.54	2.33	2.19	2.10	2.04	1.98	1.94	1.91
	.95	4.26	3.40	3.01	2.78	2.62	2.51	2.42	2.36	2.30
	.975	5.72	4.32	3.72	3.38	3.15	2.99	2.87	2.78	2.70
	.99	7.82	5.61	4.72	4.22	3.90	3.67	3.50	3.36	3.26
	.995	9.55	6.66	5.52	4.89	4.49	4.20	3.99	3.83	3.69
	.999	14.0	9.34	7.55	6.59	5.98	5.55	5.23	4.99	4.80

TABLE B.7 | *(continued)*

r_2	$1-\alpha$	10	12	15	20	24	30	60	120	∞
						r_1				
8	.50	1.02	1.03	1.04	1.05	1.06	1.07	1.08	1.08	1.09
	.90	2.54	2.50	2.46	2.42	2.40	2.38	2.34	2.32	2.29
	.95	3.35	3.28	3.22	3.15	3.12	3.08	3.01	2.97	2.93
	.975	4.30	4.20	4.10	4.00	3.95	3.89	3.78	3.73	3.67
	.99	5.81	5.67	5.52	5.36	5.28	5.20	5.03	4.95	4.86
	.995	7.21	7.01	6.81	6.61	6.50	6.40	6.18	6.06	5.95
	.999	11.5	11.2	10.8	10.5	10.3	10.1	9.73	9.53	9.33
9	.50	1.01	1.02	1.03	1.04	1.05	1.05	1.07	1.07	1.08
	.90	2.42	2.38	2.34	2.30	2.28	2.25	2.21	2.18	2.16
	.95	3.14	3.07	3.01	2.94	2.90	2.86	2.79	2.75	2.71
	.975	3.96	3.87	3.77	3.67	3.61	3.56	3.45	3.39	3.33
	.99	5.26	5.11	4.96	4.81	4.73	4.65	4.48	4.40	4.31
	.995	6.42	6.23	6.03	5.83	5.73	5.62	5.41	5.30	5.19
	.999	9.89	9.57	9.24	8.90	8.72	8.55	8.19	8.00	7.81
10	.50	1.00	1.01	1.02	1.03	1.04	1.05	1.06	1.06	1.07
	.90	2.32	2.28	2.24	2.20	2.18	2.16	2.11	2.08	2.06
	.95	2.98	2.91	2.84	2.77	2.74	2.70	2.62	2.58	2.54
	.975	3.72	3.62	3.52	3.42	3.37	3.31	3.20	3.14	3.08
	.99	4.85	4.71	4.56	4.41	4.33	4.25	4.08	4.00	3.91
	.995	5.85	5.66	5.47	5.27	5.17	5.07	4.86	4.75	4.64
	.999	8.75	8.45	8.13	7.80	7.64	7.47	7.12	6.94	6.76
12	.50	0.989	1.00	1.01	1.02	1.03	1.03	1.05	1.05	1.06
	.90	2.19	2.15	2.10	2.06	2.04	2.01	1.96	1.93	1.90
	.95	2.75	2.69	2.62	2.54	2.51	2.47	2.38	2.34	2.30
	.975	3.37	3.28	3.18	3.07	3.02	2.96	2.85	2.79	2.72
	.99	4.30	4.16	4.01	3.86	3.78	3.70	3.54	3.45	3.36
	.995	5.09	4.91	4.72	4.53	4.43	4.33	4.12	4.01	3.90
	.999	7.29	7.00	6.71	6.40	6.25	6.09	5.76	5.59	5.42
15	.50	0.977	0.989	1.00	1.01	1.02	1.02	1.03	1.04	1.05
	.90	2.06	2.02	1.97	1.92	1.90	1.87	1.82	1.79	1.76
	.95	2.54	2.48	2.40	2.33	2.29	2.25	2.16	2.11	2.07
	.975	3.06	2.96	2.86	2.76	2.70	2.64	2.52	2.46	2.40
	.99	3.80	3.67	3.52	3.37	3.29	3.21	3.05	2.96	2.87
	.995	4.42	4.25	4.07	3.88	3.79	3.69	3.48	3.37	3.26
	.999	6.08	5.81	5.54	5.25	5.10	4.95	4.64	4.48	4.31
20	.50	0.966	0.977	0.989	1.00	1.01	1.01	1.02	1.03	1.03
	.90	1.94	1.89	1.84	1.79	1.77	1.74	1.68	1.64	1.61
	.95	2.35	2.28	2.20	2.12	2.08	2.04	1.95	1.90	1.84
	.975	2.77	2.68	2.57	2.46	2.41	2.35	2.22	2.16	2.09
	.99	3.37	3.23	3.09	2.94	2.86	2.78	2.61	2.52	2.42
	.995	3.85	3.68	3.50	3.32	3.22	3.12	2.92	2.81	2.69
	.999	5.08	4.82	4.56	4.29	4.15	4.00	3.70	3.54	3.38
24	.50	0.961	0.972	0.983	0.994	1.00	1.01	1.02	1.02	1.03
	.90	1.88	1.83	1.78	1.73	1.70	1.67	1.61	1.57	1.53
	.95	2.25	2.18	2.11	2.03	1.98	1.94	1.84	1.79	1.73
	.975	2.64	2.54	2.44	2.33	2.27	2.21	2.08	2.01	1.94
	.99	3.17	3.03	2.89	2.74	2.66	2.58	2.40	2.31	2.21
	.995	3.59	3.42	3.25	3.06	2.97	2.87	2.66	2.55	2.43
	.999	4.64	4.39	4.14	3.87	3.74	3.59	3.29	3.14	2.97

TABLE B.7 | *(continued)*

r_2	$1 - \alpha$	1	2	3	4	5	6	7	8	9
						r_1				
30	.50	0.466	0.709	0.807	0.858	0.890	0.912	0.927	0.939	0.948
	.90	2.88	2.49	2.28	2.14	2.05	1.98	1.93	1.88	1.85
	.95	4.17	3.32	2.92	2.69	2.53	2.42	2.33	2.27	2.21
	.975	5.57	4.18	3.59	3.25	3.03	2.87	2.75	2.65	2.57
	.99	7.56	5.39	4.51	4.02	3.70	3.47	3.30	3.17	3.07
	.995	9.18	6.35	5.24	4.62	4.23	3.95	3.74	3.58	3.45
	.999	13.3	8.77	7.05	6.12	5.53	5.12	4.82	4.58	4.39
60	.50	0.461	0.701	0.798	0.849	0.880	0.901	0.917	0.928	0.937
	.90	2.79	2.39	2.18	2.04	1.95	1.87	1.82	1.77	1.74
	.95	4.00	3.15	2.76	2.53	2.37	2.25	2.17	2.10	2.04
	.975	5.29	3.93	3.34	3.01	2.79	2.63	2.51	2.41	2.33
	.99	7.08	4.98	4.13	3.65	3.34	3.12	2.95	2.82	2.72
	.995	8.49	5.80	4.73	4.14	3.76	3.49	3.29	3.13	3.01
	.999	12.0	7.77	6.17	5.31	4.76	4.37	4.09	3.86	3.69
120	.50	0.458	0.697	0.793	0.844	0.875	0.896	0.912	0.923	0.932
	.90	2.75	2.35	2.13	1.99	1.90	1.82	1.77	1.72	1.68
	.95	3.92	3.07	2.68	2.45	2.29	2.18	2.09	2.02	1.96
	.975	5.15	3.80	3.23	2.89	2.67	2.52	2.39	2.30	2.22
	.99	6.85	4.79	3.95	3.48	3.17	2.96	2.79	2.66	2.56
	.995	8.18	5.54	4.50	3.92	3.55	3.28	3.09	2.93	2.81
	.999	11.4	7.32	5.78	4.95	4.42	4.04	3.77	3.55	3.38
∞	.50	0.455	0.693	0.789	0.839	0.870	0.891	0.907	0.918	0.927
	.90	2.71	2.30	2.08	1.94	1.85	1.77	1.72	1.67	1.63
	.95	3.84	3.00	2.60	2.37	2.21	2.10	2.01	1.94	1.88
	.975	5.02	3.69	3.12	2.79	2.57	2.41	2.29	2.19	2.11
	.99	6.63	4.61	3.78	3.32	3.02	2.80	2.64	2.51	2.41
	.995	7.88	5.30	4.28	3.72	3.35	3.09	2.90	2.74	2.62
	.999	10.8	6.91	5.42	4.62	4.10	3.74	3.47	3.27	3.10

TABLE B.7 | *(concluded)*

r_2	$1 - \alpha$	r_1								
		10	12	15	20	24	30	60	120	∞
30	.50	0.955	0.966	0.978	0.989	0.994	1.00	1.01	1.02	1.02
	.90	1.82	1.77	1.72	1.67	1.64	1.61	1.54	1.50	1.46
	.95	2.16	2.09	2.01	1.93	1.89	1.84	1.74	1.68	1.62
	.975	2.51	2.41	2.31	2.20	2.14	2.07	1.94	1.87	1.79
	.99	2.98	2.84	2.70	2.55	2.47	2.39	2.21	2.11	2.01
	.995	3.34	3.18	3.01	2.82	2.73	2.63	2.42	2.30	2.18
	.999	4.24	4.00	3.75	3.49	3.36	3.22	2.92	2.76	2.59
60	.50	0.945	0.956	0.967	0.978	0.983	0.989	1.00	1.01	1.01
	.90	1.71	1.66	1.60	1.54	1.51	1.48	1.40	1.35	1.29
	.95	1.99	1.92	1.84	1.75	1.70	1.65	1.53	1.47	1.39
	.975	2.27	2.17	2.06	1.94	1.88	1.82	1.67	1.58	1.48
	.99	2.63	2.50	2.35	2.20	2.12	2.03	1.84	1.73	1.60
	.995	2.90	2.74	2.57	2.39	2.29	2.19	1.96	1.83	1.69
	.999	3.54	3.32	3.08	2.83	2.69	2.55	2.25	2.08	1.89
120	.50	0.939	0.950	0.961	0.972	0.978	0.983	0.994	1.00	1.01
	.90	1.65	1.60	1.55	1.48	1.45	1.41	1.32	1.26	1.19
	.95	1.91	1.83	1.75	1.66	1.61	1.55	1.43	1.35	1.25
	.975	2.16	2.05	1.95	1.82	1.76	1.69	1.53	1.43	1.31
	.99	2.47	2.34	2.19	2.03	1.95	1.86	1.66	1.53	1.38
	.995	2.71	2.54	2.37	2.19	2.09	1.98	1.75	1.61	1.43
	.999	3.24	3.02	2.78	2.53	2.40	2.26	1.95	1.77	1.54
∞	.50	0.934	0.945	0.956	0.967	0.972	0.978	0.989	0.994	1.00
	.90	1.60	1.55	1.49	1.42	1.38	1.34	1.24	1.17	1.00
	.95	1.83	1.75	1.67	1.57	1.52	1.46	1.32	1.22	1.00
	.975	2.05	1.94	1.83	1.71	1.64	1.57	1.39	1.27	1.00
	.99	2.32	2.18	2.04	1.88	1.79	1.70	1.47	1.32	1.00
	.995	2.52	2.36	2.19	2.00	1.90	1.79	1.53	1.36	1.00
	.999	2.96	2.74	2.51	2.27	2.13	1.99	1.66	1.45	1.00

Source: Adapted from table 5 of Pearson and Hartley, *Biometrika Tables for Statisticians,* volume 2, 1972, published by the Cambridge University Press, on behalf of The Biometrika Society, by permission of the authors and publishers.

TABLE B.8 | Five-place logarithms

N	0	1	2	3	4	5	6	7	8	9
0	− ∞	00000	30103	47712	60206	69897	77815	84510	90309	95424
10	00000	00432	00860	01284	01703	02119	02531	02938	03342	03743
11	04139	04532	04922	05308	05690	06070	06446	06819	07188	07555
12	07918	08279	08636	08991	09342	09691	10037	10380	10721	11059
13	11394	11727	12057	12385	12710	13033	13354	13672	13988	14301
14	14613	14922	15229	15534	15836	16137	16435	16732	17026	17319
15	17609	17898	18184	18469	18752	19033	19312	19590	19866	20140
16	20412	20683	20952	21219	21484	21748	22011	22272	22531	22789
17	23045	23300	23533	23805	24055	24304	24551	24797	25042	25285
18	25527	25768	26007	26245	26482	26717	26951	27184	27416	27646
19	27875	28103	28330	28556	28780	29003	29226	29447	29667	29885
20	30103	30320	30535	30750	30963	31175	31387	31597	31806	32015
21	32222	32428	32634	32838	33041	33244	33445	33646	33846	34044
22	34242	34439	34635	34830	35025	35218	35411	35603	35793	35984
23	36173	36361	36549	36736	36922	37107	37291	37475	37658	37840
24	38021	38202	38382	38561	38739	38917	39094	39270	39445	39620
25	39794	39967	40140	40312	40483	40654	40824	40993	41162	41330
26	41497	41664	41830	41996	42160	42325	42488	42651	42813	42975
27	43136	43297	43457	43616	43775	43933	44091	44248	44404	44560
28	44716	44871	45025	45179	45332	45484	45637	45788	45939	46090
29	46240	46389	46538	46687	46835	46982	47129	47276	47422	47567
30	47712	47857	48001	48144	48287	48430	48572	48714	48855	48996
31	49136	49276	49415	49554	49693	49831	49969	50106	50243	50379
32	50515	50651	50786	50920	51055	51188	51322	51455	51587	51720
33	51851	51983	52114	52244	52375	52504	52634	52763	52892	53020
34	53148	53275	53403	53529	53656	53782	53908	54033	54158	54283
35	54407	54531	54654	54777	54900	55023	55145	55267	55388	55509
36	55630	55751	55871	55991	56110	56229	56348	56467	56585	56703
37	56820	56937	57054	57171	57287	57403	57519	57634	57749	57864
38	57978	58092	58206	58320	58433	58546	58659	58771	58883	58995
39	59106	59218	59329	59439	59550	59660	59770	59879	59988	60097
40	60206	60314	60423	60531	60638	60746	60853	60959	61066	61172
41	61278	61384	61490	61595	61700	61805	61900	62014	62118	62221
42	62325	62428	62531	62634	62737	62839	62941	63043	63144	63246
43	63347	63448	63548	63649	63749	63849	63949	64048	64147	64246
44	64345	64444	64542	64640	64738	64836	64933	65031	65128	65225
45	65321	65418	65514	65610	65706	65801	65896	65992	66087	66181
46	66276	66370	66464	66558	66652	66745	66839	66932	67025	67117
47	67210	67302	67394	67486	67578	67669	67761	67852	67943	68034
48	68124	68215	68305	68395	68485	68574	68664	68753	68842	68931
49	69020	69108	69197	69285	69373	69461	69548	69636	69723	69810

TABLE B.8 | (concluded)

N	0	1	2	3	4	5	6	7	8	9
50	69897	69984	70070	70157	70243	70329	70415	70501	70586	70672
51	70757	70842	70927	71012	71096	71181	71265	71349	71433	71517
52	71600	71684	71767	71850	71933	72016	72099	72181	72263	72346
53	72428	72509	72591	72673	72754	72835	72916	72997	73078	73159
54	73239	73320	73400	73480	73560	73640	73719	73799	73878	73957
55	74036	74115	74194	74273	74351	74429	74507	74586	74663	74741
56	74819	74896	74974	75051	75128	75205	75282	75358	75435	75511
57	75587	75664	75740	75815	75891	75967	76042	76118	76193	76268
58	76343	76418	76492	76567	76641	76716	76790	76864	76938	77012
59	77085	77159	77232	77305	77379	77452	77525	77597	77670	77743
60	77815	77887	77960	78032	78104	78176	78247	78319	78390	78462
61	78533	78604	78675	78746	78817	78888	78958	79029	79099	79169
62	79239	79309	79379	79449	79518	79588	79657	79727	79796	79865
63	79934	80003	80072	80140	80209	80277	80346	80414	80482	80550
64	80618	80686	80754	80821	80889	80956	81023	81090	81158	81224
65	81291	81358	81425	81491	81558	81624	81690	81757	81823	81889
66	81954	82020	82086	82151	82217	82282	82347	82413	82478	82543
67	82607	82672	82737	82802	82866	82930	82995	83059	83123	83187
68	83251	83315	83378	83442	83506	83569	83632	83696	83759	83822
69	83885	83948	84011	84073	84136	84198	84261	84323	84386	84448
70	84510	84572	84634	84696	84757	84819	84880	84942	85003	85065
71	85126	85187	85248	85309	85370	85431	85491	85552	85612	85673
72	85733	85794	85854	85914	85974	86034	86094	86153	86213	86273
73	86332	86392	86451	86510	86570	86629	86688	86747	86806	86864
74	86923	86982	87040	87099	87157	87216	87274	87332	87390	87448
75	87506	87564	87622	87679	87737	87795	87852	87910	87967	88024
76	88081	88138	88195	88252	88309	88366	88423	88480	88536	88593
77	88649	88705	88762	88818	88874	88930	88986	89042	89098	89154
78	89209	89265	89321	89376	89432	89487	89542	89597	89653	89708
79	89763	89818	89873	89927	89982	90037	90091	90146	90200	90255
80	90309	90363	90417	90472	90526	90580	90634	90687	90741	90795
81	90849	90902	90956	91009	91062	91116	91169	91222	91275	91328
82	91381	91434	91487	91540	91593	91645	91698	91751	91803	91855
83	91908	91960	92012	92065	92117	92169	92221	92273	92324	92376
84	92428	92480	92531	92583	92634	92686	92737	92788	92840	92891
85	92942	92993	93044	93095	93146	93197	93247	93298	93349	93399
86	93450	93500	93551	93601	93651	93702	93752	93802	93852	93902
87	93952	94002	94052	94101	94151	94201	94250	94300	94349	94399
88	94448	94498	94547	94596	94645	94694	94743	94792	94841	94890
89	94939	94988	95036	95085	95134	95182	95231	95279	95328	95376
90	95424	95472	95521	95569	95617	95665	95713	95761	95809	95856
91	95904	95952	95999	96047	96095	96142	96190	96237	96284	96332
92	96379	96426	96473	96520	96567	96614	96661	96708	96755	96802
93	96848	96895	96942	96988	97035	97081	97128	97174	97220	97267
94	97313	97359	97405	97451	97497	97543	97589	97635	97681	97727
95	97772	97818	97864	97909	97955	98000	98046	98091	98137	98182
96	98227	98272	98318	98363	98408	98453	98498	98543	98588	98632
97	98677	98722	98767	98811	98856	98900	98945	98989	99034	99078
98	99123	99167	99211	99255	99300	99344	99388	99432	99476	99520
99	99564	99607	99651	99695	99739	99782	99826	99870	99913	99957

Source: John Neter, William Wasserman, and G. A. Whitmore, *Fundamental Statistics for Business and Economics,* 4th ed. (Boston: Allyn & Bacon, Inc., 1972 ©).

TABLE B.9 | Critical values for Hartley's *H*-statistic

$1 - \alpha = .95$

						r					
n	2	3	4	5	6	7	8	9	10	11	12
3	39.0	87.5	142	202	266	333	403	475	550	626	704
4	15.4	27.8	39.2	50.7	62.0	72.9	83.5	93.9	104	114	124
5	9.60	15.5	20.6	25.2	29.5	33.6	37.5	41.1	44.6	48.0	51.4
6	7.15	10.8	13.7	16.3	18.7	20.8	22.9	24.7	26.5	28.2	29.9
7	5.82	8.38	10.4	12.1	13.7	15.0	16.3	17.5	18.6	19.7	20.7
8	4.99	6.94	8.44	9.70	10.8	11.8	12.7	13.5	14.3	15.1	15.8
9	4.43	6.00	7.18	8.12	9.03	9.78	10.5	11.1	11.7	12.2	12.7
10	4.03	5.34	6.31	7.11	7.80	8.41	8.95	9.45	9.91	10.3	10.7
11	3.72	4.85	5.67	6.34	6.92	7.42	7.87	8.28	8.66	9.01	9.34
13	3.28	4.16	4.79	5.30	5.72	6.09	6.42	6.72	7.00	7.25	7.48
16	2.86	3.54	4.01	4.37	4.68	4.95	5.19	5.40	5.59	5.77	5.93
21	2.46	2.95	3.29	3.54	3.76	3.94	4.10	4.24	4.37	4.49	4.59
31	2.07	2.40	2.61	2.78	2.91	3.02	3.12	3.21	3.29	3.36	3.39
61	1.67	1.85	1.96	2.04	2.11	2.17	2.22	2.26	2.30	2.33	2.36
∞	1.00	1.00	1.00	1.00	1.00	1.00	1.00	1.00	1.00	1.00	1.00

$1 - \alpha = .99$

						r					
n	2	3	4	5	6	7	8	9	10	11	12
3	199	448	729	1,036	1,362	1,705	2,063	2,432	2,813	3,204	3,605
4	47.5	85	120	151	184	216	249	281	310	337	361
5	23.2	37	49	59	69	79	89	97	106	113	120
6	14.9	22	28	33	38	42	46	50	54	57	60
7	11.1	15.5	19.1	22	25	27	30	32	34	36	37
8	8.89	12.1	14.5	16.5	18.4	20	22	23	24	26	27
9	7.50	9.9	11.7	13.2	14.5	15.8	16.9	17.9	18.9	19.8	21
10	6.54	8.5	9.9	11.1	12.1	13.1	13.9	14.7	15.3	16.0	16.6
11	5.85	7.4	8.6	9.6	10.4	11.1	11.8	12.4	12.9	13.4	13.9
13	4.91	6.1	6.9	7.6	8.2	8.7	9.1	9.5	9.9	10.2	10.6
16	4.07	4.9	5.5	6.0	6.4	6.7	7.1	7.3	7.5	7.8	8.0
21	3.32	3.8	4.3	4.6	4.9	5.1	5.3	5.5	5.6	5.8	5.9
31	2.63	3.0	3.3	3.4	3.6	3.7	3.8	3.9	4.0	4.1	4.2
61	1.96	2.2	2.3	2.4	2.4	2.5	2.5	2.6	2.6	2.7	2.7
∞	1.00	1.0	1.0	1.0	1.0	1.0	1.0	1.0	1.0	1.0	1.0

Source: Reprinted, with permission, from H. A. David, "Upper 5 and 1% Points of the Maximum *F*-Ratio," *Biometrika*, vol. 39 (1952), pp. 422–24.

TABLE B.10 | Unit normal losses

D	.00	.01	.02	.03	.04	.05	.06	.07	.08	.09
.0	.3989	.3940	.3890	.3841	.3793	.3744	.3697	.3649	.3602	.3556
.1	.3509	.3464	.3418	.3373	.3328	.3284	.3240	.3197	.3154	.3111
.2	.3069	.3027	.2986	.2944	.2904	.2863	.2824	.2784	.2745	.2706
.3	.2668	.2630	.2592	.2555	.2518	.2481	.2445	.2409	.2374	.2339
.4	.2304	.2270	.2236	.2203	.2169	.2137	.2104	.2072	.2040	.2009
.5	.1978	.1947	.1917	.1887	.1857	.1828	.1799	.1771	.1742	.1714
.6	.1687	.1659	.1633	.1606	.1580	.1554	.1528	.1503	.1478	.1453
.7	.1429	.1405	.1381	.1358	.1334	.1312	.1289	.1267	.1245	.1223
.8	.1202	.1181	.1160	.1140	.1120	.1100	.1080	.1061	.1042	.1023
.9	.1004	.09860	.09680	.09503	.09328	.09156	.08986	.08819	.08654	.08491
1.0	.08332	.08174	.08019	.07866	.07716	.07568	.07422	.07279	.07138	.06999
1.1	.06862	.06727	.06595	.06465	.06336	.06210	.06086	.05964	.05844	.05726
1.2	.05610	.05496	.05384	.05274	.05165	.05059	.04954	.04851	.04750	.04650
1.3	.04553	.04457	.04363	.04270	.04179	.04090	.04002	.03916	.03831	.03748
1.4	.03667	.03587	.03508	.03431	.03356	.03281	.03208	.03137	.03067	.02998
1.5	.02931	.02865	.02800	.02736	.02674	.02612	.02552	.02494	.02436	.02380
1.6	.02324	.02270	.02217	.02165	.02114	.02064	.02015	.01967	.01920	.01874
1.7	.01829	.01785	.01742	.01699	.01658	.01617	.01578	.01539	.01501	.01464
1.8	.01428	.01392	.01357	.01323	.01290	.01257	.01226	.01195	.01164	.01134
1.9	.01105	.01077	.01049	.01022	$.0^29957$	$.0^29698$	$.0^29445$	$.0^29198$	$.0^28957$	$.0^28721$
2.0	$.0^28491$	$.0^28266$	$.0^28046$	$.0^27832$	$.0^27623$	$.0^27418$	$.0^27219$	$.0^27024$	$.0^26835$	$.0^26649$
2.1	$.0^26468$	$.0^26292$	$.0^26120$	$.0^25952$	$.0^25788$	$.0^25628$	$.0^25472$	$.0^25320$	$.0^25172$	$.0^25028$
2.2	$.0^24887$	$.0^24750$	$.0^24616$	$.0^24486$	$.0^24358$	$.0^24235$	$.0^24114$	$.0^23996$	$.0^23882$	$.0^23770$
2.3	$.0^23662$	$.0^23556$	$.0^23453$	$.0^23352$	$.0^23255$	$.0^23159$	$.0^23067$	$.0^22977$	$.0^22889$	$.0^22804$
2.4	$.0^22720$	$.0^22640$	$.0^22561$	$.0^22484$	$.0^22410$	$.0^22337$	$.0^22267$	$.0^22199$	$.0^22132$	$.0^22067$
2.5	$.0^22004$	$.0^21943$	$.0^21883$	$.0^21826$	$.0^21769$	$.0^21715$	$.0^21662$	$.0^21610$	$.0^21560$	$.0^21511$
3.0	$.0^33822$	$.0^33689$	$.0^33560$	$.0^33436$	$.0^33316$	$.0^33199$	$.0^33087$	$.0^32978$	$.0^32873$	$.0^32771$
3.5	$.0^45848$	$.0^45620$	$.0^45400$	$.0^45188$	$.0^44984$	$.0^44788$	$.0^44599$	$.0^44417$	$.0^44242$	$.0^44073$
4.0	$.0^57145$	$.0^56835$	$.0^56538$	$.0^56253$	$.0^55980$	$.0^55718$	$.0^55468$	$.0^55227$	$.0^54997$	$.0^54777$

Illustration: The value of $L(D)$ for $D = 3.01$ is $0.0^33689 = 0.0003689$.

Source: Reproduced from Robert Schlaifer, *Introduction to Statistics for Business Decisions,* published by McGraw-Hill Book Company, 1961, by permission from the copyright holder, the President and Fellows of Harvard College.

TABLE B.11 | Critical values of the Mann-Whitney test statistic

n	α	m = 2	3	4	5	6	7	8	9	10	11	12	13	14	15	16	17	18	19	20
2	.001	0	0	0	0	0	0	0	0	0	0	0	0	0	0	0	0	0	0	0
	.005	0	0	0	0	0	0	0	0	0	0	0	0	0	0	0	0	0	1	1
	.01	0	0	0	0	0	0	0	0	0	0	0	1	1	1	1	1	1	2	2
	.025	0	0	0	0	1	1	1	1	1	1	2	2	2	2	3	3	3	3	3
	.05	0	0	1	1	1	1	1	2	2	2	3	3	4	4	4	4	5	5	5
	.10	0	1	1	2	2	2	3	3	4	4	5	5	5	6	6	7	7	8	8
3	.001	0	0	0	0	0	0	0	0	0	0	0	0	0	0	0	1	1	1	1
	.005	0	0	0	0	0	0	0	1	1	1	2	2	2	3	3	3	3	4	4
	.01	0	0	0	0	1	1	1	2	2	2	3	3	3	4	4	5	5	5	6
	.025	0	0	0	1	2	2	3	3	4	4	5	5	6	6	7	7	8	8	9
	.05	0	1	1	2	3	3	4	5	5	6	6	7	8	8	9	10	10	11	12
	.10	1	2	2	3	4	5	6	6	7	8	9	10	11	11	12	13	14	15	16
4	.001	0	0	0	0	0	0	0	0	1	1	1	2	2	2	3	3	4	4	4
	.005	0	0	0	0	1	1	2	2	3	3	4	4	5	6	6	7	7	8	9
	.01	0	0	0	1	2	2	3	4	4	5	6	6	7	8	9	9	10	10	11
	.025	0	0	1	2	3	4	5	5	6	7	8	9	10	11	12	12	13	14	15
	.05	0	1	2	3	4	5	6	7	8	9	10	11	12	13	15	16	17	18	19
	.10	1	2	4	5	6	7	8	10	11	12	13	14	16	17	18	19	21	22	23
5	.001	0	0	0	0	0	0	1	2	2	3	3	4	4	5	6	6	7	8	8
	.005	0	0	0	1	2	2	3	4	5	6	7	8	8	9	10	11	12	13	14
	.01	0	0	1	2	3	4	5	6	7	8	9	10	11	12	13	14	15	16	17
	.025	0	1	2	3	4	6	7	8	9	10	12	13	14	15	16	18	19	20	21
	.05	1	2	3	5	6	7	9	10	12	13	14	16	17	19	20	21	23	24	26
	.10	2	3	5	6	8	9	11	13	14	16	18	19	21	23	24	26	28	29	31
6	.001	0	0	0	0	0	0	2	3	4	5	5	6	7	8	9	10	11	12	13
	.005	0	0	1	2	3	4	5	6	7	8	10	11	12	13	14	16	17	18	19
	.01	0	0	2	3	4	5	7	8	9	10	12	13	14	16	17	19	20	21	23
	.025	0	2	3	4	6	7	9	11	12	14	15	17	18	20	22	23	25	26	28
	.05	1	3	4	6	8	9	11	13	15	17	18	20	22	24	26	27	29	31	33
	.10	2	4	6	8	10	12	14	16	18	20	22	24	26	28	30	32	35	37	39

Source: Adapted from L. R. Verdooren, "Extended Tables of Critical Values for Wilcoxon's Test Statistic," *Biometrika*, vol. 50 (1963), pp. 177–86.

TABLE B.11 (continued)

n	α	m=2	3	4	5	6	7	8	9	10	11	12	13	14	15	16	17	18	19	20
7	.001	0	0	0	0	1	2	3	4	6	7	8	9	10	11	12	14	15	16	17
	.005	0	0	1	2	4	5	7	8	10	11	13	14	16	17	19	20	22	23	25
	.01	0	1	2	4	5	7	8	10	12	13	15	17	18	20	22	24	25	27	29
	.025	1	2	4	6	7	9	11	13	15	17	19	21	23	25	27	29	31	33	35
	.05	1	3	5	7	9	12	14	16	18	20	22	25	27	29	31	34	36	38	40
	.10	2	5	7	9	12	14	17	19	22	24	27	29	32	34	37	39	42	44	47
8	.001	0	0	0	1	2	3	5	6	7	9	10	12	13	15	16	18	19	21	22
	.005	0	0	2	3	5	7	8	10	12	14	16	18	19	21	23	25	27	29	31
	.01	0	1	3	5	7	8	10	12	14	16	18	21	23	25	27	29	31	33	35
	.025	0	3	5	7	9	11	14	16	18	20	23	25	27	30	32	35	37	39	42
	.05	1	4	6	9	11	14	16	19	21	24	27	29	32	34	37	40	42	45	48
	.10	3	6	8	11	14	17	20	23	25	28	31	34	37	40	43	46	49	52	55
9	.001	0	0	0	2	3	4	6	8	9	11	13	15	16	18	20	22	24	26	27
	.005	0	1	2	4	6	8	10	12	14	17	19	21	23	25	28	30	32	34	37
	.01	0	2	4	6	8	10	12	15	17	19	22	24	27	29	32	34	37	39	41
	.025	1	3	5	8	11	13	16	18	21	24	27	29	32	35	38	40	43	46	49
	.05	2	5	7	10	13	16	19	22	25	28	31	34	37	40	43	46	49	52	55
	.10	3	6	10	13	16	19	23	26	29	32	36	39	42	46	49	53	56	59	63
10	.001	0	0	1	2	4	6	7	9	11	13	15	18	20	22	24	26	28	30	33
	.005	0	1	3	5	7	10	12	14	17	19	22	25	27	30	32	35	38	40	43
	.01	0	2	4	7	9	12	14	17	20	23	25	28	31	34	37	39	42	45	48
	.025	1	4	6	9	12	15	18	21	24	27	30	34	37	40	43	46	49	53	56
	.05	2	5	8	12	15	18	21	25	28	32	35	38	42	45	49	52	56	59	63
	.10	4	7	11	14	18	22	25	29	33	37	40	44	48	52	55	59	63	67	71
11	.001	0	0	1	3	5	7	9	11	13	16	18	21	23	25	28	30	33	35	38
	.005	0	1	3	6	8	11	14	17	19	22	25	28	31	34	37	40	43	46	49
	.01	0	2	5	8	10	13	16	19	23	26	29	32	35	38	42	45	48	51	54
	.025	1	4	7	10	14	17	20	24	27	31	34	38	41	45	48	52	56	59	63
	.05	2	6	9	13	17	20	24	28	32	35	39	43	47	51	55	58	62	66	70
	.10	4	8	12	16	20	24	28	32	37	41	45	49	53	58	62	66	70	74	79

TABLE B.11 | (continued)

n	α	m = 2	3	4	5	6	7	8	9	10	11	12	13	14	15	16	17	18	19	20
12	.001	0	0	1	3	5	8	10	13	15	18	21	24	26	29	32	35	38	41	43
	.005	0	2	4	7	10	13	16	19	22	25	28	32	35	38	42	45	48	52	55
	.01	0	3	6	9	12	15	18	22	25	29	32	36	39	43	47	50	54	57	61
	.025	2	5	8	12	15	19	23	27	30	34	38	42	46	50	54	58	62	66	70
	.05	3	6	10	14	18	22	27	31	35	39	43	48	52	56	61	65	69	73	78
	.10	5	9	13	18	22	27	31	36	40	45	50	54	59	64	68	73	78	82	87
13	.001	0	0	2	4	6	9	12	15	18	21	24	27	30	33	36	39	43	46	49
	.005	0	2	4	8	11	14	18	21	25	28	32	35	39	43	46	50	54	58	61
	.01	1	3	6	10	13	17	21	24	28	32	36	40	44	48	52	56	60	64	68
	.025	2	5	9	13	17	21	25	29	34	38	42	46	51	55	60	64	68	73	77
	.05	3	7	11	16	20	25	29	34	38	43	48	52	57	62	66	71	76	81	85
	.10	5	10	14	19	24	29	34	39	44	49	54	59	64	69	75	80	85	90	95
14	.001	0	0	2	4	7	10	13	16	20	23	26	30	33	37	40	44	47	51	55
	.005	0	2	5	8	12	16	19	23	27	31	35	39	43	47	51	55	59	64	68
	.01	1	3	7	11	14	18	23	27	31	35	39	44	48	52	57	61	66	70	74
	.025	2	6	10	14	18	23	27	32	37	42	46	51	56	60	65	70	75	79	84
	.05	4	8	12	17	22	27	32	37	42	48	52	57	62	67	72	78	83	88	93
	.10	5	11	16	21	26	32	37	42	48	53	59	64	70	75	81	86	92	98	103
15	.001	0	0	2	5	8	11	15	18	22	25	29	33	37	41	44	48	52	56	60
	.005	0	3	6	9	13	17	21	25	30	34	38	43	47	52	56	61	65	70	74
	.01	1	4	8	12	16	20	25	29	34	38	43	48	52	57	62	67	71	76	81
	.025	2	6	11	15	20	25	30	35	40	45	50	55	60	65	71	76	81	86	91
	.05	4	8	13	19	24	29	34	40	45	51	56	62	67	73	78	84	89	95	101
	.10	6	11	17	23	28	34	40	46	52	58	64	69	75	81	87	93	99	105	111
16	.001	0	0	3	6	9	12	16	20	24	28	32	36	40	44	49	53	57	61	66
	.005	0	3	6	10	14	19	23	28	32	37	42	46	51	56	61	66	71	75	80
	.01	1	4	8	13	17	22	27	32	37	42	47	52	57	62	67	72	77	83	88
	.025	2	7	12	16	22	27	32	38	43	48	54	60	65	71	76	82	87	93	99
	.05	4	9	15	20	26	31	37	43	49	55	61	66	72	78	84	90	96	102	108
	.10	6	12	18	24	30	37	43	49	55	62	68	75	81	87	94	100	107	113	120

TABLE B.11 | (concluded)

n	α	m = 2	3	4	5	6	7	8	9	10	11	12	13	14	15	16	17	18	19	20
17	.001	0	1	3	6	10	14	18	22	26	30	35	39	44	48	53	58	62	67	71
	.005	0	3	7	11	16	20	25	30	35	40	45	50	55	61	66	71	76	82	87
	.01	1	5	9	14	19	24	29	34	39	45	50	56	61	67	72	78	83	89	94
	.025	3	7	12	18	23	29	35	40	46	52	58	64	70	76	82	88	94	100	106
	.05	4	10	16	21	27	34	40	46	52	58	65	71	78	84	90	97	103	110	116
	.10	7	13	19	26	32	39	46	53	59	66	73	80	86	93	100	107	114	121	128
18	.001	0	1	4	7	11	15	19	24	28	33	38	43	47	52	57	62	67	72	77
	.005	0	3	7	12	17	22	27	32	38	43	48	54	59	65	71	76	82	88	93
	.01	1	5	10	15	20	25	31	37	42	48	54	60	66	71	77	83	89	95	101
	.025	3	8	13	19	25	31	37	43	49	56	62	68	75	81	87	94	100	107	113
	.05	5	10	17	23	29	36	42	49	56	62	69	76	83	89	96	103	110	117	124
	.10	7	14	21	28	35	42	49	56	63	70	78	85	92	99	107	114	121	129	136
19	.001	0	1	4	8	12	16	21	26	30	35	41	46	51	56	61	67	72	78	83
	.005	1	4	8	13	18	23	29	34	40	46	52	58	64	70	75	82	88	94	100
	.01	2	5	10	16	21	27	33	39	45	51	57	64	70	76	83	89	95	102	108
	.025	3	8	14	20	26	33	39	46	53	59	66	73	79	86	93	100	107	114	120
	.05	5	11	18	24	31	38	45	52	59	66	73	81	88	95	102	110	117	124	131
	.10	8	15	22	29	37	44	52	59	67	74	82	90	98	105	113	121	129	136	144
20	.001	0	1	4	8	13	17	22	27	33	38	43	49	55	60	66	71	77	83	89
	.005	1	4	9	14	19	25	31	37	43	49	55	61	68	74	80	87	93	100	106
	.001	2	6	11	17	23	29	35	41	48	54	61	68	74	81	88	94	101	108	115
	.025	3	9	15	21	28	35	42	49	56	63	70	77	84	91	99	106	113	120	128
	.05	5	12	19	26	33	40	48	55	63	70	78	85	93	101	108	116	124	131	139
	.10	8	16	23	31	39	47	55	63	71	79	87	95	103	111	120	128	136	144	152

Table entries are upper-tail critical values W_α such that $P(\mathbf{W} \geq W_\alpha) = \alpha$.

For n or m greater than 20, the critical values W_α may be approximated by

$$W_\alpha = \frac{nm}{2} + z_\alpha \sqrt{\frac{nm(n + m + 1)}{12}}$$

where z_α is the standard normal variate such that a proportion α of the area is to the right of z_α.

TABLE B.12 | Critical values of the Wilcoxon signed rank test statistic

Sample size n	$W_{.005}$	$W_{.01}$	$W_{.025}$	$W_{.05}$	$W_{.10}$	$W_{.20}$	$n(n+1)/2$
4	0	0	0	0	1	3	10
5	0	0	0	1	3	4	15
6	0	0	1	3	4	6	21
7	0	1	3	4	6	9	28
8	1	2	4	6	9	12	36
9	2	4	6	9	11	15	45
10	4	6	9	11	15	19	55
11	6	8	11	14	18	23	66
12	8	10	14	18	22	28	78
13	10	13	18	22	27	33	91
14	13	16	22	26	32	39	105
15	16	20	26	31	37	45	120
16	20	24	30	36	43	51	136
17	24	28	35	42	49	58	153
18	28	33	41	48	56	66	171
19	33	38	47	54	63	74	190
20	38	44	53	61	70	82	210

Table entries are lower-tail critical values: W_α such that $P(\mathbf{W} \leq W_\alpha) = \alpha$.

The upper-tail critical values may be determined by using the relationship:

$$W_\alpha = [n(n+1)/2] - W_{1-\alpha}, \qquad \alpha > 0.50$$

For example, if $n = 10$, $\alpha = 0.05$, and the test is two-tailed, $W_{\alpha/2} = W_{0.025} = 9$ from the table;

$$W_{1-\alpha/2} = W_{0.975} = [10(10+1)/2] - W_{0.025} = 55 - 9 = 46$$

For $n > 20$, the critical value W_α may be approximated from:

$$W_\alpha = [n(n+1)/4] + z_\alpha \sqrt{n(n+1)(2n+1)/24}$$

where z_α is the standard normal variate such that a proportion α of the area is to the right of z_α.

Source: Adapted from R. L. McCornack, "Extended Tables of the Wilcoxon Matched Pairs Signed Rank Statistics," *Journal of the American Statistical Association* 60 (1965), pp. 864–71.

TABLE B.13 | Critical values of the Kruskal-Wallis test statistic for three samples and small sample size

n_1	n_2	n_3	Critical value	α	n_1	n_2	n_3	Critical value	α
2	1	1	2.7000	.500	4	3	2	6.4444	.009
2	2	1	3.6000	.267				6.3000	.011
2	2	2	4.5714	.067				5.4444	.046
			3.7143	.200				5.4000	.051
3	1	1	3.2000	.300				4.5111	.098
								4.4444	.102
3	2	1	4.2857	.100	4	3	3	6.7455	.010
			3.8571	.133				6.7091	.013
3	2	2	5.3572	.029				5.7909	.046
			4.7143	.048				5.7273	.050
			4.5000	.067				4.7091	.092
			4.4643	.105				4.7000	.101
3	3	1	5.1429	.043	4	4	1	6.6667	.010
			4.5714	.100				6.1667	.022
			4.0000	.129				4.9667	.048
3	3	2	6.2500	.011				4.8667	.054
			5.3611	.032				4.1667	.082
			5.1389	.061				4.0667	.102
			4.5556	.100	4	4	2	7.0364	.006
			4.2500	.121				6.8727	.011
3	3	3	7.2000	.004				5.4545	.046
			6.4889	.001				5.2364	.052
			5.6889	.029				4.5545	.098
			5.6000	.050				4.4455	.103
			5.0667	.086	4	4	3	7.1439	.010
			4.6222	.100				7.1364	.011
4	1	1	3.5714	.200				5.5985	.049
4	2	1	4.8214	.057				5.5758	.051
			4.5000	.076				4.5455	.099
			4.0179	.114				4.4773	.102
4	2	2	6.0000	.014	4	4	4	7.6538	.008
			5.3333	.033				7.5385	.011
			5.1250	.052				5.6923	.049
			4.3750	.100				5.6538	.054
			4.1667	.105				4.6539	.097
4	3	1	5.8333	.021				4.5001	.104
			5.2083	.050	5	1	1	3.8571	.143
			5.0000	.057	5	2	1	5.2500	.036
			4.0556	.093				5.0000	.048
			3.8889	.129				4.4500	.071
								4.2000	.095
								4.0500	.119

Source: Adapted from W. H. Kruskal and W. A. Wallis, "Use of Ranks on One-Criterion Variance Analysis," *Journal of the American Statistical Association* 47 (1952), pp. 583–621.

TABLE B.13 | *(concluded)*

Sample sizes			Critical		Sample sizes			Critical	
n_1	n_2	n_3	value	α	n_1	n_2	n_3	value	α
5	2	2	6.5333	.008	5	4	4	7.7604	.009
			6.1333	.013				7.7440	.011
			5.1600	.034				5.6571	.049
			5.0400	.056				5.6176	.050
			4.3733	.090				4.6187	.100
			4.2933	.112				4.5527	.102
5	3	1	6.4000	.012	5	5	1	7.3091	.009
			4.9600	.048				6.8364	.011
			4.8711	.052				5.1273	.046
			4.0178	.095				4.9091	.053
			3.8400	.123				4.1091	.086
5	3	2	6.9091	.009				4.0364	.105
			6.8281	.010	5	5	2	7.3385	.010
			5.2509	.049				7.2692	.010
			5.1055	.052				5.3385	.047
			4.6509	.091				5.2462	.051
			4.4945	.101				4.6231	.097
5	3	3	7.0788	.009				4.5077	.100
			6.9818	.011	5	5	3	7.5780	.010
			5.6485	.049				7.5429	.010
			5.5152	.051				5.7055	.046
			4.5333	.097				5.6264	.051
			4.4121	.109				4.5451	.100
5	4	1	6.9545	.008				4.5363	.102
			6.8400	.011	5	5	4	7.8229	.010
			4.9855	.044				7.7914	.010
			4.8600	.056				5.6657	.049
			3.9873	.098				5.6429	.050
			3.9600	.102				4.5229	.100
5	4	2	7.2045	.009				4.5200	.101
			7.1182	.010	5	5	5	8.0000	.009
			5.2727	.049				7.9800	.010
			5.2682	.050				5.7800	.049
			4.5409	.098				5.6600	.051
			4.5182	.101				4.5600	.100
5	4	3	7.4449	.010				4.5000	.102
			7.3949	.011					
			5.6564	.049					
			5.6308	.050					
			4.5487	.099					
			4.5231	.103					

TABLE   B.14 | Critical values of the Spearman test statistic

n	α = .100	.050	.025	.010	.005	.001
4	.8000	.8000				
5	.7000	.8000	.9000	.9000		
6	.6000	.7714	.8286	.8857	.9429	
7	.5357	.6786	.7450	.8571	.8929	.9643
8	.5000	.6190	.7143	.8095	.8571	.9286
9	.4667	.5833	.6833	.7667	.8167	.9000
10	.4424	.5515	.6364	.7333	.7818	.8667
11	.4182	.5273	.6091	.7000	.7455	.8364
12	.3986	.4965	.5804	.6713	.7273	.8182
13	.3791	.4780	.5549	.6429	.6978	.7912
14	.3626	.4593	.5341	.6220	.6747	.7670
15	.3500	.4429	.5179	.6000	.6536	.7464
16	.3382	.4265	.5000	.5824	.6324	.7265
17	.3260	.4118	.4853	.5637	.6152	.7083
18	.3148	.3994	.4716	.5480	.5975	.6904
19	.3070	.3895	.4579	.5333	.5825	.6737
20	.2977	.3789	.4451	.5203	.5684	.6586
21	.2909	.3688	.4351	.5078	.5545	.6455
22	.2829	.3597	.4241	.4963	.5426	.6318
23	.2767	.3518	.4150	.4852	.5306	.6186
24	.2704	.3435	.4061	.4748	.5200	.6070
25	.2646	.3362	.3977	.4654	.5100	.5962
26	.2588	.3299	.3894	.4564	.5002	.5856
27	.2540	.3236	.3822	.4481	.4915	.5757
28	.2490	.3175	.3749	.4401	.4828	.5660
29	.2443	.3113	.3685	.4320	.4744	.5567
30	.2400	.3059	.3620	.4251	.4665	.5479

Table entries are critical values ρ_α such that the $P(\rho \geq \rho_\alpha) = \alpha$.

If $n > 30$, the approximate critical values ρ_α may be obtained from

$$\rho_\alpha = \frac{z_\alpha}{\sqrt{n-1}}$$

where z_α is the standard normal variate such that a proportion α of the area is to the right of z_α. The lower percentiles may be obtained from the equation,

$$\rho_\alpha = -\rho_{1-\alpha}, \qquad \rho < 0.50$$

Source: Adapted from G. J. Glasser and R. F. Winter, "Critical Values of the Coefficient of Rank Correlation for Testing the Hypothesis of Independence," *Biometrika,* vol. 48 (1961), pp. 444–48.

TABLE B.15 | Critical values of the Kolmogorov test statistic: Two-sided test

	$\alpha = .20$	.10	.05	.02	.01		$\alpha = .20$	.10	.05	.02	.01
n = 1	.900	.950	.975	.990	.995	n = 21	.226	.259	.287	.321	.344
2	.684	.776	.842	.900	.929	22	.221	.253	.281	.314	.337
3	.565	.636	.708	.785	.829	23	.216	.247	.275	.307	.330
4	.493	.565	.624	.689	.734	24	.212	.242	.269	.301	.323
5	.447	.509	.563	.627	.669	25	.208	.238	.264	.295	.317
6	.410	.468	.519	.577	.617	26	.204	.233	.259	.290	.311
7	.381	.436	.483	.538	.576	27	.200	.229	.254	.284	.305
8	.358	.410	.454	.507	.542	28	.197	.225	.250	.279	.300
9	.339	.387	.430	.480	.513	29	.193	.221	.246	.275	.295
10	.323	.369	.409	.457	.489	30	.190	.218	.242	.270	.290
11	.308	.352	.391	.437	.468	31	.187	.214	.238	.266	.285
12	.296	.338	.375	.419	.449	32	.184	.211	.234	.262	.281
13	.285	.325	.361	.404	.432	33	.182	.208	.231	.258	.277
14	.275	.314	.349	.390	.418	34	.179	.205	.227	.254	.273
15	.266	.304	.338	.377	.404	35	.177	.202	.224	.251	.269
16	.258	.295	.327	.366	.392	36	.174	.199	.221	.247	.265
17	.250	.286	.318	.355	.381	37	.172	.196	.218	.244	.262
18	.244	.279	.309	.346	.371	38	.170	.194	.215	.241	.258
19	.237	.271	.301	.337	.361	39	.168	.191	.213	.238	.255
20	.232	.265	.294	.329	.352	40	.165	.189	.210	.235	.252
				Approximation for n > 40			$\dfrac{1.07}{\sqrt{n}}$	$\dfrac{1.22}{\sqrt{n}}$	$\dfrac{1.36}{\sqrt{n}}$	$\dfrac{1.52}{\sqrt{n}}$	$\dfrac{1.63}{\sqrt{n}}$

Table entries are critical values W_α such that $P(\mathbf{W} \geq W_\alpha) = \alpha$.

Source: Adapted from L. H. Miller, "Tables of Percentage Points of Kolmogorov Statistic," *Journal of the American Statistical Association*, vol. 51 (1956), pp. 111–21.

TABLE B.16 | Critical values of the Lilliefors test statistic

	$\alpha = .20$	.15	.10	.05	.01
Sample size n = 4	.300	.319	.352	.381	.417
5	.285	.299	.315	.337	.405
6	.265	.277	.294	.319	.364
7	.247	.258	.276	.300	.348
8	.233	.244	.261	.285	.331
9	.223	.233	.249	.271	.311
10	.215	.224	.239	.258	.294
11	.206	.217	.230	.249	.284
12	.199	.212	.223	.242	.275
13	.190	.202	.214	.234	.268
14	.183	.194	.207	.227	.261
15	.177	.187	.201	.220	.257
16	.173	.182	.195	.213	.250
17	.169	.177	.189	.206	.245
18	.166	.173	.184	.200	.239
19	.163	.169	.179	.195	.235
20	.160	.166	.174	.190	.231
25	.149	.153	.165	.180	.203
30	.131	.136	.144	.161	.187
Over 30	$\dfrac{.736}{\sqrt{n}}$	$\dfrac{.768}{\sqrt{n}}$	$\dfrac{.805}{\sqrt{n}}$	$\dfrac{.886}{\sqrt{n}}$	$\dfrac{1.031}{\sqrt{n}}$

Table entries are critical values W_α such that $P(\mathbf{W} \geq W_\alpha) = \alpha$.

Source: Adapted from H. W. Lilliefors, "On the Kolmogorov-Smirnov Test for Normality with Mean and Variance Unknown," *Journal of the American Statistical Association*, vol. 62 (1967), pp. 399–402.

TABLE B.17 | Critical values of the runs statistic

		W_α					$W_{1-\alpha}$				
N_1	N_2	$W_{.005}$	$W_{.01}$	$W_{.025}$	$W_{.05}$	$W_{.10}$	$W_{.90}$	$W_{.95}$	$W_{.975}$	$W_{.99}$	$W_{.995}$
2	5	—	—	—	—	3	—	—	—	—	—
	8	—	—	—	3	3	—	—	—	—	—
	11	—	—	—	3	3	—	—	—	—	—
	14	—	—	3	3	3	—	—	—	—	—
	17	—	—	3	3	3	—	—	—	—	—
	20	—	3	3	3	4	—	—	—	—	—
5	5	—	3	3	4	4	8	8	9	9	—
	8	3	3	4	4	5	9	10	10	—	—
	11	4	4	5	5	6	10	—	—	—	—
	14	4	4	5	6	6	—	—	—	—	—
	17	4	5	5	6	7	—	—	—	—	—
	20	5	5	6	6	7	—	—	—	—	—
8	8	4	5	5	6	6	12	12	13	13	14
	11	5	6	6	7	8	13	14	14	15	15
	14	6	6	7	8	8	14	15	15	16	16
	17	6	7	8	8	9	15	15	16	—	—
	20	7	7	8	9	10	15	16	16	—	—
11	11	6	7	8	8	9	15	16	16	17	18
	14	7	8	9	9	10	16	17	18	19	19
	17	8	9	10	10	11	17	18	19	20	21
	20	9	9	10	11	12	18	19	20	21	21
14	14	8	9	10	11	12	18	19	20	21	22
	17	9	10	11	12	13	20	21	22	23	23
	20	10	11	12	13	14	21	22	23	24	24
17	17	11	11	12	13	14	22	23	24	25	25
	20	12	12	14	14	16	23	24	25	26	27
20	20	13	14	15	16	17	25	26	27	28	29

Table entries are critical values W_p such that $P(\mathbf{W} \le W_p) = p$, $p = \alpha$ if $\alpha \le 0.10$ or $p = 1 - \alpha$ if $\alpha \ge 0.90$.

For n or m greater than 20, the critical value $W_p (p = \alpha$ or $p = 1 - \alpha)$ may be approximated by:

$$W_\rho = \frac{2nm}{n+m} + 1 + z_\rho \sqrt{\frac{2nm(2nm - n - m)}{(n+m)^2(n+m+1)}}$$

where z_p is the standard normal variate such that a proportion p of the area is to the right of z_p.

To use the table: Let N_1 be the smaller sample size and N_2 the larger. If the exact values of N_1 and N_2 are not listed, use the nearest values given as an approximation. Reject H_0 if T is less than W_α (or greater than $W_{1-\alpha}$) for the one-tailed test at the significance level. For the two-tailed test, reject H_0 if either $T > W_{1-\alpha/2}$ or $T < W_{\alpha/2}$ at the α significance level.

Source: Adapted from F. S. Swed and C. Eisenhart, "Tables for Testing Randomness of Grouping in a Sequence of Alternatives," *The Annals of Mathematical Statistics,* vol. 14 (1943), pp. 66–87.

TABLE B.18 | Relationship between z and r

z	.00	.01	.02	.03	.04	.05	.06	.07	.08	.09
.0	.0000	.0100	.0200	.0300	.0400	.0500	.0599	.0699	.0798	.0898
.1	.0997	.1096	.1194	.1293	.1391	.1489	.1587	.1684	.1781	.1878
.2	.1974	.2070	.2165	.2260	.2355	.2449	.2543	.2636	.2729	.2821
.3	.2913	.3004	.3095	.3185	.3275	.3364	.3452	.3540	.3627	.3714
.4	.3800	.3885	.3969	.4053	.4136	.4219	.4301	.4382	.4462	.4542
.5	.4621	.4700	.4777	.4854	.4930	.5005	.5080	.5154	.5227	.5299
.6	.5370	.5441	.5511	.5581	.5649	.5717	.5784	.5850	.5915	.5980
.7	.6044	.6107	.6169	.6231	.6291	.6352	.6411	.6469	.6527	.6584
.8	.6640	.6696	.6751	.6805	.6858	.6911	.6963	.7014	.7064	.7114
.9	.7163	.7211	.7259	.7306	.7352	.7398	.7443	.7487	.7531	.7574
1.0	.7616	.7658	.7699	.7739	.7779	.7818	.7857	.7895	.7932	.7969
1.1	.8005	.8041	.8076	.8110	.8144	.8178	.8210	.8243	.8275	.8306
1.2	.8337	.8367	.8397	.8426	.8455	.8483	.8511	.8538	.8565	.8591
1.3	.8617	.8643	.8668	.8693	.8717	.8741	.8764	.8787	.8810	.8832
1.4	.8854	.8875	.8896	.8917	.8937	.8957	.8977	.8996	.9015	.9033
1.5	.9052	.9069	.9087	.9104	.9121	.9138	.9154	.9170	.9186	.9202
1.6	.9217	.9232	.9246	.9261	.9275	.9289	.9302	.9316	.9329	.9342
1.7	.9354	.9367	.9379	.9391	.9402	.9414	.9425	.9436	.9447	.9458
1.8	.9468	.9478	.9488	.9498	.9508	.9518	.9527	.9536	.9545	.9554
1.9	.9562	.9571	.9579	.9587	.9595	.9603	.9611	.9619	.9626	.9633
2.0	.9640	.9647	.9654	.9661	.9668	.9674	.9680	.9687	.9693	.9699
2.1	.9705	.9710	.9716	.9722	.9727	.9732	.9738	.9743	.9748	.9753
2.2	.9757	.9762	.9767	.9771	.9776	.9780	.9785	.9789	.9793	.9797
2.3	.9801	.9805	.9809	.9812	.9816	.9820	.9823	.9827	.9830	.9834
2.4	.9837	.9840	.9843	.9846	.9849	.9852	.9855	.9858	.9861	.9863
2.5	.9866	.9869	.9871	.9874	.9876	.9879	.9881	.9884	.9886	.9888
2.6	.9890	.9892	.9895	.9897	.9899	.9901	.9903	.9905	.9906	.9908
2.7	.9910	.9912	.9914	.9915	.9917	.9919	.9920	.9922	.9923	.9925
2.8	.9926	.9928	.9929	.9931	.9932	.9933	.9935	.9936	.9937	.9938
2.9	.9940	.9941	.9942	.9943	.9944	.9945	.9946	.9947	.9949	.9950
3.0	.9951									
4.0	.9993									
5.0	.9999									

The z-values appear in the scales at the left and above the table; the r-values appear in the body of the table.

Source: Huntsberger, Bellingsley, and Croft, *Statistical Inference for Management and Economics* (Boston: Allyn & Bacon, Inc., 1975©).

TABLE B.19 | Random digits

Line	(1)–(5)	(6)–(10)	(11)–(15)	(16)–(20)	(21)–(25)	(26)–(30)	(31)–(35)
101	13284	16834	74151	92027	24670	36665	00770
102	21224	00370	30420	03883	94648	89428	41583
103	99052	47887	81085	64933	66279	80432	65793
104	00199	50993	98603	38452	87890	94624	69721
105	60578	06483	28733	37867	07936	98710	98539
106	91240	18312	17441	01929	18163	69201	31211
107	97458	14229	12063	59611	32249	90466	33216
108	35249	38646	34475	72417	60514	69257	12489
109	38980	46600	11759	11900	46743	27860	77940
110	10750	52745	38749	87365	58959	53731	89295
111	36247	27850	73958	20673	37800	63835	71051
112	70994	66986	99744	72438	01174	42159	11392
113	99638	94702	11463	18148	81386	80431	90628
114	72055	15774	43857	99805	10419	76939	25993
115	24038	65541	85788	55835	38835	59399	13790
116	74976	14631	35908	28221	39470	91548	12854
117	35553	71628	70189	26436	63407	91178	90348
118	35676	12797	51434	82976	42010	26344	92920
119	74815	67523	72985	23183	02446	63594	98924
120	45246	88048	65173	50989	91060	89894	36036
121	76509	47069	86378	41797	11910	49672	88575
122	19689	90332	04315	21358	97248	11188	39062
123	42751	35318	97513	61537	54955	08159	00337
124	11946	22681	45045	13964	57517	59419	58045
125	96518	48688	20996	11090	48396	57177	83867
126	35726	58643	76869	84622	39098	36083	72505
127	39737	42750	48968	70536	84864	64952	38404
128	97025	66492	56177	04049	80312	48028	26408
129	62814	08075	09788	56350	76787	51591	54509
130	25578	22950	15227	83291	41737	59599	96191
131	68763	69576	88991	49662	46704	63362	56625
132	17900	00813	64361	60725	88974	61005	99709
133	71944	60227	63551	71109	05624	43836	58254
134	54684	93691	85132	64399	29182	44324	14491
135	25946	27623	11258	65204	52832	50880	22273
136	01353	39318	44961	44972	91766	90262	56073
137	99083	88191	27662	99113	57174	35571	99884
138	52021	45406	37945	75234	24327	86978	22644
139	78755	47744	43776	83098	03225	14281	83637
140	25282	69106	59180	16257	22810	43609	12224
141	11959	94202	02743	86847	79725	51811	12998
142	11644	13792	98190	01424	30078	28197	55583
143	06307	97912	68110	59812	95448	43244	31262
144	76285	75714	89585	99296	52640	46518	55486
145	55322	07598	39600	60866	63007	20007	66819
146	78017	90928	90220	92503	83375	26986	74399
147	44768	43342	20696	26331	43140	69744	82928
148	25100	19336	14605	86603	51680	97678	24261
149	83612	46623	62876	85197	07824	91392	58317
150	41347	81666	82961	60413	71020	83658	02415

Source: *Table of 105,000 Random Decimal Digits*, Interstate Commerce Commission, Bureau of Transport Economics and Statistics, 1949.

TABLE B.20 | Durbin-Watson test bounds

Level of significance $\alpha = .05$

	Number of predictors									
	1		2		3		4		5	
n	d_L	d_U	d_L	d_U	d_L	d_U	d_L	d_U	d_L	d_U
15	1.08	1.36	0.95	1.54	0.82	1.75	0.69	1.97	0.56	2.21
16	1.10	1.37	0.98	1.54	0.86	1.73	0.74	1.93	0.62	2.15
17	1.13	1.38	1.02	1.54	0.90	1.71	0.78	1.90	0.67	2.10
18	1.16	1.39	1.05	1.53	0.93	1.69	0.82	1.87	0.71	2.06
19	1.18	1.40	1.08	1.53	0.97	1.68	0.86	1.85	0.75	2.02
20	1.20	1.41	1.10	1.54	1.00	1.68	0.90	1.83	0.79	1.99
21	1.22	1.42	1.13	1.54	1.03	1.67	0.93	1.81	0.83	1.96
22	1.24	1.43	1.15	1.54	1.05	1.66	0.96	1.80	0.86	1.94
23	1.26	1.44	1.17	1.54	1.08	1.66	0.99	1.79	0.90	1.92
24	1.27	1.45	1.19	1.55	1.10	1.66	1.01	1.78	0.93	1.90
25	1.29	1.45	1.21	1.55	1.12	1.66	1.04	1.77	0.95	1.89
26	1.30	1.46	1.22	1.55	1.14	1.65	1.06	1.76	0.98	1.88
27	1.32	1.47	1.24	1.56	1.16	1.65	1.08	1.76	1.01	1.86
28	1.33	1.48	1.26	1.56	1.18	1.65	1.10	1.75	1.03	1.85
29	1.34	1.48	1.27	1.56	1.20	1.65	1.12	1.74	1.05	1.84
30	1.35	1.49	1.28	1.57	1.21	1.65	1.14	1.74	1.07	1.83
31	1.36	1.50	1.30	1.57	1.23	1.65	1.16	1.74	1.09	1.83
32	1.37	1.50	1.31	1.57	1.24	1.65	1.18	1.73	1.11	1.82
33	1.38	1.51	1.32	1.58	1.26	1.65	1.19	1.73	1.13	1.81
34	1.39	1.51	1.33	1.58	1.27	1.65	1.21	1.73	1.15	1.81
35	1.40	1.52	1.34	1.58	1.28	1.65	1.22	1.73	1.16	1.80
36	1.41	1.52	1.35	1.59	1.29	1.65	1.24	1.73	1.18	1.80
37	1.42	1.53	1.36	1.59	1.31	1.66	1.25	1.72	1.19	1.80
38	1.43	1.54	1.37	1.59	1.32	1.66	1.26	1.72	1.21	1.79
39	1.43	1.54	1.38	1.60	1.33	1.66	1.27	1.72	1.22	1.79
40	1.44	1.54	1.39	1.60	1.34	1.66	1.29	1.72	1.23	1.79
45	1.48	1.57	1.43	1.62	1.38	1.67	1.34	1.72	1.29	1.78
50	1.50	1.59	1.46	1.63	1.42	1.67	1.38	1.72	1.34	1.77
55	1.53	1.60	1.49	1.64	1.45	1.68	1.41	1.72	1.38	1.77
60	1.55	1.62	1.51	1.65	1.48	1.69	1.44	1.73	1.41	1.77
65	1.57	1.63	1.54	1.66	1.50	1.70	1.47	1.73	1.44	1.77
70	1.58	1.64	1.55	1.67	1.52	1.70	1.49	1.74	1.46	1.77
75	1.60	1.65	1.57	1.68	1.54	1.71	1.51	1.74	1.49	1.77
80	1.61	1.66	1.59	1.69	1.56	1.72	1.53	1.74	1.51	1.77
85	1.62	1.67	1.60	1.70	1.57	1.72	1.55	1.75	1.52	1.77
90	1.63	1.68	1.61	1.70	1.59	1.73	1.57	1.75	1.54	1.78
95	1.64	1.69	1.62	1.71	1.60	1.73	1.58	1.75	1.56	1.78
100	1.65	1.69	1.63	1.72	1.61	1.74	1.59	1.76	1.57	1.78

TABLE B.20 | *(concluded)*

Level of significance $\alpha = .01$

	Number of predictors									
	1		2		3		4		5	
n	d_L	d_U	d_L	d_U	d_L	d_U	d_L	d_U	d_L	d_U
15	0.81	1.07	0.70	1.25	0.59	1.46	0.49	1.70	0.39	1.96
16	0.84	1.09	0.74	1.25	0.63	1.44	0.53	1.66	0.44	1.90
17	0.87	1.10	0.77	1.25	0.67	1.43	0.57	1.63	0.48	1.85
18	0.90	1.12	0.80	1.26	0.71	1.42	0.61	1.60	0.52	1.80
19	0.93	1.13	0.83	1.26	0.74	1.41	0.65	1.58	0.56	1.77
20	0.95	1.15	0.86	1.27	0.77	1.41	0.68	1.57	0.60	1.74
21	0.97	1.16	0.89	1.27	0.80	1.41	0.72	1.55	0.63	1.71
22	1.00	1.17	0.91	1.28	0.83	1.40	0.75	1.54	0.66	1.69
23	1.02	1.19	0.94	1.29	0.86	1.40	0.77	1.53	0.70	1.67
24	1.04	1.20	0.96	1.30	0.88	1.41	0.80	1.53	0.72	1.66
25	1.05	1.21	0.98	1.30	0.90	1.41	0.83	1.52	0.75	1.65
26	1.07	1.22	1.00	1.31	0.93	1.41	0.85	1.52	0.78	1.64
27	1.09	1.23	1.02	1.32	0.95	1.41	0.88	1.51	0.81	1.63
28	1.10	1.24	1.04	1.32	0.97	1.41	0.90	1.51	0.83	1.62
29	1.12	1.25	1.05	1.33	0.99	1.42	0.92	1.51	0.85	1.61
30	1.13	1.26	1.07	1.34	1.01	1.42	0.94	1.51	0.88	1.61
31	1.15	1.27	1.08	1.34	1.02	1.42	0.96	1.51	0.90	1.60
32	1.16	1.28	1.10	1.35	1.04	1.43	0.98	1.51	0.92	1.60
33	1.17	1.29	1.11	1.36	1.05	1.43	1.00	1.51	0.94	1.59
34	1.18	1.30	1.13	1.36	1.07	1.43	1.01	1.51	0.95	1.59
35	1.19	1.31	1.14	1.37	1.08	1.44	1.03	1.51	0.97	1.59
36	1.21	1.32	1.15	1.38	1.10	1.44	1.04	1.51	0.99	1.59
37	1.22	1.32	1.16	1.38	1.11	1.45	1.06	1.51	1.00	1.59
38	1.23	1.33	1.18	1.39	1.12	1.45	1.07	1.52	1.02	1.58
39	1.24	1.34	1.19	1.39	1.14	1.45	1.09	1.52	1.03	1.58
40	1.25	1.34	1.20	1.40	1.15	1.46	1.10	1.52	1.05	1.58
45	1.29	1.38	1.24	1.42	1.20	1.48	1.16	1.53	1.11	1.58
50	1.32	1.40	1.28	1.45	1.24	1.49	1.20	1.54	1.16	1.59
55	1.36	1.43	1.32	1.47	1.28	1.51	1.25	1.55	1.21	1.59
60	1.38	1.45	1.35	1.48	1.32	1.52	1.28	1.56	1.25	1.60
65	1.41	1.47	1.38	1.50	1.35	1.53	1.31	1.57	1.28	1.61
70	1.43	1.49	1.40	1.52	1.37	1.55	1.34	1.58	1.31	1.61
75	1.45	1.50	1.42	1.53	1.39	1.56	1.37	1.59	1.34	1.62
80	1.47	1.52	1.44	1.54	1.42	1.57	1.39	1.60	1.36	1.62
85	1.48	1.53	1.46	1.55	1.43	1.58	1.41	1.60	1.39	1.63
90	1.50	1.54	1.47	1.56	1.45	1.59	1.43	1.61	1.41	1.64
95	1.51	1.55	1.49	1.57	1.47	1.60	1.45	1.62	1.42	1.64
100	1.52	1.56	1.50	1.58	1.48	1.60	1.46	1.63	1.44	1.65

Source: Reprinted, with permission, from J. Durbin and G. S. Watson, "Testing for Serial Correlation in Least Squares Regression. II," *Biometrika* 38 (1951), pp. 159–78.

TABLE B.21 | Normal scores table of r-values

Reject H_0: *Variable is normally distributed* if r, the correlation between the variable and its *n* scores, is *less* than the tabulated value.

		α	
n	0.10	0.05	0.01
4	0.8951	0.8734	0.8318
5	0.9033	0.8804	0.8320
10	0.9347	0.9180	0.8804
15	0.9506	0.9383	0.9110
20	0.9600	0.9503	0.9290
25	0.9662	0.9582	0.9408
30	0.9707	0.9639	0.9490
40	0.9767	0.9715	0.9597
50	0.9807	0.9764	0.9664
60	0.9835	0.9799	0.9710
75	0.9865	0.9835	0.9757

Printed with permission from: MINITAB Reference Manual, T. A. Ryan, B. L. Joiner and B. F. Ryan; MINITAB, Inc. State College, Pennsylvania, 1982 (page 49).

Appendix C
Data sets

■

Information from a Sample of 100 Employees' Files

C1 = Sex (1 = male; 0 = female)
C2 = Age
C3 = Number of dependents
C4 = Years with the firm
C5 = Most recent employee evaluation (0 to 100)
C6 = College degree (1 = yes; 0 = no)
C7 = Participate in profit-sharing program (1 = yes; 0 = no)
C8 = Employee classification (1 = laborer; 2 = sales; 3 = executive)
C9 = Annual wage (not including bonuses or profit sharing)

Employee Number	C1	C2	C3	C4	C5	C6	C7	C8	C9
1	1	34	2	8	85	1	1	2	36250
2	1	27	1	2	80	1	1	3	28760
3	0	25	0	4	95	1	0	2	22540
4	1	22	0	1	90	1	0	2	23215
5	1	36	3	16	90	0	1	1	31600
6	1	41	4	20	75	0	1	1	38610
7	0	32	2	8	100	1	1	3	39800
8	1	48	1	16	90	0	0	1	44675
9	1	37	2	12	70	0	0	1	34525
10	1	30	4	7	80	0	1	1	28960
11	1	34	3	11	90	0	1	1	33665
12	0	35	2	8	95	0	1	1	29815
13	1	52	1	30	90	1	1	2	58800
14	1	41	2	21	100	0	1	1	40050
15	1	46	1	19	90	0	0	1	40890
16	1	37	0	20	70	1	1	2	38125
17	1	48	1	9	90	0	1	1	41445
18	1	43	1	13	95	0	1	1	38990
19	0	22	0	1	100	1	1	2	24000
20	1	24	0	2	95	1	1	2	26130
21	1	27	0	4	90	1	0	1	26890
22	1	36	4	10	85	0	1	1	30250
23	1	26	2	6	65	0	0	1	24670

TABLE C.1 | *(continued)*

Employee Number	C1	C2	C3	C4	C5	C6	C7	C8	C9
24	1	33	2	9	75	0	1	1	27890
25	1	31	1	11	80	0	1	1	28760
26	0	28	1	6	65	0	0	1	25585
27	0	31	3	8	70	0	1	1	27440
28	1	24	2	8	80	0	0	1	25430
29	1	52	1	26	95	1	1	3	50780
30	1	46	1	18	90	1	1	3	47500
31	1	22	0	4	60	0	0	1	22230
32	1	25	1	5	80	0	1	1	26615
33	1	36	3	11	95	1	1	2	38760
34	1	23	0	4	90	0	1	1	24470
35	1	26	0	2	100	0	1	1	24845
36	1	23	0	3	55	0	0	1	21780
37	1	45	0	16	90	1	1	3	44240
38	0	22	0	2	95	0	1	1	21900
39	0	26	0	6	90	0	1	1	24200
40	1	28	1	4	70	0	0	1	25560
41	1	33	3	9	95	1	1	2	36200
42	1	25	0	3	75	0	1	1	23550
43	1	25	0	5	85	0	1	1	25000
44	1	32	2	8	100	1	1	3	35910
45	0	29	3	8	95	0	1	1	25780
46	1	24	2	2	90	0	0	1	22115
47	1	38	2	12	70	0	1	1	36195
48	0	22	1	2	90	0	1	1	21550
49	1	56	1	24	95	1	1	3	54660
50	1	52	1	20	90	1	1	2	48300
51	1	30	3	10	70	0	1	1	27600
52	1	40	2	12	90	1	1	2	40400
53	1	27	0	7	65	0	0	1	24800
54	0	26	0	5	90	0	1	1	24000
55	1	24	3	3	95	0	0	1	23555
56	1	33	2	7	70	0	1	2	31560
57	1	24	0	1	55	0	0	1	22420
58	1	26	0	5	80	0	1	1	24110
59	0	28	1	3	100	1	1	2	29800
60	1	46	1	19	100	1	1	3	56900
61	1	38	4	10	95	1	1	2	46800
62	1	25	0	2	60	0	0	1	24000
63	1	24	1	3	75	0	0	1	24675
64	1	29	3	8	80	1	0	1	29760
65	1	36	2	6	90	1	1	2	36420
66	1	28	0	6	55	0	1	1	24000
67	1	34	2	4	90	1	1	3	39200
68	0	29	1	4	95	1	1	2	36815
69	1	24	0	6	100	0	1	1	26750
70	1	22	0	1	50	0	0	1	20000
71	1	19	0	1	80	0	0	1	18960
72	1	52	1	26	95	1	1	3	60875
73	1	20	0	2	65	0	0	1	20675

TABLE C.1 | (concluded)

Employee Number	C1	C2	C3	C4	C5	C6	C7	C8	C9
74	0	21	0	2	85	0	0	1	20600
75	1	36	5	12	90	1	1	2	38000
76	0	23	0	3	85	0	0	1	24990
77	1	28	0	2	60	0	0	1	23875
78	1	40	4	13	100	1	1	3	48000
79	1	36	2	9	95	1	1	2	38990
80	1	24	2	2	60	0	0	1	24500
81	0	28	3	5	75	0	1	1	24000
82	1	20	0	2	80	0	0	1	22290
83	1	27	2	4	70	0	1	1	25445
84	1	46	1	20	95	1	1	2	58000
85	1	38	3	7	90	1	1	3	40400
86	1	25	1	4	100	0	1	1	27690
87	0	36	3	5	100	0	1	2	30250
88	1	26	0	2	60	0	0	1	25000
89	1	43	1	12	100	1	1	3	47500
90	1	50	1	18	95	1	1	3	62500
91	1	56	1	24	100	1	1	3	66000
92	1	48	1	22	100	1	1	2	55425
93	1	52	1	23	100	1	1	3	57500
94	1	44	3	17	100	1	1	2	54950
95	1	55	1	21	95	1	1	3	60000
96	1	40	3	14	100	1	1	2	48000
97	1	49	1	21	100	1	1	3	60500
98	1	60	1	34	100	1	1	3	82500
99	1	58	1	27	100	1	1	3	75000
100	1	54	0	18	100	1	1	3	72475

TABLE C.2 | Miles per gallon data for the case in Chapter 15 (multiple linear regression analysis)

Make	NOCYL	MPG	TANK	ENGS	HP	WT	DUMMY
Chevrolet Chevette	4	28	13	1.6	65	2060	0
Chevrolet Chevette (D)	4	36	13	1.8	51	2225	1
Dodge Challenger Coupe	4	23	16	2.6	100	2725	0
Dodge Colt	4	39	11	1.4	64	1729	0
Dodge Omni	4	30	13	1.7	63	2118	0
Dodge Omni	4	25	13	2.2	84	2230	0
Dodge 024	4	30	13	1.7	63	2147	0
Ford Escort	4	28	10	1.6	70	2001	0
Ford Exp	4	26	11	1.6	70	2135	0
Mercury LN7	4	26	11	1.6	70	2146	0
Mercury Lynx	4	29	10	1.6	70	2017	0
Plymouth Champ	4	39	11	1.4	64	1865	0
Plymouth Horizon	4	35	13	1.7	63	2118	0

TABLE C.2 | *(continued)*

Make	NOCYL	MPG	TANK	ENGS	HP	WT	DUMMY
Plymouth Horizon	4	25	13	2.2	84	2230	0
Plymouth TC3	4	34	13	1.7	63	2207	0
Plymouth Sapparo	4	23	16	2.5	100	2756	0
Pontiac T1000	4	28	13	1.6	65	2065	0
Buick Skylark	4	25	14	2.5	90	2531	0
Buick Skylark	6	22	14	2.8	112	2593	0
Chevrolet Camaro	4	23	16	2.5	90	2850	0
Chevrolet Cavalier	4	25	14	1.8	51	2368	0
Chevrolet Celebrity	4	25	16	2.5	90	2734	0
Chevrolet Citation	4	25	14	2.5	90	2483	0
Chevrolet Citation	6	22	14	2.8	112	2538	0
Chevrolet Citation X-11	6	22	14	2.8	135	2538	0
Dodge Aries	4	25	13	2.2	84	2328	0
Dodge Aries	4	23	13	2.6	92	2438	0
Ford Fairmont Futura	4	21	16	2.3	86	2727	0
Ford Fairmont Futura	6	20	16	3.3	87	2781	0
Ford Granada	4	21	16	2.3	86	2801	0
Ford Granada	6	19	16	3.8	112	2897	0
Ford Mustang	4	21	15	2.3	86	2644	0
Ford Mustang	6	20	15	3.3	97	2739	0
Ford Mustang	8	19	15	4.2	111	2847	0
Mercury Capri	4	21	15	2.3	86	2698	0
Mercury Capri	6	21	15	3.3	87	2750	0
Mercury Capri	8	19	15	4.2	111	2799	0
Mercury Zephyr	4	21	16	2.3	86	2759	0
Mercury Zephyr	6	20	16	3.3	87	2813	0
Oldsmobile Ciera	4	25	16	2.5	90	2659	0
Oldsmobile Ciera	6	28	17	4.3	85	2887	0
Oldsmobile Omega	4	25	14	2.5	90	2506	0
Oldsmobile Omega	6	22	14	2.8	112	2563	0
Plymouth Reliant	4	23	13	2.2	84	2328	0
Plymouth Reliant	4	23	13	2.6	92	2439	0
Pontiac Phoenix	4	25	14	2.5	90	2518	0
Buick Century	4	25	16	2.5	90	2683	0
Buick Regal	6	21	17	3.7	119	3245	0
Buick Regal (D)	8	23	18	5.7	105	3605	1
Cadillac Cimarron	4	25	14	1.8	88	3591	0
Chevrolet Malibu	6	21	18	3.8	110	3190	0
Chevrolet Malibu	8	18	18	4.4	115	3319	0
Chevrolet Malibu (D)	8	23	20	5.7	105	3535	1
Chevrolet Monte Carlo	6	21	18	3.8	110	3209	0
Chevrolet Monte Carlo	8	18	18	4.4	115	3338	0
Chevrolet Monte Carlo (D)	8	23	20	5.7	105	3552	1
Dodge Diplomat	6	18	18	3.7	90	3364	0
Dodge Diplomat	8	17	18	5.2	130	3570	0
Dodge 400	4	25	13	2.2	84	2483	0
Dodge Mirada	6	18	18	3.7	90	3395	0
Dodge Mirada	8	17	18	5.2	130	3544	0
Ford Thunderbird	6	19	21	3.8	112	3180	0
Mercury Cougar XR7	6	19	21	3.8	112	3163	0
Oldsmobile Cutlass Supreme	6	21	18	3.8	110	3231	0

TABLE C.2 | *(concluded)*

Make	NOCYL	MPG	TANK	ENGS	HP	WT	DUMMY
Oldsmobile Cutlass Supreme	6	25	18	4.3	85	3341	0
Oldsmobile Cutlass Supreme	8	19	18	4.3	100	3401	0
Oldsmobile Cutlass Supreme (D)	8	23	18	5.7	105	3535	0
Pontiac Grand Prix	6	21	18	3.8	110	3335	1
Buick Electra	6	18	25	4.1	125	3803	0
Buick Electra	8	16	22	5.0	140	3982	0
Buick LeSabre	6	19	25	3.8	110	3620	0
Buick LeSabre	6	18	25	4.1	125	3613	0
Buick LeSabre	8	17	22	5.0	140	3811	0
Buick Riviera	8	16	21	5.0	140	3868	0
Buick Riviera (D)	8	20	21	5.7	105	3988	1
Chevrolet Caprice	8	17	25	4.4	115	3640	0
Chevrolet Caprice (D)	8	22	25	5.7	105	3895	1
Chevrolet Impala	6	19	25	3.8	110	3503	0
Chevrolet Impala	8	17	25	4.4	115	3626	0
Chevrolet Impala (D)	8	22	25	5.7	105	3882	1
Chrysler Cordoba	6	18	18	3.7	90	3460	0
Chrysler Cordoba	8	17	18	5.2	130	3610	0
Chrysler LeBaron	4	25	13	2.2	84	2481	0
Chrysler LeBaron	4	23	13	2.6	92	2582	0
Chrysler New Yorker	6	18	18	3.7	90	3602	0
Chrysler New Yorker	8	17	18	5.2	130	3740	0
Ford LTD "S"	8	18	20	4.2	122	3677	0
Ford LTD "S"	8	17	20	5.0	131	3690	0
Mercury Marquis	8	18	20	4.2	122	3710	0
Mercury Marquis	8	17	20	5.0	132	3723	0
Oldsmobile Delta 88	6	19	25	3.8	110	3557	0
Oldsmobile Delta 88	8	17	25	4.3	100	3683	0
Oldsmobile Delta 88 (D)	8	23	27	5.7	105	3891	1
Oldsmobile Ninety-eight	6	18	25	4.1	125	3785	0
Oldsmobile Ninety-eight	8	17	25	5.0	140	3964	0
Oldsmobile Ninety-eight (D)	8	22	27	5.7	105	4107	1
Oldsmobile Toronado (D)	8	21	23	5.7	105	3932	1
Plymouth Gran Fury	6	18	18	3.7	90	3364	0
Plymouth Gran Fury	8	17	18	5.2	130	3507	0
Pontiac Bonneville	6	21	18	3.8	110	3257	0
Cadillac DeVille	8	17	25	4.1	125	3923	0
Cadillac DeVille	6	18	25	4.1	125	3929	0
Cadillac DeVille (D)	8	22	27	5.7	105	4305	1
Cadillac Eldorado	8	17	20	4.1	125	3733	0
Cadillac Eldorado (D)	8	22	23	5.7	105	4108	1
Cadillac Fleetwood Brougham	8	17	25	4.1	125	3965	0
Cadillac Fleetwood Brougham (D)	8	22	27	5.7	105	4305	1
Cadillac Seville	8	17	20	4.1	125	3814	0
Cadillac Seville (D)	8	20	23	5.7	105	4196	1
Chrysler Imperial	8	16	18	5.2	140	4036	0
Lincoln Continental	6	18	20	3.8	112	3555	0
Lincoln Continental	8	17	23	5.0	134	3699	0
Lincoln Mark IV	8	17	18	5.0	134	4039	0
Lincoln Town Car	8	17	18	5.0	134	4006	0
Chevrolet Corvette	8	15	24	5.7	200	3342	0

Appendix D
Answers to selected problems

Chapter 1

1.2. *Primary data*—data obtained from the organization that collected them. *Secondary data*—data obtained from a source other than the organization that collected them.

Primary data are usually more reliable. The more times the data are presented, the greater the chance for errors (typographical errors, misinterpretations, and so forth). In addition, the primary source will often clearly state the nature of the survey or experiment that produced the data, and the secondary sources typically will not.

1.4. *a.* *Population*—the totality of units under study.

b. *Population characteristic*—an attribute of a population unit.

c. *Census*—the evaluation of every unit in the population under study.

d. *Sample*—a part of a population in which the population characteristic is studied so that inferences may be made from the sample study about the entire population.

e. *Sampling error*—the difference between (1) studying a sample and inferring a *result* about a population characteristic and (2) determining the *result* by taking a census of the population and evaluating the population characteristic.

1.6. *a.* If the manufacturer and supplier both selected random samples fairly, then the difference between the two sample results could be ascribed to sampling error.

b. Yes—by arguing that the manufacturer's sample was not representative of the population; that is, the sample was subject to sampling error.

1.8. *a.* Census; *b.* sample; *c.* sample (although computerization of tax returns may make a census possible).

1.10. *a.* *Statistical survey*—a process of collecting data from existing population units, with no particular control over factors that may affect the population characteristics of interest in the study.

b. *Statistical experiment*—a process of collecting data about a population characteristic when control is exercised over some or all factors that may affect the characteristics of interest in the study.

1.12. *a.* An experiment—the factors that affect gas mileage (precision of recording equipment, type of tires, type of drivers, and so forth) need to be closely controlled.

b. A survey—there is no reason to control factors that may affect the number of sick days taken per month.

c. A survey—again, we are not interested in controlling factors that may affect the interest rate.

1.14. *Self-enumeration*—a questionnaire is used to gather the information, and the respondent completes the questionnaire himself or herself. The most common form of this method of information acquisition is the mail questionnaire. *Personal interview*—an interviewer personally contacts individuals selected to participate in the survey or in the experiment. The interviewer records the responses of each selected individual on a *schedule* (a questionnaire form filled out by the interviewer). *Telephone interview*—an interviewer contacts individuals selected to participate in a survey or in an experiment by telephone. The interviewer records the responses of each selected individual on a schedule.

1.16. *a. Dichotomous question*—binary response, usually yes or no. Its advantage is being simple and straightforward. Its disadvantage is that it is useless with questions that cannot be answered with a simple "yes" or "no." For these questions, we would like to qualify the response or indicate a degree of conviction in the response.

b. Multiple choice—a list of responses is given, from which the respondent usually must select one response. The responses generally provide a degree of strength in one's belief about a specific issue, but the responses can simply be alternative answers to a question, such as a multiple-choice test question. The advantage is that a greater range of responses is provided than with the dichotomous question. The disadvantage is that the range of responses may be too "fine" for the respondent. That is, the difference between specific pairs of responses in a question may not have much meaning to the respondent.

c. Free form—an answer to a question is provided in essay form. The advantage is allowing the respondent to state in his or her own words the answer to the question. The disadvantages are the difficulty evaluating the response and the time-consuming reading and recording of the answers for each question.

1.18. Pretesting identifies ambiguous questions, inappropriate ordering of questions, and questions that need to be stated in different forms. Further, the information gathered in the pretest may be used to estimate quantities required for the proper planning of the survey or the experiment (for example, the determi-

nation of the sample size required to meet specified precision requirements).

1.20. *a. Variable*—the characteristic of the population under study.

b. Dependent variable—the variable whose properties are of primary interest.

c. Independent variable—a variable that may affect the properties of a dependent variable. In an experiment, we wish to control the independent variables to better understand their effects on the dependent variable.

d. Quantitative variable—a variable that can be measured numerically, such as age.

e. Qualitative variable—a variable that is nonnumeric, such as eye color.

1.22. *a.* nominal; *b.* ordinal; *c.* ratio; *d.* ratio; *e.* ratio; *f.* ordinal.

1.24. *a.* The target population is the set of *potential* customers of the drugstore.

b. Yes, the respondents are those who *currently* use the drugstore, and this sample excludes those who have never used the store or who used it at one time but are presently not customers.

c. By household interview or mail survey of residents located within the market area of the drugstore.

1.26. *a.* The residents of Seal Point and, perhaps, surrounding communities.

b. Residents with telephones, a possible source of bias.

c. 183 yes votes, 307 no votes, and 24 no-opinion votes.

d. Ratio—120 votes is "twice" as many as 60.

e. 183, 307, 24.

1.28. Any example is suitable that uses descriptive statistics for describing data sets and inferential statistics to induce a result about a population from a sample.

1.30. No, not usually. The proportion who would typically respond to a first mailing will usually not change much as the sample size is increased. It is better to use follow-up mailings to the nonrespondents than to increase the sample size.

1.32. *a.* It is a leading question—it asks for

agreement. "The amount of money spent on national defense is (*a*) about right, (*b*) too much, (*c*) too little."

b. Few people will be able to recall the number over an entire year. The length of the period should be shortened to one month or two weeks.

c. This statement may "place" the name in some people's minds. "List the five (or so) brand names that come to mind when hi-fi equipment is mentioned."

d. It is a leading question. "The money allocated to send people into space is (*a*) well spent, (*b*) not well spent."

Chapter 2

2.2.

Class number	Class limit	Frequency	Relative frequency	Cumulative frequency	Cumulative relative frequency
1	10 but less than 25	5	5/30 = 0.167	5	0.167
2	25 but less than 40	8	8/30 = 0.267	13	0.434
3	40 but less than 55	10	10/30 = 0.333	23	0.767
4	55 but less than 70	4	4/30 = 0.133	27	0.900
5	70 but less than 85	3	3/30 = 0.100	30	1.000
		30	1.000		

2.4. *a.* Nautilus equipment:

Class number	Class limit	Frequency	Relative frequency	Cumulative frequency	Cumulative relative frequency
1	0 but less than 5	39	0.39	39	0.39
2	5 but less than 10	18	0.18	57	0.57
3	10 but less than 15	17	0.17	74	0.74
4	15 but less than 20	9	0.09	83	0.83
5	20 but less than 25	10	0.10	93	0.93
6	25 but less than 30	4	0.04	97	0.97
7	30 but less than 35	3	0.03	100	1.00
		100	1.00		

b. Aerobic exercise:

Class number	Class limit	Frequency	Relative frequency	Cumulative frequency	Cumulative relative frequency
1	0 but less than 5	36	0.36	36	0.36
2	5 but less than 10	11	0.11	47	0.47
3	10 but less than 15	22	0.22	69	0.69
4	15 but less than 20	9	0.09	78	0.78
5	20 but less than 25	16	0.16	94	0.94
6	25 but less than 30	3	0.03	97	0.97
7	30 but less than 35	3	0.03	100	1.00
		100	1.00		

c. Jogging track:

Class number	Class limit	Frequency	Relative frequency	Cumulative frequency	Cumulative relative frequency
1	0 but less than 5	40	0.40	40	0.40
2	5 but less than 10	16	0.16	56	0.56
3	10 but less than 15	26	0.26	82	0.82
4	15 but less than 20	6	0.06	88	0.88
5	20 but less than 25	11	0.11	99	0.99
6	25 but less than 30	1	0.01	100	1.00
7	30 but less than 35	0	0.00	100	1.00
		100	1.00		

2.6.

a.

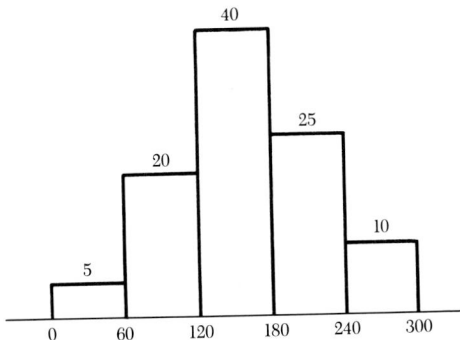

b.

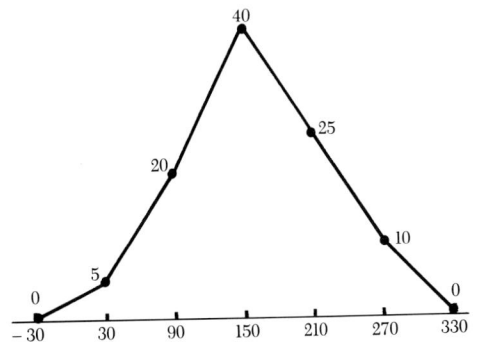

2.8.

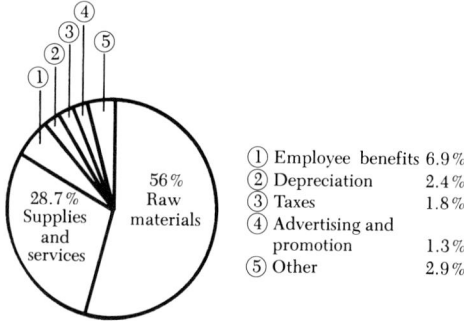

① Employee benefits 6.9%
② Depreciation 2.4%
③ Taxes 1.8%
④ Advertising and
promotion 1.3%
⑤ Other 2.9%

56% Raw materials

28.7% Supplies and services

c. and *d.* Histogram and polygon.

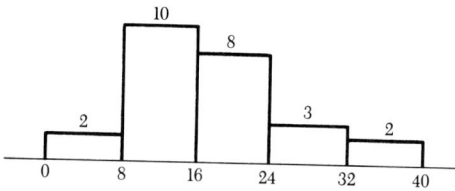

e.

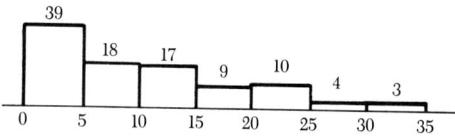

2.10.

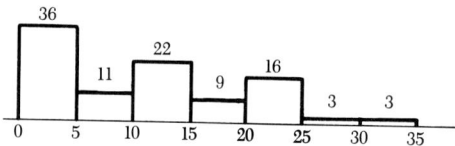

2.12.

a. Nautilus:

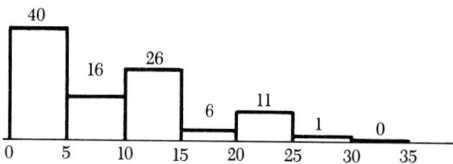

b. Aerobic exercise:

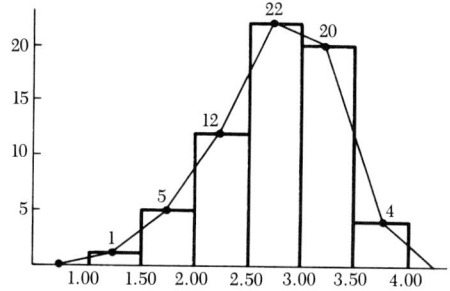

c. Jogging track:

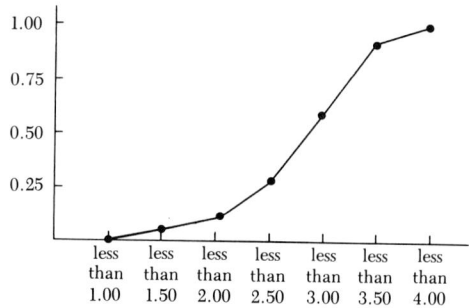

2.14. *a.* 62.50 percent.
b. 71.88 percent.

2.16. *a.* 16.4; 28.4; mean = 21.936; mode: trimodal (18.0, 20.0, 25.50); median = 21.0.
b. 13.9; 23.5; mean = 16.890; mode: four (13.9, 15.0, 16.0, 20.0); median = 16.0.
c. Yes. Mean > median. Skewed to the right.

2.18. *a.* 16.68.
b. 16.
c. 12.

2.20. *a.* 5.44.
b. 5.685.

2.22. *a.* 10.820.
b. 10.75.
c. 10.75.

2.24. *a.* 20 percent.
b. 59.9997.
c. 59.9995.
d. 0.0150.
e. 0.0045.

2.26. *a.* 60.
b. 307.0851.
c. 17.5238.

2.28. Variance = 53.6475; standard deviation = 7.3244.

2.30. *a.* (9 + 10)/2 = 9.5. For either 9 or 10, there are at least 20 percent at or below these numbers and at least 80 percent at or above these numbers.

b. 20. For 20, there are at least 70 percent at or below 20 and at least 30 percent at or above 20 in the ordered array.

2.32. Mean = 3.2; median = 2; skewness measure = 0.7927.

2.34. *a.* 107.735.
b. (104.4 + 102.1)/2 = 103.25.
c. 106.616.
d. Geometric mean or median; distribution is skewed to the right.

2.36. $\mu = \$20,000$; $\sigma = \$2,000$; $0.84 = [1 - (1/k^2)]$, $k = 2.5$.
$\mu \pm 2.5\sigma \rightarrow \$15,000$ to $\$25,000$.

2.38. *a.* $\mu = 18,642$; $\sigma = 3,613$; $M = 16,021$;

$$\gamma = \frac{3(\mu - M)}{\sigma}$$
$$= \frac{3(18,642 - 16,021)}{3,613} = 2.176$$

skewed right.

b. $\mu \pm 2\sigma$ contains at least 75 percent → 11,416 to 25,868.
$\mu \pm 3\sigma$ contains at least 89 percent → 7,803 to 29,481.

2.40.

2.42. *a.* Nautilus:

0*	00000000000000000000000000000000002344
0	555666888888889999
1*	00001222222222444
1	555556668
2*	0000024444
2	5558
3*	000

b. Aerobic exercise:

0*	00000000000000000000000000000000000004
0	56688888888
1*	0000000022222222222444
1	555568888
2*	0000000000444444
2	555
3*	000

c. Jogging:

0*	0000000000000000000000000000000000012244
0	5555666888888899
1*	00000000000222222222222244
1	555566
2*	00000004444
2	8

2.44. *a.* 1st quartile = 2.5.
2nd quartile = 7.5.
3rd quartile = 10.5.

b.

Yes, skewed right.

2.46. *a.* Yes, mean > median, skewed right.

b.

2.48. *a.* No. The heights are relatively in the correct proportion, but the areas are not. The area of the 1986 circle is much too big.

b. Area (1982 circle) $= \pi r^2 = (3.14)(0.033)^2 = 0.0034$.
Area (1986 circle) $= \pi r^2 = (3.14)(1)^2 = 3.14$.
Percent increase in area $= [(3.14 - 0.0034)/0.0034](100\%) = 92,252.94$ percent.
Percent increase in amount $= [(600 - 20)/20](100\%) = 2900$ percent.
Lie factor $= 92,252.94/2,900 = 31.81$.

2.50. Mean $= 1.3$; median $= 3$; mode $= 0, 4$ (bimodal); range $= 20$; variance $= 34.41$.

2.52. $\mu = 0$; $\sigma = 7.30$ (treating the data set as a population). $\bar{x} = 0$; $s = 8.00$ (treating the data set as a sample).

2.54. Not necessarily. It depends on how many shares there are in total. For example, 1.46 percent hold over 500 shares, but suppose one individual in this group holds 1 million shares. Shares are not "widely held" in this case. Rather, they are concentrated in this person's portfolio!

2.56. $\mu = -6.0$; median $= 7.5$; mode $= 10$; range $= 95$; $\sigma^2 = 999.00$ (treating the data set as a population). $\bar{x} = -6.0$; median $= 7.5$; mode $= 10$; range $= 95$; $s^2 = 1,110.00$ (treating the data set as a sample).

2.58. Mean $= 4.9$; geometric mean $= 3.73$; median $= 4.5$ (skewed right because median $<$ mean).

2.60. *a.* $k = 2$; $15.60 \pm 2(6.62) \Rightarrow \$2.36$ to $\$28.84$.
b. $k = \sqrt{10} = 3.1623$; $15.60 \pm 3.1623(6.62) \Rightarrow -\$5.33$ to $\$36.53$ (can truncate lower range at $\$0 \Rightarrow \$0$ to $\$36.53$).
c. $\$31.82 - \$15.60 = 16.22$; $16.22/6.62 = 2.45$ standard deviations. Thus, $k = 2.45$; $1 - 1/k^2 = 1 - 1/(2.45)^2 = 0.8334 \Rightarrow 83.34$ percent. We must assume that we can truncate at $\$0$, since $\$15.60 - 16.22 = -\$0.62$.

2.62. *a.* Mean $= \$62,687.50$.
b. Median $= \$64,000$.

c.

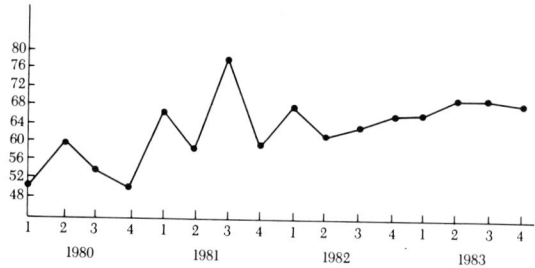

1980 1981 1982 1983

2.64. *a.* $\bar{x} = 510.05$; median $= 500$; 1st quartile $= 476$; 3rd quartile $= 532$; range $= 164$; $s = 48.37$.

b. $\bar{x} = 3.088$; median $= 3.10$; 1st quartile $= 2.75$; 3rd quartile $= 3.38$; range $= 1.56$; $s = 0.456$.

c.

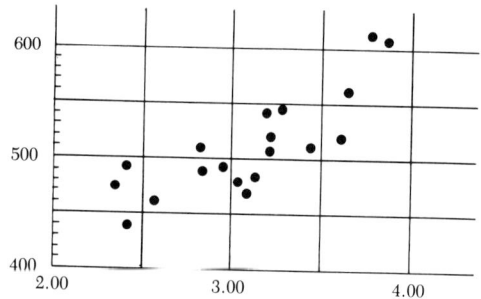

Yes, as the GPA increases, so does the GMAT score.

d.

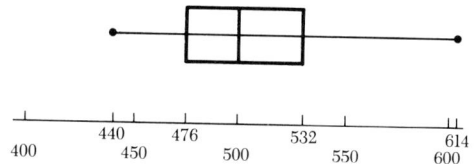

e.

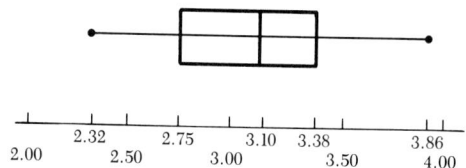

f. $\bar{x} = 24.9$; median $= 24$; mode $= 23, 24$ (bimodal).

g. $\bar{x} = 3.4$ years.

h.

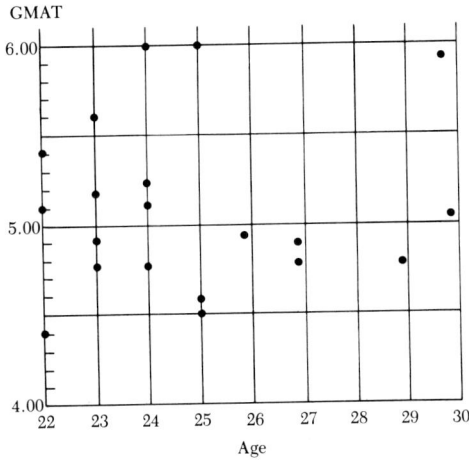

GMAT vs Age scatter plot

There does not appear to be a relationship between age and GMAT score.

i.

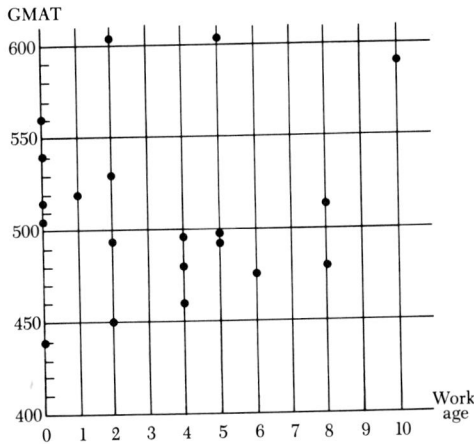

GMAT vs Work age scatter plot

There does not appear to be any relationship between work experience and GMAT score.

j.

Males (12)	Females (8)
GMAT $\bar{x} = 491.08$	GMAT $\bar{x} = 538.5$
GPA $\bar{x} = 2.95$	GPA $\bar{x} = 3.29$

Yes, the females in the study do have higher GMAT scores and GPAs.

Chapter 3

3.2. Let M_1 = Younger man, M_2 = Older man, F_1 = Younger woman, F_2 = Older woman. (See table below.)

a.

Simple events	Chairman	Secretary
E_1	M_1	M_2
E_2	M_1	F_1
E_3	M_1	F_2
E_4	M_2	M_1
E_5	M_2	F_1
E_6	M_2	F_2
E_7	F_1	M_1
E_8	F_1	M_2
E_9	F_1	F_2
E_{10}	F_2	M_1
E_{11}	F_2	M_2
E_{12}	F_2	F_1

b. A: E_1, E_2, E_3
c. B: E_1, E_2, E_3, E_4, E_5, E_6
d. C: E_2, E_3, E_5, E_6, E_9, E_{12}
e. D: E_2, E_3
f. E: E_1, E_2, E_3, E_4, E_5, E_6, E_9, E_{12}

3.4. *a.* e_1: A+; e_2: A; e_3: A−; e_4: B+; e_5: B; e_6: B−; e_7: C; e_8: F; or

e_1: A+, A, A−; e_2: B+, B, B−; e_3: C, F; or

e_1: A+, A, A−,B+, B, B−; e_2: C, F.

3.6.

Number of students
Bill hires

Number of students Herman hires		0	1	2	3
	0	e_{00}	e_{01}	e_{02}	e_{03}
	1	e_{10}	e_{11}	e_{12}	
	2	e_{20}	e_{21}		
	3	e_{30}			

S: $e_{ij} \rightarrow$ event, Herman hires i students; Bill hires j students; $i, j = 0, 1, 2, 3$ with restriction $i + j \leq 3$.
There are 10 points in the sample space: $P(e_{ij}) = 1/10$:
a. $P(A) = P(e_{20}) + P(e_{21}) + P(e_{30}) = 3/10$.
b. $P(B) = P(e_{03}) = 1/10$.

c. $P(C) = P(e_{01}) + P(e_{11}) + P(e_{21}) = 3/10$.

d. $P(D) = P(O) = 0$.

e. $P(E) = P(e_{20}) + P(e_{21}) + P(e_{30}) + P(e_{01}) + P(e_{11}) = 5/10$.

f. $P(F) = P(e_{21}) = 1/10$.

3.8.
a. Subjective.
b. Objective or subjective.
c. Objective (if all have been tested).
d. Subjective.
e. Subjective.

3.10. Product rule: $(26)(26)(26)(10)(10)(10) = 17,576,000$.

3.12.
a. Product rule: $(6)(5)(4)(3)(2)(1) = 720$ (or $P_6^6 = 720$).
b. Product rule: $(A)(5)(4)(3)(2)(1) = 120$.

3.14.
a. Product rule: $(5)(5)(5)(5) = 625$.
b. $(3)(5)(5)(5) = 375$.
c. $(1)(1)(1)(5) = 5$, if he knows in which two courses he will receive an A.

3.16. $C_7^{10} = C_3^{10} = 120$.

3.18. Product rule: Number of combinations $= (40)(40)(40) = 64,000$. Probability $= 1/64,000 = 0.0000156$.

3.20. $10! = 3,628,800$ ways to match:
For the 1st president, one of 10 vice presidents can be chosen to match.
For the 2nd president, one of nine remaining vice presidents can be chosen to match.
For the 3rd president, one of eight remaining vice presidents can be chosen to match.
.
.
.
For the 10th president, one remaining vice president is chosen to match.
By the product rule, $(10)(9)(8) \ldots (1) = 10!$
Probability of correct match by random $= 1/3,628,800 = 0.0000003$.

3.22. Product rule: $(5)(4)(5)(8)(6) = 4,800$.

3.24. Product rule: $(5)(7)(10) = 350$.

3.26. Product rule: $\boxed{W}\boxed{26}\boxed{26} + \boxed{W}\boxed{26}$ $\boxed{26}\boxed{26} = 26^2 + 26^3 = 18,252$.

3.28. $P_5^5 = 5! = 120$.

3.30. Hypergeometric:

$$C_5^{10} \cdot C_5^6 \cdot C_5^8 \cdot C_5^{12} = \frac{10!}{5!5!} \cdot \frac{6!}{5!1!} \cdot \frac{8!}{5!3!} \cdot \frac{12!}{5!7!}$$
$$= 67,060,224$$

3.32.
a. Objective probability is based on the relative frequency definition of probability—if the experiment is repeated N times and event A occurs n times, then the probability that event A occurs in any one of these N trials is n/N. Subjective probability, on the other hand, reflects a personal degree of belief that a certain event will occur.

b. Yes. We can certainly set both objective and subjective probabilities in a coin-flip experiment. Objectively, we may say that Prob(II) = 1/2, arguing on the basis of symmetry or relative frequency, and an individual may set $P(II) = 3/4$, believing, therefore, that the coin is unfair.

c. Whenever the experiment is not repeatable a sufficient number of times to appeal to the relative frequency definition of probability or when other arguments fail (such as those based on symmetry).

3.34. Using the sample space in the text: 36 points in sample space.

a. Total points in event space: $e_{61}, e_{52}, e_{43}, e_{34}, e_{25}, e_{16}, e_{56}, e_{65} = 8$ total points.
$P(7 \text{ or } 11) = 8/36 = 2/9$.

b. Total points in event space: $e_{11}, e_{66} = 2$ total points.
$P(2 \text{ or } 12) = 2/36 = 1/18$.

3.36.
a.

	Male	Female	
Older than 30	2	1	3
Age 30 or younger	0	1	1
	2	2	4

b. Probability (man and older than 30) = 2/4 = 0.50.

3.38. Product rule: $(1)(1)(10)(12)(3)(15) = 5,400$.

3.40.
a. $P_4^{16} = \dfrac{16!}{(16 - 4)!}$
$= (16)(15)(14)(13) = 43,680$.

b. $\dfrac{C_1^1 \cdot C_3^{15}}{C_4^{16}} = \dfrac{(1)(455)}{1820} = 1/4 = 0.25$.

c. Selected second: $\dfrac{P_1^1 P_3^{15}}{P_4^{16}} = \dfrac{2,730}{43,680} = 0.0625$.

Selected third: $\dfrac{P_1^1 P_3^{15}}{P_4^{16}} = 0.0625$.

Selected fourth: $\dfrac{P_1^1 P_3^{15}}{P_4^{16}} = \dfrac{0.0625}{0.1875}$.

Chapter 4

4.2. *a.* 5/6
  b. 1/6
  c. 0
  d. 4/6

4.4. *a.* $P(A \cup E) = 10/36$
  b. $P(A \cup D) = 8/36$
  c. $P(D \cup E) = 4/36$
  d. $P(D \cap E) = 2/36$
  e. $P(A \cap D) = 0$
  f. $P(A \cap E) = 0$

4.6. *a.* Prob [used car] $= (25 + 25)/150 = 1/3$.
  b. Prob [installments/new car] $= 95/100 = 0.95$.

4.8. *a.* $P(A_3) = 260/1,900 = 0.1368$
  b. $P(A_1 \cap B_4) = 300/1,900 = 0.1579$
  c. $P(A_4/B_1) = 150/500 = 0.30$
  d. $P(A_3 \cup B_4) = (260 + 800 - 150)/1,900 = 0.4789$
  e. $P[(B_3 \cup B_4)/A_1] = (50 + 300)/710 = 0.4930$
  f. $P[(A_1 \cup A_3)/B_5] = (10 + 10)/150 = 0.1333$
  g. $P(A_1 \cup A_2 \cup A_3) = (710 + 390 + 260)/1,900 = 0.7158$
  h. $P[B_1/(A_3 \cup A_4)] = (50 + 150)/(260 + 540) = 0.25$
  i. $P[A_3/(B_3 \cup B_5)] = (20 + 10)/(150 + 150) = 0.10$

4.10.

	B_1	B_2	
A_1	0.25	0.25	0.50
A_2	0.15	0.35	0.50
	0.40	0.60	1.00

  a. $(A_1, A_2), (B_1, B_2)$.
  b. $(A_1, A_2), (A_1, B_1), (A_1, B_2), (A_2, B_1),$ $(A_2, B_2), (B_1, B_2)$.
  c. $P(B_2/A_1) = 0.25/0.50 = 0.50$.
  d. $P(A_1 \cap B_2) + P(A_2 \cap B_1) = 0.25 + 0.15 = 0.40$.

4.12. *a.* $P(A/B) = \dfrac{P(A \cap B)}{P(B)} = \dfrac{0.05}{0.40} = 0.125$.

  b. $P(B/A) = \dfrac{P(A \cap B)}{P(A)} = \dfrac{0.05}{0.10} = 0.50$.

4.14. Yes. If two events are mutually exclusive, they cannot jointly occur. Thus if one occurs, the other cannot. This is the strongest form of dependence. The occurrence of one event *precludes* the occurrence of the other. [Note: $P(A \cap B) = 0 \neq P(A)P(B)$ unless either $P(A)$ or $P(B)$ or both is zero, a trivial case.]
No. All we can say is that they may or may not be independent.

4.16. *a.* Yes: $P(A \cap B) = P(A)P(B)$.
  b. No: $P(A \cap C) \neq P(A)P(C)$.
  c. No: $P(A \cap B) \neq 0$.
  d. No: $P(A \cap C) \neq 0$.
  e. No: $P(B \cup C) = P(B) + P(C) - P(B \cap C)$.
Thus $P(B \cap C) = P(B) + P(C) - P(B \cup C) = 0.3 + 0.4 - 0.0 = 0.70$, since $P(B \cap C) \neq 0$, B and C are not mutually exclusive.

4.18. P(booster does not fail)
$$= 1 - P(\text{booster fails})$$
$$= 1 - 0.001 = 0.999.$$

4.20. Let $D_i = i$th item is defective, where $i = 1, 2, 3$.
  a. $P(D_1 \cap D_2 \cap D_3)$
$$= P(D_1)P(D_2)P(D_3) = (0.10)^3$$
$$= 0.001.$$
  b. P(at least one defective) $= 1 - P$(none defective) $= 1 - (0.90)^3 = 0.271$.

4.22. $P(A \cup B) = P(A) + P(B) - P(A \cap B)$.
$P(A \cup B)$ will be less when $P(A \cap B) > 0$.

4.24. *a.* $(1/2)^5 = 1/32$.
  b. $1/2$.

4.26.

Mechanic A   Mechanic B P(Defective found) $=$
$P(D_A) + P(\overline{D}_A \cap D_B)$
$= 0.25 + 0.25 = \underline{0.50}$

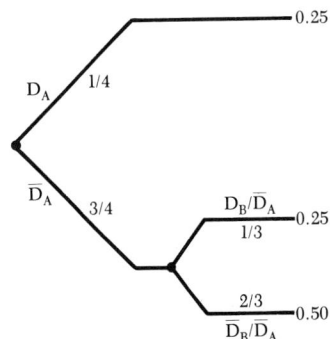

4.28. *a.* $1/P_5^5 = 1/120 = 0.08333$

b.

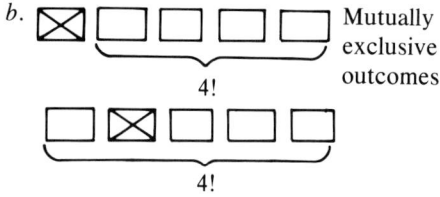

Mutually exclusive outcomes

$$\text{Prob} = \frac{4! + 4!}{5!} = 48/120 = 0.40.$$

4.30. Let S_i = plane shot down at ith battery.

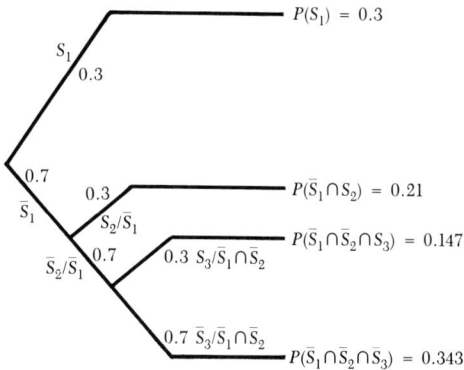

$P(S_1) = 0.3$

$P(\bar{S}_1 \cap S_2) = 0.21$

$P(\bar{S}_1 \cap \bar{S}_2 \cap S_3) = 0.147$

$P(\bar{S}_1 \cap \bar{S}_2 \cap \bar{S}_3) = 0.343$

$P(\text{shoot down}) = 1 - 0.343 = 0.657.$

4.32. $\text{Prob [all four work]} = \left[\dfrac{C_2^4 C_0^2}{C_2^6}\right]\left[\dfrac{C_2^4 C_0^2}{C_2^6}\right]$

$$\left[\frac{(6)(1)}{15}\right]\left[\frac{(6)(1)}{15}\right] = \frac{36}{225} = 0.16.$$

4.34.

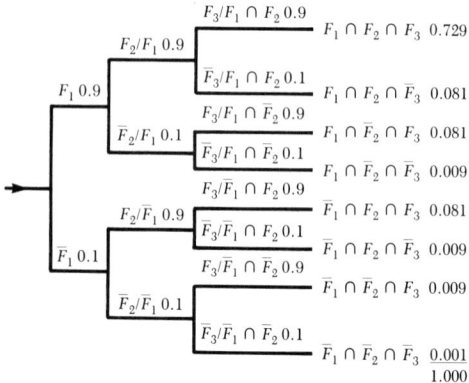

$P[\text{at least one detector signals fire}]$
$\qquad = 1 - P[\text{no detectors signal fire}]$
$\qquad = 1 - 0.001 = 0.999$

4.36.

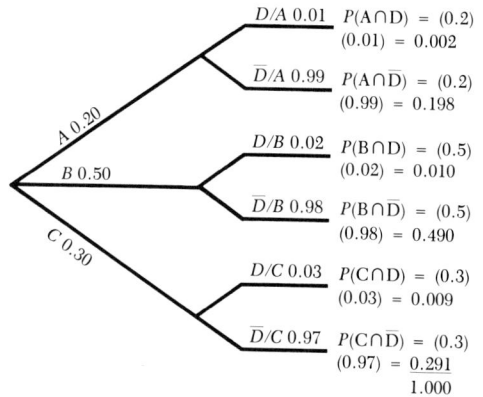

D/A 0.01	$P(A \cap D) = (0.2)(0.01) = 0.002$
$\bar{D}/A$ 0.99	$P(A \cap \bar{D}) = (0.2)(0.99) = 0.198$
D/B 0.02	$P(B \cap D) = (0.5)(0.02) = 0.010$
$\bar{D}/B$ 0.98	$P(B \cap \bar{D}) = (0.5)(0.98) = 0.490$
D/C 0.03	$P(C \cap D) = (0.3)(0.03) = 0.009$
$\bar{D}/C$ 0.97	$P(C \cap \bar{D}) = (0.3)(0.97) = \underline{0.291}$

1.000

$$P(B/D) = \frac{P(B \cap D)}{P(D)} = \frac{0.010}{0.002 + 0.010 + 0.009}$$

$$= \frac{10}{21} = 0.476.$$

4.38. *a.* $P(\text{no pages}) = (0.99)^6 = 0.94148.$

$P(\text{at least one page}) = 1 - P(\text{no pages}) = 1 - 0.94148 = 0.0585.$

b. $P(\text{three pages}) = C_3^6 (0.01)^3 (0.99)^3 = 0.000019406.$

4.40. Define A as trip within continental United States; define B as fraudulent expense statement. $P(A) = 0.75$; $P(B/A) = 0.05$; $P(B/\bar{A}) = 0.25$.

$P(B) = P(A \cap B) + P(\bar{A} \cap B)$
$\qquad = P(A)P(B/A) + P(\bar{A})P(B/\bar{A})$
$\qquad = (0.75)(0.05) + (0.25)(0.25)$
$\qquad = 0.0375 + 0.0625 = 0.10$

4.42.

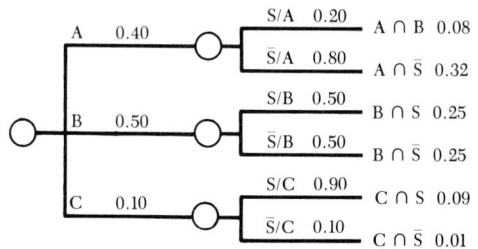

S/A 0.20	$A \cap B$ 0.08
$\bar{S}/A$ 0.80	$A \cap \bar{S}$ 0.32
S/B 0.50	$B \cap S$ 0.25
$\bar{S}/B$ 0.50	$B \cap \bar{S}$ 0.25
S/C 0.90	$C \cap S$ 0.09
$\bar{S}/C$ 0.10	$C \cap \bar{S}$ 0.01

$$P(C/S) = \frac{P(C \cap S)}{P(S)} = \frac{0.09}{0.08 + 0.25 + 0.09} = \frac{0.09}{0.42}$$

$$= 0.2143$$

4.44. $P(A \cup B \cup C) = P(A) + P(B) + P(C) - P(A \cap B) - P(A \cap C) - P(B \cap C) + P(A \cap B \cap C).$

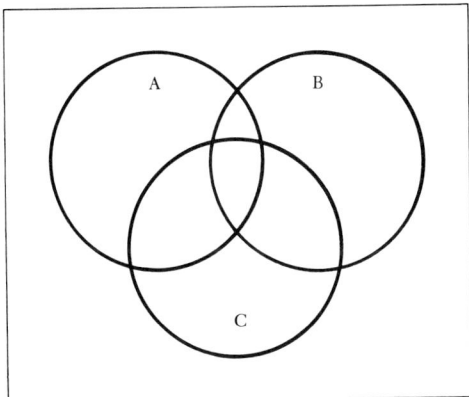

4.46. $P(A) = \frac{1}{3}$; $P(B) = \frac{3}{4}$; $P(A \cap B) = \frac{1}{6}$.

a. $P(A \cup B) = P(A) + P(B) - P(A \cap B) =$
$\frac{1}{3} + \frac{3}{4} - \frac{1}{6} = \frac{4}{12} + \frac{9}{12} - \frac{2}{12} = \frac{11}{12}$.

b. $P(B/A) = \dfrac{P(A \cap B)}{P(A)} = \dfrac{\frac{1}{6}}{\frac{1}{3}} = \frac{1}{2}$.

c. $P(B/A) = \frac{1}{2}$; $P(B) = \frac{3}{4}$; since $P(B/A) \neq P(B)$, no.

4.48. Let A = Town house burglarized, and let B = Beach house burglarized.

$$P(A \cap B) = P(A)P(B) \quad \text{(since events are independent)}$$
$$= (0.01)(0.05)$$
$$= 0.0005.$$

4.50.

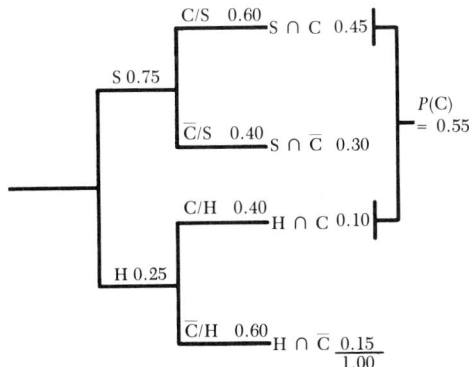

4.52.
a. $P(B_3) = \frac{30}{200} = 0.15$.
b. $P(A_3 \cap B_2) = \frac{15}{200} = 0.075$.
c. $P[(A_1 \cup A_2)/B_3] = (10 + 10)/30 = 0.6667$.
d. $P[(A_1 \cup A_3)/B_2] = (15 + 15)/70 = 0.4286$.
e. $P[A_1/(B_1 \cup B_2)] = (25 + 15)/(100 + 70)$
$= \frac{40}{170} = 0.2353$.
f. $P(A_1/B_3) = \frac{10}{30} = 0.3333$.
g. $P(A_1/B_1) = \frac{25}{100} = 0.25$; $P(A_1) = \frac{50}{200}$
$= 0.25$; $0.25 = 0.25$; yes.

4.54.
a. $P(A_1) = \frac{70}{100} = 0.70$.
b. $P(A_1 \cap A_2) = 0$.
c. $P(A_1/B_1) = \frac{42}{60} = 0.70$.
d. $P(B_3/A_1 \cup A_2) = (0 + 5)/(70 + 25) = \frac{5}{95}$
$= 0.0526$.
e. $P(B_1 \cup B_2/A_2) = (18 + 2)/25 = \frac{20}{25}$
$= 0.80$.
f. $P(\bar{A}_3) = \frac{95}{100} = 0.95$.
g. Yes: $P(A_1/B_1) = P(A_1)$.
h. No: $P(A_1 \cap B_3) \neq P(A_1)P(B_3)$.
i. No: $P(A_1 \cap B_1) \neq 0$.
j. Yes: $P(A_1 \cap B_3) = 0$.
k. Yes: $P(B_1 \cap B_3) = 0$.

4.56.
a. A: Individual has AIDS. B: Test is positive. $P(A) = 0.20$; $P(B) = 0.10$; $P(B/A) = 0.25$.
b. $P(A \cap B) = P(B/A)P(A) = (0.25)(0.20)$
$= 0.05$.
c. $P(A/B) = P(A \cap B)/P(B) = 0.05/0.10$
$= 0.50$.
d. No. $P(A \cap B) \neq P(A)P(B)$.
e. No. $P(A \cap B) \neq 0$.

4.58. Define A as one or more adults in household watch the show; define B as children in household watch show. $P(A) = 0.50$; $P(B) = 0.60$; $P(A/B) = 0.80$.
a. $P(A \cap B) = P(A/B)P(B) = (0.80)(0.60)$
$= 0.48$.
b. $P(A \cup B) = P(A) + P(B) - P(A \cap B)$
$= 0.5 + 0.6 - 0.48 = 0.62$.
c. $1 - P(A \cup B) = 1 - 0.62 = 0.38$.
d. $P(A/\bar{B}) = P(A \cap \bar{B})/P(\bar{B})$.
$P(A) = P(A \cap B) + P(A \cap \bar{B})$: $P(A \cap \bar{B})$
$= P(A) - P(A \cap B) = 0.50 - 0.48$
$= 0.02$.
$P(A/\bar{B}) = 0.02/0.40 = 0.05$.

Chapter 5

5.2.
a. Yes. It establishes a functional relationship between elementary outcomes of an experiment and numerical values.

b.

x	$P(x)$
0	0.20
1	0.30 + 0.10 = 0.40
2	0.40
	1.00

5.4. A discrete random variable with a finite set of values has a finite number of values that the random variable can assume. A discrete random variable with a countably infinite number of values has a set of values that can be placed in a one-to-one correspondence with the positive integers $(1, 2, 3, \ldots)$. In the former case, there are a finite number of values of the random variable, and in the latter case there are an infinite number of values of the random variable.

5.6. The PMF gives the probability that the random variable assumes a value x; the CMF gives the probability that the random variable assumes a value less than or equal to x.

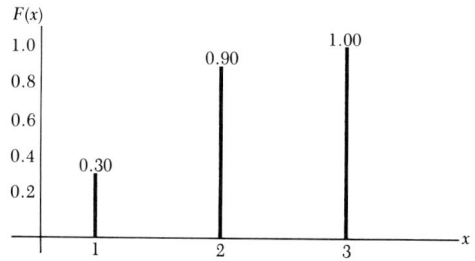

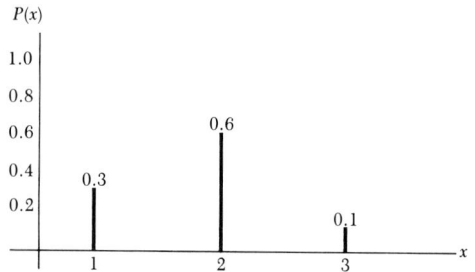

5.8.

x	$P(x)$
0	0.0611
1	0.2894
2	0.4379
3	0.2116
	1.0000

The trials are not independent. Therefore, the hypergeometric rule must be used. For example, $P[\mathbf{X} = 0] =$

$$\frac{C_0^{60} C_3^{40}}{C_4^{100}} = \frac{(1)\dfrac{(40)(39)(38)}{(3)(2)(1)}}{\dfrac{(100)(99)(98)}{(3)(2)(1)}}$$

$$= \frac{9,880}{161,700} = 0.0611.$$

5.10.

x	$P(x)$	$F(x)$
1	.30	.30
2	.60	.90
3	.10	1.00
	1.00	

$$P(3) = F(3) - F(2)$$
$$= 1.00 - 0.90 = 0.10$$
$$P(2) = F(2) - F(1)$$
$$= 0.90 - 0.30 = 0.60$$
$$P(1) = F(1) = 0.30$$

5.12.

x	$P(x)$
0	0.1667
1	0.5000
2	0.3000
3	0.0333
	1.0000

5.14.

x	$P(x)$
2	1/36
3	2/36
4	3/36
5	4/36
6	5/36
7	6/36
8	5/36
9	4/36
10	3/36
11	2/36
12	1/36
	36/36

5.16. 0.8; 0.64.

5.18. 0.0925; 0.

5.20. Yes.

x	$P(x)$
0	0.30
1	0.60
2	0.10
	1.00

$$P(0) = \frac{C_0^2 C_2^3}{C_2^5} = \frac{(1)(3)}{10} = 0.30; \; E(\mathbf{X}) = 0.80.$$

$$P(1) = \frac{C_1^2 C_1^3}{C_2^5} = \frac{(2)(3)}{10} = 0.60.$$

$$P(2) = \frac{C_2^2 C_0^3}{C_2^5} = \frac{(1)(1)}{10} = 0.10.$$

5.22.

x	$P(x)$	$xP(x)$	x^2	$x^2P(x)$
0	0.064	0	0	0
1	0.288	0.288	1	0.288
2	0.432	0.864	4	1.728
3	0.216	0.648	9	1.944
	1.000	1.800		3.960

$E(\mathbf{X}) = 1.800.$
$V(\mathbf{X}) = \Sigma x^2 P(x) - [E(\mathbf{X})]^2$
$\quad 3.96 - (1.8)^2 = 0.72.$

5.24. *a.*

x	$P(x)$
0	0.7737809
1	0.2036266
2	0.0214344
3	0.0011281
4	0.0000297
5	0.0000003
	1.0000000

b.

x	$F(x)$
0	0.7737809
1	0.9774075
2	0.9988419
3	0.9999700
4	0.9999997
5	1.0000000

c. $E(\mathbf{X}) = 0.25.$
d. $V(\mathbf{X}) = 0.30 - (0.25)^2$
$\quad = 0.2375.$

5.26. 0.

5.28. *a.*

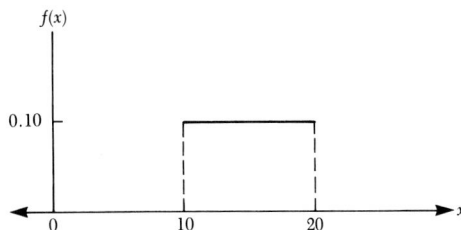

b. Yes. $f(x) \geq 0$; area under $f(x) = 1$.
c. $(6)(0.10) = 0.60.$
d. 0.
e. $(2)(0.10) = 0.20.$

5.30. *a.* $\int_{-\infty}^{\infty} f(x)dx$

$$= \int_{-\infty}^{0} (0)dx + \int_{0}^{3} (1/9)x^2 dx + \int_{3}^{+\infty} (0)dx$$

$$= 0 + \frac{x^3}{27} \Big|_0^3 + 0$$

$$= 27/27 - 0/27 = 1.$$
Also, $f(x) \geq 0$ for all x.

b. $E(\mathbf{X}) = \int_0^3 x(1/9)x^2 dx$

$$= \frac{x^4}{36} \Big|_0^3 = 81/36 - 0 = 81/36 = 2.25.$$

c. $V(\mathbf{X}) = E(\mathbf{X}^2) - [E(\mathbf{X})]^2.$

$$E(\mathbf{X}^2) = \int_0^3 x^2(1/9)x^2 dx$$

$$= \frac{x^5}{45} \Big|_0^3 = \frac{243}{45} - 0 = \frac{243}{45} = 5.4.$$

$V(\mathbf{X}) = 5.4 - (2.25)^2 = 0.3375.$

d. $\max f(x) = 1$ at $x = 3$; thus mode = 3.

e. $\int_0^x f(x)dx = 0.50$; solve for x.

$$\int_0^x (1/9)x^2 dx = x^3/27 \Big|_0^x$$

$$= x^3/27 - 0 = x^3/27.$$

$x^3/27 = 0.50$; $x^3 = (1/2)(27) = 13.5$;
$$x = \sqrt[3]{13.5} = 2.38.$$

f. $F(x) = \int_0^x (1/9)x^2 dx = x^3/27 \Big|_0^x$

$$= x^3/27 - 0 = x^3/27.$$

$$F(x) = \begin{cases} 0 & x < 0 \\ x^3/27 & 0 \leq x \leq 3. \\ 1 & x \geq 3 \end{cases}$$

5.32. $F(x) = \int_{10}^{x} 0.10\,dx = 0.10x \Big|_{10}^{x} = 0.1x - 1.$

$$F(x) = \begin{cases} 0 & x < 10 \\ 0.1x - 1 & 10 \le x \le 20. \\ 1 & x > 20 \end{cases}$$

5.34.

x	P(x)	xP(x)	2x	2xP(x)	(x − 1)²	(x − 1)²P(x)	[x − E(X)]	[x − E(X)]P(x)
0	0.10	0.00	0	0.00	1	0.10	−2.10	−.21
1	0.10	0.10	2	0.20	0	0.00	−1.10	−.11
2	0.40	0.80	4	1.60	1	0.40	−0.10	−.04
3	0.40	1.20	6	2.40	4	1.60	+0.90	+.36
	1.00	2.10		4.20		2.10		0.00

a. $E(\mathbf{X}) = 2.10.$
b. $E(2\mathbf{X}) = 4.20.$
c. $E(\mathbf{X} - 1)^2 = 2.10.$
d. $E[\mathbf{X} - E(\mathbf{X})] = 0.00.$

5.36.

x	P(x)	xP(x)
0	0.80	0.000
1	0.16	0.160
2	0.02	0.040
3	0.008	0.024
4	0.006	0.024
5	0.006	0.030
		0.278

Let $\mathbf{X}$ = The number of defectives per day.
a. $E(\mathbf{X}) = \Sigma xP(x) = 0.278.$
b. $E(\mathbf{S}) = \$50 + \$30 \cdot 0.278 = \$58.34.$
c. Mode is 0; hence most likely salary is $\$50 + \$30(0) = \$50.$

5.38. a.

		y			
		0	1	2	
	0	0.2	0.2	0.2	0.6
x	1	0.1	0.1	0.2	0.4
		0.3	0.3	0.4	1.0

b. No. $P[\mathbf{X} = 0 \cap \mathbf{Y} = 0] \ne P[\mathbf{X} = 0]P[\mathbf{Y} = 0].$

5.40. a. 7.60, 3.00, 4.64, 0.
b. 22.80.
c. 0; yes.

5.42. *Theoretical distribution:*

x	P(x)
0	20/56 = 0.357
1	30/56 = 0.536
2	6/56 = 0.107
	1.000

An example of simulated distribution (do one yourself!):

x	P(x)
0	41/100 = 0.41
1	48/100 = 0.48
2	11/100 = 0.11
	1.00

5.44. No. Probabilities do not sum to one.

5.46.

x	P(x)
0	0.3164
1	0.4219
2	0.2109
3	0.0469
4	0.0039
	1.0000

$P(x) = C_x^4 (0.25)^x (0.75)^{4-x}$.

$x = 0, 1, 2, 3, 4.$

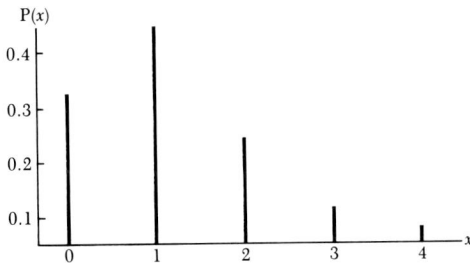

5.48. *a.* 0.60.

b.

x	F(x)
0	0.75
1	0.80
2	0.90
3	0.95
4	1.00

c. $P[X \geq 1] = P(1) + P(2) + P(3) + P(4)$
$= 0.25.$

d. Let Y = Time required to complete report. Then $Y = (1)X$ hour. Thus $E(Y) = E[(1)(X)] = (1)E(X) = E(X) = 0.60$ hour.

5.50. *a.* \$36.50.
b. \$3,650,000.
c. Slow production.

5.52. *a.*

x	P(x)	xP(x)	x²P(x)
0	.48	0	0
1	.20	.20	.20
2	.15	.30	.60
3	.08	.24	.72
4	.05	.20	.80
5	.03	.15	.75
6	.01	.06	.36
	1.00	1.15	3.43

$E(X) = 1.15.$
$V(X) = E(X^2) - [E(X)]^2$
$= 3.43 - (1.15)^2 = 2.1075.$

b.

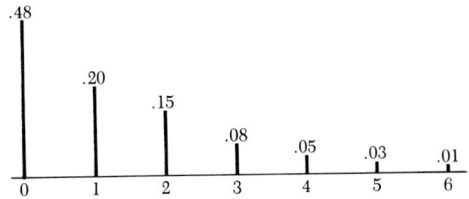

Yes—skewed right.

c. $P[Y = 1/X > 0]$

$= \dfrac{P[Y = 1 \cap X > 0]}{P[X > 0]} = \dfrac{P[X = 1]}{P[X > 0]}$

$= \dfrac{0.20}{0.52} = 0.38.$

Similarly, $P[Y = k/X > 0]$

$= \dfrac{P[X > k]}{P[X > 0]}, \quad k = 2, 3, 4, 5, 6.$

d.

y	P(y)	yP(y)	y²P(y)
1	.38	.38	0.38
2	.29	.58	1.16
3	.15	.45	1.35
4	.10	.40	1.60
5	.06	.30	1.50
6	.02	.12	0.72
		2.23	6.71

$E(Y) = 2.23; \quad V(Y) = 6.71 - (2.23)^2 = 1.74.$

e. $E(Y)$ is a better measure of "family size"; it is the conditional mean, given there is at least one child.

f. $Z = Y + 2; E(Z) = E(Y) + 2 = 4.43;$ $V(Z) = V(Y) = 1.74.$ The random variable Z includes the parents. This is probably the best measure yet if we mean by family the inclusion of children and parents.

g. $\mu_X \pm 2\sigma_X; 1.15 \pm (2)(1.4517); -1.75$ to 4.05 contains at least 3/4 of values of X. Actually, 96 percent are contained in this interval.

5.54. *a.* 1.15 units.
b. 1.4925 units.
c. 0.

d. Let $\mathbf{Y}$ = Cost to IBM. $\mathbf{Y}$ = \$250$\mathbf{X}$. $E(\mathbf{Y})$ = $E(250\mathbf{X})$ = $250E(\mathbf{X})$ = $250(1.15)$ = \$287.50.

5.56. *a.* $E(\mathbf{XY})$ = $E(\mathbf{X})E(\mathbf{Y})$ = $(10)(8)$ = 80.
b. $V(\mathbf{X} + \mathbf{Y})$ = $V(\mathbf{X}) + V(\mathbf{Y})$ = $5 + 4$ = 9.
c. $V(\mathbf{X} - \mathbf{Y})$ = $V(\mathbf{X}) + V(\mathbf{Y})$ = $5 + 4$ = 9.
d. $COV(\mathbf{X}, \mathbf{Y})$ = $E(\mathbf{XY}) - E(\mathbf{X})E(\mathbf{Y})$ = $80 - (10)(8)$ = 0.

Chapter 6

6.2. 0.17 to 5.83; 100 percent.

6.4. $2\frac{1}{6}$; $9\frac{1}{6} - (2\frac{1}{6})^2$.

6.6. *a.* 0.054; *b.* 0.073; *c.* 0.981; *d.* 0.191; *e.* 6; 3.

6.8. 0.432.

6.10. *a.* 0.969; *b.* 0.938.

6.12. *a.* 0.048; *b.* 0.140; *c.* 0.031.

6.14. *a.* 0.098; *b.* 0.266; *c.* 2.5.

6.16. *a.* 1; *b.* 0.774; *c.* 0.109; *d.* 0.349; *e.* 0.736; *f.* 0.001; *g.* 0.019; *h.* 0.098; *i.* 0.677; *j.* 0.001.

6.18. 0.1316.

6.20. 0.128.

6.22. *a.* 0.011; *b.* 2; *c.* 2.

6.24. *a.* $N = 10$; $N_1 = 2$; $N_2 = 8$; $n = 3$. $P(\mathbf{X} = 2)$ = $P(2)$ = $C_2^2 C_1^8 / C_3^{10}$ = $(1)(8)/120$ = 0.0667.
b. $n = 3$; $\pi = 0.20$. $P(\mathbf{X} = 2)$ = $C_2^3 (0.20)^2 (0.80)^1$ = 0.096.
c. Hypergeometric—trials are dependent.

6.26. *a.* $\dfrac{C_5^{10} C_0^{10}}{C_5^{20}}$ = 0.01626.
b. $\dfrac{C_3^{10} C_2^{10} + C_4^{10} C_1^{10} + C_5^{10} C_0^{10}}{C_5^{20}}$ = 0.5.

6.28. *a.* 0.1804; *b.* 0.3233; *c.* yes.

6.30. 0.9084; Poisson distribution requirements are met.

6.32. 0.406.

6.34. 400.

6.36. 0.098.

6.38. *a.* 0.616.
b. 5.
c. 0.185.
d. 5.

6.40. *a.* Successes and failures only, with independent trials.

b. The population of successes and failures is divided between two subpopulations, and trials are not independent.
c. Independence versus nonindependence.

6.42. *a.* $(0.95)^2$ = 0.7738; *b.* $1 - (0.05)^5$ = 0.9999997; *c.* $(5)(0.95)$ = 4.75.

6.44. *a.* Bernoulli.

b.

x	$P(x)$
0	0.05
1	0.95
	1.00

c. $E(\mathbf{X})$ = $0(0.05) + 1(0.95)$ = 0.95. It represents the probability that the switch operates correctly.

6.46. *a.* 0.013; *b.* 0.358; *c.* $(20)(0.05)$ = 1.

6.48. *a.* Hypergeometric.
b. $\dfrac{C_6^6 C_0^4}{C_6^{10}} = \dfrac{(1)(1)}{210}$ = 0.00476.

6.50. *a.* 0.0463; *b.* $1 - (0.0111 + 0.0500)$ = 0.9389; *c.* 0.0016; *d.* Approximately 1.

Chapter 7

7.2. *a.* 67.5 minutes; *b.* 4.33 minutes; *c.* $\frac{1}{3}$; *d.* $\frac{1}{3}$.

7.4. *a.* 0.4938; *b.* 0.1554; *c.* 0.3944; *d.* 0.2681; *e.* 0.0198; *f.* 0.0475; *g.* 0.2119; *h.* 0.9713.

7.6. *a.* 0.3849; *b.* 0.3159; *c.* 0.4265; *d.* 0.4251; *e.* 0.3227; *f.* 0.0224; *g.* 0.9516; *h.* 0.5070; *i.* 0.0062; *j.* 0.7257; *k.* 0.1112; *l.* 0.9515.

7.8. *a.* 482; *b.* 62.9; *c.* 159.80.

7.10. 0.0228.

7.12. 78.32 inches.

7.14. 0.8399.

7.16. \$32,800.

7.18. *a.* \$38,840; *b.* \$37,010; *c.* \$32,990.

7.20. *a.* 0.50; *b.* 0.7123; *c.* 0.4246.

7.22. 0.0228.

7.24. *a.* 0.578; *b.* 0.5752 (with correction).

7.26. 0.9946.

7.28. 0.6757.

7.30. *a.* $f(x) = \lambda e^{-\lambda x} = (0.80)e^{-0.80(x)}$

x	$f(x)$
0	0.80
1	0.36
2	0.16
3	0.07
4	0.03
5	0.01

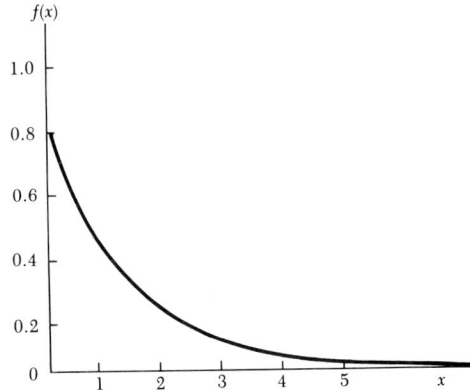

b. $E(\mathbf{X}) = 1/0.8 = 1.25$. $V(\mathbf{X}) = 1/(0.8)^2 = 1.56$.
$\sqrt{1.56} = 1.25$.

c. Within one standard deviation:
$P[1.25 - 1.25 \le \mathbf{X} \le 1.25 + 1.25] = P[0 \le \mathbf{X} \le 2.5] = F(2.5); F(2.5) = 1 - e^{-(0.80)(2.5)} = 1 - e^{-2} = 1 - 0.135337 = 0.865$.
Within two standard deviations:
$P[1.25 - 2(1.25) \le \mathbf{X} \le 1.25 + 2(1.25)] = P[0 \le \mathbf{X} \le 3.75] = F(3.75); F(3.75) = 1 - e^{-(0.80)(3.75)} = 1 - e^{-3} = 1 - 0.049789 = 0.950$.

7.32. *a.* $\lambda = 550; F(375) = 1 - e^{-375/550} = 1 - e^{-0.68}; e^{-0.65} = 0.522; e^{-0.70} = 0.497$. By interpolation, $e^{-0.68} = 0.507; F(375) = 1 - 0.507 = 0.493$.

b. $F(1,000) = 1 - e^{-1000/550} = 1 - e^{-1.82} = 1 - 0.162$ (by interpolation) $= 0.838$.

c. $F(1,200) - F(500)$
$= [1 - e^{-1,200/550}] - [1 - e^{-500/550}]$
$= [1 - e^{-2.18}] - [1 - e^{-0.91}]$
$= 0.887 - 0.597$ (by interpolation)
$= 0.290$.

7.34. One day = Unit of measure; $\lambda = 1/2$ ship/day; $P[\mathbf{X} \ge 4] = 1 - F(4) = 1 - [1 - e^{-1/2(4)}] = e^{-2} = 0.135$.

7.36. $z = (120 - 100)/20 = 1$.
Let $\mathbf{A}_i$, $i = 1, 2$, be the time required to complete job i.
$P(\mathbf{A}_1 \le 120) = P(\mathbf{Z} \le 1) = 0.5000 + .3413 = .8413$;
$P(\mathbf{A}_2 \le 120) = P(\mathbf{Z} \le 1) = .8413$;
$P(\mathbf{A}_1 \cap \mathbf{A}_2) = P(\mathbf{A}_1)P(\mathbf{A}_2) = (0.8413)(0.8413) = 0.7078$ if the events are independent.

7.38. 0.2877.

7.40. $\lambda = 1/10,000$.
a. $1 - F(30,000) = 1 - [1 - e^{-30,000/10,000}] = e^{-3} = 0.0498$.
b. No, the true distribution should probably assign a small probability in the interval between 0 and 5,000—the exponential will assign a large probability in this interval.

7.42. $\lambda = 1/3$.
a. $1 - F(5) = 0.189$ (by interpolation).
b. $F(2) = 0.487$ (by interpolation).
c. $F(4) - F(2) = 0.250$ (by interpolation).

7.44. *a.* 0.0475; *b.* 0.1488; *c.* 0.0099; *d.* 571.

7.46. *a.* $\mu = 5; \sigma = 2.18; P[\mathbf{X} \ge 8] = P[\mathbf{Z} \ge (8 - 5)/2.18]$ (without correction); $P[\mathbf{Z} \ge (7.5 - 5)/2.18]$ (with correction); 0.0838 (without correction); 0.1251 (with correction).
b. 0.4984 (without correction); 0.6298 (with correction).

7.48. *a.*
$$f(x) = \begin{cases} \dfrac{1}{0.5 - (-0.5)} = 1 & -0.5 \le x \le 0.5 \\ 0 & x < -0.5, x > 0.5 \end{cases}$$
b. 0; *c.* $\sqrt{1/12} = 0.2887$; *d.* 0.43.

7.50. *a.* 1; *b.* 1/6; *c.* 1.

7.52. $\lambda = 1/10 = 0.10$; *a.* 0.777; *b.* 0.148; *c.* 6.93 hours.

Chapter 8

8.2. *Parameter*—numerical measure of a population characteristic. *Statistic*—numerical measure calculated from a sample set of observations.

8.4. Each possible subset (sample) of the population of a specified size has an equal probability of being selected.

8.6. Number the items from 1 to 100. Place numbers on circular disks, and mix disks thoroughly in a box. Select one disk from the box, and record the number. Mix the remaining disks thoroughly, and draw a second disk from the box. Record its number. Repeat this process three more times.

8.8. *Census*—procedure or study that includes (measures) every element in the population. Examples: Determining the average annual salary of all employees in a small firm; the proportion of Delta flights that take off on time from the Dallas/Fort Worth Regional Airport on a given day.

8.10. Sampling error is the difference between the numerical measure of a characteristic in a population and the analogous numerical measure of the characteristic in the sample. Sampling error can be *controlled*, but is *unavoidable*. Typically, the amount of this error can be reduced by taking larger samples and carefully selecting samples according to definitions of randomness. Nonsampling errors (e.g., measurement errors, computational errors) are *avoidable* by careful attention to the process of acquiring, recording, and tabulating statistical data.

8.12. *a.* Mean = 3; median = 2.5; standard deviation = 2.16; proportion greater than or equal to 2 = 5/6 = 0.83.

b. There are 15 possible samples. The following table shows the samples and the values of the three sample statistics in each sample.

Item	Values	Mean	Median	P
AB	(0,2)	1	1	0.50
AC	(0,7)	3.5	3.5	0.50
AD	(0,4)	2	2	0.50
AE	(0,3)	1.5	1.5	0.50
AF	(0,2)	1	1	0.50
BC	(2,7)	4.5	4.5	1.00
BD	(2,4)	3	3	1.00
BE	(2,3)	2.5	2.5	1.00

Item	Values	Mean	Median	P
BF	(2,2)	2	2	1.00
CD	(7,4)	5.5	5.5	1.00
CE	(7,3)	5	5	1.00
CF	(7,2)	4.5	4.5	1.00
DE	(4,3)	3.5	3.5	1.00
DF	(4,2)	3	3	1.00
EF	(3,2)	2.5	2.5	1.00

c. The sampling distribution of the sample mean. Its distribution is based on 15 values of $\bar{X}$. The larger the sample, the smaller the variability of $\bar{X}$ will become.

8.14. *a.* $\pi = 1/5 = 0.20$.

b.

Item	Values	P
AB	0,0	0
AC	0,0	0
AD	0,1	0.5
AE	0,0	0
BC	0,0	0
BD	0,1	0.5
BE	0,0	0
CD	0,1	0.5
CE	0,0	0
DE	1,0	0.5

8.16. *a.* $\mu = 1.622$; $\sigma = 0.2715$.

b.

Students	Values	$\bar{x}$	s
B,W,D	1.20, 1.85, 1.66	1.57	0.334
B,W,S	1.20, 1.85, 1.95	1.67	0.407
B,W,C	1.20, 1.85, 1.45	1.50	0.328
W,D,S	1.85, 1.66, 1.95	1.82	0.147
W,D,C	1.85, 1.66, 1.45	1.65	0.200
B,D,S	1.20, 1.66, 1.95	1.60	0.378
B,D,C	1.20, 1.66, 1.45	1.44	0.230
W,S,C	1.85, 1.95, 1.45	1.75	0.265
D,S,C	1.66, 1.95, 1.45	1.69	0.251
B,S,C	1.20, 1.95, 1.45	1.53	0.382

c. $\mu_{\bar{x}} = 1.622$; $\sigma_{\bar{x}} = 0.1109$.

8.18. μ = Average of all population values; $\bar{X}$ = Average of all sample values; $\mu_{\bar{x}}$ = Average of $\bar{X}$ values from repeated samples of a fixed size $[E(\overline{\mathbf{X}}) = \mu_{\overline{\mathbf{x}}}]$; μ and $\overline{\mathbf{X}}$ should be of approximately the same magnitude; $\mu_{\overline{\mathbf{x}}} = \mu$.

8.20. *a.* 0.1056; *b.* 0.9392.

8.22. *a.* 0.1587; *b.* 0.8664.

8.24. *a.* 0.0228; *b.* no—sample size is too small for the central limit theorem to apply. Note: Use finite population correction factor.

8.26. *a.* 0.5962 (without interpolation); *b.* 0.1118 (without interpolation); *c.* 0 ($\sigma_{\overline{\mathbf{x}}} = 0$).

8.28. Binomial distribution; the normal distribution can be used as a model if n is "large."

8.30. *a.* $z = 1.25$; 0.1056. *b.* $z = -2.50, +2.50$; 0.9876. Part *a* probability is decreased; part *b* probability is increased.

8.32. *a.* $z = -2.04, +2.04$; 0.9586. *b.* (1) Population size is sufficiently large for binomial distribution to be used; (2) sample size is sufficiently large for normal approximation to be used: $n\pi = (100)(0.6) = 60 > 5$, and $n(1 - \pi) = (100)(.4) = 40 > 5$.

8.34. *a.* Mean = 0.10; standard deviation = 0.03; approximately normal; *b.* 0.0475; *c.* 0.0918.

8.36. $n/N = 25/500 = 0.05$. Since n/N is not less than 0.05, use the FPC factor.

8.38. *a.* Simple random sampling: *Finite population*—each possible sample set of size n has an equal probability of being selected. *Infinite population*—all n sampled observations are statistically independent. Example: A simple random sample of 100 students at a university for the purpose of responding to a questionnaire on the issue of a tuition increase.

   b. *Cluster sampling*—a set of clusters is selected from a population on the basis of simple random sampling, and a sample is formed by taking a census of each cluster. Example: Airplane flights from Washington, D.C., to Los Angeles daily. Cluster is each flight.

   c. *Stratified sampling*—the population is stratified, and from each strata a simple random sample is taken. These simple random samples are consolidated to form the sample. Example: Estimating residential insurance rates—stratify by type of residence.

   d. *Systematic sampling*—selecting every kth unit, $k > 1$. Example: Sampling every 10th item from a production process.

   e. *Judgment sample*—an expert familiar with the population characteristic forms the sample. Example: An auditor decides which invoices to select in an audit of invoices.

   f. *Convenience sample*—"convenient" population units are selected to form the sample. Example: Editorial page in daily newspaper—a convenience sample consists of letters sent to the editor each day.

8.40. *a.* Use 1,000 discs or chips to designate the households, and randomly select 50 "out of the hat." Most computer software packages contain algorithms for randomly selecting n numbers from N.

   b. Identify the households that fall into each of the three strata—IRS returns could be used for this purpose. From each strata, randomly sample in proportion to the strata size so that 50 households in total are sampled.

   c. Select the 50 most easily accessed.

   d. If the purpose is to choose a representative sample of 50 households from the 1,000, use an expert familiar with the neighborhood to select 50 that suitably reflect the typical household and the "variability" among households in the neighborhood.

8.42. *a.* $C_3^7 = \dfrac{7!}{4!3!} = 35.$

b.

Sample	$\bar{x}$	Mx	Sample	$\bar{x}$	Mx	Sample	$\bar{x}$	Mx
1,2,3	2.00	2	1,5,6	4.00	5	2,6,7	5.00	6
1,2,4	2.33	2	1,5,7	4.33	5	3,4,5	4.00	4
1,2,5	2.67	2	1,6,7	4.67	6	3,4,6	4.33	4
1,2,6	3.00	2	2,3,4	3.00	3	3,4,7	4.67	4
1,2,7	3.33	2	2,3,5	3.33	3	3,5,6	4.67	5
1,3,4	2.67	3	2,3,6	3.67	3	3,5,7	5.00	5
1,3,5	3.00	3	2,3,7	4.00	3	3,6,7	5.33	6
1,3,6	3.33	3	2,4,5	3.67	4	4,5,6	5.00	5
1,3,7	3.67	3	2,4,6	4.00	4	4,5,7	5.33	5
1,4,5	3.33	4	2,4,7	4.33	4	4,6,7	5.67	6
1,4,6	3.67	4	2,5,6	4.33	5	5,6,7	6.00	6
1,4,7	4.00	4	2,5,7	4.67	5			

c.

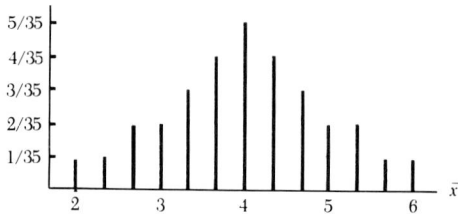

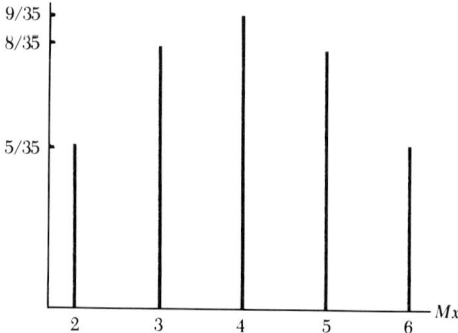

d. $\mu_{\bar{x}} = \dfrac{2 + 2.33 + \cdots + 6.00}{35} = 4.00$

$\sigma_{\bar{x}} =$

$\sqrt{\dfrac{(2-4)^2 + (2.33-4)^2 + \cdots + (6-4)^2}{35}}$

$\qquad = 0.898$

$\mu_{M_x} = 4$

$\sigma_{M_x} =$

$\sqrt{\dfrac{(2-4)^2 + (2-4)^2 + \cdots + (6-4)^2}{35}}$

$\qquad = 1.260$

e. Use sample mean: $\sigma_{\bar{x}}$ is smaller than σ_{M_x}.

8.44. σ^2 is the variance of the set of population values; s^2 is the variance of the set of sample values; and $\sigma_{\bar{x}}$ is the variance of the sample mean $\bar{X}$. σ^2 and s^2 should be approximately the same and $\sigma_{\bar{x}}$ should be much smaller than either as the same size increases.

8.46. *a.* $z = 1.42$; 0.0778. *b.* $z = -1.42, 2.83$; 0.9199.

8.48. *a.* Use FPC, since $n/N = 0.16 > 0.05$. $z = -1.90, +1.90$; 0.9426. *b.* 0.0574.

8.50. *a.* $z = -2.50$; 0.0062. *b.* $z = -3.00$; 0.00135. *c.* No, too unlikely to happen (probability < 0.00135) if μ is really 40 mpg.

8.52. *a.* Mean = \$40,000; standard deviation = \$800; approximately normal. *b.* $z = -1.25$; 0.1056.

8.54.   *a.* $\mu = 205$; $\sigma = 28.81$. *b.* $\pi = 2/5 = 0.40$.

	Executives	Salaries	MBAs	$\bar{x}$	p
c.	O,N,S	215, 185, 240	1, 0, 0	213.33	0.33
	O,N,M	215, 185, 160	1, 0, 0	186.67	0.33
	O,N,C	215, 185, 225	1, 0, 1	208.33	0.67
	O,S,M	215, 240, 160	1, 0, 0	205.00	0.33
	O,S,C	215, 240, 225	1, 0, 1	226.67	0.67
	O,M,C	215, 160, 225	1, 0, 1	200.00	0.67
	N,S,M	185, 240, 160	0, 0, 0	195.00	0.00
	N,S,C	185, 240, 225	0, 0, 1	216.67	0.33
	N,M,C	185, 160, 225	0, 0, 1	190.00	0.33
	S,M,C	240, 160, 225	0, 0, 1	208.33	0.37

d. $\mu_{\bar{X}} = 205$; $\sigma_{\bar{X}} = 11.76$. *e.* $\mu_P = 0.40$; $\sigma_P = 0.200$.

Chapter 9

9.2.   Estimation and hypothesis testing. In estimation, a population parameter value is point estimated by a sample statistic or is estimated by a confidence interval, based on the sample statistics. In hypothesis testing, a value or set of values of a population parameter is conjectured, and the sample information is used to decide whether or not the null hypothesis should be rejected.

9.4.   A single number computed from the sample values is used to estimate the value of a population parameter.

9.6.   *Estimator*—a random variable that has a sampling distribution (e.g., $\mathbf{X}$ is an estimator of μ). *Estimate*—a value of an estimator in a specific sample (e.g., $\bar{x}$ is an estimate of μ).

9.8.   *Unbiasedness*—in repeated samples of a fixed size, the average of the point estimator is equal to the population parameter value. That is $E(\hat{\theta}) = \theta$. *Minimum variance*—in repeated samples of a fixed size, the variance of the point estimator is minimum when compared with competing point estimators. That is, $V(\hat{\theta})$ is minimum among all point estimators of the population parameter. *Minimum mean square error*—in repeated samples of a fixed size, the average squared error of the estimator is minimum when compared with competing point estimators.

9.10.   An estimator is biased if its expectation (average value) is not equal to the value of the population parameter.

9.12.   *a.* $1,452. *b.* $88.44 (using the FPC).

9.14.   *a.*   0.40; 0.041 (using the FPC), 0.049 without FPC. *b.* 0.50, 0.042 (using the FPC); 0.050 without FPC.

9.16.   *a.* $35,000. *b.* $4,000.

9.18.   *a.*   15.56. *b.* 5.42. *c.* 6/25 = 0.24.

9.20.   The confidence interval width gives an indication of the quality of the point estimate. The narrower the width, the better the point estimate.

9.22.   *a.* $1.45 \pm (1.96)(0.1/\sqrt{16})$; 1.402 to 1.499. *b.* No. *c.* Do not produce the part; the confidence interval suggests that $\mu < 1.50$.

9.24. $0.10 \pm (1.96)\sqrt{0.10(0.90)/100}$; 0.0412 to 0.1588.

9.26. $50 \pm 1.75(2.6/\sqrt{100})$; 49.545 to 50.455.

9.28. $154.30 \pm (2.262)(8.097/\sqrt{10})$; 148.51 to 160.09.

9.30. $6 \pm (1.984)(2/\sqrt{100})$; 5.60 to 6.40.

9.32.   *a.* $2.8 + (1.690)0.6/\sqrt{36}$; 2.969. *b.* Since upper confidence bound is 2.969, the manufacturer's claim of three months is not reasonable.

9.34.   *a.* $0.125 + (1.28)\sqrt{0.125(0.875)/200}$; 0.155. *b.* Since we are 90 percent confident that π

does not exceed 0.155, the company's contention that at least 20 percent of the customers prefer its cereal to other brands is not supported by the sample data.

9.36. Using 70 df from table (instead of $df = 71$),

$$\frac{(72-1)(80)}{95} \le \sigma^2 \le \frac{(72-1)(80)}{48.8}; 59.79$$

to 116.39.

9.38. $a.$ 0.10966. $b.$ $\dfrac{(25-1)(0.10966)^2}{39.4} \le \sigma^2 \le$

$\dfrac{(25-1)(0.10966)^2}{12.4}; \sqrt{0.00732}$ to $\sqrt{0.02327}$;

0.086 to 0.153.

9.40. 2,401.

9.42. 246.

9.44. 25.

9.46. $a.$ 385. $b.$ 139.

9.48. $a.$ $50 \pm (2.131)(10/\sqrt{16})$; 44.67 to 55.33.

 $b.$ Ninety-five percent confident that the mean is between 44.67 and 55.33. That is, in forming this interval, we took a 5 percent risk that the resulting interval would not contain the mean. The interval from 44.67 to 55.33 either does or doesn't contain the mean.

9.50. Sample size is directly related to the precision of the estimation process.

9.52. If n is doubled, width decreases by a factor of $\sqrt{2}$.

9.54. $a.$ 246. $b.$ 97.

Chapter 10

10.2. H_0: $\theta_1 = \theta_2$ H_0: $\theta_1 \ge \theta_2$ H_0: $\theta_1 \le \theta_2$
 H_A: $\theta_1 \ne \theta_2$ H_A: $\theta_1 < \theta_2$ H_A: $\theta_1 > \theta_2$

10.4. Type I error: rejecting H_0 when H_0 is true. Type II error: not rejecting H_0 when H_0 is false.

10.6. The action limit or limits set the point or points that separate the *do not reject region* from the *rejection region* in the classical method decision rule.

10.8. Test statistic is the sample statistic whose sampling distribution is used to decide whether or not to reject the null hypothesis.

10.10. If the sample size is large, the approximate sampling distribution of a test statistic is often taken to be the normal distribution via the central limit theorem.

10.12. No. A Type I error can only be committed when the null hypothesis is rejected.

10.14. $a.$ $\alpha = P[\overline{X} > 55] = P[Z > 3.00] = 0.00135$.
 $b.$ $\beta = 0.00135$. By increasing the sample size, the probabilities of committing a Type I and a Type II error have been reduced substantially.

10.16. $a.$ $z = -2.5, +2.5$; 0.0124. $b.$ $z = -5, 0$; approximately 0.5000.

10.18. $a.$ $z = -2.00$; 0.0228. $b.$ $z = -4.00$; approximately 0.0000.

10.20. When the population random variable is normally distributed (or when n is sufficiently large to appeal to the Central Limit Theorem) *and* when s is used to estimate σ in constructing the decision rule.

10.22. H_0: $\mu \ge 54$; H_A: $\mu < 54$; $z = 1.64$.
Reject H_0 if $\bar{x} < 54 - 1.64(6/\sqrt{36}) = 52.36$.
Since $\bar{x} = 51 < 52.36$, *reject* H_0.

10.24. Reject H_0 if $\bar{x} > 5,000 + 1.860(50/\sqrt{9}) = 5,031$.
Note: use t with $df = 8$ ($t = 1.860$).
Since $\bar{x} = 5,060 > 5,031$; *reject* H_0.

10.26. Since $n > 100$, use $z = 1.64$.
Reject H_0 if $\bar{x} < 20,000 - 1.64(156/\sqrt{145}) = 19,978.75$.
Since $\bar{x} = 19,953 < 19,978.75$, *reject* H_0.

10.28. $t = 2.010$. Action limits: $\mu_0 - ts/\sqrt{n}$ and $\mu_0 + ts/\sqrt{n}$, where $\mu_0 = 70$, $s = 15$, and $n = 50$. Action limits: 65.74 and 74.26. Since $\bar{x} = 68$ falls between action limits, *do not reject* H_0.

10.30. $a.$ Action limits: $100 - (2.064)(20/\sqrt{25}) = 91.74$; $100 + (2.064)(20/\sqrt{25}) = 108.26$. Since $\bar{x} = 90 < 91.74$, *reject* H_0.
 $b.$ $(20/\sqrt{25})\sqrt{475/499} = 3.903$. Action limits: $100 - (2.064)(3.903) = 91.94$. $100 + (2.064)(3.903) = 108.06$. Since $\bar{x} = 90 < 91.94$, *reject* H_0.
 $c.$ No.

10.32. $a.$ H_0: $\pi \le 0.02$. H_A: $\pi > 0.02$.
 $b.$ $z = 2.33$; action limit $= 0.02 + 2.33\sqrt{0.02(0.98)/225} = 0.04175$. Since $p = 6/225 = 0.02667 < 0.04175$, *do not reject* H_0.

cision rule: Reject H_0 if $t < -2.101$ or if $t > 2.101$. Since $t = -3.51 < -2.101$, *reject* H_0.

   c. p-value $= 2P[t < -3.51]$. Since $P[t < -2.878] = 0.005$ and $P[t < -3.610] = 0.001$, p-value is between $2(0.005) = 0.01$ and $2(0.001) = 0.002$.

11.14. Power is lost—the test is not as efficient due to the failure to capitalize on the matched pairs which control unwanted sources of variability.

11.16. *a.* Paired, two observations on same sales-person.

   b. $t = 1.76$. Since $t = 1.76$ is between -2.093 and $+2.093$, *do not reject* H_0.

   c. Approximately $2(0.05) = 0.10$.

11.18. *a.* $(0.60 - 0.3333) \pm 1.645(0.0582)$; 0.171 to 0.36.

   b. Independent samples; normal approximation to binomial distribution.

11.20. *a.* H_0: $\pi_A = \pi_B$; H_A: $\pi_A \neq \pi_B$.

   b. $z = \dfrac{(0.5 - 0.667) - 0}{0.066} = -2.53$. Since $z = -2.53 < -1.96$, *reject* H_0.

   c. Since $z < D_0 = 0$, compute $2P[Z \leq -2.53] = 2(0.0057) = 0.0114$. Since p-value $= 0.0114 < \alpha = 0.05$, *reject* H_0.

11.22. *a.* Classical method: Action limits $= D_0 \pm$
$$z\sqrt{\frac{p_1(1 - p_1)}{n_1} + \frac{p_2(1 - p_2)}{n_2}} \text{ with } D_0 = 0$$
and $z = 1.96$.

    Decision rule: Reject H_0 is $(p_1 - p_2) <$ Lower action limit or $(p_1 - p_2) >$ Upper action limit; do not reject H_0 if $(p_1 - p_2)$ is between action limits.

    Standardized testing: Compute $z =$
$$\frac{(p_1 - p_2) - D_0}{\sqrt{\dfrac{p_1(1 - p_1)}{n_1} + \dfrac{p_2(1 - p_2)}{n_2}}}, D_0 = 0.$$ Deci-

sion rule: Reject H_0 if $z < -1.96$ or if $z > 1.96$; do not reject if $-1.96 \leq z \leq 1.96$.

   b. Standardized test:
$$z = \frac{(0.26 - 0.31) - 0}{0.06374} = -0.784.$$

Since -0.784 is between -1.96 and 1.96, do not reject H_0.

   c. p-value $= 2P(Z < -0.784) = 2(0.2177) = 0.4354$.

11.24. $f = 2.78$; $F(0.95;23,17) = 2.24$ (by interpolation); reject H_0. $0.01 < $ p-value < 0.05.

11.26. $f = 100/25 = 4$; $F(0.975; 24, 24) = 2.27$. Since $f = 4 > 2.27$, *reject* H_0.

11.28. Whenever there is a basis for pairing observations, such as the response of an individual to some stimulus prior to treatment and after treatment. By pairing, unwanted sources of variation that may adversely affect treatment comparisons are controlled.

11.30. The null hypothesis should not be rejected, since the confidence interval contained the hypothesized difference (0.30); $\alpha = 0.01$.

11.32. The paired experiment is more efficient than the independent samples experiment—fewer sample values are needed.

11.34. Assuming population variances are equal:
$$z = \frac{(496 - 454) - (0)}{(85.147)(0.1414)} = 3.488; \text{ reject } H_0$$
since $z = 3.488 > 1.645$. Note: Use z, since $n_1 + n_2 - 2 > 100$. Not assuming population variances are equal: $z = \dfrac{(496 - 454) - (0)}{12.042} = 3.488$; $\Delta = 166.49 > 100$; use $z = 1.645$. Since $z = 3.488 > 1.645$, *reject* H_0.

11.36. *a.* H_0: $\mu_A \geq \mu_B$; H_A: $\mu_A < \mu_B$.

   b. Assuming population variances are equal: $t = \dfrac{(400 - 430) - (0)}{(45.277)(0.2828)} = -2.343$; $df = 48$; action limit $= -1.677$ (t). Since $t = -2.343 < -1.677$, *reject* H_0.

   c. Between 0.025 and 0.01.

   d. No. It doesn't take into account different sizes in population of the two cities. They should consider the average number of robberies per 10,000 people.

11.38. *a.* H_0: $\pi_D - \pi_A = 0$; H_A: $\pi_D - \pi_A \neq 0$.

   b. $(0.75 - 0.90) \pm 1.96(0.05268)$; -0.253 to -0.047.

   c. Reject H_0, since confidence interval does not contain the hypothesized difference of 0.

11.40. *a.* $(0.80 - 0.60) \pm 2.575(0.1265)$; -0.126 to 0.526.

   b. Do not reject H_0, since confidence interval contains the hypothesized difference of 0.

11.42. *a.* H_0: $\pi_B - \pi_T = 0$; H_A: $\pi_B - \pi_T \neq 0$.

Test statistic value:

$$z = \frac{(0.15 - 0.09375) - (0)}{\sqrt{(0.11786)(0.88214)(0.01458)}}$$

$$= 1.44.$$

Do not reject H_0, since $-1.96 < z = 1.44 < 1.96$.

b. $(0.15 - 0.09375) \pm 1.96(0.03992); -0.022$ to 0.134.

11.44. Let μ_B = Average before program; μ_A = Average after program. Two independent samples; assume variances equal.

a. H_0: $\mu_B \le \mu_A$; H_A: $\mu_B > \mu_A$.

b. $z = \dfrac{(1.24 - 0.87) - (0)}{(0.21213)(0.1414)} = 12.33$.

Since $z = 12.33 > 1.96$, reject H_0; the program is effective.

Chapter 12

12.2.   a.   SST = 4,374.95, SSTR = 2,668.05; $f = 28.14$; reject H_0; p-value < 0.001

b. $s_P = 9.74$; $t = -5.30$; since $t = -5.30 < -1.734$, *reject* H_0.

c.   Yes; Yes; $t^2 = f$.

12.4.   SST = 73,800, SSTR = 50,138; $f = 9.54$; reject H_0; p-value between 0.005 and 0.01.

12.6.   SSTR = 1.20; SSE = 0.60; $f = 12.5$; reject H_0 at $\alpha = 0.05$ significance level; p-value < 0.001.

12.8.   a.   SST = 442; SSTR = 372.50; SSB = 16.50; $f = 33.83$; *reject* H_0.

b.   Much less than 0.001.

c.   Yes; the p-value is so small that we should *reject* H_0.

12.10.   a. H_0: $\mu_1 = \mu_2 = \mu_3 = \mu_4$ (all machines produce the same average output); H_A: at least one mean is different than the rest.

b.   SST = 18,302; SSTR = 13,410; SSB = 2,203; $f = 19.96$; *reject* H_0.

c. p-value < 0.001.

12.12.   a. $X_{ijk} = \mu + \alpha_i + \beta_j + (\alpha\beta)_{ij} + \varepsilon_{ijk}$; $i = 1, 2, 3, 4$; $j = 1, 2, 3$; $k = 1, 2$. α_i = Fertilizer effect; β_j = Corn variety; $(\alpha\beta)_{ij}$ = Interaction effect.

b.   SST = 12,543.8; SSA = 11,404.8; SSB = 238.6; SSAB = 587.4. Test for interaction effect: $f = 97.9/26.1 = 3.75$. $F(0.975; 6, 12) = 3.73$; thus p-value = 0.025. Since null hypothesis would not be rejected at the $\alpha = 0.05$ and $\alpha = 0.10$ levels, do not perform the hypothesis tests for the main effects. If α were set at 0.01, then you would perform the hypothesis tests for the main effects.

12.14.   B = $t_{0.05/12;6} = t_{0.004;6} = 4.00$ (by interpolation); $s^2 = (0.67)(2)/3 = 0.447$; $s = 0.67$; Bs = 2.68.

(A,B): -4.00 ± 2.68 (A,C): -1.00 ± 2.68
(A,D): 2.00 ± 2.68 (B,C): 3.00 ± 2.68
(B,D): 6.00 ± 2.68 (C,D): 3.00 ± 2.68

12.16.   B = $t_{.05/20; 20} = t_{0.0025; 20} = 3.29$ (by interpolation); $s^2 = 6.86(2)/5 = 2.74$; $s = 1.66$; Bs = 5.46.

(A,B): -1.20 ± 5.46 (A,C): 9.8 ± 5.46
(A,D): 2.8 ± 5.46 (A,E): 10.2 ± 5.46
(B,C): 11 ± 5.46 (B,D): 4.00 ± 5.46
(B,E): 11.4 ± 5.46 (C,D): -7.0 ± 5.46
(C,E): 0.4 ± 5.46 (D,E): 7.4 ± 5.46

12.18.   B = $t_{0.05/6; 6} = t_{0.008; 6} = 3.37$ (by interpolation); $s^2 = 92.1(2)/4 = 46.05$; $s = 6.79$, Bs = 22.88. (1,2): 43.75 ± 22.88; (1,3): 18 ± 22.88; (2,3): -25.75 ± 22.88.

12.20.   B = $t_{0.05/6; 6} = t_{0.008; 6} = 3.37$ (by interpolation); $s^2 = 4.75(2)/4 = 2.375$; $s = 1.54$; Bs = 5.19. (1,2): -10.75 ± 5.19; (1,3): -18 ± 5.19; (2,3): -7.25 ± 5.19.

12.22.   (1) Normally distributed, (2) equal variance, (3) independence; independence.

12.24.   The procedure compares variances to test the hypothesis of equal means.

12.26.   a.   *Homogeneity of variances*—the t population variances must be equal. *Normality*—for the purposes of hypothesis testing, the t population distributions must be normal. *Additivity*—the model must be additive—no terms excluded that explain a significant amount of variability in the dependent variable. *Independence*—the error terms must be uncorrelated.

b.   Same as in *a* with these additions: (1) Treatment and block effects fixed. (2) No treatment by block interaction.

12.28. By controlling the source of variation represented by the block variable.

12.30. The second model is a reparameterization of the first model; $\mu_i = \mu + \tau_i$. The term τ_i represents the treatment effect. It is the difference between the ith treatment mean (μ_i) and the overall mean (μ).

12.32. A contrast is a linear combination of two or more means; $L = c_1\mu_1 + c_2\mu_2 + \cdots + c_k\mu_k$, such that the coefficients sum to zero: $c_1 + c_2 + \cdots + c_k = 0$. Contrasts provide a means for comparing treatment means once it is determined that the treatment effects significantly differ.

12.34. By using transformations or other techniques, such as nonparametric methods (Chapter 20).

12.36. *a.*

SV	df	SS	MS	F
Treatments	4	10	2.50	5
Error	15	7.5	0.50	
Total	19			

b. $0.005 < p$-value < 0.01.

c. $B = t_{0.05/6;\ 15} = t_{0.008;\ 15} = 2.74$ (by interpolation). (*i*) $s^2 = (0.50)(2)/4 = 0.25$; $s = 0.5$; $-0.50 \pm (2.74)(0.5)$. (*ii*) $s^2 = (0.50)(4)/4 = 0.50$; $s = 0.707$; $-3.0 \pm (2.74)(0.707)$. (*iii*) $s^2 = (0.50)(6)/4 = 0.75$; $s = 0.866$; $1.5 \pm (2.74)(0.866)$.

12.38. *a.*

SV	df	SS	MS	F
Oil	2	280	140	4.94
Error	12	340	28.33	
Total	14	620		

b. $B = t_{0.05/6;\ 12} = t_{0.008;\ 12} = 2.83$ (by interpolation); $s^2 = 28.33(2)/5 = 11.332$; $s = 3.37$; $Bs = 9.54$. (A,B): 2 ± 9.54; (A,C): 10 ± 9.5; (B,C): 8 ± 9.5.

c. $0.025 < p$-value < 0.05.

12.40. *a.* $F(0.95; 3, 16) = 3.25$ (by interpolation); since $f = 83.02 > 3.25$, *reject* H_0.

b. p-value is *much less than* 0.001.

12.42. *a.* $F(0.95; 2, 27) = 3.36$ (by interpolation); since $f = 15.44 > 3.36$, *reject* H_0.

b. p-value is less than 0.001.

c. $B = t_{0.05/6;\ 27} = t_{0.008;\ 27} = 2.59$ (by interpolation); $s^2 = 16(2)/10 = 3.20$, $s = 1.79$; $Bs = 8.29$. (1,2): -5.6 ± 8.29; (1,3): -9.9 ± 8.29; (2,3): -4.3 ± 8.29.

d. $H = 24.9/9.36 = 2.66$, $H_{0.05;\ 3,\ 10} = 5.34$. Since $H = 2.66 < 5.34$, the hypothesis of equal variances *cannot be rejected*.

12.44. *b.* $F(0.95; 4, 18) = 2.95$ (by interpolation). Since $f = 7.98 > 2.95$, *reject* H_0.

c. $F(0.95; 2, 18) = 3.57$ (by interpolation). Since $f = 48.26$, *reject* H_0.

d. $F(0.95; 2, 18) = 3.75$ (by interpolation). Since $f = 7.98$, *reject* H_0.

e. $B = t_{0.05/6;\ 18} = t_{0.008;\ 18} = 2.68$ by interpolation; $s^2 = 4.78(2)/9 = 1.06$; $s = 1.03$; $Bs = 2.76$. (1,2): 3 ± 2.76; (1,3): 4.55 ± 2.76; (2,3): 1.55 ± 2.76.

f. $H = 9.36/1 = 9.36$; $H(0.95, 9, 27) = 3.62$ (by interpolation). Since $H = 9.36 > 3.62$, *reject* H_0.

12.46. *b.* $F(0.95; 6, 36) = 2.39$ (by interpolation). Since $f = 52.14 > 2.39$, reject H_0.

c. Pressure: $F(0.95, 2, 36) = 3.29$ (by interpolation). Since $f = 120.25 > 3.29$, *reject* H_0. Temperature: $F(0.95; 3, 36) = 2.89$ (by interpolation). Since $f = 422.08 > 2.89$, *reject* H_0.

d. $H = (5.38)^2/(2.06)^2 = 6.82$; $H(0.95; 12, 48) = 2.77$ (by interpolation). Since $H = 6.82 > 2.77$, *reject* H_0.

e. $B = t_{0.05(2);\ 47} = t_{0.025;\ 47} = 2.012$; $s^2 = 9.95(2)/4 = 4.975$; $s = 2.23$; $Bs = 4.49$; $(87.00 - 84.25) \pm 4.49$; no.

Chapter 13

13.2. *a* and *b.*

13.4. *a.* $y = 100 + 3x$. *b.* linear statistical. *c.* $y = 100 + x^2$. *d.* $y = (x - 1)^2$.

13.6. In using the simple linear regression model, we must assume that:

1. The mean or the expected value of the conditional probability distribution of **Y** given a value of x is equal to $\beta_0 + \beta_1 x$. Or, $E[Y_i] = \beta_0 + \beta_1 x_i$.

2. The variance of the conditional probability distribution of **Y** given a value of x is $\sigma^2 = V[\epsilon_i]$. We further assume that each

conditional probability distribution has the *same* variance, σ^2.

3. For constructing hypothesis testing decision rules or calculating confidence intervals for the regression parameters, we must assume that the error components of each conditional distribution are normally distributed. This assumption is *not* required, however, in obtaining point estimates of the regression parameters.

13.8. By conducting an experiment or taking a survey.

13.10. For *any* value of x, the expected value of **Y** is β_0.

13.12. The least-squares criterion in a regression model requires that we find regression coefficient estimates, denoted by b_0 and b_1, that minimize the sum of squared *vertical* deviations between the determined regression line and the scatter of points. That is, we seek estimates b_0 and b_1 of β_0 and β_1, respectively, which minimize the following equation:

$$LS = \sum_{i=1}^{n} (y_i - \beta_0 - \beta_1 x_i)^2 = \sum_{i=1}^{n} \epsilon_i^2.$$

13.14. The estimators are unbiased and are minimum variance among all linear unbiased estimators.

13.16. *a.* Scatterplot suggests a strong linear relationship.
b. $\hat{y} = 7.77 - 0.084x$.
c. For a one unit increase in average daily temperature, the average consumption of fuel oil *decreases* by 0.084 gallon.
d. 3.15 gallons.

13.18. 25.833.

13.20. $200/98 = 2.041$.

13.22. *b.* 1. *c.* Yes, all points fit exactly on a line.

13.24. The amount of variation in the dependent variable **Y** that is explained by the fitted regression equation.

13.26. It is the ratio of the explained to the total variation.

13.28. It is a scatter plot of the residuals (or standardized residuals) on the vertical axis and the values of the independent variable (x) or the fitted values of the dependent variable ($\hat{y}$) on the horizontal axis. It is used to provide qual-

itative information regarding how well the fitted regression function fits and satisfies the regression model assumptions.

13.30. *b.* No. The residuals appear to have greater variability for larger values of x.

13.32. The advantage of linear correlation is that the x-values are determined by random sampling. Thus there is greater likelihood of a broader coverage of the independent variable. If we desire to determine the *strength* of the linear relationship, correlation may be preferred because of the broader coverage of the independent variable. If we are interested in determining the form of the functional relationship, then linear regression may be preferred.

13.34. Nonsense correlations exist when two variables are correlated through a third variable. In this instance, a more meaningful measure would be the partial correlation between the two variables, *adjusted* for the third variable.

13.36. In simple linear regression, a random sample of **Y** values is taken for *each* level of x. In correlation analysis, a random sample of size n is drawn from the joint distribution of **X** and **Y**.

13.38. The correlation coefficient measures the *linear* relationship between **X** and **Y**; if $r = 0$, there is no linear relationship; if $r = +1$, perfect linear relationship with positive slope exists in the sample values; if $r = -1$, perfect linear relationship exists with negative slope.

13.40. The covariance of two random variables **X** and **Y**, denoted $C(\mathbf{X},\mathbf{Y})$, is a measure of how the two random variables tend to "co-vary." That is, if, for example, large values of **Y** tend to be associated with large values of **X**, and low values tend to be associated with low values, then the covariance of **X** and **Y** will be a large positive number. If, on the other hand, low values of **Y** tend to be associated with high values of **X**, then the covariance of **X** and **Y** will be a large negative number. If the two random variables are independent, then $C(\mathbf{X},\mathbf{Y}) = 0$. Unfortunately, the covariance measure is seriously affected by the units of measurement of the variables involved. This is why correlation is often thought of as a more useful measure of the linear association between random variables.

13.42. *b.* The plot tends to curve like a quadratic function, although a line may provide a good fit.

c. $\hat{y} = 9.77 + 0.00065485x$.

d. No, assumptions are not met. The residual plot has a distinctive parabola form resulting in nonindependent residuals. A line does not provide a satisfactory fit in part *b*.

13.44. *b.* $\hat{y} = 4.00 + 2.075x$. *c.* \$12.30. *d.* 0.88. *e.* Appears to be appropriate. *f.* Since $r^2 = 0.88$, the model may be fairly good. To confirm this, we would have to validate the model by demonstrating that the regression model assumptions are satisfied.

13.46. *b.* $\hat{y} = 2.98 + 6.06x$. *c.* 0.72. *d.* (*i*) 13.90; (*ii*) 19.96; (*iii*) 18.14. *e.* 2.70.

13.48. *a.* Very weak linear relationship. *b.* $\hat{y} = 3.02 + 0.001x$. *c.* 0.041. *d.* 3.62. *e.* Very little, since r^2 is so low.

13.50. *a.* Positive.
b. Positive.
c. Negative.
d. Positive.
e. Probably negative today.
f. Zero or slightly positive.

13.52. *a.* $r = 0$. *b.* $r = 1$. *c.* $r = 1$. *d.* $r = 1$.

13.54. *b.* $r_{X,Y} = .51865$: Any linear relationship is very slight.

c. $r_{X,Z} = .41593$: Since r is closer to 0 than to +1, it is increasingly doubtful whether a linear relationship exists.

$r_{Y,Z} = .97468$: Because r is so close to +1, there is a very strong linear relationship indicated.

d. $r_{X,Y}^2 = .269$: $r_{X,Z}^2 = .173$; $r_{Y,Z}^2 = .95$. It can be inferred that 95 percent of the variation in salary can be explained by age.

13.56. *a.* Sales and price. *b.* $\hat{y} = 84.3 - 26.1x$. *c.* $\hat{y} = -30.6 + 21.8x$. *d.* $\hat{y} = 21.7 - 26.4x$. *e.* $r_{price}^2 = 0.723$; $r_{average\ price}^2 = 0.145$; $r_{new\ variable}^2 = 0.904$. The regression based on the new independent variable. *f.* 23.02 ($x = 2.30 - 2.35 + -0.05$).

13.58. *a.* $\hat{y} = 11.7 - 0.0000x$; the slope is 0 to four

decimal places. *b.* 0.00. *c.* No. A parabola. *d.* No. The residual plot also looks like a parabola, indicating that the model has been misspecified.

Chapter 14

14.2. A forecast interval is for an individual value, whereas a confidence interval is constructed for a mean value.

14.4. *a.* *t*-test: Calculate $t = b_1/s_{b_1}$; reject H_0 if $t > -t_{\alpha/2;n-2}$ or if $t > t_{\alpha/2;n-2}$. *F*-test: Calculate $f = MSR/MSE$; reject H_0 if $f > F(1 - \alpha; 1, n - 2)$. Confidence interval: Calculate the interval given by $b_1 \pm t_{\alpha/2;n-2}s_{b_1}$; reject H_0 if the confidence interval docs not contain 0.

b. Yes; $t^2 = f$ and the confidence interval decision will be the same as the *t*-test decision (and therefore the *F*-test decision).

14.6. $5.67 \pm 1.708(0.943)$.

14.8. *a.* Yes. *b.* $\hat{y} = 221.86 + 10.508x$. *c.* \$6,758. *d.* \$6,521 to \$6,995. *e.* \$6,216 to \$7,300. *f.* $t = 19.13$; reject H_0. *g.* $f = 366.35$; reject H_0. *h.* Yes, since the tests are equivalent.

14.10. *a.* $\hat{y} = 8342 + 4789x$. *b.* 0.621. *c.* $22709 \pm 2.306(2149.52)$. *d.* 21,751. *e.* $21,751 \pm 2.306(648.72)$.

14.12. Since $f = 6.31 > 4.49$, reject H_0.

14.14. 0.0281.

14.16. Since $t = 19.1 > 2.265$, reject H_0; *p*-value < 0.01; average daily temperature is significant in predicting fuel use.

14.18. *a.* Yes, reject H_0: $\beta_1 = 0$ beyond the 0.01 significance level. *b.* \$8.37 to \$16.22.

14.20. *a.* Since $t = -7.01 < -2.069$, reject H_0; *p*-value < 0.001. *b.* 15.31 to 20.40.

14.22. The *Z*-transformation is undefined: $Z = (1/2)ln[(1 + r)/(1 - r)]$. If $r = +1$, we have $ln(\infty)$; and if $r = -1$, we have $ln(0)$.

14.24. $t = 5.16$; reject H_0.

14.26. $l = 0.58 - 2.58(0.06363) = 0.42$; $u = 0.58 + 2.58(0.6363) = 0.74$; 0.397 to 0.629.

14.28. $l = 0.38 - 1.64(0.1667) = 0.11$; $u = 0.38 + 1.64(0.1667) = 0.65$; 0.110 to 0.572.

14.30. $l = 0.39 - 1.96(0.1459) = 0.10$; $u = 0.39 + 1.96(0.1459) = 0.68$; 0.100 to 0.592. Since the confidence interval contains the hypothesized value (0.50), do not reject H_0.

14.32. *a.* 0.585 to 0.790. *b.* A strong relationship does not exist; the confidence interval contains the upper part of the interval defined by -0.60 to 0.60.

14.34. $r = 0.2025$; $t = 0.4136$. Since $t = 0.4136 < 2.776$, do not reject H_0.

14.36. *a.* $\hat{y} = 1.72 + 0.0028$ (GMAT score); $\hat{y} = 2.15 + 0.331$ (undergrad GPA). *b.* GMAT, since $r^2 = 0.137$ is bigger than $r^2 = 0.096$ for undergrad GPA. *c.* $t = 2.70 > 2.013$; reject H_0. *d.* $t = 2.22 > 2.013$; reject H_0. *e.* 2.57671; 2.57671 $\pm$ 2.013(0.232521). *f.* 2.98106; 2.98106 $\pm$ 2.013(0.116170). *g.* 3.71654; 3.71654 $\pm$ 1.679(0.581217). *h.* 3.39443; 3.39443 $\pm$ 1.679(0.566514). *i.* Yes, though the plot does suggest unequal variances may exist in the population.

14.38. *a.* 0.9477.
   b. $t = 5.94 > 2.776$; reject H_0. Yes, H_0: $\beta_1 = 0$ is rejected and $r^2 = 0.898$.
   c. The largest ad expenditure in the sample is 35; 65 represents an extrapolation of our fitted linear regression function. The function may no longer hold when $x = 65$.

14.40. *a.* $\hat{y} = -310.62 + 7.0679(65) = 148.79$ lbs. *b.* $\hat{y} = 46.842 + 0.12446(148.79) = 65$ inches. *c.* The prediction in part *b* is the input in part *a*. *d.* $r^2 = 0.880$ in both models; No. *e.* No. Residual plots show a distinct pattern indicating that the error terms may not be independent.

14.42. -0.623 to -0.766.

14.44. $r = 0.3801$; $t = 1.16 < 1.860$; do not reject H_0.

14.46. $r = 0.993$; $t = 20.59 > 2.447$; reject H_0.

14.48. *a.* 0.840. *c.* 0.840 $\pm$ 2.179(0.1899); 0.426 to 1.254. *d.* Yes, since the confidence interval contains 1. *e.* 0.620; 62 percent of the variation in the rate of return of the electronics security is explained by the fitted linear function of the rate of return of the average market security.

14.50. *a.* $b_1 = 0.461$; for every 1 percent increase in percentage, job satisfaction increases by 0.461 units on the 100-point scale. *b.* $t = 0.461/0.05998 = 7.68 > 2.306$; reject H_0. *c.* $\hat{y} = 52.656 + 0.461(80) = 89.54$.

Chapter 15

15.2. The error terms have zero means, equal variances, and are normally distributed and independent.

15.4. The solution to the normal equations produces the estimates of the regression model coefficients.

15.6. To answer this question, we must consider the effect both on the numerator and on the denominator in the expression for estimating σ^2. In the numerator is found SSE, and in the denominator is found the degrees of freedom, $(n - k - 1)$. By excluding one variable from the analysis, we reduce k by 1, thereby increasing the value of the denominator and reducing the value of the point estimate of σ^2. However, we also increase the value of the numerator, since SSE is going to increase. Therefore, there is a trade-off when we drop a variable from the model. It is possible that the point estimate of σ^2 will actually decrease, since the gain in one degree of freedom more than offsets the increase in the value of SSE. It is also possible that the estimate will increase in value because of the loss in the explanatory "power" of the variable being dropped from the regression equation. The point is that the value of the estimate will not necessarily get larger as variables are dropped from the regression equation. As more and more variables are added, for example, there is simply less unexplained variation to explain, and the loss of degrees of freedom more than offsets any gain in reduction of SSE in estimating σ^2.

15.8. *a.* $396/486 = 0.8148$. *b.* 81.48 percent of the variation in $\mathbf{Y}$ is explained by the fitted regression function.

15.10. *a.* 6. *b.* 45. *c.* 51. *d.* 6,410.60. *e.* 763.93. *f.* 11.807.

15.12. β_1: $t = 1.72 < 2.093$; do not reject H_0. β_2: $t = 4.00 > 2.093$; reject H_0.

15.14. We cannot claim that the overall significance of the *model* is 0.05 because this is a *joint* test of significance. That is, it is a test on *both* β_1 and β_2. In performing the t-test, we are performing the test independently on each term, rather than jointly. The significance level of the joint test is greater than 0.05.

15.16. Since $f = 6.84 > 2.40$, reject H_0.

15.18. *a.* 29.24.
 b. $f = 91.21$ and p-value $= 0.0001$. Since p-value < 0.05, reject H_0.
 c. β_1: $t = 6.83 > 2.201$; reject H_0. β_2: $t = -1.63$ is between -2.201 and 2.201; do not reject H_0.
 d. p-value $= 0.0001$. If the null hypothesis is true ($\beta_1 = \beta_2 = 0$), then it is very unlikely to get a value of the test statistic equal to or larger than $f = 92.21$ (the probability is less than 0.0001).

15.20. It is a joint test that controls the family-wise probability of committing a Type I error (α).

15.22. It is wider since the prediction of an individual value is more difficult (more variable) than estimating a mean value.

15.24. Four, or k-1 binary variables are required. These could be defined as follows,

$$x_{i,1} = \begin{cases} 1 & \text{if the } i\text{th person is single} \\ 0 & \text{otherwise} \end{cases}$$

$$x_{i,2} = \begin{cases} 1 & \text{if the } i\text{th person is married} \\ 0 & \text{otherwise} \end{cases}$$

$$x_{i,3} = \begin{cases} 1 & \text{if the } i\text{th person is separated} \\ 0 & \text{otherwise} \end{cases}$$

$$x_{i,4} = \begin{cases} 1 & \text{if the } i\text{th person is divorced} \\ 0 & \text{otherwise} \end{cases}$$

where $x_{i,1} = x_{i,2} = x_{i,3} = x_{i,4} = 0$ would denote a widowed individual.

15.26. (1) normally distributed, (2) independent, (3) equal variances; independence. If the error terms are not independent, then no hypothesis tests using the t or F-distribution can be used, and no confidence intervals based on the t-distribution can be computed.

15.28. Plotting the residuals (or standardized residuals) against the predicted values and plotting the residuals (or standardized residuals) against each independent variable.

15.30. DW $= 8.60238/14.0459 = 0.612$; $d_L = 1.08$; $d_U = 1.36$. Since DW $= 0.612 < 1.08$, reject

H_0. There does appear to be evidence of positive first-order autocorrelation among the error terms.

15.32. Multicollinearity exists in a multiple regression model when two or more of the independent variables are themselves highly interrelated, and explains the variability in the random variable **Y** in overlapping ways. One of the effects of the presence of multicollinearity in a regression model is that the variances of the b_i terms will be larger than they otherwise would be. It is then possible to eliminate a variable x_i from the model because its relationship with the dependent variable is not judged to be important, when it really is.

15.34. $VIF_1 = 5$; $VIF_2 = 1.82$; $VIF_3 = 2.86$. Maximum VIF $= 5 < 10$, and average VIF is 3.23, which is not considerably larger than 1. Thus multicollinearity is not a problem.

15.36. $VIF_1 = 200$; $VIF_2 = 250$; $VIF_3 = 21.74$; $VIF_4 = 58.82$; $VIF_5 = 333.33$. Maximum VIF $= 333.33$, and average VIF $= 172.78$.

15.38. Although each of these three variables is useful, the contribution over the variable 1/HP or 1/WT is minimal. Due to the added complexity of the resulting model, they are not worth using.

15.40. The estimates of the regression coefficients may be far-removed from the actual values.

15.42. $16/18 = 0.889$.

15.44. 0.89.

15.46. *a.* 0.8683. *b.* 0.260.

15.48. *a.* 0.918; possible severe multicollinearity problems. *b.* 0.241. *c.* 0.014.

15.50. The coefficient of determination gives the proportion of variation in a dependent random variable explained by or accounted for by relating the dependent variable to an independent variable. The coefficient of multiple determination gives the proportion of variation in a dependent random variable explained by or accounted for by more than one independent variable. The coefficient of partial determination gives the proportion of variation explained by an independent variable of that variation remaining after the initial variable has already been introduced into the model. The coefficient of determination between the two independent variables is simply a measure of the linear relationship between variables, this time between the two independent variables.

15.52. *a.* For GPA, $\hat{y} = -13.41 + 15.15$GPA, GPA is significant for all alpha-values greater than 0.0128; for HOURS, $\hat{y} = 18.777 + 1.224$HOURS, HOURS is significant for all alpha-values greater than 0.0046; for GPA and HOURS, $\hat{y} = -3.302 + 8.837$GPA $+ 0.8744$HOURS, the model is significant for alpha-values greater than 0.0055. *b.* combined: \$32,139, GPA: \$33,858, HOURS: \$29,743; the combined model is probably the best. *c.* $r_{X_1 X_2} = 0.5343$.

15.54. Since $f = 120.98 > 3.03$, reject H_0.

15.56. *a.* 5.09. *b.* 5.17.

15.58. *a.* 0.88. *b.* 0.938. *c.* 0.938; 0.536. *d.* 0.5613. *e.* $r^2_{Y/X_2 \cdot X_1} = 0.001$. *f.* $r^2_{Y/X_1 \cdot X_2} = 0.832$.

15.60. *a.* 0.8797.
  b. 0.938.
  c. $r^2_{Y/X_1 \cdot X_2} = 0.105$; $r^2_{Y/X_2 \cdot X_1} = 0.85$.
  d. The coefficient of partial determination for sex gives the proportion of variation explained, which remains after a portion of the variation has already been explained by length of service in the company. For the data in question, $r^2 = 0.8656$, or 86.56 percent of the variation in wages is explained statistically by relating this variable to length of service in the company. Of that which can still be explained $[(1 - 0.8656)]$, sex explains about 10.5 percent of it. This is more evidence to support the contention that a sex bias with regard to wages does not exist in this division of the firm.

15.62. $r^2_{Y/X_1 \cdot X_2} = 0.809$; $r^2_{Y/X_2 \cdot X_1} = 0.195$; $r^2 = 0.943$. $r^2_{Y/X_1 \cdot X_2}$ measures the proportion of previously unexplained variation that is explained by adding x_1 (TEMP) to the model based on x_2 (CAT) alone. $r^2_{Y/X_2 \cdot X_1}$ measures the proportion of previously unexplained variation by adding x_2 (CAT) to the model based on x_1 (TEMP) alone. These two coefficients of partial determination demonstrate the impor-

tance of the temperature variable (see also Problem 15.48).

15.64. Based on the plot of weight versus height, the relationship looks very linear. There is a hint of a quadratic (parabola) form. The regression model, $\mathbf{Y} = \beta_0 + \beta_1 x_1 + \beta_2 x_2 + \boldsymbol{\varepsilon}$, where $\mathbf{Y}$ = Weight, x_1 = Height, and $x_2 = 0,1$ (1 = male), explains 52.5 percent of the variability in $\mathbf{Y}$. Try the model $\mathbf{Y} = \beta_0 + \beta_1 x_1 + \beta_2 x_2 + \beta_3 x_3 + \boldsymbol{\varepsilon}$, where $x_3 = x_1^2$.

15.66. *a.* $\hat{y} = 0.940 + 0.0026x_1 + 0.292x_2$. *b.* 0.211; 21.1 percent of the variation in first year MBA grade point averages is explained by the fitted model. *c.* β_1: $t = 2.56 > 2.014$; reject H_0. β_2: $t = 2.06 > 2.014$; reject H_0.

15.68. *a.* No. *b.* There is little correlation between GMAT and UGPA ($r = 0.105$). *c.* VIF = 1.01. No. VIF < 10, and average VIF = VIF = 1.01 is not considerably larger than 1.

15.70. The following steps should be taken:
  (1) Form the scatterplot of the values of $\mathbf{Y}$ against each independent variable set of values. This provides information about how useful each independent variable is *and* in what functional form it is related to $\mathbf{Y}$.
  (2) Produce the 10 simple linear regression equations. This will determine how strongly each independent variable is *linearly* related to the dependent variable $\mathbf{Y}$.
  (3) Produce the correlation matrix for the 11 variables (the dependent variable $\mathbf{Y}$ and the 10 independent variables). This provides information on the degree of linear relationships among the independent variables. Thus information is provided on the potential for multicollinearity problems.
  (4) With this information, select a subset of the independent variables or functions of the independent variables that explains as much variation in $\mathbf{Y}$ as possible, while keeping the model parsimonious—simple and understandable. Check the model for multicollinearity problems by computing the VIFs.

Chapter 16

16.2. The fundamental pattern of the time series will not change in the future.

16.4. Because the techniques used in the analysis are the classical statistics techniques of linear regression and moving averages.

16.6. The additive model is given by $\mathbf{Y} = T + S + C + I$, where T = Trend, S = Seasonal, C = Cyclical, and I = Irregular. In the additive model, all components that compose the model enter additively.

16.8. *Trend* or *secular trend* represents the smooth, long-term movement or growth pattern of the series. *Cycle* represents the irregular, oscillating movement in the series, where each cycle is more than one year in duration. *Seasonal* movements in a time series are of less than one year's duration and reflect working days in each month, holidays (Christmas, Easter, etc.), the weather, and so on. The seasonal movements in a time series repeat themselves (or approximately so) each year. *Irregular* movements in a time series are those movements that cannot be attributed to trend, cyclical, or seasonal factors.

16.10. Growth curves approach a finite limit.

16.12. *a.*

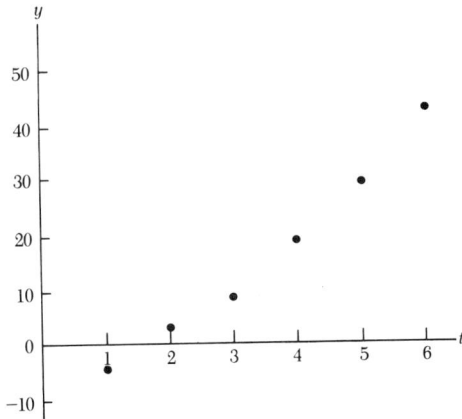

b. Parabola.

16.14. *a.*

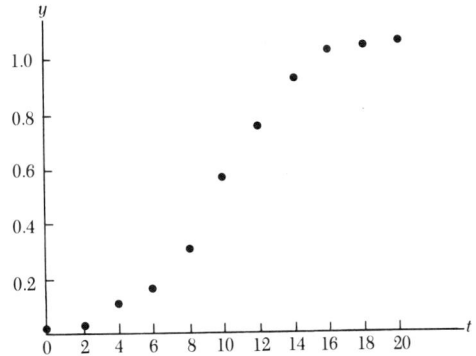

b. Growth curve.

16.16. 14.2687.

16.18. $\hat{y} = 2{,}579.57 + 21.32t + 6.75t^2$; $t = 0$ at 1980; t is in years. The model appears to be appropriate; a very pronounced cyclical component is present.

16.20. 17.623.

16.22. $\hat{y} = 87.856 + 4.530t$; $r^2 = 0.471$. The linear fit is only fair. From the scatterplot, there is a very obvious cyclical pattern.

16.24. $\hat{y} = 9.85 + 5.033t$; $r^2 = 1.00$. Based on the scatterplot and r^2, the fit is excellent.

16.26. $\log y = 0.9881 + 0.0808t$; $r^2 = 0.999$. There is an excellent semilogarithmic fit.

16.28. $1{,}043/1.12 = 931$.

16.30. 456.12.

16.32. From the scatterplot, it is evident that a cyclical (alternating above and below the parabola) does exist.

16.34. $\hat{y} = 94.271 + 4.3313t$; $r^2 = 0.967$. The scatterplot does indicate a cyclical pattern.

16.36. $\hat{y} = 100.770 + 10.270t$; $t = 0$ at 1983; t is in years.

16.38. *a.* Company A: $\hat{y} = 155.764 + 4.8260t$; $t = 0$ at 1975; t is in years; y is in $000, or $\log \hat{y} = 2.196 + 0.012t$; $t = 0$ at 1975; t is in years; y is in $000. Company B: $\hat{y} = 98.803 + 3.465t$ (same conditions), or $\log \hat{y} = 1.999 + 0.013t$ (same conditions). *b.* Based on the semilogarithmic model for both firms, the cyclical pattern is present in both series, and they are similar. *c.* Using the semilogarithmic model for both firms, estimated sales are 221.50 and 147.23 (in $000), respectively.

16.40. Based on the median of specific relatives: 81.86, 118.74, 106.33, and 93.07.

16.42. $t = 0$ at 1969. $\log y = 1.7607 + 0.06158t$; $r^2 = 0.951$. Based on r^2, the fit is quite good. However, the scatterplot indicates that the semi-logarithmic pattern is not linear over the last two or three periods (the pattern is falling below a line). The estimate for 1985 is given by: $\log y = 1.7607 + 0.06158(16) = 2.74598$, and $10^{2.74598} = 557.16$. From the logarithmic scatterplot, this may be an overestimate.

16.44. 45,708 units; 41,088 units; the first estimate should be the more accurate.

Chapter 17

17.2. 12; 1975.

17.4. 187.40.

17.6. 4.

17.8. *a.* $\hat{y} = 41.9811 + 2.0873t$; $r^2 = 0.973$.

b. The residual plot shows several large negative residuals, and the positive residuals tend to oscillate closely about 0.80. The variance of the error terms appears to be larger for smaller values of $\hat{y}$ than for larger values of $\hat{y}$.

	Least Squares	
t	$\hat{y}_t$	e_t
1	45.2	44.07	1.13
2	47.1	46.16	0.94
3	45.3	48.24	−2.94
4	51.3	50.33	0.97
5	51.9	52.42	−0.52
6	55.5	54.50	1.00
7	56.1	56.59	−0.49
8	60.1	58.68	1.42
9	58.2	60.77	−2.57
10	64.3	62.85	1.45
11	63.6	64.94	−1.34
12	66.3	67.03	−0.73
13	69.9	69.12	0.78
14	72.1	71.20	0.90

MAD = 1.227   MSE = 1.980

c. See the tables below for the exponential smoothing forecasts.

$\alpha = 0.5$

y_t	$\hat{y}_t$	$(y_t - \hat{y}_t)$	$(y_t - \hat{y}_t)^2$
45.200	39.800	5.400	29.160
47.100	42.500	4.600	21.160
45.300	44.800	.500	.250
51.300	45.050	6.250	39.063
51.900	48.175	3.725	13.876
55.500	50.037	5.463	29.839
56.100	52.769	3.331	11.097
60.100	54.434	5.666	32.099
58.200	57.267	.933	.870
64.300	57.734	6.566	43.118
63.600	61.017	2.583	6.673
66.300	62.308	3.992	15.933
69.900	64.304	5.596	31.313
72.100	67.102	4.998	24.979

MAD   4.257
MSE   21.388

$\alpha = 0.9$

y_t	$\hat{y}_t$	$(y_t - \hat{y}_t)$	$(y_t - \hat{y}_t)^2$
45.200	39.800	5.400	29.160
47.100	44.660	2.440	5.954
45.300	46.856	−1.556	2.421
51.300	45.436	5.884	34.157
51.900	50.716	1.184	1.403
55.500	51.782	3.718	13.827
56.100	55.128	.972	.944
60.100	56.003	4.097	16.787
58.200	59.690	−1.490	2.221
64.300	58.349	5.951	35.414
63.600	63.705	− .105	.011
66.300	63.610	2.690	7.233
69.900	66.031	3.869	14.969
72.100	29.513	2.587	6.692

MAD   2.993
MSE   12.228

d. There are far too many positive residuals, even with the smoothing constant set at 0.90. The trend line provides much better forecasts than does exponential smoothing with these two values of the smoothing constant.

17.10. *a.* $\alpha = 0.10$.

t	y_t	S_t	$S_t^{(2)}$	$\hat{y}_t$	e_t
1	147.000	84.900	13.890	158.000	−11.000
2	160.000	92.410	21.742	163.800	−3.800
3	152.000	98.369	29.405	170.930	−18.930
4	195.000	108.032	37.267	174.996	20.004
5	190.000	116.229	45.164	186.659	3.341
6	220.000	126.606	53.308	195.190	24.810
7	256.000	139.545	61.932	208.048	47.952
8	212.000	146.791	70.417	225.783	−13.783
9	210.000	153.112	78.687	231.650	−21.650
10	205.000	158.301	86.648	235.806	−30.806
11	210.000	163.470	94.330	237.914	−27.914
12	240.000	171.123	102.010	240.293	−0.293

$$\text{MAD} = 18.6901 \qquad \text{MSE} = 517.2473$$

b. $\alpha = 0.20$.

t	y_t	S_t	$S_t^{(2)}$	$\hat{y}_t$	e_t
1	147.000	123.800	93.560	158.000	−11.000
2	160.000	131.040	101.056	161.600	−1.600
3	152.000	135.232	107.891	168.520	−16.520
4	195.000	147.186	115.750	169.408	25.592
5	190.000	155.748	123.750	186.480	3.520
6	220.000	168.599	132.719	195.747	24.253
7	256.000	186.079	143.391	213.448	42.552
8	212.000	191.263	152.966	239.438	−27.438
9	210.000	195.010	161.375	239.135	−29.135
10	205.000	197.008	168.501	237.055	−32.055
11	210.000	199.607	174.722	232.642	−22.642
12	240.000	207.685	181.315	230.712	9.288

$$\text{MAD} = 20.4663 \qquad \text{MSE} = 557.5728$$

c. Model with $\alpha = 0.10$.

17.12. $\hat{y}_t = 2.942 + 0.985 y_{t-1}$; $r = 0.95$; a good fit.

17.14. $\hat{y}_t = 1846.1 + 0.0672 y_{t-1}$; $r^2 = 0.005$; a very poor fit.

17.16. Step one in the Box-Jenkins forecasting methodology is to tentatively identify the forecasting model to use. This is accomplished by examining the sample autocorrelations and sample partial autocorrelation of the series. The second step consists of estimating the parameters of the ARIMA model using a nonlinear least squares curve-fitting procedure. In the third step, diagnostic checking of the model developed takes place to assure that the correct model has been specified. This is accomplished by analyzing the sample autocorrelations of the residual terms of the model that has been specified. The model specified is usually correct if the sample residuals are normally distributed with a mean or expected value of 0, and are also independent.

17.18. 3; the model could not be fitted due to collinearity among the four dummy variables.

17.20. $t_0 = 3$; $b_2 = \cos 90° = 0$; $b_3 = \sin 90° = -1$; $\hat{y}_t = 200 - 50 \cos 30 t°$.

17.22. 197.467; 263.741; all appear to be significant.

17.24. *a.* A (MAD = 1.6 < 2.8). *b.* A (MSE = 4.4 < 9.6). *c.* Residuals are not independent—they form a line.

17.26.

Period	y_t	$\hat{y}_t$	$(y_t - \hat{y}_t)$	$(y_t - \hat{y}_t)^2$
1	2,105	2,125	−20	400
2	2,000	2,119	−119	14,161
3	1,905	2,083	−178	31,684
4	1,995	2,030	−35	1,225
5	2,090	2,020	−70	4,900
6	2,005	2,041	−36	1,296
7	1,895	2,030	−135	18,225
			−453	71,891

$\hat{y}_{t+1} = \alpha(y_t - \hat{y}_t) + \hat{y}_t$.
$\alpha = 0.3 \quad \text{MAD} = 593/7 = 84.71$.
$\text{MSE} = 71,891/7 = 10,270$.

17.28. As the accompanying graph shows, some type of a model that incorporates seasonal factors should be investigated, especially given the rather large MSEs and MADs reported in the previous two problems. Although the number of observations is low (7), one still might be tempted to fit a four-term sinusoidal model to this data. There does not appear to be a significant trend component in this series, so the terms that might be included are: (1) level; (2) one or both of the sinusoidal functions described (sine or cosine). The method used to obtain a forecasting equation would likely be multiple linear regression.

Sales

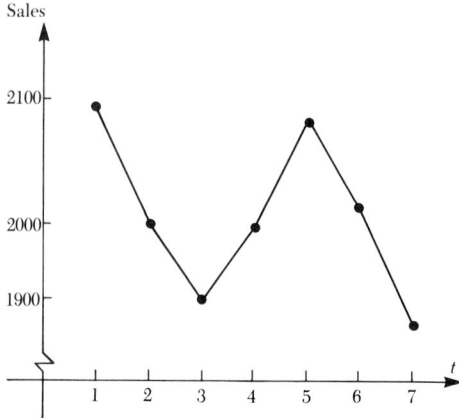

17.30. $\hat{y} = 44.069 + 2.087t$ (starting with $t = 0$).

17.32. Mean is not changing over time; differencing.

Chapter 18

18.2. It is the starting point for later comparisons to it from the developed index. It should not be in the too distant past and should be during a period when goods and prices are not subject to unusual fluctuations.

18.4. They are reported as proportions or percentages. The most common method of reporting is as a percentage.

18.6. $Q_{1984/1980} = 84.62$.

18.8. $P_{1982/1981} = 114.83$; $P_{1983/1981} = 119.45$; $P_{1984/1981} = 147.22$; $P_{1985/1981} = 159.21$; $P_{1986/1981} = 138.71$.

18.10. *a.* $P_{n/o} = 17.78$. *b.* $\Sigma P_{n/o} = 37.87$. *c.* $\bar{P}_{n/o} = 33.78$.

18.12. According to the law of supply and demand, people tend to purchase less of a commodity when the price rises and tend to purchase more when the price falls. The denominator of the Paasche index, $\Sigma p_0 q_n$, will be somewhat larger than it should be (under increasing prices), because as prices increase from the base year to the given year, people will tend to purchase fewer of the high-priced commodities and more of the lower-priced commodities, so that total costs would be less than predicted by $\Sigma p_0 q_n$. As a result, the Paasche price index will appear to be lower than it should be.

18.14. Time Reversal Test:

$$F^*_{n/0} F^*_{0/n}$$
$$= \sqrt{\left(\frac{\Sigma p_n q_0}{\Sigma p_0 q_0}\right)\left(\frac{\Sigma p_n q_n}{\Sigma p_0 q_n}\right)\left(\frac{\Sigma p_0 q_n}{\Sigma p_n q_n}\right)\left(\frac{\Sigma p_0 q_0}{\Sigma p_n q_0}\right)}$$
$$= \sqrt{\left(\frac{14.64}{12.59}\right)\left(\frac{11.4975}{9.8925}\right)\left(\frac{9.8925}{11.4975}\right)\left(\frac{12.59}{14.64}\right)}$$
$$= \sqrt{1} = 1$$

Factor Reversal Test:
The factor reversal test is satisfied if: (Price index)(Quantity index) = Value index.

$$\sqrt{\left(\frac{\Sigma p_n q_0}{\Sigma p_0 q_0}\right)\left(\frac{\Sigma p_n q_n}{\Sigma p_0 q_n}\right)}\sqrt{\left(\frac{\Sigma q_n p_0}{\Sigma q_0 p_0}\right)\left(\frac{\Sigma q_n p_n}{\Sigma q_0 p_n}\right)}$$
$$= \sqrt{\left(\frac{14.64}{12.59}\right)\left(\frac{11.4975}{9.8925}\right)}\sqrt{\left(\frac{9.895}{12.59}\right)\left(\frac{11.4975}{14.64}\right)}$$
$$= \sqrt{\frac{(11.4975)^2}{(12.59)^2}}$$
$$= \frac{11.4975}{12.49} = \frac{p_n q_n}{p_0 q_0} = \text{Value index.}$$

18.16. *a.* $L^*_{n/0} = 36.05$. *b.* $P^*_{n/0} = 35.30$. *c.* $F^*_{n/0} = 35.67$.

18.18. *a.*

1977	1978	1979	1980	1981	1982	1983	1984
$3.27	3.20	2.98	2.69	2.57	2.54	2.61	2.67

b. 39.97 percent. *c.* −18.35 percent.

18.20. *a.* $53.53. *b.* 0.043; purchasing power has increased by 4.3 percent.

18.22. 11.16 percent.

18.24. The CPI for 1981 to 1984 was 272.4, 289.1, 298.4, and 311.1. Find the CPI for 1985 and 1986 at your library to deflate the closing prices.

18.26. *a.* $P^*_{n/0} = 133.52$. *b.* $L^*_{n/0} = 134.33$. *c.* $F^*_{n/0} = 133.92$. *d.* $F^*_{n/0}$ is a geometric mean.

Chapter 19

19.2. $\chi^2 = 15.87$; since $\chi^2 = 15.87 > 9.49$, reject H_0. *p*-value is between 0.005 and 0.001 (0.004 by interpolation).

19.4. $\chi^2 = 42.60$; *p*-value is much less than 0.001, since $\chi^2_{0.001;5} = 20.5$. Therefore reject H_0.

19.6. $\chi^2 = 35.48$; $df = 8 - 1 - 2 = 5$; p-value is much less than 0.001, since $\chi^2_{0.001;5} = 20.5$. Therefore, reject H_0; data are not normal.

19.8. *a.* $\chi^2 = 23.53$; $\chi^2_{0.05;6} = 12.6$; reject H_0. $\chi^2 = 13.36$; $\chi^2_{0.05;2} = 5.99$; reject H_0.

19.10. -0.2813 to 0.0146.

19.12. $\chi^2 = 31.37$; $df = 9$. Since $\chi^2_{0.001;9} = 27.9$, p-value is less than 0.001; reject H_0.

19.14. $\chi^2 = 8.39$; $df = 4$. Since $\chi^2 = 8.39 < \chi^2_{0.05;4} = 9.49$, do not reject H_0.

19.16. $\chi^2 = 49.06$; $df = 8$. $\chi^2_{0.05;8} = 15.5$; reject H_0. Since $\chi^2_{0.001;8} = 26.1$, p-value is much less than 0.001.

19.18. $\bar{x} = 71.85$; $s = 12.62$.

Class	O_i		z	$prob$	E_i		$(O_i - E_i)$	$(O_i - E_i)^2$	$(O_i - E_i)^2/E_i$
Less than 40	0		-2.52	0.0059	0.236				
40 but less than 52	3	} 9	-1.57	0.0523	2.092	} 10.704	-1.704	2.9036	0.271
52 but less than 64	6		-0.62	0.2094	8.376				
64 but less than 76	14		0.33	0.3617	14.468		-0.468	0.2190	0.015
76 but less than 88	13		1.28	0.2704	10.816		2.184	4.7699	0.441
88 but less than 100	3	} 4	2.23	0.0874	3.496	} 4.012	0.012	0.0001	0.000
100 or more	1			0.0129	0.516				
	40								0.727

$df = 4 - 1 - 2 = 1$; $\chi^2_{0.05;1} = 3.84$. Since $\chi^2 = 0.727 < 3.84$, do not reject H_0.

19.20. Die number one: $\chi^2 = 1.70$. Since $\chi^2 = 1.70 < \chi^2_{0.05,5} = 11.1$, do not reject H_0. Die number two: $\chi^2 = 34.4$. Since $\chi^2 = 34.4 > \chi^2_{0.05,5} = 11.1$, reject H_0.

19.22. $\chi^2 = 10.40$; $\chi^2_{0.05;3} = 7.81$. Since $\chi^2 = 10.40 > 7.81$, reject H_0.

19.24. $\chi^2 = 10.273$; $\chi^2_{0.05;4} = 9.49$. Since $\chi^2 = 10.273 > 9.49$, reject H_0.

Chapter 20

20.2. $r^+ = 14$; $r^- = 41$; $r = 14$. $W_{0.05} = 11$ and $W_{0.95} = 44$; do not reject H_0. Approximate p-value (by interpolation) $= 2(0.09) = 0.18$.

20.4. $\bar{x}_1 = 65.00$, $s_1 = 9.08$; $\bar{x}_2 = 78.00$, $s_2 = 12.99$.

a. $f = s_2^2/s_1^2 = 2.047$. Since $f = 2.047 < F(0.975, 9, 9) = 4.03$, do not reject H_0.

b. $s_p = 11.21$; $t = -2.59$. Since $t = -2.59 < t_{0.025;18} = 2.101$, reject H_0. Since $t_{0.01;18} = 2.552$, the p-value is approximately $2(0.01) = 0.02$.

20.6. $\bar{d} = -785/10 = -78.5$; $s_d = 121.08$; $t = -2.05$. Since $t = -2.05$ is between -2.262 and 2.262, do not reject H_0.

20.8. $r^+ = 14$; $r^- = 31$; $r = 14$ using (before-after) and eliminating $d_4 = 0$ so that $n = 9$. Since 14 is between 6 and 39, do not reject H_0.

20.10. $R_1 = 15$; $R_2 = 40$; and $R_3 = 23$. $T = 6.269$. Since $T = 6.269 > T_{0.049} = 5.6923$, reject H_0. p-value $= 0.037$ (by interpolation).

20.12. *a.* 0.756. *b.* 0.753; bivariate normal distribution; linear relationship. *c.* Since $\rho = 0.756 > \rho_{0.95} = 0.619$, reject H_0. p-value $= 0.018$ (by interpolation).

20.14.

x	Sample CDF $S(x_i)$	Pop CDF $F(x_i)$	$\|F(x_i) - S(x_i)\|$	$\|F(x_i) - S(x_{i-1})\|$
5.6	$1/12 = 0.0833$	0.267	0.1837	0.2670
12.2	$2/12 = 0.1667$	0.492	0.3253	0.4087
14.8	$3/12 = 0.2500$	0.561	0.3110	0.3943
16.4	$4/12 = 0.3333$	0.598	0.2647	0.3480
16.8	$5/12 = 0.4167$	0.607	0.1903	0.2737
19.0	$6/12 = 0.5000$	0.652	0.1520	0.2353
19.4	$7/12 = 0.5833$	0.660	0.0767	0.1600
21.0	$8/12 = 0.6667$	0.688	0.0213	0.1047
22.4	$9/12 = 0.7500$	0.712	0.0380	0.0453
23.5	$10/12 = 0.8333$	0.729	0.1043	0.0210
23.7	$11/12 = 0.9167$	0.732	0.1847	0.1013
24.9	$12/12 = 1.0000$	0.749	0.2510	0.1677

$Y = X/18$, $F(y) = 1 - e^{-y}$; use Table B.4 to find e^{-y}. For example, with $x = 5.6$, $y = 5.6/18 = 0.311$; $1 - e^{-0.311} = 0.267$. $t = 0.4087$. Since $t = 0.4087 > W_{0.95} = 0.375$, reject H_0. p-value $= 0.027$ (by interpolation).

20.16. *a.* $t = 7$; since $W_{0.025} = 3$, and $W_{0.975} = 9$, and $t = 7$ falls between the action limits, do not reject H_0.

b. $t = 0.256$; since $t = 0.256 < W_{0.95} = 0.258$, do not reject H_0.

20.18. $S = 71$; $t = 16$. Since $T = 16 < W_{0.025} = 24$, reject H_0. p-value $= 0.008$ (approximate).

20.20. *a.* $s = 396$; $t = 71$. $W_{0.025} = 211.48$ using the z-approximation with $z_{0.025} = -1.96$. Since $t = 71 < 211.48$, reject H_0.

b. The two distributions of tire wear are symmetric.

c. $71 = \dfrac{(25)(25)}{2} - z \sqrt{\dfrac{(25)(25)(51)}{12}}$;

$z = 4.68$.

p-value $= 2P[\mathbf{Z} > 4.68] = 0.0000$ (zero to 4 decimal places).

20.22. *a.* $r^{+} = 39.5$; $r^{-} = 170.5$; $r = 39.5$. $W_{0.025} = 53$; $W_{0.975} = 157$. Since $r = 39.5 < W_{0.025}$, reject H_0.

b. p-value is between $2(0.005) = 0.01$ and $2(0.01) = 0.02$; 0.012 by interpolation.

20.24. *a.* $\rho = 0.862$; $\rho_{0.975} = 0.4451$; $\rho_{0.025} = -0.4451$. Since $\rho = 0.862 > 0.4451$, reject H_0.

b. p-value $< 2(0.001) = 0.002$, since $\rho_{0.999} = 0.6586$.

Chapter 21

21.2.

Act \ State of nature	1	2	3	4	5
1	20	0	6	1	11
2	0	4	4	7	8
3	11	2	18	0	6
4	12	10	7	2	0
5	6	8	0	10	6

21.4. $\Omega_{ij} = \begin{cases} \$3a_i & \text{if } \theta_j \geq a_i \\ \$3\theta_j - \$2(a_i - \theta_j) & \text{if } \theta_j < a_i \end{cases}$
$= \$5\theta_j - \$2a_i$

21.6.

Act (order quantity)	State of nature (demand) 0	1	2	3	4	5
0	$ 0	$ 0	$ 0	$ 0	$ 0	$ 0
1	−200	1,000	1,000	1,000	1,000	1,000
2	−400	800	2,000	2,000	2,000	2,000
3	−600	600	1,800	3,000	3,000	3,000
4	−800	400	1,600	2,800	4,000	4,000
5	−1,000	200	1,400	2,600	3,800	5,000

21.8. Take your umbrella with you or not on departing from home this morning. Risk could be reduced by listening to weather forecast before leaving home.

21.10.

Act	State of nature 1	2	3	4	5
1	100	90	10	160	100
2	120	20	40	0	50
3	0	120	80	10	0
4	150	0	0	110	100

21.12. Yes. a_1 dominates a_7: For every state of nature, the payoff for a_1 is at least as high as the payoff for a_7. a_5 dominates a_2; thus acts a_2 and a_7 may be dropped from the table.

21.14. a. a_3. b. a_4. c. a_3.

21.16. a. Don't buy any roses. b. Buy six dozen roses. c. Buy four dozen roses. d. Buy four dozen roses.

21.18.

Act	State of nature 0	1	2	3	4
0	0	30	60	90	120
1	10	0	30	60	90
2	20	10	0	30	60
3	30	20	10	0	30
4	40	30	20	10	0

Minimax loss strategy is order three cases.

21.20.

State of nature / Act	0	1	2	3	4	5
0	0	1,000	2,000	3,000	4,000	5,000
1	200	0	1,000	2,000	3,000	4,000
2	400	200	0	1,000	2,000	3,000
3	600	400	200	0	1,000	2,000
4	800	600	400	200	0	1,000
5	1,000	800	600	400	200	0

Minimax strategy is to order either 4 or 5 microcomputers.

21.22. *a.* a_2 or a_4. *b.* a_1. *c.* a_2.

21.24.

State of nature / Act	1	2	3	4
1	6,000	1,000	1,000	0
2	0	3,000	0	3,000
3	8,000	0	3,000	2,500
4	4,000	3,000	9,000	10,000

f.

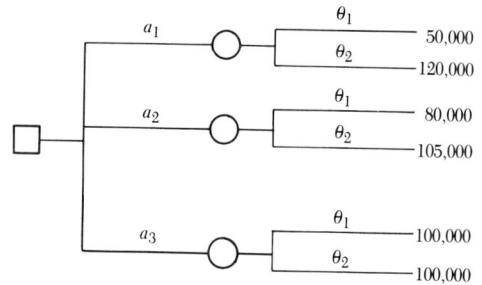

Minimax loss strategy is to select act a_2.

21.26. *a.* a_3. *b.* a_1. *c.* a_3.

21.28. *a.* a_3. *b.* a_1. *c.* a_1.

d.

State of nature / Act	1	2
1	0	20,000
2	30,000	5,000
3	50,000	0

e. a_1.

Chapter 22

22.2. Profit/dozen = 12($2 − $1 − $0.25) =
12($0.75) = $9.
Cost of overage/dozen = 12($1 − $0.50) =
12($0.50) = $6.

Payoff matrix

Act \ State of nature	10	11	12	13	14	15	16	EMV
10	90	90	90	90	90	90	90	90
11	84	99	99	99	99	99	99	97.67
12	78	93	108	108	108	108	108	103.67
13	72	87	102	117	117	117	117	107.33
14	66	81	96	111	126	126	126	108.11
15	60	75	90	105	120	135	135	106.11
16	54	69	84	99	114	129	144	101.44
P(S)	0.089	0.111	0.156	0.200	0.178	0.178	0.089	

Loss matrix

Act \ State of nature	10	11	12	13	14	15	16	EOL
10	0	9	18	27	36	45	54	28.43
11	6	0	9	18	27	36	45	20.76
12	12	6	0	9	18	27	36	14.75
13	18	12	6	0	9	18	27	11.08
14	24	18	12	6	0	9	18	10.41
15	30	24	18	12	6	0	9	12.41
16	36	30	24	18	12	6	0	17.08
P(S)	0.089	0.111	0.156	0.200	0.178	0.178	0.089	

Optimal act is to buy 14 dozen hot dogs (EMV = $107.99). EVPI = $10.41.

22.4. *a.*

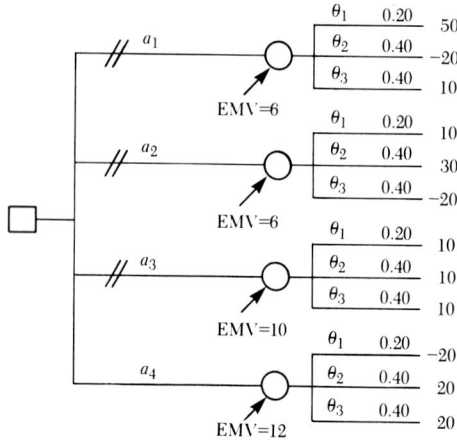

b.

Act \ State of nature	1	2	3
1	0	50	10
2	40	0	40
3	40	20	10
4	70	10	0
P(S)	0.20	0.40	0.40

c. EPPP (EVUC) = $(0.20)(50) + (0.40)(30) + (0.40)(20) = 30$.

d. EVPI
$$= \begin{cases} \text{EPPP} - \text{MAX EMV} = 30 - 12 = 18 \\ \text{or} \\ \text{MIN EOL} \qquad\qquad = 18 \end{cases}$$

22.6. Payoff function:

$$\Omega_{ij} = \begin{cases} \$3a_i & \text{if } \theta_j \geq a_i \\ \$3\theta_j - \$2(a_i - \theta_j) & \text{if } \theta_j < a_i \end{cases}$$
$$= \$5\theta_j - \$2a_i$$

Loss function:

$$l(a_i, \theta_j) = \begin{cases} \$3(\theta_j - a_i) & \text{if } \theta_j > a_i \\ 0 & \text{if } \theta_j = a_i \\ \$2(a_i - \theta_j) & \text{if } \theta_j < a_i \end{cases}$$

22.8.

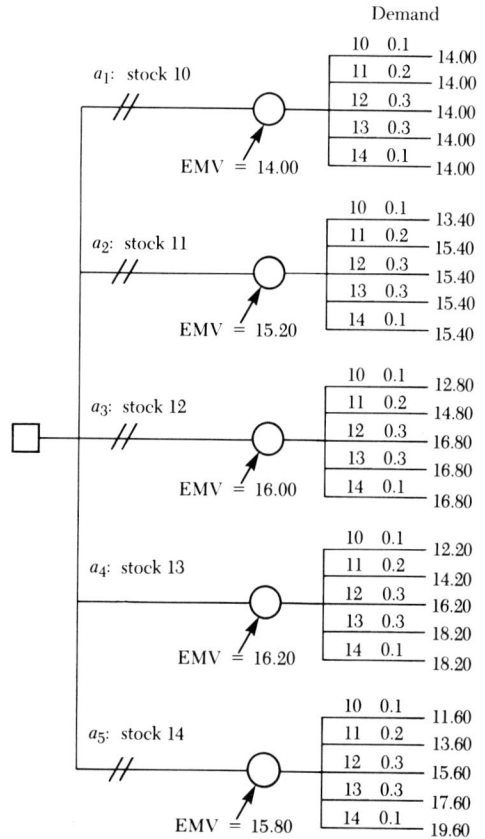

22.10. Either PCA or RCC (EMV = $1,675,000). EVPI = 1,693,000 − 1,675,000 = $18,000.

22.12. *a.* Yes, decision alternative A is dominated by both alternatives B and C.

b. EMV (B) = 66,500. EMV (C) = 61,000. Therefore, choose the money market fund.

c. EVPI = 68,500 − 66,500 = $2,000. It is the maximum amount she would be willing to pay to determine which state of nature regarding the economy will prevail in the future.

22.14. Consideration of nonmonetary consequences; utility is not linear in dollars.

22.16. Since $E(\theta) = 1,500 < \theta_b$, do not purchase equipment.

$$\text{EVPI} = (5)(150)\text{N}\left[\frac{|2000 - 1500|}{150}\right]$$
$$= (750)\text{N}(3.33)$$
$$= (750)(0.0001685448) = 0.126.$$

N(3.33) determined by interpolation.

$$\text{CVPI} = \begin{cases} 0 & \text{if } \theta < \theta_b. \\ 5(\theta - \theta_b) & \text{if } \theta \geq \theta_b. \end{cases}$$

22.18. 103.

22.20. $\theta_b = 15,000/6.10 = 2.459.$ EVPI $= (6.10)(500)(0.3602) = \$1,098.61.$

22.22. EMV$(a_1) = 4,250.$ EMV$(a_2) = 700.$ EMV$(a_3) = 8,000.$ EMV$(a_4) = 4,350.$ EOL$(a_1) = 5,700.$ EOL$(a_2) = 9,250.$ EOL$(a_3) = 1,950.$ EOL$(a_4) = 5,600.$ Optimal act is a_3: The \$8,000 for certain. EVPI $= \$1,950.$ CVPI$(\theta_1) = 2,000.$ CVPI$(\theta_2) = 400.$ CVPI$(\theta_3) = 4,500.$

22.24. *a.* EMV$(a_1) = -\$50,000.$ EMV$(a_2) = -\$35,000.$ EMV$(a_3) = -\$42,500.$ a_2: Drill the well is the optimal act.
b. EVPI $= \$77,500 - (-\$35,000) = \$112,500.$
c. EMV$(a_1) = -\$50,000.$ EMV$(a_2) = -\$80,000.$ EMV$(a_3) = -\$65,000.$ Now, a_1 is the optimal act.

Chapter 23

23.2. *a.* 0.28. *b.* $P(\theta_1/B) = 0.429;$ $P(\theta_2/B) = 0.571.$

23.4. *a.* 0.515. *b.* $P(\theta_1/B) = 0;$ $P(\theta_2/B) = 0;$ $P(\theta_3/B) = 0.243;$ $P(\theta_4/B) = 0.699;$ $P(\theta_5/B) = 0.058.$

23.6. *a.* 0.385. *b.* $P(I/B) = 0;$ $P(II/B) = 0;$ $P(III/B) = 0.034;$ $P(IV/B) = 0.748;$ $P(V/B) = 0.218.$ *c.* a_3; EMV $= 6.516.$

23.8. *a.* a_1; EMV $= \$400,000.$ *b.* EVPI $= \$600,000.$ *c.* a_1; EMV $= \$2,024,000.$ *d.* a_2; EMV $= \$0.$

23.10. Yes. EMV(prior) $= -\$35,000$ for act a_2
$$\text{EMV(take survey)} = \text{EMV(posterior)}$$
$$- \$25,000$$
$$= \$16,727.50 - \$25,000$$
$$= -\$8,272.50.$$
$$\text{EVSI} = \text{EMV(posterior)} - \text{EMV(prior)} =$$
$$\$16,727.50 - (-\$35,000) = \$51,727.50.$$

$$\text{ESI} = \frac{\text{EVSI}}{\text{EVPI}}(100\%)$$
$$= [\$51,727.50/\$112,500](100\%) = 45.98\%.$$

23.12. As n increases, the amount of information increases by $1/s^2$, or by the reciprocal of the variance.

23.14. Substitute $1/\sigma_1^2$ for I_1, etc; solve for σ_1 and simplify.

23.16. Since the sample size n is less than 30, we should not use the formulas presented in Theorem 23.1 to obtain estimates of the parameters of the revised distribution. In fact, for small n ($n < 30$), the revised distribution is not normal. Therefore we cannot find the revised probability distribution.

23.18. $H_0: \pi \leq 0.10;$ $H_A: \pi > 0.10;$ where $\pi = $ Proportion of defectives. Due to the discrete distribution with $n = 10$, it may not be possible to set α at exactly 0.05. The classical approach ignores the consequence (opportunity loss) if a Type I or a Type II error is made.

23.20.

θ	Posterior Probability
0.01	0.0159
0.02	0.1060
0.03	0.3965
0.04	0.2924
0.05	0.1892
	1.0000

23.22. There is often no way to conduct an experiment or a survey to determine frequencies corresponding to the states of nature to use for setting the prior probabilities.

23.24. *a.* 0.60. *b.* 0.40. *c.* $P(\theta_1/B) = 0.300;$ $P(\theta_2/B) = 0.350;$ $P(\theta_3/B) = 0.333;$ $P(\theta_4/B) = 0.017.$

23.26. *a.* $P(E_1) = 0.380;$ $P(E_2) = 0.295;$ $P(E_3) = 0.325.$ *b.* EVSI $= 53,599.70 - 47,500 = \$6,099.70.$ *c.* ESI $= [\$6,099.70/\$10,500](100\%) = 58.09\%.$ *d.* \$6,099.70. *e.* a_2.

23.28. *a.* a_1, EMV $= \$38.50.$ *b.* \$11.50. *c, d, e:*

State	Posterior Distribution B_1	Posterior Distribution B_2	Posterior Distribution B_3
θ_1	0.424	0.091	0.087
θ_2	0.273	0.341	0.261
θ_3	0.303	0.568	0.652

e. EMV(prior) = \$38.50; EMV(posterior) = \$38.50 (you may not get this exactly due to roundoff error). EVSI = \$0. EVSI is zero, since the prior act is a_1, and the posterior act is also a_1, *regardless of the survey outcome*. Thus the survey should not be taken. It adds no new information.

23.30. 349; highly unlikely.

23.32. Normal with $\mu_1 = 199.69$ and $\sigma_1 = 2.495$.

Index

*This book has been set Linotron in 10 and 9 point
Times Roman, leaded 2 points. Chapter titles are 16
point Helvetica Bold. The size of the type page is 36
picas by 50 picas.*

The *t*-distribution

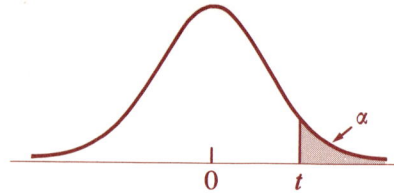

d.f. \ α	.10	.05	.025	.01	.005
1	3.078	6.314	12.706	31.821	63.657
2	1.886	2.920	4.303	6.965	9.925
3	1.638	2.353	3.182	4.541	5.841
4	1.533	2.132	2.776	3.747	4.604
5	1.476	2.015	2.571	3.365	4.032
6	1.440	1.943	2.447	3.143	3.707
7	1.415	1.895	2.365	2.998	3.499
8	1.397	1.860	2.306	2.896	3.355
9	1.383	1.833	2.262	2.821	3.250
10	1.372	1.812	2.228	2.764	3.169
11	1.363	1.796	2.201	2.718	3.106
12	1.356	1.782	2.179	2.681	3.055
13	1.350	1.771	2.160	2.650	3.012
14	1.345	1.761	2.145	2.624	2.977
15	1.341	1.753	2.131	2.602	2.947
16	1.337	1.746	2.120	2.583	2.921
17	1.333	1.740	2.110	2.567	2.898
18	1.330	1.734	2.101	2.552	2.878
19	1.328	1.729	2.093	2.539	2.861
20	1.325	1.725	2.086	2.528	2.845
21	1.323	1.721	2.080	2.518	2.831
22	1.321	1.717	2.074	2.508	2.819
23	1.319	1.714	2.069	2.500	2.807
24	1.318	1.711	2.064	2.492	2.797
25	1.316	1.708	2.060	2.485	2.787
26	1.315	1.706	2.056	2.479	2.779
27	1.314	1.703	2.052	2.473	2.771
28	1.313	1.701	2.048	2.467	2.763
29	1.311	1.699	2.045	2.462	2.756
30	1.310	1.697	2.042	2.457	2.750
40	1.303	1.684	2.021	2.423	2.704
60	1.296	1.671	2.000	2.390	2.660
120	1.289	1.658	1.980	2.358	2.617
∞	1.282	1.645	1.960	2.326	2.576

Source: Hoel, *Elementary Statistics*, 3d ed. (New York: John Wiley & Sons, ©1971).